Sensory Polymers

From their Design to Practical Applications

Sensory Polymers
From their Design to Practical Applications

Edited by

José Miguel García
Department of Chemistry, Universidad de Burgos, Burgos, Spain

Saúl Vallejos
Department of Chemistry, Universidad de Burgos, Burgos, Spain

Miriam Trigo-López
Department of Chemistry, Universidad de Burgos, Burgos, Spain

ELSEVIER

Elsevier
Radarweg 29, PO Box 211, 1000 AE Amsterdam, Netherlands
125 London Wall, London EC2Y 5AS, United Kingdom
50 Hampshire Street, 5th Floor, Cambridge, MA 02139, United States

ISBN: 978-0-443-13394-7

For Information on all Elsevier publications
visit our website at https://www.elsevier.com/books-and-journals

Publisher: Matthew Deans
Acquisitions Editor: Ana Claudia Garcia
Editorial Project Manager: Rafael Guilherme Trombaco
Production Project Manager: Sharmila Kirouchenadassou
Cover Designer: Mark Rogers

Typeset by MPS Limited, Chennai, India

Contents

Section 3 Research trends and challenges in polymer sensors

List of contributors

Laxmi Raman Adil
Department of Chemistry, Indian Institute of Technology Guwahati, Guwahati, Assam, India

Roberto J. Aguado
LEPAMAP-PRODIS Research Group, University of Girona, Girona, Girona (Catalonia), Spain

Giancarla Alberti
Department of Chemistry, University of Pavia, Pavia, Italy

Luis E. Alonso-Pastor
Cellular Materials Laboratory (CellMat), Department of Condensed Material Physics, Faculty of Science, University of Valladolid, Valladolid (Valladolid), Spain

Ana Arnaiz
Facultad de Ciencias, Departamento de Química Orgánica, Universidad de Burgos, Burgos, Spain; Universidad Politécnica de Madrid, Madrid, Spain

Pakize Aslan
Department of Analytical Chemistry, Faculty of Pharmacy, Ankara University, Ankara, Turkey

Isaí Barboza-Ramos
Department of Chemistry, University of Texas at San Antonio, San Antonio, TX, United States

Debika Barman
Department of Chemistry, Indian Institute of Technology Guwahati, Guwahati, Assam, India

Raffaela Biesuz
INSTM, Pavia Research Unit, Firenze, Italy; Department of Chemistry, University of Pavia, Pavia, Italy

Umut Bulut
Faculty of Pharmacy, Department of Analytical Chemistry, Acıbadem Mehmet Ali Aydınlar University, Ataşehir, Istanbul, Turkey

Ana S. Castro
Centre of Chemistry, University of Minho, Braga, Portugal

Sevki Can Cevher
Institute of Computational Physics, Zurich University of Applied Sciences, ZHAW, Winterthur, Zurich, Switzerland

Moirangthem Anita Chanu
Department of Chemistry, Indian Institute of Technology Guwahati, Guwahati, Assam, India

Daniela M. Correia
Centre of Chemistry, University of Minho, Braga, Portugal

Samar Damiati
Department of Chemistry, University of Sharjah, Sharjah, United Arab Emirates

Liliana C. Fernandes
Physics Centre of Minho and Porto Universities (CF-UM-UP), and Laboratory of Physics for Materials and Emergent Technologies, LapMET, University of Minho, Braga, Portugal

Manuela F. Frasco
BioMark, CEMMPRE, ARISE, Department of Chemical Engineering, Faculty of Sciences and Technology, University of Coimbra, Coimbra, Portugal

José M. García
Departamento de Química, Facultad de Ciencias, Universidad de Burgos, Burgos, Spain

Arup K. Ghosh
Department of Chemistry, Sardar Vallabhbhai National Institute of Technology (SVNIT), Surat, Gujarat, India

Priyam Ghosh
Department of Chemistry, Indian Institute of Technology Guwahati, Guwahati, Assam, India

Clara Pérez González
Materials Science Department, University of Valladolid, Valladolid, Spain

Marta Guembe-García
Facultad de Ciencias, Departamento de Química Orgánica, Universidad de Burgos, Burgos, Spain; Department of Chemistry, University of Pavia, Pavia, Italy

Manuel Herrero
Cellular Materials Laboratory (CellMat), Department of Condensed Material Physics, Faculty of Science, University of Valladolid, Valladolid (Valladolid), Spain

Xiao Huang
Institute of Advanced Materials (IAM), Nanjing Tech University (Nanjing Tech), Nanjing, P. R. China

Parameswar Krishnan Iyer
Department of Chemistry, Indian Institute of Technology Guwahati, Guwahati, Assam, India; Centre for Nanotechnology and Centre for Drone Technology, Indian Institute of Technology Guwahati, Guwahati, Assam, India; School for Health Science and Technology, Indian Institute of Technology Guwahati, Guwahati, Assam, India

Sukriye Nihan Karuk Elmas
The Faculty of Pharmacy, The Department of Analytical Chemistry, Istanbul University-Cerrahpaşa, Istanbul, Turkey

Gulsu Keles
Department of Analytical Chemistry, Faculty of Pharmacy, Ankara University, Ankara, Turkey

Mst Nasima Khatun
Department of Chemistry, Indian Institute of Technology Guwahati, Guwahati, Assam, India

Rungthip Kunthom
International Center for Interdisciplinary Research and Innovation of Silsesquioxane Science, Key Laboratory of Special Functional Aggregated Materials, Ministry of Education, School of Chemistry and Chemical Engineering, Shandong University, Jinan, Shandong, P. R. China

Sevinc Kurbanoglu
Department of Analytical Chemistry, Faculty of Pharmacy, Ankara University, Ankara, Turkey

Senentxu Lanceros-Méndez
Physics Centre of Minho and Porto Universities (CF-UM-UP), and Laboratory of Physics for Materials and Emergent Technologies, LapMET, University of Minho, Braga, Portugal; BCMaterials, Basque Centre for Materials and Applications, UPV/EHU Science Park, Leioa, Spain; IKERBASQUE, Basque Foundation for Science, Bilbao, Spain

Hongzhi Liu
International Center for Interdisciplinary Research and Innovation of Silsesquioxane Science, Key Laboratory of Special Functional Aggregated Materials, Ministry of Education, School of Chemistry and Chemical Engineering, Shandong University, Jinan, Shandong, P. R. China

Lisa Rita Magnaghi
INSTM, Pavia Research Unit, Firenze, Italy; Department of Chemistry, University of Pavia, Pavia, Italy

Miguel Manzano
Departamento de Química en Ciencias Farmacéuticas, Instituto de Investigación Sanitaria Hospital 12 de Octubre i + 12, Universidad Complutense de Madrid, Plaza Ramón y Cajal, Madrid, Spain; Networking Research Center on Bioengineering, Biomaterials and Nanomedicine (CIBER-BBN), Madrid, Spain

Magdalena Mieloszyk
Institute of Fluid-Flow Machinery, Polish Academy of Sciences, Gdansk, Poland

Álvaro Miguel
Department of Chemistry, Universidad de Burgos, Burgos, Spain; Universidad Autónoma de Madrid, Madrid, Spain

Yosuke Mizuno
Faculty of Engineering, Yokohama National University, Yokohama, Japan

Karina C. Núñez-Carrero
Cellular Materials Laboratory (CellMat), Department of Condensed Material Physics, Faculty of Science, University of Valladolid, Valladolid (Valladolid), Spain

Aysel Oktay
Molecular Biotechnology, Faculty of Science, Turkish-German University, Istanbul, Turkey

Jesus L. Pablos
Departamento de Química en Ciencias Farmacéuticas, Instituto de Investigación Sanitaria Hospital 12 de Octubre i + 12, Universidad Complutense de Madrid, Plaza Ramón y Cajal, Madrid, Spain

María Luz Rodríguez Méndez
Inorganic and Physical Chemistry Department, University of Valladolid, Valladolid, Spain

Suban K. Sahoo
Department of Chemistry, Sardar Vallabhbhai National Institute of Technology (SVNIT), Surat, Gujarat, India

Anuj Saini
Department of Chemistry, Sardar Vallabhbhai National Institute of Technology (SVNIT), Surat, Gujarat, India

M. Goreti F. Sales
BioMark, CEMMPRE, ARISE, Department of Chemical Engineering, Faculty of Sciences and Technology, University of Coimbra, Coimbra, Portugal

Coral Salvo Comino
Inorganic and Physical Chemistry Department, University of Valladolid, Valladolid, Spain

Tapashi Sarmah
Department of Chemistry, Indian Institute of Technology Guwahati, Guwahati, Assam, India

Kirk S. Schanze
Department of Chemistry, University of Texas at San Antonio, San Antonio, TX, United States

Carlos Sedano
Departamento de Química, Facultad de Ciencias, Universidad de Burgos, Burgos, Spain

Saniye Soylemez
Faculty of Engineering, Department of Biomedical Engineering, Necmettin Erbakan University, Meram, Konya, Turkey

Dilek Soyler
Faculty of Engineering, Department of Biomedical Engineering, Necmettin Erbakan University, Meram, Konya, Turkey

Miriam Trigo-López
Department of Chemistry, Universidad de Burgos, Burgos, Spain

Aditi Tripathi
Department of Chemistry, Sardar Vallabhbhai National Institute of Technology (SVNIT), Surat, Gujarat, India

Carmen R. Tubio
BCMaterials, Basque Centre for Materials and Applications, UPV/EHU Science Park, Leioa, Spain

Saúl Vallejos
Department of Chemistry, Universidad de Burgos, Burgos, Spain

María Vallet-Regí
Departamento de Química en Ciencias Farmacéuticas, Instituto de Investigación Sanitaria Hospital 12 de Octubre i + 12, Universidad Complutense de Madrid, Plaza Ramón y Cajal, Madrid, Spain; Networking Research Center on Bioengineering, Biomaterials and Nanomedicine (CIBER-BBN), Madrid, Spain

Raquel Vaz
BioMark, CEMMPRE, ARISE, Department of Chemical Engineering, Faculty of Sciences and Technology, University of Coimbra, Coimbra, Portugal

Aysu Yarman
Molecular Biotechnology, Faculty of Science, Turkish-German University, Istanbul, Turkey

Camilla Zanoni
Department of Chemistry, University of Pavia, Pavia, Italy

Jian Zhang
Institute of Advanced Materials (IAM), Nanjing Tech University (Nanjing Tech), Nanjing, P. R. China

CHAPTER

Foundation of sensory polymers

1

Miriam Trigo-López[1] and Carlos Sedano[2]

[1]Department of Chemistry, Universidad de Burgos, Burgos, Spain [2]Departamento de Química, Facultad de Ciencias, Universidad de Burgos, Burgos, Spain

1.1 Introduction

Polymers are defined as macromolecules composed of the repetitive assembly of distinct subunits called monomers, which are covalently bonded to one another. Polymers are materials that are ubiquitous in our daily lives. Society's habits are ever-changing, and the evolution of polymers has driven a portion of these shifts as they gradually replaced traditional materials. Many technological innovations in our society stem from novel inventions that either discovered a new application or unveiled a property of polymers (Sperling, 2005).

Polymeric materials are a vigorously active field of research. From a practical application standpoint, they can be categorized based on their general purpose/everyday use, their utilization as engineering polymers, or their application as high-value-added polymers. General-purpose polymers encompass a broad range of properties that render them ideal for myriad applications, including packaging, wrapping, and construction materials, among many others. There are numerous commonly used polymers, with notable examples including polyethylene, polypropylene, poly(ethylene terephthalate), polyvinyl chloride, and polystyrene. Engineering polymers are a group of polymers with the capability to maintain their dimensional, mechanical, and chemical stability under demanding environmental conditions, such as temperatures above 100 °C and below 0 °C, corrosive surroundings, or high pressures. The development of these types of polymers surged after World War II as existing materials did not meet the demands of the time, exacerbated by supply issues. These challenges led to increased interest and research into new materials for weaponry and the aerospace race, particularly for improving aircraft performance (Vora et al., 2022). Examples of these polymers include aliphatic and aromatic polyamides, polyimides, polycarbonates, polytetrafluoroethylene, polyether ketones, or polysulfones (Rusli et al., 2022). High-value-added or functional polymers are those intentionally designed for specific applications or purposes, resulting in their global annual production being below 1000 tons. Among these polymers, notable examples include conductive, antimicrobial, antioxidant, intelligent polymers, and biopolymers (Hosseini et al., 2019).

Among high-value-added polymers, smart or intelligent polymers stand out. These materials rely on the molecular recognition principle and possess the ability to respond with a change in their physicochemical properties (specific signal or action) to physical or chemical stimuli in their

Sensory Polymers. DOI: https://doi.org/10.1016/B978-0-443-13394-7.00001-X

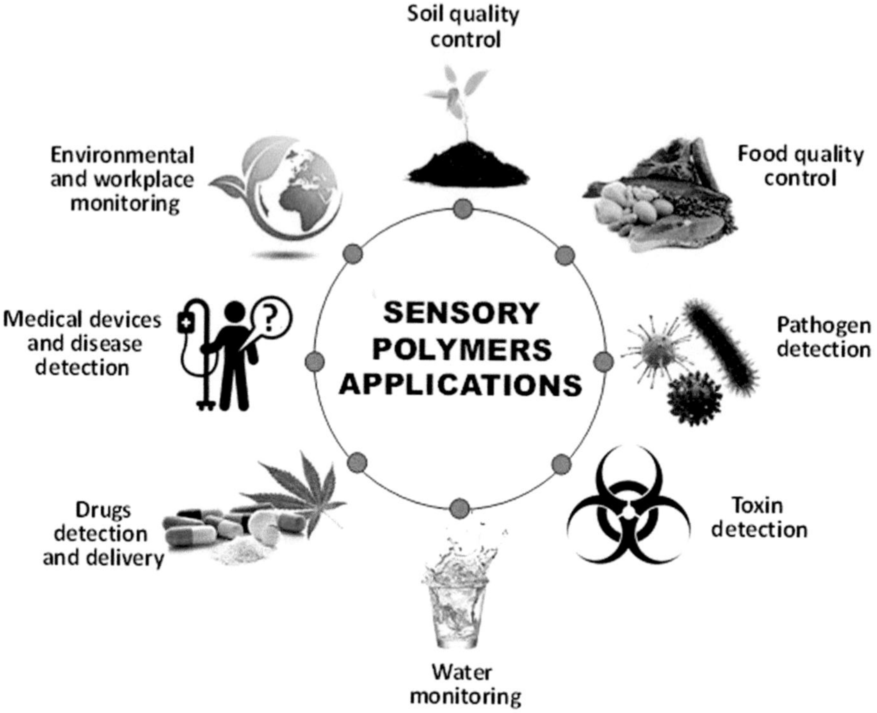

FIGURE 1.1 Applications of sensory polymers.

Various applications of sensory polymers in different fields.

environment, such as changes in the electromagnetic field, pH, pressure, temperature, or the presence of specific analytes. They are engineered to exhibit a wide range of sensory responses, including solubility, color changes, fluorescence, electrical conductivity, mechanical deformations, solvation, or volume variations, among others, in response to various stimuli. The applications of these materials span from food quality control, tissue regeneration, or controlled drug release to the development of photodegradable substances. Moreover, smart polymers can selectively react to a particular analyte (chemical and/or biological stimulus), such as the presence of heavy metals, organic molecules, or even viruses and bacteria (García et al., 2022), in different ways (Fig. 1.1), producing a macroscopic, measurable signal.

1.1.1 Sensory polymers

Polymers that provide valuable analytical information in response to the stimulus are referred to as sensory polymers, acting like an alert system to detect and signal the presence of specific stimuli. The classification of these materials can be approached in different ways. In this book, we sort them by three different criteria: their composition or structure; the mechanism of transduction, based on the nature of the response they elicit; and the stimulus to which they react (Fig. 1.2).

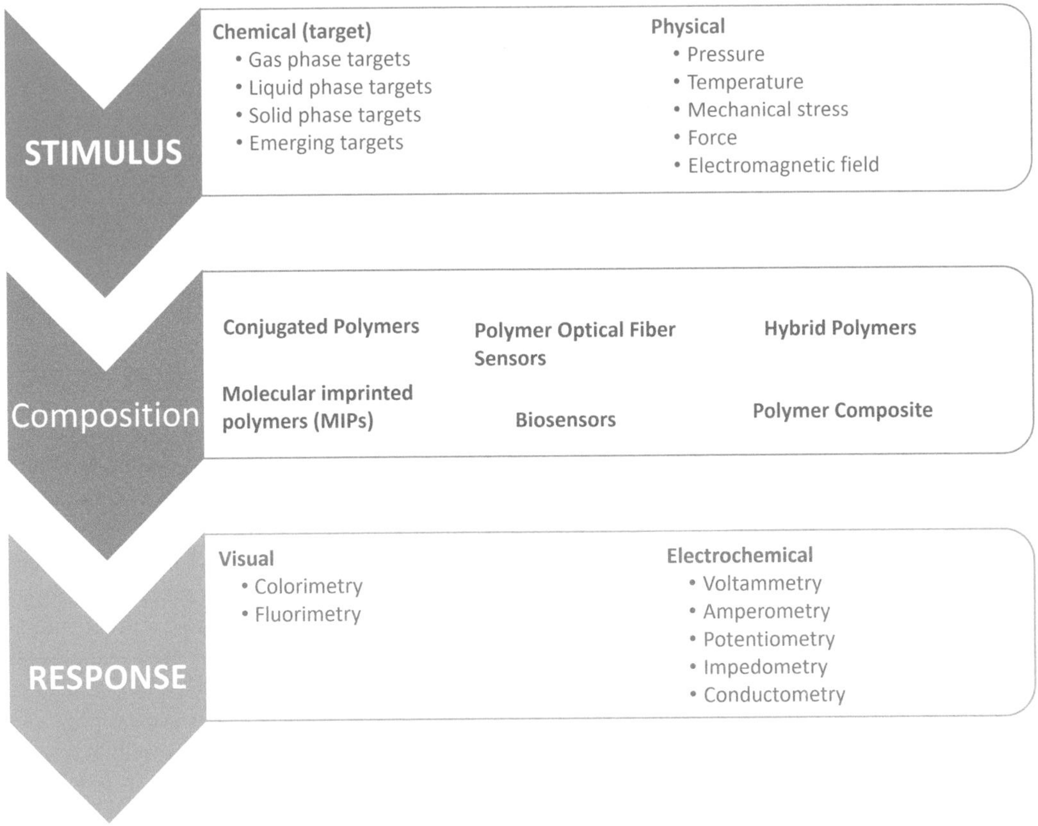

FIGURE 1.2 Classification of sensory polymers.

Classification of sensory by stimulus, composition and response.

In this sense, smart polymers respond to various types of stimuli, which can be categorized into chemical (or biological) or physical stimuli. Chemical or biological stimuli encompass discrete molecules (gases, solids, or liquids), macromolecules (enzymes, proteins, etc.), and even microorganisms like viruses or bacteria. Conversely, physical stimuli include factors such as pressure, temperature, light, electromagnetic fields, and more (García et al., 2022).

According to the type of response, smart polymers can be divided into stimuli-responsive polymers and sensory polymers. The former reacts to the external stimuli with an action, such as a variation in the interaction with solvents, in the volume or shape, the break or formation of chemical bonds, isomerization, etc. Sensory polymers react to the stimulus with an alert, a useful analytical signal, which can be a color or fluorescence change or an electrochemical response (García et al., 2022). The field of stimuli-responsive polymers is expanding and revolves around meticulous control and manipulation of both their chemical and physical properties. Sensory polymers are macromolecules characterized by the inclusion of binding sites (receptor motifs) for guest molecules (or target molecules) as well as signaling units within their structure (García et al., 2011).

Lastly, these materials can be specific sensors for different stimuli, such as gases, humidity, pH, temperature, nitroaromatic explosives, cationic and anionic species, proteins, drugs, and other neutral species.

Taking into account all these possibilities, the applications of sensory polymers are widespread, finding utility in sensing and detection systems such as environmental monitoring, biomedical diagnostics, and food safety. Throughout the book, we will study in more detail the intricacies of sensory polymers, investigating their mechanisms of response, design principles, and diverse applications across fields. By offering a comprehensive understanding of sensory polymers, this book aims to contribute to the advancement of this fascinating area of research.

1.1.2 Designing sensory polymers

Life is fundamentally structured around selective interactions between different chemical species, with biomacromolecules playing pivotal roles in essential molecular interactions. Among these biomacromolecules, proteins stand out for their remarkable versatility, serving as receptors, enzymes, antibodies, and transporters. On the other hand, nucleic acids, such as DNA and RNA, serve as the foundation of life, engaging in molecular recognition processes through base pairing interactions. The accurate replication and transmission of genetic information are facilitated by complementary base pairing, which also holds significant importance in the process of protein synthesis. Carbohydrates are also key participants in recognition processes, involved in cell-cell recognition, immune responses, and signal transduction. Lipids, as vital components of cell membranes, contribute to molecular recognition processes as well. In addition, small molecular-mass biomolecules, including neurotransmitters, saccharides, and specific substrates recognized by enzymes, play vital roles in various biochemical reactions, further contributing to the complexity of molecular interactions in living systems. These selective interactions and recognition processes form the very foundation of life's intricate and diverse mechanisms.

The remarkably specific and complex recognition processes observed in biological systems have served as a profound inspiration for the field of supramolecular chemistry. This field was pioneered by Cram, Lehn, and Pedersen, who were awarded the Nobel Prize in Chemistry in 1987 for their groundbreaking contributions. Supramolecular chemistry strives to replicate the remarkable efficiency of nature in achieving selective and effective interactions. Drawing inspiration from the fundamental principles observed in biological recognition processes, this area of study of sensory polymers seeks to harness the power of molecular self-assembly and non-covalent interactions to create sophisticated and highly functional supramolecular structures.

On the other hand, Staudinger's description of the macromolecular characteristics of polymers in the 1920s marked a significant turning point, laying the groundwork for the emergence of polymer science and the birth of the polymer age (Staudinger, 1920; Staudinger & Fritschi, 1922; Sutton, 2020). Since then, an abundance of modified polymeric structures, both synthetic and natural, has emerged, revolutionizing the production of goods and profoundly impacting advanced societies' well-being. These innovative materials have found their way into various applications, including synthetic textiles, medications, adhesives, coatings and paints, smart packaging, toys, and lightweight components for the automotive and aerospace industries. The properties of synthetic polymers as materials are intrinsically linked to their chemical structure and size, much like biomacromolecules' biological characteristics and properties rely on their chemical composition, especially concerning molecular recognition. This recognition is closely tied to the sequences of monomers, such as amino acids in proteins, as well as the conformation that the macromolecular structure adopts in its primarily aqueous biological environment. Within this environment, specific partially hydrophobic interactions have a relevant role in various functions, such as catalysis occurring in active sites. Similarly, in synthetic polymer sensors, the chemical structure is essential, dictating their performance and capabilities.

Accordingly, sensory polymers are designed by incorporating specific groups into the polymer structure known as host units, binding sites, or receptor subunits that can interact selectively with guests or target species, such as cations, anions, neutral molecules, or even stimuli, and generate a measurable response. Usually, these groups are incorporated in a monomer that is subsequently polymerized with other majority comonomers, usually commercially available, which will conform to the polymer matrix, which commonly also plays a key role in the sensing procedure, though lacking binding sites are partially responsible for the specificity and selectivity of the interaction, in the same way that the tertiary and quaternary structures play in the enzyme activity of proteins. Thus, the design process involves two key steps: (1) selection of polymer matrix, a suitable polymer matrix that can provide the desired properties such as flexibility, stability under ambient or hazardous conditions, hydrophilic/hydrophobic balance, physical shape, and compatibility with the target application; and (2) introduction of receptor moieties by chemical incorporation of functional groups or moieties into the polymer backbone or pendant chains, groups that should be understood broadly, for example, specific binding sites, irreversible reactive groups, or responsive units such as chromophores or fluorophores.

In this sense, sensory polymers are designed to recognize analytes or target species through various mechanisms, such as specific binding, molecular imprinting, or selective chemical reactions. Upon recognition, signal transduction is the measurable response that can be achieved by designing the polymer to change physical, chemical, or optical properties. Useful signals are changes in fluorescence, color, shape, conductivity, solubility, solvent affinity, or any other measurable chemical or physical property.

After design, preparation, and initial assessment of performance, the sensory polymers are optimized to enhance their sensitivity, selectivity, and response time. This can involve fine-tuning the polymer structure, optimizing the ratio of receptor groups, or incorporating additional components to amplify the signal. It also requires testing for their performance in detecting the target species in vitro and in an operational environment, something that usually involves further laboratory experiments, calibration curves, and comparison with known applications, systems, reference materials, or previously published results. Finally, if the technology readiness level and the commercial interest are high enough, the sensory polymers can be integrated into practical applications such as sensors, detectors, or diagnostic devices, where they can provide real-time monitoring or detection of the targets.

1.2 Molecular recognition

Molecular recognition is a naturally occurring phenomenon where two or more molecules are chemically, geometrically, and/or structurally complementary. This process usually occurs in aqueous environments with natural macromolecules or complex systems, such as cells. Examples of these interactions include enzyme/substrate, antigen/antibody, drug/biological target, and synthesis of mRNA from DNA templates. (Chen et al., 2002).

Supramolecular chemistry, also known as host-guest chemistry, is a field closely linked to molecular recognition in natural systems, aiming to emulate the effectiveness, simplicity, and outcomes of recognition processes observed in biological environments. The term "supramolecular" refers to the highly selective and reversible interaction between two molecules through relatively weak forces, including hydrogen bonding, van der Waals interactions, electrostatic interactions, and hydrophobic effects. In this interaction, one molecule serves as the host or receptor, while the other acts as the guest or target. The concept of recognition is specifically applied when the intermolecular host-guest interaction exhibits exceptional specificity. The host or receptor, in these interactions, can be utilized in developing various technological devices, including permselective membranes, catalysts, and sensors, among others (Dong et al., 2014). By understanding and harnessing these interactions, researchers can design and produce synthetic receptors capable of selectively bind specific target molecules (Aguilar et al., 2014).

In this regard, chemical sensors are devices that are specifically designed to detect and measure the presence and concentration of various chemical species or analytes in a given environment. If the recognition system utilizes a biochemical mechanism, it is categorized as a biosensor. They provide a means to monitor and analyze chemical processes, identify substances, and ensure safety and quality control in numerous industry applications. Chemical sensors operate based on the principle of selective interaction between the target analyte and a sensing element, which produces a measurable signal indicative of the analyte's presence or concentration. The sensing element in a chemical sensor can be a specific material, such as a molecule, a polymer, a metal oxide, or a biological component, such as an enzyme or an antibody (García et al., 2011).

Hence, a chemical sensor is inherently a tool that converts chemical information into a signal that proves analytically valuable (Hulanicki et al., 1991). This chemical information may arise from either the chemical reaction of the analyte or a physical characteristic of the studied system.

Chemical sensors consist of two interconnected subunits, each playing a specific role in the detection process (Vallejos et al., 2022). The receptor subunit serves as the recognition element capable of interacting with the target analyte or molecule of interest. On the other hand, the indicator subunit, or transducer, is responsible for generating a readable signal in response to the interaction with the analyte (Suksai & Tuntulani, 2005). Together, these subunits work in tandem to facilitate accurate and sensitive detection in sensor systems. The detection process depends on the indicator subunit or transducer, which converts the interaction between the receptor and the target analyte into a measurable property, such as color, fluorescence, conductivity, or other detectable signals. The indicator subunit plays a crucial role in translating the recognition event into a quantifiable response, allowing for the accurate and sensitive detection of the target analyte in sensor systems. In general, it can be stated that the forces governing the interactions between coordinating units and analytes can be electrostatic interactions, interactions through hydrogen bond formation, or covalent interactions with metal cores.

Based on the nature of the bond between the coordinating unit and the indicator unit, there are different types of molecular chemosensor designs (Fig. 1.3). If the bond is covalent, it is referred to as the binding site-signaling subunit approach; if not, it is referred to as the displacement assay approach. In addition to molecular chemosensors, there is another approach to

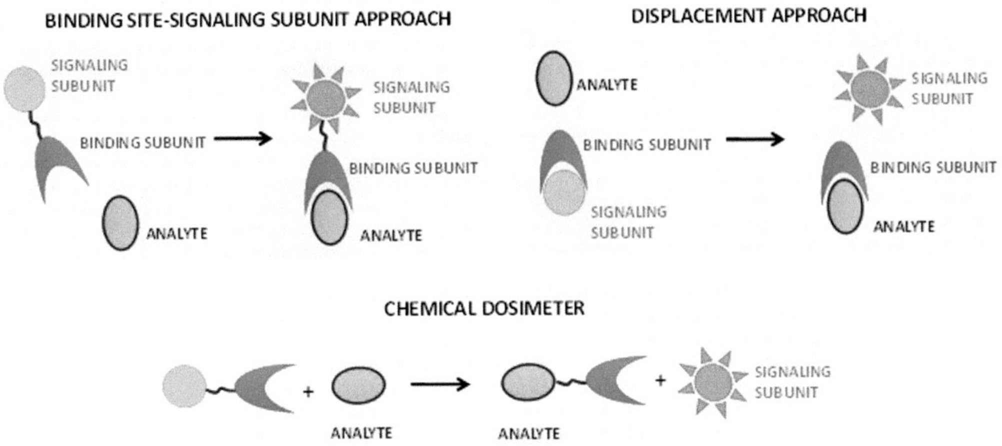

FIGURE 1.3 Chemical sensing mechanisms.

Schematic representation of different sensing approaches in chemical sensors.

selective chemical detection known as chemical dosimeters or chemodosimeters. In the chemical dosimeter approach, the ultimate idea is not to employ selective coordination systems, as in previous cases, but rather highly specific and typically irreversible chemical reactions induced by a particular analyte.

On the other hand, sensory polymers offer numerous benefits over individual or low-molecular-mass chemosensors. Most basic research conducted in the field of sensors is carried out using discrete molecules, which typically have low molecular weight, limited thermal and chemical resistance, and laborious and expensive separation and recovery processes. Additionally, these molecules are often insoluble in water, and their application in analyte detection is mostly restricted to organic or organic-aqueous media, limiting their potential for future clinical, biological, and environmental applications. In this regard, one of the main objectives of recent research in this field is focused on physically or chemically (covalently attached, especially) supporting sensors within polymer matrices (Guembe-García et al., 2020; Vallejos, Estévez, et al., 2011; Vallejos, El Kaoutit, et al., 2011). This allows them to be utilized as solid-state sensor materials, immobilizing selective receptors to prevent their migration, promoting reusability, and enhancing their mechanical properties so that they can be used in any way, including aqueous and non-aqueous environments, air, soil, and food (Fig. 1.4) (Arnaiz et al., 2023; González-Ceballos et al., 2020; Trigo-López et al., 2015; Vallejos et al., 2022).

Some other advantages of smart polymers compared with other systems and low-mass sensors are that they lack migration of the sensory motifs, can be easily processed into different shapes such as beads, coatings, films, soluble linear polymers, and fibers, and are biocompatible. Additionally, they can be designed or tuned through appropriate synthetic or modification techniques to meet the application requirements and the hydrophilic/hydrophobic balance needs (Adhikari & Majumdar, 2004; Jiang et al., 2020; Shahbazi & Jäger, 2021; Song et al., 2018). Moreover,

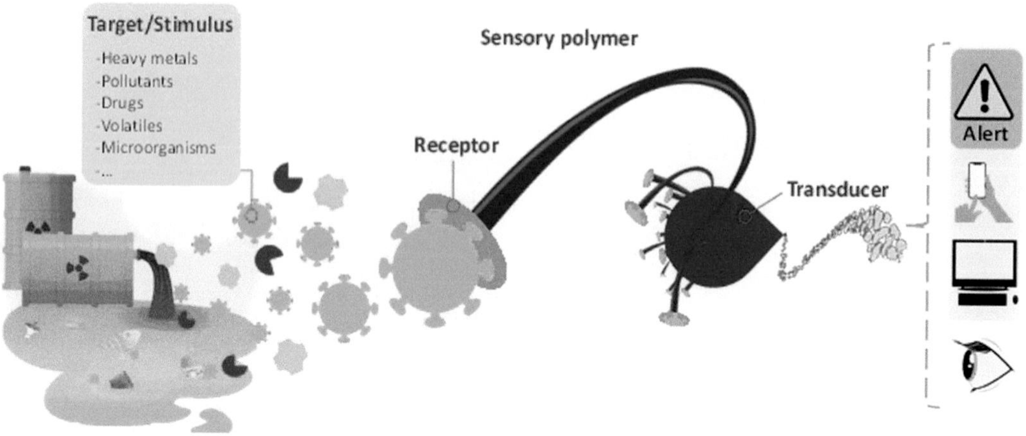

FIGURE 1.4 Sensory polymer.

Target detection and transduction by a sensory polymer.

compared to traditional sensory evidence, polymers provide a protective environment in which receptors interact with target molecules more efficiently, as demonstrated in many studies, and may show an allosteric effect and other measurable properties that are sensitive to minimal perturbations (Arnaiz et al., 2022; Bustamante et al., 2019).

The constant advancement of polymer sensors, aimed at achieving higher reliability and enabling continuous monitoring, remains an ongoing and significant challenge within the scientific community. Meeting this challenge is essential for various applications, such as measuring bioanalytes in the medical domain, overseeing numerous industrial processes, and monitoring air and water quality in environmental settings. To address these requirements, the development of cutting-edge sensor technologies is imperative.

1.3 Classification of sensory polymers

In addition to the classification by stimuli, sensing mechanism, and type of response, sensory polymers can be classified by the type of transduction mechanism, which refers to how they convert the sensory input (stimulus) into a measurable output (response). The transduction mechanism of sensory polymers determines the type of signal they generate in response to specific stimuli. On the other hand, sensory polymers can also be classified according to their composition and/or structure, which influences their sensing properties and capabilities.

The choice of transduction mechanism and the composition and structure of the sensory polymer depend on the specific application and the nature of the target analyte or stimulus. Sensory polymers are designed and tailored to respond to specific stimuli and provide a reliable and selective response, making them valuable tools in various sensing technologies and smart materials.

1.3.1 Type of transduction mechanism

Sensory polymers are designed to interact with specific stimuli and exhibit detectable changes in their properties or behavior due to this interaction. The transduction mechanism is essential for the functioning of sensory polymers as it allows them to sense and respond to various environmental cues. Advancements in instrumentation, microelectronics, and computers have enabled the design of sensors by leveraging well-established chemical, physical, and biological principles from the field of chemistry. Common transduction mechanisms in sensory polymers are summarized in Fig. 1.5 (Hulanicki et al., 1991).

1.3.1.1 Optical sensors

Optical sensory polymers change light absorption, emission, or scattering properties in response to the stimulus. These changes can manifest as color variations, fluorescence, luminescence, IR shifts, or shifts in the optical properties of the polymer, which can be easily measured using optical instruments (Sanjuán et al., 2018).

In simple terms, a small, discrete sensing molecule is a chemical structure composed of at least two structural motifs. One motif serves as a binding site, responsible for selectively interacting or reacting with a specific analyte, while the other motif acts as the signaling subunit. The detection capability of this molecule depends on the combined action of these two motifs.

The binding site allows the molecule to interact with the target analyte through feeble interactions, such as hydrogen bonds, electrostatic interactions, or metal-analyte interactions. This selective binding event creates a chemical interaction between the sensing molecule and the analyte. The signaling subunit plays a crucial role in translating this chemical interaction into a macroscopic signal. This signal can manifest as variations in the color and/or fluorescence of the sensing

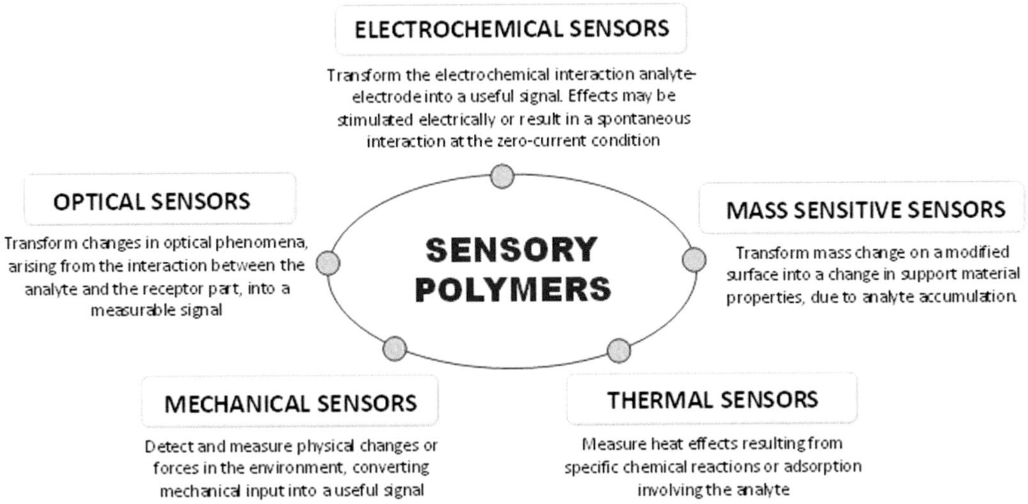

FIGURE 1.5 Classification. Transduction mechanism.

Classification of sensory polymers based on the operating principle of the transducer.

molecule. These changes in the macroscopic properties serve as a detectable indication of the presence of the analyte (García et al., 2011).

- *Fluorogenic sensor principles*: Photoluminescence can be described as the spontaneous emission of radiation, or light, from an electronically excited state of any substance. Regarding the nature of the excited state, it is divided into two categories: fluorescence and phosphorescence. In fluorescence, the excited molecule has the same multiplicity as the molecule in its ground state. After being irradiated at a specific wavelength, a molecule emits radiation at longer wavelengths than the one used for excitation (Lakowicz, 2006; Lemke & Schultz, 2011). A substructure of the molecule responsible for fluorescence is referred to as a fluorophore.
- *Chromogenic sensor principles*: Colored systems refer to conjugated structures with specific energy differences between the highest occupied molecular orbital and the lowest unoccupied molecular orbital that correspond to visible light. When the conjugate system is extended, these energy differences decrease, resulting in a red shift, also known as a bathochromic shift, of the absorption band. This shift causes the absorption energy to decrease.

In addition to controlling the length or design of the conjugation system, the absorption wavelength can also be manipulated by introducing electron-donating groups (e.g., NR_2, OR, and O^-) or electron-withdrawing groups (NO_2, CN, SO_3H, SO_3^-, COR, etc.) into the conjugated system. These electron-donating or electron-withdrawing groups alter the electronic distribution within the structure, influencing the energy levels and consequently affecting the absorption wavelength (El Sayed, 2023).

By strategically modifying the conjugation system and introducing specific electron-donating or electron-withdrawing groups, researchers can tailor the absorption properties of colored systems. This level of control enables the design of materials with desired optical properties, making them valuable in various applications, including organic electronics, sensors, and photonic devices (García et al., 2011).

The presence of an analyte induces changes in the absorption or emission spectrum of the polymer with sensor motifs, which could manifest as a spectral shift, intensity variation, or the emergence of new absorption or emission bands. Monitoring these alterations can be accomplished using laboratory UV/visible equipment, a fluorimeter, or portable diode array spectrometers connected to a laptop or even to a mobile phone (Guembe-García et al., 2021; Guembe-García et al., 2022), enabling quantitative determination of the analyte concentration. The detection limit is influenced by the probe's sensitivity and instrument constraints (Evans et al., 2013). In instances where the absorbance change results in a shift in wavelength within the visible region, it can lead to a noticeable alteration in color or fluorescence, allowing for qualitative or semi-quantitative detection of the analyte through visual observation alone. In these cases, the colorimetric polymer chemosensors can be film-shaped, prepared or coated into fabrics to prepare portable detection systems, or even integrated into garments that can be cost-effective and used in situ. This capability proves advantageous in situations in which quick evaluation of analyte's presence is required.

In addition to visual detection, using a smartphone's digital camera to record color changes is becoming an increasingly reliable method for finely quantifying species of interest, where an automatic autocalibration app has even been reported (Vallejos et al., 2021a, 2021b; Guembe-García et al., 2022).

In numerous instances, conventional laboratory-based techniques for sampling and analyzing substances fail to meet expectations, leading to a preference for continuous and *in situ* measurements. Innovative sensors capable of providing real-time data directly at the source are sought after to overcome the limitations of traditional instrumentation. Within this framework, optical fiber sensors can play a fundamental role (Baldini & Bracci, 2000). To date, using silica-glass optical fibers, a variety of distributed and quasi-distributed sensing techniques have been developed. In general, their performance, as for spatial resolution, measurement range, and sampling rate, has been considerably improved over time. Nevertheless, these enhanced properties have been demonstrated with silica glass optical fibers, which are fragile and can break at not especially high strains (Mizuno et al., 2021). Their development is based on combining optical fiber technology and polymeric materials to create highly sensitive and versatile sensors (Peters, 2011). These sensors use optical fibers made of special polymeric materials as the core, which are capable of transmitting light over long distances with minimal signal loss.

An optical fiber (Fig. 1.6) sensor comprises three primary components:

- An optoelectronic system, encompassing both hardware and software elements, which is responsible for interrogating the probe and processing the signals obtained.
- An optical link serves as a conduit to transmit the optical signal between the optoelectronic instrumentation and the probe and vice versa, and it is made of silica or polymer fibers.
- The probe is referred to as an optode or optrode, where modulation of the optical signal occurs. This is the sensing element of the sensor and is responsible for detecting and responding to the target analyte or parameter being measured. The probe's design allows for modulation of the

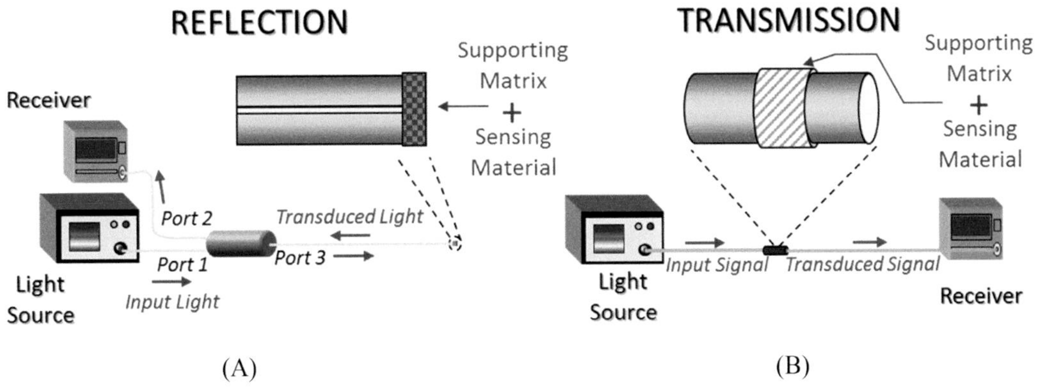

FIGURE 1.6 Polymeric optical fiber sensors.

Schematic representation of two different configurations for optical fiber sensors, such as reflection mode (A) and transmission mode (B). (Elosua et al., 2017). Figure under Creative Commons license.

From Elosua, C., Arregui, F. J., Del Villar, I., Ruiz-Zamarreño, C., Corres, J. M., Bariain, C., Goicoechea, J., Hernaez, M., Rivero, P. J., Socorro, A. B., Urrutia, A., Sanchez, P., Zubiate, P., Lopez-Torres, D., De Acha, N., Ascorbe, J., Ozcariz, A., & Matias, I. (2017). Micro and nanostructured materials for the development of optical fibre sensors. Sensors, *17(10). Available from https://doi.org/10.3390/s17102312.*

optical signal based on the changes induced by the analyte, facilitating the conversion of the physical or chemical measurement into an optical signal.

Polymers are extensively utilized in optical fiber chemical sensors, serving as both optode constituents and materials for the optical fibers themselves. Polymers can show three different functions in an optical fiber sensor: they can act as a solid support containing the chemical transducer, they can act as the selective element, or they can be the chemical transducer.

Undoubtedly, the most critical component of an optical fiber sensor is the optode. It must possess the necessary sensitivity and accuracy required for the specific application while also demonstrating a long lifetime, ease of manufacturing, and seamless connection with the optical link. The optode is responsible for modulating one of the optical properties of the light carried by the fiber, such as intensity, phase, or wavelength, in response to the analyte under study. This modulation of light can occur either directly or indirectly (Baldini & Bracci, 2000).

In the case of direct modulation, the analyte itself possesses optical properties that can be detected, such as absorption, fluorescence, or Raman scattering. Creating these types of optodes is relatively straightforward, often requiring an optimized photometric cell connected to the fiber. In contrast, indirect modulation involves an interaction between the analyte and a chemical transducer housed within the optode. In most instances, this transduction process entails a chemical reaction between the analyte and the transducer, causing changes in the optical properties of the latter with variations in analyte concentration. Chemical reactions between reagents and the parameter under investigation may be direct or, on the contrary, undergo a chemical reaction with the reagent, producing a detectable product (Bilro et al., 2012).

There are different types of sensors based on polymeric optical fibers, including:

- *Refractive index change sensors*: These sensors detect changes in the refractive index of the polymeric material coating the optical fiber. This can occur due to changes in temperature, pressure, or the concentration of certain analytes in the environment (Teng et al., 2022).
- *Absorption sensors*: In these sensors, the polymeric material is impregnated with substances that can absorb light at specific wavelengths. When the target analyte is present, light absorption occurs and can be detected and quantified.
- *Emission sensors*: These sensors use polymeric materials that can emit light in response to specific stimuli. When the analyte interacts with the material, it triggers light emission, which can be measured to determine the presence or quantity of the analyte.

Polymeric optical fiber sensors offer numerous advantages, including high sensitivity, the ability to measure parameters in harsh or hard-to-reach environments, and the capability to detect multiple analytes simultaneously. Additionally, polymeric materials are more flexible and cost-effective than other materials used in traditional optical sensors, making them ideal for a wide range of applications in industries such as environmental monitoring, healthcare, and security (Oliveira et al., 2019; Rivero et al., 2018).

1.3.1.2 Electrochemical sensors

Electrochemical sensory polymers respond to the stimulus by altering their electrical properties. They are highly attractive compared to other chemical sensors due to their remarkable detectability, experimental simplicity, and cost-effectiveness. These electrochemical sensors utilize the electrode as the transducer element and have found widespread use in commercial applications since the second half of the 20th century. They offer several advantages, including using electrons for signal acquisition, leading to cleaner analytical applications without waste generation. Additionally, miniaturization enables portable devices with fast analysis and low production costs, making these sensors accessible to various fields, such as commercial glucose sensors. Moreover, the development of electrochemical sensors contributes to advancing other techniques like chromatography detectors, especially when coupled with nanotechnology, enhancing their precision, selectivity, specificity, and sensitivity (Simões & Xavier, 2017).

The strength of electrochemical sensors lies in their superior sensitivity and selectivity, quick response time, operational simplicity, and miniaturization. They operate through a redox reaction that occurs between a target analyte present in an electrolyte and a working electrode. This interaction leads to changes in an electrical signal, and the magnitude of the signal is directly proportional to the concentration of the analyte. These sensors measure the electric current generated by the chemical reactions in the electrochemical system. The interface between the electrode surface, the recognition element, and the target analyte generates an electrical double layer. This potential is then measured after converting the chemical reactions into an electrochemical signal using a recognition element and a sensor transducer. An electrochemical sensor typically comprises two main components: a chemical recognition system for identifying the analyte species and a physicochemical transducer converting chemical interactions into easily detectable electrical signals (Fig. 1.7) (Murthy et al., 2022).

Traditionally, the electrode surface has been considered a robust tool for quantifying analytes in electrochemical sensors. However, bare electrodes have certain drawbacks, including the non-reproducibility of surface behaviors, slower kinetics of electrochemical reactions for certain compounds, and the requirement of higher overpotentials. These limitations restrict their capabilities and hinder their effectiveness in achieving analytical objectives. To overcome these challenges, chemical and physical modifications of electrode surfaces have been extensively explored. These modifications provide a versatile approach to tailoring the sensor's properties for specific analytical

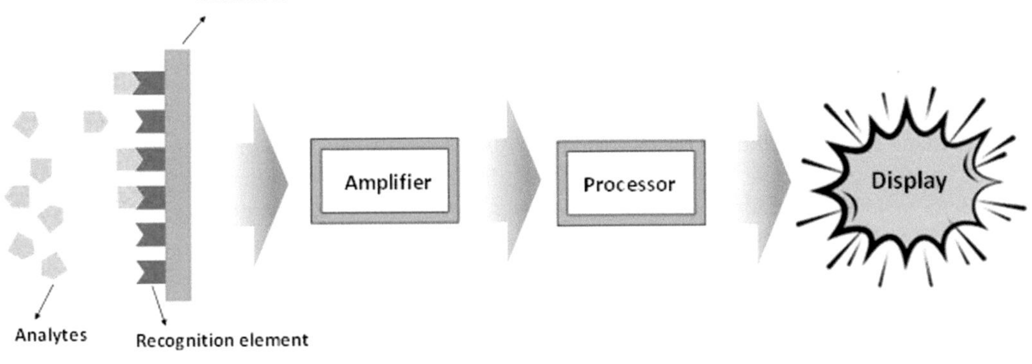

FIGURE 1.7 Electrochemical sensor.

Scheme of the working principle of an electrochemical sensor (Murthy et al., 2022).

requirements, making them invaluable tools in various applications (Chekin et al., 2016; He et al., 2019; Murthy et al., 2022).

Presently, an extensive range of materials are being employed to modify electrochemical sensors, with the primary objective of enhancing their properties. These materials are chosen to improve crucial aspects like electrical conductivity, surface area, and mechanical and chemical stability of the sensors. By carefully selecting and incorporating these materials into the sensor design, researchers can optimize the performance and sensitivity, making electrochemical sensors even more effective in diverse applications (Bonetto et al., 2018). In this context, intrinsically conducting polymers (CPs) stand out as highly relevant and extensively used materials for sensor modification due to their unique physical and chemical properties (Naseri et al., 2018). These polymers offer adjustable architecture, adaptability, versatility, room stability, and remarkable sensitivity to surface changes in their electrochemical activity, even with slight modifications (Samanta et al., 2020).

Typically, electrode surfaces are modified with thin CP films through a combination of adsorptive attraction and low solubility in the working solution, using pre-formed soluble polymers or electrochemical polymerization techniques. Among the CPs, polypyrroles (PPys), polyanilines (PAni), and polythiophenes have garnered significant attention due to their excellent film-forming properties, electrical semiconductivity, high transparency in the visible region, and exceptional thermal and environmental stability (Rattan et al., 2013; Ramanavičius et al., 2006; Runsewe et al., 2019). Furthermore, CPs can form composites with various types of metal and metal oxide nanoparticles, quantum dots, silica nanoparticles, and biomolecules, enabling the combination of desirable properties without sacrificing processability or adding excessive weight (Domínguez-Renedo et al., 2013).

These sensors are classified into different types based on the mode of detection: amperometric/voltammetric (the current generated during a redox reaction is measured), potentiometric (the potential difference between the reference and the working electrode upon interaction is measured), and conductometric (the electrical conductivity change of the polymer upon interaction is measured). They produce electronic outputs in digital signals, facilitating further analysis. Polymeric electrochemical sensors have been employed to quantitatively detect proteins, antibiotics, pesticides, heavy metal ions, bacteria, organic compounds, substances related to the central nervous system, hormones, and more. The versatility and effectiveness of these sensors have opened up promising avenues for sensitive and specific detection across various fields (Su et al., 2018).

1.3.1.3 Mechanical sensors

High mechanical forces acting upon macromolecules result in a decrease in the average molecular mass. This phenomenon is commonly observed during the transformation of polymers and is referred to as mechanical degradation. This process is attributed to the nonspecific cleavage of homolytic bonds, which is followed by indiscriminate side reactions or chain rearrangements. Nonetheless, the concept of mechanical sensory polymers stems from the notion that mechanically induced bond cleavage or rearrangement can be directed toward mechanically labile groups known as mechanophores. These mechanophores are meticulously incorporated to confer intelligent functionalities to the materials. The cleavage or rearrangement events can be detected in various settings, including solution, flow fields, ultrasound exposure, and, in the solid state, typically under conditions of compression or elongation forces (Brown & Craig, 2015; Zhang et al., 2015).

Among mechanical sensors, electromechanical sensors stand out. The operational mechanism of mechanical sensors is rooted in the physical distortion of a structure (such as the bending of a membrane or the suspension of a mass through a beam). This physical distortion is then converted into an electrical response. Various methods exist to quantify the mechanical deformation arising from mechanical forces. These methods encompass the generation of charges (Chen, Liu, et al., 2019; Khan et al., 2017), piezoelectric effects, alterations in electric resistance caused by modifications in resistor geometry or strain (Liu et al., 2017; Sun et al., 2018), shifts in electric capacitance (Chen et al., 2020; Li et al., 2020), adjustments in the resonant frequency of vibrating components within the structure, and variations in optical resonance upon mechanical loading (e.g., pressure, strain) (Okuzaki, 2019; Cochrane and Cayla, 2013). Among them, piezoelectric sensors stand out. Piezoelectric sensors utilize piezoelectric materials that generate an electric charge when subjected to mechanical stresses. These materials can be deformed under applied mechanical force, and this deformation generates an electric signal proportional to the applied force. Piezoelectric sensors find various applications, including pressure, acceleration, and vibration measurements (Smith & Kar-Narayan, 2022). These sensors are normally based on conductive polymers, such as PPy and poly (3,4-ethylenedioxythiophene) (PEDOT), together with other nanomaterials.

What sets mechanical polymeric sensors apart is their inherent flexibility and adaptability. Traditional piezoelectric materials, often ceramics, tend to be rigid and brittle. In contrast, piezoelectric polymers can be fabricated in thin, flexible layers, allowing them to conform to various shapes and surfaces. This flexibility is a game-changer, as it enables the creation of wearable sensors, conformable medical devices, and even the integration of sensors into everyday objects. The sensing mechanism relies on the specific interaction between a solid matrix, such as a supported polymer, molecularly imprinted polymer, or metal-polymer hybrid material, and a target molecule, resulting in a detectable electric charge variation. Piezoelectric behavior is a common property of several polymer families, including fluoropolymers, polyureas, polyamides, polypeptides, polysaccharides, and polyesters (Harrison & Ounaies, 2002; Usher et al., 2018).

Polymeric mechanical sensors find a multitude of applications in the medical domain. For instance, wearable health devices often integrate these sensors to monitor vital signs such as heart rate, respiratory rate, and body movement. The flexibility and comfort of polymeric materials make them an ideal choice for these applications, ensuring a seamless interaction between the sensor and the wearer's body (Wu et al., 2017). Furthermore, advancements in nanotechnology have unlocked new avenues for enhancing polymeric mechanical sensors. By embedding nanoparticles or nanofibers (NFs) within the polymer matrix, heightened sensitivity and accuracy can be achieved. This innovation has been particularly transformative in the field of prosthetics, enabling the creation of responsive artificial limbs that can closely mimic natural movements and provide amputees with an improved quality of life (Kim et al., 2018; Lee et al., 2018). The industrial landscape has also embraced polymeric mechanical sensors with open arms. By detecting minute changes in vibrations or strains, these sensors enable proactive maintenance, preventing costly breakdowns and production downtime (Li et al., 2019).

1.3.1.4 Thermal sensors

Polymer temperature sensors are devices that utilize the properties of polymers to detect and measure temperature changes. These sensors rely on how polymers respond to temperature variations, resulting in measurable changes in their physical or electrical characteristics (Niskanen & Tenhu,

2017). Polymer temperature sensors find applications in various fields, from industrial processes to medical monitoring.

The mechanism behind polymer temperature sensors involves exploiting the temperature-dependent properties of polymers. Polymers can change dimensions, electrical conductivity, or other physical properties in response to temperature fluctuations (García et al., 2022). These changes can be harnessed to create sensors that accurately sense temperature changes and provide corresponding output signals (Seuring & Agarwal, 2013).

There are various types of polymer temperature sensors, each based on different principles:

- *Thermistors*: Thermistors are temperature-sensitive resistors made from semiconductor materials. Polymer-based thermistors employ conductive polymers that exhibit significant changes in electrical resistance with temperature variations. As temperature changes, the electrical resistance of the polymer thermistor changes proportionally, providing a measurable output (Okutani et al., 2022).
- *Optical fiber sensors*: Polymer-coated optical fibers can be used to create temperature sensors based on the principle of changes in optical properties. The polymer's refractive index or fluorescence properties may vary with temperature, which alters the behavior of light passing through the fiber. This change in light behavior can be measured and correlated with temperature changes (Szczerska, 2022).
- *Piezoelectric sensors*: Some polymers, when subjected to temperature changes, exhibit piezoelectric behavior, generating electric charges when mechanically stressed. These piezoelectric polymers can be integrated into sensors that measure temperature-induced mechanical stress and convert it into an electrical signal (Jiang et al., 2014).
- *Conductive polymers*: Certain polymers exhibit changes in electrical conductivity as a response to temperature shifts. These polymers can be integrated into sensors that measure changes in their electrical properties, providing a reliable indication of temperature changes (Nitani et al., 2019).

Polymer temperature sensors offer several advantages, including cost-effectiveness, ease of fabrication, and flexibility in design. They can be tailored to specific temperature ranges and sensitivities by choosing the appropriate polymer and sensor configuration. Additionally, polymer temperature sensors can be used in various environments, including harsh conditions where traditional sensors might be less effective.

1.3.1.5 Mass sensitive sensors

Polymeric mass-sensitive sensors represent a remarkable advancement in sensor technology, offering the capability to detect minute changes in mass with precision and sensitivity. Polymeric mass-sensitive sensors are devices that detect changes in mass by utilizing polymeric materials that respond to the adsorption or interaction of specific molecules or substances on their surface (Hulanicki et al., 1991). These sensors measure alterations in the resonant frequency of the material as mass accumulates, allowing for the detection and quantification of specific analytes, such as gases or liquids. This innovative approach to sensing finds applications in various fields, ranging from environmental monitoring to biomedical research, where the ability to detect and quantify changes in mass plays a critical role.

At the heart of polymeric mass-sensitive sensors is the interaction between the analyte and the polymer material. The polymeric surface is designed to have a high affinity for certain molecules, allowing them to be adsorbed or attached upon contact. This interaction triggers a change in the mass of the sensor's surface, altering its resonant behavior. These polymers can exhibit responsive behaviors similar to those observed in biological systems, such as pH-responsive behavior, light sensitivity, and especially selective binding to biologically specific analytes. The remarkable sensitivity of these sensors lies in their ability to detect even the slightest accumulation of mass, enabling the detection of analytes at trace levels (Janata & Bezegh, 1988).

One of the key components of polymeric mass-sensitive sensors is the resonator. The resonator is often made from a piezoelectric material, such as quartz, which exhibits the piezoelectric effect, which is the ability to generate an electric change in response to mechanical stress. When mass accumulates on the resonator's surface due to analyte adsorption, it changes the resonator's mechanical properties, leading to a shift in its resonant frequency. This shift is precisely measured, allowing for the quantification of the mass change and, subsequently, the concentration of the target analyte (Uludağ et al., 2007).

Polymeric mass-sensitive sensors offer several advantages that make them highly attractive for various applications. First, their sensitivity allows for detecting analytes at extremely low concentrations, which is especially important in fields like environmental monitoring and medical diagnostics. Second, their versatility allows for customization to specific analytes of interest, tailoring the polymer's surface chemistry to promote selective interactions. This customization is performed by incorporating specific receptors in the polymer structure, frequently carried out using molecularly imprinted polymers (MIPs; which will be defined later in this chapter). This selectivity ensures accurate and reliable detection, even in complex sample matrices. Mass differences are measured using quartz crystal microbalances (QCMs), which are devices that measure changes in mass at the nanogram level by utilizing the piezoelectric properties of quartz crystals (Brimo & Serdaroğlu, 2021).

1.3.2 Classification of sensory polymers based on their structure

The composition and structure of sensory polymers play a crucial role in determining their sensing capabilities and responsiveness to specific stimuli. Different polymer compositions and structures can show different transduction mechanisms, leading to various sensory polymer systems. Table 1.1 summarizes the properties of different polymer structures. By tailoring the polymer's composition and design, researchers can create sensory polymers with unique properties, making them versatile tools for various sensing applications. The next sections provide further information about the compositions of sensory polymers.

1.3.2.1 Sensors based on conjugated polymers

Conjugated polymers are composed of repeating monomer units that are polymerized by electrochemical or chemical polymerization. They are a class of organic polymers with alternating single and multiple bonds along their polymer chain, creating a system of delocalized π-electrons. This arrangement of conjugated bonds allows for the movement of electrons across the polymer structure, giving rise to unique electronic and optical properties. Conjugated polymers exhibit semiconducting behavior, meaning they can conduct electricity to some extent but not as effectively as metals. Conjugated polymers, in general, are wide bandgap materials exhibiting semiconducting to

Table 1.1 Summary of different polymer structures for sensory polymers.

Sensory polymer	Description
Conjugated polymer sensors	They are characterized by their extended π-conjugated systems, which impart unique electronic properties. These polymers can exhibit changes in electrical conductivity, optical absorption, or emission in response to specific stimuli, making them valuable for various sensor applications.
Molecularly imprinted polymer (MIP) sensors	They are designed to have specific binding sites, or recognition cavities, for a particular target molecule. These polymers exhibit high selectivity toward the target analyte, making them ideal for sensor applications that require precise molecular recognition.
Biosensors polymers	They are designed to mimic natural systems and processes. These polymers can exhibit responsive behaviors similar to those observed in biological systems, such as pH-responsive behavior, light sensitivity, and especially selective binding to biologically specific analytes.
Nanomaterials based sensors	These systems incorporate nanomaterials, such as metal nanoparticles or carbon nanomaterials, into the polymer matrix. The presence of nanomaterials can enhance the sensitivity, selectivity, and responsiveness of the sensor, leading to improved sensing performance.
Polymer composite sensors	They are a type of sensory material that combines polymer matrices with one or more filler materials, often referred to as reinforcements or additives. These filler materials can be in the form of particles, fibers, microspheres, or other components. Incorporating these fillers into the polymer matrix enhances the sensor's properties, making it more sensitive, selective, and responsive to specific stimuli.
Electrospun polymer sensors	Electrospun polymer sensors are innovative devices that utilize electrospinning techniques to fabricate nanoscale polymer fibers. These sensors possess a high surface area-to-volume ratio, enabling sensitive detection of various physical and chemical changes. They find applications in diverse fields such as healthcare, environmental monitoring, and electronics due to their flexibility in design and exceptional sensitivity. The controlled fabrication process allows for tailoring their properties to specific sensing needs.
Sensors arrays	They consist of an assembly of multiple sensors based on polymeric materials, each designed to respond to specific stimuli. These arrays are used to detect and analyze complex mixtures of analytes, providing a more comprehensive and discriminative sensing capability than single-sensor systems. By combining sensors with varying selectivities, polymeric sensory arrays can achieve pattern recognition, allowing for the identification and quantification of multiple analytes simultaneously.

insulating conductivity levels in their pristine neutral state (Samuel et al., 1997; Smilowitz et al., 1993). They possess a direct bandgap, which allows for efficient absorption or emission of light at the band edge. Depending on the specific conjugated polymer, strong luminescence can be observed, which is often associated with the delocalization and polarization of the electronic structure.

The continuous overlapping of p-orbitals along the polymer backbone and its interaction with electron acceptors or donors make them materials with interesting optoelectronic properties (Dai et al., 2001). Their properties can be tuned through chemical modification and structural design, allowing for the development of materials with tailored functionality (Dai et al., 2002). Therefore, they have been extensively studied and utilized in various fields, including organic electronics, optoelectronics, photovoltaics (solar cells), sensors, and light-emitting devices.

As a result, sensors based on conjugated polymers have gained significant attention due to their tunable electrical and optical properties, chemical stability, and facile processability. These sensors operate on the transduction principle, wherein changes in a physical or chemical parameter are converted into an electrical or optical signal. The bandgap of conjugated polymers can be tailored by modifying their chemical structure, allowing for fine-tuning of their response to specific analytes, by modifying the polymeric chain length, the attached functional groups, the pH of the solution, doping, different oxidation states, etc (McQuade et al., 2000).

In sensory devices, conductive polymers can be part of the sensing element or the immobilization matrix. The main objective of fabricating conjugated polymer sensors is to detect a measurable signal. These polymers can be used as receptors for the binding phenomenon with the target analyte since this interaction (either covalent or physical) modifies the properties (electrostatic, optical, or chemical) of the conjugated polymer.

Conductive polymer-based sensors generate responses that can be analyzed using various techniques such as voltammetry, amperometry, potentiometry, impedometry, and conductometry. These methods are intuitive as they allow for a straightforward correlation between the electrical response and the intrinsic electrical conductivity of these sensor polymers. However, in addition to their electrical responses, these polymers can also produce visual responses, including colorimetric and fluorometric signals. Even in these cases, the visual response is a direct consequence of the conductive nature of these polymers (García et al., 2022).

Therefore, different signals can be detected with conductive polymers, developing different transduction modes, which can be classified as conjugated polymer sensors with electrical or optical transducers (Yoon, 2013).

- *Electrical transduction*: Conjugated polymers exhibit a change in electrical conductivity upon exposure to analytes. The interaction between the polymer chains and the analyte molecules alters the electronic structure, resulting in a change in conductivity. This change can be detected by measuring the resistance or current flow through the polymer, enabling the quantification of analyte concentration. This conductance modulation follows the next mechanisms: The analyte of interest either adsorbs or absorbs into the conjugated polymer film. Depending on the interaction between the analyte and the polymer, charge transfer or doping processes may occur. This can lead to an increase or decrease in the charge carrier concentration within the polymer. The change in charge carrier concentration results in a variation in the electrical conductivity of the polymer film. The electrical response of the sensor is measured and used to determine the concentration or identity of the analyte (Le et al., 2017; Verma et al., 2023). Field-effect transistors (FETs) based on conjugated polymers are commonly used for sensor conductance modulation. In FET-based sensors, the gate voltage is modulated by the charge carrier concentration changes in the polymer, leading to measurable variations in the electrical current.
- *Optical transduction*: Optical-based sensing relies on the alteration of the emission properties of conjugated polymers upon analyte interaction. Conjugated polymers often exhibit strong optical absorption in the visible or near-infrared regions of the electromagnetic spectrum. The energy levels of the polymer's excited states are modified by the analyte, leading to changes in

absorbance or emission intensity, wavelength, or lifetime. By monitoring these changes, the presence and concentration of the analyte can be determined. Fluorescence-based sensing offers several advantages, such as high sensitivity, selectivity, and the potential for real-time monitoring. Optical modulation occurs when the energy bands of the polymer change upon absorption or interaction with the analytes. As a result, the polymer's absorption or emission spectrum is modified, which can be measured and correlated with the presence or concentration of the analyte. Optical modulation in polymer sensors is valuable in various applications, including chemical sensing, fluorescence-based assays, and optoelectronic devices (Chen et al., 2021; Santos & Fahari, 2014).

Conductance and optical modulation offer unique advantages for sensing applications based on conjugated polymers. The choice between these mechanisms depends on the specific requirements of the sensor, the target analyte, and the detection method employed. Additionally, there is a third transduction mechanism that is less relevant in the field of sensory polymers, which is gravimetrical transduction. The gravimetric approach capitalizes on the alteration in weight of a conducting polymer due to its interaction with an analyte. Even slight fluctuations in the polymer's weight can be effectively tracked using a quartz crystal microbalance (Yoon, 2013).

To achieve reliable and efficient sensing performance, careful consideration must be given to the design and fabrication of conjugated polymer sensors. Factors such as polymer selection, film morphology, sensing architecture, and signal readout techniques significantly influence sensor performance. The best-known conductive polymers are PPy, PAni, and PEDOT, together with polyfluorene, poly(*p*-phenylene ethynylene), and polyacetylene derivatives (Fig. 1.8) (García et al., 2022; Namsheer and Raut, 2021). Nevertheless, many conjugated polymers with functional groups, synthetic receptors, or bio-based receptors (polyalkyl ether, crown ether, pyridyl-based ligands, protein ligands, redox-active enzymes, etc.) are also used to prepare conjugated-polymer-based sensors (McQuade et al., 2000).

Conjugated polymer sensors find application in diverse fields, including healthcare, environmental monitoring, food safety, and industrial process control. Their versatility and tunability make them suitable for detecting gases, volatile organic compounds (VOCs), heavy metals, biomolecules, pH, temperature, and humidity. Their potential integration into wearable devices, Internet of Things systems, and point-of-care diagnostics opens up new avenues for personalized health monitoring and rapid, on-site analysis (Anantha-Iyengar et al., 2019).

1.3.2.2 Molecularly imprinted polymers

Various biological recognition elements, such as microorganisms, antibodies, enzymes, tissues, DNA probes, and cells, are commonly used in biosensor technology due to their high specificity. These receptors interact with transducers to convert recognition events in complex samples. For instance, immunosensors employ antibodies as recognition elements where target analytes bind to specific antibodies. However, the practical application of such devices is challenging due to the complex protocols and the instability of biomolecules.

To address these limitations, developing synthetic recognition receptors is essential (Ye & Haupt, 2004). Molecular imprinting technology offers a promising approach by using template polymerization techniques to create artificial recognition systems. These MIPs can selectively bind target molecules, mimicking the affinity and specificity of natural receptors. This strategy

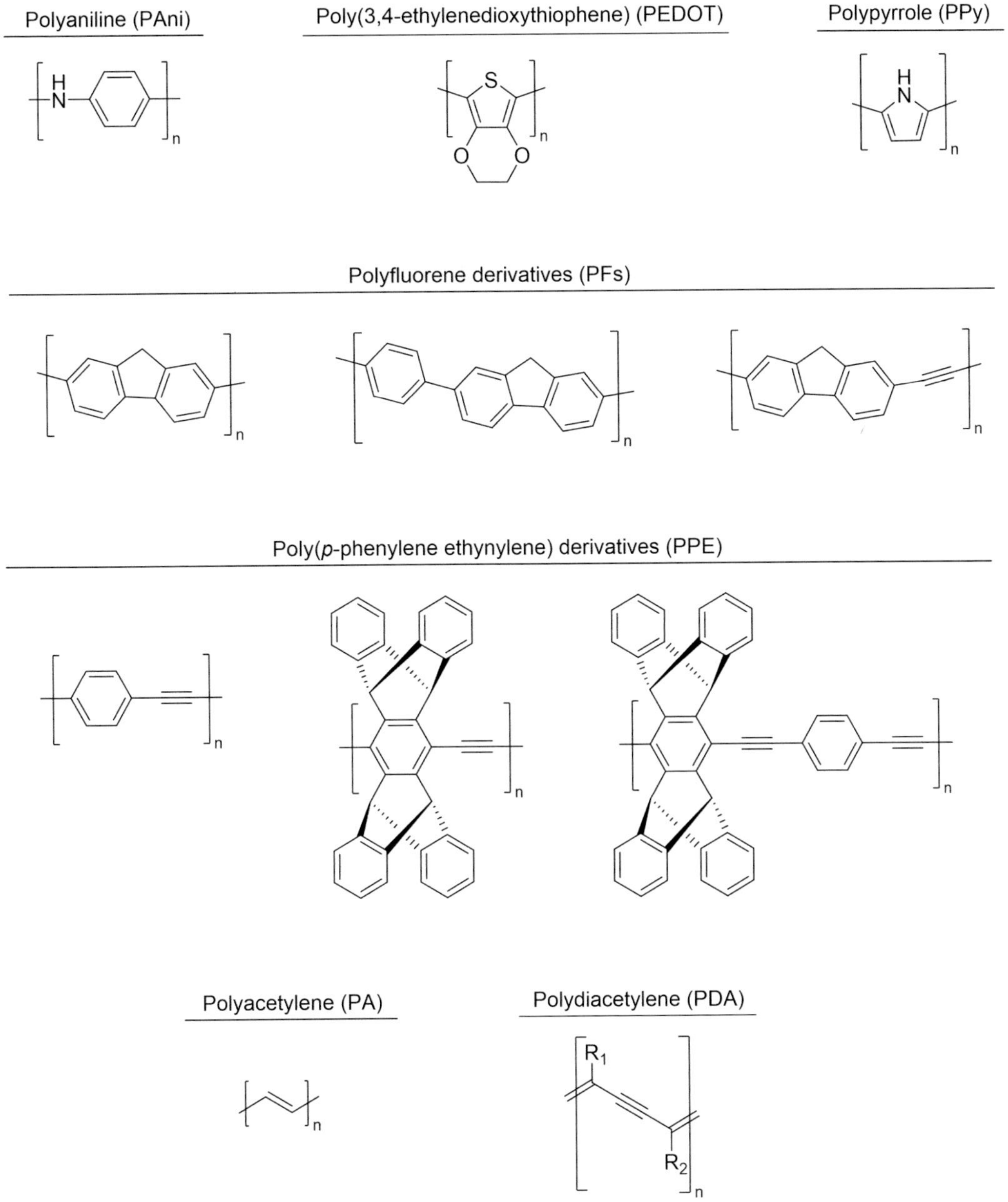

FIGURE 1.8 Conductive polymers.

Structure of common conductive polymers.

provides a synthetic route for developing artificial recognition systems and overcomes the limitations of biological recognition elements. These synthetic polymers are designed to have specific cavities or binding sites that can mimic the shape, size, and functional groups of the target analyte. As a result, MIPs offer several advantages over traditional sensing materials, including high selectivity, stability, and cost-effectiveness, as well as compatibility with sterilization processes, making them promising candidates for use in biosensors and other chemical or biochemical applications. What is more, their ability to mimic the binding properties of natural receptors opens up new possibilities and offers a versatile platform for creating specific binding sites for various templates, including amino acids, proteins, enzymes, hormones, antibodies, nucleic acids, bacteria, viruses, drugs, metal ions, toxins, antibiotics, pesticides, and more. As research and development in the field continue to progress, MIPs are expected to play an increasingly significant role in addressing current challenges in environmental monitoring, healthcare diagnostics, and other sensing applications (Uygun et al., 2015).

Regarding their preparation, the molecular imprinting technique involves the polymerization of a functional monomer and a cross-linker around a template molecule. Initially, a complex is formed between the template molecule and the functional monomer. Subsequently, polymerization takes place around this complex with the addition of an initiator and a cross-linker. After the polymerization, the template molecule is extracted, resulting in the creation of three-dimensional cavities within the polymer matrix. These cavities possess specific recognition capabilities tailored to match the template molecules (Turiel & Esteban, 2020). Achieving complete polymerization requires the use of optimal ratios of template molecules, functional monomers, cross-linkers, and initiators. It can be carried out using various methods. Fig. 1.9 illustrates several approaches, including the utilization of reversible covalent bonds (A), the attachment of polymerizable binding groups that are activated for non-covalent interactions (B), the involvement of electrostatic interactions (C), the presence of hydrophobic or van der Waals interactions (D), and the coordination with a metal center (E). These interactions are based on the complementary functional groups of the template molecules (A–E) (Saylan et al., 2019).

MIPs can be synthesized in different formats, depending on the polymerization technique. Possible morphology designs include bulk monoliths, particles, films, membranes, fibers, tubes, and nanostructured MIPs (Sellergren & Hall, 2001). A variety of functional monomers, such as methacrylic acid, 4-vinylpyrrolidine, acrylamide, and hydroxyethyl methacrylate, can be used to implement binding interactions within the imprinted sites. However, it is important to carefully select cross-linkers and solvents to ensure that the resulting MIPs have the necessary mechanical stability and processability. There are several techniques for synthesizing MIPs (Kadhem et al., 2021; Shen et al., 2023; Yang & Shen, 2022), each with its advantages and limitations.

- *Bulk polymerization*: This method involves polymerizing a monomer mixture in the presence of the target molecule. The target molecule acts as a template around which the polymer matrix forms. After polymerization, the target molecule is removed, leaving behind cavities that match its shape and size.
- *Surface imprinting*: In this approach, the target molecule is immobilized on a solid surface, such as a glass slide or silica particle. A polymer coating is then applied, and subsequent polymerization occurs. The polymer coating adopts the shape of the immobilized molecule, resulting in surface-imprinted sites that can selectively bind the target analyte. Since imprinting

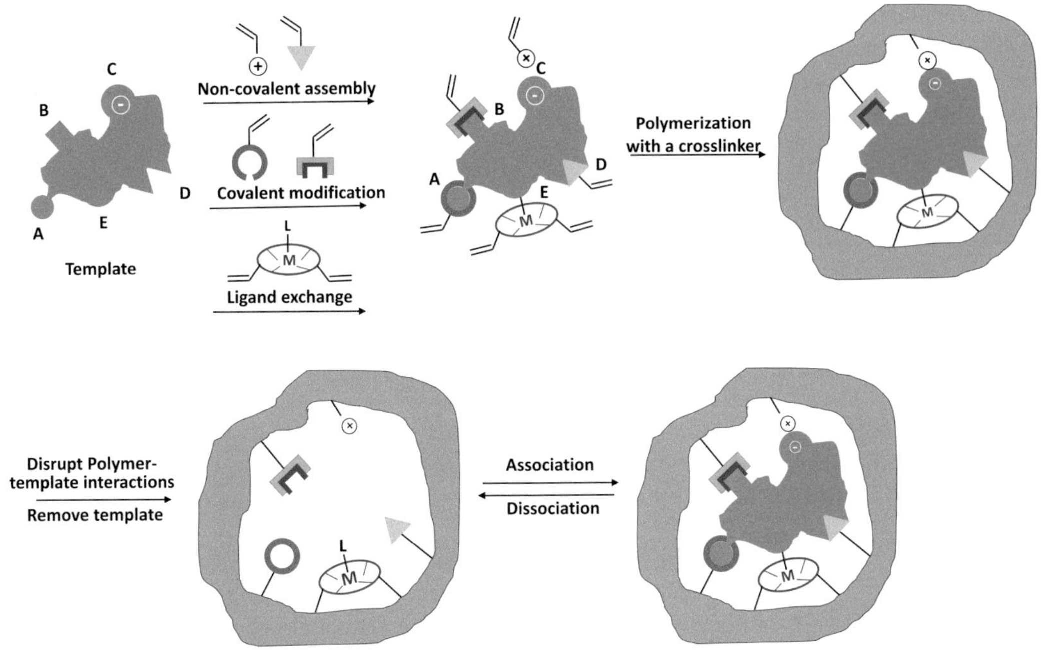

FIGURE 1.9 Molecularly imprinted polymers.

Preparation technique and principles of molecular imprinting recognition (Saylan et al., 2019).

occurs exclusively on the surface, a lesser quantity of template molecules suffices; however, sensitivity diminishes due to the decrease in available imprinting sites.

- *Emulsion polymerization*: This method involves dispersing the template molecule in an aqueous solution along with monomers, cross-linkers, and initiators. The mixture is then emulsified, and polymerization occurs within the droplets. The resulting MIP particles exhibit imprinted sites that are specific to the target molecule.

MIPs possess several key characteristics related to their preparation procedure that make them well-suited for sensory applications (Sajini & Mathew, 2021; Yan & Row, 2006). First of all, one of their major advantages is their selectivity toward target molecules. Through their careful design and synthesis, MIPs can be tailored to recognize specific analytes, even in complex sample matrices, due to their own preparation method starting from a template molecule. This selectivity allows for accurate and reliable detection in various sensing applications. Second, MIPs are characterized by their high stability and resistance to environmental factors such as temperature, pH, and solvents. This stability ensures their durability and long-term performance, making them suitable for both laboratory and field applications. Finally, the versatility of MIPs is a key feature since they can be designed for a wide range of analytes, including small organic molecules, peptides, proteins, and even metal ions.

The use of MIPs as sensory materials was first proposed in the first decade of this century by McCluskey et al. (2007), and since then, they have shown great potential for use in a wide range of fields, including environmental monitoring, food safety, biomedical diagnostics, and pharmaceutical analysis. MIPs can be utilized as sensing elements in gas sensors, electronic noses, and optical sensors to detect VOCs and hazardous gases. Their selectivity and stability make them valuable for environmental monitoring and industrial safety applications. Moreover, MIPs can be employed for the detection and removal of pollutants, such as pesticides, heavy metals, and aromatic compounds, from environmental samples, including soil, water, and air (Cao et al., 2012; Caro et al., 2006; He et al., 2007; Sanagi et al., 2013). Also, MIPs have been developed to detect food contaminants, including mycotoxins, allergens, and foodborne pathogens. These sensors provide rapid and reliable detection, contributing to improved food safety and quality control (Wang et al., 2016; Yang et al., 2014). Additionally, MIP-based sensors have shown promise in detecting biomarkers associated with various diseases, including cancer, cardiovascular disorders, and infectious diseases. The high selectivity of MIPs allows for the sensitive and specific detection of these biomarkers in patient samples (Ji et al., 2015; Xue et al., 2013; Zheng et al., 2018). Intended for biomedical applications, MIPs can also be incorporated into drug delivery systems to enhance drug targeting and controlled release. MIPs designed to recognize specific drugs or therapeutic molecules can improve drug efficacy, minimize side effects, and enable site-specific drug delivery (Kadhem et al., 2021; Kurczewska et al., 2017; Li et al., 2011).

While MIPs have shown great potential in sensory applications, further research is still required to overcome some challenges. Enhancing the sensitivity of MIPs, improving their response time, and expanding their applicability to a broader range of analytes are areas of ongoing investigation. Additionally, efforts are being made to integrate MIPs with advanced sensing platforms, such as microfluidics and nanomaterials, to enhance their performance and enable real-time monitoring.

1.3.2.3 Biosensors

The utilization of environmental triggers within biological environments, such as chemical species, pH, light, and temperature, constitutes an effort to replicate the responsive behavior of biological systems in smart polymeric materials. In the detection of target molecules by chemical sensors, the initial step involves the selective interaction between the sensor and the analyte, revealing the sensor's recognition and affinity for the analyte. When this recognition is based on biological or biochemical interactions, the term "biosensor" is employed instead of "chemosensor." An example of this is the interaction between an antigen and its corresponding antibody (Amid et al., 2019). In the field of biosensing, polymer-based biosensors are especially relevant due to the benefits they bring, including specificity, versatility, ease of design, and cost-effectiveness. Among these applications, the development of biosensors involves emulating nature's recognition processes, where the interaction between specific segments of a polymeric structure acting as a host or receptor and a target species elicits detectable changes. These changes can be measured, recorded, and analyzed to gather information about the presence and concentration of the target substance (Amid et al., 2019).

In this field, polymer biosensors have emerged as a cutting-edge field in sensing technology, combining the unique properties of polymers with the specificity and sensitivity of biomolecules. These biosensors utilize various types of polymers, including natural and synthetic polymers, to detect and quantify analytes of biological interest. With their remarkable versatility, ease of fabrication, and high biocompatibility, polymer biosensors have found widespread application in areas such as healthcare, environmental monitoring, and food safety (Davis & Higson, 2007; Spychalska et al., 2020).

Depending on the level of interaction, biosensors can be categorized into three generations. In the first generation, the biocatalyst is either tethered to or enclosed within a membrane, which is then incorporated into the transducer's surface. The second generation of biosensors involves the covalent union or adsorption of the biologically active component to the transducer's surface, enabling the removal of a semi-permeable membrane. In third-generation biosensors, it entails the direct connection of the biocatalyst to an electronic apparatus that transforms and magnifies the signal. Biosensors founded on CPs fall into this classification.

Biological constituents that operate as biochemical transducers can encompass enzymes, tissues, bacteria, yeast, antibodies/antigens, liposomes, and organelles (Foulds & Lowe, 1988; Sadik & Wallace, 1993). Inside a biosensor, the sensing biomolecule that is integrated exhibits a remarkable degree of specificity but is susceptible to extreme conditions of temperature, pH, and ionic strength (Fig. 1.10) (Karyakin et al., 1999). Additionally, most biological substances, such as enzymes, receptors, antibodies, cells, and so on, have a relatively brief lifespan when in a solution state. Thus, they need to be affixed to an appropriate matrix. The immobilization of the biological element is a means of shielding it from environmental conditions, although this can lead to a reduction in enzyme activity (Evtugyn, 1998; Schuhmann et al., 1992). The efficacy of immobilized molecules is influenced by factors like surface area, porosity, the hydrophilic nature of the immobilizing matrix, reaction conditions, and the chosen immobilization method.

Numerous techniques, including physical adsorption, cross-linking, gel entrapment, and covalent coupling, have been employed to anchor biological molecules into carrier materials. A range of

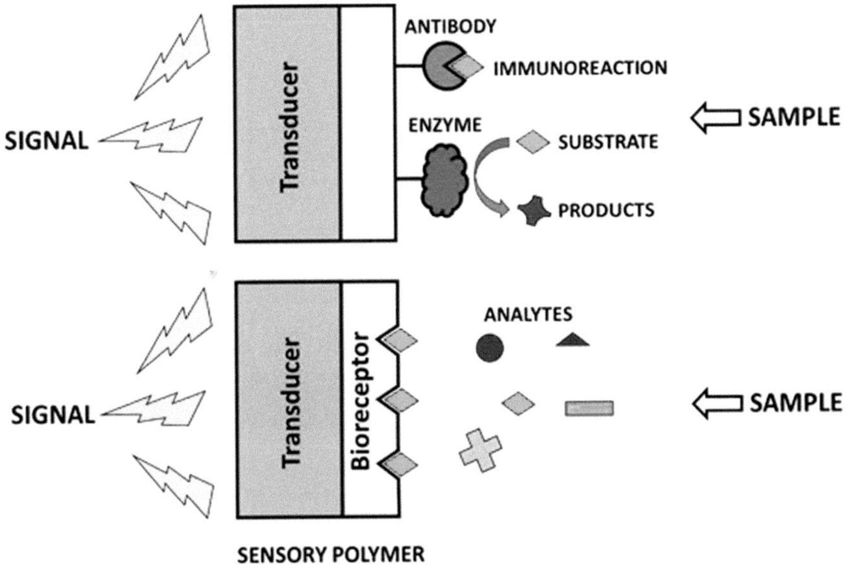

FIGURE 1.10

Biosensors general scheme of biosensors (Girigoswami et al., 2021).

matrices, such as membranes, gels, carbon, graphite, silica, and polymeric films, have been utilized for immobilizing enzymes (Alvarez-Malmagro et al., 2020; Ikeda et al., 1985; Lakard, 2020). Consequently, there's a pressing necessity to develop electrodes that harmonize with the biological component, facilitating swift electron transfer at the electrode surface. What is more, surface modification techniques, such as plasma treatment, grafting, and self-assembled monolayers, have been employed to modify the surfaces of polymer biosensors, thus facilitating the immobilization of biomolecules and enhancing the stability, selectivity, and sensitivity of the biosensor (Ramanaviciene & Plikusiene, 2021).

Regarding the matrix of the biosensor, it can be made of different types of polymers, including natural polymers (such as proteins, peptides, and nucleic acids) and synthetic polymers (such as hydrogels, conductive polymers, and MIPs). The choice of polymer will depend on the target analyte and the desired sensing mechanism (Amid et al., 2019).

Immobilizing biomolecules such as antibodies, antibody fragments, and enzymes is essential for detecting other biomolecules, such as antigens or peptide substrates. This mimicry of biological systems is harnessed across fields like chemistry, biology, biotechnology, and materials science. Immobilization techniques play a crucial role in preparing systems capable of detecting specific biomolecules related to diseases, such as antigens. Through the immobilization of proteins (antibody, antibody fragment, or enzyme) on solid supports, detection systems that recognize other proteins (antigen, peptide, etc.) can be created. The effective immobilization of biomolecules on solid supports ensures the proper construction and performance of biosensors. Preferable orientation prevents denaturation and maintains the active site's accessibility to maintain affinity for target molecules (Shen et al., 2017; Lyu et al., 2021).

This procedure can be conducted physically (adsorption, electrostatic interactions, and affinity) or chemically (via covalent bonding) (Fig. 1.11). Suitable techniques for immobilization include conventional physical adsorption utilizing electrostatic forces or hydrophobic interactions, which

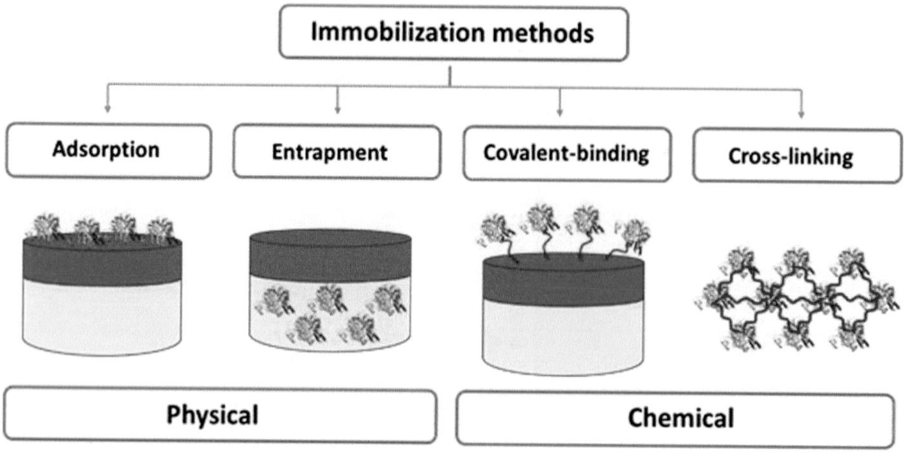

FIGURE 1.11 Enzymes immobilization.

Methods for immobilization of enzymes (Imam et al., 2021).

might exhibit instability and diminish sensitivity. Covalent fixation by means of functional groups within the biomolecule to modified substrates is irreversible and efficient, but it lacks precise orientation. To illustrate, physical approaches do not alter the structure of the immobilized biomolecule; they merely retain it through electrostatic interactions. This is a notable benefit, given that modifying the biomolecule's structure could result in denaturation or deactivation. Nevertheless, the interactions between the polymer and the protein are less robust than in the context of chemical bonding, occasionally causing the protein to detach from the substrate. Moreover, within physically linked encapsulation, the interaction between the attached antibodies and the targets (antigens) might encounter greater hindrance (Imam et al., 2021; Lyu et al., 2021).

On the other hand, chemical techniques form robust and dependable covalent bonds. However, depending on where the antibody binds to the substrate and the specificity of these bonds, the antibody might experience a loss of functionality. The interaction between antibodies and antigens is so finely tuned and delicate that any alteration in the chemical configuration of either component could lead to system inactivation. This is precisely why this kind of polymer must be meticulously designed for each application.

The design of biosensors encompasses the intriguing junction of chemistry and biology, resulting in many opportunities and applications. In the healthcare sector and clinical diagnosis, they enable the detection of biomarkers for diseases such as cancer, cardiovascular disorders, and infectious diseases. Polymer-based biosensors offer rapid and sensitive analysis, aiding in early diagnosis, personalized medicine, and point-of-care testing. In this context, another relevant issue is referred to as wearable and implantable sensors. These sensors offer continuous monitoring of physiological parameters, such as glucose levels, heart rate, and drug concentrations, enhancing patient care and disease management (Gupta et al., 2020).

Regarding environmental monitoring, polymer biosensors are invaluable tools for environmental monitoring, enabling the detection of pollutants, toxins, and pathogens in air, water, and soil. They contribute to assessing environmental quality, pollution control, and the monitoring of water resources (Madej-Kiełbik et al., 2022). Within the food safety and quality control industry, polymer biosensors find applications in detecting foodborne pathogens, allergens, and contaminants. These biosensors ensure food safety and quality control, allowing for rapid, on-site analysis and real-time monitoring of food production processes (Madej-Kiełbik et al., 2022; Vallejos et al., 2022). Last, in industrial processes, they can facilitate the optimization of fermentation, cell culture, and enzyme reactions. They provide real-time measurement of key parameters, such as metabolites, nutrients, and pH, aiding in process control and optimization (Kišija et al., 2020).

Unfortunately, there are also certain drawbacks associated with the design and application of biosensors. On occasion, the biological entities required can be prohibitively expensive or challenging to extract with sufficient purity. When these entities are immobilized, their activity might diminish, and the presence of diverse chemical components within the test solution can further lead to activity loss (for instance, heavy metals can readily disrupt enzyme function). In biological samples such as blood or saliva, there might be solutes that exhibit electrochemical activity and consequently interfere with the accurate detection of the target substance. In this sense, in physiological fluids like blood, there could be various components that adhere to the sensor's surface, resulting in fouling and subsequent loss of sensor response (Otero & Magner, 2020; Sassolas et al., 2012).

1.3.2.4 Hybrid, nanoparticle-based, and composite polymer sensors

Hybrid materials represent composite structures comprising two distinct constituents operating at the nanoscale or molecular level. Typically, this pairing involves integrating an inorganic element alongside an organic counterpart. This distinctive characteristic sets them apart from conventional composites, where constituents are amalgamated on a macroscopic scale ranging from micrometers to millimeters in the form of nanoparticles, fibers, flakes, or other microstructures (Ates et al., 2017). In both cases, as sensory systems, their sensing properties are enhanced due to an increased surface area and a more homogeneous distribution of the functional components.

This amalgamation often entails the combination of diverse polymers, resulting in a composite material. Typically, hybrid and composite polymers involve the blending of a polymer with either inorganic substances like metals, ceramics, or nanoparticles or with other organic materials such as carbon nanomaterials or other polymers, such as fibers. These materials have emerged as a relevant field of research in sensing technologies since the synergy achieved through this blending transcends the individual characteristics of each constituent, ushering forth a new set of properties that often exceed what is achievable through traditional polymer or material combinations. By thoughtfully selecting and combining different elements, researchers harness the strengths of each, mitigating weaknesses and crafting materials with a tailored balance of attributes (Andre et al., 2018).

One fascinating aspect of this type of polymer research is the diverse range of constituents that can be integrated. Inorganic nanoparticles, such as metallic or semiconductor nanoparticles, lend electrical, optical, or catalytic functionalities to polymers. Meanwhile, organic-inorganic hybrids further extend the material palette, resulting in matrices that combine the elasticity and processability of polymers with the structural integrity and properties of inorganic materials. The incorporation of reinforcing agents, be they carbon nanotubes, glass fibers, or nanoclays, endows the resulting composites with heightened mechanical strength, stiffness, and improved resistance to wear and fatigue, rendering composite polymers well-suited for applications requiring lightweight yet durable materials, like automotive components or structural elements in civil engineering.

Hybrid materials can be crafted through a diverse array of methods, each tailored to suit specific applications and desired properties. Some techniques for fashioning hybrid nanomaterials are layer-by-layer self-assembly and electrospinning methodologies, solution casting, emulsion polymerization, sol-gel processes, and covalent linkage techniques (Ghosh et al., 2015; Pang et al., 2014; Pang et al., 2016; Patil et al., 2015; Paulis & Asua, 2016; Pomogailo, 2005; Rubio-Aguinaga et al., 2023; Trigo-López et al., 2018).

On the other hand, the design and manufacturing of composite polymers are deeply intertwined with cutting-edge technologies. Advanced fabrication techniques such as 3D printing, electrospinning, and melt blending enable precise control over the spatial distribution of components, leading to intricate microstructures with optimized properties.

Both hybrid and composite polymers need to show intimate interaction among the constituent materials to result in synergistic effects that ultimately enhance the end properties of the materials (Tian et al., 2015). This is achieved through covalent or other strong secondary interactions between the polymer and the other component. In general, hybrid polymer-based sensors can be classified into two categories depending on the strength of the interaction between their different components. Hybrids require strong interactions, like covalent bonding or hydrogen bonding between components, whereas composites present weak bonding, such as van der Waals forces or weak electrostatic interactions.

Regarding hybrid polymer-based sensors, hybrids in which the receptors are inorganic species are of remarkable relevance. These materials require the presence of inorganic species, such as heavy metal complexes, in the chemical structure of the polymer. In these cases, the polymeric material acts as a support for the sensing component, making the material easy to handle and allowing a perfect interaction between the analytes and the sensory species, even in non-compatible media with organic functional components, such as aqueous or gas phases (Pablos et al., 2015; Trigo-López et al., 2018; Vallejos et al., 2012).

Due to these combined properties, hybrid materials have applications in numerous fields. In environmental monitoring, they allow for the detection of pollutants, toxins, and hazardous substances, enabling real-time monitoring of air quality, water pollution, or soil contamination, among others. In biomedicine, they offer a non-invasive and rapid method for detecting biomarkers, pathogens, and different chemical species, finding applications in disease diagnosis, drug discovery, testing, and even personalized medicine. Regarding food safety and quality control, they enable the detection of foodborne pathogens, contaminants, adulterants, food storage conditions, or freshness, ensuring the integrity of products.

On the other hand, composites combine polymers and different fillers like nanoparticles, carbon nanotubes, flakes, fibers, or inorganic substances to detect various analytes with enhanced sensing capabilities and weak interactions between their components, providing the materials with improved sensitivity, selectivity, mechanical strength, and stability. Polymers that include fillers with a dimension less than 100 nm are categorized as polymer nanocomposites. These nanocomposites are created using extremely small amounts (less than 5% by volume) of finely dispersed nanofillers. This contrasts with conventional polymer composites, which contain larger fillers in higher amounts (over 60% by volume) and are in the micrometer size range.

Regarding composite polymeric sensors based on nanomaterials, they leverage the unique properties of nanomaterials, such as nanoparticles, nanowires, and nanocomposites, to detect and quantify a wide range of analytes. Their remarkable surface-to-volume ratio, tunable properties, and high surface reactivity due to their small size allow for better interaction with analytes, leading to improved sensing performance and widespread applications (Brett & Oliveira-Brett, 2011). Sensors based on nanomaterials operate on diverse sensing mechanisms, including optical, electrical, electrochemical, and piezoelectric effects.

Nanoparticles ranging from 20 to 200 nm include single crystals, polycrystalline structures, and amorphous substances, exhibiting a variety of morphologies like spheres, cubes, and platelets. Magnetic nanoparticles find applications in targeted drug delivery, highly sensitive disease detection, gene therapy, efficient genetic screening, biochemical sensing, and rapid detoxification processes (Vaseashta, n.d.).

Carbon-based materials are the preferred choice when designing electrochemical sensors due to their exceptional reliability, low residual current, broad potential window, and renewability of the surface. Carbon nanotubes possess a strength 100 times greater than steel while only weighing one-sixth as much. Their conductivity rivals copper's, and they can even exhibit semiconductor behavior (Lines, 2008). Graphene's electrochemical properties include a broad potential window, low charge resistance, and clearly defined redox peaks. Sensors built upon graphene display electrical conductivity even at low temperatures and superior catalytic activity relative to other materials.

With electrochemically active sites uniformly distributed, the active surface of sensors is reliable (Salavagione et al., 2014). Leveraging their two-dimensional structure, these sensors excel at detecting adsorbed molecules, thereby enabling effective molecular detection with minimal electrochemical measurement noise. These sensors can be incorporated into garments, sports equipment, or wearable accessories, allowing for real-time monitoring of vital signs, body motion, or environmental factors.

1.3.2.5 Electrospun polymer sensors

Electrospun polymeric sensors have emerged as a cutting-edge area of research, blending the principles of polymer chemistry and sensor technology. These sensors, fabricated through electrospinning techniques, offer remarkable properties that make them invaluable in various fields.

Electrospinning is a versatile and scalable process that enables the production of ultrafine fibers with diameters ranging from nanometers to micrometers. The technique involves the application of an electric field to a polymer solution or melt, causing a charged jet of the polymer to be ejected from a syringe or nozzle. As the solvent evaporates during flight, solid fibers are deposited onto a collector, forming a nonwoven mat with an extremely high surface-area-to-volume ratio (Chen, Chou, et al., 2019).

The conventional setup for electrospinning involves assembling a configuration comprising key components: a syringe housing a solution containing the chosen polymer, linked to a spinneret (which can be exemplified as a needle or a pipette tip), a syringe pump, a collector connected to a ground (which could take the form of a plate or a rotary device), and a high-voltage power supply (as depicted in Fig. 1.12). This arrangement has the flexibility to be oriented either vertically or horizontally based on requirements. The process of electrospinning hinges on an electrohydrodynamic mechanism. Initially, due to surface tension effects, the polymer solution gathers at the tip of the needle, forming a droplet. Upon introducing a charge into the polymer solution, an electric field

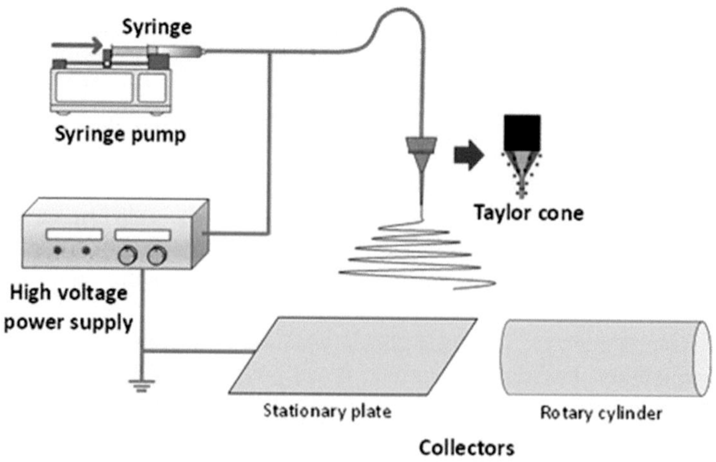

FIGURE 1.12

Electrospinning system (Chen, Chou, et al., 2019).

develops between the needle and the collector. Consequently, the droplet transforms into what is known as a Taylor cone. As the electric force surpasses the surface tension, a charged jet emerges from the cone. Initially, this jet follows a linear trajectory, subsequently undergoing vigorous whipping motions. This intricate dance leads to the solvent's gradual evaporation while the polymer is simultaneously extended and elongated. This progression culminates in the creation of solid fibers characterized by slender diameters. These fibers are then deposited onto the grounded collector, where they solidify (Bhardwaj & Kundu, 2010; Xue et al., 2019).

There are several advantages to be considered with the electrospinning technique. One of them is the possibility of tailoring polymeric sensors. Electrospinning offers the ability to control fiber morphology, diameter, and composition, which allows researchers to design sensors with enhanced sensitivity, selectivity, and response times. By choosing appropriate polymers and incorporating functional additives, the resulting sensors can be tailored to detect specific analytes or stimuli. Another interesting benefit of electrospinning is the enhanced surface area since the electrospun fibers' high surface-area-to-volume ratio is a critical feature that significantly boosts their sensing capabilities. This expanded surface area provides abundant sites for interactions between the target analyte and the sensing material, leading to improved detection limits (Chen et al. 2019). For instance, in gas sensing applications, the large surface area allows for more effective adsorption and interaction with gas molecules, leading to increased sensor responsiveness. Moreover, the versatility of electrospinning extends to the incorporation of functional groups and additives into the polymer matrix. Researchers can introduce specific receptors, such as enzymes or antibodies, onto the fibers' surfaces to impart selectivity for a particular analyte. This tailoring of surface chemistry enables the creation of sensors that can differentiate between various analytes in complex samples, a vital capability in fields like medical diagnostics and environmental monitoring (Medeiros et al., 2022).

There exist two primary methodologies for the functionalization of NFs. The first involves direct integration, wherein novel compounds are introduced into the polymer solution before the electrospinning process. This results in their effective embedding within the resulting material's structure. Conversely, the second approach, termed surface modification, entails the presentation of fresh compounds onto the surface of pre-fabricated NFs (illustrated in Fig. 1.13). The spectrum of molecules available for this modification of NFs is extensive, encompassing entities such as polymers, nanoparticles (including metallic and semiconducting varieties like quantum dots), organic compounds, dyes, carbon nanomaterials, and biomolecules such as enzymes (Balusamy et al., 2019).

The exceptional attributes of NFs, including their substantial surface area and porous configuration, contribute to an elevated capacity to immobilize biomolecules. This is in stark contrast to materials at the microscale. Consequently, the enhanced immobilization capability significantly augments sensor performance parameters like sensitivity and response time (Chen et al. 2019).

Among the advantages of sensors fabricated with electrospinning techniques is the remarkable relevance of the multi-functionality achieved with hybrid sensors. Electrospun polymeric sensors are not limited to single-analyte detection. By combining different polymers or integrating various nanomaterials, it is possible to create hybrid sensors that respond to multiple stimuli or detect multiple analytes simultaneously. This multi-functionality is particularly useful in applications where a comprehensive analysis is required, such as in monitoring water quality or assessing physiological parameters. Another important feature is the inherent flexibility of electrospun fibers, which opens

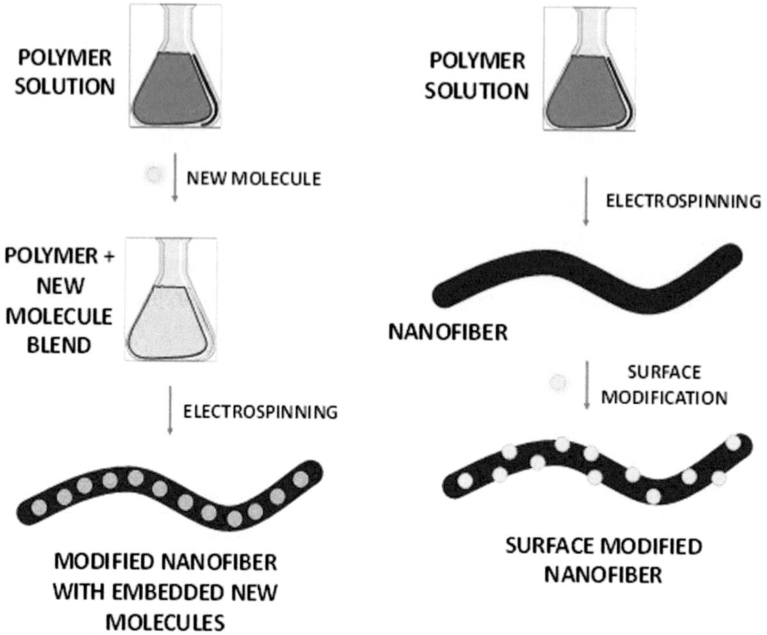

FIGURE 1.13 Functionalization of nanofibers.

Methodologies for the preparation of functionalized electrospun nanofibers (Halicka & Cabaj, 2021).

Adapted from Halicka, K., & Cabaj, J. (2021). Electrospun nanofibers for sensing and biosensing applications—A review. International Journal of Molecular Sciences, *22(12). Available from https://doi.org/10.3390/ijms22126357.*

possibilities for flexible and wearable sensor platforms. These sensors can conform to irregular surfaces, making them suitable for applications such as wearable health monitors and personalized medical devices. Moreover, the lightweight nature of the electrospun mats ensures user comfort, essential for long-term wearable applications (Al-Abduljabbar & Farooq, 2023).

There are additional applications in which NFs generated by electrospinning play an important role, such as humidity sensors, which are indispensable tools for gauging moisture levels in the air. Humidity sensors are found in diverse settings such as homes, industrial facilities, and agricultural operations (Aussawasathien et al., 2005). In this context, gas sensors can also be included since they are critical for identifying and quantifying specific atmospheric gases. Their significance spans safety applications, where they ensure early detection of hazardous gas leaks (Erdem Yilmaz & Erdem, 2020). Other environmental monitoring sensors play a pivotal role in tracking various environmental factors, from monitoring air quality and pollution levels in urban areas to assessing the health of ecosystems (Ding et al., 2010).

1.3.2.6 Sensor arrays

Traditional sensor designs have long relied on the "lock-and-key" concept, offering exceptional selectivity and specificity for detecting particular analytes. Nonetheless, this approach proves inadequate when confronted with the challenge of concurrently detecting multiple analytes in complex media. For example, reactive oxygen species, biothiols, and proteins cannot be distinguished simultaneously in metabolic processes in the human body. Leveraging pattern recognition technologies, sensor arrays exhibit a remarkable capacity to differentiate nuanced alterations arising from a multitude of analytes possessing analogous structures within intricate systems. A collection of sensing units forming a sensor array showcases diverse reactions to distinct target analytes *via* either physical or chemical interactions, enabling both qualitative and quantitative detection capabilities (Lee et al., 2020). The reactions of sensing units toward individual analytes contribute to creating distinct patterns, each akin to a unique fingerprint. These patterns are subsequently subjected to analysis through pattern recognition techniques, such as principal component analysis, hierarchical cluster analysis, linear discriminant analysis, and others. This analytical process facilitates the differentiation and categorization of analytes, enabling effective separation and identification (Fig. 1.14) (Geng et al., 2019).

Polymer sensor arrays have revolutionized the field of polymer-based sensing technologies by offering the ability to detect and differentiate multiple analytes simultaneously. These sensor arrays utilize arrays of individual sensors, each coated or functionalized with different polymers, to create a highly selective and versatile sensing platform. Polymer sensor arrays offer advantages such as enhanced sensitivity and selectivity (Li, Zhu, et al., 2023).

Regarding their configuration, polymer sensor arrays consist of multiple individual sensors arranged in a specific configuration that generates a different response pattern for each analyte. The response can be converted into one or more measurable signals, such as colorimetric, electrical, or fluorescent, which are analyzed by pattern recognition methods. These configurations can include linear arrays, two-dimensional arrays, or three-dimensional arrays, depending on the desired application.

In recent years, there have been important advancements in the fabrication techniques of sensor arrays, such as microfabrication, inkjet printing, and 3D printing (Naware et al., 2003; Sharafeldin et al., 2018; Ye et al., 2022). These advancements have enabled the high-throughput production of

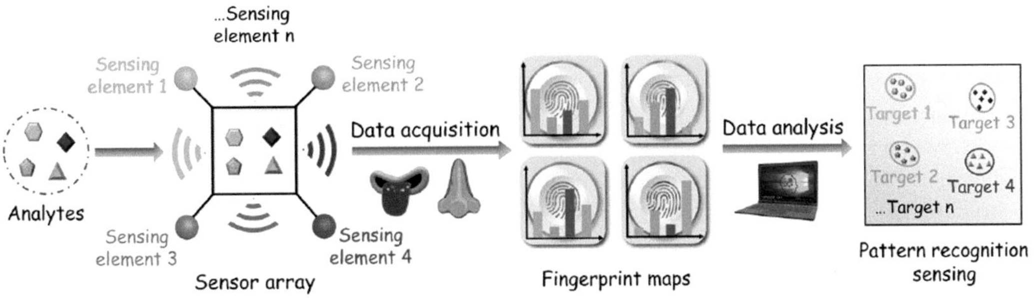

FIGURE 1.14 Sensor array.

Detection mechanism in a sensory array. Reprinted with permission from (Li, Zhu, et al., 2023).

Reprinted with permission from Li, T., Zhu, X., Hai, X., Bi, S., & Zhang, X. (2023). Recent progress in sensor arrays: From construction principles of sensing elements to applications. ACS Sensors, *8(3), 994–1016. Available from https://doi.org/10.1021/acssensors.2c02596.*

polymer sensor arrays. They permit the precise deposition of different polymers onto individual sensors, resulting in sensor arrays with uniform and reproducible sensing properties.

Another significant advantage was achieved with the miniaturization of individual sensors within the array, enabling high-density packing of sensors, leading to increased sensing capacity and improved spatial resolution. Miniaturized polymer sensor arrays find applications in portable devices, point-of-care diagnostics, and wearable technologies (Chiu & Tang, 2013; Hu et al., 2018).

An additional interesting possibility of sensory arrays is the integration of data analysis techniques, such as pattern recognition algorithms and machine learning, which have enhanced their capabilities. Sensor fusion techniques enable combining data from multiple sensors, enhancing discrimination power, and reducing false positives. Advanced data analysis techniques improve the accuracy and reliability of the sensing results (Gabrieli et al., 2021; Li et al. 2023).

1.3.3 Lab-on-a-chip and sensory devices

As previously mentioned, a diverse range of sensory materials, such as molecularly imprinted, hybrid, and conductive polymers, among others, have been extensively studied for their ability to sense, detect, and quantify various target species with a high level of selectivity. These polymers are integrated into advanced sensory devices, catering to environmental, industrial, civil, food safety, health safety, and biomedical applications. These sensory materials are adept at detecting an endless variety of sample targets, including cations and anions, explosives, radionuclides in soil solutions, and gas phases, as well as gases and microorganisms like bacteria and viruses. Moreover, they can also be utilized to detect specific biomarkers in biomedical research.

Sensory devices primarily rely on transducing the interaction between recognition receptors and target species into measurable physical properties. To effectively implement these sensory polymers, various device configurations are employed. They may be utilized as coatings, thin films, membranes, and fibers, tailored to suit specific sensing requirements. These different formats allow the enhancement of some properties of the materials, such as surface area, processability, usability, and even their sensitivity and sensibility (Sanjuán et al., 2018).

In this sense, sensory devices can appear in various formats:

- *Fibers and NFs*: These types of sensory polymers have gained a lot of interest due to the combination of a high surface area with great porosity, leading to properties such as high selectivity and sensitivity and even fast responses. The presence of a surface-to-volume ratio improves the characteristics in comparison with other polymeric sensors, which makes these structures suitable for many devices and applications. These sensors can be classified by their detection mechanisms into surface acoustic wave (SAW) sensors, resistive sensors, optical sensors, and amperometric sensors (Wang et al., 2014).
- *Films and membranes*: Due to their convenient processing characteristics, polymers can be readily transformed into films, coatings, or finished materials with specific shapes. This attribute has given rise to the development of numerous sensory devices using these materials. Typically, polymeric films act as a substrate in which the sensory monomer is dispersed or anchored in small proportions (up to 5% by weight). Subsequent processes like radical polymerization are employed to manufacture the sensory film, which is then utilized for

detection through mechanisms such as colorimetry and fluorescence. These films can be customized in terms of flexibility, hydrophilicity, or transparency, resulting in sensory materials that are easily adaptable. When it comes to polymer coatings, the sensory device consists of a solid substrate onto which a thin layer of the sensory polymer or a polymer solution is deposited, forming a coating layer of 1–10 μm in width. This deposition process is commonly carried out using techniques like spraying or sputtering a polymer solution (Guembe-García et al., 2022; Guirado-Moreno et al., 2023; Trigo-López et al., 2018).

- *Quartz crystal microbalances*: This type of system provides additional interest to polymeric sensors since they acquire many advantages, such as eliminating the preparation of samples and costly off-site analysis (Easley et al., 2022). The sensory impact arises from a shift in the polymer's mass due to analyte absorption, leading to a change in the resonance frequency of the piezoelectric crystal that can be easily measured. Therefore, the choice of suitable coating materials on QCM sensor surfaces significantly influences their overall properties and applications. As a result, QCM sensors coated with polymers have played a crucial role in detecting organic vapors and noxious gases. In this approach, diverse sensory polymers are applied to the QCM balance, interacting with various VOCs (Si et al., 2007). Additionally, there has been a growing interest in biomedical and biological applications of QCM-based sensors in recent years, particularly in the immobilization of antibodies and DNA, as well as the detection of biological hormones (Kugimiya & Takeuchi, 1999). Last, it is worth noting that polymer-coated QCM resonator devices are currently being employed for the detection of heavy metal ions (Sartore et al., 2011).
- *Polymeric sensors based on microfluidic devices*: Between 2000 and 2002, the exploration and advancement of microfluidic and nanofluidic technologies started. In the past decade, these technologies have evolved into a captivating method for creating micro- and nano-sensory devices, with a significant emphasis on the use of smart polymers (Wu & Gu, 2011). Among all possible detection methods, electrochemical measurements stand out due to their heightened selectivity and sensitivity, although others, like optical or mass spectrometry, are also effective. This approach operates by modulating electrically analyte species that are involved in redox reactions, thus being primarily employed for identifying electroactive substances through cyclic voltammetry assessments. In this context, numerous studies focusing on biological and biomedical sensing have emerged during the last few years, such as the detection of glucose with biocompatible chips (Fan et al., 2017; Thammakhet et al., 2011), identification of diverse proteins (Hajizadeh et al., 2008), or biomolecules (Lee et al., 2007), like bodily fluid testing *via* sensory chips. Additionally, there are notable applications regarding the detection of heavy metal ions, particularly Cu^{2+} and Hg^{2+}, in the field of environmental and safety applications (Qi et al., 2017; Kudr et al., 2017).
- *Polymeric sensors based on modified electrodes*: Another category of sensory devices is centered on modified electrodes, employing electrochemical measures, particularly cyclic voltammetry, to achieve the sensing effects. For this purpose, the attributes of unaltered electrodes can be enhanced through their chemical adjustment (Naveen et al., 2017). Primarily, metallic or graphene electrodes are enveloped both with sensory polymers and CPs, or polymer nanocomposites. Subsequently, the electrochemical properties of the adjusted device are measured, and the voltammetric findings are directly linked to the sensor's detection and quantification capabilities. Notably, electrochemical sensors incorporated into electrodes are

actively employed in the detection of heavy metal ions (Somerset et al., 2010a, 2010b), various medications, such as anticancer drugs (Liu et al., 2018), and even humidity, primarily utilizing CPs (Najjar & Nematdoust, 2016).

1.3.3.1 pH sensors

pH sensors are used extensively in chemical and biological fields, serving important roles in environmental monitoring, blood pH measurements, and laboratory pH tests, among other applications. The pH level of a solution has a profound impact on chemical reactions, making pH measurement and control crucial in chemistry, biochemistry, clinical chemistry (Alam et al., 2018), and environmental science (Korostynska et al., 2007). Back in the day, the earliest method of pH measurement relied on chemical indicators like litmus paper, which changed color based on the pH of the solution. Nowadays, the most common pH sensing systems utilize either amperometric or potentiometric devices. These technologies play a significant role in accurately determining pH values in various samples. Apart from pH detection, another relevant application is related to substance release, such as drug delivery in clinical science.

pH polymer sensors, also known as pH-sensitive polymer sensors, are devices that utilize pH-sensitive polymers to detect and measure changes in pH levels. These sensors respond to alterations in the acidity or alkalinity of a solution by generating a measurable signal. The working principle relies on multiple mechanisms, including chemical interactions (protonation/deprotonation, doping/de-doping, and redox reactions), physical interactions (charge depletion/accumulation), and hybrid interactions (combining chemical and physical mechanisms) (Adhikari & Majumdar, 2004; Trigo-López et al., 2015).

These mechanisms enable precise and sensitive pH measurements in various applications, producing output signals that may involve color changes (Pablos et al., 2015; Trigo-López et al., 2015) (Fig. 1.15), fluorescence (Pablos et al., 2022), and electrical or electrochemical behavior. Common devices for pH sensing encompass polymer-coated fiber optic sensors, electrodes modified with pH-sensitive polymers, fluorescent and colorimetric pH indicators, potentiometric pH sensors, and combinatory approaches for ion concentration monitoring. These pH sensors find applications in diverse fields, including environmental monitoring, biomedicine, the food and beverage industries, and chemical analysis.

FIGURE 1.15 pH sensory textile.

Reusable smart textiles polymeric coatings for the colorimetric detection of high acidity ambients.

Reproduced from Trigo-López, M., Pablos, J. L., Muñoz, A., Ibeas, S., Serna, F., García, F. C., & García, J. M. (2015). Aromatic polyamides and acrylic polymers as solid sensory materials and smart coated fibres for high acidity colorimetric sensing. Polymer Chemistry, *6(16), 3110–3120. Available from https://doi.org/10.1039/c4py01545b, with permission from the Royal Society of Chemistry.*

1.3.3.2 Temperature sensors

Temperature-sensitive polymers exhibit a wide range of physical and chemical changes in response to temperature fluctuations. These changes include an action consisting of variations in solubility, swelling behavior, mechanical properties, and interactions with other molecules or surfaces. By modifying the chemical structure and composition of the polymer, the temperature at which these changes occur can be tuned (Kim & Matsunaga, 2017). Temperature-sensitive polymers find common use in biomedical applications, particularly in drug delivery and tissue engineering (Fig. 1.16). In drug delivery, they can encapsulate drugs and deliver them to specific target sites in the body (Amin et al., 2022). By designing temperature-sensitive polymers to become more soluble at higher temperatures found in tumors, drugs can be delivered directly to cancer cells, minimizing damage to healthy tissue. In tissue engineering, these polymers create scaffolds for cell growth that can be adjusted to change their porosity or stiffness with temperature, enabling tissue-like structures with mechanical properties resembling those of natural tissues. Hydrogels can also be prepared for the controlled release of growth factors, supporting tissue regeneration. Furthermore, temperature-sensitive polymers are employed in coatings and adhesives, where their reversible changes enable adhesion to the surface or release as needed. They are also utilized in sensors and actuators, where their response to temperature changes detects or induces mechanical motion.

Accurate temperature measurements play a crucial role in numerous human endeavors, spanning industries, technology, and scientific pursuits. The necessity for sensors capable of measuring temperatures in chemical and biochemical processes is evident in both industrial manufacturing and laboratory investigations. Thus, though less exploited, variation in temperature can cause a change in measurable properties, and this response allows for the fine quantification of temperature in polymer temperature sensors. For instance, upon heating or cooling, the inner conductive pathway inside the conductive polymer composites changes in thermo-electrical dependence of the polymer and composites, thus causing the change in electrical resistance. In water environments, they can become more hydrophilic or hydrophobic at a specific temperature called the lower critical solution temperature for those becoming more soluble with increasing temperature or the upper critical solution temperature for those becoming less soluble, giving rise to cloud points that can be used as output signals (visual, light transmittance, etc.). Control of

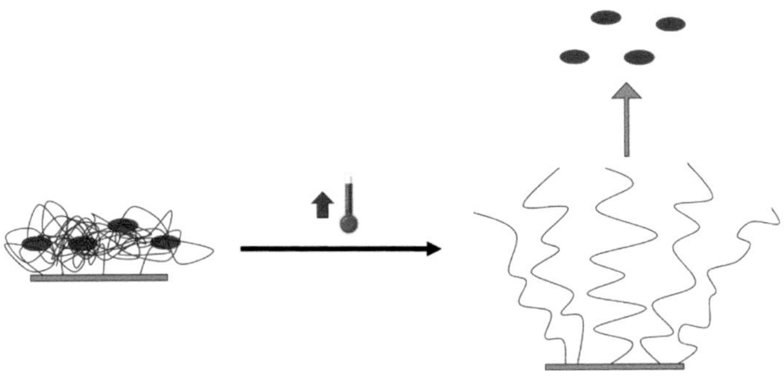

FIGURE 1.16 Thermo-responsive polymer.

Substances are released from the thermo-responsive polymer upon a temperature increase.

both temperatures can be achieved by adjusting the chemical composition of the polymer. Also, temperature polymer optical fiber sensors offer significant advantages in temperature measurement applications. They can be easily integrated into measuring systems using common and low-cost optoelectronic elements, making them accessible across various industries and research settings. With the integration of advanced sensing techniques like Fabry-Pçérot interferometry and Bragg gratings, these sensors enable continuous and real-time temperature monitoring (Szczerska, 2022). Fabry-Pérot interferometry utilizes a polymer cavity between reflective surfaces to precisely measure temperature changes based on interference patterns. On the other hand, Bragg grating-based sensors (Jin et al., 2006) reflect specific wavelengths of light, detecting temperature variations with high sensitivity and accuracy. These sensors are resilient in harsh conditions and immune to electromagnetic interference, making them suitable for aerospace, oil and gas, and structural health monitoring applications. Their cost-effectiveness and simplicity make them appealing for widespread use in temperature-sensitive fields, providing valuable insights into temperature variations for enhanced safety and efficiency (Khonina et al., 2023).

1.3.3.3 Nitroaromatic explosives detection

The detection of explosives by chemical sensors with high sensitivity holds significant importance for society, especially for security screening in demining, military operations, and protection against terrorist attacks. Additionally, it plays a crucial role in environmental monitoring. Nitroaromatic explosives, characterized by their electron-deficient nature, can interact electronically and structurally with electron-rich polymeric structures, particularly conductive polymers, making them well-suited for transducing these interactions to detect explosives (Toal & Trogler, 2006). Various sensory polymers are utilized for this purpose, including organic polymers like polyacetylenes, poly(*p*-phenylenevinylenes), poly(*p*-phenyleneethynylenes), and polymeric porphyrins, as well as inorganic polymers like polysilanes (Taniya et al., 2023). Notable nitroaromatic explosives include trinitrotoluene (TNT), trinitrophenol (picric acid, PA), and 2,4,6-trinitrophenylmethylnitramine (tetryl), and the same strategy can be applied for the detection of non-aromatic nitro explosives, such as 1,3,5-trinitro-1,3,5-triazinane (known as Research Department eXplosive, RDX), pentaerythritol tetranitrate (PETN), and nitroglycerine (Fig. 1.17). TNT, widely used in military and civil applications, requires detection for minefield clearance and security areas in conflict or post-conflict zones, as well as in groundwater solutions or soil extractions to control and remediate abandoned military sites, industrial waste, and spills related to explosive industries, due to the toxicity of nitroaromatics to blood and liver (Pablos et al., 2014a).

The quenching of luminescent conductive polymers by electron-deficient nitroaromatic explosives like TNT can be monitored for explosive detection (Nie et al., 2011). Similarly, resistive sensing using conductive polymers or conductive composites prepared with polymer matrices and conductive fillers such as carbon particles, platelets, or nanofibers can be employed for the same purpose. Additionally, SAW devices coated with specific polymers can undergo frequency changes upon adsorption and interaction with nitroaromatics, facilitating their detection. Moreover, the interaction of nitroaromatics with electron pairs of amines or other groups can lead to the formation of highly colored complexes, enabling their visual detection in both the solution and the gas phase (Pablos et al., 2014b) (Fig. 1.18). Importantly, the same strategy can be applied to the detection of non-aromatic nitro explosives, such as RDX, PETN, and nitroglycerine.

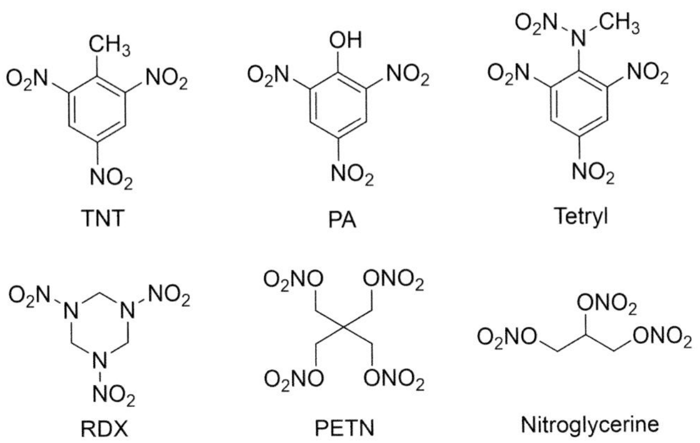

FIGURE 1.17 Explosive nitro-compounds.

Examples of nitroaromatic, nitramine, and nitrate ester explosive substances.

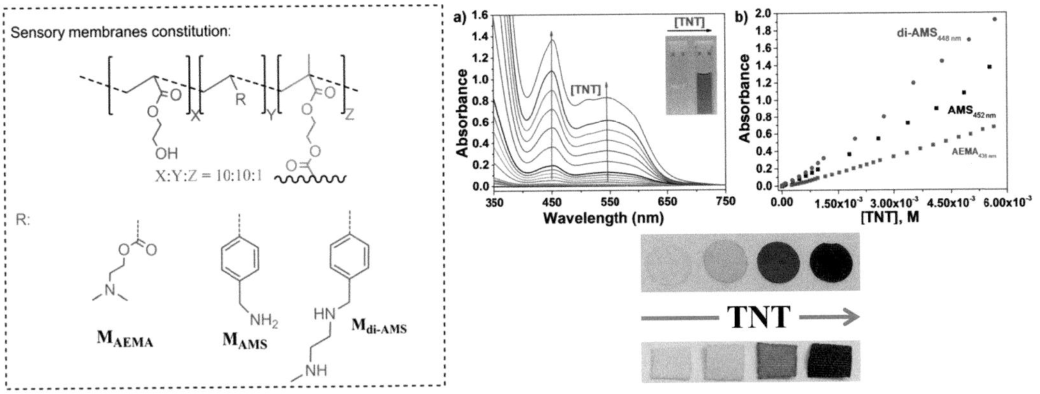

FIGURE 1.18 TNT polymer sensor.

Composition of polymers (sensory membranes) for the detection of TNT in solution and air. TNT titration by monomers containing sensory motifs with increasing amounts of TNT and visual detection of TNT vapors using membranes and coating textiles.

Reproduced from (Pablos et al., 2014b) with permission from the Royal Society of Chemistry. Reproduced from Pablos, J. L., Trigo-López, M., Serna, F., García, F. C., & García, J. M. (2014). Solid polymer substrates and smart fibres for the selective visual detection of TNT both in vapour and in aqueous media. RSC Advances, *4(49), 25562–25568. Available from https://doi.org/10.1039/C4RA02716G, with permission from the Royal Society of Chemistry.*

1.3.3.4 Cation and anion detection using fluorescent polymeric nanoparticles

Fluorescent polymer nanoparticles have garnered considerable attention due to their compelling physicochemical attributes and wide-ranging applicability across diverse fields, including photonics, electronics, sensors, catalysis, drug delivery, bioimaging, and coatings, among others.

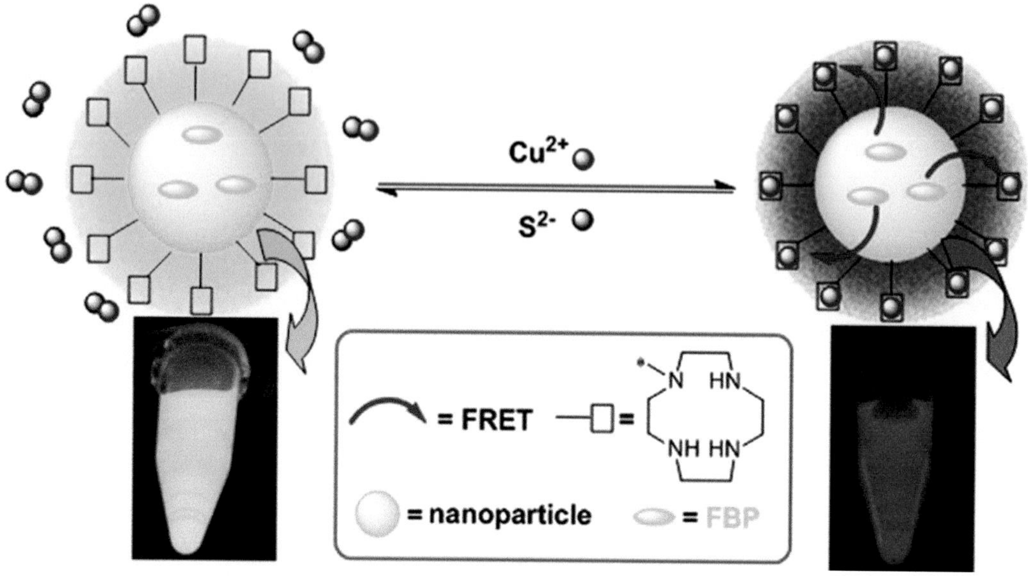

FIGURE 1.19 Sensory polymers using fluorescent polymeric nanoparticles.

Schematic illustration of cyclen-functionalized fluorescent polymeric nanoparticles sensors for Cu^{2+} and S^{2-} selective recognition.

From Chen, J., Li, Y., Zhong, W., Hou, Q., Wang, H., Sun, X., & Yi, P. (2015). Novel fluorescent polymeric nanoparticles for highly selective recognition of copper ion and sulfide anion in water. Sensors and Actuators B: Chemical, 206, *230–218. Available from https://doi.org/10.1016/j.snb.2014.09.034.*

Characterized by their diminutive size at the micro/nanoscale, substantial specific surface areas, and customizable properties, these nanoparticles offer distinct advantages over bulk polymer materials. Their synthesis encompasses various techniques, including emulsion polymerization, precipitation polymerization, dispersion polymerization, suspension polymerization, nanoprecipitation, self-assembly, and post-polymerization modification, facilitating tailoring for specific applications (Zhong et al., 2023).

Regarding chemical sensing, fluorescent polymer nanoparticles have found particular utility. Their synthesis methods, coupled with the integration of fluorescent properties into polymer matrices, have yielded materials well-suited for the detection of explosives, pH fluctuations, anions, and metal ions (Fig. 1.19). Notably, their merits for ion detection are underscored by their high sensitivity, arising from their ability to amplify fluorescence signals, and their selectivity, achieved through meticulous receptor molecule design (Chen et al., 2015). These nanoparticles find diverse applications in environmental monitoring, biological assays, and medical diagnostics, rendering them indispensable in the advancement of sensing and detection technologies across various domains (Krämer et al., 2022).

1.3.3.5 Protein sensors

Protein detection and quantification are of the utmost importance in the biomedical field, providing valuable insights into various aspects. These include evaluating protein presence and levels, obtaining clues about protein structures, analyzing protein activity, and understanding the mechanisms of protein interactions involved in biological processes.

The substances employed in protein detection and measurement platforms must adhere to specific criteria, encompassing ease of mass production, biological durability, chemical versatility, cost efficiency, and portability (Hahm, 2011). Polymers can fulfill numerous of these prerequisites and are frequently regarded as preferred materials in diverse biological detection platforms (Akgönüllü et al., 2023). In this context, polymers are studied as active and/or passive components in a wide array of protein detection methods and systems that display a diverse range of physical properties, encompassing optical, electrochemical, electrical, mass-sensitive, and magnetic characteristics (Fig. 1.20). The inherent versatility of polymers significantly enhances the throughput, sensitivity, and pursuit of miniaturization in protein detection techniques (You et al., 2007). Moreover, ideal sensor materials should possess attributes that enable easy and cost-effective large-scale production.

1.3.3.6 Humidity, gases, and volatile organic compounds

VOCs are organic chemicals characterized by their high vapor pressure at room temperature. These substances are diverse and plentiful, commonly categorized based on their functional groups, such as aliphatic hydrocarbons, simple oxygenated hydrocarbons, halogenated hydrocarbons, carbonyl compounds, and aromatic hydrocarbons, among others (Tomić et al., 2021). Many applications requiring monitoring of these compounds often involve a complex mixture of various vapors and gases. For example, the presence of thousands of VOCs has been identified in the environment, food products, and exhaled breath. Consequently, ensuring the selectivity of sensitive materials or minimizing cross-sensitivity is crucial in such scenarios.

In this sense, polymeric sensors are crucial for the detection of gases due to their high sensitivity, allowing for early and accurate identification of substances in diverse environments, such as the quantification of water (humidity), harmful gases (e.g., NH_3, NO_x, HCN, isocyanates, or chlorinated gases), as well as harmless gases like CO_2 and VOCs (e.g., alcohols, hydrocarbons, and organic amines). Their versatility permits customization, ensuring enhanced selectivity for different applications, while their lightweight and flexible nature facilitates cost-effective deployment in extensive sensor networks for real-time monitoring (Fig. 1.21). The durability of polymeric sensor materials guarantees long-term stability, making them reliable for continuous use in challenging conditions, such as industrial settings or remote locations. With rapid response times, these sensors provide timely warnings, enabling quick actions to mitigate potential hazards.

These measures can be achieved using sensory polymers with electrochemical transductions. These polymers respond piezoelectrically, with variations in resistance and impedance, or through cyclic voltammetry measurements. Additionally, spectroscopically based responses such as colorimetric, fluorogenic, or variations in IR or RAMAN responses can also be employed. A wide range of sensory polymers are available for gas phase detection and quantification, encompassing conductive (Bai & Shi, 2007), hybrid, branched, and graft polymers, as well as nanocomposites (Zegebreal et al., 2023). For instance, researchers have utilized various polymers, matrices, and fillers, such as polyethylenimine, graphene oxide, poly(acrylic acid), poly(sodium styrene sulfonate), poly(vinyl pyrrolidone), polythiophene, and poly(*p*-phenylene ethynylene).

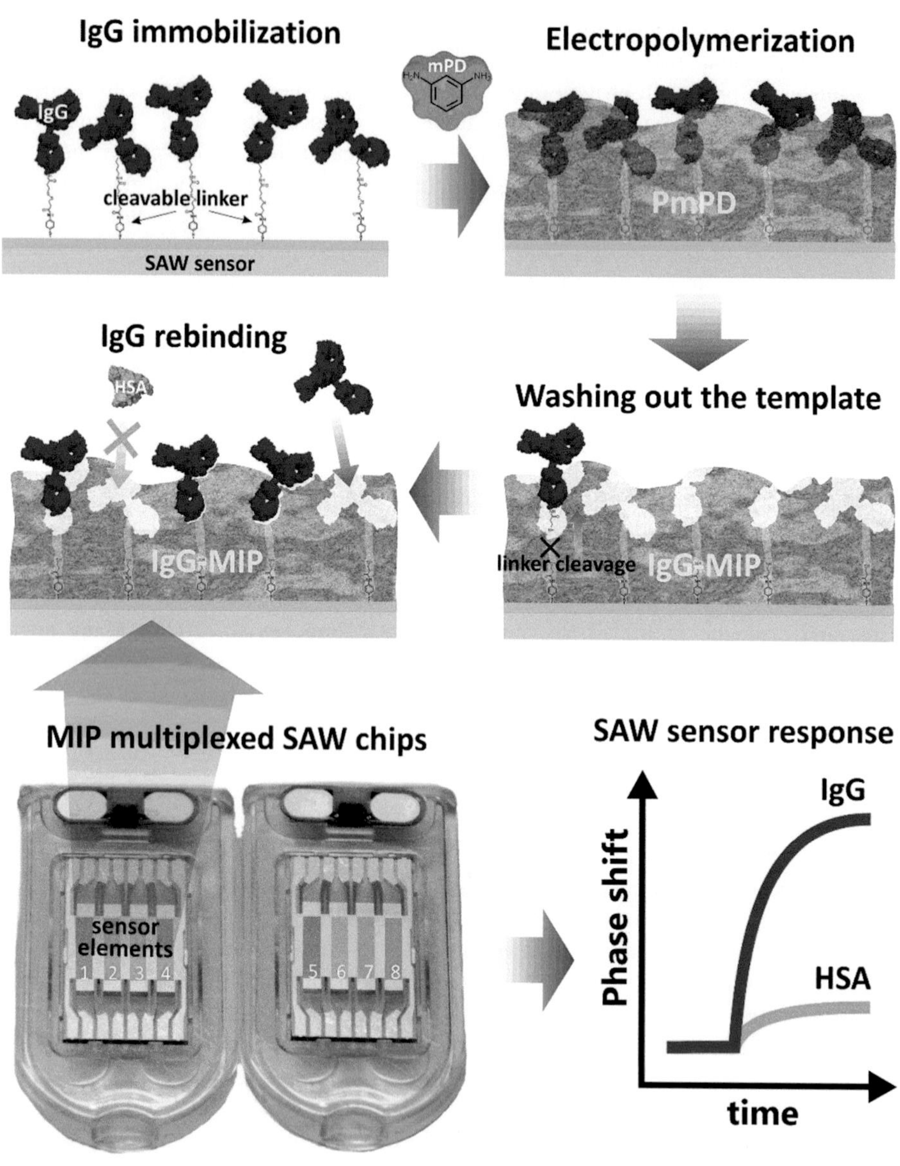

FIGURE 1.20 Protein sensor.

Schematic representation of the synthesis concept for ultrathin polymeric films with surface imprints of immunoglobulin G (IgG-MIP) sensing layer integration with a surface acoustic wave chip.

From Tretjakov, A., Syritski, V., Reut, J., Boroznjak, R., & Öpik, A. (2016). Molecularly imprinted polymer film interfaced with Surface Acoustic Wave technology as a sensing platform for label-free protein detection. Analytica Chimica Acta, 902, 182–188. Available from https://doi.org/10.1016/j.aca.2015.11.004.

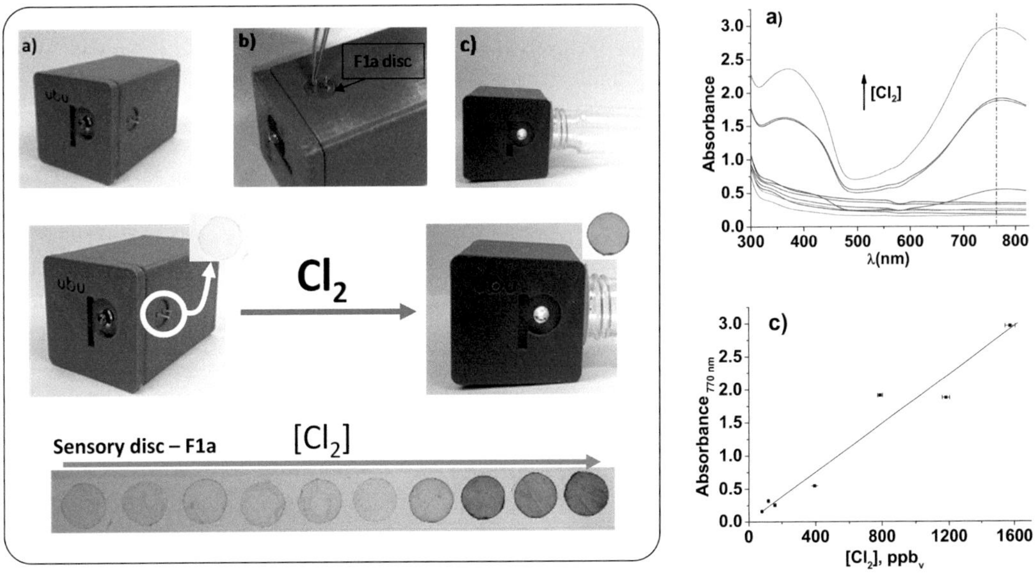

FIGURE 1.21 Chlorine sensory polymer.

Visual alarm system: (A) physical appearance; (B) setting of the sensory polymer in the shape of a film in the metallic terminals; (C) blue LED switched on due to the formation of emeraldine salt by the presence of chlorine in the environment of the sensory disc. As emeraldine is formed, the film also turns from colorless to green, which can be correlated to the [Cl_2] concentration as observed for the UV–vis spectra and the titration curve of the sensory discs after exposing them to increasing concentration of Cl_2.

From Pascual, B. S., Vallejos, S., Reglero Ruiz, J. A., Bertolín, J. C., Represa, C., García, F. C., & García, J. M. (2019). Easy and inexpensive method for the visual and electronic detection of oxidants in air by using vinylic films with embedded aniline. Journal of Hazardous Materials, 364, *238–243. Available from https://doi.org/10.1016/j.jhazmat.2018.10.039.*

1.4 Conclusions

In conclusion, the diverse landscape of sensory polymers presents an exciting frontier in materials science and sensor technology. Throughout this chapter, we have delved into a comprehensive spectrum of sensory polymer applications, ranging from molecular recognition to advanced sensor arrays. The remarkable adaptability of sensory polymers to various stimuli and specific analytes underscores their potential to revolutionize fields such as biomedicine, environmental monitoring, and electronics. The amalgamation of MIPs, nanomaterials, and conjugated polymers has paved the way for highly sensitive and selective sensors, enabling real-time monitoring and detection.

As we have explored the multifaceted world of sensory polymers, it becomes evident that their impact transcends individual sensor categories. The fusion of disciplines, such as optics, electrochemistry, and nanotechnology, has yielded sensor platforms that push the boundaries of sensitivity and accuracy. Additionally, the emergence of electrospun polymer sensors highlights the continuous innovation in fabrication techniques, offering tailored solutions for various sensing needs. The evolution toward sensor arrays underscores the quest for holistic data acquisition, enhancing the reliability and efficiency of information retrieval. In sum, sensory polymers are a testament to the remarkable synergy between materials engineering and sensor technology, promising a future where our environment can be understood, harnessed, and manipulated with unprecedented precision.

The field of polymeric sensors continues to advance, driven by ongoing research and development efforts. Further exploration of novel polymer materials, advanced fabrication techniques, and integration with emerging technologies will enhance sensor performance, increase sensitivity, and broaden analyte selectivity.

Acknowledgments

We gratefully acknowledge the financial support the Regional Government of Castilla y León (Junta de Castilla y León) and by the Ministry of Science and Innovation MICIN and the European Union NextGenerationEU PRTR.

References

Adhikari, B., & Majumdar, S. (2004). Polymers in sensor applications. *Progress in Polymer Science*, *29*(7), 699–766. Available from https://doi.org/10.1016/j.progpolymsci.2004.03.002, https://linkinghub.elsevier.com/retrieve/pii/S0079670004000383.

Aguilar, M. R., Román, S., & Smart, J. (2014). *Polymers and their applications* (first ed.). Woodhead Publishing.

Akgönüllü, S., Kılıç, S., Esen, C., & Denizli, A. (2023). Molecularly imprinted polymer-based sensors for protein detection. *Polymers*, *15*(3), 629. Available from https://doi.org/10.3390/polym15030629, https://www.mdpi.com/2073-4360/15/3/629.

Al-Abduljabbar, A., & Farooq, I. (2023). Electrospun polymer nanofibers: Processing, properties, and applications. *Polymers*, *15*(1), 65. Available from https://doi.org/10.3390/polym15010065.

Alam, A. U., Qin, Y., Nambiar, S., Yeow, J. T. W., Howlader, M. M. R., Hu, N.-X., & Deen, M. J. (2018). Polymers and organic materials-based pH sensors for healthcare applications. *Progress in Materials Science*, *96*, 174–216. Available from https://doi.org/10.1016/j.pmatsci.2018.03.008, https://linkinghub.elsevier.com/retrieve/pii/S0079642518300380.

Alvarez-Malmagro, J., García-Molina, G., & López De Lacey, A. (2020). Electrochemical biosensors based on membrane-bound enzymes in biomimetic configurations. *Sensors (Basel, Switzerland)*, *20*(12), 3393. Available from https://doi.org/10.3390/s20123393.

Amid, A., Jimat, D. N., Sulaiman, S., & Azmin, N. F. M. (2019). *Multifaceted protocol in biotechnology. Multifaceted protocol in biotechnology* (pp. 1–235). Malaysia: Springer Singapore. Available from http://www.springer.com/in/book/9789811322563, https://doi.org/10.1007/978-981-132257-0.

Amin, M., Lammers, T., & Ten Hagen, T. L. M. (2022). Temperature-sensitive polymers to promote heat-triggered drug release from liposomes: Towards bypassing EPR. *Advanced Drug Delivery Reviews*, *189*. Available from https://doi.org/10.1016/j.addr.2022.114503, https://linkinghub.elsevier.com/retrieve/pii/S0169409X22003933, 114503.

Anantha-Iyengar, G., Shanmugasundaram, K., Nallal, M., Lee, K. P., Whitcombe, M. J., Lakshmi, D., & Sai-Anand, G. (2019). Functionalized conjugated polymers for sensing and molecular imprinting applications. *Progress in Polymer Science*, *88*, 1–129. Available from https://doi.org/10.1016/j.progpolymsci.2018.08.001, http://www.sciencedirect.com/science/journal/00796700.

Andre, R. S., Sanfelice, R. C., Pavinatto, A., Mattoso, L. H. C., & Correa, D. S. (2018). Hybrid nanomaterials designed for volatile organic compounds sensors: A review. *Materials and Design, 156*, 154–166. Available from https://doi.org/10.1016/j.matdes.2018.06.041.

Arnaiz, A., Guembe-García, M., Delgado-Pinar, E., Valente, A. J. M., Ibeas, S., García, J. M., & Vallejos, S. (2022). The role of polymeric chains as a protective environment for improving the stability and efficiency of fluorogenic peptide substrates. *Nature Research, Spain Scientific Reports, 12*(1). Available from https://doi.org/10.1038/s41598-022-12848-4, http://www.nature.com/srep/index.html, 8818.

Arnaiz, A., Guirado-Moreno, J. C., Guembe-García, M., Barros, R., Tamayo-Ramos, J. A., Fernández-Pampín, N., García, J. M., & Vallejos, S. (2023). Lab-on-a-chip for the easy and visual detection of SARS-CoV-2 in saliva based on sensory polymers. *Sensors and Actuators B: Chemical., 379*. Available from https://doi.org/10.1016/j.snb.2022.133165, https://www.journals.elsevier.com/sensors-and-actuators-b-chemical, 133165.

Ates, B., Koytepe, S., Balcioglu, S., Alu, S., & Gurses, C. (2017). *Biomedical applications of hybrid polymer composite materials. Hybrid Polymer Composite Materials*. Elsevier. Available from https://doi.org/10.1016/B978-0-08-100785-3.00012-7.

Aussawasathien, D., Dong, J.-H., & Dai, L. (2005). Electrospun polymer nanofiber sensors. *Synthetic Metals, 154*(1–3), 37–40. Available from https://doi.org/10.1016/j.synthmet.2005.07.018.

Bai, H., & Shi, G. (2007). Gas sensors based on conducting polymers. *Sensors (Basel, Switzerland), 7*(3), 267–307. Available from https://doi.org/10.3390/s7030267, http://www.mdpi.com/1424-8220/7/3/267.

Baldini, F., & Bracci, S. (2000). *Polymers for optical fiber sensors* (pp. 91–107). Springer Science and Business Media LLC. Available from https://doi.org/10.1007/978-3-662-04068-3_3.

Balusamy, B., Senthamizhan, A., & Uyar, T. (2019). Functionalized electrospun nanofibers as colorimetric sensory probe for mercury detection: A review. *Sensors (Basel, Switzerland), 19*(21), 4763. Available from https://doi.org/10.3390/s19214763.

Bhardwaj, N., & Kundu, S. C. (2010). Electrospinning: A fascinating fiber fabrication technique. *Biotechnology Advances, 28*(3), 325–347. Available from https://doi.org/10.1016/j.biotechadv.2010.01.004.

Bilro, L., Alberto, N., Pinto, J. L., & Nogueira, R. (2012). Optical sensors based on plastic fibers. *Sensors (Basel, Switzerland), 12*(9), 12184–12207. Available from https://doi.org/10.3390/s120912184.

Bonetto, M. C., Muñoz, F. F., Diz, V. E., Sacco, N. J., & Cortón, E. (2018). Fused and unzipped carbon nanotubes, electrochemically treated, for selective determination of dopamine and serotonin. *Electrochimica Acta, 283*, 338–348. Available from https://doi.org/10.1016/j.electacta.2018.06.179, http://www.journals.elsevier.com/electrochimica-acta/.

Brett, C. M. A., & Oliveira-Brett, A. M. (2011). Electrochemical sensing in solution-origins, applications and future perspectives. *Journal of Solid State Electrochemistry, 15*(7–8), 1487–1494. Available from https://doi.org/10.1007/s10008-011-1447-z.

Brimo, N., & Serdaroğlu, D. Ç. (2021). *Molecular imprinted polymers for mass sensitivesensors: Comparation of performance toward immuno-sensing strategies. Molecular imprinting for nanosensors and other sensing applications* (pp. 335–365). Turkey: Elsevier. Available from https://www.sciencedirect.com/book/9780128221174, 10.1016/B978-0-12-822117-4.000137.

Brown, C. L., & Craig, S. L. (2015). Molecular engineering of mechanophore activity for stress-responsive polymeric materials. *Chemical Science, 6*(4), 2158–2165. Available from https://doi.org/10.1039/c4sc01945h, http://pubs.rsc.org/en/Journals/JournalIssues/SC.

Bustamante, S. E., Vallejos, S., Pascual-Portal, B. S., Muñoz, A., Mendia, A., Rivas, B. L., García, F. C., & García, J. M. (2019). Polymer films containing chemically anchored diazonium salts with long-term stability as colorimetric sensors. *Journal of Hazardous Materials, 365*, 725–732. Available from https://doi.org/10.1016/j.jhazmat.2018.11.066, http://www.elsevier.com/locate/jhazmat.

Cao, S., Chen, J., Sheng, W., Wu, W., Zhao, Z., & Long, F. (2012). The fabrication and development of molecularly imprinted polymer-based sensors for environmental application. *Molecularly Imprinted*

Sensors, 57–72. Available from https://doi.org/10.1016/B978-0-444-56331-6.00003-7, http://www.sciencedirect.com/science/book/9780444563316.

Caro, E., Marcé, R. M., Cormack, P. A. G., Sherrington, D. C., & Borrull, F. (2006). Direct determination of ciprofloxacin by mass spectrometry after a two-step solid-phase extraction using a molecularly imprinted polymer. *Journal of Separation Science*, *29*(9), 1230–1236. Available from https://doi.org/10.1002/jssc.200500439.

Chekin, F., Singh, S. K., Vasilescu, A., Dhavale, V. M., Kurungot, S., Boukherroub, R., & Szunerits, S. (2016). Reduced graphene oxide modified electrodes for sensitive sensing of gliadin in food samples. *ACS Sensors*, *1*(12), 1462–1470. Available from https://doi.org/10.1021/acssensors.6b00608, http://pubs.acs.org/journal/ascefj.

Chen, B., Piletsky, S., & Turner, A. P. F. (2002). Molecular recognition: Design of "keys". *United Kingdom Combinatorial Chemistry and High Throughput Screening*, *5*(6), 409–427. Available from https://doi.org/10.2174/1386207023330129, http://www.eurekaselect.com/585/journal/scombinatorial-chemistry-amp-high-throughput-screening.

Chen, J., Li, Y., Zhong, W., Hou, Q., Wang, H., Sun, X., & Yi, P. (2015). Novel fluorescent polymeric nanoparticles for highly selective recognition of copper ion and sulfide anion in water. *Sensors and Actuators B: Chemical.*, *206*, 230–238. Available from https://doi.org/10.1016/j.snb.2014.09.034, https://linkinghub.elsevier.com/retrieve/pii/S0925400514011034.

Chen, J., Liu, H., Wang, W., Nabulsi, N., Zhao, W., Yeon Kim, J., Kwon, M.-K., & Ryou, J.-H. (2019). High durable, biocompatible, and flexible piezoelectric pulse sensor using single-crystalline III-N thin film. *Advanced Functional Materials*, *29*(37), 1903162. Available from https://doi.org/10.1002/adfm.201903162.

Chen, K., Chou, W., Liu, L., Cui, Y., Xue, P., & Jia, M. (2019). Electrochemical sensors fabricated by electrospinning technology: An overview. *Sensors (Basel, Switzerland)*, *19*(17), 3676. Available from https://doi.org/10.3390/s19173676.

Chen, X., Hussain, S., Hao, Y., Tian, X., & Gao, R. (2021). Review—Recent advances of signal amplified smart conjugated polymers for optical detection on solid support. *ECS Journal of Solid State Science and Technology*, *10*(3), 037006. Available from https://doi.org/10.1149/2162-8777/abeed1.

Chen, X., Lin, X., Mo, D., Xia, X., Gong, M., Lian, H., & Luo, Y. (2020). High-sensitivity, fast-response flexible pressure sensor for electronic skin using direct writing printing. *RSC Advances*, *10*(44), 26188–26196. Available from https://doi.org/10.1039/d0ra04431h.

Chiu, S. W., & Tang, K. T. (2013). Towards a chemiresistive sensor-integrated electronic nose: A review. *MDPI AG, Taiwan Sensors (Switzerland)*, *13*(10), 14214–14247. Available from https://doi.org/10.3390/s131014214, http://www.mdpi.com/1424-8220/13/10/14214/pdf.

Cochrane, C., & Cayla, A. (2013). *Polymer-based resistive sensors for smart textiles. Multidisciplinary know-how for smart-textiles developers* (pp. 129–153). France: Elsevier Ltd.. Available from http://www.sciencedirect.com/science/book/9780857093424, https://doi.org/10.1533/9780857093530.1.129.

Dai, L., Winkler, B., Dong, L., Tong, L., & Mau, A. W. H. (2001). Conjugated polymers for light-emitting applications. *Advanced Materials*, *13*(12–13), 915–925. Available from https://doi.org/10.1002/1521-4095(200107)13:12/13 < 915::AID-ADMA915 > 3.0.CO;2-N.

Dai, L., Soundarrajan, P., & Kim, T. (2002). Sensors and sensor arrays based on conjugated polymers and carbon nanotubes. *Pure and Applied Chemistry*, *74*(9), 1753–1772. Available from https://doi.org/10.1351/pac200274091753.

Davis, F., & Higson, S. P. J. (2007). *Polymers in biosensors biomedical polymers* (pp. 174–196). United Kingdom: Elsevier Inc.. Available from http://www.sciencedirect.com/science/book/9781845690700, 10.1533/9781845693640.174.

Ding, B., Wang, M., Wang, X., Yu, J., & Sun, G. (2010). Electrospun nanomaterials for ultrasensitive sensors. *Materials Today*, *13*(11), 16–27. Available from https://doi.org/10.1016/s1369-7021(10)70200-5.

Domínguez-Renedo, O., Alonso-Lomillo, M. A., & Arcos-Martínez, M. J. (2013). Determination of metals based on electrochemical biosensors. *Critical Reviews in Environmental Science and Technology*, *43*(10), 1042–1073. Available from https://doi.org/10.1080/10934529.2011.627034.

Dong, S., Zheng, B., Wang, F., & Huang, F. (2014). Supramolecular polymers constructed from macrocycle-based host–guest molecular recognition motifs. *Accounts of Chemical Research*, *47*(7), 1982–1994. Available from https://doi.org/10.1021/ar5000456.

Easley, A. D., Ma, T., Eneh, C. I., Yun, J., Thakur, R. M., & Lutkenhaus, J. L. (2022). A practical guide to quartz crystal microbalance with dissipation monitoring of thin polymer films. *Journal of Polymer Science*, *60*(7), 1090–1107. Available from https://doi.org/10.1002/pol.20210324, https://onlinelibrary.wiley.com/doi/10.1002/pol.20210324.

Elosua, C., Arregui, F. J., Del Villar, I., Ruiz-Zamarreño, C., Corres, J. M., Bariain, C., Goicoechea, J., Hernaez, M., Rivero, P. J., Socorro, A. B., Urrutia, A., Sanchez, P., Zubiate, P., Lopez-Torres, D., De Acha, N., Ascorbe, J., Ozcariz, A., & Matias, I. (2017). Micro and nanostructured materials for the development of optical fibre sensors. *Sensors (Basel, Switzerland)*, *17*(10), 2312. Available from https://doi.org/10.3390/s17102312.

El Sayed, S. (2023). *Chromo-fluorogenic chemosensors for sensing applications. Fundamentals of sensor technology: principles and novel designs* (pp. 631–667). United Kingdom: Elsevier. Available from https://www.sciencedirect.com/book/9780323884310, 10.1016/B978-0-323-88431-0.00020-X.

Erdem Yilmaz, O., & Erdem, R. (2020). Evaluating hydrogen detection performance of an electrospun CuZnFe2O4 nanofiber sensor. *International Journal of Hydrogen Energy*, *45*(50), 26402–26412. Available from https://doi.org/10.1016/j.ijhydene.2020.06.053.

Evans, R. C., Douglas, P., & Burrow, H. D. (2013). *Applied photochemistry*. Springer Netherlands. Available from https://doi.org/10.1007/978-90-481-3830-2.

Evtugyn, G. (1998). Sensitivity and selectivity of electrochemical enzyme sensors for inhibitor determination. *Talanta*, *46*(4), 465–484. Available from https://doi.org/10.1016/s0039-9140(97)00313-5.

Fan, Y.-Q., Gao, F., Wang, M., Zhuang, J., Tang, G., & Zhang, Y.-J. (2017). Recent development of wearable microfluidics applied in body fluid testing and drug delivery. *Chinese Journal of Analytical Chemistry.*, *45* (3), 455–463. Available from https://doi.org/10.1016/S1872-2040(17)61002-8, https://linkinghub.elsevier.com/retrieve/pii/S1872204017610028.

Foulds, N. C., & Lowe, C. R. (1988). Immobilization of glucose oxidase in ferrocene-modified pyrrole polymers. *Analytical Chemistry*, *60*(22), 2473–2478. Available from https://doi.org/10.1021/ac00173a008.

Gabrieli, G., Hu, R., Matsumoto, K., Temiz, Y., Bissig, S., Cox, A., Heller, R., López, A., Barroso, J., Kaneda, K., Orii, Y., & Ruch, P. W. (2021). Combining an integrated sensor array with machine learning for the simultaneous quantification of multiple cations in aqueous mixtures. *Analytical Chemistry*, *93*(50), 16853–16861. Available from https://doi.org/10.1021/acs.analchem.1c03709.

García, J. M., García, F. C., Ruiz, J. A., Vallejos, S., & Trigo-López, M. (2022). *Smart polymers*. De Gruyter STEM.

García, J. M., García, F. C., Serna, F., & De La Peña, J. L. (2011). Fluorogenic and chromogenic polymer chemosensors. *Polymer Reviews*, *51*(4), 341–390. Available from https://doi.org/10.1080/15583724.2011.616084.

Geng, Y., Peveler, W. J., & Rotello, V. M. (2019). Array-based "chemical nose" sensing in diagnostics and drug discovery. *Angewandte Chemie – International Edition*, *58*(16), 5190–5200. Available from https://doi.org/10.1002/anie.201809607, http://onlinelibrary.wiley.com/journal/10.1002/(ISSN)1521-3773.

Ghosh, R., Nayak, A. K., Santra, S., Pradhan, D., & Guha, P. K. (2015). Enhanced ammonia sensing at room temperature with reduced graphene oxide/tin oxide hybrid films. *RSC Advances*, *5*(62), 50165–50173. Available from https://doi.org/10.1039/c5ra06696d, http://pubs.rsc.org/en/journals/journalissues.

González-Ceballos, L., Melero, B., Trigo-López, M., Vallejos, S., Muñoz, A., García, F. C., Fernandez-Muiño, M. A., Sancho, M. T., & García, J. M. (2020). Functional aromatic polyamides for the preparation of coated fibres as smart labels for the visual detection of biogenic amine vapours and fish spoilage. *Sensors and Actuators, B: Chemical, 304*. Available from https://doi.org/10.1016/j.snb.2019.127249, https://www.journals.elsevier.com/sensors-and-actuators-b-chemical, 127249.

Girigoswami, A., Ghosh, M. M., Pallavi, P., Ramesh, S., & Girigoswami, K. (2021). Nanotechnology in detection of food toxins – Focus on the dairy products. *Biointerface Research in Applied Chemistry, 11*(6), 14155–14172. Available from https://doi.org/10.33263/briac116.1415514172.

Guembe-García, M., González-Ceballos, L., Arnaiz, A., Fernández-Muiño, M. A., Sancho, M. T., Osés, S. M., Ibeas, S., Rovira, J., Melero, B., Represa, C., García, J. M., & Vallejos, S. (2022). Easy nitrite analysis of processed meat with colorimetric polymer sensors and a smartphone app. *ACS Applied Materials and Interfaces, 14*(32), 37051–37058. Available from https://doi.org/10.1021/acsami.2c09467, http://pubs.acs.org/journal/aamick.

Guembe-García, M., Peredo-Guzmán, P. D., Santaolalla-García, V., Moradillo-Renuncio, N., Ibeas, S., Mendía, A., García, F. C., García, J. M., & Vallejos, S. (2020). Why is the sensory response of organic probes within a polymer film different in solution and in the solid-state? Evidence and application to the detection of amino acids in human chronic wounds. *Polymers, 12*(6). Available from https://doi.org/10.3390/POLYM12061249, https://www.mdpi.com/2073-4360/12/6/1249, 1249.

Guembe-García, M., Santaolalla-García, V., Moradillo-Renuncio, N., Ibeas, S., Reglero, J. A., García, F. C., Pacheco, J., Casado, S., García, J. M., & Vallejos, S. (2021). Monitoring of the evolution of human chronic wounds using a ninhydrin-based sensory polymer and a smartphone. *Sensors and Actuators, B: Chemical, 335*. Available from https://doi.org/10.1016/j.snb.2021.129688, https://www.journals.elsevier.com/sensors-and-actuators-b-chemical, 129688.

Guirado-Moreno, J. C., González-Ceballos, L., Carreira-Barral, I., Ibeas, S., Fernández-Muiño, M. A., Teresa Sancho, M., García, J. M., & Vallejos, S. (2023). Smart sensory polymer for straightforward Zn(II) detection in pet food samples. *Spectrochimica Acta Part A: Molecular and Biomolecular Spectroscopy, 284*. Available from https://doi.org/10.1016/j.saa.2022.121820, https://linkinghub.elsevier.com/retrieve/pii/S1386142522009684, 121820.

Gupta, S., Sharma, A., & Verma, R. S. (2020). Polymers in biosensor devices for cardiovascular applications. *Current Opinion in Biomedical Engineering, 13*, 69–75. Available from https://doi.org/10.1016/j.cobme.2019.10.002, https://www.journals.elsevier.com/current-opinion-in-biomedical-engineering.

Hahm, J.-in (2011). Functional polymers in protein detection platforms: optical, electrochemical, electrical, mass-sensitive, and magnetic biosensors. *Sensors (Basel, Switzerland), 11*(3), 3327–3355. Available from https://doi.org/10.3390/s110303327, http://www.mdpi.com/1424-8220/11/3/3327.

Hajizadeh, S., Ivanov, A. E., Jahanshahi, M., Sanati, M. H., Zhuravleva, N. V., Mikhalovska, L. I., & Galaev, I. Y. (2008). Glucose sensors with increased sensitivity based on composite gels containing immobilized boronic acid. *Reactive and Functional Polymers, 68*(12), 1625–1635. Available from https://doi.org/10.1016/j.reactfunctpolym.2008.09.006, https://linkinghub.elsevier.com/retrieve/pii/S1381514808001545.

Halicka, K., & Cabaj, J. (2021). Electrospun nanofibers for sensing and biosensing applications—A review. *International Journal of Molecular Sciences, 22*(12), 6357. Available from https://doi.org/10.3390/ijms22126357.

Harrison, J.S., & Ounaies, Z. (2002). Polymers, Piezoelectric. In *Encyclopedia of Smart Materials*; Wiley.

He, C., Long, Y., Pan, J., Li, K., & Liu, F. (2007). Application of molecularly imprinted polymers to solid-phase extraction of analytes from real samples. *Journal of Biochemical and Biophysical Methods, 70*(2), 133–150. Available from https://doi.org/10.1016/j.jbbm.2006.07.005.

He, Q., Wu, Y., Tian, Y., Li, G., Liu, J., Deng, P., & Chen, D. (2019). Facile electrochemical sensor for nanomolar rutin detection based on magnetite nanoparticles and reduced graphene oxide decorated electrode. *Nanomaterials, 9*(1), 115. Available from https://doi.org/10.3390/nano9010115.

Hosseini, M. S., Amjadi, I., Mohajeri, M., Zubair Iqbal, M., Wu, A., & Mozafari, M. (2019). *Functional polymers: An introduction in the context of biomedical engineering. Advanced functional polymers for biomedical applications* (pp. 1–20). Iran: Elsevier. Available from https://www.sciencedirect.com/book/9780128163498, 10.1016/B978-0-12-816349-8.00001-1.

Hu, J., Qu, H., Chang, Y., Pang, W., Zhang, Q., Liu, J., & Duan, X. (2018). Miniaturized polymer coated film bulk acoustic wave resonator sensor array for quantitative gas chromatographic analysis. *Sensors and Actuators B: Chemical, 274*, 419–426. Available from https://doi.org/10.1016/j.snb.2018.07.162.

Hulanicki, A., Glab, S., & Ingman, F. (1991). Chemical sensors: Definitions and classification. *Pure and Applied Chemistry, 63*(9), 1247–1250. Available from https://doi.org/10.1351/pac199163091247.

Ikeda, T., Hamada, H., Miki, K., & Senda, M. (1985). Glucose oxidase-immobilized benzoquinone-carbon paste electrode as a glucose senso. r. *Agricultural and Biological Chemistry, 49*(2), 541–543. Available from https://doi.org/10.1080/00021369.1985.10866761.

Imam, H. T., Marr, P. C., & Marr, A. C. (2021). Enzyme entrapment, biocatalyst immobilization without covalent attachment. *Green Chemistry, 23*(14), 4980–5005. Available from https://doi.org/10.1039/d1gc01852c, http://pubs.rsc.org/en/journals/journal/gc.

Lakard, B. (2020). Electrochemical biosensors based on conducting polymers: A review. *Applied Sciences, 10* (18), 6614. Available from https://doi.org/10.3390/app10186614.

Janata, J., & Bezegh, A. (1988). Chemical sensors. *Analytical Chemistry, 60*(12), 62–74. Available from https://doi.org/10.1021/ac00163a004.

Ji, J., Zhou, Z., Zhao, X., Sun, J., & Sun, X. (2015). Electrochemical sensor based on molecularly imprinted film at Au nanoparticles-carbon nanotubes modified electrode for determination of cholesterol. *Biosensors and Bioelectronics, 66*, 590–595. Available from https://doi.org/10.1016/j.bios.2014.12.014.

Jiang, X., Kim, K., Zhang, S., Johnson, J., & Salazar, G. (2014). High-temperature piezoelectric sensing. *Sensors (Basel, Switzerland), 14*(1), 144–169. Available from https://doi.org/10.3390/s140100144.

Jiang, Z., Diggle, B., Tan, M. L., Viktorova, J., Bennett, C. W., & Connal, L. A. (2020). Extrusion 3D printing of polymeric materials with advanced properties. *Advanced Science, 7*(17). Available from https://doi.org/10.1002/advs.202001379, http://onlinelibrary.wiley.com/journal/10.1002/(ISSN)2198-3844, 2001379.

Jin, W., Lee, T. K. Y., Ho, S. L., Ho, H. L., Lau, K. T., Zhou, L. M., & Zhou, Y. (2006). *Structural strain and temperature measurements using fiber bragg grating sensors. Guided Wave Optical Components and Devices* (pp. 389–400). Elsevier. Available from https://linkinghub.elsevier.com/retrieve/pii/B9780120884810500267, https://doi.org/10.1016/B978-012088481-0/50026-7.

Kadhem, A. J., Gentile, G. J., & Fidalgo de Cortalezzi, M. M. (2021). Molecularly imprinted polymers (MIPs) in sensors for environmental and biomedical applications: A review. *Molecules, 26*(20), 6233. Available from https://doi.org/10.3390/molecules26206233.

Karyakin, A. A., Vuki, M., Lukachova, L. V., Karyakina, E. E., Orlov, A. V., Karpachova, G. P., & Wang, J. (1999). Processible polyaniline as an advanced potentiometric pH transducer. Application to biosensors. *Analytical Chemistry, 71*(13), 2534–2540. Available from https://doi.org/10.1021/ac981337a.

Khan, U., Kim, T. H., Ryu, H., Seung, W., & Kim, S. W. (2017). Graphene tribotronics for electronic skin and touch screen applications. *Advanced Materials, 29*(1). Available from https://doi.org/10.1002/adma.201603544, http://www3.interscience.wiley.com/journal/119030556/issue, 1603544.

Khonina, S. N., Voronkov, G. S., Grakhova, E. P., Kazanskiy, N. L., Kutluyarov, R. V., & Butt, M. A. (2023). Polymer waveguide-based optical sensors—Interest in bio, gas, temperature, and mechanical sensing applications. *Coatings, 13*(3). Available from https://doi.org/10.3390/coatings13030549, https://www.mdpi.com/2079-6412/13/3/549, 549.

Kim, K. H., Jang, N. S., Ha, S. H., Cho, J. H., & Kim, J. M. (2018). Highly sensitive and stretchable resistive strain sensors based on microstructured metal nanowire/elastomer composite films. *Small, 14*(14).

Available from https://doi.org/10.1002/smll.201704232, http://onlinelibrary.wiley.com/journal/10.1002/(ISSN)1613-6829, 1704232.

Kim, Y.-J., & Matsunaga, Y. T. (2017). Thermo-responsive polymers and their application as smart biomaterials. *Journal of Materials Chemistry B.*, *5*(23), 4307–4321. Available from https://doi.org/10.1039/C7TB00157F, http://xlink.rsc.org/?DOI = C7TB00157F.

Kišija, E., Osmanović, D., Nuhić, J., & Cifrić, S. (2020). Review of biosensors in industrial process control. In *1 2020/01 IFMBE proceedings*. Bosnia and Herzegovina: Springer Verlag. https:/doi.org/10.1007/978-3-030-17971-7_103, 14339277. Available from http://www.springer.com/series/7403, 73.

Korostynska, O., Arshak, K., Gill, E., & Arshak, A. (2007). State key laboratory of nonlinear mechanics (LNM), Institute of Mechanics, Chinese Academy of Sciences, Beijing 100080, China. *Sensors (Basel, Switzerland)*, *7*(12), 3027–3042. Available from https://doi.org/10.3390/s7123027, http://www.mdpi.com/1424-8220/7/12/3027.

Krämer, J., Kang, R., Grimm, L. M., De Cola, L., Picchetti, P., & Biedermann, F. (2022). Molecular probes, chemosensors, and nanosensors for optical detection of biorelevant molecules and ions in aqueous media and biofluids. *Chemical Reviews*, *122*(3), 3459–3636. Available from https://doi.org/10.1021/acs.chemrev.1c00746, https://pubs.acs.org/doi/10.1021/acs.chemrev.1c00746.

Kudr, J., Zitka, O., Klimanek, M., Vrba, R., & Adam, V. (2017). Microfluidic electrochemical devices for pollution analysis–A review. *Sensors and Actuators B: Chemical*, *246*, 578–590. Available from https://doi.org/10.1016/j.snb.2017.02.052, https://linkinghub.elsevier.com/retrieve/pii/S0925400517302721.

Kugimiya, A., & Takeuchi, T. (1999). Molecularly imprinted polymer-coated quartz crystal microbalance for detection of biological hormone. *Electroanalysis*, *11*(15), 1158–1160. Available from https://onlinelibrary.wiley.com/doi/10.1002/(SICI)1521-4109(199911)11:15%3C1158::AID-ELAN1158%3E3.0.CO;2-P, https://doi.org/10.1002/(SICI)1521-4109(199911)11:15 < 1158::AID-ELAN1158 > 3.0.CO;2-P.

Kurczewska, J., Cegłowski, M., Pecyna, P., Ratajczak, M., Gajęcka, M., & Schroeder, G. (2017). Molecularly imprinted polymer as drug delivery carrier in alginate dressing. *Materials Letters*, *201*, 46–49. Available from https://doi.org/10.1016/j.matlet.2017.05.008.

Lakowicz, J.R. (Ed.) (2006). *Principles of fluorescence spectroscopy*.Baltimore (MD): Springer.

Le, T. H., Kim, Y., & Yoon, H. (2017). Electrical and electrochemical properties of conducting polymers. *Polymers*, *9*(4). Available from https://doi.org/10.3390/polym9040150, http://www.mdpi.com/2073-4360/9/4/150/pdf, 150.

Lee, D.-S., Gil Choi, H., Hyo Chung, K., Lee, B. Y., Pyo, H.-B., & Yoon, H. C. (2007). A temperature-controllable microelectrode and its application to protein immobilization. *ETRI Journal.*, *29*(5), 667–669. Available from https://doi.org/10.4218/etrij.07.0207.0035, http://doi.wiley.com/10.4218/etrij.07.0207.0035.

Lee, J. H., Heo, J. S., Kim, Y. J., Eom, J., Jung, H. J., Kim, J. W., Kim, I., Park, H. H., Mo, H. S., Kim, Y. H., & Park, S. K. (2020). A behavior-learned cross-reactive sensor matrix for intelligent skin perception. *Advanced Materials*, *32*(22). Available from https://doi.org/10.1002/adma.202000969, http://onlinelibrary.wiley.com/journal/10.1002/(ISSN)1521-4095, 2000969.

Lee, S., Koo, J., Kang, S. K., Park, G., Lee, Y. J., Chen, Y. Y., Lim, S. A., Lee, K. M., & Rogers, J. A. (2018). Metal microparticle – Polymer composites as printable, bio/ecoresorbable conductive inks. *Materials Today*, *21*(3), 207–215. Available from https://doi.org/10.1016/j.mattod.2017.12.005, http://www.journals.elsevier.com/materials-today/.

Lemke, E. A., & Schultz, C. (2011). Principles for designing fluorescent sensors and reporters. *Germany Nature Chemical Biology*, *7*(8), 480–483. Available from https://doi.org/10.1038/nchembio.620, http://www.nature.com/nchembio.

Li, J., Fang, L., Sun, B., Li, X., & Kang, S. H. (2020). Review—Recent progress in flexible and stretchable piezoresistive sensors and their applications. *Journal of the Electrochemical Society*, *167*(3), 037561. Available from https://doi.org/10.1149/1945-7111/ab6828.

Li, S., Ge, Y., & Turner, A. P. F. (2011). A catalytic and positively thermosensitive molecularly imprinted polymer. *Advanced Functional Materials*, *21*(6), 1194–1200. Available from https://doi.org/10.1002/adfm.201001906.

Li, S., Tang, W., Chen, S., Si, Y., Liu, R., & Guo, X. (2023). Flexible organic polymer gas sensor and system integration for smart packaging. *Advanced Sensor Research*, *2*, 2300030. Available from https://doi.org/10.1002/adsr.202300030.

Li, T., Zhu, X., Hai, X., Bi, S., & Zhang, X. (2023). Recent progress in sensor arrays: from construction principles of sensing elements to applications. *ACS Sensors*, *8*(3), 994–1016. Available from https://doi.org/10.1021/acssensors.2c02596.

Li, Y., Zhang, K., Nie, M., & Wang, Q. (2019). Tubular sensor with multi-axial strain sensibility and heating capability based on bio-mimic helical networks. *Industrial & Engineering Chemistry Research*, *58*(49), 22273–22282. Available from https://doi.org/10.1021/acs.iecr.9b04783.

Lines, M. G. (2008). Nanomaterials for practical functional uses. *Journal of Alloys and Compounds*, *449* (1–2), 242–245. Available from https://doi.org/10.1016/j.jallcom.2006.02.082.

Liu, H., Dong, M., Huang, W., Gao, J., Dai, K., Guo, J., Zheng, G., Liu, C., Shen, C., & Guo, Z. (2017). Lightweight conductive graphene/thermoplastic polyurethane foams with ultrahigh compressibility for piezoresistive sensing. *Journal of Materials Chemistry C*, *5*(1), 73–83. Available from https://doi.org/10.1039/C6TC03713E.

Liu, Y., Wei, M., Hu, Y., Zhu, L., & Du, J. (2018). An electrochemical sensor based on a molecularly imprinted polymer for determination of anticancer drug Mitoxantrone. *Sensors and Actuators B: Chemical*, *255*, 544–551. Available from https://doi.org/10.1016/j.snb.2017.08.023, https://linkinghub.elsevier.com/retrieve/pii/S0925400517314442.

Lyu, X., Gonzalez, R., Horton, A., & Li, T. (2021). Immobilization of enzymes by polymeric materials. *Catalysts*, *11*(10), 1211. Available from https://doi.org/10.3390/catal11101211.

Madej-Kiełbik, L., Gzyra-Jagieła, K., Jóźwik-Pruska, J., Dziuba, R., & Bednarowicz, A. (2022). Biopolymer composites with sensors for environmental and medical applications. *Materials*, *15*(21), 7493. Available from https://doi.org/10.3390/ma15217493.

McCluskey, A., Holdsworth, C. I., & Bowyer, M. C. (2007). Molecularly imprinted polymers (MIPs): Sensing, an explosive new opportunity? *Organic & Biomolecular Chemistry*, *5*(20). Available from https://doi.org/10.1039/b708660a, http://xlink.rsc.org/?DOI = b708660a, 3233.

McQuade, D. T., Pullen, A. E., & Swager, T. M. (2000). Conjugated polymer-based chemical sensors. *Chemical Reviews*, *100*(7), 2537–2574. Available from https://doi.org/10.1021/cr9801014.

Medeiros, G. B., Lima, F. D. A., de Almeida, D. S., Guerra, V. G., & Aguiar, M. L. (2022). Modification and functionalization of fibers formed by electrospinning: A Review. *Membranes*, *12*(9). Available from https://doi.org/10.3390/membranes12090861, http://www.mdpi.com/journal/membranes, 861.

Mizuno, Y., Theodosiou, A., Kalli, K., Liehr, S., Lee, H., & Nakamura, K. (2021). Distributed polymer optical fiber sensors: A review and outlook. *Photonics Research.*, *9*(9). Available from https://doi.org/10.1364/PRJ.435143, https://opg.optica.org/abstract.cfm?URI = prj-9-9-1719, 1719.

Murthy, H. C. A., Wagassa, A. N., Ravikumar, C. R., & Nagaswarupa, H. P. (2022). *Functionalized metal and metal oxide nanomaterial-based electrochemical sensors. Functionalized nanomaterial-based electrochemical sensors: principles, fabrication methods, and applications* (pp. 369–392). Ethiopia: Elsevier. Available from https://www.sciencedirect.com/book/9780128237885, https://doi.org/10.1016/B978-0-12-823788-5.00001-6.

Najjar, R., & Nematdoust, S. (2016). A resistive-type humidity sensor based on polypyrrole and ZnO nanoparticles: Hybrid polymers vis-a-vis nanocomposites. *RSC Advances.*, *6*(113), 112129–112139. Available from https://doi.org/10.1039/C6RA24002J, http://xlink.rsc.org/?DOI = C6RA24002J.

Naseri, M., Fotouhi, L., & Ehsani, A. (2018). Recent progress in the development of conducting polymer-based nanocomposites for electrochemical biosensors applications: A mini-review. *The Chemical Record*, *18*(6), 599–618. Available from https://doi.org/10.1002/tcr.201700101.

Naveen, M. H., Ganesh Gurudatt, N., & Shim, Y.-B. (2017). Applications of conducting polymer composites to electrochemical sensors: A review. *Applied Materials Today.*, *9*, 419–433. Available from https://doi.org/10.1016/j.apmt.2017.09.001, https://linkinghub.elsevier.com/retrieve/pii/S235294071730269X.

Sellergren, B., & Hall, A. J. (2001). Fundamental aspects on the synthesis and characterisation of imprinted network polymers. In *Techniques and Instrumentation in Analytical Chemistry*, (23, pp. 21–57). Elsevier. Available from https://doi.org/10.1016/S0167-9244(01)80005-6.

Naware, M., Thakor, N.V., Orth, R.N., Murari, K., & Passeraub, P. (2003). 12 2003/12 Annual International Conference of the IEEE Engineering in Medicine and Biology – Proceedings 1952–1955 United States design and microfabrication of a polymer-modified carbon sensor array for the measurement of neurotransmitter signals 2.

Nie, H., Zhao, Y., Zhang, M., Ma, Y., Baumgarten, M., & Müllen, K. (2011). Detection of TNT explosives with a new fluorescent conjugated polycarbazole polymer. *Chemical Communications*, *47*(4), 1234–1236. Available from https://doi.org/10.1039/C0CC03659E, http://xlink.rsc.org/?DOI = C0CC03659E.

Niskanen, J., & Tenhu, H. (2017). How to manipulate the upper critical solution temperature (UCST)? *Polymer Chemistry*, *8*(1), 220–232. Available from https://doi.org/10.1039/C6PY01612J.

Nitani, M., Nakayama, K., Maeda, K., Omori, M., & Uno, M. (2019). Organic temperature sensors based on conductive polymers patterned by a selective-wetting method. *Organic Electronics*, *71*, 164–168. Available from https://doi.org/10.1016/j.orgel.2019.05.006.

Okuzaki, H. (2019). Progress and current status of materials and properties of soft actuators. In K. Asaka, & H. Okuzaki (Eds.), *Soft Actuators*. Singapore: Springer Singapore. Available from https://doi.org/10.1007/978-981-13-6850-9_1.

Okutani, C., Yokota, T., & Someya, T. (2022). Ultrathin fiber-mesh polymer thermistors. *Advanced Science*, *9* (30), 2202312. Available from https://doi.org/10.1002/advs.202202312.

Oliveira, R., Sequeira, F., Bilro, L., & Nogueira, R. (2019). *Polymer optical fiber sensors and devices. Handbook of optical fibers* (pp. 1957–1996). Portugal: Springer Singapore. Available from https://link.springer.com/book/10.1007/978-981-10-7087-7, https://doi.org/10.1007/978-981-10-7087-7_1.

Otero, F., & Magner, E. (2020). Biosensors—Recent advances and future challenges in electrode materials. *Sensors (Basel, Switzerland)*, *20*(12), 3561. Available from https://doi.org/10.3390/s20123561.

Pablos, J. L., Trigo-López, M., Serna, F., García, F. C., & García, J. M. (2014a). Water-soluble polymers, solid polymer membranes, and coated fibres as smart sensory materials for the naked eye detection and quantification of TNT in aqueous media. *Chemical Communications*, *50*(19), 2484–2487. Available from https://doi.org/10.1039/C3CC49260E, http://xlink.rsc.org/?DOI = C3CC49260E.

Pablos, J. L., Trigo-López, M., Serna, F., García, F. C., & García, J. M. (2014b). Solid polymer substrates and smart fibres for the selective visual detection of TNT both in vapour and in aqueous media. *RSC Advances*, *4*(49), 25562–25568. Available from https://doi.org/10.1039/C4RA02716G, http://xlink.rsc.org/?DOI = C4RA02716G.

Pablos, J. L., Hernández, E., Catalina, F., & Corrales, T. (2022). Solid fluorescence pH sensors based on 1,8-naphthalimide copolymers synthesized by UV curing. *Chemosensors*, *10*, 73. Available from https://doi.org/10.3390/chemosensors10020073.

Pablos, J. L., Sarabia, L. A., Ortiz, M. C., Mendía, A., Muñoz, A., Serna, F., García, F. C., & García, J. M. (2015). Selective detection and discrimination of nitro explosive vapors using an array of three luminescent sensory solid organic and hybrid polymer membranes. *Sensors and Actuators, B: Chemical*, *212*, 18–27. Available from https://doi.org/10.1016/j.snb.2015.01.103.

Pang, Z., Fu, J., Luo, L., Huang, F., & Wei, Q. (2014). Fabrication of PA6/TiO2/PANI composite nanofibers by electrospinning–electrospraying for ammonia sensor. *Colloids and Surfaces A: Physicochemical and Engineering Aspects*, *461*(1), 113–118. Available from https://doi.org/10.1016/j.colsurfa.2014.07.038.

Pang, Z., Yang, Z., Chen, Y., Zhang, J., Wang, Q., Huang, F., & Wei, Q. (2016). A room temperature ammonia gas sensor based on cellulose/TiO 2 /PANI composite nanofibers. *Colloids and Surfaces A: Physicochemical and Engineering Aspects*, *494*, 248–255. Available from https://doi.org/10.1016/j.colsurfa.2016.01.024.

Patil, U. V., Ramgir, N. S., Karmakar, N., Bhogale, A., Debnath, A. K., Aswal, D. K., Gupta, S. K., & Kothari, D. C. (2015). Room temperature ammonia sensor based on copper nanoparticleintercalated polyaniline nanocomposite thin films. *Applied Surface Science*, *339*(1), 69–74. Available from https://doi.org/10.1016/j.apsusc.2015.02.164, http://www.journals.elsevier.com/applied-surface-science/.

Paulis, M., & Asua, J. M. (2016). Knowledge-based production of waterborne hybrid polymer materials. *Macromolecular Reaction Engineering*, *10*(1), 8–21. Available from https://doi.org/10.1002/mren.201500042, http://onlinelibrary.wiley.com/journal/10.1002/(ISSN)1862-8338.

Peters, K. (2011). Polymer optical fiber sensors—a review. *Smart Materials and Structures*, *20*(1). Available from https://doi.org/10.1088/0964-1726/20/1/013002, https://iopscience.iop.org/article/10.1088/0964-1726/20/1/013002, 013002.

Pomogailo, A. D. (2005). Polymer sol-gel synthesis of hybrid nanocomposites. *Colloid Journal*, *67*(6), 658–677. Available from https://doi.org/10.1007/s10595-005-0148-7.

Qi, J., Li, B., Wang, X., Zhang, Z., Wang, Z., Han, J., & Chen, L. (2017). Three-dimensional paper-based microfluidic chip device for multiplexed fluorescence detection of Cu2 + and Hg2 + ions based on ion imprinting technology. *Sensors and Actuators B: Chemical*, *251*, 224–233. Available from https://doi.org/10.1016/j.snb.2017.05.052, https://linkinghub.elsevier.com/retrieve/pii/S0925400517308651.

Ramanaviciene, A., & Plikusiene, I. (2021). Polymers in sensor and biosensor design. *Polymers*, *13*(6), 917. Available from https://doi.org/10.3390/polym13060917.

Ramanavičius, A., Ramanavičienė, A., & Malinauskas, A. (2006). Electrochemical sensors based on conducting polymer—Polypyrrole. *Electrochimica Acta*, *51*(27), 6025–6037. Available from https://doi.org/10.1016/j.electacta.2005.11.052, https://linkinghub.elsevier.com/retrieve/pii/S0013468606003252.

Rattan, S., Singhal, P., & Verma, A. L. (2013). Synthesis of PEDOT:PSS (poly(3,4-ethylenedioxythiophene))/poly(4-styrene sulfonate)/ngps (nanographitic platelets) nanocomposites as chemiresistive sensors for detection of nitroaromatics. *Polymer Engineering & Science*, *53*(10), 2045–2052. Available from https://doi.org/10.1002/pen.23466.

Rivero, P., Goicoechea, J., & Arregui, F. (2018). Optical fiber sensors based on polymeric sensitive coatings. *Polymers*, *10*(3), 280. Available from https://doi.org/10.3390/polym10030280.

Namsheer, K., & Rout, C. S. (2021). Conducting polymers: A comprehensive review on recent advances in synthesis, properties and applications. *RSC Advances*. Available from https://doi.org/10.1039/D0RA07800J.

Rubio-Aguinaga, A., Reglero-Ruiz, J. A., García-Gómez, A., Peña Martín, E., Ando, S., Muñoz, A., García, J. M., & Trigo-López, M. (2023). Boron nitride-reinforced porous aramid composites with enhanced mechanical performance and thermal conductivity. *Composites Science and Technology*, *242*. Available from https://doi.org/10.1016/j.compscitech.2023.110211, http://www.journals.elsevier.com/composites-science-and-technology/, 110211.

Runsewe, D., Betancourt, T., & Irvin, J. A. (2019). Biomedical application of electroactive polymers in electrochemical sensors: A review. *Materials*, *12*(16). Available from https://doi.org/10.3390/ma12162629, 2629–2629, Available from, https://www.mdpi.com/1996-1944/12/16/2629, 2629.

Rusli, A., Othman, M. B. H., & Ku Marsilla, K. I. (2022). Plastics in high heat resistant applications. *Encyclopedia of Materials: Plastics and Polymers*, 1–4. Available from https://doi.org/10.1016/B978-0-12-820352-1.00073-0, https://www.sciencedirect.com/book/9780128232910.

Sadik, O. A., & Wallace, G. G. (1993). Pulse damperometric detection of proteins using antibody containing conducting polymers. *Analytica Chimica Acta*, *279*(2), 209–212. Available from https://doi.org/10.1016/0003-2670(93)80319-g.

Sajini, T., & Mathew, B. (2021). A brief overview of molecularly imprinted polymers: Highlighting computational design, nano and photo-responsive imprinting. *Talanta Open*, *4*, 100072. Available from https://doi.org/10.1016/j.talo.2021.100072.

Salavagione, H. J., Díez-Pascual, A. M., Lázaro, E., Vera, S., & Gómez-Fatou, M. A. (2014). Chemical sensors based on polymer composites with carbon nanotubes and graphene: The role of the polymer. *Journal of Materials Chemistry A*, *2*(35), 14289–14328. Available from https://doi.org/10.1039/C4TA02159B, http://xlink.rsc.org/?DOI = C4TA02159B.

Samanta, S., Roy, P., & Kar, P. (2020). Sensing of ethanol and other alcohol contaminated ethanol by conducting functional poly(o-phenylenediamine. *Materials Science and Engineering: B*, *256*, 114541. Available from https://doi.org/10.1016/j.mseb.2020.114541.

Samuel, I. D. W., Rumbles, G., Collison, C. J., Friend, R. H., Moratti, S. C., & Holmes, A. B. (1997). Picosecond time-resolved photoluminescence of PPV derivatives. *Synthetic Metals*, *84*(1–3), 497–500. Available from https://doi.org/10.1016/s0379-6779(97)80837-5.

Sanagi, M. M., Salleh, S., Ibrahim, W. A. W., Naim, A. A., Hermawan, D., Miskam, M., Hussain, I., & Aboul-Enein, H. Y. (2013). Molecularly imprinted polymer solid-phase extraction for the analysis of organophosphorus pesticides in fruit samples. *Journal of Food Composition and Analysis*, *32*(2), 155–161. Available from https://doi.org/10.1016/j.jfca.2013.09.001.

Sanjuán, A. M., Reglero Ruiz, J. A., García, F. C., & García, J. M. (2018). Recent developments in sensing devices based on polymeric systems. *Reactive and Functional Polymers*, *133*, 103–125. Available from https://doi.org/10.1016/j.reactfunctpolym.2018.10.007.

Santos, J. L., & Fahari, F. (2014). Overview of optical sensing. In F. Farahi (Ed.), *Handbook of Optical Sensors* (1st, pp. 1–10). Boca Raton: CRC Press.

Sartore, L., Barbaglio, M., Borgese, L., & Bontempi, E. (2011). Polymer-grafted QCM chemical sensor and application to heavy metal ions real time detection. *Sensors and Actuators B: Chemical.*, *155*(2), 538–544. Available from https://doi.org/10.1016/j.snb.2011.01.003, https://linkinghub.elsevier.com/retrieve/pii/S0925400511000062.

Sassolas, A., Blum, L. J., & Leca-Bouvier, B. D. (2012). Immobilization strategies to develop enzymatic biosensors. *Biotechnology Advances*, *30*(3), 489–511. Available from https://doi.org/10.1016/j.biotechadv.2011.09.003.

Saylan, Y., Akgönüllü, S., Yavuz, H., Ünal, S., & Denizli, A. (2019). Molecularly imprinted polymer based sensors for medical applications. *Sensors (Basel, Switzerland)*, *19*(6), 1279. Available from https://doi.org/10.3390/s19061279.

Schuhmann, W., Lehn, C., Schmidt, H. L., & Gründig, B. (1992). Comparison of native and chemically stabilized enzymes in amperometric enzyme electrodes. *Sensors and Actuators: B. Chemical*, *7*(1–3), 393–398. Available from https://doi.org/10.1016/0925-4005(92)80331-Q.

Seuring, J., & Agarwal, S. (2013). Polymers with upper critical solution temperature in aqueous solution: Unexpected properties from known building blocks. *ACS Macro Letters.*, *2*(7), 597–600. Available from https://doi.org/10.1021/mz400227y.

Shahbazi, M., & Jäger, H. (2021). Current status in the utilization of biobased polymers for 3D printing process: A systematic review of the materials, processes, and challenges. *ACS Applied Bio Materials*, *4*(1), 325–369. Available from https://doi.org/10.1021/acsabm.0c01379.

Sharafeldin, M., Jones, A., & Rusling, J. (2018). 3D-printed biosensor arrays for medical diagnostics. *Micromachines*, *9*(8), 394. Available from https://doi.org/10.3390/mi9080394.

Shen, M., Rusling, J. F., & Dixit, C. K. (2017). Site-selective orientated immobilization of antibodies and conjugates for immunodiagnostics development. *Methods*, *116*, 95–111. Available from https://doi.org/10.1016/j.ymeth.2016.11.010, http://www.elsevier.com/inca/publications/store/6/2/2/9/1/4/index.htt.

Shen, Y., Miao, P., Liu, S., Gao, J., Han, X., Zhao, Y., & Chen, T. (2023). Preparation and application progress of imprinted polymers. *Polymers*, *15*(10), 2344. Available from https://doi.org/10.3390/polym15102344.

Si, P., Mortensen, J., Komolov, A., Denborg, J., & Juul Møller, P. (2007). Polymer coated quartz crystal microbalance sensors for detection of volatile organic compounds in gas mixtures. *Analytica Chimica Acta*, *597*(2), 223–230. Available from https://doi.org/10.1016/j.aca.2007.06.050, https://linkinghub.elsevier.com/retrieve/pii/S0003267007010720.

Simões, F. R., & Xavier, M. G. (2017). *Electrochemical sensors nanoscience and its applications* (pp. 155–178). Brazil: Elsevier Inc.. Available from http://www.sciencedirect.com/science/book/9780323497800, 10.1016/B978-0-323-49780-0.00006-5.

Smilowitz, L., Hays, A., Heeger, A. J., Wang, G., & Bowers, J. E. (1993). Time-resolved photoluminescence from poly[2-methoxy, 5-(2′-ethyl-hexyloxy)- p -phenylene-vinylene]: Solutions, gels, films, and blends. *The Journal of Chemical Physics*, *98*(8), 6504–6509. Available from https://doi.org/10.1063/1.464790.

Smith, M., & Kar-Narayan, S. (2022). Piezoelectric polymers: Theory, challenges and opportunities. *International Materials Reviews*, *67*(1), 65–88. Available from https://doi.org/10.1080/09506608.2021.1915935.

Somerset, V., Leaner, J., Mason, R., Iwuoha, E., & Morrin, A. (2010a). Development and application of a poly (2,2′-dithiodianiline) (PDTDA)-coated screen-printed carbon electrode in inorganic mercury determination. *Electrochimica Acta*, *55*(14), 4240–4246. Available from https://doi.org/10.1016/j.electacta.2009.01.029, https://linkinghub.elsevier.com/retrieve/pii/S0013468609001066.

Somerset, V., Leaner, J., Mason, R., Iwuoha, E., & Morrin, A. (2010b). Determination of inorganic mercury using a polyaniline and polyaniline-methylene blue coated screen-printed carbon electrode. *International Journal of Environmental Analytical Chemistry*, *90*(9), 671–685. Available from https://doi.org/10.1080/03067310902962536, http://www.tandfonline.com/doi/abs/10.1080/03067310902962536.

Song, R., Murphy, M., Li, C., Ting, K., Soo, C., & Zheng, Z. (2018). Current development of biodegradable polymeric materials for biomedical applications. *Drug Design, Development and Therapy*, *12*, 3117–3145. Available from https://doi.org/10.2147/dddt.s165440.

Sperling, L. H. (2005). *Introduction to physical polymer science. Introduction to physical polymer science* (4th ed., pp. 1–845). United States: John Wiley and Sons. Available from http://onlinelibrary.wiley.com/book/10.1002/0471757128, 10.1002/0471757128.

Spychalska, K., Zając, D., Baluta, S., Halicka, K., & Cabaj, J. (2020). Functional polymers structures for (bio) sensing application—A review. *Polymers*, *12*(5), 1154. Available from https://doi.org/10.3390/polym12051154.

Staudinger, H. (1920). Über polymerisation. *Berichte der deutschen chemischen Gesellschaft (A and B Series)*, *53*(6), 1073–1085. Available from https://doi.org/10.1002/cber.19200530627, https://onlinelibrary.wiley.com/doi/10.1002/cber.19200530627.

Staudinger, H., & Fritschi, J. (1922). Über Isopren und Kautschuk. 5. Mitteilung. Über die Hydrierung des Kautschuks und über seine Konstitution. *Helvetica Chimica Acta*, *5*(5), 785–806. Available from https://doi.org/10.1002/hlca.19220050517, https://onlinelibrary.wiley.com/doi/10.1002/hlca.19220050517.

Su, S., Chen, S., & Fan, C. (2018). Recent advances in two-dimensional nanomaterials-based electrochemical sensors for environmental analysis. *Green Energy & Environment*, *3*(2), 97–106. Available from https://doi.org/10.1016/j.gee.2017.08.005.

Suksai, C., & Tuntulani, T. (2005). *Chromogenic anion sensors* (255). Springer Nature. Available from https://doi.org/10.1007/b101166.

Sun, B., McCay, R. N., Goswami, S., Xu, Y., Zhang, C., Ling, Y., Lin, J., & Yan, Z. (2018). Gas-permeable, multifunctional on-skin electronics based on laser-induced porous graphene and sugar-templated elastomer sponges. *Advanced Materials*, *30*(50), 1804327. Available from https://doi.org/10.1002/adma.201804327.

Sutton, M. (2020). The birth of the polymer age. *Chemistry World*. Available from https://www.chemistryworld.com/features/the-birth-of-the-polymer-age/4011418.article#commentsJump.

Szczerska, M. (2022). Temperature sensors based on polymer fiber optic interferometer. *Chemosensors*, *10*(6), 228. Available from https://doi.org/10.3390/chemosensors10060228.

Taniya, O. S., Khasanov, A. F., Sadieva, L. K., Santra, S., Nikonov, I. L., Al-Ithawi, W. K. A., Kovalev, I. S., Kopchuk, D. S., Zyryanov, G. V., & Ranu, B. C. (2023). Polymers and polymer-based materials for the detection of (nitro-)explosives. *Materials*, *16*(18). Available from https://doi.org/10.3390/ma16186333, https://www.mdpi.com/1996-1944/16/18/6333, 6333.

Teng, C., Min, R., Zheng, J., Deng, S., Li, M., Hou, L., & Yuan, L. (2022). Intensity-modulated polymer optical fiber-based refractive index sensor: A review. *Sensors (Basel, Switzerland)*, *22*(1), 81. Available from https://doi.org/10.3390/s22010081.

Thammakhet, C., Thavarungkul, P., & Kanatharana, P. (2011). Development of an on-column affinity smart polymer gel glucose sensor. *Analytica Chimica Acta*, *695*(1–2), 105–112. Available from https://doi.org/10.1016/j.aca.2011.03.062, https://linkinghub.elsevier.com/retrieve/pii/S000326701100465X.

Tian, L., Hu, W., Zhong, X., & Liu, B. (2015). Glucose sensing characterisations of TiO2/CuO nanofibres synthesised by electrospinning. *Materials Research Innovations*, *19*(3), 160–165. Available from https://doi.org/10.1179/1433075X14Y.0000000236, http://www.maneyonline.com/doi/pdfplus/10.1179/1433075X14Y.0000000236.

Toal, S. J., & Trogler, W. C. (2006). Polymer sensors for nitroaromatic explosives detection. *Journal of Materials Chemistry*, *16*(28). Available from https://doi.org/10.1039/b517953j, http://xlink.rsc.org/?DOI = b517953j, 2871.

Tomić, M., Šetka, M., Vojkůvka, L., & Vallejos, S. (2021). VOCs sensing by metal oxides, conductive polymers, and carbon-based materials. *Nanomaterials*, *11*(2). Available from https://doi.org/10.3390/nano11020552, https://www.mdpi.com/2079-4991/11/2/552, 552.

Trigo-López, M., Muñoz, A., Mendía, A., Ibeas, S., Serna, F., García, F. C., & García, J. M. (2018). Palladium-containing polymers as hybrid sensory materials (water-soluble polymers, films and smart textiles) for the colorimetric detection of cyanide in aqueous and gas phases. *Sensors and Actuators, B: Chemical*, *255*, 2750–2755. Available from https://doi.org/10.1016/j.snb.2017.09.089.

Trigo-López, M., Pablos, J. L., Muñoz, A., Ibeas, S., Serna, F., García, F. C., & García, J. M. (2015). Aromatic polyamides and acrylic polymers as solid sensory materials and smart coated fibres for high acidity colorimetric sensing. *Polymer Chemistry*, *6*(16), 3110–3120. Available from https://doi.org/10.1039/c4py01545b, http://pubs.rsc.org/en/Journals/JournalIssues/PY.

Turiel, E., & Esteban, A. M. (2020). *Molecularly imprinted polymers* (pp. 215–233). Elsevier BV. Available from https://doi.org/10.1016/b978-0-12-816906-3.00008-x.

Uludağ, Y., Piletsky, S. A., Turner, A. P. F., & Cooper, M. A. (2007). Piezoelectric sensors based on molecular imprinted polymers for detection of low molecular mass analytes. *FEBS Journal*, *274*(21), 5471–5480. Available from https://doi.org/10.1111/j.1742-4658.2007.06079.x.

Usher, T. D., Cousins, K. R., Zhang, R., & Ducharme, S. (2018). The promise of piezoelectric polymers. *Polymer International*, *67*(7), 790–798. Available from https://doi.org/10.1002/pi.5584, http://onlinelibrary.wiley.com/journal/10.1002/(ISSN)1097-0126.

Uygun, Z. O., Ertugrul Uygun, H. D., Ermis, N., & Canbay, E. (2015). *Molecularly imprinted sensors—New sensing technologies*. InTech. Available from 10.5772/60781.

Vallejos, S., Estévez, P., Ibeas, S., García, F. C., Serna, F., & García, J. M. (2012). An organic/inorganic hybrid membrane as a solid "Turn-On" fluorescent chemosensor for coenzyme a (CoA), cysteine (Cys), and glutathione (GSH) in aqueous media. *Sensors (Basel, Switzerland)*, *12*(3), 2969–2982. Available from https://doi.org/10.3390/s120302969Spain, http://www.mdpi.com/1424-8220/12/3/2969/pdf.

Vallejos, S., Estévez, P., Ibeas, S., Muñoz, A., García, F. C., Serna, F., & García, J. M. (2011). A selective and highly sensitive fluorescent probe of Hg2 + in organic and aqueous media: The role of a polymer network in extending the sensing phenomena to water environments. *Sensors and Actuators, B: Chemical*, *157*(2), 686–690. Available from https://doi.org/10.1016/j.snb.2011.05.041.

Vallejos, S., El Kaoutit, H., Estévez, P., García, F. C., De La Peña, J. L., Sema, F., & García, J. M. (2011). Working with water insoluble organic molecules in aqueous media: Fluorene deri vative-containing polymers as sensory materials for the colorimetric sensing of cyanide in waterf. *Polymer Chemistry*, *2*(5), 1129–1138. Available from https://doi.org/10.1039/c1py00013f.

Vallejos, S., Trigo-López, M., Arnaiz, A., Miguel, Á., & García-García, J. M. (2022). Specialty polymers for food and water quality control and safety. In R. K. Gupta (Ed.), Smart polymers. Boca Raton: CRC Press.

Vallejos, S., Marta, G.-G., José, G., Represa, M., García, C., & Clemente, F. (2021a). Colorimetric titration. Available from https://play.google.com/store/apps/details?id = es.inforapps.chameleon&gl = ES.

Vallejos, S., Marta, G.-G., José, G., Represa, M., García, C., & Clemente, F. (2021b), Colorimetric titration. Available from https://apps.apple.com/si/app/colorimetric-titration/id1533793244.

Vallejos, S., Trigo-López, M., Arnaiz, A., Miguel, Á., Muñoz, A., Mendía, A., & Miguel García, J. (2022). From classical to advanced use of polymers in food and beverage applications. *Polymers*, *14*(22), 4954. Available from https://doi.org/10.3390/polym14224954.

Vaseashta, A. (n.d.). *Nanostructured materials based next generation devices and sensors*. Springer Science and Business Media LLC. Available from https://doi.org/10.1007/1-4020-3562-4_1.

Verma, A., Gupta, R., Verma, A. S., & Kumar, T. (2023). A review of composite conducting polymer-based sensors for detection of industrial waste gases. *Sensors and Actuators Reports.*, *5*, 100143. Available from https://doi.org/10.1016/j.snr.2023.100143.

Wang, G. N., Yang, K., Liu, H. Z., Feng, M. X., & Wang, J. P. (2016). Molecularly imprinted polymer-based solid phase extraction combined high performance liquid chromatography for determination of fluoroquinolones in milk. *Analytical Methods*, *8*(27), 5511–5518. Available from https://doi.org/10.1039/c6ay00810k, http://www.rsc.org/Publishing/Journals/AY/About.asp.

Vora Rohitkumar H. & Lau, Kreisler S.Y. (2022). *Research and development of high-performance polymeric materials including polyimides and fluoro-polyimides and their industrialized products, Editor(s): Hanna Dodiuk, In Plastics Design Library, Handbook of Thermoset Plastics* (Fourth Edition), William Andrew Publishing, 266–325, ISBN 9780128216323, https://doi.org/10.1016/B978-0-12-821632-3.00036-1.

Wang, X., Li, Y., & Ding, B. (2014). *Electrospun nanofiber-based sensors.*, 267–297. Available from https://doi.org/10.1007/978-3-642-54160-5_11, https://link.springer.com/10.1007/978-3-642-54160-5_11.

Wu, J., & Gu, M. (2011). Microfluidic sensing: State of the art fabrication and detection techniques. *Journal of Biomedical Optics*, *16*(8). Available from https://doi.org/10.1117/1.3607430, http://biomedicaloptics.spiedigitallibrary.org/article.aspx?doi = 10.1117/1.3607430, 080901.

Wu, Y.-hui, Liu, H.-zhou, Chen, S., Dong, X.-chu, Wang, P.-P., Liu, S.-qi, Lin, Y., Wei, Y., & Liu, L. (2017). Channel crack-designed gold@PU sponge for highly elastic piezoresistive sensor with excellent detectability. *ACS Applied Materials & Interfaces*, *9*(23), 20098–20105. Available from https://doi.org/10.1021/acsami.7b04605.

Xue, C., Han, Q., Wang, Y., Wu, J., Wen, T., Wang, R., Hong, J., Zhou, X., & Jiang, H. (2013). Amperometric detection of dopamine in human serumby electrochemical sensor based on gold nanoparticles doped molecularly imprinted polymers. *Biosensors and Bioelectronics*, *49*, 199–203. Available from https://doi.org/10.1016/j.bios.2013.04.022.

Xue, J., Wu, T., Dai, Y., & Xia, Y. (2019). Electrospinning and electrospun nanofibers: Methods, materials, and applications. *Chemical Reviews*, *119*(8), 5298–5415. Available from https://doi.org/10.1021/acs.chemrev.8b00593.

Yan, H., & Row, K. (2006). Characteristic and synthetic approach of molecularly imprinted polymer. *International Journal of Molecular Sciences*, *7*(5), 155–178. Available from https://doi.org/10.3390/i7050155.

Yang, Y., Yu, J., Yin, J., Shao, B., & Zhang, J. (2014). Molecularly imprinted solid-phase extraction for selective extraction of bisphenol analogues in beverages and canned food. *Journal of Agricultural and Food Chemistry*, *62*(46), 11130–11137. Available from https://doi.org/10.1021/jf5037933.

Yang, Y., & Shen, X. (2022). Preparation and application of molecularly imprinted polymers for flavonoids: Review and perspective. *Molecules*, *27*(21), 7355. Available from https://doi.org/10.3390/molecules27217355.

Ye, L., & Haupt, K. (2004). Molecularly imprinted polymers as antibody and receptor mimics for assays, sensors and drug discovery. *Analytical and Bioanalytical Chemistry*, *378*(8), 1887–1897. Available from https://doi.org/10.1007/s00216-003-2450-8.

Ye, X., Ge, L., Jiang, T., Guo, H., Chen, B., Liu, C., & Hayashi, K. (2022). Fully inkjet-printed chemiresistive sensor array based on molecularly imprinted sol–gel active materials. *ACS Sensors*, *7*(7), 1819–1828. Available from https://doi.org/10.1021/acssensors.2c00093.

Yoon, H. (2013). Current trends in sensors based on conducting polymer nanomaterials. *Nanomaterials*, *3*(3), 524–549. Available from https://doi.org/10.3390/nano3030524.

You, C.-C., Miranda, O. R., Gider, B., Ghosh, P. S., Kim, I.-B., Erdogan, B., Krovi, S. A., Bunz, U. H. F., & Rotello, V. M. (2007). Detection and identification of proteins using nanoparticle–fluorescent polymer 'chemical nose' sensors. *Nature Nanotechnology*, *2*(5), 318–323. Available from https://doi.org/10.1038/nnano.2007.99, https://www.nature.com/articles/nnano.2007.99.

Zegebreal, L. T., Tegegne, N. A., & Hone, F. G. (2023). Recent progress in hybrid conducting polymers and metal oxide nanocomposite for room-temperature gas sensor applications: A review. *Sensors and Actuators A: Physical*, *359*, 114472. Available from https://doi.org/10.1016/j.sna.2023.114472, https://linkinghub.elsevier.com/retrieve/pii/S0924424723003217.

Zhang, H., Lin, Y., Xu, Y., & Weng, W. (2015). *Mechanochemistry of topological complex polymer systems* (369). Springer Science and Business Media LLC. Available from 10.1007/128_2014_617.

Zheng, W., Wu, H., Jiang, Y., Xu, J., Li, X., Zhang, W., & Qiu, F. (2018). A molecularly-imprinted-electrochemical-sensor modified with nano-carbon-dots with high sensitivity and selectivity for rapid determination of glucose. *Analytical Biochemistry*, *555*, 42–49. Available from https://doi.org/10.1016/j.ab.2018.06.004.

Zhong, H., Zhao, B., & Deng, J. (2023). Synthesis and application of fluorescent polymer micro- and nanoparticles. *Small (Weinheim an der Bergstrasse, Germany)*, *19*(26). Available from https://doi.org/10.1002/smll.202300961, https://onlinelibrary.wiley.com/doi/10.1002/smll.202300961, 2300961.

SECTION

Sensory polymers for advanced applications

1

CHAPTER

Sensors based on conjugated polymers

2

Sevki Can Cevher[1], Dilek Soyler[2], Umut Bulut[3] and Saniye Soylemez[2]

[1]*Institute of Computational Physics, Zurich University of Applied Sciences, ZHAW, Winterthur, Zurich, Switzerland*
[2]*Faculty of Engineering, Department of Biomedical Engineering, Necmettin Erbakan University, Meram, Konya, Turkey* [3]*Faculty of Pharmacy, Department of Analytical Chemistry, Acıbadem Mehmet Ali Aydınlar University, Ataşehir, Istanbul, Turkey*

2.1 Introduction

The first polymer discovered as a conductive material was polyacetylene. When polyacetylene is doped with a dopant anion, positively charged species are created. With the help of these charged species, right on the way of π-conjugation of the polymer backbone, electrons can be delocalized. As a result, electrons, or electricity, can be carried through the polymer chain. Accordingly, three leading scientists, Shirakawa, Heeger, and MacDiarmid (Chiang et al., 1977), were awarded the Nobel Prize in Chemistry in 2002 for showing "how plastic can be made to conduct electric current." The discovery of the unique charge transport of conjugated polymers (CPs) gave birth to an extraordinary opponent to metals in electronic devices. It is noteworthy here that migration of charges occurs not only along the backbone of the polymer chain but also in a π-stacked direction onto the segment of adjacent polymer chains in the crystalline regions of their thin film morphologies. From this date forward, organic materials were used to construct numerous electronic devices, which are called organic electronics. In time, the structure of conjugated organic materials was easily tailored with the help of new and existing synthetic methodologies and the availability of chemical reagents and reactants. Thus, the unique requirements of different applications can be successfully satisfied by modifying organic structures. Currently, various commercial products from organic light-emitting diodes (OLEDs), including flexible OLEDs for wearable electronics, are on the market (Gamota, 2020), and more are on the way to commercialization.

Among various fields of organic electronics, sensors are particularly prominent devices. They can both monitor essential features and detect the concentration or identity of chemical substances vital for the natural environment and/or human lives. Among sensors, bio- or chemo-sensors constructed by CPs are generally categorized depending on optical change (fluorescence or absorption) or electrical response (oxidation/reduction properties, capacitance, or conductivity) (McQuade et al., 2000). Based on transduction modes, operating principles and device configuration may change. Thus, analyte recognition *via* physical or chemical interactions and sensing mechanisms with signal processing methodologies should be carefully investigated. Pristine CPs have limited application due to the lack of strong interaction between the analyte and the CPs to detect a specific

Sensory Polymers. DOI: https://doi.org/10.1016/B978-0-443-13394-7.00002-1

target in a complex environment, such as wastewater and blood. Thus, well-configured structure-property relations between the CPs and the analyte are essential. We will further discuss, in this part, the emerging challenges correlating structure-property relations for individual sensors.

Pristine CPs built the key understanding of sensors by using early-type CPs (Rasmussen, 2020), as shown in Fig. 2.1. Although pristine CPs are still in use, new perspectives are now being implemented in designing CPs for sensors to improve their conductivity, selectivity, and detection limits. Hence, the performance criteria are all dependent on the intrinsic properties of CPs and the mechanism of generation of a measurable signal while the analyte is under inspection. Therefore, three main strategies are offered to make CPs versatile and capable, pushing forward the boundaries of sensing technology: (1) functionalization of the π-conjugated main chain; (2) functionalization of the pendant groups; and (3) producing composites of CPs with secondary or auxiliary materials. These strategies are investigated in detail for individual examples here and there. It is worthy to state here that renowned CPs such as polypyrroles, polyanilines, and polythiophenes are the choice

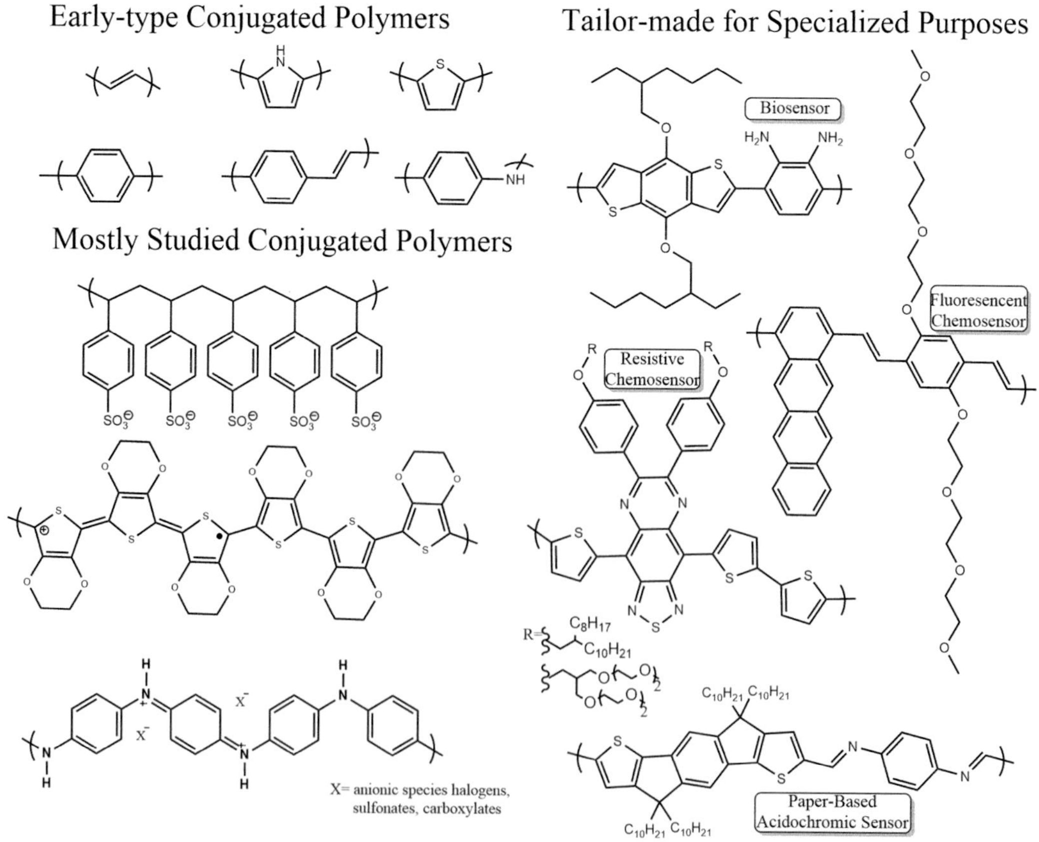

FIGURE 2.1

Representative presentation of the structures of CPs.

of materials due to their intrinsic high electrical conductivities, various processing techniques, and commercial availability. Yet, in the literature, more sophisticated molecular structures are also chosen for specific applications like fluorescence or chemiresistive sensors.

Among the very first pristine CPs, polypyrrole was the first polymer where enzyme molecules were entrapped decades ago by Leland C. Clark, who is considered "the father of biosensors" (Clark & Lyons, 1962). Biosensors, constructed by entrapping glucose oxidase, an enzyme, at the electrode surface, irreversibly detect the formation of hydrogen peroxide by the enzymatic reaction of glucose and glucose oxidase. The sensing mechanism depends on the electrochemical reduction of hydrogen peroxide, and it correlates the concentration of hydrogen peroxide to glucose. Belanger (Bélanger et al., 1989) and later Cooper et al. (Cooper & Bloor, 1993) proved the success of hydrogen peroxide scavengers to cease the response to glucose. Since then, CPs have been considered entrapment surfaces, mediators, and transduction elements.

CPs are attractive materials thanks to their controllable thickness and porosity during electropolymerization or coating because of their intrinsic structural properties and their applicability to small areas to construct microelectrodes. Moreover, enzymes can be entrapped physically or covalently to CPs films (Rahman et al., 2008), and numerous entrapped or modified CPs film-biorecognition element couples can be found in the literature (McQuade et al., 2000; Nguyen et al., 2023; Rahman et al., 2008).

The development of the biosensor field aligned with the health concerns of individuals facilitates the rise in the medical trend of patient self-monitoring of health. Point-of-care (POC) diagnostics are classified as analytical assays outside the laboratory, ensuring rapid testing of target analytes with the same or almost similar accuracy as laboratory tests. The recent COVID-19 pandemic had an insightful impact on the development and use of POC diagnosis with rapid test kits. Thanks to their excellent photophysical and electrochemical properties as well as their biocompatibility with facile modifications, CPs have great potential for POC devices: naked-eye detection of pH1N1 virus by polydiacetylene (PDA) (**polymer A**) (see in Fig. 2.2), CPs-based paper chips through a color change (Son et al., 2019), paper-based POC assay containing luminescent CPs, poly[{2-methoxy-5-(2-ethylhexyloxy)-1,4-(1-cyanovinylenephenylene)}-co-{2,5-bis(N,N′-diphenylamino)-1,4-phenylene}] (**Polymer B**) (Fig. 2.2), for detection of blood phenylalanine levels for phenylketonuria screening (Chen, Yu, et al., 2021), colorimetric and fluorometric immunochromatographic test strips based on gold-CP dots (prepared by **Polymer C** and **Polymer D**) (Fig. 2.2), nanohybrids for the determination of carcinoembryonic antigen (CEA), and cytokeratin 19 fragment (CYFRA 21−1) (Yang et al., 2021), and for quantitative rapid screening of prostate-specific antigen, α-fetoprotein, and CEA CP dot-based (**Polymer C**, **Polymer D**, and **Polymer E**) (Fig. 2.2) immunochromatographic test strips were developed (Fang et al., 2018).

The development of CP-based biosensor devices, particularly POC systems, involves several key steps.

2.1.1 Design and synthesis of conjugated polymers

Developing and synthesizing CPs with specific properties as discussed previously, that is, fluorescent or containing a functional group for incorporation of biorecognition elements suitable for targeted biosensing applications. These polymers are designed to have a conjugated backbone, which allows for efficient charge or energy transfer upon interaction with the target biomolecule. It is

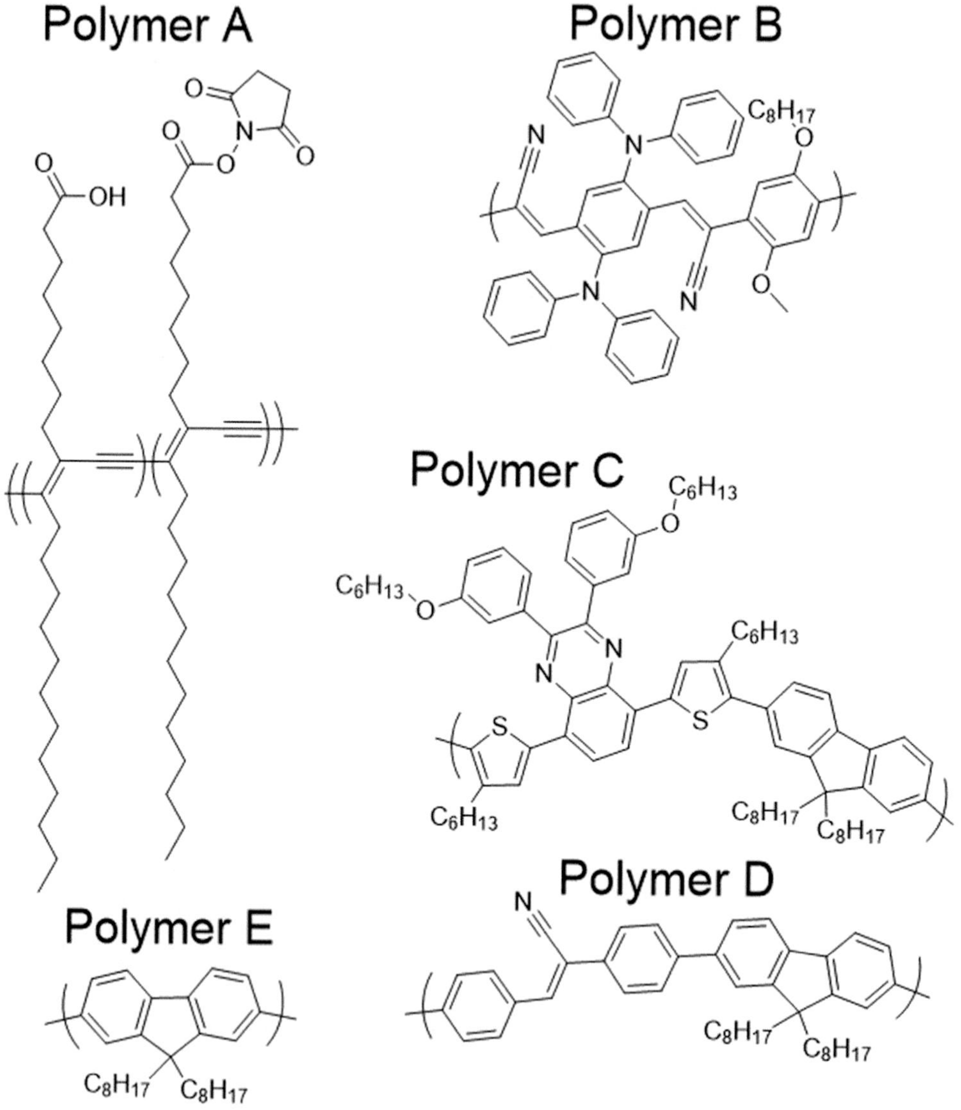

FIGURE 2.2

Examples: CP structures used for point-of-care diagnosis.

important to note that the scientific approach may vary depending on the specific design and application of the CPs-based biosensors. Researchers continuously explore new materials, recognition elements, and detection techniques to improve the performance and expand the capabilities of these biosensors, thanks to the vast availability of tuning chemical structures for CPs. This tunability allows for the design of sensors with specific selectivity toward target analytes. By tailoring the optical and electronic properties of polymers, it becomes possible to create sensors that are highly sensitive to a particular analyte while being less responsive to interfering substances.

2.1.2 Functionalization and immobilization

The CPs are functionalized with specific recognition elements, such as antibodies, aptamers, or enzymes, that can selectively bind to the target biomolecule. These recognition elements enhance the sensitivity and selectivity of the biosensor. The functionalized polymers are then immobilized onto a solid support, such as a substrate or an electrode surface. In the case of electrochemical polymerization, the functionalized CPs are generated on the surface of the electrode. Even more, molecular imprinting techniques cover the *in situ* generation of an electrochemically polymerized CPs electrode substrate containing an imprinted biorecognition element (Pilvenyte et al., 2023).

2.1.3 Transduction mechanism

Depending on the interaction between the target biomolecule and the immobilized CPs with the biorecognition element, qualitative or quantitative information about the target biomolecule is gathered. These interactions can lead to changes in the electrical or optical properties of the CPs. This transduction mechanism can be based on various principles, such as fluorescence, electrochemical, or conductance changes. Sensors based on their transduction mechanisms are further discussed in the next section.

The choice of transduction mechanism depends on the specific application, device configuration, desired sensitivity, and detection limits. These options, on the other hand, may restrict the choice of materials and applications. For example, light in the visible or ultraviolet region has a penetration limit through the skin, and scattering of individual layers of skin (Finlayson et al., 2022) also influences the choice of light that is examined.

2.1.4 Signal detection and analysis

The changes in the electrical or optical properties of CPs are measured using appropriate detection techniques. For example, fluorescence-based biosensors may utilize spectroscopic methods to measure changes in emission intensity or wavelength. Electrochemical biosensors may employ techniques such as voltammetry or impedance spectroscopy to measure changes in current or impedance. These signals can be informative in both qualitative and quantitative ways, depending on the system. Moreover, the complementary analysis system can be all in one integrated circuit that leads to wearable, implantable, and even wireless devices (Kim et al., 2023).

2.1.5 Calibration and validation

To ensure accurate and reliable results, calibration curves are established using known concentrations of the target biomolecule. These curves help correlate the measured signal changes with the concentration of the analyte. Validation studies are conducted to assess the biosensor's performance, including sensitivity, specificity, reproducibility, and stability. The interference parameters are important factors that should be taken into consideration thoroughly to avoid any false negative or positive results and incorrect reporting.

2.1.6 Point-of-care integration

CPs-based biosensors are integrated into portable and user-friendly devices suitable for POC applications. These devices may include microfluidic systems for sample handling, miniaturized electronics for signal processing, and user interfaces for result interpretation. At this stage, extensive experimental studies are necessary. Each stage has its own challenging environment that needs to be tackled to improve the suitability of the device for being biocompatible, non-toxic, implantable, or wearable and operating without external expert support for friendly use for general use purposes.

POC biosensor devices are small, portable, and user-friendly diagnostic tools that integrate biological sensing elements with electronic components. They are designed to provide real-time analysis of biological samples, such as blood, saliva, or urine, at the POC, which can be a hospital, clinic, or even a patient's home. POC biosensor devices enable healthcare professionals to quickly and accurately diagnose various medical conditions. They can detect biomarkers or specific molecules in bodily fluids, such as blood or urine, allowing for timely and targeted treatment decisions. These devices enable early detection and monitoring of various medical conditions, including infectious diseases, chronic diseases, and metabolic disorders. Healthcare providers can make immediate treatment decisions based on real-time diagnostic results. This can lead to faster initiation of appropriate therapies, reducing the risk of complications, and improving patient outcomes. They eliminate the need for sending samples to a central laboratory for analysis, which can be time-consuming and inconvenient for patients. POC biosensor devices can potentially reduce healthcare costs by minimizing the need for expensive laboratory tests and associated infrastructure. They can also help prevent unnecessary hospital admissions or visits, leading to cost savings for both patients and healthcare systems. These devices are particularly valuable in remote or resource-limited settings where access to laboratory facilities may be limited. POC biosensor devices enable healthcare providers to deliver timely and accurate diagnostics even in challenging environments, improving healthcare delivery in underserved areas. It is important to note that while POC biosensor devices offer numerous advantages, they also have limitations. These devices may have lower sensitivity or specificity compared to laboratory-based tests, and their accuracy can vary depending on the specific device and the expertise of the user. Therefore, proper training, quality control measures, and adherence to regulatory standards are essential for their effective and reliable use in healthcare settings.

2.2 Transducers based on conjugated polymers

The most common transducers used for the construction of sensors in which CPs are used can be broadly classified as electrochemical, optical, field-effect transistor (FET), and piezoresistive sensors.

2.2.1 Electrochemical sensors

Changes in the electrical conductivity or charge carrier mobility of the CPs upon interaction with the target analyte are screened. These changes can arise from charge transfer or doping processes, as the polymer's electronic structure is modified by interacting with the analyte. CPs, such as polyaniline or polythiophene, are commonly employed in electronic transduction-based sensors for their high sensitivity and ease of integration into electronic devices.

Electrochemical sensors play a crucial role in detecting and quantifying various analytes in a variety of applications, including environmental monitoring, biomedical diagnostics, and industrial processes. The fundamental principles, advantages, and applications of electrochemical sensors based on CPs are explored in this section, highlighting their significant contributions to advancing analytical chemistry.

Electrochemical sensors operate by converting chemical information into electrical signals. CPs, with their extended π-conjugation and delocalized electrons, exhibit distinctive electrochemical behavior. The detection process relies on the alteration of the polymer's electrical conductivity or electrochemical properties in response to the presence of specific analytes. This can be achieved through several mechanisms, including charge transfer interactions, doping or dedoping processes, or conformational changes induced by analyte binding (McQuade et al., 2000).

Electrochemical sensors can be classified according to the parameters measured, such as amperometric, voltammetric, conductometric, and potentiometric. Amperometric sensors depend on measuring the current signal generated by the reduction or oxidation of the redox species in the medium. This species can be an analyte or an electrochemically related material. The concentration of the analyte under investigation in amperometric sensors is directly correlated with the current measured and strongly depends on the electron transfer between the electrode/CPs surface and the catalytic molecule. In voltammetric sensors, the current is quantified based on the applied varying potential, which is related to the analyte concentration. The basis for conductometric sensing is variations in the amount of ions that can be produced or consumed as a result of biological events in bioreceptors. Conductometric electronic noses based on CPs have applications in smoke and fire detectors, as well as electronic tongues in the analysis of fluids such as wine or juice (Lange et al., 2008). The potentiometric sensors monitor the difference in potential between a reference electrode and a working electrode modified with CPs in the presence of a negligible current flow, which is directly related to the ion activity.

Glucose sensors utilizing electrochemical transduction systems have achieved remarkable success due to their low cost, ease of use, and reliability, which are especially beneficial for individuals with diabetes. Different mechanisms observed in the catalytic cycle of the amperometric devices employing glucose oxidase have led to the classification of these devices into four distinct categories, which are schematically represented in Fig. 2.3 (Adeel et al., 2021).

The first-generation biosensors cover devices where biorecognition elements consuming oxygen or producing hydrogen peroxide in their catalytic cycles are monitored by applying a fixed potential to monitor their change. Oxidase-based biosensors are monitored by the production of hydrogen peroxide by applying a fixed anodic potential at +0.7 V versus Ag/AgCl or oxygen consumption by applying a fixed cathodic potential at −0.7 V versus Ag/AgCl. However, the need for molecular oxygen in the catalytic cycle of oxidase enzymes limits the applicability of first-generation oxidase biosensors. Similarly, first-generation dehydrogenase-based biosensors need a cofactor for the

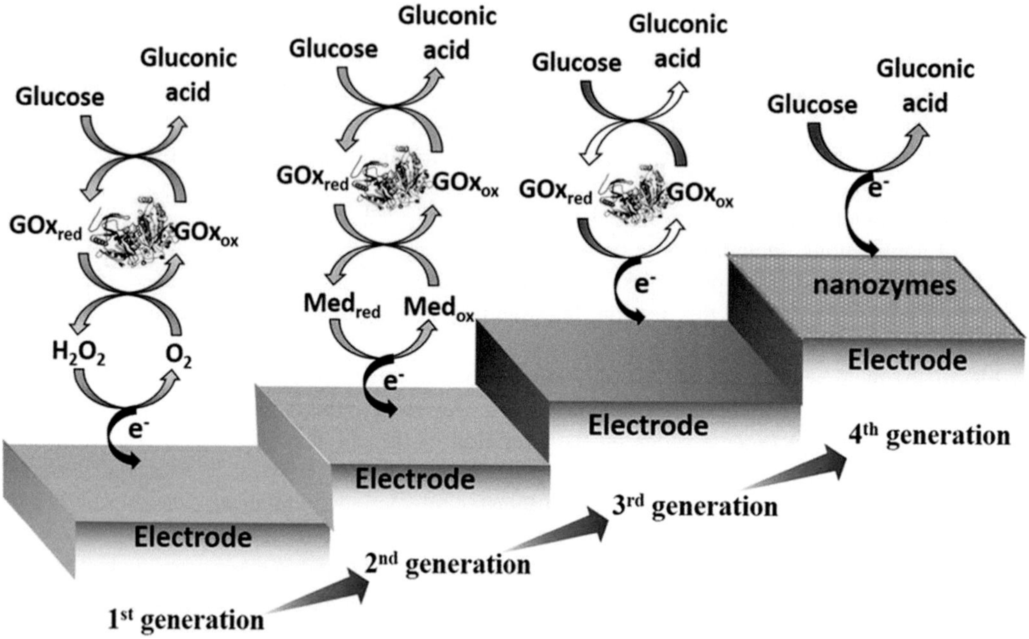

FIGURE 2.3

Structures and mechanisms of different generations of electrochemical glucose sensors.

Reproduced from Copyright 2021 Wiley-VCH GmbH: Adeel, M., Asif, K., Rahman, Md. M., Daniele, S., Canzonieri, V., & Rizzolio, F. (2021). Glucose detection devices and methods based on metal–organic frameworks and related materials. Advanced Functional Materials, 31*(52). Available from https://doi.org/10.1002/adfm.202106023.*

functional working of enzymes, which limits their use. Although first-generation biosensors provide high sensitivity and fast response times, the necessity of pretreatment of the electrode surface and corrections for matrix effects often limits their usage.

The second-generation biosensors have a mediator element that promotes electron transfer between the electrode and enzyme. This allows for reducing the applied potential and eliminating the oxygen dependency; however, there is a risk of mediator leakage into the medium that decreases biosensor performance over time. Thus, second-generation biosensors are less commonly used than the previous version.

The third-generation biosensors are independent of oxygen and cofactors. Due to the direct communication between enzyme and electrode, electrons are considered a secondary analyte for the catalytic reaction, resulting in the generation of current, which leads to improved sensing capability. CPs offer superior properties as third-generation biosensor electrode materials due to the electrical conductivity provided by their π-conjugated backbones. Their porous surfaces provide a suitable place for enzyme immobilization. Moreover, the ease of the introduction of functional groups for immobilization or altering the surface characteristics with diverse optic or electronic properties makes them a great candidate for third-generation biosensor platforms.

The fourth-generation biosensors do not contain enzymes, unlike the other three generations. Thus, disadvantages resulting from utilizing enzymes in biosensors, such as the low stability of enzymes with chemical deformation during the manufacturing process, storage, and use, are eliminated (Adeel et al., 2020; Chen et al., 2013; Lee et al., 2018). For example, a conjugated polymer/bimetallic metal oxide (polyaniline/MnBaO) composite electrode was electrochemically prepared for non-enzymatic glucose sensing (Khan et al., 2022). In this study, the amperometric determination of glucose was investigated, whereas in another study, a potentiometric glucose sensor was designed. A molecularly imprinted sensor was developed for the potentiometric analysis of glucose and showed the synergetic effect of individual compartments (Kim et al., 2017) (metal nanoparticles, CPs, and molecular imprinting layer). On the other hand, there are a tremendous number of papers that use metal/metal alloys/metal oxides as electrode materials (Huo et al., 2021; Jafarian et al., 2009; Verma, 2018; Wang et al., 2016). Yet there is a great potential for CPs to advance and optimize their catalytic properties for unique analytes.

2.2.2 Optical sensors

The alterations in the optical properties of the polymers, such as absorption or emission spectra, upon exposure to the target analyte are monitored. CPs can undergo a range of processes, including fluorescence, phosphorescence, and energy transfer. By integrating receptors specific to the analyte into the polymer structure, the presence of the target analyte leads to alterations in the fluorescence intensity of the polymer, enabling sensitive detection.

CPs exhibit high fluorescence quantum yields, and they can be readily synthesized, modified, and incorporated into various matrices, making them highly convenient materials for the development of custom-designed sensors for specific target analytes. Although the literature mainly covers composite materials of (semi)pristine and functionalized CPs for biosensors, chemical sensors based on fluorometric measurements and gas sensors can be composed solely of CPs. Here, the important criterion is to correlate the proper relationship between the polymer and the analyte. For example, Lewis basic lone pairs of pyridines interact well with Lewis acids like carbon dioxide, as well as boron centers with the lone pair electrons of nitrogen in molecules like pyridine and ammonia, two or more pyridine structures with lone pairs that can coordinate the metals. Particular attention should be given to the fact that the analyte fits the empty space within the 3-dimensional structure of the CPs. In the case of the pedant crown ethers group attached to CPs, the size of the crowns and the typical constituent elements of crowns are specific for the size of cations like Li^+, Na^+, and K^+. Unique chemical reactions or structural modifications can be triggered by specific chemicals (analytes), and fluorescence can be either turned on or off according to the chemical and/or structural change. Such changes can also be correlated with colorimetric changes, and color changes can give either qualitative or quantitative information about the analyte.

The first demonstration of fluorescent sensing was provided by Swager and coworkers in 1995 with a study that revealed the attenuated emission intensity by producing trapping sites as the analyte methyl viologen binds to the CPs (Zhou & Swager, 1995). Since then, CPs-based sensing has gained dramatic interest. Plentiful CPs-based fluorescent sensors have been developed. Thanks to the innumerable options for the structure or arrangement of backbone and functional side chains with diverse conformations, CPs and specific receptor components have shown significant promise in sensing and monitoring biological and environmental parameters. These encompass metal ions, anions, biomolecules, explosive materials, volatile organic compounds, gases like oxygen/carbon dioxide, etc. (Dai et al., 2002; McQuade et al., 2000; Wang et al., 2020). Swager and coworkers (Zhou & Swager, 1995) stated that CPs have the ability to migrate excited-state electrons through their chains or even diffuse certain distances in the bulk solid. Space dipolar couplings and strong mixing of electronic states make CPs behave like molecular wires. Hence, the interaction, electronic excitation, or energy transfer of an analyte and the receptor moiety of the CPs is sensed by the whole CPs backbone. Consequently, an amplified signal with enhanced sensitivity can be achieved.

In the field of fluorescent sensors, there are three common approaches: fluorescence turn-on, fluorescence turn-off, and fluorescence shift. The detection process relies on how the analyte (e.g., biological molecules, metal ion solution, gases, and liquids) interacts with the CP. The turn-on mechanism can be explained by aggregation-induced emission enhancement (Cai & Liu, 2020; Suzuki et al., 2020), the crosslink-enhanced emission effect (Tao et al., 2020; Yang, Liu, et al., 2020), or Förster resonance energy transfer (FRET) (Geng et al., 2021; Schneckenburger, 2020). Fluorescence shifting results in a bathochromic shift (higher wavelength than original emission) or a hypsochromic shift (lower wavelength than original emission), as seen in Fig. 2.4. Changing the solvent, the acidity of the solvent, or the presence of different metal cations can shift the fluorescence spectrum and alter the color of the emission. The fluorescence turn-off mechanism in total quenches the fluorescence. Here, the analyte can act either as a dynamic quencher (electron/energy

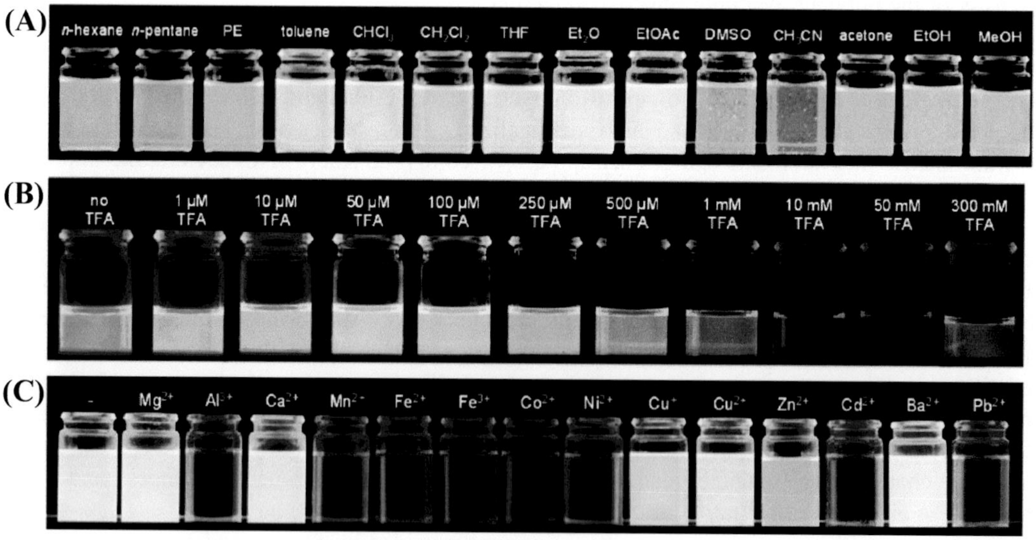

FIGURE 2.4

Fluorescence response of CPs (A) in different organic solvents, (B) at different acid concentrations, and (C) combined with different metal cations.

Reproduced from Copyright 2020 Royal Society of Chemistry: Wang, T., Zhang, N., Bai, W., & Bao, Y. (2020). Fluorescent chemosensors based on conjugated polymers with N-heterocyclic moieties: Two decades of progress. Polymer Chemistry, 11*(18), 3095–3114. Available from https://doi.org/10.1039/d0py00336k.*

transfer from the excited state of the CPs to the analyte) or a static quencher (fluorescent CPs forms non-fluorescent complexes or combinations with the analyte). Dynamic quenching mechanisms can be considered as fluorescence resonance energy transfer or photoinduced energy transfer. Static quenching mechanisms can be considered an inner filter effect or dexter energy transfer, for which further and comparative explanations can be found in a recent publication (The Huy et al., 2022).

Chemiluminescence (CL) is another useful optical detection method where the light emission is triggered by chemical reactions. The lack of an external source for excitation leads to high signal-to-noise ratios and prevents photobleaching and light damage. To improve the CL, CPs are currently employed as energy acceptors in CL nanoparticles (Zhu et al., 2018).

2.2.3 Field-effect transistor sensors

FET-sensors are advantageous for their fast response, ease of miniaturization, and high-throughput screening. The last one specifically demonstrates great potential for biomarker detection platforms. Principally, when the sensitive probe in FET-sensors interacts or binds with the target analyte, charged species are generated. These charged species in the device induce a change in carrier concentration in a channel. This output change, such as mobility, source-drain currents, and the on/off ratio of the threshold voltage, can be correlated quantitatively for the detection of biological substances. Several device architectures are used to construct FET-sensors. Four main and basic structures can be found in Fig. 2.5. FET sensors contain three electrodes: source, drain, and gate (Zaumseil & Sirringhaus, 2007). When a threshold voltage is reached, the device is considered to be "energized," and the carriers in the device start to flow along the channel between the source and drain. Eventually, the on/off state of the device is related to the bias voltage applied to the gate electrode.

The electric field generated due to the contact of metal and semiconductor electrodes with different work functions causes a shift in band edges. When an additional electric field is applied, insufficient shielding by the low charge carrier concentrations causes a generation of the electric field near the surface of the semiconductor, which is called field-effect-induced band bending (Zhang & Yates, 2012). Furthermore, charge transport within the device strongly depends on the presence of charged molecules adhered to the surface of a semiconductor. When an analyte nears the surface of a semiconductor, the potential energy gradient of charge carriers in the near-surface region changes accordingly by forming Helmholtz layers and causing band bending. Thus, this band-bending effect can be correlated with the effectiveness of transferring charge from the surface of the semiconductor to the approaching molecule (Zhang & Yates, 2010).

Recently, researchers have modified the FET devices to expand their application fields by replacing traditional insulators with electrolytes that allow contact with the gate electrode. These device structures, as seen in Fig. 2.6, cause an increase in gate capacitance and a decrease in operational voltage to less than 1 V.

An interesting example of the FET-sensors done by Zhu's group (Shen et al., 2018) covers the chemical modification of the CPs surface by oxygen plasma to generate binding sites for direct interaction between the CPs in the conductive channel and the target adenosine triphosphate in solution (see Fig. 2.7). Plasma-assisted interface-grafting generates binding sites for molecular antennas on the surface of a CPs-coated transistor that has a detection limit of 0.1 nM.

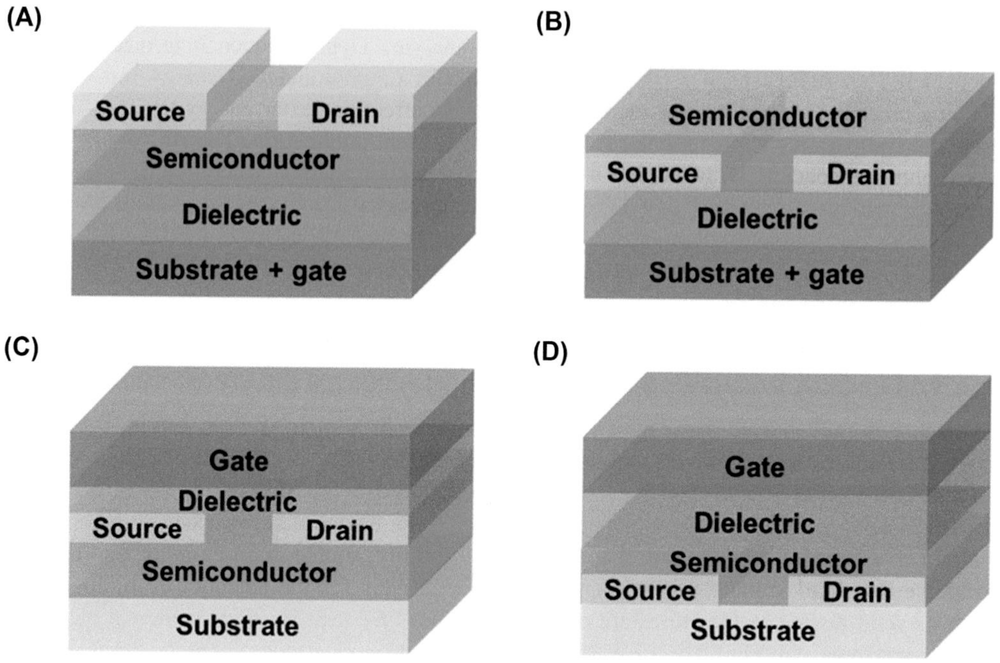

FIGURE 2.5

Graphic depiction of four device architectures: (A) bottom-gate top contact, (B) bottom-gate bottom contact, (C) top-gate top contact, (D) top-gate bottom contact.

Reproduced from Copyright 2023 MDPI, Basel, Switzerland: Hao, R., Liu, L., Yuan, J., Wu, L., & Lei, S. (2023). Recent advances in field effect transistor biosensors: Designing strategies and applications for sensitive assay. Biosensors, 13*(4). Available from https://doi.org/10.3390/bios13040426.*

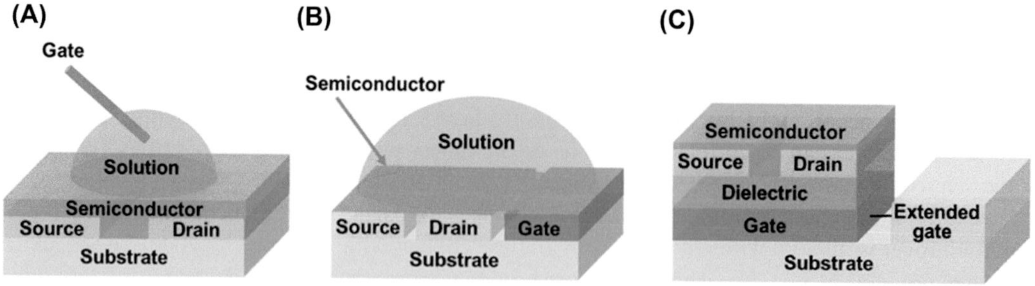

FIGURE 2.6

Schematic representation of three electrolyte-modified FET devices: (A) top-gate structure, (B) side-gate architecture, and (C) extended-gate structure (Hao et al., 2023).

Reproduced from Copyright 2023 MDPI, Basel, Switzerland: Hao, R., Liu, L., Yuan, J., Wu, L., & Lei, S. (2023). Recent advances in field effect transistor biosensors: Designing strategies and applications for sensitive assay. Biosensors, 13*(4). Available from https://doi.org/10.3390/bios13040426.*

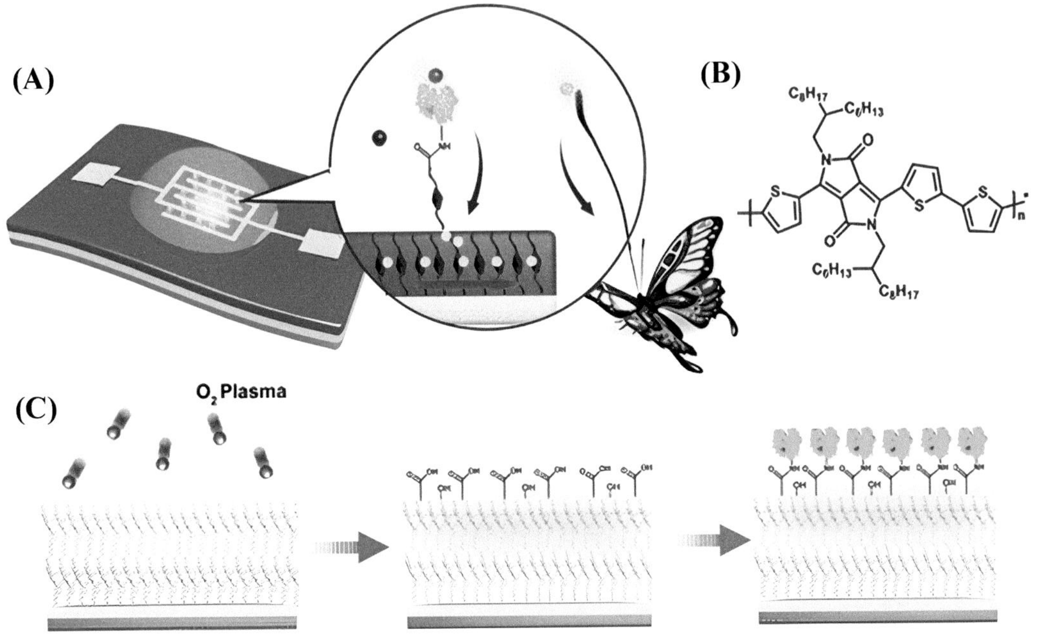

FIGURE 2.7

Schematic representation of (A) FET-sensors, (B) CPs used to fabricate FET-sensors, and (C) plasma-assisted functionalization of the CPs surface.

Reproduced from Copyright 2018 Royal Society of Chemistry: Shen, H., Zou, Y., Zang, Y., Huang, D., Jin, W., Di, C., & Zhu, D. (2018). Molecular antenna tailored organic thin-film transistors for sensing application. Materials Horizons, 5*(2), 240–247. Available from https://doi.org/10.1039/C7MH00887B.*

2.2.4 Piezoresistive sensors

Flexible sensors rely on different transduction mechanisms based on piezoresistivity, piezoelectricity, capacitance, and triboelectricity (Wang et al., 2021). Piezoresistive transducers are highly attractive for strain and pressure sensing applications due to their simple structure, improved sensitivity, low cost, and ease of processing (Gupta et al., 2023; Huang et al., 2022). In the presence of applied or external mechanical stimuli, the mechanical deformation signal of the substrate is transformed into an electrical signal change.

CPs have received a lot of attention in the construction of flexible sensors because of their aforementioned electrical properties as well as their superior mechanical flexibility, stretchability, lightweight, and biocompatibility (Georgopoulou & Clemens, 2020; Lai et al., 2023; Yang et al., 2020), and they are commonly used in the fabrication of piezoresistive sensors, in which changes in the resistance of a substrate upon mechanical deformation are monitored. Examples of piezoresistive sensors based on CPs are provided in detail in Section 2.3.

2.3 Conjugated polymer-based sensors by product

2.3.1 Wearable sensors

Wearable application of chemosensors can be utilized either by direct application of sensory materials on textiles or incorporation of sensory electrodes onto the textile or directly onto the skin, as seen in Fig. 2.8. Vital electrolytes, metabolites, and nutrients, as well as toxic gases and heavy metals, can be analyzed directly in alternative body fluids, including tears, sweat, saliva, interstitial fluid (ISF), or even exhausted breath (Singh et al., 2022). Moreover, biomarker tracking, which gained enormous attention with the COVID-19 pandemic, can give powerful and dynamic information about health conditions as well as assist in the prediction, therapy, diagnosis, and screening of diseases. Not only chemical but also physical parameters can be monitored, such as awake-sleep body motion, skin and body temperature, blood pressure, and heart activity (Shuvo et al., 2022). CP-based sensors have never fulfilled their potential due to the lack of real-time monitoring technologies, such as identifying the roles of biomarkers promptly. Thus, it has time to deal with all these limiting factors for the availability of fully functional operational sensors for cost-effective

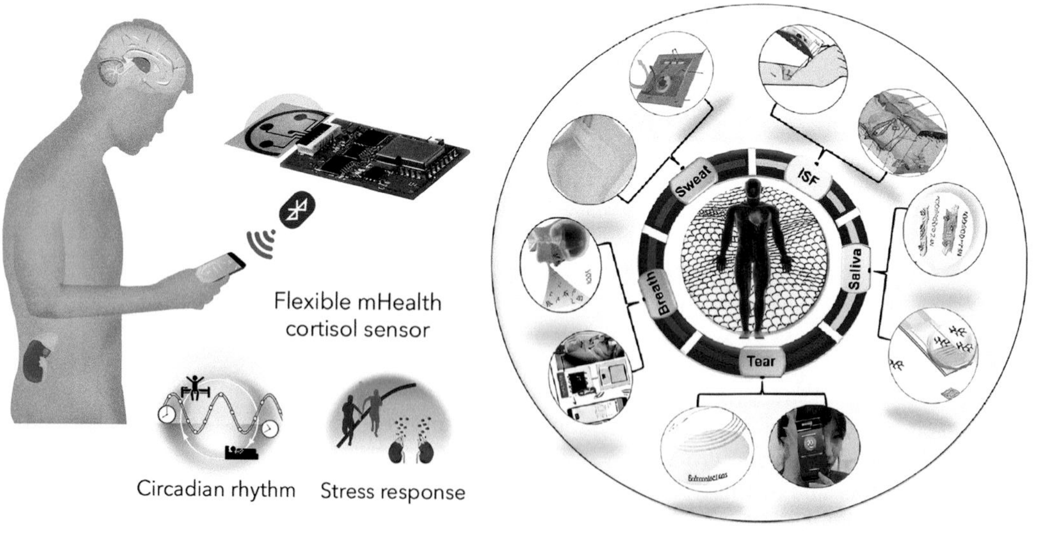

FIGURE 2.8

Typical wearable sensors and conceptual illustration of sensor devices (Torrente-Rodríguez et al., 2020).

Reproduced from Copyright 2022 Springer: Singh, S. U., Chatterjee, S., Lone, S. A., Ho, H.-H., Kaswan, K., Peringeth, K., Khan, A., Chiang, Y.-W., Lee, S., & Lin, Z.-H. (2022). Advanced wearable biosensors for the detection of body fluids and exhaled breath by graphene. Microchimica Acta, 189*(6), 236. Available from https://doi.org/10.1007/s00604-022-05317-2 and Copyright 2020 Elsevier: Torrente-Rodríguez, R. M., Tu, J., Yang, Y., Min, J., Wang, M., Song, Y., Yu, Y., Xu, C., Ye, C., IsHak, W. W., & Gao, W. (2020). Investigation of cortisol dynamics in human sweat using a graphene-based wireless mHealth system.* Matter, 2*(4), 921–937. Available from https://doi.org/10.1016/j.matt.2020.01.021.*

production with supply and demand chains available in the market due to their respective intrinsic restrictions. Wearable sensor studies are summarized in Tables 2.1 and 2.2.

One criterion that wearable sensors should satisfy is being flexible and stretchable. Hence, sensors can maintain their reliability under deformation actions that the human body can withstand, such as stretching, twisting, and bending. To fulfill this, Xu et al. confined CPs inside an elastomer that can retain its electrical conductivity even under large deformations (Xu et al., 2017). Here, the nanoconfinement helps to control glass transition temperature, mechanical modulus, and mechanical ductility. They eventually made fully stretchable CPs-based transistors that exhibited biaxial stretchability.

Coating/depositing CPs-based sensors directly onto skin or onto/into a material like fabric is a good choice for the inclusion of CPs in wearable applications. Noninvasive classes of sensors can be embedded into t-shirts, bands, underwear, contact lenses, and tattoos. Simple amperometric or potentiometric noninvasive sensors can qualitatively and quantitatively collect tremendous data without the need for complicated device compartments like lamps, monochromators, and light detectors. Fabric-embedded sensors should allow the flow of heat, air, and humidity as usual in garments and be fully functional. Moreover, they should be light and comfortable. They should operate at low power and exhibit simple and informative readout electronics to be effectively portable. Thus, CPs-based wearable sensors exhibit such a wide portfolio that conventional conductor materials fail to meet all the requirements of wearable sensor technology.

By using an ion-induced self-assembly process, PEDOT:PSS fibers with array microstructures were created by Wang et al., similar to the villi of spiders (Wang et al., 2021). Specific surface size improved from 2.2 to 10.8 m^2/g with adequate active points, which results in 0.34 kPa^{-1} for pressure sensitivity and detection limits of 82 Pa. PEDOT:PSS fibers were demonstrated in several impressive wearable applications, which include gravity/pressure sensing, real-time reporting, and airflow detection.

Chang et al. acquired elastic and conductive spines by in situ polymerizing PANI on a melamine sponge and then operating the freeze-drying method to deposition monolayer MXene nanoflakes on the conductive sponge (Chang et al., 2021). Utilizing a piezoresistive pressure sensor, it exhibits outstanding sensing achievement with a high sensitivity, a wide working range of 0–23 kPa, and a low detection limit of 256.3 Pa through the combination of the excellent elasticity and compressibility of the polymeric sponge structure, the 3D interconnected conducting PANI structures, and the tight electrical connection between constructed MXene flakes. To demonstrate the MXene/PANI sponge sensor's significant potential for flexible/wearable electrics, including health-care monitoring, electronic skin, and human-machine interaction, a variety of human movements were observed in real-time.

The structure of the roof's layered tiles was the inspiration for the creation of the MXene/PANIF nanocomposite sensing layer, which was produced by layering PANIF and MXene sheets over an elastic rubber substrate. The obtained strain sensor by Chao et al. shows exceptional sensing capability with a low detection limit (0.1538% strain), an extensive sensing range (up to 80% strain), an extremely high gauge factor (up to 2369.1), and incredible cycle stability (Chao et al., 2020). The strain sensor can be used to observe large-scale human muscular actions, such as bending the fingers and elbows, with a wireless transmitter (Fig. 2.9). It can also wirelessly detect small vital signals, for example, facial expressions, phonation, and human wrist pulses.

Table 2.1 Some studies of wearable conjugated polymer-based sensors.

Modified sensor	Method	Analyte	Linear range	Limit of detection	Sensitivity	Application	References
PEDOT:PSS-Cu^{2+} fiber-based tactile sensor	Piezoresistive	Pressure	82–600 Pa	~ 82 Pa	0.34 kPa^{-1}	Oscilloscope	Wang et al. (2021)
MXene/PANI composite sponge sensor	Piezoresistive	Pressure	0–23 kPa	256.3 Pa	0.3106 kPa^{-1}	A circuit with a Bluetooth module to wirelessly detect on-body various human motions and body activities	Chang et al. (2021)
MXene/PANIF strain sensor	Piezoresistive	Strain	Up to 80% strain	0.1538% strain	Up to 2369.1 for the gauge factor	Wirelessly human motion monitoring	Chao et al. (2020)
PU/PEDOT:PSS fibers	Piezoresistive	Strain	Up to 160% strain	5% strain	−1.0 for the gauge factor	The wireless strain sensing in a knee sleeve prototype	Seyedin et al. (2015)
PPy-coated fabrics	Piezoresistive	Strain	10% strain	6% strain	−12.5 for the gauge factor	The tracking of upper body movements using a sensorized leotard and the location and motion of fingers in relation to the palm	Scilingo et al. (2003)
MIPs-AgNWs biosensor	Voltammetric	Lactate	10^{-6}–0.1 M	0.22 μM	NR	Noninvasive skin lactate monitoring in humans during physical activity	Zhang et al. (2020)
PDMS/Graphene/AuNP/GOx microfluidic chip	Amperometry	Glucose	0–162 mg/dl	1.44 mg/dl	10 mA/cm^2	Continuous glucose monitoring with ISF extraction chip on human skin	Pu et al. (2016)
Nafion/GOx/rGO/SiO_2/C-PDMS biosensor	Amperometry	Glucose	0.1–9 mM	3.7 μM	60.8 μA mM^{-1} cm^{-2}	Sweat sample	Xu, Zhu, et al. (2021)

PEDOT, Poly (3,4-ethylenedioxythiophene); *PSS*, poly (styrene sulfonate); *PANI*, polyaniline; *PANIF*, polyaniline fiber; *PU*, polyurethane; *PPy*, polypyrrole; *MIP*, molecularly imprinted polymers; *AgNW*, Ag Nanowires; *PDMS*, polydimethylsiloxane; *AuNP*, gold nanoparticles; *GOx*, glucose oxidase; *ISF*, interstitial fluid; *rGO*, reduced graphene oxide; *SiO_2*, silicon dioxide; *NR*, not reported.

Table 2.2 Some studies of wearable conducting polymer hydrogel-based sensors.

Modified sensor	Method	Analyte	Linear range	Limit of detection	Sensitivity	Application	References
SF/TA@PPy conductive hydrogel strain sensor	Piezoresistive	Strain	>500% strain	NR	0.7154 for the gauge factor	Detection of various human motions in air and underwater	Zheng et al. (2022)
PAM/Gelatin/PEDOT:PSS conductive hydrogel strain sensor	Piezoresistive	Strain	2850% strain	NR	1.58 for the gauge factor	Detection of various human movements	Sun et al. (2020)
PPy–rGO–PPy nanosheets conductive hydrogel	Piezoresistive	Strain	405% strain	≤ 0.1% strain	1.98 for the gauge factor	Detection of human body movements and simulation of human body thermoperception	Yang, Cao, et al. (2020)
TOCNF/PAA-PPy composite hydrogel	Piezoresistive	Strain	800% strain	NR	7.3 for the gauge factor	Identification of repetitive finger motions at various angles	Chen et al. (2019)
P(AA-co-DMC)/PANI conductive hydrogel	Piezoresistive	Strain	576% strain	NR		Detection of human movements	Chen, Hao, et al. (2021)
PA-doped PPy hydrogel strain sensor	Piezoresistive	Strain	500% strain	NR	2.90 for the gauge factor	Detection of human movements under extreme temperature scenarios	Liu et al. (2022)
PEDOT:PSS/DF/PB/GOx noninvasive glucose sensor	Amperometry	Glucose	1–243 μM 243–3243 μM	0.85 μM	340.1 μA mM^{-1} cm^{-2} 184.3 μA mM^{-1} cm^{-2}	On-body glucose monitoring	Xu et al. (2022)
PEDOT:PSS/CPE uric acid sensor	Amperometry	Uric acid	2.0–250 μM	1.2 μM	0.875 μA μM^{-1} cm^{-2}	On-body uric acid monitoring	Xu, Song, et al. (2021)
PNIPAM/CMCS/MWCNT/PANI hydrogel	Piezoresistive	Strain	225% strain	NR	3.603 for the gauge factor	Detection of human movements	Zhan et al. (2021)

SF, Silk fibroin; *TA*, tannic acid; *PAM*, polyacrylamide; *TOCNF*, TEMPO-oxidized cellulose nanofibers; *PAA*, polyacrylic acid; *AA*, acrylic acid; *DMC*, methyl acryloyl oxygen ethyl trimethyl ammonium chloride; *PANI*, polyaniline; *PA*, phytic acid; *DF*, DMSO and Zonyl FS-300; *PB*, Prussian blue; *CPE*, carbon screen-printed electrodes; *CMCS*, carboxymethyl chitosan; *MWCNT*, multiwall carbon nanotubes; *PNIPAM*, poly(N-isopropylacrylamide); *NR*, not reported.

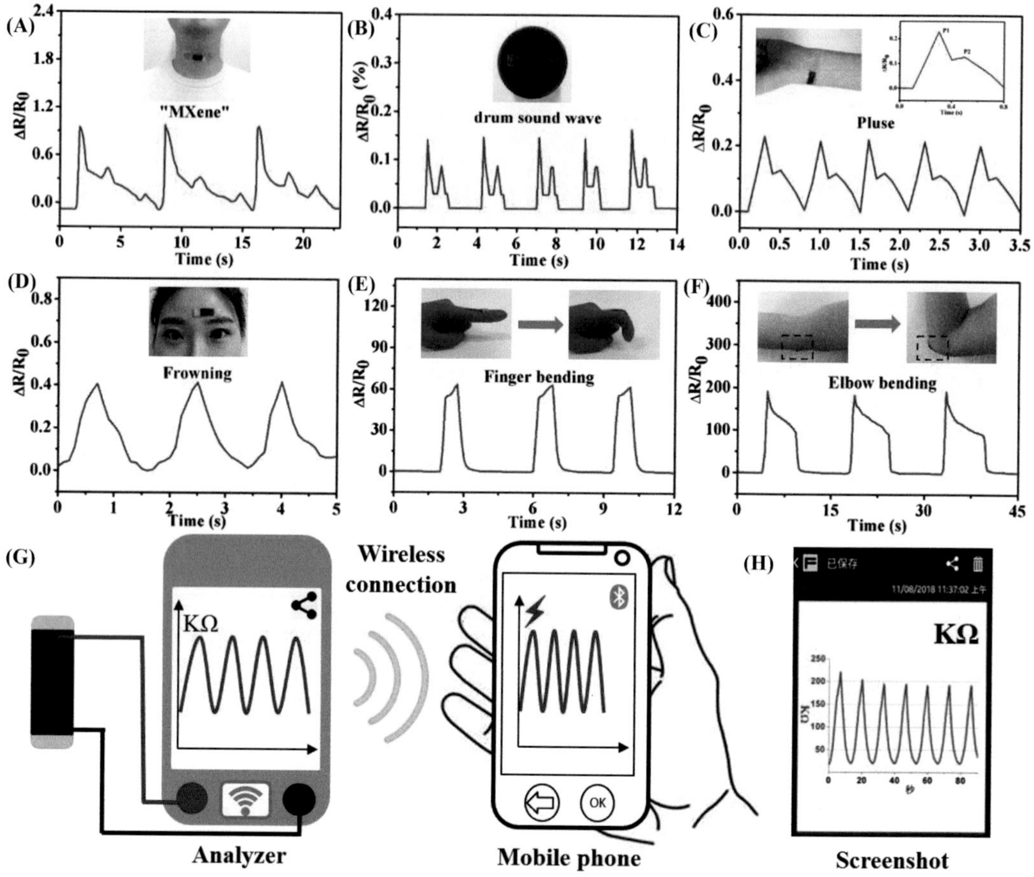

FIGURE 2.9

Wireless human movement observation with an MXene/PANIF strain sensor.

Reproduced from Copyright 2020 Elsevier: Chao, M., Wang, Y., Ma, D., Wu, X., Zhang, W., Zhang, L., & Wan, P. (2020). Wearable MXene nanocomposites-based strain sensor with tile-like stacked hierarchical microstructure for broad-range ultrasensitive sensing. Nano Energy, 78, *105187. Available from https://doi.org/10.1016/J.NANOEN.2020.105187.*

A wearable strain sensor with extra use underwater was created by Zheng et al. (2022). Without any chemical cross-linkers, the hydrogel-based sensor was constructed of PPy, tannic acid (TA), and silk fibroin (SF). SF/TA@PPy has a fast resistance response, strain sensitivity (500%, GF = 0.7154), and a broad strain range (> 500%). It can quickly detect large-strain body motions such as knee joints, elbows, wrists, and fingers in both air and water. It is also possible to accurately record small-strain physiological signal changes like respiration, frowning, smiling, and coughing (Fig. 2.10).

Sun et al. designed a triboelectric nanogenerator by sandwiching the conducting polymer hydrogels, which were built on PAM, gelatin, and PEDOT:PSS, between polyurethane (PU) and silicone

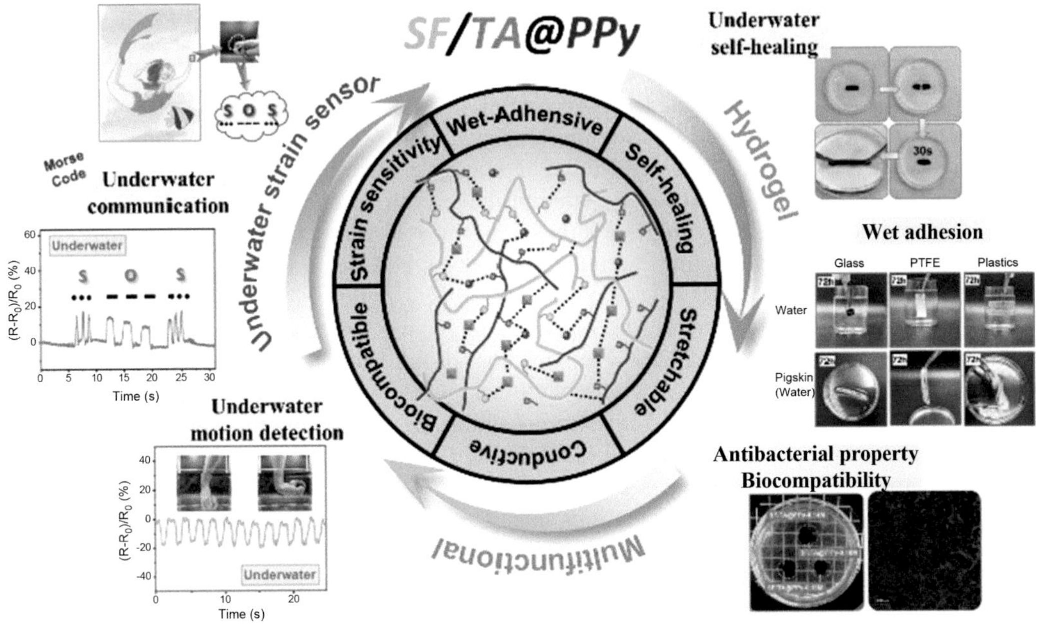

FIGURE 2.10

Applications of a SF/TA@PPy conductive hydrogel strain sensor.

Reproduced from Copyright 2022 Elsevier: Zheng, H., Chen, M., Sun, Y., & Zuo, B. (2022). Self-healing, wet-adhesion silk fibroin conductive hydrogel as a wearable strain sensor for underwater applications. Chemical Engineering Journal, 446, *136931. Available from https://doi.org/10.1016/J.CEJ.2022.136931.*

rubber (Sun et al., 2020). When bonded to the knee joint, face, and fingers, the strain sensor was able to recognize a variety of human motions and showed a gauge factor of up to 1.58 at strains of up to 2850%.

A skin-compliant hydrogel constructed from gelatin and sandwich panels made of reduced graphene oxide (rGO) and PPy was created by Yang et al. (Yang, Cao, et al., 2020). Solvothermal synthesis was used to create the sandwich layer, which was made up of three layers (PPy−rGO−PPy). It was subsequently integrated into gelatin hydrogels that had been chemically and physically cross-linked. The physical connection of the sheets that make up the matrix not only provided the hydrogel with conductivity but also enhanced its mechanical qualities. At the maximum PPy−rGO−PPy range, the designed hybrid hydrogel had up to 405% and a tensile strength of up to 110 kPa. The hydrogel's performance as a sensor included a pressure sensitivity of 3.57% kPa^{-1} at under 10 kPa, conductivity up to 0.8 S/cm, a strain gauge factor of 1.98 at 0%−200% strain, and a temperature resistance of 1.29%$°C^{-1}$ within human body temperature.

Through a simple integrated two-step preparation process, Chen et al. effectively synthesized an unusual form of electro-conductive, self-healing, and skin-inspired TOCNF/PAA-PPy hybrid hydrogel shaped like a triple network (Chen et al., 2019). With an acceptable gauge factor of 7.3, the hydrogel-based strain sensors showed quick, sensitive, and stable electrical responses. They could also successively observe small- and large-scale human movements in real time, presenting a significant opportunity for the development of stretchable, skin-like, self-healing sensing electronic instruments.

Chen et al. successfully designed P(AA-co-DMC)/PANI conductive hydrogels with a semi-interpenetrating network form. Superior mechanical qualities are provided by the conductive hydrogel's electrostatic force and semi-interpenetrating network structure (Chen, Hao, et al., 2021). The hydrogels' strain and tensile strength have both reached ideal levels of 576% and 0.173 MPa, respectively. The hydrogel exhibits superior real-time monitoring of human motion by rapidly and repeatedly detecting the telescopic waves and flexural movements of the throat and fingers.

A conductive hydrogel was created by Liu et al. employing first L-composited PAAm and secondly PA-doped PPy (Liu et al., 2022). The stretching of the hydrogels increased with the L concentration, peaking at 1896% with 18 wt.% L. Because PA was present, the hydrogels showed effective flame retardancy. They had high stretchability at 30 s of burning via an alcohol lamp (~600°C) and a low surface temperature (71.2°C) after heating (200°C) for 1200 s. With a wide strain range from 0% to 500% and a constant GF up to 2.90, the hydrogel-based strain sensor showed suitable sensing actions for monitoring human motion in burning and cold conditions.

Zhan et al. developed a CPH that exhibited pressure and pH sensitivity involving PNIPAm, which is the basis of thermosensitivity (Zhan et al., 2021). PANI was specifically chosen as the conductivity and pH-sensitive component, and MWCNTs were included to increase the hydrogel's stability and conductivity. The hybrid hydrogel as a strain sensor exhibits excellent linearity and superior sensitivity (gauge factor 3.603) to the strain. As a result of circuit disconnection brought on by volume shrinking at high temperatures, they were utilized for extremely sensitive temperature alerts. In addition, the hybrid hydrogels were used as pH sensors to detect differences between alkali and acid in the medium.

Existing studies demonstrate the materials' use not only in the detection of biomolecules in human samples but also as an operational wearable sensor. Nevertheless, the fortification of this instrument should lead to additional research that includes in vivo tests, long-term stability, and sensor effects on the skin in the near future.

For the noninvasive detection of lactate in human sweat, Zhang et al. presented a wearable electrochemical biosensor constructed from molecularly imprinted polymers (MIPs) and Ag nanowires (AgNWs) generated on a screen-printed electrode (Zhang et al., 2020). The MIPs-AgNWs biosensor has detected lactate in the 10^{-6} M–0.1 M linear range, with a detection limit of 0.22 μM, demonstrating exceptional specificity and sensitivity. In addition, the biosensors demonstrated remarkable repeatability and stability, with 99.8% ± 1.7% recovery over the storage period of seven months at ambient temperature in the dark. In vivo studies of MIPs-AgNWs biosensors were performed by affixing them to participants' skin to enable noninvasive lactate detection. In future years, analysis of human sweat in biological areas, the military, and sports will be feasible thanks to wearable electrochemical biosensors.

Pu et al. fabricated a continued glucose monitoring microsystem merged into a microfluidic chip that consisted of five polydimethylsiloxane (PDMS) layers utilizing micro-molding methods (Pu et al., 2016). With a detection limit of 1.44 mg/dl, the testing findings showed that the suggested sensor could accurately detect glucose in the linear range of 0–162 mg/dl. The sensor confirmed its ability to identify hypoglycemia in medical facilities.

Xu et al. fabricated sweat analysis using a 3D flexible glucose biosensor, which was produced by first creating the rGO/SiO_2 nanocomposites via electrostatic self-assembly and then immobilizing glucose oxidase (Fig. 2.11) (Xu, Zhu, et al., 2021). The designed biosensor in the 3D form showed enhanced sensing performance over the 2D one, showing a low detection limit (3.7 μM) and high sensitivity (60.8 $\mu A/mM \cdot cm^2$) for glucose detection in a range of 0.1–9 mM. The 3D flexible biosensor demonstrated interference-free performance, good stability, and high trueness in the detection of glucose from sweat samples.

Wearable biosensors can also utilize CPH. For instance, by Xu et al., a wearable hydrogel-based glucose sensor was constructed of PEDOT:PSS with Prussian blue NPs and GOx, then mixed with DMSO and Zonyl FS-300 (ZFS) on screen-printed carbon electrodes (Fig. 2.12) (Xu et al., 2022). Due to differences in the enzymatic reaction, the fabricated wearable sensor showed

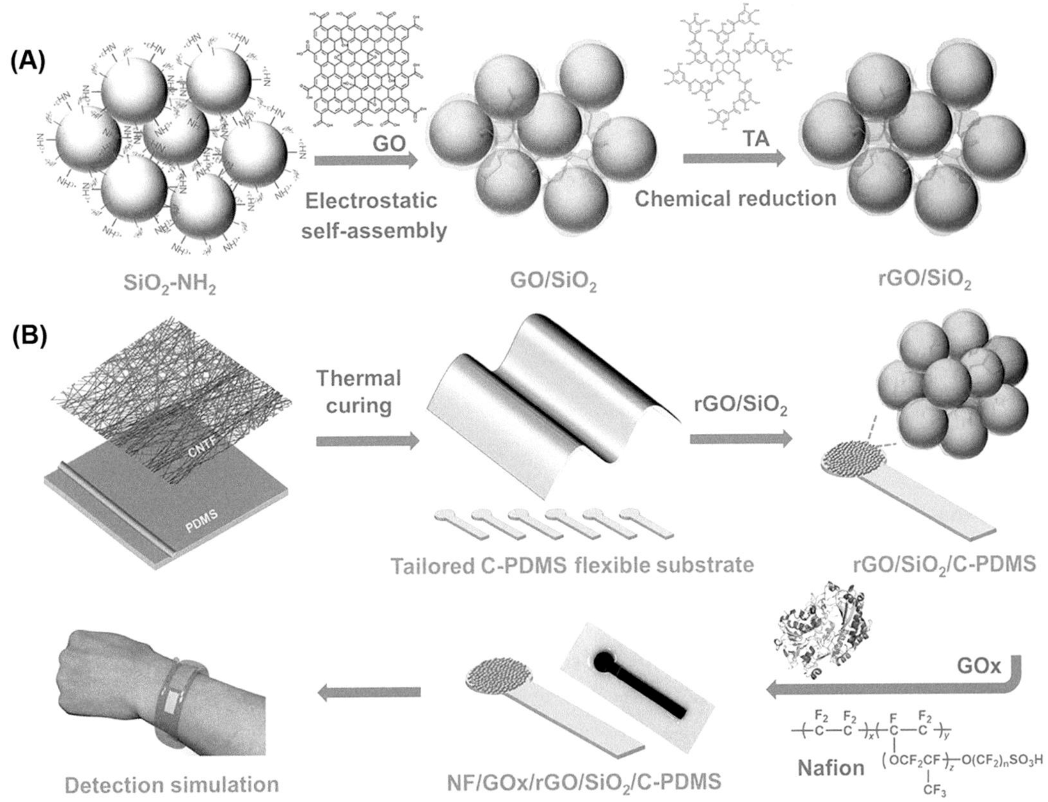

FIGURE 2.11

Schematic demonstration of the design of a 3D flexible glucose biosensor.

Reproduced from Copyright 2021 American Chemical Society (ACS): Xu, M., Zhu, Y., Gao, S., Zhang, Z., Gu, Y., & Liu, X. (2021). Reduced graphene oxide-coated silica nanospheres as flexible enzymatic biosensors for detection of glucose in sweat. ACS Applied Nano Materials, 4*(11), 12442–12452. Available from https://doi.org/10.1021/acsanm.1c02887.*

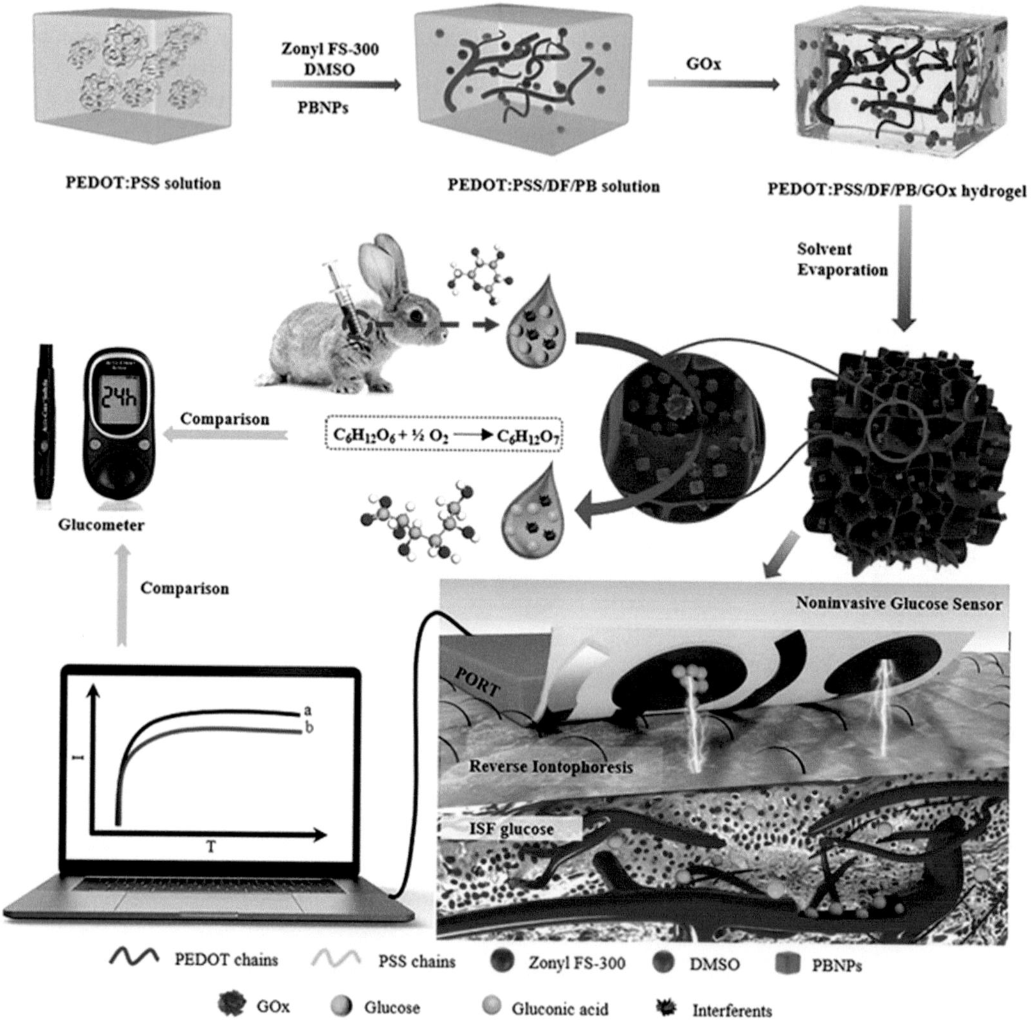

FIGURE 2.12

The schema of the PEDOT:PSS/DF/PB/GOx sensor for glucose monitoring on human skin and in serum.

Reproduced from Copyright 2022, Elsevier: Xu, C., Jiang, D., Ge, Y., Huang, L., Xiao, Y., Ren, X., Liu, X., Zhang, Q., & Wang, Y. (2022). A PEDOT:PSS conductive hydrogel incorporated with Prussian blue nanoparticles for wearable and noninvasive monitoring of glucose. Chemical Engineering Journal, 431, *134109. Available from https://doi.org/10.1016/J.CEJ.2021.134109.*

different sensitivities of 340.1 μA/mM·cm^2 at lower levels of glucose and 184.3 μA/mM·cm^2 at higher levels. In the present research, the sensor was applied to the skin of the two volunteer individuals, and glucose was collected using iontophoresis in reverse from the ISF, which demonstrated an excellent match with the results of the glucometer's finger-stick test for blood.

A wearable CPH-based microfluidic sensor has been produced by a different study team to detect uric acid in sweat. In this study, copper was used as a sacrificial anode during electropolymerization to generate PEDOT:PSS on screen-printed carbon electrodes, and PDMS was used to create a microfluidic device to collect sweat. The device's electrochemical signal to uric acid was linear, having a sensitivity of 0.875 $\mu A/\mu M \cdot cm^2$ in the 2–250 μM range. The microfluidic system was also analyzed with human sweat and correlated well with uric acid readings from an ELISA. In addition, the device kept 95% of its initial reaction after 25 days and displayed no appreciable changes in its amperometric response (Xu, Song, et al., 2021).

2.3.2 Non-wearable sensors

Biosensors for POC testing that are provided in real-time greatly benefit from the superior electroactive properties of CPs. These sensors may be advantageous due to their inexpensive cost and wide range of detection, which facilitates better access to therapy (Runsewe et al., 2019). Non-wearable sensor and POC studies are summarized in Tables 2.3 and 2.4.

Cevik et al. developed an electrochemical biosensor using a conducting (DTP(aryl)aniline) polymer and cholesterol oxidase (ChOx). The bio-catalytic activity successfully transferred electrons, and the unique conformation of the polymer provided successful immobilization of the ChOx. The biosensor exhibited outstanding performance for the biocompatibility of the polymer, reproducibility, long-time stability, and a lower detection limit in the 2.0–23.7 μM linear range. The real human serum sample analyses demonstrate that biosensors can be utilized for measuring cholesterol (Cevik et al., 2018).

Dervisevic et al. reported an amperometric urea biosensor that was designed by electropolymerization of SNS-Aniline on PGE, then using DAFc as a mediator to modify the polymer-coated electrode surface, and finally coating the urease enzyme *via* crosslinking. The values of the urea biosensor were determined to have a short response time of 2 s, a detection limit of 12 μM, and a sensitivity of 0.54 μA/mM in the 0.1–8.5 mM range. The fabricated biosensor showed superior analytical performance in real human urine and blood samples (Dervisevic et al., 2017).

Ohayon et al. have shown how to combine a redox enzyme with an n-type conjugated polymer to generate power and detect glucose from body fluids. In an enzymatic fuel cell, this n-type polymer has also been used as an anode in conjunction with a polymeric cathode to transform the chemical energy of oxygen and glucose into electrical power. The performance of the all-polymer biofuel cell has increased with the amount of glucose present in the solution, and its stability has lasted longer than thirty days. Furthermore, it produces sufficient energy to run an organic electrochemical transistor from fluids like human saliva at physiologically acceptable glucose levels. This can advance the development of self-powered micrometer-scale sensors and actuators that operate on metabolites produced in the body (Ohayon et al., 2020).

Al-Ogaidi et al. used resonant energy transmission from CL to graphene quantum dots to create an immunoassay sensor for recording the biomarker CA-125 for ovarian cancer. This biosensor exhibited a detection limit of 0.05 U/mL in a wide linear range of 0.1–600 U/mL (Ogaidi et al., 2014).

Son et al. have created a colorimetric biosensor based on PDA for POC that uses color change to visually identify highly infectious pH1N1 viruses among influenza A viruses. The PDA-PVDF membrane was first created by immobilizing PDA onto the PVDF membrane *via* photopolymerization. Subsequently, this membrane was coated with an antibody specific to influenza viruses, and the color

Table 2.3 Some studies of non-wearable conjugated polymer-based sensors.

Modified sensor	Method	Analyte	Linear range	Limit of detection	Sensitivity	References
The Polyisoprene-MWCNT composite strain sensor	Piezoresistive	Strain	Up to 40% strain	NR	NR	Knite et al. (2007)
PMMA/MWNT films	Piezoresistive	Strain	NR	NR	15.32 for the gauge factor	Pham et al. (2008)
TPU/CNT conductive polymer composite	Piezoresistive	Strain	30% strain	NR	2.3 for the gauge factor	Zhang et al. (2012)
EPDM/MWNTs nanocomposites	Piezoresistive	Strain	10% strain	NR	NR	Ciselli et al. (2010)
PPy-coated PA6 fibers	Piezoresistive	Strain	43% strain	NR	2.0 for the gauge factor	Xue et al. (2004)
PPy-coated knitted fabric	Piezoresistive	Strain	50% strain	NR	300 for the gauge factor	Wang et al. (2014)
PET/CNF/P-BDT-BTz:BDA/GOx Biosensor	Amperometry	Glucose	20−500 μM	8.5 μM	98.192 μA mM^{-1} cm^{-2}	Bulut et al. (2023)
GE/poly(EDOT-TPD)/GOx Biosensor	Amperometry	Glucose	0.1−0.5 mM	0.018 mM	65.765 μA mM^{-1} cm^{-2}	Karabag et al. (2023)
GE/P(DTFBT)/Lac Biosensor	Amperometry	Catechol	25−400 μM	14 μM	166.74 μA mM^{-1} cm^{-2}	Udum et al. (2023)

PMMA, Poly(methyl methacrylate)*; TPU*, thermoplastic polyurethane*; CNT*, carbon nanotube*; EPDM*, ethylene-propylene-diene-monomer*; PA6*, polyamide 6*; PET*, polyethylene terephthalate*; CNF*, carbon nanofiber*; P-BDT-BTz,BDA*, benzotriazole; benzodithiophene; and benzenediamine*; GE*, graphite electrode*; TPD*, thieno[3,4-c]pyrrole-4,6-dione*; DTFBT*, 5-fluoro-4,7-bis (4-hexylthiophen-2 yl) benzo [c][1,2,5]thiadiazole*; Lac*, laccase*; NR*, not reported.

Table 2.4 Some point-of-care studies of non-wearable conjugated polymer based sensors.

Modified sensor	Method	Analyte	Linear range	Limit of detection	Sensitivity	Application	References
GCE/ P(DTP(aryl) aniline)/ ChOx biosensor	Amperometry	Cholesterol	2–27.6 μM	0.174 μM	11.246 μA/μM	Human serum	Cevik et al. (2018)
PGE/SNS-Anilin/DAFc/ Urease biosensor	Amperometry	Urea	0.1–8.5 mM	12 μM	0.54 μA/mM	Human blood serum and human urine	Dervisevic et al. (2017)
P-90/GOxOECT biosensor	Amperometry	Glucose	10 nM–100 μM	10 nM		Human saliva	Ohayon et al. (2020)
PDMS/GQDs/cAb/CA-125/Ab–HRP	Chemiluminescence	Ovarian cancer biomarker CA-125	0.1–600 U/mL	0.05 U/mL	NR	Human blood plasma	Ogaidi et al. (2014)
FMICP NFs biosensor	Fluorescence	Cancer biomarkers α-fetoprotein (AFP) Carcinoembryonic antigen (CEA)	0.01–100 ng mL^{-1} 0.001–200 ng mL^{-1}	15 fg mL^{-1} 3.5 fg mL^{-1}	NR	Human saliva and serum samples	Tawfik et al. (2020)
PF–DBT–Im/SDS	Fluorescence	HAS BSA	0–0.5 mg mL^{-1} 0–0.5 mg mL^{-1}	0.81 μg/mL 0.50 μg/mL	NR	Human serum	Hussain et al. (2022)
DDNBLateral flow immunoassay strips	Colorimetric	Methamphetamine Morphine	0.5 –16 ng/mL 0.5 –4 ng/mL		8.0 ng/mL	Human hair	Fan et al. (2022)
PDA-paper chips	Colorimetric	Influenza A (pH1N1) virus		5×10^3 $TCID_{50}$		Smartphone-based Virus sensing	Son et al. (2019)
Pdot Immunochromatographic test strips	Fluorescence	PSA AFP CEA	3–15 ng/mL	2.05 pg/mL 3.30 pg/mL 4.92 pg/mL	NR	Human serum	Fang et al. (2018)
GNP-CeO_2-PANI hydrogel/SPE	Voltammetric	Peroxynitrite anion	5–100 μM	0.14 μM	29.35 ± 1.4 μA μM^{-1}	The microfluidic device	Kumar et al. (2021)
PNA-C*/CP biosensor	Fluorescence	ssDNA		10 pM			Gaylord et al. (2002)

(Continued)

Table 2.4 Some point-of-care studies of non-wearable conjugated polymer based sensors. *Continued*

Modified sensor	Method	Analyte	Linear range	Limit of detection	Sensitivity	Application	References
PMNT-based DNA biosensor	Fluorescence	ssDNA	1−10 nM	1 nM			Zhang et al. (2022)
DPA-CNPPV/PSMA Pdot /PheDH paper-based biosensor	Fluorescence	Phenylalanine	0−2400 μM	3.5 μM		Human blood plasma	Chen, Yu, et al. (2021)
PFP-FB/AChE/ChOx	Fluorescence	Acetylcholine	5−100 μM			Cancer cells	Wang et al. (2015)
PF-Lac	Fluorescence	Cholera toxin B subunits	$0-8.0 \times 10^{-8}$ M				Feng et al. (2019)
MOF-Apt@MIP paper-based fluorescence sensor	Fluorescence	Malachite green (MG)	0.1−10 ng/mL	0.06 ng/mL	NR	Fish and water samples	Duan et al. (2023)

GCE, Glassy carbon electrode; *DTP(aryl)aniline,* 4-(4H-dithienol[3,2-b : 2′,3′-d]pyrrole-4)aniline polymer; *ChOx,* cholesterol oxidase; *PGE,* pencil graphite electrode; *SNS-Anilin,* 4-(2,5-Di (thiophen-2-yl)-1H-pyrrol-1-yl)aniline; *DAFc,* di-amino-Ferrocene; *GQDs,* graphene quantum dots; *cAb,* capture antibody; *Ab−HRP,* Horseradish peroxidase-labeled antibody; *FMICPs,* fluorescent molecularly imprinted conjugated polythiophene; *NF,* nanofiber; *DBT,* 1,4-dithienylbenzothiadiazole; *SDS,* sodium dodecyl sulfate; *HSA,* human serum albumin; *BSA,* Bovine serum albümin; *DDNB,* deeply dyed nanobead; *Pdot,* polymer dot; *GNP-CeO$_2$,* graphene nanoplatelets-cerium oxide nanocomposite; *PSA,* prostate-specific antigen; *MOF,* metal-organic frameworks; *Apt,* aptamer; *MIP,* molecularly imprinted polymer; *NR,* not reported.

shift of this paper chip was observed when the pH1N1 virus was present. Additionally, using both paper chips and programs (App.), PDA-paper chips-based POC testing systems have been created that enable the visual detection and analysis of viruses at low and high concentrations (Son et al., 2019).

An assay for the identification of cancer biomarkers that uses fluorescent molecularly imprinted conjugated polythiophenes (FMICPs) has been developed by Tawfik et al. (2020) and is enzyme-free, economical, and simple to use. It successfully demonstrated how printed-paper technology could be utilized in combination with a sensitive and affordable smartphone and a movable prototype testing device for quick POC cancer diagnosis. With this biosensor, the detection limits for CEA and AFP could be reduced to 3.5 fg mL^{-1} and 15 fg mL^{-1}, respectively, which are three times lower than those of the standard enzymatic biosensors. The designed sensors are successfully used to quickly diagnose AFP in patients with liver cancer, and the results of the FMICP and FMICP NFs are in perfect accordance with those of the ELISA (Tawfik et al., 2020).

Hussain et al. described a successful technique based on polymer-surfactant self-assembly for the super-amplified and target detection of serum albumins (SAs) in complicated biological conditions. The developed cationic CP (CCP) PF-DBT-Im with violet emission showed powerful interchain FRET when linked with the anionic surfactant SDS and created red-emitting polymer-surfactant groups. A portable diagnostic device integrated with a smartphone is designed for the quick POC detection of SAs using modern mobile health (mHealth) technologies (Fig. 2.13).

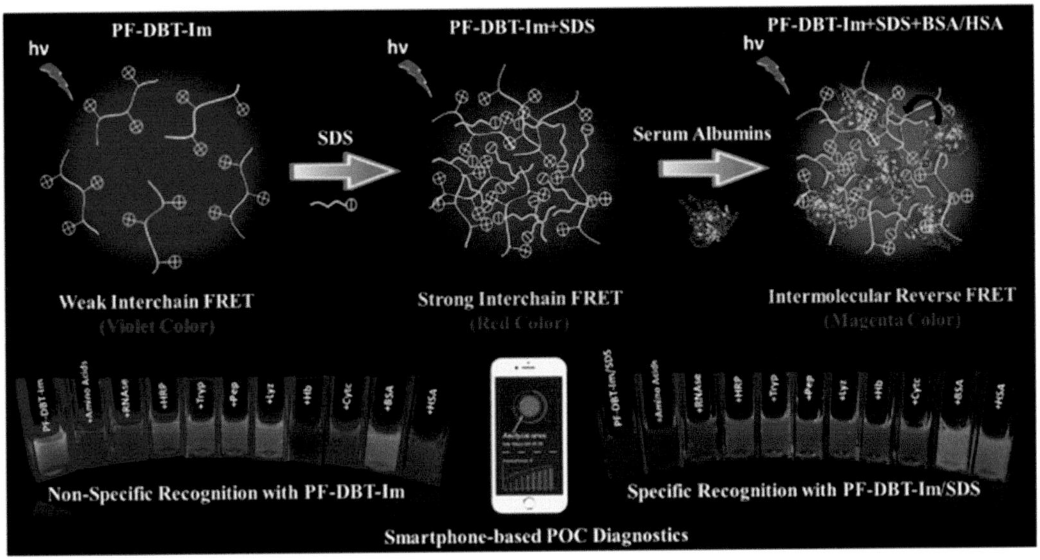

FIGURE 2.13

Albumin detection of the PF-DBT-Im/SDS biosensor under fluorescence light.

Reproduced from Copyright 2022, American Chemical Society (ACS): Hussain, S., Chen, X., Wang, C., Hao, Y., Tian, X., He, Y., Li, J., Shahid, M., Krishnan Iyer, P., & Gao, R. (2022). Aggregation and binding-directed fret modulation of conjugated polymer materials for selective and point-of-care monitoring of serum albumins. Analytical Chemistry, *94(30), 10685–10694. Available from https://doi.org/10.1021/acs.analchem.2c00984.*

The current biosensor not only reveals a new material for using the polymer-surfactant combination as a practical fluorescent instrument for target protein detection but also creates a strong, dependable, and cost-effective technique for the clinical diagnosis of disorders associated with albumin at preferred locations (Hussain et al., 2022).

Duan et al. fabricated a paper fluorescence sensor using metal-organic frameworks coated with aptamer and molecularly imprinted polymer (MOF-Apt@MIP) for malachite green (MG) and sensitively detected *via* digital image colorimetry with a smartphone. The green fluorescence terbium MOF-76 was utilized as a carrier MG-specific aptamer. The detection limits were 0.06 ng/mL for fluorescence sensors and 0.1 ng/mL for paper-based detection. This POC technique was used to detect MG in fish and water samples with rates of 94.0%−105.6% (Duan et al., 2023).

Water-soluble CPs exhibit remarkable fluorescence quenching efficiency when opposing-charged acceptors are present. In real-time DNA detection, CCPs have been employed to activate the dye emission on a particular peptide nucleic acid (PNA) sequence by Gaylord et al. The hybridization of the negative DNA target with the neutral PNA probe regulates the transmission of signals. Thus, a straightforward and very sensitive technique utilizing optical amplification and the enhanced hybridization action of PNA/DNA complexes has been presented (Gaylord et al., 2002).

CCPs, particularly polythiophene, have been widely employed as probes in the development of biosensors based on DNA and aptamers. To comprehend the interactions between DNA and poly (3-(3′-N,N,N-triethylamino-1′-propyloxy)-4-methyl-2,5-thiophene) (PMNT), Zhang et al. conducted systematic binding experiments. cDNA has been added to the PMNT−ssDNA complex to evaluate the analytical performance of the CCP-based biosensor for DNA quantification. The fluorescence ratio dropped as the cDNA concentration rose, and cDNA concentrations as low as 1 nM may be detected with high sensitivity. The fluorescence ratio has shown a nice linear range between 1 and 10 nM (Zhang et al., 2022).

An NAD(P)H-sensitive polymer dot (Pdot) biosensor has been introduced by Chen et al. for POC detection of metabolites. Using a previously published approach, the luminous polymer DPA-CNPPV and the amphiphilic polymer poly(styrene-co-maleic anhydride) (PSMA) were nanoprecipitated to create a NAD(P)H-sensitive Pdot. Accurate ratiometric detection of metabolite levels has been made possible by the selective oxidation of the analyte in a NAD(P)H-dependent enzyme-catalyzed process and the change in sensor emission from red to blue with higher NAD(P)H levels. Phenylalanine biosensing in human plasma has been used for a POC paper-based technique with smartphone imaging and a phenylketonuria screening test. It can also be used to quantify various other disease-related metabolites (Chen, Yu, et al., 2021).

Wang et al. have designed and produced a novel water-soluble CP, poly(fluorene-co-phenylene) (PFP-FB), with a boronate-protected fluoran pendent. A novel optical test for choline and acetylcholine (ACh) has been created based on the effective fluorescence signal of PFP-FB to H_2O_2 that has emerged from choline and ACh under cascade enzymatic processes of AChE/ChOx coupling. The effective utilization of PFP-FB has allowed for the ratiometric detection of choline and ACh, as well as the ongoing, continuous monitoring of the choline consumption process in cancer cells. As a result, this novel polymer probe facilitates the advancement of enzymatic biosensors and offers a more straightforward and efficient method of identifying the chemical messenger of live cells (Wang et al., 2015).

Feng et al. created fluorescently conjugated polyfluorene with a side-chain lactopyranosyl ligand (PF-Lac) in a single pot using Yamamoto C-C coupling polymerization and triazole-linked glycoconjugate synthesis. This well-defined fluorescent glycoconjugate polymer was utilized to examine the interaction of CTB subunits by fluorescence spectrophotometric titration using fluorescent fluorene scaffolds and sugar-binding ligands. When the lactose group was added to the conjugated polymer's side chain, the water solubility and biocompatibility of the polymer were improved. Fluorescence spectrophotometric titration was used to investigate the interaction between the pyranogalactose group and the cholera toxin B component in a fluorescent multivalent system of glycoconjugates containing pyranogalactose groups. There are many potential uses for PF-Lac in the investigation of glycoprotein interactions and cholera toxin detection (Feng et al., 2019).

In the study of Bulut et al., a novel conjugated polymer comprising three moieties of benzotriazole, benzodithiophene, and benzenediamine (P-BDT-BTz:BDA) was used as a platform to construct a selective and reliable biosensor for glucose employing a flexible, modified electrode with a carbon nanofiber (CNF). The electrode responded to glucose in the 20 μM–500 μM linear range under ideal circumstances with a detection limit of 8.5 μM. Furthermore, it was noted that the biosensor's ability to detect glucose in beverages was remarkable. This study's findings demonstrate that the suggested amperometric glucose oxidase (GOx)-based biosensor for glucose has a wide linear range, a low detection limit, and good sensitivity (Bulut et al., 2023).

2.4 Conclusions and perspectives

The adaptability of conjugated polymer-based biosensors has made them appropriate for a range of bioanalytical applications, such as POC diagnostics for clinical use and even disposable personal monitoring devices (glucose, urea, cholesterol, and strain sensors). By reorienting the emphasis from conventional laboratory analysis to on-the-spot identification and analysis, their adoption by both patients and physicians has the potential to change the present medical profession and the healthcare infrastructure. Significant chemical diversity is provided by CPs, which include a range of backbone compositions with unique behaviors depending on the target analysis. Furthermore, CPs have functional groups that allow them to be put into operation with different kinds of biorecognition molecules.

POC devices offer regular and real-time monitoring of physiological data about patient health through dynamic and noninvasive investigation of various biological mediums such as sweat, saliva, blood, exhaled breath, and ISF. The current growing demand for health concerns encourages researchers to develop enhanced analytical tools and devices like POC sensors that exhibit advanced selectivity, sensitivity, and rapid response. Here, CPs take the stage with their unique optical and electrical properties. Ease of synthesis, tremendous modification alternatives to tune their electrical and optical features, and applicability to various platforms, from solid devices to flexible and even wearable and implantable ones, make CPs an essentially important material for POC sensors and devices. Although there are promising wearable and implantable examples of POC sensors and devices, translation from lab to clinical practice to become a solid device on the market, there are still several obstacles to be surmounted, such as sensing device fabrication at the micro or nanoscale with biocompatibility, implantation, sensor calibration, data sharing, and data analytics. At this point, the versatility of newly designed CPs will open a new door to exploring different systems with high sensitivity and selectivity. Forthcoming advancement: integrating the use of the Internet of Things offers great advantages in self-healthcare by using wearable integrated POC systems. Chemists, physicists, biologists, electrical engineers, biomedical engineers, clinicians, and more bring their efforts continuously to overcome these obstacles. Providing research and development platforms that bring everything together in one place is an efficient way to explore the best possible approaches.

Acknowledgments

S. Soylemez thanks to the Scientific and Technological Research Council of Turkey (TÜBİTAK-122N510).

References

Adeel, M., Rahman, M. M., Caligiuri, I., Canzonieri, V., Rizzolio, F., & Daniele, S. (2020). Recent advances of electrochemical and optical enzyme-free glucose sensors operating at physiological conditions. *Biosensors and Bioelectronics*, *165*, 112331. Available from https://doi.org/10.1016/j.bios.2020.112331.

Adeel, M., Asif, K., Rahman, M. M., Daniele, S., Canzonieri, V., & Rizzolio, F. (2021). Glucose detection devices and methods based on metal–organic frameworks and related materials. *Advanced Functional Materials*, *31*(52). Available from https://doi.org/10.1002/adfm.202106023.

Bulut, U., Sayin, V. O., Altin, Y., Cevher, S. C., Cirpan, A., Bedeloglu, A. C., & Soylemez, S. (2023). A flexible carbon nanofiber and conjugated polymer-based electrode for glucose sensing. *Microchemical Journal*, *184*. Available from https://doi.org/10.1016/j.microc.2022.108148.

Bélanger, D., Nadreau, J., & Fortier, G. (1989). Electrochemistry of the polypyrrole glucose oxidase electrode. *Journal of Electroanalytical Chemistry and Interfacial Electrochemistry*, *274*(1–2), 143–155. Available from https://doi.org/10.1016/0022-0728(89)87036-6.

Cai, X., & Liu, B. (2020). Aggregation-induced emission: Recent advances in materials and biomedical applications. *Angewandte Chemie International Edition*, *59*(25), 9868–9886. Available from https://doi.org/10.1002/anie.202000845.

Cevik, E., Cerit, A., Gazel, N., & Yildiz, H. B. (2018). Construction of an amperometric cholesterol biosensor based on DTP(aryl)aniline conducting polymer bound cholesterol oxidase. *Electroanalysis*, *30*(10), 2445–2453. Available from https://doi.org/10.1002/elan.201800248, http://onlinelibrary.wiley.com/journal/10.1002/(ISSN)1521-4109.

Chang, K., Li, L., Zhang, C., Ma, P., Dong, W., Huang, Y., & Liu, T. (2021). Compressible and robust PANI sponge anchored with erected MXene flakes for human motion detection. *Composites Part A: Applied Science and Manufacturing*, *151*. Available from https://doi.org/10.1016/j.compositesa.2021.106671.

Chao, M., Wang, Y., Ma, D., Wu, X., Zhang, W., Zhang, L., & Wan, P. (2020). Wearable MXene nanocomposites-based strain sensor with tile-like stacked hierarchical microstructure for broad-range ultrasensitive sensing. *Nano Energy*, *78*. Available from https://doi.org/10.1016/j.nanoen.2020.105187.

Chen, C., Xie, Q., Yang, D., Xiao, H., Fu, Y., Tan, Y., & Yao, S. (2013). Recent advances in electrochemical glucose biosensors: A review. *RSC Advances*, *3*(14), 4473. Available from https://doi.org/10.1039/c2ra22351a.

Chen, H., Yu, J., Zhang, J., Sun, K., Ding, Z., Jiang, Y., Hu, Q., Wu, C., & Chiu, D. T. (2021). Monitoring metabolites using an NAD(P)H-sensitive polymer dot and a metabolite-specific enzyme. *Angewandte Chemie – International Edition*, *60*(35), 19331–19336. Available from https://doi.org/10.1002/anie.202106156, http://onlinelibrary.wiley.com/journal/10.1002/(ISSN)1521-3773.

Chen, X., Hao, W., Lu, T., Wang, T., Shi, C., Zhao, Y., & Liu, Y. (2021). Stretchable zwitterionic conductive hydrogels with semi-interpenetrating network based on polyaniline for flexible strain sensors. *Macromolecular Chemistry and Physics*, *222*(24). Available from https://doi.org/10.1002/macp.202100165.

Chen, Y., Lu, K., Song, Y., Han, J., Yue, Y., Biswas, S. K., Wu, Q., & Xiao, H. (2019). A skin-inspired stretchable, self-healing and electro-conductive hydrogel with a synergistic triple network for wearable strain sensors applied in human-motion detection. *Nanomaterials*, *9*(12). Available from https://doi.org/10.3390/nano9121737.

Chiang, C. K., Fincher, C. R., Park, Y. W., Heeger, A. J., Shirakawa, H., Louis, E. J., Gau, S. C., & MacDiarmid, A. G. (1977). Electrical conductivity in doped polyacetylene. *Physical Review Letters*, *39*(17), 1098–1101. Available from https://doi.org/10.1103/PhysRevLett.39.1098.

Ciselli, P., Lu, L., Busfield, J. J. C., & Peijs, T. (2010). Piezoresistive polymer composites based on EPDM and MWNTs for strain sensing applications. *E-Polymers*. Available from https://doi.org/10.1515/epoly.2010.10.10.125, http://www.e-polymers.org/journal/papers/tpeijs_170210.pdf.

Clark, L. C., & Lyons, C. (1962). Electrode systems for continuous monitoring in cardiovascular surgery. *Annals of the New York Academy of Sciences*, *102*(1), 29–45. Available from https://doi.org/10.1111/j.1749-6632.1962.tb13623.x.

Cooper, J. M., & Bloor, D. (1993). Evidence for the functional mechanism of a polypyrrole glucose oxidase electrode. *Electroanalysis*, *5*(9-10), 883–886. Available from https://doi.org/10.1002/elan.1140050925.

Dai, L., Soundarrajan, P., & Kim, T. (2002). Sensors and sensor arrays based on conjugated polymers and carbon nanotubes. *Pure and Applied Chemistry*, *74*(9), 1753–1772. Available from https://doi.org/10.1351/pac200274091753.

Dervisevic, M., Dervisevic, E., Senel, M., Cevik, E., Yildiz, H. B., & Camurlu, P. (2017). Construction of ferrocene modified conducting polymer based amperometric urea biosensor. *Enzyme and Microbial Technology*, *102*, 53–59. Available from https://doi.org/10.1016/j.enzmictec.2017.040.002, http://www.elsevier.com/locate/enzmictec.

Duan, N., Chen, X., Lin, X., Ying, D., Wang, Z., Yuan, W., & Wu, S. (2023). Paper-based fluorometric sensing of malachite green using synergistic recognition of aptamer-molecularly imprinted polymers and luminescent metal–organic frameworks. *Sensors and Actuators B: Chemical*, *384*. Available from https://doi.org/10.1016/j.snb.2023.133665.

Fan, L., Yang, J., Wu, J., Li, F., Yan, W., Tan, F., Zhang, M., Draz, M. S., Han, H., & Zhang, P. (2022). Deeply-dyed nanobead system for rapid lateral flow assay testing of drugs at point-of-care. *Sensors and Actuators B: Chemical*, *362*. Available from https://doi.org/10.1016/j.snb.2022.131829.

Fang, C. C., Chou, C. C., Yang, Y. Q., Wei-Kai, T., Wang, Y. T., & Chan, Y. H. (2018). Multiplexed detection of tumor markers with multicolor polymer dot-based immunochromatography test strip. *Analytical Chemistry*, *90*(3), 2134–2140. Available from https://doi.org/10.1021/acs.analchem.7b04411, http://pubs.acs.org/journal/ancham.

Feng, L., Zhong, M., Zhang, S., Wang, M., Sun, Z.-Y., & Chen, Q. (2019). Synthesis of water-soluble fluorescent polymeric glycoconjugate for the detection of cholera toxin. *Designed Monomers and Polymers*, *22*(1), 150–158. Available from https://doi.org/10.1080/15685551.2019.1654695.

Finlayson, L., Barnard, I. R. M., McMillan, L., Ibbotson, S. H., Brown, C. T. A., Eadie, E., & Wood, K. (2022). Depth penetration of light into skin as a function of wavelength from 200 to 1000 nm. *Photochemistry and Photobiology*, *98*(4), 974–981. Available from https://doi.org/10.1111/php.13550.

Gamota, D. (2020). Organic electronics: Ingredients for innovation. *Forbes Technology Council*.

Gaylord, B. S., Heeger, A. J., & Bazan, G. C. (2002). DNA detection using water-soluble conjugated polymers and peptide nucleic acid probes. *Proceedings of the National Academy of Sciences*, *99*(17), 10954–10957. Available from https://doi.org/10.1073/pnas.162375999.

Geng, W. C., Ye, Z., Zheng, Z., Gao, J., Li, J. J., Shah, M. R., Xiao, L., & Guo, D. S. (2021). Supramolecular bioimaging through signal amplification by combining indicator displacement assay with förster resonance energy transfer. *Angewandte Chemie – International Edition*, *60*(36), 19614–19619. Available from https://doi.org/10.1002/anie.202104358, http://onlinelibrary.wiley.com/journal/10.1002/(ISSN)1521-3773.

Georgopoulou, A., & Clemens, F. (2020). Piezoresistive elastomer-based composite strain sensors and their applications. *ACS Applied Electronic Materials*, *2*(7), 1826–1842. Available from https://doi.org/10.1021/acsaelm.0c00278.

Gupta, N., Adepu, V., Tathacharya, M., Siraj, S., Pal, S., Sahatiya, P., & Kuila, B. K. (2023). Piezoresistive pressure sensor based on conjugated polymer framework for pedometer and smart tactile glove applications. *Sensors and Actuators A: Physical*, *350*. Available from https://doi.org/10.1016/j.sna.2022.114139.

Hao, R., Liu, L., Yuan, J., Wu, L., & Lei, S. (2023). Recent advances in field effect transistor biosensors: Designing strategies and applications for sensitive assay. *Biosensors*, *13*(4), 426. Available from https://doi.org/10.3390/bios13040426.

Huang, H., Zhong, J., Ye, Y., Wu, R., Luo, B., Ning, H., Qiu, T., Luo, D., Yao, R., & Peng, J. (2022). Research Progresses in microstructure designs of flexible pressure sensors. *Polymers*, *14*(17). Available from https://doi.org/10.3390/polym14173670.

Huo, J., Lu, L., Shen, Z., Gao, H., & Liu, H. (2021). Rational design of CoNi alloy and atomic Co/Ni composite as an efficient electrocatalyst. *Surface Innovations*, *9*(1), 37–48. Available from https://doi.org/10.1680/jsuin.20.00028.

Hussain, S., Chen, X., Wang, C., Hao, Y., Tian, X., He, Y., Li, J., Shahid, M., Iyer, P. K., & Gao, R. (2022). Aggregation and binding-directed FRET modulation of conjugated polymer materials for selective and point-of-care monitoring of serum albumins. *Analytical Chemistry*, *94*(30), 10685–10694. Available from https://doi.org/10.1021/acs.analchem.2c00984.

Jafarian, M., Forouzandeh, F., Danaee, I., Gobal, F., & Mahjani, M. G. (2009). Electrocatalytic oxidation of glucose on Ni and NiCu alloy modified glassy carbon electrode. *Journal of Solid State Electrochemistry*, *13*(8), 1171–1179. Available from https://doi.org/10.1007/s10008-008-0632-1.

Karabag, A., Soyler, D., Udum, Y. A., Toppare, L., Gunbas, G., & Soylemez, S. (2023). Building block engineering toward realizing high-performance electrochromic materials and glucose biosensing platform. *Biosensors*, *13*(7). Available from https://doi.org/10.3390/bios13070677.

Khan, A., Khan, A. A. P., Marwani, H. M., Alotaibi, M. M., Asiri, A. M., Manikandan, A., Siengchin, S., & Rangappa, S. M. (2022). Sensitive non-enzymatic glucose electrochemical sensor based on electrochemically synthesized PANI/bimetallic oxide composite. *Polymers*, *14*(15), 3047. Available from https://doi.org/10.3390/polym14153047.

Kim, D.-M., Moon, J.-M., Lee, W.-C., Yoon, J.-H., Choi, C.S., Shim, Y.-B. (2017). A potentiometric non-enzymatic glucose sensor using a molecularly imprinted layer bonded on a conducting polymer. *Biosensors and Bioelectronics, 91*, 276–283. Available from https://doi.org/10.1016/j.bios.2016.120.046.

Kim, E. R., Joe, C., Mitchell, R. J., & Gu, M. B. (2023). Biosensors for healthcare: Current and future perspectives. *Trends in Biotechnology*, *41*(3), 374–395. Available from https://doi.org/10.1016/j.tibtech.2022.120.005.

Knite, M., Tupureina, V., Fuith, A., Zavickis, J., & Teteris, V. (2007). Polyisoprene—Multi-wall carbon nanotube composites for sensing strain. *Materials Science and Engineering: C*, *27*(5-8), 1125–1128. Available from https://doi.org/10.1016/j.msec.2006.080.016.

Kumar, V., Matai, I., Kumar, A., & Sachdev, A. (2021). GNP-CeO_2- polyaniline hybrid hydrogel for electrochemical detection of peroxynitrite anion and its integration in a microfluidic platform. *Microchimica Acta*, *188*(12). Available from https://doi.org/10.1007/s00604-021-05105-4.

Lai, Q. T., Sun, Q. J., Tang, Z., Tang, X. G., & Zhao, X. H. (2023). Conjugated polymer-based nanocomposites for pressure sensors. *Molecules*, *28*(4). Available from https://doi.org/10.3390/molecules28041627, http://www.mdpi.com/journal/molecules.

Lange, U., Roznyatovskaya, N. V., & Mirsky, V. M. (2008). Conducting polymers in chemical sensors and arrays. *Analytica Chimica Acta*, *614*(1), 1–26. Available from https://doi.org/10.1016/j.aca.2008.02.068.

Lee, H., Hong, Y. J., Baik, S., Hyeon, T., & Kim, D. -H. (2018). Enzyme-based glucose sensor: From invasive to wearable device. *Advanced Healthcare Materials*, *7*(8). Available from https://doi.org/10.1002/adhm.201701150.

Liu, C., Zhang, R., Li, P., Qu, J., Chao, P., Mo, Z., Yang, T., Qing, N., & Tang, L. (2022). Conductive hydrogels with ultrastretchability and adhesiveness for flame- and cold-tolerant strain sensors. *ACS Applied Materials & Interfaces*, *14*(22), 26088–26098. Available from https://doi.org/10.1021/acsami.2c07501.

McQuade, D. T., Pullen, A. E., & Swager, T. M. (2000). Conjugated polymer-based chemical sensors. *Chemical Reviews*, *100*(7), 2537–2574. Available from https://doi.org/10.1021/cr9801014.

Nguyen, T. N., Phung, V. D., & Tran, V. V. (2023). Recent advances in conjugated polymer-based biosensors for virus detection. *Biosensors*, *13*(6). Available from https://doi.org/10.3390/bios13060586, http://www.mdpi.com/journal/biosensors/.

Ogaidi, I. A., Gou, H., Aguilar, Z. P., Guo, S., Melconian, A. K., Al-Kazaz, A. K. A., Meng, F., & Wu, N. (2014). Detection of the ovarian cancer biomarker CA-125 using chemiluminescence resonance energy transfer to graphene quantum dots. *Chemical Communications*, *50*(11), 1344–1346. Available from https://doi.org/10.1039/c3cc47701k.

Ohayon, D., Nikiforidis, G., Savva, A., Giugni, A., Wustoni, S., Palanisamy, T., Chen, X., Maria, I. P., Fabrizio, E. D., Costa, P. M. F. J., McCulloch, I., & Inal, S. (2020). Biofuel powered glucose detection in bodily fluids with an n-type conjugated polymer. *Nature Materials*, *19*(4), 456–463. Available from https://doi.org/10.1038/s41563-019-0556-4.

Pham, G. T., Park, Y.-B., Liang, Z., Zhang, C., & Wang, B. (2008). Processing and modeling of conductive thermoplastic/carbon nanotube films for strain sensing. *Composites Part B: Engineering*, *39*(1), 209–216. Available from https://doi.org/10.1016/j.compositesb.2007.020.024.

Pilvenyte, G., Ratautaite, V., Boguzaite, R., Ramanavicius, S., Chen, C.-F., Viter, R., & Ramanavicius, A. (2023). Molecularly imprinted polymer-based electrochemical sensors for the diagnosis of infectious diseases. *Biosensors*, *13*(6), 620. Available from https://doi.org/10.3390/bios13060620.

Pu, Z., Zou, C., Wang, R., Lai, X., Yu, H., Xu, K., & Li, D. (2016). A continuous glucose monitoring device by graphene modified electrochemical sensor in microfluidic system. *Biomicrofluidics*, *10*(1). Available from https://doi.org/10.1063/1.4942437.

Rahman, M. A., Kumar, P., Park, D. S., & Shim, Y. B. (2008). Electrochemical sensors based on organic conjugated polymers. *Sensors*, *8*(1), 118–141. Available from https://doi.org/10.3390/s8010118, http://www.mdpi.org/sensors/papers/s8010118.pdf.

Rasmussen, S. C. (2020). Conjugated and conducting organic polymers: The first 150 years. *ChemPlusChem*, *85*(7), 1412–1429. Available from https://doi.org/10.1002/cplu.202000325, http://onlinelibrary.wiley.com/journal/10.1002/(ISSN)2192-6506.

Runsewe, D., Betancourt, T., & Irvin, J. A. (2019). Biomedical application of electroactive polymers in electrochemical sensors: A review. *Materials*, *12*(16). Available from https://doi.org/10.3390/ma12162629.

Schneckenburger, H. (2020). Förster resonance energy transfer—what can we learn and how can we use it? *Methods and Applications in Fluorescence*, *8*(1). Available from https://doi.org/10.1088/2050-6120/ab56e1.

Scilingo, E. P., Lorussi, F., Mazzoldi, A., & De Rossi, D. (2003). Strain-sensing fabrics for wearable kinaesthetic-like systems. *IEEE Sensors Journal*, *3*(4), 460–467. Available from https://doi.org/10.1109/JSEN.2003.815771.

Seyedin, S., Razal, J. M., Innis, P. C., Jeiranikhameneh, A., Beirne, S., & Wallace, G. G. (2015). Knitted strain sensor textiles of highly conductive all-polymeric fibers. *ACS Applied Materials and Interfaces*, *7*(38), 21150–21158. Available from https://doi.org/10.1021/acsami.5b04892, http://pubs.acs.org/journal/aamick.

Shen, H., Zou, Y., Zang, Y., Huang, D., Jin, W., Di, C.-an, & Zhu, D. (2018). Molecular antenna tailored organic thin-film transistors for sensing application. *Materials Horizons*, *5*(2), 240–247. Available from https://doi.org/10.1039/C7MH00887B.

Shuvo, I. I., Shah, A., & Dagdeviren, C. (2022). Electronic textile sensors for decoding vital body signals: State-of-the-art review on characterizations and recommendations. *Advanced Intelligent Systems, 4*(4). Available from https://doi.org/10.1002/aisy.202100223.

Singh, S. U., Chatterjee, S., Lone, S. A., Ho, H. H., Kaswan, K., Peringeth, K., Khan, A., Chiang, Y. W., Lee, S., & Lin, Z. H. (2022). Advanced wearable biosensors for the detection of body fluids and exhaled breath by graphene. *Microchimica Acta, 189*(6). Available from https://doi.org/10.1007/s00604-022-05317-2, http://www.springer/at/mca.

Son, S. U., Seo, S. B., Jang, S., Choi, J., Lim, Jw, Lee, D. K., Kim, H., Seo, S., Kang, T., Jung, J., & Lim, E. K. (2019). *Naked-eye detection of pandemic influenza a (pH1N1) virus by polydiacetylene (PDA)-based paper sensor as a point-of-care diagnostic platform. Sensors and Actuators, B: Chemical* (291, pp. 257–265). South Korea: Elsevier B.V. Available from https://www.journals.elsevier.com/sensors-and-actuators-b-chemical, https://doi.org/10.1016/j.snb.2019.040.081.

Sun, H., Zhao, Y., Wang, C., Zhou, K., Yan, C., Zheng, G., Huang, J., Dai, K., Liu, C., & Shen, C. (2020). Ultra-Stretchable, durable and conductive hydrogel with hybrid double network as high performance strain sensor and stretchable triboelectric nanogenerator. *Nano Energy, 76*. Available from https://doi.org/10.1016/j.nanoen.2020.105035.

Suzuki, S., Sasaki, S., Sairi, A. S., Iwai, R., Tang, B. Z., & Konishi, Gi (2020). Principles of aggregation-induced emission: Design of deactivation pathways for advanced AIEgens and applications. *Angewandte Chemie - International Edition, 59*(25), 9856–9867. Available from https://doi.org/10.1002/anie.202000940, http://onlinelibrary.wiley.com/journal/10.1002/(ISSN)1521-3773.

Tao, S., Zhu, S., Feng, T., Zheng, C., & Yang, B. (2020). Crosslink-enhanced emission effect on luminescence in polymers: Advances and perspectives. *Angewandte Chemie International Edition, 59*(25), 9826–9840. Available from https://doi.org/10.1002/anie.201916591.

Tawfik, S. M., Elmasry, M. R., Sharipov, M., Azizov, S., Lee, C. H., & Lee, Y. I. (2020). *Dual emission non-ionic molecular imprinting conjugated polythiophenes-based paper devices and their nanofibers for point-of-care biomarkers detection. Biosensors and Bioelectronics* (160). South Korea: Elsevier Ltd. Available from http://www.elsevier.com/locate/bios, https://doi.org/10.1016/j.bios.2020.112211.

The Huy, B., Thangadurai, D. T., Sharipov, M., Ngoc Nghia, N., Van Cuong, N., & Lee, Y. I. (2022). Recent advances in turn off-on fluorescence sensing strategies for sensitive biochemical analysis – A mechanistic approach. *Microchemical Journal, 179*. Available from https://doi.org/10.1016/j.microc.2022.107511, http://www.elsevier.com/inca/publications/store/6/2/0/3/9/1.

Torrente-Rodríguez, R. M., Tu, J., Yang, Y., Min, J., Wang, M., Song, Y., Yu, Y., Xu, C., Ye, C., IsHak, W. W., & Gao, W. (2020). Investigation of cortisol dynamics in human sweat using a graphene-based wireless mHealth system. *Matter, 2*(4), 921–937. Available from https://doi.org/10.1016/j.matt.2020.010.021, http://www.cell.com/matter.

Udum, Y. A., Aktas Gemci, M., Cevher, D., Soylemez, S., Cirpan, A., & Toppare, L. (2023). D-A-D type functional conducting polymer: Development of its electrochromic properties and laccase biosensor. *Journal of Applied Polymer Science, 140*(11). Available from https://doi.org/10.1002/app.53614, http://onlinelibrary.wiley.com/journal/10.1002/(ISSN)1097-4628.

Verma, N. (2018). A green synthetic approach for size tunable nanoporous gold nanoparticles and its glucose sensing application. *Applied Surface Science, 462*, 753–759. Available from https://doi.org/10.1016/j.apsusc.2018.080.175.

Wang, J., Zhao, D., & Xu, C. (2016). Nonenzymatic electrochemical sensor for glucose based on nanoporous platinum-gold alloy. *Journal of Nanoscience and Nanotechnology, 16*(7), 7145–7150. Available from https://doi.org/10.1166/jnn.2016.11377.

Wang, J., Xue, P., Tao, X., & Yu, T. (2014). Strain sensing behavior and its mechanisms of electrically conductive PPy-coated Fabric. *Advanced Engineering Materials, 16*(5), 565–570. Available from https://doi.org/10.1002/adem.201300407.

Wang, P., Wang, M., Zhu, J., Wang, Y., Gao, J., Gao, C., & Gao, Q. (2021). Surface engineering via self-assembly on PEDOT: PSS fibers: Biomimetic fluff-like morphology and sensing application. *Chemical Engineering Journal, 425*, 131551. Available from https://doi.org/10.1016/J.CEJ.2021.131551.

Wang, T., Zhang, N., Bai, W., & Bao, Y. (2020). Fluorescent chemosensors based on conjugated polymers with N-heterocyclic moieties: Two decades of progress. *Polymer Chemistry, 11*(18), 3095–3114. Available from https://doi.org/10.1039/D0PY00336K.

Wang, W., Yang, S., Ding, K., Jiao, L., Yan, J., Zhao, W., Ma, Y., Wang, T., Cheng, B., & Ni, Y. (2021). Biomaterials- and biostructures Inspired high-performance flexible stretchable strain sensors: A review. *Chemical Engineering Journal, 425*, 129949. Available from https://doi.org/10.1016/j.cej.2021.129949.

Wang, Y., Li, S., Feng, L., Nie, C., Liu, L., Lv, F., & Wang, S. (2015). Fluorescence ratiometric assay strategy for chemical transmitter of living cells using H_2O_2 – Sensitive conjugated polymers. *ACS Applied Materials & Interfaces, 7*(43), 24110–24118. Available from https://doi.org/10.1021/acsami.5b07172.

Xu, C., Jiang, D., Ge, Y., Huang, L., Xiao, Y., Ren, X., Liu, X., Zhang, Q., & Wang, Y. (2022). A PEDOT: PSS conductive hydrogel incorporated with Prussian blue nanoparticles for wearable and noninvasive monitoring of glucose. *Chemical Engineering Journal, 431*, 134109. Available from https://doi.org/10.1016/j.cej.2021.134109.

Xu, J., Wang, S., Wang, G. J. N., Zhu, C., Luo, S., Jin, L., Gu, X., Chen, S., Feig, V. R., To, J. W. F., Rondeau-Gagné, S., Park, J., Schroeder, B. C., Lu, C., Oh, J. Y., Wang, Y., Kim, Y. H., Yan, H., Sinclair, R., ... Bao, Z. (2017). Highly stretchable polymer semiconductor films through the nanoconfinement effect. *Science (New York, N.Y.), 355*(6320). Available from https://doi.org/10.1126/science.aah4496, http://science.sciencemag.org/content/sci/355/6320/59.full.pdf.

Xu, M., Zhu, Y., Gao, S., Zhang, Z., Gu, Y., & Liu, X. (2021). Reduced graphene oxide-coated silica nanospheres as flexible enzymatic biosensors for detection of glucose in sweat. *ACS Applied Nano Materials, 4*(11), 12442–12452. Available from https://doi.org/10.1021/acsanm.1c02887.

Xu, Z., Song, J., Liu, B., Lv, S., Gao, F., Luo, X., & Wang, P. (2021). A conducting polymer PEDOT:PSS hydrogel based wearable sensor for accurate uric acid detection in human sweat. *Sensors and Actuators B: Chemical, 348*, 130674. Available from https://doi.org/10.1016/J.SNB.2021.130674.

Xue, P., Tao, X. M., Kwok, K. Wy, Leung, M. Y., & yu, T. X. (2004). Electromechanical behavior of fibers coated with an electrically conductive polymer. *Textile Research Journal, 74*(10), 929–936. Available from https://doi.org/10.1177/004051750407401013.

Yang, M., Liu, M., Wu, Z., He, Y., Ge, Y., Song, G., & Zhou, J. (2020). Fluorescence enhanced detection of water in organic solvents by one-pot synthesis of orange-red emissive polymer carbon dots based on 1,8-naphthalenediol. *Micro & Nano Letters, 15*(7), 469–473. Available from https://doi.org/10.1049/mnl.2019.0170.

Yang, X., Cao, L., Wang, J., & Chen, L. (2020). Sandwich-like polypyrrole/reduced graphene oxide nanosheets integrated gelatin hydrogel as mechanically and thermally sensitive skinlike bioelectronics. *ACS Sustainable Chemistry and Engineering, 8*(29), 10726–10739. Available from https://doi.org/10.1021/acssuschemeng.0c01998, http://pubs.acs.org/journal/ascecg.

Yang, Y. C., Liu, M. H., Yang, S. M., & Chan, Y. H. (2021). Bimodal multiplexed detection of tumor markers in non-small cell lung cancer with polymer dot-based immunoassay. *American Chemical Society, Taiwan ACS Sensors, 6*(11), 4255–4264. Available from https://doi.org/10.1021/acssensors.1c02025, http://pubs.acs.org/journal/ascefj.

Yang, Y., Deng, H., & Fu, Q. (2020). Recent progress on PEDOT:PSS based polymer blends and composites for flexible electronics and thermoelectric devices. *Materials Chemistry Frontiers, 4*(11), 3130–3152. Available from https://doi.org/10.1039/D0QM00308E.

Zaumseil, J., & Sirringhaus, H. (2007). Electron and ambipolar transport in organic field-effect transistors. *Chemical Reviews, 107*(4), 1296–1323. Available from https://doi.org/10.1021/cr0501543.

Zhan, T., Xie, H., Mao, J., Wang, S., Hu, Y., & Guo, Z. (2021). Conductive PNIPAM/CMCS/MWCNT/PANI hydrogel with temperature, pressure and pH sensitivity. *ChemistrySelect*, *6*(17), 4229–4237. Available from https://doi.org/10.1002/slct.202101003.

Zhang, P., Lu, C., Niu, C., Wang, X., Li, Z., & Liu, J. (2022). Binding studies of cationic conjugated polymers and DNA for label-free fluorescent biosensors. *ACS Applied Polymer Materials*, *4*(8), 6211–6218. Available from https://doi.org/10.1021/acsapm.2c00986.

Zhang, Q., Jiang, D., Xu, C., Ge, Y., Liu, X., Wei, Q., Huang, L., Ren, X., Wang, C., & Wang, Y. (2020). Wearable electrochemical biosensor based on molecularly imprinted Ag nanowires for noninvasive monitoring lactate in human sweat. *Sensors and Actuators B: Chemical*, *320*, 128325. Available from https://doi.org/10.1016/j.snb.2020.128325, https://linkinghub.elsevier.com/retrieve/pii/S0925400520306705.

Zhang, R., Deng, H., Valenca, R., Jin, J., Fu, Q., Bilotti, E., & Peijs, T. (2012). Carbon nanotube polymer coatings for textile yarns with good strain sensing capability. *Sensors and Actuators A: Physical*, *179*, 83–91. Available from https://doi.org/10.1016/j.sna.2012.030.029.

Zhang, Z., & Yates, J. T. (2010). Effect of adsorbed donor and acceptor molecules on electron stimulated desorption: O_2/TiO_2(110. *Journal of Physical Chemistry Letters*, *1*(14), 2185–2188. Available from https://doi.org/10.1021/jz1007559.

Zhang, Z., & Yates, J. T. (2012). Band bending in semiconductors: Chemical and physical consequences at surfaces and interfaces. *Chemical Reviews*, *112*(10), 5520–5551. Available from https://doi.org/10.1021/cr3000626.

Zheng, H., Chen, M., Sun, Y., & Zuo, B. (2022). Self-healing, wet-adhesion silk fibroin conductive hydrogel as a wearable strain sensor for underwater applications. *Chemical Engineering Journal*, *446*. Available from https://doi.org/10.1016/j.cej.2022.136931.

Zhou, Q., & Swager, T. M. (1995). Fluorescent chemosensors based on energy migration in conjugated polymers: The molecular wire approach to increased sensitivity. *Journal of the American Chemical Society*, *117*(50), 12593–12602. Available from https://doi.org/10.1021/ja00155a023.

Zhu, B., Tang, W., Ren, Y., & Duan, X. (2018). Chemiluminescence of conjugated-polymer nanoparticles by direct oxidation with hypochlorite. *Analytical Chemistry*, *90*(22), 13714–13722. Available from https://doi.org/10.1021/acs.analchem.8b04109.

CHAPTER

Molecularly Imprinted Polymers (MIPs)

3

Giancarla Alberti, Camilla Zanoni, Lisa Rita Magnaghi and Raffaela Biesuz
Department of Chemistry, University of Pavia, Pavia, Italy

3.1 Introduction

3.1.1 What are molecularly imprinted polymers?

The specific recognition mechanism of antibodies for antigens, receptors for hormones, and enzymes for substrates has encouraged the development of polymers able to mimic this natural process, that is, the selective capture of chemical species from complex matrixes (Haupt et al., 2011). Such materials are called Molecularly Imprinted Polymers (MIPs). The molecular imprinting process involves the polymerization of functional monomers and a crosslinker in the presence of a target analyte that acts as a template. After the polymerization, the subsequent template removal leaves specific cavities in the polymer that are complementary in size and shape to the target molecule. Therefore, when contacted with a sample solution, the empty MIP can specifically rebind the analyte or closely related molecules (Alexander et al., 2006; Beltran et al., 2010; Kadhem et al., 2021; Kupai et al., 2016).

MIPs are appealing for their recognition ability, similar to natural receptors, but also because they are available for several target analytes and possess higher physical and chemical stability as compared to bioreceptors (Chen et al., 2016; Leibl et al., 2021).

In detail, MIPs are polymers, that is, macromolecules composed of monomers belonging to the class of homopolymers or heteropolymers.

MIPs are characterized by selective recognition cavities interacting with the analyte through different mechanisms, such as non-covalent, covalent, and semi-covalent (Belbruno, 2019). Monomers are polymerized around a template to obtain selective cavities. The template must be removed after the polymerization to obtain cavities complementary in size and shape to itself. Generally, the template is the analyte, or a molecule with a structure and reactivity very similar to the analyte. The nature of the interaction with the active site, monomers, crosslinkers, solvents, and type of polymerization must be chosen to obtain a suitable MIP for a specific analyte. Based on that choice, the obtained polymer will be implemented in sensors with the most suitable signal transduction method (Belbruno, 2019).

3.1.1.1 Non-covalent interactions

In the non-covalent approach, the analyte molecules interact with polymer residues in the active sites through weak bounds like van der Waals and hydrophobic forces, hydrogen bonding, and ionic

Sensory Polymers. DOI: https://doi.org/10.1016/B978-0-443-13394-7.00003-3

interactions. This interaction occurs in a solution, the so-called prepolymeric mixture, composed of the template, monomers, crosslinkers, and a suitable solvent. The most commonly exploited is free radical polymerization, which might be thermal, photochemical, or electrochemically induced (Zhang et al., 2006). MIPs prepared by non-covalent interaction coupled with free radical polymerization are mainly faced with electrochemical transduction.

To maximize the interactions between the analyte and the MIP's cavities, the appropriate monomer must be chosen. For example, suppose the analyte structure contains an acidic residue. In that case, it might be appropriate to use a basic monomer (for example, a monomer with aminic residues) to establish a charge-charge interaction (Yan & Ramström, 2004). Some of the most commonly employed monomers for the non-covalent approach are methacrylic acid, trifluoro-methacrylic acid (MAA), hydroxyethylmethacrilate, 4-vinylpyridine, 4-vinylimidazole, 4-vynylbenzoic acid, acrylamide, acrylonitrile, aniline, pyrrole, thiophene, and 3,4-ethylenedioxythiophene.

In recent years, scientists have focused their attention on investigation lines that are more sustainable, environmentally friendly, and not harmful to human health; for this reason, MIPs based on biocompatible and biodegradable monomers, for example, proteins, dopamine, polycaprolactone, polyhydroxybutyrate, or alginate, are among the most recently studied.

Crosslinkers and solvents, as well as monomers, are fundamental for the MIPs' performances because they can affect the porosity, rigidity, and selectivity of the polymer.

The crosslinker provides rigidity to the polymer structure and gives a more rigid packing to the monomer, making the polymer more selective. To obtain these features, the amount of crosslinker in the prepolymeric mixture must be much higher than the monomer (Yan & Ramström, 2004). Some of the most commonly used crosslinkers for MIP prepared by a non-covalent approach are divinylbenzene, *N,N′*-methylene-bisacylamide, ethyleneglycol dimethacrylate, and trimethylpropane trimethacrylate.

Last but not least is the solvent. Solvent has fundamental roles: it dissolves all the reagents, directs the interactions between the monomers and the template molecules, and affects the porosity of the polymer (if the polymer is too porous, non-specific interactions of the analyte or interfering molecules occur). The solvent does not interact with the analyte as the monomer, that is, if the analyte molecule interacts with the monomer through hydrogen bonds, protic solvents or polar aprotic solvents are not recommended since they might interfere with van Der Waals interactions.

Once monomers, crosslinkers, and solvents are chosen, the radical initiator must be added to the prepolymeric solution. The most commonly used thermal initiators are peroxides and azo-compounds, while photochemical initiators are peroxides, azo-compounds, disulfides, and ketones (AIBN, BPO, phenyl disulfide, and benzoin) (Yan & Ramström, 2004).

When the prepolymeric components are chosen, the last step is to decide the molar ratio of each building block. A widespread molar ratio for preparing an MIP with good performance is 1:4:20 = template, functional monomer, crosslinker, and 4:3 between the solvent and the monomers. These ratios are not a fixed rule; performing a design of experiments might be advisable to reach the optimal prepolymeric mixture (Yan & Ramström, 2004).

The non-covalent approach is the most widespread technique for MIP development thanks to the easy procedures for the polymer design; it does not need extensive knowledge in polymer chemistry, and the chemicals commonly used are usually low-cost and easy to handle. Unfortunately, this approach has a consistent drawback, such as the excess monomer required, which leads to the formation of several nonspecific binding sites affecting the polymer selectivity.

Other approaches, such as covalent or semi-covalent, might be employed to enhance the selectivity of the polymer (Hashim et al., 2014).

3.1.1.2 Covalent interactions

In the covalent approach, the analyte molecule interacts with residues in the polymer cavity through covalent bonds. As mentioned before, the covalent approach allows for more selective polymers due to the limited nonspecific interactions between the analyte and the binding sites (Yan & Ramström, 2004). In this approach, the binding site monomer and the template molecule are ideally present in a stoichiometric ratio; thus, each analyte molecule corresponds to one recognition cavity. To obtain high-performance MIPs, the interaction between the template and the monomer must be stable during the polymerization process; otherwise, the active site may not have the size and shape complementary to the template. Once the polymer is formed, the analyte must be removed from the scaffold in mild conditions to preserve the structure of the MIP; the formation of the covalent bond between the analyte and the binding site during the equilibration must be thermodynamically and kinetically preferred (Yan & Ramström, 2004).

The covalent approach is slightly more complicated than the non-covalent one since it requires a deep understanding of the chemical interactions between molecules. The crucial step to obtaining a high-performance polymer is to choose the binding site monomer. Some of the most commonly used monomers for covalent MIP synthesis are 4-vinyl benzene boric acid, 4-vinyl benzaldehyde, 4-vinyl aniline, and tert-butyl *p*-vinylphenylcarbonate.

Since the monomer is in a stoichiometric ratio with the analyte, the presence of a crosslinker is crucial because it stabilizes the cavities created by the template molecule. The crosslinker should be added between 50% and 66% to obtain a good balance between the polymer's stability and the cavities' selectivity. The most commonly used crosslinkers are ethylene glycol dimethacrylate (EGDMA) and divinylbenzene.

The crucial step for the formation of cavities with high selectivity is the removal of the template. In the covalent approach, the removal consists in breaking the bond; for example, if the binding site contains a boronic acid, the template might be removed between 85% and 95% using a water/alcohol solution; otherwise, if the binding site forms a Shiff base with the analyte, the template might be displaced using an acidic water/alcohol solution. As stated above, the template's release occurs in mild conditions that preserve the polymeric structure.

The covalent approach is advantageous regarding selectivity against a target molecule since it removes the problem of the unspecific interactions that can occur with non-covalent MIPs. Conversely, the weakness of the covalent approach is that the condensation reaction between the analyte and the active site is not always reversible; this phenomenon reduces the efficiency of the polymer. Moreover, the molecule for the imprinting still linked to the polymer might be released in the sample. The condensation reaction involves the loss or addition of water molecules; these steps reduce the interchange rate between the target molecule and the binding sites. The hindrance of the molecule at the site even influences the rate. To avoid some of these problems, the semi-covalent approach might be considered for MIP development.

3.1.1.3 Semi-covalent interactions

The semi-covalent approach is a perfect mix between the non-covalent and covalent approaches; indeed, it takes the qualities of the two methods. In the semi-covalent strategy, the imprinting is performed according to the covalent approach's methods, while the target molecule's rebinding follows the non-covalent one (Yan & Ramström, 2004). A MIP designed using a semi-covalent approach does not suffer from nonspecific interactions precisely as for the MIP synthesized by the covalent method, and the rebinding is influenced only by diffusion and not by the reaction kinetics. The semi-covalent approach can be divided into two families: (1) the template is the analyte itself, and (2) the template is modified with a sacrificial spacer that will be eliminated to allow the formation of the selective site.

3.1.2 Why and how to incorporate molecularly imprinted polymers in sensing devices?

As described before, MIPs are polymers characterized by the presence of cavities able to selectively sorb an analyte from a class of similar compounds through weak or covalent interactions. MIPs are widely applied in sensing devices like optical, electrochemical, or thermal sensors. Immobilizing the MIP on the sensors' surface is a critical issue. The most common used strategy is to let the prepolymeric mixture polymerize thermally, photochemically, or electrochemically directly on the sensor's surface. The polymer can be obtained for precipitation as a powder or as micro- or nanobeads and incorporated into the sensor embedded in pastes or inks (this last technique is often used in developing electrochemical sensors).

A key point is defining the best method to deposit the prepolymeric mixture to obtain the appropriate film thickness. Drop coating is the most widely used method for the deposition of the prepolymeric mixture, but it is not easy to control the film's thickness and uniformity. The polymer's thickness and homogeneity change depending on the quantity of prepolymeric mixture deposited on the sensor's surface. For example, Alberti et al. developed a MIP-modified screen-printed cell for the potentiometric atrazine determination using drop-coating to modify the working electrode surface with the acrylic prepolymeric mixture; the polymerization was carried out thermally (Alberti et al., 2022).

Spin coating is a simple and fast method to deposit prepolymeric mixtures, obtaining MIPs with small thicknesses depending on the time and the rotation speed; moreover, the centrifugal force can favor solvent evaporation. This strategy was adopted by Zeni and Cennamo's group in developing several surface plasmon resonance (SPR) sensors based on plastic optical fiber modified with MIPs for the detection of perfluorooctanoate, perfluorooctanesulfonate, perfluorinated alkylated substances (Cennamo et al., 2018), 2-furaldehyde (Cennamo et al., 2015; Pesavento, Zeni, et al., 2021), 2,4,6-trinitrotoluene (Cennamo et al., 2013), L-nicotine (Cennamo et al., 2014), dibenzyl disulfide (Cennamo et al., 2016), and SARS-CoV-2 (Cennamo, D'agostino, et al., 2021). Different thicknesses might be obtained, varying the time and the rotation speed of the platform; the disadvantage is that the obtained film is not perfectly uniform: the thickness at the central point of the surface is higher than the margins of the platform.

Microcontact imprinting is a helpful method used to control polymeric film thickness. In this case, the prepolymeric mixture is deposited on a surface while a second transparent surface is functionalized with the target molecule (this method is used to imprint high-weight molecules such as microorganisms or biomolecules) (Unger & Lieberzeit, 2021). The surfaces are put in contact under pressure, and depending on the force imposed on the sandwich, the film thickness changes and only the surface of the polymer are imprinted. The polymerization can be thermal, photoinitiated (thanks to the transparent supports), or electrochemical. Once polymerization is performed, the surface used as mold must be removed to obtain the superficial sites of the MIP (Ertürk & Mattiasson, 2016; Gavrilă et al., 2022).

The Yang et al. group developed an impedimetric sensor for neutrophil gelatinase-associated lipocalin using the microcontact surface imprinting method to modify the sensing surface. The mold surface was composed of a polydimethylsiloxane (PDMS) hemispherical concave pore array modified by a neutrophil gelatinase-associated lipocalin binding peptide (NGAL B1). The mold surface was covered with the prepolymeric mixture of acrylamide, *N,N′*-methylenebisacrylamide, diallyamine, and ammonium persulfate; the sandwich was closed using a quartz crystal electrode as a lid. The photopolymerization was performed using a UV lamp and applying a certain pressure to the electrode surface for 50 minutes. After the microcontact printing, the modified QC electrode was detached from the PDMS mold and washed with sodium dodecyl sulfate solution and acetic acid to remove the template (Yang et al., 2022).

Another method used to modify conducting sensing surfaces is electropolymerization. Electrochemical polymerization is a method to obtain conducting thin films on electrode surfaces. The polymerization can be performed by oxidation or reduction, depending on the electroactive monomer employed. The anodic method is the most widespread, consisting of the monomers' oxidation and forming cationic radicals that lead to polymer production.

Another method used to modify conducting sensing surfaces is electropolymerization. Electrochemical polymerization is a method to obtain conducting thin films on electrode surfaces. The polymerization can be performed by oxidation or reduction, depending on the electroactive monomer employed. The anodic method is the most widespread, consisting of the monomers' oxidation and forming cationic radicals that lead to polymer production (Gavrilă et al., 2022). To obtain an MIP using this technique, the template molecule is not electroactive in the potential range applied during the electropolymerization. The monomers start polymerizing around the template molecule by applying a suitable potential to the cell, generally using cyclic voltammetry (CV) or amperometric techniques. Once the polymerization is performed, the selective recognition cavities are formed after the template removal. Electrochemical polymerization has several advantages; for example, it can be performed in water easily and quickly, and the film thickness can be modulated by changing the monomer concentration and the number of cycles in CV.

Several examples of this approach can be found in the recent literature (e.g., see the review of L.M. Gonçalves (Moreira Gonçalves, 2021)).

Another interesting method to modify the working carbon paste electrode with an MIP is mixing the polymer powder (obtained by bulk synthesis and ground) with graphite power and adding a binder (such as paraffin oil or n-eicosane) to form a homogeneous paste. Some recent papers have proposed this strategy to develop electrochemical sensors for emerging contaminants and drugs (Akhoundian et al., 2018; Alizadeh & Azizi, 2016; Antuña-Jiménez et al., 2012).

3.2 Molecularly imprinted polymer-based sensors

The core of a chemical sensor or biosensor is the recognition element, which is in close contact with an interrogative transducer and selectively recognizes and interacts with the target analyte. The transducer then translates the chemical signal generated upon analyte interaction into an output signal.

In recent years, chemical sensors and biosensors have been increasingly interested in analytical chemistry, as evident from the growing number of published papers. New demands have been manifesting in clinical diagnostics, environmental analysis, and food analysis.

Recently, a significant effort has been made to synthesize artificial receptors that bind a target analyte with similar affinity and selectivity to natural antibodies or enzymes. MIPs can be used for this aim thanks to their long-term stability, chemical inertness, and insolubility in water and several organic solvents. MIPs have been effectively used with different transducers, and several methods have been exploited in developing sensors (Vasapollo et al., 2011).

3.2.1 Electrochemical sensors

One of the most common uses of MIPs is in electrochemical sensors, judging by the number of publications in the last few years (2360 results only in 2023 in Google Scholar). Why are electrochemical sensors so popular? Because they are easy to handle, the procedures are not complicated, and the devices can be miniaturized and included in automatic systems; besides, the instrumentation is cost-effective, and the sensors have a short response time and high mechanical strength. The main advantage of integrating MIPs with electrochemical transducers is their improved selectivity, allowing the detection of the target analyte even in the presence of interferents or complex matrixes. In contrast, the sensitivity is similar to that of the unmodified electrodes, depending on the detection technique.

The following sections discuss the different types of electrochemical sensors coupled with MIPs.

3.2.1.1 Potentiometric sensors

Potentiometry is a widespread technique thanks to the non-destructive nature and rapidity of measurements.

The analytes to be detected with potentiometry must be ions; for this reason, the pH at which the measurement is performed is crucial. The electrochemical potential generated is proportional to the analyte concentration. Generally, a Nernstian response occurs, following Equation 4.1:

$$E(\mathrm{mV}) = \mathrm{K}' \pm \frac{59.2}{n} \cdot \log[\mathrm{A}] \tag{3.1}$$

K′ is a constant characteristic of the particular electrode, accounting for the sum of the potential differences at each interface membrane/solution, the activity coefficients, and the liquid junction potential, and n depends on the charge of the detected ion (Scholz, 2010). Plotting E vs log [A], the ideal Nernstian slope for a single charge ion is 59.2 mV.

MIP-based potentiometric sensors have undergone rapid development in recent years.

In the classical design, defined as a "symmetrical arrangement," the permselective membrane incorporating MIP separates two solutions' compartments (inner and sample solutions) at different ion concentrations. Generally, MIP's particles are dispersed in a PVC matrix (around 30% wt.); the plasticizer (solvent) is added (about 60%–70%). The solvent establishes the dielectric medium and provides proper mechanical resistance.

MIPs can be incorporated into the PVC membrane by exploiting different approaches: traditional bulk polymers must be ground and sieved before being mixed into PVC and plasticizer; on the other hand, MIPs can be prepared by suspension polymerization or the multistep-swelling method, allowing the production of spherical monodispersed polymeric particles (Vishnuvardhan et al., 2007).

MIP-based potentiometric electrodes based on this design have been developed for biomedical target analytes or pesticides, that is, organic molecules generally in their cationic form (Abass & Rzaij, 2020; Qi et al., 2020; Wang et al., 2020, 2021).

These sensors have very stable potential but are not easily miniaturized. Another possible design avoids using an inner solution. In this case, an asymmetrical cell is created: one side of the ion-selective membrane is directly contacted with a solid phase (such as the electrode surface), while the other is exposed to the sample.

Some examples of this kind of potentiometric sensor are summarized in Table 3.1.

Table 3.1 Example of MIP-based potentiometric sensors.

Analyte	MIP composition	Type of electrode	LOD(M)	Ref.
Atrazine	Atrazine, methacrylic acid, ethylene glycol dimethacrylate, and AIBN	Graphite screen-printed electrode	4×10^{-7}	Alberti et al. (2022)
Phenoxy herbicides	(4-chloro-2-methyl phenoxy) acetic acid, methacrylic acid, ethylene glycol dimethacrylate, and AIBN	Graphite screen-printed electrode	1×10^{-8}	Zanoni, Spina, et al. (2022)
Triclosan	Triclosan, acrylamide, methyl methacrylate, divinylbenzene, trimethylolpropane trimethacrylate, and AIBN	Rotating Ag disk	2×10^{-9}	Liang et al. (2013)
Sulfadiazine	Sulfadiazine,2-hydroxyethyl methacrylate, divinylbenzene, and AIBN	Glassy carbon	7×10^{-6}	G. Cui et al. (2023)
Bisphenol A	Bisphenol A, methyl methacrylate, divinylbenzene, and AIBN	Glassy carbon	3×10^{-8}	Liu, Song, et al. (2020)
Phosphate	Diphenyl phosphate, pyrrole, and ethylene glycol dimethacrylate	Gold wire	1×10^{-7}	Hikmat and Tasfiyati (2023)
Aminophylline	Aminophylline, methacrylic acid, acrylamide, ethylene glycol dimethacrylate, and potassium persulfate	Glassy carbon	1×10^{-10}	Saher et al. (2023)
Melamine	Melamine, methacrylic acid, ethylene glycol dimethacrylate, and AIBN	Rotating Ag disk	2×10^{-6}	Liang et al. (2009)
Clozapine	Clozapine, methacrylic acid, ethylene glycol dimethacrylate, and AIBN	Carbon paste	1×10^{-6}	Ganjali et al. (2012)
Hydroxyzine	hydroxyzine dihydrochloride, methacrylic acid, ethylene glycol dimethacrylate, and AIBN	Carbon paste	7×10^{-7}	Javanbakht et al. (2008)

For example, Alberti et al. developed two potentiometric sensors for the pesticides atrazine (Alberti et al., 2022) and MCPA (2-methyl-4-chlorophenoxyacetic acid) (Zanoni, Spina, et al., 2022). In both cases, acrylic-based prepolymeric mixtures of MIPs are drop-coated on the graphite-ink screen-printed electrodes. Thermal polymerization is carried out. Both sensors permitted the detection of the pesticides at sub-μM concentrations, with the advantages of simple realization and quick responses.

Alternatively, the MIP-based ion-selective membrane can be cast over the electrode surface, such as carbon material (graphite or glassy carbon) or noble metal electrodes (Pt and Au). With this approach, electrodes are used as solid contacts in PVC membrane-based ion-selective electrodes to detect several organic analytes (Cui et al., 2023; Hikmat & Tasfiyati, 2023; Liang et al., 2009; Liang et al., 2013; Liu, Song, et al., 2020; Saher et al., 2023).

In other studies, the MIP was mixed with carbon paste; in this design, a wider surface and very stable interfaces can be obtained (Ganjali et al., 2012; Javanbakht et al., 2008).

MIP-based potentiometric sensors based on field-effect devices are another group of solid-contact sensors; in this case, the ion-selective membrane is deposited over ion-sensitive field-effect transistors (ISFETs). Willner's group has pioneered MIP-modified ISFETs exploiting TiO_2/sol-gel as the polymeric structure for molecular imprinting for the determination of different acidic organic compounds such as chloroaromatic acids, benzylphosphonic acids, and thiophenols (Lahav, Kharitonov, & Willner, 2001; Lahav, Kharitonov, Katz, et al., 2001; Pogorelova et al., 2004; Zayats et al., 2002). These sensors are of great interest because of their ease of miniaturization and suitability for mass-production processes.

3.2.1.2 Voltammetric and amperometric sensors

For these electrochemical sensors, measurements are performed in a three-electrode configuration electrochemical cell by applying a potential difference (fixed in amperometry and variable in voltammetry) between the working (sensor) and the reference (the most commonly used are the Ag/AgCl and saturated calomel electrodes) electrodes. The current, flowing through the counter electrode (generally an inert metal with a large surface) and the working electrode, is constantly monitored and constitutes the analytical signal. Due to oxidation or reduction reactions at the electrode surface, only electroactive species give rise to a voltammetric signal, that is, the so-called faradic current. Therefore, these sensors are generally used to detect electroactive analytes, although indirect methods for non-electroactive species have also been exploited, in which electroactive probes are used to obtain the electrical signal.

Voltammetric techniques differ in how the potential between the working electrode and the reference electrode is applied. If the potential scan is linear in time, the technique is called linear voltammetry; this is a classical and rarely used approach since the high detection limits due to capacitive and eddy currents mask the analytical signal. The most diffused techniques are Differential Pulse Voltammetry (DPV), in which potential impulses are overimposed on a linear potential scan, and Square Wave Voltammetry (SWV), which involves the application of a potential staircase overlaid with a symmetrical square wave pulse, one in the forward and one in the reverse directions. The main advantage of DPV and SWV is the low detection limits that can be achieved; in particular, SWV is the fastest and most sensitive pulse voltammetry technique, allowing detection limits comparable with those of spectroscopic and chromatographic methods. The voltammetric method proper to investigate the electrochemical reactions at the electrode surface is CV, in which the potential is applied following a triangular pattern (a scan in oxidation is then followed by a scan in reduction, and conversely).

Voltammetric and amperometric sensors using MIP as the recognition element require the diffusion of the electroactive species to the electrode surface. Accordingly, the recognition layer should possess high electrical conductivity. Therefore, different approaches to providing conductivity to the MIP layer have been exploited.

A fundamental aspect of designing a MIP-based voltammetric/amperometric sensor is how to face the recognition element and the transducer.

One classical approach consists of a two-step procedure: the MIP synthesis, followed by the interfacing of the MIP with the transducer. Traditional polymerization techniques (precipitation, bulk) have been employed to synthesize the MIP. The integration step could be a simple drop-coating of the prepolymeric mixture on the electrode surface or entrapping MIP particles into different matrixes, such as membranes, gels, electropolymerized materials, or carbon, gold, and silver nanoparticles (NPs). Some examples of MIP-based voltammetric sensors developed by this two-step approach are presented below.

Voltammetric sensors obtained by a simple drop coating of the MIP's prepolymeric solution onto graphite screen-printed electrodes were proposed to detect environmental and forensic interest analytes (Pesavento et al., 2013; Pesavento, Merli, et al., 2021; Zanoni, Rovida, et al., 2022). In all three studies, MAA was the functional monomer, EGDMA, or divinyl benzene (DVB), was the crosslinker, and 2,2-azobisisobutyronitrile (AIBN), was the radical initiator. The polymerization was carried out thermally at 60 °C–80 °C after dropping a few μL (2–7 μL) of the prepolymeric solution onto the clean electrode surface. DPV or SWV were the techniques employed for the analytes' quantification. The high selectivity was proved by analyzing real samples, and the detection limits are comparable to those of the corresponding unmodified screen-printed cells.

Another recent example of electrode modification with MIP by the two-step procedure is proposed by Limthin et al. (2022). A magnetic MIP-modified screen-printed electrode was developed for amperometric gluten detection. A mixture of gluten (template), methyl methacrylate (MMA) (functional monomer), EGDMA (crosslinker), 2,2-azobisisobutyronitrile (AIBN, radical initiator), and superparamagnetic iron oxide nanoparticles (SPIONs) was spin-coated on the graphite-ink screen-printed electrode, and the thermal polymerization was performed at 60 °C. The analytical performance of the modified electrode was assessed by amperometric detection. The sensor's good sensitivity, selectivity, and reproducibility were proven, making it promising for gluten detection in real samples.

An advantage of using MIP in sensing devices is that they can mimic bioreceptors, so they can be efficiently employed in developing biosensors.

For example, using a two-step approach, Altintas et al. developed an MIP-modified electrochemical sensor for cancer biomarker detection in human urine samples (Sheydaei et al., 2020). In particular, uniform nano-MIP particles were prepared by the sol-gel technique using sarcosine (SAR, template), MAA (functional monomer), EGDMA (crosslinker), AIBN (radical initiator), and a mixture of acetonitrile/water as porogen. The nanostructured MIP was mixed with graphite powder and paraffin oil to obtain the modified carbon paste electrode. DPV was the detection technique.

Thanks to the excellent performance of the sensors, their possible application for the detection of SAR in urine samples of cancer patients and healthy humans was proposed.

These and some other examples of two-step-approaches MIP-modified voltammetric/amperometric sensors are summarized in Table 3.2.

Table 3.2 Examples of two-step-approach MIP-modified voltammetric/amperometric sensors.

Analyte	MIP composition/sensor preparation	Electrochemical technique/electrode	LOD(M)	Ref.
2,4,6-trinitrotoluene (TNT)	TNT:MAA:EGDMA = 1:4:40 + AIBN/Drop-coating of the prepolymeric mixture	DPV/graphite screen-printed electrode	4×10^{-7}	Pesavento et al. (2013)
Sarcosine (SAR)	SAR:MAA:EGDMA = 1:8:32 + AIBN/Graphite powder: nanoMIP particles: paraffin oil = 55:15:30 = modified carbon paste electrode	DPV/carbon paste electrode	4×10^{-7}	Sheydaei et al. (2020)
Carbofuran (CRBF)	CRBF + *N,N′*-methylene diacrylamide + MAA/Casting of the prepolymeric mixture and Aptamer-AuNPs	DPV/gold electrode	7×10^{-11}	Li et al. (2018)
Trinitroperhydro-1,3,5-triazine (RDX)	RDX:MAA:EGDMA = 1:3:12 + AIBN/Drop-coating of a solution of MIP particles + MWCNTs in DMF-H_2O	DPV/glassy carbon electrode	2×10^{-11}	Alizadeh et al. (2019)
2-furaldehyde (2-FAL)	2-FAL:MAA:DVB = 1:4:40 + AIBN/Drop-coating of the prepolymeric mixture	SWV/graphite screen-printed electrode	5×10^{-5}	Pesavento, Merli, et al. (2021)
Irbesartan (IRB)	IRB:MAA:EGDMA = 1:4:10 + AIBN/Drop-coating of the prepolymeric mixture	SWV/graphite screen-printed electrode	1×10^{-8}	Zanoni, Rovida, et al. (2022)
SARS-CoV-2	spike protein subunit S1 + 3-aminophenyl boronic acid/Surface imprinting technique	SWV/Au-TFME	15×10^{-15}	Ayankojo et al. (2022)
Gluten (GLU)	GLU + MMA + EGDMA + AIBN + SPIONs/Spin-coating of the prepolymeric mixture	Amperometry/ graphite screen-printed electrode	1.5 ppm	Limthin et al. (2022)
Tetracycline (TC)	TC:MAA:EGDMA = 1:x:y + AIBN Immersion of the electrode in the prepolymeric mixture	Amperometry/Pt/Ti electrode	6×10^{-8}	Zhao et al. (2013)
Creatinine	Creatinine, methacrylic acid, *N,N′*-(1,2-dihydroxyethylene) bis (acrylamide), 2,2′-azobis (2-methylpropionitrile)	Amperometry/ CuO@carbon paste electrode	8.3×10^{-8}	Nontawong et al. (2019)

Another favorable approach for integrating MIPs with electrochemical transducers is synthesizing the recognition element directly onto the electrode surface. In particular, electropolymerization is an advantageous technique for MIP-based sensors' development. This method allows the deposition of uniform MIP on the transducer surface by controlling the thickness of the film layer by optimizing the electrodeposition experimental conditions. Electropolymerization is generally carried out by CV in a solution containing the template molecule and the functional monomer. Pyrrole, o-phenylenediamine, aniline, and o-aminophenol are the monomers usually employed for MIPs' electropolymerization onto different electrodes.

Some examples of this kind of sensor are summarized in Table 3.3.

3.2.1.3 Impedimetric sensors

Electrochemical Impedance Spectroscopy (EIS) combines the analysis of a material's resistive and capacitive properties based on the system's perturbation by a small-amplitude sinusoidal excitation signal (Macdonald & Johnson, 2018). Generally, the applied AC potential frequencies are scanned, and the impedance is collected for each frequency. The Randles equivalent circuit model helps interpret the impedance spectrum, and it is an excellent technique for characterizing batteries, coatings, and fuel cells; it has also been used for investigating conducting polymers, electrode kinetics, and sensors. The Nyquist plot is the most commonly used graphical output to display how impedance changes with frequency. It is obtained by plotting the negative imaginary impedance vs the real impedance; depending on the shape of the obtained curve, different phenomena at the electrode surface can be determined.

The ideal Nyquist plot of an MIP-modified electrode, that is, a semi-circle, followed by a straight line, is shown in Fig. 3.1. The R_i is the resistance at the electrode/electrolyte interface, the R_{ct} is the resistance at charge transfer due to the presence of the MIP layer onto the working electrode surface, which makes the electrode less accessible to the electrochemical probe, and D is the diffusion of the probe in the bulk solution.

As reported above, all the phenomena represented in the Nyquist plot can be schematized in a Randless equivalent circuit (see the circuit at the bottom of Fig. 3.1). Each chemical-physical phenomenon occurring at the electrode/electrolyte interface is assigned an electrical circuit element; for example, the resistance at the electrode/electrolyte interface and that due to the charge transfer is represented by resistors R_i and R_{ct}, respectively. The capacitor represents the double layer capacitance (C_{dl}), while the Warburg element (W, typical of impedance spectroscopy) corresponds to the diffusion (Simic et al., 2023).

Fig. 3.2 represents Fig. 3.1.

When a MIP-modified electrode has to be characterized by EIS, the most used setup involves a three-electrode cell in a solution containing an electrochemical probe such as $K_3Fe(CN)_6/K_4Fe(CN)_6$. The probe reaches the electrode surface (crossing the MIP coating) and reacts at the working electrode surface, gaining or losing an electron.

In MIP-modified electrodes, R_{ct} depends on the characteristics of the polymer, for example, the thickness. If the polymer is very thick, the diffusion from the bulk solution to the electrode surface is prevented, and R_{ct} is absent in the Nyquist plot. As is well known, MIPs can interact selectively with the target analyte; accordingly, the R_{ct} value increased from the empty recognition sites to the saturated cavities of the polymer. The measure of R_{ct} after known quantities of the target analyte to

Table 3.3 Examples of electropolymerized MIP-based sensors.

Analyte	MIP composition	Electrochemical technique/electrode	LOD(M)	Ref.
Ascorbic Acid (AA)	Pyrrole and AA in $LiClO_4$	DPV/graphite screen-printed electrode	2×10^{-6}	Alberti, Zanoni, Magnaghi, et al. (2023)
Adenine (AD) and Guanine (GU)	Phenol, AD, or GU in phosphate buffer solution (PBS)	DPV/gold screen-printed electrode	2×10^{-10} (AD) 4×10^{-10} (GU)	Zolfaghari Asl et al. (2023)
Lactate	3-aminophenyl-boronic acid and Lactate in NaCl/PBS	DPV/Ag-nanowires-coated carbon electrode	2×10^{-7}	Zhang, Jiang, et al. (2020)
Cholesterol	Aminothiophenol, Cholesterol in $LiClO_4$	DPV/AuNPs/polydopamine/graphene-coated glassy carbon electrode	3×10^{-19}	Yang et al. (2017)
5-hydroxyindole-3-acetic acid (5-HIAA)	Pyrrole and 5-HIAA in $LiClO_4$	DPV/glassy carbon electrode	2×10^{-11}	Moncer et al. (2021)
Glucose	*o*-Phenylendiamine, Glucose, and AuNPs in PBS	SWV/gold electrode	1×10^{-9}	Sehit et al. (2020)
Norfloxacin	Pyrrole and Norfloxacin in H_2SO_4	SWV/glassy carbon electrode	5×10^{-8}	Da Silva et al. (2015)
Danazol	(Vinylbenzyl) trimethylammonium, divinylbenzene, and Azo-bis-isobutyronitrile Danazol	Amperometry/CuO-graphite screen-printed electrode	3×10^{-13}	Wang and Li (2022)
Carbofuran	Carbofuran, 4-tert-butylcalix [8] arene, and Phenylenediamine.	Amperometry/carbon-paste electrode decorated with carbon nanotubes and gold-coated magnetite	4×10^{-9}	Amatatongchai et al. (2018)
Dopamine	Dopamine and *p*-aminobenzenethiol	Amperometry/AnNPs-modified gold electrode	8×10^{-9}	Xue et al. (2013)
Salicylic acid	Salicylic acid and *o*-phenylenediamine	Amperometry/glassy carbon electrode	2×10^{-5}	Kang et al. (2009)
Benzophenone	Benzophenone and *o*-phenylenediamine	Amperometry/glassy carbon electrode	1×10^{-8}	Li et al. (2012)

an MIP-based electrode is the approach used to perform quantitative analysis by the EIS technique. EIS is also an advantageous method for detecting non-electroactive analytes.

Table 3.4 summarizes some examples of MIP-based impedimetric sensors.

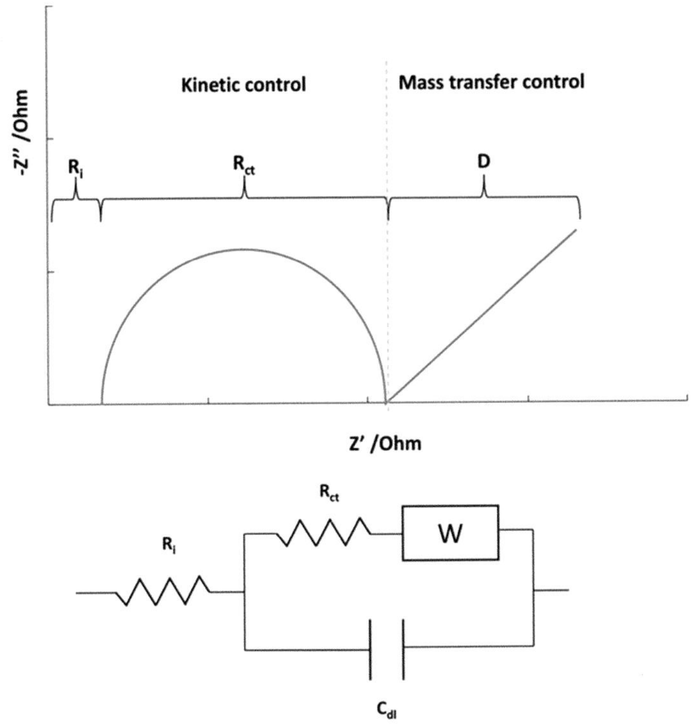

FIGURE 3.1 Nyquist plot and the corresponding Randless equivalent circuit.

The ideal Nyquist plot and the corresponding Randless equivalent circuit for a MIP modified electrode. R_i is the resistance at the interface, R_{ct} is the resistance at the charge transfer, and D is the diffusion coefficient.

3.2.2 Optical platforms

Optical MIP sensors are a trendy research field. Due to their excellent properties, MIPs can be quickly used with optical devices. Several strategies have been proposed for MIP-based optical sensors' design, depending on the optical properties of the target analyte.

In optical sensors, the information on the binding events on the MIP surface is related to the change in optical properties, such as UV-vis or IR absorption, reflection, or fluorescence of the MIP layer, that is recorded upon the interaction of the analyte with the recognition cavities.

A critical limitation to a broader spread of optical sensors, particularly those based on fluorescence, is the still restricted number of fluorescent templates. A tentative solution is the development of sensors operating in a competitive mode, in which the template and a fluorescently labeled analog compete for binding with the MIP's sites (Haupt et al., 1998).

In various applications, the target analyte has been detected by evanescent field sensors, measuring the refractive index variation; in particular, with MIP-based Surface Plasmon Resonance (SPR) sensors, very low detection limits (nM or pM levels) can be achieved.

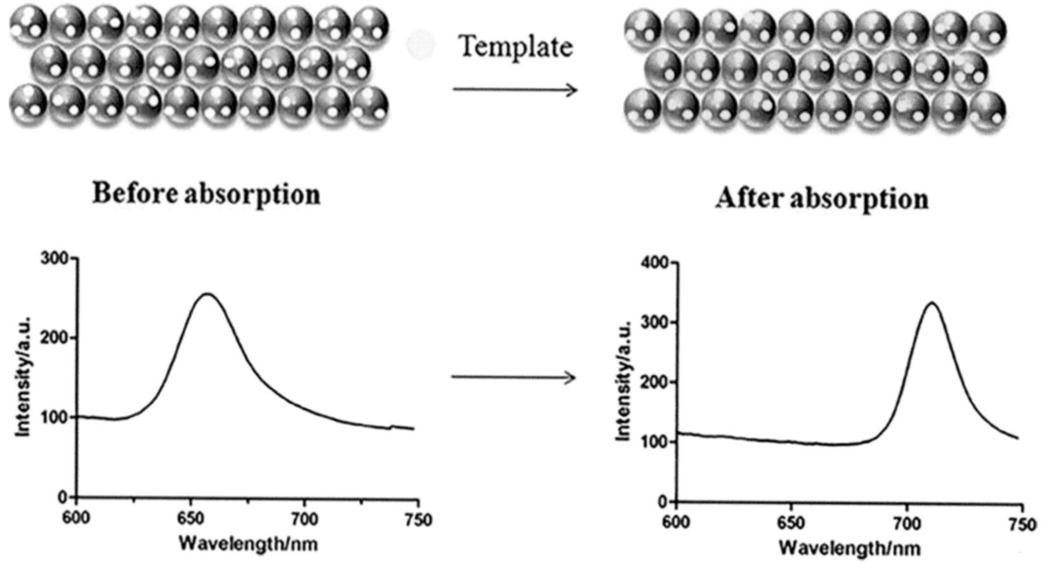

FIGURE 3.2 MIP@PhC-based sensors.

The response mechanism of MIP@PhC-based sensors.

Reproduced with permission from Fan, J., Qiu, L., Qiao, Y., Xue, M., Dong, X., & Meng, Z. (2021). Recent advances in sensing applications of molecularly imprinted photonic crystals. Frontiers in Chemistry, 9. *Available from https://doi.org/10.3389/fchem.2021.665119. Open access Creative Common CC BY 4.0, Frontiers.*

MIP-based optical sensors are versatile tools adaptable to environmental monitoring, food controls, and biomedical analyses. The current challenge in developing these kinds of sensors is the possible minimization of the interferences derived from the complex matrixes of environmental and biological samples.

The last 10-year trend in the development of optical sensors interfaced with MIPs is discussed in the following paragraphs.

3.2.2.1 Colorimetric and fluorimetric sensors

MIP-based colorimetric sensors enable selective and fast qualitative and quantitative determination of several analytes. Colorimetric sensors have attracted significant attention for their simple preparation, rapid response, and low cost. Moreover, single-step detection without complicated instrumentation is another favorable point (Piriya VS et al., 2017). Compared with other analytical methods, these sensors based on color changes evident simply by naked eyes are more suitable for daily life and in situ monitoring. On the contrary, the interference problems restrict their applicability to the analysis of liquids of simple composition or samples where the color response is not susceptible to interference.

The possibility of integrating colorimetric sensors with MIP can significantly improve selectivity, reducing interference problems; besides, MIP-based colorimetric sensors can be efficiently applied to more complex sample matrices (Wang et al., 2023).

Table 3.4 Examples of MIP-based impedimetric sensors.

Analyte	MIP composition	Electrode/probe	LOD	Ref.
A549-derived exosomes	A549-derived exosomes ortho-aniline boronic acid and NaF	GCE/$K_3Fe(CN)_6$-$K_4Fe(CN)_6$ in PBS	2×10^3 particles/mL	Zhang et al. (2023)
Sulfamethizole	Sulfamethizole, thiophene-3-ethanol	GCE/$K_3Fe(CN)_6$-$K_4Fe(CN)_6$	1.8×10^{-10} M	Kong et al. (2023)
Carbendazim	Carbendazinm, *o*-aminophenol	GCE/$K_3Fe(CN)_6$-$K_4Fe(CN)_6$	2×10^{-10} M	Li et al. (2023)
Aflatoxin-B1	Aflatoxin-B1, aniline, and Dioctyl sulfosuccinate sodium salt	Graphite screen-printed electrode/ $K_3Fe(CN)_6$/$K_4Fe(CN)_6$	1×10^{-11} M	Ong et al. (2023)
Cytokine IL-1β	Cytokine IL-1β and *o*-phenylenediamine	Graphite screen-printed electrode/ $K_3Fe(CN)_6$/$K_4Fe(CN)_6$	2×10^{-13} g/mL	Choi et al. (2022)
SARS-CoV-2-RBD	SARS-CoV-2-RBD and *o*-phenylenediamine	Gold screen-printed electrode/$K_3Fe(CN)_6$/ $K_4Fe(CN)_6$	7×10^{-13} g/mL	Amouzadeh Tabrizi et al. (2022)
Testosterone	Testosterone, *N,O*-bismethacryloyl ethanolamine, and 2,2-dimethoxy-2-phenylacetophenone	Amorphous carbon (diamond)-ethanol/ PBS (pH = 7.4)	5×10^{-7} M	Kellens et al. (2018)
Tramadol	Tramadol, 2-aminothiophenol,	Au-SPE/$K_3Fe(CN)_6$/ $K_4Fe(CN)_6$ in PBS	4×10^{-5} M	Diouf et al. (2021)
Alpha-fetoprotein and Carcinoembryonic antigen	Alpha-fetoprotein, carcinoembryonic antigen, methyl orange, and pyrrole	Fluorine-doped tin oxide/$K_3Fe(CN)_6$/ $K_4Fe(CN)_6$ in PBS	2×10^{-12} 3×10^{-12} g/mL	Taheri et al. (2022)
Dengue fever biomarker	Dengue NS1 protein and dopamine	Carbon-SPE $K_3Fe(CN)_6$/$K_4Fe(CN)_6$ in PBS	3×10^{-10} g/mL	Arshad et al. (2020)

The substrates for these sensors are carbon nanomaterials, magnetic NPs, and paper. Carbon nanomaterials have some advantages, such as high porosity, a large and modifiable surface, a hollow structure, and high stability. Magnetic NPs, for example, those of Fe_3O_4, provide magnetic properties to the MIPs, simplifying the separation and recovery of the analyte from the sensor. The most promising substrate is paper, showing several advantages such as abundance, easy functionalization, and low cost (Wang et al., 2023).

The pioneering approach to developing MIP-based colorimetric sensors has been based on the competitive displacement of a template-dye conjugate from the MIP's recognition cavities by the analyte (Greene & Shimizu, 2005; Levi et al., 1997; McNiven et al., 1998). The same strategy was recently applied in a few other studies (Lowdon et al., 2020; Silverio et al., 2017) because of its quite low sensitivity and reproducibility.

Different approaches have recently been proposed for improving the selectivity and sensitivity of colorimetric sensors in conjunction with MIPs, and they are mainly applied in biochemical assays. Most of them involved NPs; in a few cases, NPs were incorporated into the bulk of MIP, as, for example, in the sensor developed by Matsui et al., in which a MIP was synthesized as functional support for an AuNP. The sensing mechanism is based on the proximity of the AuNPs immobilized in the imprinted polymer, which exhibits selective recognition of the target analyte that caused the swelling with a consequent blue shift in the plasmon absorption of the NPs (Matsui et al., 2004). Most other studies employed NPs as separate sensing elements for colorimetric detection. For example, Wu et al. reported the application of NPs after MIP separation for the colorimetric detection of the insecticide Cartap in tea (Wu et al., 2018). Cartap was extracted from the tea samples by magnetic molecularly imprinted microspheres (Fe_3O_4-$mSiO_2$-MIPs). After elution of the analyte from the MIPs, the AgNPs were added, and a color change from yellow to gray, dependent on the presence of Cartap, was observed (Wu et al., 2018).

Kong et al. developed a microfluidic paper-based colorimetric sensor based on MIPs and AgNPs. The device consisted of a MIP membrane imprinted with bisphenol A (BPA) (the outer layer), $ZnFe_2O_4$ NPs (the second layer), and cellulose paper (the inner layer). Adding 3,3′,5,5′-tetramethylbenzidine (TMB) and peroxide to the system provoked the oxidation of peroxide by the NPs, with the consequent production of OH species able to discolor the cellulose paper. The binding of BPA to the MIP layer prevents this reaction, keeping the paper's color unchanged (Kong et al., 2017).

Another colorimetric transduction method is based on photonic crystals (PhCs). PhCs are materials with an ordered structure obtained by a periodic arrangement of materials with different refractive indices. PhCs follow Bragg's law and possess a photonic band gap thanks to their periodic structure. The photonic band gap allows PhCs to have the function of wavelength selectors. Indeed, when a light beam hits the PhCs, a bright color appears on the surface of the PhCs (structural color) due to the Bragg diffracted light within the visible range (Fan et al., 2021).

Combining MIPs with PhCs is a helpful strategy for developing colorimetric sensors.

Indeed, in the molecular recognition process, if the refractive index of the target analyte is different from that of the molecularly imprinted PhC, the average refractive index of PhC, n_a, will change. In particular, the interaction of the MIP cavities with the target analyte will cause the particle to shrink or swell due to the change in osmotic pressure, resulting in a change in the particles' diameter that causes the shift of the maximum diffraction light wavelength (λ_{max}) and consequently, the color variation if the λ_{max} is included in the visible/near-infrared light region (see Fig. 3.2).

Examples of this kind of sensor were reported in the last 10-year literature, and they are mainly applied for environmental monitoring and human health; some of the most recent and promising are described below.

Wang et al. developed a MIP@PhC two-dimensional hydrogel sensor for tetracyclines antibiotic screening in milk. The hydrogel was molecularly imprinted with oxytetracycline. An increase in the particle spacing of the MIP@PhC sensor was observed by increasing the tetracycline concentration, with a red shifting of the λ_{max}. The sensor allowed selective tetracycline quantification in a concentration range of 0.04–0.24 μM (Wang et al., 2012). A similar approach was exploited by Qin et al., who developed a sensor for the pesticide 2,4-dichlorophenol. A change in the MIP@PhC color occurred by increasing the analyte concentration, varying from red to green, allowing naked-eye detection (Qin et al., 2020).

Resende et al. proposed a MIP-based 3D silica PhC for sensing fibrinopeptide B (FPB), a biomarker of venous thromboembolism. Differently from what was previously described, this sensor showed small shifts in the λmax, increasing the analyte concentration. However, a significant decrease in the intensity of the reflectance peak linearly proportional to the FPB concentration was verified due to the refractive index change in the presence of the target analyte (Resende et al., 2020).

α-Amanitin is one of the most toxic amanitas in mushrooms (the lethal dose to humans is around 0.1 mg/kg), so it is crucial for its detection for human health. Qiu et al. (2020) developed a MIP@PhC sensor for the detection of this toxin. The excellent performances of the sensor were highlighted, such as the very low LOD (10^{-10} mg/L), the short response time (about 2 min), and the five-cycle reusability. A linear redshift proportional to the increase in α-Amanitin quantity was verified in the analyte concentration range of 10^{-9}–10^{-3} mg/L (Qiu et al., 2020).

Despite the good figures of merit, such as the high selectivity and sensitivity, the low detection limit, and the short response time, the MIP@PhC sensors present some criticalities that prevent their large-scale production and commercialization. Particularly challenging is the development of these MIP@PhC sensors using less-toxic reagents and high-throughput methods for simultaneous detection of multiple analytes.

One of the most frequent optical transductors interfaced with MIPs is fluorescence. Fluorescence-based sensing is an alternative to conventional strategies thanks to several advantages such as high sensitivity, short response time, and low cost; moreover, integrating an MIP layer in fluorescent sensors enhances their selectivity (Yang et al., 2018).

In MIP-based fluorescent sensors, the selective interaction between the recognition cavities and the target analyte affects the fluorescence characteristics of the sensor, including fluorescence quenching or enhancement, shift of the characteristic wavelength (blue or red shift), anisotropy and lifetime through different signal transduction systems. These signal variations can be related to the target analyte concentration, allowing a quantitative determination (Yang et al., 2018).

The MIP-based fluorimetric sensors are prepared for emissive target analytes using the common MIPs' synthesis. The fluorescent molecule acts as the template so that fluorescence signals can be registered in the rebinding process for the analyte quantification (Liu & Ko, 2023; Yang et al., 2018). Since only a few analytes are intrinsically fluorescent, a limited number of studies have reported on this type of MIP-based sensor (see, for example, the sensors developed by Ton et al. for enrofloxacin and fluoroquinolone analogs (Ton et al., 2012) and that by Carrasco et al. for rhodamine derivatives (Carrasco et al., 2014)).

Competitive assays have been proposed for non-fluorescent analytes. By this approach, the MIP's cavities are saturated with a fluorescent indicator, that is, a fluorescent analog of the target analyte or the fluorescent-labeled analyte. When a sample is put in contact with the MIP, the displacement of the fluorescent indicator by the analyte molecules occurs due to competitive binding; thus, a fluorescent signal proportional to the quantity of the indicator replaced can be measured. The competitive assay was exploited in very few cases since this indirect method presents some drawbacks. First, labeling the target analyte or synthesizing an analog fluorescent compound is difficult and time-consuming. Moreover, controlling the complete saturation of the MIP's recognition cavities with the fluorescent indicator molecules is problematic; indeed, an excess of free fluorescent indicator (unlinked to the MIP's sites) in the sample solution could produce interferent background fluorescence. On the other hand, the sensor's sensitivity decreases if there is no complete saturation of the recognition sites (Yang et al., 2018).

An alternative and promising approach for sensing non-fluorescent analytes with MIP-based sensors is the employment of fluorescent monomers or comonomers in the MIP synthesis. This method has been extensively exploited in the last ten years; some examples are reported below.

A MIP-based sensor for caffeine detection, prepared with 1,8-naphthalimide dye as a florescence functional monomer, was proposed by Rouhani et al. (Rouhani & Nahavandifard, 2014). In particular, an on-off probe based on the fluorescence quenching of the MIP was developed. Fig. 3.3 reports the scheme of the process.

Other MIP-based sensors developed using fluorescent functional monomers have also been proposed for the detection of carboxylate-containing bioanalytes (Z-L-phenylalanine, Z-L-glutamic acid, and penicillin G) (Wagner et al., 2013), diquat and paraquat herbicides (Chen et al., 2014), tetracycline in biological samples (Niu et al., 2015), beta-cyfluthrin in agricultural products (Qiu et al., 2017), and p-nitroaniline in wastewater (Lu et al., 2018).

The main drawbacks of this approach are the need for complex synthesis procedures for derivatizing fluorophores or providing non-fluorescent functional monomers with fluorophores, the employment of several different methods for preparing suitable fluorescent functional monomers for diverse templates, and the poor photostability and photobleaching.

Another more efficient and straightforward strategy consists of embedding fluorescent substances, such as fluorescent dyes, quantum dots (QDs), metal NPs, and lanthanide chelates, into the MIP-based sensors.

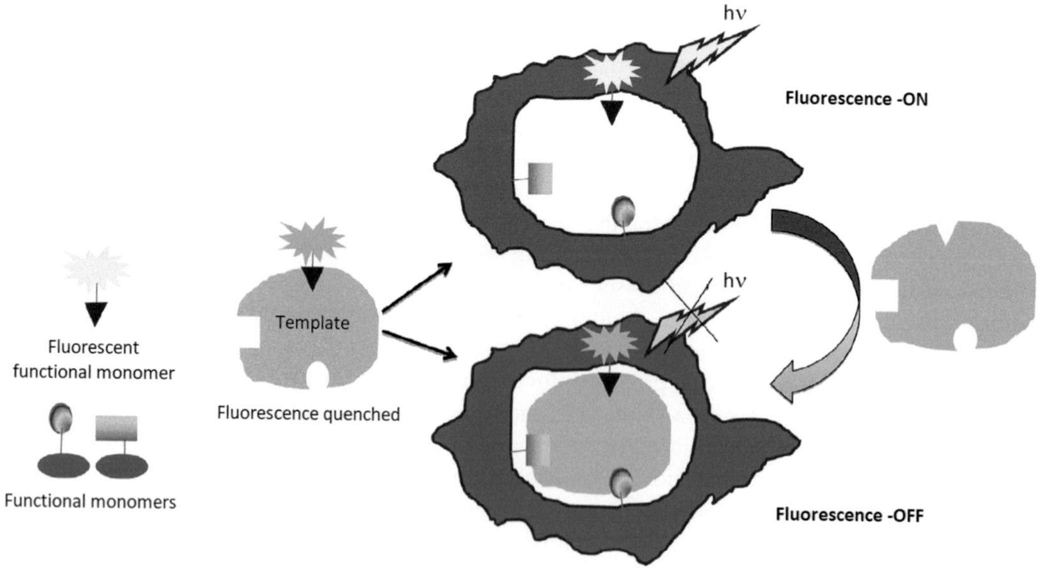

FIGURE 3.3 MIP-based fluorimetric sensor for caffeine detection.

Preparation and functioning scheme of the MIP-based fluorimetric sensor for caffeine detection.

Reproduced with permission from Rouhani, S., & Nahavandifard, F. (2014). Molecular imprinting-based fluorescent optosensor using a polymerizable 1,8-naphthalimide dye as a florescence functional monomer. Sensors and Actuators, B: Chemical, 197, 185–192. Available from https://doi.org/10.1016/j.snb.2014.02.082. Copyright © 2014 Elsevier B.V.

Classical and widely used fluorescent dyes are ATTO 647N (Cheubong et al., 2021; Tsutsumi et al., 2021), fluorescein isothiocyanate (Ashley, Feng, et al., 2018; Ming et al., 2017; Sunayama et al., 2010; Wang et al., 2015; Xu et al., 2017), rhodamine (Kunath et al., 2015; Luo et al., 2022), luminol (Duan et al., 2016), coumarin (Dai et al., 2020; Ray et al., 2016), and nitrobenzoxadiazole (Li et al., 2020; Shinde et al., 2015; Zhang, Ma, et al., 2020), generally introduced in the polymers by post-imprinting procedures. The advantages of this strategy are the wide variety of colors, the pretty good biocompatibility of the dyes, and their relatively high fluorescent intensity. On the contrary, these fluorescent substances have close excitation peaks, tailing and broad emission peaks, scarce stability, and possible photobleaching (Yang et al., 2018).

Inorganic QDs, particularly those based on Cd and Zn, are promising nanomaterials for developing MIP-based fluorimetric sensors. Compared to fluorescent dyes, QDs have a more stable photoluminescence, a broad range of excitation and emission bands, and their emission properties can be modulated by size. Several fluorescent MIP-based sensors with QDs-MIP core-shell designs have been developed (Abha et al., 2019; Assi et al., 2023; Chantada-Vázquez et al., 2016; Cui et al., 2020; Ensafi et al., 2018; Huang et al., 2018; Masteri-Farahani et al., 2020; Sistani & Shekarchizadeh, 2021; Wang, Fang, et al., 2022; Yang et al., 2016). In this case, the principal drawback is the toxicity of the heavy metal-based QDs; moreover, they sometimes present intermittent fluorescence (Liu & Ko, 2023; Yang et al., 2018).

To overcome the toxicity problems of QDs, metal nanoclusters, mainly those of Ag and Au, were proposed, thanks to their biocompatibility, low cytotoxicity, and environmental friendliness. In particular, gold nanoclusters (AuNCs) exhibit more durable and intense fluorescence signals due to harder aggregation in aqueous solutions and lower surface activation energy (Yang et al., 2018). Surface imprinting polymerization is the technique generally used to incorporate metal nanoclusters in MIP-based sensors (Liu & Ko, 2023). Examples of this kind of sensor are those proposed for detecting BPA using AuNCs (Wu et al., 2015) or AgNCs (Deng et al., 2016). Some attempts have been made using copper nanoclusters, but despite being cheap, they suffer from easy air oxidation that can induce aggregation and quench their fluorescence (Li, Zhang, et al., 2016; Liu et al., 2015).

Lanthanide chelates, mainly those of europium (III) and terbium (III), have also been exploited as fluorophores for developing MIP-based sensors using different strategies, such as physical incorporation into the polymer network (Zheng et al., 2015), post-functionalization of the MIP surface by chemical modification (Horikawa et al., 2016; Li, Kamra, et al., 2016; Mori et al., 2019; Zhao et al., 2017), and the use of lanthanide chelates with ligand-functionalized alkene monomers (Özgür et al., 2020; Quílez-Alburquerque et al., 2021; Rico-Yuste et al., 2021). Europium(III) chelates have been the most generally studied due to their emission characteristics showing visible-light fluorescence, long emission lifetimes, and significant Stokes, which can be well distinguished from interfering fluorescence or background derived from the MIP network or other interfering substances commonly present in real samples.

MIP-based colorimetric and fluorimetric sensors have attracted interest as promising biosensor alternatives thanks to their easy preparation methods, long-term stability, and relatively low cost.

The number of applications of MIP-based colorimetric and fluorimetric sensors has grown progressively over the last decade. However, despite their potentiality, MIPs have not yet replaced biological receptors in commercial sensor production. Indeed, some challenges remain, such as the controlled synthesis of MIP materials for improving reproducibility and increasing the sensitivity and selectivity of MIP-based sensors to be efficiently applied to complex systems and for trace analysis.

3.2.2.2 Plasmonic sensors

Plasmonic sensors have been exploited recently in various fields, such as environmental monitoring, medical diagnosis, pharmaceutical analysis, food quality assessment, and forensics. Incorporating MIP into plasmonic sensors is a significant challenge in developing easy-to-use, selective, portable devices suitable even for trace analysis.

SPR and localized SPR (LSPR) are the most promising transducing methods in MIP-based sensors since they possess appealing properties such as the possibility for label-free analysis, the suitability for automation, and simple surface modification (Alberti, Zanoni, Spina, et al., 2023) (Fig. 3.4).

The figure shows a scheme of SPR and LSPR mechanisms (Liu, Jalali, et al., 2020).

Briefly, SPR sensors exploit the evanescent field of a surface plasmon propagating through a metallic surface (generally a gold or silver film) to detect changes in the dielectric constant of a sample around 100 nm of the plasmonic material. The interaction of the evanescent field with a target analyte results in a shift of the wavelength (or angle) of the transmitted excitation light or a change in the light intensity proportional to the variation of the sample's refractive index. This change enables the detection of the analyte (Chen & Ming, 2012; Daghestani & Day, 2010).

When the surface plasmon is instead confined to a NP of dimensions in the same order as the light's wavelength, the free electrons of the NPs contribute to the coherent oscillation; this phenomenon is defined as LSPR. LSPR has two significant effects: the electric field near NPs' surfaces is noticeably enhanced; besides, the optical transmission spectrum of the NPs shows a maximum peak at the plasmon resonant frequency in the UV-vis region dependent on the surrounding medium's

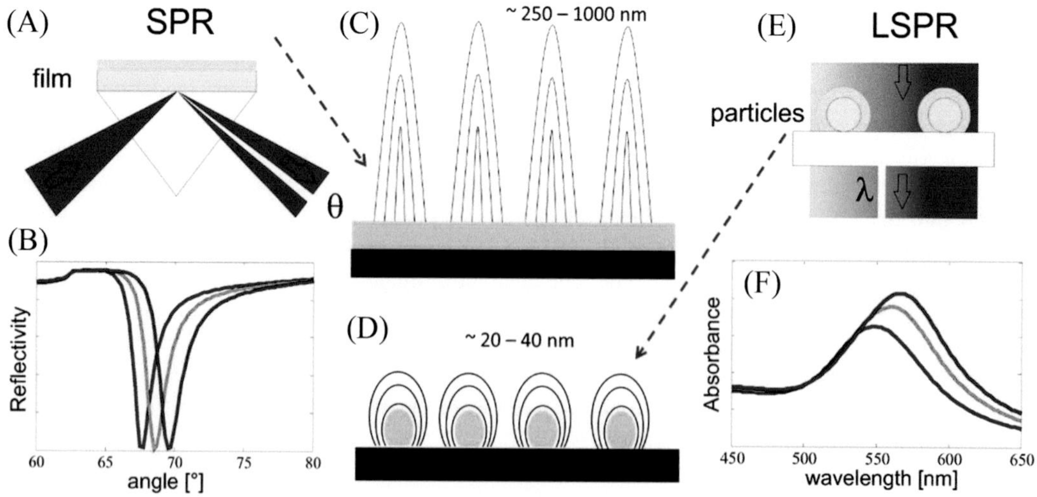

FIGURE 3.4 SPR and LSPR mechanisms.

SPR (A–C) and LSPR (D–F) mechanisms.

Reproduced with permission from Liu, J., Jalali, M., Mahshid, S., & Wachsmann-Hogiu, S. (2020). Are plasmonic optical biosensors ready for use in point-of-need applications? Analyst, 145*(2), 364–384. Available from https://doi.org/10.1039/c9an02149c. Open access Creative Common CC licensed 3.0, Royal Chemical Society – RSC.*

refractive index. A change in the dielectric properties of the NPs' surrounding medium results in the plasmon resonance spectral shift; this is the sensing principle for LSPR sensors (Mayer & Hafner, 2011; Verellen et al., 2011).

A considerable increase in fiber optic sensors has appeared in the last thirty years thanks to their peculiar characteristics, which enable online or remote monitoring and the fabrication of miniaturized devices suitable for point-of-care analysis. The devices are called extrinsic sensors if the fiber optic is used as a waveguide to transport light from the source to the sensing part. Conversely, the optical fiber is used as a sensing waveguide in intrinsic sensors. In this context, fiber optic-based SPR/LSPR platforms were exploited, intending to obtain small-size sensors with high sensitivity. Besides, very selective sensors can be obtained by modifying the SPR/LSPR surfaces with receptors, such as MIPs (Chiappini et al., 2020; Fang et al., 2021; Qu et al., 2020).

MIP-based SPR sensors are generally obtained by depositing a polymer's film layer on the plasmonic surface. The usual techniques are drop-coating and spin-coating of the prepolymeric mixture; the spin-coating approach is the greatest for the higher control of the MIP film thickness (Rico-Yuste & Carrasco, 2019).

The Zeni and Cennamo groups developed various SPR sensors based on D-shaped plastic optical fibers (POFs) functionalized with MIPs. For manufacturing these devices, the optical fiber's cladding and, partially, the core were removed; then, a photoresist buffer layer was spin-coated on the fiber's exposed core, and a gold film was sputtered over the buffer. Ultimately, the MIP's prepolymeric mixture was drop-coated or spin-coated on the gold sensing surface and left to polymerize overnight thermally. Fig. 3.5. shows a scheme for this kind of sensor.

The same strategy was adopted for several sensors targeting different analytes, such as perfluorinated compounds, 2-furaldehyde, L-nicotine, 2,4,6-trinitrotoluene (TNT), dibenzyl disulfide, and recently SARS-CoV-2 (Cennamo et al., 2011, 2014, 2015, 2016, 2018; Cennamo, D'agostino, et al., 2021; Cennamo, Pesavento, et al., 2020; Pesavento, Zeni, et al., 2021).

Microcontact imprinting is another approach to producing and controlling the layer thickness in MIP film-based SPR sensors, particularly for imprinting high molecular-weight molecules, such as biomolecules and microorganisms (Ertürk Bergdahl et al., 2019; Kidakova et al., 2018; Perçin et al., 2017).

Electropolymerization can also be employed for developing MIP-based plasmonic sensors, although this strategy is still scarcely used (Bartold et al., 2018; Choi et al., 2009; Pernites et al., 2010; Pernites et al., 2011).

In recent years, efforts have been devoted to immobilizing nanoMIPs (MIP NPs) onto the gold surface of SPR devices (Altintas, 2018; Arcadio et al., 2022; Ashley, Shukor, et al., 2018; Cennamo, Bossi, et al., 2021; Cennamo, Maniglio, et al., 2020; Erdem et al., 2019; Rahtuvanoğlu et al., 2020; Sari et al., 2016; Yılmaz et al., 2017). In most cases, MIP NPs were synthesized using the two-phase mini-emulsion polymerization method (Erdem et al., 2019; Rahtuvanoğlu et al., 2020; Sari et al., 2016; Yılmaz et al., 2017); otherwise, solid-state (Altintas, 2018; Ashley, Shukor, et al., 2018) or hydrogel (Arcadio et al., 2022; Cennamo, Bossi, et al., 2021; Cennamo, Maniglio, et al., 2020) syntheses were exploited.

The recent progress in nano-optics has permitted the development of LSPR-based sensors. Thanks to their notable properties, AuNPs have been widely studied for developing sensors. Although AuNPs are often inappropriate for sensing applications, their surface modification with organic or inorganic functionalities is fundamental for improving stability and selectivity. Among

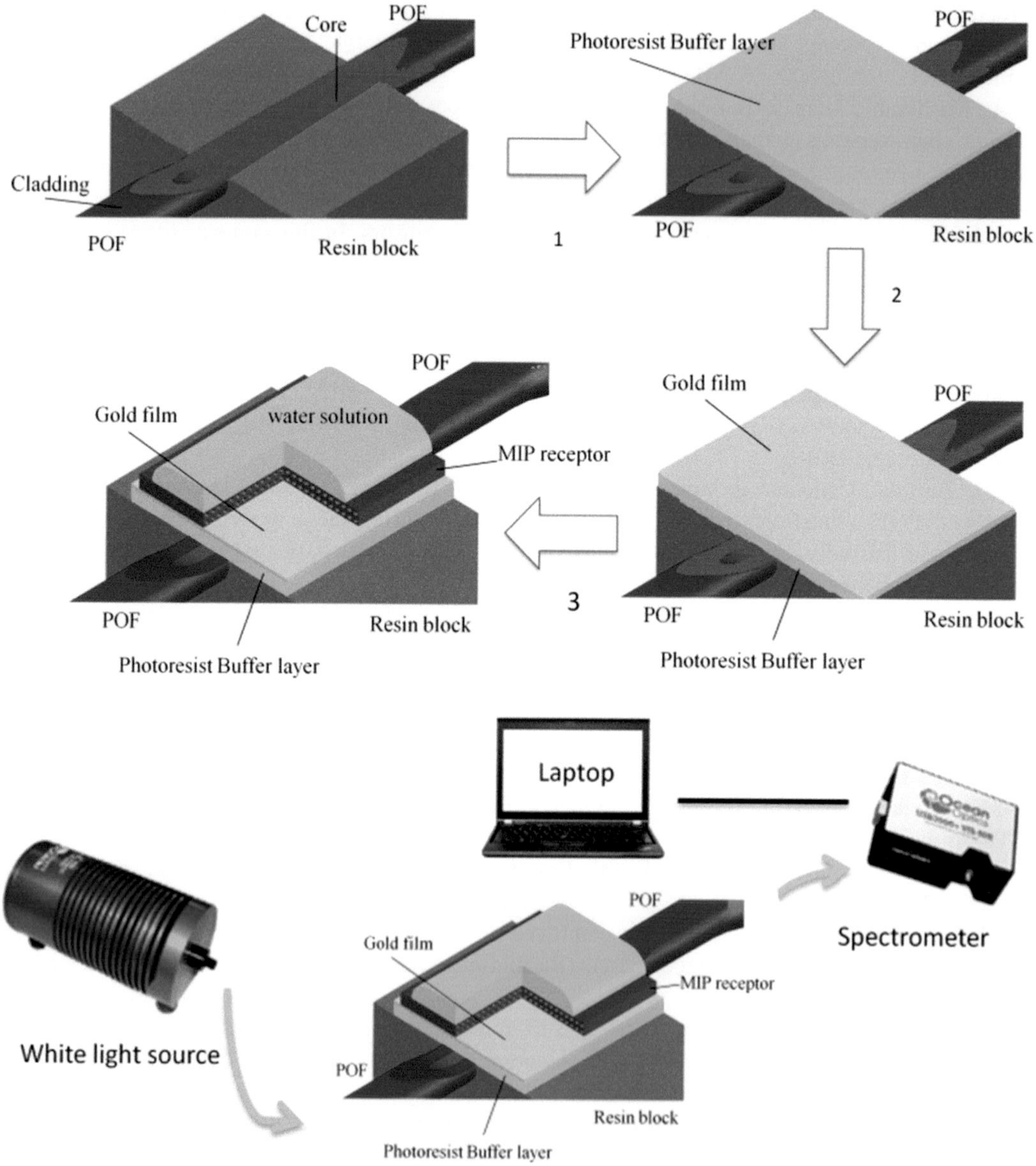

FIGURE 3.5 MIP-based POF-SPR sensors.

(A) Steps for the realization of a MIP-based SPR sensor using a D-shaped POF; (B) outline of the experimental setup.

Reproduced with permission from Cennamo N., D'Agostino G., Porto G., Biasiolo A., Perri C., Arcadio F., & Zeni L. (2018). A molecularly imprinted polymer on a plasmonic plastic optical fiber to detect perfluorinated compounds in water. Sensors (Switzerland), 18*(6). Available from https://doi.org/10.3390/s18061836. Open access Creative Common CC licensed 4.0, MDPI.*

different receptors, MIPs are remarkably interesting thanks to their figures of merit, including easy synthesis and high stability. Plasmonic sensors with high sensitivity and selectivity can be obtained by hybrid structures achieved by linking AuNPs with MIPs (Wu et al., 2022). Besides, MIP-modified LSPR sensors are compatible with microfluidic systems, thus becoming suitable for in situ and point-of-care analyses (Abbas et al., 2013).

For example, gold nanoobjects such as NPs, nanorods, and nanodisks, generally prepared by sol-gel synthesis, were efficiently imprinted for detecting target proteins (Hu et al., 2017), tear biomarkers (Culver et al., 2018), volatile cis-jasmone in plants (Shang et al., 2018), and explosives (Chegel et al., 2020).

The MIP-based plasmonic sensors discussed in this paragraph emphasized the demand for high-throughput analysis, preventing time-consuming sample preparation, and the use of expensive instrumental methods.

In recent years, the development of MIP-based plasmonic sensors for various applications has massively increased thanks to the low detection limits achieved, their high sensitivity and selectivity, rapid responses, and low-cost instrumentations.

In any case, some drawbacks must be solved to enhance reproducibility, improve sensitivity, and expand the fields of MIP-based plasmonic sensors' applications.

3.2.3 Thermal readout

The thermal readout method for MIP-based sensors has gained ground in the last ten years. This technique, also called the "Heat Transfer Method" (HTM), is based on analyzing the thermal energy transport over the MIP layer. The thermal resistance (R_{th}) at the interface MIP layer/test solution can change due to the binding of the target analyte to the receptor layer.

The potential of HTM to detect small organic molecules by MIP-based sensors was first proposed in 2013 by Peeters et al. for l-nicotine, histamine, and serotonin sensing (Peeters et al., 2013). MIP particles (synthesized by bulk polymerization) were partially sunk in a polyphenylene-vinylene adhesive layer, pre-coating an aluminum chip, after heating at a temperature higher than 120°C (above the glass transition temperature). The MIP-functionalized aluminum substrate was installed into a flow cell filled with PBS at pH 7.4. The thermal resistance R_{th} was determined by keeping the temperature of the copper block at 37 °C (by the heating power of the adjustable heat source, P), and T_2 was measured after increasing additions of the target analyte. The interaction of the analyte with the recognition cavities of the MIP produced a reduction in heat transport in that direction with a corresponding increase in heat transfer resistance.

After this pioneering work, other strategies have been proposed, in particular, to avoid employing additional adhesive layers and reduce the preparation time. Screen-printing technology is efficiently proposed; in this case, MIP particles are mixed with ink and printed onto screen-printed electrodes.

Due to the simplicity of realization, low cost, and transportability of the setup, screen-printed electrodes faced with a thermal readout seem like a promising choice for on-site screening in diverse areas such as environmental, biological, and food analysis.

Table 3.5 reports some examples of thermal detection with MIP-based sensors.

Table 3.5 Examples of MIP-based sensors with thermal readout.

Analyte	MIP composition	Sensing Chip	LOD(M)	Ref.
L-nicotine, Histamine, and Serotonin	L-nicotine, Histamine and Serotonin, methacrylic acid, ethylene glycol dimethacrylate, and AIBN	Aluminum chip	1×10^{-7} 3×10^{-8} 2×10^{-8}	Peeters et al. (2013)
Glucose	D-glucuronic acid, methacrylic acid, ethylene glycol dimethacrylate, and AIBN	Aluminum chip	2×10^{-5}	Caldara et al. (2021)
Histamine	Histamine, methacrylic acid, ethylene glycol dimethacrylate, and AIBN	Silicon electrodes	2×10^{-8}	Peeters et al. (2014)
Dopamine	Dopamine, methacrylic acid, ethylene glycol dimethacrylate, and AIBN	Graphite screen-printed electrodes	3×10^{-7}	Peeters et al. (2016)
Melamine	Melamine, methacrylic acid, ethylene glycol dimethacrylate, and AIBN	Aluminum chip	3×10^{-6}	Caldara et al. (2022)
Pseudomonas aruginosa	Phenazine, (vinylbenzyl) trimethylammonium chloride, ethylene glycol dimethacrylate, and AIBN	Aluminum chip	3×10^{-7}	Frigoli et al. (2023)
2-methoxphenidine	2-methoxphenidine, methacrylic acid, ethylene glycol dimethacrylate, and AIBN	Aluminum chip	7×10^{-7}	Lowdon et al. (2019)
Noradrenaline	Noradrenaline, itaconic acid, ethylene glycol dimethacrylate, and AIBN	Graphite screen printed electrode	3×10^{-4}	Casadio et al. (2017)
Nafcillin	Fluorescein methacrylate or methacrylic acid trimethylolpropane trimethacrylate, and AIBN	Glass slides	5×10^{-6}	Hudson et al. (2021)
Serotonin	Serotonin hydrochloride, acrylamide, methacrylic acid, ethylene glycol dimethacrylate, and AIBN	Aluminum chip	1×10^{-7}	Vandenryt et al. (2017)

3.2.4 Mass-sensitive transduction

The intrinsic properties of an analyte are the crucial point in every sensing device. Mass is the innate property of any analyte and can be directly measured. Therefore, a MIP coating on a mass-sensitive transducer makes selective detection possible. These devices' sensing mechanism is based on piezoelectric phenomena (Dickert & Lieberzeit, 2007).

Piezoelectricity is a property of particular materials that, when subjected to mechanical stress, their molecules are polarized, producing a voltage. Inversely, mechanical deformation occurs by applying an electric potential to these materials. Natural piezoelectric materials include quartz and Rochelle salt, while lithium tantalate and niobate are synthetically derived. Since they have the highest stability, quartz crystals are the most widely used in developing sensors.

Most MIP-based mass-sensitive sensors are developed with MIPs facing quartz crystal microbalance (QCM). These sensors show excellent performances, such as high sensitivity and low cost; besides, they can be automated and miniaturized easily (Ávila et al., 2008).

Moreover, integrating MIPs with QCM sensors provides significant advantages such as long-term storage stability, reusability, and resistance to experimental conditions, that is, pH and temperature changes.

MIPs can be integrated into piezoelectric sensors by different strategies, including the entrapment of MIP in spin-coated polymeric films, grafting, covalent, or physical immobilization of microparticles or NPs, and in situ polymerization.

Some examples of MIP-modified QCM sensors are below discussed.

One of the first MIP-based piezoelectric sensors was developed by Percival et al. for L-menthol detection (Percival et al., 2001). The in situ casting of a thin-film MIP on the gold-coated QCM surface was the approach adopted. The frequency shift proportional to the analyte concentration was observed in the 0.2–1 mg/L range.

Li et al. modified the surface of a quartz microbalance with a spin-coated mixture of polyvinyl chloride and microspheres of MIP for a highly sensitive and selective detection of the pesticide endosulfan in milk and drinking water samples. The simple realization, the low cost of the analysis, the sensor's reusability, and the pretty good analytical performances make the sensor promising for environmental monitoring and food control (Liu et al., 2013).

MIP-based QCM devices were also proposed for gas detection. For example, Matsuguchi et al. developed a sensor for monitoring volatile organic compounds (Matsuguchi & Uno, 2006). Firstly, a cross-linked polymer was synthesized using MMA as the functional monomer, DVB as the cross-linker, and toluene or p-xylene as the template. Then, MIP particles were blended in a soluble matrix polymer; in this case, poly(methyl methacrylate) (PMMA) was selected thanks to its good miscibility with MIP. The dispersion of MIP particles with PMMA in acetone was spin-coated onto the QCM surface. The sensor responded reversibly towards toluene p-xylene vapor but with a low response time, probably due to the matrix polymer around the MIP particles.

Another MIP-based QCM sensor was developed by Hussain et al. for formaldehyde detection in air streams (Hussain et al., 2016). In this case, a copolymer thin film obtained by copolymerization of styrene, MAA, and EGDMA and using formaldehyde as the template was directly spin-coated on one electrode of a QCM sensor. Formaldehyde concentrations down to 500 ppb in the gas phase were detected with high selectivity.

MIP-modified piezoelectric sensors were also developed for detecting antibiotic residuals in complex-matrix samples. For example, residuals of amoxicillin in eggs (Bereli et al., 2020), penicillins and methimazole in meat products (Karaseva et al., 2016; Zhao et al., 2020), lovastatin in red yeast rice (Eren et al., 2015), ractopamine in swine and feed products (Pan et al., 2018), amantadine (Yun et al., 2018), and tobramycin and enrofloxacin in animal-derived food samples (Yola et al., 2014) were determined by MIP-coated QCM sensors.

The detection of macromolecules of interest in biology, medicine, and diagnostics is an important research field. For example, human serum albumin (HSA) is an essential component and the most abundant plasma. A variation in its concentration can indicate coronary heart disease or multiple myeloma; moreover, low levels of this molecule can signal chronic hepatitis, cirrhosis, and liver failure. Consequently, the detection of HSA is of great significance. For this purpose, Ma et al. developed an MIP-based QCM sensor exploiting epitope imprinting. The synthesis was performed using as the template the C-terminus epitope of HSA and zinc acrylate as the functional monomer in DMF as a porogenic solvent. The sensor showed good analytical performances; indeed, the linear range of the calibration curve was between 0.05 and 0.5 μg/mL, and the detection limit was about 0.03 μg/mL (Ma et al., 2017).

Recently, nanoMIP technology has been exploited in preparing a QCM-modified sensor for *N*-hexanoyl-L-homoserine lactone, that is, a biomarker of bacterial infection (Guha et al., 2020). MIP NPs were prepared according to the solid-phase technique involving the covalent immobilization of the template molecule on the glass beads' surface, followed by polymerization and the final affinity separation step to remove unreacted binders. UV polymerization was carried out using MAA as a functional monomer, trimethylolpropane trimethacrylate and EGDMA as crosslinkers, and acetonitrile as a porogenic solvent. The nanoMIP particles were immobilized onto the QCM gold surface through a covalent bond with the cysteamine self-assembled monolayer previously deposited on the sensor's gold surface.

NanoMIPs can also be deposited onto the QCM sensor surface by electropolymerization; for example, this approach was adopted by Eslami et al. for naproxen detection (Eslami & Alizadeh, 2016). The polypyrrole-based MIP was electrodeposited by CV directly onto the QCM gold electrode, and the overoxidation of the polypyrrole was undertaken to increase the sensor selectivity. The low detection limit (40 nM), good selectivity, rapid response, and low cost are good prerequisites, making the sensor promising for naproxen analysis.

Mass-sensitive sensors are potentially suitable for all analytes since mass is a property of all compounds. These transducers improve their selectivity and sensitivity when interfaced with MIPs. The growing number of studies on MIP-based QCM sensors report the efficient use of these devices for various applications spanning from environmental monitoring and volatile compound detection to biomedical analysis. The excellent performances of MIP-coated QCM sensors make them potentially usable for process control and monitoring and in assisting the development of new products.

3.2.5 Molecularly implemented polymer-based sensor arrays

The sensor arrays represent a powerful tool to exploit the full potentiality of MIPs in analytical applications. Indeed, the array format faces one of the main drawbacks of MIP-based sensors, that is, their high cross-reactivity and poor specificity. The single sensor in an array does not need high levels of selectivity; it only needs different selectivity compared to the other sensing elements. Molecular imprinting technology efficiently addresses the array's aim.

In a MIP-based array, multiple sensors are modified with different MIPs, each synthesized with a diverse template molecule. A unique response pattern, resulting when every analyte is analyzed by the MIP-based array, can be used to differentiate and classify the analytes. Chemometric pattern recognition methods have to be employed for the data treatment.

The success of MIP-based arrays has been demonstrated by the increasing number of studies reported in the last two decades of literature. Some recent examples are described below.

A voltammetric array for the simultaneous detection of ciprofloxacin, levofloxacin, and moxifloxacin in pharmaceutical and biological samples was proposed by Wang, Cetó, et al. (2022). The array comprised four graphite–epoxy composite electrodes: three modified with as many customized MIPs and one with a not-imprinted polymer (NIP). The polymers were integrated with multiwall carbon nanotubes decorated with gold NPs acting as supporting substrates. MIP and NIP films were deposited by electropolymerization of pyrrole in the presence of the three different target fluoroquinolone antibiotics. The quantitative electrochemical analysis was performed by DPV, and for the data treatment, principal component analysis (PCA) and artificial neural networks were employed. Under the optimal conditions, the array showed good performances with detection limits of about 1 μM and a wide detection range (up to 0.3 mM) for the three fluoroquinolones.

Lin et al. proposed a MIP-based photonic crystal sensor array for sulfonamide discrimination and detection (Lin et al., 2020). The array comprised four units: three MIP-modified using sulfathiazole, sulfaguanidine, or sulfamethazine as the template; the fourth was a not-imprinted (NIP)-based sensing unit. The Bragg diffraction patterns of the target sulfonamides at different concentrations were measured, and linear discrimination analysis and PCA were used for the data treatment. The array was used to detect six sulfonamides in fish samples, and the high recovery percentage (higher than 90%) suggested the promising application of the array for the identification of sulfonamides in food samples.

A three-electrode QCM sensor array for the simultaneous and selective detection of the two atherosclerosis biomarkers, that is, low-density lipoprotein (LDL) and high-density lipoprotein (HDL), was proposed by Chunta et al. (2019). Two QCM gold electrodes were modified with as many MIPs, one using LDL and the other HDL as a template molecule; the third electrode was covered by a film of NIP.

The sensor signals, that is, the frequency shift (Hz) vs time (min), were recorded in 10 mM phosphate buffer at pH 7.4. PCA was the selected chemometric tool for the data treatment. This QCM sensor array allowed the simultaneous quantification of LDL and HDL in less than 10 min. Moreover, it provides a strategy to develop a non-fasting blood test suitable for clinical purposes since LDL and HDL can be directly detected without the requirement of total cholesterol and triglyceride measurements otherwise necessary for indirect assays.

Turmeric (*Curcuma longa L.*) has been a popular phytomedicine for years; nowadays, its employment for medicinal purposes has spread widely due to the properties of some of its bioactive compounds. In a recent study, a MIP-based QCM gas sensor array was developed for identifying four bioactive compounds in turmeric, that is, ar-turmerone, curlone, ethyl-*p*-methoxycinnamate, and tumerone (Hardoyono & Windhani, 2021). Four MIPs were synthesized by a polyacrylic acid polymer and one of the four target volatile compounds. Each prepolymeric mixture was drop-coated on the QCM electrode. The so-prepared MIP-based QCM array was used to identify and quantify the target compounds in the vapor of turmeric essential oils. The frequency shift due to the interaction of each target analyte with the MIP layer was used as the sensor response, and for pattern recognition, the PCA and back-propagation neural network chemometric tools were employed. Based on the good performance obtained, the developed approach could also be expanded to detect other volatile bioactive compounds in different phytomedicines.

MIP-based sensor arrays can incorporate other imprinted materials such as NPs, hydrogels, colloidal crystals, and porous organic matrices. A notable trend is the development of molecularly imprinted microstructured matrices; these hybrid materials, thanks to their high surface area, can improve the binding performance of MIPs. Besides, they can provide MIPs with optical and electrical transduction mechanisms. With the advancements in this direction, it could be expected that MIP-based sensor arrays can effectively challenge classical analytical techniques in terms of accuracy, detection limits, and costs.

3.3 Conclusion and perspectives

In recent decades, MIPs have achieved an outstanding role as receptors in sensors, mainly because of the improved selectivity of the analysis. The possibility of developing sensors able to detect different analytes outlines their wide range of applications, including environmental and industrial monitoring, medical, toxicological, and forensic analysis.

Nevertheless, the commercialization of MIP-based sensors is confined to a niche market due to significant obstacles. The first problem is the interference of analog compounds with the target analyte, and some of the synthesis processes proposed to solve this issue are not feasible for large-scale production due to manufacturing restrictions and extra costs. Second, clinical trials of MIP-based sensors are necessary for regulatory approval and validation for their future applications as biomedical tools.

Despite the significant challenges, MIP technology continues to attract researchers working towards achieving the full potential of MIP-based sensors.

References

Abass, A. M., & Rzaij, J. M. (2020). A review on: Molecularly imprinting polymers by ion selective electrodes for determination drugs. *Journal of Chemical Reviews*, *2*, 148–156. Available from https://doi.org/10.33945/SAMI/JCR.2020.3.2.

Abbas, A., Tian, L., Morrissey, J. J., Kharasch, E. D., & Singamaneni, S. (2013). Hot spot-localized artificial antibodies for label-free plasmonic biosensing. *Advanced Functional Materials.*, *23*(14), 1789–1797. Available from https://doi.org/10.1002/adfm.201202370.

Abha, K., Nebu, J., Anjali Devi, J. S., Aparna, R. S., Anjana, R. R., Aswathy, A. O., & George, S. (2019). Photoluminescence sensing of bilirubin in human serum using L-cysteine tailored manganese doped zinc sulphide quantum dots. *Sensors and Actuators, B: Chemical*, *282*, 300–308. Available from https://doi.org/10.1016/j.snb.2018.11.063, https://www.journals.elsevier.com/sensors-and-actuators-b-chemical.

Akhoundian, M., Alizadeh, T., Ganjali, M. R., & Rafiei, F. (2018). A new carbon paste electrode modified with MWCNTs and nano-structured molecularly imprinted polymer for ultratrace determination of trimipramine: The crucial effect of electrode components mixing on its performance. *Biosensors and Bioelectronics*, *111*, 27–33. Available from https://doi.org/10.1016/j.bios.2018.03.061, http://www.elsevier.com/locate/bios.

Alberti, G., Zanoni, C., Magnaghi, L. R., & Biesuz, R. (2023). Ascorbic acid sensing by molecularly imprinted electrosynthesized polymer (e-MIP) on screen-printed electrodes. *Chemosensors.*, *11*(6), 348. Available from https://doi.org/10.3390/chemosensors11060348.

Alberti, G., Zanoni, C., Spina, S., Magnaghi, L. R., & Biesuz, R. (2022). MIP-based screen-printed potentiometric cell for atrazine sensing. *Chemosensors.*, *10*(8), 339. Available from https://doi.org/10.3390/chemosensors10080339.

Alberti, G., Zanoni, C., Spina, S., Magnaghi, L. R., & Biesuz, R. (2023). Trends in molecularly imprinted polymers (MIPs)-based plasmonic sensors. *Chemosensors.*, *11*(2), 144. Available from https://doi.org/10.3390/chemosensors11020144, http://www.mdpi.com/journal/chemosensors.

Alexander, C., Andersson, H. S., Andersson, L. I., Ansell, R. J., Kirsch, N., Nicholls, I. A., O'Mahony, J., & Whitcombe, M. J. (2006). Molecular imprinting science and technology: A survey of the literature for the years up to and including 2003. *Journal of Molecular Recognition.*, *19*(2), 106–180. Available from https://doi.org/10.1002/jmr.760.

Alizadeh, T., Atashi, F., & Ganjali, M. R. (2019). Molecularly imprinted polymer nano-sphere/multi-walled carbon nanotube coated glassy carbon electrode as an ultra-sensitive voltammetric sensor for picomolar level determination of RDX. *Talanta*, *194*, 415–421. Available from https://doi.org/10.1016/j.talanta.2018.10.040, https://www.journals.elsevier.com/talanta.

Alizadeh, T., & Azizi, S. (2016). Graphene/graphite paste electrode incorporated with molecularly imprinted polymer nanoparticles as a novel sensor for differential pulse voltammetry determination of fluoxetine. *Biosensors and Bioelectronics*, *81*, 198–206. Available from https://doi.org/10.1016/j.bios.2016.02.052, http://www.elsevier.com/locate/bios.

Altintas, Z. (2018). Surface plasmon resonance based sensor for the detection of glycopeptide antibiotics in milk using rationally designed nanoMIPs. *Scientific Reports.*, *8*(1), 111222. Available from https://doi.org/10.1038/s41598-018-29585-2, http://www.nature.com/srep/index.html.

Amatatongchai, M., Sroysee, W., Jarujamrus, P., Nacapricha, D., & Lieberzeit, P. A. (2018). Selective amperometric flow-injection analysis of carbofuran using a molecularly-imprinted polymer and gold-coated-magnetite modified carbon nanotube-paste electrode. *Talanta*, *179*, 700–709. Available from https://doi.org/10.1016/j.talanta.2017.11.064, https://www.journals.elsevier.com/talanta.

Amouzadeh Tabrizi, M., Fernández-Blázquez, J. P., Medina, D. M., & Acedo, P. (2022). An ultrasensitive molecularly imprinted polymer-based electrochemical sensor for the determination of SARS-CoV-2-RBD by using macroporous gold screen-printed electrode. *Biosensors and Bioelectronics*, *196*, 113729. Available from https://doi.org/10.1016/j.bios.2021.113729, http://www.elsevier.com/locate/bios.

Antuña-Jiménez, D., Díaz-Díaz, G., Blanco-López, M. C., Lobo-Castañón, M. J., Miranda-Ordieres, A. J., & Tuñón-Blanco, P. (2012). *Molecularly imprinted electrochemical sensors: Past, present, and future. Molecularly Imprinted Sensors* (pp. 1–34). Spain: Elsevier. Available from http://www.sciencedirect.com/science/book/9780444563316, 10.1016/B978-0-444-56331-6.00001-3.

Arcadio, F., Seggio, M., Prete., Buonanno, G., Mendes, J., Coelho, L. C. C., Jorge, P. A. S., Zeni, L., Bossi, A. M., & Cennamo, N. (2022). A plasmonic biosensor based on light-diffusing fibers functionalized with molecularly imprinted nanoparticles for ultralow sensing of proteins. *Nanomaterials.*, *12*, 1400. Available from https://doi.org/10.3390/nano12091400.

Arshad, R., Rhouati, A., Hayat, A., Nawaz, M. H., Yameen, M. A., Mujahid, A., & Latif, U. (2020). MIP-based impedimetric sensor for detecting dengue fever biomarker. *Applied Biochemistry and Biotechnology*, *191*(4), 1384–1394. Available from https://doi.org/10.1007/s12010-020-03285-y, http://www.springer.com/humana + press/journal/12010.

Ashley, J., Feng, X. T., & Sun, Y. (2018). A multifunctional molecularly imprinted polymer-based biosensor for direct detection of doxycycline in food samples. *Talanta*, *182*, 49–54. Available from https://doi.org/10.1016/j.talanta.2018.01.056, https://www.journals.elsevier.com/talanta.

Ashley, J., Shukor, Y., D'Aurelio, R., Trinh, L., Rodgers, T. L., Temblay, J., Pleasants, M., & Tothill, I. E. (2018). Synthesis of molecularly imprinted polymer nanoparticles for α-casein detection using surface plasmon resonance as a milk allergen sensor. *ACS Sensors.*, *3*(2), 418–424. Available from https://doi.org/10.1021/acssensors.7b00850, http://pubs.acs.org/journal/ascefj.

Assi, N., Rypar, T., Macka, M., Adam, V., & Vaculovicova, M. (2023). Microfluidic paper-based fluorescence sensor for L-homocysteine using a molecularly imprinted polymer and in situ-formed fluorescent quantum dots. *Talanta*, *255*, 124185. Available from https://doi.org/10.1016/j.talanta.2022.124185.

Ávila, M., Zougagh, M., Ríos, A., & Escarpa, A. (2008). Molecularly imprinted polymers for selective piezoelectric sensing of small molecules. *TrAC – Trends in Analytical Chemistry*, *27*(1), 54–65. Available from https://doi.org/10.1016/j.trac.2007.10.009.

Ayankojo, A. G., Boroznjak, R., Reut, J., Öpik, A., & Syritski, V. (2022). Molecularly imprinted polymer based electrochemical sensor for quantitative detection of SARS-CoV-2 spike protein. *Sensors and*

Actuators B: Chemical., *353*, 131160. Available from https://doi.org/10.1016/j.snb.2021.131160, https://www.journals.elsevier.com/sensors-and-actuators-b-chemical.

Bartold, K., Pietrzyk-Le, A., Golebiewska, K., Lisowski, W., Cauteruccio, S., Licandro, E., D'Souza, F., & Kutner, W. (2018). Oligonucleotide determination via peptide nucleic acid macromolecular imprinting in an electropolymerized CG-rich artificial oligomer analogue. *ACS Applied Materials and Interfaces*, *10*(33), 27562–27569. Available from https://doi.org/10.1021/acsami.8b09296, http://pubs.acs.org/journal/aamick.

Belbruno, J. J. (2019). Molecularly imprinted polymers. *Chemical Reviews*, *119*(1), 94–119. Available from https://doi.org/10.1021/acs.chemrev.8b00171, http://pubs.acs.org/journal/chreay.

Beltran, A., Borrull, F., Marcé, R. M., & Cormack, P. A. G. (2010). Molecularly-imprinted polymers: Useful sorbents for selective extractions. *TrAC – Trends in Analytical Chemistry*, *29*(11), 1363–1375. Available from https://doi.org/10.1016/j.trac.2010.07.020.

Bereli, N., Çimen, D., Hüseynli, S., & Denizli, A. (2020). Detection of amoxicillin residues in egg extract with a molecularly imprinted polymer on gold microchip using surface plasmon resonance and quartz crystal microbalance methods. *Journal of Food Science*, *85*(12), 4152–4160. Available from https://doi.org/10.1111/1750-3841.15529, http://onlinelibrary.wiley.com/journal/10.1111/(ISSN)1750-3841.

Caldara, M., Lowdon, J. W., Rogosic, R., Arreguin-Campos, R., Jimenez-Monroy, K. L., Heidt, B., Tschulik, K., Cleij, T. J., Diliën, H., Eersels, K., & van Grinsven, B. (2021). Thermal detection of glucose in urine using a molecularly imprinted polymer as a recognition element. *ACS Sensors.*, *6*(12), 4515–4525. Available from https://doi.org/10.1021/acssensors.1c02223, http://pubs.acs.org/journal/ascefj.

Caldara, M., Lowdon, J. W., Royakkers, J., Peeters, M., Cleij, T. J., Diliën, H., Eersels, K., & van Grinsven, B. (2022). A molecularly imprinted polymer-based thermal sensor for the selective detection of melamine in milk samples. *Foods.*, *11*(18), 2906. Available from https://doi.org/10.3390/foods11182906, http://www.mdpi.com/journal/foods.

Carrasco, S., Canalejas-Tejero, V., Navarro-Villoslada, F., Barrios, C. A., & Moreno-Bondi, M. C. (2014). Cross-linkable linear copolymer with double functionality: Resist for electron beam nanolithography and molecular imprinting. *Journal of Materials Chemistry C.*, *2*(8), 1400–1403. Available from https://doi.org/10.1039/c3tc31499e.

Casadio, S., Lowdon, J. W., Betlem, K., Ueta, J. T., Foster, C. W., Cleij, T. J., van Grinsven, B., Sutcliffe, O. B., Banks, C. E., & Peeters, M. (2017). Development of a novel flexible polymer-based biosensor platform for the thermal detection of noradrenaline in aqueous solutions. *Chemical Engineering Journal.*, *315*, 459–468. Available from https://doi.org/10.1016/j.cej.2017.01.050, http://www.elsevier.com/inca/publications/store/6/0/1/2/7/3/index.htt.

Cennamo, N., Maniglio, D., Tatti, R., Zeni, L., & Bossi, A. M. (2020). Deformable molecularly imprinted nanogels permit sensitivity-gain in plasmonic sensing. *Biosensors and Bioelectronics.*, *156*, 112126. Available from https://doi.org/10.1016/j.bios.2020.112126.

Cennamo, N., Bossi, A. M., Arcadio, F., Maniglio, D., & Zeni, L. (2021). On the effect of soft molecularly imprinted nanoparticles receptors combined to nanoplasmonic probes for biomedical applications. *Frontiers in Bioengineering and Biotechnology*, *9*, 801489. Available from https://doi.org/10.3389/fbioe.2021.801489.

Cennamo, N., D'agostino, G., Perri, C., Arcadio, F., Chiaretti, G., Parisio, E. M., Camarlinghi, G., Vettori, C., Di Marzo, F., Cennamo, R., Porto, G., & Zeni, L. (2021). Proof of concept for a quick and highly sensitive on-site detection of sars-cov-2 by plasmonic optical fibers and molecularly imprinted polymers. *Sensors*, *21*(5), 1–17. Available from https://doi.org/10.3390/s21051681, https://www.mdpi.com/1424-8220/21/5/1681/pdf.

Cennamo, N., D'Agostino, G., Porto, G., Biasiolo, A., Perri, C., Arcadio, F., & Zeni, L. (2018). A molecularly imprinted polymer on a plasmonic plastic optical fiber to detect perfluorinated compounds in water. *Sensors (Switzerland)*, *18*(6), 1836. Available from https://doi.org/10.3390/s18061836, http://www.mdpi.com/1424-8220/18/6/1836/pdf.

Cennamo, N., D'Agostino, G., Galatus, R., Bibbò, L., Pesavento, M., & Zeni, L. (2013). Sensors based on surface plasmon resonance in a plastic optical fiber for the detection of trinitrotoluene. *Sensors and Actuators, B: Chemical*, *188*, 221–226. Available from https://doi.org/10.1016/j.snb.2013.07.005.

Cennamo, N., D'Agostino, G., Pesavento, M., & Zeni, L. (2014). High selectivity and sensitivity sensor based on MIP and SPR in tapered plastic optical fibers for the detection of L-nicotine. *Sensors and Actuators, B: Chemical*, *191*, 529–536. Available from https://doi.org/10.1016/j.snb.2013.10.067.

Cennamo, N., De Maria, L., Chemelli, C., Profumo, A., Zeni, L., & Pesavento, M. (2016). Markers detection in transformer oil by plasmonic chemical sensor system based on POF and MIPs. *IEEE Sensors Journal*, *16*(21), 7663–7670. Available from https://doi.org/10.1109/JSEN.2016.2603168, http://ieeexplore.ieee.org/xpl/RecentIssue.jsp?punumber = 7361.

Cennamo, N., De Maria, L., D'Agostino, G., Zeni, L., & Pesavento, M. (2015). Monitoring of low levels of furfural in power transformer oil with a sensor system based on a POF-MIP platform. *Sensors (Switzerland)*, *15*(4), 8499–8511. Available from https://doi.org/10.3390/s150408499, http://www.mdpi.com/1424-8220/15/4/8499/pdf.

Cennamo, N., Massarotti, D., Conte, L., & Zeni, L. (2011). Low cost sensors based on SPR in a plastic optical fiber forbiosensor implementation. *Sensors.*, *11*(12), 11752–11760. Available from https://doi.org/10.3390/s111211752Italy, http://www.mdpi.com/1424-8220/11/12/11752/pdf.

Cennamo, N., Pesavento, M., Marchetti, S., & Zeni, L. (2020). Molecularly imprinted polymers and optical fiber sensors for security applications. *Springer Proceedings in Materials.*, *4*, 17–24. Available from https://doi.org/10.1007/978-3-030-34123-7_2, http://www.springer.com/series/16157.

Chantada-Vázquez, M. P., Sánchez-González, J., Peña-Vázquez, E., Tabernero, M. J., Bermejo, A. M., Bermejo-Barrera, P., & Moreda-Piñeiro, A. (2016). Simple and sensitive molecularly imprinted polymer – Mn-doped ZnS quantum dots based fluorescence probe for cocaine and metabolites determination in urine. *Analytical Chemistry*, *88*(5), 2734–2741. Available from https://doi.org/10.1021/acs.analchem.5b04250, http://pubs.acs.org/journal/ancham.

Chegel, V. I., Lopatynskyi, A. M., Lytvyn, V. K., Demydov, P. V., Martínez-Pastor, J. P., Abargues, R., Gadea, E. A., & Piletsky, S. A. (2020). Localized surface plasmon resonance nanochips with molecularly imprinted polymer coating for explosives sensing, National Academy of Sciences of Ukraine - Institute of Semiconductor Physics, Ukraine Semiconductor Physics*Quantum Electronics and Optoelectronics*, *23*(4), 431–436. Available from https://doi.org/10.15407/spqeo23.04.431, http://journal-spqeo.org.ua/n4_2020/v23n4-p437-441.pdf.

Chen, L., Wang, X., Lu, W., Wu, X., & Li, J. (2016). Molecular imprinting: Perspectives and applications. *Chemical Society Reviews*, *45*(8), 2137–2211. Available from https://doi.org/10.1039/c6cs00061d, http://pubs.rsc.org/en/journals/journal/cs.

Chen, Y., & Ming, H. (2012). Review of surface plasmon resonance and localized surface plasmon resonance sensor? *Photonic Sensors.*, *2*(1), 37–49. Available from https://doi.org/10.1007/s13320-011-0051-2.

Chen, Z., Álvarez-Pérez, M., Navarro-Villoslada, F., Moreno-Bondi, M. C., & Orellana, G. (2014). Fluorescent sensing of "quat" herbicides with a multifunctional pyrene-labeled monomer and molecular imprinting. *Sensors and Actuators B: Chemical*, *191*, 137–142. Available from https://doi.org/10.1016/j.snb.2013.09.097.

Cheubong, C., Takano, E., Kitayama, Y., Sunayama, H., Minamoto, K., Takeuchi, R., Furutani, S., & Takeuchi, T. (2021). Molecularly imprinted polymer nanogel-based fluorescence sensing of pork contamination in halal meat extracts. *Biosensors and Bioelectronics.*, *172*, 112775. Available from https://doi.org/10.1016/j.bios.2020.112775.

Chiappini, A., Pasquardini, L., & Bossi, A. M. (2020). Molecular imprinted polymers coupled to photonic structures in biosensors: The state of art. *Sensors.*, *20*, 5069. Available from https://doi.org/10.3390/s20185069.

Choi, D. Y., Yang, J. C., Hong, S. W., & Park, J. (2022). Molecularly imprinted polymer-based electrochemical impedimetric sensors on screen-printed carbon electrodes for the detection of trace cytokine IL-1β. *Biosensors and Bioelectronics*, *204*, 114073. Available from https://doi.org/10.1016/j.bios.2022.114073, http://www.elsevier.com/locate/bios.

Choi, S. W., Chang, H. J., Lee, N., Kim, J. H., & Chun, H. S. (2009). Detection of mycoestrogen zearalenone by a molecularly imprinted polypyrrole-based surface plasmon resonance (SPR)sensor. *Journal of Agricultural and Food Chemistry*, *57*(4), 1113–1118. Available from https://doi.org/10.1021/jf804022pSouth, http://pubs.acs.org/doi/pdfplus/10.1021/jf804022p.

Chunta, S., Suedee, R., Singsanan, S., & Lieberzeit, P. A. (2019). Sensing array based on molecularly imprinted polymers for simultaneous assessment of lipoproteins. *Sensors and Actuators, B: Chemical*, *298*, 126828. Available from https://doi.org/10.1016/j.snb.2019.126828, https://www.journals.elsevier.com/sensors-and-actuators-b-chemical.

Cui, G., Liang, R., & Qin, W. (2023). Potentiometric sensor based on a computationally designed molecularly imprinted receptor. *Analytica Chimica Acta*, *1239*, 340720. Available from https://doi.org/10.1016/j.aca.2022.340720.

Cui, Z., Li, Z., Jin, Y., Ren, T., Chen, J., Wang, X., Zhong, K., Tang, L., Tang, Y., & Cao, M. (2020). Novel magnetic fluorescence probe based on carbon quantum dots-doped molecularly imprinted polymer for AHLs signaling molecules sensing in fish juice and milk. *Food Chemistry*, *328*, 127063. Available from https://doi.org/10.1016/j.foodchem.2020.127063.

Culver, H. R., Wechsler, M. E., & Peppas, N. A. (2018). Label-free detection of tear biomarkers using hydrogel-coated gold nanoshells in a localized surface plasmon resonance-based biosensor. *ACS Nano*, *12*(9), 9342–9354. Available from https://doi.org/10.1021/acsnano.8b04348, http://pubs.acs.org/journal/ancac3.

Da Silva, H., Pacheco, J., Silva, J., Viswanathan, S., & Delerue-Matos, C. (2015). Molecularly imprinted sensor for voltammetric detection of norfloxacin. *Sensors and Actuators, B: Chemical*, *219*, 301–307. Available from https://doi.org/10.1016/j.snb.2015.04.125.

Daghestani, H. N., & Day, B. W. (2010). Theory and applications of surface plasmon resonance, resonant mirror, resonant waveguide grating, and dual polarization interferometry biosensors. *Sensors.*, *10*(11), 9630–9646. Available from https://doi.org/10.3390/s101109630, http://www.mdpi.com/1424-8220/10/11/9630/pdf, United States.

Dai, H., Deng, Z., Zeng, Y., Zhang, J., Yang, Y., Ma, Q., Hu, W., Guo, L., Li, L., Wan, S., et al. (2020). Highly sensitive determination of 4-nitrophenol with coumarin-based fluorescent molecularly imprinted poly (ionic liquid). *Journal of Hazardous Materials*, *398*, 122854. Available from https://doi.org/10.1016/j.jhazmat.2020.122854.

Deng, C., Zhong, Y., He, Y., Ge, Y., & Song, G. (2016). Selective determination of trace bisphenol a using molecularly imprinted silica nanoparticles containing quenchable fluorescent silver nanoclusters. *Microchimica Acta.*, *183*(1), 431–439. Available from https://doi.org/10.1007/s00604-015-1662-x, http://www.springer/at/mca.

Dickert, F. L., & Lieberzeit, P. A. (2007). *Imprinted polymers in chemical recognition for mass-sensitive devices* (5). Springer Science and Business Media LLC. Available from https://doi.org/10.1007/5346_027.

Diouf, A., Aghoutane, Y., Burhan, H., Sen, F., Bouchikhi, B., & El Bari, N. (2021). Tramadol sensing in non-invasive biological fluids using a voltammetric electronic tongue and an electrochemical sensor based on biomimetic recognition. *International Journal of Pharmaceutics*, *593*, 120114. Available from https://doi.org/10.1016/j.ijpharm.2020.120114.

Duan, H., Li, L., Wang, X., Wang, Y., Li, J., & Luo, C. (2016). CdTe quantum dots@luminol as signal amplification system for chrysoidine with chemiluminescence-chitosan/graphene oxide-magnetite-molecularly imprinting sensor. *Spectrochimica Acta – Part A: Molecular and Biomolecular Spectroscopy*, *153*, 535–541. Available from https://doi.org/10.1016/j.saa.2015.09.016.

Ensafi, A. A., Nasr-Esfahani, P., & Rezaei, B. (2018). Synthesis of molecularly imprinted polymer on carbon quantum dots as an optical sensor for selective fluorescent determination of promethazine hydrochloride. *Sensors and Actuators, B: Chemical*, *257*, 889–896. Available from https://doi.org/10.1016/j.snb.2017.11.050.

Erdem, Ö., Saylan, Y., Cihangir, N., & Denizli, A. (2019). Molecularly imprinted nanoparticles based plasmonic sensors for real-time Enterococcus faecalis detection. *Biosensors and Bioelectronics*, *126*, 608–614. Available from https://doi.org/10.1016/j.bios.2018.11.030, http://www.elsevier.com/locate/bios.

Eren, T., Atar, N., Yola, M. L., & Karimi-Maleh, H. (2015). A sensitive molecularly imprinted polymer based quartz crystal microbalance nanosensor for selective determination of lovastatin in red yeast rice. *Food Chemistry*, *185*, 430–436. Available from https://doi.org/10.1016/j.foodchem.2015.03.153, http://www.elsevier.com/locate/foodchem.

Ertürk Bergdahl, G., Andersson, T., Allhorn, M., Yngman, S., Timm, R., & Lood, R. (2019). In vivo detection and absolute quantification of a secreted bacterial factor from skin using molecularly imprinted polymers in a surface plasmon resonance biosensor for improved diagnostic abilities. *ACS Sensors.*, *4*(3), 717–725. Available from https://doi.org/10.1021/acssensors.8b01642.

Ertürk, G., & Mattiasson, B. (2016). From imprinting to microcontact imprinting – A new tool to increase selectivity in analytical devices. *Journal of Chromatography B: Analytical Technologies in the Biomedical and Life Sciences*, *1021*, 30–44. Available from https://doi.org/10.1016/j.jchromb.2015.12.025, http://www.elsevier.com/inca/publications/store/5/0/2/6/8/9.

Eslami, M. R., & Alizadeh, N. (2016). Nanostructured conducting molecularly imprinted polypyrrole based quartz crystal microbalance sensor for naproxen determination and its electrochemical impedance study. *RSC Advances.*, *6*(12), 9387–9395. Available from https://doi.org/10.1039/c5ra21489k, http://pubs.rsc.org/en/journals/journalissues.

Fan, J., Qiu, L., Qiao, Y., Xue, M., Dong, X., & Meng, Z. (2021). Recent advances in sensing applications of molecularly imprinted photonic crystals. *Frontiers in Chemistry*, *9*, 665119. Available from https://doi.org/10.3389/fchem.2021.665119.

Fang, L., Jia, M., Zhao, H., Kang, L., Shi, L., Zhou, L., & Kong, W. (2021). Molecularly imprinted polymer-based optical sensors for pesticides in foods: Recent advances and future trends. *Trends in Food Science and Technology*, *116*, 387–404. Available from https://doi.org/10.1016/j.tifs.2021.07.039, http://www.elsevier.com/wps/find/journaldescription.cws_home/601278/description#description.

Frigoli, M., Lowdon, J. W., Caldara, M., Arreguin-Campos, R., Sewall, J., Cleij, T. J., Diliën, H., Eersels, K., & Van Grinsven, B. (2023). Thermal pyocyanin sensor based on molecularly imprinted polymers for the indirect detection of pseudomonas aeruginosa. *ACS Sensors.*, *8*(1), 353–362. Available from https://doi.org/10.1021/acssensors.2c02345, http://pubs.acs.org/journal/ascefj.

Ganjali, M. R., Alizade, T., Larijani, B., Faridbod, F., & Norouzi, P. (2012). Nano-composite clozapine potentiometric carbon paste sensor based on biomimetic molecular imprinted polymer. *International Journal of Electrochemical Science*, *2012*(5), 4756–4765. Available from https://doi.org/10.1016/s1452-3981(23)19579-4.

Gavrilă, A. M., Stoica, E. B., Iordache, T. V., & Sârbu, A. (2022). Modern and dedicated methods for producing molecularly imprinted polymer layers in sensing applications. *Applied Sciences (Switzerland).*, *12*(6), 3038. Available from https://doi.org/10.3390/app12063080, https://www.mdpi.com/2076-3417/12/6/3080/pdf.

Greene, N. T., & Shimizu, K. D. (2005). Colorimetric molecularly imprinted polymer sensor array using dye displacement. *Journal of the American Chemical Society*, *127*(15), 5695–5700. Available from https://doi.org/10.1021/ja0468022.

Guha, A., Ahmad, O. S., Guerreiro, A., Karim, K., Sandström, N., Ostanin, V. P., van der Wijngaart, W., Piletsky, S. A., & Ghosh, S. K. (2020). Direct detection of small molecules using a nano-molecular imprinted polymer receptor and a quartz crystal resonator driven at a fixed frequency and amplitude. *Biosensors and Bioelectronics*, *158*, 112176. Available from https://doi.org/10.1016/j.bios.2020.112176, http://www.elsevier.com/locate/bios.

Hardoyono, F., & Windhani, K. (2021). Identification and detection of bioactive compounds in turmeric (Curcuma longa L.) using a gas sensor array based on molecularly imprinted polymer quartz crystal microbalance. *New Journal of Chemistry.*, *45*(38), 17930–17940. Available from https://doi.org/10.1039/d1nj03640h, http://pubs.rsc.org/en/journals/journal/nj.

Hashim, S. N. N. S., Boysen, R. I., Schwarz, L. J., Danylec, B., & Hearn, M. T. W. (2014). A comparison of covalent and non-covalent imprinting strategies for the synthesis of stigmasterol imprinted polymers. *Journal of Chromatography. A*, *1359*, 35–43. Available from https://doi.org/10.1016/j.chroma.2014.07.034, http://www.elsevier.com/locate/chroma.

Haupt, K., Dzgoev, A., & Mosbach, K. (1998). Assay system for the herbicide 2,4-dichlorophenoxyacetic acid using a molecularly imprinted polymer as an artificial recognition element. *Analytical Chemistry*, *70*(3), 628–631. Available from https://doi.org/10.1021/ac9711549.

Haupt, K., Linares, A. V., Bompart, M., & Bui, B. T. S. (2011). Molecularly imprinted polymers. *Topics in Current Chemistry*, *325*, 1–28. Available from https://doi.org/10.1007/128_2011_307.

Hikmat, H., & Tasfiyati, A. N. (2023). Synthesis and characterization of a polypyrrole-based molecularly imprinted polymer electrochemical sensor for the selective detection of phosphate ion. *Journal of Analytical Chemistry.*, *78*(1), 117–124. Available from https://doi.org/10.1134/S1061934823010057, https://www.springer.com/journal/10809.

Horikawa, R., Sunayama, H., Kitayama, Y., Takano, E., & Takeuchi, T. (2016). A programmable signaling molecular recognition nanocavity prepared by molecular imprinting and post-imprinting modifications. *Angewandte Chemie – International Edition.*, *55*(42), 13023–13027. Available from https://doi.org/10.1002/anie.201605992, http://onlinelibrary.wiley.com/journal/10.1002/(ISSN)1521-3773.

Hu, R., Luan, J., Kharasch, E. D., Singamaneni, S., & Morrissey, J. J. (2017). Aromatic functionality of target proteins influences monomer selection for creating artificial antibodies on plasmonic biosensors. *ACS Applied Materials & Interfaces*, *9*(1), 145–151. Available from https://doi.org/10.1021/acsami.6b12505.

Huang, S., Guo, M., Tan, J., Geng, Y., Wu, J., Tang, Y., Su, C., Lin, C. C., & Liang, Y. (2018). Novel fluorescence sensor based on all-inorganic perovskite quantum dots coated with molecularly imprinted polymers for highly selective and sensitive detection of omethoate. *ACS Applied Materials and Interfaces*, *10*(45), 39056–39063. Available from https://doi.org/10.1021/acsami.8b14472, http://pubs.acs.org/journal/aamick.

Hudson, A. D., Jamieson, O., Crapnell, R. D., Rurack, K., Soares, T. C. C., Mecozzi, F., Laude, A., Gruber, J., Novakovic, K., & Peeters, M. (2021). Dual detection of nafcillin using a molecularly imprinted polymer-based platform coupled to thermal and fluorescence read-out. *Materials Advances.*, *2*(15), 5105–5115. Available from https://doi.org/10.1039/d1ma00192b, http://www.rsc.org/journals-books-databases/about-journals/materials-advances/.

Hussain, M., Kotova, K., & Lieberzeit, P. (2016). Molecularly imprinted polymer nanoparticles for formaldehyde sensing with QCM. *Sensors.*, *16*(7), 1011. Available from https://doi.org/10.3390/s16071011.

Javanbakht, M., Fard, S. E., Mohammadi, A., Abdouss, M., Ganjali, M. R., Norouzi, P., & Safaraliee, L. (2008). Molecularly imprinted polymer based potentiometric sensor for the determination of hydroxyzine in tablets and biological fluids. *Analytica Chimica Acta*, *612*(1), 65–74. Available from https://doi.org/10.1016/j.aca.2008.01.085.

Kadhem, A. J., Gentile, G. J., & Fidalgo de Cortalezzi, M. M. (2021). Molecularly imprinted polymers (MIPs) in sensors for environmental and biomedical applications: A review. *Molecules (Basel, Switzerland)*, *26*(20), 6233. Available from https://doi.org/10.3390/molecules26206233.

Kang, J., Zhang, H., Wang, Z., Wu, G., & Lu, X. (2009). A novel amperometric sensor for salicylic acid based on molecularly imprinted polymer-modified electrodes. *Polymer – Plastics Technology and Engineering*, *48*(6), 639–645. Available from https://doi.org/10.1080/03602550902824499.

Karaseva, N., Ermolaeva, T., & Mizaikoff, B. (2016). Piezoelectric sensors using molecularly imprinted nanospheres for the detection of antibiotics. *Sensors and Actuators, B: Chemical*, *225*, 199–208. Available from https://doi.org/10.1016/j.snb.2015.11.045.

Kellens, E., Bové, H., Vandenryt, T., Lambrichts, J., Dekens, J., Drijkoningen, S., D'Haen, J., De Ceuninck, W., Thoelen, R., Junkers, T., Haenen, K., & Ethirajan, A. (2018). Micro-patterned molecularly imprinted polymer structures on functionalized diamond-coated substrates for testosterone detection. *Biosensors and Bioelectronics.*, *118*, 58–65. Available from https://doi.org/10.1016/j.bios.2018.07.032.

Kidakova, A., Reut, J., Rappich, J., Öpik, A., & Syritski, V. (2018). Preparation of a surface-grafted protein-selective polymer film by combined use of controlled/living radical photopolymerization and microcontact imprinting. *Reactive and Functional Polymers.*, *125*, 47–56. Available from https://doi.org/10.1016/j.reactfunctpolym.2018.02.004.

Kong, J., Xu, X., Ma, Y., Miao, J., & Bian, X. (2023). Rapid and sensitive detection of sulfamethizole using a reusable molecularly imprinted electrochemical sensor. *Foods.*, *12*(8), 1693. Available from https://doi.org/10.3390/foods12081693, http://www.mdpi.com/journal/foods.

Kong, Q., Wang, Y., Zhang, L., Ge, S., & Yu, J. (2017). A novel microfluidic paper-based colorimetric sensor based on molecularly imprinted polymer membranes for highly selective and sensitive detection of bisphenol A. *Sensors and Actuators, B: Chemical*, *243*, 130–136. Available from https://doi.org/10.1016/j.snb.2016.11.146.

Kunath, S., Panagiotopoulou, M., Maximilien, J., Marchyk, N., Sänger, J., & Haupt, K. (2015). Cell and tissue imaging with molecularly imprinted polymers as plastic antibody mimics. *Advanced Healthcare Materials.*, *4*(9), 1322–1326. Available from https://doi.org/10.1002/adhm.201500145, http://onlinelibrary.wiley.com/journal/10.1002/(ISSN)2192-2659.

Kupai, J., Razali, M., Buyuktiryaki, S., Kecili, R., & Szekely, G. (2016). Long-term stability and reusability of molecularly imprinted polymers. *Polymer Chemistry.*, *8*(4), 666–673. Available from https://doi.org/10.1039/c6py01853j, http://pubs.rsc.org/en/Journals/JournalIssues/PY.

Lahav, M., Kharitonov, A. B., Katz, O., Kunitake, T., & Willner, I. (2001). Tailored chemosensors for chloroaromatic acids using molecular imprinted TiO2 thin films on ion-sensitive field-effect transistors. *Analytical Chemistry*, *73*(3), 720–723. Available from https://doi.org/10.1021/ac000751j.

Lahav, M., Kharitonov, A. B., & Willner, I. (2001). Imprinting of chiral molecular recognition sites in thin TiO2 films associated with field-effect transistors: Novel functionalized devices for chiroselective and chirospecific analyses. *Chemistry – A European Journal*, *7*(18), 3992–3997. Available from 10.1002/1521-3765(20010917)7:18 < 3992::AID-CHEM3992 > 3.0.CO;2-G.

Leibl, N., Haupt, K., Gonzato, C., & Duma, L. (2021). Molecularly imprinted polymers for chemical sensing: A tutorial review. *Chemosensors.*, *9*(6), 123. Available from https://doi.org/10.3390/chemosensors9060123.

Levi, R., McNiven, S., Piletsky, S. A., Cheong, S. H., Yano, K., & Karube, I. (1997). Optical detection of chloramphenicol using molecularly imprinted polymers. *Analytical Chemistry*, *69*(11), 2017–2021. Available from https://doi.org/10.1021/ac960983b.

Li, H., Guan, H., Dai, H., Tong, Y., Zhao, X., Qi, W., Majeed, S., & Xu, G. (2012). An amperometric sensor for the determination of benzophenone in food packaging materials based on the electropolymerized molecularly imprinted poly-o-phenylenediamine film. *Talanta*, *99*, 811–815. Available from https://doi.org/10.1016/j.talanta.2012.07.033, https://www.journals.elsevier.com/talanta.

Li, Q., Kamra, T., & Ye, L. (2016). A modular approach for assembling turn-on fluorescence sensors using molecularly imprinted nanoparticles. *Chemical Communications*, *52*(82), 12237–12240. Available from https://doi.org/10.1039/c6cc06628c, http://pubs.rsc.org/en/journals/journal/cc.

Li, Q., Shinde, S., Grasso, G., Caroli, A., Abouhany, R., Lanzillotta, M., Pan, G., Wan, W., Rurack, K., & Sellergren, B. (2020). Selective detection of phospholipids using molecularly imprinted fluorescent sensory core-shell particles. *Nature Research, Sweden Scientific Reports.*, *10*(1), 9924. Available from https://doi.org/10.1038/s41598-020-66802-3, http://www.nature.com/srep/index.html.

Li, S., Li, J., Luo, J., Xu, Z., & Ma, X. (2018). A microfluidic chip containing a molecularly imprinted polymer and a DNA aptamer for voltammetric determination of carbofuran. *Microchimica Acta.*, *185*(6). Available from https://doi.org/10.1007/s00604-018-2835-1, http://www.springer/at/mca.

Li, T., Zhang, X., Gao, X., Lin, J., Zhao, F., & Zeng, B. (2023). Sensitive dual-mode detection of carbendazim by molecularly imprinted electrochemical sensor based on biomass-derived carbon-loaded gold nanoparticles. *Microchimica Acta.*, *190*(6), 236. Available from https://doi.org/10.1007/s00604-023-05821-z, https://www.springer.com/journal/604.

Li, X. G., Zhang, F., Gao, Y., Zhou, Q. M., Zhao, Y., Li, Y., Huo, J. Z., & Zhao, X. J. (2016). Facile synthesis of red emitting 3-aminophenylboronic acid functionalized copper nanoclusters for rapid, selective and highly sensitive detection of glycoproteins. *Biosensors and Bioelectronics*, *86*, 270–276. Available from https://doi.org/10.1016/j.bios.2016.06.054, http://www.elsevier.com/locate/bios.

Liang, R., Kou, L., Chen, Z., & Qin, W. (2013). Molecularly imprinted nanoparticles based potentiometric sensor with a nanomolar detection limit. *Sensors and Actuators, B: Chemical.*, *188*, 972–977. Available from https://doi.org/10.1016/j.snb.2013.07.110.

Liang, R., Zhang, R., & Qin, W. (2009). Potentiometric sensor based on molecularly imprinted polymer for determination of melamine in milk. *Sensors and Actuators, B: Chemical*, *141*(2), 544–550. Available from https://doi.org/10.1016/j.snb.2009.05.024.

Limthin, D., Leepheng, P., Klamchuen, A., & Phromyothin, D. (2022). Enhancement of electrochemical detection of gluten with surface modification based on molecularly imprinted polymers combined with superparamagnetic iron oxide nanoparticles. *Polymers.*, *14*(1), 91. Available from https://doi.org/10.3390/polym14010091.

Lin, Z. Z., Li, L., Fu, G. Y., Lai, Z. Z., Peng, A. H., & Huang, Zy (2020). Molecularly imprinted polymer-based photonic crystal sensor array for the discrimination of sulfonamides. *Analytica Chimica Acta*, *1101*, 32–40. Available from https://doi.org/10.1016/j.aca.2019.12.032, http://www.journals.elsevier.com/analytica-chimica-acta/.

Liu, J., Jalali, M., Mahshid, S., & Wachsmann-Hogiu, S. (2020). Are plasmonic optical biosensors ready for use in point-of-need applications? *Analyst.*, *145*(2), 364–384. Available from https://doi.org/10.1039/c9an02149c, http://pubs.rsc.org/en/journals/journal/an.

Liu, K., Song, Y., Song, D., & Liang, R. (2020). Plasticizer-free polymer membrane potentiometric sensors based on molecularly imprinted polymers for determination of neutral phenols. *Analytica Chimica Acta*, *1121*, 50–56. Available from https://doi.org/10.1016/j.aca.2020.04.074, http://www.journals.elsevier.com/analytica-chimica-acta/.

Liu, N., Han, J., Liu, Z., Qu, L., & Gao, Z. (2013). Rapid detection of endosulfan by a molecularly imprinted polymer microsphere modified quartz crystal microbalance. *Analytical Methods.*, *5*(17), 4442–4447. Available from https://doi.org/10.1039/c3ay40697k, http://www.rsc.org/Publishing/Journals/AY/About.asp.

Liu, R., & Ko, C. C. (2023). Molecularly imprinted polymer-based luminescent chemosensors. *Biosensors*, *13*(2), 295. Available from https://doi.org/10.3390/bios13020295, http://www.mdpi.com/journal/biosensors/.

Liu, Z. C., Qi, J. W., Hu, C., Zhang, L., Song, W., Liang, R. P., & Qiu, J. D. (2015). Cu nanoclusters-based ratiometric fluorescence probe for ratiometric and visualization detection of copper ions. *Analytica Chimica Acta*, *895*, 95–103. Available from https://doi.org/10.1016/j.aca.2015.09.002, http://www.journals.elsevier.com/analytica-chimica-acta/.

Lowdon, J. W., Diliën, H., Singla, P., Peeters, M., Cleij, T. J., van Grinsven, B., & Eersels, K. (2020). MIPs for commercial application in low-cost sensors and assays – An overview of the current status quo.

Sensors and Actuators, B: Chemical, 325, 128973. Available from https://doi.org/10.1016/j.snb.2020.128973, https://www.journals.elsevier.com/sensors-and-actuators-b-chemical.

Lowdon, J. W., Eersels, K., Rogosic, R., Boonen, T., Heidt, B., Diliën, H., van Grinsven, B., & Cleij, T. J. (2019). Surface grafted molecularly imprinted polymeric receptor layers for thermal detection of the new psychoactive substance 2-methoxphenidine. *Sensors and Actuators, A: Physical, 295*, 586–595. Available from https://doi.org/10.1016/j.sna.2019.06.029.

Lu, X., Yang, Y., Zeng, Y., Li, L., & Wu, X. (2018). Rapid and reliable determination of p-nitroaniline in wastewater by molecularly imprinted fluorescent polymeric ionic liquid microspheres. *Biosensors and Bioelectronics, 99*, 47–55. Available from https://doi.org/10.1016/j.bios.2017.07.041, http://www.elsevier.com/locate/bios.

Luo, J., Tan, H., Yang, B., Chen, D., & Fei, J. (2022). A mesoporous silica-based probe with a molecularly imprinted polymer recognition and Mn:ZnS QDs@rhodamine B ratiometric fluorescence sensing strategy for the analysis of 4-nitrophenol. *Analytical Methods., 14*(39), 3881–3889. Available from https://doi.org/10.1039/d2ay01147f.

Ma, X. T., He, X. W., Li, W. Y., & Zhang, Y. K. (2017). Epitope molecularly imprinted polymer coated quartz crystal microbalance sensor for the determination of human serum albumin. *Sensors and Actuators, B: Chemical, 246*, 879–886. Available from https://doi.org/10.1016/j.snb.2017.02.137.

Macdonald, J. R., & Johnson, W. B. (2018). *Fundamentals of impedance spectroscopy* (pp. 1–20). Wiley. Available from https://doi.org/10.1002/9781119381860.ch1

Masteri-Farahani, M., Mashhadi-Ramezani, S., & Mosleh, N. (2020). Molecularly imprinted polymer containing fluorescent graphene quantum dots as a new fluorescent nanosensor for detection of methamphetamine. *Spectrochimica Acta Part A: Molecular and Biomolecular Spectroscopy, 229*, 118021. Available from https://doi.org/10.1016/j.saa.2019.118021.

Matsuguchi, M., & Uno, T. (2006). Molecular imprinting strategy for solvent molecules and its application for QCM-based VOC vapor sensing. *Sensors and Actuators, B: Chemical, 113*(1), 94–99. Available from https://doi.org/10.1016/j.snb.2005.02.028.

Matsui, J., Akamatsu, K., Nishiguchi, S., Miyoshi, D., Nawafune, H., Tamaki, K., & Sugimoto, N. (2004). Composite of Au nanoparticles and molecularly imprinted polymer as a sensing material. *Analytical Chemistry, 76*(5), 1310–1315. Available from https://doi.org/10.1021/ac034788q.

Mayer, K. M., & Hafner, J. H. (2011). Localized surface plasmon resonance sensors. *Chemical Reviews, 111*(6), 3828–3857. Available from https://doi.org/10.1021/cr100313v.

McNiven, S., Kato, M., Levi, R., Yano, K., & Karube, I. (1998). Chloramphenicol sensor based on an in situ imprinted polymer. *Analytica Chimica Acta, 365*(1-3), 69–74. Available from https://doi.org/10.1016/S0003-2670(98)00096-8.

Ming, W., Wang, X., Lu, W., Zhang, Z., Song, X., Li, J., & Chen, L. (2017). Magnetic molecularly imprinted polymers for the fluorescent detection of trace 17β-estradiol in environmental water. *Sensors and Actuators, B: Chemical, 238*, 1309–1315. Available from https://doi.org/10.1016/j.snb.2016.09.111.

Moncer, F., Adhoum, N., Catak, D., & Monser, L. (2021). Electrochemical sensor based on MIP for highly sensitive detection of 5-hydroxyindole-3-acetic acid carcinoid cancer biomarker in human biological fluids. *Analytica Chimica Acta, 1181*, 338925. Available from https://doi.org/10.1016/j.aca.2021.338925.

Moreira Gonçalves, L. (2021). Electropolymerized molecularly imprinted polymers: Perceptions based on recent literature for soon-to-be world-class scientists. *Current Opinion in Electrochemistry., 25*, 100640. Available from https://doi.org/10.1016/j.coelec.2020.09.007.

Mori, K., Hirase, M., Morishige, T., Takano, E., Sunayama, H., Kitayama, Y., Inubushi, S., Sasaki, R., Yashiro, M., Takeuchi, T., & Pretreatment-Free, A. (2019). Polymer-based platform prepared by molecular imprinting and post-imprinting modifications for sensing intact exosomes. *Angewandte Chemie –*

International Edition., *58*(6), 1612–1615. Available from https://doi.org/10.1002/anie.201811142, http://onlinelibrary.wiley.com/journal/10.1002/(ISSN)1521-3773.

Niu, H., Yang, Y., & Zhang, H. (2015). Efficient one-pot synthesis of hydrophilic and fluorescent molecularly imprinted polymer nanoparticles for direct drug quantification in real biological samples. *Biosensors and Bioelectronics*, *74*, 440–446. Available from https://doi.org/10.1016/j.bios.2015.06.071, http://www.elsevier.com/locate/bios.

Nontawong, N., Amatatongchai, M., Thimoonnee, S., Laosing, S., Jarujamrus, P., Karuwan, C., & Chairam, S. (2019). Novel amperometric flow-injection analysis of creatinine using a molecularly-imprinted polymer coated copper oxide nanoparticle-modified carbon-paste-electrode. *Journal of Pharmaceutical and Biomedical Analysis*, *175*, 112770. Available from https://doi.org/10.1016/j.jpba.2019.07.018.

Ong, J. Y., Phang, S. W., Goh, C. T., Pike, A., & Tan, L. L. (2023). Impedimetric polyaniline-based aptasensor for aflatoxin B1 determination in agricultural products. *Foods.*, *12*(8), 1698. Available from https://doi.org/10.3390/foods12081698, http://www.mdpi.com/journal/foods.

Özgür, E., Patra, H. K., Turner, A. P. F., Denizli, A., & Uzun, L. (2020). Lanthanide [terbium(III)]-doped molecularly imprinted nanoarchitectures for the fluorimetric detection of melatonin. *Industrial and Engineering Chemistry Research*, *59*(36), 16068–16076. Available from https://doi.org/10.1021/acs.iecr.0c02387, http://pubs.acs.org/journal/iecred.

Pan, M., Li, R., Xu, L., Yang, J., Cui, X., & Wang, S. (2018). Reproducible molecularly imprinted piezoelectric sensor for accurate and sensitive detection of ractopamine in swine and feed products. *Sensors*, *18*(6), 1870. Available from https://doi.org/10.3390/s18061870.

Peeters, M., Csipai, P., Geerets, B., Weustenraed, A., Van Grinsven, B., Thoelen, R., Gruber, J., De Ceuninck, W., Cleij, T. J., Troost, F. J., & Wagner, P. (2013). Heat-transfer-based detection of L-nicotine, histamine, and serotonin using molecularly imprinted polymers as biomimetic receptors. *Analytical and Bioanalytical Chemistry*, *405*(20), 6453–6460. Available from https://doi.org/10.1007/s00216-013-7024-9.

Peeters, M., Kobben, S., Jiménez-Monroy, K. L., Modesto, L., Kraus, M., Vandenryt, T., Gaulke, A., Van Grinsven, B., Ingebrandt, S., Junkers, T., & Wagner, P. (2014). Thermal detection of histamine with a graphene oxide based molecularly imprinted polymer platform prepared by reversible addition-fragmentation chain transfer polymerization. *Sensors and Actuators, B: Chemical.*, *203*, 527–535. Available from https://doi.org/10.1016/j.snb.2014.07.013.

Peeters, M. M., Van Grinsven, B., Foster, C. W., Cleij, T. J., & Banks, C. E. (2016). Introducing thermal wave transport analysis (TWTA): A thermal technique for dopamine detection by screen-printed electrodes functionalized with molecularly imprinted polymer (MIP) particles. *Molecules (Basel, Switzerland)*, *21*(5), 552. Available from https://doi.org/10.3390/molecules21050552, http://www.mdpi.com/1420-3049/21/5/552/pdf.

Perçin, I., Idil, N., Bakhshpour, M., Yılmaz, E., Mattiasson, B., & Denizli, A. (2017). Microcontact imprinted plasmonic nanosensors: Powerful tools in the detection of salmonella paratyphi. *Sensors*, *17*(6), 1375. Available from https://doi.org/10.3390/s17061375.

Percival, C. J., Stanley, S., Galle, M., Braithwaite, A., Newton, M. I., McHale, G., & Hayes, W. (2001). Molecular-imprinted, polymer-coated quartz crystal microbalances for the detection of terpenes. *Analytical Chemistry*, *73*(17), 4225–4228. Available from https://doi.org/10.1021/ac0155198.

Pernites, R., Ponnapati, R., Felipe, M. J., & Advincula, R. (2011). Electropolymerization molecularly imprinted polymer (E-MIP) SPR sensing of drug molecules: Pre-polymerization complexed terthiophene and carbazole electroactive monomers. *Biosensors and Bioelectronics.*, *26*(5), 2766–2771. Available from https://doi.org/10.1016/j.bios.2010.10.027.

Pernites, R. B., Ponnapati, R. R., & Advincula, R. C. (2010). Surface plasmon resonance (SPR) detection of theophylline via electropolymerized molecularly imprinted polythiophenes. *Macromolecules*, *43*(23), 9724–9735. Available from https://doi.org/10.1021/ma101868y.

Pesavento, M., Zeni, L., De Maria, L., Alberti, G., & Cennamo, N. (2021). SPR-optical fiber-molecularly imprinted polymer sensor for the detection of furfural in wine. *Biosensors*, *11*(3), 72. Available from https://doi.org/10.3390/bios11030072.

Pesavento, M., D'Agostino, G., Alberti, G., Biesuz, R., & Merli, D. (2013). Voltammetric platform for detection of 2,4,6-trinitrotoluene based on a molecularly imprinted polymer. *Analytical and Bioanalytical Chemistry*, *405*(11), 3559–3570. Available from https://doi.org/10.1007/s00216-012-6553-y.

Pesavento, M., Merli, D., Biesuz, R., Alberti, G., Marchetti, S., & Milanese, C. (2021). A MIP-based low-cost electrochemical sensor for 2-furaldehyde detection in beverages. *Analytica Chimica Acta*, *1142*, 201–210. Available from https://doi.org/10.1016/j.aca.2020.10.059, http://www.journals.elsevier.com/analytica-chimica-acta/.

Piriya VS, A., Joseph, P., Daniel SCG, K., Lakshmanan, S., Kinoshita, T., & Muthusamy, S. (2017). Colorimetric sensors for rapid detection of various analytes. *Materials Science and Engineering C.*, *78*, 1231–1245. Available from https://doi.org/10.1016/j.msec.2017.05.018.

Pogorelova, S. P., Kharitonov, A. B., Willner, I., Sukenik, C. N., Pizem, H., & Bayer, T. (2004). Development of ion-sensitive field-effect transistor-based sensors for benzylphosphonic acids and thiophenols using molecularly imprinted TiO2 films. *Analytica Chimica Acta*, *504*(1), 113–122. Available from https://doi.org/10.1016/S0003-2670(03)00532-4, http://www.journals.elsevier.com/analytica-chimica-acta/.

Qi, L., Liang, R., & Qin, W. (2020). Stimulus-responsive imprinted polymer-based potentiometric sensor for reversible detection of neutral phenols. *Analytical Chemistry*, *92*(6), 4284–4291. Available from https://doi.org/10.1021/acs.analchem.9b04911, http://pubs.acs.org/journal/ancham.

Qin, X., Liu, W., Liu, G., Ren, C., Liu, C., Li, H., & Cao, Y. (2020). 2,4-Dichlorophenol molecularly imprinted two-dimensional photonic crystal hydrogels. *Journal of Applied Polymer Science*, *137*(42), 49299. Available from https://doi.org/10.1002/app.49299.

Qiu, H., Gao, L., Wang, J., Pan, J., Yan, Y., & Zhang, X. (2017). A precise and efficient detection of Beta-Cyfluthrin via fluorescent molecularly imprinted polymers with ally fluorescein as functional monomer in agricultural products. *Food Chemistry*, *217*, 620–627. Available from https://doi.org/10.1016/j.foodchem.2016.09.028, http://www.elsevier.com/locate/foodchem.

Qiu, X., Chen, W., Luo, Y., Wang, Y., Wang, Y., & Guo, H. (2020). Highly sensitive α-amanitin sensor based on molecularly imprinted photonic crystals. *Analytica Chimica Acta*, *1093*, 142–149. Available from https://doi.org/10.1016/j.aca.2019.09.066, http://www.journals.elsevier.com/analytica-chimica-acta/.

Qu, J. H., Dillen, A., Saeys, W., Lammertyn, J., & Spasic, D. (2020). Advancements in SPR biosensing technology: An overview of recent trends in smart layers design, multiplexing concepts, continuous monitoring and in vivo sensing. *Analytica Chimica Acta*, *1104*, 10–27. Available from https://doi.org/10.1016/j.aca.2019.12.067, http://www.journals.elsevier.com/analytica-chimica-acta/.

Quílez-Alburquerque, J., Descalzo, A. B., Moreno-Bondi, M. C., & Orellana, G. (2021). Luminescent molecularly imprinted polymer nanocomposites for emission intensity and lifetime rapid sensing of tenuazonic acid mycotoxin. *Polymer*, *230*, 124041. Available from https://doi.org/10.1016/j.polymer.2021.124041, http://www.journals.elsevier.com/polymer.

Rahtuvanoğlu, A., Akgönüllü, S., Karacan, S., & Denizli, A. (2020). Biomimetic nanoparticles based surface plasmon resonance biosensors for histamine detection in foods. *ChemistrySelect.*, *5*(19), 5683–5692. Available from https://doi.org/10.1002/slct.202000440, http://onlinelibrary.wiley.com/journal/10.1002/(ISSN)2365-6549.

Ray, J. V., Mirata, F., Pérollier, C., Arotcarena, M., Bayoudh, S., & Resmini, M. (2016). Smart coumarin-tagged imprinted polymers for the rapid detection of tamoxifen. *Analytical and Bioanalytical Chemistry*, *408*(7), 1855–1861. Available from https://doi.org/10.1007/s00216-015-9296-8, http://link.springer.de/link/service/journals/00216/index.htm.

Resende, S., Frasco, M. F., & Sales, M. G. F. (2020). A biomimetic photonic crystal sensor for label-free detection of urinary venous thromboembolism biomarker. *Sensors and Actuators, B: Chemical, 312*, 127947. Available from https://doi.org/10.1016/j.snb.2020.127947, https://www.journals.elsevier.com/sensors-and-actuators-b-chemical.

Rico-Yuste, A., & Carrasco, S. (2019). Molecularly imprinted polymer-based hybrid materials for the development of optical sensors. *Polymers, 11*(7), 1173. Available from https://doi.org/10.3390/polym11071173.

Rico-Yuste, A., Abouhany, R., Urraca, J. L., Descalzo, A. B., Orellana, G., & Moreno-Bondi, M. C. (2021). Eu(III)-templated molecularly imprinted polymer used as a luminescent sensor for the determination of tenuazonic acid mycotoxin in food samples. *Sensors Actuators B Chem., 329*, 129256. Available from https://doi.org/10.1016/j.snb.2020.129256.

Rouhani, S., & Nahavandifard, F. (2014). Molecular imprinting-based fluorescent optosensor using a polymerizable 1,8-naphthalimide dye as a florescence functional monomer. *Sensors and Actuators, B: Chemical, 197*, 185–192. Available from https://doi.org/10.1016/j.snb.2014.02.082.

Saher, A., Bahgat, A., Molouk, A., Mortada, W., & Khalifa, M. (2023). MIP/GO/GCE sensor for the determination of aminophylline in pharmaceutical ingredients and urine samples. *Analytical and Bioanalytical Chemistry Research., 10*(4), 435–443. Available from https://doi.org/10.22036/abcr.2023.392413.1915.

Sari, E., Üzek, R., Duman, M., & Denizli, A. (2016). Fabrication of surface plasmon resonance nanosensor for the selective determination of erythromycin via molecular imprinted nanoparticles. *Talanta, 150*, 607–614. Available from https://doi.org/10.1016/j.talanta.2015.12.043, https://www.journals.elsevier.com/talanta.

Scholz, F. (2010). *Electroanalytical methods: Guide to experiments and applications. Electroanalytical Methods: Guide to Experiments and Applications* (pp. 1–359). Germany: Springer Berlin Heidelberg. Available from http://www.springerlink.com/openurl.asp?genre = book&isbn = 978-3-642-02914-1, 10.1007/978-3-642-02915-8.

Sehit, E., Drzazgowska, J., Buchenau, D., Yesildag, C., Lensen, M., & Altintas, Z. (2020). Ultrasensitive nonenzymatic electrochemical glucose sensor based on gold nanoparticles and molecularly imprinted polymers. *Biosensors and Bioelectronics., 165*, 112432. Available from https://doi.org/10.1016/j.bios.2020.112432.

Shang, L., Liu, C., Chen, B., & Hayashi, K. (2018). Development of molecular imprinted sol-gel based LSPR sensor for detection of volatile cis-jasmone in plant. *Sensors and Actuators, B: Chemical, 260*, 617–626. Available from https://doi.org/10.1016/j.snb.2017.12.123.

Sheydaei, O., Khajehsharifi, H., & Rajabi, H. R. (2020). Rapid and selective diagnose of Sarcosine in urine samples as prostate cancer biomarker by mesoporous imprinted polymeric nanobeads modified electrode. *Sensors and Actuators B: Chemical., 309*, 127559. Available from https://doi.org/10.1016/j.snb.2019.127559.

Shinde, S., El-Schich, Z., Malakpour, A., Wan, W., Dizeyi, N., Mohammadi, R., Rurack, K., Gjörloff Wingren, A., & Sellergren, B. (2015). Sialic acid-imprinted fluorescent core-shell particles for selective labeling of cell surface glycans. *Journal of the American Chemical Society, 137*(43), 13908–13912. Available from https://doi.org/10.1021/jacs.5b08482, http://pubs.acs.org/journal/jacsat.

Silverio, O. V., So, R. C., Elnar, K. J. S., Malapit, C. A., & Nepomuceno, M. C. M. (2017). Development of dieldrin, endosulfan, and hexachlorobenzene-imprinted polymers for dye-displacement array sensing. *Journal of Applied Polymer Science, 134*(2), 44401. Available from https://doi.org/10.1002/app.44401, http://onlinelibrary.wiley.com/journal/10.1002/(ISSN)1097-4628.

Simic, M., Stavrakis, A. K., Kojic, T., Jeoti, V., & Stojanovic, G. M. (2023). Parameter estimation of the randles equivalent electrical circuit using only real part of the impedance. *IEEE Sensors Journal, 23*(5), 4922–4929. Available from https://doi.org/10.1109/JSEN.2023.3238074, http://ieeexplore.ieee.org/xpl/RecentIssue.jsp?punumber = 7361.

Sistani, S., & Shekarchizadeh, H. (2021). Fabrication of fluorescence sensor based on molecularly imprinted polymer on amine-modified carbon quantum dots for fast and highly sensitive and selective detection of tannic acid in food samples. *Analytica Chimica Acta, 1186*, 339122. Available from https://doi.org/10.1016/j.aca.2021.339122.

Sunayama, H., Ooya, T., & Takeuchi, T. (2010). Fluorescent protein recognition polymer thin films capable of selective signal transduction of target binding events prepared by molecular imprinting with a post-imprinting treatment. *Biosensors and Bioelectronics., 26*(2), 458–462. Available from https://doi.org/10.1016/j.bios.2010.07.091.

Taheri, N., Khoshsafar, H., Ghanei, M., Ghazvini, A., & Bagheri, H. (2022). Dual-template rectangular nanotube molecularly imprinted polypyrrole for label-free impedimetric sensing of AFP and CEA as lung cancer biomarkers. *Talanta, 239*, 123146. Available from https://doi.org/10.1016/j.talanta.2021.123146.

Ton, X. A., Acha, V., Haupt, K., & Tse Sum Bui, B. (2012). Direct fluorimetric sensing of UV-excited analytes in biological and environmental samples using molecularly imprinted polymer nanoparticles and fluorescence polarization. *Biosensors and Bioelectronics., 36*(1), 22–28. Available from https://doi.org/10.1016/j.bios.2012.03.033.

Tsutsumi, K., Sunayama, H., Kitayama, Y., Takano, E., Nakamachi, Y., Sasaki, R., & Takeuchi, T. (2021). Fluorescent signaling of molecularly imprinted nanogels prepared via postimprinting modifications for specific protein detection. *Advanced NanoBiomed Research., 1*(4), 2000079. Available from https://doi.org/10.1002/anbr.202000079.

Unger, C., & Lieberzeit, P. A. (2021). Molecularly imprinted thin film surfaces in sensing: Chances and challenges. *Reactive and Functional Polymers., 161*, 104855. Available from https://doi.org/10.1016/j.reactfunctpolym.2021.104855.

Vandenryt, T., Van Grinsven, B., Eersels, K., Cornelis, P., Kholwadia, S., Cleij, T. J., Thoelen, R., De Ceuninck, W., Peeters, M., & Wagner, P. (2017). Single-shot detection of neurotransmitters in whole-blood samples by means of the heat-transfer method in combination with synthetic receptors. *Sensors (Switzerland)., 17*(12), 2701. Available from https://doi.org/10.3390/s17122701, http://www.mdpi.com/1424-8220/17/12/2701/pdf.

Vasapollo, G., Sole, R. D., Mergola, L., Lazzoi, M. R., Scardino, A., Scorrano, S., & Mele, G. (2011). Molecularly imprinted polymers: Present and future prospective. *International Journal of Molecular Sciences., 12*(9), 5908–5945. Available from https://doi.org/10.3390/ijms12095908Italy, http://www.mdpi.com/1422-0067/12/9/5908/pdf.

Verellen, N., Van Dorpe, P., Huang, C., Lodewijks, K., Vandenbosch, G. A. E., Lagae, L., & Moshchalkov, V. V. (2011). Plasmon line shaping using nanocrosses for high sensitivity localized surface plasmon resonance sensing. *Nano Letters, 11*(2), 391–397. Available from https://doi.org/10.1021/nl102991v.

Vishnuvardhan, V., Prathish, K. P., Naidu, G. R. K., & Prasada Rao, T. (2007). Fabrication and topographical analysis of non-covalently imprinted polymer inclusion membranes for the selective sensing of pinacolyl methylphosphonate-A simulant of Soman. *Electrochimica Acta, 52*(24), 6922–6928. Available from https://doi.org/10.1016/j.electacta.2007.05.005.

Wagner, R., Wan, W., Biyikal, M., Benito-Peña, E., Moreno-Bondi, M. C., Lazraq, I., Rurack, K., & Sellergren, B. (2013). Synthesis, spectroscopic, and analyte-responsive behavior of a polymerizable naphthalimide-based carboxylate probe and molecularly imprinted polymers prepared thereof. *Journal of Organic Chemistry., 78*(4), 1377–1389. Available from https://doi.org/10.1021/jo3019522.

Wang, C., Qi, L., & Liang, R. (2021). A molecularly imprinted polymer-based potentiometric sensor based on covalent recognition for the determination of dopamine. *Analytical Methods., 13*(5), 620–625. Available from https://doi.org/10.1039/d0ay02100h, http://pubs.rsc.org/en/journals/journal/ay.

Wang, J., Liang, R., & Qin, W. (2020). Molecularly imprinted polymer-based potentiometric sensors. *TrAC Trends in Analytical Chemistry, 130*, 115980. Available from https://doi.org/10.1016/j.trac.2020.115980.

Wang, J., Gao, L., Han, D., Pan, J., Qiu, H., Li, H., Wei, X., Dai, J., Yang, J., Yao, H., & Yan, Y. (2015). Optical detection of λ-cyhalothrin by core-shell fluorescent molecularly imprinted polymers in Chinese spirits. *Journal of Agricultural and Food Chemistry*, *63*(9), 2392–2399. Available from https://doi.org/10.1021/jf5043823, http://pubs.acs.org/journal/jafcau.

Wang, L., & Li, Y. (2022). A sensitive amperometric sensor based on CuO and molecularly imprinted polymer composite for determination of danazol in human urine. *International Journal of Electrochemical Science.*, *17*(11), 221178. Available from https://doi.org/10.20964/2022.11.72.

Wang, L. Q., Lin, F. Y., & Yu, L. P. (2012). A molecularly imprinted photonic polymer sensor with high selectivity for tetracyclines analysis in food. *Analyst.*, *137*(15), 3502–3509. Available from https://doi.org/10.1039/c2an35460h, http://pubs.rsc.org/en/journals/journal/an.

Wang, M., Cetó, X., & del Valle, M. (2022). A novel electronic tongue using electropolymerized molecularly imprinted polymers for the simultaneous determination of active pharmaceutical ingredients. *Biosensors and Bioelectronics.*, *198*, 113807. Available from https://doi.org/10.1016/j.bios.2021.113807.

Wang, S., Xu, S., Zhou, Q., Liu, Z., & Xu, Z. (2023). State-of-the-art molecular imprinted colorimetric sensors and their on-site inspecting applications. *Journal of Separation Science*, *46*(12), e2201059. Available from https://doi.org/10.1002/jssc.202201059, http://onlinelibrary.wiley.com/journal/10.1002/(ISSN)1615-9314.

Wang, Y. Q., Fang, Z., Min, H., Xu, X. Y., & Li, Y. (2022). Sensitive determination of ofloxacin by molecularly imprinted polymers containing ionic liquid functionalized carbon quantum dots and europium ion. *ACS Applied Nano Materials.*, *5*(6), 8467–8474. Available from https://doi.org/10.1021/acsanm.2c01583, https://pubs.acs.org/journal/aanmf6.

Wu, L., Li, X., Miao, H., Xu, J., & Pan, G. (2022). State of the art in development of molecularly imprinted biosensors. *View.*, *3*(3), 20200170. Available from https://doi.org/10.1002/VIW.20200170.

Wu, M., Deng, H., Fan, Y., Hu, Y., Guo, Y., & Xie, L. (2018). Rapid colorimetric detection of cartap residues by AgNP sensor with magnetic molecularly imprinted microspheres as recognition elements. *Molecules (Basel, Switzerland)*, *23*(6), 1443. Available from https://doi.org/10.3390/molecules23061443.

Wu, X., Zhang, Z., Li, J., You, H., Li, Y., & Chen, L. (2015). Molecularly imprinted polymers-coated gold nanoclusters for fluorescent detection of bisphenol A. *Sensors and Actuators, B: Chemical*, *211*, 507–514. Available from https://doi.org/10.1016/j.snb.2015.01.115.

Xu, J., Haupt, K., & Tse Sum Bui, B. (2017). Core-shell molecularly imprinted polymer nanoparticles as synthetic antibodies in a sandwich fluoroimmunoassay for trypsin determination in human serum. *ACS Applied Materials and Interfaces.*, *9*(29), 24476–24483. Available from https://doi.org/10.1021/acsami.7b05844, http://pubs.acs.org/journal/aamick.

Xue, C., Han, Q., Wang, Y., Wu, J., Wen, T., Wang, R., Hong, J., Zhou, X., & Jiang, H. (2013). Amperometric detection of dopamine in human serum by electrochemical sensor based on gold nanoparticles doped molecularly imprinted polymers. *Biosensors and Bioelectronics.*, *49*, 199–203. Available from https://doi.org/10.1016/j.bios.2013.04.022.

Yan, M., & Ramström, O. (2004). *Molecularly imprinted materials: Science and technology. Molecularly imprinted materials: Science and technology* (pp. 1–734). United States: CRC Press. Available from http://www.tandfebooks.com/doi/book/10.1201/9781420030303, 10.1201/9781420030303.

Yang, H., Li, L., Ding, Y., Ye, D., Wang, Y., Cui, S., & Liao, L. (2017). Molecularly imprinted electrochemical sensor based on bioinspired Au microflowers for ultra-trace cholesterol assay. *Biosensors and Bioelectronics*, *92*, 748–754. Available from https://doi.org/10.1016/j.bios.2016.09.081, http://www.elsevier.com/locate/bios.

Yang, J. C., Cho, C. H., Choi, D. Y., Park, J. P., & Park, J. (2022). Microcontact surface imprinting of affinity peptide for electrochemical impedimetric detection of neutrophil gelatinase-associated lipocalin. *Sensors and Actuators B: Chemical.*, *364*, 131916. Available from https://doi.org/10.1016/j.snb.2022.131916, https://www.journals.elsevier.com/sensors-and-actuators-b-chemical.

Yang, Q., Li, J., Wang, X., Peng, H., Xiong, H., & Chen, L. (2018). Strategies of molecular imprinting-based fluorescence sensors for chemical and biological analysis. *Biosensors and Bioelectronics*, *112*, 54–71. Available from https://doi.org/10.1016/j.bios.2018.04.028, http://www.elsevier.com/locate/bios.

Yang, Y., Niu, H., & Zhang, H. (2016). Direct and highly selective drug optosensing in real, undiluted biological samples with quantum-dot-labeled hydrophilic molecularly imprinted polymer microparticles. *ACS Applied Materials and Interfaces*, *8*(24), 15741–15749. Available from https://doi.org/10.1021/acsami.6b04176, http://pubs.acs.org/journal/aamick.

Yılmaz, E., Özgür, E., Bereli, N., Türkmen, D., & Denizli, A. (2017). Plastic antibody based surface plasmon resonance nanosensors for selective atrazine detection. *Materials Science and Engineering: C.*, *73*, 603–610. Available from https://doi.org/10.1016/j.msec.2016.12.090.

Yola, M. L., Uzun, L., Özaltın, N., & Denizli, A. (2014). Development of molecular imprinted nanosensor for determination of tobramycin in pharmaceuticals and foods. *Talanta*, *120*, 318–324. Available from https://doi.org/10.1016/j.talanta.2013.10.064.

Yun, Y., Pan, M., Fang, G., Gu, Y., Wen, W., Xue, R., & Wang, S. (2018). An electrodeposited molecularly imprinted quartz crystal microbalance sensor sensitized with AuNPs and rGO material for highly selective and sensitive detection of amantadine. *RSC Advances.*, *8*(12), 6600–6607. Available from https://doi.org/10.1039/c7ra09958d, http://pubs.rsc.org/en/journals/journal/ra.

Zanoni, C., Rovida, R., Magnaghi, L. R., Biesuz, R., & Alberti, G. (2022). Voltammetric detection of irbesartan by molecularly imprinted polymer (MIP)-modified screen-printed electrodes. *Chemosensors.*, *10*(12), 517. Available from https://doi.org/10.3390/chemosensors10120517, http://www.mdpi.com/journal/chemosensors.

Zanoni, C., Spina, S., Magnaghi, L. R., Guembe-Garcia, M., Biesuz, R., & Alberti, G. (2022). Potentiometric MIP-modified screen-printed cell for phenoxy herbicides detection. *International Journal of Environmental Research and Public Health*, *19*(24), 16488. Available from https://doi.org/10.3390/ijerph192416488.

Zayats, M., Lahav, M., Kharitonov, A. B., & Willner, I. (2002). Imprinting of specific molecular recognition sites in inorganic and organic thin layer membranes associated with ion-sensitive field-effect transistors. *Tetrahedron*, *58*(4), 815–824. Available from https://doi.org/10.1016/S0040-4020(01)01112-7.

Zhang, H., Ye, L., & Mosbach, K. (2006). Non-covalent molecular imprinting with emphasis on its application in separation and drug development. *Journal of Molecular Recognition*, *19*(4), 248–259. Available from https://doi.org/10.1002/jmr.793.

Zhang, J., Chen, Q., Gao, X., Lin, Q., Suo, Z., Wu, D., Wu, X., & Chen, Q. (2023). A label-free and antibody-free molecularly imprinted polymer-based impedimetric sensor for NSCLC-cells-derived exosomes detection. *Biosensors*, *13*(6), 647. Available from https://doi.org/10.3390/bios13060647.

Zhang, Q., Jiang, D., Xu, C., Ge, Y., Liu, X., Wei, Q., Huang, L., Ren, X., Wang, C., & Wang, Y. (2020). Wearable electrochemical biosensor based on molecularly imprinted Ag nanowires for noninvasive monitoring lactate in human sweat. *Sensors and Actuators B: Chemical.*, *320*, 128325. Available from https://doi.org/10.1016/j.snb.2020.128325.

Zhang, Z., Ma, X., Li, B., Zhao, J., Qi, J., Hao, G., Jianhui, R., & Yang, X. (2020). Fluorescence detection of 2,4-dichlorophenoxyacetic acid by ratiometric fluorescence imaging on paper-based microfluidic chips. *Analyst.*, *145*(3), 963–974. Available from https://doi.org/10.1039/c9an01798d, http://pubs.rsc.org/en/journals/journal/an.

Zhao, H., Wang, H., Quan, X., & Tan, F. (2013). Amperometric sensor for tetracycline determination based on molecularly imprinted technique. *Procedia Environmental Sciences.*, *18*, 249–257. Available from https://doi.org/10.1016/j.proenv.2013.04.032.

Zhao, T., Wang, J., He, J., Deng, Q., & Wang, S. (2017). One-step post-imprint modification achieve dual-function of glycoprotein fluorescent sensor by "Click Chemistry.". *Biosensors and Bioelectronics*, *91*, 756–761. Available from https://doi.org/10.1016/j.bios.2017.01.046, http://www.elsevier.com/locate/bios.

Zhao, X., He, Y., Wang, Y., Wang, S., & Wang, J. (2020). Hollow molecularly imprinted polymer based quartz crystal microbalance sensor for rapid detection of methimazole in food samples. *Food Chemistry*, *309*, 125787. Available from https://doi.org/10.1016/j.foodchem.2019.125787.

Zheng, X., Pan, J., Gao, L., Wei, X., Dai, J. D., Shi, W. D., & Yan, Y. (2015). Silica nanoparticles doped with a europium(III) complex and coated with an ion imprinted polymer for rapid determination of copper(II). *Microchimica Acta*, *182*(3-4), 753–761. Available from https://doi.org/10.1007/s00604-014-1382-7, http://www.springer/at/mca.

Zolfaghari Asl, A., Rafati, A. A., & Khazalpour, S. (2023). Highly sensitive molecularly imprinted polymer-based electrochemical sensor for voltammetric determination of Adenine and Guanine in real samples using gold screen-printed electrode. *Journal of Molecular Liquids.*, *369*, 120942. Available from https://doi.org/10.1016/j.molliq.2022.120942.

CHAPTER

Colorimetric sensors

4

Raquel Vaz, M. Goreti F. Sales and Manuela F. Frasco

BioMark, CEMMPRE, ARISE, Department of Chemical Engineering, Faculty of Sciences and Technology, University of Coimbra, Coimbra, Portugal

4.1 Introduction

The importance of monitoring the environment, health, food, pharmaceuticals, and safety systems has led to the rapid development of biosensors. A biosensor is a device that converts the chemical or biological reaction that occurs when the analyte is detected into a measurable signal. A biosensor consists of a biorecognition element that is immobilized on the sensor substrate and has a selective affinity for the desired target; the transducer, which converts the binding between the biorecognition element and the analyte into a measurable signal; and the output system, which allows the signal to be amplified and displayed (Kawamura & Miyata, 2016). The biorecognition element is based on enzymes, nucleic acid probes, antibodies, cells, tissues, or biomimetic materials. Biosensors can be categorized according to the type of signal transduction, with electrochemical, optical, thermal, and piezoelectric being the most common (Asal et al., 2019).

Although the first biosensor dates back to 1962, biosensors only came onto the market more than twenty years later (Vaz et al., 2022). Optical transducers use light as a physical parameter that changes with the binding of the target analyte and has experienced significant growth in recent years. This is due to the fact that they provide fast, reliable, and sensitive signals and also allow easy expansion toward multiplex detection and miniaturized systems (Kawamura & Miyata, 2016). In particular, naturally occurring colors, which are used for mating, hunting, and fighting, have stimulated the curiosity of the research community and serve as bioinspiration for the development of new materials whose design principles originate from nature. Colorimetric sensors use or mimic natural colors to create simple sensing devices whose results can be easily visible to the naked eye. Colorimetric detection has been used for the detection of pollutants (e.g., heavy metal ions), biomolecules, and physical properties such as temperature and pressure, as this type of detection is not only cost-effective but also has high sensitivity (Wu et al., 2023). The improvement of the interface between receptor and transducer for higher sensitivity and selectivity, as well as the incorporation of nanomaterials, catalysts, and other nature-inspired materials, has only occurred in the last twenty years (Vaz et al., 2022).

Colorimetric strategies are generally characterized by the materials used to manufacture them. For example, colorimetric sensors can use catalysts or nanozyme-like materials to directly or indirectly catalyze chemical reactions with chromogenic substrates that act as labeling agents because

Sensory Polymers. DOI: https://doi.org/10.1016/B978-0-443-13394-7.00004-5

these reactions lead to color changes. Metal nanoparticles (NPs) are also used as a colorimetric transduction element or as a localized surface plasmon resonance (LSPR) amplification element. Indeed, colorimetric approaches have explored the unique optical properties of these NPs in the visible wavelength range, especially of gold (Au) and silver (Ag), as they can be easily synthesized in different sizes and shapes, as well as functionalized and thus modulating the final composition. Most of the developed colorimetric designs rely on the strong variation of the LSPR bands with interparticle distance, as the plasmon coupling that occurs between aggregated NPs causes a pronounced color change (Kaleeswaran et al., 2016). Graphene or other carbon-based materials enhance the sensitivity of the sensors due to high-density surface-active sites, and there are also organic dye-incorporated assays (Wu et al., 2023) (Fig. 4.1A).

Currently, the use of polymers in sensors is also of great interest, as polymers can be easily customized for specific applications. They exhibit a range of mechanical, thermal, and chemical resistances and are also capable of reversibly or irreversibly changing their properties in response to

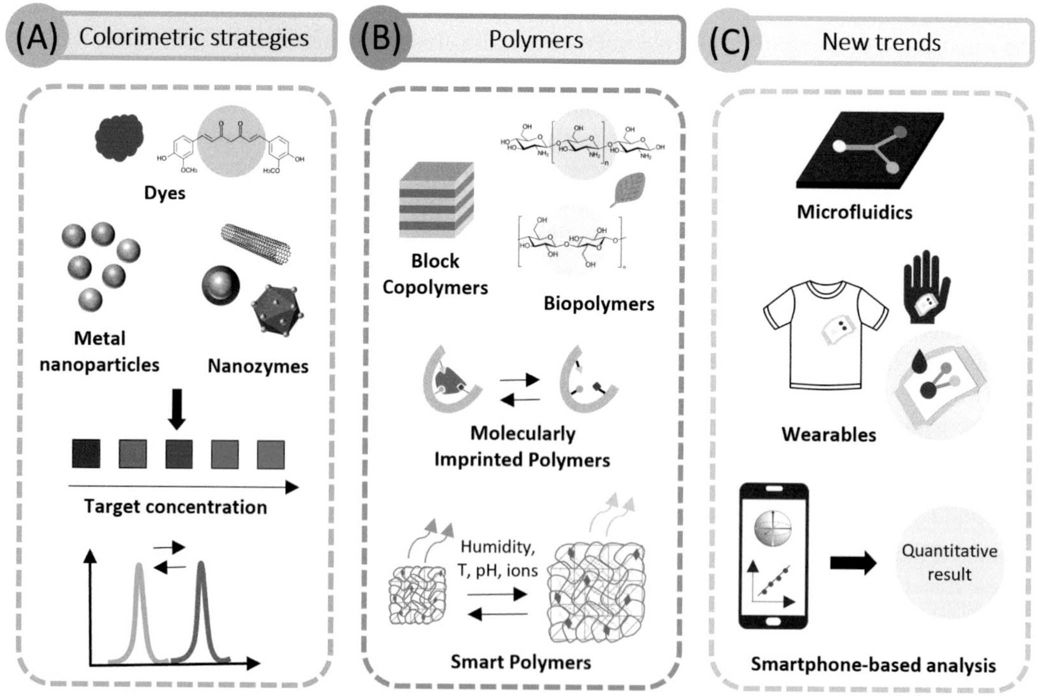

FIGURE 4.1

Schematic illustration of the progress of colorimetric polymer sensors, considering: (A) the colorimetric strategies characterized by the materials used to fabricate the sensors and ways to obtain the colorimetric response; (B) the incorporation of polymers for easy adaptability of the sensor for specific applications; and (C) recent trends for the improvement of colorimetric systems, taking into consideration their miniaturization, wearability, and sensitivity.

No Permission Required.

external inputs, such as electric or magnetic fields (Alberti et al., 2021) (Fig. 4.1B). In addition, solvent affinity and diffusion can be modulated to facilitate the colorimetric detection process (Urbano et al., 2021). The polymer-based materials used in sensors frequently improve the recognition step, act as supports for the immobilization of dyes, NPs, and others, or change their physical or chemical properties, enabling the transduction of the recognition reaction (Alberti et al., 2021). For example, the most successful synthetic biorecognition elements are molecularly imprinted polymers (MIPs), also known as plastic antibodies. These are produced by mixing monomers, crosslinkers, and the template molecule. The template is removed after polymerization, leading to a polymeric matrix with sites of similar size, shape, and functional groups complementary to the template molecule (Wang et al., 2023). MIPs have clear advantages over other biorecognition elements, such as antibodies and proteins, as they have higher thermal, chemical, and mechanical tolerance and are also a more cost-effective option (Wang et al., 2023).

The need to develop low-cost and miniaturized sensing systems with high accuracy and low power intake has recently led to efforts to develop microfluidic colorimetric devices that allow multiplexing in simple designs and easy interpretation of the measured data (Ma et al., 2023). In addition, with the introduction of smartphones as colorimeters, portability, accessibility to any user, high resolution and good accuracy of color analysis, and the possibility of integration with other functions, such as storing images and voice notes along with the measurement results, have been achieved (Ciaccheri et al., 2023) (Fig. 4.1C).

As there has been a rapid development of polymer-based colorimetric sensors in recent years, this book chapter presents the latest advances in these sensors. The sensors are characterized according to the integration of polymers with metal NPs, nanomaterials, including those with enzyme-like behavior, and dyes. Emphasis is also given to sensors based on structural color polymers and smart polymers. The discussion of colorimetric sensors shows application examples for the detection of physical properties, such as temperature and strain, and environmental or health-related analytes. New trends, including the miniaturization of systems into microfluidic colorimetric devices and smartphone-aided color visualization, are also explored. Finally, current polymer-based colorimetric sensors are discussed, and future perspectives are identified.

4.2 Colorimetric polymer sensors

4.2.1 Incorporation of metal nanoparticles

Metal NPs, particularly AuNPs and AgNPs, have unique optical properties that change with the surrounding dielectric medium and the interparticle distance, size, and shape. Metal NPs-based colorimetric methods have already been developed to detect ions and small molecules in environmental and health monitoring (Bamrungsap et al., 2019; Kaleeswaran et al., 2016; Liu et al., 2018; Oziri et al., 2022; Raja et al., 2020, 2023; Wikantyasning et al., 2023). Metal NPs are popular because their interaction with the analyte triggers aggregation or stabilization when their surface is functionalized, leading to color changes (Raja et al., 2020).

AuNPs range in size from 1.0 to 100 nm and are often used as molecular probes for bioactive compounds, such as drugs, toxic metals, and pesticides, among others (Raja et al., 2020). Bamrungsap et al. (2019) reported a visual colorimetric system based on Au nanorods and graphene oxide bonded with the polycationic polydiallyldimethylammonium chloride polymer, which prevents the absorption of Au nanorods into the graphene oxide. In this way, the Au is well dispersed, and the solution shows a red color. If polyanionic heparin is present, the polymer binds to it, and the Au nanorods self-assemble on the surface of graphene oxide, causing the particles to aggregate and change in color from red to purple (Bamrungsap et al., 2019). Oziri et al. (2022) developed AuNPs coated with cyclic poly(ethylene glycol), a polymer that interacts with bovine serum albumin (BSA). Thus, when BSA is present in the sample under analysis, it leads to aggregation of the NPs, changing their interparticle space and, consequently, the color of the solution (Oziri et al., 2022). Another study reported the use of AuNPs combined with the polymer poly(acrylic acid) conjugated to 3-aminophenyl boronic acid for the qualitative detection of bacteria. This was possible because the link between the boronic acid and the polyols of saccharides on the outer membranes of *Escherichia coli* and *Staphylococcus aureus* led to a decrease in the distance between the AuNPs and thus to a color change from red to blue (Wikantyasning et al., 2023) (Fig. 4.2).

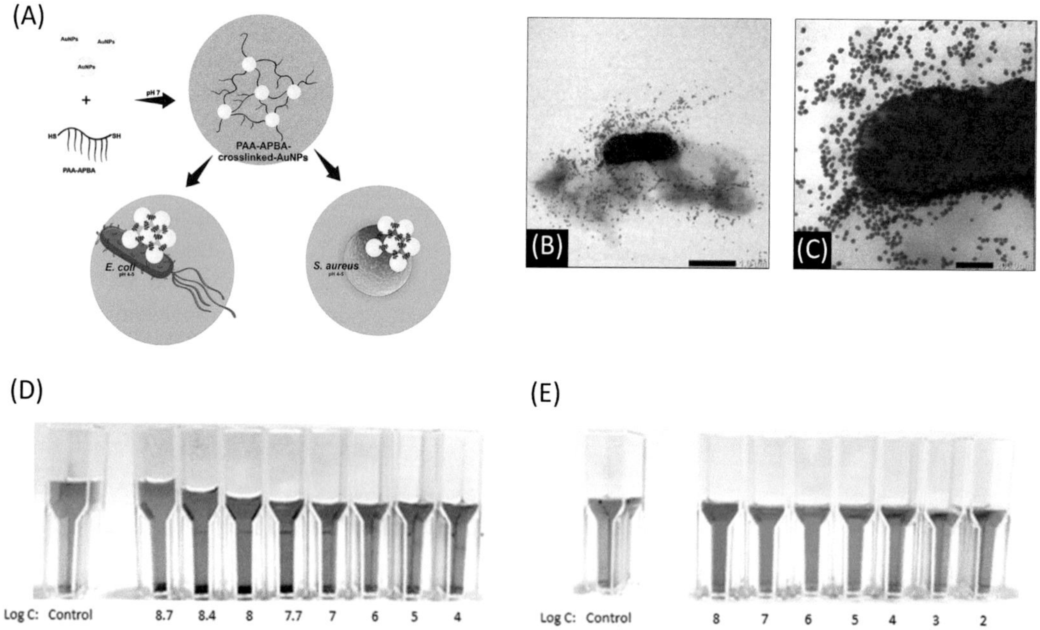

FIGURE 4.2

(A) Schematic illustration of the colorimetric polymer sensor based on AuNPs for bacteria detection. Scanning electron microscopy (SEM) of the aggregation of AuNPs around *E. coli* at a (B) magnification of 5000 ×, and (C) magnification of 20,000 ×. Colorimetric recognition at different concentrations of (D) *E. coli* and (E) *S. aureus*.

Reproduced under the terms and conditions of the Creative Commons Attribution (CC BY-NC-SA 4.0) License. Wikantyasning, E. R., Da'i, M., Cholisoh, Z., & Kalsum, U. (2023). 3-Aminophenylboronic acid conjugation on responsive polymer and gold nanoparticles for qualitative bacterial detection. Journal of Pharmacy and Bioallied Sciences, 15*(2), 81–87. Available from https://doi.org/10.4103/jpbs.jpbs_646_22. Copyright 2023, Medknow Publications.*

Regarding AgNPs, they have advantages over AuNPs, such as lower preparation costs (Kaleeswaran et al., 2016). Recently, AgNPs capped with polypropylene glycol were produced for colorimetric sensing of the carbaryl insecticide. The chosen polymer allows stabilization of the NPs, biocompatibility, and specific binding to this insecticide. The biosensor changed color from yellow to brown in the presence of carbaryl (Raja et al., 2023).

4.2.2 Incorporation of nanozymes

The use of natural catalysts for bioanalytical purposes has been shown to be very efficient, such as the detection of glucose with glucose oxidase (GOx) or horseradish peroxidase (HRP) to amplify the signals from optical assays used for analytes as diverse as proteins, nucleic acids, and polysaccharides (Cardoso et al., 2021). However, natural enzymes undergo easy denaturation, and their activity is limited by external properties such as pH and temperature. Therefore, in recent years, research has been conducted on artificial enzymes based on nanomaterials, also known as nanozymes. Nanozymes are nanomaterials with enzyme-like activities that are highly stable and inexpensive (Cardoso et al., 2021). The nanomaterials used for the development of nanozymes are usually metals, metal oxides, and carbon-based materials, and their activities are catalase-, peroxidase-, and oxidase-like activities (Cardoso et al., 2021). Several examples of applications for the detection of health-related biomarkers (Bhaduri et al., 2023; Kang et al., 2023; Sengupta et al., 2020; Wang et al., 2019; Yang et al., 2021; Yin et al., 2019; Zhang et al., 2022; Zhang et al., 2020; Zhao et al., 2019), ions (Akhond et al., 2020; Ko et al., 2021; Nguyen et al., 2022; Preman et al., 2023), toxins and pollutants (Adegoke & Daeid, 2021; Amirzehni et al., 2020; Housaindokht et al., 2022; Komal et al., 2023), and other targets (Guo et al., 2020; Nsuamani et al., 2022; Wang & Kan, 2020; Zhou et al., 2023) can already be found in the literature.

Within metal and metal oxide nanomaterials, artificial enzyme mimics generally include AuNPs, platinum nanomaterials, iron oxide (Fe_3O_4) NPs, and cerium (IV) oxide (CeO_2) NPs (Cardoso et al., 2021), but other metal-based materials can also be used. A cost-effective approach was developed for the specific recognition of metallothioneins, which are intracellular cysteine-rich metal-binding proteins and thus can be used as a biomarker for heavy metal poisoning. This work was based on the catalytic activity of polymer-caged AuNPs for sensitive colorimetric analysis. The polymer used was a polyhedral oligomeric silsesquioxane, whose cubic geometry led to a higher area per volume and exposed functional groups. Overall, the nanozyme was responsible for the reduction of the yellow 4-nitrophenol to a colorless product. The presence and binding of metallothioneins to the nanozyme inhibited its catalytic activity, thereby maintaining the yellow color (Zhang et al., 2020) (Fig. 4.3). A similarly simple strategy was also based on AuNPs for the detection of cancer cells with the naked eye. The AuNPs were coated with polydopamine and functionalized with an antinucleolin AS1411 probe to specifically detect breast cancer cells overexpressing the nucleolin protein on their surface. The nanozyme can catalyze the reduction of the blue-colored methylene blue to a colorless product. When cancer cells are present, this reaction is inhibited (Yang et al., 2021). In another work, a mimic of peroxidase activity was developed using an iron-doped ceria and Au hybrid coated with a polymer. Then, the hybrid nanozyme was functionalized for selective recognition of the explosive 2,4,6-trinitrotoluene. When the explosive target is present, it tunes the nanozyme-based oxidation of 3,3′,5,5′-tetramethylbenzidine (TMB) by hydrogen peroxide (H_2O_2), leading to the appearance of a green product (Adegoke & Daeid, 2021). Efficient

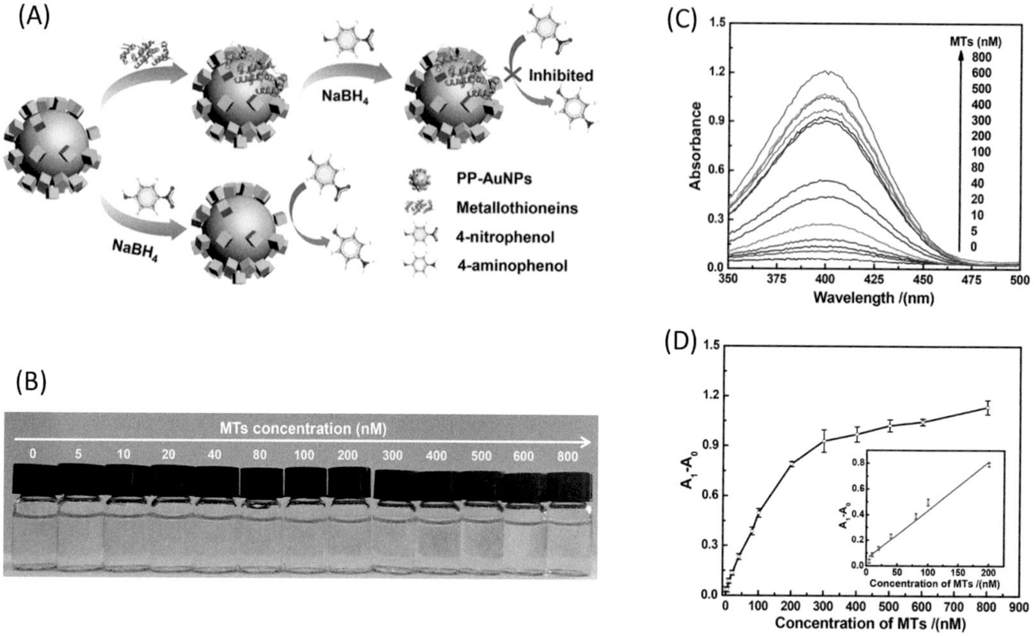

FIGURE 4.3

(A) Colorimetric principle of the nanozyme polymer-caged AuNPs for specific recognition of metallothioneins. The nanozyme was responsible for the reduction of the yellow 4-nitrophenol to a colorless product. (B) The presence and binding of metallothioneins to the nanozyme inhibited its catalytic activity, thereby maintaining the yellow color. (C) UV – vis absorption spectra at increasing concentrations of metallothioneins, and (D) calibration curve.

Reproduced with permission from Zhang, Y., Hao, J., Xu, X., Chen, X., & Wang, J. (2020). Protein corona-triggered catalytic inhibition of insufficient POSS polymer-caged gold nanoparticles for sensitive colorimetric detection of metallothioneins. Analytical Chemistry, 92*(2), 2080–2087. Available from https://doi.org/10.1021/acs.analchem.9b04593. Copyright 2019, American Chemical Society.*

peroxidase-mimicking catalysts have been successfully developed using different structures, such as a bimetallic structure of cobalt and zinc metal-organic framework (MOF) coated with a MIP layer for the selective detection of the pesticide dimethoate (Amirzehni et al., 2020) or a nanozyme constituted by iron-incorporated porphyrin-based conjugated polymer for the detection of glucose in a sensing system containing also GOx (Bhaduri et al., 2023).

However, other materials, namely covalent organic frameworks (COFs), can also be used for the efficient construction of nanozymes. This is the case of rhodium NPs carried on COFs, which are highly-ordered porous organic polymers. The composite formed showed peroxidase-like activity. Cysteine, an important amino acid in the human body, interacts with rhodamine through its thiol group and inhibits the peroxidase activity. In the presence of TMB, it is possible to visually follow the presence of cysteines in the tested sample. In addition to its inhibiting properties of peroxidase activity, cysteine is also able to reduce the oxidized TMB substrate. Considering these two mechanisms, it was possible to develop a sensitive colorimetric method for the detection of cysteine (Zhang et al., 2023) (Fig. 4.4). In a different sensing strategy, a competitive reaction was the detection principle to obtain a colorimetric signal in a nanozyme-based lateral flow assay for the detection of aflatoxin B1. For this purpose, copper-cobalt NPs were developed and coated with the polymer polydopamine (CuCo@PDA), resulting in a nanozyme with peroxidase-like activity and a dark color, enriched with functional groups for further functionalization with the aptamer biorecognition probes. In the absence of aflatoxin B1, the aptamer probe can hybridize with complementary DNA immobilized on the control and test lines of the strip, resulting in a visually dark color. When aflatoxin B1 is present, the probe first binds the target with high affinity, and the hybridization is inhibited. The colorimetric signal was amplified by introducing the chromogenic substrate TMB

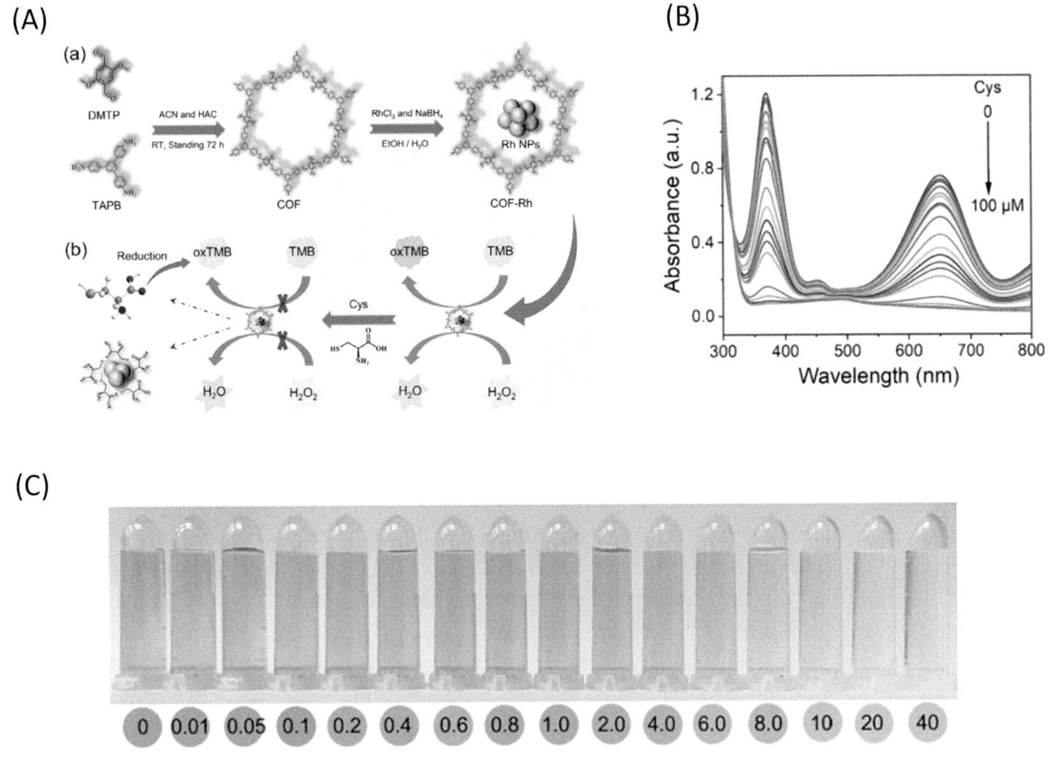

FIGURE 4.4

(A) Schematic illustration of the (a) synthesis of the COF with rhodium NPs (COF-Rh), and (b) strategy for cysteine detection based on inhibition of the nanozyme peroxidase-like activity and the reduction of the oxidized TMB substrate. Colorimetric detection of cysteine accompanied by (B) UV–vis absorption spectra of the solution, and (C) naked eye detection.

Reproduced with permission from Zhang, L., Zhang, W., Nie, Y., Wang, Y., & Zhang, P. (2023). Covalent organic framework-supported ultrasmall Rh nanoparticles as peroxidase mimics for colorimetric sensing of cysteine. Journal of Colloid and Interface Science, 636, *568–576. Available from https://doi.org/10.1016/j.jcis.2023.01.020. Copyright 2023, Elsevier Inc.*

since its oxidation by the CuCo@PDA nanozyme with peroxidase-like activity generated a blue color, which decreased with increasing concentrations of the target aflatoxin B1 (Zhu et al., 2024).

Graphene oxide, carbon nanotubes, fullerene derivates, and other carbon-based materials are also known for their catalytic activity (Cardoso et al., 2021). A metal-free nanozyme hybrid of chemically modified graphitic carbon nitride, chitin, and acetic acid showed oxidase-mimicking activity to detect glucose, also exhibiting peroxidase activity. In the presence of glucose, the graphitic carbon nitride of the bifunctional nanozyme hybrid oxidizes it to gluconic acid and H_2O_2, and then the chitin-acetic acid mimicking peroxidase oxidizes the TMB substrate in the presence of the H_2O_2. Thus, the observed color changes enabled the detection of peroxide and glucose (Sengupta et al., 2020). The combination of MIPs with these materials has been demonstrated by a study where a MIP incorporated into graphitic carbon nitride with peroxide-like activity was developed for sensing the pollutant bisphenol A. In this study, the colorless TMB turns into a blue product when oxidized, but the introduction of the analyte decreases the catalytic ability of the nanozyme. Thus, the increasing concentration of bisphenol A was followed by a disappearance of the blue color (Komal et al., 2023).

Nanozymes have been recently produced with other materials, such as $CsPbX_3$ perovskite nanocrystals, due to their optoelectrical properties, namely, wide visible light coverage, high fluorescence quantum yield, and biocatalytic properties (Ye et al., 2023). In this study, octylamine-modified polyacrylic acid-capped $CsPbBr_3$ nanocrystals were prepared, which are not only stable in an aqueous solution but also exhibit oxidase-like and ascorbate oxidase-like activities. MOFs were also combined with the nanocrystals to enhance their optical properties. The nanozyme was able to generate oxide-free radicals from oxygen *via* the oxidase-like pathway and oxidize ascorbic acid to produce H_2O_2. The addition of Prussian blue with peroxidase-like properties made it possible to generate a colored product when TMB was added. The colorimetric system was then constructed on a portable paper-based device. In this way, the authors successfully obtained a colorimetric point-of-care device for the detection of ascorbic acid (Ye et al., 2023). The hydrolase activities of COF nanozymes have also been explored, including the possibility of photo-enhanced catalytic efficiency. The degradation of hydrolase substrates, such as *p*-nitrophenyl acetate, under light irradiation resulted in color revelation. In addition, organophosphorus nerve agents are known inhibitors of hydrolase activity. Therefore, their presence led to a change in the visible color that depended on the concentration of nerve agents and enabled the analysis with the naked eye (Xiao et al., 2023).

As mentioned, the possibility of combining MIPs with nanozymes for selective and sensitive colorimetric sensors is also interesting (Amirzehni et al., 2020; Chen, Chen, et al., 2023; Guo et al., 2020; Wang & Kan, 2020; Wang et al., 2019; Xu et al., 2023; Zhang et al., 2022; Zhou et al., 2023). MIPs are considered plastic antibodies in the sense that they form cavities within a polymer matrix that specifically recognize the target. Molecular imprinting of high-molecular-weight proteins is challenging, but it has already proven successful with certain techniques. To detect the protein thyroglobulin, a dimeric protein of 660 kDa and a biomarker for thyroid carcinoma, hemin-graphene nanosheets with peroxidase-like activity were developed and placed on filter paper. An acrylamide-based MIP hydrogel film was then produced using thyroglobulin as a template and deposited on the surface of the nanozyme. The working principle of the colorimetric polymer sensor was based on the catalysis of the oxidation of TMB by the nanozyme, resulting in a blue color. When thyroglobulin was detected by the MIP layer, the catalytic capacity of the nanozyme decreased. Therefore, there was a color fading with increasing concentrations of the target (Wang et al., 2019). The same blocking effect of peroxidase-like activity in the presence of the target molecule was adopted in a different study for the detection of aloe-emodin. Aloe-emodin is a compound of Aloe vera used in extracts due to its antiinflammatory and anti-oxidative effects, but its dose is important in order not to have the opposite effect, that is, cytotoxicity. In this work, the nanozyme was composed of Fe_3O_4 microparticles synthesized directly on a multilayer exfoliated graphite paper, and the MIP layer was prepared from pyrrole as a monomer and aloe-emodin as the template. Due to the peroxidase-like activity of the Fe_3O_4 microparticles, TMB could be oxidized to deep blue by the prepared sensor in the presence of H_2O_2. When the imprinted cavities of the MIP were bound with the target molecule, the nanozyme activity was blocked, resulting in a light blue color. In addition, the sensor could also be used as an electrochemical interface due to the electrical properties of pyrrole (Wang & Kan, 2020). Another interesting application of nanozyme-based colorimetric assays was presented for the detection of the allergen ovalbumin in vaccines. First, magnetic silicon dioxide MIP particles were prepared, and then different concentrations of ovalbumin were added to the system. The separation between linked ovalbumin in the MIP particles and free ovalbumin was achieved by magnetic separation. Subsequently, Ni-Fe-MOF nanozymes with peroxidase activity were added. Since the nanozymes were only bound to the particles through the ovalbumin, a colored product was only generated in its presence, enabling a sandwich assay, and by adding TMB and H_2O_2. Therefore, the high sensitivity of this strategy depended on the sandwich structure of MIP NPs, ovalbumin, and nanozymes, and the intensity of the color increased with increasing concentrations of ovalbumin (Xu et al., 2023) (Fig. 4.5).

4.2.3 Incorporation of DNA nanostructures

DNA is a highly stable and programmable biomolecule that is often used as a probe in sensing devices. The mechanisms of DNA nanomaterials on colorimetric sensors are based on the affinity of DNA with the target and the subsequent alteration of its structure (Wu et al., 2023). Typically, G-quadruplexes formed by metal ions, ion-bridged complexes, and DNA nanozymes are reported for colorimetric purposes (Bhadra et al., 2014; Kan et al., 2014; Wang et al., 2017). However, to our knowledge, there are only a few examples in the literature that also include polymeric structures (Gu et al., 2018; Sha et al., 2021; Wang et al., 2010).

One colorimetric strategy for pH sensing was developed based on DNA i-motif and water-soluble conjugated polymers. The DNA i-motif changed its conformation with pH changes, and the polymer sensed this structural variation, resulting in a change in its optical properties. The water-soluble conjugated polymer chosen was cationic polythiophene. At pH below 6.5, the cytosines of the DNA were protonated, and it adopted a closely packed four-stranded i-motif structure. The polymer interacted with it and formed a complex with a yellow color. However, at a pH above 6.5, the DNA appeared in a random-coiled state and formed a different conjugate with the polymer, which was red in color (Wang et al., 2010). In another work, a DNA probe with two catalytic functions, namely a Cu^{2+}-mediated DNAzyme, that is, a DNA Cu^{2+}-induced self-cleaving domain, and a G-rich sequence, was used for the detection of L-histidine. The core of this assay mechanism was the step-by-step activation of the catalytic domains, which enabled the correlation of the colorimetric signal with the quantity of L-histidine present. The colorimetric signal was generated from the H_2O_2-mediated oxidation of 2,2′-azino-bis(3-ethylbenzothiazoline-6-sulfonic acid) diammonium salt, catalyzed by the G-quadruplex–hemin complex. Moreover, the G-quadruplex formation only occurs in the presence of the metal ion, as it induces the self-cleaving domain. However, when

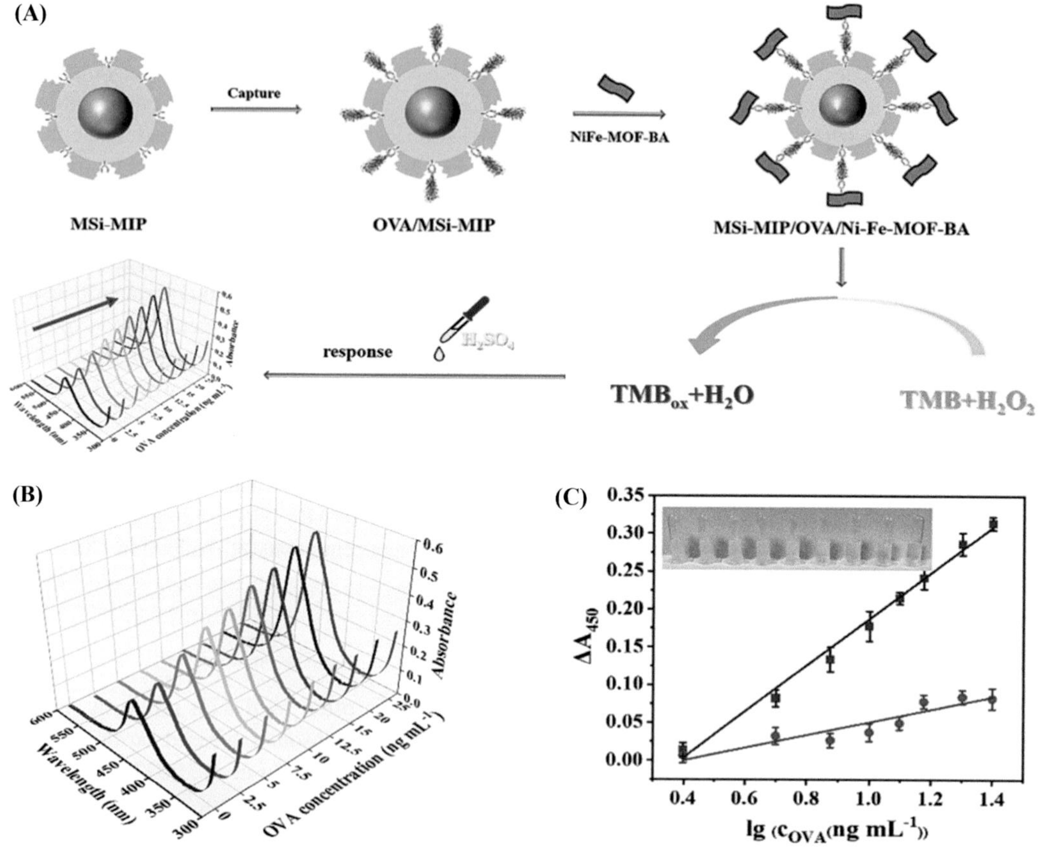

FIGURE 4.5

(A) Scheme of the colorimetric detection of ovalbumin based on a sandwich assay between MIP particles and Ni-Fe-MOF nanozymes with peroxidase-like activity. (B) UV-vis absorption spectra with increasing concentrations of ovalbumin. (C) Calibration curve of the MIP (*black* line) and the non-imprinted control (*red* line), as well as photographs of the colorimetric response to increasing concentrations of ovalbumin.

Reproduced with permission from Xu, Z., Jin, D., Lu, X., Zhang, Y., & Liu, W. (2023). Sandwich-type colorimetric assay based on molecularly imprinted polymers and boronic acid functionalized Ni-Fe-MOF nanozyme for sensitive detection of allergen ovalbumin. Microchemical Journal, 194, *109349. Available from https://doi.org/10.1016/j.microc.2023.109349. Copyright 2023, Elsevier B. V.*

L-histidine is present, it binds to copper ions and inhibits G-quadruplex formation and color development (Gu et al., 2018). The design of novel MOF colloid nanorods by grafting carcinoembryonic antigen aptamer-incorporated DNA tetrahedral nanostructures resulted in a sensitive colorimetric detection of this tumor biomarker (Sha et al., 2021). In the MOF material, the iron porphyrin ring exhibited strong HRP mimicking activity, while the tetravalent zirconium ions on the surface of the material were used as ligands to establish coordination assembly with the phosphorylated DNA tetrahedral structure. Then, the colorimetric immunosensor was based on a functionalized surface with an antibody against the carcinoembryonic antigen. Once the target molecule was recognized by the antibody, the DNA nanostructures with the MOF mimicking peroxidase catalytic activity were also able to detect it efficiently. Upon addition of the substrate TMB, the nanozyme oxidized it, resulting in a color change (Sha et al., 2021) (Fig. 4.6).

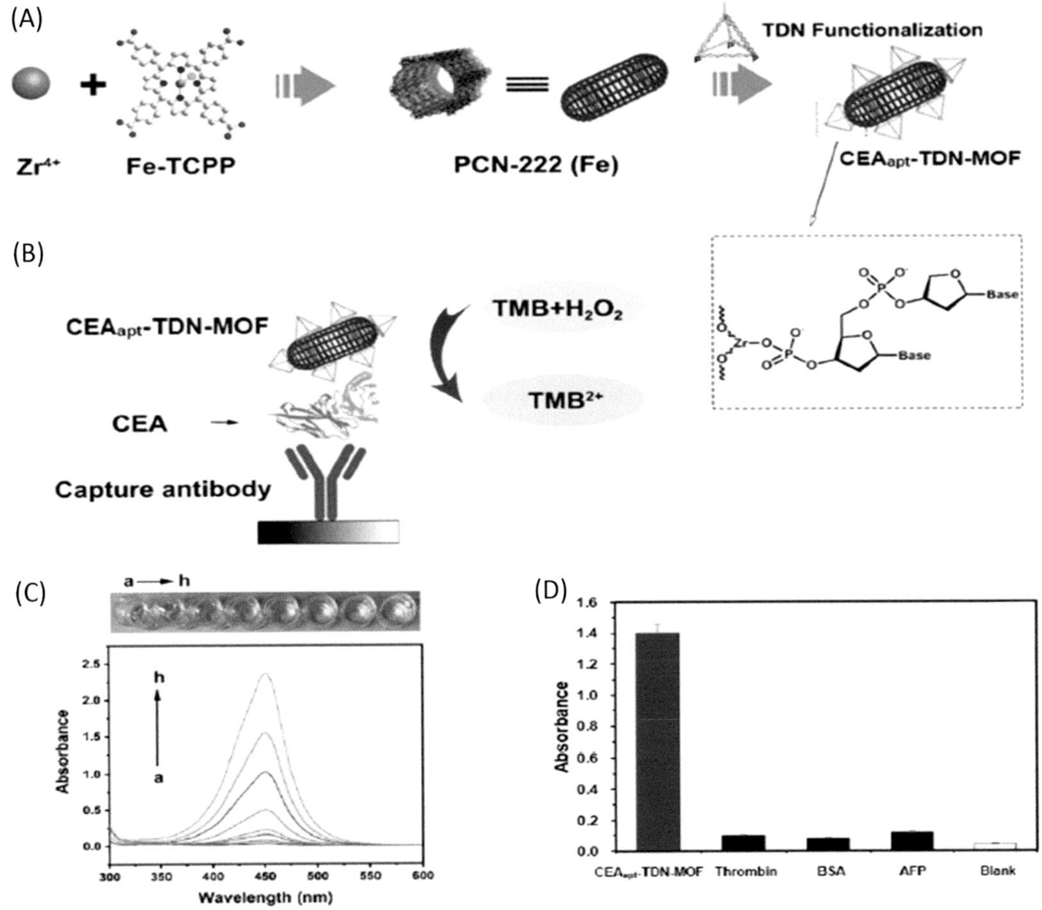

FIGURE 4.6

(A) Diagram of the preparation process of the DNA tetrahedron nanostructure functionalized with MOF colloidal nanorods, and (B) its combination with a surface functionalized with an antibody against the carcinoembryonic antigen. The detection system showed (C) a color response to increasing concentrations of the antigen after stopping the reaction with sulfuric acid, and (D) specificity when other proteins were analyzed.

4.2.4 Incorporation of natural and synthetic dyes

Colorimetric sensors based on a pigment or dye immobilized on a polymer matrix have broad applications. Among the natural dyes most commonly used in the development of these films, polyphenols are the most important due to their known ability to change color with pH or through complexation with ions (Ahmad et al., 2019; Chumee et al., 2022; Matthew et al., 2022; Nascimento et al., 2023; Tavassoli et al., 2023). For example, for the quantification of ions in beverages and environmental samples, a film of starch and poly(vinyl alcohol) (PVA) containing quercetin was developed. When copper ions were present, complexation with quercetin occurred, resulting in a color change of this natural dye (Nascimento et al., 2023). A common strategy is also the incorporation of anthocyanins, a broad class of polyphenols, as pH indicators to monitor the freshness of food. Since a biocompatible sensor system is required for use with food, the colorimetric assay based on these natural pH indicators also includes natural polymers. For example, a colorimetric biofilm was prepared from natural polymers, namely agar, sodium alginate, and PVA, incorporating anthocyanin extracts from red cabbage as a pH indicator. This pH sensor was used to monitor the freshness of pork, as the pH changes with the formation of ammonia during decomposition (Chumee et al., 2022). Furthermore, a pH-sensitive smart packaging film was fabricated using carboxymethyl cellulose as the polymer matrix, anthocyanins as the pH indicator, and graphene oxide to enhance the colorimetric signal. In addition to the excellent biodegradability of the film, it enabled the visual detection of ammonia formation (Zhao et al., 2022). Other natural polymers with colorimetric properties can also be efficient for application in smart packaging, as shown by the work developed using the soluble soybean polysaccharide instead of carboxymethyl cellulose. A pH-responsive film was obtained and tested to monitor in real-time the freshness/spoilage of salmon fish fillets. The obtained results demonstrated distinguishable color variations of the film in response to changes in the pH values (Kafashan et al., 2023) (Fig. 4.7).

Nevertheless, there are currently few studies on the use of natural dyes for the fabrication of colorimetric polymer sensors. Most studies rely on synthetic pigments (Fabregat et al., 2017; Mishra et al., 2024; Park et al., 2020; Sutthasupa et al., 2021; Wu & Selvaganapathy, 2023; Xie et al., 2014). For instance, poly(styrene-*co*-maleic anhydride)/polyethyleneimine core-shell NPs were produced and then impregnated with methyl red, bromocresol purple, or 4-nitrophenol for the detection of formaldehyde. In the presence of this gas, the dyes changed color to yellow, red, and gray, respectively, demonstrating the versatility of pigments that can be used for the same purpose (Park et al., 2020). Another example is the alginate-methyl cellulose hydrogel that was fabricated for pH sensing by introducing the dye bromothymol blue into the hydrogel and thus can be used as a spoilage indicator in smart packaging, particularly tested in this study to monitor minced pork spoilage under different storage conditions (Sutthasupa et al., 2021). This type of approach in colorimetric sensors can also be adapted for ion detection. For example, a polyvinylpyrrolidone-capped rhodamine 6G dye probe was developed to respond to different ions. In the presence of sulfide ions, it changed from orange to pale yellow, while in the presence of bicarbonate ions, it changed from orange to pink (Mishra et al., 2024). Another example was the development of a paper-based device aiming for a fast, low-cost, and portable analysis of cadmium ions, which are environmental pollutants of concern. To this end, a paper substrate was modified so that the developed ion-imprinted polymer could be produced on its surface using a reversible addition-fragmentation chain transfer polymerization strategy. The imprinted polymer was based on polyethyleneimine and methacrylic acid monomers, and cadmium

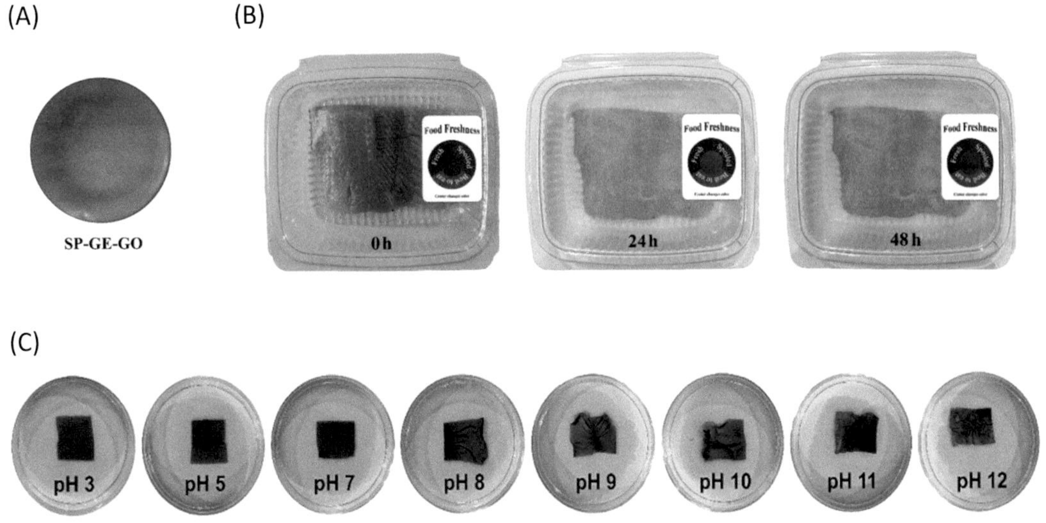

FIGURE 4.7

(A) Photograph of the colorimetric polymer film based on soybean polysaccharide (SP), grape skin anthocyanin extract (GE), and graphene oxide (GO) for colorimetric detection in smart packaging. (B) Real-time freshness and spoilage monitoring through differences in color due to pH changes. (C) Color variations of the film at various pH values.

Reproduced with permission from Kafashan, A., Joze-Majidi, H., Kazemi-Pasarvi, S., Babaei, A., & Jafari, S. M. (2023). Nanocomposites of soluble soybean polysaccharides with grape skin anthocyanins and graphene oxide as an efficient halochromic smart packaging. Sustainable Materials and Technologies, 38, *e00755. Available from https://doi.org/10.1016/j.susmat.2023.e00755. Copyright 2023, Elsevier B. V.*

ions were used as the template. The detection was based on a pink-colored complex formed between dithizone and cadmium ions (Huang et al., 2017) (Fig. 4.8).

An interesting technique used in colorimetric sensors is dye displacement. One strategy is to use MIPs, where the binding of the target molecule to the specific cavity leads to the dissociation of the previous link between the polymer and the dye. This technique has already proven successful in the identification of different aromatic amines (Greene & Shimizu, 2005) and antibiotics (Lowdon et al., 2021). However, MIPs are not obligatory but only a reversible link between the dye and the polymer. For example, the addition of the competing analyte to displace the dye alizarin red S from the boronic acid-rich benzoxaborole hydrogel proved to be effective for the detection of saccharides (Lampard et al., 2018).

4.3 Structural polymer-based colorimetric sensors

4.3.1 Block copolymers

There are three different forms of coloration in nature based on absorbing pigments, structural coloration, and bioluminescence. As the name suggests, pigment coloration comes from pigments that

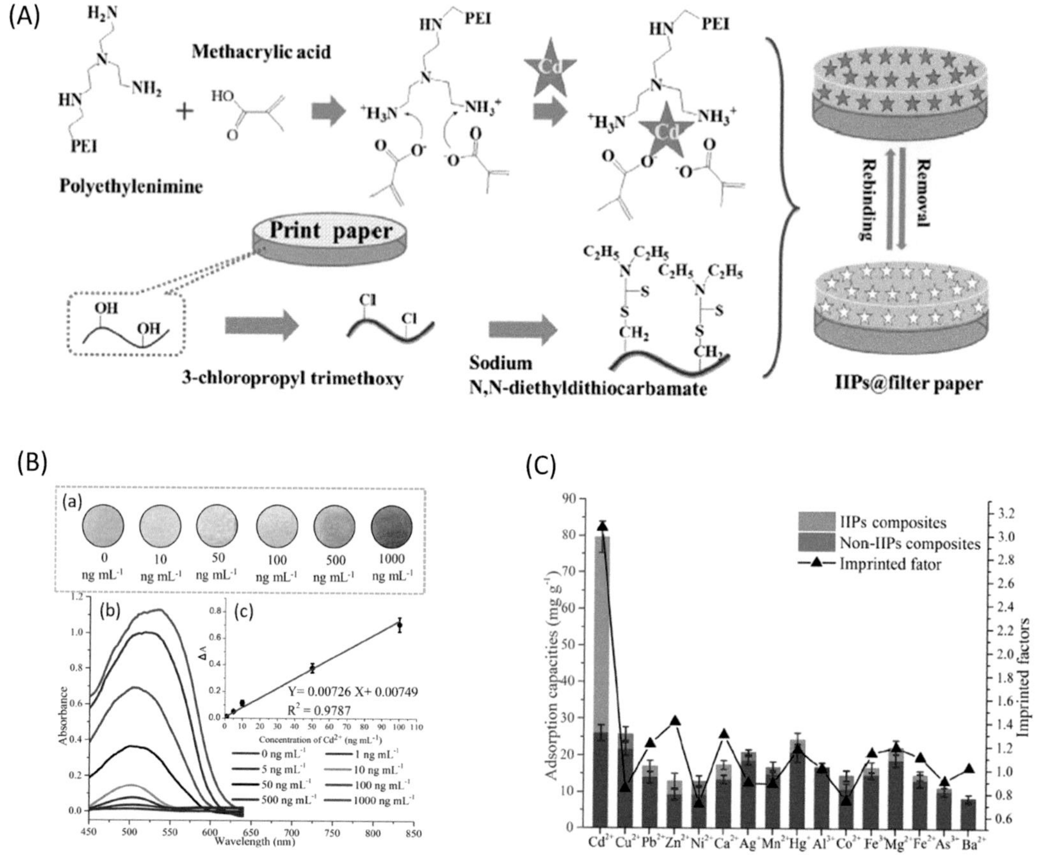

FIGURE 4.8

(A) Preparation procedure of the portable paper-based colorimetric device for detection of cadmium ions using an ion imprinted polymer (IIP) (IIPs@filter paper). (B) Colorimetric response of the sensor based on (a) naked eye detection, (b) absorption curves and (c) calibration curve. (C) Binding capacity of the IIP-based sensor and its non-imprinted control, and specificity to cadmium ions.

Reproduced with permission from Huang, K., Chen, Y., Zhou, F., Zhao, X., Liu, J., Mei, S., Zhou, Y., & Jing, T. (2017). Integrated ion imprinted polymers-paper composites for selective and sensitive detection of Cd(II) ions. Journal of Hazardous Materials, 333, *137–143. Available from https://doi.org/10.1016/j.jhazmat.2017.03.035. Copyright 2017, Elsevier B. V.*

absorb specific wavelengths and reflect others, and it is usually angle-independent (Sun et al., 2013). In contrast, bioluminescence involves the production of light by an exergonic reaction of oxygen with certain substrates and enzymes (Chatterjee, 2022). Structural coloration originates from different mechanisms, such as thin-film interference, diffraction grating, or photonic crystals (PCs) (Vaz et al., 2020). PCs consist of arrays with different refractive indexes and a periodic spatial arrangement in one, two, or three dimensions. Their color is often angle-dependent and derives from the periodic modulation of the refractive index, which prevents certain wavelengths from propagating through the PC. If the reflected light is in the visible range, the PC has an iridescent color (Vaz et al., 2020).

Block copolymers can be precisely tuned to control their domain spacing and refractive index contrast, so a variety of photonic nanostructures, especially multilayer films and particles, have been obtained from self-assembled block copolymers (Liu et al., 2020; Wang et al., 2022). The ordered morphologies (e.g., lamellar, spherical, gyroid, and cylindrical) resulting from phase separation between different blocks can reach hundreds of nanometers and offer exciting possibilities for reflective optics in the visible spectral range (Lin et al., 2018; Sveinbjörnsson et al., 2012). These periodic structures can be modulated by various factors, including the molecular weight and molecular structure of the polymers, as well as additives such as small molecules and inorganic NPs, which influence the final optical properties of the materials (Liu et al., 2020; Wang et al., 2022).

Among the linear block copolymers most commonly used to produce photonic films, polystyrene-*b*-poly(2-vinylpyridine) (PS-*b*-P2VP) dominates, forming well-defined lamellae by easily controlling the block ratio to achieve a planar interface. However, in addition to the intrinsic parameters of domain spacing and average refractive index, there are also extrinsic factors such as the casting of annealing steps that influence the properties of the lamellae and determine the wavelength of the photonic multilayer (Wang et al., 2022). The possibility of observing color changes due to domain swelling has been explored for sensing purposes. The PS-*b*-P2VP thin films were prepared as 1D photonic lamellar gels for sensitive detection of pH and the type of acid based on the protonation of the pyridyl groups in the P2VP domain (Kim et al., 2021; Noro et al., 2016). For example, when a protic ionic liquid was added to the PS-*b*-P2VP film, the formation of hydrogen bonds between the pyridyl groups in the P2VP block and the hydrogen groups of the protic ionic liquid led to swelling of the P2VP phase, and the film exhibits photonic properties in the visible light spectrum (Noro et al., 2016). In the presence of increasing concentrations of a disulfonic acid, whose molecules protonate the pyridyl groups, the degree of swelling is controlled and can be accompanied by a color change from blue to red (Noro et al., 2016) (Fig. 4.9). In addition, the versatility of soft PS-*b*-P2VP photonic films for the detection of solvents was demonstrated (Lim et al., 2012). In this study, color tunability was achieved by a sequential process of self-assembly of the PS-*b*-P2VP block copolymer, which comprises a hydrophobic block and a hydrophilic polyelectrolyte block, followed by quaternization of the P2VP layers with 1-bromoethane (Lim et al., 2012). The sensing ability was based on a direct exchange of counterions, leading to selective swelling of the hydrophilic quaternized P2VP block. The increase in both the thickness and refractive index of this domain led to a color change in the photonic gel films. Moreover, since the swelling depends on the hydration characteristics of the counterions, selectivity was achieved as observed by different shifts in the wavelength of the reflected light (Lim et al., 2012). Another interesting application is the use of quaternized PS-*b*-P2VP films for the detection of environmentally harmful surfactants, namely sodium dodecyl sulfate (SDS) and sodium methyl sulfate (SMS) (Lin et al., 2023). In the case of SDS, two types of behavior were observed. At lower concentrations (<1 mM), the mechanism of counter-anionic exchange with the quaternized P2VP chains occurs, leading to a blue shift in the observed color. At higher concentrations (>1 mM), a red shift in the reflectance originated from the increased periodicity caused by the swelling of the quaternized P2VP microdomain. Regarding SMS, only red shifts in reflectance were observed with increasing concentrations of this surfactant due to the weaker hydrophobicity (Lin et al., 2023).

In the case of biomolecules, an example is the detection of fructose using PS-*b*-P2VP films functionalized with boronic acid moieties that bind reversibly to sugars. The increasing

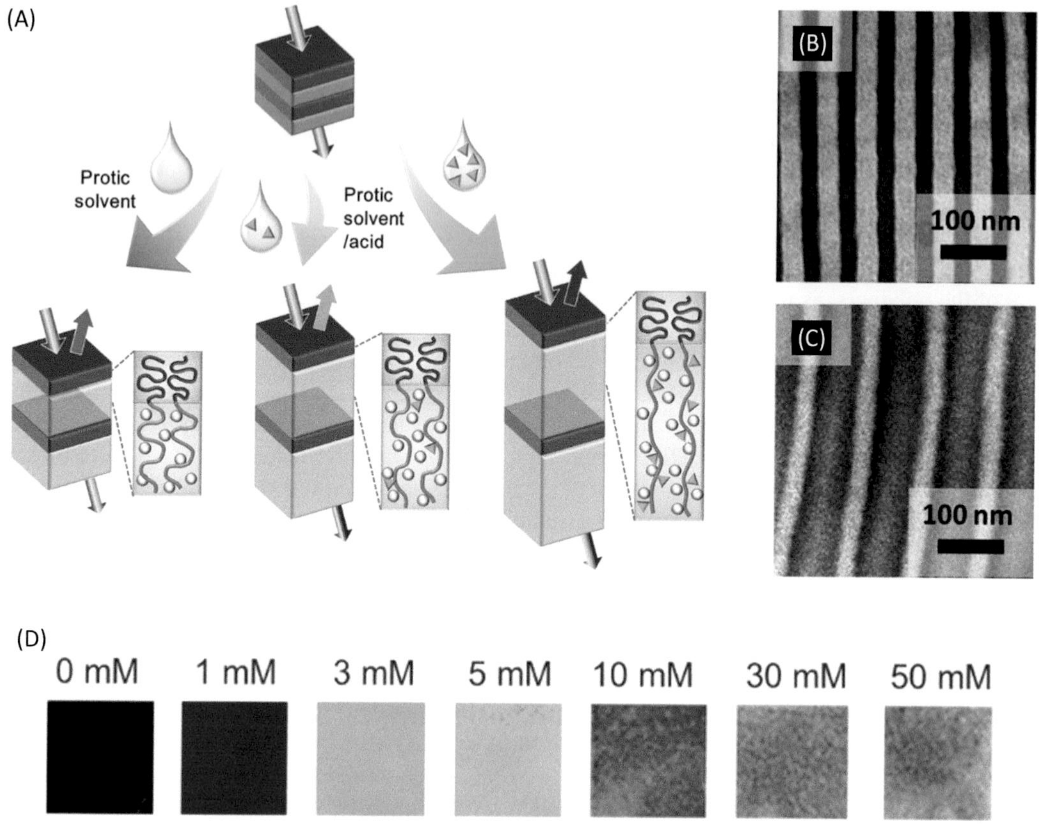

FIGURE 4.9

(A) Schematic representation of the fabrication and mode of action of a photonic block copolymer for the colorimetric recognition of a disulfonic acid. Transmission electron microscopy images of (B) the photonic block copolymer without any stimulus, and (C) after swelling due to the solvent tetraethylene glycol. (D) Photographs of the colorimetric recognition signal derived from different concentrations of a disulfonic acid.

Reproduced with permission from Noro, A., Tomita, Y., Matsushita, Y., & Thomas, E. L. (2016). Enthalpy-driven swelling of photonic block polymer films. Macromolecules, *49(23), 8971–8979. Available from https://doi.org/10.1021/acs.macromol.6b01867.*

concentration of fructose led to an increase in the volume of the film and, thus, a shift of the reflected wavelength from blue to orange (Ayyub et al., 2013). Furthermore, using the ionic interactions of polyelectrolytes with proteins was the basis of the method applied to differentiate proteins with lamellar PS-*b*-P2VP films. As previously described, the P2VP block was quaternized with primary bromides, and its swelling ratios varied with the size and pH-dependent charge of the proteins, resulting in blue or red shifts (Fan et al., 2014). For example, in the presence of BSA, a blue shift was observed due to a coacervation protein-polymer mechanism, causing a decrease in the hydration of the quaternized P2VP and thus a decrease in the domain spacing and a change in the reflectance peak to lower wavelengths (Fan et al., 2014). Other sensing capabilities include detecting strain involved in human motion by embedding lamellar PS-*b*-P2VP in an elastomeric film and following strain-dependent capacitances through color changes (Park et al., 2018), as well as detecting humidity in a simultaneously responsive self-powered motion-sensing device based on structural color readouts (Kim et al., 2022).

Brush block copolymers, a class of comb polymers in which polymeric side chains are densely grafted to a linear backbone (Liberman-Martin et al., 2017), are increasingly used for the production of photonic materials because they have a high molecular weight, and present more rigid molecular structures and reduced entanglement during self-assembly, which allows them to self-assemble rapidly with larger domain spacings (Wang et al., 2022). Recent strategies have focused on the preparation of brush copolymers with precise structures by controlled polymerization techniques (Liberman-Martin et al., 2017; Qiao et al., 2018; Seo et al., 2019) and on achieving tunable photonic properties by incorporating short linear homopolymers (Macfarlane et al., 2014), or NPs such as Au (Song et al., 2015) and zirconium oxide (Song et al., 2016). Moreover, photonic pigments were developed using bottlebrush block copolymers and confining their self-assembly into highly ordered concentric lamellar structures within spherical emulsified microdroplets. These photonic pigments presented a selective and reversible response to different solvents, namely, ethanol, cyclohexanol, and *tert*-butanol, by color change due to differential swelling correlated with the polarity of the alcohols compared to water (Song et al., 2019) (Fig. 4.10). In another study, a dendronized brush block copolymer of poly(norbornene alkyl monomer-r-norbornene dodecyl-methacrylamide)-*b*-poly(norbornene benzyl monomer) was reported to distinguish various alkyl alcohols, alcohol isomers, and alcohol mixtures. The increase in alkyl chain length in alkyl alcohols leads to a decrease in the polarity of the alcohols, which in turn leads to a different ability to be absorbed by the 1D photonic film. In this way, the domain size of the photonic films with different alcohols increased differently, resulting in various color changes (Zhao, Guo, et al., 2019).

4.3.2 Cholesteric liquid crystals

Cholesteric liquid crystals (CLCs) are a type of liquid crystal that self-assembles into a helicoidal structure that is chiral and exhibits selective reflection of light based on the Bragg relationship (Zhou et al., 2022). CLCs are also referred to as chiral nematic liquid crystals because the cholesteric phase is a special state of the nematic phase. While CLCs exhibit a spontaneous helicoidal structure with the twist axis perpendicular to the local orientation, in the nematic phase, the crystals behave as an oriented, sequentially elongated molecule. Due to the unique helical structure of the CLCs, they show not only selective light reflection but also circular dichroism and rotational properties (Zhou et al., 2022). The wavelength of reflection in the visible region leads to their use in colorimetric applications.

Polymeric CLCs can be obtained through polymers such as cellulose. Cellulose nanocrystals (CNCs) can be prepared by acid-catalyzed hydrolysis of bulk cellulose, and the introduction of sulfate ester groups enables the formation of chiral nematic liquid crystals exhibiting iridescent colors (He et al., 2018). An interesting property of CNCs is their ability to preserve the obtained ordering upon drying, and the periodicity of the photonic structure can be adjusted through fabrication

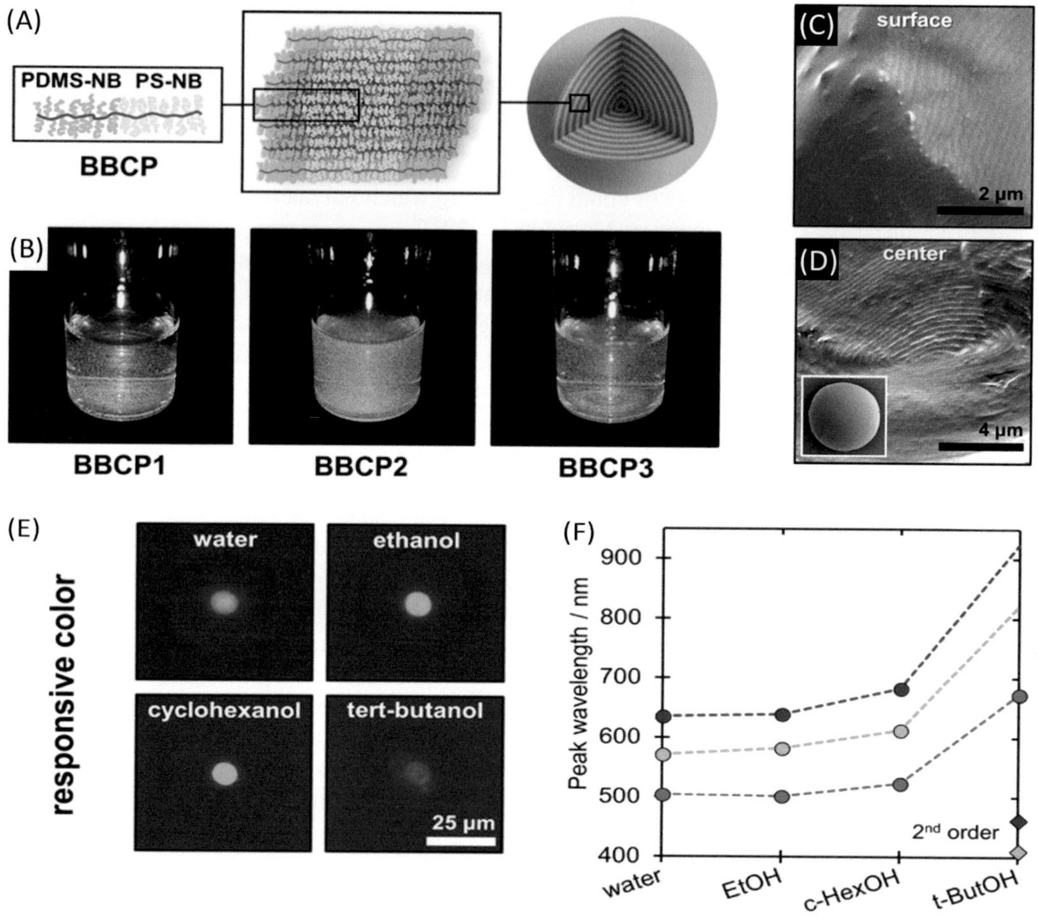

FIGURE 4.10

(A) Schematic illustration of the photonic pigments created through lamellar bottlebrush block copolymers (BBCPs) within emulsified microdroplets, and (B) photographs of different BBCPs showing different colors. BBCP1 microspheres were analyzed through SEM, where it can be observed their cross-section and lamellar structure from (C) the surface and (D) the center. The pigments presented a selective response to different solvents, namely ethanol, cyclohexanol and *tert*-butanol, exemplified by (E) color change of BBCP1, and (F) shifts in peak wavelength of all the tested pigments.

Reproduced with permission from Song, D. P., Zhao, T. H., Guidetti, G., Vignolini, S. & Parker, R. M. (2019). Hierarchical photonic pigments via the confined self-assembly of bottlebrush block copolymers. ACS Nano, *13(2), 1764–1771. Available from https://doi.org/10.1021/acsnano.8b07845. Copyright 2019, American Chemical Society.*

techniques (Zhao, Parker, et al., 2019). Due to these properties, advances in 3D printing of cholesteric hydroxypropyl CNCs within a photo-crosslinkable system made it possible to obtain solid objects with predefined structural colors. Using this method, filaments with an internal helicoidal nanoarchitecture were produced, enabling the direct ink writing of objects with structural color. This promises environmentally friendly future photonic inks with many applications in colorimetric polymer sensing (Chan et al., 2022).

There are several reports in the literature on the use of CNCs in colorimetric sensors for the detection of environmental properties, such as temperature (Yi et al., 2022), pH (Yang et al., 2019), humidity (Bumbudsanpharoke et al., 2019), and gases (Dai et al., 2017; Song et al., 2018), as well as diverse chemicals (An et al., 2023; Hu et al., 2022). In addition, multi-responsive biosensors based on CNCs already exist and are a great alternative for anticounterfeiting technologies, environmental sensing, food quality testing, and health monitoring, among others (Chen, Ling, et al., 2023; He et al., 2018; Zhao et al., 2020).

A simple colorimetric sensor based on films with Cu(II)-doped CNCs was fabricated for the monitoring of ammonia gas. The incorporation of copper ions into the CNCs enabled the tuning and modulation of their color and increased the sensing capacity of ammonia gas, as the copper ions have a strong chelation affinity for ammonia. Therefore, when ammonia gas entered the nematic structure, the CNC structure changed, and a visible red shift occurred (Dai et al., 2017). In a different work, hydroxypropyl cellulose was used for temperature sensing due to its colorimetric properties. The hydroxypropyl cellulose self-assembled into a temperature-responsive cholesteric liquid crystalline mesophase, and ethylene glycol was added as the coolant, preventing the mesophase from freezing at low temperatures. The sensor could distinguish temperatures from -20 °C to 25 °C through an increase in the helical pitch and a subsequent color shift from blue to orange (Yi et al., 2022). CNCs were also used for methanol discrimination, and in this case, the CNCs were embedded in a water-soluble polymer of β-cyclodextrin. The sensor showed different colors for methanol and ethanol, two homologous alcohols and, thus, it could differentiate them. Also, there was a red shift from blue-greenish to orange with the increasing concentration of methanol, so the colorimetric sensor enabled quantitative analysis (Hu et al., 2022) (Fig. 4.11).

CLCs can also be made from polymers other than cellulose for environmental and health monitoring (Foelen et al., 2023; Lee et al., 2018; Pan et al., 2023; Sutarlie & Yang, 2011). For example, a polymer-dispersed CLC layer was developed and easily integrated in microfluidic devices for monitoring volatile organic compounds (Sutarlie & Yang, 2011). In a different study, a CLC made from acrylates was used both as a decontamination device for poisonous organophosphates and as a sensing device for the same compounds, described as a photonic absorbent of the vapor molecules. The photonic CLC showed a blue color but changed to green in the presence of dimethyl methylphosphonate, which was used as a nerve agent simulant. The color shift was due to hydrogen bonding between the target and the carboxylic groups of the CLC (Foelen et al., 2023). There are several techniques that can be used to fabricate polymer CLC microcapsules, such as bulk emulsification, phase separation, interfacial polymerization, or coacervation (Pan et al., 2023). In a study to develop thermochromic microcapsules, monodisperse CLC-in-water emulsion drops were produced by microfluidics, and they consisted of a commercial CLC core and a thin transparent polyurethane shell layer. Then, the core-shell particles were placed on a hole-patterned flexible film, which enabled microwriting and the development of a smart color display. The photonic sensor based on CLCs was responsive to temperature, with changes in color that could be observed with the naked eye (Lee et al., 2018) (Fig. 4.12).

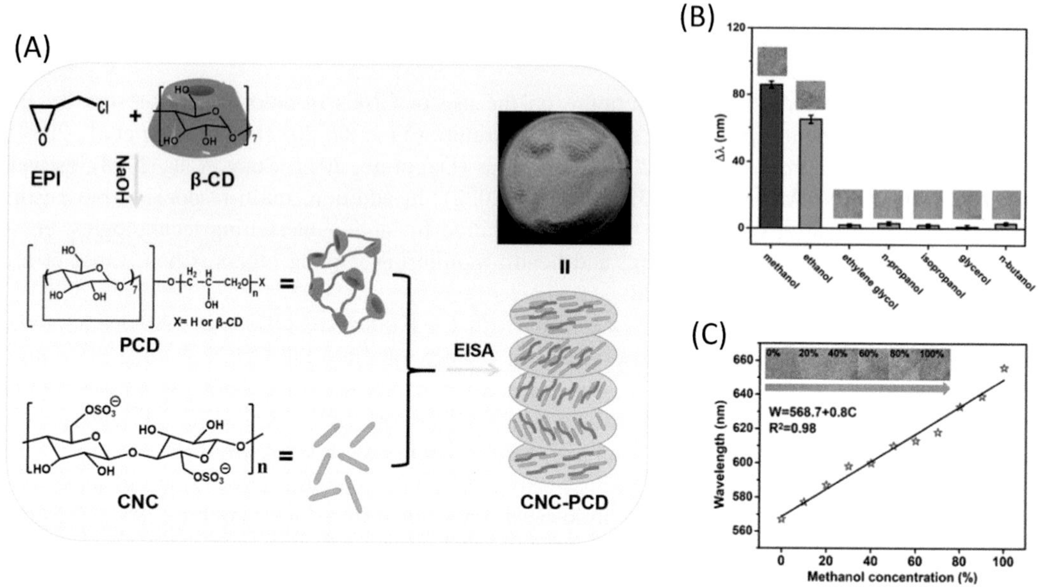

FIGURE 4.11

(A) Design of a colorimetric polymer sensor based on CNCs photonic structure. (B) Shift of color according to the solvent used, showing response to both methanol and ethanol, but still being able to differentiate them. (C) Color changes with increasing concentrations of methanol.

Reproduced with permission from Hu, C. Y., Bai, L., Song, F., Wang, Y. L., & Wang, Y. Z. (2022). Cellulose nanocrystal and beta-cyclodextrin chiral nematic composite films as selective sensor for methanol discrimination. Carbohydrate Polymers, 296, *119929. Available from https://doi.org/10.1016/j.carbpol.2022.119929. Copyright 2022, Elsevier Ltd.*

4.4 Smart polymer-based colorimetric sensors

The maintenance of biological systems demands sensitivity toward the environment and a reaction toward its changes. Mimicking these systems has led to the development of stimulus-responsive polymers, also known as smart polymers. Smart polymers undergo reversible changes due to chemical or physical events in their environment as a direct effect of the interaction between the polymer chains and the solvent molecules (Kuckling et al., 2012). Depending on their chemical composition, smart polymers can respond to temperature, solvents, humidity, pH, mechanical stress, or ionic strength. Common mechanisms that lead to changes in smart polymers include sol-gel transitions (due to the transition of linear and solubilized macromolecules from a monophasic to a biphasic state), swelling/shrinking of crosslinked networks, or changes in the polymer hydrophilicity (Aguilar & San Román, 2014).

Temperature-sensitive polymers have a low critical solution temperature (LCST) or an upper critical solution temperature (UCST). This means that they undergo a transition from monophasic to biphasic when the temperature is raised above a critical point, or vice versa, respectively. A biphasic system exists when an aqueous phase with nearly no polymer and a polymer-enriched

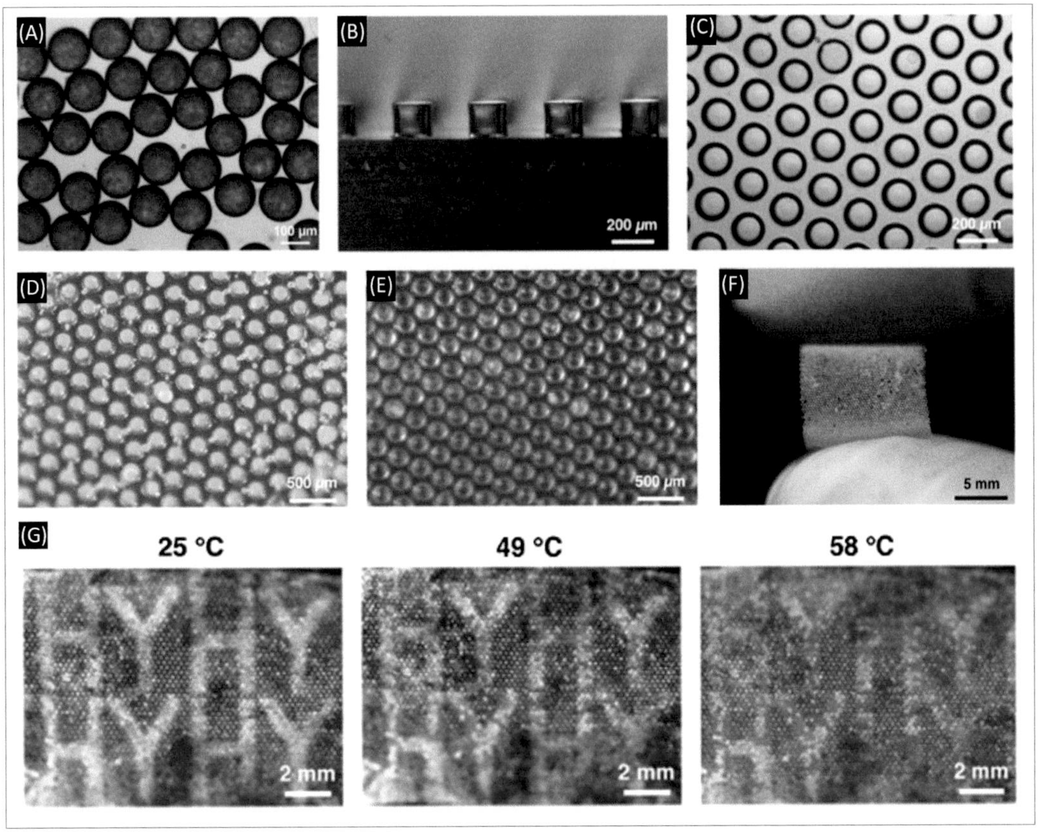

FIGURE 4.12

Photonic sensor based on CLCs responsive to temperature. Optical microscopic image of (A) the CLC emulsion drops, (B) lateral view and (C) top view of the holes-patterned flexible film, where the CLC particles were placed for microwriting, as well as (D) color at T = 25 °C, and (E) lack of color at T = 58 °C. (F) Photograph of the colorimetric sensor. (G) Colorimetric thermometer display of letters "HY" by positioning two types of CLC microcapsules on the flexible film.

Reproduced with permission from Lee, W. J., Kim, B., Han, S. W., Seo, M., Choi, S.-E., Yang, H., Kim, S.-H., Jeong, S., & Kim, J. W. (2018). 2-Dimensional colloidal micropatterning of cholesteric liquid crystal microcapsules for temperature-responsive color displays. Journal of Industrial and Engineering Chemistry, 68, *393–398. Available from https://doi.org/10.1016/j.jiec.2018.08.014. Copyright 2018, Elsevier B. V.*

phase are formed (Chakraborty et al., 2018). LCST polymers have attracted more interest than UCST, with the families of poly(*N*-substituted acrylamide), poly(vinyl amide), and poly(oligoethylene glycol (meth)acrylate) being the most commonly used (Aguilar & San Román, 2014). The pH-sensitive polymers incorporate poly(acrylic acid), poly(methacrylic acid), poly(*N*,*N*-dimethylaminoethyl methacrylate), poly(*N*,*N*-diethylaminoethyl methacrylate), or poly(*N*-ethyl pyrrolidine methacrylate) moieties. Depending on the pH values, these enable protonation or deprotonation, which causes a shift from a soluble to an insoluble state of the polymer (Chakraborty et al., 2018).

Polymers with shape memory are also commonly used because they can recover their predefined shape when stimulated. They consist of a stable network and a reversible switch transition. While the former is responsible for the original shape, the switching transition allows the polymer to hold a temporary shape. The switch can occur through crystallization/melting, redox reactions of mercapto groups, hydrogen bonding, and metal-ligand coordination, among others. Light, pH, mechanical stress, humidity, and electric and magnetic fields can also trigger the shape memory effect (Aguilar & San Román, 2014). However, in a world where environmental awareness is increasing, the use of polymers of natural origin or bioinspired materials is a major objective. Natural polymers have great advantages, such as their inherent biocompatibility and biodegradability, which enable applications in the fields of therapy, diagnostics, sensing, and imaging. To fully achieve their potential, their availability in the ecosystem, their chemical and biological properties, and their mechanical properties must be considered (Prianka et al., 2018).

Smart polymer-based colorimetric sensors have been widely studied in the last fifteen years as they provide valuable naked-eye detection in various fields such as environmental monitoring, clinical applications, and the agrifood industry. Several examples can be found in the literature, especially for humidity (Dong et al., 2022; Mills et al., 2017), temperature (Son et al., 2022; Wibowo et al., 2022), volatile organic compounds (Wang et al., 2013), heavy metal ions (Huang et al., 2017; Veedu et al., 2023), and mechanical properties (Goni-Lizoain et al., 2023; Han et al., 2014; Kim et al., 2023; Park et al., 2019). Moreover, multi-response colorimetric biosensors have been studied to improve their application and translation from the laboratory to real-world settings (Liu et al., 2011; Liu et al., 2010).

A colorimetric temperature-activated humidity indicator was produced based on the aggregation of thiazine dyes encapsulated in a hydroxypropyl cellulose polymer. Initially, the polymer showed a purple color due to aggregation but turned blue with heat activation. Then, the sensor was ready to respond to humidity by returning to its initial purple color within seconds. This change is reversible and makes the sensor reusable (Mills et al., 2017). A different colorimetric reversible sensor for temperature control was also fabricated using the LCST polymer poly(*N*-isopropylacrylamide) and AuNPs embedded in the polymeric matrix. The response of the prepared hydrogel to temperature led to changes in the interparticle spacing of the AuNPs, demonstrating a color change from red to pink between 25 °C and 50 °C (Son et al., 2022). This change in color presented by AuNPs has been extensively explored, including obtaining a stress-responsive colorimetric film. In this case, the application of force resulted in the deformation of the polyvinylpyrrolidone film in which a 1D AuNPs chain was incorporated, thus leading to the separation of the NPs. The color alteration from blue to red was visible to the naked eye (Han et al., 2014). A different strategy, based on a porous mechanochromic polymer comprising poly(dimethyl siloxane), the mechanophore spiropyran, and silica NPs, was presented. Upon application of external forces, the polymer matrix changed and led to spiropyran undergoing a ring-opening process, which in turn originated the formation of a colored product. The presence of hydrophilic silica NPs in the composite was demonstrated to enhance the mechanochromic sensitivity and stretchability (Park et al., 2019). A simple strategy to measure strain relied on preparing colloidal nanocapsules of polystyrene-grafted AuNPs. These were then embedded in an elastomer matrix. Upon application of a tensile tension, increased inter-nanoparticle distance was observed through a color change from blue to red. The sift in color could be reversed as the tension was released, making it a reusable colorimetric sensor (Kim et al., 2023).

4.5 New trends in colorimetric polymer sensors

4.5.1 Miniaturization

The miniaturization of colorimetric sensors into microfluidic systems makes it possible to reduce the sample volume, concentrate, and analyze it in the same device (Ma et al., 2023). However, to our knowledge, there are only a few examples of microfluidics adapted to colorimetric polymer sensors. Some of these studies have combined MIPs with nanozymes for colorimetric detection of chemical substances (Amatatongchai et al., 2023; Kong et al., 2017), while others used dyes for analyte recognition (Catalan-Carrio et al., 2022; Godino et al., 2014; Krauss et al., 2017; Punnoy et al., 2021) or natural enzymes (He et al., 2019).

A paper-based microfluidics to constrain the biorecognition event in a small area was developed. That area was modified with a MIP for bisphenol A and a nanozyme with peroxidase-like behavior based on zinc ferrite. In the absence of the target analyte, the nanozyme reacted with added H_2O_2, generating hydroxyl radicals that oxidized the TMB, resulting in a blue color. In the presence of the target analyte, this reaction was inhibited, and the blue color faded with increasing analyte concentrations (Kong et al., 2017). A similar approach was used to detect the insecticide carbaryl. The device was a 3D-microfluidic paper origami device constituted by two foldable layers: the bottom layer consisted of a circular area for sample loading, connected by channels to three reaction zones, and a lower circular control reaction zone; the upper layer consisted of three circular reservoirs, forming three detection zones, and a lower control zone. This means that the microfluidic device can analyze multiple tests with the same sample when folded. Furthermore, the nanozyme consisted of mesoporous silica-platinum NPs, and the colorimetric reaction was based on the "turn-off" effect on the enzymatic activity in the presence of the insecticide (Amatatongchai et al., 2023). The detection of analytes in sweat has been attracting intensive research, considering it is a non-invasive method. For colorimetric detection of glucose, a thermoresponsive textile and paper-based microfluidic device was proposed. For this purpose, polyurethane was used in cotton fabric for the ability to have temperature-dependent shape memory, for patterning of channels, and for thermal-triggered sweat transport. Then, a chitosan-modified paper-based colorimetric sensor combined with GOx formed the detection area (He et al., 2019) (Fig. 4.13).

Regarding dye-incorporation methodologies, a polyester-paper microfluidic device was reported that had a reagent storage chamber and could be integrated into a multiplex design. It was successfully tested for analytes as different as total serum protein, cocaine, an explosive, and Fe(III) ions, showing a different color change for each analyte (Krauss et al., 2017). More recently, natural polymers were used to construct a microfluidic device for the colorimetric detection of glucose in tears. In this research, PVA and starch were combined and modified by the incorporation of GOx, HRP, and potassium iodine. Glucose is oxidized by the enzyme, producing H_2O_2, which then induces the oxidation of potassium iodine *via* the action of HRP. This last reaction leads to color development that can be seen with the naked eye. This system was incorporated into cotton threads, which are known to have twisted strands of cellulosic fibers and gaps between the strands, which in turn allow liquid samples to flow by capillary force. Finally, the described system was inserted into a 3D microfluidic cassette to obtain a miniaturized device for diabetes screening (Punnoy et al., 2021). In another work, a microfluidic platform based on cellulose paper to monitor cooking oil

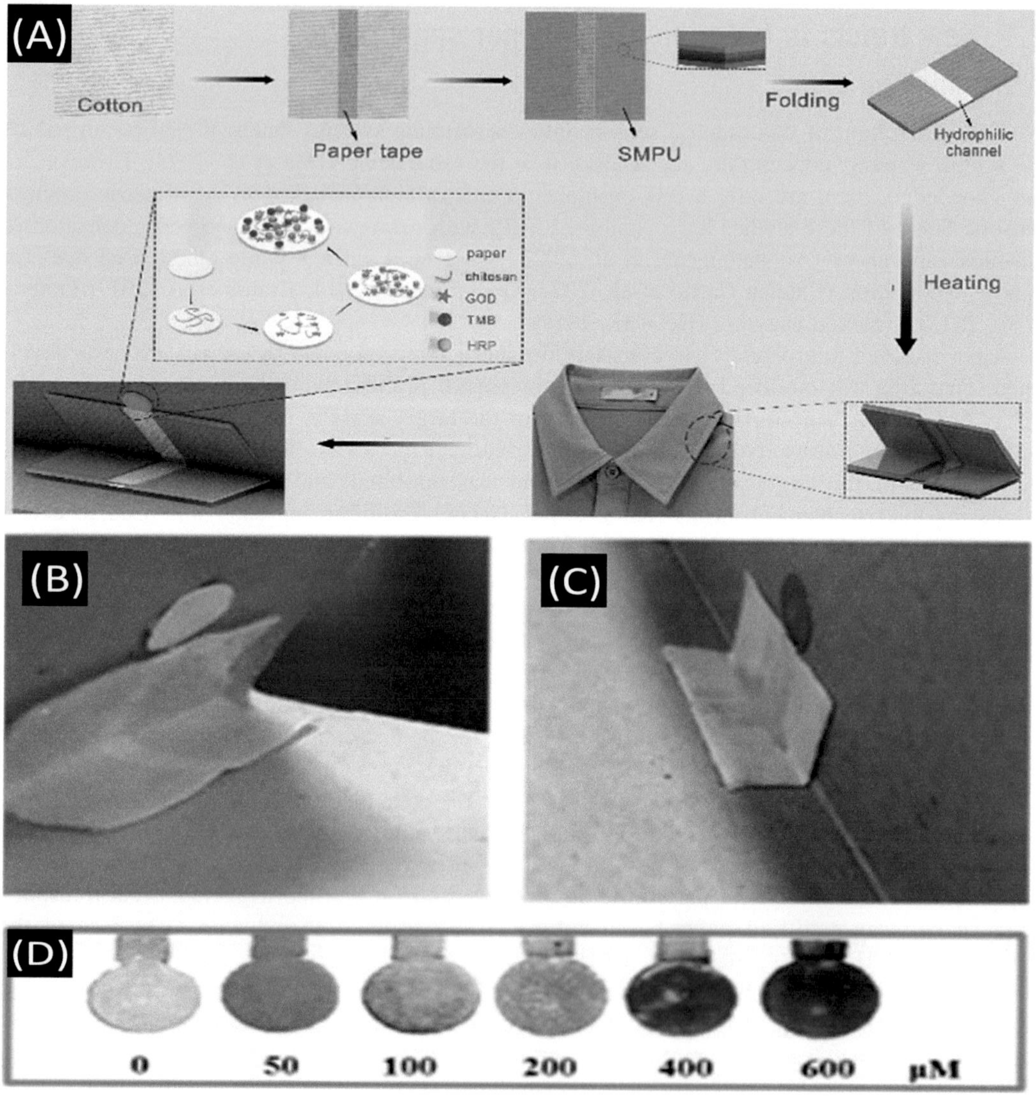

FIGURE 4.13

(A) Scheme of the fabrication process of the thermoresponsive textile/paper-based microfluidic sensing system, and photographs of the (B) final microfluidic sensor without glucose, and (C) in the presence of glucose. (D) Colorimetric analysis of artificial sweat samples containing 0, 50, 100, 200, 400, and 600 μM of glucose.

quality by detecting and determining the concentration of oleic acid molecules was presented. For this purpose, the cellulose paper was modified with graphene oxide, not only to increase its hydrophobicity but also to improve color visualization. Then, phenolphthalein and sodium carbonate were added at each spot of the microfluidic device for colorimetric acid-base titration. Therefore, when oleic acid was present, the pH was neutralized, and the initial pink color of phenolphthalein faded. The color disappearance was gradual with increasing concentrations of oleic acid (Saedan et al., 2023) (Fig. 4.14).

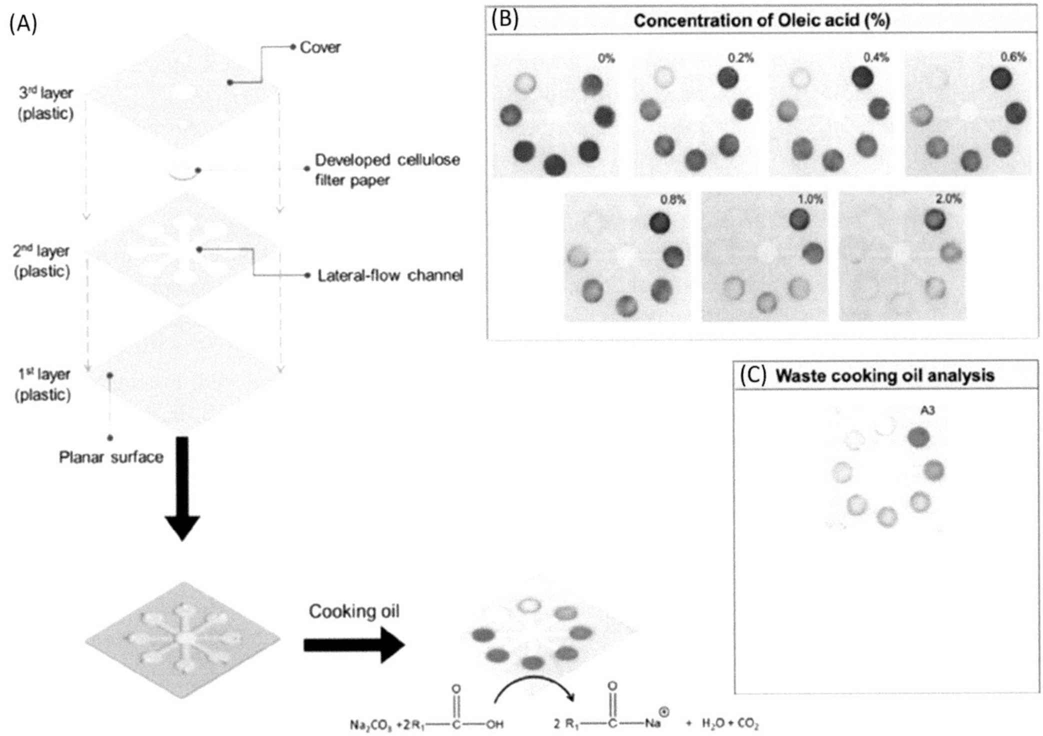

FIGURE 4.14

(A) Schematic illustration of the microfluidic platform for cooking oil quality monitoring. A cellulose paper was modified with graphene oxide, and phenolphthalein and sodium carbonate were added at different concentrations in each spot for acid-base titration and color revealing. (B) Photographs of the response of the developed colorimetric sensor using oleic acid as a testing molecule at different concentrations. (C) Photograph of real waste cooking oil analysis.

Reproduced with permission from Saedan, A., Yukird, J., Rodthongkum, N., & Ummartyotin, S. (2023). Graphene oxide modified cellulose paper-based device: A novel platform for cooking oil quality evaluation. Food Control, 148, *109675. Available from https://doi.org/10.1016/j.foodcont.2023.109675. Copyright 2023, Elsevier Ltd.*

4.5.2 Smartphone-based technologies

Although colorimetric analysis can be assessed with the naked eye for qualitative or semi-quantitative observation, quantitative analysis is usually accompanied by spectroscopic methods. Portable instrumentation that integrates optical fiber probes and miniaturized spectrometers has attracted much attention for in situ analysis due to time-saving result analysis, versatility, and simplicity (Martínez-Aviño et al., 2022). Thus, the colorimetric polymer sensors described in the previous sections have recently been coupled with smartphone-based technologies as analytical devices (Akhoundian & Alizadeh, 2023; Bustamante et al., 2019; El Hani et al., 2023; Guembe-García et al., 2022; Guirado-Moreno et al., 2023; Han et al., 2022; Hosu et al., 2019; Ko et al., 2021; Kuşçuoğlu et al., 2019; Li et al., 2022; Vallejos et al., 2013; Zhang et al., 2022; Zhang et al., 2020; Zhou et al., 2024). Basically, the smartphone is used as a camera to capture images for further color analysis by an installed software that provides color parameters such as red-green-blue coordinates (RGB system) or CIELAB parameters. Smartphones have facilitated the development of new colorimetric sensors and the processing of images, improving the linearity and sensitivity of colorimetric methods (Martínez-Aviño et al., 2022).

Some examples of smartphone-assisted colorimetric analysis include polymer sensors that rely on enzymatic activity for color development. For example, the enzymes HRP, GOx, and tyrosinase were incorporated into a poly(aniline-*co*-anthranilic acid) film. In the presence of the analytes, H_2O_2, glucose, or catechol, the enzymes catalyzed a reversible redox color change of the polymer from green to blue. The same device thus enabled multiplex analysis. Subsequently, the color change was captured by a smartphone with the integrated ColorLab® application for optical analysis and simple, quantitative sensing (Hosu et al., 2019). A different portable device for monitoring uric acid in patients with gout was presented. For this purpose, a wearable microneedle colorimetric patch based on a PVA microneedle with the enzyme uricase embedded for the oxidation of uric acid was developed. In addition, polypyrrole NPs were incorporated due to their peroxidase-like activity, as well as TMB. In this way, color variation related to the H_2O_2 concentration and indirectly to the uric acid concentration was observed. Correct quantification was performed using a smartphone (Zhang et al., 2022). Another example is the study that used a magnetite chitosan hydrogel, described as a peroxidase nanozyme, for the colorimetric detection of the pesticide thiabendazole using a smartphone. Chitosan was used because it is polycationic, which is similar to the microenvironment of the enzyme HRP. In the presence of the pesticide, the oxidation of TMB was inhibited as a function of concentration. The smartphone-based measurements were not as sensitive as those of a spectrometer, but the smartphone outperforms the spectrometer due to its ease of use and applicability in point-of-care (Ariti et al., 2023).

MIP nanozymes can also be easily allied with smartphone-assisted colorimetric sensors. One strategy relied on a paper-based microfluidic chip, which consisted of a reaction area incorporating a MIP for microcystin recognition as well as a nanozyme based on $ZnFe_2O_4$, H_2O_2, and TMB to trigger color development. The use of a smartphone to scan the color progression enabled the analysis of complex water samples (Han et al., 2022). A naked-eye visual colorimetric sensing device for glucose by incorporating a natural enzyme and a nanozyme into the same system was recently presented. The nanozyme consisted of Fe_3O_4 covered with a MIP constructed with acrylamide as the monomer and TMB as the template. It was found that the MIP nanozyme strategy resulted in a 9-fold higher catalytic activity than the free nanozyme. Regarding the process of glucose determination, when glucose was present, GOx led to the formation of gluconic acid and H_2O_2. TMB was then added, which was specifically recognized by the MIP nanozyme, and oxidized, whose degree

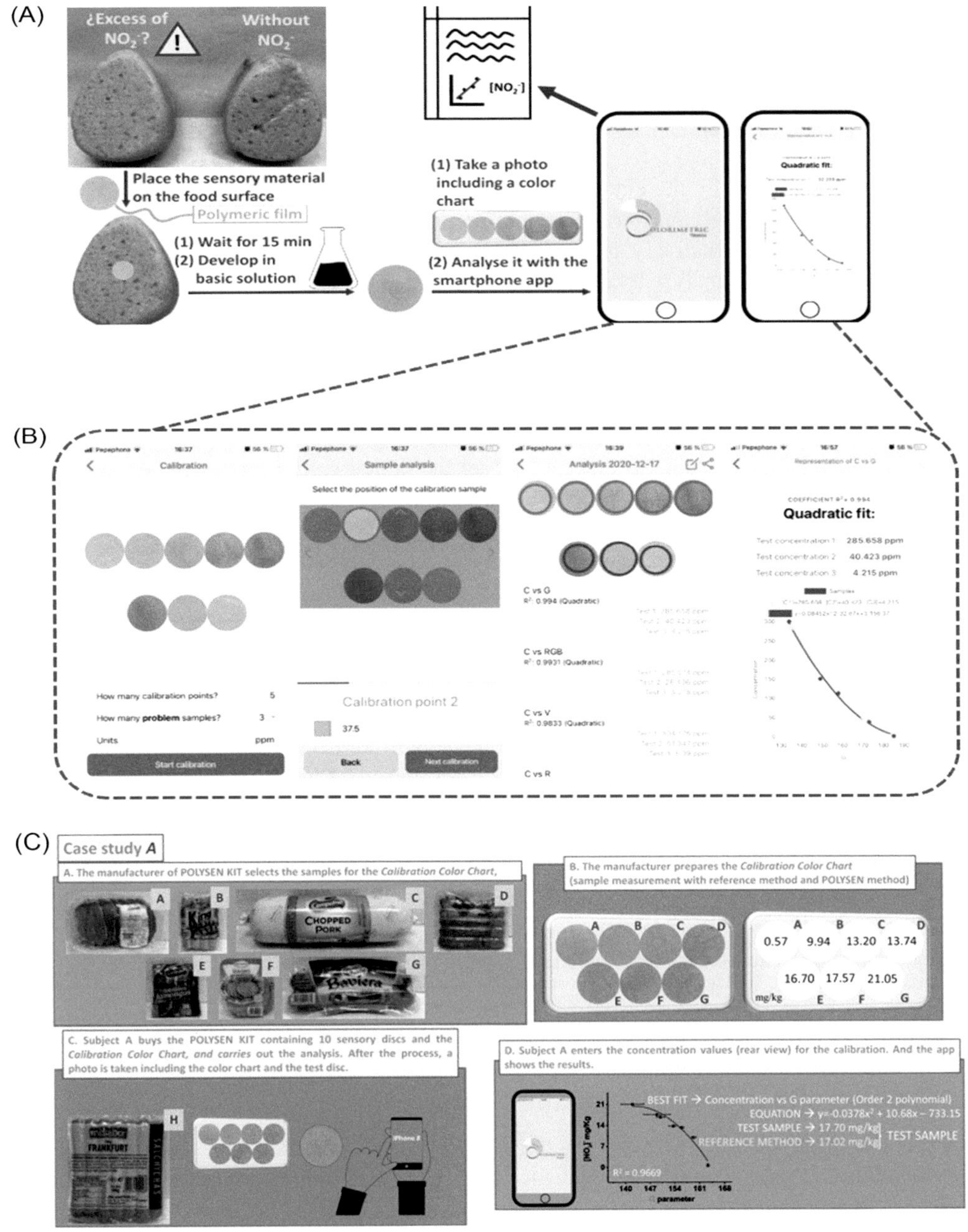

FIGURE 4.15

(A) Illustration of the smartphone-assisted one-pot determination of the nitrite concentration in food samples. (B) The colorimetric signal from nitrite recognition is captured by the camera of the smartphone, and calibration was performed in the developed app after performing 16 linear and 16 quadratic fits of concentration versus different digital color parameters. The best fit was chosen as the best option for further data analysis. (C) Case study as proof of concept, with an example of the measurement of a problem sample with a smartphone.

of oxidation is determined by the amount of H_2O_2 formed as a result of glucose oxidation. In this way, the increasing glucose concentration could be detected with the naked eye by a color shift from pink to blue and quantitatively determined using a smartphone (Li et al., 2024).

Another work aimed to analyze food quality in situ by detecting nitrite formation in meat samples. A colorimetric film-shaped polymeric sensor based on *N*-vinyl-2-pyrrolidone and methylmethacrylate was obtained to ensure that the sensing film could withstand handling and be absorbent. Then, the patch was placed on the meat samples for the absorption of nitrite, followed by immersing it in an HCl solution. The acid allowed the nitrosyl cation to form in the presence of nitrite, which in turn attacked the polymer matrix, resulting in structural changes and a yellow-brownish color. The smartphone with a mobile application was able to analyze the digital color parameters and automatically calculate the nitrite concentration (Guembe-García et al., 2022) (Fig. 4.15). A different sensor for food quality control was constructed to determine the copper content of grape must. For this purpose, a smart polymeric film constituted by commercially available monomers (1-vinyl-2-pyrrolidone and methyl methacrylate crosslinked by ethylene glycol dimethacrylate) and the copper chelating agent bicinchoninic acid was developed. When copper ions were present, they interacted with the chelating agent, which led to a structural change in the polymer film. A color was then generated and examined using a smartphone to obtain the RGB parameters (Guirado-Moreno et al., 2023).

A very interesting approach to this new trend was designed for the recognition of bacteria and the differentiation between different bacteria. An array membrane chip was composed of biomimetic lipid bilayers, also called liposomes, assembled with polydiacetylene. Due to the backbone distortion of polydiacetylene when the toxins interacted with the lipid bilayers, these liposomes presented chromatic transitions that differed depending on the mode of action of the toxins released by the bacteria. Thus, the liposomes were immobilized on a paper-like substrate and tested with *S. aureus* and *Pseudomonas aeruginosa* as model bacteria. The color analysis was performed with a smartphone application, which could determine how many bacterial cells were present within 4 hours, which greatly shortened the reaction time compared to conventional methods (Zhou et al., 2024).

4.6 Conclusions and future perspectives

Color arises from physical or chemical processes, both of which can be adapted to sensing devices using naturally derived colors or synthetic materials that mimic their optical properties. Colorimetric sensors are based on the optical properties of a variety of materials, such as the intrinsic properties of metallic NPs that change color according to interparticle distance; the catalytic activity of enzymes or enzyme-like materials, which catalyze chemical reactions involving labeling agents, that is, chromogenic substrates or products for signal amplification; DNA-based materials whose structure changes upon recognition of the target molecule; dyes or natural pigments to develop a color that transduces the recognition of target analytes; and natural or synthetic polymeric assemblies that exhibit structural color, which can be tuned and stimuli-responsive. The diversity of colorimetric strategies, their ease of use, simplicity, and cost-effective performance have led to the development of numerous technologies in the environmental, health, and food sectors. Moreover, in comparison to other optical technologies, colorimetric devices are highly advantageous because they enable qualitative or semi-quantitative analysis with the naked eye.

Currently, the integration of polymers in sensors is also of great interest, as polymers can be easily customized for specific applications and have a range of mechanical, thermal, and chemical properties. Polymers can be used in colorimetric sensors for several purposes, which shows their versatility: as supports for the immobilization of revealing elements such as nanomaterials, enzymes, and dyes; to improve the affinity and diffusion of the sample through the sensor; as the structural color element, such as 1D photonic block copolymers and CLCs, with the capacity of tuning their color through the fabrication process; and MIPs, synthetic biorecognition elements with multiple applications.

The explored examples reflect the increasing interest in colorimetric polymer sensors for simple, easy, and low-cost recognition of the most varied molecules and physical properties. The use of metal NPs and dyes excels in simplicity and low-cost approaches without losing sensitivity and selectivity. DNA nanomaterials, such as G-quadruplex and other structures, are highly stable and programmable but are not representative in the literature when considering their assembly with polymers. Recently, the incorporation of nanozymes has overtaken other techniques. This is due to their greater design flexibility, which makes it possible to incorporate materials as diverse as metal NPs, carbon-based materials, and inorganic nanocrystals, among others, as well as their combination to achieve peroxidase-, oxidase-, or catalase-like activities. In addition, nanozymes show high stability and constitute a less expensive option than natural enzymes. Moreover, using polymers for obtaining structural color through self-assembly of block copolymers, CNCs, or other CLCs allows the reflected light to be modulated during the fabrication process and pursues greater color diversity systems, and a wider range of wavelength shifts upon target recognition. Moreover, engineering smart polymer networks has led to major advances in creating responsive soft materials. The ability of smart polymers to recognize different physicochemical events is based on their functional groups, and copolymerization expands the possibility for multi-responsive sensors. Future advances aim at improving the functionalities of polymer materials and the implementation of natural biopolymers to develop easily biodegradable disposable products. Mastering polymer manipulation and design for optical purposes will, therefore, enable market realization.

Advances in microfluidics and the combination of colorimetric sensors with wearable applications, artificial intelligence, and portable instrumentation have evolved into innovative approaches that are expected to have a major impact on smart healthcare, environmental monitoring, and the agri-food industry. In the area of portability, the inclusion of smartphones for accurate and sensitive quantification of analytes is a topic of great interest. However, color analysis is highly dependent on lighting conditions and camera quality. Therefore, work has been done in recent years to improve camera lenses and provide new camera settings that can be used to adjust parameters such as light sensitivity, exposure time, and contrast ratio. It is also important to systematize and create standardized camera and analysis applications that can be set up on smartphones to reduce the number of errors in the analysis of color parameters and to be independent of the smartphone or user.

Acknowledgments

The authors acknowledge the financial support from the European Commission/H2020, through MindGAP/FET-Open/GA829040 project. The author RV also acknowledges Fundação para a Ciência e a Tecnologia her PhD grant (2020.09673.BD).

References

Adegoke, O., & Daeid, N. N. (2021). Polymeric-coated Fe-doped ceria/gold hybrid nanocomposite as an aptasensor for the catalytic enhanced colorimetric detection of 2,4-dinitrophenol. *Colloids and Surfaces A: Physicochemical and Engineering Aspects*, *627*, 127194. Available from https://doi.org/10.1016/j.colsurfa.2021.127194, http://www.elsevier.com/locate/colsurfa.

Aguilar, M. R., & San Román, J. (2014). *Smart polymers and their applications*. Woodhead Publishing. Available from https://doi.org/10.1533/9780857097026.1.

Ahmad, N. A., Yook Heng, L., Salam, F., Mat Zaid, M. H., & Abu Hanifah, S. (2019). A colorimetric pH sensor based on Clitoria sp and Brassica sp for monitoring of food spoilage using chromametry. *Sensors*, *19* (21), 4813. Available from https://doi.org/10.3390/s19214813.

Akhond, M., Hormozi Jangi, S. R., Barzegar, S., & Absalan, G. (2020). Introducing a nanozyme-based sensor for selective and sensitive detection of mercury(II) using its inhibiting effect on production of an indamine polymer through a stable n-electron irreversible system. *Chemical Papers*, *74*(4), 1321–1330. Available from https://doi.org/10.1007/s11696-019-00981-w, https://www.springer.com/journal/11696.

Akhoundian, M., & Alizadeh, T. (2023). Enzyme-free colorimetric sensor based on molecularly imprinted polymer and ninhydrin for methamphetamine detection. *Spectrochimica Acta Part A: Molecular and Biomolecular Spectroscopy*, *285*, 121866. Available from https://doi.org/10.1016/j.saa.2022.121866.

Alberti, G., Zanoni, C., Losi, V., Magnaghi, L. R., & Biesuz, R. (2021). Current trends in polymer based sensors. *Chemosensors*, *9*(5), 108. Available from https://doi.org/10.3390/chemosensors9050108, https://www.mdpi.com/2227-9040/9/5/108/pdf.

Amatatongchai, M., Thimoonnee, S., Somnet, K., Chairam, S., Jarujamrus, P., Nacapricha, D., & Lieberzeit, P. A. (2023). Origami 3D-microfluidic paper-based analytical device for detecting carbaryl using mesoporous silica-platinum nanoparticles with a molecularly imprinted polymer shell. *Talanta*, *254*, 124202. Available from https://doi.org/10.1016/j.talanta.2022.124202.

Amirzehni, M., Hassanzadeh, J., & Vahid, B. (2020). Surface imprinted CoZn-bimetalic MOFs as selective colorimetric probe: Application for detection of dimethoate. *Sensors and Actuators B: Chemical*, *325*, 128768. Available from https://doi.org/10.1016/j.snb.2020.128768.

An, B., Xu, M., Sun, J., Sun, W., Miao, Y., Ma, C., Luo, S., Li, J., Li, W., & Liu, S. (2023). Cellulose nanocrystals-based bio-composite optical materials for reversible colorimetric responsive films and coatings. *International Journal of Biological Macromolecules*, *233*, 123600. Available from https://doi.org/10.1016/j.ijbiomac.2023.123600.

Ariti, A. M., Geleto, S. A., Gutema, B. T., Mekonnen, E. G., Workie, Y. A., Abda, E. M., & Mekonnen, M. L. (2023). Magnetite chitosan hydrogel nanozyme with intrinsic peroxidase activity for smartphone-assisted colorimetric sensing of thiabendazole. *Sensing and Bio-Sensing Research*, *42*, 100595. Available from https://doi.org/10.1016/j.sbsr.2023.100595.

Asal, M., Özen, Ö., Şahinler, M., Baysal, H. T., & Polatoğlu, İ. (2019). An overview of biomolecules, immobilization methods and support materials of biosensors. *Sensor Review*, *39*(3), 377–386. Available from https://doi.org/10.1108/SR-04-2018-0084, http://www.emeraldinsight.com/info/journals/sr/sr.jsp.

Ayyub, O. B., Ibrahim, M. B., Briber, R. M., & Kofinas, P. (2013). Self-assembled block copolymer photonic crystal for selective fructose detection. *Biosensors and Bioelectronics*, *46*, 124–129. Available from https://doi.org/10.1016/j.bios.2013.02.025.

Bamrungsap, S., Cherngsuwanwong, J., Srisurat, P., Chonirat, J., Sangsing, N., & Wiriyachaiporn, N. (2019). Visual colorimetric sensing system based on the self-assembly of gold nanorods and graphene oxide for heparin detection using a polycationic polymer as a molecular probe. *Analytical Methods*, *11*(10), 1387–1392. Available from https://doi.org/10.1039/C8AY02129E.

Bhadra, S., Codrea, V., & Ellington, A. D. (2014). G-quadruplex-generating polymerase chain reaction for visual colorimetric detection of amplicons. *Analytical Biochemistry*, *445*, 38–40. Available from https://doi.org/10.1016/j.ab.2013.10.010.

Bhaduri, S. N., Ghosh, D., Chatterjee, S., Biswas, R., Bhaumik, A., & Biswas, P. (2023). Fe(III)-incorporated porphyrin-based conjugated organic polymer as a peroxidase mimic for the sensitive determination of glucose and H2O2. *Journal of Materials Chemistry B*, *11*(37), 8956–8965. Available from https://doi.org/10.1039/d3tb00977g.

Bumbudsanpharoke, N., Kwon, S., Lee, W., & Ko, S. (2019). Optical response of photonic cellulose nanocrystal film for a novel humidity indicator. *International Journal of Biological Macromolecules*, *140*, 91–97. Available from https://doi.org/10.1016/j.ijbiomac.2019.08.055.

Bustamante, S. E., Vallejos, S., Pascual-Portal, B. S., Muñoz, A., Mendia, A., Rivas, B. L., García, F. C., & García, J. M. (2019). Polymer films containing chemically anchored diazonium salts with long-term stability as colorimetric sensors. *Journal of Hazardous Materials*, *365*, 725–732. Available from https://doi.org/10.1016/j.jhazmat.2018.11.066, http://www.elsevier.com/locate/jhazmat.

Cardoso, A. R., Frasco, M. F., Serrano, V., Fortunato, E., & Sales, M. G. F. (2021). Molecular imprinting on nanozymes for sensing applications. *Biosensors*, *11*(5), 152. Available from https://doi.org/10.3390/bios11050152, https://www.mdpi.com/2079-6374/11/5/152/pdf.

Catalan-Carrio, R., Saez, J., Fernández Cuadrado, L. Á., Arana, G., Basabe-Desmonts, L., & Benito-Lopez, F. (2022). Ionogel-based hybrid polymer-paper handheld platform for nitrite and nitrate determination in water samples. *Analytica Chimica Acta*, *1205*, 339753. Available from https://doi.org/10.1016/j.aca.2022.339753.

Chakraborty, D. D., Nath, L. K., & Chakraborty, P. (2018). Recent progress in smart polymers: Behavior, mechanistic understanding and application. *Polymer – Plastics Technology and Engineering*, *57*(10), 945–957. Available from https://doi.org/10.1080/03602559.2017.1364383, http://www.tandf.co.uk/journals/titles/03602559.asp.

Chan, C. L. C., Lei, I. M., van de Kerkhof, G. T., Parker, R. M., Richards, K. D., Evans, R. C., Huang, Y. Y. S., & Vignolini, S. (2022). 3D printing of liquid crystalline hydroxypropyl cellulose—Toward tunable and sustainable volumetric photonic structures. *Advanced Functional Materials*, *32*(15), 2108566. Available from https://doi.org/10.1002/adfm.202108566, http://onlinelibrary.wiley.com/journal/10.1002/(ISSN)1616-3028.

Chatterjee, A. (2022). At the intersection of natural structural coloration and bioengineering. *Biomimetics*, *7*(2), 66. Available from https://doi.org/10.3390/biomimetics7020066.

Chen, G.-Y., Chen, L.-X., Gao, J., Chen, C., Guan, J., Cao, Z., Hu, Y., & Yang, F.-Q. (2023). A novel molecularly imprinted sensor based on CuO Nanoparticles with peroxidase-like activity for the selective determination of astragaloside-IV. *Biosensors*, *13*(11), 959. Available from https://doi.org/10.3390/bios13110959.

Chen, J., Ling, Z., Wang, X., Ping, X., Xie, Y., Ma, H., Guo, J., & Yong, Q. (2023). All bio-based chiral nematic cellulose nanocrystals films under supramolecular tuning by chitosan/deacetylated chitin nanofibers for reversible multi-response and sensor application. *Chemical Engineering Journal*, *466*, 143148. Available from https://doi.org/10.1016/j.cej.2023.143148.

Chumee, J., Kumpun, S., Nimanong, N., Banditaubol, N., & Ohama, P. (2022). Colorimetric biofilm sensor with anthocyanin for monitoring fresh pork spoilage. *Materials Today: Proceedings*, *65*, 2467–2472. Available from https://doi.org/10.1016/j.matpr.2022.07.103.

Ciaccheri, L., Adinolfi, B., Mencaglia, A. A., & Mignani, A. G. (2023). Smartphone-enabled colorimetry. *Sensors*, *23*(12), 5559. Available from https://doi.org/10.3390/s23125559, http://www.mdpi.com/journal/sensors.

Dai, S., Prempeh, N., Liu, D., Fan, Y., Gu, M., & Chang, Y. (2017). Cholesteric film of Cu(II)-doped cellulose nanocrystals for colorimetric sensing of ammonia gas. *Carbohydrate Polymers*, *174*, 531–539. Available from https://doi.org/10.1016/j.carbpol.2017.06.098.

Dong, X., Zhang, Z. L., Zhao, Y. Y., Li, D., Wang, Z. L., Wang, C., Song, F., Wang, X. L., & Wang, Y. Z. (2022). Bio-inspired non-iridescent structural coloration enabled by self-assembled cellulose nanocrystal composite films with balanced ordered/disordered arrays. *Composites Part B: Engineering*, *229*, 109456. Available from https://doi.org/10.1016/j.compositesb.2021.109456, https://www.journals.elsevier.com/composites-part-b-engineering.

El Hani, O., Karrat, A., Digua, K., & Amine, A. (2023). Advanced molecularly imprinted polymer-based paper analytical device for selective and sensitive detection of Bisphenol-A in water samples. *Microchemical Journal*, *184*, 108157. Available from https://doi.org/10.1016/j.microc.2022.108157.

Fabregat, V., Burguete, M. I., Galindo, F., & Luis, S. V. (2017). Influence of polymer composition on the sensitivity towards nitrite and nitric oxide of colorimetric disposable test strips. *Environmental Science and Pollution Research*, *24*(4), 3448–3455. Available from https://doi.org/10.1007/s11356-016-8068-0, http://www.springerlink.com/content/0944-1344.

Fan, Y., Tang, S., Thomas, E. L., & Olsen, B. D. (2014). Responsive block copolymer photonics triggered by protein-polyelectrolyte coacervation. *ACS Nano*, *8*(11), 11467–11473. Available from https://doi.org/10.1021/nn504565r, http://pubs.acs.org/journal/ancac3.

Foelen, Y., Puglisi, R., Debije, M. G., & Schenning, A. P. H. J. (2023). Photonic liquid crystal polymer absorbent for immobilization and detection of gaseous nerve agent simulants. *ACS Applied Optical Materials*, *1* (1), 107–114. Available from https://doi.org/10.1021/acsaom.2c00014.

Godino, N., Vereshchagina, E., Gorkin, R., & Ducrée, J. (2014). Centrifugal automation of a triglyceride bioassay on a low-cost hybrid paper-polymer device. *Microfluidics and Nanofluidics*, *16*(5), 895–905. Available from https://doi.org/10.1007/s10404-013-1283-9.

Goni-Lizoain, C., Bonnaire, R., Fontanier, J. C., Copin, E., Gilblas, R., Aubry-Meneveau, C., Rumeau, P., & Le Maoult, Y. (2023). Polypropylene/bis(benzoxazolyl)stilbene mechanochromic blends, an attractive feature for colorimetric strain detection. *Sensors and Actuators A: Physical*, *355*, 114310. Available from https://doi.org/10.1016/j.sna.2023.114310, https://www.journals.elsevier.com/sensors-and-actuators-a-physical.

Greene, N. T., & Shimizu, K. D. (2005). Colorimetric molecularly imprinted polymer sensor array using dye displacement. *Journal of the American Chemical Society*, *127*(15), 5695–5700. Available from https://doi.org/10.1021/ja0468022.

Gu, P., Zhang, G., Deng, Z., Tang, Z., Zhang, H., Khusbu, F. Y., Wu, K., Chen, M., & Ma, C. (2018). A novel label-free colorimetric detection of l-histidine using Cu2 + -modulated G-quadruplex-based DNAzymes. *Spectrochimica Acta Part A: Molecular and Biomolecular Spectroscopy*, *203*, 195–200. Available from https://doi.org/10.1016/j.saa.2018.05.084.

Guembe-García, M., González-Ceballos, L., Arnaiz, A., Fernández-Muiño, M. A., Sancho, M. T., Osés, S. M., Ibeas, S., Rovira, J., Melero, B., Represa, C., García, J. M., & Vallejos, S. (2022). Easy nitrite analysis of processed meat with colorimetric polymer sensors and a smartphone app. *ACS Applied Materials & Interfaces*, *14*(32), 37051–37058. Available from https://doi.org/10.1021/acsami.2c09467.

Guirado-Moreno, J. C., Carreira-Barral, I., Ibeas, S., García, J. M., Granès, D., Marchet, N., & Vallejos, S. (2023). Democratization of copper analysis in grape must following a polymer-based lab-on-a-chip approach. *Applied Materials and Interfaces*, *15*(12), 16055–16062. Available from https://doi.org/10.1021/acsami.3c00395, http://pubs.acs.org/journal/aamick.

Guo, L., Zheng, H., Zhang, C., Qu, L., & Yu, L. (2020). A novel molecularly imprinted sensor based on PtCu bimetallic nanoparticle deposited on PSS functionalized graphene with peroxidase-like activity for selective determination of puerarin. *Talanta*, *210*, 120621. Available from https://doi.org/10.1016/j.talanta.2019.120621.

Han, J., Liu, F., Qi, J., Arabi, M., Li, W., Wang, G., Chen, L., & Li, B. (2022). A ZnFe2O4-catalyzed segment imprinted polymer on a three-dimensional origami paper-based microfluidic chip for the detection of microcystin. *The Analyst*, *147*(6), 1060–1065. Available from https://doi.org/10.1039/d2an00032f.

Han, X., Liu, Y., & Yin, Y. (2014). Colorimetric stress memory sensor based on disassembly of gold nanoparticle chains. *Nano Letters*, *14*(5), 2466–2470. Available from https://doi.org/10.1021/nl500144k.

He, J., Xiao, G., Chen, X., Qiao, Y., Xu, D., & Lu, Z. (2019). A thermoresponsive microfluidic system integrating a shape memory polymer-modified textile and a paper-based colorimetric sensor for the detection of glucose in human sweat. *RSC Advances*, *9*(41), 23957–23963. Available from https://doi.org/10.1039/c9ra02831e.

He, Y. D., Zhang, Z. L., Xue, J., Wang, X. H., Song, F., Wang, X. L., Zhu, L. L., & Wang, Y. Z. (2018). Biomimetic optical cellulose nanocrystal films with controllable iridescent color and environmental stimuli-responsive chromism. *Applied Materials and Interfaces*, *10*(6), 5805–5811. Available from https://doi.org/10.1021/acsami.7b18440, http://pubs.acs.org/journal/aamick.

Hosu, O., Lettieri, M., Papara, N., Ravalli, A., Sandulescu, R., Cristea, C., & Marrazza, G. (2019). Colorimetric multienzymatic smart sensors for hydrogen peroxide, glucose and catechol screening analysis. *Talanta*, *204*, 525–532. Available from https://doi.org/10.1016/j.talanta.2019.06.041.

Housaindokht, M. R., Jamshidi, A., Zonoz, F. M., & Firouzi, M. (2022). A novel nanocomposite (g-C3N4/Fe3O4@P2W15V3) with dual function in organic dyes degradation and cysteine sensing. *Chemosphere*, *304*, 135305. Available from https://doi.org/10.1016/j.chemosphere.2022.135305.

Hu, C. Y., Bai, L., Song, F., Wang, Y. L., & Wang, Y. Z. (2022). Cellulose nanocrystal and beta-cyclodextrin chiral nematic composite films as selective sensor for methanol discrimination. *Carbohydrate Polymers*, *296*, 119929. Available from https://doi.org/10.1016/j.carbpol.2022.119929.

Huang, K., Chen, Y., Zhou, F., Zhao, X., Liu, J., Mei, S., Zhou, Y., & Jing, T. (2017). Integrated ion imprinted polymers-paper composites for selective and sensitive detection of Cd(II) ions. *Journal of Hazardous Materials*, *333*, 137–143. Available from https://doi.org/10.1016/j.jhazmat.2017.030.035.

Kafashan, A., Joze-Majidi, H., Kazemi-Pasarvi, S., Babaei, A., & Jafari, S. M. (2023). Nanocomposites of soluble soybean polysaccharides with grape skin anthocyanins and graphene oxide as an efficient halochromic smart packaging. *Sustainable Materials and Technologies*, *38*, e00755. Available from https://doi.org/10.1016/j.susmat.2023.e00755.

Kaleeswaran, P., Nandhini, T., & Pitchumani, K. (2016). Naked eye sensing of melamine: Aggregation induced recognition by sodium d-gluconate stabilised silver nanoparticles. *New Journal of Chemistry*, *40*(4), 3869–3874. Available from https://doi.org/10.1039/C5NJ03083H.

Kan, Y., Jiang, C., Xi, Q., Wang, X., Peng, L., Jiang, J., & Yu, R. (2014). A simple, sensitive colorimetric assay for coralyne based on target induced split G-quadruplex formation. *Analytical Sciences*, *30*(5), 561–568. Available from https://doi.org/10.2116/analsci.30.561.

Kang, Z.-W., Li, Z.-Z., Kankala, R. K., Wang, S.-B., & Chen, A.-Z. (2023). Supercritical fluid-assisted fabrication of Pt-modified cerium oxide nanozyme based on polymer nanoreactors for peroxidase-like and glucose detection characteristics. *The Journal of Supercritical Fluids*, *198*, 105915. Available from https://doi.org/10.1016/j.supflu.2023.105915.

Kawamura, A., & Miyata, T. (2016). *Biosensors. Biomaterials Nanoarchitectonics* (pp. 157–176). Japan: Elsevier Inc. Available from http://www.sciencedirect.com/science/book/9780323371278, https://doi.org/10.1016/B978-0-323-37127-8.00010-8.

Kim, D., Hwang, K. S., Kim, J. H., Lee, C., & Lee, J. Y. (2021). Multituning of structural color by protonation and conjugate bases. *ACS Applied Polymer Materials*, *3*(6), 2902–2910. Available from https://doi.org/10.1021/acsapm.0c01382, http://pubs.acs.org/journal/aapmcd.

Kim, J. H., Rosenfeld, J., Kim, Y. C., Choe, S., Composto, R. J., Lee, D., & Dreyfus, R. (2023). Polymer-grafted, gold nanoparticle-based nano-capsules as reversible colorimetric tensile strain sensors. *Small (Weinheim an der Bergstrasse, Germany)*, *19*(36), 2300361. Available from https://doi.org/10.1002/smll.202300361, http://onlinelibrary.wiley.com/journal/10.1002/(ISSN)1613-6829.

Kim, T., Lee, J. W., Park, C., Lee, K., Lee, C. E., Lee, S., Kim, Y., Kim, S., Jeon, S., Ryu, D. Y., Koh, W. G., & Park, C. (2022). Self-powered finger motion-sensing structural color display enabled by block copolymer photonic crystal. *Nano Energy*, *92*, 106688. Available from https://doi.org/10.1016/j.nanoen.2021.106688, http://www.journals.elsevier.com/nano-energy/.

Ko, E., Hur, W., Son, S. E., Seong, G. H., & Han, D. K. (2021). Au nanoparticle-hydrogel nanozyme-based colorimetric detection for on-site monitoring of mercury in river water. *Microchimica Acta*, *188*(11), 382. Available from https://doi.org/10.1007/s00604-021-05032-4, http://www.springer/at/mca.

Komal, M., Vinoth Kumar, J., Arulmozhi, R., Sherlin Nivetha, M., Pavithra, S., & Abirami, N. (2023). Selective and sensitive on-site colorimetric detection of 4,4′-isopropylidenediphenol using non-enzymatic molecularly imprinted graphitic carbon nitride hybrids in milk and water samples. *New Journal of Chemistry*, *47*(19), 9087–9100. Available from https://doi.org/10.1039/d3nj01241g.

Kong, Q., Wang, Y., Zhang, L., Ge, S., & Yu, J. (2017). A novel microfluidic paper-based colorimetric sensor based on molecularly imprinted polymer membranes for highly selective and sensitive detection of bisphenol A. *Sensors and Actuators B: Chemical*, *243*, 130–136. Available from https://doi.org/10.1016/j.snb.2016.11.146.

Krauss, S. T., Holt, V. C., & Landers, J. P. (2017). Simple reagent storage in polyester-paper hybrid microdevices for colorimetric detection. *Sensors and Actuators, B: Chemical*, *246*, 740–747. Available from https://doi.org/10.1016/j.snb.2017.020.018.

Kuckling, D., Doering, A., Krahl, F., & Arndt, K.-F. (2012). *Polymer science: A comprehensive reference*. *Stimuli-Responsive Polymer Systems* (8). Elsevier B.V. Available from https://doi.org/10.1016/B978-0-444-53349-4.00214-4.

Kuşçuoğlu, C. K., Güner, H., Söylemez, M. A., Güven, O., & Barsbay, M. (2019). A smartphone-based colorimetric PET sensor platform with molecular recognition via thermally initiated RAFT-mediated graft copolymerization. *Sensors and Actuators, B: Chemical*, *296*, 126653. Available from https://doi.org/10.1016/j.snb.2019.126653, https://www.journals.elsevier.com/sensors-and-actuators-b-chemical.

Lampard, E. V., Sedgwick, A. C., Sombuttan, T., Williams, G. T., Wannalerse, B., Jenkins, A. T. A., Bull, S. D., & James, T. D. (2018). Dye displacement assay for saccharides using benzoxaborole hydrogels. *ChemistryOpen*, *7*(3), 266–268. Available from https://doi.org/10.1002/open.201700193, http://onlinelibrary.wiley.com/journal/10.1002/(ISSN)2191-1363.

Lee, W. J., Kim, B., Han, S. W., Seo, M., Choi, S.-E., Yang, H., Kim, S.-H., Jeong, S., & Kim, J. W. (2018). 2-Dimensional colloidal micropatterning of cholesteric liquid crystal microcapsules for temperature-responsive color displays. *Journal of Industrial and Engineering Chemistry*, *68*, 393–398. Available from https://doi.org/10.1016/j.jiec.2018.08.014.

Li, Q., Sun, T., Salentijn, G. I. J., Ning, B., Han, D., Bai, J., Peng, Y., Gao, Z., & Wang, Z. (2022). Bifunctional ligand-mediated amplification of polydiacetylene response to biorecognition of diethylstilbestrol for on-site smartphone detection. *Journal of Hazardous Materials*, *432*, 128692. Available from https://doi.org/10.1016/j.jhazmat.2022.128692.

Li, T., Bu, J., Yang, Y., & Zhong, S. (2024). A smartphone-assisted one-step bicolor colorimetric detection of glucose in neutral environment based on molecularly imprinted polymer nanozymes. *Talanta*, *267*, 125256. Available from https://doi.org/10.1016/j.talanta.2023.125256.

Liberman-Martin, A. L., Chu, C. K., & Grubbs, R. H. (2017). Application of bottlebrush block copolymers as photonic crystals. *Macromolecular Rapid Communications*, *38*(13), 1700058. Available from https://doi.org/10.1002/marc.201700058, http://www3.interscience.wiley.com/journal/117932056/grouphome.

Lim, H. S., Lee, J.-H., Walish, J. J., & Thomas, E. L. (2012). Dynamic swelling of tunable full-color block copolymer photonic gels via counterion exchange. *ACS Nano*, *6*(10), 8933–8939. Available from https://doi.org/10.1021/nn302949n.

Lin, E.-L., Hsu, W.-L., & Chiang, Y.-W. (2018). Trapping structural coloration by a bioinspired gyroid microstructure in solid state. *ACS Nano*, *12*(1), 485–493. Available from https://doi.org/10.1021/acsnano.7b07017.

Lin, I. M., Tsai, R. S., Chou, Y. T., & Chiang, Y. W. (2023). Photonic crystal reflectors with ultrahigh sensitivity and discriminability for detecting extremely low-concentration surfactants. *Applied Materials and*

Interfaces, *15*(38), 45249–45259. Available from https://doi.org/10.1021/acsami.3c06946, http://pubs.acs.org/journal/aamick.

Liu, L., Li, W., Liu, K., Yan, J., Hu, G., & Zhang, A. (2011). Comblike thermoresponsive polymers with sharp transitions: Synthesis, characterization, and their use as sensitive colorimetric sensors. *Macromolecules*, *44* (21), 8614–8621. Available from https://doi.org/10.1021/ma201874c.

Liu, S., Yang, Y., Zhang, L., Xu, J., & Zhu, J. (2020). Recent progress in responsive photonic crystals of block copolymers. *Journal of Materials Chemistry C*, *8*, 16633–16647. Available from https://doi.org/10.1039/d0tc04561f.

Liu, X., Zhu, C., Xu, L., Dai, Y., Liu, Y., & Liu, Y. (2018). Green and facile synthesis of highly stable gold nanoparticles via hyperbranched polymer in-situ reduction and their application in Ag + detection and separation. *Polymers*, *10*(1), 42. Available from https://doi.org/10.3390/polym10010042.

Liu, X.-Y., Cheng, F., Liu, Y., Li, W.-G., Chen, Y., Pan, H., & Liu, H.-J. (2010). Thermoresponsive gold nanoparticles with adjustable lower critical solution temperature as colorimetric sensors for temperature, pH and salt concentration. *Journal of Materials Chemistry*, *20*(2), 278–284. Available from https://doi.org/10.1039/B916125B.

Lowdon, J. W., Diliën, H., van Grinsven, B., Eersels, K., & Cleij, T. J. (2021). Colorimetric sensing of amoxicillin facilitated by molecularly imprinted polymers. *Polymers*, *13*(13), 2221. Available from https://doi.org/10.3390/polym13132221.

Ma, X., Guo, G., Wu, X., Wu, Q., Liu, F., Zhang, H., Shi, N., & Guan, Y. (2023). Advances in integration, wearable applications, and artificial intelligence of biomedical microfluidics systems. *Micromachines*, *14* (5), 972. Available from https://doi.org/10.3390/mi14050972.

Macfarlane, R. J., Kim, B., Lee, B., Weitekamp, R. A., Bates, C. M., Lee, S. F., Chang, A. B., Delaney, K. T., Fredrickson, G. H., Atwater, H. A., & Grubbs, R. H. (2014). Improving brush polymer infrared one-dimensional photonic crystals via linear polymer additives. *Journal of the American Chemical Society*, *136* (50), 17374–17377. Available from https://doi.org/10.1021/ja5093562, http://pubs.acs.org/journal/jacsat.

Martínez-Aviño, A., de Diego-Llorente-Luque, M., Molins-Legua, C., & Campíns-Falcó, P. (2022). Advances in the measurement of polymeric colorimetric sensors using portable instrumentation: Testing the light influence. *Polymers*, *14*(20), 4285. Available from https://doi.org/10.3390/polym14204285.

Matthew, S. A. L., Egan, G., Witte, K., Kaewchuchuen, J., Phuagkhaopong, S., Totten, J. D., & Seib, F. P. (2022). Smart silk origami as eco-sensors for environmental pollution. *ACS Applied Bio Materials*, *5*(8), 3658–3666. Available from https://doi.org/10.1021/acsabm.2c00023, http://pubs.acs.org/journal/aabmcb.

Mills, A., Hawthorne, D., Burns, L., & Hazafy, D. (2017). Novel temperature-activated humidity-sensitive optical sensor. *Sensors and Actuators B: Chemical*, *240*, 1009–1015. Available from https://doi.org/10.1016/j.snb.2016.08.182.

Mishra, A., Kushwaha, A., Maurya, P., & Verma, R. (2024). Colorimetric and absorbance based sensor for sulfide and bicarbonate ions by dye doped polymer composite. *Spectrochimica Acta Part A: Molecular and Biomolecular Spectroscopy*, *305*, 123554. Available from https://doi.org/10.1016/j.saa.2023.123554.

Nascimento, L. L. B. S., Mageste, A. B., Ferreira, G. M. D., Patrício, P. d. R., Rezende, S. d. S., de Oliveira, J. E., Cardoso, M. d. G., & Ferreira, G. M. D. (2023). Flavonoid-incorporated starch and poly(vinyl alcohol) film: Sensitive and selective colorimetric sensor for copper identification and quantification in beverages and environmental samples. *Colloids and Surfaces A: Physicochemical and Engineering Aspects*, *679*, 132574. Available from: http://www.elsevier.com/locate/colsurfa, https://doi.org/10.1016/j.colsurfa.2023.132574.

Nguyen, T. H. A., Le, T. T. V., Huynh, B. A., Nguyen, N. V., Le, V. T., Doan, V. D., Tran, V. A., Nguyen, A. T., Cao, X. T., & Vasseghian, Y. (2022). Novel biogenic gold nanoparticles stabilized on poly(styrene-co-maleic anhydride) as an effective material for reduction of nitrophenols and colorimetric detection of Pb(II. *Environmental Research*, *212*, 113281. Available from https://doi.org/10.1016/j.envres.2022.113281, http://www.elsevier.com/inca/publications/store/6/2/2/8/2/1/index.htt.

Noro, A., Tomita, Y., Matsushita, Y., & Thomas, E. L. (2016). Enthalpy-driven swelling of photonic block polymer films. *Macromolecules*, *49*(23), 8971–8979. Available from https://doi.org/10.1021/acs.macromol.6b01867, http://pubs.acs.org/journal/mamobx.

Nsuamani, M. L., Zolotovskaya, S., Abdolvand, A., Daeid, N. N., & Adegoke, O. (2022). Thiolated gamma-cyclodextrin-polymer-functionalized CeFe(3)O(4) magnetic nanocomposite as an intrinsic nanocatalyst for the selective and ultrasensitive colorimetric detection of triacetone triperoxide. *Chemosphere*, *307*, 136108. Available from https://doi.org/10.1016/j.chemosphere.2022.136108.

Oziri, O. J., Maeki, M., Tokeshi, M., Isono, T., Tajima, K., Satoh, T., Sato, S. I., & Yamamoto, T. (2022). Topology-dependent interaction of cyclic poly(ethylene glycol) complexed with gold nanoparticles against bovine serum albumin for a colorimetric change. *Langmuir: The ACS Journal of Surfaces and Colloids*, *38*(17), 5286–5295. Available from https://doi.org/10.1021/acs.langmuir.1c03027, http://pubs.acs.org/journal/langd5.

Pan, Y., Xie, S., Wang, H., Huang, L., Shen, S., Deng, Y., Ma, Q., Liu, Z., Zhang, M., Jin, M., & Shui, L. (2023). Microfluidic construction of responsive photonic microcapsules of cholesteric liquid crystal for colorimetric temperature microsensors. *Advanced Optical Materials*, *11*, 2202141. Available from https://doi.org/10.1002/adom.202202141.

Park, J., Lee, Y., Barbee, M. H., Cho, S., Cho, S., Shanker, R., Kim, J., Myoung, J., Kim, M. P., Baig, C., Craig, S. L., & Ko, H. (2019). A hierarchical nanoparticle-in-micropore architecture for enhanced mechanosensitivity and stretchability in mechanochromic electronic skins. *Advanced Materials*, *31*(25), 1808148. Available from https://doi.org/10.1002/adma.201808148, http://onlinelibrary.wiley.com/journal/10.1002/(ISSN)1521-4095.

Park, J. J., Kim, Y., Lee, C., Kook, J. W., Kim, D., Kim, J. H., Hwang, K. S., & Lee, J. Y. (2020). Colorimetric visualization using polymeric core-shell nanoparticles: Enhanced sensitivity for formaldehyde gas sensors. *Polymers*, *12*(5), 998. Available from https://doi.org/10.3390/POLYM12050998, https://www.mdpi.com/2073-4360/12/5/998.

Park, T. H., Yu, S., Cho, S. H., Kang, H. S., Kim, Y., Kim, M. J., Eoh, H., Park, C., Jeong, B., Lee, S. W., Ryu, D. Y., Huh, J., & Park, C. (2018). Block copolymer structural color strain sensor. *NPG Asia Materials*, *10*(4), 328–339. Available from https://doi.org/10.1038/s41427-018-0036-3, http://www.nature.com/am/index.html.

Preman, N. K., Jain, S., Antony, A., Shetty, D. M., Fathima, N., Prasad, K. S., & Johnson, R. P. (2023). Stimuli-responsive copolymer-mediated synthesis of gold nanoparticles for nanozyme-based colorimetric detection of mercury(II) ions. *ACS Applied Polymer Materials*, *5*(8), 6377–6389. Available from https://doi.org/10.1021/acsapm.3c00977, http://pubs.acs.org/journal/aapmcd.

Prianka, T. R., Subhan, N., Reza, H. M., Hosain, M. K., Rahman, M. A., Lee, H., & Sharker, S. M. (2018). Recent exploration of bio-mimetic nanomaterial for potential biomedical applications. *Materials Science and Engineering C*, *93*, 1104–1115. Available from https://doi.org/10.1016/j.msec.2018.09.012.

Punnoy, P., Preechakasedkit, P., Aumnate, C., Rodthongkum, N., Potiyaraj, P., & Ruecha, N. (2021). Polyvinyl alcohol/starch modified cotton thread surface as a novel colorimetric glucose sensor. *Materials Letters*, *299*, 130076. Available from https://doi.org/10.1016/j.matlet.2021.130076.

Qiao, Y., Zhao, Y., Yuan, X., Zhao, Y., & Ren, L. (2018). One-dimensional photonic crystals prepared by self-assembly of brush block copolymers with broad PDI. *Journal of Materials Science*, *53*(23), 16160–16168. Available from https://doi.org/10.1007/s10853-018-2754-x.

Raja, D. A., Musharraf, S. G., Shah, M. R., Jabbar, A., Bhanger, M. I., & Malik, M. I. (2020). Poly(propylene glycol) stabilized gold nanoparticles: An efficient colorimetric assay for ceftriaxone. *Journal of Industrial and Engineering Chemistry*, *87*, 180–186. Available from https://doi.org/10.1016/j.jiec.2020.03.041, http://www.sciencedirect.com/science/journal/1226086X.

Raja, D. A., Rahim, S., Shah, M. R., Bhanger, M. I., & Malik, M. I. (2023). Silver nanoparticle based efficient colorimetric assay for carbaryl – An insecticide. *Journal of Molecular Liquids*, *372*, 121200. Available from https://doi.org/10.1016/j.molliq.2023.121200, https://www.journals.elsevier.com/journal-of-molecular-liquids.

Saedan, A., Yukird, J., Rodthongkum, N., & Ummartyotin, S. (2023). Graphene oxide modified cellulose paper-based device: A novel platform for cooking oil quality evaluation. *Food Control*, *148*, 109675. Available from https://doi.org/10.1016/j.foodcont.2023.109675.

Sengupta, P., Pramanik, K., Datta, P., & Sarkar, P. (2020). Chemically modified carbon nitride-chitin-acetic acid hybrid as a metal-free bifunctional nanozyme cascade of glucose oxidase-peroxidase for "click off" colorimetric detection of peroxide and glucose. *Biosensors and Bioelectronics*, *154*, 112072. Available from https://doi.org/10.1016/j.bios.2020.112072.

Seo, H. B., Yu, Y. G., Chae, C. G., Kim, M. J., & Lee, J. S. (2019). Synthesis of ultrahigh molecular weight bottlebrush block copolymers of ω-end-norbornyl polystyrene and polymethacrylate macromonomers. *Polymer*, *177*, 241–249. Available from https://doi.org/10.1016/j.polymer.2019.06.009, http://www.journals.elsevier.com/polymer/.

Sha, L., Zhu, M., Lin, F., Yu, X., Dong, L., Wu, L., Ding, R., Wu, S., & Xu, J. (2021). Stable DNA aptamer–metal–organic framework as horseradish peroxidase mimic for ultra-sensitive detection of carcinoembryonic antigen in serum. *Gels*, *7*, 181. Available from https://doi.org/10.3390/gels7040181.

Son, H., Shin, H., & Kim, Y. (2022). Thermo-responsive and reversible colorimetric sensors as fever-checkers using low critical solution temperature polymer–Au nanocomposites. *Colloids and Surfaces A: Physicochemical and Engineering Aspects*, *651*, 129794. Available from https://doi.org/10.1016/j.colsurfa.2022.129794.

Song, D.-P., Li, C., Colella, N. S., Lu, X., Lee, J.-H., & Watkins, J. J. (2015). Thermally tunable metallodielectric photonic crystals from the self-assembly of brush block copolymers and gold nanoparticles. *Advanced Optical Materials*, *3*(9), 1169–1175. Available from https://doi.org/10.1002/adom.201500116.

Song, D. P., Li, C., Li, W., & Watkins, J. J. (2016). Block copolymer nanocomposites with high refractive index contrast for one-step photonics. *ACS Nano*, *10*(1), 1216–1223. Available from https://doi.org/10.1021/acsnano.5b06525, http://pubs.acs.org/journal/ancac3.

Song, D. P., Zhao, T. H., Guidetti, G., Vignolini, S., & Parker, R. M. (2019). Hierarchical photonic pigments via the confined self-assembly of bottlebrush block copolymers. *ACS Nano*, *13*(2), 1764–1771. Available from https://doi.org/10.1021/acsnano.8b07845, http://pubs.acs.org/journal/ancac3.

Song, W., Lee, J. K., Gong, M. S., Heo, K., Chung, W. J., & Lee, B. Y. (2018). Cellulose nanocrystal-based colored thin films for colorimetric detection of aldehyde gases. *Applied Materials and Interfaces*, *10*(12), 10353–10361. Available from https://doi.org/10.1021/acsami.7b19738, http://pubs.acs.org/journal/aamick.

Sun, J., Bhushan, B., & Tong, J. (2013). Structural coloration in nature. *RSC Advances*, *3*(35), 14862–14889. Available from https://doi.org/10.1039/c3ra41096j.

Sutarlie, L., & Yang, K.-L. (2011). Monitoring spatial distribution of ethanol in microfluidic channels by using a thin layer of cholesteric liquid crystal. *Lab on a Chip*, *11*(23), 4093–4098. Available from https://doi.org/10.1039/C1LC20460B.

Sutthasupa, S., Padungkit, C., & Suriyong, S. (2021). Colorimetric ammonia (NH3) sensor based on an alginate-methylcellulose blend hydrogel and the potential opportunity for the development of a minced pork spoilage indicator. *Food Chemistry*, *362*, 130151. Available from https://doi.org/10.1016/j.foodchem.2021.130151.

Sveinbjörnsson, B. R., Weitekamp, R. A., Miyake, G. M., Xia, Y., Atwater, H. A., & Grubbs, R. H. (2012). Rapid self-assembly of brush block copolymers to photonic crystals. *Proceedings of the National Academy of Sciences of the United States of America*, *109*(36), 14332–14336. Available from https://doi.org/10.1073/pnas.1213055109UnitedStates, http://www.pnas.org/content/109/36/14332.full.pdf + html.

Tavassoli, M., Khezerlou, A., Firoozy, S., Ehsani, A., & Punia Bangar, S. (2023). Chitosan-based film incorporated with anthocyanins of red poppy (Papaver rhoeas L.) as a colorimetric sensor for the detection of shrimp freshness. *International Journal of Food Science & Technology*, *58*(6), 3050–3057. Available from https://doi.org/10.1111/ijfs.16432.

Urbano, B. F., Bustamante, S., Palacio, D. A., Vera, M., & Rivas, B. L. (2021). Polymer-based chromogenic sensors for the detection of compounds of environmental interest. *Polymer International*, *70*(9), 1202–1208. Available from http://onlinelibrary.wiley.com/journal/10.1002/(ISSN)1097-0126, https://doi.org/10.1002/pi.6223.

Vallejos, S., Muñoz, A., Ibeas, S., Serna, F., García, F. C., & García, J. M. (2013). Solid sensory polymer substrates for the quantification of iron in blood, wine and water by a scalable RGB technique. *Journal of Materials Chemistry A*, *1*(48), 15435–15441. Available from https://doi.org/10.1039/c3ta12703f.

Vaz, R., Frasco, M. F., & Sales, M. G. F. (2020). Photonics in nature and bioinspired designs: Sustainable approaches for a colourful world. *Nanoscale Advances*, *2*(11), 5106–5129. Available from https://doi.org/10.1039/d0na00445f, http://pubs.rsc.org/en/journals/journalissues/na?_ga2.190536939.1555337663.1552312502-1364180372.1550481316#!issueidna001002&typecurrent&issnonline2516-0230.

Vaz, R., Frasco, M. F., & Sales, M. G. F. (2022). *Chapter 4 – Biosensors: Concept and importance in point-of-care disease diagnosis. Biosensor Based Advanced Cancer Diagnostics* (pp. 59–84). Academic Press. Available from https://doi.org/10.1016/B978-0-12-823424-2.00001-6.

Veedu, A. P., Kuppusamy, S., Mohan, A. M., & Deivasigamani, P. (2023). Chromogenic probe adhered porous polymer monolith as real-time solid-state sensor for the detection of ultra-trace toxic mercury ions. *Environmental Research*, *239*, 117399. Available from https://doi.org/10.1016/j.envres.2023.117399.

Wang, L., Liu, X., Yang, Q., Fan, Q., Song, S., Fan, C., & Huang, W. (2010). A colorimetric strategy based on a water-soluble conjugated polymer for sensing pH-driven conformational conversion of DNA i-motif structure. *Biosensors and Bioelectronics*, *25*(7), 1838–1842. Available from https://doi.org/10.1016/j.bios.2009.120.016.

Wang, M., & Kan, X. (2020). Imprinted polymer/Fe3O4 micro-particles decorated multi-layer graphite paper: Electrochemical and colorimetric dual-modal sensing interface for aloe-emodin assay. *Sensors and Actuators B: Chemical*, *323*, 128672. Available from https://doi.org/10.1016/j.snb.2020.128672.

Wang, S., Xu, S., Zhou, Q., Liu, Z., & Xu, Z. (2023). State-of-the-art molecular imprinted colorimetric sensors and their on-site inspecting applications. *Journal of Separation Science*, *46*(12), 2201059. Available from https://doi.org/10.1002/jssc.202201059, http://onlinelibrary.wiley.com/journal/10.1002/(ISSN)1615-9314.

Wang, X., Sun, X., Hu, P. A., Zhang, J., Wang, L., Feng, W., Lei, S., Yang, B., & Cao, W. (2013). Colorimetric sensor based on self-assembled polydiacetylene/graphene-stacked composite film for vapor-phase volatile organic compounds. *Advanced Functional Materials*, *23*(48), 6044–6050. Available from https://doi.org/10.1002/adfm.201301044.

Wang, X., Huang, K., Zhang, H., Zeng, L., Zhou, Y., & Jing, T. (2019). Preparation of molecularly imprinted polymers on hemin-graphene surface for recognition of high molecular weight protein. *Materials Science and Engineering: C*, *105*, 110141. Available from https://doi.org/10.1016/j.msec.2019.110141.

Wang, Z., Zhao, J., Li, Z., Bao, J., & Dai, Z. (2017). Sequence and structure dual-dependent interaction between small molecules and DNA for the detection of residual silver ions in As-prepared silver nanomaterials. *Analytical Chemistry*, *89*(12), 6815–6820. Available from https://doi.org/10.1021/acs.analchem.7b01238.

Wang, Z., Chan, C. L. C., Parker, R. M., & Vignolini, S. (2022). The limited palette for photonic block-copolymer materials: A historical problem or a practical limitation? *Angewandte Chemie International Edition*, *61*, e202117275. Available from https://doi.org/10.1002/anie.202117275.

Wibowo, A. F., Han, J. W., Kim, J. H., Prameswati, A., Park, J., Aisyah, S., Entifar, N., Lee, J., Kim, S., Lim, D. C., Moon, M. W., Kim, M. S., & Kim, Y. H. (2022). Multiple functionalities of highly conductive and flexible photo- and thermal-responsive colorimetric cellulose films. *Materials Research Letters*, *10*(1), 36–44. Available from https://doi.org/10.1080/21663831.2021.2013330, http://www.tandfonline.com/loi/tmrl20#.VyirCfl9670.

Wikantyasning, E. R., Da'i, M., Cholisoh, Z., & Kalsum, U. (2023). 3-Aminophenylboronic acid conjugation on responsive polymer and gold nanoparticles for qualitative bacterial detection. *Journal of Pharmacy and Bioallied Sciences*, *15*(2), 81–87. Available from https://doi.org/10.4103/jpbs.jpbs_646_22, http://www.jpbsonline.org.

Wu, R., & Selvaganapathy, P. R. (2023). Porous biocompatible colorimetric nanofiber-based sensor for selective ammonia detection on personal wearable protective equipment. *Sensors and Actuators B: Chemical*, *393*, 134270. Available from https://doi.org/10.1016/j.snb.2023.134270.

Wu, Y., Feng, J., Hu, G., Zhang, E., & Yu, H. H. (2023). Colorimetric sensors for chemical and biological sensing applications. *Sensors*, *3*(5), 2749. Available from https://doi.org/10.3390/s23052749, http://www.mdpi.com/journal/sensors.

Xiao, S.-J., Yuan, M.-Y., Shi, Y.-D., Wang, M.-P., Li, H.-H., Zhang, L., & Qiu, J.-D. (2023). Construction of covalent organic framework nanozymes with photo-enhanced hydrolase activities for colorimetric sensing of organophosphorus nerve agents. *Analytica Chimica Acta*, *1278*, 341706. Available from https://doi.org/10.1016/j.aca.2023.341706.

Xie, K., Gao, A., Li, C., & Li, M. (2014). Highly water-soluble and pH-sensitive colorimetric sensors based on a D–π–A heterocyclic azo chromosphere. *Sensors and Actuators B: Chemical*, *204*, 167–174. Available from https://doi.org/10.1016/j.snb.2014.07.090.

Xu, Z., Jin, D., Lu, X., Zhang, Y., & Liu, W. (2023). Sandwich-type colorimetric assay based on molecularly imprinted polymers and boronic acid functionalized Ni-Fe-MOF nanozyme for sensitive detection of allergen ovalbumin. *Microchemical Journal*, *194*, 109349. Available from https://doi.org/10.1016/j.microc.2023.109349.

Yang, B. Z., Su, Z. Y., & Jou, A. F. J. (2021). Exploiting the catalytic ability of polydopamine-remodeling gold nanoparticles toward the naked-eye detection of cancer cells at a single-cell level. *ACS Applied Bio Materials*, *4*(3), 2821–2828. Available from https://doi.org/10.1021/acsabm.1c00041, http://pubs.acs.org/journal/aabmcb.

Yang, H., Choi, S. E., Kim, D., Park, D., Lee, D., Choi, S., Nam, Y. S., & Kim, J. W. (2019). Color-spectrum-broadened ductile cellulose films for vapor-pH-responsive colorimetric sensors. *Journal of Industrial and Engineering Chemistry*, *80*, 590–596. Available from https://doi.org/10.1016/j.jiec.2019.08.039, http://www.sciencedirect.com/science/journal/1226086X.

Ye, Q., Yuan, E., Shen, J., Ye, M., Xu, Q., Hu, X., Shu, Y., & Pang, H. (2023). Amphiphilic polymer capped perovskite compositing with nano Zr-MOF for nanozyme-involved biomimetic cascade catalysis. *Advanced Science*, *10*, 2304149. Available from https://doi.org/10.1002/advs.202304149.

Yi, H., Lee, S. H., Kim, D., Jeong, H. E., & Jeong, C. (2022). Colorimetric sensor based on hydroxypropyl cellulose for wide temperature sensing range. *Sensors*, *22*(3), 886. Available from https://doi.org/10.3390/s22030886, https://www.mdpi.com/1424-8220/22/3/886/pdf.

Yin, M., Duan, Z., Zhang, C., Feng, L., Wan, Y., Cai, Y., Liu, H., Li, S., & Wang, H. (2019). A visualized colorimetric detection strategy for heparin in serum using a metal-free polymer nanozyme. *Microchemical Journal*, *145*, 864–871. Available from https://doi.org/10.1016/j.microc.2018.11.059.

Zhang, L., Zhang, W., Nie, Y., Wang, Y., & Zhang, P. (2023). Covalent organic framework-supported ultra-small Rh nanoparticles as peroxidase mimics for colorimetric sensing of cysteine. *Journal of Colloid and Interface Science*, *636*, 568–576. Available from https://doi.org/10.1016/j.jcis.2023.01.020.

Zhang, P., Wu, X., Xue, H., Wang, Y., Luo, X., & Wang, L. (2022). Wearable transdermal colorimetric microneedle patch for uric acid monitoring based on peroxidase-like polypyrrole nanoparticles. *Analytica Chimica Acta*, *1212*, 339911. Available from https://doi.org/10.1016/j.aca.2022.339911.

Zhang, Q., Wang, X., Decker, V., & Meyerhoff, M. E. (2020). Plasticizer-free thin-film sodium-selective optodes inkjet-printed on transparent plastic for sweat analysis. *Applied Materials and Interfaces*, *12*(23), 25616–25624. Available from https://doi.org/10.1021/acsami.0c05379, http://pubs.acs.org/journal/aamick.

Zhang, X., Peng, J., Xi, L., Lu, Z., Yu, L., Liu, M., Huo, D., & He, H. (2022). Molecularly imprinted polymers enhanced peroxidase-like activity of AuNPs for determination of glutathione. *Microchimica Acta*, *189*(12), 457. Available from https://doi.org/10.1007/s00604-022-05576-z.

Zhang, Y., Hao, J., Xu, X., Chen, X., & Wang, J. (2020). Protein corona-triggered catalytic inhibition of insufficient POSS polymer-caged gold nanoparticles for sensitive colorimetric detection of metallothioneins. *Analytical Chemistry*, *92*(2), 2080–2087. Available from https://doi.org/10.1021/acs.analchem.9b04593.

Zhao, G., Zhang, Y., Zhai, S., Sugiyama, J., Pan, M., Shi, J., & Lu, H. (2020). Dual response of photonic films with chiral nematic cellulose nanocrystals: Humidity and formaldehyde. *ACS Applied Materials & Interfaces*, *12*(15), 17833–17844. Available from https://doi.org/10.1021/acsami.0c00591.

Zhao, T. H., Parker, R. M., Williams, C. A., Lim, K. T. P., Frka-Petesic, B., & Vignolini, S. (2019). Printing of responsive photonic cellulose nanocrystal microfilm arrays. *Advanced Functional Materials*, *29*(21), 1804531. Available from https://doi.org/10.1002/adfm.201804531, http://onlinelibrary.wiley.com/journal/10.1002/(ISSN)1616-3028.

Zhao, Y., Guo, T., Yang, J., Li, Y., Yuan, X., Zhao, Y., & Ren, L. (2019). Alcohols responsive photonic crystals prepared by self-assembly of dendronized block copolymers. *Reactive and Functional Polymers*, *139*, 162–169. Available from https://doi.org/10.1016/j.reactfunctpolym.2019.04.001.

Zhao, Y., Du, J., Zhou, H., Zhou, S., Lv, Y., Cheng, Y., Tao, Y., Lu, J., & Wang, H. (2022). Biodegradable intelligent film for food preservation and real-time visual detection of food freshness. *Food Hydrocolloids*, *129*, 107665. Available from https://doi.org/10.1016/j.foodhyd.2022.107665.

Zhao, Z., Lin, T., Liu, W., Hou, L., Ye, F., & Zhao, S. (2019). Colorimetric detection of blood glucose based on GOx@ZIF-8@Fe-polydopamine cascade reaction. *Spectrochimica Acta Part A: Molecular and Biomolecular Spectroscopy*, *219*, 240–247. Available from https://doi.org/10.1016/j.saa.2019.04.061.

Zhou, H., Wang, H., He, W., Yang, Z., Cao, H., Wang, D., & Li, Y. (2022). Research progress of cholesteric liquid crystals with broadband reflection. *Molecules (Basel, Switzerland)*, *27*(14), 4427. Available from https://doi.org/10.3390/molecules27144427.

Zhou, Y., Liu, A., Li, Y., & Liu, S. (2023). Magnetic molecular imprinted polymers-based nanozyme for specific colorimetric detection of protocatechuic acid. *Coatings*, *13*(8), 1374. Available from https://doi.org/10.3390/coatings13081374.

Zhou, Y., Xue, Y., Lin, X., Duan, M., Hong, W., Geng, L., Zhou, J., & Fan, Y. (2024). Smartphone-based polydiacetylene colorimetric sensor for point-of-care diagnosis of bacterial infections. *Smart Materials in Medicine*, *5*(1), 140–152. Available from https://doi.org/10.1016/j.smaim.2023.10.002.

Zhu, X., Tang, J., Ouyang, X., Liao, Y., Feng, H., Yu, J., Chen, L., Lu, Y., Yi, Y., & Tang, L. (2024). A versatile CuCo@PDA nanozyme-based aptamer-mediated lateral flow assay for highly sensitive, on-site and dual-readout detection of Aflatoxin B1. *Journal of Hazardous Materials*, *465*, 133178. Available from https://doi.org/10.1016/j.jhazmat.2023.133178.

CHAPTER

Fluorogenic sensors

5

Isaí Barboza-Ramos[1], Sukriye Nihan Karuk Elmas[2] and Kirk S. Schanze[1]

[1]Department of Chemistry, University of Texas at San Antonio, San Antonio, TX, United States [2]The Faculty of Pharmacy, The Department of Analytical Chemistry, Istanbul University-Cerrahpaşa, Istanbul, Turkey

5.1 Introduction

The emergence of fluorescent molecular materials throughout the past few decades has favored advancements and improvements to develop highly sensitive fluorogenic probes for the detection and identification of relevant analytes in chemical and biological-related samples (Bazylevich et al., 2017; Nadler & Schultz, 2013). Among the efforts for the creation of chemo- and bio-sensors, developing fluorogenic probes with superior light harvesting properties, fluorescence emission tunability, suitable photostability, and appropriate signal modulation has acquired noteworthy importance (Qi et al., 2015; Mayr et al., 2009; Squeo et al., 2017; Zhang et al., 2018). Particularly, identifying fluorescence-lacking analytes with fluorescent probes is an essential approach to gathering evidence about the presence of those nonfluorescent molecules in a sample, representing a practical tool to obtain additional information about the localization and quantification of analytes (Chan & Wu, 2015; Li, Qiao, et al., 2018). Chromo-fluorogenic sensory probes have been viable for the rapid recognition of cations, small organic and inorganic molecules, volatile organic compounds, and macromolecules, including proteins and nucleic acids (Shan et al., 2018; Shanmugaraju et al., 2019; Wang et al., 2021). Studies have been oriented to increase the diversity of fluorogenic probes, modifying the structural nature of the fluorogenic probe and the mechanism of action behind the signal modulation (Kumar et al., 2019; Reineck & Gibson, 2017; Wei et al., 2021).

The structural features of polymeric materials as fluorogenic probes offer several advances as sensory components in optical recognition platforms (Chakraborty et al., 2022; Samanta et al., 2021). These probes provide direct or indirect efficient detectability by a range of fluorescence modulation schemes and the nature of interaction toward the various target analytes (Roy & Bandyopadhyay, 2018). The structural versatility and tunability of polymer-based materials allow the design of unique electronic structures, simultaneously serving as charge or energy transfer scaffolds to successfully trace other molecules (Gu et al., 2017; Jia et al., 2019). The integration of diverse chromophores into the backbone or as pendant groups gives rise to unique optoelectronic properties and selective interactions between the polymer chains and analyte that induce specific morphology, distribution, and arrangement of the polymeric chains (Anantha-Iyengar et al., 2019; Li, Acharya, et al., 2018). Such modifications mediate processes of energy or charge transport, affecting the absorption and emission spectral patterns with changes transduced into detectable signals to monitor target analytes (Hasegawa & Kitagawa, 2022; Sun & Schanze, 2022; Wang et al., 2014).

Sensory Polymers. DOI: https://doi.org/10.1016/B978-0-443-13394-7.00005-7

Recent advances in the field of polymeric materials as fluorescent sensors have pursued a variety of molecular architectures by modifying the monomeric repeating units, either backbone or pendant groups (Sanjuán et al., 2018; Zhou, Chua, et al., 2019). In practice, the performance of polymeric fluorogenic probes is determined by the resulting optical properties originating from the electronic states or energy band gaps that arise from the delocalized electronic structure of the polymer backbone or the conjugated moieties attached to the main chain (Pangeni & Nesterov, 2013; Zhao et al., 2006). Newly developed chromo-fluorogenic chemosensors will exhibit unique absorption and fluorescence emission spectral bands that enable the generation of measurable signals and particular detection mechanisms (Hu et al., 2019; Liu, Luo, et al., 2020). The objective of this chapter is to overview existing developments in the field of polymeric-based fluorogenic probes and the increasing availability and diversity of sensory materials (Fig. 5.1). The chapter starts with theoretical foundations on chemical sensing, which is followed by a comparison of polymers and small fluorophore molecules that are utilized in sensory probes. The next section includes the fluorescence modulation approaches for detection, including fluorescence sensing based on aggregation-induced emission (AIE), aggregation-caused quenching (ACQ), and charge and energy transfer processes. In closing, the chapter includes a discussion of specific examples of polymeric fluorogenic probes to detect a wide range of target species, providing an overview of relevant applicability in the subject under consideration.

5.2 Principles of fluorescence-based chemical sensing

The use of artificial sensory systems has extrapolated over a variety of disciplines, continuously growing into the quest for new fluorogenic materials (Bazylevich et al., 2017; Collot et al., 2018; Usama et al., 2021). A fluorogenic probe (P) is a molecule containing a moiety that can absorb photons of light at certain wavelengths to reach a higher molecular electronic state, which is called an excited state,

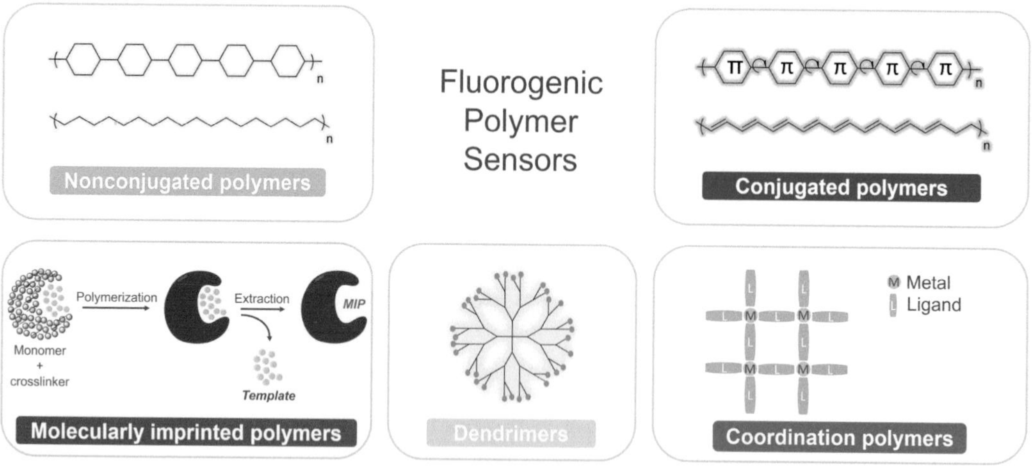

FIGURE 5.1

Types of fluorogenic sensory polymers presented in this chapter. This figure shows the 5 types of fluorogenic sensory polymers presented in this chapter: non-conjugated polymers, conjugated polymers, molecularly-imprinted polymers, dendrimers, and coordination polymers.

throughout a process called excitation (Fig. 5.2). Upon excitation, the excited molecules can follow a relaxation pathway in which the excess of energy can be released by radiative decay known as fluorescence, in which photons are released at lower energy wavelengths. When employed as a fluorogenic probe signal, the fluorescence emission delivers a real-time readout of a molecule's state and environment (Fu et al., 2018; Zhao, Takano, et al., 2022). Fluorescence emission is a fast process that usually falls into the time scale of 10^{-10} to 10^{-8} seconds. Even though fluorescence is a typical observable relaxation pathway upon excitation, there exist other competitive processes occurring as well, including intersystem crossing, phosphorescence emission, and nonradiative processes that occur by the release of thermal energy to the environment or collision with neighboring molecules (Singh et al., 2022; Sun & Liu, 2020). Those alternative events are favored based on the chemical nature of the polymer. However, for fluorogenic sensory probes, fluorescence modulation is considered the key component and contributor to developing detection methods (Kumar et al., 2017).

Most suitable fluorogenic probes are desired to have a high quantum yield and molar absorptivity (Marín-Hernández et al., 2016). Ideally, the fluorescence emission bands should be separated enough from the excitation wavelength so the signal can be isolated without the interference of the scattering of the excitation source. Upon excitation, the fluorophore emits light at longer wavelengths compared to the shorter emission wavelength of the excitation. For conjugated systems, as the conjugation length increases, the fluorescence emission experiences a redshift. As expected, the quantum yield and molar absorptivity of such systems increase due the enhanced conjugation and stabilization (Mao et al., 2021). The range of the concentration of the probe has potential effects on the efficacy of detecting an analyte, particularly at high concentrations where inner filter effects (IFEs) exist and attenuate the photon absorption and emission. An increase in the IFE causes the

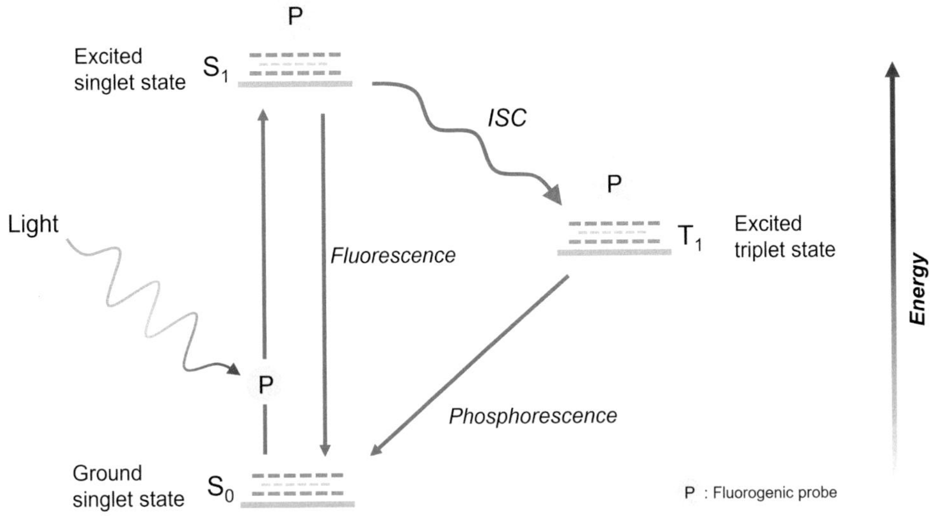

FIGURE 5.2

Illustration of the Jablonski diagram related to the electronic and vibrational states on a molecule upon the light excitation process. The image includes a general representation of the Jablonski diagram that is related to the common electronic and vibrational states on a molecule upon light absorption.

self-quenching that generates the occurrence of interferent charge or energy transfer events (Panigrahi & Mishra, 2019). However, either working at low concentrations of the probe or incorporating well-solubilizing groups into the fluorophore can provide an approach to enhance the solubility properties and intermolecular interactions and, therefore, a reduction of the interfering IFEs.

5.3 Polymer versus small molecules in fluorescence sensing

By comparing polymer-based fluorescence sensors with small molecule-based fluorescence sensors, it is known that polymeric fluorescent sensory materials have noticeable advantages due to their superior and distinctive features (Adhikari & Majumdar, 2004). Sensitivity, selectivity, stability, and cost are some key factors that differentiate the effectiveness between small molecule-based fluorescence sensors and polymer-based fluorescence sensors (Jiang et al., 2009; Zhang, Wang, et al., 2016; Zhou & Swager, 1995). For example, the performance of small molecule-based chemical sensors in terms of selectivity and sensitivity mostly depends on the interaction of the host-guest. This kind of interaction is based on weak forces, therefore, the sensor is highly affected by the nature of the solvent in the system. The use of polymeric materials for sensing applications have been previously reviewed by García et al. (2011). For chemical sensors built from small molecules, the host-guest interaction is a major determinant for the sensitivity and selectivity towards specific analytes. Especially, the kind of interactions involved in the sensing mechanism may be affected by secondary interactions with solvent molecules, therefore, the sensor is highly affected by the nature of the solvent in the system. In the case of water-soluble chemical sensors, the sensing capacity could be influenced by additional sensor-solvent interaction processes, including secondary electrostatic interactions and presence of hydrogen bonding. One advantage of sensory polymeric materials centers on the various geometry, morphology, and structural functionality parameters that open the versatility and availability of diverse polymer chemical structures for sensing platforms. The properties of macromolecular chemical sensors generally show much higher sensitivity and selectivity than small molecule sensors by the possibility to conduct modifications of the polymer structure. (García et al., 2011) On the one hand, by the incorporation of different functional groups on the side chains linked to the polymer backbone, the selectivity for specific analytes can be tuned. On the other hand, by modifying the hydrophobic or hydrophilic sections of the polymer chains, the intrachain and interchain polymer interactions and polymer-solvent interactions can be modified to enhance the performance of the sensing platform. Also, tuning the density of cross-links of the polymer represents an advantage to modulate and enhance the sensing capacity of these materials. Synthetic polymers are high molecular weight structures consisting of repeating monomers that lead to enhancement of the analyte-receptor interactions given the multiple binding sites for the target analytes. Due to the high molecular weights of the polymers and the several binding moieties, these materials display intrinsic properties that are sensitive to minor changes in the chemical environment. In the case of fluorescent probes based on conjugated polymers (CPs), a single recognition event can quench the fluorescence emission through several repeat units on the polymer main chain, occurring a notable signal amplification; however, sensing platforms based on small molecules with the same receptor units, this signal amplification is limited. Another advantage is the ability of polymers to be prepared in solution, solid state, and gel form, giving numerous alternatives for their processability (Adhikari & Majumdar, 2004; Ansari & Masoum, 2021; García et al., 2011; Hillberg et al., 2005; Huang et al., 2010; Jiang et al., 2009; Long et al., 2012; Zhou et al., 2014). In summary, there are strengths and weaknesses of fluorescence sensors using both polymers and small molecules. However, the most important aspect to note is that the design of the sensor depends on the special requirements and specific applications for fluorescence determination of the target analytes.

5.4 Modes of fluorescence modulation in polymeric materials

5.4.1 Fluorescence sensing via aggregation-caused quenching

Fluorescence emission quenching is characteristic of conventional organic fluorophores that occur upon aggregate formation and increase of fluorophore concentration in solution (Cho et al., 2020; Klymchenko, 2017; Wang, Zhang, et al., 2023). The phenomenon of ACQ is a turn-off mechanism in which a highly emissive fluorophore becomes nonemissive when interacting with a quencher analyte that induces the aggregation that suppresses the fluorescence (Fig. 5.3). The ACQ property of organic molecules has enabled them to serve as sensing probes (Elabd & Elhefnawy, 2022). At the aggregate level, the quenching effect is derived from the restriction of the fluorescence radiative decay, favoring nonradiative pathways and possibly inducing changes in the fluorescence lifetime and fluorophore quantum yield based on the quenching mechanism (Noh et al., 2010; Tan et al.,

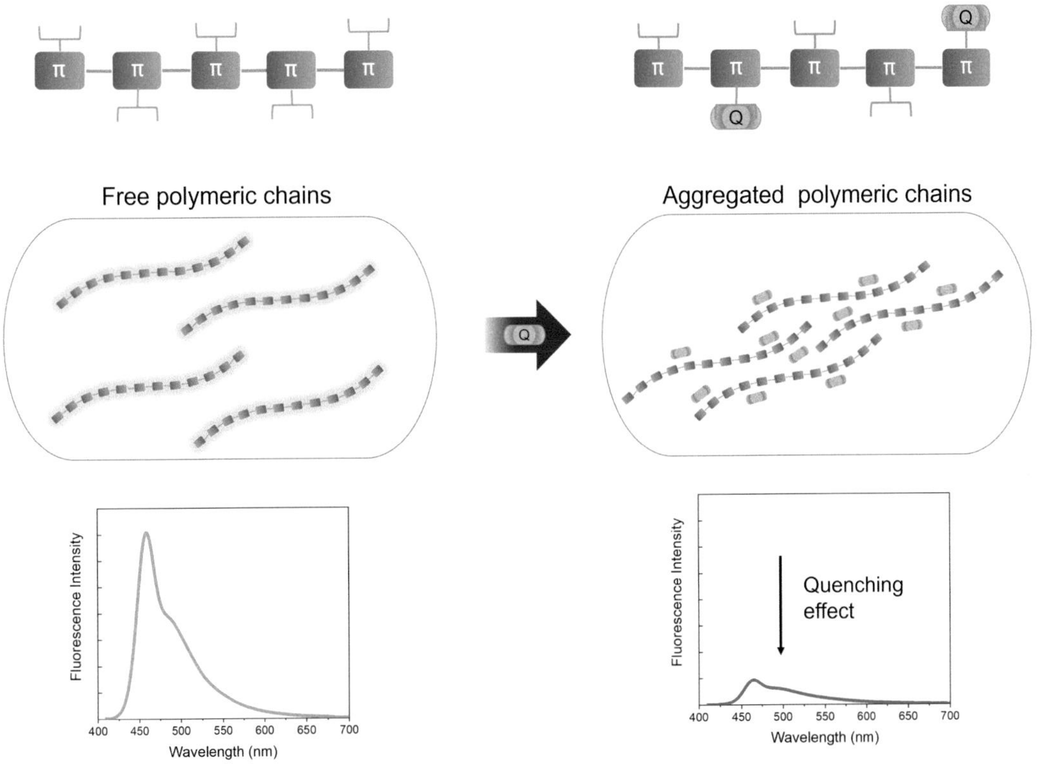

FIGURE 5.3

Illustration of the amplified quenching effect of the polymeric conjugated system upon the addition of a quencher analyte (Q). The figure illustrates the amplified quenching effect of the polymeric conjugated system upon the addition of a quencher analyte (Q). The quencher induces the aggregation of the polymer chains, suppressing the fluorescence emission.

2002). By experimentation, the ACQ is studied to monitor the fluorescence emission reduction with continuous addition and increment of the quencher molecule concentration (Liu et al., 2012). The acquired quenching data and the derived Stern-Volmer analysis allows to understand the interactions between the fluorescent molecules and quencher, and aditionally, to charactize the quenching process generated by static ground state-complex formation. In theory, a quenching process can occur by static quenching mechanism or dynamic quenching mechanism, however, for the case of ACQ the quenching mechanism is distinctly static. The static quenching mechanism is an instantaneous process generated from the fluorophore-quencher stable complex assemblage, which activates electron or excitation energy transfer processes, and therefore, the fluorescence emission pathway is quenched (Cabarcos & Carter, 2005; Ciotta et al., 2019; Gehlen, 2020). While it is true that certain systems are governed by static quenching, a detailed analysis of the relative contribution of this mechanism is required to fully understand the interactions behind the sensory response.

Sensing via quenching of the fluorescence of polymeric systems is also feasible for the detection of different analytes (Dong et al., 2018). Polymeric materials, which design incorporate conjugated backbones, can harvest light efficiently and to respond rapidly to structural or morphological changes induced by other molecules in the system. The exceptional optical properties and stability of light-harvesting polymeric systems enable their usage in sensors and other optoelectronic devices. Particularly, the versatility of the large delocalized system allows the fine-tuning of the energy bandgap to improve the sensitivity and detectability of the fluorogenic probe (Anantha-Iyengar et al., 2019; Wang et al., 2018). In addition, the incorporation of ionic side chains has improved processability aspects in polar solvents and expanded prospects for sensing diverse targets with biological and chemical relevance, such as ions, proteins, DNAs, and other molecules (Kumaraswamy et al., 2004; Cui et al., 2015; Liu et al., 2011). One key factor of the enhanced sensing capacity of polymeric systems centers on the amplified quenching effect, especially for conjugated backbones. In conjugated systems, as opposed to individual chromophores, the fluorescence emission is more efficiently quenched due to the enhanced complexation of the polymer chains and quencher. The quenching amplification is additionally favored by the exciton-transporting capacity of the conjugated backbone, known as the molecular wire effect (Jiang et al., 2007). Employing polymeric systems with intrinsic quenching amplification, coupled with aggregation formation, has enabled obtaining fluorogenic probes with an observable analytical response at nanomolar scale concentration.

5.4.2 Fluorescence sensing via aggregation-induced emission

AIE materials have recently offered exceptional prospects for diverse sensing applications, which take advantage of the outstanding photostability, enhanced signal-to-noise ratio, and generation of significant Stokes shifts to improve the sensitivity (Chang et al., 2018; Gao et al., 2016; Zhou, Xu, et al., 2019). Given that conventional fluorophores lead to limitations with self-quenching in the aggregated state, AIE-based fluorogenic probes have provided the potential to overcome ACQ constraints such as minimal photostability, reduced signal-to-noise ratio, and minor Stokes shifts (Wang & Tang, 2019). AIE-based sensing probes utilize fluorophores that are nonemissive when in dilute solution; however, upon aggregation, the chromophores become strongly emissive (Fig. 5.4) (Wang et al., 2020; Würthner, 2020). Experimental data has suggested that vibrational and rotational factors dissipate the excited state energy through nonradiative pathways in the solution state;

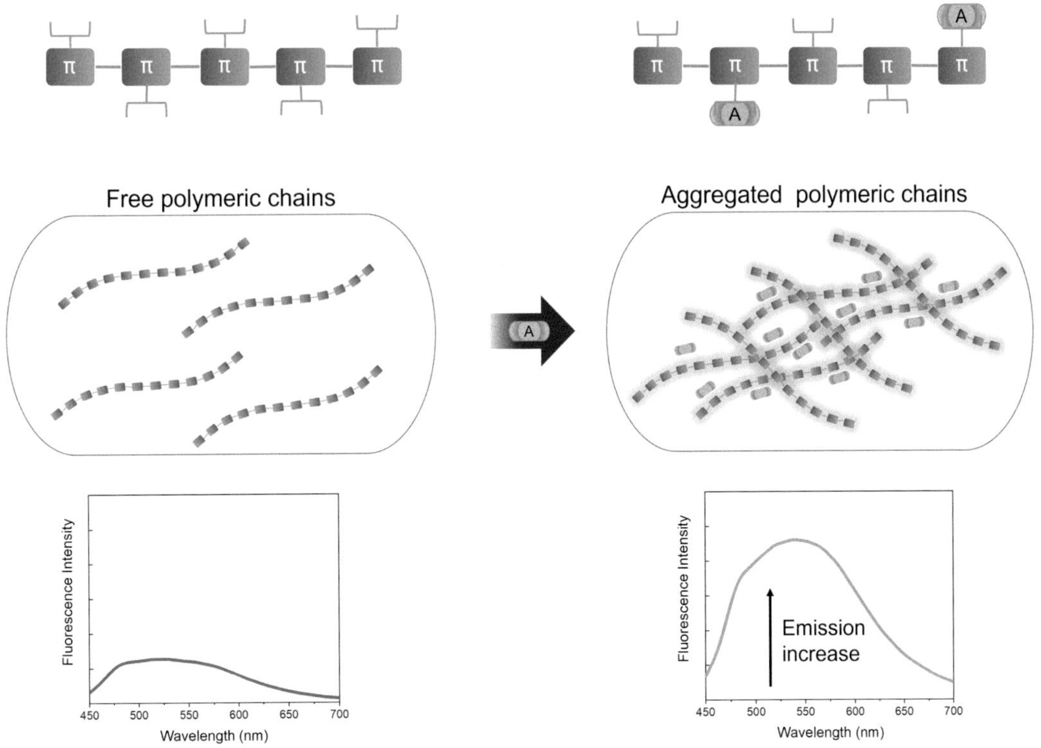

FIGURE 5.4

Illustration of the aggregation-induced emission effect of the polymeric conjugated system upon the addition of analyte (A). The figure represents the general mechanism of aggregation-induced emission of the polymeric conjugated system upon the addition of an analyte (A). The aggregation-induced emission is produced upon the formation of the polymer aggregates, occurring enhancing in the fluorescence emission.

however, fluorescence radiative decay is favored in the aggregate state due to the restriction of the intramolecular motion that gives rise to a decrease in the nonradiative decay rate (Liu et al., 2022; Zhang et al., 2023).

The development of the AIE concept has expanded to materials science, especially by merging the advantages of polymers and AIE fluorophores to build new sensing platforms with superior performance compared to low molecular weight AIE molecules (Li et al., 2012; Zhang et al., 2020). Notably, AIE-based polymers provide remarkable benefits of structural versatility and enhanced processability in solution, deriving into varied and unique optical features for the detection of numerous analytes. The design of AIE polymers involves the covalent linking of low-mass AIE moieties into the polymeric backbone as pendant groups, following a variety of polymerization strategies to obtain linear and nonlinear polymers. Alternatively, AIE-active fluorophores can be incorporated as part of the polymer backbone (Dong et al., 2019; Wei et al., 2018). In this case, the extended polymeric chain contributes to the restriction of the intermolecular rotation, hence, higher emission is achieved. The importance of fine-tuning the structural and morphological characteristics of polymeric AIE materials dwells in the adaptability to fulfill desirable features for selective and efficient sensory probes.

5.4.3 Fluorescence sensing via energy transfer

Another mechanism in which polymer-based materials serve for the effective detection of analytes consists of fluorescence resonance energy transfer (FRET) processes (Fang et al., 2022; Murphy et al., 2004). The FRET process occurs by the transfer of energy of a donor chromophore in the excited state to another chromophore with lower excited state energy. Energy transfer is favored by the proximity of the donor and acceptor that allows dipole-dipole coupling. A conjugated polymeric backbone can harvest light in a polymeric-based chemosensor; the system is then promoted to the excited state, and the energy is transmitted to a detected molecule. In the last step, the new excited molecule will emit photons at longer wavelengths (Hussain et al., 2022; Zhao et al., 2017). Chromophores can also be located as pendant groups and exhibit energy or charge transfer processes (Kumar et al., 2020; Taya et al., 2018). The FRET is delimited in distance by the Förster radius, which in turn depends on factors such as the spectral overlap of the donor emission and absorbance of the acceptor, the dipoles alignment, and the index of refraction of the solvent. The Förster radius is the donor-acceptor separation distance at which the energy transfer reaches 50% of efficiency (Aneja et al., 2008). Ideally, the light source must selectively excite the donor to avoid undesired direct excitation and emission from the reporter acceptor. Overall, detection strategies based on FRET rely on the interactions between the donor and acceptor (Wu et al., 2020). Commonly, the resulting conformational arrangement of polymer chains and target analytes results in complex aggregate states where pairwise interactions are unlikely to occur.

After reviewing the background information, in the remaining portions of the chapter, we summarize a variety of unique fluorogenic probes that rely on improved sensory systems. The focus of the discussion is on polymeric sensory materials that use analytical responses based on ACQ, AIQ, and FRET modes of fluorescence modulation to detect a variety of analytes, such as ions, proteins, and other biologically relevant targets.

5.5 Fluorogenic sensors based on linear polymers

Linear condensation polymers are obtained from monomers with different reactive side groups by the polymerization process described as a condensation reaction. This type of polymers has typically been created using traditional methods of reaction, such as the condensation reaction between monomers that possess two different functional groups (A-B) or between two monomers that have two functional groups each (A-A + B-B). Additionally, copolymers can be synthesized by copolymerization of more than two monomers (Fakirov, 2019). Linear condensation polymers are designed as linear structures with finite molecular weights and good solubility in appropriate solvents (Flory, 1946).

In applications, linear condensation polymers are used as chemical sensors, and the monomeric units are designed to have features for the determination of specific analytes. Regarding the applicability of these types of polymers as chemical sensing materials, the analyte recognition unit might be incorporated into backbones or lateral chains, originating the possibility of many detection strategies based on the nature of the various monomeric units. Although they possess certain properties that make them favorable for use in sensory systems, this class of polymers is typically not soluble in water and may even be difficult to dissolve in common organic solvents. Depending on the specific requirements of the sensory system, this lack of solubility could either be beneficial or detrimental. One possible advantage is that the sensor can be utilized as a solid system for detecting analytes in a solvent medium without the dissolution of the polymeric material. Another advantage is the easy recovery and purification of the polymer using simple techniques such as solvent washing and filtration (García et al., 2011).

The following are some drawbacks of linear condensation polymers that restrict sensor application. On the one hand, these polymers are soluble in a small number of available solvents, especially for developing the application of the sensor in a single-phase solution system. On the other hand, the polymeric materials in the solid state display low wettability with certain solvents, water being a particular example (Miller-Chou & Koenig, 2003). This effect can result in a poor fluorogenic signal of the material to solvated analytes.

5.5.1 Fluorogenic sensors based on nonconjugated polymers

5.5.1.1 Nonconjugated linear polymers

Nonconjugated polymers (NCPs) are a class of organic polymers that lack alternating delocalized π-bonding in their backbone. NCPs are characterized to have an all-carbon backbone with only single bonds, unlike the CP congeners. NCPs are electrically insulating materials because of the absence of π-type bonds. NCPs can be synthesized utilizing various methods such as condensation polymerization, cationic polymerization, anionic polymerization, and radical polymerization. The characteristics of NCPs can be tailored by controlling factors, including their chemical composition, molecular weight, and degree of branching (Shrivastava, 2018). Packaging supplies represent one of the most widespread uses for NCPs due to their ability to impart mechanical strength and barrier features. Due to their biocompatibility and capacity for customization with specific functionalities, these polymers find applications in the biomedical field, particularly in drug delivery and sensing systems (Lai, 2020).

Fluorogenic sensors based on NCPs have been reported for several applications because of the ability to selectively sense target analytes with high-sensitive responses (Fig. 5.5). NCPs have been designed to incorporate fluorophores as pendant groups, generating the potential to absorb and emit light when exposed to specific excitation wavelengths. NCPs with inserted fluorophores are promising candidates because of their notable optical signal amplification and light harvesting behavior, displaying an advantage over small fluorescent molecules employed in conventional sensing platforms (Zhu et al., 2012). NCPs could be synthesized by binding nonconjugated polymeric backbones with the conjugated chromophores, preserving the fluorescence features of conjugated motifs (Zheng et al., 2000). Wang and coworkers synthesized and characterized a new water-soluble **NCP1** obtained by the Cu-catalyzed "Click" coupling reaction between diazido triethyleneglycol and fluorescent diethynyl-fluorene. The sensor was utilized to conduct the optical discrimination and determination of three different microbial pathogens by using fluorescence microscopy. The

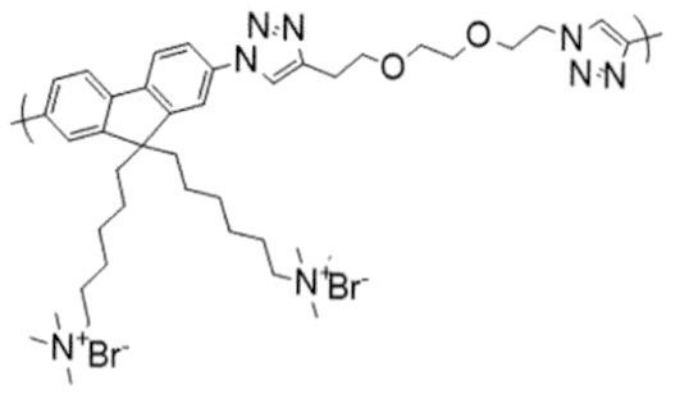

NCP1

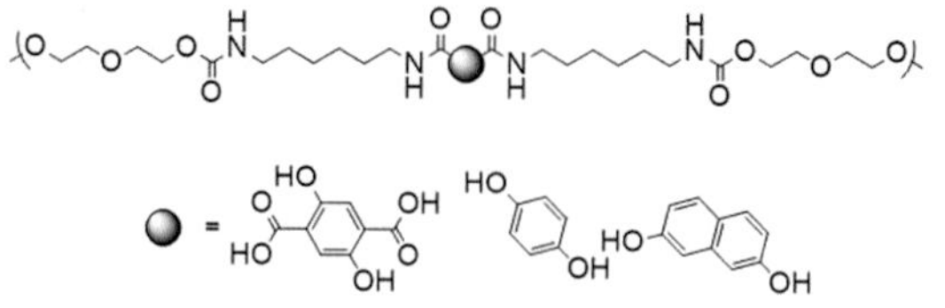

NCP2

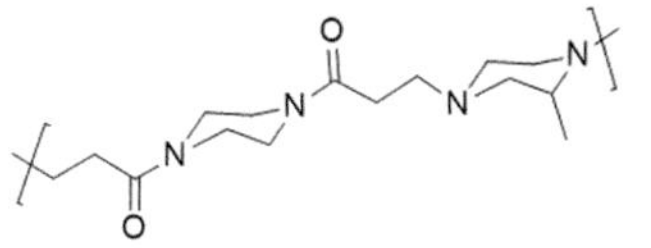

NCP3

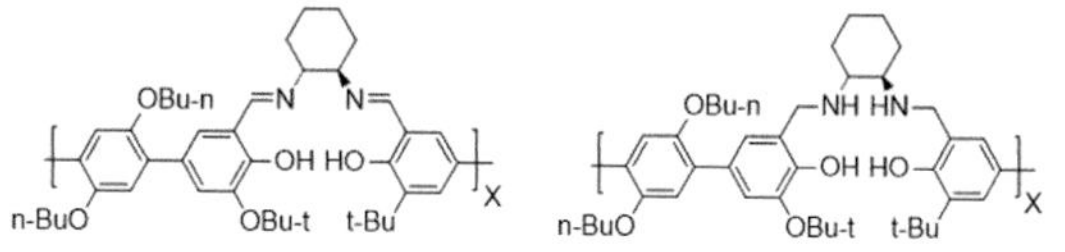

NCP4

FIGURE 5.5

Structure of nonconjugated linear polymers used as sensors for the detection of different biological and chemical targets. The figure includes examples of nonconjugated linear polymers used as fluorogenic sensors for the detection of biological and chemical analytes.

method was rapid and simple, and the pathogen determination was performed without the use of a complex instrument (Wang et al., 2016).

Luminescent nonconjugated polyurethane sensors (**NCP2**) were synthesized by Jiang et al. All of the obtained polymers demonstrate highly sensitive and selective signals for the turn-off fluorescence detection of 2,4,6-trinitrophenol (TNP) in both solid and solution states. The polymers could determine TNP via different mechanisms such as photoinduced electron transfer (PET), fluorescence resonance energy transfer (FRET), and inner filter effect (IFE). Performing filter paper-based assays allowed to achieve a robust determination of TNP in H_2O with a good detection limit value (10^{-10} M) (Jiang et al., 2018).

A series of polyamidoamines (PAAs) for the reversible switching process between glutathione (GSH) and Hg^{2+} were improved by Zhao et al. The polymers (**NCP3**) were obtained using a Michael addition reaction of piperazine compounds with N, N-bisacrylamide piperazine. A noticeable AIE effect was detected for PAAs with emission at 425–525 nm, and the fluorescence mechanism was found to be affected by temperature, excitation wavelength, and conjugated variations. The amide groups and piperazine rings located on PAA backbones allow the AIE process through the clusteration-triggered emission (CTE) mechanism, consistent with the redshift of theoretical calculations. The coordination with Hg^{2+} ions successfully quenched the emissions of PAAs, which presented an enriched quenching effect with increasing Hg^{2+} concentrations. As a reducing agent, GSH was able to restore the emission of the polymer when mixed with Hg^{2+}. By reversibly switching between fluorescence-on and fluorescence-off processes, mercury and GSH concentrations were determined with accuracy (Zhao, Zheng, et al., 2022).

Enantioselective fluorescent sensors based on chiral polymers represent another practical option because of their high sensitivity and potential in analytical and biological applications (Hembury et al., 2008; Xu & McCarroll, 2005). Interestingly, these polymers have been demonstrated to successfully ensure real-time analytical sensing for chiral molecules (Gokel et al., 2004).The use of these polymers not only significantly facilitates the fast detection of the enantiomeric contents of chiral molecular mixtures but also allows the monitoring of high-throughput catalysts for asymmetric synthesis routes. In this perspective, the enantioselective determination of phenylglycinol was performed by Xu et al. Two chiral polymers containing (*R*, *R*)-Salen-type units (**NCP4**) were prepared using the nucleophilic addition–elimination process. These chiral polymers presented a gradual increase of fluorescence in the presence of (*R*)- or (*S*)-phenylglycinol. Particularly, these polymers showed a sensitive response in the presence of (*S*)-phenylglicynol (Xu et al., 2010).

5.5.1.2 Nonconjugated polymer dots

Non-conjugated polymer dots (NCPDs), which are built on aggregated structures, are potentially useful for sensing; hence study in this area is crucial for developing new detection approaches (Zhu et al., 2015). This type of polymeric material has attracted great interest in fluorescence sensing platforms due to inherent features such as outstanding optical performance, good biocompatibility, and stability (Fan et al., 2021). Hence, some NCPDs have successfully been designed by implementing methods that include hydrothermal synthesis, cross-linking techniques, and polymerization methods. Water soluble fluorescent nanoparticles have been reported by following these sinthetic methods (Zhu et al., 2015). NCPDs do not usually comprise conjugated fluorescent molecules but instead feature many chemical groups (e.g., N = O and C = N) with intrinsically weak fluorescence characteristics. These groups are immobilized to polymeric networks that cause fluorescence generation from a cross-linked-enhanced emission.

Chemical detection of pesticides is vital to human health and environmental protection, as the use of pesticides has increased significantly in agricultural processes. As is known, changes in fluorescence intensity in the presence of pesticides are mostly “turn-off” mechanisms, resulting in a decrease in the detected fluorescence signal, and this phenomenon, the IFE effect, is very important in facilitating pesticide detection (Sharma et al., 2020). Fernandes et al. reported NCPDs based on a β-glucan material, prepared using polysaccharide from the Usnea lichen biomass. The material could detect methyl parathion and dichloro pesticides with the quenching mechanism caused by the IFE. The sensor NCPDs obtained from renewable resources could be used in applications of carbon-based nanomaterials, particularly to sense environmentally and biologically significant analytes. The emission properties of NCPDs proved a significant advantage for monitoring the analytes, including excellent photostability, fast response, and high sensitivity (Fig. 5.6) (Fernandes et al., 2023).

Qi et al. created a fluorescence “turn-on” sensor for GSH determination using chitosan-based glutaraldehyde nonconjugated polymers (GCPF) through the FRET mechanism. Because of the overlapped absorbance spectrum of MnO_2 nanostructures and the emission spectrum of GCPF, the fluorescence signal was quenched in the presence of MnO_2 nanostructures according to the FRET mechanism. In the presence of GSH, MnO_2 nanosheets were decreased and decomposed into Mn^{2+} ions because of the reducing features of GSH. Consequently, the fluorescence signal was monitored again after the reduction process, confirming the decomposition process. Hence, this strategy allowed the detection of GSH from 0.5 to 50 μM with a LOD of 84 nM. It presented high

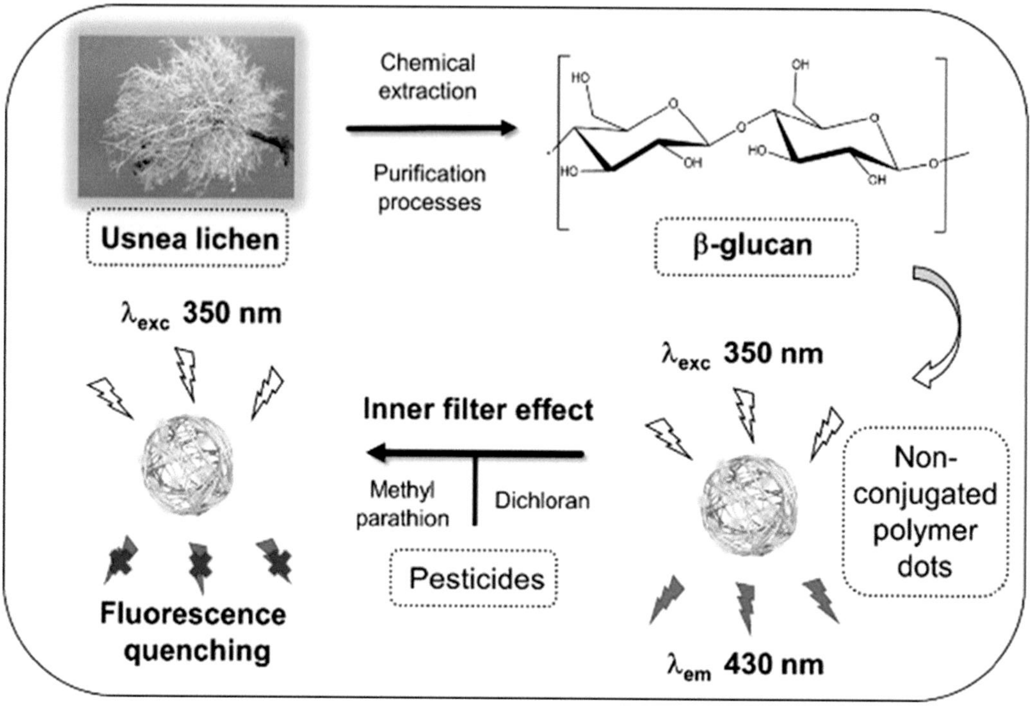

FIGURE 5.6

Nonconjugated polymer dot-based fluorogenic sensors for the detection of pesticides. This figure includes the description of nonconjugated polymer dot-based fluorogenic sensors for the detection of methyl parathion and dichloro pesticides.

Source: *Reproduced with permission from Fernandes, R. F., Atvars, T. D. Z., & Temperini, M. L. A. (2023). Exploring the non-traditional fluorescence emission of non-conjugated polymers dots for sensing pesticides.* Reactive and Functional Polymers, 182, *105483. https://doi.org/10.1016/j.reactfunctpolym.2022.105483.*

sensitivity and selectivity in GSH determination in the presence of other analytes such as L-valine, DL-aspartic acid, L-lysine, L-glutamic acid, and L-threonine (Qi et al., 2022).

Liu and coworkers reported the synthesis of a polymer nanoparticle prepared with the interaction of a small nonfluorescence electron-rich molecule and a large nonfluorescent polymer through a hydrothermal method. The small molecule causes a decrease in bond rotations and vibrations, resulting in the delocalization of n and π electrons. The polymer nanoparticles with strong fluorescence signals were assembled using D-glucose and hyperbranched poly(ethylenimine) (PEI-G) following the Schiff base reaction. The preparation process for the polymer nanoparticles was considered a simple and eco-friendly process. PEI-G nanoparticles displayed a good fluorescence quantum yield and unique emission features with high sensitivity toward changes in concentration and solvent polarity. For this study, PEI-G polymeric nanoparticles were utilized as a fluorescent sensor for robust, sensitive, and selective determination of picric acid (PA) in an aqueous solution.

Among the nitroaromatic compounds (NACs), 2,4,6-trinitrotoluene (TNT) is the most commonly known nitro-explosive compound; however, PA is also considered an explosive compound with high detonation velocity and low safety coefficient. Regarding the sensing study, PA showed an important degree of fluorescence quenching, while nitroaromatic explosives caused almost no change in the fluorescence signal. The quenching of fluorescence upon adding PA was attributed to the effect of resonance energy transfer, electron transfer, and IFE between PEI-G polymeric nanoparticles and PA. A linear range and a detection limit for PA were found to be 0.05–70 μM and 26 nM, respectively. Also, the improved chemosensor was effectively performed to determine PA in environmental samples (Liu et al., 2016).

Quercetin (Qc) is one type of natural flavonoid that exists in most vegetables, fruits, and plants. These natural derivatives have exclusive biological features such as antibacterial, antiviral, antioxidant, antiproliferative, antiinflammatory, and cardiovascular protection effects. Because of the importance of Qc, a novel NCPD was fabricated by Fan et al., employing L-threonine and hyperbranched poly(ethylenimine) (PEI). The sensor NCPD showed unique optical characteristics and stability. In the presence of Qc, the fluorescence signal of sensor NCPD is enhanced through the mechanisms of the inner filter and the synergistic effect of the static quenching mechanism. The wide linear range for the determination of Qc was found, and the Qc detection strategy has offered a novel, effective, sensitive, and selective sensing platform (Fig. 5.7) (Fan et al., 2021).

Other NCPDs were prepared by Yang and coworkers through the pyrolytic method utilizing polyethyleneimine. Obtained NCPDs display outstanding colloidal stability caused by the surface-distributed polymer chains. The use of NCPDs can enable the precise and accurate identification of metronidazole facilitated by the electrostatic affinity between the sensory dots and the antibiotic. Metronidazole, a synthetic nitroimidazole-based antibiotic, is widely employed for treating bacterial infections in both animals and humans. Additionally, it is utilized as a feed supplement to accelerate the growth of animals (Yang et al., 2019; Gholivand & Torkashvand, 2011). Accumulation of metronidazole in the human body might cause some poisoning reactions, such as ataxia, neuropathy, and peripheral disorders. The emission of the sensory material was quenched by metronidazole as a result of IFE and static quenching mechanism. The improved technique was successfully performed for monitoring metronidazole in milk samples with good results. By using non-conjugated polymer carbon dots, an easy-to-prepare, fast, and practical method has been reported for food quality monitoring (Yang et al., 2019).

Recently, a series of polymer dot-based were reported for sensing heavy metal ions such as Pb^{2+} and Zn^{2+}, including effective discrimination of metal ions in multivariate mixtures. The differential sensing array used the fluorescence response profiles and recognition fingerprint patterns to develop linear discrimination analysis (LDA) based on a single polymer dots system or "all-in-one" configuration. The sensing mechanism of the multiple polymer-dot-based platform is reported to work through the nanoprecipitation method of the polymers and monitoring of the fluorescence responses in which energy transfer mechanisms are involved. The sensor array was demonstrated to be efficient in differentiating binary mixtures of Pb^{2+}/Zn^{2+} metal ions in real lake water samples, including the presence of other ions (Chen et al., 2023).

In brief, NCPDs have been utilized in several fluorescence sensor applications, with the ability to selectively determine specific analytes and provide high-sensitivity responses. By modifying the polymer backbone to incorporate fluorophores, it is possible to design sensors that respond to specific stimuli, including pH, temperature, or specific analytes.

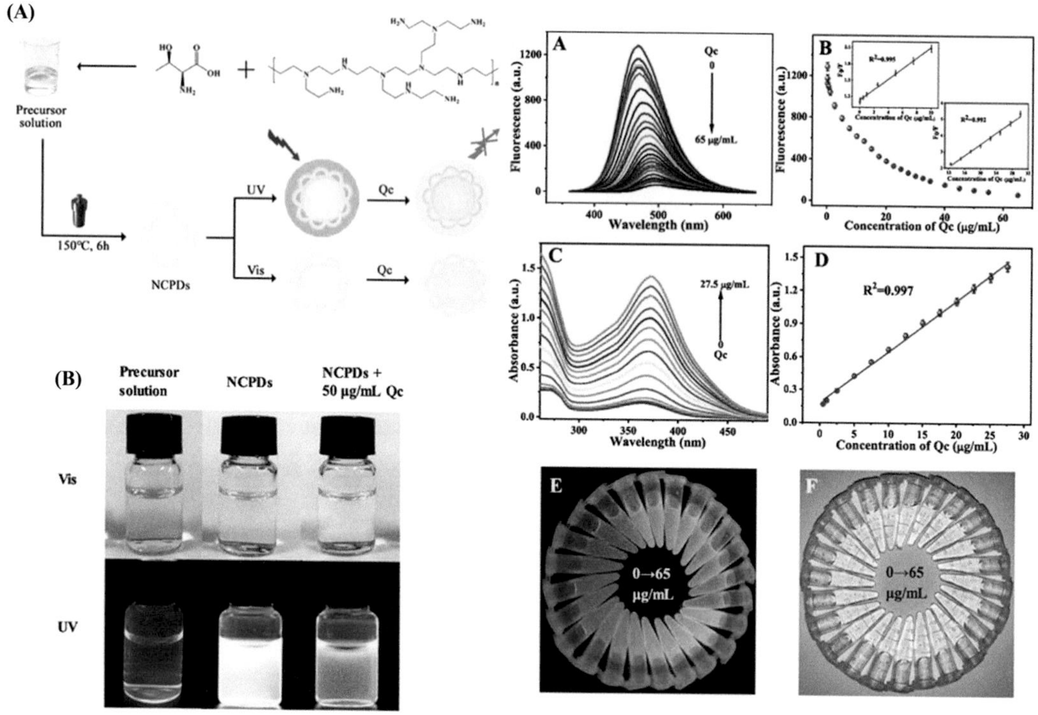

FIGURE 5.7

Nonconjugated polymer dot-based fluorogenic sensors for the detection of quercetin. The image describes nonconjugated polymer dot-based fluorogenic sensors for the detection of the natural flavonoid quercetin.

Source: *Reproduced with permission from Fan, Y., Yao, J., Huang, M., Linghu, C., Guo, J., & Li, Y. (2021). Non-conjugated polymer dots for fluorometric and colorimetric dual-mode detection of quercetin.* Food Chemistry, 359, *129962. https://doi.org/10.1016/j.foodchem.2021.129962.*

5.5.2 Fluorogenic sensors based on conjugated polymers

Conjugated polymers (CPs) are a class of macromolecular materials built with large π-electronic delocalized backbones (Qiu et al., 2020). Due to the extended conjugated system that enables distinctive tunable photophysical and optoelectronic properties, CPs have received considerable attention for the development of advantageous polymeric materials (Morin et al., 2016). The intrinsic features comprise light-harvesting capability, highly efficient quantum yields, conductive properties, and enhanced energy and charge transfer behavior (Vezie et al., 2016). The structural controllability of the polymeric chains has allowed the expansion of CP diversity, either varying the nature of the conjugated backbone or fine-tuning the various pendant groups attached to the polymer main chain. Some examples of CPs are presented in Fig. 5.8. In particular, the facile synthesis, coupled with the versatile structural tunability, has launched these polymers as platforms for applications related to light-emitting diodes, polymeric solar cells, chemosensory probes, and biosensors (Bulut et al., 2022; Haldar et al., 2020; Li et al., 2017; Liu, Hua, et al., 2020; Tan et al., 2024).

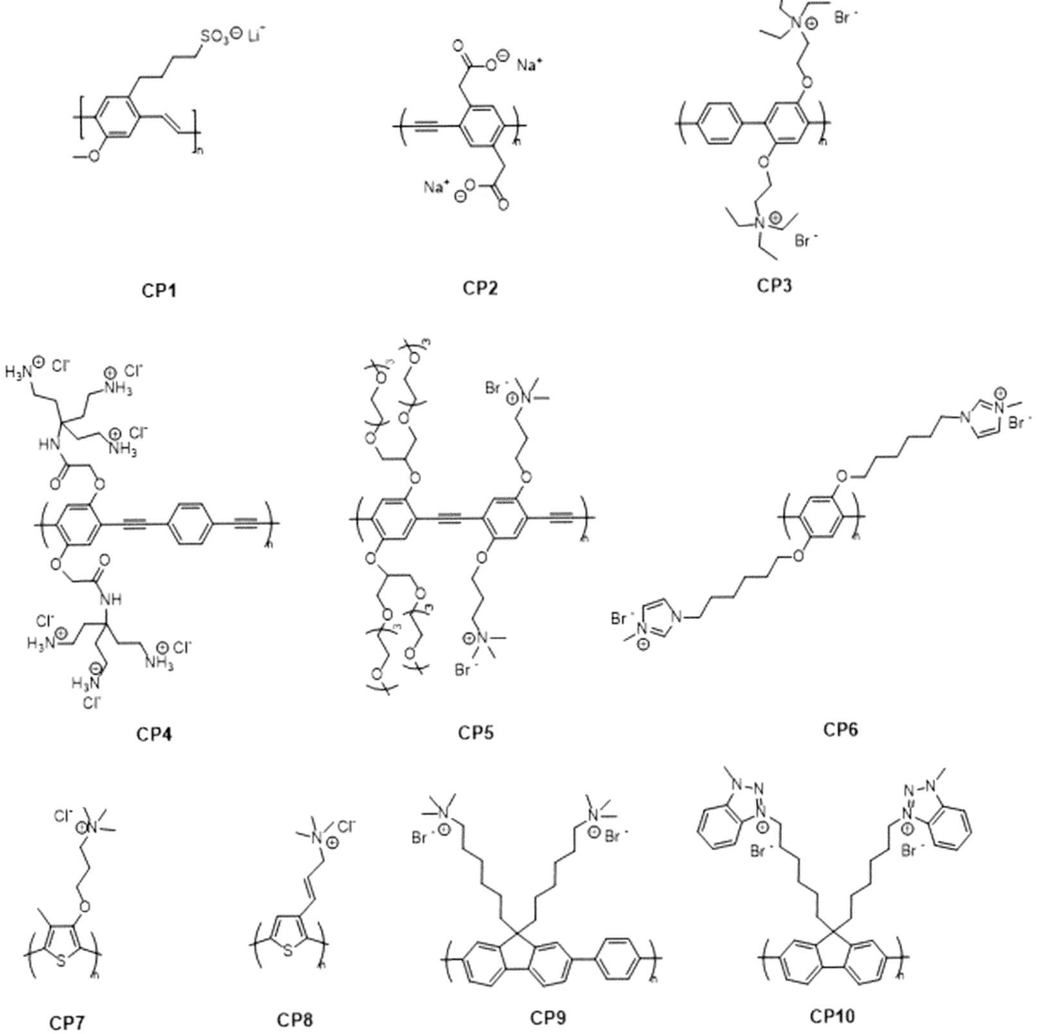

FIGURE 5.8

Charged conjugated polyelectrolytes used for sensing applications. This figure includes examples of charged conjugated polyelectrolytes with different pendant groups used for various sensing applications.

CPs can act as coupling receptors to modulate specific interactions into transducing responses to detect the presence of different analytes. As CPs have the potential to exhibit an amplified quenching effect, the sensitivity of the system is significantly increased given that minor perturbation originating from the CPs-analyte triggers an augmented signaling response compared to small molecule chemosensors (Tan et al., 2002; Thomas et al., 2007). As the electronic localization results in a wide energy band gap, CPs display strong absorption features in the UV-visible absorption region and well-structured emission bands optimal for the detection of analytes. The optical properties are affected by the nature of the delocalization, polarization, and twisting effects exerted along the conjugated polymeric chains (Bolinger et al., 2012; Huang et al., 2006).

A CP-based fluorescent probe comprises mechanisms in which the fluorescence emission is reduced or enhanced by the complexation of the polymer chains with the analyte. For CPs with neutral receptors incorporated into the polymer backbone, the sensory response is originated from the host-guest interactions that facilitate efficient energy or electron transfer processes and, therefore, bimolecular quenching via static or dynamic quenching pathways (Tanwar et al., 2022; Wosnick & Swager, 2004; Wang et al., 2000). In charged CPs, the conjugated backbone of CPs allows the rapid exciton transport along the conjugated system and the charged pendant groups provide the binding points for interactions with oppositely charged molecules. Upon binding of the analyte, the polymer aggregation is induced, and the sensory-signal amplification is triggered by a static quenching mechanism (Harrison et al., 2000; Tan et al., 2004).

5.5.2.1 Anionic Conjugated Polymers

Water-soluble CPs with negatively charged pendant groups have demonstrated fluorescence turn-off features upon binding with a variety of cations (Chen et al., 2009; Ihde et al., 2022). The photophysical properties of anionic CPs are also delimited by the structural nature of the conjugated backbone. Early efforts led to the design of sulfonate-containing CPs to study the reversible fluorescence quenching effect in the presence of cationic electron acceptors. **CP1** initially served as a highly sensitive and selective probe based on reversible fluorescence quenching. Upon addition of viologen (MV^{2+}) at low concentrations of 10^{-7} M, a significant quenching of 95% was observed. Further recovery of the polymer's strong fluorescence was achieved in the presence of the protein avidin. The weak complex formation between the polymer and viologen-type quencher allowed the new complexation of quencher-avidin, reversing and preventing the quenching effect (Chen et al., 1999). **CP1** was also noted to exhibit static quenching through ultrafast PET with proteins cytochrome c in an aqueous solution (Fan et al., 2002).

Amplified fluorescence quenching of polyanionic materials was previously explored by incorporating carboxyl-side groups into the PPE backbone. Particularly, this study shed light on the aggregational, and conformational changes induced by charged quenchers and the influence on the degree of quenching (Jiang et al., 2006). Branched carboxylate-functionalized CPs with a different backbone were first investigated by Schanze and coworkers. The main findings indicated suppression of aggregation and enhanced quantum yields, suggesting potential uses of polyanionic-based sensory materials (Lee, Kömürlü, et al., 2011). To optimize the selectivity and sensitivity of polymer-based fluorescence sensory systems, the reduction of the aggregate state of polymeric chains is desired. **CP2** is a simple polymer designed to avoid aggregation with improved optical features such as quantum yield enhancement and spectrally narrowed band shape (Koenen et al., 2014).

5.5.2.2 Cationic Conjugated Polymers

Within the structural diversity of cationic CPs, common solubilizing groups consist of ammonium, imidazolium, bis-imidazolium, and phosphonium moieties (Hu et al., 2018; Jagadesan et al., 2020; Lee, Liu, et al., 2011; Zhu et al., 2018). The unique photophysical properties derived from the different backbones have provided alternative and promising paths for the development of fluorescence chemosensors. One of the first reported sensing cationic CP systems in the aqueous and solid state with cationic moieties (**CP3**) displayed high amplified fluorescence quenching efficiency in the presence of anionic quenchers such as $Ru(phen')_3^{4-}$ and $Fe(CN)_6^{4-}$, inspiring the exploration of new cationic sensing CP systems (Harrison et al., 2000). The PPE-based CPs with ammonium solubilizing groups have been recently studied for the ability for the detection of biologically important phosphate ions involved in cellular processes. The water-soluble **CP4** has been reported to display an effective chemosensory response toward PPi and ATP ions with K_{SV} values in the order 10^5 M^{-1}. The main findings suggest that the phosphate analytes induce aggregation on the polymeric chains and the fluorescence quenching effect, accompanied by shifted emission to longer wavelengths. The multiple cationic charged groups of **CP4** allow an enhanced solubility and great equilibrium between the free chains and the aggregate state of the polymer, which improves the sensory response. Wu and coworkers described the fluorescent sensor array of **CP4** and a series of other cationic CP derivatives for sensing and differentiation of NACs that act as electron acceptors. The diverse fluorescence quenching response patterns allowed the classification of the combination of NACs by linear discriminant analysis (LDA). Results confirmed an accurate differentiation method of NACs into two-dimensional discrimination plots at low concentrations of 1.0 μM (Wu et al., 2016). Similarly, **CP5** with a quantum yield of 37% and a series of other PPE-type charged polymers were designed as sensor arrays for aromatic carboxylic acid derivatives in a water solution. The main finding suggests there exists a modulation of the fluorescence and static quenching, generating a reliable classification by LDA with reliability and accuracy at different pH (Han et al., 2016).

Imidazolium-containing CPs have also been reported in the literature. Parthasarathy et al. reported PPE-based CPs (PIM-2 and PIM-4) that demonstrated to have effective fluorescence quenching toward the addition of the quenchers anthraquinone (AQS) and pyrophosphate (PPi), especially in methanol solution. Methanol has been considered a suitable polar solvent for the photophysical characterization of charged CPs. The main finding suggests that electrostatic interactions facilitate the complexation of the AQS with the electron-rich polymeric chains, and the fluorescence is quenched by PET. The quenching effect with PPi is mainly attributed to the strong association and formation of polymer aggregates in methanol solution, causing 90% of quenching of PIM-2 and PIM-4 upon quencher concentrations of 1 μM and 3 μM, respectively. The K_{SV} for all cases was found to be $> 10^5$ M^{-1} (Parthasarathy et al., 2015). Hussain and coworkers also reported the synthesis of the conjugated polyelectrolyte **CP6** for the detection of PA. Due to the favorable electrostatic interaction between the polymer chains and PA, and the presence of a ground-state charge transfer complex, this polymer demonstrates extraordinary selectivity and sensitivity. Interfering agents such as NACs were shown to have no effect on the ultrasensitive detection function (Hussain et al., 2015).

The polythiophene (PT) backbone offers distinct optical features for the development of fluorogenic probes for molecular sensing. For instance, the fluorescence sensory properties of **CP7** containing ammonium pendant groups demonstrated the selective chemical sensing toward the ATP nucleotide (Li et al., 2005). Further research revealed that this polymer also serves as an effective probe for sensing other relevant molecules such as cysteine, homocysteine, taurine, and glutathione (Yao et al., 2009). Most recently, **CP8** showed selective sensing of ATP over a variety of nucleotides such as AMP, ADP, UTP, and coenzyme A (CoA) (Wang et al., 2014). Other backbones, such as polyfluorene (PF) backbone, have been utilized for synthesizing water-soluble CPs with various cationic moieties for sensory materials. Within this framework, polymer **CP9** was designed for the detection of the DNA hybridization process, where the energy transfer mechanism from the conjugated backbone to the DNA strands modulates the recognition signal (Gaylord et al., 2003). Similarly, the polymer **CP10** was developed for the ultrasensitive detection of nitro explosive PA mediated by PET, IFE, and resonance energy transfer processes (Tanwar et al., 2018).

5.5.2.3 Neutral Conjugated Polymers

Sensing materials derived from neutral CPs have been alternatively utilized to detect specific analytes based on changes in the structural conformation and optoelectronic properties induced by hydrogen bonding or π-π interactions (Zhou, Chua, et al., 2019). Giovannitti et al. described the development of neutral CPs with crown ether bithiophene units for the selective optical sensing of alkali metal ions. This sensory system exhibits the backbone twisting in the presence of sodium and potassium alkali ions, showing a selectivity dependence on the crown ether ring size (Giovannitti et al., 2016). Another report by Hu and coworkers includes the synthesis of a neutral polymer for the recognition of uranyl ions (UO_2^{2+}). The neutral polymer demonstrated to have an enhanced turn-off response with a low detection limit of 7.4×10^{-9} M, without observable interference of lanthanide or actinide ions (Hu et al., 2021). On another note, Shin et al. reported the self-assembly and polymerization of an imidazole-containing diacetylene called polymer PDA-Im. After polymerization and immobilization on filter paper, PDA-Im showed changes in the fluorescence emission when exposed to Fe(III) ions with a colorimetric response from blue to red and a fluorescence turn-on effect (Shin et al., 2022).

Novel approaches to polymer probes have explored macromolecular engineering areas that render unconventional polymeric materials. Bridged multicyclic CP features multiple ring-like structures as part of the polymer backbone and emissive properties in an aggregate state that serve as sensory materials. A recent report details the design of a series of neutral polymers with conjugated macrocycles with AIE properties. The series of polymers demonstrated to form highly emissive nanoaggregates in solvent mixtures with 90% water content. The presence of explosive molecules was demonstrated to induce static quenching and PET upon interaction with the nanoaggregates, existing selectivity of the systems toward PA compared to other nitroaromatic compounds (Liu et al., 2023).

CPs also can be designed by incorporating aromantic pendant groups cross-conjugated with the main chain. Wang et al. recently reported the synthesis of two series of six CPs that display the cross-conjugated ligand triazacryptan (TAC) on the side chains. The cross-conjugated term refers to the conjugation of the pendant group and the main chain is interrupted by a nonconjugated segment. The polymer PP3 stands as an optical probe for tracking potassium ion (K^+) in intracellular conditions with rapid and selective sensitivity even in the presence of other interferent ions. Remarkably, PP3 demonstrated the ability to the dynamic balance of potassium ion concentration in living cell models in a real-time manner giving insights into the influx and efflux dynamic of intracellular potassium ions (Wang, Pan, et al., 2023).

5.6 Fluorogenic Sensors Based on Molecularly Imprinted Polymers

The molecular imprinting technique is known as an appealing method to design functional synthetic receptors, in particular, to develop molecularly imprinted polymers (MIPs) (Gao et al., 2014;

Huang et al., 2021). MIPs refer to networks of polymers that are created through the copolymerization of cross-linkers and functional monomers containing various reactive groups mediated by non-reactive template molecules (Kan et al., 2012; Wang et al., 2010). A large variety of analytes can be determined with this approach. Following polymerization, templates are identified as MIPs sensing materials suitable for detecting particular target analytes in terms of form, size, and functional groups (Alexander et al., 2006; Anirudhan et al., 2013). As a result, MIPs are thought to have "molecular memory" and exhibit specific binding features as a molecular template with sensory dependence on the structural nature of the analytes (Gao et al., 2014). MIPs show unique characteristics such as simple preparation, good reusability, selective adsorption, outstanding mechanical/chemical/thermal stability, and low cost. Together, these factors make it possible to increase the sensitivity of practical analytical determination. The improved stability has been observed upon various conditions of high pressure, temperature, presence of acids and bases, extreme pH, and organic solvent medium. According to these advantages, MIPs have received increasing attention in diverse areas, including drug delivery, separation techniques, mimicking enzymes, and chemical/biomimetic sensing technology (Carrasco-Correa et al., 2015; Wang et al., 2010). The combination of the recognition characteristics of MIPs and various transducers is extensively exploited for various sensors, including quartz crystal microbalance-based sensors, electrode-type sensors, surface plasmon resonance, ellipsometry, and optical/fluorescence sensors (Ansari & Karimi, 2017; Baltrons et al., 2013; Chen et al., 2012; Fuchs et al., 2012; Kong et al., 2010; Lucci et al., 2010). MIPs are conceivably viewed as a potential and strong optical detection tool with excellent selectivity.

Traditional MIPs have distinct recognition behavior, but they are limited in their ability to provide signals during the analysis and determination of target analytes. However, the development of molecularly imprinted fluorescence sensors (MIFs) involves effective recognition and output signaling. MIFs can be obtained directly using intrinsic fluorescence for the recognition of target species (Huang et al., 2021). In addition, MIFs can be prepared indirectly by the addition of fluorescence-labeled analogs. Fluorescence-specific functional monomers in polymer imprinting processes also allow the diversity of fluorophore substances embedded in chemosensors, allowing their applicability in fields related to the environment, medicine, and the analysis of food safety.

The application of MIFs for the robust determination of environmental organic pollutants, heavy metals, and agricultural and veterinary drug residues is presented in this section (Ansari & Karimi, 2017). 3-monochloropropane-1,2-diol (3-MCPD) is a carcinogenic substance that frequently forms as a chemical contaminant during food processing with the possibility of ending in final products (Martin et al., 2021; Yaman et al., 2021). Fang et al. have presented a new approach to fluorescence sensing that involves using a CD-infused filter paper in combination with an MIP film. The CDs are embedded into the filter paper through electrostatic attraction, producing high fluorescence signals, while MIP films are added to the CD-infused papers to enhance the detection efficacy. The sensor showed a good adsorption capacity, a low detection limit ($0.6\ ng \cdot mL^{-1}$), and a high selectivity with an imprinting factor of 4.5 (Fig. 5.9). This research provides a powerful and new strategy for the fast, simple, rapid, and sensitive detection of 3-MCPD (Fang et al., 2019).

On another note, phenobarbital is commonly utilized for the treatment of various types of neurological diseases, such as cerebral palsy and epilepsy migraines, to name a few. The excess phenobarbital level could lead to hypnosis, sedation, and some other severe symptoms (Huang et al., 2011). A fluorogenic chemosensor was prepared for the determination of phenobarbital utilizing

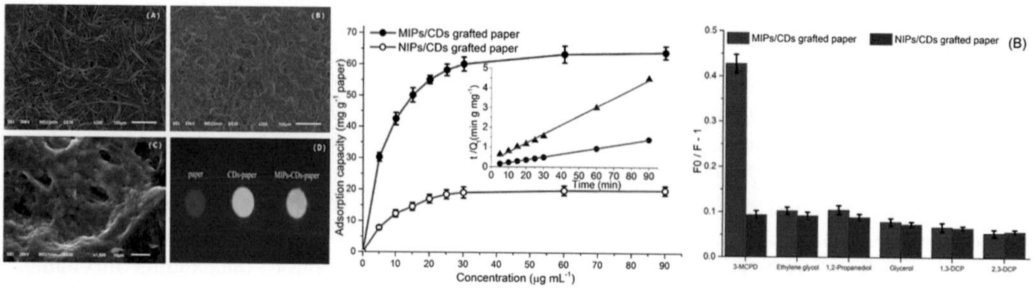

FIGURE 5.9

Characterization and emission spectra of CD-infused papers. This image displays characterization and emission spectral data of carbon dot (CD)-infused filter paper in combination with a molecularly imprinted fluorescence film for the detection of a carcinogenic substance (3-MCPD).

Source: *Reproduced with permission from Fang, M., Zhou, L., Zhang, H., Liu, L., & Gong, Z.-Y. (2019). A molecularly imprinted polymers/carbon dots-grafted paper sensor for 3-monochloropropane-1,2-diol determination.* Food Chemistry, 274, *156–161. https://doi.org/10.1016/j.foodchem.2018.08.133.*

MIP-coated GSCDs composite (GSCDs) by Shariati et al. A hydrothermal method was utilized to obtain GSCDs from Cedrus as a carbon source. Tetraethoxysilane (TEOS), 3-aminopropyltriethoxysilane (APTES), and phenobarbital were used to design the material. The fluorescence intensity of MIP-GSCD was decreased during the determination of phenobarbital. GSCDs are demonstrated to display a great selectivity to phenobarbital with a detection limit of 0.01 nM and a linear range of 0.04–3.45 nM. As part of the sensor capability evaluation, phenobarbital detection in human blood plasma samples was successfully performed (Shariati et al., 2019). Another fluorometric sensor based on MIP@N-doped carbon dots for phenobarbital sensing is reported by Nejhad et al. The assay consisted of the phenobarbital determination of the exhaled breath condensate obtained from the expiratory circuit of a mechanical ventilator (MVEBC). The sensing mechanism was determined to occur based on the adsorption of the analyte on the MIP and the charge transfer effect between phenobarbital and N-CDs. The decrease in emission was found to be dependent on the phenobarbital concentrations. This technique has important usage for the phenobarbital quantification from the MVEBC apparatus, allowing to monitoring of the phenobarbital administered as a drug for the treatment of seizure disorders in infants (Nasehi Nejhad et al., 2022).

A MIP-based fluorescence probe for the determination of Japanese encephalitis virus (JEV) was first published by Liang and coworkers. A FRET-based strategy for virus recognition was performed. Pyrene-1-carboxaldehyde (PC) acted as an energy acceptor, and the virus as an energy donor. Highly sensitive detection of JEV in water was achieved with a wide linear range (Fig. 5.10). The results evidenced the fluorescence signal increased upon the determination of the virus, reaching a detection limit of 9.6 pM (Liang et al., 2016).

N-acyl homoserine lactones (AHLs) are signaling molecules in bacterial quorum sensing. With the help of AHLs, bacterial cells can identify factors such as population density and upregulate specific phenotypes, including biofilm formation, food spoilage, and virulence factor production (Huang et al., 2016). Because AHLs display an important role in foodborne disorders and food

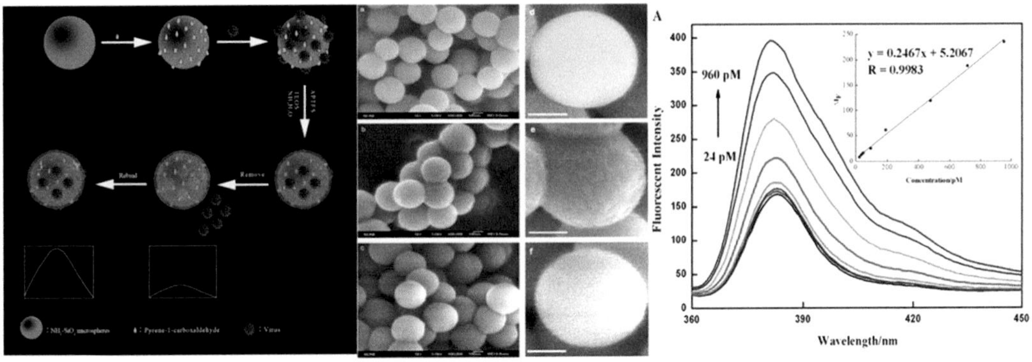

FIGURE 5.10

Characterization and emission spectra of MIP-based fluorescence probe for the determination of Japanese encephalitis virus. The figure describes the characterization and emission spectral data of molecularly-imprinted based fluorescence probe for the determination of Japanese encephalitis virus.

Source: *Reproduced with permission from Liang, C., Wang, H., He, K., Chen, C., Chen, X., Gong, H., & Cai, C. (2016). A virus-MIPs fluorescent sensor based on FRET for highly sensitive detection of JEV.* Talanta, 160, *360–366. https://doi.org/10.1016/j.talanta.2016.06.010.*

spoilage, the determination of AHLs as a universal biomarker of Gram-negative bacteria is vital. A magnetic fluorescent sensor was prepared by a CD-doped MIP layer on the surface of Fe_3O_4 particles (MFMP) for the determination of AHL signaling molecules. N-Z-L-homoserine lactone, MAA, and HEMA were utilized as templates and functional monomers for the preparation of AHLs MIP layers. The improved sensor had a rapid response, good selectivity, and superparamagnetism characteristics. AHLs were detected by the MFMP material with a decreasing effect in fluorescence intensity (Fig. 5.11). Most importantly, the sensor was efficient in sensing AHLs in milk and fish juice as real samples. The improved fluorescent sensor represents a great potential tool for a variety of applications, such as early determination of food contaminants in agri-food products (Cui et al., 2020).

In recent years, natural contamination in the atmosphere has expanded more widely, potentially generating serious threats to humanity. Thus, the identification of toxic substances in the atmosphere is vital for the safety and well-being of human health (Kulawik et al., 2013; Wang et al., 2018). 2,4,6-trinitrophenol (TNP) and other NACs are known to be ecological toxins with poor biodegradability, high water dissolvability, and significant toxicity (Rodgers & Bunce, 2001). Therefore, Murugan and coworkers proposed a novel concept of a fluorescence chemosensor based on nitrogen-passivated carbon dots infused with an MIP (N-CDs@MIP) for the determination of TNP. The sensor was utilized for sensitive and selective determination of trace amounts of TNP. The fluorescence signal of the sensor was quenched upon the determination of TNP. The N-CDs@MIP displayed a linear range from 0.5 nM to 2.5 nM with a lower detection limit (0.15 nM). For the practical applicability of the sensor, analyses of TNP in regular tap and lake water samples were successfully performed (Murugan et al., 2022).

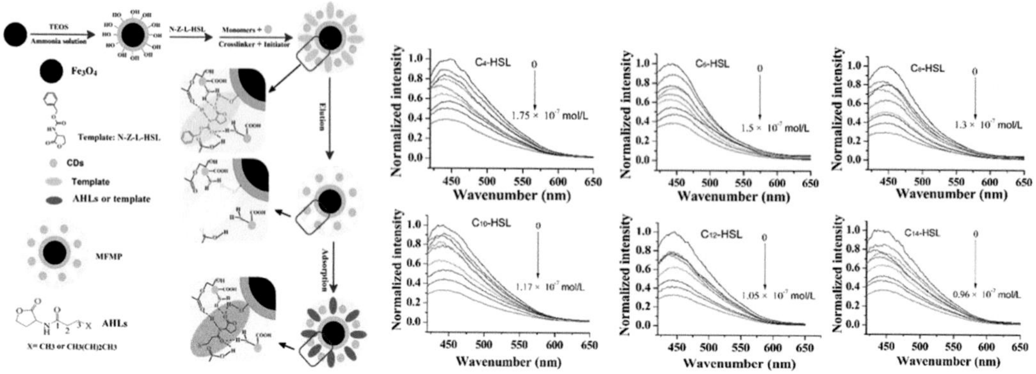

FIGURE 5.11

Characterization studies and emission spectra of MFMP. The image includes the sensing properties characterization and emission spectral study of carbon dots doped molecularly-imprinted polymer (MFMP) for the determination of N-acyl homoserine lactones.

Source*: Reproduced with permission from reference Cui, Z., Li, Z., Jin, Y., Ren, T., Chen, J., Wang, X., Zhong, K., Tang, L., Tang, Y., & Cao, M. (2020). Novel magnetic fluorescence probe based on carbon quantum dots-doped molecularly imprinted polymer for AHLs signaling molecules sensing in fish juice and milk.* Food Chemistry, 328, *127063. https://doi.org/10.1016/j.foodchem.2020.127063.*

5.7 Dendrimers

Fluorescent sensors based on polymeric dendrimers are chemical sensors that utilize fluorescence signals to determine the presence of target analytes in various samples. Polymeric dendrimers are highly branched and symmetrically organized macromolecules with well-defined structures and sizes. They contain a central core, repeated branching units, and a surface functionalized with specific groups that could interact with the target analytes (Avudaiappan et al., 2019; Klajnert & Bryszewska, 2001) These polymers can be obtained from ionic chain polymerization, emulsion polymerization, and radical chain polymerization. The unique structures and features of polymeric dendrimers make them ideal platforms for improving highly selective and sensitive fluorescence sensors. The nature of dendrimers allows for incorporating multiple fluorescence molecules or receptor groups, which can improve the selectivity and sensitivity of the sensor. Moreover, the dense surface of dendrimers could provide a large surface area for interaction with the target molecule, resulting in improved signal amplification and binding affinity (Klajnert & Bryszewska, 2001; Takahashi et al., 2003)

Fluorescence sensors based on polymeric dendrimers have been developed for numerous applications, including environmental monitoring, drug discovery, and biomedical diagnostics (Avudaiappan et al., 2019; Krasteva et al., 2003). Antibiotics have become more prevalent in the environment as contaminants due to their overuse (Zhu et al., 2017). To evaluate the presence of such contaminants, Xu's research group designed fifth-generation poly(amidoamine)s (PAMAMs) (PF1-PF3) to achieve rapid detection of antibiotics. The most common methods for determining antibiotics are time-consuming and expensive to use. A series of 19 different antibiotics were

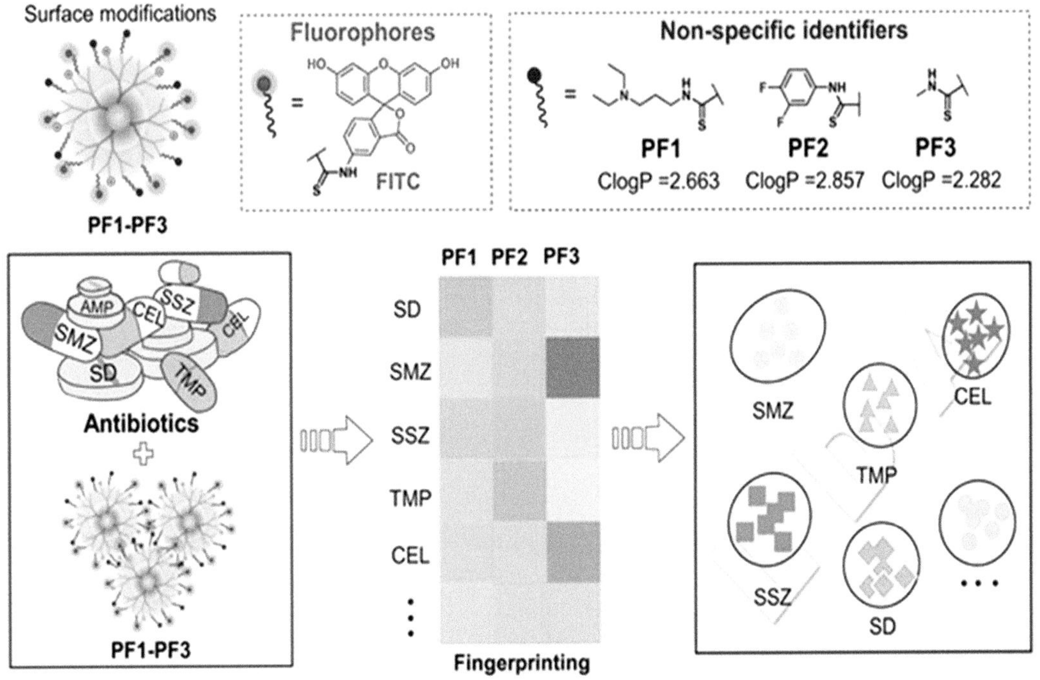

FIGURE 5.12

Structures of PF1-PF3 dendrimers and classification analysis of antibiotics. The figure includes the structures of the fifth-generation dendrimers used for the detection of antibiotics. In addition, the classification analysis of the antibiotics is included.

Source: *Reproduced with permission from Xu, L., Wang, H., Xiao, W., Zhang, W., Stewart, C., Huang, H., Li, F. & Han, J. (2023). PAMAM dendrimer-based tongue rapidly identifies multiple antibiotics. Sensors and Actuators B: Chemical, 382, 133519. https://doi.org/10.1016/j.snb.2023.133519.*

accurately and quickly discriminated in H_2O with 97% accuracy. Also, the sensors allowed the determination of multiple sulfonamide computation residues in milk samples with 100% accuracy (Fig. 5.12). The improved sensors show the robust real-sample determination of antibiotics (Xu et al., 2023).

Within the same research group, Xu et al. utilized a machine learning-assisted array of dendrimers for the detection of amyloid-beta (Aβ) peptides related to Alzheimer's disease. As shown in Fig. 5.13, the sensor array is composed of three pyrene-substituted dendrimers, which accurately differentiate a series of metalloproteins and charged nonmetalloproteins by excimer emission quenching. Similarly, the system evidences the efficient differentiation of Aβ-40 and Aβ-42 fibril aggregates in basic conditions. The identification of Aβ aggregates was evaluated in the presence of interferent ions and proteins, displaying significant antiinterference capacity and distinction of different Aβ40/Aβ42 monomers, oligomers, and fibrils. The sensor array was demonstrated to

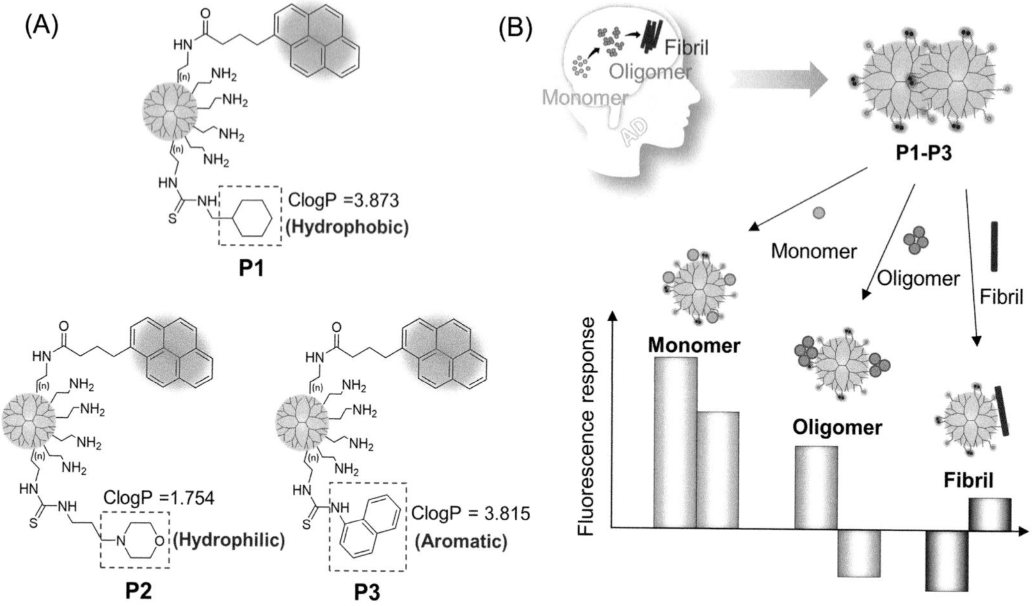

FIGURE 5.13

(A) Structure of the synthesized fluorescent dendrimers. (B) Illustration of the fluorescence response of the dendrimers in the presence of various types of Aβ40/Aβ42 monomers, oligomers, and fibrils. The figure shows the structure of the series of three fluorescent dendrimers. In addition, the image depicts the fluorescence response of the sensor array in the presence of amyloid beta peptides with monomeric, oligomeric, and fibril configurations

Source: *Reproduced with permission from Xu, L., Wang, H., Xu, Y., Cui, W., Ni, W., Chen, M., Huang, H., Stewart, C., Li, L., Li, F., & Han, J. (2022).* ACS Sensors, *7(5), 1315–1322. https://doi.org/10.1021/acssensors.2c00132.*

identify with accuracy Aβ40/Aβ42 more complex biological samples, including serum and cerebrospinal fluid (Xu et al., 2022).

Wang et al. synthesized a novel hybrid ratiometric chemosensor composed of a nanoaggregate polymer of tetraphenylethylene functionalized hyperbranched polyamidoamine and rhodamine B for the rapid, selective, and sensitive determination of free bilirubin in urine and water. Sensing free bilirubin for bedside assessment is vital, especially for liver dysfunction patients and newborns (Wang et al., 2022; Huber et al., 2012). The sensor exhibited two emission signals at 465 nm and 577 nm. The emission at 465 nm could be related to the Förster resonance energy transfer mechanism between the polymer and bilirubin. Meanwhile, the emission at 577 nm caused by the structure of rhodamine B did not change during the determination of bilirubin. In addition, there was a noticeable shift from blue to orange during the determination of bilirubin with the predicted mechanism (Fig. 5.14). A moveable hydrogel-based in-situ testing device was also improved for the bilirubin test, which makes the sensor portable and suitable for household use (Wang et al., 2022).

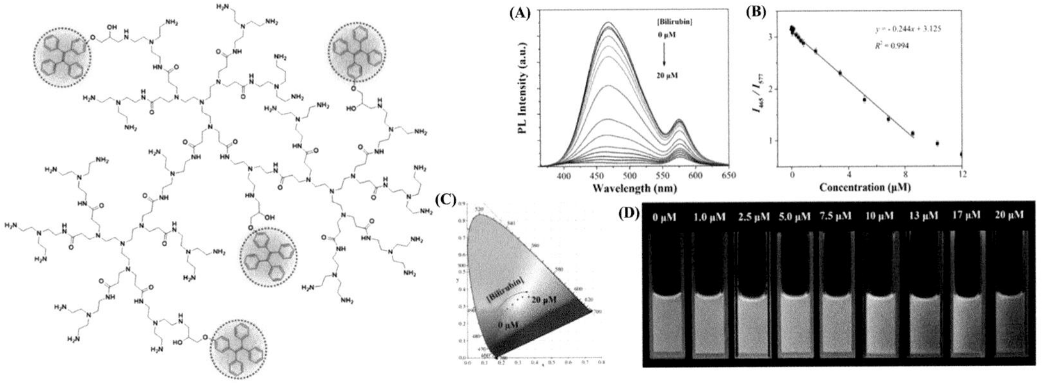

FIGURE 5.14

Structure and emission spectra of the tetraphenylethylene functionalized polymer for the determination of bilirubin. The image shows the structure and emission spectra of the tetraphenylethylene functionalized polymer (dendrimer) for the determination of bilirubin.

Source: *Reproduced with permission from Wang, B., Zhou, X.-Q., Li, L., Li, Y.-X., Yu, L.-P., & Chen, Y. (2022). Ratiometric fluorescence sensor for point-of-care testing of bilirubin based on tetraphenylethylene functionalized polymer nanoaggregate and rhodamine B.* Sensors and Actuators B: Chemical, 369, *132392. https://doi.org/10.1016/j.snb.2022.132392.*

The dendritic polymer of porphyrin-cored poly epichlorohydrin was prepared as a sensor for the cyanide ion determination by Avudaiappan et al. This dendrimer presented red emission at 656 and 718 nm, which was gradually decreased and quenched during the determination of cyanide ion due to the self-assembling of the porphyrin-cored dendritic chains. The quenching mechanism was also supported by density functional theory studies, field emission scanning electron microscopy images, and time-resolved fluorescence lifetime measurements. The dendrimeric polymer was highly sensitive (LOD, 1.2 μM) and selective, and it had a quick response to cyanide ions (CN^-) (Fig. 5.15; Avudaiappan et al., 2019).

Overall, fluorescent sensors based on polymeric dendrimers offer a potential approach for the improvement of highly selective and sensitive chemical sensors for a wide range of applications. The unique features of polymeric dendrimers provide a versatile platform for developing customized sensors with enhanced functionality and performance.

5.8 Coordination polymers

Coordination polymers are coordinated hybrid systems of repeating units self-assembled by coordinate bonds of various metallic ions as inorganic centers and ligands as organic components. By selecting numerous metal ions and ligands of diverse chemical nature, researchers have accessed structural tunability to develop a wide range of coordination networks (Du et al., 2013; Zhang, Liu, et al., 2016). Generally, the photoluminescence of coordination polymers originates from the excitation of the metallic clusters and the energy transfer process between the metal center and ligands,

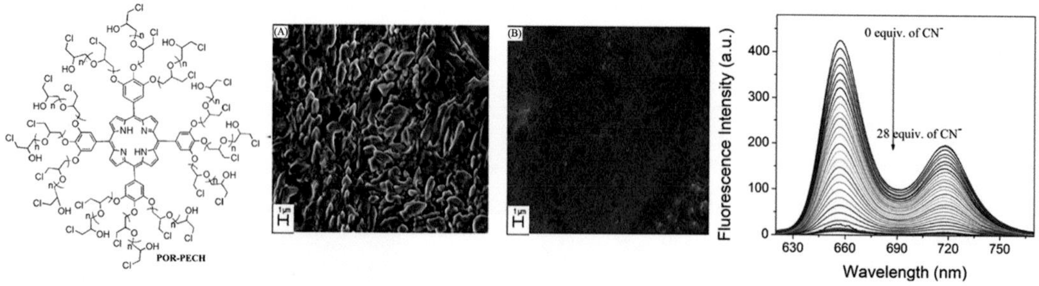

FIGURE 5.15

Characterization and fluorescence studies of the porphyrin-cored poly epichlorohydrin dendrimer. This figure displays the characterization data and fluorescence studies of the porphyrin-cored poly epichlorohydrin dendrimer.

Source: *Reproduced with permission from Avudaiappan, G., Anjaly Jacob, K., Theresa, L. V., Shebitha, A. M., Hiba, K., Shenoi, P. K., Unnikrishnan, V., & Sreekumar, K. (2019). A novel dendritic polymer based turn- off fluorescence sensor for the selective detection of cyanide ion in aqueous medium.* Reactive and Functional Polymers, 137, *71–78. https://doi.org/10.1016/j.reactfunctpolym.2019.01.018.*

in which the emission properties depend on defined metal-ligand interactions. In coordination polymers, the immobilization of ligands within the networks induces an enhancement in the photoluminescence intensity (Alexandrov et al., 2022; Bünzli, 2014; Liu, Luo, et al., 2020). Quantum chemical calculations usually enable the optimization of the polymer structures and prediction of optical properties for target applications. Therefore, the theoretical methods aid in the precise selection of metals and ligands needed to build well-suited coordination polymers. The chemical structural adaptability of coordination polymers has contributed to adjusting the optical and luminescence properties for sensing various analytes (Kansız & Dege, 2018; Sorg et al., 2018). The coordination polymer sensing mechanism often involves interactions between the analyte and binding sites in the metal cluster or ligand moieties, altering the excited state energy of the polymeric host and generating new radiative relaxation pathways. Changes in the emission spectral patterns, the appearance of pronounced stock shifts, and the activation of nonradiative decay deactivation modulate the sensory signals (Alsharabasy et al., 2021; Zhang et al., 2015).

The structure of luminescent coordination polymers is often derived from rare-earth metal, transition metal, and alkaline-earth metal ions. A relevant example was reported by Tan et al., in which a lanthanide coordination polymer was designed based on adenine, dipicolinic acid, and terbium ion (Tb^{3+}). For this system, as shown in Fig. 5.16, the fluorescence emission is quenched due to the PET occurring between the adenine rings and di-picolinic acid moieties. Consequently, the intramolecular energy transfer to the metal center is suppressed. However, greater fluorescence emission is observed when increasing the concentration of mercury ions (Hg^{2+}) in the system in the range of 0–200 nM. The strong interaction between adenine and Hg^{2+} deactivates the PET process and favors the intramolecular energy transfer to the Tb^{3+} ion generating fluorescence. The limit of detection was reported as 0.2 nM, indicating the sensor is ultra sensitive. This first report on lanthanide-based coordination polymers and PET process has opened the route to further investigate the applicability of these materials in sensing various relevant analytes (Tan et al., 2012).

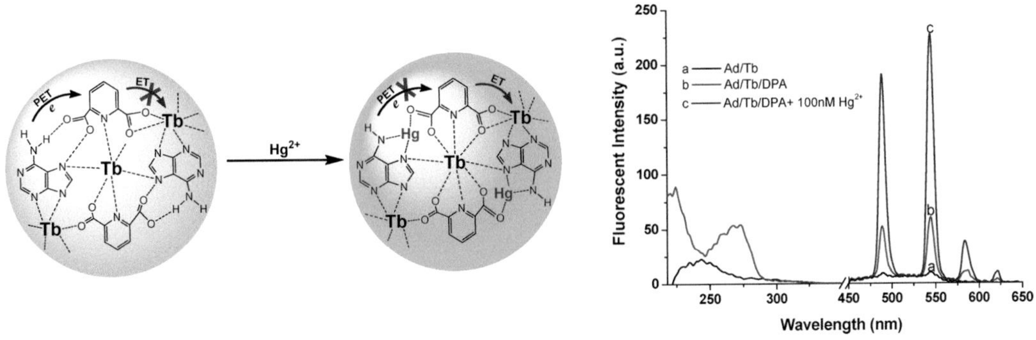

FIGURE 5.16

Spherical coordination polymer nanoparticles that enhance fluorescence in the presence of Hg^{2+}. The red spectral bands show the coordination of polymer nanoparticles in the absence of Hg^{2+} ions, and the blue spectral bands show nanoparticle emission in the presence of Hg^{2+} ions. This image represents the formation of the highly emissive coordination polymer nanoparticles in the presence of the mercury (II) ions. The figure includes the spectral bands of the nanoparticles in the absence and presence of the mercury (II) ions. In the presence of mercury (II) ions, there exists a significant enhancement in the emission spectral bands of the nanoparticles.

Source: *Reproduced with permission from Tan, H., Liu, B., & Chen, Y. (2012).* ACS Nano, 6*(12), 10505–10511. https://doi.org/10.1021/nn304469j.*

Most recent reports include the incorporation of transition metals such as cobalt as metal centers for developing coordination polymers as fluorogenic probes. Luo et al. described a set of four cobalt (II) coordination polymers with anthracene-containing dicarboxylic acid with high selectivity in water solution. The mode of fluorescence-modulated signal derives from the chemical nature of the auxiliary ligand. For instance, the sensing of Cr^{3+}, Al^{3+}, and Pb^{2+} ions follows the fluorescence turn-on approach, in which the energy transfer process from ligand to Co(II) is responsible for the fluorescence increase (Luo et al., 2023). In another study, Wang et al. reported a cobalt(II)-based coordination polymer built with amide-featuring tetracarboxylic acid, which displays sensing capacity toward Fe^{3+} ions by quenching mechanism, and sensing of Tb^{3+} by fluorescence turn-on effect. The sensory system was able to detect other ions such as Pb^{2+} and Al^{3+} by radiometric fluorescence response (Wang, Sun, et al., 2023).

Fig. 5.17 represents the coordination polymer system Cd(II)-CP containing the host ligand H_2L and the auxiliary ligand 1,4-pbyb. The fluorogenic polymer probe was utilized for the detection of NACs and sulfide ions (S^{2-}) as analytes. Upon addition of PA, the polymer exhibits turn-off fluorescence dictated by quenching through energy transfer and PET processes. On the contrary, the presence of S^{2-} ions induces the fluorescence to intensify by dissociating the polymer networks into emissive species (Dou et al., 2023). Guo and coworkers also developed the coordination polymer chemosensor for sensing PA and sulfide S^{2-} ions. The polymer was tailored with a different auxiliary ligand (4,4′-bipy) incorporated into the host ligand H_2L. The sensory response follows the same sensing turn-off in the presence of PA, and turn-on mechanism in the presence of S^{2-} ions (Guo et al., 2023).

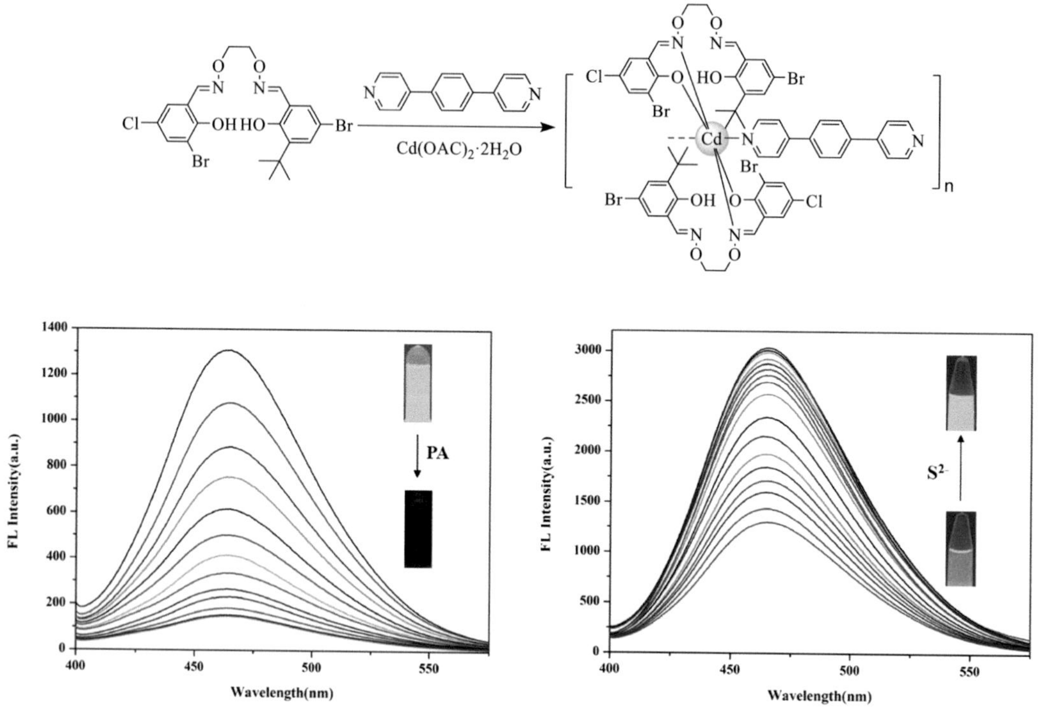

FIGURE 5.17

The synthetic reaction of the coordination polymer Cd(II)-CP and changes in the emission intensity of the polymer in the presence of picric acid (PA) and sulfide ions (S^{2-}). The PA concentration was increased in the range of 0–110 μM. The S^{2-} concentration was increased in the range 0–160 μM. The figure includes the synthetic route of the cadmium-containing coordination polymer. In addition, the titration graphs evidence the quenching effect on the emission band of the polymer upon addition of the picric acid concentration. A second graph shows the enhancement of emission of the sensor when the concentration of sulfide ions is increased.

Source: *Reproduced with permission from Dou, L., Tong, L., Ma, C.-Y., Dong, W.-K., & Ding, Y.-J. (2023).* Journal of Molecular Structure, 1292, *136162. https://doi.org/10.1016/j.molstruc.2023.136162.*

An electrochemiluminescence (ECL) sensor with a series of binuclear coordination polymers is presented in Fig. 5.18. The polymers were designed with different molar ratios of transition metal (Zn^{2+}) and lanthanide (Tb^{3+}) ions for electrochemiluminescent sensing of the antibiotic molecule tetracycline in $K_2S_2O_8$. The coordination polymer with a ratio of metals Tb:Zn of 9:3 adopts a flower-like morphology that allows the efficient ligand antenna effect and sensitization of the Zn^{2+} ions. The organic ligands serve as antenna groups to increase the luminescence by energy transfer from the ligand to the Tb^{3+} ions, and the Zn^{2+} metal centers enhance the intramolecular energy transfer by inducing better matching of the ligand and metal energy levels. In the presence of tetracycline, ECL resonance energy transfer from the coordinated network to the tetracycline promotes

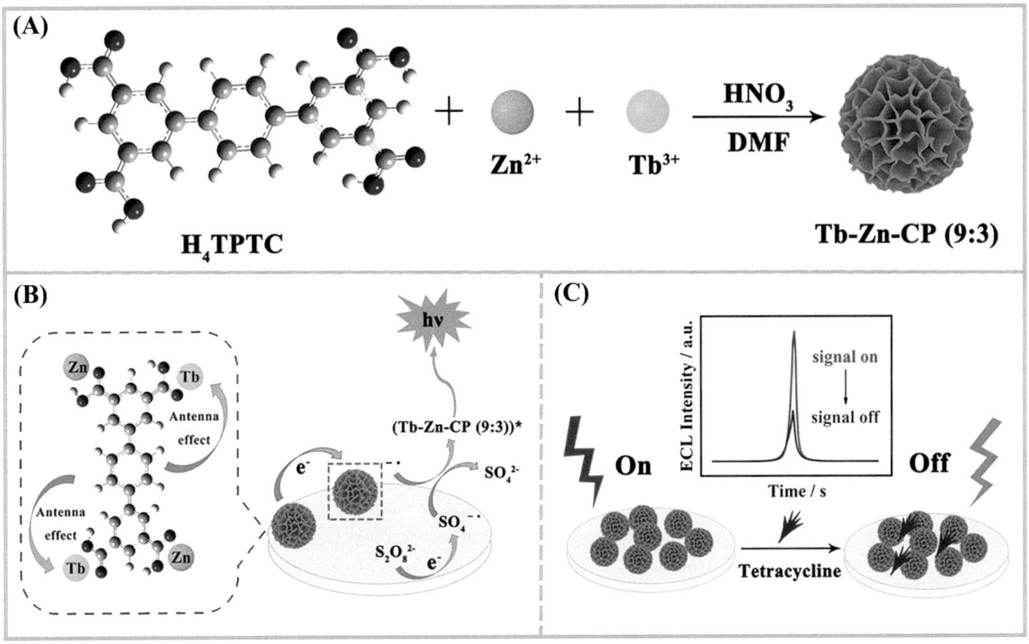

FIGURE 5.18

(A) Synthesis process of the coordination polymer Tb-Zn-CP (9:3). (B) Schematic representation of the luminescence mechanism of the lanthanide-based sensor. (C) Illustration of the turn-off sensing mode of the Tb-Zn-CP (9:3) system in the presence of tetracycline. The image illustrates the synthesis of the binuclear coordination polymer with a flower-like morphology. In addition, a schematic representation of the electrochemiluminescence mechanism of the lanthanide-based coordination polymer is included. The illustration also includes the turn-off sensing mode of the chemosensor in the presence of the antibiotic tetracycline.

Source: *Reproduced with permission from Kuang, G., Wang, C., Song, L., Zhang, G., Yang, Y., & Fu, Y. (2023).* Food Chemistry, 403, *134376. https://doi.org/10.1016/j.foodchem.2022.134376.*

the emission quenching effect. This lanthanide-based sensor was reported to be selective and effective for the detection of antibiotic molecules in dairy products (Kuang et al., 2023).

5.9 Conclusions and Perspectives

In this chapter, an overview of the structure, features, and usages of fluorogenic polymers has been reported. It is evident from research on fluorogenic sensing with polymers that the range of polymer structures and applications have become quite widespread. By using the high sensitivity of fluorescent polymers and the ability to optically respond to changes in the chemical environment, polymer-based fluorogenic sensors have been developed for a variety of analytes, including small ions, such as cations and anions, as well as biological cells, including pathogens. Essentially, fluorogenic polymers are materials that are responsive to various stimuli, with the sensing response being a change of the fluorescence of the materials. This chapter aims to provide a general introduction to the field of fluorogenic polymers. While several specific applications are described, readers are encouraged to explore the numerous excellent references cited for further insight.

Acknowledgments

KSS and IBR acknowledge the Welch Foundation for support of this work through the Welch Chair (Grant No. AX-0045–20110629).

References

Adhikari, B., & Majumdar, S. (2004). Polymers in sensor applications. *Progress in Polymer Science.*, *29*(7), 699–766. Available from https://doi.org/10.1016/j.progpolymsci.2004.03.002.

Alexander, C., Andersson, H. S., Andersson, L. I., Ansell, R. J., Kirsch, N., Nicholls, I. A., O'Mahony, J., & Whitcombe, M. J. (2006). Molecular imprinting science and technology: a survey of the literature for the years up to and including 2003. *Journal of Molecular Recognition*, *19*(2), 106–180. Available from https://doi.org/10.1002/jmr.760.

Alexandrov, E. V., Shevchenko, A. P., Nekrasova, N. A., & Blatov, V. A. (2022). Topological methods for analysis and design of coordination polymers. *Russian Chemical Reviews*, *91*(4), RCR5032. Available from https://doi.org/10.1070/RCR5032.

Alsharabasy, A. M., Pandit, A., & Farràs, P. (2021). Recent Advances in the Design and Sensing Applications of Hemin/Coordination Polymer-Based Nanocomposites. *Advanced Materials*, *33*(2), 2003883. Available from https://doi.org/10.1002/adma.202003883.

Anantha-Iyengar, G., Shanmugasundaram, K., Nallal, M., Lee, K. P., Whitcombe, M. J., Lakshmi, D., & Sai-Anand, G. (2019). Functionalized conjugated polymers for sensing and molecular imprinting applications. *Progress in Polymer Science*, *88*, 1–129. Available from https://doi.org/10.1016/j.progpolymsci.2018.08.001.

Aneja, A., Mathur, N., Bhatnagar, P. K., & Mathur, P. C. (2008). Triple-FRET technique for energy transfer between conjugated polymer and TAMRA dye with possible applications in medical diagnostics. *Journal of Biological Physics.*, *34*(5), 487–493. Available from https://doi.org/10.1007/s10867-008-9107-y.

Anirudhan, T. S., Divya, P. L., & Nima, J. (2013). Silylated montmorillonite based molecularly imprinted polymer for the selective binding and controlled release of thiamine hydrochloride. *Reactive and Functional Polymers.*, *73*(8), 1144–1155. Available from https://doi.org/10.1016/j.reactfunctpolym.2013.05.004.

Ansari, S., & Karimi, M. (2017). Novel developments and trends of analytical methods for drug analysis in biological and environmental samples by molecularly imprinted polymers. *TrAC Trends in Analytical Chemistry*, *89*, 146–162. Available from https://doi.org/10.1016/j.trac.2017.02.002.

Ansari, S., & Masoum, S. (2021). Recent advances and future trends on molecularly imprinted polymer-based fluorescence sensors with luminescent carbon dots. *Talanta*, *223*, 121411. Available from https://doi.org/10.1016/j.talanta.2020.121411.

Avudaiappan, G., Jacob, K. A., Theresa, L. V., Shebitha, A. M., Hiba, K., Shenoi, P. K., Unnikrishnan, V., & Sreekumar, K. (2019). A novel dendritic polymer based turn- off fluorescence sensor for the selective detection of cyanide ion in aqueous medium. *Reactive and Functional Polymers.*, *137*, 71–78. Available from https://doi.org/10.1016/j.reactfunctpolym.2019.01.018.

Baltrons, O., López-Mesas, M., Palet, C., Le Derf, F., & Portet-Koltalo, F. (2013). Molecularly imprinted polymer-liquid chromatography/fluorescence for the selective clean-up of hydroxylated polycyclic aromatic hydrocarbons in soils. *Analytical Methods.*, *5*(22), 6297. Available from https://doi.org/10.1039/c3ay41227j.

Bazylevich, A., Patsenker, L. D., & Gellerman, G. (2017). Exploiting fluorescein based drug conjugates for fluorescent monitoring in drug delivery. *Dyes and Pigments.*, *139*, 460–472. Available from https://doi.org/10.1016/j.dyepig.2016.11.057.

Bolinger, J. C., Traub, M. C., Brazard, J., Adachi, T., Barbara, P. F., & Vanden Bout, D. A. (2012). Conformation and Energy Transfer in Single Conjugated Polymers. *Accounts of Chemical Research*, *45*(11), 1992–2001. Available from https://doi.org/10.1021/ar300012k.

Bulut, U., Sayin, V. O., Cevher, S. C., Cirpan, A., & Soylemez, S. (2022). Benzodithiophene bearing conjugated polymer-based surface anchoring for sensitive electrochemical glucose detection. *Express Polymer Letters.*, *16*(10), 1012–1021. Available from https://doi.org/10.3144/expresspolymlett.2022.74.

Bünzli, J. C. G. (2014). Review: Lanthanide coordination chemistry: from old concepts to coordination polymers. *Journal of Coordination Chemistry*, *67*(23–24), 3706–3733. Available from https://doi.org/10.1080/00958972.2014.957201, https://doi.org/10.1080/00958972.2014.957201.

Cabarcos, E. L., & Carter, S. A. (2005). Characterization of the photoluminescence quenching of mixed water-soluble conjugated polymers for potential use as biosensor materials. *Macromolecules*, *38*(10), 4409–4415. Available from https://doi.org/10.1021/ma050153c.

Carrasco-Correa, E. J., Ramis-Ramos, G., & Herrero-Martínez, J. M. (2015). Hybrid methacrylate monolithic columns containing magnetic nanoparticles for capillary electrochromatography. *Journal of Chromatography. A*, *1385*, 77–84. Available from https://doi.org/10.1016/j.chroma.2015.01.044.

Chakraborty, G., Chittela, R. K., Jonnalgadda, P. N., & Pal, H. (2022). Polyanionic amphiphilic polymer based supramolecular dye-host assembly: Highly selective turn–on probe for protamine sensing. *Sensors and Actuators B: Chemical.*, *371*, 132582. Available from https://doi.org/10.1016/j.snb.2022.132582.

Chan, Y. H., & Wu, P. J. (2015). Semiconducting Polymer Nanoparticles as Fluorescent Probes for Biological Imaging and Sensing. *Particle & Particle Systems Characterization*, *32*(1), 11–28. Available from https://doi.org/10.1002/ppsc.201400123.

Chang, J., Li, H., Hou, T., Duan, W., & Li, F. (2018). Paper-based fluorescent sensor via aggregation induced emission fluorogen for facile and sensitive visual detection of hydrogen peroxide and glucose. *Biosensors and Bioelectronics.*, *104*, 152–157. Available from https://doi.org/10.1016/j.bios.2018.01.007.

Chen, C., Zhang, X., Long, Z., Zhang, J., & Zheng, C. (2012). Molecularly imprinted dispersive solid-phase microextraction for determination of sulfamethazine by capillary electrophoresis. *Microchimica Acta.*, *178*(3–4), 293–299. Available from https://doi.org/10.1007/s00604-012-0833-2.

Chen, H., Zhu, L., Jiang, W., Ji, H., Zhou, X., Qin, Y., & Wu, L. (2023). Multiple fluorescence polymer dots-based differential array sensors for highly efficient heavy metal ions detection. *Environmental Research*, *232*, 116278. Available from https://doi.org/10.1016/j.envres.2023.116278.

Chen, L., McBranch, D. W., Wang, H. L., Helgeson, R., Wudl, F., & Whitten, D. G. (1999). Highly sensitive biological and chemical sensors based on reversible fluorescence quenching in a conjugated polymer. *Proceedings of the National Academy of Sciences*, *96*(22), 12287–12292. Available from https://doi.org/10.1073/pnas.96.22.12287.

Chen, Y., Pu, K. Y., Fan, Q. L., Qi, X. Y., Huang, Y. Q., Lu, X. M., & Huang, W. (2009). Water-soluble anionic conjugated polymers for metal ion sensing: Effect of interchain aggregation. *Journal of Polymer Science Part A: Polymer Chemistry.*, *47*(19), 5057–5067. Available from https://doi.org/10.1002/pola.23558.

Cho, J., Keum, C., Lee, S. G., & Lee, S. Y. (2020). Aggregation-driven fluorescence quenching of imidazole-functionalized perylene diimide for urea sensing. *The Analyst*, *145*(22), 7312–7319. Available from https://doi.org/10.1039/d0an01252a.

Ciotta, E., Prosposito, P., & Pizzoferrato, R. (2019). Positive curvature in Stern–Volmer plot described by a generalized model for static quenching. *Journal of Luminescence.*, *206*, 518–522. Available from https://doi.org/10.1016/j.jlumin.2018.10.106.

Collot, M., Fam, T. K., Ashokkumar, P., Faklaris, O., Galli, T., Danglot, L., & Klymchenko, A. S. (2018). Ultrabright and fluorogenic probes for multicolor imaging and tracking of lipid droplets in cells and tissues. *Journal of the American Chemical Society*, *140*(16), 5401–5411. Available from https://doi.org/10.1021/jacs.7b12817.

Cui, W., Wang, L., Xiang, G., Zhou, L., An, X., & Cao, D. (2015). A colorimetric and fluorescence "turn-off" chemosensor for the detection of silver ion based on a conjugated polymer containing 2,3-di(pyridin-2-yl) quinoxaline. *Sensors and Actuators B: Chemical*, *207*, 281–290. Available from https://doi.org/10.1016/j.snb.2014.10.072.

Cui, Z., Li, Z., Jin, Y., Ren, T., Chen, J., Wang, X., Zhong, K., Tang, L., Tang, Y., & Cao, M. (2020). Novel magnetic fluorescence probe based on carbon quantum dots-doped molecularly imprinted polymer for AHLs signaling molecules sensing in fish juice and milk. *Food Chemistry*, *328*, 127063. Available from https://doi.org/10.1016/j.foodchem.2020.127063.

Dong, W., Fei, T., & Scherf, U. (2018). Conjugated polymers containing tetraphenylethylene in the backbones and side-chains for highly sensitive TNT detection. *RSC Advances*, *8*(11), 5760–5767. Available from https://doi.org/10.1039/c7ra13536j.

Dong, W., Ma, Z., Chen, P., & Duan, Q. (2019). Carbazole and tetraphenylethylene based AIE-active conjugated polymer for highly sensitive TNT detection. *Materials Letters*, *236*, 480–482. Available from https://doi.org/10.1016/j.matlet.2018.10.162.

Dou, L., Tong, L., Ma, C. Y., Dong, W. K., & Ding, Y. J. (2023). Inserting auxiliary ligand to construct a Cd(II)-based salamo-like coordination polymer as bifunctional chemosensor for detecting picric acid and S2 − . *Journal of Molecular Structure*, *1292*, 136162. Available from https://doi.org/10.1016/j.molstruc.2023.136162.

Du, M., Li, C. P., Liu, C. S., & Fang, S. M. (2013). Design and construction of coordination polymers with mixed-ligand synthetic strategy. *Coordination Chemistry Reviews*, *257*(7), 1282–1305. Available from https://doi.org/10.1016/j.ccr.2012.10.002.

Elabd, A. A., & Elhefnawy, O. A. (2022). A new benzeneacetic acid derivative-based sensor for assessing thorium (IV) in aqueous solution based on aggregation caused quenching (ACQ) and aggregation induced emission (AIE. *Journal of Photochemistry and Photobiology A: Chemistry.*, *428*, 113866. Available from https://doi.org/10.1016/j.jphotochem.2022.113866.

Fakirov, S. (2019). Condensation polymers: their chemical peculiarities offer great opportunities. *Progress in Polymer Science.*, *89*, 1–18. Available from https://doi.org/10.1016/j.progpolymsci.2018.09.003.

Fan, C., Plaxco, K. W., & Heeger, A. J. (2002). High-efficiency fluorescence quenching of conjugated polymers by proteins. *Journal of the American Chemical Society*, *124*(20), 5642–5643. Available from https://doi.org/10.1021/ja025899u.

Fan, Y., Yao, J., Huang, M., Linghu, C., Guo, J., & Li, Y. (2021). Non-conjugated polymer dots for fluorometric and colorimetric dual-mode detection of quercetin. *Food Chemistry*, *359*, 129962. Available from https://doi.org/10.1016/j.foodchem.2021.129962.

Fang, B., Shen, Y., Peng, B., Bai, H., Wang, L., Zhang, J., Hu, W., Fu, L., Zhang, W., Li, L., & Huang, W. (2022). Small-molecule quenchers for förster resonance energy transfer: structure, mechanism, and applications. *Angewandte Chemie*, *134*(41), 116278. Available from https://doi.org/10.1002/ange.202207188.

Fang, M., Zhou, L., Zhang, H., Liu, L., & Gong, Z. Y. (2019). A molecularly imprinted polymers/carbon dots-grafted paper sensor for 3-monochloropropane-1,2-diol determination. *Food Chemistry*, *274*, 156–161. Available from https://doi.org/10.1016/j.foodchem.2018.08.133.

Fernandes, R. F., Atvars, T. D. Z., & Temperini, M. L. A. (2023). Exploring the non-traditional fluorescence emission of non-conjugated polymers dots for sensing pesticides. *Reactive and Functional Polymers.*, *182*, 105483. Available from https://doi.org/10.1016/j.reactfunctpolym.2022.105483.

Flory, P. J. (1946). Fundamental principles of condensation polymerization. *Chemical Reviews*, *39*(1), 137–197. Available from https://doi.org/10.1021/cr60122a003.

Fu, Y., Yu, J., Wang, K., Liu, H., Yu, Y., Liu, A., Peng, X., He, Q., Cao, H., & Cheng, J. (2018). Simple and efficient chromophoric-fluorogenic probes for diethylchlorophosphate vapor. *ACS Sensors.*, *3*(8), 1445–1450. Available from https://doi.org/10.1021/acssensors.8b00313.

Fuchs, Y., Soppera, O., & Haupt, K. (2012). Photopolymerization and photostructuring of molecularly imprinted polymers for sensor applications—A review. *Analytica Chimica Acta*, *717*, 7–20. Available from https://doi.org/10.1016/j.aca.2011.12.026.

Gao, M., Li, S., Lin, Y., Geng, Y., Ling, X., Wang, L., Qin, A., & Tang, B. Z. (2016). Fluorescent light-up detection of amine vapors based on aggregation-induced emission. *ACS Sensors.*, *1*(2), 179–184. Available from https://doi.org/10.1021/acssensors.5b00182.

Gao, R., Mu, X., Hao, Y., Zhang, L., Zhang, J., & Tang, Y. (2014). Combination of surface imprinting and immobilized template techniques for preparation of core–shell molecularly imprinted polymers based on directly amino-modified Fe_3O_4 nanoparticles for specific recognition of bovine hemoglobin. *Journal of Materials Chemistry B*, *2*(12), 1733–1741. Available from https://doi.org/10.1039/C3TB21684E.

García, J. M., García, F. C., Serna, F., & de la Peña, J. L. (2011). Fluorogenic and chromogenic polymer chemosensors. *Polymer Reviews.*, *51*(4), 341–390. Available from https://doi.org/10.1080/15583724.2011.616084.

Gaylord, B. S., Heeger, A. J., & Bazan, G. C. (2003). DNA hybridization detection with water-soluble conjugated polymers and chromophore-labeled single-stranded DNA. *Journal of the American Chemical Society*, *125*(4), 896–900. Available from https://doi.org/10.1021/ja027152 + .

Gehlen, M. H. (2020). The centenary of the Stern-Volmer equation of fluorescence quenching: From the single line plot to the SV quenching map. *Journal of Photochemistry and Photobiology C: Photochemistry Reviews.*, *42*, 100338. Available from https://doi.org/10.1016/j.jphotochemrev.2019.100338.

Gholivand, M. B., & Torkashvand, M. (2011). A novel high selective and sensitive metronidazole voltammetric sensor based on a molecularly imprinted polymer-carbon paste electrode. *Talanta*, *84*(3), 905–912. Available from https://doi.org/10.1016/j.talanta.2011.02.022.

Giovannitti, A., Nielsen, C. B., Rivnay, J., Kirkus, M., Harkin, D. J., White, A. J. P., Sirringhaus, H., Malliaras, G. G., & McCulloch, I. (2016). Sodium and potassium ion selective conjugated polymers for optical ion detection in solution and solid state. *Advanced Functional Materials.*, *26*(4), 514–523. Available from https://doi.org/10.1002/adfm.201503791.

Gokel, G. W., Leevy, W. M., & Weber, M. E. (2004). Crown ethers: sensors for ions and molecular scaffolds for materials and biological models. *Chem Rev*, *104*(5), 2723–2750. Available from https://doi.org/10.1021/cr020080k.

Gu, J. Z., Liang, X. X., Cui, Y. H., Wu, J., Shi, Z. F., & Kirillov, A. M. (2017). Introducing 2-(2-carboxyphenoxy)terephthalic acid as a new versatile building block for design of diverse coordination polymers: synthesis, structural features, luminescence sensing, and magnetism. *CrystEngComm.*, *19*(18), 2570–2588. Available from https://doi.org/10.1039/c7ce00219j.

Guo, W. T., Ding, Y. F., Li, X., Tong, L., Dou, L., & Dong, W. K. (2023). Highly efficient and selective detection of sulfur ions and picric acid through salamo-Cd(II) coordination polymer chemosensor. *Inorganica Chimica Acta*, *557*, 121704. Available from https://doi.org/10.1016/j.ica.2023.121704.

Haldar, U., Sharma, R., Kim, E. J., & Lee, H. I. (2020). Azobenzene–hemicyanine conjugated polymeric chemosensor for the rapid and selective detection of cyanide in pure aqueous media. *Journal of Polymer Science*, *58*(1), 124–131. Available from https://doi.org/10.1002/pol.20190285.

Han, J., Wang, B., Bender, M., Seehafer, K., & Bunz, U. H. F. (2016). Water-soluble poly(p-aryleneethynylene)s: a sensor array discriminates aromatic carboxylic acids. *ACS Applied Materials & Interfaces*, *8*(31), 20415–20421. Available from https://doi.org/10.1021/acsami.6b06462.

Harrison, B. S., Ramey, M. B., Reynolds, J. R., & Schanze, K. S. (2000). Amplified fluorescence quenching in a poly(p-phenylene)-based cationic polyelectrolyte. *Journal of the American Chemical Society*, *122*(35), 8561–8562. Available from https://doi.org/10.1021/ja000819c.

Hasegawa, Y., & Kitagawa, Y. (2022). Luminescent lanthanide coordination polymers with transformative energy transfer processes for physical and chemical sensing applications. *Journal of Photochemistry and Photobiology C: Photochemistry Reviews.*, *51*, 100485. Available from https://doi.org/10.1016/j.jphotochemrev.2022.100485.

Hembury, G. A., Borovkov, V. V., & Inoue, Y. (2008). Chirality-Sensing Supramolecular Systems. *Chemical Reviews*, *108*(1), 1–73. Available from https://doi.org/10.1021/cr050005k.

Hillberg, A. L., Brain, K. R., & Allender, C. J. (2005). Molecular imprinted polymer sensors: Implications for therapeutics. *Molecularly Imprinted Polymers: Technology and Applications*, *57*(12), 1875–1889. Available from https://doi.org/10.1016/j.addr.2005.07.016.

Hu, L., Zhang, Q., Li, X., & Serpe, M. J. (2019). Stimuli-responsive polymers for sensing and actuation. *Materials Horizons*, *6*(9), 1774–1793. Available from https://doi.org/10.1039/c9mh00490d.

Hu, Q., Zhang, W., Yin, Q., Wang, Y., & Wang, H. (2021). A conjugated fluorescent polymer sensor with amidoxime and polyfluorene entities for effective detection of uranyl ion in real samples. *Spectrochimica Acta Part A: Molecular and Biomolecular Spectroscopy*, *244*, 118864. Available from https://doi.org/10.1016/j.saa.2020.118864.

Hu, Z., Zheng, N., Dong, S., Liu, X., Chen, Z., Ying, L., Duan, C., Huang, F., & Cao, Y. (2018). Phosphonium conjugated polyelectrolytes as interface materials for efficient polymer solar cells. *Organic Electronics.*, *57*, 151–157. Available from https://doi.org/10.1016/j.orgel.2018.03.006.

Huang, C., Wang, H., Ma, S., Bo, C., Ou, J., & Gong, B. (2021). Recent application of molecular imprinting technique in food safety. *Journal of Chromatography. A*, *1657*, 462579. Available from https://doi.org/10.1016/j.chroma.2021.462579.

Huang, J., Shi, Y., Zeng, G., Gu, Y., Chen, G., Shi, L., Hu, Y., Tang, B., & Zhou, J. (2016). Acyl-homoserine lactone-based quorum sensing and quorum quenching hold promise to determine the performance of biological wastewater treatments: An overview. *Chemosphere*, *157*, 137–151. Available from https://doi.org/10.1016/j.chemosphere.2016.05.032.

Huang, X., Meng, J., Dong, Y., Cheng, Y., & Zhu, C. (2010). Polymer-based fluorescence sensor incorporating triazole moieties for Hg2 + detection via click reaction. *Polymer*, *51*(14), 3064–3067. Available from https://doi.org/10.1016/j.polymer.2010.05.001.

Huang, Y. Q., Fan, Q. L., Lu, X. M., Fang, C., Liu, S. J., Yu-Wen, L. H., Wang, L. H., & Huang, W. (2006). Cationic, water-soluble, fluorene-containing poly(arylene ethynylene)s: Effects of water solubility on aggregation, photoluminescence efficiency, and amplified fluorescence quenching in aqueous solutions. *Journal of Polymer Science Part A: Polymer Chemistry*, *44*(19), 5778–5794. Available from https://doi.org/10.1002/pola.21628.

Huang, Y., Zhao, S., Shi, M., Liu, J., & Liang, H. (2011). Competitive immunoassay of phenobarbital by microchip electrophoresis with laser induced fluorescence detection. *Anal Chim Acta*, *694*(1-2), 162–166. Available from https://doi.org/10.1016/j.aca.2011.03.036.

Huber, A. H., Zhu, B., Kwan, T., Kampf, J. P., Hegyi, T., & Kleinfeld, A. M. (2012). Fluorescence sensor for the quantification of unbound bilirubin concentrations. *Clinical Chemistry*, *58*(5), 869–876. Available from https://doi.org/10.1373/clinchem.2011.176412.

Hussain, S., Chen, X., Wang, C., Hao, Y., Tian, X., He, Y., Li, J., Shahid, M., Iyer, P. K., & Gao, R. (2022). Aggregation and binding-directed FRET modulation of conjugated polymer materials for selective and point-of-care monitoring of serum albumins. *Analytical Chemistry*, *94*(30), 10685–10694. Available from https://doi.org/10.1021/acs.analchem.2c00984.

Hussain, S., Malik, A. H., Afroz, M. A., & Iyer, P. K. (2015). Ultrasensitive detection of nitroexplosive – picric acid via a conjugated polyelectrolyte in aqueous media and solid support. *Chemical Communications*, *51*(33), 7207–7210. Available from https://doi.org/10.1039/C5CC02194D.

Ihde, M. H., Tropp, J., Diaz, M., Shiller, A. M., Azoulay, J. D., & Bonizzoni, M. (2022). A Sensor Array for the Ultrasensitive Discrimination of Heavy Metal Pollutants in Seawater. *Advanced Functional Materials.*, *32*(33), 2112634. Available from https://doi.org/10.1002/adfm.202112634.

Jagadesan, P., Yu, Z., Barboza-Ramos, I., Lara, H. H., Vazquez-Munoz, R., López-Ribot, J. L., & Schanze, K. S. (2020). Light-activated antifungal properties of imidazolium-functionalized cationic conjugated polymers. *Chemistry of Materials.*, *32*(14), 6186–6196. Available from https://doi.org/10.1021/acs.chemmater.0c02076.

Jia, X., Ge, Y., Shao, L., Wang, C., & Wallace, G. G. (2019). Tunable conducting polymers: toward sustainable and versatile batteries. *ACS Sustainable Chemistry & Engineering*, *7*(17), 14321–14340. Available from https://doi.org/10.1021/acssuschemeng.9b02315.

Jiang, H., Taranekar, P., Reynolds, J. R., & Schanze, K. S. (2009). Conjugated polyelectrolytes: synthesis, photophysics, and applications. *Angewandte Chemie International Edition.*, *48*(24), 4300–4316. Available from https://doi.org/10.1002/anie.200805456.

Jiang, H., Zhao, X., & Schanze, K. S. (2006). Amplified fluorescence quenching of a conjugated polyelectrolyte mediated by Ca^{2+}. *Langmuir: the ACS Journal of Surfaces and Colloids*, *22*(13), 5541–5543. Available from https://doi.org/10.1021/la060429p.

Jiang, H., Zhao, X., & Schanze, K. S. (2007). Effects of polymer aggregation and quencher size on amplified fluorescence quenching of conjugated polyelectrolytes. *Langmuir: the ACS Journal of Surfaces and Colloids*, *23*(18), 9481–9486. Available from https://doi.org/10.1021/la701192t.

Jiang, N., Li, G., Che, W., Zhu, D., Su, Z., & Bryce, M. R. (2018). Polyurethane derivatives for highly sensitive and selective fluorescence detection of 2,4,6-trinitrophenol (TNP). *Journal of Materials Chemistry C*, *6*(42), 11287–11291. Available from https://doi.org/10.1039/c8tc04250k.

Kan, X., Xing, Z., Zhu, A., Zhao, Z., Xu, G., Li, C., & Zhou, H. (2012). Molecularly imprinted polymers based electrochemical sensor for bovine hemoglobin recognition. *Sensors and Actuators B: Chemical.*, *168*, 395–401. Available from https://doi.org/10.1016/j.snb.2012.04.043.

Kansız, S., & Dege, N. (2018). Synthesis, crystallographic structure, DFT calculations and Hirshfeld surface analysis of a fumarate bridged Co(II) coordination polymer. *Journal of Molecular Structure*, *1173*, 42–51. Available from https://doi.org/10.1016/j.molstruc.2018.06.071.

Klajnert, B., & Bryszewska, M. (2001). Dendrimers: properties and applications. *Acta Biochim Pol*, *48*(1), 199–208.

Klymchenko, A. S. (2017). Solvatochromic and fluorogenic dyes as environment-sensitive probes: design and biological applications. *Accounts of Chemical Research*, *50*(2), 366–375. Available from https://doi.org/10.1021/acs.accounts.6b00517.

Koenen, J. M., Zhu, X., Pan, Z., Feng, F., Yang, J., & Schanze, K. S. (2014). Enhanced fluorescence properties of poly(phenylene ethynylene)-conjugated polyelectrolytes designed to avoid aggregation. *ACS Macro Letters.*, *3*(5), 405–409. Available from https://doi.org/10.1021/mz500067k.

Kong, Y., Zhao, W., Yao, S., Xu, J., Wang, W., & Chen, Z. (2010). Molecularly imprinted polypyrrole prepared by electrodeposition for the selective recognition of tryptophan enantiomers. *Journal of Applied Polymer Science*, *115*(4), 1952–1957. Available from https://doi.org/10.1002/app.31165.

Krasteva, N., Guse, B., Besnard, I., Yasuda, A., & Vossmeyer, T. (2003). Gold nanoparticle/PPI-dendrimer based chemiresistors—Vapor-sensing properties as a function of the dendrimer size. *Sensors and Actuators, B: Chemical*, *92*(1–2), 137–143. Available from https://doi.org/10.1016/S0925-4005(03)00250-8.

Kuang, G., Wang, C., Song, L., Zhang, G., Yang, Y., & Fu, Y. (2023). Novel electrochemiluminescence luminophore based on flower-like binuclear coordination polymer for high-sensitivity detection of tetracycline in food products. *Food Chemistry*, *403*, 134376. Available from https://doi.org/10.1016/j.foodchem.2022.134376.

Kulawik, P., Özogul, F., Glew, R., & Özogul, Y. (2013). Significance of antioxidants for seafood safety and human health. *Journal of Agricultural and Food Chemistry*, *61*(3), 475–491. Available from https://doi.org/10.1021/jf304266s.

Kumar, K., Tarai, M., & Mishra, A. K. (2017). Unconventional steady-state fluorescence spectroscopy as an analytical technique for analyses of complex-multifluorophoric mixtures. *TrAC Trends in Analytical Chemistry.*, *97*, 216–243. Available from https://doi.org/10.1016/j.trac.2017.09.004.

Kumar, V., Choudhury, N., Kumar, A., De, P., & Satapathi, S. (2020). Poly-tryptophan/carbazole based FRET-system for sensitive detection of nitroaromatic explosives. *Optical Materials.*, *100*, 109710. Available from https://doi.org/10.1016/j.optmat.2020.109710.

Kumar, V., Maiti, B., Chini, M. K., De, P., & Satapathi, S. (2019). Multimodal fluorescent polymer sensor for highly sensitive detection of nitroaromatics. *Scientific Reports*, *9*(1), 7269. Available from https://doi.org/10.1038/s41598-019-43836-w.

Kumaraswamy, S., Bergstedt, T., Shi, X., Rininsland, F., Kushon, S., Xia, W., Ley, K., Achyuthan, K., McBranch, D., & Whitten, D. (2004). Fluorescent-conjugated polymer superquenching facilitates highly sensitive detection of proteases. *Proceedings of the National Academy of Sciences*, *101*(20), 7511–7515. Available from https://doi.org/10.1073/pnas.0402367101.

Lai, W. F. (2020). Non-conjugated polymers with intrinsic luminescence for drug delivery. *Journal of Drug Delivery Science and Technology*, *59*, 101916. Available from https://doi.org/10.1016/j.jddst.2020.101916.

Lee, R. H., Liu, J. K., Ho, J. H., Chang, J. W., Liu, B. T., Wang, H. J., & Jeng, R. J. (2011). Synthesis of quaternized ammonium iodide-containing conjugated polymer electrolytes and their application in dye-sensitized solar cells. *Polymer International*, *60*(3), 483–492. Available from https://doi.org/10.1002/pi.2972.

Lee, S. H., Kömürlü, S., Zhao, X., Jiang, H., Moriena, G., Kleiman, V. D., & Schanze, K. S. (2011). Water-soluble conjugated polyelectrolytes with branched polyionic side chains. *Macromolecules*, *44*(12), 4742–4751. Available from https://doi.org/10.1021/ma200574d.

Li, C., Numata, M., Takeuchi, M., & Shinkai, S. (2005). A Sensitive colorimetric and fluorescent probe based on a polythiophene derivative for the detection of ATP. *Angewandte Chemie*, *117*(39), 6529–6532. Available from https://doi.org/10.1002/ange.200501823.

Li, C., Wu, T., Hong, C., Zhang, G., & Liu, S. (2012). A General strategy to construct fluorogenic probes from charge-generation polymers (CGPs) and AIE-active fluorogens through triggered complexation. *Angewandte Chemie International Edition.*, *51*(2), 455–459. Available from https://doi.org/10.1002/anie.201105735.

Li, G., Chang, W. H., & Yang, Y. (2017). Low-bandgap conjugated polymers enabling solution-processable tandem solar cells. *Nature Reviews Materials*, *2*(8), 17043. Available from https://doi.org/10.1038/natrevmats.2017.43.

Li, X., Qiao, Y., Guo, S., Xu, Z., Zhu, H., Zhang, X., Yuan, Y., He, P., Ishida, M., & Zhou, H. (2018). Direct visualization of the reversible $O_2 - /O -$ redox process in Li-rich cathode materials. *Advanced Materials.*, *30*(14), 1705197. Available from https://doi.org/10.1002/adma.201705197.

Li, Z., Acharya, R., Wang, S., & Schanze, K. S. (2018). Photophysics and phosphate fluorescence sensing by poly(phenylene ethynylene) conjugated polyelectrolytes with branched ammonium side groups. *Journal of Materials Chemistry C.*, *6*(14), 3722–3730. Available from https://doi.org/10.1039/c7tc05081j.

Liang, C., Wang, H., He, K., Chen, C., Chen, X., Gong, H., & Cai, C. (2016). A virus-MIPs fluorescent sensor based on FRET for highly sensitive detection of JEV. *Talanta*, *160*, 360–366. Available from https://doi.org/10.1016/j.talanta.2016.06.010.

Liu, B., Bao, Y., Wang, H., Du, F., Tian, J., Li, Q., Wang, T., & Bai, R. (2012). An efficient conjugated polymer sensor based on the aggregation-induced fluorescence quenching mechanism for the specific detection of palladium and platinum ions. *Journal of Materials Chemistry*, *22*(8), 3555. Available from https://doi.org/10.1039/c2jm15651b.

Liu, C., Yang, J. C., Lam, J. W. Y., Feng, H. T., & Tang, B. Z. (2022). Chiral assembly of organic luminogens with aggregation-induced emission. *Chemical Science.*, *13*(3), 611–632. Available from https://doi.org/10.1039/d1sc02305e.

Liu, J. Q., Luo, Z. D., Pan, Y., Singh, A. K., Trivedi, M., & Kumar, A. (2020). Recent developments in luminescent coordination polymers: Designing strategies, sensing application and theoretical evidences. *Coordination Chemistry Reviews*, *406*, 213145. Available from https://doi.org/10.1016/j.ccr.2019.213145.

Liu, Y., Hua, L., Yan, S., & Ren, Z. (2020). Halogenated π-conjugated polymeric emitters with thermally activated delayed fluorescence for highly efficient polymer light emitting diodes. *Nano Energy.*, *73*, 104800. Available from https://doi.org/10.1016/j.nanoen.2020.104800.

Liu, S. G., Luo, D., Li, N., Zhang, W., Lei, J. L., Li, N. B., & Luo, H. Q. (2016). Water-soluble nonconjugated polymer nanoparticles with strong fluorescence emission for selective and sensitive detection of nitro-explosive picric acid in aqueous medium. *ACS Applied Materials & Interfaces*, *8*(33), 21700–21709. Available from https://doi.org/10.1021/acsami.6b07407.

Liu, X., Fan, Q., & Huang, W. (2011). DNA biosensors based on water-soluble conjugated polymers. *Biosensors and Bioelectronics.*, *26*(5), 2154–2164. Available from https://doi.org/10.1016/j.bios.2010.09.025.

Liu, X., Lei, P., Liu, X., Li, Y., Wang, Y., Wang, L., Zeng, Q. D., & Liu, Y. (2023). From luminescent π-conjugated macrocycles to bridged multi-cyclic -conjugated polymers: cyclic topology, aggregation-induced emission, and explosive sensing. *Polymer Chemistry*, *14*(25), 2979–2986. Available from https://doi.org/10.1039/D3PY00298E.

Long, Y., Chen, H., Wang, H., Peng, Z., Yang, Y., Zhang, G., Li, N., Liu, F., & Pei, J. (2012). Highly sensitive detection of nitroaromatic explosives using an electrospun nanofibrous sensor based on a novel fluorescent conjugated polymer. *Analytica Chimica Acta*, *744*, 82–91. Available from https://doi.org/10.1016/j.aca.2012.07.028.

Lucci, P., Derrien, D., Alix, F., Pérollier, C., & Bayoudh, S. (2010). Molecularly imprinted polymer solid-phase extraction for detection of zearalenone in cereal sample extracts. *Analytica Chimica Acta*, *672*(1–2), 15–19. Available from https://doi.org/10.1016/j.aca.2010.03.010.

Luo, R., Xu, C., Chen, G., Xie, C. Z., Chen, P., Jiang, N., Zhang, D. M., Fan, Y., & Shao, F. (2023). Four novel cobalt(II) coordination polymers based on anthracene-derived dicarboxylic acid as multi-functional fluorescent sensors toward different inorganic ions. *Crystal Growth & Design*, *23*(4), 2395–2405. Available from https://doi.org/10.1021/acs.cgd.2c01374.

Mao, W., Tang, J., Dai, L., He, X., Li, J., Cai, L., Liao, P., Jiang, R., Zhou, J., & Wu, H. (2021). A general strategy to design highly fluorogenic far-red and near-infrared tetrazine bioorthogonal probes. *Angewandte Chemie International Edition*, *60*(5), 2393–2397. Available from https://doi.org/10.1002/anie.202011544.

Martin, A. A., Fodjo, E. K., Marc, G. B. I., Albert, T., & Kong, C. (2021). Simple and rapid detection of free 3-monochloropropane-1,2-diol based on cysteine modified silver nanoparticles. *Food Chemistry*, *338*, 127787. Available from https://doi.org/10.1016/j.foodchem.2020.127787.

Marín-Hernández, C., Toscani, A., Sancenón, F., Wilton-Ely, J. D. E. T., & Martínez-Máñez, R. (2016). Chromo-fluorogenic probes for carbon monoxide detection. *Chemical Communications*, *52*(35), 5902–5911. Available from https://doi.org/10.1039/c6cc01335j.

Mayr, T., Borisov, S. M., Abel, T., Enko, B., Waich, K., Mistlberger, G., & Klimant, I. (2009). Light harvesting as a simple and versatile way to enhance brightness of luminescent sensors. *Analytical Chemistry*, *81*(15), 6541–6545. Available from https://doi.org/10.1021/ac900662x.

Miller-Chou, B. A., & Koenig, J. L. (2003). A review of polymer dissolution. *Progress in Polymer Science*, *28*(8), 1223–1270. Available from https://doi.org/10.1016/S0079-6700(03)00045-5.

Morin, P.-O., Bura, T., & Leclerc, M. (2016). Realizing the full potential of conjugated polymers: innovation in polymer synthesis. *Materials Horizons*, *3*(1), 11–20. Available from https://doi.org/10.1039/C5MH00164A.

Murphy, C. B., Zhang, Y., Troxler, T., Ferry, V., Martin, J. J., & Jones, W. E. (2004). Probing Förster and Dexter energy-transfer mechanisms in fluorescent conjugated polymer chemosensors. *The Journal of Physical Chemistry. B*, *108*(5), 1537–1543. Available from https://doi.org/10.1021/jp0301406.

Murugan, K., Jothi, V. K., Rajaram, A., & Natarajan, A. (2022). Novel metal-free fluorescent sensor based on molecularly imprinted polymer N-CDs@MIP for highly selective detection of TNP. *ACS Omega.*, *7*(1), 1368–1379. Available from https://doi.org/10.1021/acsomega.1c05985.

Nadler, A., & Schultz, C. (2013). The power of fluorogenic probes. *Angewandte Chemie International Edition.*, *52*(9), 2408–2410. Available from https://doi.org/10.1002/anie.201209733.

Nasehi Nejhad, P., Jouyban, A., Khoubnasabjafari, M., Jouyban-Gharamaleki, V., Hosseini, M., & Rahimpour, E. (2022). Development of a fluoremetric probe based on molecularly imprinted polymers for determination of phenobarbital in exhaled breath condensate. *Chemical Papers*, *76*(6), 3447–3457. Available from https://doi.org/10.1007/s11696-022-02105-3.

Noh, M., Kim, T., Lee, H., Kim, C. K., Joo, S. W., & Lee, K. (2010). Fluorescence quenching caused by aggregation of water-soluble CdSe quantum dots. *Colloids and Surfaces A: Physicochemical and Engineering Aspects*, *359*(1–3), 39–44. Available from https://doi.org/10.1016/j.colsurfa.2010.01.059.

Pangeni, D., & Nesterov, E. E. (2013). "Higher energy gap" control in fluorescent conjugated polymers: turn-on amplified detection of organophosphorous agents. *Macromolecules*, *46*(18), 7266–7273. Available from https://doi.org/10.1021/ma4016278.

Panigrahi, S. K., & Mishra, A. K. (2019). Inner filter effect in fluorescence spectroscopy: As a problem and as a solution. *Journal of Photochemistry and Photobiology C: Photochemistry Reviews.*, *41*, 100318. Available from https://doi.org/10.1016/j.jphotochemrev.2019.100318.

Parthasarathy, A., Pappas, H. C., Hill, E. H., Huang, Y., Whitten, D. G., & Schanze, K. S. (2015). Conjugated polyelectrolytes with imidazolium solubilizing groups. Properties and application to photodynamic inactivation of bacteria. *ACS Applied Materials & Interfaces*, *7*(51), 28027–28034. Available from https://doi.org/10.1021/acsami.5b02771.

Qi, J., Han, J., Zhou, X., Yang, D., Zhang, J., Qiao, W., Ma, D., & Wang, Z. Y. (2015). Optimization of broad-response and high-detectivity polymer photodetectors by bandgap engineering of weak donor–strong acceptor polymers. *Macromolecules*, *48*(12), 3941–3948. Available from https://doi.org/10.1021/acs.macromol.5b00859.

Qi, W. J., He, H. K., Tian, X., Song, Y., Li, X. N., Li, R., Hu, P. P., & Huang, X. M. (2022). Fluorescence "turn-on" sensing platform for glutathione detection using chitosan-based glutaraldehyde non-conjugated polymers. *Journal of Analysis and Testing*, *6*(3), 327–334. Available from https://doi.org/10.1007/s41664-021-00173-0.

Qiu, Z. J., Hammer, B., & Müllen, K. (2020). Conjugated polymers – problems and promises. *Progress in Polymer Science*, *100*, 101179. Available from https://doi.org/10.1016/j.progpolymsci.2019.101179.

Reineck, P., & Gibson, B. C. (2017). Near-infrared fluorescent nanomaterials for bioimaging and sensing. *Advanced Optical Materials.*, *5*(2), 1600446. Available from https://doi.org/10.1002/adom.201600446.

Rodgers, J. D., & Bunce, N. J. (2001). Treatment methods for the remediation of nitroaromatic explosives. *Water Research*, *35*(9), 2101–2111. Available from https://doi.org/10.1016/s0043-1354(00)00505-4.

Roy, B., & Bandyopadhyay, S. (2018). The design strategies and mechanisms of fluorogenic and chromogenic probes for the detection of hydrazine. *Analytical Methods*, *10*(10), 1117–1139. Available from https://doi.org/10.1039/C7AY02866K.

Samanta, T., Das, N., Patra, D., Kumar, P., Sharmistha, B., & Shunmugam, R. (2021). Reaction-Triggered ESIPT Active Water-Soluble Polymeric Probe for Potential Detection of Hg^{2+}/CH_3 Hg^+ in Both Environmental and Biological Systems. *ACS Sustainable Chemistry & Engineering*, *9*(14), 5196–5203. Available from https://doi.org/10.1021/acssuschemeng.1c00437.

Sanjuán, A. M., Reglero Ruiz, J. A., García, F. C., & García, J. M. (2018). Recent developments in sensing devices based on polymeric systems. *Reactive and Functional Polymers.*, *133*, 103–125. Available from https://doi.org/10.1016/j.reactfunctpolym.2018.10.007.

Shan, Y., Yao, W., Liang, Z., Zhu, L., Yang, S., & Ruan, Z. (2018). Reaction-based AIEE-active conjugated polymer as fluorescent turn on probe for mercury ions with good sensing performance. *Dyes and Pigments.*, *156*, 1–7. Available from https://doi.org/10.1016/j.dyepig.2018.03.060.

Shanmugaraju, S., Umadevi, D., González-Barcia, L. M., Delente, J. M., Byrne, K., Schmitt, W., Watson, G. W., & Gunnlaugsson, T. (2019). "Turn-on" fluorescence sensing of volatile organic compounds using a

4-amino-1,8-naphthalimide Tröger's base functionalised triazine organic polymer. *Chemical Communications*, *55*(81), 12140–12143. Available from https://doi.org/10.1039/c9cc05585a.

Shariati, R., Rezaei, B., Jamei, H. R., & Ensafi, A. A. (2019). Application of coated green source carbon dots with silica molecularly imprinted polymers as a fluorescence probe for selective and sensitive determination of phenobarbital. *Talanta*, *194*, 143–149. Available from https://doi.org/10.1016/j.talanta.2018.09.069.

Sharma, P., Kumar, M., & Bhalla, V. (2020). "Metal-free" fluorescent supramolecular assemblies for distinct detection of organophosphate/organochlorine pesticides. *ACS Omega*, *5*(31), 19654–19660. Available from https://doi.org/10.1021/acsomega.0c02315.

Shin, H., Jannah, F., Yoo, E. J., & Kim, J. M. (2022). A colorimetric and fluorescence "turn-on" sensor for Fe (III) ion based on imidazole-functionalized polydiacetylene. *Sensors and Actuators B: Chemical*, *350*, 130885. Available from https://doi.org/10.1016/j.snb.2021.130885.

Shrivastava, A. (2018). Polymerization (pp. 17–48). Elsevier BV. Available from https://doi.org/10.1016/b978-0-323-39500-7.00002-2.

Singh, A. K., Nair, A. V., & Singh, N. D. P. (2022). Small two-photon organic fluorogenic probes: sensing and bioimaging of cancer relevant biomarkers. *Analytical Chemistry*, *94*(1), 177–192. Available from https://doi.org/10.1021/acs.analchem.1c04306.

Sorg, J. R., Wehner, T., Matthes, P. R., Sure, R., Grimme, S., Heine, J., & Müller-Buschbaum, K. (2018). Bismuth as a versatile cation for luminescence in coordination polymers from BiX3/4,4′-bipy: understanding of photophysics by quantum chemical calculations and structural parallels to lanthanides. *Dalton Transactions (Cambridge, England: 2003)*, *47*(23), 7669–7681. Available from http://doi.org/10.1039/C8DT00642C.

Squeo, B. M., Gregoriou, V. G., Avgeropoulos, A., Baysec, S., Allard, S., Scherf, U., & Chochos, C. L. (2017). BODIPY-based polymeric dyes as emerging horizon materials for biological sensing and organic electronic applications. *Progress in Polymer Science.*, *71*, 26–52. Available from https://doi.org/10.1016/j.progpolymsci.2017.02.003.

Sun, B., & Liu, L. (2020). Co-effects of the electron transfer and intersystem crossing on the photophysics of a phenothiazine based Hg^{2+} sensor. *Spectrochimica Acta Part A: Molecular and Biomolecular Spectroscopy*, *229*, 117939. Available from https://doi.org/10.1016/j.saa.2019.117939.

Sun, H., & Schanze, K. S. (2022). Functionalization of water-soluble conjugated polymers for bioapplications. *ACS Applied Materials & Interfaces*, *14*(18), 20506–20519. Available from https://doi.org/10.1021/acsami.2c02475.

Takahashi, S., Balzani, V., & Kawa, M. (2003). Dendrimers V: Functional and hyperbranched building blocks, photophysical properties, applications in materials and life sciences. In C. A. Schalley, & F. Vögtle (Eds.), *Topics in Current Chemistry* (228). Heidelberg: Springer Berlin. Available from https://doi.org/10.1007/BFb0121312.

Tan, C., Atas, E., Müller, J. G., Pinto, M. R., Kleiman, V. D., & Schanze, K. S. (2004). Amplified quenching of a conjugated polyelectrolyte by cyanine dyes. *Journal of the American Chemical Society*, *126*(42), 13685–13694. Available from https://doi.org/10.1021/ja046856b.

Tan, C., Pinto, M. R., & Schanze, K. S. (2002). Photophysics, aggregation and amplified quenching of a water-soluble poly(phenylene ethynylene). *Chemical Communications.*, *2*(5), 446–447. Available from https://doi.org/10.1039/b109630c.

Tan, H., Liu, B., & Chen, Y. (2012). Lanthanide coordination polymer nanoparticles for sensing of mercury (II) by photoinduced electron transfer. *ACS Nano.*, *6*(12), 10505–10511. Available from https://doi.org/10.1021/nn304469j.

Tan, C., Wang, S., Barboza-Ramos, I., & Schanze, K. S. (2024). A perspective looking backward and forward on the 25th anniversary of conjugated polyelectrolytes. *ACS Appl Mater Interfaces*. Available from https://doi.org/10.1021/acsami.4c02617.

Tanwar, A. S., Adil, L. R., Afroz, M. A., & Iyer, P. K. (2018). Inner filter effect and resonance energy transfer based attogram level detection of nitroexplosive picric acid using dual emitting cationic conjugated polyfluorene. *ACS Sensors*, *3*(8), 1451–1461. Available from https://doi.org/10.1021/acssensors.8b00093.

Tanwar, A. S., Parui, R., Garai, R., Chanu, M. A., & Iyer, P. K. (2022). Dual "static and dynamic" fluorescence quenching mechanisms based detection of TNT via a cationic conjugated polymer. *ACS Measurement Science AU*, *2*(1), 23–30. Available from https://doi.org/10.1021/acsmeasuresciau.1c00023.

Taya, P., Maiti, B., Kumar, V., De, P., & Satapathi, S. (2018). Design of a novel FRET based fluorescent chemosensor and their application for highly sensitive detection of nitroaromatics. *Sensors and Actuators B: Chemical.*, *255*, 2628–2634. Available from https://doi.org/10.1016/j.snb.2017.09.073.

Thomas, S. W., Joly, G. D., & Swager, T. M. (2007). Chemical sensors based on amplifying fluorescent conjugated polymers. *Chemical Reviews*, *107*(4), 1339–1386. Available from https://doi.org/10.1021/cr0501339.

Usama, S. M., Inagaki, F., Kobayashi, H., & Schnermann, M. J. (2021). Norcyanine-carbamates are versatile near-infrared fluorogenic probes. *Journal of the American Chemical Society*, *143*(15), 5674–5679. Available from https://doi.org/10.1021/jacs.1c02112.

Vezie, M. S., Few, S., Meager, I., Pieridou, G., Dörling, B., Ashraf, R. S., Goñi, A. R., Bronstein, H., McCulloch, I., Hayes, S. C., Campoy-Quiles, M., & Nelson, J. (2016). Exploring the origin of high optical absorption in conjugated polymers. *Nat Mater*, *15*(7), 746–753. Available from https://doi.org/10.1038/nmat4645.

Wang, B., Zhou, X. Q., Li, L., Li, Y. X., Yu, L. P., & Chen, Y. (2022). Ratiometric fluorescence sensor for point-of-care testing of bilirubin based on tetraphenylethylene functionalized polymer nanoaggregate and rhodamine B. *Sensors and Actuators B: Chemical.*, *369*, 132392. Available from https://doi.org/10.1016/j.snb.2022.132392.

Wang, C., Javadi, A., Ghaffari, M., & Gong, S. (2010). A pH-sensitive molecularly imprinted nanospheres/hydrogel composite as a coating for implantable biosensors. *Biomaterials*, *31*(18), 4944–4951. Available from https://doi.org/10.1016/j.biomaterials.2010.02.073.

Wang, C., Tang, Y., Liu, Y., & Guo, Y. (2014). Water-soluble conjugated polymer as a platform for adenosine deaminase sensing based on fluorescence resonance energy transfer technique. *Analytical Chemistry*, *86*(13), 6433–6438. Available from https://doi.org/10.1021/ac500837f.

Wang, D., & Tang, B. Z. (2019). Aggregation-induced emission luminogens for activity-based sensing. *Accounts of Chemical Research*, *52*(9), 2559–2570. Available from https://doi.org/10.1021/acs.accounts.9b00305.

Wang, H., Zhang, C., Hu, L., Tang, F., Wang, Y., Ding, F., Lu, J., & Ding, A. (2023). Red-emissive dual-state fluorogenic probe for wash-free imaging of lipid droplets in living cells and fatty liver tissues. *Chemistry – An Asian Journal*, *18*(7), e202201291. Available from https://doi.org/10.1002/asia.202201291.

Wang, L., Sun, X., Cheng, J., Lu, J., Tian, H., Li, Y., Dou, J., & Wang, S. (2023). A multiresponsive luminescent Co(II) coordination polymer assembled from amide-functionalized organic units for effective pH and cation sensing. *J. Mater. Chem. C*, *11*(5), 1812–1823. Available from http://doi.org/10.1039/D2TC03973G.

Wang, Z., Pan, T., Shen, M., Liao, J., & Tian, Y. (2023). Cross-conjugated polymers as fluorescent probes for intracellular potassium ion detection. *Sensors and Actuators B: Chemical*, *390*, 134008. Available from https://doi.org/10.1016/j.snb.2023.134008.

Wang, J., Lv, F., Liu, L., Ma, Y., & Wang, S. (2018). Strategies to design conjugated polymer based materials for biological sensing and imaging. *Coordination Chemistry Reviews*, *354*, 135–154. Available from https://doi.org/10.1016/j.ccr.2017.06.023.

Wang, J., Wang, D., Miller, E. K., Moses, D., Bazan, G. C., & Heeger, A. J. (2000). Photoluminescence of water-soluble conjugated polymers: origin of enhanced quenching by charge transfer. *Macromolecules*, *33*(14), 5153–5158. Available from https://doi.org/10.1021/ma000081j.

Wang, Y., Chen, H., Li, M., Hu, R., Lv, F., Liu, L., & Wang, S. (2016). Synthesis of a new cationic nonconjugated polymer for discrimination of microbial pathogens. *Polymer Chemistry.*, *7*(44), 6699–6702. Available from https://doi.org/10.1039/c6py01532h.

Wang, Y., Hao, X., Liang, L., Gao, L., Ren, X., Wu, Y., & Zhao, H. (2020). A coumarin-containing Schiff base fluorescent probe with AIE effect for the copper(II) ion. *RSC Advances.*, *10*(10), 6109–6113. Available from https://doi.org/10.1039/c9ra10632d.

Wang, Y. N., Wang, S. D., Cao, K. Z., & Zou, G. D. (2021). A dual-functional fluorescent Co(II) coordination polymer sensor for the selective sensing of ascorbic acid and acetylacetone. *Journal of Photochemistry and Photobiology A: Chemistry.*, *411*, 113204. Available from https://doi.org/10.1016/j.jphotochem.2021.113204.

Wei, G., Jiang, Y., & Wang, F. (2018). A new click reaction generated AIE-active polymer sensor for Hg^{2+} detection in aqueous solution. *Tetrahedron Letters*, *59*(15), 1476–1479. Available from https://doi.org/10.1016/j.tetlet.2018.03.008.

Wei, S., Li, Z., Lu, W., Liu, H., Zhang, J., Chen, T., & Tang, B. Z. (2021). Multicolor fluorescent polymeric hydrogels. *Angewandte Chemie International Edition*, *60*(16), 8608–8624. Available from https://doi.org/10.1002/anie.202007506.

Wosnick, J. H., & Swager, T. M. (2004). Enhanced fluorescence quenching in receptor-containing conjugated polymers: a calix[4]arene-containing poly(phenylene ethynylene). *Chemical Communications* (23), 2744. Available from https://doi.org/10.1039/b411489b.

Wu, J., Tan, C., Chen, Z., Chen, Y. Z., Tan, Y., & Jiang, Y. (2016). Fluorescence array-based sensing of nitroaromatics using conjugated polyelectrolytes. *The Analyst*, *141*(11), 3242–3245. Available from https://doi.org/10.1039/C6AN00678G.

Wu, L., Huang, C., Emery, B. P., Sedgwick, A. C., Bull, S. D., He, X. P., Tian, H., Yoon, J., Sessler, J. L., & James, T. D. (2020). Förster resonance energy transfer (FRET)-based small-molecule sensors and imaging agents. *Chemical Society Reviews*, *49*(15), 5110–5139. Available from https://doi.org/10.1039/c9cs00318e.

Würthner, F. (2020). Aggregation-induced emission (AIE): a historical perspective. *Angewandte Chemie International Edition*, *59*(34), 14192–14196. Available from https://doi.org/10.1002/anie.202007525.

Xu, L., Wang, H., Xiao, W., Zhang, W., Stewart, C., Huang, H., Li, F., & Han, J. (2023). PAMAM dendrimer-based tongue rapidly identifies multiple antibiotics. *Sensors and Actuators B: Chemical.*, *382*, 133519. Available from https://doi.org/10.1016/j.snb.2023.133519.

Xu, L., Wang, H., Xu, Y., Cui, W., Ni, W., Chen, M., Huang, H., Stewart, C., Li, L., Li, F., & Han, J. (2022). Machine learning-assisted sensor array based on poly(amidoamine) (PAMAM) dendrimers for diagnosing Alzheimer's disease. *ACS Sensors*, *7*(5), 1315–1322. Available from https://doi.org/10.1021/acssensors.2c00132.

Xu, Y., & McCarroll, M. E. (2005). Fluorescence anisotropy as a method to examine the thermodynamics of enantioselectivity. *The Journal of Physical Chemistry. B*, *109*(16), 8144–8152. Available from https://doi.org/10.1021/jp044380c.

Xu, Y., Zheng, L., Huang, X., Cheng, Y., & Zhu, C. (2010). Fluorescence sensors based on chiral polymer for highly enantioselective recognition of phenylglycinol. *Polymer*, *51*(5), 994–997. Available from https://doi.org/10.1016/j.polymer.2010.01.038.

Yaman, Y. T., Bolat, G., Saygin, T. B., & Abaci, S. (2021). Molecularly imprinted label-free sensor platform for impedimetric detection of 3-monochloropropane-1,2-diol. *Sensors and Actuators B: Chemical*, *328*, 128986. Available from https://doi.org/10.1016/j.snb.2020.128986.

Yang, S., Wang, L., Zuo, L., Zhao, C., Li, H., & Ding, L. (2019). Non-conjugated polymer carbon dots for fluorometric determination of metronidazole. *Microchimica Acta*, *186*(9), 652. Available from https://doi.org/10.1007/s00604-019-3746-5.

Yao, Z., Feng, X., Li, C., & Shi, G. (2009). Conjugated polyelectrolyte as a colorimetric and fluorescent probe for the detection of glutathione. *Chemical Communications*, *39*, 5886. Available from https://doi.org/10.1039/b912811e.

Zhang, E., Hou, X., Han, X., Wang, S., Li, H., Jiang, D., Zhao, Y., Xiang, M., Qu, F., & Ju, P. (2023). A series of AIE dyes derived from shikimic acid: the thermo/pressure dependent fluorescence and application in the sensing of arginine. *Dyes and Pigments*, *210*, 111041. Available from https://doi.org/10.1016/j.dyepig.2022.111041.

Zhang, H., Liu, G., Shi, L., Liu, H., Wang, T., & Ye, J. (2016). Engineering coordination polymers for photocatalysis. *Nano Energy*, *22*, 149–168. Available from https://doi.org/10.1016/j.nanoen.2016.01.029.

Zhang, X., Wang, C., Wang, P., Du, J., Zhang, G., & Pu, L. (2016). Conjugated polymer-enhanced enantioselectivity in fluorescent sensing. *Chemical Science.*, *7*(6), 3614–3620. Available from https://doi.org/10.1039/C6SC00266H.

Zhang, J., Xia, H., Ren, S., Jia, W., & Zhang, C. (2020). Three AIE-ligand-based Cu (I) coordination polymers: synthesis, structures and luminescence sensing of TNP. *New Journal of Chemistry.*, *44*(14), 5285–5292. Available from https://doi.org/10.1039/d0nj00207k.

Zhang, W. D., Li, G., Xu, L., Zhuo, Y., Wan, W. M., Yan, N., & He, G. (2018). 9,10-Azaboraphenanthrene-Containing Small Molecules and Conjugated Polymers: Synthesis and Their Application in Chemodosimeters for the Ratiometric Detection of Fluoride Ions. *Chemical Science*, *9*(19), 4444–4450. Available from https://doi.org/10.1039/C8SC00688A.

Zhang, X., Wang, W., Hu, Z., Wang, G., & Uvdal, K. (2015). Coordination polymers for energy transfer: Preparations, properties, sensing applications, and perspectives. *Coordination Chemistry Reviews*, *284*, 206–235. Available from https://doi.org/10.1016/j.ccr.2014.10.006.

Zhao, H., Takano, Y., Sasikumar, D., Miyatake, Y., & Biju, V. (2022). Excitation-wavelength-dependent functionalities of temporally controlled sensing and generation of singlet oxygen by a photoexcited state engineered rhodamine 6G-anthracene conjugate. *Chemistry – A European Journal*, *28*(71), e202202014. Available from https://doi.org/10.1002/chem.202202014.

Zhao, R., Zheng, J., Chen, Z., Wang, M., Zhang, D., Ding, L., Fu, C., Zhang, C., & Deng, K. (2022). Synthesis and aggregation-induced emission of polyamide-amines as fluorescent switch controlled by Hg^{2+}-glutathione. *ChemistrySelect*, *7*(13), e202103562. Available from https://doi.org/10.1002/slct.202103562.

Zhao, Q., Zhang, Z., & Tang, Y. (2017). A new conjugated polymer-based combination probe for ATP detection using a multisite-binding and FRET strategy. *Chemical Communications.*, *53*(68), 9414–9417. Available from https://doi.org/10.1039/C7CC04293K.

Zhao, X., Pinto, M. R., Hardison, L. M., Mwaura, J., Müller, J., Jiang, H., Witker, D., Kleiman, V. D., Reynolds, J. R., & Schanze, K. S. (2006). Variable band gap poly(arylene ethynylene) conjugated polyelectrolytes. *Macromolecules*, *39*(19), 6355–6366. Available from https://doi.org/10.1021/ma0611523.

Zheng, S., Shi, J., & Mateu, R. (2000). Novel blue light emitting polymer containing an adamantane moiety. *Chemistry of Materials*, *12*(7), 1814–1817. Available from https://doi.org/10.1021/cm000228l.

Zhou, C., Xu, W., Zhang, P., Jiang, M., Chen, Y., Kwok, R. T. K., Lee, M. M. S., Shan, G., Qi, R., Zhou, X., Lam, J. W. Y., Wang, S., & Tang, B. Z. (2019). Engineering sensor arrays using aggregation-induced emission luminogens for pathogen identification. *Advanced Functional Materials.*, *29*(4), 1805986. Available from https://doi.org/10.1002/adfm.201805986.

Zhou, H., Chua, M. H., Tang, B. Z., & Xu, J. (2019). Aggregation-induced emission (AIE)-active polymers for explosive detection. *Polymer Chemistry*, *10*(28), 3822–3840. Available from https://doi.org/10.1039/c9py00322c.

Zhou, Q., & Swager, T. M. (1995). Method for enhancing the sensitivity of fluorescent chemosensors: energy migration in conjugated polymers. *Journal of the American Chemical Society*, *117*(26), 7017–7018. Available from https://doi.org/10.1021/ja00131a031.

Zhou, Y., Qu, Z., Zeng, Y., Zhou, T., & Shi, G. (2014). A novel composite of graphene quantum dots and molecularly imprinted polymer for fluorescent detection of paranitrophenol. *Biosensors and Bioelectronics.*, *52*, 317–323. Available from https://doi.org/10.1016/j.bios.2013.09.022.

Zhu, C., Liu, L., Yang, Q., Lv, F., & Wang, S. (2012). Water-soluble conjugated polymers for imaging, diagnosis, and therapy. *Chemical Reviews*, *112*(8), 4687–4735. Available from https://doi.org/10.1021/cr200263w.

Zhu, S., Song, Y., Shao, J., Zhao, X., & Yang, B. (2015). Non-conjugated polymer dots with crosslink-enhanced emission in the absence of fluorophore units. *Angewandte Chemie International Edition*, *54*(49), 14626–14637. Available from https://doi.org/10.1002/anie.201504951.

Zhu, S., Wang, X., Yang, Y., Bai, H., Cui, Q., Sun, H., Li, L., & Wang, S. (2018). Conjugated polymer with aggregation-directed intramolecular förster resonance energy transfer enabling efficient discrimination and killing of microbial pathogens. *Chemistry of Materials*, *30*(10), 3244–3253. Available from https://doi.org/10.1021/acs.chemmater.8b00164.

Zhu, Y. G., Zhao, Y., Li, B., Huang, C. L., Zhang, S. Y., Yu, S., Chen, Y. S., Zhang, T., Gillings, M. R., & Su, J. Q. (2017). Continental-scale pollution of estuaries with antibiotic resistance genes. *Nature Microbiology.*, *2*(4), 16270. Available from https://doi.org/10.1038/nmicrobiol.2016.270.

CHAPTER

Electrochemical sensors

6

Gulsu Keles[1], Aysel Oktay[2], Pakize Aslan[1], Aysu Yarman[2] and Sevinc Kurbanoglu[1]

[1]*Department of Analytical Chemistry, Faculty of Pharmacy, Ankara University, Ankara, Turkey* [2]*Molecular Biotechnology, Faculty of Science, Turkish-German University, Istanbul, Turkey*

6.1 What is a sensor?

The term sensor is derived from the term sense, which refers "to feel". Sensors are devices that collect signals to identify source/environment changes and respond accordingly. More broadly, a sensor is a compact apparatus that transforms an input signal from a given stimulus into a measurable output signal, where the input signal can be any quantifiable property, such as amount or physical change, while the output is finally an electrical pulse. There are many types of sensors, such as chemical, optical, electrical, magnetic, mass-sensitive, thermal, and electrochemical sensors (Kimmel et al., 2012; O' Riordan & Barry, 2016).

Chemical sensors have attracted considerable interest due to their capacity to facilitate data gathering and facilitate information generation with minimal impact on the observed system. This allows for the analysis of results and their correlation with other variables in the same environment. These devices possess distinct characteristics that set them apart from mainstream instrumental methods, exhibiting increased sensitivity, responsiveness, and selectivity. Moreover, chemical sensors offer advantages such as portability, user-friendly operation, potential for automation, opportunities for miniaturization, and cost-effectiveness. In essence, chemical sensors play a crucial role in detecting minute quantities of chemical vapors. Various sensing elements, including zinc oxide nanowires, carbon nanotubes (CNTs), and palladium nanoparticles (NPs), can be employed as chemical sensors (Kimmel et al., 2012; Kurbanoglu et al., 2023).

Fig. 6.1 shows that optical sensors operate based on light and biosensing elements, measuring variations after the analyte reacts. Electrical sensors encompass metal oxide, organic semiconductive sensors, and electrolytic permittivity sensors (Bozal-Palabiyik et al., 2021; Kapoor & Rajput, 2021; Karimi-Maleh et al., 2020; Wang et al., 2008; Wang, 1991).

A magnetic sensor is a detection apparatus that operates based on the presence of a magnetic field. Magnetic sensor applications utilize both flexible and solid magnetic materials. Sensors can gauge temperature, pressure, and light through their magnetic properties (Nejad et al., 2021; Wang et al., 2014).

Electrochemical sensors employ electrodes as the primary sensing elements. Originating in the latter part of the 20th century, these instruments are now widely adopted in commercial applications. Notable features of electrochemical sensors include utilizing electrons for signal acquisition and

Sensory Polymers. DOI: https://doi.org/10.1016/B978-0-443-13394-7.00006-9

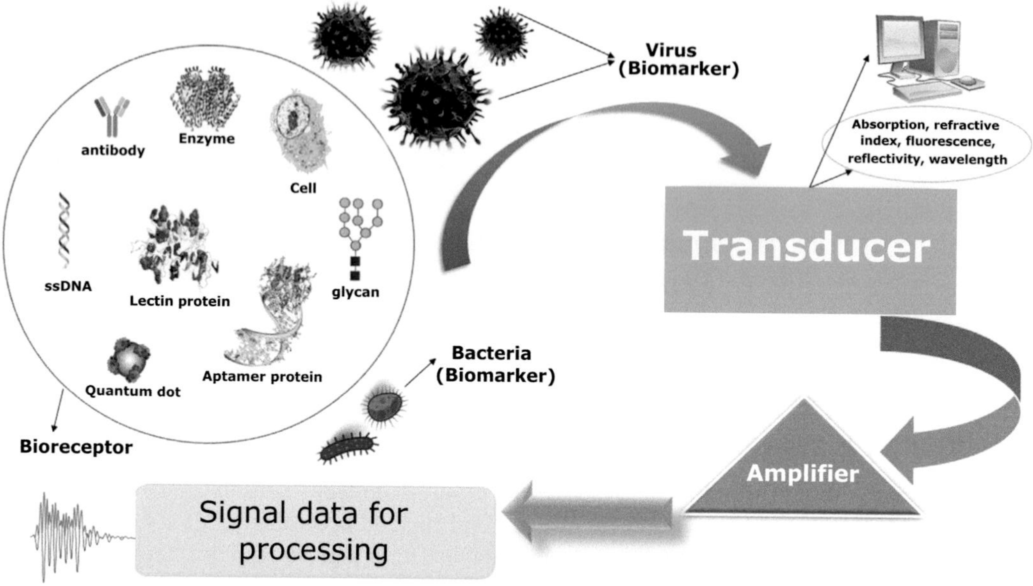

FIGURE 6.1

A schematic diagram of a nanosensor.

Modified from Kapoor, A. & Rajput, J. K. (2021). Advanced nanosensors for virus detection. In Nanosensors for Smart Agriculture (pp. 609–640). Elsevier. https://doi.org/10.1016/B978-0-12-824554-5.00024-0.

providing a clean model for analytical applications without generating waste. These sensors excel in miniaturization for portable devices, enabling analyses with micro-volumes of samples. They offer swift analysis and are cost-effective, contributing to their popularity, as seen in commercial glucose sensors. Furthermore, the advancement of electrochemical sensors contributes to enhancing other technologies, such as chromatographic detectors (Kissinger, 2005; Xueji et al., 2011).

6.2 Electrochemical sensors

Developed in 1950, an electrochemical sensor is a tool designed to identify the concentration of analytes through the processes of oxidation or reduction occurring at an electrode, measuring the subsequent current flow. These sensors hold great significance in both chemistry and medicine for identifying and sensing the presence of chemical species. Within the realm of electrochemical sensing, carbon-based nanomaterials find utility as electrodes or supporting matrices. They serve specific functions, such as enhancing electrocatalytic properties, facilitating improved electron transfer across interfaces, and providing an excellent interfacial environment for signal amplification (Erkmen, Selcuk, et al., 2022; Erkmen, Unal, et al., 2022; Hammond et al., 2016; Islam & Channon, 2019).

Contemporary electrochemical sensors employ diverse features to identify varied parameters in our daily lives, such as chemical, physical, or biological aspects. Instances include sensors for healthcare applications and instrumentation, environmental monitoring, as well as those integrated into machines such as mobile phones, cars, airplanes, and various technological devices. In the preceding decades, progress in microelectronics and microengineering has significantly benefited modern sensor systems, allowing for the creation of even more miniaturized sensors characterized by heightened selectivity and sensitivity, along with reduced production and operational costs. Electrochemical sensors primarily fall into three categories: potentiometric, conductometric, and amperometric/voltammetric (Davis, 1985; Malhotra et al., 2005).

Electrochemical sensors are able to generate electronic outputs in the form of digital signals far beyond analysis. In general, the response of these sensors is due to the chemical and electrical interactions that result from the potentiometric, amperometric, and conductometric readings. Electrochemical transducers are the most widely employed transducers in the construction of novel sensors, encompassing biosensors (Xueji et al., 2011). They are at the forefront of market availability and have already proven their actual advantages. Electrochemical measurement stands out among other measurement systems due to its speed and simplicity. Amperometry and voltammetry-based electrochemical sensor systems provide economical, rapid, user-friendly, dependable, and sensitive approaches for identifying various pollutive agents such as pharmaceutical compounds, pesticides, herbicides, heavy metals, colorants, and in intricate environmental samples. In most instances, users can effectively utilize electrochemical sensor systems with straightforward instructions, requiring no prior training. This capability allows for the downsizing of sensor systems, making them easy, speedy, and efficient for on-the-spot specimen analysis (Becker et al., 2006; Malhotra et al., 2005; Schubert & Scheller, 1992; Švorc et al., 1997; Zare & Shekari, 2019).

6.2.1 Electrochemical methods

Electrochemical sensors are able to generate electronic outputs in the form of digital signals for advanced analysis. In general, the response of these sensors is due to the chemical and electrical interactions that result from the conductometric, amperometric, and potentiometric readings (Machado & Cincotto, 2021). The electrochemical process is a proven and effective technique for producing coatings of the desired thickness, achieved by regulating the voltage, electrodeposition time, and the concentration of the electrolyte and monomers. Electrochemical methods are employed to eliminate and recover heavy metals, functioning based on the concept that metals undergo deposition in their elemental state on solid electrodes when subjected to an electric potential. The fundamental reaction entails reducing metals in several oxidation states to a zero-oxidation state at the electrode, facilitated by the flow of electrons via the circuit from the anode. Three reduction types distinguished by their operation, include electrocoagulation (EC), electroflotation, and electrodeposition. In EC, electrodes release coagulant ions that interact with metals, and the resulting coagulates can be subsequently removed through centrifugation (Manna et al., 2022; Stefan, 2001; Yarman, Kurbanoglu, Erkmen et al., 2021; Yarman, Kurbanoglu, Zebger et al., 2021).

Electrodeposition involves the simultaneous removal and recovery processes induced by the application of potential or current. Heavy metals in various oxidation states undergo reduction at the cathode and are subsequently deposited onto it. The utilization of electrochemical processes for the removal of copper (Cu), chromium (Cr), nickel (Ni), zinc (Zn), arsenic (As), manganese (Mn), and silver (Ag) has been well-documented (Compton & Banks, 2018; Sun et al., 2023).

Electrochemical techniques, for example, cyclic voltammetry (CV) and linear sweep voltammetry (LSV), which necessitate a potentiostat, provide enhanced insights into the electron transport mechanism. These techniques involve monitoring the current passing via a working electrode while executing linear sweeps (back and forth for CV) at a designated scanning rate. Valuable details regarding the electron transfer procedure, midpoint potentials of catalytic sites, and more can be obtained by recording CVs at various scan rates, solution convections, or diverse substrate conditions (Aristov & Habekost, 2015; Elgrishi et al., 2018; Isaev et al., 2018).

CV has been frequently employed in electroanalytical applications recently. Originating in the 1960s through the work of Nicholson and Shain, it was created as a more straightforward investigative technique. This method allows for the analysis of the current–voltage curve produced in voltammetric analysis, eliminating the necessity for intricate examination (Paulsen, 1982). This method has a wide application in the study of reduction/oxidation reactions, the determination of reactants, and monitoring the in-progress reactions of the formed products. The measurement of current involves assessing potential changes in one direction and then evaluating it once more as it shifts in the opposite direction. In other words, current is measured in terms of voltage. As the potential change takes place, the current utilized in CV increases and decreases linearly. CV is widely used and offers advantages in areas as diverse as analyzing solids, biological systems, enzymes, and bacterial cultures. Furthermore, it offers crucial information for comprehending the thermodynamics involved in oxidation/reduction operation and the kinetics governing heterogeneous electron transfer reactions (Aristov & Habekost, 2015; Compton & Banks, 2018; Elgrishi et al., 2018; Isaev et al., 2018).

Differential pulse voltammetry (DPV) is an electroanalytical technique for the quantitation of organic and inorganic compounds. In this method, current measurements are made before and after the pulse, and the distinctness between these currents is plotted toward the potential. With this method, very small concentrations can be measured, and analytes with similar redox potentials can be quantified. It also provides better measurement by eliminating capacitive/background current. The differential pulse voltammogram consists of current peaks that vary depending on the analyte concentration. This method is a sensitive and efficient tool for chemical analysis and has many applications (Laborda et al., 2014; Machado & Cincotto, 2021).

LSV is a technique in which the scan rate is kept constant by applying a linear potential wave. Voltammograms obtained with this method provide information on redox systems, such as analyte concentrations, diffusion constants, and rate constants. While this technique has continued to evolve over the years, it has been used for the determination of various types of compounds in water. An important point in terms of application is the new concept of working electrodes. For example, single-use electrodes, such as a basic screen-printed carbon electrode (SPCE) modified with CNTs, can be applied for assessing the antidepressant vortioxetine in pharmaceutical formulations and biological fluids. This method allows rapid and sensitive analytes to be determined in a diversity of applications (Elgrishi et al., 2018; Majeed et al., 2022).

Stripping voltammetry is a sensitive electrochemical technique for the assessment of trace metals. It is based on the preliminary deposition of metal ions and the subsequent stripping from the dissolved phase. The stripping analysis used with mercury electrodes is implemented as a two-stage technique. In the first stage, some of the metal ions are deposited on the anode, followed by the stripping stage. The most common form of this analytical method is anodic stripping voltammetry. In this technique, the working electrode serves as the cathode during the precipitation operation and as the anode throughout the stripping process. Stripping analysis can be applied following different stripping processes, and the predeposition operation is applied under controlled voltage and time. The deposition voltage is set to be more negative than the potential of the most challenging reducing metal ion. Stripping analysis is a method in which the metal is oxidized and reduced, and the analyte is stripped from the electrode surface as an insoluble film during the measurement. Cathodic stripping voltammetry is the reverse of anodic stripping voltammetry (Zuhri, 1998).

Square wave voltammetry (SWV) is an electroanalytical method that allows the measurement process to be completed in a short time using a fast voltage scan rate. In this technique, voltage is supplied to the working electrode using large rectangle-shaped pulses. Two current readings are taken in every rectangular pulse cycle, and the difference between these readings provides a square wave voltammogram against the potential of the working electrode. Since this method is fast, a distinctive current signal can be obtained in the presence of air, and rapid measurements can be made in flow systems. However, fast measurements can reduce sensitivity because shorter pulse handling intervals lead to a lower iF/iC ratio. Therefore, while the analysis time decreases, the sensitivity can also decrease (Frew et al., 1987; Gómez Avila et al., 2019; Machado & Cincotto, 2021).

Electrochemical impedance spectroscopy (EIS) is a powerful technique utilized for the specification of electrochemical systems that is widely employed in diverse fields such as energy conversion and storage technologies, electrocatalysis, corrosion studies, chemical sensing and biosensing, semiconductor science, and noninvasive diagnostics (Ciucci, 2019). Moreover, EIS is the most significant electrochemical technique for measuring impedance in a circuit, utilizing ohms as the unit of measurement for resistance (Magar et al., 2021). EIS is especially attractive for two main reasons. First, the EIS data can be correlated with various essential physical characteristics, including diffusion and reaction rates, as well as microstructural features of the electrochemical system. Second, performing an EIS experiment is relatively simple. Furthermore, the EIS method is commonly employed in the development of various biosensors, including immunosensors and aptasensors. EIS-based biosensors are preferred for their simplicity in modification, rapid response, cost efficiency, and the ability to detect samples with low concentrations (Bonora et al., 1996; Lazanas & Prodromidis, 2023).

6.2.2 Design of electrochemical sensors

The electrochemical polymerization method is commonly utilized in the application of conductive polymers. This method involves applying an external potential and oxidizing the monomer dissolved in the supporting electrolyte solution to form radical cations. This method was first successfully employed by Letheby and has since become commonly used in the synthesis of conductive polymers. During electrochemical polymerization, a single or double-compartment cell is used, which contains a triple electrode system and a supporting electrolyte (Goyal & Bishnoi, 2012).

The three-electrode system comprises a working electrode, a reference electrode, and a counter electrode. The working electrode must be a nonoxidizable substance such as platinum, gold, tin oxide, or indium tin oxide (Kant et al., 2022). The counter electrode can be made of platinum, gold, nickel, etc., while the reference electrode is typically a calibrated calomel or silver/silver chloride electrode. By utilizing the electrochemical polymerization method and the triple electrode system, conductive polymer films can be successfully formed, making this method an important technique in the field. R4NX as a supporting electrolyte can be expressed by the general formula

(R: alkyl, aryl radical, and X = Cl^-, Br^-, BF_4^-, I^-, ClO_4^-, $CF_3SO_3^-$, PF_6^-, $CH_3C_6H_4SO_3^-$) quaternary ammonium salts are used as they are easily soluble in many solvents. As a solvent, many solvents can be used, depending on the solubility and inertness of the monomer and the salt used as a supporting electrolyte (Dutta, 2008; Erkmen, Unal, et al., 2022).

The layer-by-layer technique is a simple and cost-effective process for creating multilayered nanocomposites on a solid base. It relies on electrostatic interactions to bind the layers together. Alternating layers can be formed through charge reversal by immersing the substrate in solutions of positively and negatively charged molecules. There are different techniques that can be used for layer-by-layer assembly, such as dipping, spinning, spraying, electromagnetism, or fluids. However, the dipping process can be time-consuming, and the spinning process is faster and produces more uniform layers. The spray method allows for coating large surfaces and can approach an industrial level of production. It is possible to vary the number and order of the layers by changing the order of the adsorptive processes. Furthermore, the layer-by-layer method is not restricted to electrostatic interactions and can include other interactions such as hydrogen bonding, biospecific interplays, metal coordination, and charge transfer. Technical aspects of this method include manual and automated processes, as well as spin coating (Arduini et al., 2017; Borisova et al., 2015; Erkmen, Selcuk, et al., 2022; Weng et al., 2013).

6.3 Polymers in electrochemical sensor

Conventionally, polymers are known as insulators due to their dielectric properties, such as charge transport and electrochemical redox efficiency. They are considered uninteresting from an electronic perspective. Furthermore, the saturated polymers (all carbon valance electrons that the polymer had tied up in covalent bindings) show the same insulating behavior. On the contrary, conjugated polymers have sp^2 p_z hybridization, which means that each carbon atom has one unpaired electron in its structure. This unique electronic structure is responsible for its ability to conduct electricity, strong electron affinity, and low ionization potentials. As a result, conjugated polymers can be expected to show semiconducting or even better metallic properties. In 1892, Letheby reported the first electrochemical application of polyaniline. Polyaniline occurred through the oxidation of aniline onto a platinum electrode in dilute sulfuric acid. The appearance of polyaniline, organic π-conjugated polymers, gained much attention from researchers (Letheby, 1862; Naveen et al., 2017). In 1988, Heeger et al. examined polyacetylene as a significant example of this matter. They reported that polyacetylene can possibly have metallic-like conductivity features. Since this discovery, organic conjugated polymer structures have taken significant roles in several different applications, such as organic light-emitting diodes, optical detectors, organic solar cells, transistors, supercapacitors, batteries, drug release systems electrochemical sensors, biosensors, and more (Heeger et al., 1988; Kurbanoglu et al., 2023; Peng et al., 2009).

6.3.1 Conducting polymers

Conducting polymers (CPs) have become a recent class of functional substances, and they have been extensively employed in electroanalytical utilizations as signal-enhancing elements due to their compatibility, corrosion resistance, low cost, lightweight, charge transfer, electrical, physical, and optical properties, and high capacity conducting abilities. CPs' polymer backbone chain is characterized by alternating single and double bond configurations, which allow CPs to conduct electricity up to a certain limit, and this provides them to possess attractive intrinsic features advantageous in electroanalytic applications. The significant potential of CPs has made them candidates for the fabrication of electrochemical sensors. The discovery of CPs applications has been a critical moment of analytical research (Dakshayini et al., 2019; Naveen et al., 2017; Shrivastava et al., 2016). The notable electrical conductivity of CPs was observed in halogen-treated polyacetylene by Shirakawa et al. in 1977 (Shirakawa et al., 1977). In addition, polypyrrole (PPy), polyaniline (PANI), polythiophene (PTh), poly(*o*-toluidine) (PoT), polyaminonaphthalenes, polyfluorene (PF), polyazulene, and their varieties take place in CPs. Particularly, CPs attracted considerable notice from academic and industrial societies because Heeger, MacDiarmid, and Shirakawa collaboratively won the Nobel Prize in Chemistry in 2000 with their leading work on CPs (Shirakawa et al., 1977; Soylemez et al., 2021). CPs are polyconjugated polymers with electronic features similar to metals while still preserving the features of traditional organic polymers (Peng et al., 2009). Conventionally, they are known as semiconductive conjugated polymers when they are in an unspoiled state. At this point, CPs can increase their conductivity via the process of doping. The term "doping" refers to the fact that the π-conjugated systems' oxidation and reduction reactions generate p-doped and n-doped CPs, respectively, and that the doping can be regulated by chemical and electrochemical techniques. Furthermore, the adjustable ability to conduct electricity of CPs results from their extended π-conjugated molecular backbone, which can be modified by the presence of dopants. CPs are typically synthesized by using chemical and electrochemical techniques primarily utilizing oxidative polymerization. Following polymerization, the oxidized polymer backbone typically acquires a positive charge. There should be a negatively charged counter-ion to balance the charge; these negatively charged molecules or materials, known as dopants, become integrated into the resulting CP (Wang et al., 2018).

With the recent progress in nanoscience and nanotechnology and to overcome some disadvantages of CPs, researchers have extensively studied CP composites with nanomaterials, which can provide CP applications in diverse fields. The primary purpose of utilizing nanomaterials is their ability to carry negative charges, making them valuable as novel dopants in the oxidative polymerization of CPs. For instance, CPs can constitute a collaboration with polymer nanocomposites with carbon quantum dots (CQDs), with CNTs, graphene, metal or metal oxide NPs, such as Cu, Au, Pd, Pt, Ni, ZnO, TiO_2, MnO_2, and Fe_3O_4, and alternative nanomaterials. This incorporation between CPs and nanomaterials provides them widespread usage in electrochemical studies such as voltammetric sensors, immunosensors, aptamer sensors, biosensors such as DNA biosensors (Naveen et al., 2017; Peng et al., 2009; Shrivastava et al., 2016; Wang et al., 2018). Especially in designing electrochemical enzyme-based biosensors, CPs have found a place because the electrode can be conjugated with the CPs and the enzyme active sites. As is well known, poly(3,4-ethylenedioxythiophene) (PEDOT) is the most widely used and investigated polythiophenes derivative, and the combinations of PEDOT and nanomaterials attracted the interest of researchers with their promising potential in sensing. For example, GR and CNTs have been employed as dopants in PEDOT. In consideration of this combination's sensing capabilities, a range of sensors has been developed (Wang et al., 2018).

6.3.2 Molecularly imprinted polymers

The origins of the concept of molecular imprinting can be traced back to 1931 when Polyakov illustrated the unique adsorption characteristics of silica gel, which recognizes its target compound, methyl orange (Polyakov, 1931). In 1972, according to Wulff et al., the initial paper on the imprinting of organic polymers was dedicated to them (Wulff & Sarhan, 1972; Wulff, 1995). Basically, researchers have developed molecularly imprinted polymers (MIPs) to replicate the functional sites found in biopolymers (Kurbanoglu et al., 2019). According to one of the pioneers of these researchers, Monarch et al. reported a basic template strategy for substrate-specific polymer synthesis that is based on complementary interactions. They named this strategy "host-guest polymerization" (Arshady & Mosbach, 1981).

MIPs serve as strong molecular identification components skilled of mimicking natural recognition elements such as antibodies, enzymes, and biological receptors (Vasapollo et al., 2011). Furthermore, the initiation of MIP utilization provides a promising alternative in many areas, such as chromatography, biosensor technology, catalysis research, and the manufacture of artificial antibodies. Therewithal, it has found widespread use in analytical applications, which are MIP-based sensors (Wulff, 1995; Yano & Karube, 1999). In the past few years, MIPs have been frequently utilized for the detection and separation of a specific, diverse range of analytes, such as low-molecular-weight molecules, including pharmaceuticals, food additives, toxins, steroids, sugars, pesticides, as well as biomacromolecules for instance hormones, nucleic acids, proteins, bacteria, and viruses. The major advantages of MIPs are their excellent affinity and selectivity toward the target molecule utilized in the imprinting process and their stability in challenging conditions, for example, extreme pH levels, organic solvents, and high temperatures (Yano & Karube, 1999; Yarman & Kurbanoglu, 2022).

Molecular imprinting of synthetic polymers is a method that incorporates selective recognition sites for detecting the molecule of affinity. MIPs involve the process of copolymerizing functional monomers, crosslinking monomers (in electropolymerization, it is unnecessary to use cross-linkers), and the target analyte, or in other words the template. After the template is removed, binding places reveal that they are characterized by size and shape complementarity with the analyte (Fig. 6.2). In this process, the polymer acquires a "molecular memory" that enables it to specifically recognize and bind to the analyte (template). There are two main approaches for molecular imprinting. One method involves establishing a prepolymerization complex between the imprint molecules and functional monomers through noncovalent interactions, essentially through self-assembly. Conversely, monomers can be covalently attached to the imprint molecule by creating a polymerizable derivative of the imprint molecule. Due to the increased constancy of covalent bonding, it is conceivable that covalent imprinting procedures would result in a more consistent population of binding places (Haupt & Mosbach, 2000; Yarman et al., 2021). On the contrary, the noncovalent imprinting method, initially developed by Mosbach et al., provides increased flexibility in choosing functional monomers, potential target molecules, and the application of the imprinted material. Additionally, hybrid strategies have been proposed to integrate the benefits of both covalent and noncovalent imprinting approaches (Haupt & Mosbach, 2000; Mosbach & Ramström, 1996).

The selection of the monomer plays a pivotal role in creating specialized cavities that are designed for the template molecule. Commonly employed functional monomers are such as carboxylic acids (methacrylic acid [MAA], acrylic acid, and vinylbenzoic acid), heteroaromatic bases

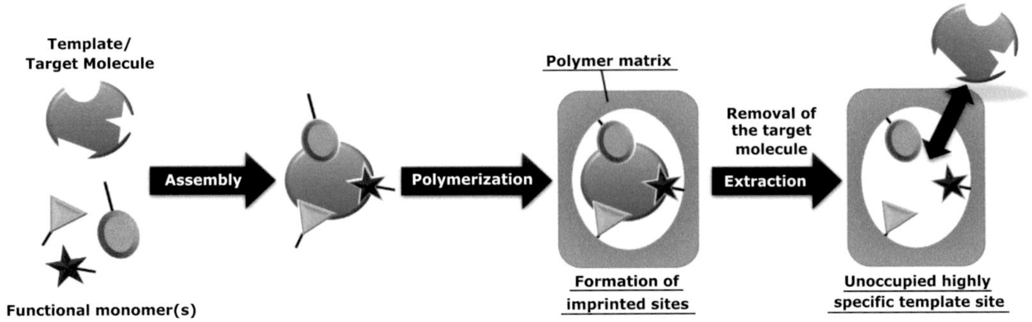

FIGURE 6.2

Schematic representation of the synthesis and the extraction of a MIP. *MIP*, molecularly imprinted polymer.

Modified from Charlier, H., David, M., Lahem, D. & Debliquy, M. (2022). Electrochemical detection of penicillin G using molecularly imprinted conductive co-polymer sensor. Applied Sciences (Switzerland), 12(15). https://doi.org/10.3390/app12157914.

(vinylpyridine and vinylimidazole), and sulfonic acids (2-acrylamido-2-methylpropane sulfonic acid). In the noncovalent approach, they are generally utilized in larger quantities than the template to support the formation of assemblies between the template and monomers. Indeed, there is an equilibrium between the monomer and the template, and the functional monomers typically have to be added more, correlative to the number of moles in the template, to support the creation of the template, the monomer assemblies (Cormack & Elorza, 2004; Vasapollo et al., 2011).

Electrochemical techniques not only enable the sophisticated development of MIPs but also serve as potent means for electrode-based signal generation. MIP-based sensors, employing full electrochemistry throughout MIP synthesis and readout, are more prevalent than alternative methods such as Raman and Infrared (IR) spectroscopy, quartz crystal microbalance (QCM), and surface plasmon resonance (SPR). The popularity of fully electronic MIPs is attributed to their cost-effectiveness and specificity (Scheller et al., 2019). These MIPs rely on three fundamental electrochemical measurement approaches. The first approach involves following the permeability of a redox marker owing to thin MIP layers. It offers a direct means for MIP-based sensors, allowing for the characterization of synthesis steps and the evaluation of target-rebinding. However, precision can be challenging due to small signal changes after template removal, and nonspecific sample components may affect the signal (Menger et al., 2016). The second approach involves analyzing enzyme-based MIP sensors based on measuring enzymatic activity or catalytic active analytes. Electrochemical detection through enzymatic activity quantifies rebinding directly on the sensor surface despite the potential for nonspecific enzyme adsorption (Yarman, 2018). This method has been studied by Ozcelikay et al. for the recognition of butyrylcholinesterase rebound to the MIP, which used *o*-phenylenediamine (*o*-PD) as a monomer on a glassy carbon electrode (GCE) by the anodic oxidation of thiocholine (Ozcelikay et al., 2019). The last approach utilizes electroactive analytes via direct electron transfer, where the faradaic current of prosthetic groups provides a direct indication of the target protein throughout the synthesis and rebinding stages (Yarman & Scheller, 2020). Using this approach, Bosserdt et al. demonstrated the determination of cytochrome c by a MIP that used scopoletin as a monomer and was prepared via a self-assembled monolayer of mercaptoundecanoic acid on a gold electrode (Bosserdt et al., 2013).

6.4 Applications of sensory polymers in electrochemical sensor

In the literature, there are many researches on the utilization of CP-based nanosensors and biosensors using electrochemical techniques. To illustrate, in our research group, Kurbanoglu et al. developed an electrochemical biosensor for detecting the tyrosinase enzyme by using the active compound, indomethacin, using a unique combination of materials. They utilized a conjugated polymer called poly[BDT-alt-(TP;BO)], which incorporates benzoxadiazole, thienopyrroledione, and benzodithiophene moieties, along with fullerene, to enhance catechol detection. This novel combination resulted in an exceedingly sensitive and rapid-response biosensor for catechol. The modified electrodes were thoroughly characterized by following several techniques, including CV, EIS, scanning electron microscope (SEM), atomic force microscopy, and chronoamperometric (CA) techniques. After optimization, the biosensor detected catechol within a range of 0.5–62.5 μM, with a detection limit of 0.11 μM. Additionally, the study revealed that indomethacin exhibited mixed-type enzyme kinetics and inhibited tyrosinase with an I_{50} value of 15.11 μM (Kurbanoglu et al., 2022).

In a study by Al-Kadhi et al., a modified GCE, combining polyaniline and NiO nanoflowers, was developed for the sensitive detection of nitrite ions in drinking water. NiO nanoflowers were characterized via X-ray diffraction (XRD) analysis, X-ray photoelectron spectroscopy (XPS), Fourier-transform infrared spectroscopy (FT-IR), and thermal gravimetric analysis to analyze their structure and surface properties. The electrode's sensitivity to nitrite sensing was examined across a pH range of 2–10. Amperometry revealed a linear detection range of 0.1–1 μM and 1–500 μM, with a low detection limit of 9.7 nM and a high diffusion coefficient of 4.91×10^{-1} cm^2/s. Nitrite detection was effective across various pH levels, with neutral and alkaline conditions producing higher oxidation currents. It displayed excellent resistance to interference from various metal ions. Moreover, in Nile River water spiked with nitrite, the modified electrode exhibited good recovery, affirming its effectiveness with a response time of approximately 4 seconds (Al-Kadhi et al., 2023).

A unique electrochemical sensor was fabricated by Deller et al. to detect the pesticide Pirimicarb. This sensor utilized a screen-printed electrode designed with poly(3,4-ethylenedioxythiophene) and gold nanoparticles (AuNPs) to increase its electrochemical performance. The characterization of the developed electrode was observed by employing SEM and CV. When Pirimicarb was present, a new peak at 1.0 V emerged in the electrode, indicating Pirimicarb oxidation. Gas chromatography-mass spectrometry analysis revealed two main peaks, suggesting that the sensor detects and degrades Pirimicarb through electroxidation. This sensor demonstrated a detection range from 93.81 to 750 μM, with limits of quantification (LOQ) and limits of detection (LOD) of 93.91 μM and 28.34 μM, respectively. Additionally, it demonstrated strong selectivity against various interferents (Deller et al., 2023).

In a study by Eswaran et al., a versatile, flexible electrode composed of polyimide, AuNP, polyaniline, and palladium (Pd) nanocomposite was crafted, boasting exceptional electrochemical attributes (Fig. 6.3). The nanocomposite's chemical structure and surface properties on the electrode were characterized by XRD, field emission scanning electron microscopy (FESEM), and FT-IR, while the electrochemical demonstration was determined by CV, DPV, and EIS. The synergistic interactions between polyaniline and Pd created heightened conductivity, swift responsiveness, and an elevated electron transfer rate. The modified electrode exhibited impressive multifunctionality,

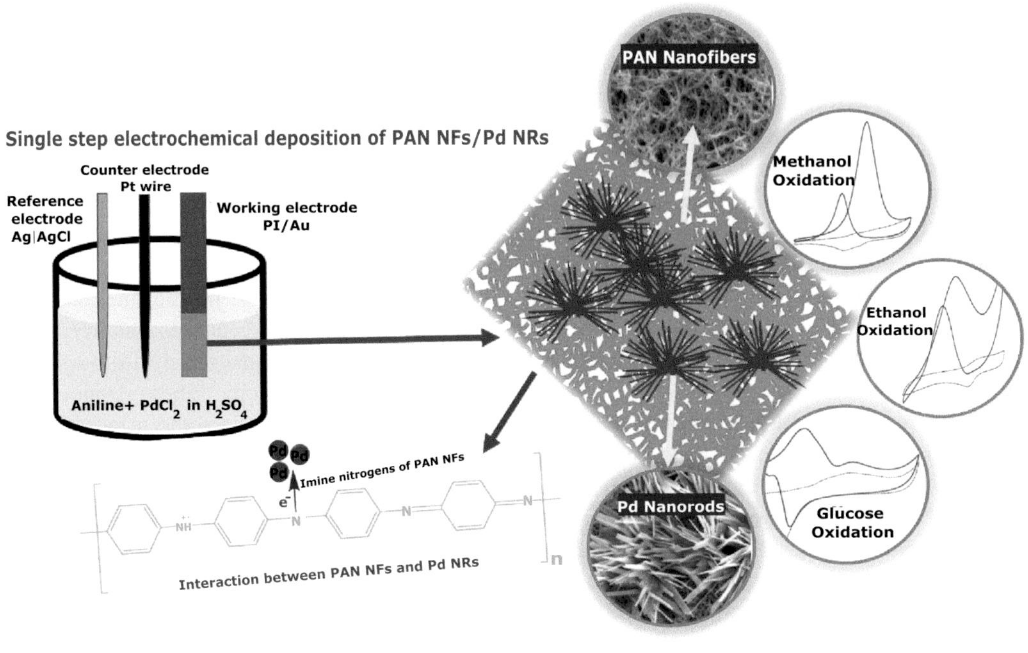

FIGURE 6.3

Overall scheme from electrochemical deposition to applications.

Modified from Eswaran, M., Rahimi, S., Pandit, S., Chokkiah, B., & Mijakovic, I. (2023). A flexible multifunctional electrode based on conducting PANI/Pd composite for non-enzymatic glucose sensor and direct alcohol fuel cell applications. Fuel, 345, 128182. https://doi.org/10.1016/j.fuel.2023.128182.

excelling in biomedical glucose sensing and energy conversion applications. Its electrochemical analysis revealed an extraordinary sensitivity of 2140 μA/μM cm^2, coupled with a low glucose detection limit of 0.3 μM. Moreover, the electrode displayed excellent electrocatalytic capabilities for methanol and ethanol oxidation in an alkaline medium, yielding current densities of 3 mA/cm^2 and 0.96 mA/cm^2, respectively (Eswaran et al., 2023).

Isotactic polypropylene surgical meshes were enhanced by Fontana-Escartin et al. with a CP layer such as 3,4-phenylenedioxythiophene (PHEDOT) and PEDOT to detect bacterial adhesion and growth. This modification allowed the sensing of the oxidation of nicotinamide adenine dinucleotide (NADH), a metabolite produced during bacterial respiration, to NAD$^+$. The process involved plasma treatment, functionalization with CP NPs, and the application of a uniform electropolymerized CP layer. The surgical meshes were characterized by SEM, FT-IR, Raman, and CV. The CV results demonstrated NADH detection with detection limits of 0.35 mM and 0.48 mM within the range found in bacterial cultures. Real-time monitoring of NADH from bacterial respiration was successfully demonstrated using *Staphylococcus aureus*, biofilm-positive, and biofilm-negative *Escherichia coli* cultures (Fontana-Escartín et al., 2023).

In a study by Gupta et al., an electrochemical paper-based sensor was devised to detect the bacterial infection biomarker procalcitonin. This sensor utilized a combination of reduced graphene oxide (RGO), AuNPs, and poly(3,4-ethylenedioxythiophene):poly(styrene sulfonate) (PEDOT:PSS) and was applied to a low-cost cellulose fiber paper substrate. The following characterization of the fabricated electrode was conducted using FESEM, FT-IR, UV-Vis spectroscopy (UV-Vis), XRD, CV, EIS, and chronoamperometry techniques. Extensive characterization confirmed the presence of the nanocomposite on the substrate. The sensor displayed a linear detection range for procalcitonin from 1×10^3 to 6×10^7 fg/mL and demonstrated good reproducibility with an approximately 3.7% relative standard deviation (Gupta et al., 2023).

Kargari et al. created an electrochemical sensor utilizing a composite of polyaniline and poly (diallyldimethylammonium chloride) combined with graphene oxide nanosheets functionalized with acrylic acid. The surface topology and chemical structure of the nanocomposite were followed and characterized by FESEM, FT-IR, and energy-dispersive X-ray spectroscopy. These components were assembled onto a GCE to craft an electrochemical sensor for detecting arsenic using CV and DPV. Notably, graphene oxide nanosheets enhanced surface area through nanoscale effects. Poly (diallyldimethylammonium chloride) augmented analyte adsorption by utilizing electrostatic interactions, while polyaniline enhanced charge transfer rates due to its excellent conductivity. The ensued electrode represented a sensitivity of 1.79 A/M with a detection limit of 0.12 μM (Hamid Kargari et al., 2023).

An electrochemical sensor for detecting phloroglucinol was developed by Keerthana et al. using a cellulose fiber paper electrode coated with CQDs derived from biomass and doped with PEDOT through an electrodeposition process (Fig. 6.4). The modified electrode underwent characterization via FT-IR, XPS, transmission electron spectroscopy (TEM), CV, and DPV. The electrocatalytic properties of biomass-derived quantum dots and the excellent conductivity of PEDOT improved the oxidation of phloroglucinol. Compared to other electrodes, this designed electrode demonstrated a higher response in terms of the oxidation peak current of phloroglucinol. The sensor displayed a good linear detection range of 36–360 nM and a low detection limit of 11 nM. Moreover, it was effectively utilized for phloroglucinol detection in industrial effluents, with a relative standard deviation of 0.84%–1.02% and recovery rates of 98.5%–101.2% (Keerthana et al., 2023).

In a study by Kummari et al., a disposable electrochemical sensor was designed for detecting the antiviral drug valganciclovir. The sensor involved the direct electrodeposition of silver nanoparticles (AgNPs) onto a SPCE modified with acid-functionalized CNTs and CP 2,6-diaminopyridine (p-DAP). The modified sensor was characterized by XRD, XPS, FESEM, CV, EIS, and square-wave voltammetry. The results demonstrated valganciclovir detection with a detection limit of 5–350 nM in phosphate-buffered saline (PBS) solution. Additionally, it effectively underwent testing using simulated urine/serum and industrial water samples, underscoring its suitability for applications in point-of-care, industrial, and environmental applications when screening antiviral drugs (Kummari et al., 2023).

Liaqat et al. investigated the suitability of PANI and its derived form, poly *o*-toluidine (POT), as electrochemical sensors for detecting the toxic compound hydrazine. The characterization of both CPs was carried out by FT-IR, XRD, and UV-Vis. Electrochemical studies such as CV and DPV showed that hydrazine strongly interacted with these polymers, resulting in irreversible charge transfer and improved oxidation current with higher hydrazine concentrations. PANI exhibited better sensitivity with a detection limit of 1 μM compared to POT's 5.13 mM in the linear

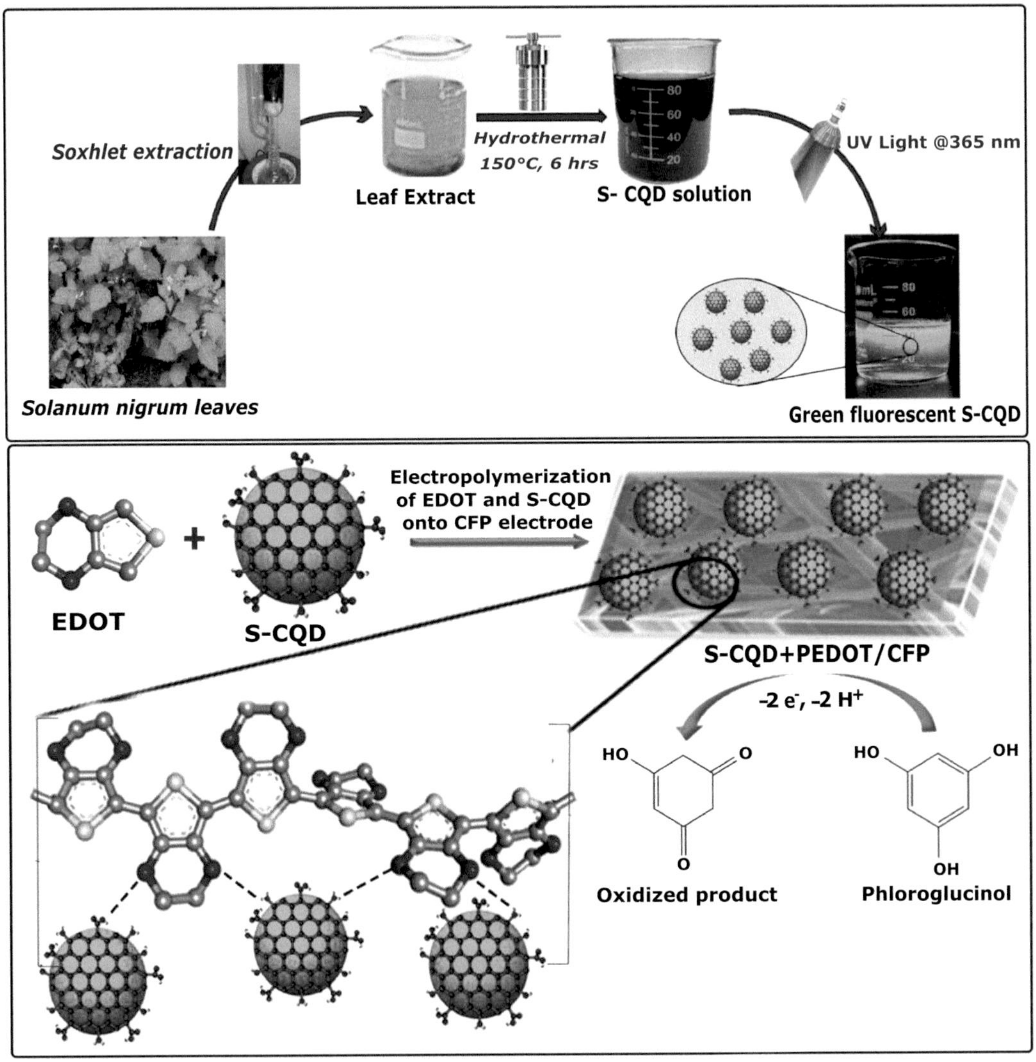

FIGURE 6.4

The synthesis of biomass-derived carbon quantum dots (S-CQDs) and the design of CFP using biomass-derived carbon quantum dots (S-CQDs) and EDOT monomer.

Modified from Keerthana, P., George, A., Benny, L., & Varghese, A. (2023). Biomass derived carbon quantum dots embedded PEDOT/CFP electrode for the electrochemical detection of phloroglucinol. Electrochimica Acta, 448, 142184. https://doi.org/10.1016/j.electacta.2023.142184.

concentration range of 1–9 mM. The diffusion coefficients for PANI and POT were correspondingly determined to be 1.89×10^{-5} cm^2/s and 2.89×10^{-6} cm^2/s (Liaqat et al., 2023).

In a study by Sierra-Padilla et al., a novel electrochemical sensor was crafted by enhancing a PANI-silicon oxide network with carbon black, improving the ability to conduct electricity and antifouling features. Characterization techniques such as FT-IR, XPS, and SEM were employed to assess the modified sensor's structure. CV and DPV were employed to study their electrochemical behavior and response to various chlorophenols, common water pollutants. The modified sensor demonstrated superior antifouling properties and achieved a sensitivity of 5.48×10^3 μA/mM cm^2, with a low detection limit of 0.83 μM for 4-chloro-3 methylphenol. Moreover, it demonstrated excellent reproducibility and repeatability, making it suitable for real-world water sample analysis and yielding high recovery values (97%–104%) (Sierra-Padilla et al., 2023).

Zhou et al. created an eco-friendly electrochemical sensor for detecting aflatoxin B1 in food and herbal medication. This sensor utilized two biological probes, an aptamer and an aflatoxin B1 polyclonal antibody with a high affinity for specific aflatoxin B1 recognition. The electrode's surface was modified via eRAFT polymerization by in situ grafting to attach numerous ferrocene polymers, significantly increasing the electrical signal (Fig. 6.5). The sensor achieved a remarkable aflatoxin B1 detection limit of 37.34 fg/mL. In addition, it demonstrated excellent reliability with recovery rates between 95.69% and 107.65% and a low relative standard deviation ranging from 0.84% to 4.92%. High-performance liquid chromatography (HPLC) with a fluorescence detector was used to validate the method's reliability (Zhou et al. 2023).

Zhou et al. introduced an innovative biosensor designed for the extremely sensitive determination of acetamiprid (ACE). This biosensor utilizes a CP composite known as phaseoloidin-doped poly(3,4-ethyloxythiophene) (PL/PEDOT). The composite possesses significant surface area, excellent conductivity, and stability. The creation of composites consisting of PL and PEDOT with a porous structure is accomplished through a single-step electrochemical copolymerization process. The resulting PAMAM/PL/PEDOT composite effectively combines the conductive ability of PEDOT, the biocompatibility of PL, and the superior surface-to-volume ratio of PAMAM dendrimers. Subsequently, the material is functionalized with third-generation poly(amidoamine) (PAMAM), and platforms (PL/PEDOT and PAMAM/PL/PEDOT) were employed for the immobilization of ACE aptamers, leading to the development of ACE aptasensors. The detection of ACE involved the assessment of signal changes in both DPV and CA, arising from the catalyzed hydrogen peroxide reduction by PL/PEDOT. It is noteworthy that the sensitivity of the DNA/PAMAM/PL/PEDOT/GCE response was determined to be 3.11 times advanced compared to that of DNA/PL/PEDOT/GCE, suggesting a significant improvement attributed to the presence of PAMAM dendrimers. Subsequently, PAMAM dendrimers were attached to the PL/PEDOT composite, creating optimal platforms for immobilizing ACE aptamers (DNA) and facilitating the improvement of biosensors. The dilution of ACE can be gauged by monitoring changes in both DPV and CA signals. DNA/PL/PEDOT/GCE and DNA/PAMAM/PL/PEDOT/GCE demonstrate effective approaches for detecting ACE in actual samples. SEM technique was employed to analyze the microstructure and surface topology of PL/PEDOT and PAMAM/PL/PEDOT composites. Furthermore, FT-IR was employed to confirm the formation of PL/PEDOT and PAMAM/PL/PEDOT composites, thereby validating the efficient fabrication of the aptasensor. Electrochemical methods such as CV, EIS, and DPV were utilized to identify the setting procedure of aptasensors and the detection of ACE. The biosensor displayed a consistent linear range for detecting ACE in both PBS and actual samples, demonstrating its reliability. The assessment of selectivity and stability, crucial factors

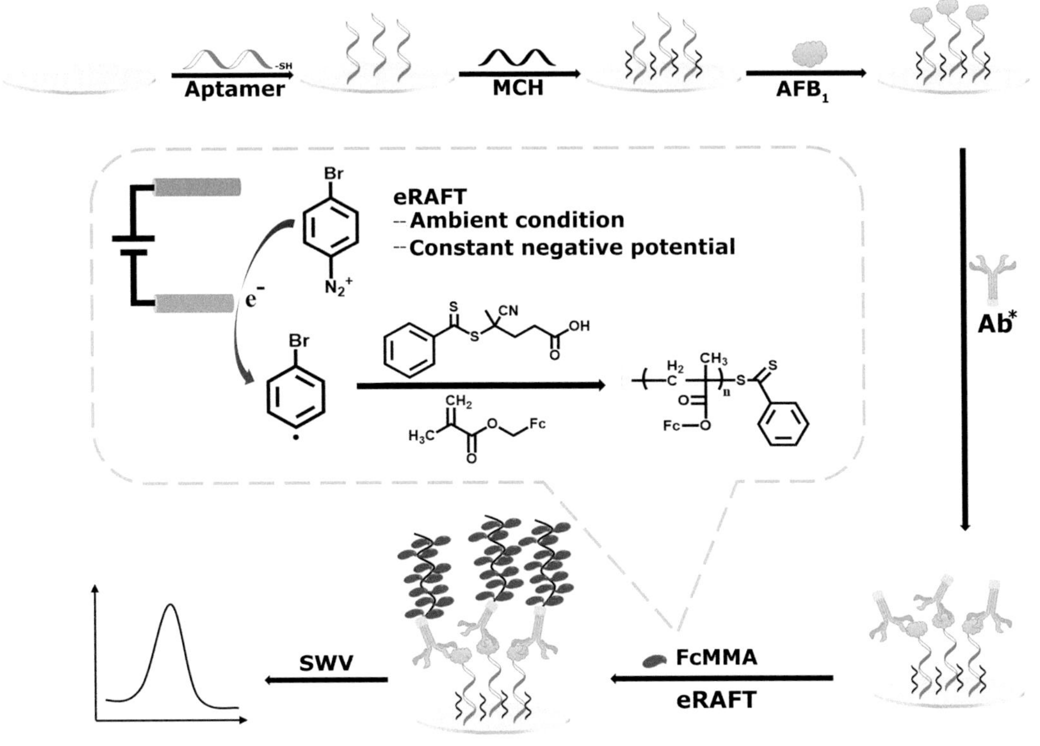

FIGURE 6.5

Highly sensitive electrochemical detection schematic of AFB1 based on eRAFT polymerization signal amplification strategy and the reaction mechanism of eRAFT.

Modified from Zhou, S., Guo, L., Shi, X., Ma, L., Yang, H., & Miao, M. (2023). In situ synthesized eRAFT polymers for highly sensitive electrochemical determination of AFB1 in foods and herbs. Food Chemistry, 421, 136176. https://doi.org/10.1016/j.foodchem.2023.136176.

for aptasensor evaluation, was conducted by exposing them to potential interfering substrates. The electrochemical signal responses of the aptasensor to both ACE and potential interferents demonstrated high selectivity, particularly in the context of ACE pesticide detection. In summary, the successful development of ACE detection aptasensors was achieved using PL/PEDOT and PAMAM/PL/PEDOT nanocomposites through DPV and CA techniques. The ACE aptasensor based on PAMAM/PL/PEDOT displayed linear ranges of 0.1 pg/mL to 10 ng/mL and 0.1 fg to 1.0 pg/mL, with impressively low detection limits of 0.0117 pg/mL and 0.0355 fg/mL through DPV and CA, respectively. These aptasensors exhibited low detection limits and exceptional sensing performance characterized by broad linear ranges and superior sensitivity, attributed to the wide specific surface area and effective immobilization of ACE aptamers. Importantly, the aptasensors were effectively utilized to detect ACE in cabbage, demonstrating encouraging practical applications. This methodology provides a robust approach for constructing aptasensors designed for pesticide residue detection (Zhou et al. 2023).

Another study reported with compositing nanofibers of polyacrylonitrile (PAN) and multiwalled carbon nanotubes (MWCNTs) NFs, which were covered with CPs (polypyrrole [PPy] or poly(3,4-ethylenedioxythiophene) [PEDOT]) through chemical vapor deposition, aiming for sensitive glucose determination. The primary goal was to enhance glucose determination efficiency. The research investigated the capability of nanofibrous formations and nanoconducting components in biosensing by utilizing both pretreatment of NFs with MWCNTs and posttreatment with CPs. These "core-shell" conducting NFs were utilized as platforms for immobilizing glucose oxidase (GOx) to enable enzymatic detection of glucose. The purpose of the CP layer was to enhance the conductivity of the matrix and enable the electron transfer essential for signal transduction. PPy and PEDOT were selected as ideal choices, with PPy selected due to its biocompatibility in biodevices and PEDOT for its excellent electrochemical properties. GOx was chosen as the "model enzyme," and careful attention was given to its concentration. Offering a detailed comparison with previous studies that primarily focused on either prefunctionalization of PAN NFs with MWCNTs or postfunctionalization with the corresponding CPs. The study revealed that enzyme concentration played a pivotal role in biosensor performance, with optimum concentrations of ~92.5 U for MWCNTs/PEDOT NFs and ~46.25 U for MWCNTs/PPy NFs. Increased enzyme concentrations from ~255 U to ~370 U led to reduced sensitivity and total effectivity of detection. The integration of MWCNTs into the NFs enhanced sensitivity and limit of detection, even for nonoptimum enzyme concentrations, with the most noticeable increase in analytical efficiency at the optimum enzyme concentration. PEDOT-based biosensors displayed modestly improved activity to PPy-based biosensors, both in the existence and absence of MWNCTs. The attained analytical factors in this study rank among the most exceptional reported to date, coupled with robust stability, rendering the practical implementation of this method highly viable for detecting various other target analytes. The stability findings demonstrated outstanding results, maintaining over 90% activity across successive measurements for both sensors. A slight decrease in activity was followed, with the PEDOT-coated NF biosensor reaching ~80% by the 30th measurement and the PPy-coated NF biosensor experiencing a more sudden decrease down to ~78% after the 25th measurement. Their research addressed a significant limitation in the practical application of biosensors, demonstrating the impressiveness of the suggested approach in detecting glucose. Furthermore, the sensors' selectivity was assessed against prevalent interfering substances encountered in beverages and biological samples. Both PEDOT- and PPy-based biosensors displayed high performance even in the presence of interferents. To showcase practical utility, both sensors were employed in detecting glucose in actual blood serum samples, yielding recoveries close to 100%. The biosensors exhibited notable sensitivities, registering 92.94 and 81.72 μA/mM/cm^2 for (PAN-MWCNTs)/PEDOT and (PAN-MWCNTs)/PPy, correspondingly, coupled with minimal LOD values of 2.30 and 2.38 μM, respectively. Applicational stability was sustained through 25 successive measurements, with both sensors retaining activity above 90% (Çetin et al., 2023).

In the investigation, the researchers described a process involving the at the same time electroreduction and deposition of graphene oxide (GO) to generate electrochemically reduced graphene oxide (ERGO) on a GCE, referred to as GCE-ERGO. Subsequently, cobalt (II) tetra-amino phthalocyanine was electropolymerized onto GCE-ERGO, forming a stable sensing surface known as GCE-ERGO/polyCoTAPc. This modified electrode, GCE-ERGO/polyCoTAPc, was employed for the highly sensitive simultaneous determination and quantification of paracetamol (PA) and dopamine (DA). Characterization of the designed surface was conducted utilizing CV and EIS to verify its conducting properties and electron transfer. The modified surface's functional groups and composition were verified using IR and EDX spectroscopy. The resultant sensing electrode demonstrated increased electrocatalytic efficiency for the redox probe ferri/ferrocyanide $\{[Fe(CN)_6]^{3-/4-}\}$. Notably, the electrocatalytic peak currents for DA and PA were considerably increased, featuring an oxidation potential difference of 264 mV, enabling their determination at the same time. DPV was used to show the linear range between the electrocatalytic oxidation peak currents of PA and DA at GCE-ERGO/polyCoTAPc, revealing concentration changes up to 100 μM for DA and up to 90 μM for PA. The detection limit values were determined as 0.095 μM for DA and 0.10 μM for PA. GCE-ERGO/polyCoTAPc exhibited good sensitivity of 1.32 $\mu A/\mu M/cm^2$ for PA and 8.39 $\mu A/\mu M/cm^2$ for DA. In an electrochemical cell containing an analytical solution and a typical three-electrode system: GCE or modified GCE as a working electrode, Pt as a counter electrode, and Ag|AgCl as a a pseudo-reference electrode. DPV, CV, and EIS techniques were performed at room temperature, and the detection of DA and PA was followed at a pH of 7.4. PBS saturated with N_2 gas. The electrochemical behavior of PA and DA was examined in 0.10 M PBS with various pH (3.0–9.0) using CV. The findings revealed a decrease in electrocatalytic oxidation potentials for both DA and PA as the pH values increased. The oxidation and reduction peak currents exhibited a proportional enhancement with the rising scan rate for 0.10 mM of PA and DA in PBS, with a scan rate of mV/s. Differential pulse voltammograms were obtained by varying the concentration of DA while maintaining a constant concentration of PA. The electrooxidation peak currents for both DA and PA were comparable, considering that the concentrations of uric acid and ascorbic acid (AA) were significantly higher than those of PA and DA. The developed sensor was used for the detection of DA and PA in urine samples spiked with these substances, demonstrating its practical usability with recoveries close to 100% (Luhana & Mashazi, 2023).

Another study focused on creating an effective electrochemical analysis by combining exceptionally conductive polymers with heterojunction materials to develop a remarkable conductive polymer by incorporating metal-free catalyst g-C_3N_4 with polypyrrole (PPy) nanocomposite. The synthesis involved in situ polymerization and a facile sonochemical technique to create a robust g-C_3N_4/PPy nanocomposite. Nitrofurantoin (NFT) was chosen as the target analyte for extremely sensitive voltammetric sensing using the modified electrode. The nanocomposite's crystallinity, functionality, and structural morphology were thoroughly examined through various spectroscopic analyses. The unique surface area of the g-C_3N_4/PPy increased due to the synergistic interaction between PPy and g-C_3N_4, enhancing electrocatalytic performance. Real human samples, including urine and serum, were successfully tested using the g-C_3N_4/PPy-modified electrode, demonstrating its applicability for practical use. Different spectroscopic methods, such as FT-IR, XRD, Brunauer–Emmett–Teller, and microscopy analyses, confirmed the successful construction of the nanocomposite. CV and EIS were employed for electrochemical studies using a conventional three-electrode method. Real-time analysis of human urine and serum samples using the g-C_3N_4/PPy-modified electrode, coupled with DPV, revealed accurate detection of spiked NFT concentrations. The results exhibited extensive recovery rates and low relative standard deviation values, indicating the sensor's suitability for real-time detection applications. The electrode's stability, repeatability, and reproducibility were evaluated through CV, demonstrating consistent performance over successive cycles. The nanocomposite's long-term stability was assessed over 25 days, showcasing only a minor reduction in peak current. Reproducibility and repeatability analyses further confirmed the electrode's reliability. Electrochemical studies demonstrated the effective reduction of NFT, with the modified electrode showing a large linear range of

0.04–585.2 μM, a low detection limit of 0.005 μM, and an impressive sensitivity of 7.813 μA/μM/cm^2 compared to previous sensors (Vinothkumar et al., 2021).

In another study, they designed a thoroughly sensitive and novel biosensor for detecting prostate-specific antigen (PSA) in human serum, utilizing antifouling peptides and a signal amplification strategy. The sensor featured a low fouling layer of poly(ethylene glycol) (PEG) doped into the conducting polymer PEDOT, followed by streptavidin immobilization and subsequent attachment of biotin-labeled peptides. These peptides were customized to incorporate a PSA-specific recognition domain (HSSKLQK) and an antifouling domain (PPPPEKEKEKE), with a terminal -SH group. Gold nanorods functionalized with DNA (DNA/AuNRs) were affixed to the electrode, and methylene blue (MB) substances were utilized as signal amplifiers. In the existence of PSA, the peptide underwent targeted cleavage, leading to the loss of AuNRs, DNA, and MB, resulting in a substantial reduction in the current signal. The SEM characterization of the PEG/PEDOT surface morphology revealed a three-dimensional porous network microstructure, significantly expanding the electroactive surface area. This larger surface area facilitated the attachment of biomolecules and electron transfer. The biosensor assembly process was monitored through CV. The successful immobilization of streptavidin and subsequent steps were verified by changes in redox peak currents. XPS analysis provided further confirmation of biosensor construction. The antifouling ability of these modified interfaces was demonstrated by minimal fluorescence signal after incubation with FITC-BSA. The biosensor's antifouling property was further validated by testing its response to various proteins, for example, hemoglobin, bovine serum albumin, lysozyme, and human serum albümin. DPV responses indicated significantly reduced nonspecific protein adsorption on PEG/PEDOT, Pep/PEG/PEDOT, and MB/DNA/AuNRs/Pep/PEG/PEDOT fabricated electrodes comparison with bare and $LiClO_4$/PEDOT-developed surfaces. Moreover, the biosensor's antifouling ability extended to complex biological samples, such as fetal bovine serum (FBS). Insomuch as after incubation in undiluted FBS, the MB/DNA/AuNRs/Pep/PEG/PEDOT-modified electrode retained 74.02% of its initial value, affirming the biosensor's effectiveness in complex sample environments. The biosensor's selectivity was validated against potential interfering proteins, showcasing a notable decrease in response change rate for PSA compared to other proteins. Long-term stability studies indicated excellent stability of DNA/AuNRs and the fabricated biosensor. Intra- and inter-assay reproducibility evaluations yielded low relative standard deviation (RSD) values. To assess accuracy, the biosensor was employed in detecting PSA concentrations in real human serum samples using the standard addition technique. The biosensor exhibited high detection recoveries (101.6%–107.0%) with acceptable RSD values (3.2%–6.1%), affirming its potential for practical determination of PSA in serum. The biosensor demonstrated a broad linear dynamic range from 0.10 pg/mL to 10.0 ng/mL and a low LOD of 0.035 pg/mL. Notably, the biosensor successfully detected PSA in real human serum, showcasing its practical potential (Hui et al., 2022).

Other related studies about the applications of CPs in electrochemical sensors are tabulated in Table 6.1.

Besides, MIPs are also crucial in applications based on nanosensors and biomimetic sensors using electrochemical techniques. The latest studies of electrochemical MIP sensors based on conducting and nonconducting polymers will be presented herein.

In a study by Salimonnafs et al., MIP consisting of MAA and ethylene glycol dimethacrylate (EGDMA) was employed to design a GCE, while polydiphenylamine served as an immobilizer through electropolymerization (Fig. 6.6). An electrochemically synthesized RGO layer was

Table 6.1 Selected studies about the applications of conducting polymers in electrochemical sensors.

Sensor	Analyte	Method	Medium	Linear range	LOD	Application	Reference
GRE/poly[BDT-*alt*-(TP;BO)]-C60/Tyr	Catechol	AMP	PBS (pH 7.0)	0.5–62.5 μM	0.11 μM	Inhibition	Kurbanoglu et al. (2022)
GCE/PANI/NiOnF	Nitrite	AMP	0.1 M PBS (pH 7.0)	0.1–1 μM and 1–500 μM	9.7 nM and 64 nM	Nile River water	Al-Kadhi et al. (2023)
SPCE/PEDOT:PSS/AuNPs	Pirimicarb	CV	ABS (pH 5)	93.81–750 μM	28.34 μM	NS	Deller et al. (2023)
PI/Au-PANI/Pd	Glucose	DPV	0.1 M NaOH	10–30 μM	0.3 μM	Real sample	Eswaran et al. (2023)
$OMLP_f$ /PHEDOT/PEDOT OME_f/PHEDOT/PEDOT	Bacterial infections	CV	0.1 M PBS	0–1 mM and 1–6 mM	0.35 mM and 0.48 mM	B + *E. coli*, B − *E. coli* bacteria, *S. aureus*	Fontana-Escartín et al. (2023)
rGO-AuNP-PEDOT:PSS	Procalcitonin	CA	0.1 M KCl solution containing 10 mM $[Fe(CN)_6]^{3-/4-}$ (pH 7.4)	1×10^3 to 6×10^7 fg/mL	0.28×10^3 fg/mL	NS	Gupta et al. (2023)
AAGO-PDDA-PA/GCE	Arsenate	DPV	0.1 M ABS	0–30 μM	0.12 μM	Tap water and rice four sample	Hamid Kargari et al. (2023)
S-CQD + PEDOT/CFP	Phloroglucinol	DPV	PBS (pH 7.0)	36–360 nM	11 nM	Pharmaceutical industrial effluent water samples, firework industrial effluent water samples	Keerthana et al. (2023)
SPCE/*f*-CNT/p-DAP-AgNPs	Valganciclovir	SWV	PBS (pH 7.0)	5 − 350 nM	2.9 nM	Urine, human serum, and tap water samples	Kummari et al. (2023)
PANI/GCE POT/GCE	Hydrazine	DPV	PBS (pH 6)	1–9 mM	1 μM and 5.13 mM	NS	Liaqat et al. (2023)
SNG-C/CB-PANI	4-Chloro-3-methylphenol 4-Chlorophenol 2,4-Dichloropheno 2,4,6-Trichlorophenol	CV	ABS (pH 4)	2–10 μM	0.83 μM 0.75 μM 0.92 μM 1.06 μM	Natural water samples	Sierra-Padilla et al. (2023)

(Continued)

Table 6.1 Selected studies about the applications of conducting polymers in electrochemical sensors. ***Continued***

Sensor	Analyte	Method	Medium	Linear range	LOD	Application	Reference
nAu/pAMT/*f*-CNT/GCE	Serotonin	SWV AMP	PBS (pH 7.0)	0.015–5 μM 0.05–2.15 μM	7.8 nM 18 nM	Human serum	Shekher et al. (2023)
AuNPs/poly-AHP/CPE	Valacyclovir	DPV	PBS (pH 7.0)	5–80 nM	1.89 nM	Pharmaceutical tablet, artificial urine, and diluted human serum sample	Kummari et al. (2022)
PEG/PEDOT/GCE	Prostate specific antigen	DPV	PBS (pH 7.4)	0.10 pg/mL to 10.0 ng/mL	0.035 pg/mL	Real human serum	Hui et al. (2022)
*o*PPy-β-CD-PMo_{12}/PGE	Propylparaben	DPV	0.04 M BRB (pH 6.0)	0.2–10 μM 10–100 μM	0.04 μM	Commercial cleansing micellar solution	Hatami et al. (2021)
g-C_3N_4/PPy/GCE	Nitrofurantoin	CV	0.05 M PB (pH 7.0)	0.04–585.2 μM	0.005 μM	Human urine and serum samples	Vinothkumar et al. (2021)
PABA/PSS/FTO	Dopamine	CV	HEPES buffer (pH 7.0)	0.1–1.0 μM	0.0628 μM	NS	Panapimonlawat et al. (2021)
POA@Ag/GCE	Dopamine	DPV CA	0.1 M PBS (6.0)	10–130 mM 5–45 μM	2.8 μM 0.83 μM	Blood and urine	Pandian et al. (2021)
PEDOT: PSS/MXene-PdAu/PEDOT:PSS	Shikonin	DPV	PBS (pH 7.0)	0.001–35 μM	0.33 nM	*Lithospermum erythrorhizon*	Huang et al. (2022)
CS/PPyNWs/GCE	Acetamiprid pesticide	CC CA	5 mL of PBS (PH 6.86)	1.0 pg/mL to 0.1 mg/mL 1.0 fg/mL–0.1 ng/mL	0.347 pg/mL 0.065 fg/mL	Real samples	Zhang, Lang, et al. (2022)
PPy-CO_2@PGE	Pb^{2+} and Cd^{2+}	DPASV	0.1 M ABS (pH 6)	0.1–1 nM	0.018 nM 0.023 nM	Natural water samples	Poudel et al. (2022)
GCE-ERGO/polyCoTAPc	Dopamine Paracetamol	DPV	0.10 M PBS (pH 7.4)	2.0–100 μM 7.0–90 μM	0.095 μM 0.104 μM	Synthetic urine samples	Luhana and Mashazi (2023)
AuNPs/p(Py-EDOT)/PGE	Carcinoembryonic antigen	DPV	0.1 MPBS (pH 7.4)	0.001–100 ng/mL	0.741 pg/mL	Human serum	Dokur et al. (2022)
NG/PPy/GCE	Acetamiprid	DPV CA	PBS (pH 7.4)	10^{-12} to 10^{-7} g/mL	1.15×10^{-13} g/mL 7.32×10^{-13} g/mL	Vegetables	Wang et al. (2022)
PrP^C /AuNPs-E-PPy-3-COOH	Amyloid-β oligomer	CV	0.01 M PBS	10^{-9} to 10^3 nM	10^{-9} nM	Ex vivo real sample	Zhao et al. (2022)

AuNPs/P(PyAmn)/ITO	Cytokeratin 19 fragment 21−1	CV	PBS (pH 7.4)	0.015−90 pg/mL	4.59 fg/mL	Real human serum samples	Aydın et al. (2023)
PPy/GOx/DGNs/GRE PANI/GOx/DGNs/GRE	Glucose	CPA	SA buffer (pH 6.0)	0.10−19.9 mM 0.30−19.9 mM	0.070 mM 0.18 mM	Real samples (human serum, saliva, wine, milk, and juice)	German et al. (2022)
(PAN-MWCNTs)/PEDOT/ GOx-2 (PAN-MWCNTs)/PPy/ GOx-1	Glucose	AMP	0.1 M PBS (pH 7.0.)	0.01−1.2 mM 0.01−2.0 mM	2.30 μM 2.38 μM	Real samples	Çetin V et al. (2023)
SiNPs/PBSeThTh/laccase SiNPs/PBSeThTh/MWCNTs/laccase	Catechol	CV	50 mM SA (pH 5.5)	50−350 μM 10−400 μM	7.2 μM 1.11 μM	Tap water	Deniz et al. (2022)
PA-TaCoPc@ZnO	Nitrite	CA	0.1 M PBS(pH 7)	1−10 μM	21 nm	NS	Sudhakara et al. (2023)
Pth-CuO/GCE	Hydrogen peroxide	AMP	0.1 M PBS (pH 7.0)	20−3300 μM	3.86 μM	Milk and tap water	Rashed et al. (2022)
DNA/PAMAM/PL/PEDOT/GCE	Acetamiprid	DPV	PBS (pH 7.4)	0.1 pg/mL −10 ng/mL	0.0117 pg/mL	Cabbage samples	Zhou, Guo, et al. (2023)
poly 1,8-DAN/*f*-MWCNT/CPE	Nitrite	AMP	0.1 M PBS (pH 7.2)	300−6500 nM	75 nM	Tap water	Salhi et al. (2022)
Co/PPy/GCE	Nitrite	AMP	0.2 M PBS (pH 8.0)	2−3318 μM	0.35 μM	Pickled Chinese cabbage and water samples	Lü et al. (2022)
Au@PPy-C/g-C_3N_4 NCs/GCE	Nitrite	DPV	PBS (pH 7.0.)	1.5−22.5 μM	1.11 μM	Underground water, tap water, industrial wastewater, and seawater behavior	Faisal et al. (2023)
SNG−C−2% PANI	Chlorophenols	DPV	0.1 M ABS (pH 4)	0.7−7.0 μM	0.69 μM	Spiked water samples	Calatayud-Macías et al. (2023)

(Continued)

Table 6.1 Selected studies about the applications of conducting polymers in electrochemical sensors. ***Continued***

Sensor	Analyte	Method	Medium	Linear range	LOD	Application	Reference
Ag-PEDOT/GCE	Caffeic acid	SWV	0.1 M PBS (pH 7)	2–100 μM	1.9 μM	Pear and orange juice	García-Guzmán et al. (2021)
Magnetite MOF@CNT-pC/GCE	Adenine and guanine	DPV	PBS (pH 7.0)	0.5–30 μM 0.5–25 μM	285.0 nM 244.0 nM	Real sample	Şenocak et al. (2022)
Ni-NPs/ $PPy_{(1)}$/GRE	Glucose	AMP	0.1 M PBS	1.0–1000 μM	0.4 μM	Real samples	Emir et al. (2021)
poly (DL-met)/AuNPs-GCE	Paroxetine	DPV	0.1 M PBS (pH 7)	50 nM –100 μM	0.01 nM	Pharmaceutical formulations	Al-Mhyawi et al. (2021)

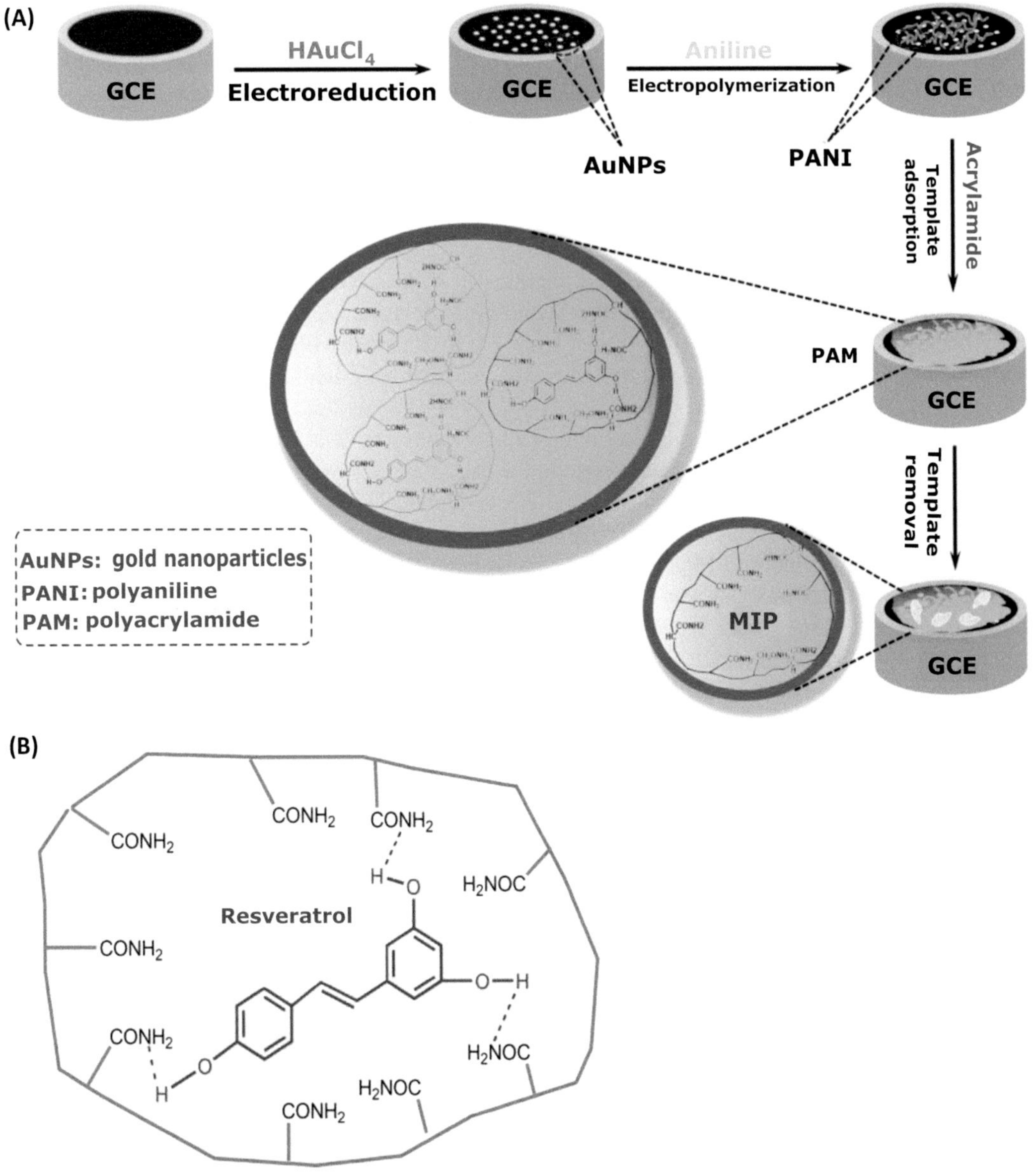

FIGURE 6.6

(A) Workflow for the preparation of PAM/PANI/AuNPs molecular imprinted film (MIP) on glassy carbon electrode (GCE). (B) Schematic of the recognition behavior between PAM with resveratrol molecule via hydrogen bond interaction.

Modified from Huang, S., Yang, J., Li, S., Qin, Y., Mo, Q., Chen, L., & Li, X. (2021). Highly sensitive molecular imprinted voltammetric sensor for resveratrol assay in wine via polyaniline/gold nanoparticles signal enhancement and polyacrylamide recognition. Journal of Electroanalytical Chemistry, 895, 115455. https://doi.org/10.1016/j.jelechem.2021.115455.

positioned beneath the polymer layer, enhancing sensitivity. Consequently, in this study, the designed electrode enabled the quantification of trace levels of ceftriaxone, exhibiting a linear range of 0.025–3.047 μM and for LOD value of 0.008 μM (Salimonnafs et al., 2023). The study by Huang et al. suggested an exceedingly sensitive electrochemical sensor for detecting resveratrol following a molecularly imprinted approach. AuNPs and PANI film were deposited on a GCE using electrodeposition and electropolymerization methods. The resulting porous PANI/AuNPs composite served as a conducting matrix for electron transfer and anchoring the imprinted substrates, polyacrylamide (PAM), which was polymerized chemically with acrylamide and resveratrol to form a specific MIP capable of electrochemically responding to resveratrol. The developed sensor exhibited a linear response in detecting resveratrol from 1.0 to 200 μM, with a detection limit of 87 nM. The sensor-enabled it to detect resveratrol directly in wines, eliminating the necessity for sample pretreatment (Huang et al., 2021).

Balciunas et al. utilized molecular imprinting technology to create a glyphosate-sensitive layer utilizing a CP, PPy, deposited onto a gold electrode for use as an electrochemical surface plasmon resonance sensor. The impact of glyphosate on the electrochemical deposition of PPy on the electrode was observed through EIS, CV, and SPR. The study revealed that the interaction of glyphosate with the bare gold electrode blocked PPy deposition. To overcome this, a self-assembled monolayer of 11-(1*H*-pyrrol-1-yl)undecane-1-thiol was introduced before depositing PPy. The dissociation constant and Gibbs free energy values were calculated as $38.18 \pm 2.33 \times 10^{-5}$ and -19.51 ± 0.15 kJ/mol, respectively (Balciunas et al., 2022).

Charlier et al. fabricated an MIP to detect penicillin G in the liquid phase. Pyrrole/pyrrole-3-carboxylic acid copolymer synthesized by oxidative polymerization was deposited on a homemade interdigitated electrode, on which a thick gold layer was sputtered. Surface characterization was observed by QCM, SEM, and EIS. The study demonstrated the viability of using pyrrole/pyrrole-3-carboxylic acid MIP for impedimetric penicillin G sensors. The authors claimed these sensors exhibited good sensitivity, reproducibility, and selectivity. The sensors provided nearly linear detection performance within the specified range of 12.5–100 ppb, meeting the required specifications (Charlier et al., 2022).

In another study where penicillin G was utilized as a template, a nanosized molecularly imprinted polymer (nanoMIP) was synthesized by Rahim et al. using miniemulsion polymerization to selectively recognize penicillin G. NanoMIP was designed using MAA as a functional monomer and EDGMA as a crosslinking substance. Through sonication, the monomer-template mixture was emulsified into a miniemulsion during prepolymerization, resulting in nanoMIP. Characterization included FTIR, FESEM-EDX, and particle diameter analysis. As a result, nanoMIP 2 (6 mmol MAA) exhibited the highest binding capacity and selectivity for penicillin G compared to nanoMIP 1 (4 mmol MAA) and nanoMIP 3 (10 mmol MAA) (Rahim et al., 2023).

In a study by Çorman et al., a novel MIP sensor was developed for detecting teriflunomide (TER) using pyrrole-histidine as a functional monomer. The sensor's thin film was formed through electropolymerization of pyrrole-histidine onto a GCE with TER as a template. The chemical structure and surface topology of the created MIP sensor were analyzed using FT-IR and SEM, while its electrochemical characteristics were determined by CV and EIS. The resulting sensor exhibited a linear TER detection range of 0.1–1.0 pM and a detection limit of 11.38 fM. The current approach for the electrochemical sensor has also exhibited remarkable recuperation results in simulated serum samples and real tablet formulations, correspondingly, yielding recoveries of 97.56% and 100.35% (Çorman et al., 2022).

In a study by Elmalahany et al., Saxagliptin, a novel antidiabetic compound for type 2 diabetes, prompted an electrochemical MIP. Through computational calculations and synthesis, MIPs were developed with Saxagliptin as a template, itaconic acid as a functional monomer, and various ratios of crosslinking agents. Noncovalent interaction analysis further aided sensor design. The optimized MIP was integrated with MWCNTs into a carbon paste electrode, exhibiting excellent linearity within the range of 1×10^{-9} to 1×10^{-15} M and low detection limits of 8×10^{-16} and 2×10^{-16} M. The sensor exhibited successful application in pharmaceutical formulations, as well as in urine and human serum samples, showcasing a recovery range of 97.45%–100.64% (Elmalahany et al., 2023).

Ratautaite et al. presented a PPy-based sensor for detecting the SARS-CoV-2-S spike glycoprotein, a vital element of the COVID-19-causing coronavirus. MIP technology was used to create the electrochemical sensor, where Py was electrodeposited on a platinum electrode. Pulsed amperometric detection demonstrated that MIP-based sensors were more sensitive and selective toward SARS-CoV-2-S glycoprotein compared to nonimprinted polymer (NIP)-based electrodes, which were prepared similarly to MIP but in the lack of template molecules. Moreover, the MIP exhibited higher specificity toward SARS-CoV-2-S glycoprotein over bovine serum albumin, showing potential for detecting SARS-CoV-2 virus proteins (Ratautaite et al. 2022).

In another study by Ratautaite et al., a comparison of two CPs, PPy and PANI, was carried out to create the detection of L-tryptophan. MIP sensors prepared with pyrrole exhibit a greater affinity to L-tryptophan. The PPy-MIP sensor responded linearly between 50 and 100 μM, with an LOD and LOQ of 16.6 μM and of 49.8 μM, respectively (Ratautaite et al. 2022).

Micro-sized MIP sensors modified by MWCNT were fabricated by Shagri et al. for potentiometric determination of the barbital. The sensors utilized a synthetic barbital sodium MIP as a recognition receptor integrated into a plasticized polyvinyl chloride membrane. MAA was used as a functional monomer. Operating in the concentration range 1.0×10^{-3} to 2.0×10^{-7} M, the sensors exhibited a Nernstian slope of -56.8 ± 0.9 mV/decade. The incorporation of MWCNTs enhanced interfacial capacitance and eliminated water layer interference. The sensors demonstrated strong selectivity for barbital detection in spiked urine samples (Al Shagri et al., 2022).

Jara-Cornejo et al. built up an electrochemical MIP sensor for the immensely sensitive determination of methotrexate, which is an antimetabolite widely employed for cancer and autoimmune disease treatment and poses environmental concerns due to its extensive use and disposal. MIP was electrodeposited onto a GCE that had been previously modified with MWCNT (Fig. 6.7). The sensor displayed a LOD of 2.7×10^{-9} M, a linear range of 0.01–125 μM, and extreme selectivity, making it promising for methotrexate quantification in environmental samples (Jara-Cornejo et al., 2023).

Mulyasuryani et al. developed a novel electrochemical sensor for the detection of *p*-aminophenol (*p*-AP) by designing a SPCE with a MIP based on chitosan. The MIP, synthesized using *p*-AP as a template and chitosan as the base polymer, was characterized for the morphology of the surface through FT-IR, and electrochemical analyses were also employed on the modified electrode. The MIP sensor displayed superior selectivity with an excellent accuracy of $94.11 \pm 0.01\%$. The anodic peak current displayed a linear increase within the concentration range of 0.5–35 μM *p*-AP, with a sensitivity of 3.6 ± 0.1 μA/μM, a LOD of 2.1 ± 0.1 μM, and a LOQ of 7.5 ± 0.1 μM (Mulyasuryani et al., 2023).

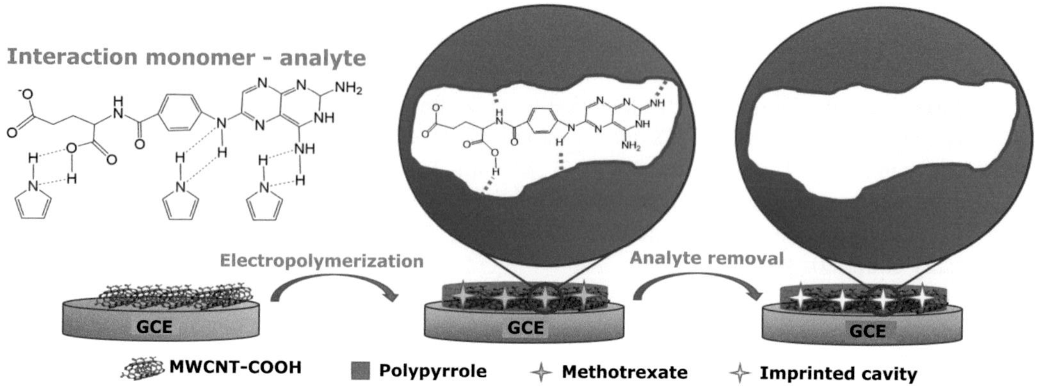

FIGURE 6.7

Schematic illustration of the preparation of electropolymerized MIP film and its application for the quantification of methotrexate.

Modified from Jara-Cornejo, E., Khan, S., Vega-Chacón, J., Wong, A., da Silva Neres, L. C., Picasso, G., & Sotomayor, M. D. P. T. (2023). Biomimetic material for quantification of methotrexate using sensor based on molecularly imprinted polypyrrole film and MWCNT/GCE. Biomimetics, 8(1). https://doi.org/10.3390/biomimetics8010077.

The study by Koçak and Gürler Akyüz focused on fabricating an electrochemical sensor for DA using an MIP on GCE and extremely oriented pyrolytic graphite with edge and basal plane orientations. A thiophene acetic acid-DA thin film was formed through electropolymerization, followed by platinum (Pt) NP deposition and elution-based DA template extraction. CV, chronoamperometry, and DPV analyses confirmed sensor behavior. The MIP sensor presented a detection limit of 14.40 nM, 42.50 nM, and 0.671 μM for glassy carbon, edge, and basal plane electrodes, respectively. The sensor demonstrated notable selectivity toward interferents and successfully analyzed DA in human serum with stability and reproducibility (Koçak & Akyüz, 2023).

Another study conducted a thorough exploration of different parameters involved in fabricating a MIP sensor for the fundamental understanding of cortisol detection. Cortisol detection was effectively carried out utilizing a bio-inspired MIP optimized via the widely employed monomer *o*-PD. Changeable parameters, including electropolymerization cycles, monomer–template ratio, diverse *o*-PD concentrations, rebinding time, pH values, and various template removal techniques, were systematically optimized to enhance the consistency and efficacy of the MIP sensor fabrication process. Under optimized conditions, the MIP sensor was benchmarked against a NIP to verify its notable sensitivity. Additionally, analogous biomolecules (corticosterone, cortisone, triamcinolone, and progesterone) were tested utilizing the under-optimized conditions of MIP to provide selectivity. Experiments testing the selectivity of a NIP sensor validated the distinctive binding preference of the MIP sensor for cortisol, successfully discerning it from various other steroid hormones. The SPCEs underwent preoxidization using 0.1 M PBS pH of 7.0 in a potential range of 0–1.3 V at a scanning rate of 0.1 V/s for 20 cycles to electrochemically cleanse the modified electrode surface. The impact of *o*-PD concentration was assessed through CV responses of the MIP electrode before template removal, in the blank solution, and after rebinding of cortisol. A reduction in the concentration of *o*-PD from 2–0.5 mM led to an elevation in the CV signal followed at the MIP electrode before the removal of the template. Electropolymerization designed a nonconductive poly *o*-PD (p*o*-PD) layer, with higher *o*-PD concentrations yielding a thicker layer, inhibiting electronic conductivity. After template removal, cavities formed within the p*o*-PD layer, facilitating redox reactions. Unexpectedly, the current in the CV was increased upon the exposure of the MIP electrode to cortisol. This unforeseen rise could be attributed to parameters such as the electropolymerized layer being too thin to create suitable cavities for cortisol rebinding or the detachment of some p*o*-PD layers throughout washing. To address this, the optimization of the number of CV cycles was carried out. Under these optimized conditions, the DPV response of the MIP sensor to cortisol was evaluated in an 80 μL solution containing PBS and $[Fe(CN)_6]^{3-/4-}$. The DPV signal for the MIP sensor exhibited a proportional decrease of around 0.8 V with increasing cortisol concentration. To assess sensor reproducibility, five diverse MIP sensors were developed, every measuring a concentration of 100 pM cortisol. The RSD was calculated, resulting in a value of 3.84%, indicating consistent and reliable results across replicas of multiple sensors. Moreover, the sensor's stability was evaluated by storing developed electrodes at room temperature for 3 days, with the sensor's response remaining within an admissible range of 10.06% during the storage term. Once again, this highlights the stability of the sensor and its capacity to maintain consistent activity over time. Quantification of peak current against cortisol concentration demonstrated exceptional sensitivity with an LOD of 0.036 nM. Under the specified optimal conditions, the MIP sensor displayed a correlated reduction in peak currents observed through DPV with enhancing cortisol concentrations spanning from 0.1 to 100 nM (Kim et al., 2023).

Yan et al. developed a TiO_2-based molecularly imprinted photoelectrochemical sensor (MIPES) for the sensitive determination of ciprofloxacin (CIP). The photoactive sites of TiO_2 were identified through NiOOH photo-deposition, and the imprinted places were anchored on these photoactive places through complexation between NiOOH and CIP. By carefully controlling the imprinted places, the photocurrent distinction before and after CIP addition was increased, enhancing the detection sensitivity of CIP. The introduction of a photo-deposition layer between TiO_2 and MIP ensured consistency between imprinted sites and photoactive sites, resulting in an increased photocurrent distinction with/without CIP. The quantity of photo-deposited NiOOH was meticulously fine-tuned, and the correlation between photoactive sites and NiOOH was confirmed owing to photocurrent assessments. Additionally, the CIP' amount as the template molecule was adjusted, and the interaction between CIP and NiOOH was verified through FT-IR and XPS tests, successfully anchoring imprinted sites on photoactive sites. The preparation process, characterization followed by FT-IR and XRD, involved TEM and SEM tests to observe the morphology of the samples. High-resolution transmission electron microscopy images revealed a TiO_2 NRs coating with an amorphous OPOD layer, lacking lattice stripes. EDS mapping confirmed the existence of C, N, O, Ni, and Ti elements, validating the presence of the NiOOH photo-deposition layer. XPS, photocurrent, and EIS tests over-characterized the TiO_2 NRs/NiOOH/rMIP preparation procedure. Photocurrent tests indicated an increase after NiOOH photo-deposition, attributed to the formation of a PN heterojunction reducing the recombination of photoexcited cavities and electrons. However, the photocurrent experienced a decline following the complexation with CIP and underwent a further reduction after the electropolymerization of OPOD. Subsequent CIP photodegradation resulted in an increased photocurrent due to the remaining imprinted cavities providing a novel pathway for electron transfer. Photocurrent tests were performed in PBS with various CIP concentrations, such as from 0.1 to 2500 μM, using TiO_2 NRs/NiOOH/rMIP as the working electrode. To establish the interaction between photocurrent and CIP concentration, a real-time amperometric photocurrent signal study was conducted in PBS solution under illuminated conditions. Comparative analysis with TiO_2 NRs/NiOOH/eMIP sensors prepared through traditional methods indicated an extensive linear detection range, superior sensitivity, and a lower LOD for TiO_2 NRs/NiOOH/rMIP. Furthermore, TiO_2 NRs/NiOOH/rMIP demonstrated well stability, selectivity, and recovery rates, showcasing its hopeful application in the actual determination of antibiotics. Selectivity was confirmed by introducing various interference agents with similar structures, such as enrofloxacin, norfloxacin, levofloxacin, and tetracycline, where the photocurrent response of CIP surpassed that of the interferences, indicating good selectivity. The study utilized TiO_2 as a photoactive matrix for MIPES (TiO_2 NRs/NiOOH/rMIP) construction through electro-polymerization. Sequential addition experiments utilizing TiO_2 NRs/NiOOH/eMIP as the working electrode revealed linear equations for different CIP concentration ranges. The LOD was found to be 0.097 μM. In the assessment of practical application feasibility, varying concentrations of CIP were introduced into clarified lake and lap water. Detection procedures were executed utilizing the TiO_2 NRs/NiOOH/rMIP. The recovery rate fell within the range of 98%–106%, exhibiting an RSD lower than 4%. This confirmed the sensor's good accuracy and practical utility for detection. The MIP photoelectrochemical sensor TiO_2 NRs/NiOOH/rMIP, highly sensitive and developed in this study, manifested consistent imprinted and photoactive sites (Yan et al., 2023).

Sunon and Ngamchuea developed a novel, rapid, and sensitive electrochemical sensor for the detection of melatonin, utilizing an MIP with melatonin as the template and MB as the functional monomer. The synthesis of the MIP was accomplished through a simple electropolymerization process, eliminating the requirement for an initiating reagent. MB, a readily available and budget-friendly compound, served as a good nominee for MIP synthesis due to its extremely conductive polymer-forming properties. Its electrochemical characteristics were adjustable to optimize melatonin detection. During electropolymerization, the positively charged MB formed strong electrostatic attractions with the negatively charged melatonin, and its aromatic rings participated in $\pi-\pi$ stacking interactions with the aromatic rings in melatonin. These types of interactions facilitated the development of precisely characterized binding places in the MIP, improving the sensitivity and selectivity of the sensor. The MIP's morphology was examined using a field-emission SEM, and its functional groups were analyzed by FT-IR. In studies related to application, samples of urine and saliva were gathered from five individuals healthy volunteers. Samples were then centrifuged, spiked with 30 μM melatonin, and subjected to DPV measurements. The sensor's electrochemical properties were examined at a bare GCE before fabricating the melatonin-templated MB using a MIP electrode (MIP/GCE). The MIP/GCE exhibited improved sensitivity, lower LOD, and higher selectivity compared to a bare GCE and a nonimprinted polymer electrode (NIP/GCE). The sensor's analytical performance was further enhanced using DPV to eliminate capacitive background charging currents. Reproducibility testing for intra and inter-electrode measurements demonstrated excellent %RSD values (≤3%). The sensor's sensitivity for melatonin measurement and its repeatability within a single MIP/GCE were confirmed. However, when stored under atmospheric and room conditions, the sensor's response decreased to 82% after 1 day but remained stable at $81.0 \pm 1.4\%$ for at least 7 days. The sensor's selectivity against potential interferences such as AA, tryptophan, serotonin, cytosine, lactate, cytidine, creatinine, creatine, and urea was evident. Percentage recoveries of $103 \pm 1\%$ and $102 \pm 1\%$ for saliva and urine samples, respectively, validated the sensor's applicability for melatonin analysis in biological fluids. The sensor exhibited notable selectivity for melatonin over various interferences such as urea, lactate, cytidine, creatinine, cytosine, tryptophan, serotonin, creatine, and AA. Detection of melatonin was achieved at a potential of 0.60 V versus Ag/AgCl, demonstrating a sensitivity of 138.8 ± 4.7 μA/μM within the linear range of 0.097–200 μM and a LOD of 29 nM. Despite the challenges associated with controlling MIP formation, our sensor presented several advantages, including simplicity, affordability, portability, rapid analysis, and cost-effectiveness. The sensor demonstrated excellent reusability, biomimetic properties, and straightforward electrode modification using the affordable MB compound through a simple electropolymerization technique. These characteristics collectively position our sensor as a promising and reliable method for melatonin detection in biological fluids (Sunon & Ngamchuea, 2023).

A recently developed SPCE designed for the potentiometric assessment of nalbuphine (NAL) is outlined in their study. The electrode incorporates carboxylated MWCNTs/polyaniline (*f*-MWCNTs/PANI) nanocomposite as an onto-electron transducer, covered with a selective polyvinyl chloride (PVC) membrane containing molecularly imprinted drug-polymer beads to act as a recognition receptor for the potentiometric detection of the synthetic narcotic, NAL. The study employed FT-IR with an attenuated total reflection for IR spectral measurements and SEM and an XRD for characterization of the designed morphology of developed *f*-MWCNTs/PANI nanocomposite surface. The FT-IR spectra of MIP, including with and without NAL drug NIP sensor, were evaluated and compared with the spectrum of pure NAL. Additionally, SEM was utilized to investigate the surface characterizations of NIP and MIP after NAL removal. The study examined the crystal behavior and orientation of the nanocomposite consisting of functionalized (*f*-MWCNTs) and PANI by observing XRD patterns for *f*-MWCNTs, PANI, and the *f*-MWCNTs/PANI nanocomposite. The potentiometric response time of the developed SPE was assessed using sequential additions of NAL solutions. The interface capacitance of the developed SPCE was measured using chronopotentiometry (CP) and EIS, revealing an improved capacitance of 52.5 μF with the utilization of the developed nanocomposite layer. This combined layer successfully inhibited the development of an undesirable thin water layer, improving both membrane and potential stability. Potential stability investigations were conducted using two different concentrations of NAL HCl such as 5.0×10^{-4} and 5.0×10^{-3} M in a PBS of pH 6.5. The pH adjustment was achieved using a limited volume of 0.1 M HCl or NaOH. The suggested SPCE maintained a consistent potential within the pH range of 4–8.5. A minor positive potential drift occurred at pH levels below 4, attributed to the interference of H^+ ions. The significant decline in the potential response observed at pH exceeding 8.5 was associated with interference from hydroxide ions and the deprotonation of the hydroxyl group in the NAL drug. As a result, all measurements were conducted in a 0.01 M PBS with a pH of 6.5. The selectivity of the developed SPCE/(*f*-MWCNTs/PANI)/MIP-ISM sensor was checked against various interfering agents and components widely used in drug formulations or present as associated matrices in biological fluids. Calibration curves were generated for various substances, including inorganic cations, simple drugs, amino acids, and NAL HCl, using both the designed SPCE/(*f*-MWCNTs/PANI)/MIP-ISM sensor and the unmodified sensor (SPCE/MIP-ISM). The sensor with the modified transducer layer exhibited significantly enhanced selectivity, demonstrated by higher selectivity coefficient values compared to the bare sensor. The absence of a transducer layer in the unmodified sensor resulted in the creation of a water layer, causing potential instability and considerable potential drift. Conversely, the introduction of a hydrophobic transducer layer made of the nanocomposite improved and stabilized the potential response by minimizing the creation of a water layer between the conducting substrate and the ion-selective membrane. The detection of NAL in a complex biological fluid for the fast diagnosis of overdose cases. Various amounts of NAL HCl were spiked into urine and diluted with PBS in a 1:5 ratio. Potentiometric studies were conducted utilizing the modified SPCE, demonstrating good recoveries of NAL and advisable no matrix effect. The developed SPCE exhibited superior performance, suggesting that known/unknown species in human urine did not interfere. The electrode, modified with nanocomposite layer as a transducing material and MIP beads embedded in a plasticized membrane as a recognition sensing layer, features a linear range of 1.1×10^{-7} M. Significantly, the created sensor provides superior sensitivity with a low LOD (as low as 40 ng/mL), enhanced potential stability and longevity (maintaining performance for a minimum of 6 months), rapid response, and notable selectivity over various common related substances (Hassan et al., 2022).

In another study, the technique of combining CPs with MWCNT was utilized to design an electrochemical sensor for the determination of chlorpyrifos (CPF) in real solutions of vegetable samples. The primary objective was to synthesize a copolymer of 3-thiophene acetic acid and PEDOT on the MWCNT surface. Subsequently, a polymer drop-casted GCE sensor was constructed for the sensitive determination of CPF. Cyclic voltammetric studies with various concentrations of ICP@MWCNT showed that 0.02 g of ICP@MWCNT/GCE provided an optimal demonstration and was thus selected for beyond examples. Electroanalytical studies of the proposed sensor were investigated utilizing CPF solutions in a pH of 7.0 buffer at different concentrations ranging from 1000 to 0.02 nM. Feasibility studies of the developed sensor involved detecting CPF spiked in cucumbers utilizing DPV and HPLC. The conventional three-electrode technique was utilized, including a bare/modified GCE, a calomel reference electrode, and a platinum wire counter electrode. Characterization methods, for instance, FT-IR, FESEM, and TEM were employed to assess the prepared materials' surface topology. EIS was chosen to study the modified electrode's surface properties, using Nyquist diagrams measured in a solution comprising $K_3Fe(CN)_6$ and 0.1 M KCl. The electrochemical responses of both bare/modified electrodes were investigated through CV. DPV was employed as an extremely sensitive electrochemical technique for the detection of CPF using the ICP@MWCNT electrode. A typical DPV profile was obtained above a concentration range of 0.02–1000 nM, showing a sharp oxidation peak for CPF at −62.63 mV. The developed sensor's applicability was revealed by detecting CPF in real samples, and validation was performed using HPLC. DPV analysis of tap water and cucumber extract demonstrated efficient CPF analysis with good repeatability. The sensor's selectivity was characterized using CPF, organophosphate compounds, and reference compounds, demonstrating good selectivity of the fabricated sensor. The reproducibility of the fabricated sensor was confirmed by preparing five individual sensors, and the response current was obtained for a 1.0×10^{-7} M CPF solution, yielding a calculated RSD of 2.3%. Repeatability was assessed by evaluating the electrochemical response of 1.0×10^{-7} M CPF for five replicate measurements utilizing the same sensor, resulting in an RSD value of 2.1%. These findings highlight the reproducibility and repeatability of the developed sensor. Under optimal conditions, the sensor demonstrated an exceptionally low LOD of 4.0×10^{-12} M for CPF. The remarkable reusability of the materials suggests potential applications in pesticide monitoring in vegetable samples (Anirudhan et al., 2022). Other related studies about the applications of MIP in electrochemical sensors are tabulated in Table 6.2.

Table 6.2 Selected studies about the Applications of Molecularly Polymers in the electrochemical sensor.

Sensor	Monomer	Template	Method	Medium	Linear range	LOD	Application	Reference
MIP/PAM/PANI/AuNPs/GCE	Acrylamide	Resveratrol	DPV	PBS (pH 4.0)	1.0–200 μM	87 nM	Red wine	Huang et al. (2021)
GCE/RGO/MIP/PDPA	Methacrylic acid	Ceftriaxone	DPV	BRB (pH 2)	0.025–3.047 μM	0.008 μM	Urine, blood serum, powder	Salimonnafs et al. (2023)
MICPs/PPy/PyCOOH	Pyrrole, pyrrole-3-carboxylic acid	Penicillin G	EIS	PBS	12.5–100 ppb	15 ppb	Milk	Charlier et al. (2022)
[Poly (Py-coPyHis)]@MIP/GCE	Pyrrole-histidine	Teriflunomide	CV	ABS (pH 5.2)	0.1–1.0 pM	11.38 fM	Synthetic serum samples and tablet dosage form	Çorman et al. (2022)
MIP/MwCNT/CPE	Itaconic acid	Saxagliptin	DPV	BRB (pH 3)	1 to 1×10^{-6} nM	8×10^{-7} nM	Spiked human serum and urine samples	Elmalahany et al. (2023)
MIP-PPy/Pt electrode	Pyrrole	SARS-CoV-2-S spike glycoprotein.	DPV	PBS (pH 7.4)	0–25 μg/mL	NS	NS	Ratautaite, Boguzaite, et al. (2022)
$MI_{Ppy!}GE$	Pyrrole and aniline	L-Tryptophan	DPV	40 mM BRB (pH 2.5)	50 –100 μM	16.6 μM	NS	Ratautaite, Boguzaite, et al. (2022)
MIP/MWCNT/GCE	Pyrrole	Methotrexate	DPV	PBS (pH 3)	0.01–125 μM	2.7 nM	Environmental and pharmaceutical samples	Jara-Cornejo et al. (2023)
SPCE-M-4G SPCE-M-4S	Chitosan–PVA	p-Aminophenol		PBS (pH 6)	0–35 μM 0–50 μM	(2.1 ± 0.1) μM 6.0 ± 0.1 μM	Paracetamol drug samples	Mulyasuryani et al. (2023)
MIP@GC@TAA-DA@PtNP MIP@EPPG@TAA-DA@PtNP MIP@BPPG@TAA-DA@PtNP	Thiophene acetic acid	Dopamine	DPV	PBS (pH 7.0)	2–100 nM 2–100 nM 0.2–100 μM	42.50 nM 14.40 nM 0.671 mM	Human serum	Koçak and Akyüz (2023)
MIP-PAT/CFP	2-Aminothiazole	4-Hexylresorcinol	CV	PBS (pH 7)	30–300 nM	6.03 nM	Real shrimp samples	George et al. (2023)

(Continued)

Table 6.2 Selected studies about the Applications of Molecularly Polymers in the electrochemical sensor. *Continued*

Sensor	Monomer	Template	Method	Medium	Linear range	LOD	Application	Reference
SPE/(*f*-MWCNTs-PANI) /MIP-ISM	Acrylamide	Nalbuphine	POT	PBS (pH 6.5)	0.24–50000 μM	0.11 μM	Pharmaceuticals formulations and spiked human urine sample	Hassan et al. (2022)
AuNP@PMD/PGE	*o*-Phenylenediamine, dopamine, and methyldopa	Darifenacin	DPV	PBS (pH 7.0)	3–50 μM	0.94 μM	Dosage form and spiked human plasma.	Wadie et al. (2022)
MIP/MAA/DVB/GCE	Methacrylic acid 3-vinyl aniline	Bisphenol A	POT	2 mM PBS (pH 7.0)	0.5–20 μM	0.23 μM	Spiked lake and river water samples	Wang et al. (2022)
DTMIP -N@Fe-MOF@C/GCE	*o*-Phenylenediamine	Clenbuterol hydrochloride and ractopamine	CV	0.1 M NaOH	10^{-3} to 8 nM	3.03×10^{-4} nM	Clenbuterol hydrochloride tablets, human urine and raw pork	Li, Li, et al. (2021)
MICP@MWCNT/GCE	*p*-Phenylenediamine	Chlorpyrifos	DPV	pH 7.0 buffer solution	0.02–1000 nM	4.0×10^{-3} nM	Tap water and cucumber extract	Anirudhan et al. (2022)
E-MIP/Cu/Au-AT-SPE	4-Aminophenol	Dioctyl phthalate	CV	PBS		9 μM	Freshwater	El-Sharif et al. (2022)
TiO_2 NRs/NiOOH/Rmip	*O*-Phenylenediamine	Ciprofloxacin	EIS	PBS (pH 7.0)	0.1–1 mM	0.032 mM	Tap water and Lake water	Yan et al. (2023)
MB-MIP/GCE	Methylene blue	Melatonin	DPV	PBS (pH 7.0)	0.097–200 μM	29 nM	Authentic urine and saliva samples	Sunon and Ngamchuea (2023)
MIP/SPCE	*O*-Phenylenediamine	Cortisol	DPV	PBS (pH 5.2)	0.1–100 nM	0.036 nM	NS	Kim et al. (2023)
MIP/MAA/SPCE	Methacrylic acid	Caffeic acid	CV	0.05 M PBS (pH 3)	0–1.11 mM	0.13 mM	Wine sample	Elhachem et al. (2023)
MIP@ERGO/GCE	Pyrrole	Domperidone	DPV	PBS (pH 7.0)	0.5–17.2 μM	3.8 nM	Human urine sample	Kumar et al. (2021)
MIP/Ppy/GCE	Pyrrole	Polyaromatic hydrocarbons	CV	PBS (pH 7.0)	0.62–7.4 μM	0.228 μM	NS	Ngwanya et al. (2021)

MIP/$CsPbBr_3$-QDs/GCE	Silica layer	Prometryn	CV	PBS (pH 7.5)	0.1−500.0 μg/L	0.010 μg/kg 0.050 μg/L	Fish seawater samples	Zhang, Gan, et al. (2022)
MIP/sol-gelMWCNTs/GCE MWCNTs/NIPs	APTES	Enrofoxacin	CV	0.1 M PBS (pH 8)	2.8 pM to 28 mM	0.9 pM	Spiked natural seawater, fish, and shrimp samples.	Chen et al. (2022)
MIP-AuNPs@SiO_2	APTES	Glutathione	UV−Vis	PBS (pH 4.0)	5−40 mM	1.16 mM	Serum sample	Zhang, Peng, et al. (2022)
MIP/BPNS-AuNP/GCE	Pyrrole	Norfloxacin	LSV	0.1 M PBS (pH 6.0)	0.1 nM to 10 mM	0.012 nM	Pharmaceutical, milk	Li et al. (2022)
MIP-AF/SPE	Aniline	Tryptophan	EIS	PBS (pH 7.4)	10 pM−80 μM	8 pM	Milk and cancer cell media	Alam et al. (2022)
AuNPs/MIP/GCE	Pyrrole	Ethanethiol	DPV	BRB (pH 2)	6.1−32.4 mg/L	7.2 mg/L	Spiked wine samples	Alonso-Lomillo and Domínguez-Renedo (2023)
GCE-MIP	*O*-phenylenediamine	1-Naphthol	DPV	PBS (pH 6.5)	5.0−25 nM	15 nM	Oilfield produced water	Bartilotti et al. (2021)
CPE-Mnp-MIP	Acrylic acid	Tetracycline	SWV	0.1 M PBS (pH 7.0)	0.5−40 μM	0.15 μM	Market and Natural milk	Zeb et al. (2021)
MIP-2 film-coated	*p*-Bis(2,20 -bithien-5-yl)- methylbenzo-18-crown-6	Tyramine	DPV	0.1 M PBS (pH 7.4)	290 μM to 2.64 mM	159 μM	Mozzarella cheese whey samples	Ayerdurai et al. (2021)
MIP/Co_3O_4@MOF-74/cycle4/GCE	Pyrrole	Fenamiphos	DPV	PBS (pH 4.0)	0.01−1.0 nM	3.0 pM	Orange juice	Karimi-Maleh et al. (2021)
MIP/DA/SPCE	Dopamine	Diclofenac	DPV	0.1 M PBS (pH 7)	0.1−10 μM	70 nM	River and tap water	Seguro et al. (2021)
GCE/MWCNT/P-Cys@MIP	L-Cysteine	Ceftizoxime	DPV	0.01 M PBS (pH 4.5)	2×10^7 to 1×10^{-4} M	$\times 10^{-10}$M 0.1 nM	Blood and urine samples	Ali et al. (2021)
Fe_3O_4@MIP/RGO/GCE	Methacrylic acid	Luteolin	DPV	PBS (pH 7.0)	2.5 pM to 0.1 μM	1 pM	Lotus leaves sample	You et al. (2021)

(Continued)

Table 6.2 Selected studies about the Applications of Molecularly Polymers in the electrochemical sensor. ***Continued***

Sensor	Monomer	Template	Method	Medium	Linear range	LOD	Application	Reference
MIP-DBP-CTS/F-CC_3/GCE	Chitosan	Dibutyl phthalate	DPV	0.01 M PBS (pH 7.0)	0−1.8 μM	0.0026 μM	Rice wine sample	Zhou et al. (2021)
MIP/L-cysteine/AuNPs/GCE	*p*-Aminobenzene sulfonic acid	Histamine	DPV	PBS (pH 7.0)	1−107 μM	0.6 μM	Spiked beer and wine samples	Li, Zhong, et al. (2021)
Nafion-MWCNTs/GCE	Tetraethoxysilane Phenylmethoxysilane Methyl methoxy silane	Ephedrine hydrochloride	DPV	PBS (pH 7.0)	0.18−75 μM	72 nM	Spiked saliva samples	Jia et al. (2021)
MIP∣ERGO∣GCE	Pyrrole	Bisphenol A	DPV	0.1 M PBS (pH 7.0)	750−0.5 nM	0.2 nM	Tap water and milk samples	Karthika et al. (2021)
CeO_2NPs/RGO−Ru $(bpy)_3^{2+}$−MIP−GCE	Vinyl benzene Methacrylic acid	Trimipramine	ECL	0.1 M PBS (pH 7.4)	0.2−100 pM	0.025 pM	Human serum and urine samples	Khonsari and Sun (2021)

6.5 Conclusion

Electrochemical sensors have played a pivotal role in modern science and technology, revolutionizing the ability to detect and quantify analytes in various fields, from environmental monitoring to medical diagnostics. These sensors, with their ability to convert chemical information into electronic signals, have become indispensable tools. They offer rapid signal acquisition, miniaturization, and clean and cost-effective solutions for a wide range of applications by enabling reliable and sensitive approaches for detecting various pollutants and analytes. They operate on the principles of conductometry, amperometry, and potentiometry, easily measuring physical, chemical, and biological parameters. Techniques such as CV, DPV, LSV, and SWV have further enriched the electrochemical sensor toolbox, offering a better understanding of electron transfer mechanisms.

CPs have introduced a new dimension to electrochemical sensing with their remarkable electrical conductivity and compatibility with various substrates. Furthermore, the synergy between CPs and nanomaterials has expanded the horizons of electrochemical sensors. CPs such as PEDOT have been coupled with materials such as graphene and CNTs, enhancing sensor performance.

Another breakthrough in analytical science is the development of MIPs. MIPs have opened new frontiers in molecular recognition, allowing for the creation of synthetic materials with high specificity and stability. MIP-based sensors have been found to be useful in detecting a diverse array of analytes, from small molecules to biomacromolecules. In recent years, improvements in nanoscience and nanotechnology have expanded the possibilities of MIP sensors. Nanomaterials such as CQDs, metal or metal oxide NPs, and other nanocomposites have been integrated with MIPs to enhance sensor performance further. This chapter primarily focuses on electrochemical sensors, their various types and applications, and related topics such as CPs, molecular imprinting, and the use of nanomaterials in sensor technology.

In summary, from the fundamental principles of detection to cutting-edge innovations in electrochemical sensing and molecular imprinting, the sensor landscape continues to evolve, driving advancements in science, healthcare, industry, and our daily lives. The integration of CPs and MIPs has further enhanced their specificity and sensitivity, promising a future filled with innovative sensing solutions for complex analytical challenges.

Acknowledgment

As women in science, we express our gratitude to our great leader M. Kemal ATATÜRK, on the 100th anniversary of the Republic of Türkiye.

References

Al Shagri, L. M. S., Kamel, A. H., Abd-Rabboh, H. S. M., & Bajaber, M. A. (2022). Molecularly imprinted polymer modified with an MWCNT nanocomposite for the fabrication of a barbital solid-contact ion-selective electrode. *ACS Omega*, *7*(37), 32988–32995. Available from https://doi.org/10.1021/acsomega.2c02250, http://pubs.acs.org/journal/acsodf.

Alam, I., Lertanantawong, B., Sutthibutpong, T., Punnakitikashem, P., & Asanithi, P. (2022). Molecularly imprinted polymer-amyloid fibril-based electrochemical biosensor for ultrasensitive detection of tryptophan. *Biosensors*, *12*(5). Available from https://doi.org/10.3390/bios12050291, https://www.mdpi.com/2079-6374/12/5/291/pdf?version = 1651486400.

Ali, M. R., Bacchu, M. S., Al-Mamun, M. R., Rahman, M. M., Ahommed, M. S., Aly, M. A. S., & Khan, M. Z. H. (2021). Sensitive MWCNT/P-Cys@MIP sensor for selective electrochemical detection of ceftizoxime. *Journal of Materials Science*, *56*(22), 12803–12813. Available from https://doi.org/10.1007/s10853-021-06115-6, http://www.springer.com/journal/10853.

Al-Kadhi, N. S., Hefnawy, M. A., Alamro, F. S., Pashameah, R. A., Ahmed, H. A., & Medany, S. S. (2023). Polyaniline-supported nickel oxide flower for efficient nitrite electrochemical detection in water. *Polymers*, *15*(7). Available from https://doi.org/10.3390/polym15071804, http://www.mdpi.com/journal/polymers.

Al-Mhyawi, S. R., Ahmed, R. K., & Nashar, R. M. E. (2021). Application of a conducting poly-methionine/gold nanoparticles-modified sensor for the electrochemical detection of paroxetine. *Polymers*, *13*(22). Available from https://doi.org/10.3390/polym13223981, https://www.mdpi.com/2073-4360/13/22/3981/pdf.

Alonso-Lomillo, M. A., & Domínguez-Renedo, O. (2023). Molecularly imprinted polypyrrole based electrochemical sensor for selective determination of ethanethiol. *Talanta*, *253*. Available from https://doi.org/10.1016/j.talanta.2022.123936, https://www.journals.elsevier.com/talanta.

Anirudhan, T. S., Athira, V. S., & Nair, S. S. (2022). Detection of chlorpyrifos based on molecular imprinting with a conducting polythiophene copolymer loaded on multi-walled carbon nanotubes. *Food Chemistry*, *381*, 132010. Available from https://doi.org/10.1016/j.foodchem.2021.132010.

Arduini, F., Cinti, S., Scognamiglio, V., Moscone, D., & Palleschi, G. (2017). How cutting-edge technologies impact the design of electrochemical (bio)sensors for environmental analysis. A review. *Analytica Chimica Acta*, *959*, 15–42. Available from https://doi.org/10.1016/j.aca.2016.12.035, http://www.journals.elsevier.com/analytica-chimica-acta/.

Aristov, N., & Habekost, A. (2015). Cyclic voltammetry—A versatile electrochemical method investigating electron transfer processes. *World Journal of Chemical Education*, *3*, 115–119. Available from https://doi.org/10.12691/wjce-3-5-2.

Arshady, R., & Mosbach, K. (1981). Synthesis of substrate-selective polymers by host-guest polymerization. *Die Makromolekulare Chemie*, *182*(2), 687–692. Available from https://doi.org/10.1002/macp.1981.021820240.

Aydın, E. B., Aydın, M., & Sezgintürk, M. K. (2023). Novel electrochemical biosensing platform based on conductive multilayer for sensitive and selective detection of CYFRA 21-1. *Sensors and Actuators B: Chemical*, *378*. Available from https://doi.org/10.1016/j.snb.2022.133208, https://www.journals.elsevier.com/sensors-and-actuators-b-chemical.

Ayerdurai, V., Cieplak, M., Noworyta, K. R., Gajda, M., Ziminska, A., Sosnowska, M., Piechowska, J., Borowicz, P., Lisowski, W., Shao, S., D'Souza, F., & Kutner, W. (2021). Electrochemical sensor for selective tyramine determination, amplified by a molecularly imprinted polymer film. *Bioelectrochemistry*, *138*. Available from https://doi.org/10.1016/j.bioelechem.2020.107695, http://www.elsevier.com/locate/bioelechem.

Balciunas, D., Plausinaitis, D., Ratautaite, V., Ramanaviciene, A., & Ramanavicius, A. (2022). Towards electrochemical surface plasmon resonance sensor based on the molecularly imprinted polypyrrole for glyphosate sensing. *Talanta*, *241*, 123252. Available from https://doi.org/10.1016/j.talanta.2022.123252, 00399140.

Bartilotti, M., Beluomini, M. A., & Zanoni, M. V. B. (2021). Using an electrochemical MIP sensor for selective determination of 1-naphthol in oilfield produced water. *Electroanalysis*, *33*(5), 1346–1355. Available from https://doi.org/10.1002/elan.202060545, http://onlinelibrary.wiley.com/journal/10.1002/(ISSN)1521-4109.

Becker, T., Hitzmann, B., Muffler, K., Pörtner, R., Reardon, K. F., Stahl, F., & Ulber, R. (2006). Future aspects of bioprocess monitoring. *Advances in Biochemical Engineering/Biotechnology*, *105*, 249–293. Available from https://doi.org/10.1007/10_2006_036.

Bonora, P. L., Deflorian, F., & Fedrizzi, L. (1996). Electrochemical impedance spectroscopy as a tool for investigating underpaint corrosion. *Electrochimica Acta*, *41*(7–8), 1073–1082. Available from https://doi.org/10.1016/0013-4686(95)00440-8, http://www.journals.elsevier.com/electrochimica-acta/.

Borisova, B., Ramos, J., Díez, P., Sánchez, A., Parrado, C., Araque, E., Villalonga, R., Pingarrón, J. M., & Layer-by-Layer Biosensing, A. (2015). Architecture based on polyamidoamine dendrimer and carboxymethylcellulose-modified graphene oxide. *Electroanalysis*, *27*(9), 2131–2138. Available from https://doi.org/10.1002/elan.201500098, http://onlinelibrary.wiley.com/journal/10.1002/(ISSN)1521-4109.

Bosserdt, M., Gajovic-Eichelman, N., & Scheller, F. W. (2013). Modulation of direct electron transfer of cytochrome c by use of a molecularly imprinted thin film. *Analytical and Bioanalytical Chemistry*, *405*(20), 6437–6444. Available from https://doi.org/10.1007/s00216-013-7009-8.

Bozal-Palabiyik, B., Kurbanoglu, S., Erkmen, C., & Uslu, B. (2021). *Future prospects and concluding remarks for electroanalytical applications of quantum dots. Electroanalytical Applications of Quantum Dot-Based Biosensors* (pp. 427–450). Turkey: Elsevier, Turkey Elsevier. Available from https://www.sciencedirect.com/book/9780128216705, https://doi.org/10.1016/B978-0-12-821670-5.00008-7.

Calatayud-Macías, P., López-Iglesias, D., Sierra-Padilla, A., Cubillana-Aguilera, L., Palacios-Santander, J. M., & García-Guzmán, J. J. (2023). Bulk modification of sonogel–carbon with polyaniline: A suitable redox mediator for chlorophenols detection. *Chemosensors*, *11*(1). Available from https://doi.org/10.3390/chemosensors11010063, http://www.mdpi.com/journal/chemosensors.

Charlier, H., David, M., Lahem, D., & Debliquy, M. (2022). Electrochemical detection of Penicillin G using molecularly imprinted conductive Co-polymer sensor. *Applied Sciences (Switzerland)*, *12*(15). Available from https://doi.org/10.3390/app12157914, http://www.mdpi.com/journal/applsci/.

Chen, J., Tan, L., Qu, K., Cui, Z., & Wang, J. (2022). Novel electrochemical sensor modified with molecularly imprinted polymers for determination of enrofloxacin in marine environment. *Microchimica Acta*, *189*(3). Available from https://doi.org/10.1007/s00604-022-05205-9, http://www.springer/at/mca.

Ciucci, F. (2019). Modeling electrochemical impedance spectroscopy. *Current Opinion in Electrochemistry*, *13*, 132–139. Available from https://doi.org/10.1016/j.coelec.2018.12.003, http://www.journals.elsevier.com/current-opinion-in-electrochemistry.

Compton, R. G., & Banks, C. E. (2018). *Understanding voltammetry* (Third Edition, pp. 1–456). United Kingdom: World Scientific Publishing Co. Ltd. Available from https://www.worldscientific.com/worldscibooks/10.1142/q0155#t = aboutBook.

Cormack, P. A. G., & Elorza, A. Z. (2004). Molecularly imprinted polymers: Synthesis and characterisation. *Journal of Chromatography B: Analytical Technologies in the Biomedical and Life Sciences*, *804*(1), 173–182. Available from https://doi.org/10.1016/j.jchromb.2004.02.013.

Çorman, M. E., Cetinkaya, A., Armutcu, C., Bellur Atici, E., Uzun, L., & Ozkan, S. A. (2022). A sensitive and selective electrochemical sensor based on molecularly imprinted polymer for the assay of teriflunomide. *Talanta*, *249*. Available from https://doi.org/10.1016/j.talanta.2022.123689, https://www.journals.elsevier.com/talanta.

Dakshayini, B. S., Reddy, K. R., Mishra, A., Shetti, N. P., Malode, S. J., Basu, S., Naveen, S., & Raghu, A. V. (2019). Role of conducting polymer and metal oxide-based hybrids for applications in ampereometric sensors and biosensors. *Microchemical Journal*, *147*, 7–24. Available from https://doi.org/10.1016/j.microc.2019.02.061, http://www.elsevier.com/inca/publications/store/6/2/0/3/9/1.

Davis, G. (1985). Electrochemical techniques for the development of amperometric biosensors. *Biosensors*, *1*(2), 161–178. Available from https://doi.org/10.1016/0265-928X(85)80002-X.

Deller, A. E., Hryniewicz, B. M., Pesqueira, C., Horta, R. P., da Silva, B. J. G., Weheabby, S., Al-Hamry, A., Kanoun, O., & Vidotti, M. (2023). PEDOT: PSS/AuNPs-based composite as voltammetric sensor for the detection of Pirimicarb. *Polymers, 15*(3). Available from https://doi.org/10.3390/polym15030739, http://www.mdpi.com/journal/polymers.

Deniz, S. A., Goker, S., Toppare, L., & Soylemez, S. (2022). Fabrication of D-A-D type conducting polymer, carbon nanotubes and silica nanoparticle-based laccase biosensor for catechol detection. *New Journal of Chemistry, 46*(32), 15521–15529. Available from https://doi.org/10.1039/d2nj02147a, http://pubs.rsc.org/en/journals/journal/nj.

Dokur, E., Uruc, S., Gorduk, O., & Sahin, Y. (2022). Ultrasensitive electrochemical detection of carcinoembryonic antigen with a label-free immunosensor using gold nanoparticle-decorated poly(pyrrole-co-3,4-ethylenedioxythiophene). *ChemElectroChem, 9*(15). Available from https://doi.org/10.1002/celc.202200121, http://onlinelibrary.wiley.com/journal/10.1002/(ISSN)2196-0216.

Dutta, R. (2008). *Fundamentals of biochemical engineering*. Available from https://doi.org/10.1007/978-3-540-77901-8.

Elgrishi, N., Rountree, K. J., McCarthy, B. D., Rountree, E. S., Eisenhart, T. T., & Dempsey, J. L. (2018). A practical beginner's guide to cyclic voltammetry. *Journal of Chemical Education, 95*(2), 197–206. Available from http://pubs.acs.org/loi/jceda8, https://doi.org/10.1021/acs.jchemed.7b00361.

Elhachem, M., Bou-Maroun, E., Abboud, M., Cayot, P., & Maroun, R. G. (2023). Optimization of a molecularly imprinted polymer synthesis for a rapid detection of caffeic acid in wine. *Foods, 12*(8). Available from https://doi.org/10.3390/foods12081660, http://www.mdpi.com/journal/foods.

Elmalahany, N. S., Abdel-Tawab, M. A. H., Elwy, H. M., Fahmy, H. M., & Nashar, R. M. E. (2023). Design and application of molecularly imprinted electrochemical sensor for the new generation antidiabetic drug saxagliptin. *Electroanalysis, 35*(5). Available from https://doi.org/10.1002/elan.202200313, http://onlinelibrary.wiley.com/journal/10.1002/(ISSN)1521-4109.

El-Sharif, H. F., Patel, S., Ndunda, E. N., & Reddy, S. M. (2022). Electrochemical detection of dioctyl phthalate using molecularly imprinted polymer modified screen-printed electrodes. *Analytica Chimica Acta, 1196*, 339547. Available from https://doi.org/10.1016/j.aca.2022.339547.

Emir, G., Dilgin, Y., Ramanaviciene, A., & Ramanavicius, A. (2021). Amperometric nonenzymatic glucose biosensor based on graphite rod electrode modified by Ni-nanoparticle/polypyrrole composite. *Microchemical Journal, 161*, 105751. Available from https://doi.org/10.1016/j.microc.2020.105751.

Erkmen, C., Selcuk, O., Unal, D. N., Kurbanoglu, S., & Uslu, B. (2022). Layer-by-layer modification strategies for electrochemical detection of biomarkers. *Biosensors and Bioelectronics: X, 12*, 100270. Available from https://doi.org/10.1016/j.biosx.2022.100270.

Erkmen, C., Unal, D. N., Kurbanoglu, S., & Uslu, B. (2022). *Basics of Electrochemical Sensors Engineering Materials* (pp. 81–99). Turkey: Springer Science and Business Media Deutschland GmbH. Available from http://springer.com/series/4288, https://doi.org/10.1007/978-3-030-98021-4_5.

Eswaran, M., Rahimi, S., Pandit, S., Chokkiah, B., & Mijakovic, I. (2023). A flexible multifunctional electrode based on conducting PANI/Pd composite for non-enzymatic glucose sensor and direct alcohol fuel cell applications. *Fuel, 345*, 128182. Available from https://doi.org/10.1016/j.fuel.2023.128182.

Faisal, M., Alam, M. M., Ahmed, J., Asiri, A. M., Algethami, J. S., Alkorbi, A. S., Madkhali, O., Aljabri, M. D., Rahman, M. M., & Harraz, F. A. (2023). Electrochemical detection of nitrite (NO2) with PEDOT: PSS modified gold/PPy-C/carbon nitride nanocomposites by electrochemical approach. *Journal of Industrial and Engineering Chemistry, 121*, 519–528. Available from https://doi.org/10.1016/j.jiec.2023.02.007, http:http://www.sciencedirect.com/science/journal/1226086X.

Fontana-Escartín, A., El Hauadi, K., Lanzalaco, S., Pérez-Madrigal, M. M., Armelin, E., Turon, P., & Alemán, C. (2023). Preparation and characterization of functionalized surgical meshes for early detection of bacterial infections. *Biomaterials Science and Engineering*, *9*(2), 1104–1115. Available from https://doi.org/10.1021/acsbiomaterials.2c01319, http://pubs.acs.org/journal/abseba.

Frew, J. E., Allen, H., & Hill, O. (1987). Electrochemical biosensors. *Analytical Chemistry*, *59*(15), 933A–944A. Available from https://doi.org/10.1021/ac00142a001.

García-Guzmán, J. J., López-Iglesias, D., Cubillana-Aguilera, L., Bellido-Milla, D., Palacios-Santander, J. M., Marin, M., Grigorescu, S. D., Lete, C., & Lupu, S. (2021). Silver nanostructures - poly(3,4-ethylenedioxythiophene) sensing material prepared by sinusoidal voltage procedure for detection of antioxidants. *Electrochimica Acta*, *393*. Available from https://doi.org/10.1016/j.electacta.2021.139082, http://www.journals.elsevier.com/electrochimica-acta/.

George, A., Cherian, A. R., Benny, L., Varghese, A., & Hegde, G. (2023). Surface-engineering of carbon fibre paper electrode through molecular imprinting technique towards electrochemical sensing of food additive in shrimps. *Microchemical Journal*, *184*, 108155. Available from https://doi.org/10.1016/j.microc.2022.108155.

German, N., Popov, A., Ramanavicius, A., & Ramanaviciene, A. (2022). Development and practical application of glucose biosensor based on dendritic gold nanostructures modified by conducting polymers. *Biosensors*, *12*(8). Available from https://doi.org/10.3390/bios12080641, http://www.mdpi.com/journal/biosensors/.

Gómez Avila, J., Heredia, A. C., Crivello, M. E., & Garay, F. (2019). Theory of square-wave voltammetry for the analysis of an EC reaction mechanism complicated by the adsorption of the reagent. *Journal of Electroanalytical Chemistry*, *840*, 117–124. Available from https://doi.org/10.1016/j.jelechem.2019.03.065.

Goyal, R. N., & Bishnoi, S. (2012). Surface modification in electroanalysis: Past, present and future. *Indian Journal of Chemistry—Section A Inorganic, Physical, Theoretical and Analytical Chemistry*, *51*(1–2), 205–225. Available from http://nopr.niscair.res.in/bitstream/123456789/13357/1/IJCA%2051A%2801-02%29%20205-225.pdf.

Gupta, Y., Pandey, C. M., & Ghrera, A. S. (2023). Development of conducting cellulose paper for electrochemical sensing of procalcitonin. *Microchimica Acta*, *190*(1). Available from https://doi.org/10.1007/s00604-022-05596-9, https://www.springer.com/journal/604.

Hamid Kargari, S., Ahour, F., & Mahmoudian, M. (2023). An electrochemical sensor for the detection of arsenic using nanocomposite-modified electrode. *Nature Research*, *13*(1). Available from https://doi.org/10.1038/s41598-023-36103-6, https://www.nature.com/srep/.

Hammond, J. L., Formisano, N., Estrela, P., Carrara, S., & Tkac, J. (2016). Electrochemical biosensors and nanobiosensors. *Essays in Biochemistry*, *60*(1), 69–80. Available from https://doi.org/10.1042/EBC20150008, http://essays.biochemistry.org/content/ppebio/60/1/69.full.pdf.

Hassan, S. S. M., Kamel, A. H., & Fathy, M. A. (2022). A novel screen-printed potentiometric electrode with carbon nanotubes/polyaniline transducer and molecularly imprinted polymer for the determination of nalbuphine in pharmaceuticals and biological fluids. *Analytica Chimica Acta*, *1227*. Available from https://doi.org/10.1016/j.aca.2022.340239, http://www.journals.elsevier.com/analytica-chimica-acta/.

Hatami, E., Ashraf, N., & Arbab-Zavar, M. H. (2021). Construction of β-cyclodextrin-phosphomolybdate grafted polypyrrole composite: Application as a disposable electrochemical sensor for detection of propylparaben. *Microchemical Journal*, *168*, 106451. Available from https://doi.org/10.1016/j.microc.2021.106451.

Haupt, K., & Mosbach, K. (2000). Molecularly imprinted polymers and their use in biomimetic sensors. *Chemical Reviews*, *100*(7), 2495–2504. Available from https://doi.org/10.1021/cr990099w.

Heeger, A. J., Kivelson, S., Schrieffer, J. R., & Su, W. P. (1988). Solitons in conducting polymers. *Reviews of Modern Physics*, *60*(3), 781–850. Available from https://doi.org/10.1103/RevModPhys.60.781.

Huang, H., Deng, L., Xie, S., Li, J., You, X., Yue, R., & Xu, J. (2022). Sandwich-structured PEDOT:PSS/MXene-PdAu/PEDOT:PSS film for highly sensitive detection of shikonin in lithospermum erythrorhizon. *Analytica Chimica Acta*, *1221*, 340127. Available from https://doi.org/10.1016/j.aca.2022.340127.

Huang, S., Yang, J., Li, S., Qin, Y., Mo, Q., Chen, L., & Li, X. (2021). Highly sensitive molecular imprinted voltammetric sensor for resveratrol assay in wine via polyaniline/gold nanoparticles signal enhancement and polyacrylamide recognition. *Journal of Electroanalytical Chemistry*, *895*, 115455. Available from https://doi.org/10.1016/j.jelechem.2021.115455.

Hui, N., Wang, J., Wang, D., Wang, P., Luo, X., & Lv, S. (2022). An ultrasensitive biosensor for prostate specific antigen detection in complex serum based on functional signal amplifier and designed peptides with both antifouling and recognizing capabilities. *Biosensors and Bioelectronics*, *200*, 113921. Available from https://doi.org/10.1016/j.bios.2021.113921.

Isaev, V. A., Grishenkova, O. V., & Zaykov, Y. P. (2018). Theory of cyclic voltammetry for electrochemical nucleation and growth. *Journal of Solid State Electrochemistry*, *22*(9), 2775–2778. Available from https://doi.org/10.1007/s10008-018-3989-9, http://link.springer-ny.com/link/service/journals/10008/index.htm.

Islam, M. N., & Channon, R. B. (2019). *Electrochemical sensors. Bioengineering Innovative Solutions for Cancer* (pp. 47–71). United Kingdom: Elsevier. Available from http://www.sciencedirect.com/science/book/9780128138861.

Jara-Cornejo, E., Khan, S., Vega-Chacón, J., Wong, A., da Silva Neres, L. C., Picasso, G., & Sotomayor, M. D. P. T. (2023). Biomimetic material for quantification of methotrexate using sensor based on molecularly imprinted polypyrrole film and MWCNT/GCE. *Biomimetics*, *8*(1). Available from https://doi.org/10.3390/biomimetics8010077, http://www.mdpi.com/journal/biomimetics.

Jia, L., Mao, Y., Zhang, S., Li, H., Qian, M., Liu, D., & Qi, B. (2021). Electrochemical switch sensor toward ephedrine hydrochloride determination based on molecularly imprinted polymer/nafion-MWCNTs modified electrode. *Microchemical Journal.*, *164*, 105981. Available from https://doi.org/10.1016/j.microc.2021.105981.

Kant, T., Shrivas, K., Dewangan, K., Kumar, A., Jaiswal, N. K., Deb, M. K., & Pervez, S. (2022). Design and development of conductive nanomaterials for electrochemical sensors: a modern approach. *Materials Today Chemistry*, *24*, 100769. Available from https://doi.org/10.1016/j.mtchem.2021.100769.

Kapoor, A., & Rajput, J. K. (2021). *Advanced nanosensors for virus detection. Nanosensors for Smart Agriculture* (pp. 609–640). India: Elsevier. Available from https://www.sciencedirect.com/book/9780128245545.

Karimi-Maleh, H., Karimi, F., Alizadeh, M., & Sanati, A. L. (2020). Electrochemical sensors, a bright future in the fabrication of portable kits in analytical systems. *Chemical Record*, *20*(7), 682–692. Available from https://doi.org/10.1002/tcr.201900092, http://onlinelibrary.wiley.com/journal/10.1002/(ISSN)1528-0691.

Karimi-Maleh, H., Yola, M. L., Atar, N., Orooji, Y., Karimi, F., Senthil Kumar, P., Rouhi, J., & Baghayeri, M. (2021). A novel detection method for organophosphorus insecticide fenamiphos: Molecularly imprinted electrochemical sensor based on core-shell Co3O4@MOF-74 nanocomposite. *Journal of Colloid and Interface Science*, *592*, 174–185. Available from https://doi.org/10.1016/j.jcis.2021.02.066, http://www.elsevier.com/inca/publications/store/6/2/2/8/6/1/index.htt.

Karthika, P., Shanmuganathan, S., Viswanathan, S., & Delerue-Matos, C. (2021). Molecularly imprinted polymer-based electrochemical sensor for the determination of endocrine disruptor bisphenol-A in bovine milk. *Food Chemistry*, *363*, 130287. Available from https://doi.org/10.1016/j.foodchem.2021.130287.

Keerthana, P., George, A., Benny, L., & Varghese, A. (2023). Biomass derived carbon quantum dots embedded PEDOT/CFP electrode for the electrochemical detection of phloroglucinol. *Electrochimica Acta*, *448*, 142184. Available from https://doi.org/10.1016/j.electacta.2023.142184.

Khonsari, Y. N., & Sun, S. (2021). A novel MIP-ECL sensor based on RGO-CeO2NP/Ru(bpy)32 + -chitosan for ultratrace determination of trimipramine. *Journal of Materials Chemistry B*, *9*(2), 471–478. Available from https://doi.org/10.1039/d0tb01666g, http://pubs.rsc.org/en/journals/journal/tb.

Kim, M., Park, D., Park, J., & Park, J. (2023). Bio-inspired molecularly imprinted polymer electrochemical sensor for cortisol detection based on O-phenylenediamine optimization. *Biomimetics*, *8*(3), 282. Available from https://doi.org/10.3390/biomimetics8030282.

Kimmel, D. W., Leblanc, G., Meschievitz, M. E., & Cliffel, D. E. (2012). Electrochemical sensors and biosensors. *Analytical Chemistry*, *84*(2), 685–707. Available from https://doi.org/10.1021/ac202878q.

Kissinger, P. T. (2005). Biosensors—A perspective. *Biosensors and Bioelectronics*, *20*(12), 2512–2516. Available from https://doi.org/10.1016/j.bios.2004.10.004, http://www.elsevier.com/locate/bios.

Koçak, İ., & Akyüz, B. G. (2023). Dopamine electrochemical sensor based on molecularly imprinted polymer on carbon electrodes with platinum nanoparticles. *Electrocatalysis*, *14*(5), 763–775. Available from https://doi.org/10.1007/s12678-023-00833-y, https://www.springer.com/journal/12678.

Kumar, D. R., Dhakal, G., Nguyen, V. Q., & Shim, J. J. (2021). Molecularly imprinted hornlike polymer@-electrochemically reduced graphene oxide electrode for the highly selective determination of an antiemetic drug. *Analytica Chimica Acta*, *1141*, 71–82. Available from https://doi.org/10.1016/j.aca.2020.10.014, http://www.journals.elsevier.com/analytica-chimica-acta/.

Kummari, S., Kumar, V. S., Goud, K. Y., & Gobi, K. V. (2022). Nano-Au particle decorated poly-(3-amino-5-hydroxypyrazole) coated carbon paste electrode for in-vitro detection of valacyclovir. *Journal of Electroanalytical Chemistry*, *904*, 115859. Available from https://doi.org/10.1016/j.jelechem.2021.115859.

Kummari, S., Panicker, L., Gobi, K. V., Narayan, R., & Kotagiri, Y. G. (2023). Electrochemical strip sensor based on a silver nanoparticle-embedded conducting polymer for sensitive in vitro detection of an antiviral drug. *ACS Applied Nano Materials*, *6*(13), 12381–12392. Available from https://doi.org/10.1021/acsanm.3c02080, https://pubs.acs.org/journal/aanmf6.

Kurbanoglu, S., Cevher, Ş. C., Gurban, A.-M., Doni, S., & Soylemez. (2023). *Conjugated polymers in enzyme-based electrochemical biosensors. Advences in materials science research* (pp. 1–49). Nova Science Publishers.

Kurbanoglu, S., Cevher, S. C., Toppare, L., Cirpan, A., & Soylemez, S. (2022). Electrochemical biosensor based on three components random conjugated polymer with fullerene (C60. *Bioelectrochemistry*, *147*, 108219. Available from https://doi.org/10.1016/j.bioelechem.2022.108219.

Kurbanoglu, S., Yarman, A., Scheller, F. W., & Ozkan, S. A. (2019). *Molecularly imprinted polymer-based nanosensors for pharmaceutical analysis. New developments in nanosensors for pharmaceutical analysis* (pp. 231–271). Turkey: Elsevier. Available from http://www.sciencedirect.com/science/book/9780128161449.

Laborda, E., González, J., & Molina, A. (2014). Recent advances on the theory of pulse techniques: A mini review. *Electrochemistry Communications*, *43*, 25–30. Available from https://doi.org/10.1016/j.elecom.2014.03.004.

Lazanas, A. C., & Prodromidis, M. I. (2023). Electrochemical impedance spectroscopy—A tutorial. *ACS Measurement Science Au*, *3*(3), 162–193. Available from https://doi.org/10.1021/acsmeasuresciau.2c00070, https://pubs.acs.org/page/amachv/about.html.

Letheby, H. (1862). XXIX—On the production of a blue substance by the electrolysis of sulphate of aniline. *Journal of the Chemical Society*, *15*(0), 161–163. Available from https://doi.org/10.1039/JS8621500161.

Li, G., Qi, X., Wu, J., Xu, L., Wan, X., Liu, Y., Chen, Y., & Li, Q. (2022). Ultrasensitive, label-free voltammetric determination of norfloxacin based on molecularly imprinted polymers and Au nanoparticle-functionalized black phosphorus nanosheet nanocomposite. *Journal of Hazardous Materials*, *436*, 129107. Available from https://doi.org/10.1016/j.jhazmat.2022.129107.

Li, S., Zhong, T., Long, Q., Huang, C., Chen, L., Lu, D., Li, X., Zhang, Z., Shen, G., & Hou, X. (2021). A gold nanoparticles-based molecularly imprinted electrochemical sensor for histamine specific-recognition and determination. *Microchemical Journal*, *171*, 106844. Available from https://doi.org/10.1016/j.microc.2021.106844, 0026265X.

Li, X., Li, Y., Yu, P., Tong, Y., & Ye, B. C. (2021). A high sensitivity electrochemical sensor based on a dual-template molecularly imprinted polymer for simultaneous determination of clenbuterol hydrochloride and ractopamine. *Analyst*, *146*(20), 6323–6332. Available from https://doi.org/10.1039/d1an01413g, http://pubs.rsc.org/en/journals/journal/an.

Liaqat, F., Jamil, M., Tariq, I., Haider, A., & Khan, M. A. (2023). Comparative electroanalytical performance of poly o-toulidine and polyaniline towards hydrazine oxidation supplemented by DFT. *Journal of Applied Polymer Science*, *140*(31). Available from https://doi.org/10.1002/app.54112, http://onlinelibrary.wiley.com/journal/10.1002/(ISSN)1097-4628.

Lü, H., Wang, H., Yang, L., Zhou, Y., Xu, L., Hui, N., & Wang, D. (2022). A sensitive electrochemical sensor based on metal cobalt wrapped conducting polymer polypyrrole nanocone arrays for the assay of nitrite. *Microchimica Acta*, *189*(1). Available from https://doi.org/10.1007/s00604-021-05131-2, http://www.springer/at/mca.

Luhana, C., & Mashazi, P. (2023). Simultaneous detection of dopamine and paracetamol on electroreduced graphene oxide−cobalt phthalocyanine polymer nanocomposite electrode. *Electrocatalysis*, *14*(3), 406−417. Available from https://doi.org/10.1007/s12678-022-00806-7, https://www.springer.com/journal/12678.

Çetin V, M. Z., Guven, N., Apetrei, R. M., & Camurlu, P. (2023). Highly sensitive detection of glucose via glucose oxidase immobilization onto conducting polymer-coated composite polyacrylonitrile nanofibers. *Enzyme and Microbial Technology*, *164*. Available from https://doi.org/10.1016/j.enzmictec.2022.110178, http://www.elsevier.com/locate/enzmictec.

Machado, S. A. S., & Cincotto, Fernando Henrique (2021). *Electrochemical methods applied for bioanalysis: Differential pulse voltammetry and square wave voltammetry* (pp. 273−282). Springer Science and Business Media LLC.

Magar, H. S., Hassan, R. Y. A., & Mulchandani, A. (2021). Electrochemical impedance spectroscopy (Eis): Principles, construction, and biosensing applications. *Sensors*, *21*(19). Available from https://doi.org/10.3390/s21196578, https://www.mdpi.com/1424-8220/21/19/6609.

Majeed, S., Naqvi, S. T. R., ul Haq, M. N., & Ashiq, M. N. (2022). Electroanalytical techniques in biosciences: Conductometry, coulometry, voltammetry, and electrochemical sensors. *Analytical techniques in biosciences: from basics to applications* (pp. 157−178). Pakistan: Elsevier. Available from https://www.sciencedirect.com/book/9780128226544.

Malhotra, B. D., Singhal, R., Chaubey, A., Sharma, S. K., & Kumar, A. (2005). Recent trends in biosensors. *Current Applied Physics*, *5*(2), 92−97. Available from https://doi.org/10.1016/j.cap.2004.06.021, http://www.elsevier.com/.

Manna, S., Sharma, A., & Satpati, A. K. (2022). Electrochemical methods in understanding the redox processes of drugs and biomolecules and their sensing. *Current Opinion in Electrochemistry*, *32*, 100886. Available from https://doi.org/10.1016/j.coelec.2021.100886.

Menger, M., Yarman, A., Erdossy, J., Yildiz, H. B., Gyurcsányi, R. E., & Scheller, F. W. (2016). MIPs and aptamers for recognition of proteins in biomimetic sensing. *Biosensors*, *6*(3). Available from https://doi.org/10.3390/bios6030035, http://www.mdpi.com/2079-6374/6/3/35/pdf.

Mosbach, K., & Ramström, O. (1996). The emerging technique of molecular imprinting and its future impact on biotechnology. *Bio/Technology*, *14*(2), 163−170. Available from https://doi.org/10.1038/nbt0296-163.

Mulyasuryani, A., Prananto, Y. P., Fardiyah, Q., Widwiastuti, H., & Darjito, D. (2023). Application of chitosan-based molecularly imprinted polymer in development of electrochemical sensor for p-aminophenol determination. *Polymers*, *15*(8). Available from https://doi.org/10.3390/polym15081818, http://www.mdpi.com/journal/polymers.

Naveen, M. H., Gurudatt, N. G., & Shim, Y. B. (2017). Applications of conducting polymer composites to electrochemical sensors: A review. *Applied Materials Today*, *9*, 419−433. Available from https://doi.org/10.1016/j.apmt.2017.09.001, http://www.journals.elsevier.com/applied-materials-today/.

Nejad, F. G., Tajik, S, Beitollahi, H, & Sheikhshoaie, I (2021). Magnetic nanomaterials based electrochemical (bio) sensors for food analysis. *Talanta*, *228*, 122075.

Ngwanya, O. W., Ward, M., & Baker, P. G. L. (2021). Molecularly imprinted polypyrrole sensors for the detection of pyrene in aqueous solutions. *Electrocatalysis*, *12*(2), 165–175. Available from https://doi.org/10.1007/s12678-020-00638-3, http://www.springer.com/chemistry/electrochemistry/journal/12678.

O' Riordan, A., & Barry, S. (2016). Electrochemical nanosensors: advances and applications. *Reports in Electrochemistry*, 1. Available from https://doi.org/10.2147/RIE.S80550.

Ozcelikay, G., Kurbanoglu, S., Zhang, X., Soz., Wollenberger, U., Ozkan, S. A., & Scheller, F. W. (2019). Electrochemical MIP sensor for butyrylcholinesterase. *Polymers.*, *11*, 1970.

Panapimonlawat, T., Phanichphant, S., & Sriwichai, S. (2021). Electrochemical dopamine biosensor based on poly(3-aminobenzylamine) layer-by-layer self-assembled multilayer thin film. *Polymers*, *13*(9). Available from https://doi.org/10.3390/polym13091488, https://www.mdpi.com/2073-4360/13/9/1488/pdf.

Pandian, P., Kalimuthu, R., Arumugam, S., & Kannaiyan, P. (2021). Solid phase mechanochemical synthesis of Poly(o-anisidine) protected silver nanoparticles for electrochemical dopamine sensor. *Materials Today Communications*, *26*. Available from https://doi.org/10.1016/j.mtcomm.2021.102191, http://www.journals.elsevier.com/materials-today-communications/.

Paulsen, H. (1982). Advances in selective chemical syntheses of complex oligosaccharides. *Angewandte Chemie International Edition in English.*, *21*(3), 155–173. Available from https://doi.org/10.1002/anie.198201553.

Peng, H., Zhang, L., Soeller, C., & Travas-Sejdic, J. (2009). Conducting polymers for electrochemical DNA sensing. *Biomaterials*, *30*(11), 2132–2148. Available from https://doi.org/10.1016/j.biomaterials.2008.12.065.

Polyakov, M. V. (1931). Adsorption properties and structure of silica gel. *Zhur Fiz Khim*, *2*, 709–805.

Poudel, A., Shyam Sunder, G. S., Rohanifar, A., Adhikari, S., & Kirchhoff, J. R. (2022). Electrochemical determination of Pb2 + and Cd2 + with a poly(pyrrole-1-carboxylic acid) modified electrode. *Journal of Electroanalytical Chemistry*, *911*. Available from https://doi.org/10.1016/j.jelechem.2022.116221, https://www.journals.elsevier.com/journal-of-electroanalytical-chemistry.

Rahim, Z. A., Yusof, N. A., Ismail, S., Mohammad, F., Abdullah, J., Rahman, N. A., Abubakar, L., & Soleiman, A. A. (2023). Functional nano molecularly imprinted polymer for the detection of Penicillin G in pharmaceutical samples. *Journal of Polymer Research*, *30*(3). Available from https://doi.org/10.1007/s10965-023-03496-x, https://www.springer.com/journal/10965.

Rashed, M. A., Ahmed, J., Faisal, M., Alsareii, S. A., Jalalah, M., Tirth, V., & Harraz, F. A. (2022). Surface modification of CuO nanoparticles with conducting polythiophene as a non-enzymatic amperometric sensor for sensitive and selective determination of hydrogen peroxide. *Surfaces and Interfaces*, *31*. Available from https://doi.org/10.1016/j.surfin.2022.101998, http://www.journals.elsevier.com/surfaces-and-interfaces.

Ratautaite, V., Boguzaite, R., Brazys, E., Ramanaviciene, A., Ciplys, E., Juozapaitis, M., Slibinskas, R., Bechelany, M., & Ramanavicius, A. (2022). Molecularly imprinted polypyrrole based sensor for the detection of SARS-CoV-2 spike glycoprotein. *Electrochimica Acta*, *403*, 139581. Available from https://doi.org/10.1016/j.electacta.2021.139581.

Ratautaite, V., Brazys, E., Ramanaviciene, A., & Ramanavicius, A. (2022). Electrochemical sensors based on L-tryptophan molecularly imprinted polypyrrole and polyaniline. *Journal of Electroanalytical Chemistry*, *917*. Available from https://doi.org/10.1016/j.jelechem.2022.116389, https://www.journals.elsevier.com/journal-of-electroanalytical-chemistry.

Salhi, O., Ez-zine, T., Oularbi, L., & El Rhazi, M. (2022). Electrochemical sensing of nitrite ions using modified electrode by poly 1,8-diaminonaphthalene/functionalized multi-walled carbon nanotubes. *Frontiers in Chemistry*, *10*. Available from https://doi.org/10.3389/fchem.2022.870393, http://journal.frontiersin.org/journal/chemistry.

Salimonnafs, Y., MemarMaher, B., Amirkhani, L., & Derakhshanfard, F. (2023). Fabrication of a molecular imprinted composite and its application in the measurement of ceftriaxone in an electrochemical sensor. *International Journal of Polymeric Materials and Polymeric Biomaterials*, *72*(5), 366–375. Available from https://doi.org/10.1080/00914037.2021.2014485, http://www.tandf.co.uk/journals/titles/00914037.asp.

Scheller, F. W., Zhang, X., Yarman, A., Wollenberger, U., & Gyurcsányi, R. E. (2019). Molecularly imprinted polymer-based electrochemical sensors for biopolymers. *Current Opinion in Electrochemistry*, *14*, 53–59. Available from https://doi.org/10.1016/j.coelec.2018.12.005, http://www.journals.elsevier.com/current-opinion-in-electrochemistry.

Schubert, F., & Scheller, F. (1992). Biosensors. *Acta Horticulturae*, *304*, 71–78. Available from https://doi.org/10.17660/ActaHortic.1992.304.7.

Seguro, I., Pacheco, J. G., & Delerue-Matos, C. (2021). Low cost, easy to prepare and disposable electrochemical molecularly imprinted sensor for diclofenac detection. *Sensors*, *21*(6), 1–11. Available from https://doi.org/10.3390/s21061975, https://www.mdpi.com/1424-8220/21/6/1975/pdf.

Şenocak, A., Tümay, S. O., Ömeroğlu, İ., & Şanko, V. (2022). Crosslinker polycarbazole supported magnetite MOF@CNT hybrid material for synergetic and selective voltammetric determination of adenine and guanine. *Journal of Electroanalytical Chemistry*, *905*. Available from https://doi.org/10.1016/j.jelechem.2021.115963, https://www.journals.elsevier.com/journal-of-electroanalytical-chemistry.

Shekher, K., Sampath, K., Vandini, Shiv, Satyanarayana, M., & Gobi, K. V. (2023). Gold nanoparticle assimilated polymer layer on carbon nanotube matrices for sensitive detection of serotonin in presence of dopamine in-vitro. *Inorganica Chimica Acta*, *549*, 121399. Available from https://doi.org/10.1016/j.ica.2023.121399.

Shirakawa, H., Louis, E. J., MacDiarmid, A. G., Chiang, C. K., & Heeger, A. J. (1977). Synthesis of electrically conducting organic polymers: Halogen derivatives of polyacetylene, (CH)x. *Journal of the Chemical Society, Chemical Communications*, *16*, 578–580. Available from https://doi.org/10.1039/C39770000578.

Shrivastava, S., Jadon, N., & Jain, R. (2016). Next-generation polymer nanocomposite-based electrochemical sensors and biosensors: A review. *TrAC - Trends in Analytical Chemistry*, *82*, 55–67. Available from https://doi.org/10.1016/j.trac.2016.04.005, http://www.elsevier.com/locate/trac.

Sierra-Padilla, A., López-Iglesias, D., Calatayud-Macías, P., García-Guzmán, J. J., Palacios-Santander, J. M., & Cubillana-Aguilera, L. (2023). Incorporation of carbon black into a sonogel matrix: Improving antifouling properties of a conducting polymer ceramic nanocomposite. *Microchimica Acta*, *190*(5). Available from https://doi.org/10.1007/s00604-023-05740-z, https://www.springer.com/journal/604.

Soylemez, S., Erkmen, C., Kurbanoglu, S., Toppare, L., & Uslu, B. (2021). *Fabrication of quantum dot-polymer composites and their electroanalytical applications. Electroanalytical applications of quantum dot-based biosensors* (pp. 271–306). Turkey: Elsevier. Available from https://www.sciencedirect.com/book/9780128216705.

Stefan, R.-I. (2001). *Electrochemical sensors in bioanalysis*. CRC Press.

Sudhakara, S. M., Devendrachari, M. C., Khan, F., Thippeshappa, S., & Kotresh, H. M. N. (2023). Highly sensitive and selective detection of nitrite by polyaniline linked tetra amino cobalt (II) phthalocyanine surface functionalized ZnO hybrid electrocatalyst. *Surfaces and Interfaces*, *36*. Available from https://doi.org/10.1016/j.surfin.2022.102565, http://www.journals.elsevier.com/surfaces-and-interfaces.

Sun, S., Chen, J., & Wong, W., (2023). Based Electrical Sensors for Aqueous Heavy Metal Ion Detections.

Sunon, P., & Ngamchuea, K. (2023). Methylene blue molecularly imprinted polymer for melatonin determination in urine and saliva samples. *Microchimica Acta*, *190*(9). Available from https://doi.org/10.1007/s00604-023-05930-9, https://www.springer.com/journal/604.

Švorc, J., Miertuš, J., Stred'ansky, M., & Stred'anský, M. (1997). Composite transducers for amperometric biosensors. The glucose sensor. *Analytical Chemistry*, *69*(12), 2086–2090. Available from https://doi.org/10.1021/ac9609485, http://pubs.acs.org/journal/ancham.

Vasapollo, G., Sole, R. D., Mergola, L., Lazzoi, M. R., Scardino, A., Scorrano, S., & Mele, G. (2011). Molecularly imprinted polymers: Present and future prospective. *International Journal of Molecular Sciences*, *12*(9), 5908–5945. Available from https://doi.org/10.3390/ijms12095908Italy, http://www.mdpi.com/1422-0067/12/9/5908/pdf.

Vinothkumar, V., Kesavan, G., & Chen, S.-M. (2021). Graphitic carbon nitride nanosheets incorporated with polypyrrole nanocomposite: A sensitive metal-free electrocatalyst for determination of antibiotic drug nitrofurantoin. *Colloids and Surfaces A: Physicochemical and Engineering Aspects*, *629*, 127433. Available from https://doi.org/10.1016/j.colsurfa.2021.127433.

Wadie, M., Abdel-Moety, E. M., Rezk, M. R., Mahmoud, A. M., & Marzouk, H. M. (2022). Electropolymerized poly-methyldopa as a novel synthetic mussel-inspired molecularly imprinted polymeric sensor for darifenacin: Computational and experimental study. *Applied Materials Today*, *29*. Available from https://doi.org/10.1016/j.apmt.2022.101595, http://www.journals.elsevier.com/applied-materials-today/.

Wang, C., Qi, L., Liang, R., & Qin, W. (2022). Multifunctional molecularly imprinted receptor-based polymeric membrane potentiometric sensor for sensitive detection of bisphenol A. *Analytical Chemistry*, *94* (22), 7795–7803. Available from https://doi.org/10.1021/acs.analchem.1c05444.

Wang, Y, Li, J, & Viehland, D (2014). Magnetoelectrics for magnetic sensor applications: status, challenges and perspectives. *Materials Today*, *17*(6), 269–275.

Wang, G., Morrin, A., Li, M., Liu, N., & Luo, X. (2018). Nanomaterial-doped conducting polymers for electrochemical sensors and biosensors. *Journal of Materials Chemistry B.*, *6*(25), 4173–4190. Available from https://doi.org/10.1039/c8tb00817e, http://pubs.rsc.org/en/journals/journal/tb.

Wang, J. (1991). Modified electrodes for electrochemical sensors. *Electroanalysis*, *3*(4–5), 255–259. Available from https://doi.org/10.1002/elan.1140030404, 15214109.

Wang, J., Zhang, D., Xu, K., Hui, N., & Wang, D. (2022). Electrochemical assay of acetamiprid in vegetables based on nitrogen-doped graphene/polypyrrole nanocomposites. *Microchimica Acta*, *189*(10). Available from https://doi.org/10.1007/s00604-022-05490-4, http://www.springer/at/mca.

Wang, Y., Xu, H., Zhang, J., & Li, G. (2008). Electrochemical sensors for clinic analysis. *Sensors*, *8*(4), 2043–2081. Available from https://doi.org/10.3390/s8042043China, http://www.mdpi.org/sensors/papers/s8042043.pdf.

Weng, S., Chen, M., Zhao, C., Liu, A., Lin, L., Liu, Q., Lin, J., & Lin, X. (2013). Label-free electrochemical immunosensor based on K3[Fe(CN) 6] as signal for facile and sensitive determination of tumor necrosis factor-alpha. *Sensors and Actuators, B: Chemical.*, *184*, 1–7. Available from https://doi.org/10.1016/j.snb.2013.03.141.

Wulff, G. (1995). Molecular imprinting in cross-linked materials with the aid of molecular templates—A way towards artificial antibodies. *Angewandte Chemie International Edition in English*, *34*(17), 1812–1832. Available from https://doi.org/10.1002/anie.199518121.

Wulff, G., & Sarhan, A. (1972). Macromelecular colloquium. *Angewandte Chemie International Edition in English*, *11*, 344.

Xueji, Z., Huangxian, J., Joseph, W., Zhang, X., Ju, H., & Wang, J. (2011). Electrochemical sensors, biosensors and their biomedical applications. *Biosensors*.

Yan, G., Han, Z., Hou, X., Yi, S., Zhang, Z., Zhou, Y., & Zhang, L. (2023). A highly sensitive TiO_2-based molecularly imprinted photoelectrochemical sensor with regulation of imprinted sites by Photo-deposition. *Journal of Colloid and Interface Science*, *650*, 1319–1326. Available from https://doi.org/10.1016/j.jcis.2023.07.105, http://www.elsevier.com/inca/publications/store/6/2/2/8/6/1/index.htt.

Yano, K., & Karube, I. (1999). Molecularly imprinted polymers for biosensor applications. *TrAC - Trends in Analytical Chemistry*, *18*(3), 199–204. Available from https://doi.org/10.1016/S0165-9936(98)00119-8.

Yarman, A. (2018). Development of a molecularly imprinted polymer-based electrochemical sensor for tyrosinase. *Turkish Journal of Chemistry*, *42*(2), 346–354. Available from https://doi.org/10.3906/kim-1708-68, http://journals.tubitak.gov.tr/chem/issues/kim-18-42-2/kim-42-2-12-1708-68.pdf.

Yarman, A., Kurbanoglu, S., Erkmen, C., Uslu, B., & Scheller, F. W. (2021). *Quantum dot-based electrochemical molecularly imprinted polymer sensors: Potentials and challenges. Electroanalytical Applications of Quantum Dot-Based Biosensors* (pp. 121–153). Germany: Elsevier. Available from https://www.sciencedirect.com/book/9780128216705.

Yarman, A., & Kurbanoglu, S. (2022). Molecularly imprinted polymer-based sensors for SARS-CoV-2: Where are we now? *Biomimetics*, *7*(2). Available from https://doi.org/10.3390/biomimetics7020058, https://www.mdpi.com/2313-7673/7/2/58/pdf?version = 1651820372.

Yarman, A., Kurbanoglu, S., Zebger, I., & Scheller, F. W. (2021). Simple and robust: The claims of protein sensing by molecularly imprinted polymers. *Sensors and Actuators B: Chemical*, *330*, 129369. Available from https://doi.org/10.1016/j.snb.2020.129369.

Yarman, A., & Scheller, F. W. (2020). How reliable is the electrochemical readout of MIP sensors? *Sensors (Switzerland)*, *20*(9). Available from https://doi.org/10.3390/s20092677, https://www.mdpi.com/1424-8220/20/9/2677/pdf.

You, Z., Fu, Y., Xiao, A., Liu, L., & Huang, S. (2021). Magnetic molecularly imprinting polymers and reduced graphene oxide modified electrochemical sensor for the selective and sensitive determination of luteolin in natural extract. *Arabian Journal of Chemistry*, *14*(3), 102990. Available from https://doi.org/10.1016/j.arabjc.2021.102990.

Zare, H. R., & Shekari, Z. (2019). Types of monitoring biosensor signals. *Electrochemical biosensors* (pp. 135–166). Iran: Elsevier. Available from http://www.sciencedirect.com/science/book/9780128164914.

Zeb, S., Wong, A., Khan, S., Hussain, S., & Sotomayor, M. D. P. T. (2021). Using magnetic nanoparticles/MIP-based electrochemical sensor for quantification of tetracycline in milk samples. *Journal of Electroanalytical Chemistry*, *900*. Available from https://doi.org/10.1016/j.jelechem.2021.115713, https://www.journals.elsevier.com/journal-of-electroanalytical-chemistry.

Zhang, D., Lang, X., Hui, N., & Wang, J. (2022). Dual-mode electrochemical biosensors based on chondroitin sulfate functionalized polypyrrole nanowires for ultrafast and ultratrace detection of acetamiprid pesticide. *Microchemical Journal*, *179*, 107530. Available from https://doi.org/10.1016/j.microc.2022.107530.

Zhang, R. R., Gan, X. T., Xu, J. J., Pan, Q. F., Liu, H., Sun, A. L., Shi, X. Z., & Zhang, Z. M. (2022). Ultrasensitive electrochemiluminescence sensor based on perovskite quantum dots coated with molecularly imprinted polymer for prometryn determination. *Food Chemistry*, *370*. Available from https://doi.org/10.1016/j.foodchem.2021.131353, http://www.elsevier.com/locate/foodchem.

Zhang, X., Peng, J., Xi, L., Lu, Z., Yu, L., Liu, M., Huo, D., & He, H. (2022). Molecularly imprinted polymers enhanced peroxidase-like activity of AuNPs for determination of glutathione. *Microchimica Acta*, *189*(12). Available from https://doi.org/10.1007/s00604-022-05576-z, https://www.springer.com/journal/604.

Zhao, C., Wang, A., Tang, X., & Qin, J. (2022). Electrochemical sensitive detection of amyloid-β oligomer harnessing cellular prion protein on AuNPs embedded poly (pyrrole-3-carboxylic acid) matrix. *Materials Today Advances*, *14*, 100250. Available from https://doi.org/10.1016/j.mtadv.2022.100250.

Zhou, Q., Guo, M., Wu, S., Fornara, D., Sarkar, B., Sun, L., & Wang, H. (2021). Electrochemical sensor based on corncob biochar layer supported chitosan-MIPs for determination of dibutyl phthalate (DBP. *Journal of Electroanalytical Chemistry*, *897*, 115549. Available from https://doi.org/10.1016/j.jelechem.2021.115549.

Zhou, S., Guo, L., Shi, X., Ma, L., Yang, H., & Miao, M. (2023). In situ synthesized eRAFT polymers for highly sensitive electrochemical determination of AFB1 in foods and herbs. *Food Chemistry*, *421*, 136176. Available from https://doi.org/10.1016/j.foodchem.2023.136176.

Zhou, Y., Lü, H., Zhang, D., Xu, K., Hui, N., & Wang, J. (2023). Electrochemical biosensors based on conducting polymer composite and PAMAM dendrimer for the ultrasensitive detection of acetamiprid in vegetables. *Microchemical Journal*, *185*, 108284. Available from https://doi.org/10.1016/j.microc.2022.108284.

Zuhri, V. (1998). Abt Phys Biochem des Physiol. *Fresenius' Journal of Analytical Chemistry*.

CHAPTER

Biosensors

7

Jesus L. Pablos[1], Miguel Manzano[1,2] and María Vallet-Regí[1,2]

[1]*Departamento de Química en Ciencias Farmacéuticas, Instituto de Investigación Sanitaria Hospital 12 de Octubre i + 12, Universidad Complutense de Madrid, Plaza Ramón y Cajal, Madrid, Spain* [2]*Networking Research Center on Bioengineering, Biomaterials and Nanomedicine (CIBER-BBN), Madrid, Spain*

7.1 Introduction

Talking about the history of biosensors, the field of research focused on the development of these systems had its beginnings more than 50 years ago. The first biosensor was described by Clark in 1956, which in 1975 became the first commercial launch of a glucose analyzer when Clark and Lyons elucidated the basic concept of a biosensor with the description of the first enzyme-electrode (Clark & Lyons, 1962). The invention of this "Clark oxygen electrode" allowed us to reason that the electrochemical detection of oxygen or hydrogen peroxide could be applied as a solid base in a wide range of bioanalytical instruments by integrating the right enzyme for the desired application.

Since then, biosensors have quickly gained a great deal of interest in research over the last few decades because of the great focus on the progress of all kinds of analytical devices for the detection and quantification of several specific chemical species of interest. Some important examples include metabolites such as glucose, urea, or cholesterol, species of major importance in clinical diagnosis. For this, it must be highlighted that biosensors are a strong new trend that is emerging in the field of diagnostic therapy. This translates into an increasing number of published papers reporting on the development of a biosensor. As for the industry, although the start was slower, nowadays it represents a market of several billion US dollars (Haleem et al., 2021; Turner, 2013).

In relation to the applications of biosensors, these are multiple and versatile. They range from applications related to food safety and environment monitoring to civil defense, although the main field is the one related to medical diagnosis. It is very helpful to be able to follow up on body chemistry continually. In this way, as described above, it should be highlighted the glucose biosensor, designed to detect diabetes early, which is the first commercial and the most well-known biosensor. From this point, numerous biosensors have been developed in recent years depending on the analyte (enzymes, proteins, glucose level) or diseases (such as cancer, diabetes, and cardiac disease) to be detected (Haleem et al., 2021).

One of the most important elements in biosensor development is the biological (biomolecules such as enzymes or proteins) together with the non-biological (polymers or carbon-based nanomaterials together with their composites) material immobilization on the electrode surface or specifically on the transducer layer, with improved properties of biocompatibility, low toxicity, and high

Sensory Polymers. DOI: https://doi.org/10.1016/B978-0-443-13394-7.00007-0

absorptive capacity (Wang & Uchiyam, 2013). This is a key aspect since the resulting architecture of materials from this immobilization will determine the performance of any biosensor and will greatly affect its sensitivity and stability. In this sense, although biomarkers and enzymes are the primary elements used in biosensors to diagnose, it must be considered that biological receptors have certain limitations (high price and instability), so in biosensor design, it is also a key aspect to consider how tailor-made chemical receptors (polymers) that are biocompatible and have improved sensitivity can be used in these devices to extend their bioanalytical application. In this sense, polymeric materials have high variability, and their relatively inexpensive nature, together with the advantage that they can be synthesized and processed on a large scale, makes them very attractive for their application in the field of biosensors.

Therefore, this chapter aims to provide a general overview of the concepts of biosensors. This will be done in several parts. First, it is very important to define the concept of biosensor, its main characteristics, and its usual composition, as well as the main features it must have, to understand the importance of this technology and its evolution. Subsequently, we will discuss the classification of biosensors from the point of view of the nature of bioreceptors and biorecognition principles (enzymatic biosensors, immunosensors, aptamer/nucleic acid biosensors, and microbial/whole-cell biosensors) and the transduction method (electrochemical, electronic, optical, and mass-based/gravimetric).

Next, we will describe the role of incorporating polymers in biosensors to increase selectivity and biocompatibility and, in general, improve the overall performance of these devices. In this sense, this chapter will be focused on one side, on the description of how polymeric materials can be used in the electrode fabrication as a transductor layer to achieve a simple and fast electron transferring process between the analyte and the electrode or transducer surface. For this, first, it is important to describe the use of membrane polymers as coatings to improve biocompatibility and performance in vivo of biosensors; on the other hand, conductive polymers are the most used as transducer materials due to their optimal chemical, electrical, and physical properties. Moreover, close attention should be paid to molecularly imprinted polymers (MIPs), where the target is replicated in 3D negative form into the polymer and can specifically identify target molecules with excellent biocompatibility with application in vivo and in vitro biosensing and a great potential to replace biological molecules as active components of biosensors. Polymeric chains formed from monomers are self-assembled around the target template through polymerization. After the removal of the target template, the 3D imprint polymer remains, which will be the recognition element of the biosensor.

Finally, this chapter will describe the main applications of biosensors in several fields, such as healthcare, environmental applications, drug development, food security, and agriculture. Furthermore, this chapter will discuss biosensors' present and future.

7.2 What is a biosensor?

The first question that needs to be raised is: What is a biosensor? Overall, a sensor can be defined as a device that helps in detecting the presence of an analyte or changes in physical amounts, such as pressure changes, heat, humidity, movement, and an electrical signal like current, and therefore converts these changes and signals into parameters that can be detected and analyzed (Ensafi, 2019). The appearance of biosensors has been made possible thanks to the growing interest in the development of analytical devices for the detection, quantification, and monitoring of specific chemical and biological analytes, so that biosensors allow non-invasive detection when the physical conditions do not allow extraction of blood, serum, or fluid. Biosensors can be presented in several different sizes and shapes and will be able to detect and measure even low concentrations of specific analytes, such as pathogens, toxic chemicals, or pH variations.

Therefore, when we talk about biosensors, in general terms, they will be defined as one analytical device whose function is to measure biological processes, biological signals, and the concentration of biological molecules or other molecules of interest, and then they can turn these changes into an electrical signal. In other words, it can transform a biochemical interaction at the biosensor surface into a quantifiable physical (optical or electric) response. In relation to these biological processes and signals, these can be any biological element or synthetic material, such as enzymes, tissues, microorganisms, cells, and acids.

Regarding the main components of a biosensor, the following parts intimately associated can be highlighted: analyte, bioreceptor, transducer, electronics, and display reader, as can be observed in Fig. 7.1, so that these devices will work with different forms of this (Wu et al., 2022).

Concerning the analyte, it can be defined as the substance of interest whose constituents will be identified or detected (e.g., glucose, ammonia, alcohol, and lactose).

Next, we can observe the bio-receptor layer, which is built of recognition material and specifically recognizes the desired analyte to produce a measurable signal. These receptor species are usually of biological origin (enzymes, DNA, and antibodies) and have high selectivity through the binding event between the receptor and the analyte. However, it should be noted that artificial receptors themselves can also be used. The process by which the signal is produced during the

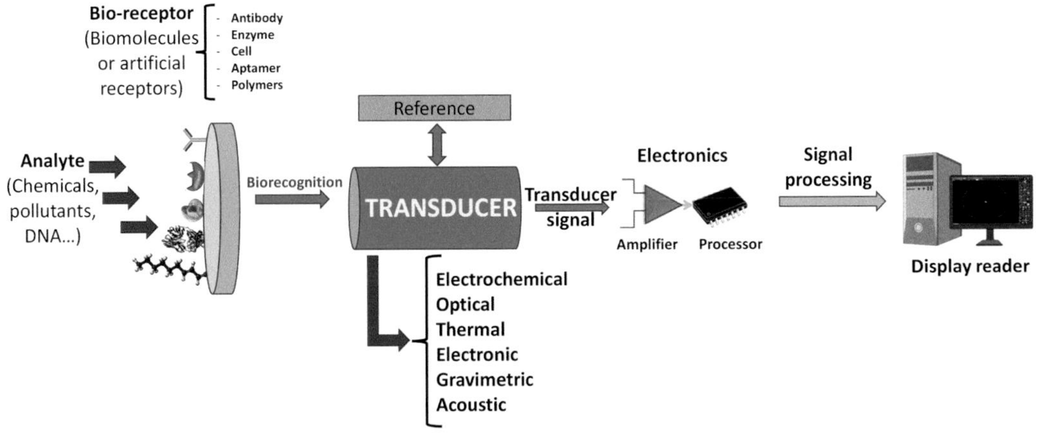

FIGURE 7.1

General scheme of a typical biosensor consisting of an analyte to be detected, the bioreceptor, a transducer together with the corresponding reference, an electronic system (composed of an amplifier and processor), and a display reader.

interaction process between the analyte and bioreceptor is called biorecognition. In the biorecognition elements, we can include antibodies, enzymes, cells, aptamers, and synthetic recognition bioelements such as conducting polymers (CPs) or MIPs (Purohit et al., 2020).

A transducer must be present, which logs the data and converts the chemical/biochemical signal of the binding event due to the interaction analyte/receptor into a measurable physical change to be quantified as output data. The process that implies this energy conversion is known as signalization. Transducers will be able to generate optical or electrical signals, which should be proportional to analyte–bioreceptor interactions. In relation to the signal generated, transducers are categorized as electrochemical, optical, thermal, or electronic, depending on the type of generated species, such as electrons, protons, electrochemically active species, or changes in conductivity, optical absorbance, or fluorescence.

Finally, a method of measuring the changes detected by the transducer should be included to turn these changes into useful information for the display reader. These devices are connected to the corresponding electronics and signal processing system to obtain the results in the form of numerical, graphical, or spreadsheet data values. Usually, the displays are customized and designed a la carte according to the application and working principles of biosensors, and therefore this part is usually the most expensive in the development of the final device.

In addition to the above, the use of different biocompatible polymers in the manufacture of these devices should be highlighted, although, in this chapter, we will only describe some of the most common polymers used in biosensor design, due to their excellent properties such as insulating properties or dielectric properties (Namsheer & Rout, 2021).

7.3 Characteristics of biosensors

In the process of developing an ideal, highly effective, and fully functional biosensor, there are certain features that every good biosensor should possess. If we meet these criteria, it will be possible to optimize these devices for commercial use (Turner, 2013).

7.3.1 Selectivity and sensitivity

Two crucial properties when selecting a bioreceptor or biosensor are selectivity and sensitivity. Regarding selectivity, an adequate bioreceptor must detect the target analyte in samples with mixtures of different species, undesired contaminants, or interferent molecules. Biosensors that generate false positives will have very limited application. The fact that a bioreceptor can generate a response only in the presence of a target molecule because of their interaction is critical for the development of reliable biosensors. Examples of these, which will be described throughout this chapter, are antibodies; these species are composed of arms with a "Y" conformational shape, which provides selectivity and high specificity (Gilles-Gonzalez & Gonzalez, 2005). Enzymes are another example; these have very specific biorecognition systems in the form of selective hydrogen bonding, which provides high specificity (Golub et al., 2015).

In relation to synthetic biosensor elements, it is worth highlighting the MIPs, which show ease of development, low cost, and high chemical stability in contrast with their lower selectivity (Wackerlig & Lieberzeit, 2015; Wackerlig & Schirhagl, 2015).

On the other hand, sensitivity can be defined as the minimum amount of target analyte that can be detected to confirm the presence of the interest analyte, even in trace form, in the sample. High-sensitivity biosensors will generate an adequate response where the desired ranges are on the scale from picomolar to nanomolar (Madsen et al., 2010).

7.3.2 Response time and reproducibility

The time necessary to obtain results is called the response time. In relation to reproducibility, this parameter is defined as the capacity to obtain similar results by measuring the sample precisely and accurately with replies.

7.3.3 Stability

This is a key parameter in biosensor applications where ongoing monitoring is needed so that these remain stable in the face of environmental disturbances around biosensing devices. Among the factors that affect the stability of biosensors, it is necessary to highlight the affinity of the bioreceptor and the degradation of this over time.

In general, when we talk about a commercial application of biosensors, they must meet a series of requirements, such as accuracy and reproducibility, good sensitivity and resolution, adequate speed of response, stability in temperature changes, and electrical or environmental interferences. Furthermore, these devices must be physically robust, reliable, and have self-control capacity.

7.4 The importance of understanding biosensor technology and its evolution

In the search for a better insight into the concept and role of biosensors, we can look for similar systems and compare them with each other. What similar systems could we find? The human body and its sensory organs. The human body undergoes multiple changes and reactions to external stimuli, such as the body's response to temperature changes or the presence of an unpleasant smell. All this is because of the receptors and their response to environmental stimuli from interaction with the body/surroundings. Our body is filled with numerous receptors that can detect many very specific stimuli, similar to the sensory activity of a biosensor. It can be described as an artificial receptor for a very specific molecule or stimuli, in which a biological recognition element that can be found in different organisms is used to target different molecules. These require high specificity even at very low targeted analyte concentrations. After analyte identification, a transducer will be responsible for generating an amplified signal for further analysis.

Over the years, biosensors have evolved according to the different integration methods between bioreceptors and transducers. Thus, in relation to the mode of attachment of the components of

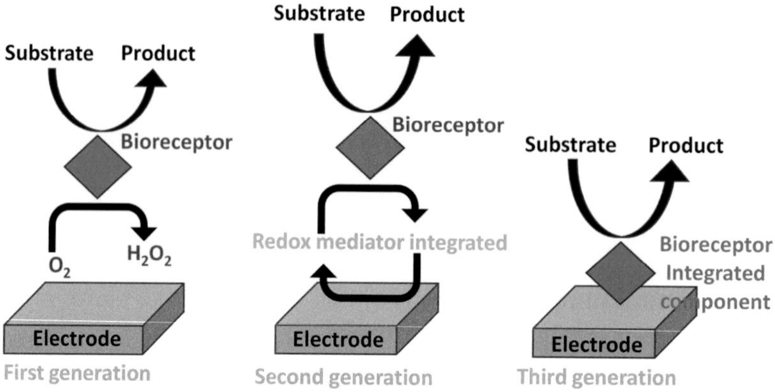

FIGURE 7.2

Three generations of biosensors related to the mode of attachment of their components.

biosensors, these can be classified into three different generations (Fernández et al., 2017). A scheme of these three generations of biosensors is shown in Fig. 7.2.

The first generation, called mediator-less amperometric biosensors, is based on the electroactivity of the bio-receptor substrate or product, that is, the electrical response that happens by means of the diffusion of the reaction products to the transducer. It can be measured by both the analyte and the product contents of the reactions occurring in the bioreceptor. The first report of the first generation of biosensors was described by Leland Charles Clark Jr. in 1956, with an electrode that measures the oxygen concentration in blood (Heineman et al., 2006). Later, in 1962, Clark continued his work in developing biosensors and demonstrated experimentally the operation of an amperometric enzyme electrode with application in glucose detection (Clark & Lyons, 1962). Those works established the basis for developing these devices, so in 1967, Updike and Hicks, based on Clark's work, conceived the first functional electrode based on an enzyme, glucose oxidase, as an oxygen sensor (Updike, 1967). Subsequently, numerous biosensors were developed for multiple fields of interest, such as the detection of urea (Guilbault & Montalvo, 2002) or hydrogen peroxide detection in a platinum electrode with a glucose and lactate enzyme sensor-based system (Guilbault & Lubrano, 1973) and the development of biosensors for alcohol (Hooda et al., 2018). A general outline of early advances in biosensor development is shown in Fig. 7.3.

The second generation of biosensors, called mediator amperometric biosensors, is centered on the enhancement of their analytical efficiency by the introduction of individual components integrated into the biological component layer using redox mediators, either free in solution or immobilized with the biomolecule, which improves system response. An example of these devices is the lactate analyzer, presented in 1976 to provide trusted measurements of blood lactic acid from blood samples based on the electron transport from lactate dehydrogenase to an electrode (Geyssant, 1985). Last, the third generation of biosensors shows that these devices move toward the integration of the bioreceptor molecule as an integral component of the electrode base sensing element with a direct interaction bioreceptor (redox-active biomolecule)-electrode surface (Liedberg et al., 1983). In this case, the biochemistry and electrochemistry are given without the involvement of any product or mediator.

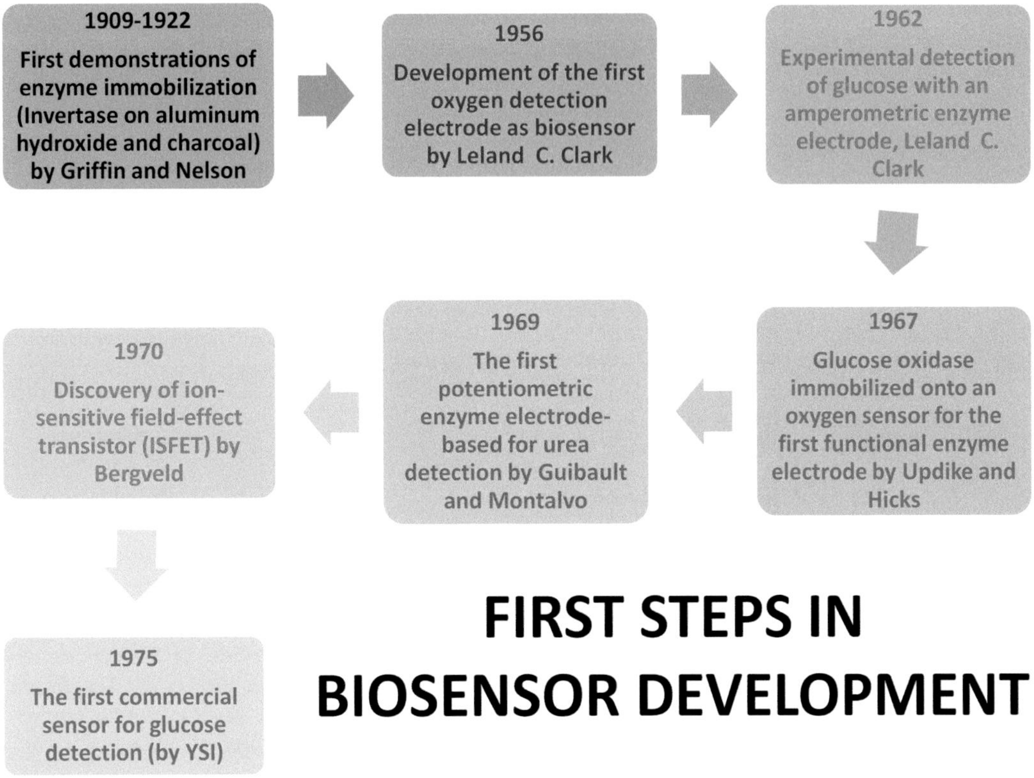

FIGURE 7.3

Graphical evolution in the biosensor development, through the description of the first steps in its development. (Bergveld, 1970; Clark & Lyons, 1962; Guilbault & Lubrano, 1973; Guilbault & Montalvo, 2002; Griffin & Nelson, 1916; Heineman et al., 2006; Nelson & Griffin, 1916; Newman & Turner, 2005).

7.5 Classification of biosensors

When talking about a biosensor classification method, we can ask different questions so that biosensors can be classified according to different criteria. In this chapter, we will discuss this classification in-depth, based on the nature of the bioreceptor and biorecognition principle (enzymatic biosensors, antibodies/immunosensors, aptamer/nucleic acid biosensors, and microbial/whole-cell biosensors) and the transduction method (electrochemical, electronic biosensor, optical, and mass-based/gravimetric). Another classification can be described depending on the detection system (optical, electrical, electronic, thermal, mechanical, and magnetic) and the technology (nanotechnology, surface plasmon resonance (SPR), biosensors-on-chip (lab-on-chip), and electrometers), although in this chapter we are not describing them. This classification is shown in Fig. 7.4.

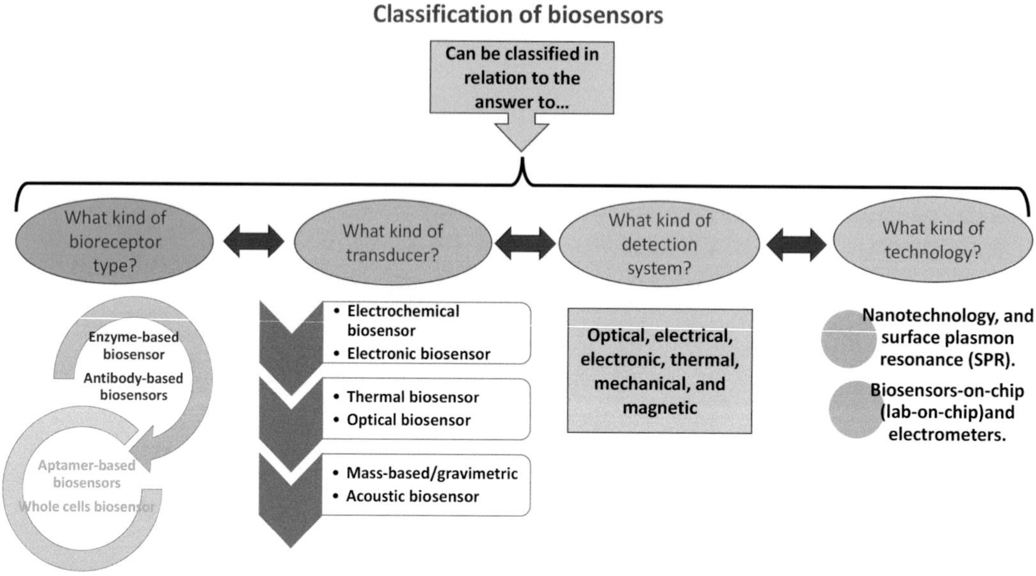

FIGURE 7.4

Classification of biosensors according to different criteria: Based on the nature of bioreceptor and biorecognition principle (enzymatic biosensors, antibodies/immunosensors, aptamer/nucleic acid biosensors, and microbial/whole-cell biosensors), on the transduction method (electrochemical, electronic biosensor, optical, and mass-based/gravimetric), on the detection system (optical, electrical, electronic, thermal, mechanical, and magnetic), and the technology (nanotechnology, surface plasmon resonance (SPR), biosensors-on-chip (lab-on-chip), and electrometers).

7.5.1 What kind of bioreceptor?

First, a bioreceptor can be defined as a biomolecule or biological element able to recognize the desired substrate or target analyte. In this process, a signal during the interaction between the bioreceptor and analyte will be produced, which is called biorecognition. In relation to the biorecognition principle, that is, the type of biological signaling mechanism and the nature of bioreceptors, biosensors can be classified according to their interaction with the analyte. In this sense, biosensors can be divided into catalytic biosensors, where this interaction between analyte and bioreceptor will result in new biochemical reaction products (enzymes, microorganisms, tissues, and whole cells), and non-catalytic biosensors. In this second type, the bound analyte-bioreceptor is irreversible without the formation of new biochemical reaction products (antibodies, cell receptors, and nucleic acids). The following describes the main ones according to the bioreceptor elements used for analyte recognition.

Moreover, concerning this part of biosensors, one of the most crucial parts in their fabrication is the process of immobilizing the biomolecules used on the sensor head. This aspect will be discussed following the description of the different types of bioreceptors.

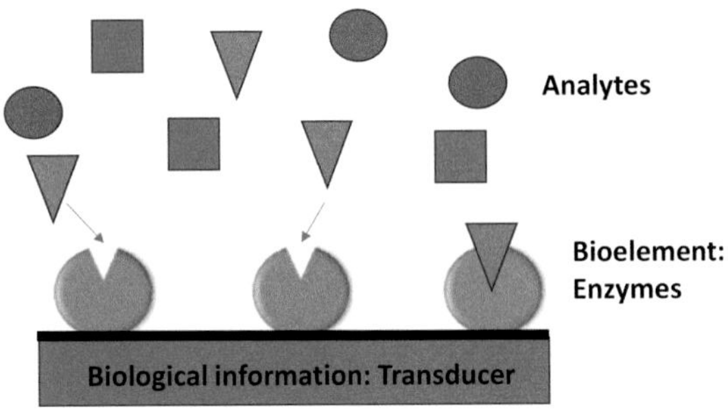

FIGURE 7.5

Working principle of an enzyme-based biosensor, where the analyte is specifically recognized by the bioelement.

7.5.1.1 Enzyme-based biosensors

One of the most widely used elements for biorecognition in the field of biosensors is enzymes, which are common biocatalysts. One of the key issues with enzymes is that they are highly specific to the target analyte or substrate and, catalyze multiple biochemical reactions. These reactions will be subsequently recorded and determined by the transducers. Therefore, the operating principle of an enzyme-based biosensor will be closely related to the catalytic reaction and the binding skills that will result in target analyte detection and the release of the enzyme for subsequent reactions (Fig. 7.5) (Sharma, 1994).

Regarding the mechanisms involved in the enzyme-analyte recognition process, there are three main ones. In the first one, the enzyme metabolizes the analyte; in this case, the enzyme concentration will be related to the catalytic transformation of the analyte by the enzyme; in the second, the enzyme can be activated or inhibited by the analyte so that the analyte concentration will be related to the enzymatic product formation; and in the third, it will be related to the follow-up of the enzyme characteristics (Perumal & Hashim, 2014). These unique properties make enzymes powerful devices for developing these analytical devices. On the other hand, it should be noted that enzyme catalytic action can be affected by factors such as temperature, the amount of the substrate, pH value, or the presence of competitive substances.

If we go back to the beginning of these biosensors, the first enzyme-based sensor was described by Updike and Hicks in 1967. These biosensors have been developed from immobilization methods of enzymes near the transducing component to producing the proportional signal for analyte detection, among which we can highlight the adsorption of enzymes by Van der Walls interactions or either ionic or covalent bonding. The different methods of enzyme loading will have a decisive influence on the performance of this kind of biosensor. However, we will discuss this aspect in more detail in a later section. In general, this kind of biosensor tends to produce or consume protons and/or electroactive species. Concerning the enzymes most used in these devices, these are usually oxidoreductases, polyphenol oxidases, peroxidases, and amino oxidases, which are usually used as diagnostic tools to determine the formation of hydrogen peroxide in blood plasma (Wilson & Hu, 2000).

Commercially and in practice, electrochemical transducers coupled to these biosensors are the most commonly used. In them, we can describe four principal types of commonly used enzyme-based biosensors: Glucose biosensor, the most widely employed and used in the diagnosis and treatment of diabetes, biotechnology, or food science; lactate biosensor, also used in biotechnology and food medicine, together with applications in sports medicine; urea biosensor, mainly used in clinical applications; and glutamate/glutamine biosensor, also employed in food science and biotechnology. We can highlight the advances developed by Cordeiro et al. (Cordeiro et al., 2018) in developing an enzyme-based sensor for glucose detection in the brain through a W-Au-based sensor, which can monitor glucose in the brain. Additionally, several studies have described the use of enzyme-based sensors to detect cholesterol, food safety, and the detection of pesticides or heavy metals (Amine et al., 2006; Soldatkin et al., 2012; Zapp et al., 2011). Furthermore, there are studies in relation to a nucleic acid biosensor with application in DNA sensing (Lin et al., 2011).

Regarding enzyme-based biosensors, it is necessary to distinguish between different types of supports that are used in their fabrication process for enzyme immobilization, the procedure whereby the enzyme molecules are entrapped onto a solid matrix that acts as a substrate to obtain the desired product. This is key since the enzyme must maintain its activity once immobilized. These can be classified into natural polymers, synthetic polymers, and inorganic materials as support.

7.5.1.1.1 Natural polymers in enzyme-based biosensors

Currently, the most used naturally derived polymers include alginate, chitosan, chitin, collagen, gelatin, cellulose, or starch, among others. Alginate and collagen are regularly used for enzyme immobilization due to their mild gelation conditions to obtain cross-linked networks with cationic cross-linkers and increase their stability. Another strategy used is to entrap enzymes in chitosan (Kapoor & Kuhad, 2007) because of the presence of functional groups (hydroxyl and amino groups) that can be easily linked to enzymes. Furthermore, a higher absorption rate makes natural polymers suitable for enzyme encapsulation, such as gelatin, a commonly used natural polymer with a very high absorption rate. And finally, cellulose, the quintessential natural polymer, is one of the most frequently used for a broad range of enzyme immobilizations, such as β-galactosidase, tyrosinase, lipase, α-amylases, or penicillin G acylase (Bryjak et al., 2007; Mislovičová et al., 2004; Namdeo & Bajpai, 2009).

7.5.1.1.2 Synthetic polymers in enzyme-based biosensors

A wide selection of synthetic polymers are used in enzyme encapsulation, with additional protective functions for the encapsulated enzymes. Examples of these include ion exchange polymers (ion exchange resins), which are insoluble porous materials with larger surface areas suitable for enzyme trapping; and amberlite and diethylaminoethyl cellulose exchange resins for α-amylase trapping (Davis et al., 2010; Kumari & Kayastha, 2011). Also, there is a need to describe other synthetic polymers, such as polyethylene glycol and glutaraldehyde. These act as a protective coating when free radicals are formed due to the enzyme immobilization of reddish peroxidases (Ashraf & Husain, 2010; Guisan et al., n.d.). Another example of protective functions is the role of polyvinyl chloride (PVC), which protects the thermal inactivation of the enzyme cyclodextrin glucosyltransferase (Sirisha et al., 2016).

7.5.1.1.3 Inorganic materials as support in enzyme-based biosensors

Another optimal support for enzymes is inorganic materials, which are inert in nature and have a porous and well-defined structure and shape, making them a good deck for enzyme anchorage. These include ceramics, zeolites, active glass, activated carbon, and charcoal. The main use of inorganic materials is in molecular adsorption; the presence of hydroxyl groups makes the formation of hydrogen bonds more likely and hence favors enzyme immobilization. This is the best support for chymotrypsin immobilization (Xing et al., 2000). If we look at some characteristic examples of the use of these inorganic materials, first we have to talk about zeolites inorganic materials with heterogeneous surfaces and with many sites susceptible to enzyme interaction and, therefore, adsorption (Zhang et al., 2021). Another example is the immobilization of lysozyme in Na-zeolite, improving its activity (Lee et al., 2019).

In addition, it should include one of the most abundant inorganic materials, silica, together with silica nanoparticles (Tripathi & Savio Melo, n.d.), with an inert nature and widely used in enzyme immobilization due to their high surface area and stability, both thermal and mechanical. One of the most used strategies is activating the surface to strengthen the interaction with the enzyme, for example, with polyvinyl alcohol. Examples of these are the immobilization of lignin, laccase, acetylcholinesterase, or tyrosinase with biocatalytic degradation properties (Suresh et al., 2023).

Finally, another widely used inorganic support type is functionalized glass, an inexpensive and tough inorganic material. Examples include the covalent immobilization of α-amylase (Kahraman et al., 2007) or urea biosensors in blood by urease immobilization on a glass pH electrode (Singh et al., 2021).

7.5.1.2 Antibody-based biosensor: immunosensors

Continuing with the description of bioelements for analyte recognition in biosensors, the second most used are antibodies (Fig. 7.6). So, what are the antibodies? These are defined as proteins that are produced by the immune systems of multiple organisms to recognize and then offset pathogens, specifically through strong antigen-antibody interactions. In terms of their structure, they present a

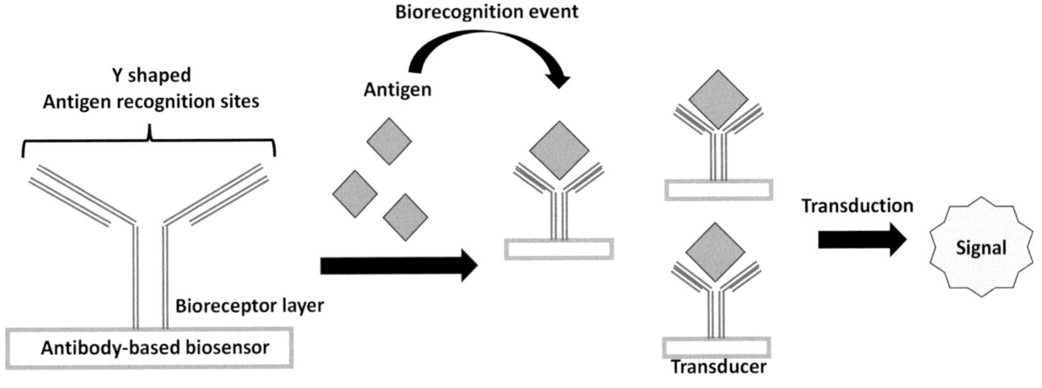

FIGURE 7.6

General scheme of antibody-based biosensor in DNA recognition events.

typical build of immunoglobulins (Ig) in the shape of a "Y," composed of two heavy and two light polypeptidic chains and with two antigen recognition sites, all connected through disulfide bonds. The usual way of producing antibodies is by immunizing an animal with the target, which after a few days will contain a high concentration of corresponding antibodies against the desired target, and all this is followed by a subsequent purification by affinity chromatography (Dzantiev & Zherdev, 2013).

The high specificity property of antibodies is a key parameter in their use in immunosensors, where only the analyte of interest will fit into the antibody binding site. Immunosensors are biosensors composed of an antibody embedded in a ligand. Biosensors based on this architecture were first applied for detection in the 1950s (Conroy et al., 2009), and their development as bioreceptors has become a valuable tool in the development of immunosensors for clinical diagnostics (D'Orazio, 2003), which are highly specific, stable, and very versatile. These can be classified as non-labeled or labeled. The first one is designed to determine the antigen-antibody complex through the physical changes produced in this complex formation, and the second one involves introducing a label that can measurably detect that antigen-antibody complex.

Regarding their scope of application, immunosensors play a significant role in public health, and significant advances have been made in multiple fields. Immunosensors for bacterial detection (Barton et al., 2009) or for use in the detection of cancer in its early stages (de Castro et al., 2020; Ushaa et al., 2011) are promising tools called to replace the traditional diagnostic methods or in environmental monitoring (Jain et al., 2019).

7.5.1.3 Aptamer-based biosensors (biomimetic)

Aptamers are synthetic short oligonucleotide sequences of DNA or RNA (around 20 or 40 bases), which are an appealing substitute for antibodies. Aptamers were reported for the first time in the early 1990s. They adopt a 2D or 3D folded structure, which allows them to show high selectivity and affinity for targets (amino acids, oligosaccharides, peptides, proteins, metal ions, cells, etc.) (Yoo et al., 2020). This will lead to a high binding capacity due to a higher surface density and less steric encumbrance (Dhiman et al., 2017; Song et al., 2012). The general aptamer-based biosensor scheme is depicted in Fig. 7.7. Compared to the antibodies described in the previous section, aptamers are smaller molecules in size and less complex; furthermore, they can be chemically synthesized, which makes it possible to easily functionalize them with alternative groups of interest or with different markers concerning detection criteria (Kumar et al., 2019). In addition, they are stable over a wide range of temperatures (they can withstand several denaturation/renaturation cycles), pH (2–12), and storage conditions. Its properties include high specificity, small size, and versatility. All this makes it possible to obtain a wide variety of different kinds of biosensor formats. However, these biosensors present challenges due to the structural and chemical simplicity of nucleic acids, which decreases their efficiency and increases their cost. This type of molecule can be isolated from oligonucleotide libraries by an in vitro selection mechanism called SELEX (Systematic Evolution of Ligands by Exponential Enrichment) (Stoltenburg et al., 2007). This in vitro selection mechanism can potentially set aptamers against non-immunogenic and toxic substances for which antibodies cannot be drawn up (Song et al., 2008). For this purpose, a random group of aptamers is incubated with the target. It is analyzed which aptamers bind to the target, and then the aptamers bound to the target are separated from the unbound aptamers. After release from the target, the bound aptamers undergo a new SELEX selection cycle. This process is repeated to

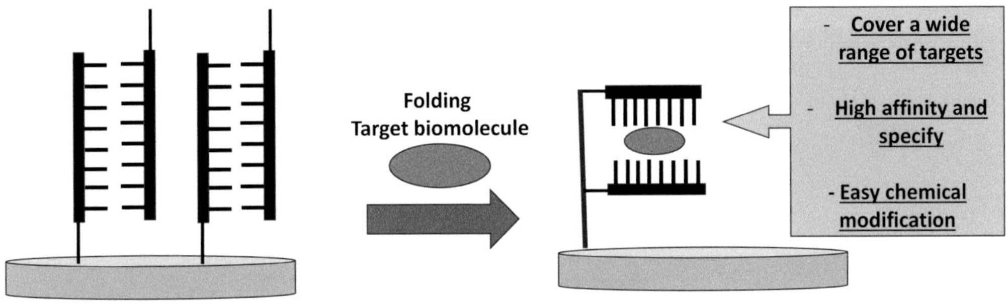

FIGURE 7.7

General aptamer-based (DNA) biosensor scheme. In this kind of device, target DNA is captured at the recognition layer, with high affinity and covering a wide range of targets. Finally, the resulting interaction is transduced into a measurable electronic signal.

obtain a sufficiently high affinity for the target for application, that is, a dissociation constant in the nanomolecular to picomolecular range. In relation to the advances in this field, the time and costs required in the adaptamer selection process have been reduced.

In relation to the most commonly used transduction techniques in this area, we have optical, electrochemical, and piezoelectric-based devices. Furthermore, these biosensors can be classified as labeled or label-free. Among the most common techniques for label-free optical sensors, SPR is long used, while in labeled optical aptamer-based biosensors, the use of fluorescent probes as markers is common (Hianik, 2018). Related to monitoring biological systems, there are great advances in the use of fluorescent NPs (such as Quantum Dots), conjugated with aptamers to identify cancer cells, pathogens, or proteins (Chu et al., 2021; Ikanovic et al., 2007).

As for advances in clinical application, they are of great practical importance: the detection of small-molecule biomarkers that are biologically active and have high significance for human health; the detection of protein biomarkers in biological fluids (blood or sweat); the detection of pathogens (whole-cell bacteria and viruses); the detection of circulating tumor cells to improve early diagnosis and therapeutic outcomes; the detection of extracellular vesicles secreted by cells such as exosomes, microvesicles, and apoptotic bodies; and tissue sample diagnostics to acquire diagnostic information through in situ immunostaining of tumor tissues (Stanciu et al., 2021).

7.5.1.4 Whole cell-based biosensors

Other bioelements to be mentioned include whole cell-based biosensors (Fig. 7.8) with application in synthetic biology (Gupta et al., 2019) that use multiple living cell types (eukaryotic cells, bacterial cells, fungi, algae, and viruses) with their array of biomolecular mechanisms, which possess various potential elements for biorecognition. It can be defined as a biosynthetic system with cellular-based components with the main purpose of converting a stimulus into a quantifiable and assessable cellular response (for example, a redox response or changes in osmotic characteristics) (Liu et al., 2022). These systems present several advantages, such as ease of management and

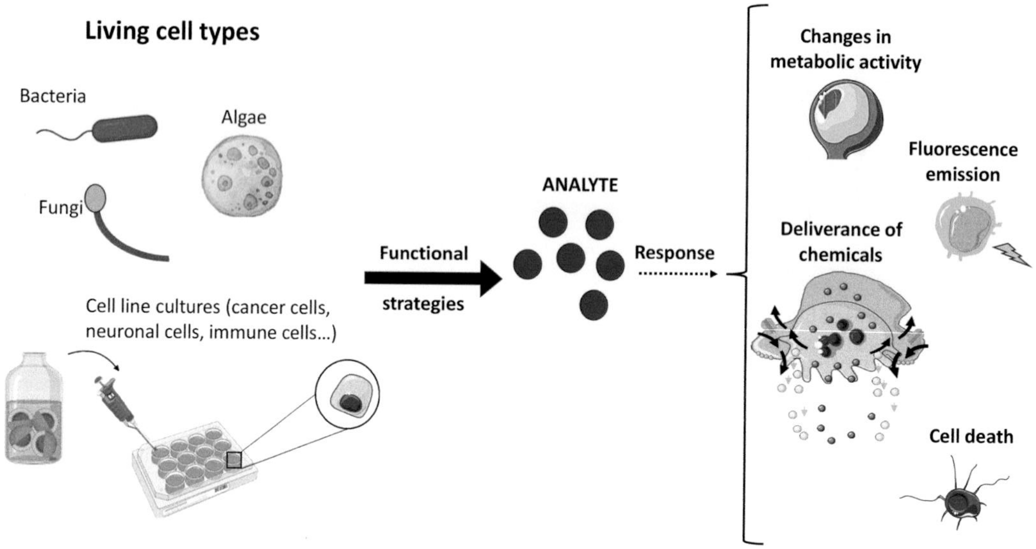

FIGURE 7.8

Scheme of the different cell types (left) and functional strategies (right) utilized in cell-based biosensors.

proliferation. Furthermore, they can generate specific recognition elements through self-replication, with the added benefit of avoiding further extraction or purification processes (Kylilis et al., 2019). Its operation principle is centered on the interaction of cells with the corresponding analyte. This interaction will result in an electrochemical response that will be registered by the transducer (Ron & Rishpon, 2010). These biosensors display high selectivity, optimal sensitivity, and a good sensing capacity. Among their common applications, these are used in multiple fields such as environmental monitoring, pharmacology, detection of organic pollutants or heavy metals, food analysis, or drug detection (Berepiki et al., 2020). Other examples of these types of biosensors to highlight are the development of a whole-cell *Escherichia coli* biosensor intended for the detection of an insecticide (pyrethroid) with a low detection limit in the order of 3 ng/mL (Riangrungroj et al., 2019) or a biosensor composed of fibroblast cells for the electrochemical measure of ATP concentration (Feng et al., 2007) and biosensors with a DNA probe placed on the transducer with application in the electrochemical detection of papillomavirus DNA that occurs when the complementary strand interacts with the probe (Nascimento et al., 2012).

7.5.1.5 Immobilization techniques of biological elements

One of the elements of biosensors, as described above, is the bioreceptor. Still, within the biosensor manufacturing process, a crucial aspect is the immobilization of the biological elements in the bioreceptor on the surface transducer and, thus, the selection of the most appropriate technique to proceed with this immobilization.

And why is this so important? First, several factors could affect their efficiency, for instance, the physical/mechanical properties and chemical environment (pH and temperature); it must also be considered that after immobilization, the biomolecule that will develop the biosensor may be inactivated or removed from the surface, thus ruining the biosensor. Therefore, after immobilization, the biomolecules must maintain their structure and properties throughout the life of the biosensor and its operation (Monošík et al., 2012).

It can be defined as the main immobilization method for biosensor construction as a function of physical and matrix entrapment, or reversible absorption and chemical or irreversible absorption, through covalent bonding or cross-linking.

In the physical absorption, a mixture of various non-covalent interactions, that is, Van der Waals forces together with hydrogen bonds, electrostatic forces, ionic bonding, and hydrophobic interactions, are used to immobilize the biomolecule to the transducer surface. That is one of the simplest methods to absorb biocomponents, although those interactions are relatively weak, so the biomolecules attached could be removed. This is the usual method employed in the case of enzyme-based sensors (Martinkova et al., 2017; Sassolas et al., 2012).

This method presents both advantages and disadvantages. Concerning their advantages, this is a simple and cheap process; moreover, it is friendly to bioreceptor activity and does not require an amendment of the biological elements of the biosensor. On the other hand, this immobilization could be subjected to changes in the media, such as pH or temperature changes, together with limited storage stability.

In connection with *physical matrix entrapping*, biomolecules are physically entrapped in a 3D organic/polymeric or inorganic matrix. The most commonly used organic matrices are polymers such as gelatin, alginate, polydimethylsiloxane (PDMS), starch, polyacrylamide, and polyvinyl alcohol, among others. On the other hand, the inorganic matrix is frequently composed of activated carbons or porous ceramic materials (Chen & Shamsi, 2017). The most used techniques to achieve this matrix entrapping are (I) electropolymerization (with the application of a potential to a solution of biomolecules and monomers to produce reactive radical species to form the polymer, which will trap the biorecognition molecules); (II) the sol-gel process (one of the most used techniques), where the main concept is based on the formation of a nanoporous material through the well-known reactions of hydrolysis and condensation in metal alkoxides. As the 3D matrix is being formed, the bioelements are trapped inside the matrix (Martinkova et al., 2017). This method allows easier synthesis in mild conditions and the possibility of entrapping high concentrations of biomolecules, along with good thermal and chemical stability; and (III) the microencapsulation, in which the biorecognition elements are contained in a semipermeable polymeric membrane that serves as a separator between the analyte and the bioelement, which is attached to the sensor by means of a simple and affordable process.

In *chemical or irreversible absorption through covalent bonding*, the sensor surface needs to be considered a reactive group through their functionalization, which will serve as a link to the binding of biological materials so that a covalent bond is formed directly between the sensor surface and the bioreceptor. This covalent binding is usually achieved by well-known reactions using common linkers such as EDC, glutaraldehyde (in materials with hydroxyl reactive groups on the surface), and bonds formed between the gold surface and thiol groups or silanes to bind the bioelements on bioactive glass/silicon surfaces. This kind of irreversible immobilization is often the most effective method of binding, as it is much more controlled and accurate. However, it is very important to keep in mind that this immobilization must not affect the recognition sites of the analyte and that it must not alter or change the functionality of the bioelement. On the other hand, this method presents a series of great advantages since it favors the uniform and homogeneous distribution of the bioelements and their stability against changes in the medium. Furthermore, it favors the reproducibility of the device. In addition, the type of bond formed between the bioelement, and the transducer surface is strong and very stable. However, it also presents some disadvantages, such as the use of aggressive chemicals and the fact that once used, the matrix can no longer be reused.

In connection with *irreversible absorption through cross-linking*, this process occurs by creating covalent cross-linking points between the biorecognition elements and inert proteins such as bovine serum albumin (BSA). This process occurs through covalent crosslinking points between the biorecognition elements and inert proteins, such as BSA (or between enzymes), by means of multifunctional reagents that behave as crosslinking agents. This mechanism helps to reduce the possibility of loss in the enzymes together with a higher bond strength, but against this, undesired crosslinking reactions may occur between the protein molecules; in addition, partial denaturation processes of the proteins used may occur, which limits their application.

7.5.2 What kind of transducer?

Another type of classification for biosensors is the one based on the transduction method (transducer), as shown in Fig. 7.1. When the target is recognized and captured by the biorecognition element (i.e., the bioreceptor), a signal will be generated. The transducer detects and converts this biological response into an electrical response. Furthermore, the biosensing process can be either qualitative to determine the presence of the interest analyte or quantitative to quantify the analyte concentration in the sample.

There are three main classes of transducers and transduction methods to obtain different types of biosensors: optical, electrochemical, and mass-based detection methods (Chen & Wang, 2020; Marazuela & Moreno-Bondi, 2002; Rubab et al., 2018).

7.5.2.1 Optical transducers (optical biosensors)

Optical transducers demand that the recognition event in biosensors can produce a signal in the form of light or color, for example, the fluorescence labeling or light produced in enzyme reactions or even the color change of the bioreceptor layer in a colorimetric assay (Liu et al., 2013). In general, optical biosensors are composed of a light source, the bioreceptor, the optical transmission device, and the optical detection system. Then, the analyte reacts with the bioreceptor, producing the biological signal (in the form of color change, light, or color), which is transformed into an electrical signal by the transducer (changes in several light parameters, such as phase, amplitude, or frequency). Among its features, it is worth mentioning its effectiveness and fast response, along with optimal selectivity, specificity, and sensitivity, in addition to not presenting electromagnetic interference in the sensing area of the sensing system. Furthermore, these are efficient analytical tools in multiple fields, such as health care, clinical diagnostics, drug and biomedical research, environmental applications, and food control. We can distinguish three types of optical biosensors: fiber optics, SPR, Raman, and FTIR (Chadha et al., 2022).

7.5.2.2 Electrochemical transducers (electrochemical biosensors)

Electrochemical transducers are usually integrated within biosensors, leading to electrochemical biosensors that are highly specific to the required substrates. These biosensors are useful in detecting biological materials such as enzymes, whole cells, tissues, specific ligands, or even non-biological matrixes. These devices imply the use of electrochemical electrodes that will oversee, measure, and transform every electrochemical change, a product of the chemical interaction with the bioreceptor sensing interface, like enzyme-substrate or antigen-antibody interaction, into an electrical signal (voltage, current, and impedance, among others). Thus, the signal can be processed and analyzed, detected by a detector, and then analyzed by a computer with the necessary software to translate that signal into a measurable parameter.

And how do you describe the performance of an electrochemical transducer in electrochemical biosensors? This one is based on changes in the electrical properties of the system on account of the binding event. One of the two electrodes should be locked together with the bioreceptor; then, the analyte will bind to the biorecognition molecule (this has been discussed in the previous section), resulting in changes in electric properties as a product of redox reactions by the biological interaction events (Srivastava et al., 2020). There are different types of devices (Pohanka & Skládal, 2008) related to the detection mechanisms: amperometric, potentiometric, conductimetric, and impedimetric biosensors. In relation to *amperometric detection*, the electroactive species are oxidized or reduced while the biochemical reaction is produced so that a current is measured. It can be divided into two different modes: if the potential remains constant, we have amperometry, and if the potential is varied, we get voltammetry (Grieshaber et al., 2008). Amperometric biosensors constantly monitor the current product of the process of oxidation-reduction reactions of electroactive species and have applications in disease diagnosis and treatment, providing quantitative analytical information.

The current produced in the reactions is proportional to the concentration of the analyte of interest to be detected. One of the most common is the Clark oxygen electrode used as a glucose amperometric biosensor (Fig. 7.9). In relation to the Clark oxygen anode, this is separated in this glucose biosensor by a structure composed of a layer of glucose permeable to oxygen. Between this layer of glucose, bio-catalyzed glucose-oxidase (GOD) is isolated from the glucose layer by another layer, permeable to oxygen and glucose. The goal is that the chemical GOD must be trapped between two different layers, one only permeable to oxygen at the top and the other permeable to both oxygen and glucose at the bottom. In summary, these devices are based on the fact that glucose reacts with GOD through an enzymatic transformation of the glucose that is oxidized by oxygen to form glucuronic acid. Furthermore, the glucose mediator reacts with surrounding oxygen to form hydrogen peroxide and GOD, which can react with more glucose. Then, the amount of oxygen consumed or the amount of hydrogen peroxide obtained can be monitored. Thus, a higher glucose content will be associated with higher oxygen consumption (Li et al., 2009).

In relation to *potentiometry*, the parameter that is measured is the potential between electrodes (working and reference) in the absence of current flow, so that the ion activity of the sample can be measured and associated with the analyte concentration (Bakker & Pretsch, 2005). The basis of its operation lies in the fact that during a biochemical reaction, there are changes in the charges of the system; for example, when there is a production of electrons during an enzymatic reaction, these changes are monitored. Thus, we can define potentiometric biosensors as devices consisting of a bioreceptor and potential transducer, monitoring the changes in potential produced at the working electrode with regard to the reference electrode. In this kind of biosensor, the electrochemical cell is related to a working electrode and a reference electrode. The bioactive molecule, namely, the bioreceptor, covers the working electrode; thus, the generated potential will be proportional to the

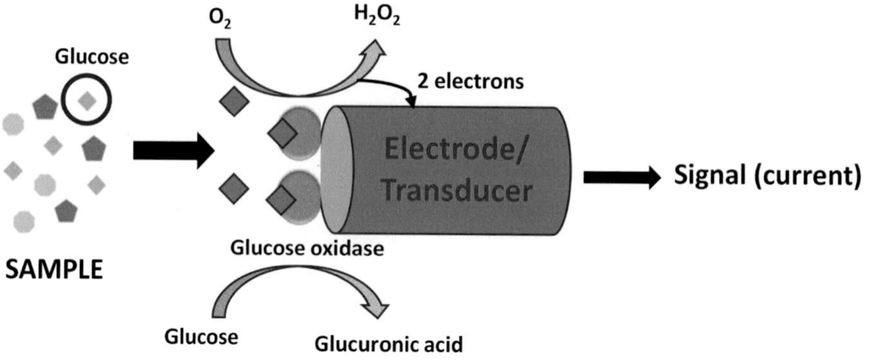

FIGURE 7.9

Scheme of operation of glucose amperometric biosensor (Clark oxygen electrode).

analyte concentration. Among the multiple advantages of these devices, their high sensitivity and selectivity and their great usefulness in the detection of low-concentration analytes in reduced sample volumes should be highlighted.

Another type of device is the *conductimetric sensor*, which evaluates the analyte's capacity to conduct electric current between electrodes, measuring changes in the electrode resistance, such as impedance changes, when the analyte interacts with the bioreceptor surface (Lisdat & Schäfer, 2008). Speaking of conductimetric biosensors, their operation is based on the synthesis and uptake of ionic species. Due to this interaction at the surface of the electrode and the fact that it occurs very close to this, aptamer-based bioreceptors, with their small size, make them an optimal option as bioelements in these electrochemical-based biosensors.

Finally, the *mass-sensitive methods* used in mass-based biosensors can be described. These are completely dependent on mass changes, so when interactions are produced between biomolecules and analytes, a mass change occurs. This change results in an alteration in oscillating frequency, which is the shift measured by these mass-based sensors. Examples of these devices include cantilever-based biosensors that can be easily functionalized with bioelements like antibodies (Datar et al., n.d.; Lang et al., 2005).

7.6 Understanding the use and applications of biosensors

In this section, the main applications of biosensors will be described, along with examples of their use in different fields of application. However, in a later section, the applications of conductive polymers in developing biosensors will be described specifically. As far as biosensor applications are concerned, they cover a wide range of applications due to their multiple advantages. Among the main applications, it is necessary to include healthcare, environmental monitoring, water analysis, food control, security, agriculture, drug detection, and civil security, as depicted in Fig. 7.10.

The development of non-intrusive and effective biosensors for medical care involves several main requirements: biomarkers that are more focused and explicit, and non-intrusive recognition

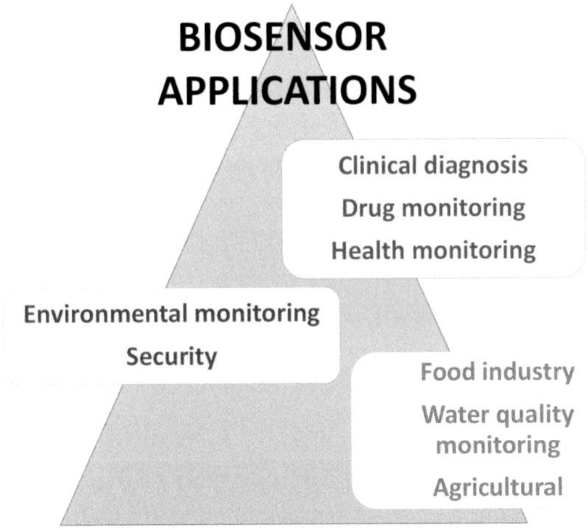

FIGURE 7.10

Scheme of the main applications of biosensors in diverse fields.

and screening. In the field of clinical diagnosis, developing simple and disposable devices with fast reaction times, ease of use, low cost, and the possibility of easy scalability in their production is critical. All this has resulted in the presence of multiple devices on the market to measure different elements of importance in human behaviors. Examples of all these are the analysis of glucose in biological samples (glucose oxidase electrode-based biosensor), clinical detection of uric acid (Kim et al., 2015), their use in cancer therapy and bio-imaging (Peng et al., 2014), or the comprehension of biological processes and molecular systems in cells (Oldach & Zhang, 2014).

In connection with environmental monitoring, the use of biosensors in this field is related to the presence of pesticides and heavy metals from industrial and agricultural sectors in the environment, which may cause many health issues and environmental alterations. Therefore, it is important to develop portable devices for monitoring and detecting all sorts of pollutants rapidly in a short time (Justino et al., 2017). Examples of biosensors in this application include immunosensors and enzymatic biosensors (McConnell et al., 2020) for pesticide detection (Verma & Bhardwaj, 2015) or even aptasensors for bacteria and virus detection (Yi et al., 2020).

Regarding water analysis, the development and applications of biosensors in this field are of great importance since water is essential for every living organism, especially humans, without which there could be no life. The importance of this field lies in the current problem of water pollution due to agriculture and industry, which strongly alter aquatic ecosystems, so today, obtaining clean water for consumption is a great challenge. Here arise the biosensors as tools capable of monitoring water toxicity and changes in it continuously, as well as all important parameters such as pH, temperature, nitrates, phosphates, dissolved oxygen, and all this with an extremely sensitive response. Among the most important applications, it is worth mentioning the detection of pesticides and heavy metals in water (Hara & Singh, 2021). Electrochemical biosensors are also of great relevance for detecting and monitoring chemical ions, pathogens, and molecules of interest (Chen et al., 2017; Su et al., 2020).

Other applications of interest for biosensors focus on the food industry, in the detection of pathogens and food nutrition labels, such as carbohydrates or amino acids, together with certain food ingredients, and all of this is aimed at food security and quality. These applications are extended to developing biosensors in the fight against bioterrorism for the detection of pathogens (bacteria, viruses, or toxins) that are dangerous for humans, wildlife, and nature (Nikolelis & Nikoleli, n.d.).

7.7 Polymers in biosensors; what about their role?

First, it is important to place the role of polymers as devices; polymers are commonly used in developing biosensors due to their relatively low cost and the fact that they can be affordably obtained on a large scale. The use of plastic materials in biosensor fabrication has gained great importance because of their properties, such as low cost and cost-effectiveness, the possibility of large-scale mass production by injection molding, and outstanding stability, thus providing inexpensive, high-performance, full-scale, and environmentally friendly devices. Although we could focus on the use of polymers in the manufacture of these devices (i.e., poly (methyl methacrylate) (PMMA) or PDMS), we will focus on the materials that are most used in the design of biosensors as electrode materials, with a more active role: conductive polymers and MIPs. Furthermore, it is important to report some aspects of the use of polymer membranes in biosensors as simple coatings to improve their selectivity and biocompatibility, together with some examples.

7.7.1 Polymer membranes in biosensors

Regarding polymer membranes as coatings in biosensors, their primary function is to prevent or at least minimize the approach of interfering molecules and compounds that could interact with elements of biosensors. There are two main problems in relation to the correct performance of a biosensor: biofouling and the presence of interfering molecules. This kind of interference is particularly serious when electrochemical measurements need to be made in physiological fluids (for example, blood). These fluids could deposit biological molecules irreversibly onto the bioreceptor surfaces, such as proteins or amino acids, which can be oxidized at electrode surfaces (in glucose sensors). Also, the biofouling formation process can decrease the response of biosensors or even deactivate them. These problems are, therefore, highly relevant for biosensors that will be used more than once and for biosensors that will be implanted in vivo (Gavalas et al., 2006). As mentioned above, polymeric permselective coatings could minimize these problems due to the interference of that electroactive species with the surface sensor as well as providing a level of biocompatibility. The first two polymeric materials used were Nafion, a fluorinated ionomer (Turner & Sherwood, 2009), and cellulose acetate (Maines et al., 1996). These are two of the most commonly used polymers that provide a good degree of biocompatibility; furthermore, cellulose acetate acts as a selective barrier in electrochemical biosensors, allowing only small molecules, like hydrogen peroxide, to meet the electrode. This allows the elimination of several electrochemically active molecules, which could interfere with the correct measurement of biosensors. The structures of the polymeric materials described are shown in Fig. 7.11.

Regarding Nafion, it is largely used as a coating polymer (Wisniewski & Reichert, 2000). It is an anionic polymer that is chemically inert. Furthermore, it is easy to form coatings by casting

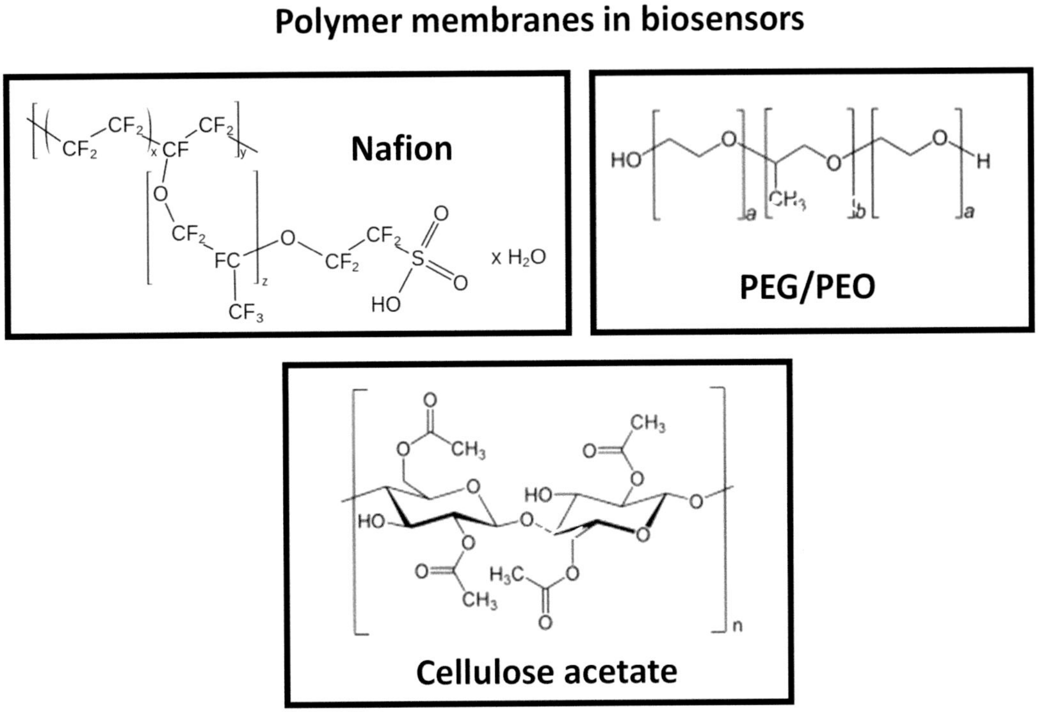

FIGURE 7.11

Structures of polymeric materials commonly used as membranes and coating polymers in biosensors.

from solutions of this, which results in hydrophilic channels inside a hydrophobic matrix. In addition, these coatings are relatively robust and can exclude anionic interferents, resulting in enhanced biocompatibility (Moussy & Harrison, 2002). This results in better results for Nafion-coated electrodes implanted in vivo.

Another problem described at the beginning of this section is the prevention of biofouling (Kingshott & Griesser, 1999). Among the most well-known materials, we have those based on polyethylene glycol/polyethylene oxide (PEG/PEO), which withstand biofouling. Their surface presents very low protein adsorption since PEG/PEO chains are solvated in aqueous systems, which means that proteins meet a surface basically composed of water, like the conditions found in biological systems, contributing to their biocompatibility.

Another important family of polymers used as coatings in sensors and biosensors are hydrogels. These are composed of water-soluble linear polymers such as poly (vinyl alcohol) or poly (acrylic acid) that, during the polymer network synthesis, are crosslinked to obtain a high hydrophilic network able to be swelled by absorbing water with high biocompatibility. Common examples include crosslinked polyethylene oxide or poly-hydroxyethyl methacrylate (PHEMA), both biocompatible and hydrophilic. Moreover, their swelling behavior can be regulated by varying the crosslinking degree so that the material's degree of hydrophilicity and stiffness can be varied. For instance, crosslinked PEO is used to steady an implanted glucose sensor (Csoeregi et al., 2002). Another usual application is as a stabilizing agent in enzyme-based biosensors to avoid denaturation and thus the loss of efficiency of enzymes through their encapsulation inside hydrogels (glucose oxidase incorporated in crosslinked PHEMA in glucose sensors improves their durability after 3 months of continuous functioning) (Doretti et al., 1996).

Finally, regarding polymers as biocompatible agents, multiple studies have been carried out to study the protein deposition on membranes and polymer coatings to minimize the absorption of blood proteins onto coating surfaces. Membranes composed of polyurethane or silicone rubber in solid-state ion sensors show less protein adsorption when exposed to blood (Cha et al., 2002), and functionalization of PVC with anionic surfactants allows an increase in biocompatibility in amperometric enzyme electrodes (Reddy & Vadgama, 1997). Another approach to minimize blood protein adsorption and, moreover, enhance biocompatibility is the synthesis of polymers containing phospholipids, as these are the main constituents of the cell wall. An example is the copolymer of methacrylate monomers and 2-methacryloyloxyethyl phosphorylcholine, a monomer based on phospholipid polar groups (Ueda et al., 1992).

7.7.2 Conducting polymer-based biosensors

The first question is defining a CP. These polymers are organic materials that are also conductive. In contrast to organic polymers, CPs possess special properties in terms of electrical conductivity, high electron affinity, and redox activity (Thompson & Reynolds, 2019). The high amount of research on CPs is related to the fact that they can be easily functionalized with a huge variety of groups to customize their properties. Among the most important developments to be highlighted are the major advances in applications of CPs in LEDs, supercapacitors, field effect transistors, and solar cells, and because this is the subject we are developing in this chapter, biosensors (Gerard et al., 2002; Peng et al., 2009). As discussed in previous sections of this chapter, bioreceptors are composed of several commonly used recognition elements, such as aptamers, whole cells, antibodies, proteins, and, more specifically, CPs, a suitable matrix for enzyme immobilization (Lakard, 2020). When talking about CP-based electrochemical biosensors, these recognition elements are immobilized on the electrode (CP electrode) so that the modification of the structures of CPs will be done by biomolecule immobilization (enzymes specifically) through compatible biological procedures.

And where does the importance of these lie? CPs are widely utilized to improve the sensitivity, versatility, and speed of detection of biosensors in measurements of analytes in diagnostics more efficiently. Moreover, electrical CPs are easily tunable to obtain the required electronic (owing to the charge mobility throughout the π electron polymer chains) and mechanical properties (flexibility, strength, plasticity, toughness, or elasticity). Another key advantage of CP is that during the electrochemical synthesis process, the polymer will be directly deposited on the electrode surface in a way that will simultaneously immobilize and trap the biological molecules, controlling the spatial distribution and the film thickness to modulate the activity of the biological molecules, for example, enzymes. Alternatively, a polymeric film can be placed electrochemically and then proceed to the adsorption or chemical grafting of the biological species to the CP film. In brief, all these aspects allow the biosensors to behave as single-use systems with optimal adaptability to multiple kinds of targets quickly and simply, biocompatible, easily synthesized, and functionalized to facilitate the immobilization of the bioelements.

One of the characteristics that a biosensor must have is a fast electron transfer process between the analyte and the electrode surface to obtain a usable reading signal through the conversion of the biorecognition event into a signal by the transducer. In the previous sections, we talked about third-generation biosensors; these have high sensitivity and selectivity, but a major barrier to their use remains the limitations in the electron-transferring process, which hinders their performance. When it comes to electrochemical systems, the main barriers are the difficulties in transmitting electrical signals between the bioanalyte, bioreceptor, and transducer. Examples of these barriers are the suppression of electroactive prosthetic groups in protein structure, or denaturation of these, which translates into the obstruction of the simple electron transfer redox proteins-transductor system. Another example is improving the matrix to entrap enzymes in enzyme-based biosensors. The incorporation of enzymes into CP films allows the integration of biologically active molecules into electrodes of all kinds of geometries or sizes.

Therefore, in this regard, CPs are a promising material to be applied in the development of analytically enhanced biosensors in terms of their biocompatibility, good stability and optimal mechanical properties, electrochemical activity, electrical conductivity, and the ability to transfer electric charges from bioreceptors to electrodes and transducers (Oztekin et al., 2011; Wang, She, et al., 2019; Wang, Yu, et al., 2019).

This section will describe the main approaches to the synthesis and applications of CPs for immobilizing biomolecules and rapid electron transfer in the biosensor fabrication field.

7.7.2.1 Conducting polymer-based structures most used in biosensors as electrode materials

Among the most commonly used conductive polymers used in biosensors, it can be highlighted polyaniline (PANI), polythiophene (PHT), poly(3,4-ethylenedioxythiophene) (PEDOT), or polypyrrole (Ppy) (Ramanavicius et al., 2014), as can be seen in Fig. 7.12. These organic polymers are defined by the presence of alternating single (σ) and double bonds (π) in a conjugated structure and, thus, by the appearance of π-bonded delocalized electrons along that structure. This implies that CPs can be oxidized or reduced, which notably increases these materials' conductivity (Le et al., 2017). Therefore, CPs can establish a direct route for electron transfer between biological molecules (enzymes and proteins, among others) and the electrode surface of biosensors, avoiding the need for a mediator to provide mobility to the electrons. Bearing all these aspects in mind, CPs are promising alternatives to be used as conventional materials for biosensors.

As for the favorable features of the most commonly used CPs that allow them to be applied as biosensors, we can highlight the following:

Polyaniline (PANI) is considered the first CP (Hush, 2003). In relation to their synthesis, PANI is obtained from the oxidation of their monomer, aniline. PANI is a semiflexible CP that possesses high conductivity, good electrochemical stability, present electro-optical properties, and, furthermore, strong interactions with biological molecules. This polymer is commonly used in the entrapment of enzymes as glucose oxidase through electropolymerization of aniline to be used as an electrochemical glucose biosensor (Dhand et al., 2011). Another example of the use of this polymer in biosensors is related to the immobilization of the enzyme horseradish peroxidase (HRP) in a layer of the conductive polymer composed of poly (aniline-co-*o*-aminobenzoic acid) through electrostatic interactions. To this end, the enzyme is mixed in the polymer (HRP-p(Ani-co-*o*-aba)) and then cast on a carbon electrode. Finally, another biopolymer, chitosan, is coated with the layer of

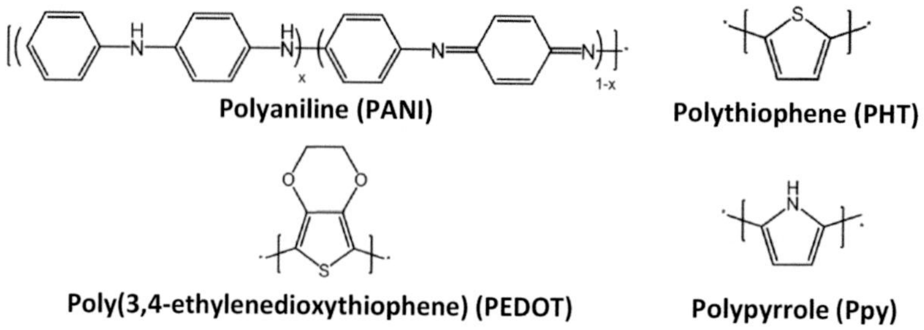

FIGURE 7.12

Structures of most commonly used conductive polymers in biosensors.

HRP-polymer, thus obtaining highly sensitive biosensors for H_2O_2 detection (Org et al., 2013). Another example is the formation of PANI and gold nanocomposites for immobilizing cholesterol oxidase in a chitosan matrix covered by a surface of indium tin oxide for cholesterol detection (Srivastava et al., 2010).

Polypyrrole and their derivates are examples of immobilizing substrates for biomolecules; moreover, they are stable under ambient conditions, present long-term environmental stability, and can be polymerized under natural and aqueous media, showing a simple one-step electrodeposition process on the electrode surface together with an easy entrapment of enzymes, which makes them an ideal material for biosensors based on enzyme immobilization (Jain et al., 2017), as well as in immunosensors or DNA biosensors.

In relation to *polythiophene*, thiophene is an aromatic ring and electron-rich. This can be oxidized to obtain the p-doped form, acquiring highly conductive PHTs. These polymeric films of PHT are formed by electrochemical oxidation. It translates into good properties in terms of good adhesion, optimal control of their thickness during the electropolymerization process, and, simultaneously, entrapment of the desired enzyme. Among its properties, their good adhesion to the substrate, biocompatibility, cell viability interface, and the ability to form blends with biodegradable materials are worth mentioning. Furthermore, these polymers can establish hydrogen bonds with complementary regions of DNA, which is used in specific DNA sequence detection (Fabregat et al., 2013), and molecules of biochemical interest can be immobilized in these electroactive polymers covalently to obtain electrically conducting materials for application in biosensors.

7.7.2.2 Advances in the application of conducting polymer-based biosensors

As stated above, CPs are considered a crucial class of functional polymers with application in biosensor fabrication because of their adjustable chemical properties along with easily definable structural and electrical properties by multiple strategies. The following are worth mentioning: chemical grafting to introduce functional groups of interest; association with other materials, such as nanoparticles, to obtain nanostructured materials; and all to improve the sensitivity, selectivity, and stability of the biosensor and their response toward the bioanalyte. This part of the chapter will be devoted to establishing the relationship between the different types of bioreceptors described above and the application of CPs in various fields.

CPs are commonly used in the fabrication of biosensors in several fields of great importance, from health care and immunosensors in medical diagnosis (biosensors for glucose, urea, lactate, and cholesterol detection) and environmental sensors to DNA sensors in the detection of genetic upheavals (inherit diseases and pathogenic infections). Furthermore, these are used in environmental monitoring and food control sensing of hazardous chemicals and analytes of interest in the food industry (glucose, fructose, starch, urea, and lactose, among others).

In relation to *health care*, the selective determination of certain analytes, such as glucose, urea, uric acid, lactate, and cholesterol, among others, both in blood and biological fluids, is crucial to monitoring and treating multiple diseases. As described earlier, there are multiple reports on the immobilization of enzymes, in this case glucose oxidase, for glucose detection and quantification. As for the immobilization of this enzyme (glucose oxidase) in CPs, we can point out several examples, such as porous devices of Ppy for the better loading of the enzyme or the covalent interaction between GOX and carboxy-functionalized PANI (Ramanathan et al., 2000). Another recent example of glucose detection is the use of composites of polyacrylonitrile and carbon nanotube nanofibers coated with another usual conducting polymer, PEDOT, through chemical vapor deposition (Çetin et al., 2023).

Regarding urea biosensors, these are based on the detection of NH_4^+ or HCO_3^- using, for this purpose, composite films of electropolymerized PANI in non-conductive hydrogels (polyacrylamide or polyvinyl alcohol) or Ppy and a polyion complex also obtained by electropolymerization. Furthermore, it is usual to entrap an enzyme, urease, or glutamate dehydrogenase, in Ppy electropolymerized films (Botewad et al., 2023).

Another biosensor of interest is that used for lactate and cholesterol detection. In the first case, lactate biosensors rely on enzymes immobilized (lactate oxidase and lactate dehydrogenase) in composites of Ppy-polyvinyl sulfonate, polyacrylate, or PANI (Rathee et al., 2016). Second, the monitoring of a very important parameter, cholesterol, must be highlighted. In the design of cholesterol biosensors, again, the immobilization of an enzyme (cholesterol oxidase) is used. For this purpose, electropolymerized Ppy films doped with dodecylbenzene sulfonate or multiple conductive polymers, such as PHT, PANI, PEDOT, or poly(carbazole), and functionalization of them are used (Soylemez et al., 2015).

As for *immunosensors*, biosensors are those in which a target analyte (an antigen) is detected due to the formation of a complex with an antibody; their importance lies in their strong affinity and in the combination between immunochemistry and electrochemistry that could allow the detection of a large set of analytes through direct electrical detection. There are multiple examples of immunosensors relying on CPs; electroplated CPs can act as antibody receptors in such a way that the polymer electrochemistry may be modulated by the reaction of antibodies against CPs or the use of Ppy films with antibodies (anti-BSA) incorporated directly for BSA detection (Lakard, 2020).

Finally, *DNA biosensors* show a great field of application in clinical diagnostic hereditary illnesses and detection of infections, or even in processes required in molecular biology, like cDNA screening, since the detection of defined sequences of DNA is of crucial importance. The simplest method of obtaining these biosensors is through the abortion method, which involves the immobilization of DNA into CPs, although it offers poor quality, stability, and response. In general, the integration of DNA with electroactive polymers provides biosensors with superior properties; the electrodeposition of Ppy films linked covalently to oligonucleotides improves the DNA interaction and detection of DNA hybridization (Wang et al., 1999).

7.7.3 Molecularly imprinted polymers as an alternative bioelement in biosensors

Throughout this chapter, we have reviewed the most important general concepts in biosensors and discussed the role of polymers in these devices, focusing on a brief description of conductive polymers. In this section, we highlight the importance of MIPs. These are of great importance nowadays due to their unique chemical and physical properties, together with an easy modification that will allow faster and more sensitive detection of analytes in multiple fields. MIPs are defined as biomimetic receptors; they are obtained by the polymerization of monomers using the target analyte as a template. The process is depicted in Fig. 7.13. First is the

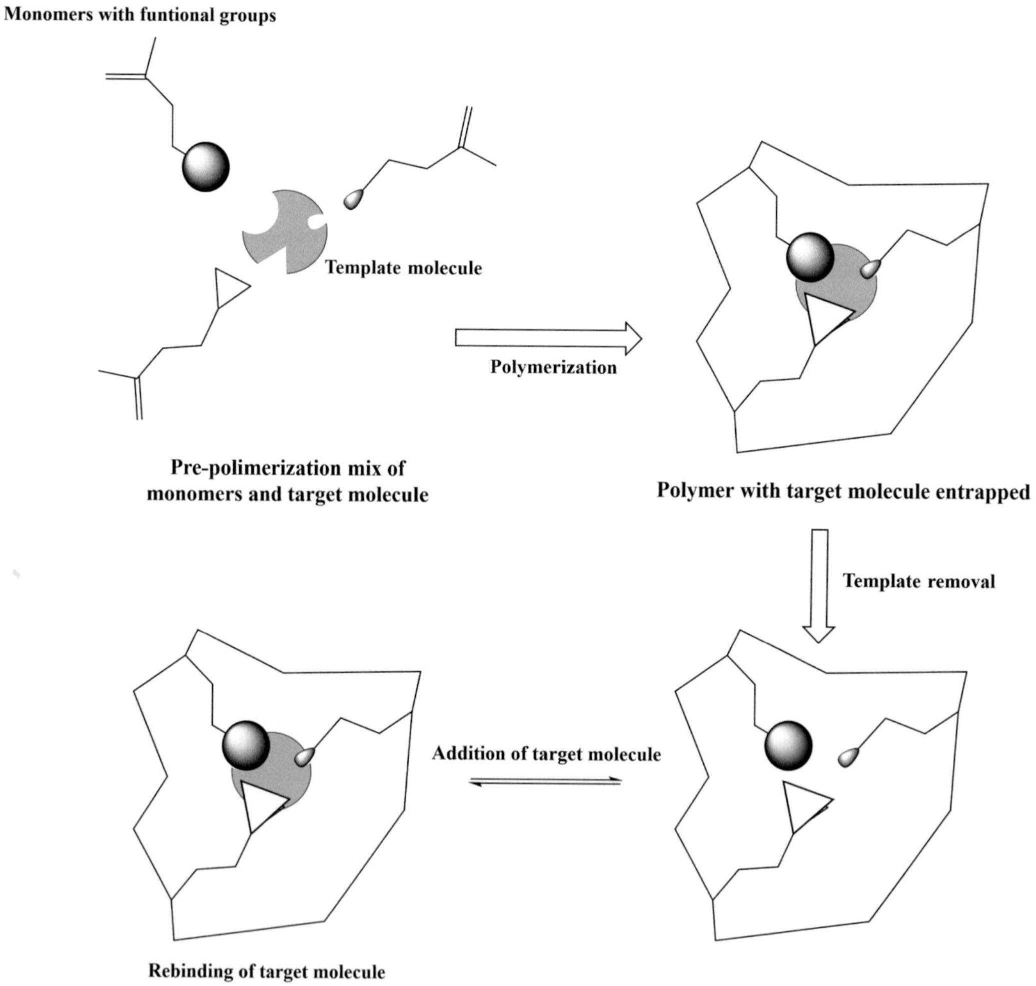

FIGURE 7.13

Scheme of general procedure of MIPs creation.

incubation of the pre-polymerization mix of monomers with the template molecule, which translates into non-covalent interactions between the monomers and template that stabilize the system; after that, the polymerization of monomers allows the polymer obtained to entrap the template into the polymer matrix. Then, when the template is removed, a three-dimensional polymer matrix is obtained with controlled cavities that will act as biomimetic receptors. The particularity of these cavities will be their size, shape, and electrostatic environment, which will be exactly the right ones to interact with the molecule target.

There are different polymerization procedures for obtaining these MIPs, such as UV or thermal polymerization and RAFT polymerization, together with the traditional bulk imprinting, a challenge in the case of proteins due to several factors like the heterogeneity of binding sites, the problems in synthetic routes in relation to solvents, or persistent trapping of the target molecule in the polymer matrix. In any case, using different synthetic routes will enable obtaining all kinds of systems for every application (Turner et al., 2006).

A very important issue in the design and production of biosensors based on MIPs is the adequate selection of the functional monomer and their subsequent polymerization since direct bonds, covalent or non-covalent, will be created between the functional monomer that will form the polymeric matrix and the target analyte molecule (Yoshikawa et al., 2016). After the process of breaking the bonds between the functional monomer forming the polymer matrix and the target molecule, new covalent bonds can be formed again in the biosensing process of the analyte molecule, which greatly increases the specificity of the biosensor. However, it may happen that the elimination of these bonds is not complete, is too slow, or that a polymeric matrix that is too rigid strengthens the bonds formed, which influences the final performance of the biosensor. Therefore, using MIPs with non-covalent interactions is very common since these types of interactions are weaker and removing the template from the polymeric matrix is easier (Wackerlig & Lieberzeit, 2015).

Although there are a huge number of examples of MPIs in biosensors with different functional monomers, in this chapter we will only show a very small part of them, which gives us an idea of the use of different monomers and polymers for a wide range of applications in the detection of analytes of interest, especially in clinical diagnostics. Some recent examples of biosensors based on MIPs are shown in the table, along with a short description of the functional monomer and template molecule used in Table 7.1.

Table 7.1 Brief description of biosensors based on MIPs for interest analyte detection.

Functional monomer	Template molecule	Reference
Aniline	Azythromicin	Jafari et al. (2018)
Pyrrole	*B. cereus*	Lahcen et al. (2018)
o-phenylenediamine	Dopamine	Teng et al. (2017)
TBA	Hemoglobin	Moon et al. (2017)
o-phenylenediamine	HIV	Babamiri et al. (2018)
Dopamine	PSA	Tamboli et al. (2016)

To conclude this section, it should be noted that developing electrochemical biosensors based on MIPs as an element of biorecognition is becoming increasingly popular as it brings many advantages. The production processes can be adapted according to the specific biosensor target; they are chemically and thermally stable. In addition, this will help to reduce the use of animals and, above all, the low cost of the polymers. Where is the biggest challenge within this field of research? In general, the opinion is that the production methods must be optimized as much as possible so that biosensors with real-world applications in the form of detection and diagnosis kits can be produced. Furthermore, these devices should display relevant detection levels from a biological point of view and be scalable for industrial production.

7.8 Future challenges of biosensor technology and conclusions

One of the key aspects of the future of biosensors will be to continue the investigation of the immense variety of options that exist for producing highly accurate drugs and diagnostic instruments. In addition, there is a great need to improve the extraction and analysis of human samples. Implantable biosensors will become increasingly important from the point of view of personalized medical treatments so that the effects and type of drug in the body can be monitored with increasing precision to determine its efficacy. The development of biosensors for the continuous monitoring of fluids such as saliva or blood in a very minimally invasive way will also be of great importance, although the useful life of these devices, as well as their sensitivity and selectivity, will have to be improved. Biosensors provide us with a detailed level of understanding of different medical aspects at the molecular level of biological structures. In addition, the study of biological processes is important in the pharmaceutical, health, and food industries, agriculture, and environmental technology, so the whole of this knowledge has favored the emergence of more and more methods of detection of biomolecules that play a decisive role in drug development, pathogen detection, gene therapy, and multiple biotechnological processes.

It should be noted that the market for biosensors for the treatment of health and medicine is becoming increasingly broad, with devices being developed in multiple areas, such as diagnosis, monitoring of important parameters in the health of patients, or the identification of diseases, which is enabling its great development. In addition, preventive treatments requiring real-time monitoring from these biosensors are becoming increasingly important, which can significantly reduce the resources and time invested. It is, therefore, crucial to develop biomolecular sensors that are increasingly sensitive and selective to the biomolecular targets of interest.

In summary, biosensors are the basis for the creation of personalized therapy through quick measurement and a highly localized system to fight against all kinds of complex medical and health problems. The sum of the combined forces of knowledge in biology, chemistry, and physics will allow us to design more and better biosensors to detect analytes in the environment, food industry, blood plasma, or urine. The objective will tend to be directed toward direct detection in body fluids or the environment, without prior sample preparation or purification, with a much faster analysis and in the absence of specialized laboratories, with the intention that the person can make the analysis himself: non-invasive monitoring and implantable biosensors.

Acknowledgments

We acknowledge the support from the "(MAD2D-CM)-UCM" project funded by Comunidad de Madrid by the Recovery Transformation and Resilience Plan and by Next-Generation EU from the European Union (Ref: PR47/21-MAD2D-CM). Line of Action 3 (LIA 3): Smart materials with advanced functionalities (LIA 3.2. Materials for biomedical applications.).

References

Amine, A., Mohammadi, H., Bourais, I., & Palleschi, G. (2006). Enzyme inhibition-based biosensors for food safety and environmental monitoring. *Biosensors and Bioelectronics*, *21*(8), 1405–1423. Available from https://doi.org/10.1016/J.BIOS.2005.07.012.

Ashraf, H., & Husain, Q. (2010). Use of DEAE cellulose adsorbed and crosslinked white radish (Raphanus sativus) peroxidase for the removal of α-naphthol in batch and continuous process. *International Biodeterioration & Biodegradation*, *64*(1), 27–31. Available from https://doi.org/10.1016/J.IBIOD.2009.10.003.

Babamiri, B., Salimi, A., & Hallaj, R. (2018). A molecularly imprinted electrochemiluminescence sensor for ultrasensitive HIV-1 gene detection using EuS nanocrystals as luminophore. *Biosensors and Bioelectronics*, *117*, 332–339. Available from https://doi.org/10.1016/J.BIOS.2018.06.003.

Bakker, E., & Pretsch, E. (2005). Potentiometric sensors for trace-level analysis. *TrAC Trends in Analytical Chemistry*, *24*(3), 199–207. Available from https://doi.org/10.1016/J.TRAC.2005.01.003.

Barton, A. C., Collyer, S. D., Davis, F., Garifallou, G. Z., Tsekenis, G., Tully, E., O'Kennedy, R., Gibson, T., Millner, P. A., & Higson, S. P. J. (2009). Labeless AC impedimetric antibody-based sensors with pg mL − 1 sensitivities for point-of-care biomedical applications. *Biosensors and Bioelectronics*, *24*(5), 1090–1095. Available from https://doi.org/10.1016/J.BIOS.2008.06.001.

Berepiki, A., Kent, R., Machado, L. F. M., & Dixon, N. (2020). Development of high-performance whole cell biosensors aided by statistical modeling. *ACS Synthetic Biology*, *9*(3), 576–589. Available from https://doi.org/10.1021/acssynbio.9b00448.

Bergveld, P. (1970). Development of an ion-sensitive solid-state device for neurophysiological measurements. *IEEE Transactions on Bio-Medical Engineering*, *17*(1), 70–71.

Botewad, S. N., Gaikwad, D. K., Girhe, N. B., Thorat, H. N., & Pawar, P. P. (2023). Urea biosensors: A comprehensive review. *Biotechnology and Applied Biochemistry*, *70*(2), 485–501. Available from https://doi.org/10.1002/bab.2168.

Bryjak, J., Aniulyte, J., & Liesiene, J. (2007). Evaluation of man-tailored cellulose-based carriers in glucoamylase immobilization. *Carbohydrate Research*, *342*(8), 1105–1109. Available from https://doi.org/10.1016/J.CARRES.2007.02.014.

de Castro, A. C. H., Alves, L. M., Silva Siquieroli, A. C., Madurro, J. M., & Brito-Madurro, A. G. (2020). Label-free electrochemical immunosensor for detection of oncomarker CA125 in serum. *Microchemical Journal*, *155*, 104746. Available from https://doi.org/10.1016/J.MICROC.2020.104746.

Çetin, M. Z., Guven, N., Apetrei, R. M., & Camurlu, P. (2023). Highly sensitive detection of glucose via glucose oxidase immobilization onto conducting polymer-coated composite polyacrylonitrile nanofibers. *Enzyme and Microbial Technology*, *164*, 110178. Available from https://doi.org/10.1016/J.ENZMICTEC.2022.110178.

Cha, G. S., Liu, D., Meyerhoff, M. E., Cantor, H. C., Midgley, A. R., Goldberg, H. D., & Brown, R. B. (2002). Electrochemical performance, biocompatibility, and adhesion of new polymer matrixes for solid-state ion sensors. *Analytical Chemistry*, *63*(17), 1666–1672. Available from https://doi.org/10.1021/ac00017a003.

Chadha, U., Bhardwaj, P., Agarwal, R., Rawat, P., Agarwal, R., Gupta, I., Panjwani, M., Singh, S., Ahuja, C., Selvaraj, S. K., Murali, B., Sonar, P., Badoni, B., & Chakravorty, A. (2022). Recent progress and growth in biosensors technology: A critical review. *Journal of Industrial and Engineering Chemistry*, *109*, 21–51. Available from https://doi.org/10.1016/j.jiec.2022.02.010.

Chen, C., & Wang, J. (2020). Optical biosensors: An exhaustive and comprehensive review. *Analyst*, *145*(5), 1605–1628. Available from https://doi.org/10.1039/c9an01998g.

Chen, I. H., Du, S., Liu, Y., Xi, J., Lu, X., Horikawa, S., Wikle, H. C., Suh, S. J., & Chin, B. A. (2017). Detection of salmonella enterica with magnetoelastic biosensors in wash water containing clorox and chlorine dioxide. *ECS Transactions*, *80*(10), 1557–1564. Available from https://doi.org/10.1149/08010.1557ecst.

Chen, S., & Shamsi, M. H. (2017). Biosensors-on-chip: A topical review. *Journal of Micromechanics and Microengineering*, *27*(8), 083001. Available from https://doi.org/10.1088/1361-6439/aa7117.

Chu, C. Y., Huang, C. C., Tseng, T. W., Tsai, P. H., Wang, C. H., & Chen, C. W. (2021). Mechanically reinforced biodegradable starch-based polyester with the specific Poly(ethylene ether carbonate). *Polymer*, *219*, 123512. Available from https://doi.org/10.1016/j.polymer.2021.123512.

Clark, L. C., & Lyons, C. (1962). Electrode systems for continuous monitoring in cardiovascular surgery. *Annals of the New York Academy of Sciences*, *102*(1), 29–45. Available from https://doi.org/10.1111/j.1749-6632.1962.tb13623.x.

Conroy, P. J., Hearty, S., Leonard, P., & O'Kennedy, R. J. (2009). Antibody production, design and use for biosensor-based applications. *Seminars in Cell & Developmental Biology*, *20*(1), 10–26. Available from https://doi.org/10.1016/J.SEMCDB.2009.01.010.

Cordeiro, C. A., Sias, A., Koster, T., Westerink, B. H. C., & Cremers, T. I. F. H. (2018). In vivo "real-time" monitoring of glucose in the brain with an amperometric enzyme-based biosensor based on gold coated tungsten (W-Au) microelectrodes. *Sensors and Actuators B: Chemical*, *263*, 605–613. Available from https://doi.org/10.1016/J.SNB.2018.02.116.

Csoeregi, E., Quinn, C. P., Schmidtke, D. W., Lindquist, S. E., Pishko, M. V., Ye, L., Katakis, I., Hubbell, J. A., & Heller, A. (2002). Design, characterization, and one-point in vivo calibration of a subcutaneously implanted glucose electrode. *Analytical Chemistry*, *66*(19), 3131–3138. Available from https://doi.org/10.1021/ac00091a022.

D'Orazio, P. (2003). Biosensors in clinical chemistry. *Clinica Chimica Acta*, *334*(1-2), 41–69. Available from https://doi.org/10.1016/S0009-8981(03)00241-9.

Datar, R., Kim, S., Jeon, S., Hesketh, P., Manalis, S., Boisen, A., & Thundat, T. (n.d.). Cantilever sensors: Nanomechanical tools for diagnostics. Available from http://www.mrs.org/bulletin.

Davis, A., Martinez, S., Nelson, D., & Middleton, K. (2010). A tubulin polymerization microassay used to compare ligand efficacy. *Methods in Cell Biology*, *95*(C), 331–351. Available from https://doi.org/10.1016/S0091-679X(10)95018-8.

Dhand, C., Das, M., Datta, M., & Malhotra, B. D. (2011). Recent advances in polyaniline based biosensors. *Biosensors and Bioelectronics*, *26*(6), 2811–2821. Available from https://doi.org/10.1016/J.BIOS.2010.10.017.

Dhiman, A., Kalra, P., Bansal, V., Bruno, J. G., & Sharma, T. K. (2017). Aptamer-based point-of-care diagnostic platforms. *Sensors and Actuators B: Chemical*, *246*, 535–553. Available from https://doi.org/10.1016/J.SNB.2017.02.060.

Doretti, L., Ferrara, D., Gattolin, P., & Lora, S. (1996). Covalently immobilized enzymes on biocompatible polymers for amperometric sensor applications. *Biosensors and Bioelectronics*, *11*(4), 365–373. Available from https://doi.org/10.1016/0956-5663(96)82732-1.

Dzantiev, B., & Zherdev, A. (2013). *Antibody-based biosensors*, 161–196. Available from https://doi.org/10.1201/b15589-7.

Ensafi, A. A. (2019). An introduction to sensors and biosensors. *Electrochemical Biosensors*, 1–10. Available from https://doi.org/10.1016/B978-0-12-816491-4.00001-2.

Fabregat, G., Ballano, G., Armelin, E., Del Valle, L. J., Cativiela, C., & Alemán, C. (2013). An electroactive and biologically responsive hybrid conjugate based on chemical similarity. *Polymer Chemistry*, *4*(5), 1412–1424. Available from https://doi.org/10.1039/c2py20894f.

Feng, X., Castracane, J., Tokranova, N., Gracias, A., Lnenicka, G., & Szaro, B. G. (2007). A living cell-based biosensor utilizing G-protein coupled receptors: Principles and detection methods. *Biosensors and Bioelectronics*, *22*(12), 3230–3237. Available from https://doi.org/10.1016/j.bios.2007.03.002.

Fernández, H., Arévalo, F. J., Granero, A. M., Robledo, S. N., Nieto, C. H. D., Riberi, W. I., & Zon, M. A. (2017). Electrochemical biosensors for the determination of toxic substances related to food safety developed in south America: Mycotoxins and herbicides. *Chemosensors*, *5*(3), 5030023. Available from https://doi.org/10.3390/chemosensors5030023.

Gavalas, V. G., Berrocal, M. J., & Bachas, L. G. (2006). Enhancing the blood compatibility of ion-selective electrodes. *Analytical and Bioanalytical Chemistry*, *384*(1), 65–72. Available from https://doi.org/10.1007/s00216-005-0039-0.

Gerard, M., Chaubey, A., & Malhotra, B. D. (2002). Application of conducting polymers to biosensors. *Biosensors & Bioelectronics*, *17*, 345–359. Available from http://www.elsevier.com/locate/bios.

Geyssant. (1985). Lactate determination with the lactate analyser LA 640: A critical study. *Scandinavian Journal of Clinical and Laboratory Investigation*, *45*(2), 145–149.

Gilles-Gonzalez, M. A., & Gonzalez, G. (2005). Heme-based sensors: Defining characteristics, recent developments, and regulatory hypotheses. *Journal of Inorganic Biochemistry*, *99*(1), 1–22. Available from https://doi.org/10.1016/J.JINORGBIO.2004.11.006.

Golub, E., Albada, H. B., Liao, W. C., Biniuri, Y., & Willner, I. (2015). Nucleoapzymes: Hemin/G-quadruplex DNAzyme – Aptamer binding site conjugates with superior enzyme-like catalytic functions. *Journal of the American Chemical Society*, *138*(1), 164–172. Available from https://doi.org/10.1021/jacs.5b09457.

Grieshaber, D., Mackenzie, R., Vörös, J., & Reimhult, E. (2008). Electrochemical biosensors-sensor principles and architectures. *Sensors*, *8*, 1400–1458. Available from http://www.mdpi.org/sensors.

Griffin, E. G., & Nelson, J. M. (1916). The influence of certain substances on the activity of invertase. *Journal of the American Chemical Society*, *38*(3), 722–730. Available from https://doi.org/10.1021/ja02260a027.

Guilbault, G. G., & Lubrano, G. J. (1973). An enzyme electrode for the amperometric determination of glucose. *Analytica Chimica Acta*, *64*(3), 439–455. Available from https://doi.org/10.1016/S0003-2670(01)82476-4.

Guilbault, G. G., & Montalvo, J. G., Jr. (2002). Urea-specific enzyme electrode. *Journal of the American Chemical Society*, *91*(8), 2164–2165. Available from https://doi.org/10.1021/ja01036a083.

Guisan, J.M., Bolivar, J.M., López-Gallego, F., & Rocha-Martín, J. (Eds.) (n.d.). Immobilization of enzymes and cells methods and protocols. Methods in molecular biology (4th ed.), 2100. Available from http://www.springer.com/series/7651.

Gupta, N., Renugopalakrishnan, V., Liepmann, D., Paulmurugan, R., & Malhotra, B. D. (2019). Cell-based biosensors: Recent trends, challenges and future perspectives. *Biosensors and Bioelectronics*, *141*, 111435. Available from https://doi.org/10.1016/j.bios.2019.111435.

Haleem, A., Javaid, M., Singh, R. P., Suman, R., & Rab, S. (2021). Biosensors applications in medical field: A brief review. *Sensors International*, *2*, 100100. Available from https://doi.org/10.1016/j.sintl.2021.100100.

Hara, T. O., & Singh, B. (2021). Electrochemical biosensors for detection of pesticides and heavy metal toxicants in water: Recent trends and progress. *ACS ES and T Water*, *1*(3), 462–478. Available from https://doi.org/10.1021/acsestwater.0c00125.

Heineman, W. R., & Jensen, W. B. (2006). Leland C. Clark Jr. (1918–2005). *Biosensors and Bioelectronics*, *21*(8), 1403–1404. Available from https://doi.org/10.1016/J.BIOS.2005.12.005.

Hianik, T. (2018). Aptamer-based biosensors. *Encyclopedia of Interfacial Chemistry: Surface Science and Electrochemistry*, 11–19. Available from https://doi.org/10.1016/B978-0-12-409547-2.13492-4.

Hooda, V., Kumar, V., Gahlaut, A., & Hooda, V. (2018). Artificial cells, alcohol quantification: Recent insights into amperometric enzyme biosensors. *Nanomedicine and Biotechnology*, *46*(2), 398–410. Available from https://doi.org/10.1080/21691401.2017.1315426.

Hush, N. S. (2003). An overview of the first half-century of molecular electronics. *Annals of the New York Academy of Sciences*, *1006*, 1–20. Available from https://doi.org/10.1196/annals.1292.016.

Ikanovic, M., Rudzinski, W. E., Bruno, J. G., Allman, A., Carrillo, M. P., Dwarakanath, S., Bhahdigadi, S., Rao, P., Kiel, J. L., & Andrews, C. J. (2007). Fluorescence assay based on aptamer-quantum dot binding to bacillus thuringiensis spores. *Journal of Fluorescence*, *17*(2), 193–199. Available from https://doi.org/10.1007/s10895-007-0158-4.

Jafari, S., Dehghani, M., Nasirizadeh, N., & Azimzadeh, M. (2018). An azithromycin electrochemical sensor based on an aniline MIP film electropolymerized on a gold nano urchins/graphene oxide modified glassy carbon electrode. *Journal of Electroanalytical Chemistry*, *829*, 27–34. Available from https://doi.org/10.1016/J.JELECHEM.2018.09.053.

Jain, R., Jadon, N., & Pawaiya, A. (2017). Polypyrrole based next generation electrochemical sensors and biosensors: A review. *TrAC Trends in Analytical Chemistry*, *97*, 363–373. Available from https://doi.org/10.1016/J.TRAC.2017.10.009.

Jain, R., Miri, S., Pachapur, V. L., & Brar, S. K. (2019). Advances in antibody-based biosensors in environmental monitoring. *Tools, Techniques and Protocols for Monitoring Environmental Contaminants*, 285–305. Available from https://doi.org/10.1016/B978-0-12-814679-8.00014-5.

Justino, C. I. L., Duarte, A. C., & Rocha-Santos, T. A. P. (2017). Recent progress in biosensors for environmental monitoring: A review. *Sensors (Switzerland)*, *17*(12), 17122918. Available from https://doi.org/10.3390/s17122918.

Kahraman, M. V., Bayramoğlu, G., Kayaman-Apohan, N., & Güngör, A. (2007). α-Amylase immobilization on functionalized glass beads by covalent attachment. *Food Chemistry*, *104*(4), 1385–1392. Available from https://doi.org/10.1016/J.FOODCHEM.2007.01.054.

Kapoor, M., & Kuhad, R. C. (2007). Immobilization of xylanase from Bacillus pumilus strain MK001 and its application in production of xylo-oligosaccharides. *Applied Biochemistry and Biotechnology*, *142*(2), 125–138. Available from https://doi.org/10.1007/s12010-007-0013-8.

Kim, J., Imani, S., de Araujo, W. R., Warchall, J., Valdés-Ramírez, G., Paixão, T. R. L. C., Mercier, P. P., & Wang, J. (2015). Wearable salivary uric acid mouthguard biosensor with integrated wireless electronics. *Biosensors and Bioelectronics*, *74*, 1061–1068. Available from https://doi.org/10.1016/J.BIOS.2015.07.039.

Kingshott, P., & Griesser, H. J. (1999). Surfaces that resist bioadhesion. *Current Opinion in Solid State and Materials Science*, *4*(4), 403–412. Available from https://doi.org/10.1016/S1359-0286(99)00018-2.

Kumar, A., Malinee, M., Dhiman, A., Kumar, A., & Sharma, T. K. (2019). Aptamer technology for the detection of foodborne pathogens and toxins. *Advanced Biosensors for Health Care Applications*, 45–69. Available from https://doi.org/10.1016/B978-0-12-815743-5.00002-0.

Kumari, A., & Kayastha, A. M. (2011). Immobilization of soybean (Glycine max) α-amylase onto Chitosan and Amberlite MB-150 beads: Optimization and characterization. *Journal of Molecular Catalysis B: Enzymatic*, *69*(1-2), 8–14. Available from https://doi.org/10.1016/J.MOLCATB.2010.12.003.

Kylilis, N., Riangrungroj, P., Lai, H. E., Salema, V., Fernández, L. Á., Stan, G. B. V., Freemont, P. S., & Polizzi, K. M. (2019). Whole-cell biosensor with tunable limit of detection enables low-cost agglutination assays for medical diagnostic applications. *ACS Sensors*, *4*(2), 370–378. Available from https://doi.org/10.1021/acssensors.8b01163.

Lahcen, A. A., Arduini, F., Lista, F., & Amine, A. (2018). Label-free electrochemical sensor based on spore-imprinted polymer for Bacillus cereus spore detection. *Sensors and Actuators B: Chemical*, *276*, 114–120. Available from https://doi.org/10.1016/J.SNB.2018.08.031.

Lakard, B. (2020). Electrochemical biosensors based on conducting polymers: A review. *Applied Sciences (Switzerland)*, *10*(18), 10186614. Available from https://doi.org/10.3390/APP10186614.

Lang, H. P., Hegner, M., & Gerber, C. (2005). Cantilever array sensors. *Materials Today*, *8*(4), 30–36. Available from https://doi.org/10.1016/S1369-7021(05)00792-3.

Le, T. H., Kim, Y., & Yoon, H. (2017). Electrical and electrochemical properties of conducting polymers. *Polymers*, *9*(4), 9040150. Available from https://doi.org/10.3390/polym9040150.

Lee, S. Y., Show, P. L., Ko, C. M., & Chang, Y. K. (2019). A simple method for cell disruption by immobilization of lysozyme on the extrudate-shaped Na-Y zeolite: Recirculating packed bed disruption process. *Biochemical Engineering Journal*, *141*, 210–216. Available from https://doi.org/10.1016/J.BEJ.2018.10.016.

Li, F., Wang, Z., & Feng, Y. (2009). Construction of bienzyme biosensors based on combination of the one-step electrodeposition and covalent-coupled sol-gel process. *Science in China, Series B: Chemistry*, *52*(12), 2269–2274. Available from https://doi.org/10.1007/s11426-009-0158-0.

Liedberg, B., Nylander, C., & Lunström, I. (1983). Surface plasmon resonance for gas detection and biosensing. *Sensors and Actuators*, *4*(C), 299–304. Available from https://doi.org/10.1016/0250-6874(83)85036-7.

Lin, L., Liu, Q., Wang, L., Liu, A., Weng, S., Lei, Y., Chen, W., Lin, X., & Chen, Y. (2011). Enzyme-amplified electrochemical biosensor for detection of PML–RARα fusion gene based on hairpin LNA probe. *Biosensors and Bioelectronics*, *28*(1), 277–283. Available from https://doi.org/10.1016/J.BIOS.2011.07.032.

Lisdat, F., & Schäfer, D. (2008). The use of electrochemical impedance spectroscopy for biosensing. *Analytical and Bioanalytical Chemistry*, *391*(5), 1555–1567. Available from https://doi.org/10.1007/s00216-008-1970-7.

Liu, C., Yu, H., Zhang, B., Liu, S., Liu, C. G., Li, F., & Song, H. (2022). Engineering whole-cell microbial biosensors: Design principles and applications in monitoring and treatment of heavy metals and organic pollutants. *Biotechnology Advances*, *60*, 108019. Available from https://doi.org/10.1016/J.BIOTECHADV.2022.108019, –108019.

Liu, G. Y., Lan, M. H., Liu, W. M., Wu, J. S., Ge, J. C., & Zhang, H. Y. (2013). Colorimetric detection of carbenicillin using cationic polythiophene derivatives. *Chinese Journal of Polymer Science (English Edition)*, *31*(11), 1484–1490. Available from https://doi.org/10.1007/s10118-013-1355-z.

Madsen, R., Lundstedt, T., & Trygg, J. (2010). Chemometrics in metabolomics—A review in human disease diagnosis. *Analytica Chimica Acta*, *659*(1-2), 23–33. Available from https://doi.org/10.1016/J.ACA.2009.11.042.

Maines, A., Ashworth, D., & Vadgama, P. (1996). Diffusion restricting outer membranes for greatly extended linearity measurements with glucose oxidase enzyme electrodes. *Analytica Chimica Acta*, *333*(3), 223–231. Available from https://doi.org/10.1016/0003-2670(96)00222-X.

Marazuela, M. D., & Moreno-Bondi, M. C. (2002). Fiber-optic biosensors – An overview. *Analytical and Bioanalytical Chemistry*, *372*(5-6), 664–682. Available from https://doi.org/10.1007/s00216-002-1235-9.

Martinkova, P., Kostelnik, A., Valek, T., & Pohanka, M. (2017). Main streams in the construction of biosensors and their applications. *International Journal of Electrochemical Science*, *12*(8), 7386–7403. Available from https://doi.org/10.20964/2017.08.02.

McConnell, E. M., Nguyen, J., & Li, Y. (2020). Aptamer-based biosensors for environmental monitoring. *Frontiers in Chemistry*, *8*, 00434. Available from https://doi.org/10.3389/fchem.2020.00434.

Mislovičová, D., Masárová, J., Vikartovská, A., Gemeiner, P., & Michalková, E. (2004). Biospecific immobilization of mannan–penicillin G acylase neoglycoenzyme on Concanavalin A-bead cellulose. *Journal of Biotechnology*, *110*(1), 11–19. Available from https://doi.org/10.1016/J.JBIOTEC.2004.01.006.

Monošík, R., Stredanský, M., & Šturdík, E. (2012). Biosensors – Classification, characterization and new trends. *Acta Chimica Slovaca*, *5*(1), 109–120. Available from https://doi.org/10.2478/v10188-012-0017-z.

Moon, J. M., Kim, D. M., Kim, M. H., Han, J. Y., Jung, D. K., & Shim, Y. B. (2017). A disposable amperometric dual-sensor for the detection of hemoglobin and glycated hemoglobin in a finger prick blood sample. *Biosensors and Bioelectronics*, *91*, 128–135. Available from https://doi.org/10.1016/J.BIOS.2016.12.038.

Moussy, F., & Harrison, D. J. (2002). Prevention of the rapid degradation of subcutaneously implanted Ag/AgCl reference electrodes using polymer coatings. *Analytical Chemistry*, *66*(5), 674–679. Available from https://doi.org/10.1021/ac00077a015.

Namdeo, M., & Bajpai, S. K. (2009). Immobilization of α-amylase onto cellulose-coated magnetite (CCM) nanoparticles and preliminary starch degradation study. *Journal of Molecular Catalysis B: Enzymatic*, *59* (1-3), 134–139. Available from https://doi.org/10.1016/J.MOLCATB.2009.02.005.

Namsheer, K., & Rout, C. S. (2021). Conducting polymers: A comprehensive review on recent advances in synthesis, properties and applications. *RSC Advances*, *11*(10), 5659–5697. Available from https://doi.org/10.1039/d0ra07800j.

Nascimento, G. A., Souza, E. V. M., Campos-Ferreira, D. S., Arruda, M. S., Castelletti, C. H. M., Wanderley, M. S. O., Ekert, M. H. F., Bruneska, D., & Lima-Filho, J. L. (2012). Electrochemical DNA biosensor for bovine papillomavirus detection using polymeric film on screen-printed electrode. *Biosensors and Bioelectronics*, *38*(1), 61–66. Available from https://doi.org/10.1016/J.BIOS.2012.04.052.

Nelson, J. M., & Griffin, E. G. (1916). Adsorption f invertase. *Journal of the American Chemical Society*, *38* (5), 1109–1115. Available from https://doi.org/10.1021/ja02262a018.

Newman, J. D., & Turner, A. P. F. (2005). Home blood glucose biosensors: A commercial perspective. *Biosensors and Bioelectronics*, *20*(12), 2435–2453. Available from https://doi.org/10.1016/J.BIOS.2004.11.012.

Nikolelis, D.P., & Nikoleli, G.P. (n.d.). Advanced sciences and technologies for security applications biosensors for security and bioterrorism applications. Available from http://www.springer.com/series/5540.

Oldach, L., & Zhang, J. (2014). Genetically encoded fluorescent biosensors for live-cell visualization of protein phosphorylation. *Chemistry & Biology*, *21*(2), 186–197. Available from https://doi.org/10.1016/J.CHEMBIOL.2013.12.012.

Org, W. E., Chairam, S., Buddhalee, P., & Amatatongchai, M. (2013). Electrochemical science: A novel hydrogen peroxide biosensor based on horseradish peroxidase immobilized on poly(aniline-co-o-aminobenzoic acid) modified glassy carbon electrode coated with Chitosan film. *International Journal of Electrochemcal Science*, *8*, 10250–10264. Available from http://www.electrochemsci.org.

Oztekin, Y., Ramanaviciene, A., Yazicigil, Z., Solak, A. O., & Ramanavicius, A. (2011). Direct electron transfer from glucose oxidase immobilized on polyphenanthroline-modified glassy carbon electrode. *Biosensors and Bioelectronics*, *26*(5), 2541–2546. Available from https://doi.org/10.1016/J.BIOS.2010.11.001.

Peng, F., Su, Y., Zhong, Y., Fan, C., Lee, S. T., & He, Y. (2014). Silicon nanomaterials platform for bioimaging, biosensing, and cancer therapy. *Accounts of Chemical Research*, *47*(2), 612–623. Available from https://doi.org/10.1021/ar400221g.

Peng, H., Zhang, L., Soeller, C., & Travas-Sejdic, J. (2009). Conducting polymers for electrochemical DNA sensing. *Biomaterials*, *30*(11), 2132–2148. Available from https://doi.org/10.1016/J.BIOMATERIALS.2008.12.065.

Perumal, V., & Hashim, U. (2014). Advances in biosensors: Principle, architecture and applications. *Journal of Applied Biomedicine*, *12*(1), 1–15. Available from https://doi.org/10.1016/j.jab.2013.02.001.

Pohanka, M., & Skládal, P. (2008). Electrochemical biosensors – Principles and applications. *Journal of Applied Biomedicine*, *6*(2), 57–64. Available from https://doi.org/10.32725/jab.2008.008.

Purohit, B., Vernekar, P. R., Shetti, N. P., & Chandra, P. (2020). Biosensor nanoengineering: Design, operation, and implementation for biomolecular analysis. *Sensors International*, *1*, 100040. Available from https://doi.org/10.1016/J.SINTL.2020.100040.

Ramanavicius, A., Oztekin, Y., & Ramanaviciene, A. (2014). Electrochemical formation of polypyrrole-based layer for immunosensor design. *Sensors and Actuators B: Chemical*, *197*, 237–243. Available from https://doi.org/10.1016/J.SNB.2014.02.072.

Rathee, K., Dhull, V., Dhull, R., & Singh, S. (2016). Biosensors based on electrochemical lactate detection: A comprehensive review. *Biochemistry and Biophysics Reports*, *5*, 35–54. Available from https://doi.org/10.1016/j.bbrep.2015.11.010.

Reddy, S. M., & Vadgama, P. M. (1997). A study of the permeability properties of surfactant modified poly (vinyl chloride) membranes. *Analytica Chimica Acta*, *350*(1-2), 67–76. Available from https://doi.org/10.1016/S0003-2670(97)00314-0.

Riangrungroj, P., Bever, C. S., Hammock, B. D., & Polizzi, K. M. (2019). A label-free optical whole-cell Escherichia coli biosensor for the detection of pyrethroid insecticide exposure. *Scientific Reports*, *9*(1), 12466. Available from https://doi.org/10.1038/s41598-019-48907-6.

Ron, E. Z., & Rishpon, J. (2010). Electrochemical cell-based sensors. *Advances in Biochemical Engineering/Biotechnology*, *117*, 77–84. Available from https://doi.org/10.1007/10_2009_17.

Rubab, M., Shahbaz, H. M., Olaimat, A. N., & Oh, D. H. (2018). Biosensors for rapid and sensitive detection of Staphylococcus aureus in food. *Biosensors and Bioelectronics*, *105*, 49–57. Available from https://doi.org/10.1016/J.BIOS.2018.01.023.

Sassolas, A., Blum, L. J., & Leca-Bouvier, B. D. (2012). Immobilization strategies to develop enzymatic biosensors. *Biotechnology Advances*, *30*(3), 489–511. Available from https://doi.org/10.1016/J.BIOTECHADV.2011.09.003.

Sharma. (1994). *Biosensors. Meas. Sci. Technol.*, *5*, 461–467.

Singh, G., Kaur, M., Kaur, H., & Kang, T. S. (2021). Synthesis and complexation of a new caffeine based surface active ionic liquid with lysozyme in aqueous medium: Physicochemical, computational and antimicrobial studies. *Journal of Molecular Liquids*, *325*, 115156. Available from https://doi.org/10.1016/j.molliq.2020.115156.

Sirisha, V. L., Jain, A., & Jain, A. (2016). Enzyme immobilization: An overview on methods, support material, and applications of immobilized enzymes. *Advances in Food and Nutrition Research*, *79*, 179–211. Available from https://doi.org/10.1016/BS.AFNR.2016.07.004.

Soldatkin, O. O., Kucherenko, I. S., Pyeshkova, V. M., Kukla, A. L., Jaffrezic-Renault, N., El'skaya, A. V., Dzyadevych, S. V., & Soldatkin, A. P. (2012). Novel conductometric biosensor based on three-enzyme system for selective determination of heavy metal ions. *Bioelectrochemistry (Amsterdam, Netherlands)*, *83* (1), 25–30. Available from https://doi.org/10.1016/J.BIOELECHEM.2011.08.001.

Song, K. M., Lee, S., & Ban, C. (2012). Aptamers and their biological applications. *Sensors*, *12*(1), 612–631. Available from https://doi.org/10.3390/s120100612.

Song, S., Wang, L., Li, J., Fan, C., & Zhao, J. (2008). Aptamer-based biosensors. *TrAC Trends in Analytical Chemistry*, *27*(2), 108–117. Available from https://doi.org/10.1016/J.TRAC.2007.12.004.

Soylemez, S., Udum, Y. A., Kesik, M., Gündoɀdu Hizliateş, C., Ergun, Y., & Toppare, L. (2015). Electrochemical and optical properties of a conducting polymer and its use in a novel biosensor for the detection of cholesterol. *Sensors and Actuators B: Chemical*, *212*, 425–433. Available from https://doi.org/10.1016/J.SNB.2015.02.045.

Srivastava, K. R., Awasthi, S., Mishra, P. K., & Srivastava, P. K. (2020). Biosensors/molecular tools for detection of waterborne pathogens. *Waterborne Pathogens: Detection and Treatment*, 237–277. Available from https://doi.org/10.1016/B978-0-12-818783-8.00013-X.

Srivastava, M., Srivastava, S.K., Nirala, N.R., & Prakash, R. (2010). 2:1;1-A Volume 2 | Number 1 | 2 (1), 1–100. Available from http://www.rsc.org/methods.

Stanciu, L.A., Wei, Q., Barui, A.K., & Mohammad, N. (2021). Annual review of biomedical engineering recent advances in aptamer-based biosensors for global health applications. Available from https://doi.org/10.1146/annurev-bioeng-082020.

Stoltenburg, R., Reinemann, C., & Strehlitz, B. (2007). SELEX—A (r)evolutionary method to generate high-affinity nucleic acid ligands. *Biomolecular Engineering*, *24*(4), 381–403. Available from https://doi.org/10.1016/J.BIOENG.2007.06.001.

Su, X., Sutarlie, L., & Loh, X. J. (2020). Sensors, biosensors, and analytical technologies for aquaculture water quality. *Research: A Journal of Science and its Applications*, *2020*, 8272705. Available from https://doi.org/10.34133/2020/8272705.

Suresh, R., Rajendran, S., Khoo, K. S., & Soto-Moscoso, M. (2023). Enzyme immobilized nanomaterials: An electrochemical bio-sensing and biocatalytic degradation properties toward organic pollutants. *Topics in Catalysis*, *66*(9–12), 691–706. Available from https://doi.org/10.1007/s11244-022-01760-w.

Tamboli, V. K., Bhalla, N., Jolly, P., Bowen, C. R., Taylor, J. T., Bowen, J. L., Allender, C. J., & Estrela, P. (2016). Hybrid synthetic receptors on MOSFET devices for detection of prostate specific antigen in human plasma. *Analytical Chemistry*, *88*(23), 11486–11490. Available from https://doi.org/10.1021/acs.analchem.6b02619.

Teng, Y., Liu, F., & Kan, X. (2017). Voltammetric dopamine sensor based on three-dimensional electrosynthesized molecularly imprinted polymers and polypyrrole nanowires. *Microchimica Acta*, *184*(8), 2515–2522. Available from https://doi.org/10.1007/s00604-017-2243-y.

Thompson, B. C., & Reynolds, J. R. (2019). *Handbook of conducting polymers. 2 Volume Set* (Fourth Edition). Boca Ratón: CRC Press.

Tripathi, A., & Savio Melo, J. (Eds.) (n.d.). Gels horizons: From science to smart materials. Available from http://www.springer.com/series/15205.

Turner, A. P. F. (2013). Biosensors: Sense and sensibility. *Chemical Society Reviews*, *42*(8), 3184–3196. Available from https://doi.org/10.1039/c3cs35528d.

Turner, N. W., Jeans, C. W., Brain, K. R., Allender, C. J., Hlady, V., & Britt, D. W. (2006). From 3D to 2D: A review of the molecular imprinting of proteins. *Biotechnology Progress*, *22*(6), 1474–1489. Available from https://doi.org/10.1021/bp060122g.

Turner, R. F. B., & Sherwood, C. S. (2009). Biocompatibility of perfluorosulfonic acid polymer membranes for biosensor applications. *ACS Symposium Series*, 211–221. Available from https://doi.org/10.1021/bk-1994-0556.ch017.

Ueda, T., Oshida, H., Kurita, K., Ishihara, K., & Nakabayashi, N. (1992). Preparation of 2-methacryloyloxyethyl phosphorylcholine copolymers with alkyl methacrylates and their blood compatibility. *Polymer Journal*, *24*, 1259–1269.

Updike, S. J. (1967). The enzyme electrode. *Nature*, *214*, 986–988.

Ushaa, S. M., Madhavilatha, M., & Rao, G. M. (2011). Design and analysis of nanowire sensor array for prostate cancer detection. *International Journal of Nano and Biomaterials*, *3*(3), 239–255.

Verma, N., & Bhardwaj, A. (2015). Biosensor technology for pesticides—A review. *Applied Biochemistry and Biotechnology*, *175*(6), 3093–3119. Available from https://doi.org/10.1007/s12010-015-1489-2.

Wackerlig, J., & Lieberzeit, P. A. (2015). Molecularly imprinted polymer nanoparticles in chemical sensing – Synthesis, characterisation and application. *Sensors and Actuators B: Chemical*, *207*(Part A), 144–157. Available from https://doi.org/10.1016/J.SNB.2014.09.094.

Wackerlig, J., & Schirhagl, R. (2015). Applications of molecularly imprinted polymer nanoparticles and their advances toward industrial use: A review. *Analytical Chemistry*, *88*(1), 250–261. Available from https://doi.org/10.1021/acs.analchem.5b03804.

Wang, J., Jiang, M., Fortes, A., & Mukherjee, B. (1999). New label-free DNA recognition based on doping nucleic-acid probes within conducting polymer films. *Analytica Chimica Acta*, *402*(1–2), 7–12. Available from https://doi.org/10.1016/S0003-2670(99)00531-0.

Wang, Q., She, W., Lu, X., Li, P., Sun, Y., Liu, X., Pan, W., & Duan, K. (2019). The interaction of hyaluronic acid and graphene tuned by functional groups: A density functional study. *Computational and Theoretical Chemistry*, *1165*, 112559. Available from https://doi.org/10.1016/j.comptc.2019.112559.

Wang, X., & Uchiyam, S. (2013). *Polymers for biosensors construction. State of the Art in Biosensors – General Aspects*. InTech. Available from https://doi.org/10.5772/54428.

Wang, Y., Yu, H., Li, Y., Wang, T., Xu, T., Chen, J., Fan, Z., Wang, Y., & Wang, B. (2019). Facile preparation of highly conductive poly(amide-imide) composite films beyond 1000 S m-1 through ternary blend strategy. *Polymers*, *11*(3). Available from https://doi.org/10.3390/polym11030546.

Wilson, G. S., & Hu, Y. (2000). Enzyme-based biosensors for in vivo measurements. *Chemical Reviews*, *100* (7), 2693–2704. Available from https://doi.org/10.1021/cr990003y.

Wisniewski, N., & Reichert, M. (2000). Methods for reducing biosensor membrane biofouling. *Colloids and Surfaces B: Biointerfaces*, *18*(3-4), 197–219. Available from https://doi.org/10.1016/S0927-7765(99)00148-4.

Wu, L., Li, X., Miao, H., Xu, J., & Pan, G. (2022). State of the art in development of molecularly imprinted biosensors. *View*, *3*(3), 20200170. Available from https://doi.org/10.1002/VIW.20200170.

Xing, G. W., Li, X. W., Tian, G. L., & Ye, Y. H. (2000). Enzymatic peptide synthesis in organic solvent with different zeolites as immobilization matrixes. *Tetrahedron*, *56*(22), 3517–3522. Available from https://doi.org/10.1016/S0040-4020(00)00261-1.

Yi, J., Xiao, W., Li, G., Wu, P., He, Y., Chen, C., He, Y., Ding, P., & Kai, T. (2020). The research of aptamer biosensor technologies for detection of microorganism. *Microbiology and Biotechnology*, *104*, 9877–9890. Available from https://doi.org/10.1007/s00253-020-10940-1.

Yoo, H., Jo, H., & Oh, S. S. (2020). Detection and beyond: Challenges and advances in aptamer-based biosensors. *Materials Advances*, *1*(8), 2663–2687. Available from https://doi.org/10.1039/d0ma00639d.

Yoshikawa, M., Tharpa, K., & Dima, Ş. O. (2016). Molecularly imprinted membranes: Past, present, and future. *Chemical Reviews*, *116*(19), 11500–11528. Available from https://doi.org/10.1021/acs.chemrev.6b00098.

Zapp, E., Brondani, D., Vieira, I. C., Scheeren, C. W., Dupont, J., Barbosa, A. M. J., & Ferreira, V. S. (2011). Biomonitoring of methomyl pesticide by laccase inhibition on sensor containing platinum nanoparticles in ionic liquid phase supported in montmorillonite. *Sensors and Actuators B: Chemical*, *155*(1), 331–339. Available from https://doi.org/10.1016/J.SNB.2011.04.015.

Zhang, C., Guo, J., Zou, X., Guo, S., Guo, Y., Shi, R., & Yan, F. (2021). Acridine-based covalent organic framework photosensitizer with broad-spectrum light absorption for antibacterial photocatalytic therapy. *Advanced Healthcare Materials*, *10*(19), 202100775. Available from https://doi.org/10.1002/adhm.202100775.

CHAPTER

Hybrid polymer-based sensors

8

Hongzhi Liu and Rungthip Kunthom

International Center for Interdisciplinary Research and Innovation of Silsesquioxane Science, Key Laboratory of Special Functional Aggregated Materials, Ministry of Education, School of Chemistry and Chemical Engineering, Shandong University, Jinan, Shandong, P. R. China

8.1 Introduction

The growth of new materials with specially designed chemical, mechanical, or physical properties becomes increasingly important in many fields of practical applications, especially sensors. Recent years have seen increased demand for hybrid polymer materials in a wide range of applications (Kalia & Pielichowski, n.d.; Yan, Yang, & Liu, 2020; Irfan, n.d.). Global research and development in the field of sensors has increased tremendously in industrial investment, as well as academic literature. The need for harmful gases and contaminants sensors for environmental monitoring has increased dramatically, leading to the development of several sensor devices for environmental, medical, industrial, and agricultural technologies. Some molecular or polymer sensors are limited in their use in chemosensors and biosensors due to their narrow chemical and physical properties. To overcome this limitation, hybrid polymers have been interestingly chosen to be the one choice in the design of sensors because they can modify their chemical and physical characteristics for specific requirements by combining two or more parts. These needs have been met with accelerated advances in polymer science and technology, taking advantage of nanophase technology geared toward better enhancements of hybrid polymer materials. To accomplish the various practical requirements, advanced material design is necessary. The nanostructure, degree of organization, and properties that can be obtained for such materials certainly depend on the chemical nature of their components, but they also rely on the synergy between these components (Wu et al., 2012).

Hybrid polymers have drawn a lot of attention because they provide outstanding features for a range of applications with the right design and synthesis. These hybrid polymers are often built utilizing functional building blocks with stiff structures employing the correct polymerization techniques to produce materials with unique features. Hybrid polymers are made up of two different types of monomers/polymers, which can be two organic moieties or an organic and inorganic mixture. Certain qualities of both components are combined in these combinations, which cannot be obtained in a single polymer alone. Epoxy-resin-based compounds, for example, offer several great qualities, such as quick drying at normal temperatures, strong adherence to most surfaces, toughness, and chemical resistance to a wide range of weak acids, alkalis, and solvents. Rees et al.

Sensory Polymers. DOI: https://doi.org/10.1016/B978-0-443-13394-7.00008-2

leveraged the unique attributes of epoxy resin and reported that incorporating a polysulfide component into epoxy resin improves some properties without altering the epoxy system's existing performance capabilities (Rees et al., n.d.). Viscosity reduction, adhesion enhancement, flexibility introduction, better impact strength, thermal stability, corrosion resistance, regulated damping properties, and improved chemical resistance are all advantages of such adjustments. Polyurethanes, alkyds, acrylics, polyesters, silicones, and other polymers can be mixed to increase performance or achieve desired outcomes. Many polymer institutions are conducting research, and new developments are being announced regularly. However, to the best of our knowledge, the use of hybrid polymers constructed from two organic monomers/polymers in sensor applications has rarely been studied (Alam et al., 2022; Cichosz et al., 2018; Park et al., 2017).

Hybrid polymers, composed of inorganic and organic components ranging in size from a few nanometers to tens of nanometers, are emerging as a highly potent and promising material category. The high stability of the inorganic part has made the construction of organic-inorganic hybrid materials a highly researched and currently popular topic. In the case of general hybrid materials, mostly the organic component holds the inorganic constituents and/or soft tissue together, while the inorganic part gives them mechanical strength and structure as a whole. These hybrid polymers are frequently the most integrated intelligent systems, which are skilled in striking trade-offs between various tasks such as mechanical behavior, density, controlled permeability, color, and hydrophobicity.

Additionally, hybrid porous polymers are of interest for use in sensor applications due to their structural characteristics, considering pore geometry, size, and surface functionality. Some porous polymers can keep their porosity even after being dissolved and processed, a unique property that other types of porous materials, such as activated carbons, zeolites, or porous silicas, do not have (Wu et al., 2012; Irfan, n.d.; Soldatov & Liu, 2021). The polymeric framework structure for porous polymers typically includes composition, topology, and functionality. Surface area is a key parameter for determining pore structure. The goal of producing more advanced porous frameworks is to develop materials with a large number of pores, a high surface area, and precisely controlled pore sizes. The porosity properties, such as geometry, surface areas, and pore volume, are derived from the Brunauer-Emmett-Teller (BET) isotherms. Using the diameter of the pore size, porous polymers can be classified as microporous polymers (<2 nm), mesoporous polymers (2–50 nm), and macroporous polymers (>50 nm) (Soldatov & Liu, 2021). Because there are many ways to make them, many hybrid porous polymers with different chemical functions (either in the porous framework or at the pore surface) have been made. The challenging preparation of hybrid porous polymers is for designing and constructing the pore structures, as well as for customizing the functionalities inside the framework or on the pore surface. Porous polymers are classified into three main categories based on their chemical makeup: Organics-organics such as polymers with intrinsic microporosity, hypercrosslinked porous polymers, conjugated microporous polymers, and covalent organic frameworks (COFs). These are included in the first class. Inorganics such as porous zeolites, porous silica, and carbon are examples of the second kind. Hybrid inorganic-organic materials fall under the third category, which includes periodic mesoporous organosilicas (Fujita & Inagaki, 2008; Croissant et al., 2016) and metal-organic frameworks (MOFs). Herein, hybrid polymers with an organic and inorganic combination are shown into three categories: metal-organic framework (MOF)-based, silsesquioxane (SQ)-based, and other-based polymers.

An overview of the design and application of hybrid polymers for sensing versatile analysts is described. The design ideas of hybrid polymers in many different types of monomers (based on MOF and SQ) are reviewed. The unique sensing of polymers is based on the functional group or moieties in those crosslinked polymers, as will be discussed.

8.2 Discussion

Researchers have been exploring different approaches to synthesizing materials with hybrid polymers to improve their nature. The development of organic-inorganic hybrid materials with distinctive structures and particular properties has been studied over the past decades. This class of materials is crucial in the creation of high-performance materials because they combine the benefits of both conventional organic and inorganic. Herein, hybrid polymers are classified into three main groups based on the linker-typed polymer, as shown below.

8.2.1 MOF-based hybrid polymer

In decades, MOFs have emerged as a new class of hybrid porous polymeric materials. The units of MOFs are made up of metals that are linked together with organic molecules to form crystal structures with defined porosity. Keeping the pore diameter in the micropore range will increase the surface area of the framework. There are a few examples of MOFs with ultrahigh porosity (BET surface areas >5000 $m^2\ g^{-1}$). Molecular design has become increasingly prevalent in porous solids due to the framework design principles outlined for MOFs. Equally important are postsynthetic modifications of the organic linkers and metal-containing units (Zhang et al., 2022). One example is when aniline polymerized in the pores of HKUST-1 on the surface of a platinum electrode. This made a fully porous, conductive polymeric monolith (Lu et al., 2014). Deposition of a thin layer of HKUST-1 onto the surface of a platinum electrode, followed by electropolymerization of aniline within the pores, created a microporous polyaniline structure with a surface area of 986 $m^2\ g^{-1}$ (Yan, Yang, & Xu, Zhang, et al., 2020), indicating a highly rigid polymer, and electrical conductivity of 0.125 S cm^{-1}, which is significantly higher than any MOF or COFs to date. Interestingly, MOFs retain their crystalline and underlying structure as compared to other frameworks, which is an advantage for sensor applications. MOF-based mixed-matrix membranes (MMMs) can be applied for toxic industrial chemical sorption. For example, Zhang et al. used Al-MIL-53-NO_2/PVDF MMMs as a novel platform for chemical sensing applications as fluorescent turn-on sensors for hydrogen sulfide. 70 wt% MMMs loaded on poly(vinylidene fluoride) (PVF) were fabricated, where the reduction of nitro groups on the MOF ligands by hydrogen sulfide resulted in a detectable fluorescence from Al-MIL-53-NH_2. It was found that the MMMs showed high selectivity and sensitivity for hydrogen sulfide over acid and halogen salts, as well as other reducing agents, with a detection limit of 92 nM, which is lower than the reported free MOF powder by 3 orders of magnitude attributed to the availability of MOF in membrane over aggregated powder (Zhang et al., 2018). Examples of hybrid polymers based on MOF for sensors are listed in Table 8.1.

Table 8.1 Examples of hybrid polymers based on MOF for sensors.

Polymer	Surface area ($m^2\ g^{-1}$)	Fields of applications/ analysts	Sensor characteristics	Ref.
IL-COOH/ Fe_3O_4@UiO-67-bpydc	438.5	Extraction and sensitive detection of fluoroquinolone antibiotics in environmental water.	The recoveries of environmental water ranged from 90.0% to 110.0%, and the detection limits were lower than 0.02 $\mu g\ L^{-1}$.	Lu et al. (2021)
TPB–HCP	717	Reversible adsorption for berberine hydrochloride (BH) and other cation organic dyes from water.	Excellent adsorption capacity for BH owing to the synergistic effects of size matching and electrostatic interaction.	Zhang et al. (2018)
MSP	N/A	Piezoresistive sensor.	High sensitivity for a broad pressure range (147 kPa^{-1} for less than 5.37 kPa region and 442 kPa^{-1} for 5.37–18.56 kPa region), a low detection limit of 9 Pa, a rapid response time, and an excellent durability over 10,000 cycles.	Yue et al. (2018)

8.2.2 SQ-based hybrid polymer

SQ is a building block for creating a new class of hybrid polymers, considerably enhancing the family of hybrid materials (Du & Liu, 2020; Dudziec et al., 2019; Lichtenhan et al., 1995; Zhang & Müller, 2013; Ye et al., 2011; Raftopoulos & Pielichowski, 2016). SQ is an organic-inorganic hybrid compound with an inner Si-O-Si core and an outside core of organic groups that are easily modified into versatile functional groups. Their structures include random, cage-like silsesquioxane (cage-SQ) or polyhedral oligomeric silsesquioxane (POSS), double-decker silsesquioxane (DDSQ), and ladder structures (Mituła et al., 2017; Xu et al., 2013; Chaiprasert et al., 2021; Chimjarn et al., 2015; Kunthom et al., 2018). Their chemical structures contain the fundamental formula $(RSiO1.5)n$ (n = 6, 8, 10, 12, 14, 16, 18, etc.) (Alves et al., 2013; Chaiprasert et al., 2021; Chen et al., 2021; Chiang et al., 2023; Chimjarn et al., 2015; Cichosz et al., 2018; Croissant et al., 2016; Dong et al., 2018; Du & Liu, 2021; Du et al., 2019; Du & Liu, 2018, 2020; Dudziec et al., 2019; Fujita & Inagaki, 2008; Ge & Liu, 2016; Ghanbari et al., 2011; He et al., 2017; Huang et al., 2018; Ishizaki et al., 2018; Jin et al., 2018; Kakuta et al., 2015; Kausar, 2017; Knauer et al., 2017; Kunthom et al., 2018; Kwon et al., n.d.; Laird et al., 2021; Lee et al., 2006; Li et al., 2019; Li et al., 2011; Li et al., 2017; Lichtenhan et al., 1995; Liu & Liu, 2017; Liu et al., 2013; Lu et al., 2014; Lu et al., 2021; Lu et al., 2019; Lv et al., 2022; Meng et al., 2020; Mituła et al., 2017; Ni et al., 2014; Noureddine et al., 2014; Oleksy & Galina, 2013; Park et al., 2017; Petit et al., 2013; Raftopoulos & Pielichowski, 2016; Raimondo et al., 2015;

Rikowski & Marsmann, 1997; Sangtrirutnugul et al., 2017; Scott, 1946; Silverstein et al., 2005; Soldatov & Liu, 2021; Suenaga et al., 2021; Sun et al., 2020; Sun, Feng, et al., 2018; Sun, Huo, et al., 2018; Wang et al., 2014; Wang et al. 2016, 2017; Wang et al., 2022; Wang et al., 2020; Wang et al., 2015; Wang et al., 2023; Wang et al., 2021; Wu et al., 2010; Xu et al., 2007; Xu et al., 2013; Yan et al., 2019; Yan, Yang, & Liu, 2020; Yang & Liu, 2020; Yang et al., 2015; Yang & Liu, 2019; Yang & Liu, 2021; Ye et al., 2011; Yue et al., 2018; Zhang et al., 2022; Zhang et al., 2018; Zhang & Müller, 2013; Zhang et al., 2012; Zhang et al., 2018; Zhao et al., 2023; Zhou et al., 2014; Zhou et al., 2017; Rikowski & Marsmann, 1997; Laird et al., 2021; Chimjarn et al., 2015). The inorganic nature and multiple reactive functionalities of SQ make these modifiable nanoparticles ideal for the construction of hybrid materials with tunable properties. SQ has distinctive features because the organic-inorganic core-shell particles undergo molecular hybridization, and the peripheral organic groups can be further functionalized as needed. Due to these properties of SQs, they are a flexible framework for creating many kinds of useful hybrid materials. Utilizing the advantages of their distinctive structures, functional SQ precursors can interact with a wide range of substrates to produce target hybrid materials with particular functions. SQs have been used as building blocks (Ge & Liu, 2016), crosslinkers (Wu et al., 2010), skeletons (Huang et al., 2018), nanofillers (Petit et al., 2013), and other components of hybrid polymers over the past few decades. They can significantly increase the stability, mechanical properties, and oxidation resistance of polymers while also satisfying the needs of both science and industry. SQ-based hybrid polymers (SQ-HPs) have considerable potential in versatile application areas (e.g., catalysis (Zhao et al., 2023), polymer science (Kausar, 2017), nanomedicine (Ghanbari et al., 2011), flame retardant (Raimondo et al., 2015), optical engineering (Zhou et al., 2017), and optoelectronic (Li et al., 2017)), including sensing applications, when combined with their finely programmable or tailored physical and chemical properties. Therefore, the creation of SQ-HPs has been a focus of research nowadays. A summary of recent developments in SQ-HPs would be extremely important because it might help materials scientists create new and better-performing SQ-HPs by designing and synthesizing them. In order to encourage future study of SQ-HPs on sensor applications, we will detail the recent advancements in SQ-HPs, ranging from functional monomer design and polymer synthesis to sensory application.

8.2.2.1 Preparation and design

SQ chemical reagents are nanostructured with sizes of 1–3 nm and can be thought of as the smallest particles of silica possible. However, unlike silica or silicones, each SQ molecule may contain nonreactive organic substituents that make the SQ nanostructure compatible with monomers or polymers. SQ molecules can be incorporated into polymer chains or networks through polymerization or grafting thanks to versatile functional groups such as olefins, acrylates, phenols, fluoroalkyls, halides, amines, sulfhydryls, azides, nitriles, carbazoles, imidazolium salts, azobenzenes, and cinnamates. A variety of SQs can be prepared *via* direct reactions such as hydrolysis and condensation, modifications, or postfunctionalization reactions. A possible method of modification is to tune the organic groups that surround the inorganic core. The following general methods can be used to create SQ-HPs: (1) physical blending to deposit or embed SQs into a polymer matrix (Li et al., 2011); (2) grafting SQs as pendants onto polymers or as blocks or terminal groups of the main chain (Liu et al., 2013); (3) specific reactions to produce polymers with various topological architectures, such as star-shaped (Zhang et al., 2012), hyperbranched (Huang et al., 2018), and dendritic

(He et al., 2017); (4) self-polymerization or copolymer (Sangtrirutnugul et al., 2017; Lu et al., 2019). Radical polymerization methods (Alves et al., 2013), atomic transfer radical polymerization (Jin et al., 2018), reversible addition-fragmentation chain transfer polymerization (RAFT) (Ishizaki et al., 2018), ionic polymerization (Du et al., 2019), ring-opening polymerization (Xu et al., 2007), click chemistry (Ni et al., 2014; Noureddine et al., 2014), Heck coupling reaction (Ge & Liu, 2016), and Friedel-Crafts reaction (Liu & Liu, 2017; Yang & Liu, 2019) are among the reactions involved. The majority of cage SQs are very soluble in common solvents (such as toluene, tetrahydrofuran, dichloromethane, and chloroform), and depending on the functional groups of the cage SQs, SQ-HPs can be produced in a variety of ways.

A variety of SQ-containing copolymers have been prepared using condensation, ring-opening metathesis, and radical, both conventional and atom-transfer, copolymerization techniques. Many varieties of substituents may be affixed to the Si atoms at the corners of the cages to optimize polymer–SQ-HPs interactions and simultaneously aid in the easy dissolution of the SQ moieties in common solvents for synthesis purposes. These groups may be incorporated into almost any conventional polymer and in a variety of different chain architectures. Thus, the inorganic nature and multiple reactive functionalities of SQ make these compounds ideal for their use in the construction of organic-inorganic hybrid nanomaterials.

Incorporation of cage-SQ into polymers often results in interesting improvements in materials' properties, including increases in use temperature, oxidation resistance, and surface hardening, resulting in improved mechanical properties, as well as reductions in flammability and heat evolution. Several works have focused on developing a repertoire of functionalized cage-SQ macromonomers as a starting point for formulating nanocomposites, with the objective of determining whether well-defined monomer/composite nanostructures and periodically placed organic/inorganic components might offer novel and predictable properties.

Many fundamental mass-producible polymers, including polyolefins (Lee et al., 2006), polyethers (Knauer et al., 2017), polyesters (Oleksy & Galina, 2013), polyamides, polystyrene, polyacrylates, polyurethanes, carbon fibers, epoxy resins, and polyethers, can be cooperated with cage SQs through copolymerization, blending, grafting, and other chemical reactions. The fabrication of such SQ-HPs by combining them with polymers that can be generated in laboratories and can be transferred to industrial-scale manufacturing nowadays.

SQ-HPs have stood out among the possibilities for sensing because of their exceptional qualities, including strong stability, quick response, high sensitivity, and emissive intensity (Qingzheng Wang et al., 2020). More importantly, the cage-SQ cage aids in preventing or lessening fluorescence quenching brought on by chromophores that are prone to aggregation or intermolecular interaction (Zhou et al., 2014; Sun et al., 2018).

Liu et al. have developed numerous cage SQ-based detection systems that can detect antibiotics, nitroaromatic compounds (NACs), and heavy metal ions (Fe^{3+}, Ru^{3+}, and Cu^{2+}) employing cross-linkers such as hexaphenylsilole, tetraphenylethene, carbazole, pyrene, and triphenylamine derivatives (Liu & Liu, 2017; Du & Liu, 2018; Yang & Liu, 2020; Wang et al., 2020; Yan et al., 2019; Wang et al., 2020; Li et al., 2019).

Wang et al. developed aminopropyl substituted T_8 SQ hydrochloride salt functionalized hybrid carbon dots for bioimaging, demonstrating excellent photoluminescence stability and photobleaching resistance in the presence of a biological sample matrix, which has been demonstrated to be suitable for cell imaging. Chiara et al.'s well-defined hetero-bifunctional cage SQs, which combine a cluster of sugar epitopes and a fluorescent dye to generate a series of derivatives, have been shown to be efficient probes for bioimaging of cell surface receptors.

Compared with small organic molecules or traditional polymers, the presence of cage SQs provides more space for capturing analytes, reduces barriers to mass transport, and offers more diffusion pathways for exciton migration. Accordingly, SQ-HPs as effective and reliable probes show promise as powerful tools in sensing applications.

SQs are introduced as an inorganic part, becoming a hybrid polymer *via* several methods. For instance, in 1995, Lichtenhan et al. described the preparation of linear polymers derived from polymerization of methacrylate-functionalized SQ monomers. This work provided new classes of monomers, polymers, and additives. Specifically, the thermal stability of obtained polymers is higher than poly(methyl methacrylate) (PMMA) due to the incorporation of SQ (Lichtenhan et al., 1995). Meanwhile, Hwang et al. presented a versatile strategy to prepare star-shaped SQ-containing polymer hybrids using RAFT polymerization and click chemistry so that the polymer's length, architecture, and graft density are controllable. Alkyne-bearing SQ is a core, and their modified structure leading to the 8CTAs-SQ used for methyl methacrylate (MMA) radical polymerization had a high surface density as compared with previous studies.

8CTAs-SQ is an excellent chain transfer agent for MMA polymerization, as indicated by its narrow polydispersity observed in contrast to that found with linear polymerization. This provides insights into the dependence of M_n and PDI on conversion for MMA polymerization. The unique confined geometry and the higher polymerization rate of the anchored intermediate macro-RAFT radical on the SQ surface may contribute to the high opportunity for star and linear chain radical termination in surface modification applications (Ye et al., 2011).

Cage-like organosiloxanes are cage-SQ and spherosilicate (Q) that are 3D organic-inorganic hybrid nanosized molecules with a silica-like core and substituents attached to each vertex of the cage. These cage components are excellent building blocks for manufacturing innovative hybrid porous polymers with a variety of intriguing applications due to their rigid core that may be linked to numerous organic moieties (Du & Liu, 2020; Soldatov & Liu, 2021) (Kalia & Pielichowski, n.d.).

By choosing suitable organic fluorophores and carrying out the appropriate chemical processes, functionalized cage-based fluorescent hybrid polymers can be effectively prepared. SQs-incorporated conjugated polymer network reduces aggregation or $\pi-\pi$ interaction in polymer, leading to an increase in fluorescence intensity (Dudziec et al., 2019; Zhou et al., 2014). SQ-HP is utilized to detect a variety of substances because they have high luminescence and stability by quenching fluorescence signals compared with fluorescent monomers.

Recently, Wang et al. produced a hybrid porous polymer called PCS-DPB, which is based on SQ (Wang et al., 2022). This hybrid polymer exhibits exceptional luminescence and a significant red shift, making it highly ideal for use in sensor applications. PCS-DPB is very good at detecting DNP and Ru^{3+} ions because it is very selective and sensitive. Nitrogen adsorption studies verified the presence of both microporous and mesoporous structures in PCS-DPB. The porosity characteristics of PCS-DPB were examined, revealing surface areas of 418 (PCS-DPB-a), 1292 $m^2 g^{-1}$ (PCS-DPB), and 973 $m^2 g^{-1}$ (PCS-DPB-b). The best molar ratio of OVS to CN-DPB in making PCS-DPB is very important for determining its surface area; at the right molar ratio, the maximum surface area is reached. The high specific surface area of PCS-DPB, resulting from its porous structure, offers advantages for applications such as adsorption and sensing. The porous structure of PCS-DPB, along with its large surface area and fluorescence properties, make it a very promising heterogeneous sensor. The report showed the advancement of covalently linked porous fluorescent polymers, which have several uses in sensing, including environmental remediation. The incorporation of a fluorescent triphenylamine derivative with a donor-acceptor-donor structure into the porous polymer improves its ability to detect contaminants and expands its possible applications.

8.2.2.2 Sensor application

Porous polymer interacts with guest molecules greater than nonporous polymer, which is beneficial in providing greater signal sensitivity. The organic monomer or functional fluorophores, for example, hexaphenylsilole, tetraphenylethene, carbazole, pyrene, and triphenylamine derivatives, have been used to introduce as fluorescent units in SQ-based fluorescent porous polymers, which are utilized as sensors to detect heavy metal ions, pH, nitroaromatic chemicals, antibiotics, etc. (Du & Liu, 2020; Soldatov & Liu, 2021). For example, azide functionalized DDSQ fluorescent polymer can be applied to metal sensing (Chiang et al., 2023).

Due to its inherent sensitivity, excellent selectivity, and ease of use, fluorescence detection is one of the most promising methods for explosive sensing. SQ-based luminous porous materials were also employed for the detection of NACs such as nitrobenzene (NB), nitrotoluene (NT), dinitrotoluene (DNT), and trinitrotoluene (TNT) (Li et al., 2019; Wang et al., 2017; Sun et al., 2018; Wang et al., 2014). For example, a hybrid porous polymer prepared by crosslinking OVS with tri(4-bromophenyl)ethane, can detect various nitroaromatic explosives such as NT, DNT, TNT, NP, DNP, and 2,4,6-trinitrophenol (TNP) by fluorescence quenching and exhibits a high selectively detecting TNT or TNP compared to other aromatic compounds such as toluene, xylene, phenol, chlorobenzene, dichlorobenzene (DCB), nitrobenzene (NB), nitrophenol (NP), and dinitrophenol (DNP) (Sun et al., 2018). It is interesting that it can be utilized to manufacture a paper sensor that can quickly and affordably detect explosives in their vapor, solid, and solution phases.

Carbazole-functionalized SQ-based fluorescent porous polymer effectively quenches NP and DNP when it is produced by the oxidative coupling process of octa[4-(9-carbazolyl)phenyl]silsesquioxane (Li et al., 2019). Another fluorescent porous polymer that had been carbazole-functionalized displayed strong selective luminous quenching for TNP ($K_{sv} = 3.98 \times 10\ M^{-1}$ (Wu et al., 2012)), and it was created by Liang et al. (Meng et al., 2020) *via* FC reaction of OVS with carbazole.

The Heck coupling of OVS with brominated distyrylpyridine (Br-DSP) produced a pyridine derivative functionalized porous polymer with a large surface area of $600\ m^2\ g^{-1}$ (Yan, Yang, & Liu, 2020) and strong pH-sensing capabilities (Yang et al., 2015). On protonation, the pH values change from 5.5 to 1.0, and the maximum emission wavelength (λ_{max}) shifts from 525 to 618 nm. The pH range of 1–4 was found to have an excellent linear correlation of pH = − 0.069 em + 44 between pH values and maximum emission wavelength (λ_{em}), with a linear correlation coefficient of 0.98. By crosslinking OVS with ACQ-active 2-(2,6-bis-4-(diphenylamino)stryryl-4H-pyranylidene)malononitrile (TPA-DCM), a typical SQ-based NIR porous nanocomposite with in situ producing aggregation-induced emission (AIE) motif was created (Yan et al., 2019). Large Stokes shift of 180 nm, NIR emission at 670 nm, good thermal stability (400°C, 5% mass loss, N_2), and high surface area of $720\ m^2\ g^{-1}$ (Yan, Yang, & Liu, 2020) were all displayed by this material. These characteristics provide it with the excellent selectivity and sensitivity needed to detect Ru^{3+}.

The size of SQ nanomaterials affects the sensory polymer characteristic. For example, Chujo et al. developed organic-inorganic hybrid gels composed of SQ-based network polymers with variable emission properties depending on the size of coexisting silica particles (SPs), as shown in Fig. 8.1 (Kakuta et al., 2015). The construction of SQ-HPs from octaammonium SQ, by connecting with bithiophene dicarboxylic acid, resulted in a series of SQ networks with various crosslinking ratios among the SQ units. The observation is that the SQ networks with relatively low crosslinking ratios showed higher affinity to SPs, with increased affinity observed from all SQ networks to small-sized SPs. The discovery that the emission properties of the SQ-HPs were varied simply by adding SPs into the sample, with red-shifted emission induced in the presence of nanoparticles and blue-shifted emission observed from the samples containing microparticles. These emission characteristics could be applicable to developing a facile optical chemosensor for discriminating the size of nanomaterials.

8.2.2.2.1 Gas sensor

Silverstein et al. reported on the combination of inorganic polysilsesquioxane with organic polystyrene to obtain hybrid polyHIPE. The PPy-coated polymer showed its potential as sensor material, resulting from reversible and repeatable changes in conductivity on exposure to acetone vapor (Silverstein et al., 2005).

Zhou et al. focus on the synthesis of acrylate monomers and hybrid polymers (Fig. 8.2). The crucial step, 4-(1,2,2-triphenylvinyl)phenol, was synthesized via Suzuki coupling from 1-bromotriphenylethylene and 4-hydroxyphenylboronic acid. Using AIBN as a radical initiator, hybrid polymers were created from a mixture of monomer 3 and acryloisobutyl SQ (4) in various ratios.

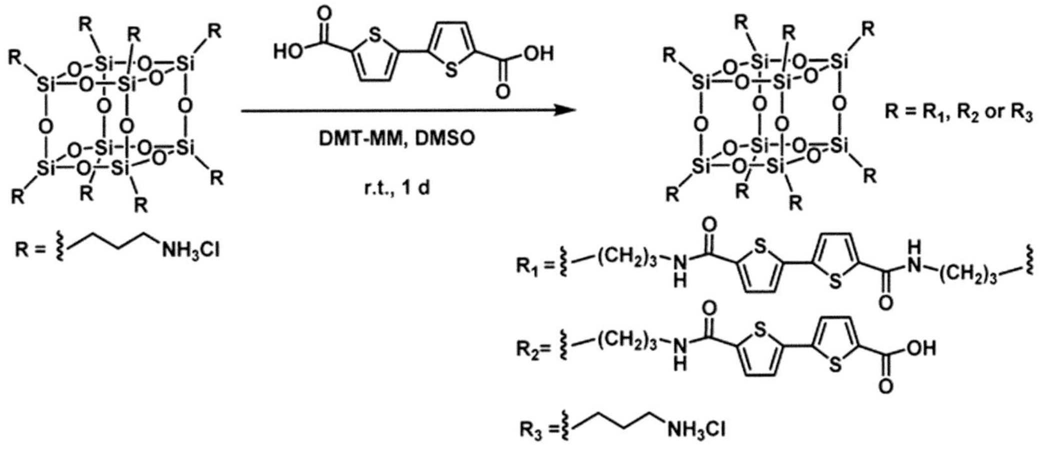

FIGURE 8.1 Synthesis of the possible chemosensor SQ-HPs.

Synthesis of the possible chemosensor SQ-HPs with bithiophene dicarboxylic acid linker.

From Kakuta, T., Tanaka, K., & Chujo, Y.. (2015). Synthesis of emissive water-soluble network polymers based on polyhedral oligomeric silsesquioxane and their application as optical sensors for discriminating the particle size. Journal of Materials Chemistry C, *3(48), 12539–12545. Available from https://doi.org/10.1039/C5TC03139G.*

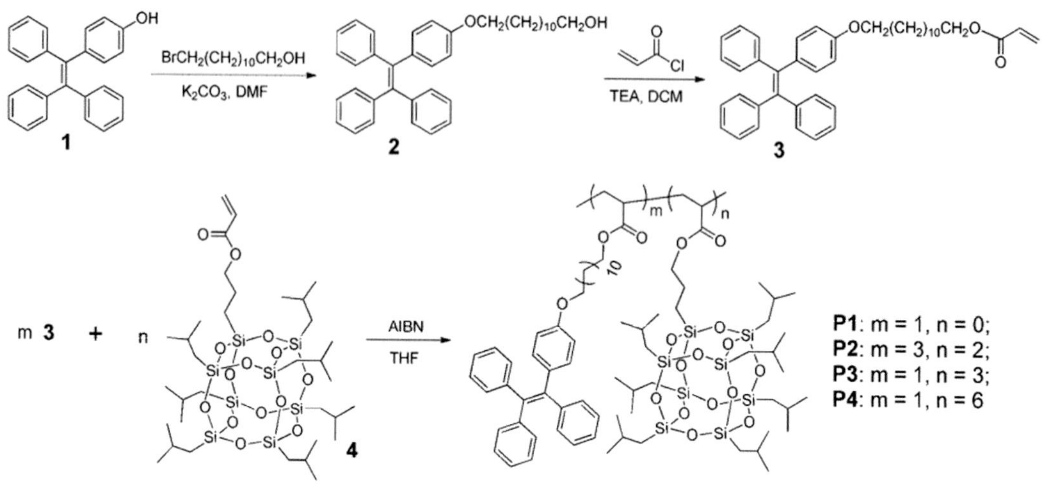

FIGURE 8.2 Synthesis of acrylate monomers and hybrid polymers.

Synthetic routes lead to monomer 3 and its corresponding polymers (P1–P4).

From Zhou, H., Ye, Q., Neo, W.T., Song, J., Yan, H., Zong, Y., Tang, B.Z., Hor, T.S.A. & Xu, J. (2014). Electrospun aggregation-induced emission active POSS-based porous copolymer films for detection of explosives. Chemical Communications, 50*(89), 13785–13788. Available from https://doi.org/10.1039/c4cc06559j.*

SQ-based AIE active copolymers display strong AIE effects in solution and solid state, and their porous films have good sensitivity and selectivity for detecting nitro-compound vapor. Despite having a thickness of 560 nm, porous copolymer films exhibit a ninefold increase in reaction to explosive vapors when compared to dense films (Zhou et al., 2014).

Wang et al. presented the preparation of a class of SQ-based porous polymers containing typical π-conjugated units, including biphenyl, tetrahedral silicon-centered units, and tetraphenylethene, by Heck reactions of OVS with the corresponding brominated monomers, as shown in Fig. 8.3A (Wang et al., 2017). The two sets of reaction conditions were employed to retrieve their original appearance, which exhibits tunable appearance with the colors from dark colors to various colors and the physical forms from coarse powders with irregular shape particles to fine powders with relatively solid spheres by altering the reaction conditions from method A to B. The variation of appearance leads to the alteration of the fluorescence from nearly no fluorescence to bright fluorescence. These polymers were afforded their original appearance by method B, which could explain the unexpected phenomenon that many conjugated porous polymers exhibit no or very low fluorescence despite their π-conjugated structures. Additionally, samples with fine powders show excellent dispersion stability in solvents, which is beneficial for the use of these materials in solution-processable methodologies such as spin-coating. Furthermore, these materials show efficient fluorescence quenching for nitrobenzene vapor in thin films, thereby indicating their potential application as sensing agents for the detection of explosives. As shown in Fig. 8.3B, fluorescent quenching was detected when these films were extremely exposed to nitrobenzene, which is explained by the donor-acceptor electron-transfer mechanism.

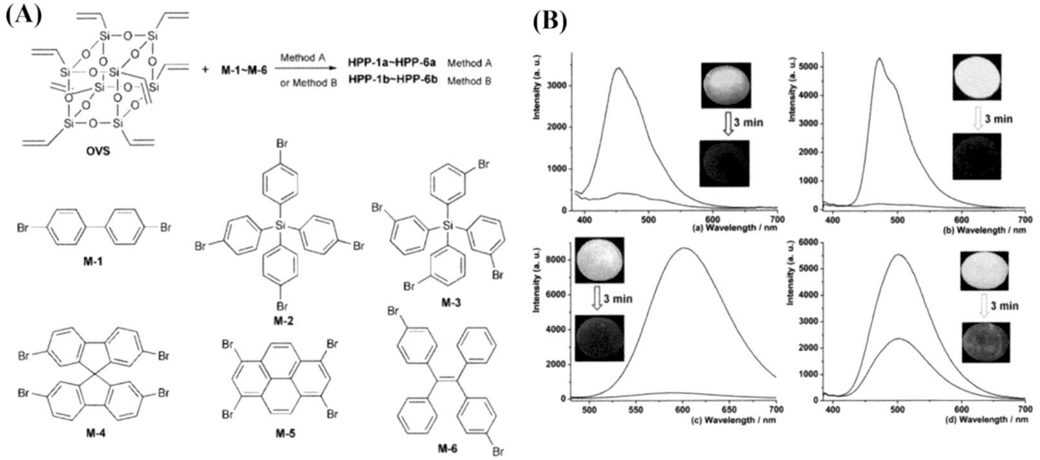

FIGURE 8.3 Synthesis of SQ-based hybrid porous polymers, and quenching of emission spectra.

(A) Synthetic routes of SQ-based hybrid porous polymers, HPP-1 to HPP-6. (B) Quenching of emission spectra of (a) HPP-1b, (b) HPP-4b, (c) HPP-5b, and (d) HPP-6b under exposure to the saturated vapor of nitrobenzene.

From Wang, D., Sun, R., Feng, S., Li, W., & Liu, H. (2017). Retrieving the original appearance of polyhedral oligomeric silsesquioxane-based porous polymers. Polymer, 130, *218–229. Available from https://doi.org/10.1016/j.polymer.2017.10.021.*

Recently, Suenaga et al. synthesized an chemosensor hybrid polymer, namely SSBKI-POSS, for oxygen detection, as shown in Fig. 8.4 (Suenaga et al., 2021). Hypoxia-selective luminescent probes based on oxygen quenching of phosphorescence can be used for an aerobic environment with positively detected by luminescence.

8.2.2.2.2 Organic sensor

Wang et al. reported that SQ-HPs have been synthesized from OVS with 2,2′,7,7′-tetrabromo-9,9′-spirobifluorene (TBrSBF) and 1,3,6,8-tetrabromopyrene (TBrPy) thought Heck reactions (HPP-1 and HPP-2), as shown in Fig. 8.5A (Wang et al., 2016). The fluorescence of these materials can be tuned by altering the reaction conditions, which could be valuable for the construction of porous polymers with tunable/controllable fluorescence. These materials could be utilized as solid absorbents for the capture and storage of CO_2 and as sensing agents for the detection of explosives. The result of the detection of explosives is that the HPP-1c material showed high sensitivity toward TNT at low concentrations (<1 ppm), and the fluorescence can be easily quenched with all the analytes. The fluorescence intensity gradually weakens with increasing concentration of analyte (Fig. 8.5B). These findings suggest that these materials could be utilized as sensing agents for detecting explosives with K_{SV} of 16344 M^{-1} for TNT.

Wang et al. synthesized luminescent hybrid porous polymers (LHPPs) using OVS and halogenated triphenylamine (TPA) as monomers *via* the Heck reaction (Fig. 8.6A) (Wang et al., 2014). Tuning of the porous and luminescent properties of LHPPs by altering TPA species and reaction conditions. Optimization of LHPP-3, which exhibits high porosity and thermal stability and emits

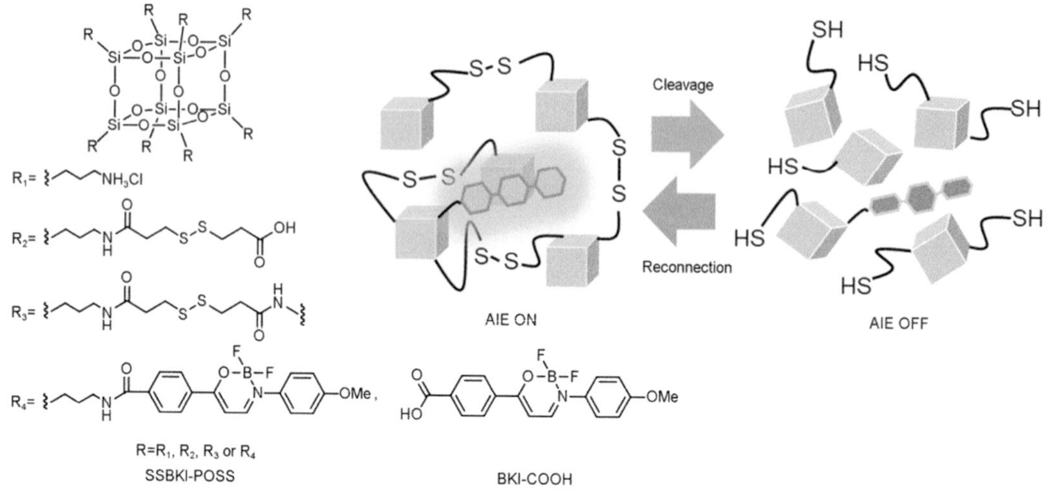

FIGURE 8.4 Chemical structure of SQ-HP, namely SSBKI-POSS.

Chemical structure of SSBKI-POSS and schematic illustration for discriminating hyperoxia and hypoxia.

From Suenaga, K., Tanaka, K., & Chujo, Y. (2021). Positive luminescent sensor for aerobic conditions based on polyhedral oligomeric silsesquioxane networks. Chemical Research in Chinese Universities, *37(1), 162–165. Available from https://doi.org/10.1007/s40242-021-0398-x.*

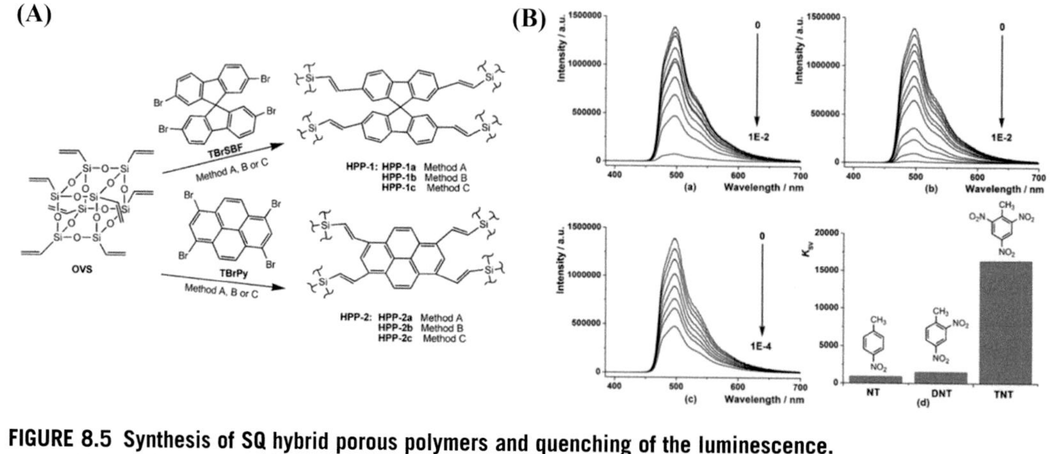

FIGURE 8.5 Synthesis of SQ hybrid porous polymers and quenching of the luminescence.

(A) Synthetic route to SQ-based fluorescent hybrid porous polymers HPP-1 and HPP-2. (B) Quenching of the luminescence of HPP-1c with various concentrations of (a) NT, (b) DNT, and (c) TNT in THF.

From Wang, D., Feng, S., & Liu, H. (2016). Fluorescence-tuned polyhedral oligomeric silsesquioxane-based porous polymers. Chemistry – A European Journal, *22(40), 14319–14327. Available from https://doi.org/10.1002/chem.201602688.*

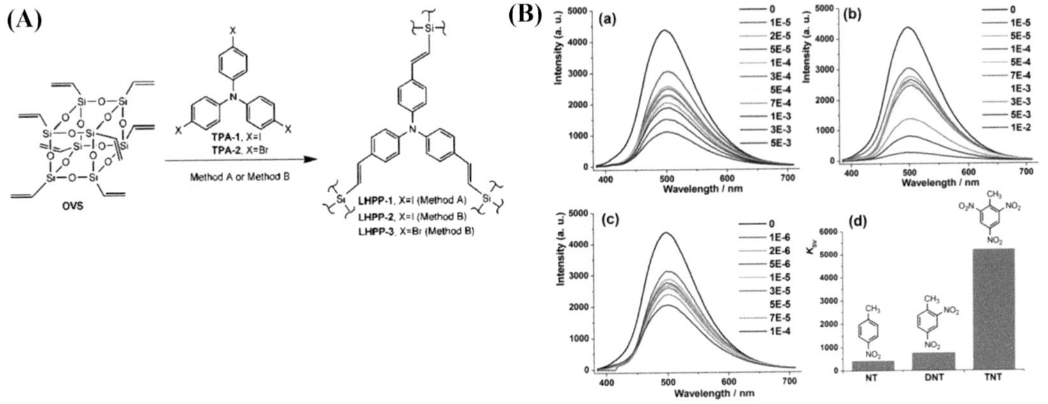

FIGURE 8.6 Hybrid porous polymers (LHPPs) using OVS and halogenated triphenylamine.

(A) Synthetic routes of luminescent hybrid porous polymers (LHPPs). (B) (a–c) Quenching of luminescent spectra of LHPP-3 with various concentrations of NT (a), DNT (b), and TNT (c) in THF. (d) The Stern–Volmer constants with NT, DNT, and TNT analytes for LHPP-3.

From Wang, D., Li, L., Yang, W., Zuo, Y., Feng, S., & Liu, H. (2014). POSS-based luminescent porous polymers for carbon dioxide sorption and nitroaromatic explosives detection. RSC Advances, *4(104), 59877–59884. Available from https://doi.org/10.1039/c4ra11069b.*

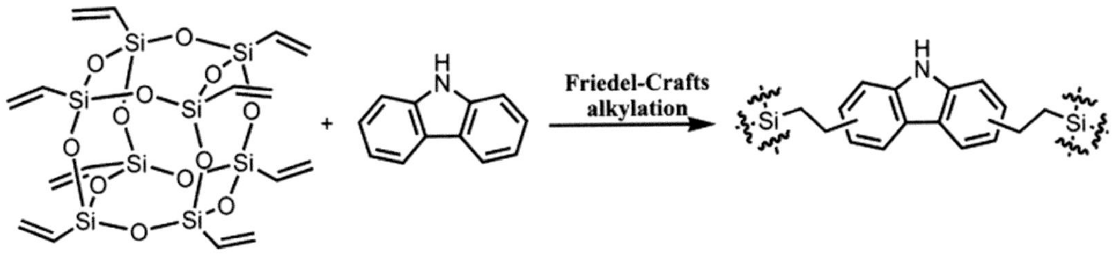

FIGURE 8.7 Synthetic route of SCHPPs.

Synthetic route of SCHPPs by varying the molar ratio of OVS to carbazole thought FC reaction.

From Meng, X., Liu, Y., Wang, S., Du, J., Ye, Y., Song, X., & Liang, Z. (2020). Silsesquioxane-carbazole-corbelled hybrid porous polymers with flexible nanopores for efficient CO2 conversion and luminescence sensing. ACS Applied Polymer Materials, *2(2), 189–197. Available from https://doi.org/10.1021/acsapm.9b00747.*

high yellow luminescence. Efficient quenching of LHPP-3's luminescence by nitroaromatic explosives (Fig. 8.6B) such as NT, DNT, and TNT, indicating its potential as a chemical sensor for explosives detection with K_{sv} of 411, 751, and 5208 M^{-1}, respectively. Additionally, LHPP-3 exhibited high sensitivity for TNT at a low concentration (<1 ppm). The increase of electron-withdrawing groups in its analyst structure leads to the enhancement of quenching efficiencies.

Meng et al. used the Friedel-Crafts reaction (Fig. 8.7) to create a series of silsesquioxane-carbazole-corbeled hybrid porous polymers (SCHPPs) with strong catalytic and luminous

Table 8.2 Other synthetic hybrid polymers for organic sensors.

Polymer	Surface area (S_{micro}) [m^3 g^{-1}]	Pore size (nm)	Pore volume (V_{micro}) [cm^3 g^{-1}]	Fields of applications/analysts	Sensor characteristics	Ref.
Py-HPP	1300	1.41 and 4.15	N/A	Detection and removal of antibiotics, namely, berberine chloride hydrate (BCH), tetracycline hydrochloride (TH), and mafenide hydrochloride (MH).	The equilibrium adsorption capacity for BCH is the highest and reaches 330 mg g^{-1}; while the values for TH and MH are 195 and 165 mg g^{-1}.	Yang and Liu. (2020)
3Ph-TSHPP	555 (322)	1.54	0.55 (0.14)	The degradation of both acidic and basic dyes (ST, CR, RB, and MB) without additional oxidation agents or pH adjustments.	Highly active and metal-free photocatalysts for environmental remediation and energy conversion.	Du and Liu. (2021)
PCS-OTS	816	N/A	N/A	Dyes adsorption and nitroaromatics detection.	The equilibrium adsorption capacity for Rhodamine B (RB), Congo red (CR), and methyl orange (MO) is selective for the nitrophenolic compounds, with the highest efficiency being seen for 2,4-dinitrophenol (DNP). Properties are 1935, 1420, and 155 mg g^{-1}. A sensitive chemical sensor detects p-nitrophenol with high sensitivity (K_{sv} = 81230 M^{-1}).	Wang et al. (2021)
PCS-CZ-DCM	954	1.7 and 2.1–7.2	N/A	The control of antibiotic pollutants in environmental applications.	Excellent photodegradation activity for antibiotics BCH without additional oxidation agents or pH adjustments.	Wang et al. (2022)

Table 8.2 Other synthetic hybrid polymers for organic sensors. ***Continued***

Polymer	Surface area (S_{micro}) [$m^3 g^{-1}$]	Pore size (nm)	Pore volume (V_{micro}) [$cm^3 g^{-1}$]	Fields of applications/analysts	Sensor characteristics	Ref.
HLPP-OTS	814 (577)	1.7 and 2.1–5.1	0.435 (0.261)	Water treatment and sensor.	Exhibits excellent luminescence properties and acts as a sensitive chemosensor to detect explosives with high sensitivity toward 5-nitrosalicylaldehyde (K_{sv} = 36560 M^{-1}).	Wang et al. (2020)
HPP	1741 (922)	1.41, 1.69, and 2.65	1.057 (0.387)	Wastewater treatment and detecting nitroaromatics.	Selective for the nitro-phenolic compounds with highest efficiency being seen for 2,4-dinitrophenol (DNP). K_{sv} of DNP and NP for HPP is 638 and 2100 M^{-1}.	Li et al. (2019)
PCS-DPBS	245	0.15	N/A	The effective capture and detection of iodine in the vapor environment.	High iodine vapor adsorption capacities, with uptake capacities up to 2.81 g g^{-1} at 70°C. Great sensitivity with K_{sv} values of 1622.4 M^{-1}.	Wang et al. (2023)

capabilities (Meng et al., 2020). By varying the molar ratio of OVS to carbazole, the SCHPPs displayed customizable surface areas, high thermal stability, and preferred luminescence. SCHPP-3 exhibits significant luminous qualities and selective sensing capability for picric acid or 2,4,6-trinitrophenol (TNP) (K_{sv} = 3.98 × 10M^{-1}) (Wu et al., 2012). TNP has high acidity, which readily and easily accepts electrons with carbazole and SQ. Other synthetic hybrid polymers for organic sensors are listed in Table 8.2.

8.2.2.2.3 Biosensors

Wang et al. created organic-inorganic hybrid carbon dots (CDs/SQ), which are functionalized with octaaminopropyl polyhedral oligomeric silsesquioxane named OA-POSS (Fig. 8.8) (Wang et al., 2015). The CDs/SQ exhibit favorable photoluminescent properties with a quantum yield of 24.0% and resistance to photobleaching, making them suitable for cell imaging in biological systems. They also have high photoluminescence stability in the presence of biological sample matrix and resistance to photobleaching, which makes it easier to image cells in biological systems. This work

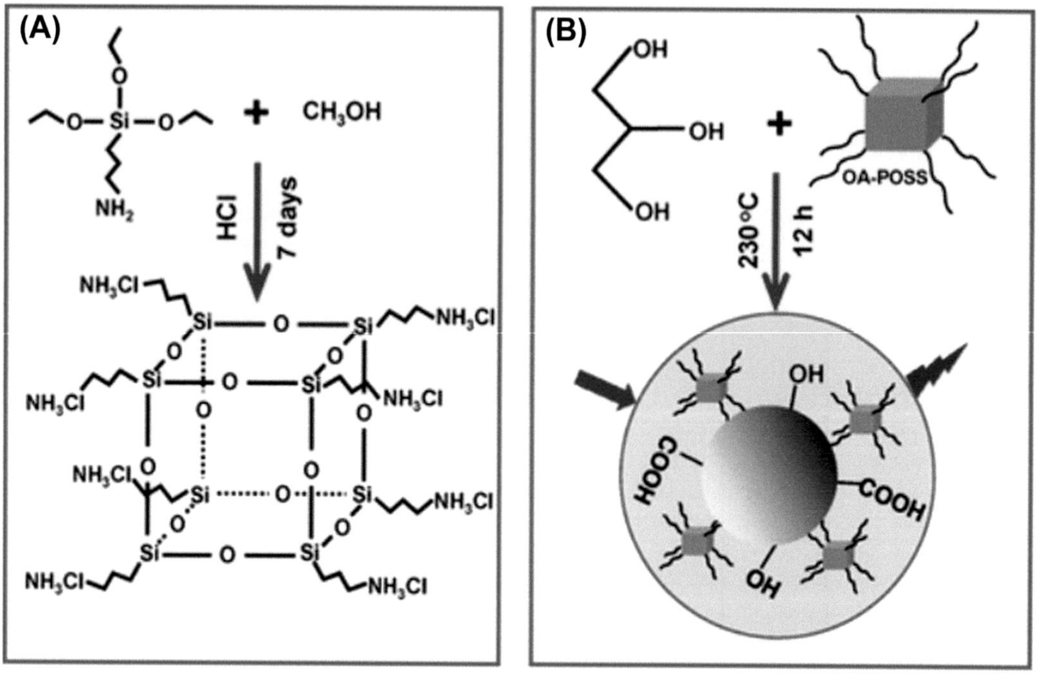

FIGURE 8.8 Organic-inorganic hybrid carbon dots (CDs/SQ), functionalized with octaaminopropyl polyhedral oligomeric silsesquioxane named OA-POSS.

(A) Diagram of the hydrolytic condensation of APTES to produce SQ named OA-POSS (B) carbon dots (CDs/SQ) preparation using glycerol as a carbon source and solvent medium, and OA-POSS as a passivation agent. *From Wang, W. J., Hai, X., Mao, Q. X., Chen, M. L. & Wang, J. H. (2015). Polyhedral oligomeric silsesquioxane functionalized carbon dots for cell imaging.* ACS Applied Materials and Interfaces, *7(30), 16609–16616. Available from https://doi.org/10.1021/acsami.5b04172.*

provides a helpful method for the creation of a photoluminescent probe for use in biological imaging. An evident for the mixture of HeLa and MCF-7 cells with CDs/SQ shows strong blue and green emissions, which are clearly captured on the fluorescent microscope with laser excitation at 340 and 495 nm, respectively, while no visible fluorescence emission is obtained in the control trials.

8.2.2.2.4 Ion-selective sensor

Lv et al. described how to make a sulfur-based fluorescent hybrid porous polymer (HPP-SH) (Fig. 8.9) that can only detect and attach to Hg^{2+} ions, as shown in Lv et al. (2022). The HPP-SH compound has high efficiency and selectivity in detecting Hg^{2+} ions. It achieves this through a fluorescence "turn-off" mechanism with a low detection limit of 4.48 ppb and demonstrates a significant ability to adsorb Hg^{2+} ions, with a maximum adsorption capacity of 900.9 mg g^{-1}. This behavior is consistent with the pseudo-second-order kinetic and Langmuir models. The material exhibits exceptional dual-functionality, surpassing the majority of previous sensors or adsorbents

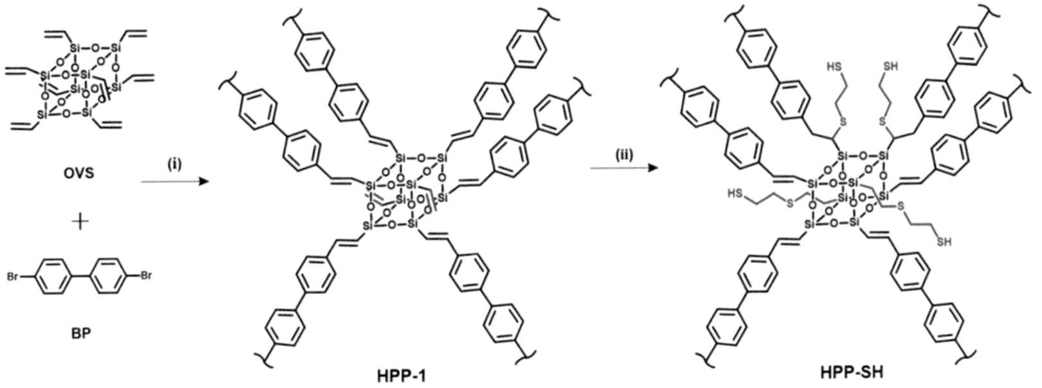

FIGURE 8.9 Synthetic route of HPP-SH.

Synthetic route of HPP-SH. Reaction conditions: (i) $Pd(PPh_3)_4$, K_2CO_3, DMF, 120°C, and 48 h; (ii) 1,2-ethanedithiol, UV, DMPA, THF, room temperature, and 10 h.

From Lv, Z., Chen, Z., Feng, S., Wang, D., & Liu, H. (2022). A sulfur-containing fluorescent hybrid porous polymer for selective detection and adsorption of Hg^{2+} *ions.* Polymer Chemistry, 13*(16), 2320–2330. Available from https://doi.org/10.1039/d2py00077f.*

for Hg^{2+}. It is highly recyclable for the repeated adsorption of Hg^{2+} ions. The sulfur content in HPP-SH is crucial for detection, as a higher sulfur content leads to a lower detection limit.

The example of hybrid polymers for ion-selective sensors is listed in Table 8.3, and the synthetic route of SQ-HP, namely POSS-2, mentioned in the table is shown in Fig. 8.10.

8.2.2.2.5 Humidity sensor

Researchers extensively studied polyelectrolytes as responsive materials for impedance-based humidity sensors. Polyelectrolytes offer adaptability, processability into solutions, and adjustability for controlling properties. Polyelectrolyte-based sensors are capable of effectively monitoring the entire range of relative humidity (RH). Nevertheless, maintaining the stability of humidity sensors based on polyelectrolytes in high RH conditions has proven to be a difficult task. The structure of the sensitive film can be changed by the strong interaction between water molecules and hydrophilic parts in the polyelectrolyte. Dong et al. enhance the advancement of a reliable humidity sensor by creating a crosslinked polyelectrolyte and enhancing its capacity to form a film by combining a hybrid ionic porous SQ-HP material, namely POSS-TPPBr, with polystyrene. Using OVS and tetraphenyphosphonium bromide (TPPBr) in a Friedel-Crafts procedure, POSS-TPPBr was obtained and employed as an impedance-type humidity sensor with negligible hysteresis and quick response over the whole RH range of 11%–95% without significantly and qualitatively degrading its impedance qualities (Dong et al., 2018). It displays micropores and mesopores, having pore sizes of around 0.7 nm and 3.8 nm, respectively. The pores serve as conduits for the adsorption and conveyance of water molecules, hence enhancing the material's ability to sense humidity. The N_2 sorption isotherms of POSS-TPPBr at low relative pressure ($P/P_0 = 0.01$) show that it is

Table 8.3 Hybrid polymers for ion-selective sensors.

Polymer	Surface area (S_{micro}) [$m^2\ g^{-1}$]	Pore size (nm)	Pore volume (V_{micro}) [$cm^3\ g^{-1}$]	Fields of applications/analysts	Sensor characteristics	Ref.
THPP	620 (170)	1.4 and 4	0.54 (0.07)	Water treatment.	The efficient detection of Ru^{3+} with high selectivity, and the detection limit (LOD) is calculated to be 5.2×10^{-6} mol L^{-1}.	Yan, Yang, and Liu. (2020)
PCS-TPPy	1236 (455)	1.69 and 2.77	0.98	Adsorption capacity of 1.53 mmol g^{-1} for 4-bromo-phenol (BP), 0.68 mmol g^{-1} for hydroquinone (HQ), and 0.50 mmol g^{-1} for phenol (PH).	Selective sensing capability for Ru^{3+} and Fe^{3+} with good linear response from 10 to 1000 μM and the corresponding detection limits are 3.12 μM and 6.78 μM.	Yang and Liu. (2021)
POSS-2	1486.5	N/A	N/A	A highly efficient dual-function material for simultaneous Au (III) detection and recovery by simply introducing abundant imidazole thione and thioether groups in one system.	Rapidly and selectively detect Au(III) with a very low limit of detection of 1.2 ppb by fluorescence quenching or a visualized color change from white to dark orange.	Chen et al. (2021)
HPP-3	1910 (177)	1.97 and 0.06	1.97	Multiple applications in chemical sensors and water treatment.	High sensitivity for Fe^{3+}, Cu^{2+}, and Ru^{3+}, especially for Fe^{3+} with K_{sv} = 140000 M^{-1}.	Liu and Liu. (2017)
SHHPP-2	1144 (320)	1.4	0.94	Multiple applications in chemical sensors and water treatment.	Serve as an fluorescence sensor and exhibited good sensitivity for Fe^{3+}, Ru^{3+}, and Cu^{2+}, especially for Fe^{3+} with observed K_{sv} = 53695 M^{-1}.	Du and Liu. (2018)
TPAIE-2	720 (320)	1.54	0.55 (0.16)	Multiple applications in chemical sensors and water treatment.	High surface areas, hierarchical pore structures, and electron-rich features allow TPAIE-2 to specifically respond to Ru^{3+} with high selectivity and sensitivity. K_{sv} was calculated to be 5300 M^{-1}.	Yan et al. (2019)

Table 8.3 Hybrid polymers for ion-selective sensors. *Continued*

Polymer	Surface area (S_{micro}) [$m^2\ g^{-1}$]	Pore size (nm)	Pore volume (V_{micro}) [$cm^3\ g^{-1}$]	Fields of applications/analysts	Sensor characteristics	Ref.
HPP-SH	639 (438)	1.7 and 2.8	0.66 (0.18)	A dual-functional material, i.e., a sulfur-containing fluorescent hybrid porous polymer (HPP-SH), for the simultaneous detection and adsorption of Hg^{2+}.	Selectively and efficiently detect Hg^{2+} ions with a low limit of detection of 4.48 ppb by a fluorescence "turn-off" mode and a maximum adsorption amount of 900.9 mg g^{-1}.	Lv et al. (2022)

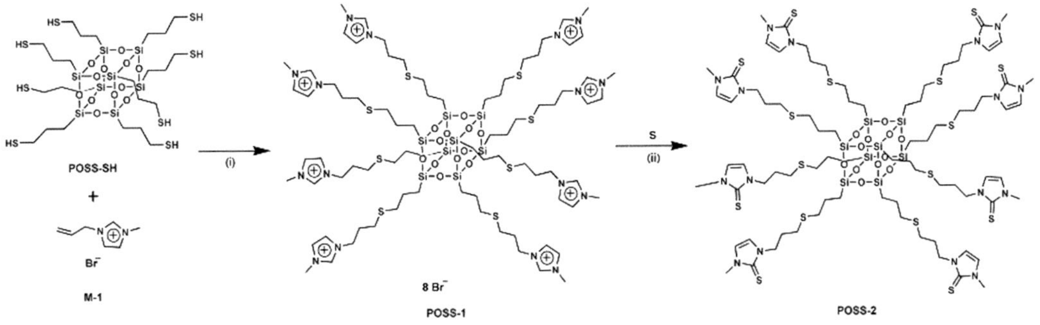

FIGURE 8.10 Synthetic Route of SQ named POSS-2.

Reaction conditions: (i) UV, DMPA, CH_2Cl_2/CH_3OH, 30 min; (ii) K_2CO_3, CH_3OH, reflux, 24 h.

From Chen, Z., Wang, D., Feng, S., & Liu, H. (2021). An imidazole thione-modified polyhedral oligomeric silsesquioxane for selective detection and adsorptive recovery of Au(III) from aqueous solutions. ACS Applied Materials & Interfaces, 13*(20), 23592–23605. Available from https://doi.org/10.1021/acsami.1c01965.*

microporous. The BET surface area of POSS-TPPBr is 810 $m^2\ g^{-1}$, which means it has a large specific surface area that helps the water molecules and polymer interact.

8.2.2.2.6 Multisensors

Some synthetic hybrid polymers based on SQ can be adjusted and used for multisensors, as shown in Table 8.4. For example, Sun et al. synthesized cyclosiloxane-linked fluorescence porous SQ-HP polymers (FPP-1, FPP-2, and FPP-3) from cyclosiloxane and three different linkers (Fig. 8.11) for detection of explosives and Fe^{3+} (Sun et al., 2020).

Table 8.4 Hybrid polymers for multisensors.

Polymer	Surface area (S_{micro}) [m^2 g^{-1}]	Pore size (nm)	Pore volume (V_{micro}) [cm^3 g^{-1}]	Fields of applications/ analysts	Sensor characteristics	Ref.
PCS-DPB	1292	0.86−1.54 and 2.65	0.796 (0.384)	Environmental remediation application.	Excellent luminescence and a large red-shift of 120 nm, which makes its vapor act as a probe to detect DNP and Ru^{3+} with high selectivity and sensitivity.	Wang et al. (2022)
FPP-3	735	N/A	0.7	On-site sensing of explosives and Fe^{3+}.	Detect the explosives and Fe^{3+} in solution, solid, and vapor phases with a rapid response (<10 s), suggesting its potential for the rapid, convenient, sensitive, and on-site sensing of explosives and Fe^{3+}.	Sun et al. (2020)

8.2.3 Other-based hybrid polymer

Apart from hybrid polymers based on MOF and SQ, there are other hybrid polymers that are useful for sensor applications. Chen et al. developed a novel strip-style electrochemical sensor using reduced graphene oxide (rGO) and poly(ethylene dioxythiophene)/poly(styrene sulfonic acid) (PEDOT/PSS) nanocomposite films for the sensitive and selective electrochemical detection of serotonin (5-hydroxytryptamine, 5-HT) neurotransmitter. The 200 μL of the rGO − PEDOT/PSS − nafion (0.5 wt%) film showed the highest sensitivity and selectivity for 5-HT with a detection limit of 0.1 μM.

Moreover, Kwon et al. prepared a hybrid polymer, namely PST-11:P3HT, from the reaction between zeolite and PST-11 (Kwon et al., n.d.). The NO_2 gas sensing performance is greater sensitivity, a faster response time, and a greater recovery rate under ambient conditions when compared with devices without zeolite.

8.3 Conclusions and perspectives

Hybrid polymers are better at trapping analytes, lower obstacles to mass transfer, and offer more diffusion paths for exciton migration when compared to tiny organic molecules or conventional polymers. hybrid polymer, as dependable and efficient probes, hold potential as effective instruments in sensing applications. We have highlighted recent developments in several SQ-HPs research areas in this perspective article, focusing primarily on functionality, synthesis

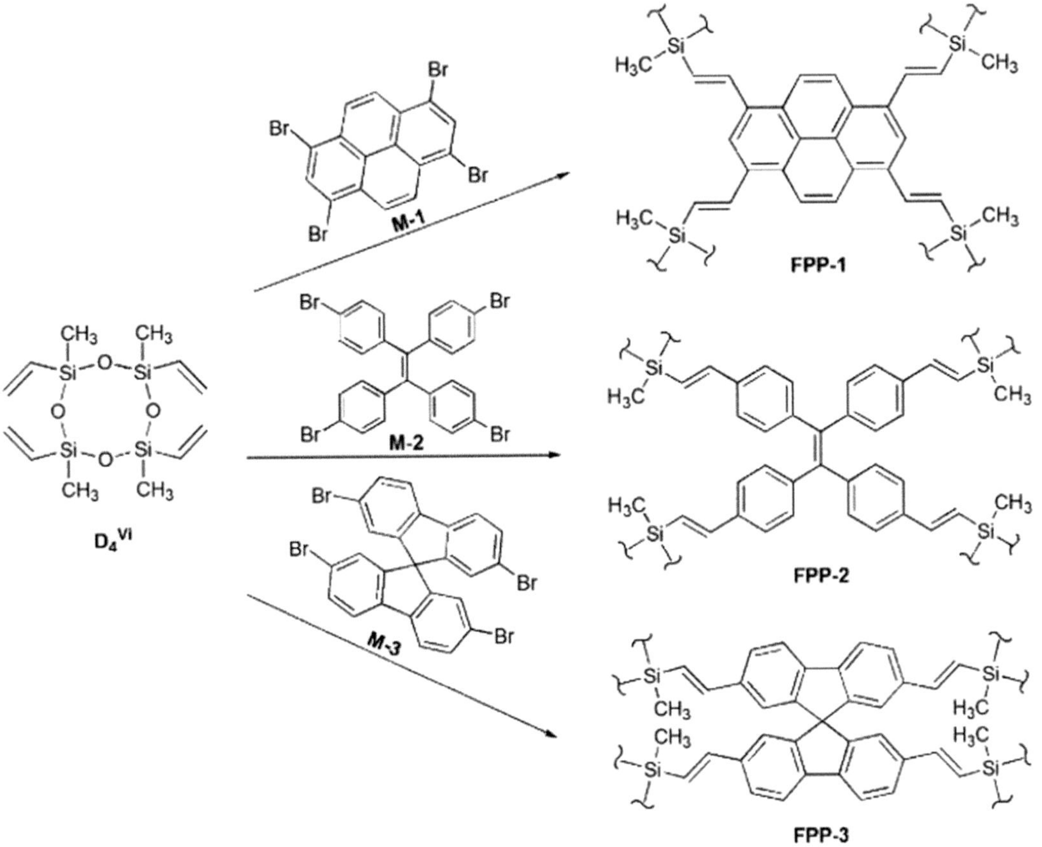

FIGURE 8.11 Synthetic routes of cyclosiloxane-linked fluorescence porous SQ-HP polymers

Synthetic routes of cyclosiloxane-linked fluorescence porous polymers (FPP-1, FPP-2, and FPP-3).

From Sun, R., Feng, S., Zhou, B., Chen, Z., Wang, D., & Liu, H. (2020). Flexible cyclosiloxane-linked fluorescent porous polymers for multifunctional chemical sensors. ACS Macro Letters, 9*(1), 43–48. Available from https://doi.org/10.1021/acsmacrolett.9b00901.*

methodology, and applications of sensors. Here, it is unquestionably clear how important SQ-HPs are in giving hybrid materials noticeably enhanced and desired performance. Despite booming research on SQ-HPs, there are still many challenges to overcome. There are still some problems with the precise or regioselective functionalization of SQs, the controllable and effective synthesis of SQ-HPs, and the processability of some SQ-HPs. The solution to these problems is conducive to promoting systematic research on SQ-HPs and their application expansion. For a larger deployment of hybrid materials with appropriate macroscopic features, additional work in various diverse areas is essential. The first is the design and synthesis of novel monomers, both in terms of structures and functional groups. The emergence of many novel monomers with different topological structures provides an idea for these problems. Hybrid materials based on these new monomers may exhibit unexpected properties, which are well worth exploring. Secondly, novel reactions are developed to further improve and optimize the present synthetic techniques. Finally, reducing the cost of synthesis and rediscovering the useful properties of the hybrid polymer can promote further development and sensory application. We believe that the underlying science and practical uses of hybrid polymers will develop further and provide novel functional materials for related applications with competitive performance.

Acknowledgments

This work was supported by the National Key R&D Program of China (2022YFE0197000) and the National Natural Science Foundation of China (Grant No. 21975144, 22111530285).

Abbreviations

AIBN azobisisobutyronitrile;
AIE aggregation-induced emission;
ACQ aggregation-caused quenching;
BCH berberine chloride hydrate;
BP 4 bromophenol;
Br-DSP brominated distyrylpyridine;
CR Congo red;
COFs covalent organic frameworks;
CMPs conjugated microporous polymers;
CV crystal violet;
DBAB 4,4′-dibromoazobenzene;
DDSQ double-decker-shaped silsesquioxane;
DMF dimethylformamide;
DNP 2,4-dinitrophenol;
DNT 2,4-dinitrotoluene;
DPPF 1,1-bis-(diphenylphosphine)ferrocene;
DPPOF 1,1-bis-(diphenylphosphineoxide)ferrocene;
DVB divinylbenzene;
EDA ethylenediamine;
FC-reaction Friedel-Crafts reaction;
HCPs hypercrosslinked polymers;
HDA hexamethylenediamine;
IMAC immobilized metal ion affinity chromatography;
I_8OPS octa(iodophenyl)silsesquioxane;
IUPAC International Union of Pure and Applied Chemistry;
MB methylene blue;
MH mafenide hydrochloride;
MO methyl orange;
MOFs metal-organic frameworks;
NACs nitroaromatic compounds;
NIR near-infrared;

NB	nitrobenzene;
NP	4-nitrophenol;
NT	nitrotoluene;
OAPS	octa(aminophenyl)silsesquioxane;
OBPS	octa(biphenyl)silsesquioxane;
OCPS	octa[4-(9-carbazolyl)phenyl]silsesquioxane;
ODHS	octa[dimethyl(hydrido)silyl]silicate;
ODVS	octa[dimethyl(vinyl)silyl]silicate;
OPS	octa(phenyl)silsesquioxane;
OTS	triphenylamine-functionalized SQ monomer;
OVS	octa(vinyl)silsesquioxane;
PA	picric acid;
p-I_8OPS	octa(p-iodophenyl)silsesquioxane;
PCSs	hybrid porous polymers based on cage-like organosiloxanes;
PCL	polycaprolactone;
PDMS	polydimethylsiloxane;
PEG	polyethylene glycol;
PIMs	polymers with intrinsic microporosity;
PMA	poly(methyl acrylate);
PMMS	poly[(mercaptopropyl)methylsiloxane];
PMOs	periodic mesoporous organosilicas;
POPs	porous organic polymers;
PPG	polypropylene glycol;
PS	polystyrene;
PVA	poly(vinyl alcohol);
PVC	polyvinyl chloride;
PVCz	poly(N-vinylcarbazole);
PVDF	polyvinylidene fluoride;
PVP	polyvinylpyrrolidone;
PXRD	powder X-ray diffraction;
Qs	spherosilicates;
RAFT	reversible addition – fragmentation chain-transfer;
RB	rhodamine B;
RH	relative humidity;
ROMP	ring-opening metathesis polymerization;
SEM	scanning electron microscopy;
SQs	silsesquioxanes;
ST	Safranine;
POSSs	polyhedral oligomeric silsesquioxanes;
T	silsesquioxane;
T_8	octasilsesquioxane;
TBB	1,3,5-tribromobenzene;
TBS	tetrabiphenylsilane;
TEA	triethylamine;
TEM	transmission electron microscopy;
TGA	thermogravimetric analysis;
TH	tetracycline hydrochloride;
THF	tetrahydrofuran;

TIC	toxic industrial chemical;
TNT	2,4,6-trinitrotoluene;
TPA	triphenylamine;
TPA-DCM	2-(2,6-bis-4-(diphenylamino)stryryl-4H-pyranylidene)malononitrile;
TPE	tetraphenylethene;
TPP	triphenylphosphine;
TPPO	triphenylphosphine oxide;
TPS	tetraphenylsilane;
UV	ultraviolet;
UV-LED	ultraviolet light-emitting diode

References

Alam, M. W., Bhat., Qahtani., Aamir, M., Amin, M. N., Farhan, M., Aldabal, S., Khan, M. S., Jeelani, I., Nawaz, A., & Souayeh, B. (2022). Recent progress, challenges, and trends in polymer-based sensors: A review. *Polymers* (11).

Alves, F., Scholder, P., & Nischang, I. (2013). Conceptual design of large surface area porous polymeric hybrid media based on polyhedral oligomeric silsesquioxane precursors: Preparation, tailoring of porous properties, and internal surface functionalization. *ACS Applied Materials & Interfaces*, *5*(7), 2517–2526. Available from https://doi.org/10.1021/am303048y.

Chaiprasert, T., Liu, Y., Intaraprecha, P., Kunthom, R., Takeda, N., & Unno, M. (2021). Synthesis of tricyclic laddersiloxane with various ring sizes (bat siloxane). *Macromolecular Rapid Communications*, *5*, 2000.

Chen, Z., Wang, D., Feng, S., & Liu, H. (2021). An imidazole thione-modified polyhedral oligomeric silsesquioxane for selective detection and adsorptive recovery of Au(III) from aqueous solutions. *ACS Applied Materials & Interfaces*, *13*(20), 23592–23605. Available from https://doi.org/10.1021/acsami.1c01965.

Chiang, C.-H., Mohamed, M.G., Chen, W.-C., Madhu, M., Tseng, W.-L., Kuo, S.-W. (2023). Construction of fluorescent conjugated polytriazole containing double-decker silsesquioxane: Click polymerization and thermal stability. *15*(2), 331, Available from https://doi.org/10.3390/polym15020331.

Chimjarn, S., Kunthom, R., Chancharone, P., Sodkhomkhum, R., Sangtrirutnugul, P., & Ervithayasuporn, V. (2015). Synthesis of aromatic functionalized cage-rearranged silsesquioxanes (T 8, T 10, and T 12) via nucleophilic substitution reactions. *Dalton Transactions*, *44*(3), 916–919. Available from https://doi.org/10.1039/c4dt02941k.

Cichosz, S., Masek, A., & Zaborski, M. (2018). Polymer-based sensors: A review. *Polymer Testing*, *67*, 342–348. Available from https://doi.org/10.1016/j.polymertesting.2018.03.024.

Croissant, J. G., Cattoën, X., Durand, J. O., Wong Chi Man, M., & Khashab, N. M. (2016). Organosilica hybrid nanomaterials with a high organic content: Syntheses and applications of silsesquioxanes. *Nanoscale*, *8*(48), 19945–19972. Available from https://doi.org/10.1039/c6nr06862f, http://www.rsc.org/publishing/journals/NR/Index.asp.

Dong, W., Ma, Z., & Duan, Q. (2018). Preparation of stable crosslinked polyelectrolyte and the application for humidity sensing. *Sensors and Actuators B: Chemical*, *272*, 14–20. Available from https://doi.org/10.1016/j.snb.2018.05.140.

Du, Y., Ge, M., & Liu, H. (2019). Porous polymers derived from octavinylsilsesquioxane by cationic polymerization. *Macromolecular Chemistry and Physics*, *220*(5). Available from https://doi.org/10.1002/macp.201800536.

Du, Y., & Liu, H. (2018). Silsesquioxane-based hexaphenylsilole-linked hybrid porous polymer as an effective fluorescent chemosensor for metal ions. *ChemistrySelect*, *3*(6), 1667–1673. Available from https://doi.org/10.1002/slct.201703133.

Du, Y., & Liu, H. (2020). Cage-like silsesquioxanes-based hybrid materials. *Dalton Transactions*, *49*(17), 5396–5405. Available from https://doi.org/10.1039/d0dt00587h.

Du, Y., & Liu, H. (2021). Triazine-functionalized silsesquioxane-based hybrid porous polymers for efficient photocatalytic degradation of both acidic and basic dyes under visible light. *ChemCatChem*, *13*(24), 5178–5190. Available from https://doi.org/10.1002/cctc.202101231, http://onlinelibrary.wiley.com/journal/10.1002/(ISSN)1867-3899.

Dudziec, B., Zak, P., & Marciniec, B. (2019). Synthetic routes to silsesquioxane-based systems as photoactive materials and their precursors. *Polymers*, *11*(3), 504. Available from https://doi.org/10.3390/polym11030504.

Fujita, S., & Inagaki, S. (2008). Self-organization of organosilica solids with molecular-scale and mesoscale periodicities. *Chemistry of Materials*, *20*(3), 891–908. Available from https://doi.org/10.1021/cm702271v.

Ge, M., & Liu, H. (2016). A silsesquioxane-based thiophene-bridged hybrid nanoporous network as a highly efficient adsorbent for wastewater treatment. *Journal of Materials Chemistry A*, *4*(42), 16714–16722. Available from https://doi.org/10.1039/C6TA06656A.

Ghanbari, H., Cousins, B. G., & Seifalian, A. M. (2011). A nanocage for nanomedicine: Polyhedral oligomeric silsesquioxane (POSS. *Macromolecular Rapid Communications*, *32*(14), 1032–1046. Available from https://doi.org/10.1002/marc.201100126.

He, H., Chen, S., Tong, X., An, Z., Ma, M., Wang, X., & Wang, X. (2017). Self-assembly of a strong polyhedral oligomeric silsesquioxane core-based aspartate derivative dendrimer supramolecular gelator in different polarity solvents. *Langmuir the ACS Journal of Surfaces and Colloids*, *33*(46), 13332–13342. Available from https://doi.org/10.1021/acs.langmuir.7b02893.

Huang, M., Yue, K., Huang, J., Liu, C., Zhou, Z., Wang, J., Wu, K., Shan, W., Shi, A. C., & Cheng, S. Z. D. (2018). Highly asymmetric phase behaviors of polyhedral oligomeric silsesquioxane-based multiheaded giant surfactants. *ACS Nano*, *12*(2), 1868–1877. Available from https://doi.org/10.1021/acsnano.7b08687, http://pubs.acs.org/journal/ancac3.

Irfan, M.H. (n.d.). *Chemistry and technology of thermosetting polymers in construction applicationshybrid polymers hybrid polymers* (Vol. 1, pp. 240–257). Dordrecht: Springer. Available from https:doi.org/10.1007/978-94-011-4954-9_10.

Ishizaki, Y., Yamamoto, S., Miyashita, T., & Mitsuishi, M. (2018). Synthesis and porous SiO2 nanofilm formation of the silsesquioxane-containing amphiphilic block copolymer. *Langmuir the ACS Journal of Surfaces and Colloids*, *34*(27), 8007–8014. Available from https://doi.org/10.1021/acs.langmuir.8b01114.

Jin, Y., Wang, P., Hou, K., Lin, Y., Li, L., Xu, S., Cheng, J., Wen, X., & Pi, P. (2018). Superhydrophobic porous surface fabricated via phase separation between polyhedral oligomeric silsesquioxane-based block copolymer and polyethylene glycol. *Thin Solid Films*, *649*, 210–218. Available from https://doi.org/10.1016/j.tsf.2018.01.030, http://www.journals.elsevier.com/journal-of-the-energy-institute.

Kakuta, T., Tanaka, K., & Chujo, Y. (2015). Synthesis of emissive water-soluble network polymers based on polyhedral oligomeric silsesquioxane and their application as optical sensors for discriminating the particle size. *Journal of Materials Chemistry C*, *3*(48), 12539–12545. Available from https://doi.org/10.1039/C5TC03139G.

Kalia, S., & Pielichowski, K. (n.d.). Polymer/POSS nanocomposites and hybrid materials. In: S. Kalia & K. Pielichowski, *Springer series on polymer and composite materials*. Springer Cham. Available from https://doi.org/10.1007/978-3-030-02327-0.

Kausar, A. (2017). State-of-the-art overview on polymer/POSS nanocomposite. *Polymer-Plastics Technology and Engineering*, *56*(13), 1401–1420. Available from https://doi.org/10.1080/03602559.2016.1276592.

Knauer, K. M., Brust, G., Carr, M., Cardona, R. J., Lichtenhan, J. D., & Morgan, S. E. (2017). Rheological and crystallization enhancement in polyphenylenesulfide and polyetheretherketone POSS nanocomposites. *Journal of Applied Polymer Science*, *134*(7), 44462. Available from https://doi.org/10.1002/app.44462.

Kunthom, R., Piyanuch, P., Wanichacheva, N., & Ervithayasuporn, V. (2018). Cage-like silsesequioxanes bearing rhodamines as fluorescence Hg2 + sensors. *Journal of Photochemistry and Photobiology A: Chemistry, 356*, 248–255. Available from https://doi.org/10.1016/j.jphotochem.2017.12.033.

Kwon, E.H., An, H., Park, M.B., Kim, M., Park, Y.D. (n.d.), Conjugated polymer–zeolite hybrids for robust gas sensors: Effect of zeolite surface area on NO2 sensing ability. *Chemical Engineering Journal, 420*, 129588. Available form https://doi.org/10.1016/j.cej.2021.129588.

Laird, M., Herrmann, N., Ramsahye, N., Totée, C., Carcel, C., Unno, M., Bartlett, J. R., & Wong Chi Man, M. (2021). Large polyhedral oligomeric silsesquioxane cages: The isolation of functionalized POSS with an unprecedented Si18O27 core. *Angewandte Chemie – International Edition, 60*(6), 3022–3027. Available from https://doi.org/10.1002/anie.202010458, http://onlinelibrary.wiley.com/journal/10.1002/(ISSN)1521-3773.

Lee, D. H., Yoon, K. B., Jung, M. S., Sung, J. K., & Noh, S. K. (2006). Preparation of ethylene/polyhedral oligomeric silsesquioxane(POSS) copolymers with rac-Et(Ind)2ZrCl2/MMAO catalyst system. *Studies in Surface Science and Catalysis, 161*, 53–58. Available from https://doi.org/10.1016/s0167-2991(06)80434-9, https://www.sciencedirect.com/bookseries/studies-in-surface-science-and-catalysis.

Li, W., Jiang, C., Liu, H., Yan, Y., & Liu, H. (2019). Octa[4-(9-carbazolyl)phenyl]silsesquioxane-based porous material for dyes adsorption and sensing of nitroaromatic compounds. *Chemistry – An Asian Journal, 14* (19), 3363–3369. Available from https://doi.org/10.1002/asia.201900951, http://onlinelibrary.wiley.com/journal/10.1002/(ISSN)1861-471X.

Li, Y. C., Mannen, S., Schulz, J., & Grunlan, J. C. (2011). Growth and fire protection behavior of POSS-based multilayer thin films. *Journal of Materials Chemistry, 21*(9), 3060–3069. Available from https://doi.org/10.1039/c0jm03752d.

Li, Z., Kong, J., Wang, F., & He, C. (2017). Polyhedral oligomeric silsesquioxanes (POSSs): An important building block for organic optoelectronic materials. *Journal of Materials Chemistry C, 5*(22), 5283–5298. Available from https://doi.org/10.1039/C7TC01327B.

Lichtenhan, J. D., Otonari, Y. A., & Carr, M. J. (1995). Linear hybrid polymer building blocks: Methacrylate-functionalized polyhedral oligomeric silsesquioxane monomers and polymers. *Macromolecules, 28*(24), 8435–8437. Available from https://doi.org/10.1021/ma00128a067.

Liu, H., & Liu, H. (2017). Selective dye adsorption and metal ion detection using multifunctional silsesquioxane-based tetraphenylethene-linked nanoporous polymers. *Journal of Materials Chemistry A, 5* (19), 9156–9162. Available from https://doi.org/10.1039/C7TA01255A.

Liu, J., Fan, J., Zhang, Z., Hu, Q., Zeng, T., & Li, B. (2013). Nano/microstructured polyhedral oligomeric silsesquioxanes-based hybrid copolymers: Morphology evolution and surface characterization. *Journal of Colloid and Interface Science, 394*(1), 386–393. Available from https://doi.org/10.1016/j.jcis.2012.11.015.

Lu, C., Ben, T., Xu, S., & Qiu, S. (2014). Electrochemical synthesis of a microporous conductive polymer based on a metal–organic framework thin film. *Angewandte Chemie International Edition, 53*(25), 6454–6458. Available from https://doi.org/10.1002/anie.201402950.

Lu, D., Qin, M., Liu, C., Deng, J., Shi, G., & Zhou, T. (2021). Ionic liquid-functionalized magnetic metal–organic framework nanocomposites for efficient extraction and sensitive detection of fluoroquinolone antibiotics in environmental water. *ACS Applied Materials & Interfaces, 13*(4), 5357–5367. Available from https://doi.org/10.1021/acsami.0c17310.

Lu, N., Lu, Y., Liu, S., Jin, C., Fang, S., Zhou, X., & Li, Z. (2019). Tailor-engineered POSS-based hybrid gels for bone regeneration. *Biomacromolecules, 20*(9), 3485–3493. Available from https://doi.org/10.1021/acs.biomac.9b00771.

Lv, Z., Chen, Z., Feng, S., Wang, D., & Liu, H. (2022). A sulfur-containing fluorescent hybrid porous polymer for selective detection and adsorption of Hg 2 + ions. *Polymer Chemistry, 13*(16), 2320–2330. Available from https://doi.org/10.1039/d2py00077f.

Meng, X., Liu, Y., Wang, S., Du, J., Ye, Y., Song, X., & Liang, Z. (2020). Silsesquioxane-carbazole-corbelled hybrid porous polymers with flexible nanopores for efficient CO2 conversion and luminescence sensing. *ACS Applied Polymer Materials*, *2*(2), 189–197. Available from https://doi.org/10.1021/acsapm.9b00747, pubs.acs.org/journal/aapmcd.

Mituła, K., Duszczak, J., Brząkalski, D., Dudziec, B., Kubicki, M., & Marciniec, B. (2017). Tetra-functional double-decker silsesquioxanes as anchors for reactive functional groups and potential synthons for hybrid materials. *Chemical Communications*, *53*(75), 10370–10373. Available from https://doi.org/10.1039/C7CC03958A.

Ni, B., Dong, X. H., Chen, Z., Lin, Z., Li, Y., Huang, M., Fu, Q., Cheng, S. Z. D., & Zhang, W. B. (2014). "Clicking" fluorinated polyhedral oligomeric silsesquioxane onto polymers: A modular approach toward shape amphiphiles with fluorous molecular clusters. *Polymer Chemistry*, *5*(11), 3588–3597. Available from https://doi.org/10.1039/c3py01670f, http://pubs.rsc.org/en/Journals/JournalIssues/PY.

Noureddine, A., Trens, P., Toquer, G., Cattoën, X., & Chi Man, M. W. (2014). Tailoring the hydrophilic/lipophilic balance of clickable mesoporous organosilicas by the copper-catalyzed azide–alkyne cycloaddition click-functionalization. *Langmuir the ACS Journal of Surfaces and Colloids*, *30*(41), 12297–12305. Available from https://doi.org/10.1021/la503151w.

Oleksy, M., & Galina, H. (2013). Unsaturated polyester resin composites containing bentonites modified with silsesquioxanes. *Industrial & Engineering Chemistry Research*, *52*(20), 6713–6721. Available from https://doi.org/10.1021/ie303433v.

Park, S. J., Park, C. S., & Yoon, H. (2017). Chemo-electrical gas sensors based on conducting polymer hybrids. *Polymers*, *9*(5), 155. Available from https://doi.org/10.3390/polym9050155.

Petit, C., Lin, K. Y. A., & Park, A. H. A. (2013). Design and characterization of liquidlike poss-based hybrid nanomaterials synthesized via ionic bonding and their interactions with CO2. *Langmuir the ACS Journal of Surfaces and Colloids*, *29*(39), 12234–12242. Available from https://doi.org/10.1021/la4007923.

Raftopoulos, K. N., & Pielichowski, K. (2016). Segmental dynamics in hybrid polymer/POSS nanomaterials. *Progress in Polymer Science*, *52*, 136–187. Available from https://doi.org/10.1016/j.progpolymsci.2015.01.003, http://www.sciencedirect.com/science/journal/00796700.

Raimondo, M., Russo, S., Guadagno, L., Longo, P., Chirico, S., Mariconda, A., Bonnaud, L., Murariu, O., & Dubois, P. (2015). Effect of incorporation of POSS compounds and phosphorous hardeners on thermal and fire resistance of nanofilled aeronautic resins. *RSC Advances*, *5*(15), 10974–10986. Available from https://doi.org/10.1039/C4RA11537F.

Rees, T. M., Thompson, N. J., & Wilford, A. (n.d.). The modern approach to modifying epoxy resins using liquid polysulphides: Part 2. *Journal of the Oil and Colour Chemists Association*, *72*(2), 0.66–71.

Rikowski, E., & Marsmann, H. C. (1997). Cage-rearrangement of silsesquioxanes. *Polyhedron*, *16*(19), 3357–3361. Available from https://doi.org/10.1016/S0277-5387(97)00092-2, http://www.journals.elsevier.com/polyhedron/.

Sangtrirutnugul, P., Chaiprasert, T., Hunsiri, W., Jitjaroendee, T., Songkhum, P., Laohhasurayotin, K., Osotchan, T., & Ervithayasuporn, V. (2017). Tunable porosity of cross-linked-polyhedral oligomeric silsesquioxane supports for palladium-catalyzed aerobic alcohol oxidation in water. *ACS Applied Materials & Interfaces*, *9*(14), 12812–12822. Available from https://doi.org/10.1021/acsami.7b03910.

Scott, D. W. (1946). Thermal rearrangement of branched-chain methylpolysiloxanes. *Journal of the American Chemical Society*, *68*(3), 356–358. Available from https://doi.org/10.1021/ja01207a003.

Silverstein, M. S., Tai, H., Sergienko, A., Lumelsky, Y., & Pavlovsky, S. (2005). PolyHIPE: IPNs, hybrids, nanoscale porosity, silica monoliths and ICP-based sensors. *Polymer*, *46*(17), 6682–6694. Available from https://doi.org/10.1016/j.polymer.2005.05.022.

Soldatov, M., & Liu, H. (2021). Hybrid porous polymers based on cage-like organosiloxanes: Synthesis, properties and applications. *Progress in Polymer Science*, *119*101419. Available from https://doi.org/10.1016/j.progpolymsci.2021.101419.

Suenaga, K., Tanaka, K., & Chujo, Y. (2021). Positive luminescent sensor for aerobic conditions based on polyhedral oligomeric silsesquioxane networks. *Chemical Research in Chinese Universities*, *37*(1), 162–165. Available from https://doi.org/10.1007/s40242-021-0398-x.

Sun, R., Feng, S., Wang, D., & Liu, H. (2018). Fluorescence-tuned silicone elastomers for multicolored ultraviolet light-emitting diodes: Realizing the processability of polyhedral oligomeric silsesquioxane-based hybrid porous polymers. *Chemistry of Materials*, *30*(18), 6370–6376. Available from https://doi.org/10.1021/acs.chemmater.8b02514.

Sun, R., Feng, S., Zhou, B., Chen, Z., Wang, D., & Liu, H. (2020). Flexible cyclosiloxane-linked fluorescent porous polymers for multifunctional chemical sensors. *ACS Macro Letters*, *9*(1), 43–48. Available from https://doi.org/10.1021/acsmacrolett.9b00901.

Sun, R., Huo, X., Lu, H., Feng, S., Wang, D., & Liu, H. (2018). Recyclable fluorescent paper sensor for visual detection of nitroaromatic explosives. *Sensors and Actuators B: Chemical*, *265*, 476–487. Available from https://doi.org/10.1016/j.snb.2018.03.072.

Wang, D., Feng, S., & Liu, H. (2016). Fluorescence-tuned polyhedral oligomeric silsesquioxane-based porous polymers. *Chemistry – A European Journal*, *22*(40), 14319–14327. Available from https://doi.org/10.1002/chem.201602688.

Wang, D., Li, L., Yang, W., Zuo, Y., Feng, S., & Liu, H. (2014). POSS-based luminescent porous polymers for carbon dioxide sorption and nitroaromatic explosives detection. *Royal Society of Chemistry, China RSC Advances*, *4*(104), 59877–59884. Available from https://doi.org/10.1039/c4ra11069b, http://pubs.rsc.org/en/journals/journalissues.

Wang, D., Sun, R., Feng, S., Li, W., & Liu, H. (2017). Retrieving the original appearance of polyhedral oligomeric silsesquioxane-based porous polymers. *Polymer*, *130*, 218–229. Available from https://doi.org/10.1016/j.polymer.2017.10.021.

Wang, Q., Liu, H., Jiang, C., & Liu, H. (2020). Silsesquioxane-based triphenylamine functionalized porous polymer for CO2, I2 capture and nitro-aromatics detection. *Polymer*, *186*. Available from https://doi.org/10.1016/j.polymer.2019.122004.

Wang, Q., Unno, M., & Liu, H. (2021). Silsesquioxane-based triphenylamine-linked fluorescent porous polymer for dyes adsorption and nitro-aromatics detection. *Materials*, *14*(14), 3851. Available from https://doi.org/10.3390/ma14143851.

Wang, Q., Unno, M., & Liu, H. (2022). Organic-inorganic hybrid near-infrared emitting porous polymer for detection and photodegradation of antibiotics. *ACS Sustainable Chemistry and Engineering*, *10*(22), 7309–7320. Available from https://doi.org/10.1021/acssuschemeng.2c00935, http://pubs.acs.org/journal/ascecg.

Wang, W. J., Hai, X., Mao, Q. X., Chen, M. L., & Wang, J. H. (2015). Polyhedral oligomeric silsesquioxane functionalized carbon dots for cell imaging. *ACS Applied Materials and Interfaces*, *7*(30), 16609–16616. Available from https://doi.org/10.1021/acsami.5b04172, http://pubs.acs.org/journal/aamick.

Wang, Z., Kunthom, R., Kostjuk, S. V., & Liu, H. (2023). Near-infrared-emitting silsesquioxane-based porous polymer containing thiophene for highly efficient adsorption and detection of iodine vapor and solution phase. *European Polymer Journal*, *192*. Available from https://doi.org/10.1016/j.eurpolymj.2023.112072.

Wang, Z., Mathew, A., & Liu, H. (2022). Silsesquioxane-based porous polymer derived from organic chromophore with AIE characteristics for selective detection of 2,4-dinitrophenol and Ru3 + . *Polymer*, *248*124788. Available from https://doi.org/10.1016/j.polymer.2022.124788, https://www.sciencedirect.com/science/article/pii/S0032386122002750.

Wu, D., Xu, F., Sun, B., Fu, R., He, H., & Matyjaszewski, K. (2012). Design and preparation of porous polymers. *Chemical Reviews*, *112*(7), 3959–4015. Available from https://doi.org/10.1021/cr200440z.

Wu, M., Wu, R., Li, R., Qin, H., Dong, J., Zhang, Z., & Zou, H. (2010). Polyhedral oligomeric silsesquioxane as a cross-linker for preparation of inorganic-organic hybrid monolithic columns. *Analytical Chemistry*, *82* (13), 5447–5454. Available from https://doi.org/10.1021/ac1003147.

Xu, W., Chung, C., & Kwon, Y. (2007). Synthesis of novel block copolymers containing polyhedral oligomeric silsesquioxane (POSS) pendent groups via ring-opening metathesis polymerization (ROMP). *Polymer*, *48*(21), 6286–6293. Available from https://doi.org/10.1016/j.polymer.2007.08.014.

Xu, Z., Zhao, Y., Wang, X., & Lin, T. (2013). A thermally healable polyhedral oligomeric silsesquioxane (POSS) nanocomposite based on Diels–Alder chemistry. *Chemical Communications*, *49*(60), 6755–6757. Available from https://doi.org/10.1039/c3cc43432j.

Yan, Y., Laine, R. M., & Liu, H. (2019). In situ methylation transforms aggregation-caused quenching into aggregation-induced emission: Functional porous silsesquioxane-based composites with enhanced near-infrared emission. *ChemPlusChem*, *84*(10), 1630–1637. Available from https://doi.org/10.1002/cplu.201900568, http://onlinelibrary.wiley.com/journal/10.1002/(ISSN)2192-6506.

Yan, Y., Yang, G., Xu, J. L., Zhang, M., Kuo, C. C., & Wang, S. D. (2020). Conducting polymer-inorganic nanocomposite-based gas sensors: A review. *Science and Technology of Advanced Materials*, *21*(1), 768–786. Available from https://doi.org/10.1080/14686996.2020.1820845, http://www.tandfonline.com/loi/tsta20#.V0KNDU1f3cs.

Yan, Y., Yang, H., & Liu, H. (2020). Silsesquioxane-based fluorescent nanoporous polymer derived from a novel AIE chromophore for concurrent detection and adsorption of Ru3. *Sensors and Actuators B: Chemical*, *319*128154. Available from https://doi.org/10.1016/j.snb.2020.128154.

Yang, H., & Liu, H. (2020). Pyrene-functionalized silsesquioxane as fluorescent nanoporous material for antibiotics detection and removal. *Microporous and Mesoporous Materials*, *300*. Available from https://doi.org/10.1016/j.micromeso.2020.110135.

Yang, N., & Liu, H. (2021). Tetraphenylpyrene-bridged silsesquioxane-based fluorescent hybrid porous polymer with selective metal ions sensing and efficient phenolic pollutants adsorption activities. *Polymer*, *230*124083. Available from https://doi.org/10.1016/j.polymer.2021.124083.

Yang, W., Jiang, X., & Liu, H. (2015). A novel pH-responsive POSS-based nanoporous luminescent material derived from brominated distyrylpyridine and octavinylsilsesquioxane. *RSC Advances*, *5*(17), 12800–12806. Available from https://doi.org/10.1039/C4RA13628D.

Yang, X., & Liu, H. (2019). Diphenylphosphine-substituted ferrocene/silsesquioxane-based hybrid porous polymers as highly efficient adsorbents for water treatment. *ACS Applied Materials & Interfaces*, *11*(29), 26474–26482. Available from https://doi.org/10.1021/acsami.9b07874.

Ye, Y. S., Shen, W. C., Tseng, C. Y., Rick, J., Huang, Y. J., Chang, F. C., & Hwang, B. J. (2011). Versatile grafting approaches to star-shaped POSS-containing hybrid polymers using RAFT polymerization and click chemistry. *Chemical Communications*, *47*(38), 10656–10658. Available from https://doi.org/10.1039/c1cc13412d, http://pubs.rsc.org/en/journals/journal/cc.

Yue, Y., Liu, N., Liu, W., Li, M., Ma, Y., Luo, C., Wang, S., Rao, J., Hu, X., Su, J., Zhang, Z., Huang, Q., & Gao, Y. (2018). 3D hybrid porous Mxene-sponge network and its application in piezoresistive sensor. *Nano Energy*, *50*, 79–87. Available from https://doi.org/10.1016/j.nanoen.2018.05.020.

Zhang, H., Chen, L., Li, Y., Hu, Y., Li, H., Xu, C. C., & Yang, S. (2022). Functionalized organic-inorganic hybrid porous coordination polymer-based catalysts for biodiesel production via trans/esterification. *Green Chemistry*, *24*(20), 7763–7786. Available from https://doi.org/10.1039/d2gc02722d, http://pubs.rsc.org/en/journals/journal/gc.

Zhang, Q. M., Wang, Z., Cheng, G., Ma, H., Zhang, Q. P., Wan, F. X., Tan, B., & Zhang, C. (2018). Efficient alkaloid capture from water using a charged porous organic polymer. *RSC Advances*, *8*(58), 33398–33402. Available from https://doi.org/10.1039/C8RA06499G, http://pubs.rsc.org/en/journals/journal/ra.

Zhang, W., & Müller, A. H. E. (2013). Architecture, self-assembly and properties of well-defined hybrid polymers based on polyhedral oligomeric silsequioxane (POSS. *Progress in Polymer Science*, *38*(8), 1121–1162. Available from https://doi.org/10.1016/j.progpolymsci.2013.03.002.

Zhang, W. B., He, J., Yue, K., Liu, C., Ni, P., Quirk, R. P., & Cheng, S. Z. D. (2012). Rapid and efficient anionic synthesis of well-defined eight-arm star polymers using octavinylPOSS and poly(styryl)lithium. *Macromolecules*, *45*(21), 8571–8579. Available from https://doi.org/10.1021/ma301597f.

Zhang, X., Zhang, Q., Yue, D., Zhang, J., Wang, J., Li, B., Yang, Y., Cui, Y., & Qian, G. (2018). Flexible metal-organic framework-based mixed-matrix membranes: A new platform for H2S sensors. *Small (Weinheim an der Bergstrasse, Germany)*, *14*(37), 1801563. Available from https://doi.org/10.1002/smll.201801563.

Zhao, X., Wang, Q., Kunthom, R., & Liu, H. (2023). Sulfonic acid-grafted hybrid porous polymer based on double-decker silsesquioxane as highly efficient acidic heterogeneous catalysts for the alcoholysis of styrene oxide. *ACS Applied Materials & Interfaces*, *15*(5), 6657–6665. Available from https://doi.org/10.1021/acsami.2c17732.

Zhou, H., Ye, Q., Neo, W. T., Song, J., Yan, H., Zong, Y., Tang, B. Z., Hor, T. S. A., & Xu, J. (2014). Electrospun aggregation-induced emission active POSS-based porous copolymer films for detection of explosives. *Chemical Communications*, *50*(89), 13785–13788. Available from https://doi.org/10.1039/c4cc06559j.

Zhou, H., Ye, Q., & Xu, J. (2017). Polyhedral oligomeric silsesquioxane-based hybrid materials and their applications. *Materials Chemistry Frontiers*, *1*(2), 212–230. Available from https://doi.org/10.1039/C6QM00062B.

CHAPTER

Polymer composite sensors

9

Karina C. Núñez-Carrero, Luis E. Alonso-Pastor and Manuel Herrero

Cellular Materials Laboratory (CellMat), Department of Condensed Material Physics, Faculty of Science, University of Valladolid, Valladolid (Valladolid), Spain

9.1 Introduction

9.1.1 Composite materials

Composite materials can be defined as the combination of two or more components with different properties (i.e., metallic, ceramic, or plastic) (Devnani & Sinha, 2019), where one of them is present in a major proportion and is known as the matrix (in polymeric composites, the matrix is polymeric) and the rest of the components act as a dispersed phase (often functional particles or fibers, reinforcements, or fillers) (Hsissou et al., 2021). It is important to note that in the case of multiple discontinuous phases of different natures, the composites are called hybrids (Ahmadijokani et al., 2020). The matrix must ensure cohesion with the reinforcement to transmit, among them, the stimulus to which the composite is subjected. In this way, the final assembly of all components will have properties superior to those of each individual constituent material on its own (Dagdag et al., 2019; Hsissou et al., 2020). The advantage of polymer composites, compared to those with metal matrices, lies in their manufacturing process, which allows the production of complex-shaped parts with lower weight and in more sustainable manufacturing processes (Tarhini & Tehrani-Bagha, 2019; Zhang et al., 2019).

Classifying composite polymeric materials is not an easy task. A multitude of factors must be considered, including the nature of components (organic, mineral, metallic, etc.), quantities, geometry, distribution of reinforcement(s), used additives, manufacturing processes, final properties, and their ultimate application field, among others. The permutations of these factors give infinite possibilities. Nevertheless, from an engineering perspective, the term "polymeric composite" is often associated with typical polymer matrix materials reinforced with glass or carbon fiber (these fibers can be continuous or discontinuous) for the transport sector (e.g., aeronautics), building, and construction applications. In these instances, the goal is to obtain high-performance materials with low densities, achieving remarkable mechanical properties to manufacture structural parts.

Beyond the mechanical properties, combining components of different natures (matrix and discontinuous phase) can confer additional advantages to the composites, such as thermal, acoustic, optical, and so on. This is the key that allows the composites to find their aptitude for sensor applications. Either component, matrix, or discontinuous phase can inherently possess the capacity to perceive and respond to alterations in their physical or chemical properties. This adaptability

Sensory Polymers. DOI: https://doi.org/10.1016/B978-0-443-13394-7.00009-4

empowers them to transmute these changes into electrical signals or data, which can be used for monitoring specific magnitudes or phenomena. As mentioned previously, in polymeric composite sensors, the discontinuous phase used to be the key factor in sensing, and it regulates the detection principles (piezoelectric, resistive, capacitive, conductive, inductive, optical, ultrasonic, magnetic field manipulation, ionization, and so on). This discontinuous phase assumes different forms and dimensions, and it could be presented as nanometric fibers, short and long fibers, or even complex fabrics. This functional phase becomes integrated within polymeric matrices, leading to different characteristics such as lightweightness, mechanical resilience, corrosion resistance, and hostile environment resistance. These properties, along with the capacity to adopt complex geometries, enable seamless integration into diverse frameworks. In line with the discussion, the overarching challenge is mainly focused on getting good distributions, dispersion, and compatibility of this discontinuous phase within the polymeric matrix, thereby enhancing the efficacy of the designed sensor.

Fig. 9.1 provides an illustrative scheme showcasing the general composition of a polymeric composite material: the polymer matrix, the discontinuous phase, and sometimes additives or fillers in very small amounts. Polymeric matrices are typically of two types: thermoplastic or thermosetting. Each type offers distinct advantages and drawbacks. Thermoplastic matrices are characterized by their capacity to shape complex geometries, high recyclability, and cost-effectiveness. On the other hand, thermosetting matrices create a crosslinked polymer network that enhances the composite material's mechanical behavior and dimensional stability (Shukla & Saxena, 2021). While this chapter primarily focuses on a review of sensor composites based on functionality and the type of

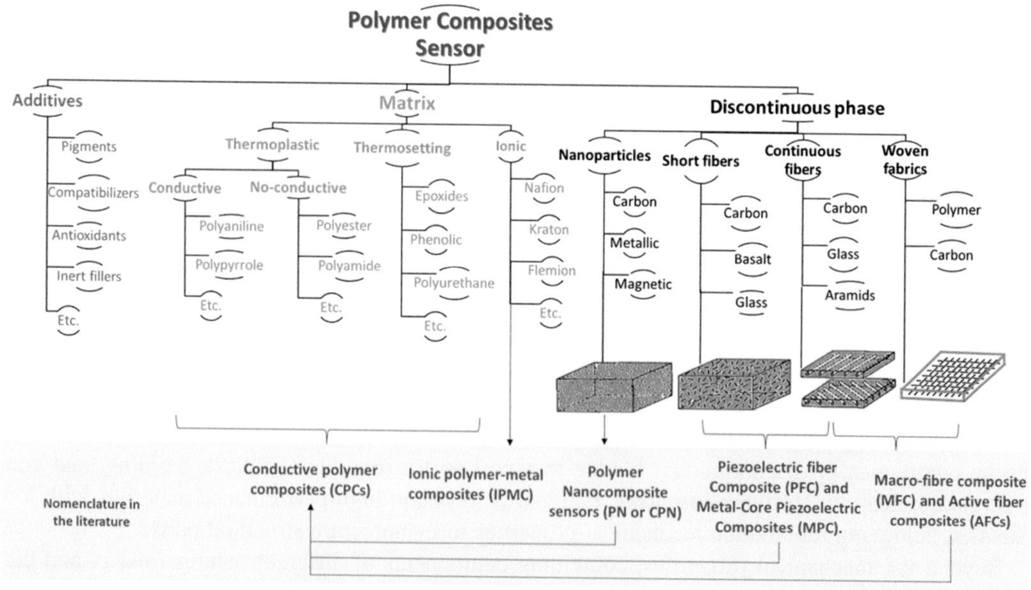

FIGURE 9.1

Schematic view of the composition of various polymeric composite sensors along with their corresponding nomenclatures in the literature.

discontinuous phase, it is noteworthy that not all polymeric composite sensors feature inert matrices. Extensive literature showcases the synergies achievable between electrically conductive polymeric matrices like polyacetylene, polypyrrole (PPy), polyaniline (PANI), and more when combined with functional fibers or particles to enhance sensor composite detection (Chen et al., 2021). Addressing limitations concerning the stability and processability of these materials to manufacture the sensor composite is pivotal in such cases. Effective composite formation, mixing, doping, and controlling the size of the functional discontinuous phase aid in mitigating these challenges. Between all these matrices, PANI has garnered the most significant attention, serving as a matrix in the design of polymeric composite sensors (Singh & Shukla, 2020).

In Fig. 9.1, the schematic representation also illustrates how, in the literature, the term *conductive polymer composites (CPCs)* can encompass sensors formed with both conductive and nonconductive polymeric matrices, as well as multiphase blends. The functional discontinuous phase refers to simple or hybrid conductive fillers (carbonaceous, metallic particles, etc.) (Al-Saleh & Sundararaj, 2009; Deng et al., 2014; Stankovich et al., 2006) of different sizes. The final composite is understood to be conductive and function as a sensor, irrespective of the type of matrix it comprises. CPCs can encompass various combinations of functionally dispersed phases, ranging from nanoparticles and micrometer-sized fibers to textiles. This gives rise to different nomenclatures such as *piezoelectric fiber composites (PFC)*, *metal-core piezoelectric composites*, *macrofiber composites (MFC)*, or *active-fiber composites (AFCs)*. The operational principles of these devices, based on the type of dispersed phase they employ, will be discussed in the following sections.

Concerning the classification based on the type of matrix employed in polymeric composite sensors, while not the primary focus of this review, it is worth mentioning *Ionic Polymer-Metal Composites (IPMCs)*, as depicted in Fig. 9.1. IPMCs are based on ionic polymers (e.g., Nafion, Flemion, Selenion-CMV, and Kraton) with a conductive medium, generally a metal. These composites can exhibit significant dynamic deformation when appropriately electroplated and subjected to a time-varying electric field (Shahinpoor & Kim, 2001). Widely studied, these composites possess a distinct sandwich-like configuration composed of a polymer matrix membrane (typically Nafion) in the center, flanked by layers of metallic electrodes on both sides (Zhao et al., 2021). In the case of Nafion, the fixed sulfonic acid ion groups bear a negative charge, and cations—typically alkali metal ions—can traverse nanochannels within the ionomer (Pugal et al., 2010; Stalbaum et al., 2017). Cations within the matrix membrane are often associated with water molecules to form hydrated cations. Upon applying a voltage to the electrodes on both sides of the IPMC, hydrated cations, along with water molecules, migrate from the anode side to the cathode side, propelled by electrostatic forces. This action leads to substantial bending and deformation in the IPMC due to asymmetrical swelling, and the reverse also occurs. The external charge manipulation alters the internal distribution of the IPMC, thus generating a potential difference on both sides of the electrode. This enables the measurement of the ionic electric response using the superficial metallic electrode (Bahramzadeh & Shahinpoor, 2011). Due to this distinctive behavior, IPMCs are regarded as transducers of intelligent materials utilized as biomimetic sensors, actuators, and artificial muscles (Amirkhani & Bakhtiarpour, 2016; Kim & Tadokoro, 2007; MohdIsa et al., 2019; Punning et al., 2007). As the IPMCs fall outside the scope of the polymer composite definition employed in this chapter (dispersed phase embedded in a polymeric matrix), this hybrid material type (polymer membrane/metal) is not within the purpose of this review.

This review will thus focus on the four most common types of composites, characterized by a continuous phase and an embedded dispersed phase, classified based on the size of the discontinuous or functional phase (which generally imparts the sensing capability): nanocomposites, short-fiber composites, continuous-fiber composites, and composites based on fabrics. To achieve this, a brief overview of the key concepts associated with these types will be provided in the following sections before delving into the discussion of operating principles and advancements in the field of sensors utilizing these composites.

9.1.2 Polymeric nanocomposites

The widely accepted definition of polymeric nanocomposites refers to materials where at least one of the phases (typically the particles or fibers of the dispersed phase) exists on a nanoscale. This typically entails physical measurements on the order of 10^{-9} meters. However, in line with convention, dimensions within the range of 1–100 nm are recognized as applicable to the nanoscale (Shukla & Saxena, 2021).

The interesting aspect of working at these dimensions lies in the fact that as the size of the dispersed particle diminishes to the nanoscale, there is an increase in the aspect ratio, consequently leading to heightened interaction between the particles and the matrix. Consequently, properties like stiffness, heat resistance, sensitivity, and electrochemical behavior, contributed by the nanoparticle, are expected to be significantly enhanced across the system. To improve the properties of the final nanocomposite, the fundamental factor is to achieve a good dispersion of the nanoparticles within the polymeric matrix, achieving a uniform distribution and minimizing agglomerates (Das et al., 2018).

Presently, nanocomposites are actively employed in developing smart sensors owing to their conductivity, mechanical properties, and chemical resistance. The nanoparticles used to fabricate polymeric sensor composites encompass carbon nanostructures such as nanotubes or graphene, metallic, magnetic, or semiconductor nanoparticles. Nanocomposites are the most extensively researched configuration in polymeric sensor composites, often referred to as *polymer nanocomposite sensors* in the literature. It is also common to use the nomenclature for *conductive polymer nanocomposites*, especially when dealing with dispersed phases based on conductive nanoparticles (see Fig. 9.1 and Section 9.2.1).

Irrespective of whether it functions as a sensor or not, the efficacy of a polymeric nanocomposite depends on several factors, such as polymer polarity, molecular weight, hydrophobicity, and reactive groups. Equally important are the type, quantity, and shape of the nanoparticle, the composite's manufacturing approach, the alignment of charges (or lack thereof), and the polymer-nanoparticle interaction when designing nanocomposites. This interaction encompasses factors like optimal dispersion, polymer-nanoparticle miscibility, electrostatic forces, and so on (Deepa et al., 2019). Numerous strategies are pursued to achieve this, including surface modification of nanoparticles and enhancement of processing and molding techniques (Alshammari, 2018; Kuilla et al., 2010). Three approaches are commonly employed to produce nanocomposites, which are later used to design sensor nanocomposites: melt-mixing, solution-mixing, and in situ polymerization. Melt-mixing is the most common method extensively used in the industry to obtain nanocomposites. This method is based on the traditional polymer processing by extrusion. The main drawback of this approach is that, depending on the matrix and the nanoparticle, it is possible that the dispersions need optimization due to the low interfacial adhesion between them. Solution-mixing promotes the dispersion of the nanoparticles in a solution of the polymer matrix. High dispersion is raised, but the main disadvantage is the large quantities of solvent used by the method. In situ polymerization is a highly effective method to obtain nanocomposites, and it is based on the dispersion of the nanoparticles during polymerization, when polymer chains are growing from monomers. This method is not currently implemented on an industrial scale (Tanasi et al., 2020).

9.1.3 Composites with short particles or fiber

These materials encompass fillers where the particles or fibers are present at the micro and/or millimeter scale. The performance of composite materials, similarly to the case of nanocomposites, predominantly depends on their constituent elements and manufacturing techniques, but in general, composites can offer exceptional properties such as high durability, stiffness, damping property, flexural strength, and resistance to corrosion, wear, impact, and fire (Rajak et al., 2019).

Depending on the type of reinforcement, a distinction between fiber-reinforced and particle-reinforced composites can be established. Within the fiber-reinforced composites, there exist subcategories encompassing those that employ synthetic and natural fibers. Synthetic fiber composites represent the more conventional, with glass, basalt, carbon, or aramid fibers being the most common. Nevertheless, the employment of natural fibers such as kenaf, jute, hemp, or flax is progressively growing (Akhil et al., 2023). Focused on particle-reinforced composites, it is usual to incorporate ceramic, inorganic, or metallic fillers to pursue specific functionalities such as fire resistance or augmented conductivity. This group is where short-fiber composites used as sensors are usually found.

Composites based on discontinuous fibers or particles are generally based on a thermoplastic polymer matrix. In this sense, the manufacturing process consists of dispersing these fibers or particles within the matrix by extrusion. During this process, the polymeric pellets and fibers are mixed rigorously and fed into the hopper of the extruder. The mixture is then carried inside the barrel with the help of a rotating screw. As the mixture travels toward the die, the pellets melt, and compounding with fibers takes place due to heating and the rotating action of the screw (Komal et al., 2019). Once the final composite has been obtained, it can be processed by different technologies, such as calendaring and blowing for films or injection molding and 3D printing for final parts. These composites find applications in various industries, including automotive, aerospace, construction, sports, electronics, marine, energy, and consumer goods. They are used to enhance properties, reduce weight, and improve durability in components such as vehicle panels, aircraft structures, building materials, sports equipment, electronic casings, marine vessels, wind turbine blades, and consumer products.

In the case of short fiber composites used as sensors, the most commonly used particles are carbon black (CB), modified glass fibers, and short fibers of basalt and aramids. The majority operate as detectors based on the piezoelectric principle, which is why they are often referred to in the literature as *PFCs* (see Fig. 9.1 and Section 9.2.2).

9.1.4 Continuous fiber composites and fabric composites

Polymer composites reinforced with long fibers and textiles are advanced materials that combine the mechanical properties of fibers with the versatility of polymers. These composites are engineered to offer exceptional strength, stiffness, and durability, making them vital components in various industrial sectors.

As mentioned previously, long fiber and textile polymer composites are also heterogeneous materials composed of a polymeric matrix, but in this case, they have embedded long fibers or textiles. When the fibers are aligned in the same direction, they are called unidirectional composites, while if it is a fabric that acts as reinforcement, they are called bidirectional composites. Furthermore, it is noteworthy that these fabrics can be stacked in different directions, earning them the name multidirectional (Rajak, Pagar, Menezes et al., 2019). As in the case of short-fiber composites, natural and synthetic fiber composites can be distinguished (Rajak, Pagar, Kumar et al., 2019). From an industrial point of view, fiberglass and carbon fiber fabrics, and aramid to a lesser extent, are the majority.

Owing to their exceptional mechanical properties and low weight, these composites stand as substitutes for metallic components. Weight reduction is particularly critical in aeronautics and the automotive industry, but its utility extends to sectors like construction, wind turbine blade manufacturing, and sports equipment.

To establish the manufacturing processes for continuous fiber or woven composites, it must be considered whether the polymeric matrix is thermoplastic or thermosetting. In the realm of thermoplastic matrices, all methodologies encompass polymer melting, followed by the application of compressive forces to facilitate the interstitial diffusion of the matrix material within fibers or textiles. In thermosetting matrices, the process is different. In these cases, the polymeric precursor is a liquid with a lower viscosity than melted polymers, and consequently, the diffusion between fibers and textiles is much easier. However, these matrices need a curing phase, typically at temperatures and/or pressures, to provide crosslinking and yield the definitive composite.

Today, available technologies are varied and depend greatly on the starting materials, both on the polymer matrix and the fiber morphology. Traditionally, the manufacturing techniques of this type of composite can be divided into two large groups: open mold technologies, in which the composites cure exposed to air, and closed mold technologies, where the resin and fibers cure inside the mold (Rajak, Pagar, Kumar et al., 2019). Typical examples of open mold technologies are hand lay-up (Elkington et al., 2015), spray-up (Zin et al., 2016), and filament winding (Azeem et al., 2022). Closed mold technologies include vacuum bag molding and vacuum infusion (Ekuase et al., 2022), resin transfer molding and reaction molding (Rudd, 2001), continuous lamination (Rajak, Pagar, Kumar et al., 2019), and pultrusion (Minchenkov et al., 2021).

In the case of sensors based on continuous long fiber or fabric composites, they typically involve textiles made of carbon fiber or optical fibers, which are referred to in the literature as *MFCs or AFCs* (see Figure 9.11 and Section 9.2.3).

It is important to mention, given the current nature of the topic, that the design of polymeric composites based on fabrics or continuous fibers used as sensors should not be confused with the technologies developed for monitoring the health of structural parts based on polymeric composite fabric (Structural Health Monitoring (SHM)). Often, these concepts are challenging to distinguish in the literature as they tend to overlap. For instance, structural components can possess intrinsic capabilities to monitor their state during operational conditions or serve as a means of quality control in the production process, utilizing the sensory abilities of the composites that constitute them or incorporating additional particles designed for such purposes. In Section 9.2.3, these aspects will be further elaborated upon.

9.2 Polymer composite sensor

9.2.1 Nanocomposites sensors

9.2.1.1 Carbon nanoparticles

Carbon nanoparticles, such as carbon nanotubes (CNTs), graphene, and carbon quantum dots, are widely used in polymer composites for sensors. These nanoparticles possess excellent electrical and mechanical properties and can improve the electrical conductivity of the polymer, as well as its sensitivity and response to different stimuli.

One of the applications of carbonaceous nanocomposite sensors is the detection of gases (carbon dioxide (CO_2), sulfur dioxide (SO_2), carbon monoxide (CO), ammonia (NH_3), methane (CH_4), nitrogen dioxide (NO_2), ozone (O_3), hydrogen (H_2), water vapor (H_2O), volatile organic compounds, etc.) (Alshammari, 2018). This application has gained significant interest recently due to the current concern for the environment and air quality (Gilbertson et al., 2014). Carbon nanomaterials, in particular, are considered excellent gas sensing elements due to their high aspect ratio, enabling high-sensitivity detection (Llobet, 2013). In this case, the mechanism is based on changes or perturbations in the electrical conductivity of the nanocomposite upon exposure to the gas, which can occur at the surface level or through internal diffusion of the gas within the polymeric matrix of the nanocomposite (Alshammari, 2018; Chen et al., 2001; Kumar et al., 2010; Li et al., 2008; Mangu et al., 2011). Upon exposure to gas molecules, the electrical properties of the composite, primarily conductivity, undergo changes according to the type of gas and its concentration, enabling gas detection. Gas molecules diffuse through the conductive composite and impact the conductive nanofiller network by disrupting electrical conduction, either through direct adsorption on the filler surface or by polymer swelling. Polymer swelling increases the spaces between conductive network elements, enhancing the carrier tunneling effect and resulting in a measurable variation in composite resistance (Alshammari, 2018; Kumar et al., 2010).

For efficient gas detection, gas sensors must exhibit high sensitivity to low gas concentrations as well as the ability to highly differentiate between different gases. Therefore, the quantity, nanoparticle shape, dispersion, and mobility, along with the molecular structure of the polymer matrix, are crucial aspects for the proper functioning of these sensors (Llobet, 2013). Additionally, their detection properties can be modified simply by introducing structural defects, attaching surface groups, and/or through chemical doping (Llobet, 2013). Generally, highly ordered carbon structures, such as CNTs and graphene, exhibit high carrier mobility and a high signal-to-noise ratio, making them suitable for detection applications compared to larger carbon structures like CB or carbon fibers (Liu et al., 2012).

As described in the introductory section, there can also exist hybrid composite materials with synergistic effects that combine the sensing ability of the dispersed phase of the composite (in this case, carbon nanostructures) with the use of conducting polymer matrices such as PANI, poly(3,4-ethylenedioxythiophene) (PEDOT), poly(styrene sulfonate) (PSS), and poly(*N*-vinyl pyrrolidone) (PVP). In the case of gas detection sensors, this synergy is widely reported (Bai et al., 2015; Ellis & Star, 2016; Yu, Cheung et al., 2017). It is worth noting that beyond the excellent detection properties of materials like CNTs, for instance, the adsorption of gas molecules affects the electrical conduction of the conducting polymer, resulting in a greater variation in the composite material's relative resistance. This, in turn, significantly enhances the sensitivity and response of the device

(Novikov et al., 2016). The synergy can be further enhanced regarding sensitivity and selectivity and result in reproducibility by working on the sensor architecture, such as aligning carbon nanofibers (Liu et al., 2012; Yang et al., 2013) and surface treatment of the nanoparticles (Alshammari, 2018; Mutuma et al., 2017).

Despite numerous advancements in this field, several researchers have demonstrated that carbon nanocomposite gas sensors are highly sensitive to the environmental conditions in which they are deployed. Temperature and humidity, for instance, can adversely impact their detection capabilities due to alterations in physisorption and chemisorption processes (Alshammari, 2018). Additionally, water vapor influences the absorption rate of the gases under detection. Mahmoud et al. present a comprehensive investigation of gas sensors based on carbon nanomaterials (Raya et al., 2022).

It is not only possible to detect gases using nanocomposites based on carbon nanoparticles, but various chemical sensors have also been developed to ascertain the presence of heavy metal contamination. This issue is of significant concern due to its detrimental effects on both human health and the environment. Different carbonaceous nanomaterials, such as singlewalled carbon nanotubes (SWCNT), multiwalled carbon nanotubes (MWCNT), or CNT, have been employed to detect toxic metal ions like Pb, Cd, Hg, or Cu (Chang et al., 2014; Huang, Chen et al., 2014; Ullah et al., 2018). The electrochemical detection of these ions relies on the alteration of current, potential electrochemical impedance, or capacitance of the carbon filler (Lee et al., 2016; Wanekaya, 2011).

Moreover, polymeric nanocomposites based on CNTs have been investigated for pH-sensing applications. The CNT network functions as an electrical conduit between the device electrodes and the transducer, typically the conducting polymer utilized as the matrix (e.g., PANI) (Kaempgen & Roth, 2006). The synergy of electronic and ionic conductivity in these conducting polymer matrices of carbon nanocomposites renders them highly promising transducers (Alshammari, 2018).

Another significant application of carbon-based polymeric nanocomposites lies in humidity sensors. Similar to gas sensors, CNT and MWCNT nanocomposites are employed (Fei et al., 2014; Li, Xu et al., 2016), achieving sensitivities within the range of (10%–90% RH) (Singh et al., 2018). It has been reported that the concentration of carbonaceous nanoparticles notably affects the humidity sensor's detection properties when CNT concentration is beyond the percolation threshold, leading to a well-linear response of the sensor across the humidity range (Yoo et al., 2010). In this application, special attention is paid to the polymeric matrix of the sensor nanocomposite, and its hydrophilicity can impact the sensor's behavior (Yang et al., 2008). Hence, it is common to find studies focused on PVP nanocomposites, which, using a hydrophilic polymer, facilitate water absorption within the sensor nanocomposite (Bhangare et al., 2018; Mallakpour & Behranvand, 2017; Zhang et al., 2010).

Carbon-based nanocomposites are also employed in optical sensors due to their unique properties associated with the interaction of light with carbon nanostructures. In the case of CNTs, their single or multi-walled cylindrical carbon structure allows the capture and emission of light across a wide range of wavelengths. The adsorption of molecules or other chemical species on the surface of nanotubes can alter their optical properties, such as absorption or fluorescence, leading to detectable changes in the light signal. This is harnessed to design sensors capable of detecting specific substances like gases (Raya et al., 2022), biomolecules (e.g., creatinine) (Babu et al., 2018), or contaminants (e.g., Arsenic As(III)) (Babu & Doble, 2018), relying on alterations in nanotube fluorescence or absorption.

In the case of graphene oxides, their two-dimensional structure composed of a single layer of carbon and oxygen atoms also presents intriguing optical properties. Alterations in the electron density on the surface of graphene oxide induced by interactions with the environment can modify its light absorbance and emissivity at various wavelengths. This characteristic is exploited to develop highly sensitive optical sensors capable of detecting changes in the surrounding environment's properties (Anju & Renuka, 2019). In recent years, research in this field has been directed toward detecting food and environmental analytes, such as additives, chemical contaminants, and microbial pollutants (Báez et al., 2021). Furthermore, these carbon nanocomposites can be hybridized using Ag or Au nanoparticles to enhance their detection capabilities (Gupta et al., 2019).

Regarding thermal sensors, carbon nanoparticles also play a significant role due to their ability to alter their electrical conductivity in response to temperature changes. Yuan et al. (2021) describe that there are two significant changes in the structure of the carbon-based nanocomposite when heat is applied. In the first one, electrons on the outer layer of carbon atoms become more active due to the increased temperature of individual nanofillers. This also leads to a decrease in composite resistance. In the second one, the expansion of the entire system is due to thermal expansion, which can subsequently reduce the number of viable pathways and ultimately increase the system's resistance.

Based on these principles, it is believed that three mechanisms govern the temperature-resistance effect of these nanocomposites: the tunneling effect mechanism (Xie & Sheng, 2009), the thermal expansion mechanism (Xi et al., 2004), and the nanofiller rearrangement mechanism (Ferrara et al., 2007). The tunneling effect mechanism proposes that an increase in the temperature could induce more tunnels, thus creating more conductive pathways. As a result, the temperature-resistance relationship exhibits negative temperature coefficient behavior. Regarding the thermal expansion mechanism, when the carbon nanocomposite is heated, the coefficient of thermal expansion of the polymers is much greater than that of the carbon nanofillers. This difference could increase the distance between neighboring nanofillers and reduce the conductive pathways, leading to higher resistance. As for the nanofiller rearrangement mechanism, it is postulated that changes occur in the collection and/or orientation of the nanofillers when the polymer matrix melts due to a temperature increase (Ferrara et al., 2007).

In recent years, possibly the most significant attention has been focused on sensors based on polymeric nanocomposites with carbon nanoparticles, particularly in applications like mechanical sensors and biosensors. In the case of mechanical sensors, the conductance of carbon nanomaterials (CNTs, graphene, etc.) can be substantially altered when subjected to mechanical strain. This is precisely why structural modifications in carbon nanomaterials lead to adjustments in their electrical properties, thus enabling effective detection (Alshammari, 2018). Consequently, incorporating carbon nanomaterials into polymers emerges as an efficient strategy for developing highly sensitive mechanical sensors with a broad measurement range, utilizing straightforward device structures. Following this principle, a multitude of carbon composite systems have been experimented with for pressure sensors (Hwang et al., 2011; Oh et al., 2022), deformation sensors (Fu et al., 2019; Hu et al., 2010), damage detection (such as structural health monitoring) (Obitayo & Liu, 2012), and motion detection sensors (Yamada et al., 2011).

The detection mechanism is rooted in the piezoresistive properties of conductive networks (Zhang et al., 2000). Mechanical deformation impacts the conductive network of the nanofiller, leading to a rise in resistivity. Applying a mechanical stimulus modifies the tunneling distance between adjacent CNTs, causing the disruption and formation of conductive networks and an increase in distances between charges within the polymer matrix. This mechanism has been corroborated through both experimental and theoretical validation, as documented in various studies (Bao et al., 2012; Liu et al., 2016; Obitayo & Liu, 2012; Park et al., 2014).

At a filler concentration just above the percolation threshold, the conductive network simply maintains ohmic conduction within the nanocomposite. Consequently, applying mechanical stress easily disrupts the conductive network and increases tunneling distance, leading to a significant variation in resistance. However, high filler concentrations hinder the disconnection of the conductive network, resulting in minimal resistance variation due to induced stress (Alshammari, 2018). This deterioration in sensor performance emphasizes the crucial role of nanoparticle quantities and types in designing mechanical sensors, as well as the careful selection of the polymer matrix for the nanocomposite. In this context, the choice of nanoparticle quantities and types is an important aspect of designing mechanical sensors, as is selecting the polymer matrix for the nanocomposite. It has been reported that the linear deformation behavior of a thermoset matrix can be advantageous compared to the elastoplastic nonlinear behavior of a thermoplastic for these applications (Avilés et al., 2016).

In this aspect, the bibliography is extensive, and multiple combinations of carbon nanocomposites have been designed, focusing on enhancing deformation-recovery cycles and reproducibility under cyclic loading. For example, (SWCNT)/poly(dimethylsiloxane) (Zhou et al., 2017), reduced graphene oxide (rGO)/CNT in poly(lactic acid) compounds (Hu et al., 2017), rGO/polyvinylidene fluoride (PVDF) (Eswaraiah et al., 2012), MWCNT/polyethylene oxide (Park et al., 2008), and functionalized graphene/PVDF (Eswaraiah et al., 2011).

Furthermore, regarding pressure sensors based on the resistance change of carbon nanoparticles, various significant applications have also been studied, including microphone design, touchscreen technology, human vital signs monitoring, and weighing scales (Bijender & Kumar, 2022; Yu, Zhang et al., 2017).

Finally, biosensors based on carbon nanocomposites have gained significant interest (Morales-Narváez et al., 2017). The capacity of carbon nanostructures to host biological molecules makes carbon nanocomposites among the most promising systems for biosensor applications (Yang, Denno et al., 2015). The potential to tailor their electrochemical detection properties to suit specific applications enhances their potential for biodetection purposes. The fundamental process involves functionalizing the carbon surface with specific biological species, such as antibodies, enzymes, or nucleic acids. These species have an affinity for the biomolecule target being detected. Essentially, the sensor is constructed by immobilizing molecules with stable bioactivity, either covalently or noncovalently, onto the surface of nanocomposite materials (Oueiny et al., 2014). When the sample containing the target biomolecules meets the functionalized nanoparticles, the biological species selectively bind to the biomolecules. This results in the formation of a biomolecule layer on the surface of the nanoparticles, causing alterations in their electrochemical properties. Consequently, these changes become a measurable and quantifiable signal (Eivazzadeh-Keihan et al., 2022) (see Fig. 9.2). The ability to achieve this controlled interaction between carbon nanocomposites and biological molecules renders them highly promising for biosensing applications.

The bibliography reports numerous applications in biodetection, such as glucose sensors (Chen et al., 2014; Mohammadpour-Haratbar et al., 2022), DNA sensors (the detection mechanism is based on the difference in electrochemical oxidation potentials of the four nucleobases: adenine, thymine, cytosine, and guanine) (Rasheed & Sandhyarani, 2017; Vikrant et al., 2019), protein markers (Hu et al., 2019), dopamine sensors (Raj et al., 2017; Tan et al., 2010), uric acid sensors (Bhambi et al., 2010), immunosensors (Tam & Hieu, 2011; Yang, Chen et al., 2015), and cholesterol sensors (Zhu et al., 2013).

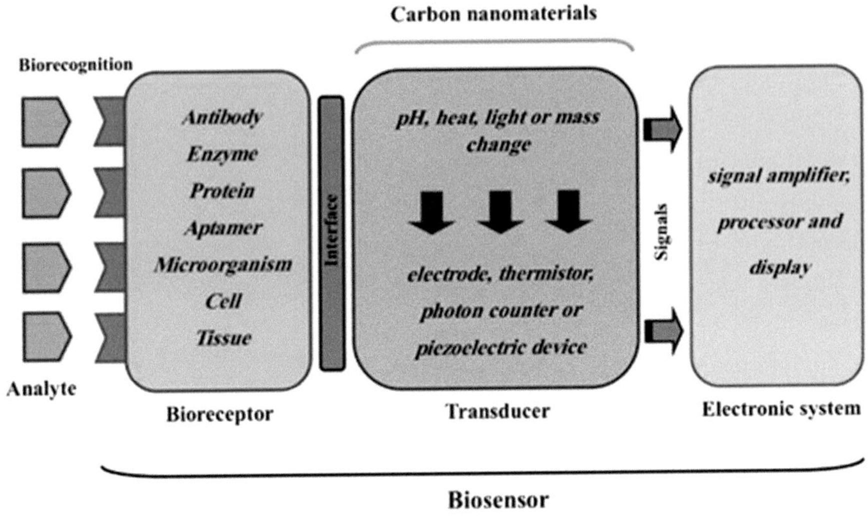

FIGURE 9.2

Schematic illustration of the components of a biosensor based on carbon nanocomposites.

From Eivazzadeh-Keihan, R., Bahojb Noruzi, E., Chidar, E., Jafari, M., Davoodi, F., Kashtiaray, A., Ghafori Gorab, M., Masoud Hashemi, S., Javanshir, S., Ahangari Cohan, R., Maleki, A. & Mahdavi, M. (2022). Applications of carbon-based conductive nanomaterials in biosensors. Chemical Engineering Journal, 442. *Available from https://doi.org/10.1016/j.cej.2022.136183.*

9.2.1.2 Metal nanoparticles

Metal nanoparticles, such as gold (AuNP), silver (AgNP), copper (CuNP), or platinum (PtNP) nanoparticles, are also employed in polymer composites for sensors. These nanoparticles can impart electrical conductivity and be used in sensors based on changes in electrical resistance or optical properties.

Particularly, polymer composites based on AgNP and AuNP have been extensively investigated in various domains due to their diverse range of properties, such as antimicrobial behavior, catalytic activity in various types of reactions, and enhanced conductivity characteristics. As a result, they find applications in water treatment, the textile industry, food packaging, sensors, and medical devices (Tamayo et al., 2018).

Regarding utilizing these polymer/metal nanocomposites as sensors, their operational principle is rooted in the fact that the bond between the recognition element and the analyte can modify the physicochemical properties of the metallic nanoparticles within the transducer composite (Rajesh et al., 2009; Saha et al., 2012). These interactions can generate a detectable response signal linked to various phenomena, such as plasmon resonance absorption, conductivity, and redox behavior (see Fig. 9.3).

The most extensively studied configuration of sensor nanocomposites with metallic nanoparticles is based on conductive polymers (CPs) where the metallic nanoparticles serve as "electron antennas," effectively channeling electrons between the electrode and the electroactive species. Consequently, these nanoparticles facilitate enhanced electron transfer between the electrode

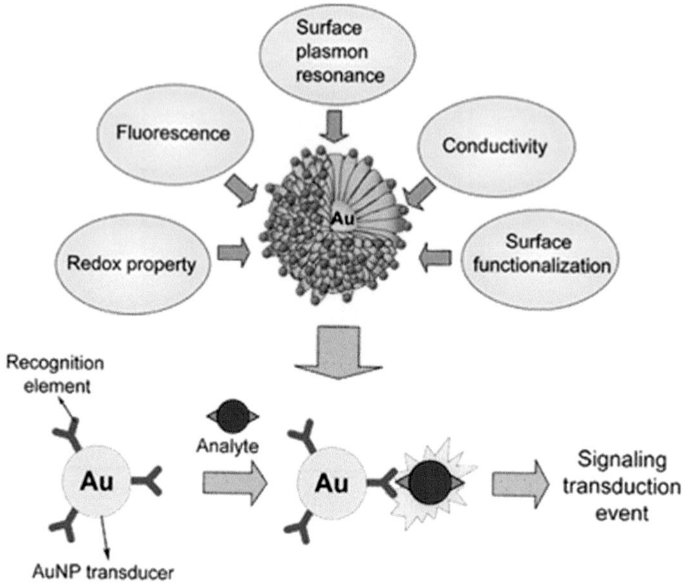

FIGURE 9.3

Physical properties of AuNPs and schematic illustration of an AuNP-based detection system.

Reprinted with permission from Saha, K., Agasti, S. S., Kim, C., Li, X., & Rotello, V. M. (2012). Gold nanoparticles in chemical and biological sensing. Chemical Reviews, 112*(5), 2739–2779. Available from 10.1021/cr2001178. Copyright 2012 American Chemical Society.*

surface and the electrolyte (Atta et al., 2012). Through this pathway, electrochemical sensors for the measurement of glucose and H_2O_2 are generated (Pingarrón et al., 2008). Examples of electropolymerized CPs with AuNPs for glucose detection include PPy and PANI (Saha et al., 2012). Furthermore, the activity of these nanocomposite sensors is often enhanced by the addition of carbon nanoparticles, such as graphene (Fang et al., 2010), CNTs (Lin et al., 2009), and MWCNT (Zhang et al., 2009).

Another common principle on which polymer/metal nanocomposite sensors are based is the plasmonic response of metal nanoparticles. This refers to the optical phenomenon that occurs when collective oscillations of electrons (plasmons) are excited on the metal surface. An example of such a sensor is the one formed by poly-2-vinylpyridine with supported AuNPs on a gold substrate used as a pH detector (Stuart et al., 2010). Another example is the AgNP/poly(vinyl alcohol) nanocomposites used for colorimetric hydrogen peroxide sensors (Filippo et al., 2009).

These metal nanoparticles can also be used to create composite sensors with CPs, employing a layer-by-layer fabrication method. Thin and durable films have been reported, comprising AuNPs and various types of organic dendrimers like polyphenylene, poly(propyleneimine), and poly(amidoamine) (Krasteva et al., 2002). In this case, the AuNPs were used to provide electrical conductivity to the composites, while the dendrimers served to crosslink the nanoparticles and offer sites for the selective sorption of analyte molecules.

Finally, the bibliography is extensive regarding applications of these nanocomposites as biosensors (Huang, Liu et al., 2014; Lin et al., 2009; Ruan et al., 2013). In this context, AuNP nanocomposites stand out, as they enable the immobilization of enzymes through amine groups and cysteine residues of macromolecules, which are known to bind with gold colloids strongly, thus maintaining their enzymatic and electrochemical activity for a significantly extended period (Phukon et al., 2014).

9.2.1.3 Magnetic nanoparticles

In general, magnetic nanocomposites can be prepared using three approaches that lead to different composite structures. Melt blending, in situ polymerization, and solution mixing gave rise to structures in which nanoparticles are dispersed in the polymer matrix (Jazzar et al., 2021). However, there are other approaches, like the coextrusion process, that allow obtaining wires based on a magnetic core and a polymer shell (D'Ambrogio et al., 2021). and the deposition of magnetic nanoparticles onto the polymer surface (Alfadhel et al., 2014).

In recent years, the magnetic nanoparticles in a polymer matrix have generated great scientific interest due to their proven ability to detect and remove various water contaminants through different mechanisms such as oil-water separation or heavy metal removal (Gu et al., 2014). Regarding oil-water separation, the magnetic nanoparticles can break the emulsions of the formed emulsion (Ali et al., 2015). In the case of heavy metal removal, the incorporation of magnetic nanoparticles into the polymer enhances the electrostatic interaction between metallic ions and the polymer, which improves adsorption capacity (Zhao et al., 2018). In both applications, the preferred nanoparticles are different iron oxides.

Another interesting field of application of magnetic nanoparticles is aircraft and aerospace, where magnetic polymer nanocomposites present interesting potential in manufacturing sensors. In this context, nanocomposites based on zirconate titanate fillers embedded in poly(dimethyl siloxane) showed outstanding piezoelectric behavior at high temperatures, which makes them suitable for sensing aircraft ball bearings (Qing et al., 2019).

9.2.1.4 Semiconductor nanoparticles

In recent years, different semiconductor nanoparticles have been used to manufacture nanocomposites. For this purpose, among the most common semiconductor nanoparticles are zinc oxide (ZnO), zinc sulfide (ZnS), and lead selenide (PbSe). Regarding their application as sensors, the nanocomposites based on these nanoparticles are the following: gas and electrochemical detection, photodetection, and optoelectronic switches.

Metal oxide semiconductor materials, such as ZnO, SnO_2, WO_3, and TiO_2, are reported to be usable as sensors because they have nonstoichiometric structures, so free electrons originating from oxygen vacancies contribute to the electronic conductivity (Skubal et al., 2002). Among all, the most extensively studied is ZnO, which is low-cost, nontoxic, and can be easily obtained through standard chemical synthesis that allows control of different morphologies, shapes, and crystallographic phases. Besides, ZnO is an n-type semiconductor with a direct bandgap at room temperature, presenting prominent importance once it has high electron mobility, thermal stability, and good electrical properties. All these characteristics make them one of the most explored semiconductive nanoparticles for chemosensitive sensors (Franco et al., 2022). The most common nanocomposites of ZnO are its combinations with PANI, PPY, and PEDOT, which have been extensively studied as different gas and electrochemical sensors (Barthwal & Singh, 2017). It is also noteworthy that it is possible to use ZnO heterostructures such as CuO (Hsu et al., 2021), NiO (Qu et al., 2016), WO_3 (Park, 2019), Fe_2O_3 (Song et al., 2019), and so on. These structures pursue different goals, like changing operating conditions and/or improving the sensitivity, selectivity, and stability of the sensor.

Zinc sulfide (ZnS) is one of the first semiconductors discovered. Its atomic structure and chemical properties are comparable to those of more popular and widely known ZnO, but ZnS presents certain advantages, such as a larger bandgap, and therefore it is more suitable for visible-blind ultraviolet light-based devices such as sensors/photodetectors (Fang et al., 2011). ZnS nanobelts have been proven to have high potential as visible-light-blind UV light photodetectors and ultrafast optoelectronic switches (Fang et al., 2009). Finally, PbSe has been extensively investigated as thin films due to their potential uses as imaging sensor devices in different applications such as night vision, missile tracking, research, medical imaging, and so on (Gupta et al., 2021). However, it is important to note that these kinds of composites traditionally use inorganic substrates such as glass, SiO_2/Si, and GaAs, which leaves them out of this review.

9.2.2 Short fibers and microparticle composites sensors

9.2.2.1 Carbon black

Important studies have been reported on polymer sensor matrices composed of CB particles dispersed within the polymer matrix for the fabrication of "electronic noses" (Carrillo et al., 2002; Chiu & Tang, 2013; Freund & Lewis, 1995; Lonergan et al., 1996; Xiao et al., 2023; Zee & Judy, 2001). The CB particles give the films electrical conductivity, while the various organic polymers are the source of chemical diversity between the sensor array elements (Lonergan et al., 1996).

These sensors are "chemiresistor" types and operate based on the change in resistance of the sensor after exposure to an analyte. The analyte diffuses into the polymer composite, promoting matrix swelling and subsequently leading to the separation of the dispersed conductive carbon particles. As a result, the resistance of the sensor increases (Lei et al., 2007).

The response mechanism of individual CB/polymer composite sensors is qualitatively explained by percolation theory, which describes the relationship between the resistivity of the composite and the CB content (Lux, 1993). When the CB concentration is very high, the CB particles pack closely in the compound and form conductive pathways, which impart a low resistance response from the sensor. As the CB concentration decreases (e.g., in a swollen polymer), the distances between the CB particles increase, and the resistance of the composite gradually increases. When the CB concentration decreases to a point where the conductive pathways are interrupted, the resistance of the composite increases abruptly (by many orders of magnitude). This point is called the percolation threshold. In CB/polymer composite sensors, the CB concentration is usually only slightly above the percolation threshold. Therefore, even a small amount of swelling can cause a dramatic change in sensor resistance. Most sensors of this nature use between 15% and 20% wt. of CB in their composition (Kim et al., 2005). Lewis and coworkers were pioneers in demonstrating the relationship of the sensory responses of these composites to percolation theory (Lonergan et al., 1996).

Removal of the stimulus leads to desorption of the vapor and decreased electrical resistance to the original value. It has been found that the effect of swelling on the conductivity of the composite depends strongly on the nature of the organic solvent, that is, at least one polymer in the composites must be able to absorb the vapor from the organic solvent by a dissolution process. For this reason, the crystallinity and molecular weight of the matrix play an important role in the efficiency of the sensor, as well as the dispersion, distribution, and amount of CB on the composite sensor (Chen & Tsubokawa, 2000; Dong et al., 2004).

Besides the use of an insulating organic polymer in chemiresistor composites (i.e., poly(vinyl acetate) (PVAc)/CB (Arshak et al., 2005) and poly(methyl methacrylate) (PMMA)/CB (Doleman et al., 1998)), the possibility of using CPC has been explored recently. This approach aims to achieve synergies through interactions and chemical manipulation of the electrical properties of the conductive polymer. For instance, polyetherimide/CB (Daneshkhah et al., 2020) has been investigated for aldehyde detection in the food and medical industry, addressing the challenge of secondary interactions with other organic compounds. Additionally, a high-performing PVP/CB (Kim, 2010) composite sensor has been studied for methanol detection, while PANI/CB (Sotzing et al., 2000) has shown promise for detecting biogenic amines.

Finally, it is important to note that increasing the surface area of carbon particles (i.e., nanoparticles) improves the efficiency of these systems (Raya et al., 2022). For this reason, research on carbon nanocomposites used as sensors is more extensive, as described in the previous section.

9.2.2.2 Glass fibers

Fiber-reinforced polymers are widely used as composite materials in the field of lightweight structures due to their high mechanical strength, exceptional fatigue behavior, low cost, and excellent heat resistance (Morampudi et al., 2020). Since the 1950s, companies such as Boeing have introduced the use of glass fiber (GF) composites for their secondary structural parts such as floors, cargo liners, air ducts, and various other parts, of the cabin of their aircraft (Das & Yokozeki, 2020). These advanced materials not only take an important position in the aerospace industry but also in other industries such as civil engineering, marine, automotive, wind turbines, and many more (Fiore et al., 2011; Friedrich & Almajid, 2013; Garcia-Espinel et al., 2015; Kensche, 2006).

In recent years, in situ monitoring of the structural health of these composites to predict damage to their structure during use because of gradual degradation or manufacturing defects has become of great interest. In fact, there is a whole new field of study dedicated to the SHM of composites (Rocha et al., 2021) and the monitoring of self-healing structures (Thostenson & Chou, 2006). Composites with self-detection of stresses and small damages can prevent catastrophic failures. These detections can be made either during the lifetime of the composite (wear damage) or as a way of validating the correct fabrication of the composite (manufacturing defects) (Yang et al., 2018; Yang, Chiesura et al., 2016). The performance of these composites as optimal sensors is one of the most current goals in the field of engineered composites.

However, in the case of GF composites, being electrically insulating materials, it is not possible to make in situ electrical resistance measurements to predict damage to their structure, unlike carbon composites (Kalashnyk et al., 2017; Wang et al., 2006; Yang, Wang et al., 2016). For this reason, the main line of research that has been carried out is the development of GF-reinforced composites with improved electrical conductivity. The general approach to overcome this challenge has been to induce the electrical conductivity of GF composites by the addition of conductive particles, such as CB, nanotubes, carbon nanofibers, and graphene (Böger et al., 2010; Gao, Zhuang et al., 2010; Hao et al., 2016; Li et al., 2014), to a polymer matrix. Other approaches include surface treatment with antistatic or metallic coatings (Thostenson & Chou, 2006, 2008) and/or fiber functionalization (Gao et al., 2007, 2008). All these approaches are applied for both short, continuous, and textile GF composites.

Fig. 9.5 shows three different approaches to GF composites used as sensors. Fig. 9.4A shows a schematic of the fabrication of GF composites by embedding carbon nanotube fiber patches during fabrication by the hand lay-up method (layer-by-layer deposition: fiber/epoxy). Rectangular samples of the composite were subjected to tensile tests according to ASTM D 3039. Mechanical response and electrical resistance data were recorded at different load cycles. Finally, a monotonic load was applied to the specimens until failure to evaluate the effectiveness of the hybrid system in detecting damage and deformation occurred. The results showed a high detection sensitivity (Khalid et al., 2018).

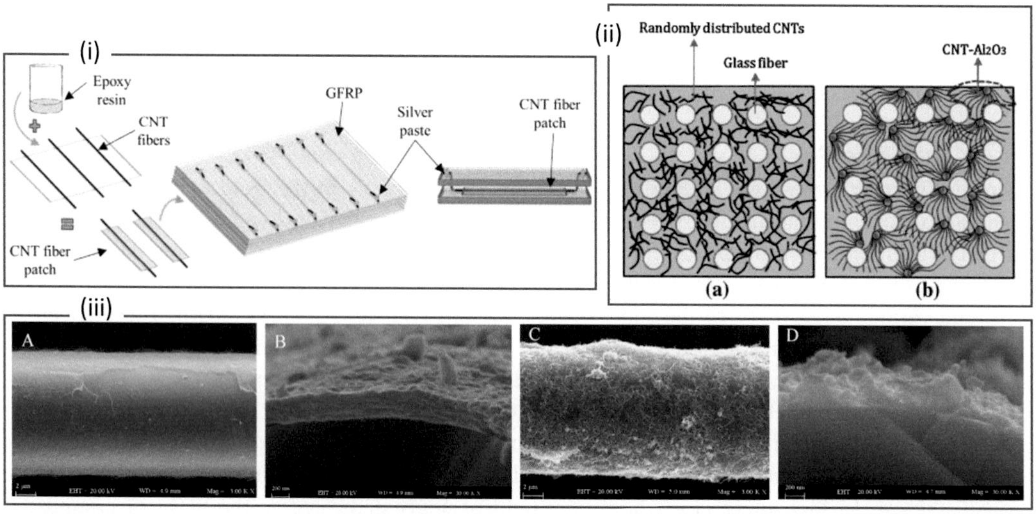

FIGURE 9.4

(A) Schematics of GFRP composites incorporating CNT fibers. (B) The preparation process of the multiscale glass fabric/epoxy/CNT–Al2O3 composites. Schematics of the dispersion of (1) randomly distributed CNTs and (2) CNT–Al2O3 hybrids in the glass fiber-reinforced composites. (C) SEM images showing the in-plane and crosssection morphologies of GFs with different coatings ((1) and (2): Graphene coating; (3) and (4): CNT coating).

(A) From Khalid, H. R., Nam, I. W., Choudhry, I., Zheng, L., & Lee, H. K. (2018). Piezoresistive characteristics of CNT fiber-incorporated GFRP composites prepared with diversified fabrication schemes. Composite Structures, 203, *835–843. Available from https://doi.org/10.1016/j.compstruct.2018.08.003; (B) From Li, W., He, D., Dang, Z., & Bai, J. (2014). In situ damage sensing in the glass fabric reinforced epoxy composites containing CNT-Al2O3 hybrids.* Composites Science and Technology, 99, *8–14. Available from https://doi.org/10.1016/j.compscitech.2014.05.005; (C) From Hao, B., Ma, Q., Yang, S., Mäder, E. & Ma, P. C. (2016). Comparative study on monitoring structural damage in fiber-reinforced polymers using glass fibers with carbon nanotubes and graphene coating.* Composites Science and Technology, 129, *38–45. Available from https://doi.org/10.1016/j.compscitech.2016.04.012.*

Another example of increasing the conductivity of GF composites is shown in Fig. 9.4B. The addition of carbon nanotube fillers with Al_2O_3 in epoxy resin/GF composites allows conductive charges to be dispersed throughout the matrix, and their increased surface area brings them into contact with each other (Li et al., 2014). This facilitates the formation of conductive networks, thus improving conductivity. Some of the studies using conductive fillers reported significant limitations with respect to resolution, which depends mainly on the density and dispersion of the conductive networks (Das & Yokozeki, 2020). Most research in this field has shown that conductive particles, when properly dispersed in the insulating polymer matrix of a fiber composite, can form electrically percolating networks that are sensitive to detecting the initiation and growth of matrix-dominated microcracks. This in situ technique has been successfully applied to detect damage in fiber composites under tensile and fatigue loading (Gao et al., 2009; Gao, Chou et al., 2010; Gao, Thostenson et al., 2010; Li & Chou, 2008).

Finally, yet another example of the approaches under investigation to enhance the electrical conductivity of GF composites for sensor applications involves fiber surface treatment. Fig. 9.4C shows the morphology of GF with different coatings. The electrophoretic deposition (EPD) process produces a uniform coating layer of graphene or CNT on the fiber surface. The graphene completely covered the GF, and the nanotubes formed networks along the direction of the fiber. These morphological differences originate from the structures of the coatings: graphene has a plate-like structure and can be easily assembled layer by layer on the GF surface, while CNTs are one-dimensional materials with a high aspect ratio. The connectivity of individual CNTs results in continuous networks on the fiber surface, which enhances the sensing capability of these GF composites (Hao et al., 2016).

In addition to the inclusion of conductive particles and the functionalization of GF, another common approach to sensor design is the use of electrical CPs, such as polyacetylene, PPy, polyindol, and PANI, which have also been used in the manufacture of GF composite sensors. The fundamental limitation of CPs is their stability and processability. However, proper compounding, blending, doping, and size help to eliminate these limitations (Singh & Shukla, 2020).

In all cases, to achieve a conductive composite GF with multifunctional sensing capability in structural damage monitoring, experimental and theoretical studies widely pointed out the need to obtain a low percolation threshold, anisotropic electrical properties, and high sensitivity to various environmental parameters (stress/strain, temperature, and humidity). Numerous investigations conclude that to obtain reasonable conductance through electrical tunneling, the typical distance between two carbon particles should be less than one nanometer (Du et al., 2004; Gao, Zhuang et al., 2010). The tunneling distance in two adjacent nanotubes immersed in the polymer matrix is complicated by many factors, such as energy levels of electronic states, contact potential barriers, dispersibility, and the polydispersity of frequently used carbon fillers (Bauhofer & Kovacs, 2009).

In addition to GF-based sensors used to monitor the structural health of composite materials, another standardized use of GF composites as sensors is known as flexible strain sensors (Abeykoon, 2019; Fu et al., 2020). Most of the flexible strain sensors reported so far are composite systems with a functional filler to provide the signal and a polymer matrix as a soft substrate. For example, several wearable devices have been developed to detect human body movements, from large scales, including joint and muscle movement, to some small behaviors, such as breathing, eye blinking, heartbeat, and pulse (Amjadi et al., 2016). By incorporating functional fillers, such as carbon fiber (Zhao et al., 2020), carbon nanoparticles (Park et al., 2016), or graphene (Ma et al., 2020), flexible GF composite sensors have demonstrated broad suitability under various conditions. Moreover, the types of sensitive sensor signals under external load can be presented by different mechanisms: electrical and optical properties, such as electrical resistance (Amjadi et al., 2014), capacitance (Mannsfeld et al., 2010), or piezochromicity (Mei et al., 2015). Developments are so promising that flexible and transparent sensors based on GF composites have already been designed for mobile electronics applications (Lim et al., 2020).

9.2.2.3 Basalt fibers

Basalt fiber (BF) is a filamentous material of micrometric size derived from volcanic rock. Since it was discovered in 1923, basalt has been widely used in military applications (Colombo et al., 2012; Pavlovski et al., 2007). In recent decades, there has been significant interest in their use as reinforcement fibers in lightweight, high-performance polymer composites for civil and infrastructure applications (Dhand et al., 2015; Khandelwal & Rhee, 2020; Sim et al., 2005). This is due to their improved mechanical properties, low water absorption, thermal insulation, marginal health risk, and excellent tolerance to temperature and environmental factors compared to GF, as well as their simple production process and low cost relative to carbon fiber (Hao et al., 2017; Jamshaid & Mishra, 2016). Like in the case of GF composites, BF composites have been studied to be used as sensors based on their electrical properties (electrical conductivity, resistivity, capacitance, etc.). Thus, BFs can be used as resistive elements that vary their electrical resistance in response to changes in the variable to be measured. These changes can be detected and quantified using appropriate electrical circuits. Like GF composites, there are numerous research efforts aimed at improving the electrical properties of these fibers to enhance their performance as sensors. Mittal et al. (Mittal & Rhee, 2018) were among the first to report on the direct synthesis of CNTs on BFs using the chemical vapor deposition (CVD) technique. Hao et al. (Hao et al., 2017) modified BF using a CVD method, aiming to enhance the functionality of the epoxy/BF composite for sensing applications. The results showed that a thin layer of pyrolytic carbon was deposited on the fiber surface, rendering the insulating fibers electrically conductive, and the composite could be calibrated for structural damage monitoring applications. Sarasini et al. (Sarasini et al., 2022) grew aligned CNTs directly on basalt fabrics through a rapid CVD process without any external catalyst addition, resulting in a highly dense coverage of the underlying substrate, which proved particularly useful for high-performance composites displaying in situ SHM capabilities.

The mechanical properties of basalt, such as high tensile strength and good flexibility, also allow these fibers to be used as deformation or strain sensors. When a force is applied or deformation occurs, BFs can change their shape, resulting in changes in their electrical properties, such as electrical resistance or capacitance. These changes can be measured and used to determine the magnitude of the applied force or deformation. For example, Wang et al. (Wang, Wang et al., 2018) designed a hybrid composite based on BF/carbon nanofibers/epoxy composites with self-sensing capabilities for deformation and damage detection. Additionally, Sun et al. (Sun et al., 2022) designed highly sensitive and biodegradable strain sensors with high mechanical strength using BF as a reinforced layer. Tang et al. (Tang et al., 2016) proposed a self-sensing intelligent basalt composite for the SHM of reinforced concrete, utilizing distributed fiber optic sensing technology. Based on the obtained strain distribution, the location of major cracks, their width, and their displacement can be evaluated. However, further research is needed to increase the accuracy of strain measurement in basalt composite smart bars.

As discussed in this section, the primary application is structural monitoring. However, due to their chemical stability and corrosion resistance, BFs are potentially interesting for pH or gas sensors. Finally, it is remarkable that the literature is scarce regarding using conductive polymer matrices with BF to manufacture sensor composites (Patel et al., 2022).

9.2.2.4 Short fibers piezoelectric: piezoelectric fiber composite, macrofiber composite, and metal-core piezoelectric fiber

Piezoelectric fibers are fibers that generate an electric charge in response to applied mechanical deformation (Jones et al., 1996). These fibers can be used to manufacture sensors that detect and quantify variables such as pressure, force, acceleration, vibration, and deformation (Scheffler & Poulin, 2022). Sensors based on piezoelectric fibers harness this property to convert mechanical energy into electrical signals. When a force or deformation is applied to the piezoelectric fiber, an electric charge is produced in the fiber, generating an electrical signal proportional to the magnitude of the applied force or deformation. This electrical signal can be measured and used to detect and measure the variable of interest.

Piezoelectricity arises in noncentrosymmetric atomic structures. The most common ones are inorganic materials (ceramics) with perovskite structures such as lead zirconate titanate (PZT), lead titanate, barium titanate, and other less popular ones like lead lanthanum zirconate titanate, and lead magnesium niobate (Hong et al., 2007). These materials exhibit high piezoelectric coefficients, which reflect the amount of charge generated in response to applied stress. High coupling coefficients also manifest the efficiency of inorganic materials (Jones et al., 1996), representing the relationship between electrical and mechanical energy conversion. It is important to note that piezoelectric fibers can also be of a polymeric nature, with fluorinated polymers being particularly noteworthy. These include PVDF and its copolymers such as polyvinylidene fluoride-trifluoroethylene, polyvinylidene fluoride-trifluoroethylene-chlorotrifluoroethylene, and polyvinylidene fluoride-hexafluoropropylene (Lang & Muensit, 2006; Lee et al., 2020; Martins et al., 2014).

In recent years, there has been significant interest in manufacturing ceramic-polymeric piezoelectric composite materials, combining the benefits of both. Ceramics are less expensive and easier to manufacture than polymers, and they have relatively high dielectric constants and good electromechanical coupling. However, they have a high acoustic impedance and, therefore, do not acoustically match well with water, which is the medium through which signals are normally transmitted or received. Additionally, due to their rigidity and brittleness, monolithic ceramics cannot be formed into curved surfaces, limiting the design flexibility of transducers. They also have a high degree of noise associated with their resonant modes (Akdogan et al., 2005).

On the other hand, piezoelectric polymers acoustically adapt well to water, are highly flexible, and have few spurious modes. However, the applications for these polymers are limited due to their low electromechanical coupling, low dielectric constant, and high manufacturing cost. Ceramic/polymer piezoelectric composites have demonstrated superior properties compared to monophasic materials. They combine high coupling, low impedance, a few spurious modes, and an intermediate dielectric constant. Additionally, they are flexible and moderately priced. The following sections outline a more detailed description of the processing and electromechanical properties of ceramic-polymer piezoelectric composites (Akdogan et al., 2005).

Additionally, it is important to note that there is considerable interest in using these piezoelectric materials in fiber form to form polymer composite sensors, as their small diameter allows for flexible fibers that can be incorporated into a wide range of structures (Chen et al., 2020; Mokhtari et al., 2020; Scheffler & Poulin, 2022). Moreover, fiber-based materials tend to have fewer defects than other structures, such as macroscopic films (Sawai et al., 2006) (i.e., thin films). Performance and properties of materials typically improve when in fiber form, as seen with widely used glass and carbon fibers known for their excellent mechanical properties. Record-breaking performances of organic polymers are achieved in synthetic fibers that can be highly drawn and aligned (Sawai et al., 2006). However, as described by Poulin et al. (Scheffler & Poulin, 2022), obtaining these fibers remains a technical and scientific challenge (see Section 9.2.3.3). In the case of inorganic materials, they cannot be melted or dissolved in common solvents. Additionally, the mechanical properties and piezoelectric performances of fibers can vary substantially depending on their composition and processing (Scheffler & Poulin, 2022). For this reason, these authors provide an excellent synthesis focused on the processing and challenges for the efficient use of different piezoelectric fibers (organic, inorganic, and composite fibers) that have been recently studied.

On the other hand, it is important to note that a fundamental variable to consider in designing piezoelectric composites is the connectivity patterns (Akdogan et al., 2005). These patterns establish the electromechanical properties based on the arrangement of phases within the composite and how individual phases are interconnected (Newnham et al., 1978). There are 10 connectivity patterns for a biphasic system, where each phase can be continuous in zero, one, two, or three dimensions. The internationally accepted nomenclature to describe such composites is $(0-0)$, $(0-1)$, $(0-2)$, $(0-3)$, $(1-1)$, $(1-2)$, $(2-2)$, $(1-3)$, $(2-3)$, and $(3-3)$ (Fernandez et al., 1995). The first digit within the parenthesis refers to the number of connectivity dimensions for the piezoelectrically active phase, and the second digit is used for the electromechanically inactive polymer phase. Based on this connectivity concept, a series of ceramic/polymer piezoelectric composites have been developed. It has been demonstrated that all these composites exhibit improved piezoelectric properties compared to monophasic piezoelectric ceramics.

In recent years, piezoelectric fiber-based composites have acted as pressure and force sensors for applications such as contact pressure monitoring, load detection, and force distribution evaluation (i.e., tactile sensors) (Wang, Sun et al., 2018; Zhu et al., 2020). They are also used as acoustic and ultrasonic transducers applied in the generation and reception of ultrasonic signals in medical devices, non-destructive inspection, and structural monitoring, among others (Grewe et al., 1990). Furthermore, they serve as energy-harvesting generators used in applications such as portable devices, autonomous sensors, and remote monitoring systems (Anton & Sodano, 2007; Khazaee et al., 2020; Narita & Fox, 2018). They are utilized as actuators and control systems for micro- and nanopositioning devices (Rocha et al., 2020), vibration control systems (Shivashankar & Gopalakrishnan, 2020), and fuel injection systems (Zeng et al., 2014). Additionally, they function as vibration and acceleration sensors to detect and measure vibration and acceleration in applications such as structural monitoring (Elvin et al., 2006), vibration assessment in machinery (Gautschi, 2002), and impact detection (Song et al., 2007), among others. In general, piezoelectric fiber-based sensors offer various advantages, including high sensitivity, rapid response, wide frequency range, the ability to operate under extreme conditions (temperature, pressure, etc.), and corrosion resistance.

Most applications of piezoelectric sensor composites have focused on the study of smart textiles based on fabrics with piezoelectric fibers embedded within a polymer matrix. This application will be described in the section active fiber composites (AFCs), dedicated to sensors based on continuous piezoelectric fibers and fabrics. In this section, some examples where short piezoelectric fibers are used in polymeric composites for sensors will be described. It is important to mention that depending on the length of these fibers, the term macrofiber composite (MFC) is coined, which is usually longer and thinner than the piezoelectric fibers used in traditional composites (PFC).

Like in GF and BF composites, short PFCs are commonly used in SHM applications (Konka et al., 2012). One of the main challenges in this field is that including short piezoelectric fibers (i.e., ceramic particles) decreases the strength of the final composite. Therefore, like any polymer composite, dispersion, distribution, and load-polymer compatibility are paramount to avoid compromising the mechanical properties. In this respect, much research has been conducted. Thus, for example, Crawley et al. (Crawley & De Luis, 1987) found that embedding piezoceramics and silicon chips within glass/epoxy and graphite/epoxy composite laminates reduces the strength of the final composite. Mall and Coleman (Mall & Coleman, 1998) studied the effects of incorporating PZT piezoelectric sensors on the tensile strength and fatigue behavior of a quasi-isotropic graphite/epoxy laminate, as well as the voltage degradation of the embedded sensor under these loading conditions. They demonstrated the significant impact on the tensile behavior that the incorporation of PZT produces in the composite. Vizzini et al. (Hansen & Vizzini, 2000) analyzed the interlaminar stress state around an interleaved active piezoceramic actuator within a unidirectional composite laminate. From their study, it can be concluded that the interleaving technique increases the strength of the composite structure. Ghezzo et al. (Ghezzo et al., 2010) observed the initiation of microcracks within glass epoxy laminates with embedded sensors under nearly static tensile loading conditions. Other more recent works, such as that of Konka et al. (Konka et al., 2012), demonstrated the effects of incorporating MFC and PFC sensors on the structural integrity of GF epoxy composite laminates, specifically in fatigue tests, concluding that both integrated MFC sensors and PZT-based composites or PFC have no significant effect on the fatigue behavior of the composite samples.

Finally, concerning SHM applications, there is another type of composite based on noncontinuous or small-sized piezoelectric fibers known as metal-core piezoelectric fibers (MPFs). These composites are based on piezoelectric fibers with a metallic core surrounded by piezoelectric material, typically ceramic, and they have significant potential to be used as structurally integrated transducers for guided wave SHM (Ricci et al., 2022).

Compared to traditional piezoelectric ceramics, the MPF exhibits unique directionality in Lamb wave detection. MFCs were designed to overcome the fragility of piezoelectric ceramics and create a new device with easily integrable shapes. MFCs are manufactured using the extrusion method and have recently undergone improvements (Qiu et al., 2010). The geometry of an MPF comprises three parts: a metal core, a surface electrode, and piezoelectric ceramic fibers. The outer surface of an MPF is coated with a metal layer serving as a surface electrode, while the core consists of a metal core surrounded by a piezoelectric ceramic layer. Thanks to the metal core, MPFs overcome the brittleness of conventional piezoelectric fibers and can comfortably function as sensors or actuators with two electrodes: the central metal core and the surface metal layer (Zhang et al., 2015). Additionally, due to the malleability of the metal core, MPFs can be processed into various shapes according to the desired structures. Typically, the metal core of the MPF is made of platinum, and the fibrous piezoelectric ceramic core is composed of slurry (PNN-PZT) mixed with an organic solvent. The length of a single MPF can vary from several millimeters to several centimeters (Epaarachchi & Kahandawa, 2016). MPFs are the least studied piezoelectric fiber-based sensing composites, and most of their applications are in the field of structural monitoring.

9.2.3 Continuous fiber composites and textile sensors

9.2.3.1 Aramid fibers

Aramid fiber was the first organic fiber with sufficient tensile modulus and strength to be used as insulation in specialized composites. In terms of weight-to-weight ratio, it exhibits significantly superior mechanical properties compared to steel and GF. Heat resistance and flame resistance are intrinsic properties of aramid fibers, which are maintained at high temperatures (Abbadi et al., 2015; Dharmavarapu & Sreekara, 2022). However, it is important to note that aramid fibers have a smooth and low chemical reactivity surface (Zhang et al., 2021), which can limit their role in sensor composites. Therefore, like other inert fibers, it is necessary to subject them to surface treatment or incorporate conductive particles onto the surface to make them suitable as sensors. For instance, Sodano et al. (Groo et al., 2021) integrated graphene into aramid fiber-reinforced composites to detect damage and deformation during mechanical loading, thus developing a multifunctional component based on aramid composites with in situ damage detection capabilities.

Based on the principle of multifunctional aramid composites, Wang et al. (Wang et al., 2021) have studied flexible sensors based on CNT/PMIA (poly(*m*-phenylene isophthalamide), aramid fiber type). PMIA fibers are combined with carboxylated CNTs to form a composite sensor with an ideal conductive network, which further acts with resistance variation in response to external changes. As a result, the CNT/PMIA sensor exhibits the ability to detect pressures generated by various human movements, as demonstrated by monitoring human movements of finger joint bending and elbow joint bending, speaking, blinking, and smiling, as well as temperature variations in the range of 30°C–90°C.

A similar idea, but based on smart textiles, has been described by Doshi and Thostenson (Doshi & Thostenson, 2018). In this case, it was used to create new textile pressure sensors based on CNTs. The EPD technique allowed for the creation of a uniform coating based on functionalized CNTs on non-conductive aramid fibers. The electrically conductive coating was firmly bonded to the fiber surface and showed piezoresistive electrical/mechanical coupling with enormous potential for use as highly flexible and sensitive pressure sensors over a wide range of pressures. In the same vein, Ehlert and Sodano (Ehlert & Sodano, 2014) developed a self-assembly method to create a thin layer of MWCNTs on aramid fibers. The fibers showed resistive behavior and had a gauge factor of approximately 1.6, which competes with current foil strain sensors. This system could be promising for developing strain sensors integrated into multifunctional composite materials.

In another vein, without the need for treatments to the aramid fiber, Parthenios et al. (Parthenios et al., 2002) have successfully developed a nondestructive multifunctional technique based on the Raman response of aramid fibers to evaluate the integrity of the interface and the overall stress distribution in a unidirectional Kevlar/epoxy composite. Aramid fibers are excellent Raman scatterers. Their spectrum includes a strong band at 1611 cm^{-1}, corresponding to the C-C stretching of phenyl. This band has good stress sensitivity (Vlattas & Galiotis, 1994) and has therefore been extensively studied as a strain sensor in aramid fiber composites (Galiotis, 1993; Jahankhani & Galiotis, 1991).

Finally, it is worth mentioning that since one of the major applications of aramid fiber composites is their use in military applications (anti-ballistics), in recent years, there have been significant studies focused on the in situ or ex situ monitoring of impact deformations of these composites. In most of these cases, the aramid composite does not participate in the sensing activity. Polymeric piezoelectric sensors (i.e., piezoelectric thin film based on vinylidene fluoride (PVDF)) (Caneva et al., 2008), piezoelectric force sensors (i.e., commercial sensor 9011A Kistler), and pressure sensors (Rubio et al., 2020) are integrated into the composite to make measurements.

9.2.3.2 Carbon fibers

Carbon fiber-reinforced polymer (CFRP) composites possess distinctive properties that set them apart: good chemical resistance, excellent mechanical properties, low density, and high mechanical strength. These properties, coupled with decreasing material costs and manufacturing expenses, have contributed to the widespread adoption of this composite, making it competitive against metallic structures (Das et al., 2019; Forintos & Czigany, 2019).

The use of carbon fibers (CFs) in composites is not limited to improving densities, prices, and mechanical properties. Their electrical properties enable the production of multifunctional structural components that are particularly noteworthy in the aerospace sector. For instance, they can be used for deicing aircraft wings (Abbas & Park, 2021; Hung et al., 1987), protecting them from lightning strikes (Li et al., 2015; Wang et al., 2016), or storing energy (Shirshova et al., 2013, 2014). Utilizing them as sensors is another example of their multifunctionality. Temperature sensors (Wang & Chung, 1999; Wang & Chung, 2002; Zhu et al., 2019), humidity sensors (Chung, 2001; Tulliani et al., 2019), strain sensors (Wang et al., 1999), and even materials that collect data throughout the product shelf life regarding its environment and structural condition (Industry 4.0) (Barbosa et al., 2023), can be additional properties incorporated into the CFRP. This eliminates the need for embedding additional sensors (e.g., Bragg grating sensors (Takeda et al., 2002) and piezoelectric sensors (Vipperman, 1999)), which could compromise mechanical properties and constrain manufacturing processes of the final products. Moreover, integrated and interconnected sensors are often expensive, not durable, and difficult to repair (Wen et al., 2011).

The electrical properties of CFs are the key to achieving this intrinsic multifunctionality and serve as the basis for designing sensors based on CF polymer composites. The conductivity of carbon stems from its highly stable, flat hexagonal grid structure and the delocalized cloud of electrons between the planes. Deformation and separation of the carbon hexagonal rings require high energy, providing the macrolevel strength of CF, while the free electrons in the electron cloud contribute to its good electrical conductivity (Forintos & Czigany, 2019; Park, 2015). Numerous factors influence the electrical properties of CFs, including the precursor or raw material used in their fabrication (Goodhew et al., 1975), manufacturing conditions (fiber orientation, oxidation processes, etc.) (Rahaman et al., 2007), and the final crystalline structure (fiber imperfections, purity, etc.) (Forintos & Czigany, 2019).

When a force or deformation is applied to the CFRP, the CFs within the polymer matrix stretch or compress, leading to a modification in the distance between them (Cesano et al., 2020; Zhao et al., 2019), changing the electrical conductivity of the composite. Generally, when the sensor is in its undeformed state, the CFs are closer together, allowing for higher electrical conductivity. However, when a force or deformation is applied to the sensor, the CFs move apart and separate, resulting in increased electrical resistivity. The change in electrical resistivity of the sensor can be measured by applying a known voltage and then measuring the resulting electrical response. This variation can be proportional to the force or deformation applied to the sensor, allowing for the quantification and monitoring of specific magnitudes.

Therefore, in the context of self-sensing electrical resistance measurements for damage detection, which stands as the primary application of these composites as sensors, addressing the matter of connecting them to the electrical circuit becomes necessary. For this purpose, direct contact or induction can be used. In the first case, the terminals of a power source (direct current (DC) or alternating current (AC)) are physically connected to the CF reinforcement of the CFRP (Abry et al., 2001). Abry et al. (Abry et al., 2001) demonstrated that DC electrical conduction allows for the detection of fiber failures, whereas AC measurements are more suitable for monitoring matrix cracks, delamination, fiber/matrix debonding, or transverse cracks in CF. In the case of the induction method, voltage is induced in a conductor placed in an alternating magnetic field, causing current to flow through the conducting material. Changes in the magnetic field can be caused by the relative movement of the magnetic field and the conductor (motion induction) or the magnitude of the field, that is, the change of flux over time (stationary induction) (Bayerl et al., 2014; Serway & Jewett, 2001). The latter method is the most extensively studied.

Based on these principles, multiple configurations have been designed (different types of matrices: conductive (Takahashi et al., 2022), nonconductive with varying degrees of crosslinking (Todoroki, 2004), addition of conductive particles (Friedrich & Breuer, 2015; Gallo & Thostenson, 2015), etc.; different arrangements of fabrics and CFs within the composite structure (Vavouliotis et al., 2011; Wang et al., 2004), etc.) to model their sensor capabilities for SHM under different conditions (cyclic, static, tensile, flexural loads (Abry et al., 2001; Prasse et al., 2001; Schulte & Baron, 1989), etc.) and to detect the type and/or size (Todoroki et al., 2005) of the occurring failure (crack, delamination, fiber breakage, etc.).

While SHM is the main focus of attention for these CFRPs as sensors, there are also highly interesting yet less explored applications in other sectors. For example, they can be used in sports applications to achieved enhanced performance, or on the Internet of Things (IoT) for sports facilities and fitness equipment to improve or add new functionalities (Ebling, 2016). Integrated sensors within sports equipment can detect speed, rotation, impact, and flight trajectory and then transmit the data to a smartphone (Wang, Chen et al., 2018). Successful studies, such as the work conducted by Narita et al. and Wang et al. (Narita et al., 2019; Wang, 2021), have explored materials based on hybrid piezoelectric CFRP laminates for energy harvesting in sports equipment, evaluating the output voltage of the composite samples due to impact loading.

Another distinctive application of these carbon composites as sensors lies in the field of medicine for gait monitoring and motion detection (Nag et al., 2017). They can be integrated into assistive devices for individuals with disabilities or prosthetics to provide precise feedback and control based on muscular activity (Nambiar & Yeow, 2011). All of this is based on the same principle of the IoT in wearable sensors, which have been extensively studied (Nag et al., 2022).

Furthermore, CFRP sensors are used in the automotive industry for various applications. They can be integrated into key parts and components such as chassis, suspension, and brakes to monitor the forces and stresses they experience while driving, addressing upcoming challenges such as automation and autonomous driving. They are also employed in safety systems, such as airbags, to detect the presence of occupants and adjust the response in the event of a collision (Joy Sequeira et al., 2019; Luo et al., 2017).

Finally, as described, CFs are commonly used in CFRP sensors as continuous fibers or textiles. When carbon structures are used in the form of particles or short fibers, nanoscale sizes are preferred to form the required conductive percolation network, as mentioned earlier. Short CFs (micrometer-sized) in composite sensors have been poorly studied (Davoodi et al., 2020; Montazerian et al., 2019) or, alternatively, conventional CB particles, as described in Section 9.2.2.1, are used instead.

9.2.3.3 Active fiber composites

As described in Section 9.2.2.4, piezoelectric materials, especially ceramics, have become popular in recent decades as materials for transducers, sensors, and microactuators. The reason for this is the wide frequency range in which devices made from these materials can operate (Chopra, 2002). However, when piezoceramic materials are used as layers, particles, or small–medium fibers (known as macrofiber composites or MFC sensors) in composite smart structures, some issues arise. The ceramic particles increase the structure's stiffness, making it difficult to conform to curved surfaces, as well as the deformation compatibility and stiffness mismatch, which can lead to delamination of the layers and vulnerability to fracture due to the material's brittleness (Kuna, 2010).

Then, a new group of sensor materials was created: AFCs. This nomenclature is specifically used to refer to uniaxially oriented piezoceramic fibers embedded in a polymer matrix and interleaved between two interdigitated electrodes (Kornmann et al., 2004), as shown in Fig. 9.5. The piezoelectric fibers are polarized in their axial direction, aligning with the direction of the electric field, making them highly efficient. Bent and Hagood initially developed this configuration (Bent & Hagood, 1997). Through the interdigitated electrodes, the piezoelectric fibers are polarized along their length. As a result, their piezoelectric properties are primarily determined by the piezoelectric strain constant (d_{33}), with an absolute value twice as large as the piezoelectric strain constant (d_{31}) exploited in most other piezoelectric devices (Bent & Hagood, 1997). When an electric voltage is applied to the electrodes, the electric field generated across the fibers causes them to elongate in

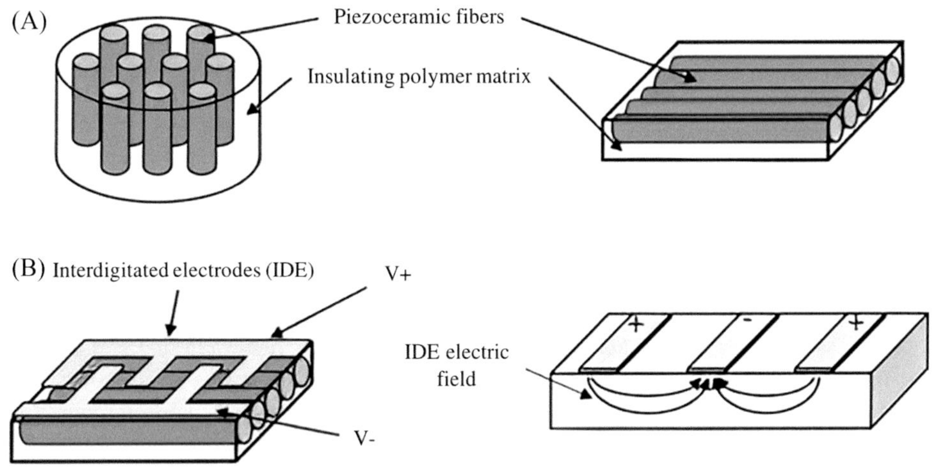

FIGURE 9.5

(A) Scheme of AFC with active piezoelectric fibers in a passive polymer matrix. (B) Schemes of AFC with interdigitated electrodes. The fibers are polarized along the direction of their long axis, so the measured piezoelectric coefficient is the d_{33} coefficient.

Reprinted with permission from Scheffler, S., & Poulin, P. (2022). Piezoelectric fibers: Processing and challenges. ACS Applied Materials and Interfaces, 14*(15), 16961–16982. Available from https://doi.org/10.1021/acsami.1c24611. Copyright 2022 American Chemical Society.*

the longitudinal direction. This continuous fiber composite configuration results in its hardness and flexibility being superior to those of monolithic piezoceramics, and its piezoelectric properties are better than those of monomaterial piezopolymers like PVDF. Authors such as Bowen et al. (Bowen et al., 2005) have studied the mechanical properties of composites based on piezoelectric fibers and a matrix, analyzing the influence of the fiber volume fraction on the mechanical properties and failure mechanism of the composite as a whole. It is crucial to highlight that the piezoelectric properties of AFCs, whether ceramic or polymeric, depend greatly on their manufacturing or spinning process. Scheffler and Poulin (Scheffler & Poulin, 2022) provide an excellent summary of all the configurations and techniques for obtaining these types of composites.

Despite their different configurations and manufacturing processes, AFCs and MFCs can be used in virtually the same applications (Nelson, 2002), including energy harvesting, SHM of composites, vibration control, damping, and actuators. However, MFCs have been developed to a more commercial level. AFCs are still in an earlier stage of development but enjoy a very promising potential considering their processing from fibers, which can be produced through low-cost and easily scalable fiber spinning processes.

There are two major applications where AFCs stand out: SHM and active and passive mechanical damping. In the first case, AFCs can be used for electromechanical (Annamdas & Radhika, 2013; Sodano et al., 2004) and acoustic emission (AE) (Li, Kong et al., 2016; Prabakar & Rao, 2005) measurements to monitor defects in a structure. The AFC is embedded within the structure or attached to the surface that needs monitoring. As with all sensor composites, good adhesion between the host structure and the piezoelectric element is required to achieve a large sensing area. The advantage of using AFC instead of monolithic piezoelectric materials is that they are more flexible and can be attached to curved surfaces. In impedance measurements, a voltage wave is generated, and its response is detected by piezoelectric actuators/sensors. The mechanical impedance of the structure is correlated with the electrical impedance of the piezoelectric device. If the structure is damaged, the electrical response is modified compared to a reference measurement. In the AE method, damage detection is continuously measured. The piezoelectric device detects vibrations associated with crack formation and defects (Scheffler & Poulin, 2022).

Piezoelectric fiber-based devices for SHM also have limitations: the quality of bonding to the structure, the maximum tension or bending the device can withstand before failure, and fatigue and efficiency loss at high temperatures (Henslee et al., 2012; Pandey & Arockiarajan, 2017). Regarding active and passive mechanical damping, it is a relatively new field that still faces significant challenges. In the case of passive damping (Duffy et al., 2013; Groo et al., 2019), mechanical vibrations are converted into electricity, and the associated energy is dissipated through the Joule effect. Piezoelectric materials enable uniform active damping, which is based on detecting vibrations associated with a closed-loop actuation system to enhance the damping effect (Duffy et al., 2013). This appealing approach has been the subject of various theoretical studies with AFCs (Ray & Pradhan, 2007; Sarangi & Ray, 2011).

Apart from SHM and damping measurement, the use of AFCs as actuators or energy harvesting devices has also been reported, but generally, MFCs are preferred (see Section 9.2.2.4) due to the rectangular sections of the active piezoelectric components, which provide higher performance in these applications (Khazaee et al., 2020). This allows for a more efficient connection to the electrodes compared to the long cylindrical fibers of AFCs. However, AFCs made of indefinitely long piezoelectric fibers could be used in larger structural parts, whereas MFCs can only be used in small devices.

With the concept in mind that AFCs are made of long piezoelectric fibers, there has been significant interest in the manufacturing of smart textiles based on AFCs in recent years, capable of converting mechanical energy into electrical energy. These textiles could find countless applications as sensors or energy harvesters in clothing, biomedicine, housing, furniture, technical textiles, etc (Chen et al., 2020; Dong et al., 2020; Mokhtari et al., 2020). However, manufacturing textiles is a major challenge as it involves issues of comfort, flexibility, and portability. Additionally, it presents technical difficulties for poling and electrical connections to conditioning circuits. In fact, it is essential to establish efficient electrical connections that can withstand repeated deformations. These connections must cover as much surface area as possible of the active elements to maximize electromechanical transduction. It is also essential to avoid short circuits and properly separate the materials from the electrodes. Despite these challenges, exciting advancements have been made in recent years. Several groups have successfully implemented piezoelectric fibers into textile structures using methods that are potentially scalable and compatible with conventional textile technologies, such as weaving and even knitting and braiding (Scheffler & Poulin, 2022).

9.2.3.4 Optical fibers and polymer optical fibers

Optical fiber (OF) utilizes the properties of light to detect and measure physical or environmental changes. OFs are employed in a diverse range of applications due to their high sensitivity, immunity to electromagnetic interference, and ability to transmit signals over long distances without significant loss of intensity. OFs are thin and flexible strands composed of transparent glass or plastic materials (polymer optical fiber (POF), with polymethylmethacrylate or PMMA being the most common material used in most cases) (Kuang et al., 2009). It refers to cylindrical waveguides made of two concentric layers, the core and the cladding, which guide light through a slight difference in refractive index between the two layers (Keiser, 2000). The fundamental principle of how OFs work as sensors is based on the change in optical properties that occurs when the fiber is exposed to a specific variable. These changes can be measured and quantified to detect the phenomenon being monitored.

Monocomponent long fibers are usually used as optical sensors. However, composite materials can also be used for this purpose. However, it is essential to mention that the high flexibility of POF allows its incorporation into soft structures, making it a suitable option for instrumentation in novel wearable robots (Leal-Junior et al., 2019). It is worth noting that 3D printing is one of the technologies that have enabled the development of these novel flexible wearable systems, which is an additive layer manufacturing process where hot or molten polymer is injected layer by layer (Gul et al., 2018). In this case, fiber optic sensors are integrated into 3D-printed structures for SHM (Fang et al., 2016) and plantar pressure detection platforms (Zhang et al., 2017).

Another example of these POF composites is deformation detection for SHM, shape determination, and space vehicle qualification testing. Friebele et al. (Friebele et al., 1999) review the results of recent work at the Naval Research Laboratory, where fiber optic strain sensors have been used in spacecraft structures and ground test hardware. The sensors have been installed on the structure's surface and embedded in fiber-reinforced polymer composites. The issue of the potential reduction in the strength of high-performance composites due to integrated OF sensors and cables is discussed, demonstrating that these are challenges to be overcome.

Another specific example where fiber optic-based sensors are used as composite materials is the well known Fiber Optic Acoustic Emission Sensors (FOAES). Conventional AE techniques employ piezoelectric sensors based on wideband or resonant glass OF to detect elastic stress waves that propagate through various types of structural materials when damage occurs. The embedding process within a polymer matrix protects FOAES against environmental stresses, thus extending their lifespan. The immunity of FOAES to electromagnetic interference makes this type of sensor attractive for monitoring purposes in a wide range of challenging operating environments. Willberry and Papaelias et al. (Willberry & Papaelias, 2020) provide a comprehensive review of these new use perspectives.

In conclusion, the most significant field of application for these OFs in the form of polymer composite materials is the structural monitoring of components made of fiberglass and CF composites. They are used both to monitor damage during their usage and to assess the quality of their manufacturing. Among the various fiber optic configurations, Fiber Bragg Gratings (FBGs) photo-inscribed in the core of an OF are the most widely used for monitoring composite materials (Othonos et al., 2006). These FBGs correspond to a modulation of the refractive index along the fiber's axis and act as wavelength-selective mirrors. They are intrinsically sensitive to temperature, pressure, and axial strain, producing a wavelength-encoded response that can be directly recorded and processed (Kersey et al., 1997). Kinet et al. (Kinet et al., 2014) provide a comprehensive review that encompasses the main demodulation techniques used to record and process the amplitude spectrum during and after the composite manufacturing process, addressing the challenges associated with the connection between embedded OF and the environment.

9.3 Conclusions, challenges, and future prospects

This chapter explores a wide range of fundamental concepts governing the use of polymer composites as sensors. These compounds are formed by combining functionally dispersed phases (with properties such as electrical conductivity and piezoelectricity) and polymeric matrices of different natures. They have shown characteristics that make them innovative solutions in the field of detection across sectors as varied as transport, sports, healthcare and medicine, manufacturing industry, defense, energy and environment, consumer electronics, and the realm of the IoT.

The primary reason behind the increasing use of polymer composites for sensor applications lies in the ease that the polymeric matrix provides in creating complex shapes, along with improvements in mechanical properties and overall durability. The design of sensor polymer composites ranges from nanocomposites, which take advantage of the exceptional surface area of nanoparticles to enhance sensor sensitivity and functionality, to large structural pieces manufactured from textile composites, which offer exceptional mechanical strength along with detection capabilities. This highlights the large number of dimensions and configurations that have been explored. Among all, it is noteworthy that nanocomposites, particularly those based on carbon nanoparticles, lead sensor research due to their ability to generate easily detectable response signals, greater versatility, low costs (in most cases), and high availability.

Despite the exciting advancements in the sensor field, it is vital to recognize the current limitations of these composite materials. The challenge of achieving a uniform distribution and dispersion of noncontinuous dispersed phases, coupled with the necessity for profound compatibility in polymer-fiber/fabric interactions, remains significant. It is envisaged that research will persist in the realm of manufacturing and modification processes for nanocomposites and other polymer composite configurations, yielding sensors that are more sensitive, selective, reproducible, durable, and cost-effective. Furthermore, with advancements in different manufacturing processes, such as 3D printing and molecular printing, combined with progress in nanotechnology, we can anticipate the creation of customized sensors seamlessly integrated into more intricate components and devices, heralding new opportunities in the era of smart technology.

Acknowledgment

The authors are grateful for the financial support provided by MCIN/AEI/10.13039/501100011033 and by the "European Union NextGenerationEU/PRTR" (ECOLAYER project TED2021-129419B-C22). The authors also acknowledge the University of Valladolid for the Postdoctoral Contract CONVOCATORIA 2020 (K.C.N.C). Finally, the authors would also like to thank the Recovery and Resilience Mechanism Funds -Next Generation EU Funds- and the Community of Castilla y León Funds. Complementary Research and Development Plans with the Autonomous Communities in R&D&I actions, of the Component 17. Investment 1.

References

Abbadi, A., Tixier, C., Gilgert, J., & Azari, Z. (2015). Experimental study on the fatigue behaviour of honeycomb sandwich panels with artificial defects. *Composite Structures*, *120*, 394–405. Available from https://doi.org/10.1016/j.compstruct.2014.10.020, http://www.elsevier.com/inca/publications/store/4/0/5/9/2/8.

Abbas, S., & Park, C. W. (2021). Frosting and defrosting assessment of carbon fiber reinforced polymer composite with surface wettability and resistive heating characteristics. *International Journal of Heat and Mass Transfer*, *169*, 120883. Available from https://doi.org/10.1016/j.ijheatmasstransfer.2020.120883, http://www.journals.elsevier.com/international-journal-of-heat-and-mass-transfer/.

Abeykoon, C. (2019). Design and applications of soft sensors in polymer processing: A review. *IEEE Sensors Journal*, *19*(8), 2801–2813. Available from https://doi.org/10.1109/JSEN.2018.2885609, http://ieeexplore.ieee.org/xpl/RecentIssue.jsp?punumber = 7361.

Abry, J. C., Choi, Y. K., Chateauminois, A., Dalloz, B., Giraud, G., & Salvia, M. (2001). In-situ monitoring of damage in CFRP laminates by means of AC and DC measurements. *Composites Science and Technology*, *61*(6), 855–864. Available from https://doi.org/10.1016/S0266-3538(00)00181-0, https://www.sciencedirect.com/science/article/pii/S0266353800001810.

Ahmadijokani, F., Shojaei, A., Dordanihaghighi, S., Jafarpour, E., Mohammadi, S., & Arjmand, M. (2020). Effects of hybrid carbon-aramid fiber on performance of non-asbestos organic brake friction composites. *Wear*, *452–453*, 203280. Available from https://doi.org/10.1016/j.wear.2020.203280.

Akdogan, E. K., Allahverdi, M., & Safari, A. (2005). Piezoelectric composites for sensor and actuator applications. *IEEE Transactions on Ultrasonics, Ferroelectrics, and Frequency Control*, *52*(5), 746–775. Available from https://doi.org/10.1109/TUFFC.2005.1503962.

Akhil, U. V., Radhika, N., Saleh, B., Aravind Krishna, S., Noble, N., & Rajeshkumar, L. (2023). A comprehensive review on plant-based natural fiber reinforced polymer composites: Fabrication, properties, and applications. *Polymer Composites*, *44*(5), 2598–2633. Available from https://doi.org/10.1002/pc.27274, http://onlinelibrary.wiley.com/journal/10.1002/(ISSN)1548-0569.

Alfadhel, A., Li, B., & Kosel, J. (2014). Magnetic polymer nanocomposites for sensing applications. *Proceedings of IEEE Sensors*, 2066–2069. Available from http://www.ieee.org/sensors, https://doi.org/10.1109/ICSENS.2014.6985442.

Ali, N., Zhang, B., Zhang, H., Zaman, W., Ali, S., Ali, Z., Li, W., & Zhang, Q. (2015). Monodispers and multifunctional magnetic composite core shell microspheres for demulsification applications. *Journal of the Chinese Chemical Society*, *62*(8), 695–702. Available from https://doi.org/10.1002/jccs.201500151, http://onlinelibrary.wiley.com/journal/10.1002/(ISSN)2192-6549/issues.

Al-Saleh, M. H., & Sundararaj, U. (2009). A review of vapor grown carbon nanofiber/polymer conductive composites. *Carbon*, *47*(1), 2–22. Available from https://doi.org/10.1016/j.carbon.2008.09.039.

Alshammari, A. S. (2018). *Carbon-based polymer nanocomposites for sensing applications. Carbon-Based Polymer Nanocomposites for Environmental and Energy Applications* (pp. 331–360). Saudi Arabia: Elsevier Inc.. Available from https://www.sciencedirect.com/book/9780128135747, https://doi.org/10.1016/B978-0-12-813574-7.00014-9.

Amirkhani, M., & Bakhtiarpour, P. (2016). Ionic polymer metal composites (IPMCs) Chapter 19, Electroactive polymers for robotic applications: Artificial muscles and sensors. *Smart Multi-Functional Materials and Artificial Muscle*, *2*. Available from http://www.springerlink.com/openurl.asp?genre = book&isbn = 978-1-84628-371-0.

Amjadi, M., Kyung, K. U., Park, I., & Sitti, M. (2016). Stretchable, skin-mountable, and wearable strain sensors and their potential applications: A review. *Advanced Functional Materials*, *26*(11), 1678–1698. Available from https://doi.org/10.1002/adfm.201504755, http://onlinelibrary.wiley.com/journal/10.1002/(ISSN)1616-3028.

Amjadi, M., Pichitpajongkit, A., Lee, S., Ryu, S., & Park, I. (2014). Highly stretchable and sensitive strain sensor based on silver nanowire-elastomer nanocomposite. *ACS Nano*, *8*(5), 5154–5163. Available from https://doi.org/10.1021/nn501204t.

Anju, M., & Renuka, N. K. (2019). Graphene–dye hybrid optical sensors. *Nano-Structures and Nano-Objects*, *17*, 194–217. Available from https://doi.org/10.1016/j.nanoso.2019.01.003, http://www.journals.elsevier.com/nano-structures-and-nano-objects/.

Annamdas, V. G. M., & Radhika, M. A. (2013). Electromechanical impedance of piezoelectric transducers for monitoring metallic and non-metallic structures: A review of wired, wireless and energy-harvesting methods. *Journal of Intelligent Material Systems and Structures*, *24*(9), 1021–1042. Available from https://doi.org/10.1177/1045389X13481254.

Anton, S. R., & Sodano, H. A. (2007). A review of power harvesting using piezoelectric materials (2003–2006). *Smart Materials and Structures*, *16*(3), R1–R21. Available from https://doi.org/10.1088/0964-1726/16/3/R01.

Arshak, K., Moore, E., Cavanagh, L., Harris, J., McConigly, B., Cunniffe, C., Lyons, G., & Clifford, S. (2005). Determination of the electrical behaviour of surfactant treated polymer/carbon black composite gas sensors. *Composites Part A: Applied Science and Manufacturing*, *36*(4), 487–491. Available from https://doi.org/10.1016/j.compositesa.2004.10.015.

Atta, N. F., Galal, A., & El-Ads, E. H. (2012). Gold nanoparticles-coated poly(3,4-ethylene-dioxythiophene) for the selective determination of sub-nano concentrations of dopamine in presence of sodium dodecyl sulfate. *Electrochimica Acta*, *69*, 102–111. Available from https://doi.org/10.1016/j.electacta.2012.02.082.

Avilés, F., May-Pat, A., Canché-Escamilla, G., Rodríguez-Uicab, O., Ku-Herrera, J. J., Duarte-Aranda, S., Uribe-Calderon, J., Gonzalez-Chi, P. I., Arronche, L., & La Saponara, V. (2016). Influence of carbon nanotube on the piezoresistive behavior of multiwall carbon nanotube/polymer composites. *Journal of Intelligent Material Systems and Structures*, *27*(1), 92–103. Available from https://doi.org/10.1177/1045389X14560367, http://jim.sagepub.com/.

Azeem, M., Ya, H. H., Kumar, M., Stabla, P., Smolnicki, M., Gemi, L., Khan, R., Ahmed, T., Ma, Q., Sadique, M. R., Mokhtar, A. A., & Mustapha, M. (2022). Application of filament winding technology in composite pressure vessels and challenges: A review. *Journal of Energy Storage*, *49*. Available from https://doi.org/10.1016/j.est.2021.103468, http://www.journals.elsevier.com/journal-of-energy-storage/.

Babu, P. J., & Doble, M. (2018). Albumin capped carbon-gold (C-Au) nanocomposite as an optical sensor for the detection of Arsenic(III). *Optical Materials*, *84*, 339–344. Available from https://doi.org/10.1016/j.optmat.2018.07.013.

Babu, P. J., Raichur, A. M., & Doble, M. (2018). Synthesis and characterization of biocompatible carbon-gold (C-Au) nanocomposites and their biomedical applications as an optical sensor for creatinine detection and cellular imaging. *Sensors and Actuators, B: Chemical*, *258*, 1267–1278. Available from https://doi.org/10.1016/j.snb.2017.11.148.

Bahramzadeh, Y., & Shahinpoor, M. (2011). Dynamic curvature sensing employing ionic-polymer-metal composite sensors. *Smart Materials and Structures*, *20*(9). Available from https://doi.org/10.1088/0964-1726/20/9/094011, http://iopscience.iop.org/0964-1726/20/9/094011/pdf/0964-1726_20_9_094011.pdf.

Bai, S., Zhao, Y., Sun, J., Tian, Y., Luo, R., Li, D., & Chen, A. (2015). Ultrasensitive room temperature NH 3 sensor based on a graphene–polyaniline hybrid loaded on PET thin film. *Chemical Communications*, *51* (35), 7524–7527. Available from https://doi.org/10.1039/C5CC01241D.

Bao, W. S., Meguid, S. A., Zhu, Z. H., & Weng, G. J. (2012). Tunneling resistance and its effect on the electrical conductivity of carbon nanotube nanocomposites. *Journal of Applied Physics*, *111*(9). Available from https://doi.org/10.1063/1.4716010.

Barbosa, G. F., Grassi, G. Z., de Andrade Bezerra, W., & Shiki, S. B. (2023). Drilling of carbon fiber parts performed by an Industry 4.0 systems-integrated technology. *International Journal of Advanced Manufacturing Technology*, *126*(11–12), 5191–5198. Available from https://doi.org/10.1007/s00170-023-11266-8, https://www.springer.com/journal/170.

Barthwal, S., & Singh, N. B. (2017). ZnO-CNT nanocomposite:A device as electrochemical sensor. *Materials Today: Proceedings*, *4*(4), 5552–5560. Available from https://doi.org/10.1016/j.matpr.2017.06.012, https://www.sciencedirect.com/journal/materials-today-proceedings.

Bauhofer, W., & Kovacs, J. Z. (2009). A review and analysis of electrical percolation in carbon nanotube polymer composites. *Composites Science and Technology*, *69*(10), 1486–1498. Available from https://doi.org/10.1016/j.compscitech.2008.06.018.

Bayerl, T., Duhovic, M., Mitschang, P., & Bhattacharyya, D. (2014). The heating of polymer composites by electromagnetic induction – A review. *Composites Part A: Applied Science and Manufacturing*, *57*, 27–40. Available from https://doi.org/10.1016/j.compositesa.2013.10.024.

Bent, A. A., & Hagood, N. W. (1997). Piezoelectric fiber composites with interdigitated electrodes. *Journal of Intelligent Material Systems and Structures*, *8*(11), 903–919. Available from https://doi.org/10.1177/1045389X9700801101, http://jim.sagepub.com/.

Bhambi, M., Sumana, G., Malhotra, B. D., & Pundir, C. S. (2010). An amperomertic uric acid biosensor based on immobilization of uricase onto polyaniline-multiwalled carbon nanotube composite film. *Artificial Cells, Blood Substitutes, and Biotechnology*, *38*(4), 178–185. Available from https://doi.org/10.3109/10731191003716344.

Bhangare, B., Jagtap, S., Ramgir, N., Waichal, R., Muthe, K. P., Gupta, S. K., Gadkari, S. C., Aswal, D. K., & Gosavi, S. (2018). Evaluation of humidity sensor based on PVP-RGO nanocomposites. *IEEE Sensors Journal*, *18*(22), 9097–9104. Available from https://doi.org/10.1109/JSEN.2018.2870324, http://ieeexplore.ieee.org/xpl/RecentIssue.jsp?punumber = 7361.

Bijender., & Kumar, A. (2022). Recent progress in the fabrication and applications of flexible capacitive and resistive pressure sensors. *Sensors and Actuators A: Physical*, *344*. Available from https://doi.org/10.1016/j.sna.2022.113770, https://www.journals.elsevier.com/sensors-and-actuators-a-physical.

Bowen, C., Dent, A., Stevens, R., Cain, M., & Steward, M. (2005). Determination of critical and minimum volume fraction for composite sensors and actuators. *Proceedings 4M*, 483–487.

Báez, D. F., Brito, T. P., Espinoza, L. C., Méndez-Torres, A. M., Sierpe, R., Sierra-Rosales, P., Venegas, C. J., Yáñez, C., & Bollo, S. (2021). Graphene-based sensors for small molecule determination in real samples. *Microchemical Journal*, *167*, 106303. Available from https://doi.org/10.1016/j.microc.2021.106303.

Böger, L., Sumfleth, J., Hedemann, H., & Schulte, K. (2010). Improvement of fatigue life by incorporation of nanoparticles in glass fibre reinforced epoxy. *Composites Part A: Applied Science and Manufacturing*, *41*(10), 1419–1424. Available from https://doi.org/10.1016/j.compositesa.2010.06.002.

Caneva, C., De Rosa, I. M., & Sarasini, F. (2008). Monitoring of impacted aramid-reinforced composites by embedded PVDF acoustic emission sensors. *Strain*, *44*(4), 308–316. Available from https://doi.org/10.1111/j.1475-1305.2007.00374.x.

Carrillo, A., Martín-Domínguez, I. R., Rosas, A., & Márquez, A. (2002). Numerical method to evaluate the influence of organic solvent absorption on the conductivity of polymeric composites. *Polymer*, *43*(23), 6307–6313. Available from https://doi.org/10.1016/S0032-3861(02)00558-X, https://www.sciencedirect.com/science/article/pii/S003238610200558X.

Cesano, F., Uddin, M. J., Lozano, K., Zanetti, M., & Scarano, D. (2020). All-carbon conductors for electronic and electrical wiring applications. *Frontiers in Materials*, 7. Available from https://doi.org/10.3389/fmats.2020.00219, http://journal.frontiersin.org/journal/materials.

Chang, J., Zhou, G., Christensen, E. R., Heideman, R., & Chen, J. (2014). Graphene-based sensors for detection of heavy metals in water: A review. Chemosensors and chemoreception. *Analytical and Bioanalytical Chemistry*, *406*(16), 3957–3975. Available from https://doi.org/10.1007/s00216-014-7804-x, http://link.springer.de/link/service/journals/00216/index.htm.

Chen, G., Li, Y., Bick, M., & Chen, J. (2020). Smart textiles for electricity generation. *Chemical Reviews*, *120*(8), 3668–3720. Available from https://doi.org/10.1021/acs.chemrev.9b00821, http://pubs.acs.org/journal/chreay.

Chen, J., & Tsubokawa, N. (2000). Novel gas sensor from polymer-grafted carbon black: Vapor response of electric resistance of conducting composites prepared from poly(ethylene-block-ethylene oxide)-grafted carbon black. *Journal of Applied Polymer Science*, *77*(11), 2437–2447. Available from https://doi.org/10.1002/1097-4628, https://doi.org/10.1002/1097-4628.

Chen, J., Zhu, Y., Huang, J., Zhang, J., Pan, D., Zhou, J., Ryu, J. E., Umar, A., & Guo, Z. (2021). Advances in responsively conductive polymer composites and sensing applications. *Polymer Reviews*, *61*(1), 157–193. Available from https://doi.org/10.1080/15583724.2020.1734818, http://www.tandf.co.uk/journals/titles/15583724.asp.

Chen, R. J., Franklin, N. R., Kong, J., Cao, J., Tombler, T. W., Zhang, Y., & Dai, H. (2001). Molecular photodesorption from single-walled carbon nanotubes. *Applied Physics Letters*, *79*(14), 2258–2260. Available from https://doi.org/10.1063/1.1408274.

Chen, X., Wu, G., Cai, Z., Oyama, M., & Chen, X. (2014). Advances in enzyme-free electrochemical sensors for hydrogen peroxide, glucose, and uric acid. *Microchimica Acta*, *181*(7-8), 689–705. Available from https://doi.org/10.1007/s00604-013-1098-0, http://www.springer/at/mca.

Chiu, S. W., & Tang, K. T. (2013). Towards a chemiresistive sensor-integrated electronic nose: A review. *Sensors (Switzerland)*, *13*(10), 14214–14247. Available from https://doi.org/10.3390/s131014214, http://www.mdpi.com/1424-8220/13/10/14214/pdf.

Chopra, I. (2002). Review of state of art of smart structures and integrated systems. *AIAA Journal*, *40*(11), 2145–2187. Available from https://doi.org/10.2514/2.1561, http://arc.aiaa.org/loi/aiaaj.

Chung, D. D. L. (2001). Continuous carbon fiber polymer-matrix composites and their joints, studied by electrical measurements. *Polymer Composites*, *22*(2), 250–270. Available from https://doi.org/10.1002/pc.10536.

Colombo, C., Vergani, L., & Burman, M. (2012). Static and fatigue characterisation of new basalt fibre reinforced composites. *Composite Structures*, *94*(3), 1165–1174. Available from https://doi.org/10.1016/j.compstruct.2011.10.007.

Crawley, E. F., & De Luis, J. (1987). Use of piezoelectric actuators as elements of intelligent structures. *AIAA Journal*, *25*(10), 1373–1385. Available from https://doi.org/10.2514/3.9792.

D'Ambrogio, G., Zahhaf, O., Hebrard, Y., Le, M. Q., Cottinet, P. J., & Capsal, J. F. (2021). Microstructuration of piezoelectric composites using dielectrophoresis: Toward application in condition monitoring of bearings. *Advanced Engineering Materials*, *23*(1). Available from https://doi.org/10.1002/adem.202000773, http://onlinelibrary.wiley.com/journal/10.1002/(ISSN)1527-2648.

Dagdag, O., Safi, Z., Hsissou, R., Erramli, H., El Bouchti, M., Wazzan, N., Guo, L., Verma, C., Ebenso, E. E., & El Harfi, A. (2019). Epoxy pre-polymers as new and effective materials for corrosion inhibition of carbon steel in acidic medium: Computational and experimental studies. *Scientific Reports*, *9*(1). Available from https://doi.org/10.1038/s41598-019-48284-0, http://www.nature.com/srep/index.html.

Daneshkhah, A., Vij, S., Siegel, A. P., & Agarwal, M. (2020). Polyetherimide/carbon black composite sensors demonstrate selective detection of medium-chain aldehydes including nonanal. *Chemical Engineering Journal*, *383*. Available from https://doi.org/10.1016/j.cej.2019.123104, http://www.elsevier.com/inca/publications/store/6/0/1/2/7/3/index.htt.

Das, R., Pattanayak, A. J., & Swain, S. K. (2018). Polymer nanocomposites for sensor devices. *Polymer-based Nanocomposites for Energy and Environmental Applications: A Volume in Woodhead Publishing Series in Composites Science and Engineering*, 206–216. Available from https://doi.org/10.1016/B978-0-08-102262-7.00007-6, https://www.sciencedirect.com/book/9780081022627/polymer-based-nanocomposites-for-energy-and-environmental-applications.

Das, S., & Yokozeki, T. (2020). Polyaniline-based multifunctional glass fiber reinforced conductive composite for strain monitoring. *Polymer Testing*, *87*. Available from https://doi.org/10.1016/j.polymertesting.2020.106547, https://www.journals.elsevier.com/polymer-testing.

Das, T. K., Ghosh, P., & Das, N. C. (2019). Preparation, development, outcomes, and application versatility of carbon fiber-based polymer composites: A review. *Advanced Composites and Hybrid Materials*, *2*(2), 214–233. Available from https://doi.org/10.1007/s42114-018-0072-z, http://springer.com/journal/42114.

Davoodi, E., Fayazfar, H., Liravi, F., Jabari, E., & Toyserkani, E. (2020). Drop-on-demand high-speed 3D printing of flexible milled carbon fiber/silicone composite sensors for wearable biomonitoring devices. *Additive Manufacturing*, *32*. Available from https://doi.org/10.1016/j.addma.2019.101016, https://www.journals.elsevier.com/additive-manufacturing.

Deepa, A. V., Murugasen, P., Muralimanohar, P., Sathyamoorthy, K., & Vinothkumar, P. (2019). A comparison on the structural and optical properties of different rare earth doped phosphate glasses. *Optik*, *181*, 361–367. Available from https://doi.org/10.1016/j.ijleo.2018.12.045, http://www.elsevier.com/journals/optik/0030-4026.

Deng, H., Lin, L., Ji, M., Zhang, S., Yang, M., & Fu, Q. (2014). Progress on the morphological control of conductive network in conductive polymer composites and the use as electroactive multifunctional materials. *Progress in Polymer Science*, *39*(4), 627–655. Available from https://doi.org/10.1016/j.progpolymsci.2013.07.007.

Devnani, G. L., & Sinha, S. (2019). Effect of nanofillers on the properties of natural fiber reinforced polymer composites. *Materials Today: Proceedings*, *18*, 647–654. Available from https://doi.org/10.1016/j.matpr.2019.06.460, https://www.sciencedirect.com/journal/materials-today-proceedings.

Dhand, V., Mittal, G., Rhee, K. Y., Park, S. J., & Hui, D. (2015). A short review on basalt fiber reinforced polymer composites. *Composites Part B: Engineering*, *73*, 166–180. Available from https://doi.org/10.1016/j.compositesb.2014.12.011.

Dharmavarapu, P., & Sreekara, S. R. (2022). Aramid fibre as potential reinforcement for polymer matrix composites: A review. *Emergent Materials*, *5*(5), 1561–1578. Available from https://doi.org/10.1007/s42247-021-00246-x, https://www.springer.com/journal/42247.

Doleman, B. J., Sanner, R. D., Severin, E. J., Grubbs, R. H., & Lewis, N. S. (1998). Use of compatible polymer blends to fabricate arrays of carbon black-polymer composite vapor detectors. *Analytical Chemistry*, *70*(13), 2560–2564. Available from https://doi.org/10.1021/ac971238h, http://pubs.acs.org/journal/ancham.

Dong, K., Peng, X., & Wang, Z. L. (2020). Fiber/fabric-based piezoelectric and triboelectric nanogenerators for flexible/stretchable and wearable electronics and artificial intelligence. *Advanced Materials*, *32*(5). Available from https://doi.org/10.1002/adma.201902549, http://onlinelibrary.wiley.com/journal/10.1002/(ISSN)1521-4095.

Dong, X. M., Fu, R. W., Zhang, M. Q., Zhang, B., & Rong, M. Z. (2004). Electrical resistance response of carbon black filled amorphous polymer composite sensors to organic vapors at low vapor concentrations. *Carbon*, *42*(12–13), 2551–2559. Available from https://doi.org/10.1016/j.carbon.2004.05.034.

Doshi, S. M., & Thostenson, E. T. (2018). Thin and flexible carbon nanotube-based pressure sensors with ultrawide sensing range. *ACS Sensors*, *3*(7), 1276–1282. Available from https://doi.org/10.1021/acssensors.8b00378, http://pubs.acs.org/journal/ascefj.

Du, F., Scogna, R. C., Zhou, W., Brand, S., Fischer, J. E., & Winey, K. I. (2004). Nanotube networks in polymer nanocomposites: Rheology and electrical conductivity. *Macromolecules*, *37*(24), 9048–9055. Available from https://doi.org/10.1021/ma049164g.

Duffy, K. P., Choi, B. B., Provenza, A. J., Min, J. B., & Kray, N. (2013). Active piezoelectric vibration control of subscale composite fan blades. *Journal of Engineering for Gas Turbines and Power*, *135*(1). Available from https://doi.org/10.1115/1.4007720.

Ebling, M. R. (2016). IoT: From sports to fashion and everything in-between. *IEEE Pervasive Computing*, *15*(4), 2–4. Available from https://doi.org/10.1109/MPRV.2016.71, http://www.computer.org/pervasive/.

Ehlert, G. J., & Sodano, H. A. (2014). Fiber strain sensors from carbon nanotubes self-assembled on aramid fibers. *Journal of Intelligent Material Systems and Structures*, *25*(17), 2117–2121. Available from https://doi.org/10.1177/1045389X13517316, http://jim.sagepub.com/.

Eivazzadeh-Keihan, R., Bahojb Noruzi, E., Chidar, E., Jafari, M., Davoodi, F., Kashtiaray, A., Ghafori Gorab, M., Masoud Hashemi, S., Javanshir, S., Ahangari Cohan, R., Maleki, A., & Mahdavi, M. (2022). Applications of carbon-based conductive nanomaterials in biosensors. *Chemical Engineering Journal*, *442*. Available from https://doi.org/10.1016/j.cej.2022.136183, http://www.elsevier.com/inca/publications/store/6/0/1/2/7/3/index.htt.

Ekuase, O. A., Anjum, N., Eze, V. O., & Okoli, O. I. (2022). A review on the out-of-autoclave process for composite manufacturing. *Journal of Composites Science*, *6*(6). Available from https://doi.org/10.3390/jcs6060172, https://www.mdpi.com/2504-477X/6/6/172/pdf?version = 1655174669.

Elkington, M., Bloom, D., Ward, C., Chatzimichali, A., & Potter, K. (2015). Hand layup: Understanding the manual process. *Advanced Manufacturing: Polymer and Composites Science*, *1*(3), 138–151. Available from https://doi.org/10.1080/20550340.2015.1114801, http://www.tandfonline.com/toc/yadm20/current.

Ellis, J. E., & Star, A. (2016). Carbon nanotube based gas sensors toward breath analysis. *ChemPlusChem*, *81*(12), 1248–1265. Available from https://doi.org/10.1002/cplu.201600478, http://onlinelibrary.wiley.com/journal/10.1002/%28ISSN%292192-6506/.

Elvin, N. G., Lajnef, N., & Elvin, A. A. (2006). Feasibility of structural monitoring with vibration powered sensors. *Smart Materials and Structures*, *15*(4), 977–986. Available from https://doi.org/10.1088/0964-1726/15/4/011.

Epaarachchi, J. A., & Kahandawa, G. C. (2016). In J. A. Epaarachchi, & G. C. Kahandawa (Eds.), *Structural health monitoring technologies and next-generation smart composite structures*. Boca Raton: CRC Press. Available from https://doi.org/10.1201/9781315373492.

Eswaraiah, V., Balasubramaniam, K., & Ramaprabhu, S. (2011). Functionalized graphene reinforced thermoplastic nanocomposites as strain sensors in structural health monitoring. *Journal of Materials Chemistry*, *21*(34), 12626–12628. Available from https://doi.org/10.1039/C1JM12302E, http://doi.org/10.1039/C1JM12302E.

Eswaraiah, V., Balasubramaniam, K., & Ramaprabhu, S. (2012). One-pot synthesis of conducting graphene–polymer composites and their strain sensing application. *Nanoscale*, *4*(4), 1258–1262. Available from https://doi.org/10.1039/C2NR11555G, http://doi.org/10.1039/C2NR11555G.

Fang, L., Chen, T., Li, R., & Liu, S. (2016). Application of embedded fiber Bragg grating (FBG) sensors in monitoring health to 3D printing structures. *IEEE Sensors Journal*, *16*(17), 6604–6610. Available from https://doi.org/10.1109/JSEN.2016.2584141, http://ieeexplore.ieee.org/xpl/RecentIssue.jsp?punumber = 7361.

Fang, X., Bando, Y., Liao, M., Gautam, U. K., Zhi, C., Dierre, B., Liu, B., Zhai, T., Sekiguchi, T., Koide, Y., & Golberg, D. (2009). Single-crystalline ZnS nanobelts as ultraviolet-light sensors. *Advanced Materials*, *21*(20), 2034–2039. Available from https://doi.org/10.1002/adma.200802441, http://www3.interscience.wiley.com/cgi-bin/fulltext/122279645/PDFSTART.

Fang, X., Zhai, T., Gautam, U. K., Li, L., Wu, L., Bando, Y., & Golberg, D. (2011). ZnS nanostructures: From synthesis to applications. *Progress in Materials Science*, *56*(2), 175–287. Available from https://doi.org/10.1016/j.pmatsci.2010.10.001.

Fang, Y., Guo, S., Zhu, C., Zhai, Y., & Wang, E. (2010). Self-assembly of cationic polyelectrolyte-functionalized graphene nanosheets and gold nanoparticles: A two-dimensional heterostructure for hydrogen peroxide sensing. *Langmuir: The ACS Journal of Surfaces and Colloids*, *26*(13), 11277–11282. Available from https://doi.org/10.1021/la100575g.

Fei, T., Jiang, K., Jiang, F., Mu, R., & Zhang, T. (2014). Humidity switching properties of sensors based on multiwalled carbon nanotubes/polyvinyl alcohol composite films. *Journal of Applied Polymer Science*, *131* (1). Available from https://doi.org/10.1002/app.39726.

Fernandez, J. F., Dogan, A., Zhang, Q. M., Tressler, J. F., & Newnham, R. E. (1995). Hollow piezoelectric composites. *Sensors and Actuators A: Physical*, *51*(2), 183–192. Available from https://doi.org/10.1016/0924-4247(95)01221-4, https://www.sciencedirect.com/science/article/pii/0924424795012214.

Ferrara, M., Neitzert, H. C., Sarno, M., Gorrasi, G., Sannino, D., Vittoria, V., & Ciambelli, P. (2007). Influence of the electrical field applied during thermal cycling on the conductivity of LLDPE/CNT composites. *Physica E: Low-Dimensional Systems and Nanostructures*, *37*(1-2), 66–71. Available from https://doi.org/10.1016/j.physe.2006.10.008.

Filippo, E., Serra, A., & Manno, D. (2009). Poly(vinyl alcohol) capped silver nanoparticles as localized surface plasmon resonance-based hydrogen peroxide sensor. *Sensors and Actuators, B: Chemical*, *138*(2), 625–630. Available from https://doi.org/10.1016/j.snb.2009.02.056.

Fiore, V., Di Bella, G., & Valenza, A. (2011). Glass-basalt/epoxy hybrid composites for marine applications. *Materials and Design*, *32*(4), 2091–2099. Available from https://doi.org/10.1016/j.matdes.2010.11.043.

Forintos, N., & Czigany, T. (2019). Multifunctional application of carbon fiber reinforced polymer composites: Electrical properties of the reinforcing carbon fibers – A short review. *Composites Part B: Engineering*, *162*, 331–343. Available from https://doi.org/10.1016/j.compositesb.2018.10.098.

Franco, M. A., Conti, P. P., Andre, R. S., & Correa, D. S. (2022). A review on chemiresistive ZnO gas sensors. *Sensors and Actuators Reports*, *4*. Available from https://doi.org/10.1016/j.snr.2022.100100, https://www.journals.elsevier.com/sensors-and-actuators-reports.

Freund, M. S., & Lewis, N. S. (1995). A chemically diverse conducting polymer-based \electronic nose\. *Proceedings of the National Academy of Sciences of the United States of America*, *92*(7), 2652–2656. Available from https://doi.org/10.1073/pnas.92.7.2652, http://www.pnas.org.

Friebele, E. J., Askins, C. G., Bosse, A. B., Kersey, A. D., Patrick, H. J., Pogue, W. R., Putnam, M. A., Simon, W. R., Tasker, F. A., Vincent, W. S., & Vohra, S. T. (1999). Optical fiber sensors for spacecraft applications. *Smart Materials and Structures*, *8*(6), 813–838. Available from https://doi.org/10.1088/0964-1726/8/6/310.

Friedrich, K., & Almajid, A. A. (2013). Manufacturing aspects of advanced polymer composites for automotive applications. *Applied Composite Materials*, *20*(2), 107–128. Available from https://doi.org/10.1007/s10443-012-9258-7.

Friedrich, K., & Breuer, U. (2015). *Multifunctionality of polymer composites: Challenges and new solutions*, 1, William Andrew.

Fu, X., Ramos, M., Al-Jumaily, A. M., Meshkinzar, A., & Huang, X. (2019). Stretchable strain sensor facilely fabricated based on multi-wall carbon nanotube composites with excellent performance. *Journal of Materials Science*, *54*(3), 2170–2180. Available from https://doi.org/10.1007/s10853-018-2954-4.

Fu, Y. F., Yi, F. L., Liu, J. R., Li, Y. Q., Wang, Z. Y., Yang, G., Huang, P., Hu, N., & Fu, S. Y. (2020). Super soft but strong E-Skin based on carbon fiber/carbon black/silicone composite: Truly mimicking tactile sensing and mechanical behavior of human skin. *Composites Science and Technology*, *186*. Available from https://doi.org/10.1016/j.compscitech.2019.107910, http://www.journals.elsevier.com/composites-science-and-technology/.

Galiotis, C. (1993). A study of mechanisms of stress transfer in continuous- and discontinuous-fibre model composites by laser Raman spectroscopy. *Composites Science and Technology*, *48*(1-4), 15–28. Available from https://doi.org/10.1016/0266-3538(93)90116-X.

Gallo, G. J., & Thostenson, E. T. (2015). Electrical characterization and modeling of carbon nanotube and carbon fiber self-sensing composites for enhanced sensing of microcracks. *Materials Today Communications*, *3*, 17–26. Available from https://doi.org/10.1016/j.mtcomm.2015.01.009, http://www.journals.elsevier.com/materials-today-communications/.

Gao, L., Chou, T. W., Thostenson, E. T., & Zhang, Z. (2010). A comparative study of damage sensing in fiber composites using uniformly and non-uniformly dispersed carbon nanotubes. *Carbon*, *48*(13), 3788–3794. Available from https://doi.org/10.1016/j.carbon.2010.06.041, https://www.sciencedirect.com/science/article/pii/S0008622310004471.

Gao, L., Thostenson, E. T., Zhang, Z., Byun, J. H., & Chou, T. W. (2010). Damage monitoring in fiber-reinforced composites under fatigue loading using carbon nanotube networks. *Philosophical Magazine*, *90*(31-32), 4085–4099. Available from https://doi.org/10.1080/14786430903352649.

Gao, S. L., Zhuang, R. C., Zhang, J., Liu, J. W., & Mäder, E. (2010). Class fibers with carbon nanotube networks as multifunctional sensors. *Advanced Functional Materials*, *20*(12), 1885–1893. Available from https://doi.org/10.1002/adfm.201000283, http://www3.interscience.wiley.com/cgi-bin/fulltext/123417501/PDFSTART.

Gao, L., Thostenson, E. T., Zhang, Z., & Chou, T. W. (2009). Sensing of damage mechanisms in fiber-reinforced composites under cyclic loading using carbon nanotubes. *Advanced Functional Materials*, *19*(1), 123–130. Available from https://doi.org/10.1002/adfm.200800865, http://www3.interscience.wiley.com/cgi-bin/fulltext/121542321/PDFSTART.

Gao, S. L., Mäder, E., & Plonka, R. (2007). Nanostructured coatings of glass fibers: Improvement of alkali resistance and mechanical properties. *Acta Materialia*, *55*(3), 1043–1052. Available from https://doi.org/10.1016/j.actamat.2006.09.020.

Gao, S. L., Mäder, E., & Plonka, R. (2008). Nanocomposite coatings for healing surface defects of glass fibers and improving interfacial adhesion. *Composites Science and Technology*, *68*(14), 2892–2901. Available from https://doi.org/10.1016/j.compscitech.2007.10.009.

Garcia-Espinel, J. D., Castro-Fresno, D., Parbole Gayo, P., & Ballester-Muñoz, F. (2015). Effects of sea water environment on glass fiber reinforced plastic materials used for marine civil engineering constructions. *Materials and Design*, *66*, 46–50. Available from https://doi.org/10.1016/j.matdes.2014.10.032.

Gautschi, G. (2002). *Background of Piezoelectric Sensors* (pp. 5–11). Springer Nature. Available from https://doi.org/10.1007/978-3-662-04732-3_2.

Ghezzo, F., Starr, A. F., & Smith, D. R. (2010). Integration of networks of sensors and electronics for structural health monitoring of composite materials. *Advances in Civil Engineering*, *2010*. Available from https://doi.org/10.1155/2010/598458.

Gilbertson, L. M., Busnaina, A. A., Isaacs, J. A., Zimmerman, J. B., & Eckelman, M. J. (2014). Life cycle impacts and benefits of a carbon nanotube-enabled chemical gas sensor. *Environmental Science and Technology*, *48*(19), 11360–11368. Available from https://doi.org/10.1021/es5006576, http://pubs.acs.org/journal/esthag.

Goodhew, P. J., Clarke, A. J., & Bailey, J. E. (1975). A review of the fabrication and properties of carbon fibres. *Materials Science and Engineering*, *17*(1), 3–30. Available from https://doi.org/10.1016/0025-5416(75)90026-9, https://www.sciencedirect.com/science/article/pii/0025541675900269.

Grewe, M. G., Gururaja, T. R., Shrout, T. R., & Newnham, R. E. (1990). Acoustic properties of particle/polymer composites for ultrasonic transducer backing applications. *IEEE Transactions on Ultrasonics, Ferroelectrics, and Frequency Control*, *37*(6), 506–514. Available from https://doi.org/10.1109/58.63106.

Groo, L., Steinke, K., Inman, D. J., & Sodano, H. A. (2019). Vibration damping mechanism of fiber-reinforced composites with integrated piezoelectric nanowires. *ACS Applied Materials and Interfaces*, *11*(50), 47373–47381. Available from https://doi.org/10.1021/acsami.9b17029, http://pubs.acs.org/journal/aamick.

Groo, L. A., Nasser, J., Inman, D., & Sodano, H. (2021). Laser induced graphene for in situ damage sensing in aramid fiber reinforced composites. *Composites Science and Technology*, *201*. Available from https://doi.org/10.1016/j.compscitech.2020.108541, http://www.journals.elsevier.com/composites-science-and-technology/.

Gu, J., Jiang, W., Wang, F., Chen, M., Mao, J., & Xie, T. (2014). Facile removal of oils from water surfaces through highly hydrophobic and magnetic polymer nanocomposites. *Applied Surface Science*, *301*, 492–499. Available from https://doi.org/10.1016/j.apsusc.2014.02.112, http://www.journals.elsevier.com/applied-surface-science/.

Gul, J. Z., Sajid, M., Rehman, M. M., Siddiqui, G. U., Shah, I., Kim, K. H., Lee, J. W., & Choi, K. H. (2018). 3D printing for soft robotics – A review. *Science and Technology of Advanced Materials*, *19*(1), 243–262. Available from https://doi.org/10.1080/14686996.2018.1431862, http://www.tandfonline.com/loi/tsta20#.V0KNDU1f3cs.

Gupta, B. D., Pathak, A., & Semwal, V. (2019). Carbon-based nanomaterials for plasmonic sensors: A review. *Sensors (Switzerland)*, *19*(16). Available from https://doi.org/10.3390/s19163536, https://www.mdpi.com/1424-8220/19/16/3536/pdf.

Gupta, M. C., Harrison, J. T., & Islam, M. T. (2021). Photoconductive PbSe thin films for infrared imaging. *Materials Advances*, *2*(10), 3133–3160. Available from https://doi.org/10.1039/D0MA00965B, http://doi.org/10.1039/D0MA00965B.

Hansen, J. P., & Vizzini, A. J. (2000). Fatigue response of a host structure with interlaced embedded devices. *Journal of Intelligent Material Systems and Structures*, *11*(11), 902–909. Available from https://doi.org/10.1177/1045389X0001101101, https://doi.org/10.1177/1045389X0001101101.

Hao, B., Förster, T., Mäder, E., & Ma, P. C. (2017). Modification of basalt fibre using pyrolytic carbon coating for sensing applications. *Composites Part A: Applied Science and Manufacturing*, *101*, 123–128. Available from https://doi.org/10.1016/j.compositesa.2017.06.010.

Hao, B., Ma, Q., Yang, S., Mäder, E., & Ma, P. C. (2016). Comparative study on monitoring structural damage in fiber-reinforced polymers using glass fibers with carbon nanotubes and graphene coating. *Composites Science and Technology*, *129*, 38–45. Available from https://doi.org/10.1016/j.compscitech.2016.04.012, http://www.journals.elsevier.com/composites-science-and-technology/.

Henslee, I. A., Miller, D. A., & Tempero, T. (2012). Fatigue life characterization for piezoelectric macrofiber composites. *Smart Materials and Structures*, *21*(10). Available from https://doi.org/10.1088/0964-1726/21/10/105037, http://iopscience.iop.org/0964-1726/21/10/105037/pdf/0964-1726_21_10_105037.pdf.

Hong, D., Yerubandi, G., Chiang, H. Q., Spiegelberg, M. C., & Wager, J. F. (2007). Electrical modeling of thin-film transistors. *Critical Reviews in Solid State and Materials Sciences*, *32*(3-4), 111–142. Available from https://doi.org/10.1080/10408430701707347.

Hsissou, R., Abbout, S., Seghiri, R., Rehioui, M., Berisha, A., Erramli, H., Assouag, M., & Elharfi, A. (2020). Evaluation of corrosion inhibition performance of phosphorus polymer for carbon steel in [1 M] HCl: Computational studies (DFT, MC and MD simulations). *Journal of Materials Research and Technology*, *9*(3), 2691–2703. Available from https://doi.org/10.1016/j.jmrt.2020.01.002, http://www.elsevier.com/journals/journal-of-materials-research-and-technology/2238-7854.

Hsissou, R., Seghiri, R., Benzekri, Z., Hilali, M., Rafik, M., & Elharfi, A. (2021). Polymer composite materials: A comprehensive review. *Composite Structures*, *262*. Available from https://doi.org/10.1016/j.compstruct.2021.113640, http://www.elsevier.com/inca/publications/store/4/0/5/9/2/8.

Hsu, K. C., Fang, T. H., Hsiao, Y. J., & Li, Z. J. (2021). Rapid detection of low concentrations of H2S using CuO-doped ZnO nanofibers. *Journal of Alloys and Compounds*, *852*. Available from https://doi.org/10.1016/j.jallcom.2020.157014, https://www.journals.elsevier.com/journal-of-alloys-and-compounds.

Hu, C., Li, Z., Wang, Y., Gao, J., Dai, K., Zheng, G., Liu, C., Shen, C., Song, H., & Guo, Z. (2017). Comparative assessment of the strain-sensing behaviors of polylactic acid nanocomposites: Reduced graphene oxide or carbon nanotubes. *Journal of Materials Chemistry C*, *5*(9), 2318–2328. Available from https://doi.org/10.1039/C6TC05261D.

Hu, M., Li, Z., Guo, C., Wang, M., He, L., & Zhang, Z. (2019). Hollow core-shell nanostructured MnO2/Fe2O3 embedded within amorphous carbon nanocomposite as sensitive bioplatform for detecting protein tyrosine kinase-7. *Applied Surface Science*, *489*, 13–24. Available from https://doi.org/10.1016/j.apsusc.2019.05.146, http://www.journals.elsevier.com/applied-surface-science/.

Hu, N., Karube, Y., Arai, M., Watanabe, T., Yan, C., Li, Y., Liu, Y., & Fukunaga, H. (2010). Investigation on sensitivity of a polymer/carbon nanotube composite strain sensor. *Carbon*, *48*(3), 680–687. Available from https://doi.org/10.1016/j.carbon.2009.10.012.

Huang, H., Chen, T., Liu, X., & Ma, H. (2014). Ultrasensitive and simultaneous detection of heavy metal ions based on three-dimensional graphene-carbon nanotubes hybrid electrode materials. *Analytica Chimica Acta*, *852*, 45–54. Available from https://doi.org/10.1016/j.aca.2014.09.010, http://www.journals.elsevier.com/analytica-chimica-acta/.

Huang, K. J., Liu, Y. J., Wang, H. B., & Wang, Y. Y. (2014). A sensitive electrochemical DNA biosensor based on silver nanoparticles-polydopamine@graphene composite. *Electrochimica Acta*, *118*, 130–137. Available from https://doi.org/10.1016/j.electacta.2013.12.019.

Hung, C. C., Dillehay, M. E., & Stahl, M. (1987). A heater made from graphite composite material for potential deicingapplication. *Journal of Aircraft*, *24*(10), 725–730. Available from https://doi.org/10.2514/3.45513.

Hwang, J., Jang, J., Hong, K., Kim, K. N., Han, J. H., Shin, K., & Park, C. E. (2011). Poly(3-hexylthiophene) wrapped carbon nanotube/poly(dimethylsiloxane) composites for use in finger-sensing piezoresistive pressure sensors. *Carbon*, *49*(1), 106–110. Available from https://doi.org/10.1016/j.carbon.2010.08.048.

Jahankhani, H., & Galiotis, C. (1991). Interfacial shear stress distribution in model composites, Part 1: A Kevlar 49® fibre in an epoxy matrix. *Journal of Composite Materials*, *25*(5), 609–631. Available from https://doi.org/10.1177/002199839102500508.

Jamshaid, H., & Mishra, R. (2016). A green material from rock: Basalt fiber – A review. *Republic Journal of the Textile Institute*, *107*(7), 923–937. Available from https://doi.org/10.1080/00405000.2015.1071940, http://www.tandf.co.uk/journals/titles/00405000.asp.

Jazzar, A., Alamri, H., Malajati, Y., Mahfouz, R., Bouhrara, M., & Fihri, A. (2021). Recent advances in the synthesis and applications of magnetic polymer nanocomposites. *Journal of Industrial and Engineering Chemistry*, *99*, 1–18. Available from https://doi.org/10.1016/j.jiec.2021.04.011, http://www.sciencedirect.com/science/journal/1226086X.

Jones, D. J., Prasad, S. E., & Wallace, J. B. (1996). Piezoelectric materials and their applications. *Key Engineering Materials*, *122-124*, 71–144. Available from https://doi.org/10.4028/http://www.scientific.net/KEM.122-124.71, https://www.scientific.net/KEM.122-124.71.

Joy Sequeira, G., Lugner, R., Jumar, U., & Brandmeier, T. (2019). A validation sensor based on carbon-fiber-reinforced plastic for early activation of automotive occupant restraint systems. *Journal of Sensors and Sensor Systems*, *8*(1), 19–35. Available from https://doi.org/10.5194/jsss-8-19-2019, http://www.journal-of-sensors-and-sensor-systems.net/home.html.

Kaempgen, M., & Roth, S. (2006). Transparent and flexible carbon nanotube/polyaniline pH sensors. *Journal of Electroanalytical Chemistry*, *586*(1), 72–76. Available from https://doi.org/10.1016/j.jelechem.2005.09.009.

Kalashnyk, N., Faulques, E., Schjødt-Thomsen, J., Jensen, L. R., Rauhe, J. C. M., & Pyrz, R. (2017). Monitoring self-sensing damage of multiple carbon fiber composites using piezoresistivity. *Synthetic Metals*, *224*, 56–62. Available from https://doi.org/10.1016/j.synthmet.2016.12.021, http://www.journals.elsevier.com/synthetic-metals/.

Keiser, G. (2000). *Optical fibre communications*. New York: McGraw-Hill.

Kensche, C. W. (2006). Fatigue of composites for wind turbines. *International Journal of Fatigue*, *28*(10), 1363–1374. Available from https://doi.org/10.1016/j.ijfatigue.2006.02.040.

Kersey, A. D., Davis, M. A., Patrick, H. J., LeBlanc, M., Koo, K. P., Askins, C. G., Putnam, M. A., & Friebele, E. J. (1997). Fiber grating sensors. *Journal of Lightwave Technology*, *15*(8), 1442–1462. Available from https://doi.org/10.1109/50.618377.

Khalid, H. R., Nam, I. W., Choudhry, I., Zheng, L., & Lee, H. K. (2018). Piezoresistive characteristics of CNT fiber-incorporatedGFRP composites prepared with diversified fabrication schemes. *Composite Structures*, *203*, 835–843. Available from https://doi.org/10.1016/j.compstruct.2018.08.003, http://www.elsevier.com/inca/publications/store/4/0/5/9/2/8.

Khandelwal, S., & Rhee, K. Y. (2020). Recent advances in basalt-fiber-reinforced composites: Tailoring the fiber-matrix interface. *Composites Part B: Engineering*, *192*. Available from https://doi.org/10.1016/j.compositesb.2020.108011, https://www.journals.elsevier.com/composites-part-b-engineering.

Khazaee, M., Rezaniakolaie, A., & Rosendahl, L. (2020). A broadband macro-fiber-composite piezoelectric energy harvester for higher energy conversion from practical wideband vibrations. *Nano Energy*, *76*. Available from https://doi.org/10.1016/j.nanoen.2020.104978, http://www.journals.elsevier.com/nano-energy/.

Kim, K. J., & Tadokoro, S. (2007). *Electroactive Polymers for Robotic Applications: Artificial Muscles and Sensors* (pp. 1–281). London, United States: Springer. Available from http://www.springerlink.com/openurl.asp?genre = book&isbn = 978-1-84628-371-0, https://doi.org/10.1007/978-1-84628-372-7.

Kim, Y. S. (2010). Fabrication of carbon black-polymer composite sensors using a position-selective and thickness-controlled electrospray method. *Sensors and Actuators, B: Chemical*, *147*(1), 137–144. Available from https://doi.org/10.1016/j.snb.2010.03.002.

Kim, Y. S., Ha, S. C., Yang, Y., Kim, Y. J., Cho, S. M., Yang, H., & Kim, Y. T. (2005). Portable electronic nose system based on the carbon black-polymer composite sensor array. *Sensors and Actuators, B: Chemical*, *108*(1-2), 285–291. Available from https://doi.org/10.1016/j.snb.2004.11.067.

Kinet, D., Mégret, P., Goossen, K. W., Qiu, L., Heider, D., & Caucheteur, C. (2014). Fiber Bragg grating sensors toward structural health monitoring in composite materials: Challenges and solutions. *Sensors (Switzerland)*, *14*(4), 7394–7419. Available from https://doi.org/10.3390/s140407394, http://www.mdpi.com/1424-8220/14/4/7394/pdf.

Komal, U. K., Lila, M. K., Chaitanya, S., & Singh, I. (2019). *Fabrication of short fiber reinforced polymer composites. Reinforced Polymer Composites: Processing, Characterization and Post Life Cycle Assessment* (pp. 21–38). India: Wiley. Available from https://onlinelibrary.wiley.com/doi/book/10.1002/9783527820979, https://doi.org/10.1002/9783527820979.ch2.

Konka, H. P., Wahab, M. A., & Lian, K. (2012). The effects of embedded piezoelectric fiber composite sensors on the structural integrity of glass-fiber-epoxy composite laminate. *Smart Materials and Structures*, *21* (1). Available from https://doi.org/10.1088/0964-1726/21/1/015016, http://iopscience.iop.org/0964-1726/21/1/015016/pdf/0964-1726_21_1_015016.pdf.

Kornmann, X., Huber, C., Barbezat, M., & Brunner, A.J. (2004). *Active fibre composites: Sensors and actuators for smart composites & structures*. In: 11th European Conference on Composite Materials ECCM-11, Proceedings (CD), Rhodes, Greece, paper B047, pp. 1–9.

Krasteva, N., Besnard, I., Guse, B., Bauer, R. E., Müllen, K., Yasuda, A., & Vossmeyer, T. (2002). Self-assembled gold nanoparticle/dendrimer composite films for vapor sensing applications. *Nano Letters*, *2*(5), 551–555. Available from https://doi.org/10.1021/nl020242s.

Kuang, K. S. C., Quek, S. T., Koh, C. G., Cantwell, W. J., & Scully, P. J. (2009). Plastic optical fibre sensors for structural health monitoring: A review of recent progress. *Journal of Sensors*, *2009*. Available from https://doi.org/10.1155/2009/312053.

Kuilla, T., Bhadra, S., Yao, D., Kim, N. H., Bose, S., & Lee, J. H. (2010). Recent advances in graphene based polymer composites. *Progress in Polymer Science (Oxford)*, *35*(11), 1350–1375. Available from https://doi.org/10.1016/j.progpolymsci.2010.07.005, http://www.sciencedirect.com/science/journal/00796700.

Kumar, B., Feller, J. F., Castro, M., & Lu, J. (2010). Conductive bio-Polymer nano-Composites (CPC): Chitosan-carbon nanotube transducers assembled via spray layer-by-layer for volatile organic compound sensing. *Talanta*, *81*(3), 908–915. Available from https://doi.org/10.1016/j.talanta.2010.01.036.

Kuna, M. (2010). Fracture mechanics of piezoelectric materials – Where are we right now? *Engineering Fracture Mechanics*, *77*(2), 309–326. Available from https://doi.org/10.1016/j.engfracmech.2009.03.016.

Lang, S. B., & Muensit, S. (2006). Review of some lesser-known applications of piezoelectric and pyroelectric polymers. *Applied Physics A: Materials Science and Processing*, *85*(2), 125–134. Available from https://doi.org/10.1007/s00339-006-3688-8.

Leal-Junior, A. G., Diaz, C. A. R., Avellar, L. M., Pontes, M. J., Marques, C., & Frizera, A. (2019). Polymer optical fiber sensors in healthcare applications: A comprehensive review. *Sensors (Switzerland)*, *19*(14). Available from https://doi.org/10.3390/s19143156, https://www.mdpi.com/1424-8220/19/14/3156/pdf.

Lee, C., Park, H., & Lee, J. H. (2020). Recent structure development of poly(vinylidene fluoride)-based piezoelectric nanogenerator for self-powered sensor. *Actuators*, *9*(3), 57. Available from https://doi.org/10.3390/act9030057.

Lee, S., Park, S. K., Choi, E., & Piao, Y. (2016). Voltammetric determination of trace heavy metals using an electrochemically deposited graphene/bismuth nanocomposite film-modified glassy carbon electrode. *Journal of Electroanalytical Chemistry*, *766*, 120–127. Available from https://doi.org/10.1016/j.jelechem.2016.02.003.

Lei, H., Pitt, W. G., McGrath, L. K., & Ho, C. K. (2007). Modeling carbon black/polymer composite sensors. *Sensors and Actuators, B: Chemical*, *125*(2), 396–407. Available from https://doi.org/10.1016/j.snb.2007.02.041.

Li, C., & Chou, T. W. (2008). Modeling of damage sensing in fiber composites using carbon nanotube networks. *Composites Science and Technology*, *68*(15-16), 3373–3379. Available from https://doi.org/10.1016/j.compscitech.2008.09.025.

Li, C., Thostenson, E. T., & Chou, T. W. (2008). Sensors and actuators based on carbon nanotubes and their composites: A review. *Composites Science and Technology*, *68*(6), 1227–1249. Available from https://doi.org/10.1016/j.compscitech.2008.01.006.

Li, W., He, D., Dang, Z., & Bai, J. (2014). In situ damage sensing in the glass fabric reinforced epoxy composites containing CNT-Al2O3 hybrids. *Composites Science and Technology*, *99*, 8–14. Available from https://doi.org/10.1016/j.compscitech.2014.05.005.

Li, W., Kong, Q., Ho, S. C. M., Lim, I., Mo, Y. L., & Song, G. (2016). Feasibility study of using smart aggregates as embedded acoustic emission sensors for health monitoring of concrete structures. *Smart Materials and Structures*, *25*(11). Available from https://doi.org/10.1088/0964-1726/25/11/115031, http://iopscience.iop.org/article/10.1088/0964-1726/25/11/115031/pdf.

Li, W., Xu, F., Sun, L., Liu, W., & Qiu, Y. (2016). A novel flexible humidity switch material based on multiwalled carbon nanotube/polyvinyl alcohol composite yarn. *Sensors and Actuators, B: Chemical*, *230*, 528–535. Available from https://doi.org/10.1016/j.snb.2016.02.108.

Li, Y., Li, R., Lu, L., & Huang, X. (2015). Experimental study of damage characteristics of carbon woven fabric/epoxy laminates subjected to lightning strike. *Composites Part A: Applied Science and Manufacturing*, *79*, 164–175. Available from https://doi.org/10.1016/j.compositesa.2015.09.019.

Lim, Y. W., Jin, J., & Bae, B. S. (2020). Optically transparent multiscale composite films for flexible and wearable electronics. *Advanced Materials*, *32*(35). Available from https://doi.org/10.1002/adma.201907143, http://onlinelibrary.wiley.com/journal/10.1002/(ISSN)1521-4095.

Lin, J., He, C., Zhao, Y., & Zhang, S. (2009). One-step synthesis of silver nanoparticles/carbon nanotubes/chitosan film and its application in glucose biosensor. *Sensors and Actuators, B: Chemical*, *137*(2), 768–773. Available from https://doi.org/10.1016/j.snb.2009.01.033.

Liu, H., Gao, J., Huang, W., Dai, K., Zheng, G., Liu, C., Shen, C., Yan, X., Guo, J., & Guo, Z. (2016). Electrically conductive strain sensing polyurethane nanocomposites with synergistic carbon nanotubes and graphene bifillers. *Nanoscale*, *8*(26), 12977–12989. Available from https://doi.org/10.1039/C6NR02216B.

Liu, Y., Dong, X., & Chen, P. (2012). Biological and chemical sensors based on graphene materials. *Chemical Society Reviews*, *41*(6), 2283–2307. Available from https://doi.org/10.1039/C1CS15270J, http://doi.org/10.1039/C1CS15270J.

Llobet, E. (2013). Gas sensors using carbon nanomaterials: A review. *Sensors and Actuators, B: Chemical*, *179*, 32–45. Available from https://doi.org/10.1016/j.snb.2012.11.014.

Lonergan, M. C., Severin, E. J., Doleman, B. J., Beaber, S. A., Grubbs, R. H., & Lewis, N. S. (1996). Array-based vapor sensing using chemically sensitive, carbon black-Polymer resistors. *Chemistry of Materials*, *8*(9), 2298–2312. Available from https://doi.org/10.1021/cm960036j, http://pubs.acs.org/journal/cmatex.

Luo, X., Du, W., Li, H., Li, P., Ma, C., Xu, S., & Zhang, J. (2017). Occupant injury response prediction prior to crash based on pre-crash systems. *SAE Technical Papers*. Available from http://papers.sae.org/2017-10.4271/2017-01-1471.

Lux, F. (1993). Models proposed to explain the electrical conductivity of mixtures made of conductive and insulating materials. *Journal of Materials Science*, *28*(2), 285–301. Available from https://doi.org/10.1007/BF00357799.

Ma, Q., Hao, B., & Ma, P. C. (2020). Modulating the sensitivity of a flexible sensor using conductive glass fibre with a controlled structure profile. *Composites Communications*, *20*. Available from https://doi.org/10.1016/j.coco.2020.100367, http://www.journals.elsevier.com/composites-communications.

Mall, S., & Coleman, J. M. (1998). Monotonic and fatigue loading behavior of quasi-isotropic graphite/epoxy laminate embedded with piezoelectric sensor. *Smart Materials and Structures*, *7*(6), 822–832. Available from https://doi.org/10.1088/0964-1726/7/6/010.

Mallakpour, S., & Behranvand, V. (2017). *Green hybrid nanocomposites from metal oxides, poly(vinyl alcohol) and poly(vinyl pyrrolidone): Structure and chemistry. Hybrid Polymer Composite Materials: Structure and Chemistry* (pp. 263–289). Iran: Elsevier Inc.. Available from http://www.sciencedirect.com/science/book/9780081007914, https://doi.org/10.1016/B978-0-08-100791-4.00010-0.

Mangu, R., Rajaputra, S., & Singh, V. P. (2011). MWCNT-polymer composites as highly sensitive and selective room temperature gas sensors. *Nanotechnology*, *22*(21). Available from https://doi.org/10.1088/0957-4484/22/21/215502, http://iopscience.iop.org/0957-4484/22/21/215502/pdf/0957-4484_22_21_215502.pdf.

Mannsfeld, S. C. B., Tee, B. C. K., Stoltenberg, R. M., Chen, C. V. H. H., Barman, S., Muir, B. V. O., Sokolov, A. N., Reese, C., & Bao, Z. (2010). Highly sensitive flexible pressure sensors with microstructured rubber dielectric layers. *Nature Materials*, *9*(10), 859–864. Available from https://doi.org/10.1038/nmat2834, http://www.nature.com/nmat/.

Martins, P., Lopes, A. C., & Lanceros-Mendez, S. (2014). Electroactive phases of poly(vinylidene fluoride): Determination, processing and applications. *Progress in Polymer Science*, *39*(4), 683–706. Available from https://doi.org/10.1016/j.progpolymsci.2013.07.006, http://www.sciencedirect.com/science/journal/00796700.

Mei, J., Leung, N. L. C., Kwok, R. T. K., Lam, J. W. Y., & Tang, B. Z. (2015). Aggregation-induced emission: Together we shine, united we soar!. *Chemical Reviews*, *115*(21), 11718–11940. Available from https://doi.org/10.1021/acs.chemrev.5b00263, http://pubs.acs.org/journal/chreay.

Minchenkov, K., Vedernikov, A., Safonov, A., & Akhatov, I. (2021). Thermoplastic pultrusion: A review. *Russian Federation Polymers*, *13*(2), 1–36. Available from https://doi.org/10.3390/polym13020180, https://www.mdpi.com/2073-4360/13/2/180/pdf.

Mittal, G., & Rhee, K. Y. (2018). Chemical vapor deposition-based grafting of CNTs onto basalt fabric and their reinforcement in epoxy-based composites. *Composites Science and Technology*, *165*, 84–94. Available from https://doi.org/10.1016/j.compscitech.2018.06.018, http://www.journals.elsevier.com/composites-science-and-technology/.

Mohammadpour-Haratbar, A., Mohammadpour-Haratbar, S., Zare, Y., Rhee, K. Y., & Park, S. J. (2022). A review on non-enzymatic electrochemical biosensors of glucose using carbon nanofiber nanocomposites. *Biosensors*, *12*(11). Available from https://doi.org/10.3390/bios12111004, http://www.mdpi.com/journal/biosensors/.

MohdIsa, W. H., Hunt, A., & HosseinNia, S. H. (2019). Sensing and self-sensing actuation methods for ionic polymer-metal composite (IPMC): A review. *Sensors (Switzerland)*, *19*(18). Available from https://doi.org/10.3390/s19183967, https://www.mdpi.com/1424-8220/19/18/3967/pdf.

Mokhtari, F., Cheng, Z., Raad, R., Xi, J., & Foroughi, J. (2020). Piezofibers to smart textiles: A review on recent advances and future outlook for wearable technology. *Journal of Materials Chemistry A*, *8*(19), 9496–9522. Available from https://doi.org/10.1039/D0TA00227E, http://doi.org/10.1039/D0TA00227E.

Montazerian, H., Dalili, A., Milani, A. S., & Hoorfar, M. (2019). Piezoresistive sensing in chopped carbon fiber embedded PDMS yarns. *Composites Part B: Engineering*, *164*, 648–658. Available from https://doi.org/10.1016/j.compositesb.2019.01.090.

Morales-Narváez, E., Baptista-Pires, L., Zamora-Gálvez, A., & Merkoçi, A. (2017). Graphene-based biosensors: Going simple. *Advanced Materials*, *29*(7). Available from https://doi.org/10.1002/adma.201604905, http://onlinelibrary.wiley.com/journal/10.1002/(ISSN)1521-4095.

Morampudi, P., Namala, K. K., Gajjela, Y. K., Barath, M., & Prudhvi, G. (2020). Review on glass fiber reinforced polymer composites. *Materials Today: Proceedings*, *43*, 314–319. Available from https://doi.org/10.1016/j.matpr.2020.11.669, https://www.sciencedirect.com/journal/materials-today-proceedings.

Mutuma, B. K., Rodrigues, R., Ranganathan, K., Matsoso, B., Wamwangi, D., Hümmelgen, I. A., & Coville, N. J. (2017). Hollow carbon spheres and a hollow carbon sphere/polyvinylpyrrolidone composite as ammonia sensors. *Journal of Materials Chemistry A*, *5*(6), 2539–2549. Available from https://doi.org/10.1039/C6TA09424D.

Nag, A., Mukhopadhyay, S. C., & Kosel, J. (2017). Wearable flexible sensors: A review. *IEEE Sensors Journal*, *17*(13), 3949–3960. Available from https://doi.org/10.1109/JSEN.2017.2705700, http://ieeexplore.ieee.org/xpl/RecentIssue.jsp?punumber = 7361.

Nag, A., Nuthalapati, S., & Mukhopadhyay, S. C. (2022). Carbon fiber/polymer-based composites for wearable sensors: A review. *IEEE Sensors Journal*, *22*(11), 10235–10245. Available from https://doi.org/10.1109/JSEN.2022.3170313, http://ieeexplore.ieee.org/xpl/RecentIssue.jsp?punumber = 7361.

Nambiar, S., & Yeow, J. T. W. (2011). Conductive polymer-based sensors for biomedical applications. *Biosensors and Bioelectronics*, *26*(5), 1825–1832. Available from https://doi.org/10.1016/j.bios.2010.09.046.

Narita, F., & Fox, M. (2018). A review on piezoelectric, magnetostrictive, and magnetoelectric materials and device technologies for energy harvesting applications. *Advanced Engineering Materials*, *20*(5). Available from https://doi.org/10.1002/adem.201700743, http://www3.interscience.wiley.com/journal/67500980/home.

Narita, F., Nagaoka, H., & Wang, Z. (2019). Fabrication and impact output voltage characteristics of carbon fiber reinforced polymer composites with lead-free piezoelectric nano-particles. *Materials Letters*, *236*, 487–490. Available from https://doi.org/10.1016/j.matlet.2018.10.174, http://www.journals.elsevier.com/materials-letters/.

Nelson, L. J. (2002). Smart piezoelectric fibre composites. *Materials Science and Technology*, *18*(11), 1245–1256. Available from https://doi.org/10.1179/026708302225007448.

Newnham, R. E., Skinner, D. P., & Cross, L. E. (1978). Connectivity and piezoelectric-pyroelectric composites. *Materials Research Bulletin*, *13*(5), 525–536. Available from https://doi.org/10.1016/0025-5408(78)90161-7, https://www.sciencedirect.com/science/article/pii/0025540878901617.

Novikov, S., Lebedeva, N., Satrapinski, A., Walden, J., Davydov, V., & Lebedev, A. (2016). Graphene based sensor for environmental monitoring of NO2. *Sensors and Actuators, B: Chemical*, *236*, 1054–1060. Available from https://doi.org/10.1016/j.snb.2016.05.114.

Obitayo, W., & Liu, T. (2012). A review: Carbon nanotube-based piezoresistive strain sensors. *Journal of Sensors*, *2012*. Available from https://doi.org/10.1155/2012/652438.

Oh, J., Kim, D. Y., Kim, H., Hur, O. N., & Park, S. H. (2022). Comparative study of carbon nanotube composites as capacitive and piezoresistive pressure sensors under varying conditions. *Materials*, *15*(21). Available from https://doi.org/10.3390/ma15217637, http://www.mdpi.com/journal/materials.

Othonos, A., Kalli, K., Pureur, D., & Mugnier, A. (2006). Fibre Bragg gratings. *Springer Series in Optical Sciences*, *123*, 189–269. Available from https://doi.org/10.1007/3-540-31770-8_6.

Oueiny, C., Berlioz, S., & Perrin, F. X. (2014). Carbon nanotube-polyaniline composites. *Progress in Polymer Science*, *39*(4), 707–748. Available from https://doi.org/10.1016/j.progpolymsci.2013.08.009, http://www.sciencedirect.com/science/journal/00796700.

Pandey, A., & Arockiarajan, A. (2017). Fatigue study on the actuation performance of macro fiber composite (MFC): Theoretical and experimental approach. *Smart Materials and Structures*, *26*(3). Available from https://doi.org/10.1088/1361-665X/aa59e9, http://iopscience.iop.org/article/10.1088/1361-665X/aa59e9/pdf.

Park, J., Lee, Y., Hong, J., Ha, M., Jung, Y. D., Lim, H., Kim, S. Y., & Ko, H. (2014). Giant tunneling piezoresistance of composite elastomers with interlocked microdome arrays for ultrasensitive and multimodal electronic skins. *ACS Nano*, *8*(5), 4689–4697. Available from https://doi.org/10.1021/nn500441k.

Park, M., Kim, H., & Youngblood, J. P. (2008). Strain-dependent electrical resistance of multi-walled carbon nanotube/polymer composite. *Nanotechnology*, *19*(5), 055705. Available from https://doi.org/10.1088/0957-4484/19/05/055705, https://doi.org/10.1088/0957-4484/19/05/055705.

Park, S. (2019). Enhancement of hydrogen sensing response of ZnO nanowires for the decoration of WO3 nanoparticles. *Materials Letters*, *234*, 315–318. Available from https://doi.org/10.1016/j.matlet.2018.09.129, http://www.journals.elsevier.com/materials-letters/.

Park, S. J. (2015). *Springer Series in Materials Science*, . *Carbon fibers* (210). Netherlands: Springer. Available from https://doi.org/10.1007/978-94-017-9478-7.

Park, S. J., Kim, J., Chu, M., & Khine, M. (2016). Highly flexible wrinkled carbon nanotube thin film strain sensor to monitor human movement. *Advanced Materials Technologies*, *1*(5). Available from https://doi.org/10.1002/admt.201600053, http://onlinelibrary.wiley.com/journal/10.1002/(ISSN)2365-709X.

Parthenios, J., Katerelos, D. G., Psarras, G. C., & Galiotis, C. (2002). Aramid fibers: A multifunctional sensor for monitoring stress/strain fields and damage development in composite materials. *Engineering Fracture Mechanics*, *69*(9), 1067–1087. Available from https://doi.org/10.1016/S0013-7944(01)00123-0, https://www.sciencedirect.com/science/article/pii/S0013794401001230.

Patel, V. K., Kant, R., Chauhan, P. S., & Bhattacharya, S. (2022). *Introduction to Applications of Polymers and Polymer Composites* (pp. 1–6). AIP Publishing. Available from https://doi.org/10.1063/9780735424555_001.

Pavlovski, D., Mislavsky, B., & Antonov, A. (2007). CNG cylinder manufacturers test basalt fibre. *Reinforced Plastics*, *51*(4), 36–39. Available from https://doi.org/10.1016/S0034-3617(07)70152-2, https://www.sciencedirect.com/science/article/pii/S0034361707701522.

Phukon, P., Radhapyari, K., Konwar, B. K., & Khan, R. (2014). Natural polyhydroxyalkanoate-gold nanocomposite based biosensor for detection of antimalarial drug artemisinin. *Materials Science and Engineering C*, *37*(1), 314–320. Available from https://doi.org/10.1016/j.msec.2014.01.019.

Pingarrón, J. M., Yáñez-Sedeño, P., & González-Cortés, A. (2008). Gold nanoparticle-based electrochemical biosensors. *Electrochimica Acta*, *53*(19), 5848–5866. Available from https://doi.org/10.1016/j.electacta.2008.03.005.

Prabakar, K., & Rao, S. P. M. (2005). Acoustic emission during phase transition in soft PZT ceramics under an applied electric field. *Ferroelectrics, Letters Section*, *32*(5–6), 99–110. Available from https://doi.org/10.1080/07315170500416579.

Prasse, T., Michel, F., Mook, G., Schulte, K., & Bauhofer, W. (2001). A comparative investigation of electrical resistance and acoustic emission during cyclic loading of CFRP laminates. *Composites Science and Technology*, *61*(6), 831–835. Available from https://doi.org/10.1016/S0266-3538(00)00179-2, http://www.journals.elsevier.com/composites-science-and-technology/.

Pugal, D., Jung, K., Aabloo, A., & Kim, K. J. (2010). Ionic polymer-metal composite mechanoelectrical transduction: Review and perspectives. *Polymer International*, *59*(3), 279–289. Available from https://doi.org/10.1002/pi.2759, http://www3.interscience.wiley.com/cgi-bin/fulltext/123238505/PDFSTART.

Punning, A., Kruusmaa, M., & Aabloo, A. (2007). A self-sensing ion conducting polymer metal composite (IPMC) actuator. *Sensors and Actuators, A: Physical*, *136*(2), 656–664. Available from https://doi.org/10.1016/j.sna.2006.12.008.

Qing, X., Li, W., Wang, Y., & Sun, H. (2019). Piezoelectric transducer-based structural health monitoring for aircraft applications. *Sensors (Switzerland)*, *19*(3). Available from https://doi.org/10.3390/s19030545, https://www.mdpi.com/1424-8220/19/3/545/pdf.

Qiu, J. H., Ji, H. L., Zhu, K. J., & Park, M. J. (2010). Response of metal core piezoelectric fibers to unsteady airflows. *Modern Physics Letters B*, *24*(13), 1453–1456. Available from https://doi.org/10.1142/S0217984910023852.

Qu, Z., Fu, Y., Yu, B., Deng, P., Xing, L., & Xue, X. (2016). High and fast H2S response of NiO/ZnO nanowire nanogenerator as a self-powered gas sensor. *Sensors and Actuators B: Chemical*, *222*, 78–86. Available from https://doi.org/10.1016/j.snb.2015.08.058, https://www.sciencedirect.com/science/article/pii/S0925400515302367.

Rahaman, M. S. A., Ismail, A. F., & Mustafa, A. (2007). A review of heat treatment on polyacrylonitrile fiber. *Polymer Degradation and Stability*, *92*(8), 1421–1432. Available from https://doi.org/10.1016/j.polymdegradstab.2007.03.023.

Raj, M., Gupta, P., Goyal, R. N., & Shim, Y. B. (2017). Graphene/conducting polymer nano-composite loaded screen printed carbon sensor for simultaneous determination of dopamine and 5-hydroxytryptamine. *Sensors and Actuators, B: Chemical*, *239*, 993–1002. Available from https://doi.org/10.1016/j.snb.2016.08.083.

Rajak, D. K., Pagar, D. D., Kumar, R., & Pruncu, C. I. (2019). Recent progress of reinforcement materials: A comprehensive overview of composite materials. *Journal of Materials Research and Technology*, *8*(6), 6354–6374. Available from https://doi.org/10.1016/j.jmrt.2019.09.068, http://www.elsevier.com/journals/journal-of-materials-research-and-technology/2238-7854.

Rajak, D. K., Pagar, D. D., Menezes, P. L., & Linul, E. (2019). Fiber-reinforced polymer composites: Manufacturing, properties, and applications. *Polymers*, *11*(10). Available from https://doi.org/10.3390/polym11101667, https://res.mdpi.com/d_attachment/polymers/polymers-11-01667/article_deploy/polymers-11-01667.pdf.

Rajesh., Ahuja, T., & Kumar, D. (2009). Recent progress in the development of nano-structured conducting polymers/nanocomposites for sensor applications. *Sensors and Actuators, B: Chemical*, *136*(1), 275–286. Available from https://doi.org/10.1016/j.snb.2008.09.014.

Rasheed, P. A., & Sandhyarani, N. (2017). Carbon nanostructures as immobilization platform for DNA: A review on current progress in electrochemical DNA sensors. *Biosensors and Bioelectronics*, *97*, 226–237. Available from https://doi.org/10.1016/j.bios.2017.06.001, http://www.elsevier.com/locate/bios.

Ray, M. C., & Pradhan, A. K. (2007). On the use of vertically reinforced 1-3 piezoelectric composites for hybrid damping of laminated composite plates. *Mechanics of Advanced Materials and Structures*, *14*(4), 245–261. Available from https://doi.org/10.1080/15376490600795683.

Raya, I., Kzar, H. H., Mahmoud, Z. H., Al Ayub Ahmed, A., Ibatova, A. Z., & Kianfar, E. (2022). A review of gas sensors based on carbon nanomaterial. *Carbon Letters*, *32*(2), 339–364. Available from https://doi.org/10.1007/s42823-021-00276-9, https://www.springer.com/journal/42823.

Ricci, F., Monaco, E., Boffa, N. D., Maio, L., & Memmolo, V. (2022). Guided waves for structural health monitoring in composites: A review and implementation strategies. *Progress in Aerospace Sciences*, *129*. Available from https://doi.org/10.1016/j.paerosci.2021.100790, https://www.journals.elsevier.com/progress-in-aerospace-sciences.

Rocha, H., Semprimoschnig, C., & Nunes, J. P. (2021). Sensors for process and structural health monitoring of aerospace composites: A review. *Engineering Structures*, *237*. Available from https://doi.org/10.1016/j.engstruct.2021.112231, http://www.journals.elsevier.com/engineering-structures/.

Rocha, R. T., Tusset, A. M., Ribeiro, M. A., Lenz, W. B., Haura Junior, R., Jarzebowska, E., & Balthazar, J. M. (2020). On the positioning of a piezoelectric material in the energy harvesting from a nonideally excited portal frame. *Journal of Computational and Nonlinear Dynamics*, *15*(12). Available from https://doi.org/10.1115/1.4048024, https://asmedigitalcollection.asme.org/computationalnonlinear.

Ruan, C., Shi, W., Jiang, H., Sun, Y., Liu, X., Zhang, X., Sun, Z., Dai, L., & Ge, D. (2013). One-pot preparation of glucose biosensor based on polydopamine-graphene composite film modified enzyme electrode. *Sensors and Actuators, B: Chemical*, *177*, 826–832. Available from https://doi.org/10.1016/j.snb.2012.12.010.

Rubio, I., Díaz-álvarez, A., Bernier, R., Rusinek, A., Loya, J. A., Miguelez, M. H., & Rodríguez-Millán, M. (2020). Postmortem analysis using different sensors and technologies on aramid composites samples after ballistic impact. *Sensors (Switzerland)*, *20*(10). Available from https://doi.org/10.3390/s20102853, https://www.mdpi.com/1424-8220/20/10/2853/pdf.

Rudd, C. D. (2001). *Resin transfer molding and structural reaction injection molding* (pp. 492–500). ASM International. Available from https://doi.org/10.31399/asm.hb.v21.a0003413.

Saha, K., Agasti, S. S., Kim, C., Li, X., & Rotello, V. M. (2012). Gold nanoparticles in chemical and biological sensing. *Chemical Reviews*, *112*(5), 2739–2779. Available from https://doi.org/10.1021/cr2001178.

Sarangi, S. K., & Ray, M. C. (2011). Active damping of geometrically nonlinear vibrations of doubly curved laminated composite shells. *Composite Structures*, *93*(12), 3216–3228. Available from https://doi.org/10.1016/j.compstruct.2011.06.005, http://www.elsevier.com/inca/publications/store/4/0/5/9/2/8.

Sarasini, F., Tirillò, J., Lilli, M., Bracciale, M. P., Vullum, P. E., Berto, F., De Bellis, G., Tamburrano, A., Cavoto, G., Pandolfi, F., & Rago, I. (2022). Highly aligned growth of carbon nanotube forests with in-situ catalyst generation: A route to multifunctional basalt fibres. *Composites Part B: Engineering*, *243*. Available from https://doi.org/10.1016/j.compositesb.2022.110136, https://www.journals.elsevier.com/composites-part-b-engineering.

Sawai, D., Fujii, Y., & Kanamoto, T. (2006). Development of oriented morphology and tensile properties upon superdawing of solution-spun fibers of ultra-high molecular weight poly(acrylonitrile). *Polymer*, *47*(12), 4445–4453. Available from https://doi.org/10.1016/j.polymer.2006.03.067, http://www.journals.elsevier.com/polymer/.

Scheffler, S., & Poulin, P. (2022). Piezoelectric fibers: Processing and challenges. *ACS Applied Materials and Interfaces*, *14*(15), 16961–16982. Available from https://doi.org/10.1021/acsami.1c24611, http://pubs.acs.org/journal/aamick.

Schulte, K., & Baron, C. (1989). Load and failure analyses of CFRP laminates by means of electrical resistivity measurements. *Composites Science and Technology*, *36*(1), 63–76. Available from https://doi.org/10.1016/0266-3538(89)90016-X.

Serway, R. A., Jewett, J. W. (2001). Physics for scientists and engineers. Cengage Learning.

Shahinpoor, M., & Kim, K. J. (2001). Ionic polymer-metal composites: I. Fundamentals. *Smart Materials and Structures*, *10*(4), 819–833. Available from https://doi.org/10.1088/0964-1726/10/4/327.

Shirshova, N., Qian, H., Houllé, M., Steinke, J. H. G., Kucernak, A. R. J., Fontana, Q. P. V., Greenhalgh, E. S., Bismarck, A., & Shaffer, M. S. P. (2014). Multifunctional structural energy storage composite supercapacitors. *Faraday Discussions, 172*, 81–103. Available from https://doi.org/10.1039/c4fd00055b, http://pubs.rsc.org/en/journals/journal/fd.

Shirshova, N., Qian, H., Shaffer, M. S. P., Steinke, J. H. G., Greenhalgh, E. S., Curtis, P. T., Kucernak, A., & Bismarck, A. (2013). Structural composite supercapacitors. *Composites Part A: Applied Science and Manufacturing, 46*, 96–107. Available from https://doi.org/10.1016/j.compositesa.2012.10.007.

Shivashankar, P., & Gopalakrishnan, S. (2020). Review on the use of piezoelectric materials for active vibration, noise, and flow control. *Smart Materials and Structures, 29*(5). Available from https://doi.org/10.1088/1361-665X/ab7541, https://iopscience.iop.org/article/10.1088/1361-665X/ab7541.

Shukla, P., & Saxena, P. (2021). Polymer nanocomposites in sensor applications: A review on present trends and future scope. *Chinese Journal of Polymer Science (English Edition), 39*(6), 665–691. Available from https://doi.org/10.1007/s10118-021-2553-8, http://www.springerlink.com/content/0256-7679.

Sim, J., Park, C., & Moon, D. Y. (2005). Characteristics of basalt fiber as a strengthening material for concrete structures. *Composites Part B: Engineering, 36*(6-7), 504–512. Available from https://doi.org/10.1016/j.compositesb.2005.02.002.

Singh, E., Kumar, U., Srivastava, R., & Yadav, B. C. (2018). Carbon nanotubes based thin films as optoelectronic moisture sensor. *Advanced Science, Engineering and Medicine, 10*(7), 785–787. Available from https://doi.org/10.1166/asem.2018.2240.

Singh, P., & Shukla, S. K. (2020). Advances in polyaniline-based nanocomposites. *Journal of Materials Science, 55*(4), 1331–1365. Available from https://doi.org/10.1007/s10853-019-04141-z, http://www.springer.com/journal/10853.

Skubal, L. R., Meshkov, N. K., & Vogt, M. C. (2002). Detection and identification of gaseous organics using a TiO2 sensor. *Semiconductor photochemistry 1 First International Conference On Semiconductor Photochemistry, 148*(1), 103–108. Available from https://doi.org/10.1016/S1010-6030(02)00079-5, https://www.sciencedirect.com/science/article/pii/S1010603002000795.

Sodano, H. A., Park, G., & Inman, D. J. (2004). An investigation into the performance of macro-fiber composites for sensing and structural vibration applications. *Mechanical Systems and Signal Processing, 18*(3), 683–697. Available from https://doi.org/10.1016/S0888-3270(03)00081-5, http://www.elsevier.com/inca/publications/store/6/2/2/9/1/2/index.htt.

Song, G., Olmi, C., & Gu, H. (2007). An overheight vehicle-bridge collision monitoring system using piezoelectric transducers. *Smart Materials and Structures, 16*(2), 462–468. Available from https://doi.org/10.1088/0964-1726/16/2/026.

Song, X., Li, L., Chen, X., Xu, Q., Song, B., Pan, Z., Liu, Y., Juan, F., Xu, F., & Cao, B. (2019). Enhanced triethylamine sensing performance of A-Fe2O3 nanoparticle/ZnO nanorod heterostructures. *Sensors and Actuators, B: Chemical, 298*. Available from https://doi.org/10.1016/j.snb.2019.126917, https://www.journals.elsevier.com/sensors-and-actuators-b-chemical.

Sotzing, G. A., Phend, J. N., Grubbs, R. H., & Lewis, N. S. (2000). Highly sensitive detection and discrimination of biogenic amines utilizing arrays of polyaniline/carbon black composite vapor detectors. *Chemistry of Materials, 12*(3), 593–595. Available from https://doi.org/10.1021/cm990694e.

Stalbaum, T., Shen, Q., & Kim, K.J. (2017). A model framework for actuation and sensing of ionic polymer-metal composites: Prospective on frequency and shear response through simulation tools. In *Proceedings of SPIE – The International Society for Optical Engineering*, 10163. Available from http://spie.org/x1848.xml. https://doi.org/10.1117/12.2258403.

Stankovich, S., Dikin, D. A., Dommett, G. H. B., Kohlhaas, K. M., Zimney, E. J., Stach, E. A., Piner, R. D., Nguyen, S. B. T., & Ruoff, R. S. (2006). Graphene-based composite materials. *Nature, 442*(7100), 282–286. Available from https://doi.org/10.1038/nature04969.

Stuart, M. A. C., Huck, W. T. S., Genzer, J., Müller, M., Ober, C., Stamm, M., Sukhorukov, G. B., Szleifer, I., Tsukruk, V. V., Urban, M., Winnik, F., Zauscher, S., Luzinov, I., & Minko, S. (2010). Emerging applications of stimuli-responsive polymer materials. *Nature Materials*, *9*(2), 101–113. Available from https://doi.org/10.1038/nmat2614, http://www.nature.com/nmat/.

Sun, T., Zhao, H., Zhang, J., Chen, Y., Gao, J., Liu, L., Niu, S., Han, Z., Ren, L., & Lin, Q. (2022). Degradable bioinspired hypersensitive strain sensor with high mechanical strength using a basalt fiber as a reinforced layer. *ACS Applied Materials and Interfaces*, *14*(37), 42723–42733. Available from https://doi.org/10.1021/acsami.2c12479, http://pubs.acs.org/journal/aamick.

Takahashi, K., Yaginuma, K., Goto, T., Yokozeki, T., Okada, T., & Takahashi, T. (2022). Electrically conductive carbon fiber reinforced plastics induced by uneven distribution of polyaniline composite micron-sized particles in thermosetting matrix. *Composites Science and Technology*, *228*. Available from https://doi.org/10.1016/j.compscitech.2022.109642, http://www.journals.elsevier.com/composites-science-and-technology/.

Takeda, S., Okabe, Y., & Takeda, N. (2002). Delamination detection in CFRP laminates with embedded small-diameter fiber Bragg grating sensors. *Composites Part A: Applied Science and Manufacturing*, *33*(7), 971–980. Available from https://doi.org/10.1016/S1359-835X(02)00036-2, https://www.sciencedirect.com/science/article/pii/S1359835X02000362.

Tam, P. D., & Hieu, N. V. (2011). Conducting polymer film-based immunosensors using carbon nanotube/antibodies doped polypyrrole. *Applied Surface Science*, *257*(23), 9817–9824. Available from https://doi.org/10.1016/j.apsusc.2011.06.028, http://www.journals.elsevier.com/applied-surface-science/.

Tamayo, L., Palza, H., Bejarano, J., & Zapata, P. A. (2018). *Polymer composites with metal nanoparticles: Synthesis, properties, and applications. synthesis, properties, and applications. Polymer Composites with Functionalized Nanoparticles: Synthesis, Properties, and Applications* (pp. 249–286). Chile: Elsevier. Available from https://www.sciencedirect.com/book/9780128140642, https://doi.org/10.1016/B978-0-12-814064-2.00008-1.

Tan, L., Zhou, K. G., Zhang, Y. H., Wang, H. X., Wang, X. D., Guo, Y. F., & Zhang, H. L. (2010). Nanomolar detection of dopamine in the presence of ascorbic acid at β-cyclodextrin/graphene nanocomposite platform. *Electrochemistry Communications*, *12*(4), 557–560. Available from https://doi.org/10.1016/j.elecom.2010.01.042.

Tanasi, P., Asensio, M., Herrero, M., Núñez, K., Cañibano, E., & Merino, J. C. (2020). Control of branches distribution in linear PE copolymers using fibrillar nanoclay as support of catalyst system. *Polymer*, *202*. Available from https://doi.org/10.1016/j.polymer.2020.122707, http://www.journals.elsevier.com/polymer/.

Tang, Y., Wang, Z., & Song, M. (2016). Self-sensing and strengthening effects of reinforced concrete structures with near-surfaced mounted smart basalt fibre-reinforced polymer bars. *Advances in Mechanical Engineering*, *8*(10), 1–19. Available from https://doi.org/10.1177/1687814016673499, http://ade.sagepub.com/.

Tarhini, A. A., & Tehrani-Bagha, A. R. (2019). Graphene-based polymer composite films with enhanced mechanical properties and ultra-high in-plane thermal conductivity. *Composites Science and Technology*, *184*. Available from https://doi.org/10.1016/j.compscitech.2019.107797, http://www.journals.elsevier.com/composites-science-and-technology/.

Thostenson, E. T., & Chou, T. W. (2006). Carbon nanotube networks: Sensing of distributed strain and damage for life prediction and self healing. *Advanced Materials*, *18*(21), 2837–2841. Available from https://doi.org/10.1002/adma.200600977.

Thostenson, E. T., & Chou, T. W. (2008). Real-time in situ sensing of damage evolution in advanced fiber composites using carbon nanotube networks. *Nanotechnology*, *19*(21). Available from https://doi.org/10.1088/0957-4484/19/21/215713.

Todoroki, A. (2004). Electric resistance change method for cure/strain/damage monitoring of CFRP laminates. *Key Engineering Materials*, *270-273*, 1812–1820. Available from https://doi.org/10.4028/http://www.scientific.net/KEM.270-273.1812, https://www.scientific.net/KEM.270-273.1812.

Todoroki, A., Tanaka, M., & Shimamura, Y. (2005). Electrical resistance change method for monitoring delaminations of CFRP laminates: Effect of spacing between electrodes. *Composites Science and Technology*, *65*(1), 37–46. Available from https://doi.org/10.1016/j.compscitech.2004.05.018.

Tulliani, J. M., Inserra, B., & Ziegler, D. (2019). Carbon-based materials for humidity sensing: A short review. *Micromachines*, *10*(4). Available from https://doi.org/10.3390/mi10040232, https://res.mdpi.com/micromachines/micromachines-10-00232/article_deploy/micromachines-10-00232-v2.pdf?filename = &attachment = 1.

Ullah, N., Mansha, M., Khan, I., & Qurashi, A. (2018). Nanomaterial-based optical chemical sensors for the detection of heavy metals in water: Recent advances and challenges. *TrAC – Trends in Analytical Chemistry*, *100*, 155–166. Available from https://doi.org/10.1016/j.trac.2018.01.002, http://www.elsevier.com/locate/trac.

Vavouliotis, A., Paipetis, A., & Kostopoulos, V. (2011). On the fatigue life prediction of CFRP laminates using the electrical resistance change method. *Composites Science and Technology*, *71*(5), 630–642. Available from https://doi.org/10.1016/j.compscitech.2011.01.003.

Vikrant, K., Bhardwaj, N., Bhardwaj, S. K., Kim, K. H., & Deep, A. (2019). Nanomaterials as efficient platforms for sensing DNA. *Biomaterials*, *214*. Available from https://doi.org/10.1016/j.biomaterials.2019.05.026, http://www.journals.elsevier.com/biomaterials/.

Vipperman, J. (1999). *Novel autonomous structural health monitoring using piezoelectrics 40th structures, structural dynamics, and materials conference and exhibit*. American Institute of Aeronautics and Astronautics, American Institute of Aeronautics and Astronautics. Available from https://doi.org/10.2514/6.1999-1507, https://doi.org/10.2514/6.1999-1507.

Vlattas, C., & Galiotis, C. (1994). Deformation behaviour of liquid crystalpolymer fibres: 1. Converting spectroscopic data into mechanical stress-strain curves in tension and compression. *Polymer*, *35*(11), 2335–2347. Available from https://doi.org/10.1016/0032-3861(94)90770-6, https://www.sciencedirect.com/science/article/pii/0032386194907706.

Wanekaya, A. K. (2011). Applications of nanoscale carbon-based materials in heavy metal sensing and detection. *Analyst*, *136*(21), 4383–4391. Available from https://doi.org/10.1039/C1AN15574A, http://doi.org/10.1039/C1AN15574A.

Wang, F. (2021). Application of new carbon fiber material in sports equipment. *IOP Conference Series: Earth and Environmental Science*, *714*(3). Available from https://doi.org/10.1088/1755-1315/714/3/032064, https://iopscience.iop.org/journal/1755-1315.

Wang, F. S., Ji, Y. Y., Yu, X. S., Chen, H., & Yue, Z. F. (2016). Ablation damage assessment of aircraft carbon fiber/epoxy composite and its protection structures suffered from lightning strike. *Composite Structures*, *145*, 226–241. Available from https://doi.org/10.1016/j.compstruct.2016.03.005, http://www.elsevier.com/inca/publications/store/4/0/5/9/2/8.

Wang, L., Zhang, M., Yang, B., Ding, X., Tan, J., Song, S., & Nie, J. (2021). Flexible, robust, and durable aramid fiber/CNT composite paper as a multifunctional sensor for wearable applications. *ACS Applied Materials and Interfaces*, *13*(4), 5486–5497. Available from https://doi.org/10.1021/acsami.0c18161, http://pubs.acs.org/journal/aamick.

Wang, S., & Chung, D. D. L. (1999). Temperature/light sensing using carbon fiber polymer-matrix composite. *Composites Part B: Engineering*, *30*(6), 591–601. Available from https://doi.org/10.1016/S1359-8368(99)00020-7, https://www.sciencedirect.com/science/article/pii/S1359836899000207.

Wang, S., & Chung, D. D. L. (2002). Mechanical damage in carbon fiber polymer-matrix composite, studied by electrical resistance measurement. *Composite Interfaces*, *9*(1), 51–60. Available from https://doi.org/10.1163/156855402753642890.

Wang, S., Kowalik, D. P., & Chung, D. D. L. (2004). Self-sensing attained in carbon-fiber-polymer-matrix structural composites by using the interlaminar interface as a sensor. *Smart Materials and Structures*, *13*(3), 570–592. Available from https://doi.org/10.1088/0964-1726/13/3/017.

Wang, S., Wang, D., Chung, D. D. L., & Chung, J. H. (2006). Method of sensing impact damage in carbon fiber polymer-matrix composite by electrical resistance measurement. *Journal of Materials Science*, *41*(8), 2281–2289. Available from https://doi.org/10.1007/s10853-006-7172-9.

Wang, X., Fu, X., & Chung, D. D. L. (1999). Strain sensing using carbon fiber. *Journal of Materials Research*, *14*(3), 790–802. Available from https://doi.org/10.1557/JMR.1999.0105, http://journals.cambridge.org/action/displayJournal?jid = JMR.

Wang, X., Sun, F., Yin, G., Wang, Y., Liu, B., & Dong, M. (2018). Tactile-sensing based on flexible PVDF nanofibers via electrospinning: A review. *MDPI AG, China Sensors (Switzerland)*, *18*(2). Available from https://doi.org/10.3390/s18020330, http://www.mdpi.com/1424-8220/18/2/330/pdf.

Wang, Y., Chen, M., Wang, X., Chan, R. H. M., & Li, W. J. (2018). IoT for next-generation racket sports training. *IEEE Internet of Things Journal*, *5*(6), 4558–4566. Available from https://doi.org/10.1109/JIOT.2018.2837347, http://ieeexplore.ieee.org/servlet/opac?punumber = 6488907.

Wang, Y., Wang, Y., Wan, B., Han, B., Cai, G., & Chang, R. (2018). Strain and damage self-sensing of basalt fiber reinforced polymer laminates fabricated with carbon nanofibers/epoxy composites under tension. *Composites Part A: Applied Science and Manufacturing*, *113*, 40–52. Available from https://doi.org/10.1016/j.compositesa.2018.07.017.

Wen, J., Xia, Z., & Choy, F. (2011). Damage detection of carbon fiber reinforced polymer composites via electrical resistance measurement. *Composites Part B: Engineering*, *42*(1), 77–86. Available from https://doi.org/10.1016/j.compositesb.2010.08.005.

Willberry, J. O., & Papaelias, M. (2020). Structural health monitoring using fibre optic acoustic emission sensors. *Sensors (Switzerland)*, *20*(21), 1–31. Available from https://doi.org/10.3390/s20216369, https://www.mdpi.com/1424-8220/20/21/6369/pdf.

Xi, Y., Ishikawa, H., Bin, Y., & Matsuo, M. (2004). Positive temperature coefficient effect of LMWPE-UHMWPE blends filled with shortcarbon fibers. *Carbon*, *42*(8-9), 1699–1706. Available from https://doi.org/10.1016/j.carbon.2004.02.027.

Xiao, H., Zhang, W., Huang, C., Xia, C., Xia, Y., He, L., & Wang, Z. (2023). A highly selective mems-based gas sensor with gelatin-carbon black composite film fabricated by the thin-film-needle-coating method. *SSRN*. Available from https://doi.org/10.2139/ssrn.4459046, https://www.ssrn.com/index.cfm/en/.

Xie, H., & Sheng, P. (2009). Fluctuation-induced tunneling conduction through nanoconstrictions. *Physical Review B – Condensed Matter and Materials Physics*, *79*(16). Available from https://doi.org/10.1103/PhysRevB.79.165419, http://oai.aps.org/oai?verb = GetRecord&Identifier = oai:aps.org:PhysRevB.79.165419&metadataPrefix = oai_apsmeta_2.

Yamada, T., Hayamizu, Y., Yamamoto, Y., Yomogida, Y., Izadi-Najafabadi, A., Futaba, D. N., & Hata, K. (2011). A stretchable carbon nanotube strain sensor for human-motion detection. *Nature Nanotechnology*, *6*(5), 296–301. Available from https://doi.org/10.1038/nnano.2011.36, http://www.nature.com/nnano/index.html.

Yang, C., Denno, M. E., Pyakurel, P., & Venton, B. J. (2015). Recent trends in carbon nanomaterial-based electrochemical sensors for biomolecules: A review. *Analytica Chimica Acta*, *887*, 17–37. Available from https://doi.org/10.1016/j.aca.2015.05.049, http://www.journals.elsevier.com/analytica-chimica-acta/.

Yang, C. Q., Wang, X. L., Jiao, Y. J., Ding, Y. L., Zhang, Y. F., & Wu, Z. S. (2016). Linear strain sensing performance of continuous high strength carbon fibre reinforced polymer composites. *Composites Part B: Engineering*, *102*, 86–93. Available from https://doi.org/10.1016/j.compositesb.2016.07.013.

Yang, G., Lee, C., Kim, J., Ren, F., & Pearton, S. J. (2013). Flexible graphene-based chemical sensors on paper substrates. *Physical Chemistry Chemical Physics: PCCP*, *15*(6), 1798–1801. Available from https://doi.org/10.1039/C2CP43717A.

Yang, N., Chen, X., Ren, T., Zhang, P., & Yang, D. (2015). Carbon nanotube based biosensors. *Sensors and Actuators, B: Chemical*, *207*, 690–715. Available from https://doi.org/10.1016/j.snb.2014.10.040.

Yang, Y., Chiesura, G., Plovie, B., Vervust, T., Luyckx, G., Degrieck, J., Sekitani, T., & Vanfleteren, J. (2018). Design and integration of flexible sensor matrix for in situ monitoring of polymer composites.

ACS Sensors, *3*(9), 1698–1705. Available from https://doi.org/10.1021/acssensors.8b00425, http://pubs.acs.org/journal/ascefj.

Yang, Y., Chiesura, G., Vervust, T., Bossuyt, F., Luyckx, G., Degrieck, J., & Vanfleteren, J. (2016). Design and fabrication of a flexible dielectric sensor system for in situ and real-time production monitoring of glass fibre reinforced composites. *Sensors and Actuators, A: Physical*, *243*, 103–110. Available from https://doi.org/10.1016/j.sna.2016.03.015.

Yang, Z., Cao, Z., Sun, H., & Li, Y. (2008). Composite films based on aligned carbon nanotube arrays and a poly(N-isopropyl acrylamide) hydrogel. *Advanced Materials*, *20*(11), 2201–2205. Available from https://doi.org/10.1002/adma.200701964China, http://www3.interscience.wiley.com/cgi-bin/fulltext/119031153/PDFSTART.

Yoo, K. P., Lim, L. T., Min, N. K., Lee, M. J., Lee, C. J., & Park, C. W. (2010). Novel resistive-type humidity sensor based on multiwall carbon nanotube/polyimide composite films. *Sensors and Actuators, B: Chemical*, *145*(1), 120–125. Available from https://doi.org/10.1016/j.snb.2009.11.041.

Yu, J., Cheung, K. W., Yan, W. H., Li, Y. X., & Ho, D. (2017). High-sensitivity low-power tungsten doped niobium oxide nanorods sensor for nitrogen dioxide air pollution monitoring. *Sensors and Actuators, B: Chemical*, *238*, 204–213. Available from https://doi.org/10.1016/j.snb.2016.07.001.

Yu, X., Zhang, W., Zhang, P., & Su, Z. (2017). Fabrication technologies and sensing applications of graphene-based composite films: Advances and challenges. *Biosensors and Bioelectronics*, *89*, 72–84. Available from https://doi.org/10.1016/j.bios.2016.01.081, http://www.elsevier.com/locate/bios.

Yuan, C., Tony, A., Yin, R., Wang, K., & Zhang, W. (2021). Tactile and thermal sensors built from carbon–polymer nanocomposites—A critical review. *Sensors (Switzerland)*, *21*(4), 1–26. Available from https://doi.org/10.3390/s21041234, https://www.mdpi.com/1424-8220/21/4/1234/pdf.

Zee, F., & Judy, J. W. (2001). Micromachined polymer-based chemical gas sensor array. *Sensors and Actuators B: Chemical*, *72*(2), 120–128. Available from https://doi.org/10.1016/S0925-4005(00)00638-9, https://www.sciencedirect.com/science/article/pii/S0925400500006389.

Zeng, W., Shu, L., Li, Q., Chen, S., Wang, F., & Tao, X. M. (2014). Fiber-based wearable electronics: A review of materials, fabrication, devices, and applications. *Advanced Materials*, *26*(31), 5310–5336. Available from https://doi.org/10.1002/adma.201400633, http://www3.interscience.wiley.com/journal/119030556/issue.

Zhang, B., Jia, L., Tian, M., Ning, N., Zhang, L., & Wang, W. (2021). Surface and interface modification of aramid fiber and its reinforcement for polymer composites: A review. *European Polymer Journal*, *147*. Available from https://doi.org/10.1016/j.eurpolymj.2021.110352, https://www.journals.elsevier.com/european-polymer-journal.

Zhang, C., Qiu, J., Ji, H., & Shan, S. (2015). An imaging method for impact localization using metal-core piezoelectric fiber rosettes. *Journal of Intelligent Material Systems and Structures*, *26*(16), 2205–2215. Available from https://doi.org/10.1177/1045389X14551432, http://jim.sagepub.com/.

Zhang, J., Shen, G., Wang, W., Zhou, X., & Guo, S. (2010). Individual nanocomposite sheets of chemically reduced graphene oxide and poly(N-vinyl pyrrolidone): Preparation and humidity sensing characteristics. *Journal of Materials Chemistry*, *20*(48), 10824–10828. Available from https://doi.org/10.1039/C0JM02440F, http://doi.org/10.1039/C0JM02440F.

Zhang, T., Sun, J., Ren, L., Yao, Y., Wang, M., Zeng, X., Sun, R., Xu, J. B., & Wong, C. P. (2019). Nacre-inspired polymer composites with high thermal conductivity and enhanced mechanical strength. *Composites Part A: Applied Science and Manufacturing*, *121*, 92–99. Available from https://doi.org/10.1016/j.compositesa.2019.03.017.

Zhang, X. W., Pan, Y., Zheng, Q., & Yi, X. S. (2000). Time dependence of piezoresistance for the conductor-filled polymer composites. *Journal of Polymer Science Part B: Polymer Physics*, *38*(21), 2739–2749. Available from https://doi.org/10.1002/1099-0488, https://doi.org/10.1002/1099-0488.

Zhang, Y., Zhang, K., & Ma, H. (2009). Electrochemical DNA biosensor based on silver nanoparticles/poly(3-(3-pyridyl) acrylic acid)/carbon nanotubes modified electrode. *Analytical Biochemistry*, *387*(1), 13–19. Available from https://doi.org/10.1016/j.ab.2008.10.043, http://www.elsevier.com/inca/publications/store/6/2/2/7/8/1/index.htt.

Zhang, Y. F., Hong, C. Y., Ahmed, R., & Ahmed, Z. (2017). A fiber Bragg grating based sensing platform fabricated by fused deposition modeling process for plantar pressure measurement. *Measurement: Journal of the International Measurement Confederation*, *112*, 74–79. Available from https://doi.org/10.1016/j.measurement.2017.08.024.

Zhao, C., Xia, Z., Wang, X., Nie, J., Huang, P., & Zhao, S. (2020). 3D-printed highly stable flexible strain sensor based on silver-coated-glass fiber-filled conductive silicon rubber. *Materials and Design*, *193*. Available from https://doi.org/10.1016/j.matdes.2020.108788, https://www.journals.elsevier.com/materials-and-design.

Zhao, G., Huang, X., Tang, Z., Huang, Q., Niu, F., & Wang, X. (2018). Polymer-based nanocomposites for heavy metal ions removal from aqueous solution: A review. *Polymer Chemistry*, *9*(26), 3562–3582. Available from https://doi.org/10.1039/C8PY00484F.

Zhao, Q., Zhang, K., Zhu, S., Xu, H., Cao, D., Zhao, L., Zhang, R., & Yin, W. (2019). Review on the electrical resistance/conductivity of carbon fiber reinforced polymer. *Applied Sciences (Switzerland)*, *9*(11). Available from https://doi.org/10.3390/app9112390, https://res.mdpi.com/applsci/applsci-09-02390/article_-deploy/applsci-09-02390-v2.pdf?filename = &attachment = 1.

Zhao, X., Hao, M., Yan, Y., Wang, J., Zhao, C., Wang, Y., & Luo, M. (2021). Modeling analysis of ionic polymer-metal composites sensors with various sizes. In *27th International Conference on Mechatronics and Machine Vision in Practice* (pp. 133–138). Available from http://ieeexplore.ieee.org/xpl/mostRecentIssue.jsp?punumber = 9664973. https://doi.org/10.1109/M2VIP49856.2021.9664985.

Zhou, J., Yu, H., Xu, X., Han, F., & Lubineau, G. (2017). Ultrasensitive, stretchable strain sensors based on fragmented carbon nanotube papers. *ACS Applied Materials and Interfaces*, *9*(5), 4835–4842. Available from https://doi.org/10.1021/acsami.6b15195, http://pubs.acs.org/journal/aamick.

Zhu, L., Xu, L., Tan, L., Tan, H., Yang, S., & Yao, S. (2013). Direct electrochemistry of cholesterol oxidase immobilized on gold nanoparticles-decorated multiwalled carbon nanotubes and cholesterol sensing. *Talanta*, *106*, 192–199. Available from https://doi.org/10.1016/j.talanta.2012.12.036, https://www.journals.elsevier.com/talanta.

Zhu, M., Lou, M., Abdalla, I., Yu, J., Li, Z., & Ding, B. (2020). Highly shape adaptive fiber based electronic skin for sensitive joint motion monitoring and tactile sensing. *Nano Energy*, *69*. Available from https://doi.org/10.1016/j.nanoen.2019.104429, http://www.journals.elsevier.com/nano-energy/.

Zhu, P., Xie, X., Sun, X., & Soto, M. A. (2019). Distributed modular temperature-strain sensor based on optical fiber embedded in laminated composites. *Composites Part B: Engineering*, *168*, 267–273. Available from https://doi.org/10.1016/j.compositesb.2018.12.078.

Zin, M. H., Razzi, M. F., Othman, S., Liew, K., Abdan, K., & Mazlan, N. (2016). A review on the fabrication method of bio-sourced hybrid composites for aerospace and automotive applications. *IOP Conference Series: Materials Science and Engineering*, *152*(1). Available from https://doi.org/10.1088/1757-899X/152/1/012041, http://www.iop.org/EJ/journal/mse.

CHAPTER

Sensors based on polymer nanomaterials

10

Mst Nasima Khatun[1], Moirangthem Anita Chanu[1], Debika Barman[1], Priyam Ghosh[1], Tapashi Sarmah[1], Laxmi Raman Adil[1] and Parameswar Krishnan Iyer[1,2,3]

[1]Department of Chemistry, Indian Institute of Technology Guwahati, Guwahati, Assam, India [2]Centre for Nanotechnology and Centre for Drone Technology, Indian Institute of Technology Guwahati, Guwahati, Assam, India [3]School for Health Science and Technology, Indian Institute of Technology Guwahati, Guwahati, Assam, India

10.1 Introduction

Sensor technology has evolved significantly over the years, becoming an integral part of modern life. From consumer electronics to industrial applications and healthcare, sensors play a crucial role in monitoring and measuring various physical, chemical, and biological parameters (Atkinson et al., 2014; Khatun et al., 2021; Lam et al., 2013; Liu et al., 2012; Orozco et al., 2013; Paleček et al., 2015; Ronkainen et al., 2010; Song et al., 2010; Suginta et al., 2013). As the demand for more sensitive, selective, and miniaturized sensors grows, researchers have turned their attention to novel materials that can address these challenges effectively (Khatun et al., 2019). Polymer nanomaterials (PNMs) have emerged as a promising class of materials with unique properties (Adhikari, 2021a; Banik et al., 2016; Parveen et al., 2012; Schadler et al., 2007) and versatile applications (Goel et al., 2022; Shifrina et al., 2020; Singh et al., 2017; Srikar et al., 2014; Wen et al., 2019), making them ideal candidates for the development of advanced sensors (Alsbaiee et al., 2016; Canfarotta et al., 2013; Malik et al., 2015). PNMs refer to nanoscale structures derived from synthetic or natural polymers, engineered to possess specific physical, chemical, and mechanical properties. They offer several advantages that make them attractive for sensor applications. A primary advantage of PNMs lies in their elevated surface-area-to-volume ratio, coupled with the molecular wire effect (Adhikari, 2021a; Banik et al., 2016; Malik et al., 2015; Parveen et al., 2012; Schadler et al., 2007). This dynamic highlighted more interactions with target analytes, leading to amplified sensitivity and improved detection limits. Additionally, their composition is amenable to customization via functionalization and modification, affording the creation of sensors with meticulously selective and precise reactions to designated analytes. Carbon-based nanomaterials, such as carbon nanotubes (CNTs) (Han et al., 2019; Schroeder et al., 2019; Zaporotskova et al., 2016), graphene (Liu et al., 2022; Verma et al., 2022), and graphene oxide (GO) (Cruz-Martínez et al., 2021; Qian et al., 2021), are among the most extensively studied nanomaterials for sensor applications. CNTs, with their exceptional electrical conductivity and mechanical strength, have demonstrated great potential in gas sensors and biosensors due to their unique adsorption properties. Graphene and GO, with their two dimensional planar structures and excellent electrical properties, have been explored in a wide range of sensors, from chemical and gas

Sensory Polymers. DOI: https://doi.org/10.1016/B978-0-443-13394-7.00010-0

sensors to biosensors and environmental monitoring devices. Conducting polymers (CPs), including polyaniline (PANI) (Al-Haidary et al., 2021; Fratoddi et al., 2015; Yang et al., 2023), polypyrrole (PPy) (Lo et al., 2020; Miah et al., 2022), and polythiophene (PTh) (Ashraf et al., 2020; Gonçalves & Balogh, 2012; Li & Shi, 2013; Mousavi et al., 2021; Sasaki et al., 2021; Shishkanova et al., 2022), are another class of PNMs that have attracted significant attention for sensor development. These polymers exhibit inherent electrical conductivity and undergo redox reactions with target analytes, leading to changes in electrical properties that can be used for sensing purposes. The tunable conductivity and facile synthesis of CPs enable the design of sensors with high sensitivity and rapid response times. CPs are macromolecules comprising luminescent units linked by covalent bonds. With elevated molar extinction coefficients and potent light absorption capabilities, CPs also exhibit remarkable photostability and biocompatibility (Gopalana et al., 2019; McQuade et al., 2000; Rahman et al., 2008). The emergence of CP nanomaterials (CPNs) has kindled fervent research interest owing to their compact size, intensified fluorescence, and facile modifiability (Kausar, 2022). Non-CPNs possess distinctive attributes stemming from their structural characteristics. Unlike CPs, they lack extended π-conjugation along their molecular backbone, resulting in unique properties like optical, electrical, biocompatibility, and chemical stability. In essence, non-CPNs possess distinct advantages and applications compared to their CP counterparts, leveraging their specific attributes to cater to various technological and scientific needs (Bhattacharya et al., 2021; Liu et al., 2016). Hybrid nanocomposites (Nugroho et al., 2019; Wei et al., 2017), which combine PNMs with other functional materials like nanoparticles, quantum dots, or metal oxides, have emerged as a versatile platform for sensor design. These composites leverage the synergistic effects of different materials, resulting in enhanced sensing performance, improved stability, and broader sensing ranges. The integration of PNMs into sensor technology has opened up new opportunities in various fields. In environmental monitoring, polymer nanomaterial–based sensors have been deployed to detect pollutants, gases, and heavy metals, offering real time data for efficient pollution control and resource management. In healthcare, these sensors have shown promise in disease diagnostics, wearable health monitoring, and drug delivery systems, revolutionizing personalized medicine and patient care. Moreover, PNM–based sensors find applications in industrial settings for process monitoring, safety control, and quality assurance. The development of PNM–based sensors is not without challenges. One of the key challenges is ensuring the reproducibility and scalability of sensor fabrication. The synthesis and functionalization processes must be well-defined and optimized to achieve consistent sensor performance across different batches. Additionally, the integration of PNMs into existing sensor platforms and their compatibility with electronic systems require careful consideration to realize practical and marketable sensor devices. Another critical aspect is the safety and environmental impact of PNMs. As these materials become more prevalent in sensor technology and other applications, it is essential to understand their potential toxicity and ecological consequences. Responsible and sustainable use of PNMs demands comprehensive investigations into their long-term effects on human health and the environment.

In this comprehensive compilation, we aim to provide a detailed overview of sensor technology based on PNMs. We will delve into the synthesis, characterization, and functionalization strategies of conjugated, nonconjugated, and hybrid PNMs. Additionally, we will explore the various sensing mechanisms employed by PNM–based sensors, including optical, electrical, and electrochemical transduction mechanisms. Moreover, we will discuss the diverse applications of these sensors in environmental monitoring, healthcare, industrial processes, and consumer electronics. The chapter will encompass case studies and examples to illustrate the potential of PNM–based sensors in real world scenarios. Lastly, we will address the challenges and future prospects in the field. We will discuss ongoing research efforts and potential directions for overcoming current limitations, such as improving sensor performance, addressing reproducibility concerns, and advancing integration into practical sensor devices. In conclusion, PNMs have emerged as a powerful and versatile class of materials for sensor technology. Their unique properties, combined with advancements in synthesis and functionalization techniques, offer tremendous potential for developing highly efficient and innovative sensors. By exploring the diverse applications, sensing mechanisms, and challenges of polymer nanomaterial–based sensors, this chapter aims to provide a comprehensive understanding of their significance in advancing sensor technology across various domains like chemosensing, biosensing, optical sensing, electrical sensing, and beyond. With continued research and responsible deployment, these sensors hold the promise of addressing critical societal challenges and shaping a more connected, sustainable, and technologically advanced future.

10.2 Classification and properties of polymer nanomaterials

PNMs are tiny particles or materials composed of polymers with dimensions ranging from 10 to 100 nm in at least one direction (Adhikari, 2021b). These polymers are composed of large molecules made up of repeating units known as monomers. These nanomaterials are well-suited for a wide range of applications due to their unique properties and structures. PNMs, synthesized at the nanoscale, possess several basic properties that make them distinctive and valuable for various applications. These properties include:

- *High surface area*: PNMs have an exceptionally high surface-area-to-volume ratio due to their small size. This property enhances their reactivity and makes them suitable for adsorption and catalytic processes.
- *Tunable properties*: The chemical and physical properties of PNMs can often be finely tuned or modified, allowing for customization to meet specific application requirements. This tunability includes properties like conductivity, porosity, and mechanical strength.
- *Enhanced mechanical properties*: Some PNMs exhibit improved mechanical properties compared to their bulk counterparts, making them attractive for applications requiring strength and durability.
- *Versatility*: PNMs can be designed to be versatile and adaptable for use in a wide range of applications, from drug delivery systems to sensors and composite materials.
- *Biocompatibility*: Many PNMs are biocompatible, making them suitable for use in biomedical applications such as drug delivery, tissue engineering, and biosensors.
- *Stability*: PNMs often demonstrate stability in different environmental conditions, including temperature, pH, and humidity, which is crucial for their functionality in various applications.
- *Controlled release*: They can be engineered to have controlled release properties, allowing for precise and sustained delivery of drugs or other active agents.
- *Electrical conductivity*: Some conductive PNMs exhibit electrical conductivity, making them useful in applications like flexible electronics and sensors.
- *Chemical functionalization*: PNMs can be easily functionalized with different chemical groups, enabling the attachment of biomolecules, drugs, or other compounds for specific applications.

- *Lightweight*: Due to their nanoscale size, PNMs are lightweight, which can be advantageous in applications where weight is a critical factor.

These basic properties of PNMs contribute to their versatility and suitability for a wide range of applications in fields such as materials science, nanotechnology, medicine, and electronics. Researchers continue to explore and harness these properties to develop innovative solutions in various industries. PNMs are categorized into three types: conjugated, nonconjugated, and hybrid, based on their chemical structures. These classifications determine their distinct properties and potential applications, which are discussed below.

10.2.1 Conjugated polymer nanomaterials

CPs are made of large organic molecules with alternating single and double bonds in their backbone chain (Burroughes et al., 1990). This bonding arrangement allows electrons from overlapping p-orbitals to spread out across adjacent orbitals, resulting in the emergence of advantageous optoelectronic features in CPs. The delocalized electron system enables charge carriers to absorb light while also producing and transporting charge carriers. As a result, the absorbed light energy can be efficiently converted into fluorescence, heat, and other forms of energy, making them optically and electronically active materials. These materials are currently of great interest, particularly in the fields of electronics, biology, and sensing, due to their wide range of advantageous properties (Elgiddawy et al., 2023). These properties include their ability to act as semiconductors, their capacity for solution processing, their flexibility, their high efficiency in fluorescence, their large absorption cross-section, their ability to adjust emission and bandgap, their photostability, their low cytotoxicity, their ease of chemical modification and surface functionalization, and their relatively low cost. Several examples of frequently employed CP backbones include conjugated polyfluorenes, polyphenylene vinylenes, PThs, and polyphenyl ethynylenes (Wang et al., 2019). Nevertheless, the majority of CPs exhibit limited solubility in water and are typically dissolved in organic solvents, which often consist of toxic chlorinated solvents that pose significant risks to both human health and the environment. In this regard, research on the environmentally friendly processing of CPs has proliferated in recent years. The synthesis of CPNs is one of these methods. Nanomaterials are defined as materials with a structure in at least one dimension smaller than 100 nm. As a result, nanomaterial morphologies vary between 1D nanofibers, which include nanorods, nanotubes, nanowires, and 2D platelets, and 3D spherical nanoparticles, which include nanospheres, nanocapsules, hyperbranched polymers, self-assembled structures, and porous materials (MacFarlane et al., 2020). The most common morphologies obtained from the current methods are 3D spherical nanoparticles, which are thermodynamically favorable for a minimum surface-to-volume ratio. Because of their small size and high surface-area-to-volume ratio, they can perform better in charge transport, light absorption, and interaction with other materials or biological systems (Fig. 10.1).

Nanoparticles of CPs can be obtained through two main methods: postpolymerization dispersion of polymers that were prepared separately or direct polymerization in dispersed heterophase systems (Elgiddawy et al., 2023). Both approaches possess their own merits and limitations. The generation of nanoparticles through postpolymerization involves separate steps for polymerization and nanoparticle formation, which can be accomplished using commercially available polymers that meet specific requirements, eliminating the need for specialized equipment and expertise in organic

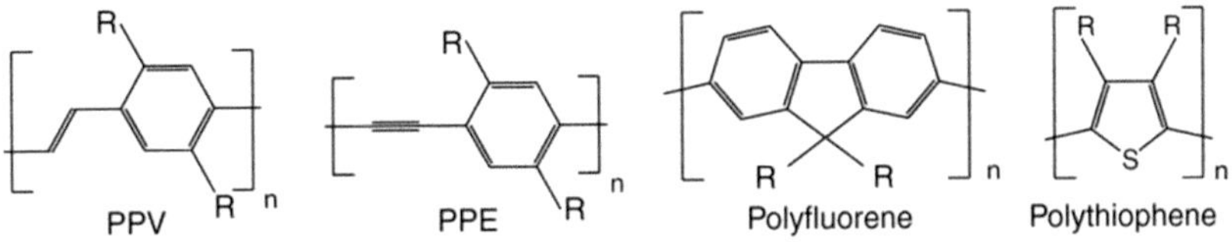

FIGURE 10.1

Four common backbones of CPs and their chemical structures.

Reproduced with permission from Wang, Y., Feng, L., & Wang, S. (2019). Conjugated polymer nanoparticles for imaging, cell activity regulation, and therapy. Advanced Functional Materials, 29, *Article 1806818.*

and polymer synthesis (Wu et al., 2006). If necessary, polymers have the potential to undergo thorough purification following the process of polymerization. In contrast, direct polymerization, where both polymerization and nanoparticle formation occur simultaneously in a single process, is not limited to polymers that can dissolve in organic solvents (Behrendt et al., 2016). As a result, it has the potential to offer a wider range of nanoparticles with greater control over size and particle structure. Direct polymerization encompasses a range of techniques, such as emulsion, microemulsion, miniemulsion, and dispersion polymerization. On the other hand, postpolymerization techniques involve miniemulsion and nanoprecipitation.

10.2.2 Nonconjugated polymer nanomaterials

Non-CPNs refer to a category of nanomaterials comprising polymers that do not possess conjugated double bonds along their primary chain (Jose et al., 2023; Liu et al., 2016; Liu et al., 2020; Zhao et al., 2023). Although most functional polymeric nanomaterials with intrinsic fluorescence emission are almost always CPs, there are still a number of challenges in producing and using these materials, including complex multistep synthetic pathways, the use of hazardous organic solvents, and the potential for fluorescence self-quenching in aqueous solutions. As a result, creating autofluorescent polymer materials that are easily prepared and have good water solubility is very difficult and is still being pursued. Recently, efforts have been made to develop water-friendly fluorescent non-CPNs by using techniques like polymerization, hydrothermal synthesis, cross-linking, and physical loading approaches (Bhattacharya et al., 2021). Typically, non-CPNs do not contain conjugated fluorescent dyes but rather possess numerous subfluorophore chemical groups (e.g., C = N and N = O) with intrinsically weak fluorescent properties. Fluorescence is generated when these groups are immobilized in a polymeric network. Traditionally, these heteroatom-containing double bonds are not considered fluorescent chromophores. A hydrothermal reaction between a large nonfluorescent polymer and a small nonfluorescent electron-rich molecule can generate a fluorescent non-CPN. This mechanism for producing fluorescence is known as cross-linked-enhanced emission (CEE) (Zhu et al., 2015). CEE operates when a small amine-rich organic molecule reacts with a large polymer under controlled hydrothermal conditions, which results in the formation of non-CPNs. Attachment of a small molecule reduces bond vibrations and rotations and the architectural confinement of electron-rich moieties (delocalized n and π electrons), resulting in fluorescence emission. Recently, a number of groups have reported the use of hydrothermal treatment of natural materials, including silk, chitosan, and natural materials modified with polyethylene imine (Bhattacharya et al., 2021). In this procedure, condensation reactions that result in fluorescent aggregated moieties cause the amine-rich precursor and large saccharide molecules to form non-CPNs. Due to their unique molecular structures and abundant terminal groups, these non-CP with fluorescent properties exhibit considerable potential as innovative materials for fluorescence-based applications in chemical and biological domains (Fig. 10.2).

10.2.3 Hybrid polymer nanomaterials

Hybrid PNMs are a class of advanced materials that combine polymers with nanoparticles at the nanoscale level. These materials often exhibit unique and enhanced properties that arise from the synergistic combination of the characteristics of both polymers and nanoparticles (Ferreira Soares et al., 2020; Jiang et al., 2016). The integration of polymers and nanoparticles allows for the tailoring of material properties such as mechanical strength, electrical conductivity, thermal stability, and optical properties, among others. Hybrid PNMs consist of a polymer matrix in which nanoparticles are dispersed. The nanoparticles can be of various types, such as metal nanoparticles (e.g., gold and silver), metal oxide nanoparticles (e.g., titanium dioxide and zinc oxide), quantum dots, carbon-based nanoparticles (e.g., CNTs, graphene), and more (Shu et al., 2020) (Fig. 10.3).

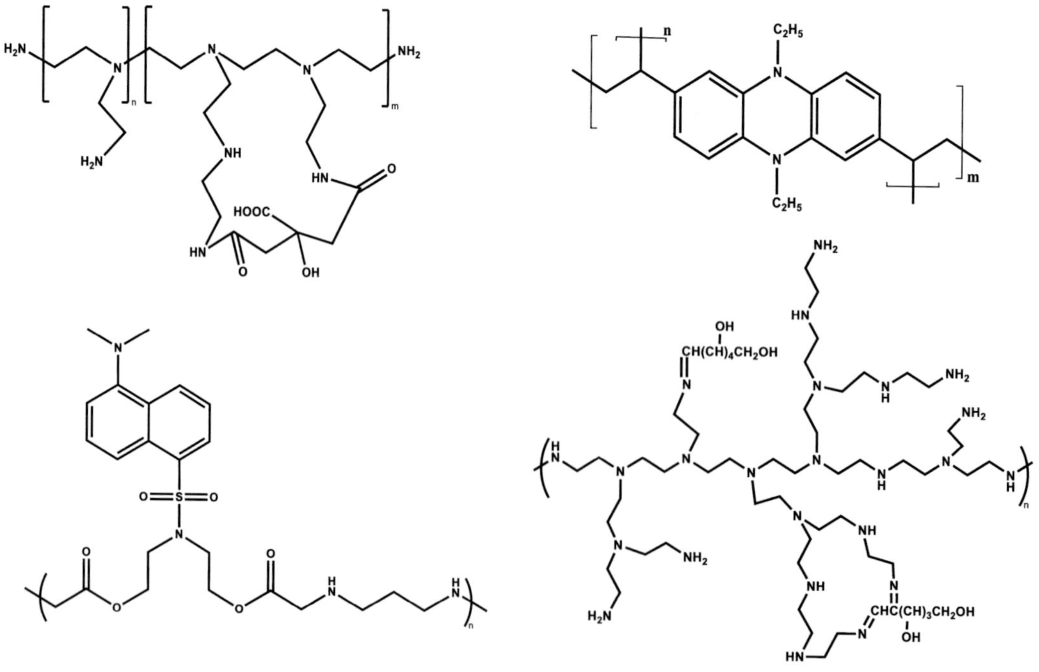

FIGURE 10.2

Some of the structures of non-CP nanomaterials.

Reproduced with permission from Jose et al. (2023); Liu et al. (2016); Liu et al. (2020); and Zhao et al. (2023).

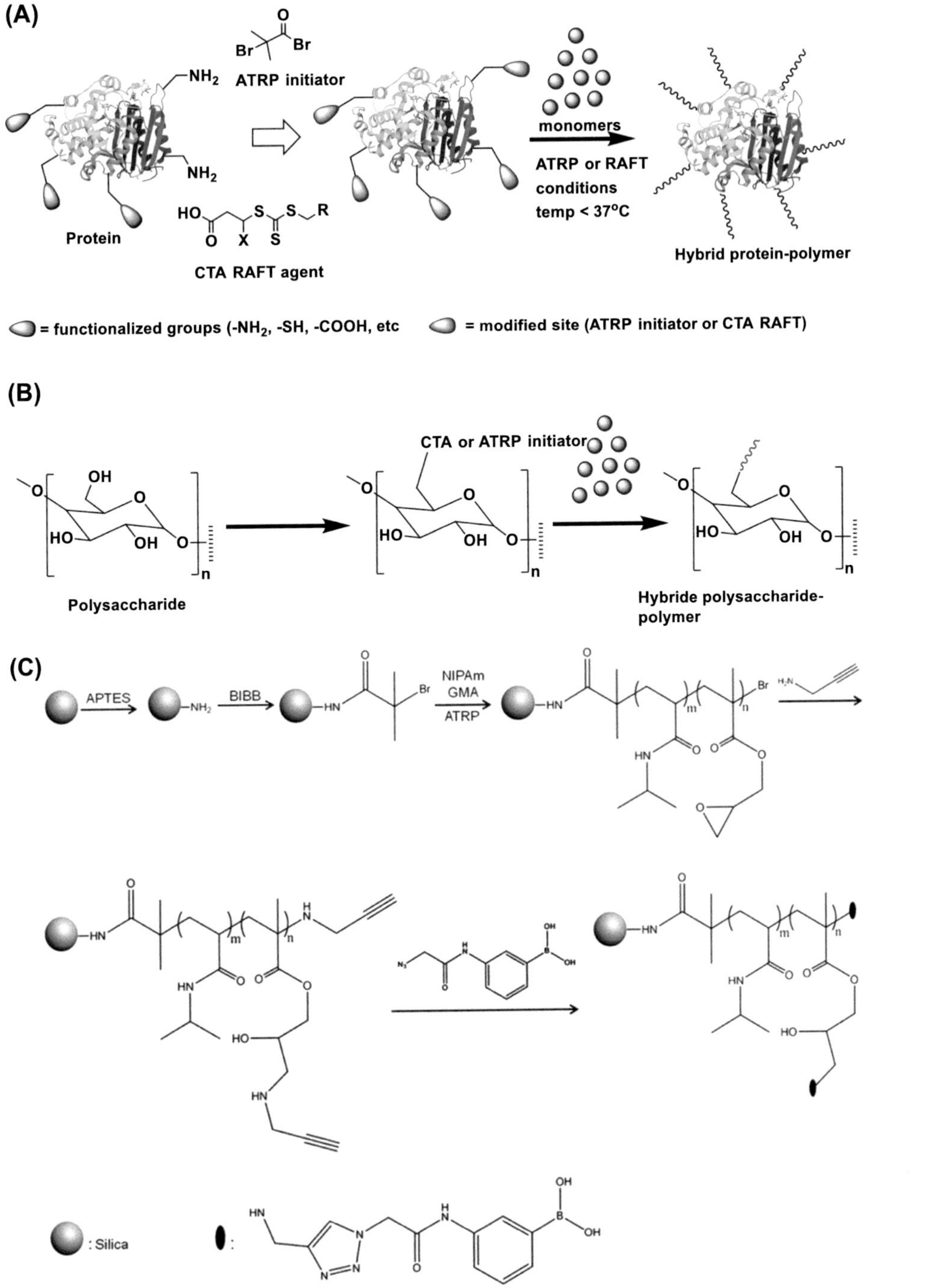

FIGURE 10.3

Si@poly(NIPAm-co-GMA)@APBA particles synthesized by combining SI-ATRP with CuAAC click reaction (Ferreira Soares et al., 2020; Jiang et al., 2016).

There are different methods to synthesize hybrid PNMs, depending on the desired properties and applications. These methods often involve the incorporation of nanoparticles into the polymer matrix during its synthesis or through postsynthesis processing. The combination of polymers and nanoparticles leads to hybrid materials with unique properties. For instance, hybrid polymer nanocomposites might have improved mechanical strength, electrical conductivity, and thermal stability compared to pure polymers. They find applications in various fields such as electronics, sensors, coatings, biomedical devices, energy storage, and more (Macchione et al., 2018). For example, they might be used in flexible electronics, drug delivery systems, and catalysis. The properties of hybrid PNMs can be tuned by adjusting the type, size, and concentration of nanoparticles, as well as the choice of polymer matrix. This tailoring allows researchers to create materials with specific characteristics suited for particular applications. While hybrid PNMs offer exciting possibilities, their development and characterization can be complex. Challenges include achieving uniform dispersion of nanoparticles within the polymer matrix, maintaining stability over time, and ensuring proper adhesion between the polymer and nanoparticles. Research in the field of hybrid PNMs is ongoing, with scientists working on novel synthesis techniques, improved characterization methods, and innovative applications.

In summary, hybrid PNMs are a promising class of materials that combine the advantages of polymers and nanoparticles, leading to materials with enhanced and tunable properties. Their diverse applications make them an exciting area of research and innovation across various industries.

10.3 Characterization of polymer nanomaterials

Characterizing polymer nanoparticles involves assessing various properties critical for their performance and applications. Among these properties, particle size and size distribution are paramount, as they significantly influence drug loading, drug release, and nanoparticle stability. Additionally, the zeta potential, which characterizes surface charge, is another crucial attribute. To delve into the details, advanced microscopic techniques such as atomic force microscopy (AFM), scanning electron microscopy (SEM), and transmission electron microscopy (TEM) are employed for in-depth characterization. These methods provide valuable insights into nanoparticle morphology and structure. For precise size measurement and distribution analysis, laser light scattering, particularly dynamic light scattering (DLS), is commonly utilized. It offers accurate data on nanoparticle dimensions. Furthermore, X-ray photoelectron spectroscopy is employed to delve into the surface chemistry of nanoparticle suspensions, enabling researchers to understand their chemical composition and functional groups. In summary, characterizing nanoparticles involves examining their crucial features, with an emphasis on size, size distribution, and zeta potential. Advanced microscopic techniques and spectroscopic methods are employed to gain comprehensive insights into their morphology, structure, and surface chemistry, all of which are fundamental for their diverse applications (Singh & Lillard, 2009).

10.3.1 Transmission electron microscopy

TEM is valuable for nanoparticle investigation, providing detailed nanoscale imaging, diffraction, and spectroscopy. It offers insights into morphology, size, structure, crystallography, and chemical composition. However, TEM requires intricate and time-consuming sample preparation because ultra-thin samples are needed to withstand the vacuum environment. The electron beam interacts with the specimen, yielding surface and structural information and allowing comprehensive nanoparticle analysis (Molpeceres et al., 2000).

10.3.2 Scanning electron microscopy

SEM operates by scanning a sample's surface with focused, high energy electrons, detecting reflected or scattered electrons to create images. It excels in morphological and dimensional analysis but has limitations in assessing size distribution and population averages. SEM uses an electron gun to generate electrons, accelerates them with an anode, and employs magnetic lenses to focus the electron beam. Electromagnetic deflector coils enable scanning of the sample's surface, yielding high-resolution images of morphology and structural features. However, SEM is not suitable for determining size distribution or statistical particle size information within a sample (Zhou & Wang, 2007).

10.3.3 Atomic force microscopy

AFM creates images by utilizing surface forces between the sample and a sharp tip attached to a scanning arm. This tip is incredibly fine, typically measuring 1–2 microns in length and less than 100 ÅngströMs (Å) in diameter. During the scanning process, the forces acting between the tip and the sample's surface cause the tip's position to deviate, which is then measured to create surface topography images (Cohen et al., 1999; Tokay & Erdem-Şenatalar, 2011; Li, 2006).

10.3.4 X-ray diffraction (XRD)

XRD is a crucial technique used for determining the crystal structure of nanoparticles. It allows scientists to identify the types of atoms present at lattice points within a crystal, as well as the arrangement of crystal planes and their respective distances from each other. Diffraction phenomena arise when incoming waves, such as X-rays or light, interact with a material's periodic structure. The formation of distinct diffraction patterns occurs due to the constructive interference of waves that are reflected from various layers within this periodic structure. This occurrence is fundamentally governed by Bragg's Law, which dictates the specific conditions under which diffraction takes place. Importantly, XRD accomplishes these tasks without causing damage to the sample. XRD is widely used in materials science, chemistry, and solid-state physics for characterizing a wide range of materials, including nanoparticles (Cullity & Strock, 2001; Katmıs et al., 2018).

10.3.5 Fourier transform infrared spectroscopy (FT-IR)

FT-IR is a valuable technique for analyzing the chemical functional groups on nanoparticle surfaces. It works by directing infrared radiation onto the sample, leading to two key processes: absorption and transmission. Absorption causes specific chemical bonds to vibrate at characteristic frequencies, while transmission reveals the wavelengths not absorbed. The resulting spectra, with peaks at specific wavenumbers, indicate vibrational frequencies of bonds and provide insights into the nanoparticle's chemical composition and functional groups' quantity. This information is crucial for understanding material properties and potential applications (Katmıs et al., 2018; Skoog et al., 1998).

10.3.6 Dynamic light spectroscopy

DLS, also known as photon correlation spectroscopy, determines nanoparticle size distribution by analyzing their Brownian motion in a liquid. It measures the hydrodynamic diameter, accounting for particle size, solvent effects, and motion, using the Stokes-Einstein equation. DLS is suitable for both uniform and varying sized nanoparticles in colloidal solutions but may not work well for very large particles. Sample preparation may involve proper dispersion. DLS instruments have a wide measuring range, covering particles from a few nanometers to several micrometers in size (Bootz et al., 2004).

10.3.7 Electrophoretic light scattering

Electrophoretic light scattering is the preferred method for determining the zeta potential of suspended particles, offering high sensitivity, accuracy, and versatility. Zeta potential characterizes particle charge in colloidal solutions and is influenced by factors like pH, ionic strength, and ion presence. It is crucial for colloidal solution stability. High zeta potential ($\pm$ 30 mV is critical) leads to particle repulsion, preventing agglomeration, while low zeta potential results in particle attraction and agglomeration. Zeta potential measurements are essential for understanding and controlling colloidal system behavior and stability (Clogston & Patri, 2011).

10.4 Synthetic strategy for the preparation of polymer nanomaterials

10.4.1 Nanoprecipitation technique

Nanoprecipitation, also known as reprecipitation, is a versatile method for modifying the bulk to produce intriguing nanostructures. It entails rapidly injecting a dilute polymeric solution into an excess amount of nonsolvent (typically water), causing a fast decline in solvent quality, generating supersaturation, and eventually leading to polymer precipitation, where it forms nanoparticles under suitable conditions (Pecher & Mecking, 2010). The process is called nucleation in nanoprecipitation. Classical nucleation theory may explain the details of nucleation and growth. As solubility decreases, ΔG changes, making phase separation more energetically favorable, resulting in the separation of the solid phase in the liquid medium.

The LaMer model illustrates how solute concentration changes over time throughout the nanoparticle fabrication procedure. In general, nucleation is induced by agitation, such as stirring, when the solute concentration rises and the system approaches the critical nucleation concentration. Nuclei are precipitated at this stage and grow by coalescence until they reach a stable size. During the process, the solute concentration decreases below the critical nucleation threshold, prohibiting any new nucleation and allowing just the current nuclei to develop continuously until the solute concentration reaches the equilibrium saturation concentration. The last stage is aggregation, whereby suspended nanoparticles aggregate with one another in the absence of a stabilizer.

The correlation between nucleation and supersaturation is given in classical nucleation theory by the equation:

$$J = A \exp\left(-\frac{16\pi\gamma^3\nu^2}{3k^3T^3(\ln S)^2}\right),$$

where A is a constant and S is the supersaturation defined by the difference between the real time concentration and the equilibrium saturation concentration. Surface tension, molar volume, Boltzmann's constant, and absolute temperature are denoted by γ, ν, k, and T, respectively. The equation demonstrates that increased supersaturation results in faster precipitation. Because the concentration drop during the second stage of the LaMer model occurs due to competition between nucleation and growth within a limited amount of solute, high supersaturation-induced rapid precipitation favors nucleation over growth, resulting in a high yield of small nanoparticles (Kang et al., 2022; Tao et al., 2019).

The reprecipitation process was used to prepare bare PTB7 CPNs. PTB7 is the abbreviated name for a polymer with a complex chemical structure (poly({4,8-bis[(2-ethylhexyl)oxy]benzo[1,2-b:4,5-b′]dithiophene-2,6-diyl}{3-fluoro-2-[(2-ethylhexyl)carbonyl]thieno[3,4-b]thiophenediyl})). The PTB7 CP was first dissolved in tetrahydrofuran (THF), which has good solubility, and then stabilized by an amphiphilic polymer. Thereafter, the prepared nanoparticle is quickly injected into water, which has poor solubility, while being sonicated and vigorously stirred. After injection, the amphiphilic copolymer forms a micelle, with the hydrophilic blocks exposed to the aqueous environment and the hydrophobic blocks flared into the PTB7-containing core (Fig. 10.4) (Zhao et al., 2021).

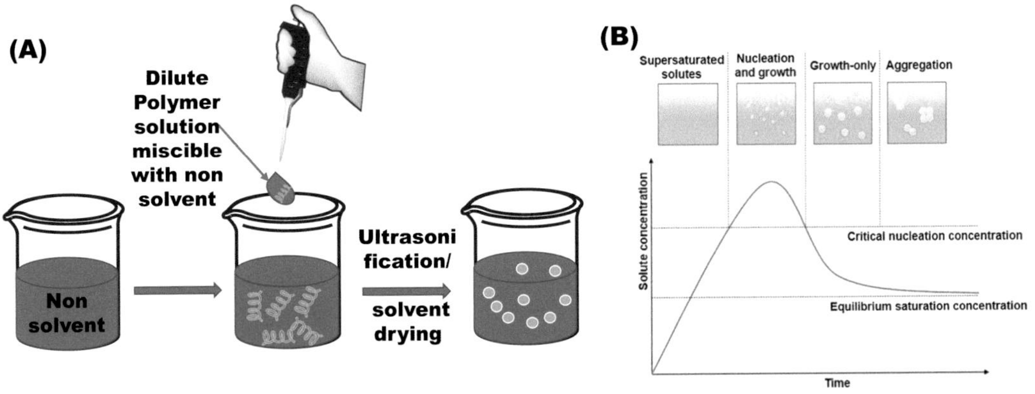

FIGURE 10.4

(A) Nanoprecipitation technique for CPN preparation. (B) Changes in solute concentration over time throughout the nanoparticle production process.

Reprinted with permission from Zhao, M., Leggett, E., Bourke, S., Poursanidou, S., Carter-Searjeant, S., Po, S., Carmo, M. P. D., Dailey, L. A., Manning, P., Ryan, S., Urbano, L., Green, M., & Rakovich, A. (2021). Theranostic near-infrared-active conjugated polymer nanoparticles. ACS Nano, 15, *8790–8802.*

10.4.2 Emulsification

Many nanomaterials, such as inorganic, organic, magnetite, or any solid material for various applications, can be enclosed in a polymer shell by employing green solvents such as water. The miniemulsion process is a versatile approach for creating a wide range of polymers and structured materials with limited geometries. Miniemulsions are continuous-phase droplets that are equally dispersed, small in size, and stable. High shear, such as high pressure or ultrasonication, is used to generate the system. The combination of a surfactant, an amphiphilic component, and a costabilizer that is soluble and homogeneously distributed in the droplet phase means that the costabilizer has a lower solubility in the continuous phase than the rest of the droplet phase and thus builds up an osmotic pressure in the droplets to counteract the Laplace pressure. Such small droplets may operate as nanocontainers for reactions to take place, either at the droplet interface or within the droplets, resulting in the generation of nanoparticles (Landfester, 2009).

The Landfester group used the emulsion approach to get the most homogenous nanoparticle size. In a 20% ethanol solution, homogenous platinum- or iron-containing particles with an exceptionally narrow size distribution were effectively generated (Schreiber et al., 2009).

Furthermore, surfactant-free emulsion polymerization for the in situ incorporation of hydrophobic iridium complexes was carried out using minidroplet-assisted mass transfer with hydrophobic solvents with a higher azeotropic boiling point than the reaction temperature for the formation of stable minidroplets in water. Monomers in the solution not only created polymer particles but also impacted the generation and movement of minidroplets. The vinyl-substituted iridium complex (IrV) copolymerized with 4-vinylpyridine (4VP) and methyl methacrylate is highly stable and has excellent photostability for more than 6 months (Fig. 10.5) (Lee et al., 2020).

10.4.3 Self-assembly method

Polymerization induced self-assembly (PISA) is an efficient and responsive approach for producing polymeric nano-objects at high concentrations. The fabrication of sophisticated polymer based bio-hybrid nanostructures (PBBNs) is becoming more feasible and achievable with three

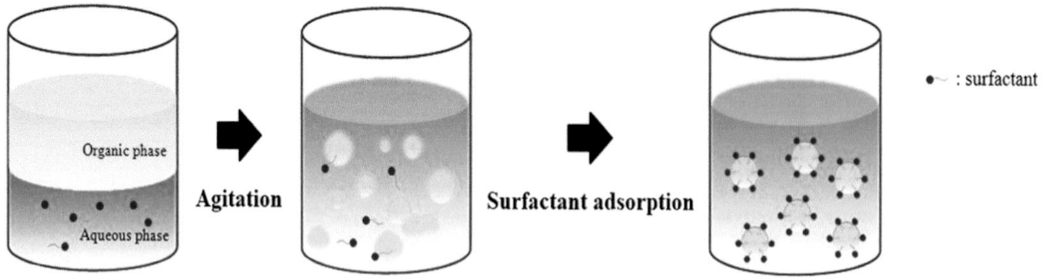

FIGURE 10.5

Emulsification polymeric nanoparticle preparation.

Reprinted with permission from Lee, S. H., Sung, J. H., & Park, T. H. (2012). Nanomaterial–based biosensor as an emerging tool for biomedical applications. Annals of Biomedical Engineering *2012,* 40, *1384–1397.*

methodologies: (1) grafting synthetic polymers to various biomacromolecules (carbohydrates, nucleic acids, and proteins) to form biomolecule polymer conjugate (BPCs), which were amphiphilic, allowing self-assembly in aqueous solution; (2) decoration of monomers with biomolecules for polymerization by grafting-through technique, followed by chain extension to amphiphilic BPCs and self-assembly in aqueous solution; and (3) developing PBBNs in aqueous solution using single or multiple coassembly processes between synthetic copolymers and biomacromolecules. With the evolution of polymer science, various polymerization techniques, in addition to ATRP-mediated PISA, have been introduced for the synthesis of PBBNs. Huang and colleagues, for example, used RAFT agent-modified BSA as the macroCTA to execute chain extension with HPMA, resulting in the synthesis of BSA-PHPMA conjugate nanostructures via RAFT-mediated PISA. Such well-dispersed BSA-PHPMA conjugate nanostructures with diameters ranging from 164 to 255 nm can be produced in a variety of solid compositions (Fig. 10.6) (Qiu et al., 2023).

10.4.4 Template-assisted technique

The template-assisted approach is a typical method for fabricating polymer nanostructures such as nanocapsules, nanofibers, nanowires, and nanotubes. It was done using electrochemical solid template-assisted methods or chemical procedures (Cho & Lee, 2008; Xiao et al., 2007). CP nanotubes have also been created using pre-prepared reactive templates, such as MnO_2 nanowires. Additionally, vertically aligned poly (3,4-ethylenedioxythiophene) (PEDOT) nanotubes were created using electrochemical deposition into anodic aluminum oxide, ZnO nanowires, and titanium nanotubes, as well as vapor deposition polymerization onto nanoporous templates.

To overcome the above limitations, we present a new straightforward method for obtaining relatively dense distributions of hollow PEDOT nanotubes using simple stainless-steel electrodes. In this alternative method, electrospun fibers are gathered onto a reasonably flat PEDOT film, which

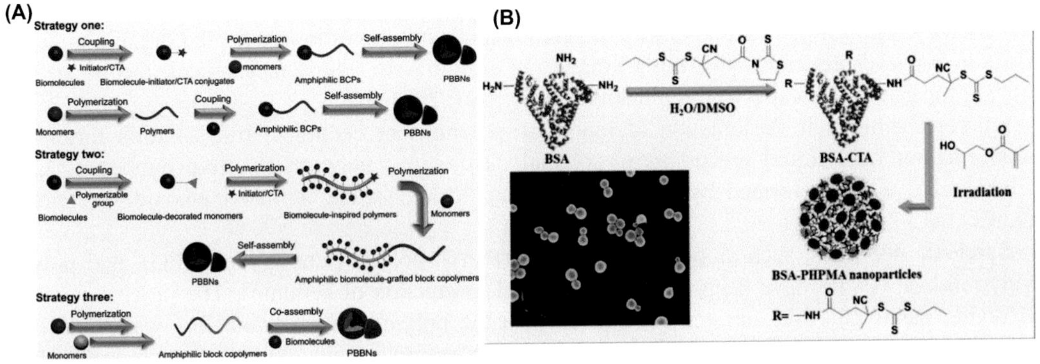

FIGURE 10.6

(A) Different strategies of self-assembly method for polymeric nanoparticle preparation by (B) BSA-PHPMA nanoparticle preparation by self-assembly.

Reprinted with permission from Qiu, L., Han, X., Xing, C., & Glebe, U. (2023). Polymerization-induced self-assembly: An emerging tool for generating polymer-based biohybrid nanostructures. Small (Weinheim an der Bergstrasse, Germany), 19, *Article 2207457.*

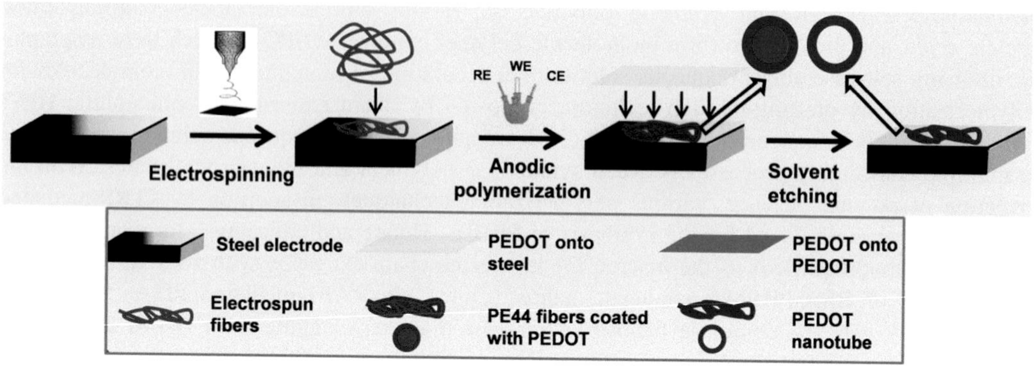

FIGURE 10.7

PEDOT nanotube preparation by template-assisted method.

Reprinted with permission from Estrany, F., Calvet, A., del Valle, L. J., Puiggalí, J., & Alemán, C. (2016). A multi-step template-assisted approach for the formation of conducting polymer nanotubes onto conducting polymer films. Polymer Chemistry, *7, 3540–3550.*

is critical for the entire coating of the template in a subsequent electropolymerization step. After removing the insulating fiber templates via solvent etching, the remaining hollow nanotubes exhibit electrochemical characteristics remarkably comparable to those obtained for films, with a considerably higher amount of electroactive surface (Fig. 10.7) (Estrany et al., 2016).

10.4.5 Sol-gel

The sol-gel method is a waste-free way of creating hybrid nanocomposites, making it environmentally friendly. The sol-gel method allows the homogenous mixing of inorganic and organic moieties in a single-phase material and affords unique opportunities to adjust mechanical, electrical, and optical properties for a variety of applications (Schottner, 2001).

Sol-gel chemistry is the creation of inorganic polymers or ceramics from solution through a transformation from liquid precursors to sol and, lastly, to a network structure dubbed a "gel." Traditionally, a sol is formed by the hydrolysis and condensation of metal alkoxide precursors (Danks et al., 2016).

Synthetic polymers, such as polyvinyl alcohol (PVA), polyethylene glycol (PEG), and polyvinylpyrrolidone (PVP), have been used in the sol-gel production of ceramics. The process involves dissolving metal salts like nitrates or acetates with the polymer in a solvent like water and then heating it to form a gel. Polymer functional groups (for example, imide on PVP) attach to dissolved metal ions to generate a homogenous gel, which limits particle nucleation and development, resulting in a nanoparticulate ceramic product (Parveen et al., 2012). The presence of the polymer guarantees that particle size is kept very tiny ($\sim$25 nm), even in the case of complex quaternary oxides such as YBCO ($YBa_2Cu_3O_7x$)0.83. This control on ceramic crystallite growth is thought to be due to metal ion complexation by hydroxyl substituents on PVA (Fig. 10.8) (Sun & Oh, 1996).

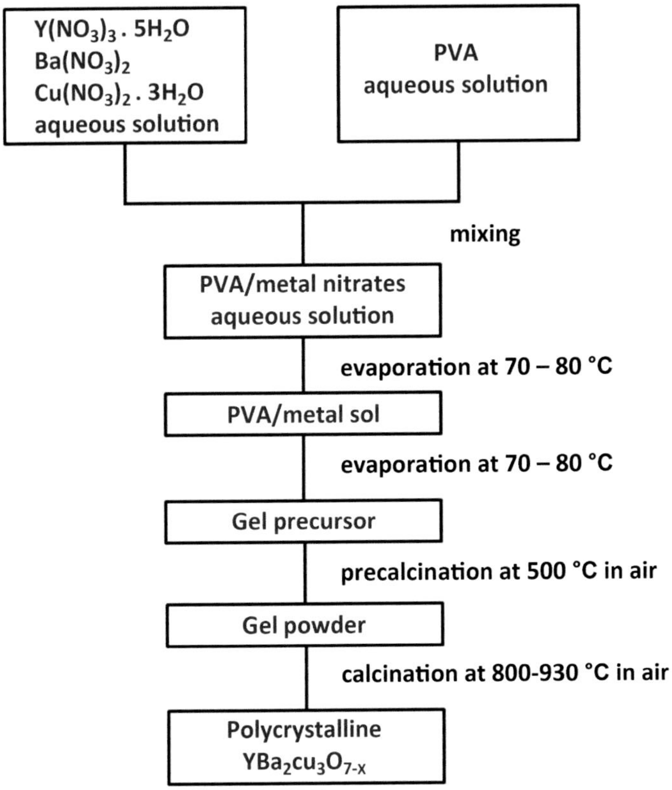

FIGURE 10.8

Flowchart showing polymeric nanoparticle preparation by sol-gel method.

Reprinted with permission from Sun, Y.-K., & Oh, I.-H. (1996). Preparation of ultrafine $YBa_2cu_3O_{7-x}$ superconductor powders by the poly(vinyl alcohol)-assisted Sol-Gel method. Industrial & Engineering Chemistry Research, 35, *4296–4300.*

10.4.6 Solvent evaporation

Solvent evaporation is the most commonly utilized approach for preparing polymeric nanoparticles for active ingredient delivery. This approach is used to create polymeric nanoparticles from biodegradable polymers, which are then employed in medication delivery systems. The polymer and medicine are dissolved in an organic solvent in this process, and emulsions are created using an aqueous solution of surfactants. Previously, the solvent for dissolving the hydrophobic medication was dichloromethane or chloroform. However, due to their toxicological profiles, ethyl acetate is now employed instead of dichloromethane and chloroform (Mohanraj & Chen, 2007; Rao & Geckeler, 2011). For the creation of emulsions, the solvent evaporation method employs two major methodologies. The first is the preparation of single-emulsions (oil-in-water [O/W]), and the second is the preparation of double-emulsions (water-in-oil-in-water [W/O/W]).

High-speed homogenization or ultrasonication is utilized to prepare the emulsion in this procedure. The organic solvent is then evaporated, and the nanoparticles are recovered using ultracentrifugation. The nanoparticles are rinsed with distilled water to remove surfactants. After completing all the processes, the nanoparticles can be lyophilized.

While the solvent evaporation approach is a simple way of preparing polymeric nanoparticles, it requires external energy. External energy power, time consumption, and probable agglomeration of nanodroplets during the evaporation process may all have an impact on the particle size and morphology of nanoparticles (Fig. 10.9) (Rao & Geckeler, 2011).

10.4.7 Dialysis

Dialysis is a simple and effective procedure for preparing tiny PNPs. The polymer is first dissolved in an organic solvent before being placed in a dialysis tube to produce nanoparticles via dialysis. It is critical that the dialysis tube is appropriate for the nanoparticle's molecular weight. As a result of displacement during dialysis, the solvent loses its solubility. Progressive polymer aggregation occurs as a result, and a homogenous dispersion of nanoparticles is formed (Fig. 10.10) (Kumar & Sharma, 2018; Rao & Geckeler, 2011).

10.4.8 Salting out

Another approach for preparing polymeric nanoparticles is salting out. This approach is based on the salting out action to separate a water miscible solvent like acetone. In other words, the salting-out technique modifies the emulsification/solvent diffusion process by starting with an emulsion (Fonseca et al., 2013; Pulingam et al., 2022). The organic phase is created by dissolving the polymer in a water miscible solvent such as acetone, THF, or ethanol. The organic phase is then mixed

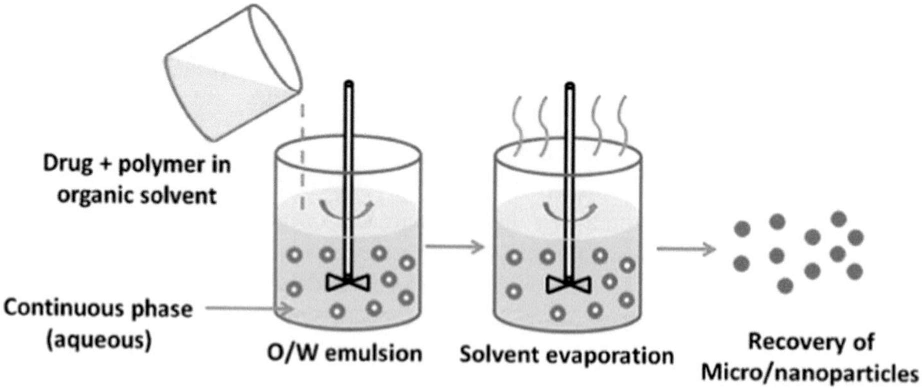

FIGURE 10.9

Solvent evaporation for polymeric nanoparticle preparation.

Reprinted with permission from Rao, J. P., & Geckeler, K. E. (2011). Polymer nanoparticles: Preparation techniques and size-control parameters. Progress in Polymer Science, 36, *887–913.*

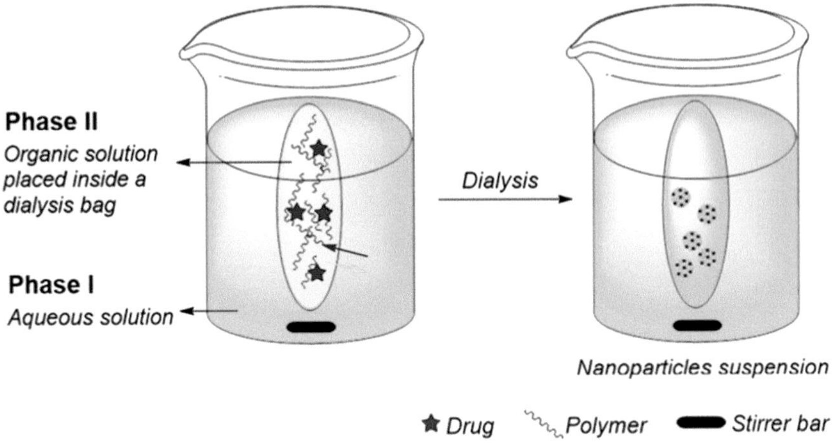

FIGURE 10.10

Dialysis method for polymeric nanoparticle preparation.

Reprinted with permission from Kumar, D., & Sharma, P. K. (2018). Nanoparticulate system for cancer therapy: An updated review. International Journal of Nanotechnology and Nanomedicine, 4, *022–034.*

with the aqueous phase, which is made up of water, the salting out agent, and a stabilizer. The choice of salting out agent, which might be electrolytes like calcium chloride or magnesium chloride or nonelectrolytes like sucrose, is critical for drug encapsulation efficiency.

Following this method, a sufficient volume of water is added to the mixture, and the resultant nanoparticles are collected by cross-flow filtration (Fonseca et al., 2013). The salting out method has some advantages, the most noteworthy of which is that it reduces stress to protein encapsulants. Another advantage is that there is no temperature increase. As a result, heat sensitive compounds can be treated using this approach. The opposite of its benefits, this approach has some drawbacks, including a focus on lipophilic medicines and significant washing processes (Fig. 10.11).

10.4.9 Supercritical fluid technology

Fluids that do not change phase despite pressure changes are known as supercritical fluids. Because it is compatible with the critical state ($T_c = 31.1$ °C, $P_c = 73.8$ bar), nontoxic, on-flammable, and affordable, supercritical CO_2 is the most commonly employed supercritical fluid (Tsai & Wang, 2019; Wang et al., 2016). Supercritical fluid technology is intended to provide an intriguing and practical particle generation process that avoids many of the drawbacks of previous technologies. The most significant advantage of using supercritical fluid technology to manufacture polymeric nanoparticles is that the precipitated product contains no solvent.

Rapid expansion of supercritical solution (RESS) and the supercritical antisolvent method (SAS) are the two most commonly employed methods. The drug ingredient is dissolved in an organic solvent and then released into the supercritical fluid using the RESS technique. The organic

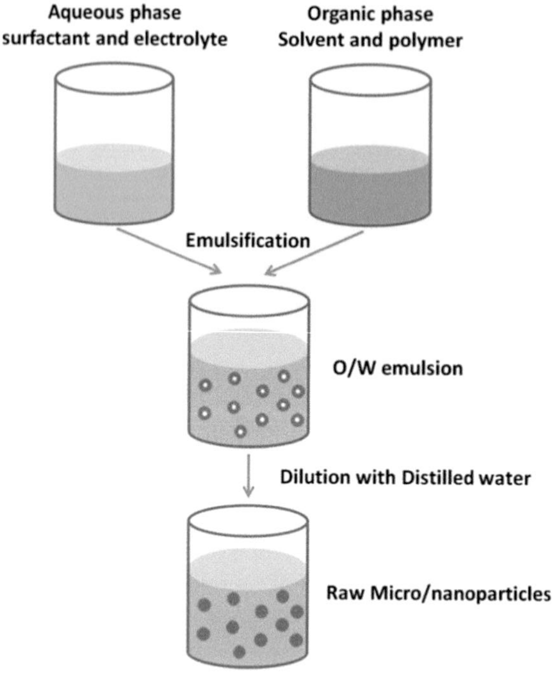

FIGURE 10.11

Salting out method for polymeric nanoparticle preparation.

Reprinted with permission from Fonseca, A. C., Ferreira, P., Cordeiro, R. A., Mendonça, P. V., Góis, J. R., & Gil, M. H. (2013). Drug delivery systems for predictive medicine: Polymers as tools for advanced applications. In New Strategies to Advance Pre/Diabetes Care: Integrative Approach by PPPM *(pp. 399–455). Dordrecht: Springer Netherlands.*

phase dissolves quickly in the supercritical solvent, leaving only nanoparticles that can be filtered back out. The active ingredient and polymer are dissolved in a supercritical solvent under high pressure in the SAS process (Fig. 10.12) (Sun et al., 2005).

10.4.10 Advanced technology

Many features of nanoparticles are determined by their shape, size, polydispersity, and surface chemistry. To be used in chemical sensing, medical diagnostics, catalysis, thermoelectrics, photovoltaics, or pharmaceuticals, nanoparticles must be synthesized with carefully regulated properties. Because nanoparticle syntheses sometimes include many chemicals and are carried out in interdependent experimental settings, this is a time-consuming, labor-intensive, and resource-intensive task. Machine learning is a promising method for the rapid creation of efficient procedures for nanoparticle synthesis and, possibly, the synthesis of new types of nanoparticles (Tao et al., 2021).

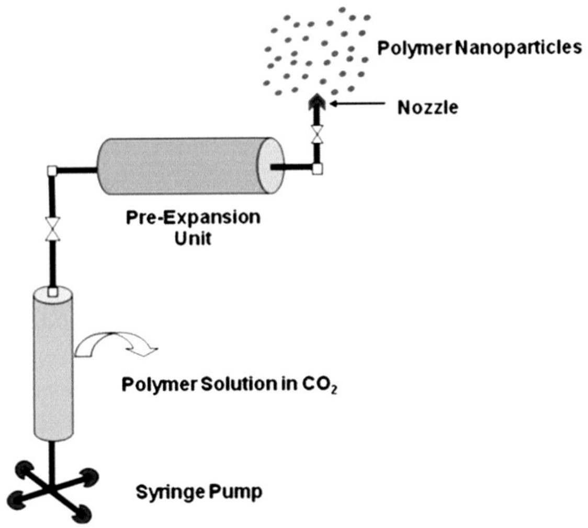

FIGURE 10.12

Supercritical fluid technology for polymeric nanoparticle preparation.

Reprinted with permission from Sun, Y.-P., Meziani, M. J., Pathak, P., & Qu, L. (2005). Polymeric nanoparticles from rapid expansion of supercritical fluid solution. Chemistry, 11, *1366–1373.*

10.5 Sensors applications of polymer nanomaterials

10.5.1 Chemosensing

The overall environmental quality is continuously declining due to the tendency toward urbanization and industrialization as the world's population slowly increases. Currently, air pollution is one of the most significant environmental risks, and it is closely linked to rising healthcare costs, a rise in the number of preventable deaths, and a decline in global economic output (Mamun & Yuce, 2020; Shafik & Bandyopadhyay, 1992). In addition, the threat of utilizing biological weapons and the use of pesticides are both growing. It is impractical for humans to entirely shield themselves from the dangers of the surrounding environment. However, due to the significance of metal ions and small chemical molecules in terms of how they affect people and the environment, extensive research has been done on the design and development of very sensitive chemical sensors. Some heavy metals like Cu^{2+}, Fe^{2+}, Pb^{2+}, Ag^{2+}, and Hg^{2+} impose extremely hazardous effects on the environment and human health due to their acute toxicity (Alvarez et al., 2011). Similarly, volatile organic compounds (VOCs) and some small molecules like nitroaromatics, which are typically toxic, explosive, and flammable, are closely correlated with the progression of some diseases like cancer, Alzheimer's disease, cardiovascular disorders, and kidney stones (Chunlei et al., 2011; Tomić et al., 2021). However, the impacts may be mitigated by early threat detection by monitoring one's surrounding microenvironment. Rapid and accurate sensors that can identify pollutants at the molecular level could improve environmental understanding and protection. If more sensitive and less expensive approaches for pollutant detection were available, manufacturing, ecosystem monitoring, compliance, process control, and environmental decision-making would also be greatly enhanced. Chemical sensors based on PNMs are becoming increasingly crucial in monitoring our environment, providing data on industrial manufacturing processes and their emissions, food, and beverage quality management, and various other uses due to their large specific surface area and good biocompatibility (Adhikari, 2021a; Goutam et al., 2011; Hussain et al., 2015; Nasir et al., 2015; Saikia et al., 2012; Simkhovich et al., 2002; Tanwar et al. 2019, 2022; Tanwar et al., 2020).

10.5.1.1 Polymer nanomaterials based on ion detection

Some heavy metal ions (Ag, Pb, Hg, Cu, Fe, and so on) are severe environmental contaminants that can cause serious concern on a global scale because of their nonbiodegradable nature. They are only necessary for our bodies up to a certain concentration limit; thus, higher consumption or exposure amounts beyond body requirements can result in both acute and chronic poisoning. As a result, there is a high demand for the precise identification and total elimination of these hazardous metal ions.

The PNMs were selectively designed to specifically chelate with ions, which will induce distortion in the architecture of the PNMs, thus leading to a change in optical properties. In 2011, a highly sensitive P1a-based PNM was introduced to detect Cu^{2+} and Fe^{2+} (Fig. 10.13). The P1a-based probe was embedded on poly(styrene-co-maleic anhydride) (PSMA) to form PNM-P1a, consisting of carboxyl groups on the surface via hydrolysis of the maleic anhydride units. The fluorescence intensity at 540 nm of the PNM-P1a probe gets quenched upon aggregation in the presence of Cu^{2+} and Fe^{2+} due to interparticle cross-linking upon coordination with metal ions. Furthermore, individual quantification of Cu^{2+} and Fe^{2+} was obtained due to the selective recovery capability of PNM-P1a toward Cu^{2+} (Chan, Jin, et al., 2011).

In 2013, another two different PNMs, i.e., P1b and P2a, were developed by doping with a photoswitchable spiropyran (pSP) for specific detection of Cu^{2+} (Fig. 10.13). Then the resulting two PNMs, pSP-PNM-P1b and pSP-PNM-P2a, were mixed to form the probe, which gets activated upon UV irradiation to form open merocyanine with the capability of chelating Cu^{2+}. This results in shortening the distance between pSP-PNM-P1b and pSP-PNM-P2a, which promotes FRET, thus allowing the detection of Cu^{2+} in the physiological range as shown in Fig. 10.13 (Wu et al., 2013).

Then an activable PNM probe for Hg^{2+} detection was developed based on P1a and a rhodamine spirolactam dye (RB-SL). As shown in Fig. 10.13, in the presence of Hg^{2+}, RB-SL changes into RB-Hg, which results in significant emission enhancement of RB-Hg at 590 nm with a decrease in emission intensity of P1a at 537 nm due to the FRET process. The PNM probe was highly sensitive to Hg^{2+}, with a detection limit as low as 0.7 parts per billion (ppb) (Childress et al., 2012).

An Ag^{2+} ion responsive PNM probe was introduced by incorporating P1 in the backbone of sulfonated polystyrene-block-poly(ethylene-ran-butylene)-block-polystyrene (PS-SO_3H), where the benzothiadiazole core of P1 acts as an active chelating or metal binding site for Ag^{2+} ions. However, the negatively charged sulfonic acid groups of amphiphilic PS-SO_3H located on the surface of nanomaterials partially took part in the metal binding process. An aggregation-induced fluorescence quenching mechanism toward the Ag^{2+} ion was observed, which can be reclaimed to its pristine state upon the addition of NaCl to sequestrate Ag + (Yang et al., 2013).

Lead ions (Pb^{2+}) are often detected using expensive equipment and laborious sample processing procedures. Therefore, an easily prepared PNM probe was developed incorporating P3 and a dye NIR695 (Fig. 10.13) with the carboxyl- and 15-crown-5-functionalized polydiacetylenes (PDAs)

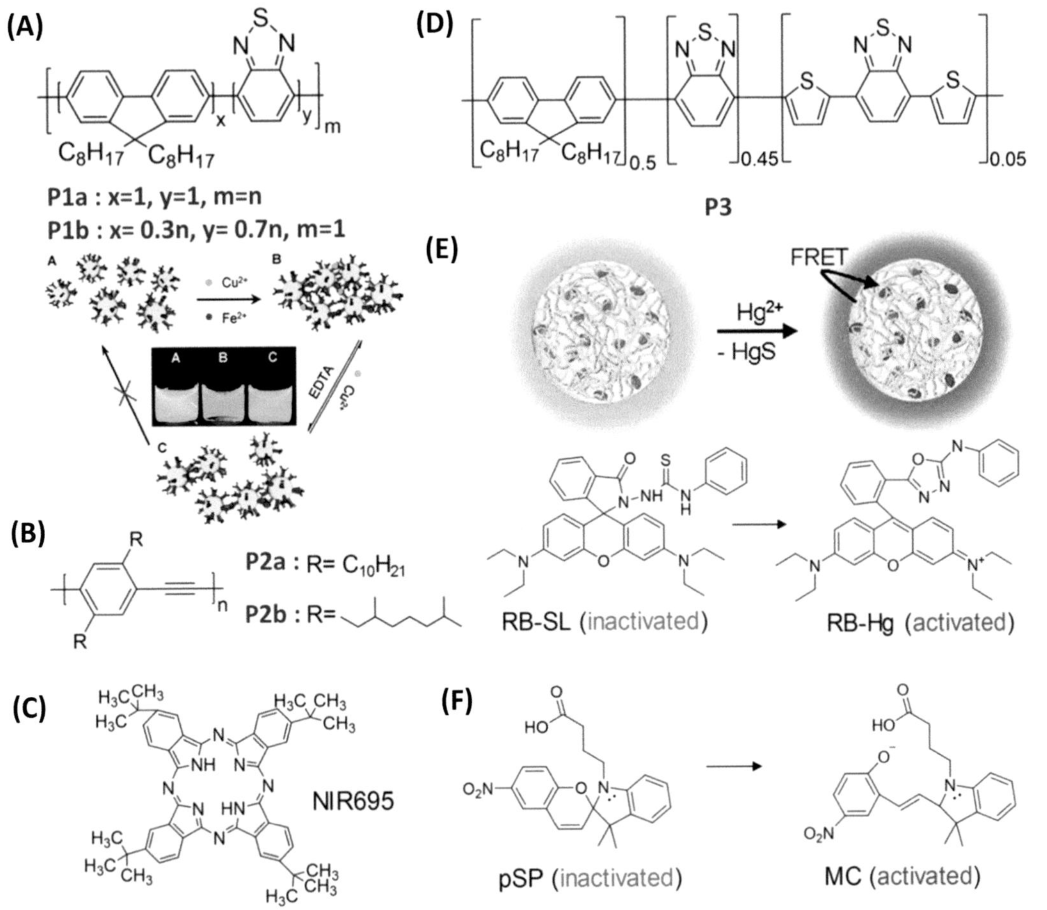

FIGURE 10.13

(A) Chemical structure of PNM and schematic representation showing Cu^{2+} and Fe^{2+} ion detection using P1a PNM. Reproduced with permission (Chan, Jin, et al., 2011). (B)–(D) Chemical structure of P2a, P2b, and P3 respectively. (E, F) Ion detection mechanism of RB-SL and pSP respectively. Reproduced with permission (Childress et al., 2012).

mixture for the detection of Pb^{2+} ions. A photochromic shift from blue to red was observed due to the PDAs chelating Pb^{2+}, which led to partial distortion or shortening of the conjugated system (Kuo et al., 2015).

10.5.1.2 Polymer nanomaterial–based volatile organic compound detection

Due to the high vapor pressure of VOCs, they can quickly pollute the air we breathe and have an impact on a number of biological functions that are essential for human health. VOCs can be generally categorized into seven kinds based on the functional groups present in the chemical structure, which include halogenated, oxygenated, aliphatic, and aromatic hydrocarbons, as well as aldehydes, esters, and terpenes. Therefore, introducing an effective and highly selective technique for VOC detection is essential. Recently, PNMs with large reactive surface areas and size-dependant characteristics now offer a new approach to dealing with such issues (Tomić et al., 2021). The sensitivity of a PNM is directly dependent on the interaction between the attached functional group to the polymer and the analyte molecule; however, mixing these PNM with other nanomaterials can enhance its sensitivity by forming an unstable nanomorphology that can smoothly reorganize upon VOC exposure (Andre et al., 2018). The absorption and adsorption processes of gas molecules may be the reason for a gas sensor's reaction. However, the swelling process of the PNM layer occurs due to gas sorption, and the swelling process is reversible as it indicates no chemical bond has formed between the polymer and the analyte. In 2004, a swelling-induced mechanochromic PNM was reported, where the conjugated polydiacetylenepolydimethylsiloxane compound (PDA-PDMS) shows a colorimetric transition from blue to red upon exposure to hydrocarbons over a certain time period depending on the analyte chain length (Fig. 10.14). The development of the red color is precisely correlated with the film's swelling, making it possible to visually distinguish between n-pentane and n-heptane. The edges of the film are significantly affected by swelling because there are weaker forces of attraction between PDA molecules there, leading to the separation of the chains from the edges to the center. The increase in the inter-chain distance brought about by this modification permits some C-C single bond-free rotation of the polymer backbone, which results in orbital distortion and, as a result, the change from blue to red (Park et al., 2014).

The ability to detect several VOCs at room temperature, including CH_4 and NH_3, was investigated using thin films of copper nanoparticles (CuNPs) embedded into PANI. The sensor

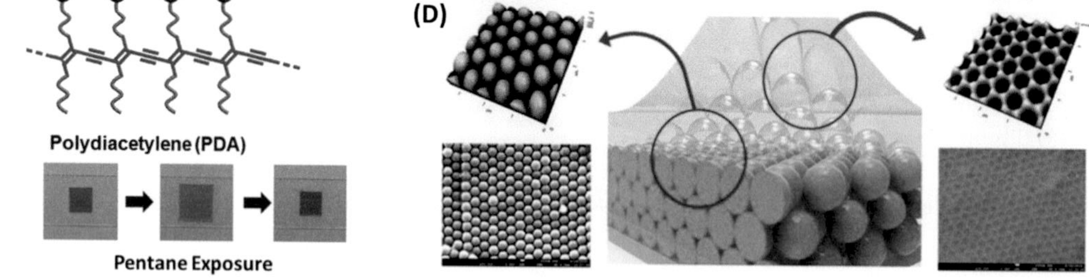

FIGURE 10.14

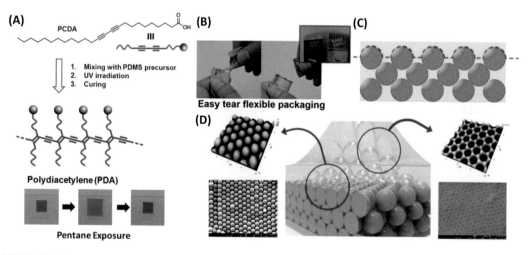

(A) Structure and Schematic of PCDA and PDA supramolecules respectively. Reproduced with permission (Park et al., 2014). (B) Representation of nanoscale easy tear (NET) process motivated by commercial easy tear packaging. (C) Colloidal crystal-PDMS composition formation. (D) AFM and SEM images of colloidal crystal-PDMS composite and upper bulk PDMS layer. Reproduced with permission (Chang et al., 2018).

demonstrated improved responsiveness and kinetics for NH_3 detection. Cyclic voltammetry was used for the detection, and an oxidation peak with a center value of +0.1 V was attributed to the electrooxidation of a copper and ammonia complex. Up to 10% of the volume, the system showed a linear relationship to ammonia content (Patil et al., 2015).

Based on the swelling effect, an ultrafast responsive photonic crystal-based sensor was developed consisting of a colloidal crystal-PDMS composite following a nanoscale easy tear process motivated by a commercially available "easy tear package" (Fig. 10.14), which exhibits distinguished and rapid color change upon exposure to VOCs. A colloidal crystal-PDMS composite was obtained through the nanoscale propagation of tears at the interface between the bulk PDMS and the outer surface of crystallized PNMs. Due to the complete removal of the residual on the colloidal crystals, the response time for VOC detection significantly decreases when VOCs are allowed to come into direct contact with the sample (Chang et al., 2018).

10.5.1.3 Polymer nanomaterial–based nitro-explosive detection

Recent years have witnessed a rapid expansion in the study of explosive material detection and precise quantification, with implications for environmental safety, public health, and homeland security. The most often used explosive substances include nitro-based compounds (2,4,6-trinitrotoluene [TNT], 2,4-dinitrotoluene [DNT], 1,3,5-trinitro-1,3,5-triazinane [RDX], and picric acid [PA]), nitrates, and peroxides like triacetone triperoxide (TATP). Nitroexplosive substances are frequently used in the leather, matchbox, dye, pharmaceutical, and fireworks industries. Reliable detection of traces of explosive chemicals is still a difficult challenge due to their high sensitivity to friction, heat, or shock and low vapor pressure. However, PNMs with the capacity to detect nitrogen-containing nitro aromatic compounds (NAC) such as explosives in both the solution and gaseous phases, with ultra-sensitivity down to picomolar and even lower concentrations, high reactive surface areas, and rapid reaction are more desirable. Numerous techniques have been reported for the minute-level detection of NAC, including surface-enhanced Raman spectroscopy, trained canines, infrared spectroscopy, X-ray imaging, mass spectrometry techniques, electrochemical techniques, gas chromatography, and others. Even though those techniques provide trustworthy results, however, a collaboration of analytical methods with PNMs would provide better onsite detection results with low cost, quick response, and high repeatability.

Since NAC are electron donating in nature, their detection relies on the electron transfer pathway from electron-rich NMs to NAC, which causes the quenching of fluorescence. The three different PNMs P4, P5, and P6 were combined to exhibit emission intensity at 614 nm via the FRET mechanism (Fig. 10.15). The presence of nitroaromatics led to photoinduced electron transfer (PET), which reduced the fluorescence emission of PNMs by moving electrons from electron-rich PNMs to electron-deficient nitroaromatics. Both test trips and solutions were used to test the ability of PNMs to detect nitroaromatics (Huang et al., 2015).

A cationic conjugated polyelectrolyte poly(3,3′-((2-phenyl-9H-fluorene-9,9-diyl)bis(hexane-6,1-diyl))bis(1-methyl-1H-imidazol-3-ium)bromide) (PFMI) was designed for selective PA detection. Additionally, contact mode detection of PA was carried out employing simple, affordable, and portable fluorescent paper strips for onsite detection. Additionally, the two-terminal PFMI-NMs sensor device was fabricated, which offers an unprecedented and outstanding platform for the vapor mode detection of PA in an ambient environment. The "molecular wire effect," PET, potential

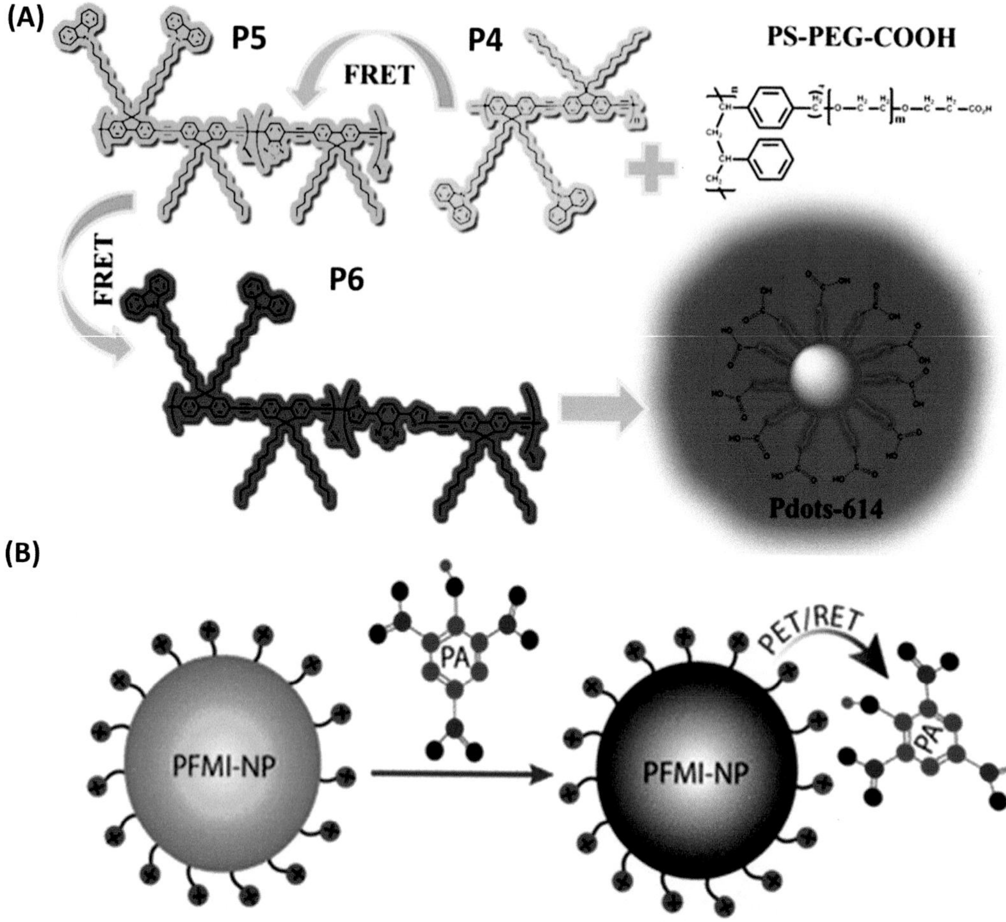

FIGURE 10.15

Representation of (A) formation of Pdots and (B) picric acid sensing mechanism using PEMI-NP.

Reproduced with permission from Huang, J., Gu, J., Meng, Z., Jia, X., & Xi, K. (2015). Signal enhancement of sensing nitroaromatics based on highly sensitive polymer dots. Nanoscale, 7, *15413–15420; Malik, A. H., Hussain, S., Kalita, A., Iyer, P. K. Conjugated polymer nanoparticles for the amplified detection of nitroexplosive picric acid on multiple platforms.* ACS Applied Materials & Interfaces, 7, *26968–26976.*

resonance energy transfer, and electrostatic interaction are all implicated in the process underlying the PFMI-NMs probe's extreme sensitivity to PA (Fig. 10.15) (Malik et al., 2015).

10.5.1.4 Polymer nanomaterials based on PH sensing

In physiological processes, pH is regarded as a crucial variable since it has the power to directly change the configuration and activity of biomolecules. By disrupting bodily homeostasis, the aberrant pH value is linked to a variety of illnesses, including Alzheimer's disease, heart ischemia, cancer, and inflammation; thus, pH sensing is crucial for life science. Based on P7 and pH-sensitive fluorescein, Chiu's team developed a PNM probe that exhibited two emission peaks due to FRET between P7 and the fluorescein dye, one from the dye at 513 nm, which was sensitive to pH, and one from P7 at 440 nm, which was inert to pH (Fig. 10.16). Thus, the PNM-P7 probe matched the standard criteria for cellular research by ratiometrically measuring pH from 5 to 8 (Chan, Wu, et al., 2011).

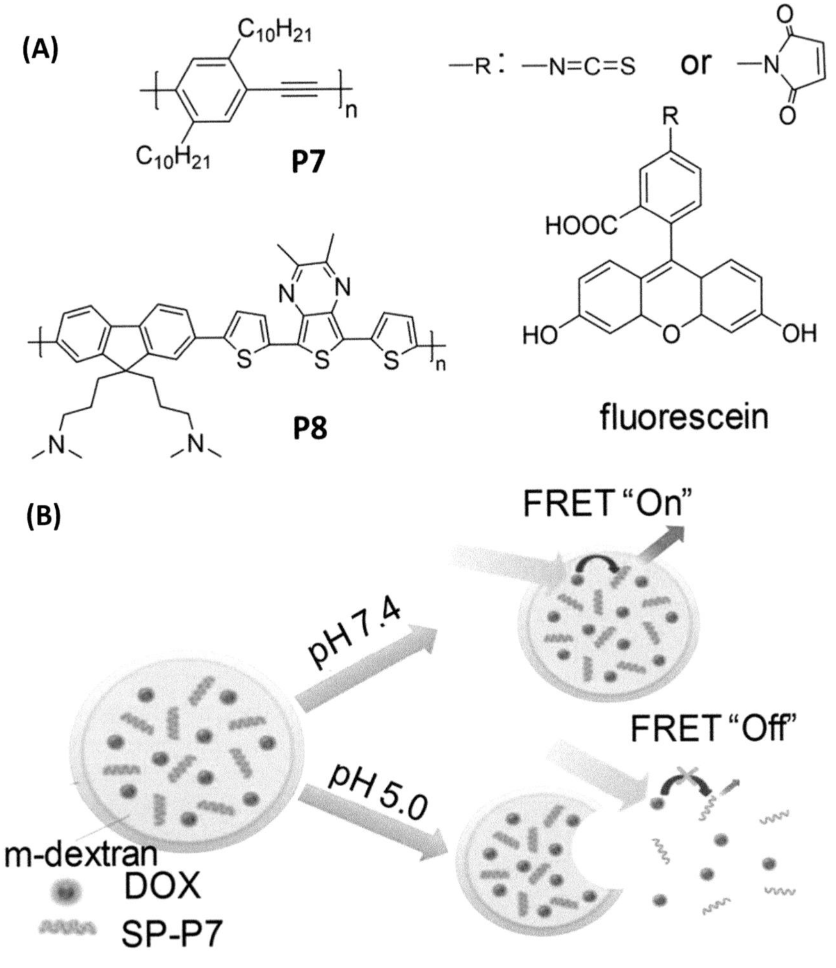

FIGURE 10.16

(A) Chemical structure of PNM-based probe and dyes utilized for fluorescence sensing of pH (Chan, Wu, et al., 2011). (B) Schematic representation illustrating the control of drug release capability and signal alteration of PNM-P8 upon different pH.

Reproduced with permission Yu, J.-C., Chen, Y.-L., Zhang, Y.-Q., Yao, X.-K., Qian, C.-G., Huang, J., Zhu, S., Jiang, X.-Q., Shen, Q.-D., & Gu, Z. (2014). PH-responsive and near-infrared-emissive polymer nanoparticles for simultaneous delivery, release, and fluorescence tracking of doxorubicin in vivo. Chemical Communications, 50, *4699–4702.*

In 2014, another PNM-based probe was synthesized utilizing poly[9,9-bis(N,N-dimethylpropan-1-amino)-2,7-fluorene-alt-5,7-bis(thiophen-2-yl)-2,3-dimethylthieno[3,4-b]pyrazine] (BTTPF) as a matrix material for pH sensing. The PNM-based probe P8 could load drugs and control their release upon pH detection. The pendant acetal-modified dextran (m-dextran) was utilized to encapsulate doxorubicin (DOX) and P8, which gave the probe its pH-sensitive properties. In a slightly acidic environment, PNMs were dissociated as a result of m-dextran being hydrolyzed. However, the FRET procedure was hampered by the separation of P8 and DOX; as a result, the DOX fluorescence was restored and the P8 emission at 690 nm diminished (Fig. 10.16) (Yu et al., 2014).

10.5.2 Biosensing

Sensors play a crucial role in modern technology, detecting temperature changes, pressure fluctuations, chemical composition variations, and more. PNMs, synthesized at the nanoscale (1–100 nm), offer unique properties like high surface area, significant porosity, and adjustable chemical and physical traits. These features make them ideal for cutting-edge sensor development, particularly in the promising field of biosensors. Biosensors are intricate apparatuses that employ biological constituents to discern precise biomolecules, such as proteins or DNA, within intricate specimens, such as blood or saliva. The electrical component is responsible for detecting, recording, and transmitting information regarding physiological alterations or the existence of various biological or chemical constituents in the environment. Biosensors manifest in diverse dimensions and configurations, enabling the discernment and quantification of minute quantities of distinct biomarkers, noxious substances, and pH levels. Polymers serve as the primary constituents for the fabrication of numerous biosensors. The facile nature of its synthesis and processing, its remarkable adaptability, and its molecular imprinting are merely a few of the fundamental constituents that contribute to its widespread utilization. Biocompatible polymers, which exhibit a high degree of compatibility with biological systems, are readily accessible, thereby establishing their status as the materials of choice for biosensors that can be injected or implanted. Diverse categories of polymers have been utilized in diverse manners to synthesize biosensors. PNMs serve as an excellent matrix for the immobilization of biologically active entities, including enzymes or antibodies, thereby facilitating the development of biosensors with exceptional sensitivity and specificity. Polymeric nanomaterials have been extensively utilized in biosensors due to their distinctive physical, chemical, and biological characteristics (Lee et al., 2012; Mondal et al., 2021; Muthuraj et al., 2013; Muthuraj et al., 2016; Zehra et al., 2018).

In recent times, numerous types of sensors implementing CP nanomaterials have been elucidated for their applications in the realm of biology. CPNs exhibit great promise as 1D nanostructured materials owing to their unique chemical-doping specificities, tunable transport properties, substantial surface areas, diverse structural variations, cost-effectiveness, and straightforward synthesis techniques (Oh et al., 2013). The oxidation state of CP materials can be readily influenced by their inherent reversible doping-dedoping (redox) mechanisms, leading to fluctuations in conductivity. Furthermore, the charge transport characteristics of CP materials are subject to the influence of structural parameters, including the diameter and aspect ratio. Henceforth, CP materials can manifest sensitive and expeditious reactions toward distinct chemical or biological entities. Electrochemical synthesis, akin to chemical polymerization, is commonly employed for the fabrication of CPNs. Synthetic strategies can be categorized as hard-template, soft-template, and template-free. The commonly employed chemical compounds for the advancement of novel biosensors encompass PPy, PANI, and PEDOT. PPy nanomaterials have been utilized in a myriad of biosensors, including DNA sensors and aptamer sensors for thrombin and vascular endothelial growth factor detection, bioelectronic nose, and anticancer agent detection. These applications are attributed to the distinctive chemical and electrical properties of PPy nanomaterials, which stem from their conjugated p-electron system. Furthermore, recent scientific literature has indicated that carbon polymer nanomaterials exhibit remarkable potential as optimal contenders for the advancement of cutting-edge biosensors with exceptional performance capabilities.

Several notable applications of PNMs based on their properties in biosensors include:

Sensing elements: PNMs exhibit remarkable potential as sensing elements in biosensors owing to their exceptional surface-area-to-volume ratio, thereby facilitating heightened sensitivity and selectivity in the detection of target molecules. As an illustration, nanoparticles composed of polymer compounds like polystyrene, PEG, and polyacrylic acid have been employed as sensing agents for the detection of glucose, cholesterol, and DNA.

Signal amplification: PNMs can serve as amplifiers of signals in biosensors. Through the conjugation of nanoparticles onto the biosensor's surface, the signal generated by the sensing element can be enhanced, thereby augmenting the biosensor's sensitivity. Gold nanoparticles, which have been adorned with polymer coatings like PEG and PVA, have been effectively employed as enhancers of signals in the realm of glucose sensing.

Biocompatibility: PNMs exhibit excellent biocompatibility, rendering them suitable for integration into biosensors without eliciting any deleterious effects on the biological milieu. As an illustration, polymeric nanoparticles, like poly(lactic-co-glycolic acid), have been effectively employed as vehicles for drug delivery and biosensors for the detection of biomolecules, including proteins and nucleic acids.

Stability and durability: PNMs exhibit commendable stability and exceptional durability, rendering them highly suitable for deployment in the realm of biosensors. These substances exhibit remarkable resilience against extreme environmental conditions, including elevated temperatures and pressures, while concurrently demonstrating notable resistance against enzymatic degradation and other biological agents. PNMs, namely PANI and PPy, have been employed as sensing elements in glucose and cholesterol sensing applications owing to their exceptional stability and durability.

Surface functionalization: PNMs can effectively alter biosensors' surface properties, thereby augmenting their sensitivity and selectivity. As an illustration, the utilization of PEG in the context of biosensors allows for surface modification, thereby enhancing biocompatibility and mitigating nonspecific binding.

Signal amplification: PNMs possess the capability to augment the signal of biosensors, thereby augmenting the detection limit and sensitivity of said biosensors. In the realm of chemical applications, it is worth noting that gold nanoparticles possess the capability to function as signal amplifiers within biosensors. This particular attribute allows for the augmentation of the surface plasmon resonance signal, thereby resulting in enhanced detection sensitivity.

Biomimetic membranes: PNMs possess the capability to fabricate biomimetic membranes intended for biosensors, thereby emulating the inherent characteristics of the cellular membrane and enhancing the discerning nature and precise targeting of biosensors. In the realm of chemistry, polymer vesicles have the potential to serve as biomimetic membranes in the context of biosensors. These membranes possess the ability to imitate the phospholipid bilayer found in cellular membranes, thereby enhancing the biocompatibility of biosensors.

Enzyme immobilization: PNMs possess the capability to effectively immobilize enzymes onto biosensors, thereby augmenting the enzymes' stability and activity. In the realm of chemistry, it is worth noting that the utilization of poly(acrylic acid) nanoparticles presents a promising avenue. These nanoparticles can effectively immobilize enzymes on biosensors. This particular application holds the potential to enhance both the stability and activity of enzymes, consequently leading to an overall improvement in the sensitivity of biosensors.

DNA sensors: PNMs possess the remarkable capability to facilitate the development of DNA sensors, enabling the detection of DNA sequences with exceptional specificity and sensitivity. In the realm of chemistry, one intriguing application involves the utilization of polymer nanowires as DNA sensors. These nanowires possess the remarkable ability to detect DNA sequences, exhibiting a detection limit as low as 10 femtomoles (Fig. 10.17).

PVP, a polymer possessing water solubility, finds extensive application across diverse sectors such as pharmaceuticals, medicine, cosmetics, and more. Within the realm of pharmaceuticals, this particular substance finds utility as a coating agent, binder, pore-former, suspending agent, solubilizer, and various other excipients. These excipients serve a crucial role in both conventional formulations and emerging controlled or targeted delivery systems, as well as in the development of biosensors (Pourmadadi et al., 2023). Some PVP-based nano-biosensors developed after 2010 are discussed in Table 10.1 (Fig. 10.18).

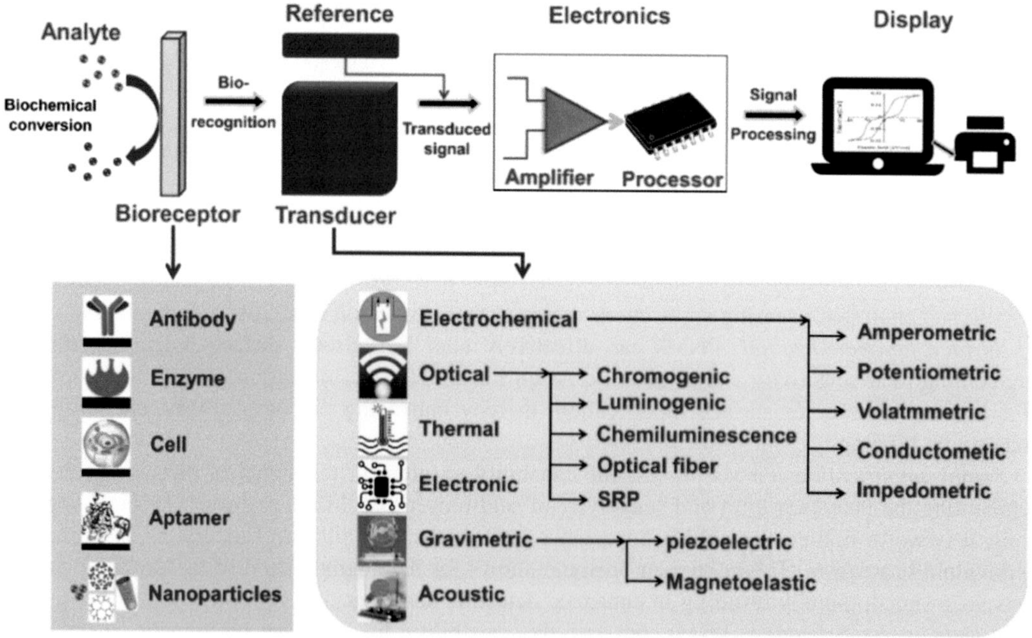

FIGURE 10.17

Schematic representation of different components of a typical biosensor.

Adapted with permission from Pourmadadi, M., Shamsabadipour, A., Aslani, A., Eshaghi, M. M., Rahdar, A., & Pandey, S. (2023). Development of polyvinylpyrrolidone-based nanomaterials for biosensors applications: A review. Inorganic Chemistry Communications, 152, *Article 110714.*

Table 10.1 Recently developed biosensors based on various miscellaneous methods.

Biosensors materials	Biorecognition elements	Analyte	Method	Sensing	Advantages	References
PVP/G	sDNA/single-strand probe DNA	A typical gene of the hepatitis B virus	Optical	LOD: 62.35 nM without PVP, 1.56 nM with PVP	Simple, rapid, sensitive, and inexpensive	Zhang et al. (2015)
CPPy NT	Aptamer	Thrombin	Electrochemical	LOD: 50 nM	Very low detection limit	Yoon et al. (2008)
PPy	Anti-CA 125 Ab	CA 125	Electrochemical	LOD: 1 U/mL	Very low detection limit, and cost-effective.	Bangar et al. (2009)
PVP/ sucrose	Enzyme/ horseradish peroxidase (HRP)	3,3′,5,5′-tetramethylbenzidine	Optical	–	Reduced operator involvement and capable of storing sensitive reagents	Dai et al. (2012)
G/PVP/ PANI	Enzyme/cholesterol oxidase	Cholesterol	Electrochemical	LOD: 1 μM Linear range: 50 μM–10 mm Sensitivity: 34.77 $\mu A/mM/cm^2$	Simple, cost-effective, disposable, and portable	Ruecha et al. (2014)
AuNPs/ PVP/PANI	Glucose oxidase	Glucose	Electrochemical	LOD: 10^{-5} M. Linear range: 0.05–2.25 mm Sensitivity: 9.62 $\mu A/\mu M/cm^2$.	High stability and reproducibility	Miao et al. (2015)
PVP capped PBNPs	Anti-human IgG antibody and anti-PTHrP	Human IgG and PTHrP	Electrochemical	LOD: 50 ng/mL	Label-free detection	Espinoza-Castañeda et al. (2015)
fMWCNT/ AuNP/PVP	Mouse monoclonal PSA antibody	Prostate specific antigen (PSA)	Electrochemical	LOD: 1 ng/mL. Linear range: 0–4 ng/mL	High sensitivity and wide linear range	Quintero-Jaime et al. (2019)
PVP NFS	HRP tagged antibody	*E. coli*; I O157	Optical	LOD: 2×10^6 CFU/mL of *E. coli*.	Capable of storing sensitive reagents	Jin et al. (2013)

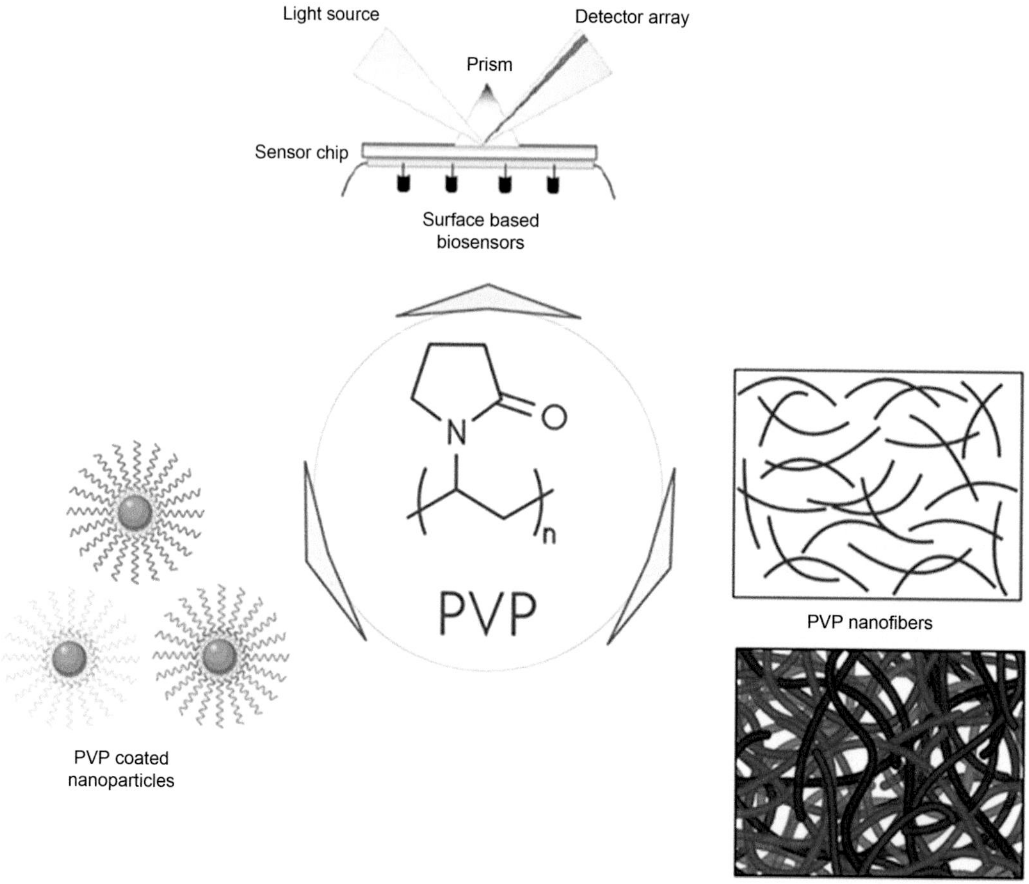

FIGURE 10.18

Schematic representation of PVP-coated nanoparticles and their application to biosensors.

Adapted with permission from Pourmadadi, M., Shamsabadipour, A., Aslani, A., Eshaghi, M. M., Rahdar, A., & Pandey, S. (2023). Development of polyvinylpyrrolidone-based nanomaterials for biosensors applications: A review. Inorganic Chemistry Communications, 152, *Article 110714.*

In general, PNMs exhibit remarkable potential for application in biosensors owing to their distinctive physical, chemical, and biological characteristics. The exceptional surface-area-to-volume ratio, remarkable biocompatibility, inherent stability, and commendable durability render them exceedingly well-suited for deployment as sensing elements, signal amplifiers, and drug delivery systems in the realm of biosensors.

10.6 Conclusion and future aspects

In conclusion, sensors based on PNMs have emerged as a cutting-edge technology with tremendous potential to revolutionize the field of sensing. The distinct attributes of PNMs—substantial surface area, adjustable physicochemical traits, biocompatibility, and the molecular wire effect—have facilitated the creation of remarkably efficient, sensitive, and specific sensors across a broad range of applications. This chapter offers an in-depth analysis of diverse polymer nanomaterial categories, spanning conjugated, nonconjugated, and hybrid variations. It further investigates essential characterization techniques and methodologies crucial for synthesizing these nanomaterials. The chapter extensively covers applications in chemosensing, biosensing, and optical and electrochemical sensing, effectively showing the manifold potential of PNMs. Hybrid nanocomposites, combining PNMs with other functional materials, have opened new avenues for sensor design. By tailoring the properties of these composites, researchers have achieved enhanced sensor performance, enabling real time monitoring of complex systems and increasing our understanding of various processes. Furthermore, the synthesis and functionalization strategies of PNMs have undergone significant refinement, allowing for precise control over their properties and sensor characteristics. Researchers are continually exploring new methods and techniques to optimize sensor performance, improve reproducibility, and scale up production to meet the demands of practical applications. Despite the remarkable progress in the field, challenges still remain. Achieving commercial viability and integrating polymer nanomaterial–based sensors into existing technologies requires further exploration and collaboration between academia and industry. Moreover, addressing the potential environmental and health impacts of these nanomaterials is crucial to ensuring their safe and sustainable use. In conclusion, the integration of PNMs into sensor technology has unlocked a realm of possibilities. From advanced environmental monitoring to personalized healthcare and beyond, these sensors hold the promise of transforming how we interact with and understand our world. With continued research, innovation, and responsible deployment, polymer nanomaterial–based sensors will undoubtedly play a pivotal role in shaping a smarter, safer, and more connected future.

Acknowledgments

The authors wish to formally acknowledge the financial support received from the following entities: The Department of Electronics & Information Technology, specifically under DeitY Project No. 5(9)/2012-NANO (Vol. II). The Department of Science and Technology (DST), denoted by Grant No. DST/SERB/EMR/2014/000034. The DST-Max Planck Society in Germany, identified by Grant No. IGSTC/MPG/PG(PKI)/2011A/48.

Conflict of interest

The authors declare that they have no conflicts of interest to disclose.

References

Adhikari, C. (2021a). Polymer nanoparticles-preparations, applications and future insights: A concise review. *Polymer-Plastics Technology and Materials*, *60*, 1996–2024.

Adhikari, C. (2021b). Polymer nanoparticles-preparations, applications and future insights: A concise review. *Polymer-Plastics Technology and Materials*, 1–29.

Al-Haidary, Q. N., Al-Mokaram, A. M., Hussein, F. M., & Ismail, A. H. (2021). Development of polyaniline for sensor applications: A review. *Journal of Physics: Conference Series*, *1853*, 012062.

Alsbaiee, A., Smith, B., Xiao, L., Ling, Y., Helbling, D., & Dichtel, W. (2016). Rapid removal of organic micropollutants from water by a porous beta-cyclodextrin polymer. *Nature*, *529*, 190–194.

Alvarez, A., Costa-Fernandez, J., Pereiro, R., Sanz-Medel, A., & Salinas-Castillo, A. (2011). Fluorescent conjugated polymers for chemical and biochemical sensing. *TrAC Trends in Analytical Chemistry*, *30*, 1513–1525.

Andre, R. S., Sanfelice, R. C., Pavinatto, A., Mattoso, L. H. C., & Correa, D. S. (2018). Hybrid nanomaterials designed for volatile organic compounds sensors: A review. *Materials & Design*, *156*, 154–166.

Ashraf, A., Farooq, U., Farooqi, B. A., & Ayub, K. (2020). Electronic structure of polythiophene gas sensors for chlorinated analytes. *Journal of Molecular Modeling*, *26*, 44.

Atkinson, M. A., Eisenbarth, G. S., & Michels, A. W. (2014). Type 1 Diabetes progress and prospects. *Lancet*, *383*, 69–82.

Bangar, M. A., Shirale, D. J., Chen, W., Myung, N. V., & Mulchandani, A. (2009). Single conducting polymer nanowire chemiresistive label-free immunosensor for cancer biomarker. *Analytical Chemistry*, *81*, 2168–2175.

Banik, B. L., Fattahi, P., & Brown, J. L. (2016). Polymeric nanoparticles: the future of nanomedicine. *WIREs Nanomedicine and Nanobiotechnology*, *8*, 271–299.

Behrendt, J. M., Esquivel Guzman, J. A., Purdie, L., Willcock, H., Morrison, J. J., Foster, A. B., O'Reilly, R. K., McCairn, M. C., & Turner, M. L. (2016). Scalable synthesis of multicolour conjugated polymer nanoparticles via Suzuki-Miyaura polymerisation in a miniemulsion and application in bioimaging. *Reactive & Functional Polymers*, *107*, 69–77.

Bhattacharya, D. S., Bapat, A., Svechkarev, D., & Mohs, A. M. (2021). Water-soluble blue fluorescent nonconjugated polymer dots from hyaluronic acid and hydrophobic amino acids. *ACS Omega*, *6*, 17890–17901.

Bootz, A., Vogel, V., Schubert, D., & Kreuter, J. (2004). Comparison of scanning electron microscopy, dynamic light scattering and analytical ultracentrifugation for the sizing of poly(butyl cyanoacrylate) nanoparticles. *European Journal of Pharmaceutics and Biopharmaceutics: Official Journal of Arbeitsgemeinschaft fur Pharmazeutische Verfahrenstechnik e.V*, *57*, 369–375.

Burroughes, J. H., Bradley, D. D. C., Brown, A. R., Marks, R. N., Mackay, K., Friend, R. H., Burn, P. L., & Holmes, A. B. (1990). Light-emitting diodes based on conjugated polymers. *Nature*, *347*, 539–541.

Canfarotta, F., Whitcombe, M. J., & Piletsky, S. A. (2013). Polymeric nanoparticles for optical sensing. *Biotechnolgy Advances*, *31*, 1585–1599.

Chan, Y.-H., Jin, Y., Wu, C., & Chiu, D. T. (2011). Copper (II) and iron (II) ion sensing with semiconducting polymer dots. *Chemical Communications*, *47*, 2820–2822.

Chan, Y.-H., Wu, C., Ye, F., Jin, Y., Smith, P. B., & Chiu, D. T. (2011). Development of ultrabright semiconducting polymer dots for ratiometric pH sensing. *Analytical Chemistry*, *83*, 1448–1455.

Chang, H.-K., Chang, G. T., Thokchom, A. K., Kim, T., & Park, J. (2018). Ultra-fast responsive colloidal-polymer composite-based volatile organic compounds (VOC) sensor using nanoscale easy tear process. *Scientific Reports*, *8*, 5291.

Childress, E. S., Roberts, C. A., Sherwood, D. Y., C. LeGuyader, L. M., & Harbron, E. J. (2012). Ratiometric fluorescence detection of mercury ions in water by conjugated polymer nanoparticles. *Analytical Chemistry*, *84*, 1235–1239.

Cho, S. I., & Lee, S. B. (2008). Fast electrochemistry of conductive polymer nanotubes: synthesis, mechanism, and application. *Accounts of Chemical Research*, *41*, 699–707.

Chunlei, Z., Qiong, Y., Libing, L., & Shu, W. (2011). Conjugated polymers for sensitive chemical Sensors. *Progress in Chemistry (Beijing)*, *23*, 1993–2002.

Clogston, J. D., & Patri, A. K. (2011). *Zeta potential measurement. Characterization of nanoparticles intended for drug delivery* (pp. 63–70). Springer.

Cohen, S. H., Lightbody, M. L., & Bray, M. T. (1999). *Atomic force microscopy/scanning tunneling microscopy 3*. Springer.

Cruz-Martínez, H., Rojas-Chávez, H., Montejo-Alvaro, F., Peña-Castañeda, Y. A., Matadamas-Ortiz, P. T., & Medina, D. I. (2021). Recent developments in graphene-based toxic gas sensors: A theoretical overview. *Sensors*, *21*, 1992.

Cullity, B. D., & Strock, S. R. (2001). *Elements of X-ray diffraction* (pp. 1–2). Addison-Wesley Publishing Company Inc.

Dai, M., Jin, S., & Nugen, S. R. (2012). Water-soluble electrospun nanofibers as a method for on-chip reagent storage. *Biosensors*, *2*, 388–395.

Danks, A. E., Hall, S. R., & Schnepp, Z. (2016). The evolution of 'sol-gel' chemistry as a technique for materials synthesis. *Materials Horizons*, *3*, 91–112.

Elgiddawy, N., Elnagar, N., Korri-Youssoufi, H., & Yassar, A. (2023). π-Conjugated polymer nanoparticles from design, synthesis to biomedical applications: Sensing, imaging, and therapy. *Microorganisms*, *11*, 2006.

Espinoza-Castañeda, M., Escosura-Muñiz, A. de la, Chamorro, A., Torres, C. de, & Merkoçi, A. (2015). Nanochannel array device operating through prussian blue nanoparticles for sensitive label-free immunodetection of a cancer biomarker. *Biosensors & Bioelectronics*, *67*, 107–114.

Estrany, F., Calvet, A., del Valle, L. J., Puiggalí, J., & Alemán, C. (2016). A multi-step template-assisted approach for the formation of conducting polymer nanotubes onto conducting polymer films. *Polymer Chemistry*, *7*, 3540–3550.

Ferreira Soares, D. C., Domingues, S. C., Viana, D. B., & Tebaldi, M. L. (2020). Polymer-hybrid nanoparticles: Current advances in biomedical applications. *Biomedicine & Pharmacotherapy = Biomedecine & Pharmacotherapie*, *131*, 110695.

Fonseca, A. C., Ferreira, P., Cordeiro, R. A., Mendonça, P. V., Góis, J. R., & Gil, M. H. (2013). *Drug delivery systems for predictive medicine: Polymers as tools for advanced applications. New Strategies to Advance Pre/Diabetes Care: Integrative Approach by PPPM* (pp. 399–455). *Dordrecht*: Springer Netherland*s*.

Fratoddi, I., Venditti, I., Cametti, C., & Russo, M. V. (2015). Chemiresistive polyaniline-based gas sensors: A mini review. *Sensors and Actuators B: Chemical*, *220*, 534–548.

Goel, H., Saini, K., Razdan, K., Khurana, R. K., Elkordy, A. A., & Singh, K. K. (2022). *In vitro* physicochemical characterization of nanocarriers: A road to optimization. *Nanoparticle Therapeutics*, 133–179.

Gonçalves, V. C., & Balogh, D. T. (2012). Optical chemical sensors using polythiophene derivatives as active layer for detection of volatile organic compounds. *Sensors and Actuators B: Chemical*, *162*, 307–312.

Gopalana, A.-I., Komathi, S., Muthuchamy, N., Lee, K.-P., Whitcombe, M. J., Dhan, L., & Sai-Anand, G. (2019). Functionalized conjugated polymers for sensing and molecular imprinting applications. *Progress in Polymer Science*, *88*, 1–129.

Goutam, P. J., Singh, D. K., Giri, P. K., & Iyer, P. K. (2011). Enhancing the photostability of poly(3-hexylthiophene) by preparing composites with multiwalled carbon nanotubes. *The Journal of Physical Chemistry. B*, *115*, 919–924.

Han, T., Nag, A., Chandra Mukhopadhyay, S., & Xu, Y. (2019). Carbon nanotubes and its gas-sensing applications: A review. *Sensors and Actuators A: Physical*, *291*, 107–143.

Huang, J., Gu, J., Meng, Z., Jia, X., & Xi, K. (2015). Signal enhancement of sensing nitroaromatics based on highly sensitive polymer dots. *Nanoscale*, *7*, 15413–15420.

Hussain, S., Malik, A. H., & Iyer, P. K. (2015). Highly precise detection, discrimination, and removal of anionic surfactants over the full pH range via cationic conjugated polymer: An efficient strategy to facilitate illicit-drug analysis. *ACS Applied Materials & Interfaces*, *7*, 3189–3198.

Jiang, L., Bagán, H., Kamra, T., Zhou, T., & Ye, L. (2016). Nanohybrid polymer brushes on silica for bioseparation. *Journal of Materials Chemistry B*, *4*, 3247–3256.

Jin, S., Dai, M., Ye, B.-C., & Nugen, S. R. (2013). Development of a capillary flow microfluidic escherichia coli biosensor with on-chip reagent delivery using water-soluble nanofibers. *Microsystem Technologies*, *19*, 2011–2015.

Jose, A., Tharayil, A., & Porel, M. (2023). Water soluble non-conjugated fluorescent polymers: aggregation induced emission, solid-state fluorescence, and sensor array applications. *Polymer Chemistry*, *14*, 3309–3316.

Kang, S., Yoon, T. W., Kim, G.-Y., & Kang, B. (2022). Review of conjugated polymer nanoparticles: From formulation to applications. *ACS Applied Nano Materials*, *5*, 17436–17460.

Katmıs, A., Fide, S., Karaismailoglu, S., & Derman, S. (2018). Synthesis and characterization methods of polymericnanoparticles. *Characterization and Application of Nanomaterials*, 1.

Kausar, A. (2022). Conjugated polymer/nanocarbon nanocomposite-sensing properties and interactions. *Journal of Macromolecular Science, Part A*, *59*, 775–785.

Khatun, M. N., Dey, A., Meher, N., & Iyer, P. K. (2021). Long alkyl chain induced OFET characteristic with low threshold voltage in an n-Type perylene monoimide semiconductor. *ACS Applied Electronic Materials*, *3*, 3575–3587.

Khatun, M. N., Tanwar, A. S., Meher, N., & Iyer, P. K. (2019). An unprecedented blueshifted naphthalimide AIEEgen for ultrasensitive detection of 4-Nitroaniline in water via "receptor-Free" IFE mechanism. *Chemistry, an Asian Journal*, *14*, 4725–4731.

Kumar, D., & Sharma, P. K. (2018). Nanoparticulate system for cancer therapy: An updated review. *International Journal of Nanotechnology and Nanomedicine*, *4*, 022–034.

Kuo, S.-Y., Li, H.-H., Wu, P.-J., Chen, C.-P., Huang, Y.-C., & Chan, Y.-H. (2015). Dual colorimetric and fluorescent sensor based on semiconducting polymer dots for ratiometric detection of lead ions in living cells. *Analytical Chemistry*, *87*, 4765–4771.

Lam, B., Das, J., Holmes, R. D., Live, L., Sage, A., Sargent, E. H., & Kelley, S. O. (2013). Solution-based circuits enable rapid and multiplexed pathogen detection. *Nature Communications*, *4*, 2001.

Landfester, K. (2009). Miniemulsion polymerization and the structure of polymer and hybrid nanoparticles. *Angewandte Chemie International Edition*, *48*, 4488–4507.

Lee, S. H., Sung, J. H., & Park, T. H. (2012). Nanomaterial-based biosensor as an emerging tool for biomedical applications. *Annals of Biomedical Engineering*, *40*, 1384–1397.

Lee, S.-J., Yu, Y.-G., Woo, H. C., Lee, C.-L., Kumar, S., Kang, N.-G., & Lee, J.-S. (2020). *In situ* incorporation of hydrophobic emissive complexes in monodisperse copolymer particles via surfactant-free emulsion polymerization. *Macromolecules*, *53*, 10097–10106.

Li, C., & Shi, G. (2013). Polythiophene-based optical sensors for small molecules. *ACS Applied Materials & Interfaces*, *5*, 4503–4510.

Li, H.-Q. (2006). *Atomic force microscopy*. Available from: http://www.chembio.uoguelph.ca/educmat/chm729/afm/firstpag.Htm.

Liu, J., Bao, S., & Wang, X. (2022). Applications of graphene-based materials in sensors: A review. *Micromachines*, *13*, 184.

Liu, J., Wu, F., Xie, A., Liu, C., & Bao, H. (2020). Preparation of non-conjugated fluorescent polymer nanoparticles for use as a fluorescent probe for detection of 2,4,6-trinitrophenol. *Analytical and Bioanalytical Chemistry*, *412*, 1235–1242.

Liu, S. G., Luo, D., Li, N., Zhang, W., Lei, J. L., Li, N. B., & Luo, H. Q. (2016). Water-soluble non-conjugated polymer nanoparticles with strong fluorescence emission for selective and sensitive detection of nitro-explosive picric acid in aqueous medium. *ACS Applied Materials & Interfaces*, *8*, 21700–21709.
Liu, Y., Dong, X., & Chen, P. (2012). Biological and chemical sensors based on graphene materials. *Chemical Society Reviews*, *41*, 2283–2307.
Lo, M., Ktari, N., Gningue-Sall, D., Madani, A., Aaron, S. E., Aaron, J.-J., Mekhalif, Z., Delhalle, J., & Chehimi, M. M. (2020). Polypyrrole: A reactive and functional conductive polymer for the selective electrochemical detection of heavy metals in water. *Emergent Materials*, *3*, 815–839.
Macchione, M. A., Biglione, C., & Strumia, M. (2018). Design, synthesis and architectures of hybrid nanomaterials for therapy and diagnosis applications. *Polymers*, *10*, 527.
MacFarlane, L. R., Shaikh, H., Garcia-Hernandez, J. D., Vespa, M., Fukui, T., & Manners, I. (2020). Functional nanoparticles through π-conjugated polymer self-assembly. *Nature Reviews Materials*, *6*, 7–26.
Malik, A. H., Hussain, S., Kalita, A., & Iyer, P. K. (2015). Conjugated polymer nanoparticles for the amplified detection of nitro-explosive picric acid on multiple platforms. *ACS Applied Materials & Interfaces*, *7*, 26968–26976.
Mamun, M. A. A., & Yuce, M. R. (2020). Recent progress in nanomaterial enabled chemical Sensors for wearable environmental monitoring applications. *Advanced Functional Materials*, *30*, 2005703.
McQuade, D. T., Pullen, A. E., & Swager, T. M. (2000). Conjugated polymer-based chemical sensors. *Chemical Reviews*, *100*, 2537–2574.
Miah, M. R., Yang, M., Khandaker, S., Bashar, M. M., Alsukaibi, A. K. D., Hassan, H. M. A., Znad, H., & Awual, M. R. (2022). Polypyrrole-based sensors for volatile organic compounds (VOCs) sensing and capturing: a comprehensive review. *Sensors and Actuators A: Physical*, *347*, 113933.
Miao, Z., Wang, P., Zhong, A., Yang, M., Xu, Q., Hao, S., & Hu, X. (2015). Development of a glucose biosensor based on electrodeposited gold nanoparticles-polyvinylpyrrolidone-polyaniline nanocomposites. *Journal of Electroanalytical Chemistry*, *756*, 153–160.
Mohanraj, V. J., & Chen, Y. (2007). Nanoparticles – A review. *Tropical Journal of Pharmaceutical Research*, *5*, 561–573.
Molpeceres, J., Aberturas, M., & Guzman, M. (2000). Biodegradable nanoparticles as a delivery system for cyclosporine: Preparation and characterization. *Journal of Microencapsulation*, *17*, 599–614.
Mondal, S., Zehra, N., Choudhury, A., & Iyer, P. K. (2021). Wearable sensing devices for point of care diagnostics. *ACS Applied Bio Materials*, *4*, 47–70.
Mousavi, S. M., Hashemi, S. A., Bahrani, S., Yousefi, K., Behbudi, G., Babapoor, A., Omidifar, N., Lai, C. W., Gholami, A., & Chiang, W.-H. (2021). Recent advancements in polythiophene-based materials and their biomedical, geno sensor and DNA detection. *International Journal of Molecular Sciences*, *22*, 6850.
Muthuraj, B., Hussain, S., & Iyer, P. K. (2013). A rapid and sensitive detection of ferritin at a nanomolar level and disruption of amyloid β fibrils using fluorescent conjugated polymer. *Polymer Chemistry*, *4*, 5096–5107.
Muthuraj, B., Mukherjee, S., Patra, C. R., & Iyer, P. K. (2016). Amplified fluorescence from polyfluorene nanoparticles with dual state emission and aggregation caused red shifted emission for live cell imaging and cancer theranostics. *ACS Applied Materials & Interfaces*, *8*, 32220–32229.
Nasir, A., Kausar, A., & Younus, A. (2015). A review on preparation, properties and applications of polymeric nanoparticle-based materials. *Polymer-Plastics Technology and Materials*, *54*, 325–341.
Nugroho, F. A. A., Darmadi, I., Cusinato, L., Susarrey-Arce, A., Schreuders, H., Bannenberg, L. J., et al. (2019). Metal-polymer hybrid nanomaterials for plasmonic ultrafast hydrogen detection. *Nature Materials*, *18*, 489–495.
Oh, W.-K., Kwon, O. S., & Jang, J. (2013). Conducting polymer nanomaterials for biomedical applications: Cellular interfacing and biosensing. *Polymer Reviews*, *53*, 407–442.

Orozco, J., García-Gradilla, V., D'Agostino, M., Gao, W., Cortés, A., & Wang, J. (2013). Artificial enzyme-powered microfish for water-quality testing. *ACS Nano*, *7*, 818–824.

Paleček, E., Tkáč, J., Bartošík, M., Bertók, T., Ostatná, V., & Paleček, J. (2015). Electrochemistry of non-conjugated proteins and glycoproteins. Toward sensors for biomedicine and glycomics. *Chemical Reviews*, *115*, 2045–2108.

Park, D.-H., Hong, J., Park, I. S., Lee, C. W., & Kim, J.-M. (2014). A Colorimetric hydrocarbon sensor employing a swelling-induced mechanochromic polydiacetylene. *Advanced Functional Materials*, *24*, 5186–5193.

Parveen, S., Misra, R., & Sahoo, S. K. (2012). Nanoparticles: a boon to drug delivery, therapeutics, diagnostics and imaging. *Nanomedicine: Nanotechnology, Biology and Medicine*, *8*, 147–166.

Patil, U. V., Ramgir, N. S., Karmakar, N., Bhogale, A., Debnath, A. K., Aswal, D. K., Gupta, S. K., & Kothari, D. C. (2015). Room temperature ammonia sensor based on copper nanoparticle intercalated polyaniline nanocomposite thin films. *Applied Surface Science*, *339*, 69–74.

Pecher, J., & Mecking, S. (2010). Nanoparticles of conjugated polymers. *Chemical Reviews*, *110*, 6260–6279.

Pourmadadi, M., Shamsabadipour, A., Aslani, A., Eshaghi, M. M., Rahdar, A., & Pandey, S. (2023). Development of polyvinylpyrrolidone-based nanomaterials for biosensors applications: A review. *Inorganic Chemistry Communications*, *152*, 110714.

Pulingam, T., Foroozandeh, P., Chuah, J.-A., & Sudesh, K. (2022). Exploring various techniques for the chemical and biological synthesis of polymeric nanoparticles. *Nanomaterials*, *12*, 576.

Qian, L., Thiruppathi, A. R., van der Zalm, J., & Chen, A. (2021). Graphene oxide-based nanomaterials for the electrochemical sensing of isoniazid. *ACS Applied Nano Materials*, *4*, 3696–3706.

Qiu, L., Han, X., Xing, C., & Glebe, U. (2023). Polymerization-induced self-assembly: An emerging tool for generating polymer-based biohybrid nanostructures. *Small (Weinheim an der Bergstrasse, Germany)*, *19*, 2207457.

Quintero-Jaime, A. F., Berenguer-Murcia, Á., Cazorla-Amorós, D., & Morallón, E. (2019). Carbon nanotubes modified with Au for electrochemical detection of prostate specific antigen: Effect of Au nanoparticle size distribution. *Frontiers in Chemistry*, 7.

Rahman, M. A., Kumar, P., Park, D.-S., & Shim, Y.-B. (2008). Electrochemical sensors based on organic conjugated polymers. *Sensors*, *8*, 118–141.

Rao, J. P., & Geckeler, K. E. (2011). Polymer nanoparticles: Preparation techniques and size-control parameters. *Progress in Polymer Science*, *36*, 887–913.

Ronkainen, N. J., Halsall, H. B., & Heineman, W. R. (2010). Electrochemical biosensors. *Chemical Society Reviews*, *39*, 1747–1763.

Ruecha, N., Rangkupan, R., Rodthongkum, N., & Chailapakul, O. (2014). Novel paper-based cholesterol biosensor using graphene/polyvinylpyrrolidone/polyaniline nanocomposite. *Biosensors & Bioelectronics*, *52*, 13–19.

Saikia, G., Dwivedi, A. K., & Iyer, P. K. (2012). Development of solution, film and membrane based fluorescent sensor for the detection of fluoride anions from water. *Analytical Methods*, *4*, 3180–3186.

Sasaki, Y., Lyu, X., Tang, W., Wu, H., & Minami, T. (2021). Polythiophene-based chemical sensors: Toward on-site supramolecular analytical devices. *Bulletin of the Chemical Society of Japan*, *94*, 2613–2622.

Schadler, L. S., Kumar, S. K., Benicewicz, B. C., Lewis, S. L., & Harton, S. E. (2007). Designed interfaces in polymer nanocomposites: A fundamental viewpoint. *MRS Bulletin/Materials Research Society*, *32*, 335–340.

Schottner, G. (2001). Hybrid sol-gel-derived polymers: applications of multifunctional materials. *Chemistry of Materials: A Publication of the American Chemical Society*, *13*, 3422–3435.

Schreiber, E., Ziener, U., Manzke, A., Plettl, A., Ziemann, P., & Landfester, K. (2009). Preparation of narrowly size distributed metal-containing polymer latexes by miniemulsion and other emulsion techniques:

Applications for nanolithography. *Chemistry of Materials: A Publication of the American Chemical Society*, *21*, 1750–1760.

Schroeder, V., Savagatrup, S., He, M., Lin, S., & Swager, T. M. (2019). Carbon nanotube chemical sensors. *Chemical Reviews*, *119*, 599–663.

Shafik, N., & Bandyopadhyay, S. (1992). *Economic growth and environmental quality: Time series and cross-country evidence*.

Shifrina, Z. B., Matveeva, V. G., & Bronstein, L. M. (2020). Role of polymer structures in catalysis by transition metal and metal oxide nanoparticle composites. *Chemical Reviews*, *120*, 1350–1396.

Shishkanova, T. V., Tobrman, T., Otta, J., Broncová, G., Fitl, P., & Vrňata, M. (2022). Substituted polythiophene-based sensor for detection of ammonia in gaseous and aqueous environment. *Journal of Materials Science*, *57*, 17870–17882.

Shu, T., Shen, Q., Zhang, X., & Serpe, M. J. (2020). Stimuli-responsive polymer/nanomaterial hybrids for sensing applications. *Analyst*, *145*, 5713–5724.

Simkhovich, L., Iyer, P., Goldberg, I., & Gross, Z. (2002). Structure and chemistry of N-substituted corroles and their rhodium(I) and zinc(II) metal-ion complexes. *Chemistry – A European Journal*, *8*, 2595–2601.

Singh, N., Joshi, A., Toor, A., & Verma, G. (2017). Drug delivery: Advancements and challenges (pp. 865–886). Elsevier.

Singh, R., & Lillard, J. W., Jr (2009). Nanoparticle-based targeted drug delivery. *Journal of Experimental and Molecular Pathology*, *86*, 215–223.

Skoog, D. A., Holler, F. J., & Nieman, T. A. (1998). *Principles of instrumental analysis*. Toronto, ON: Thomson Learning, Inc.

Song, S., Qin, Y., He, Y., Huang, Q., Fan, C., & Chen, H.-Y. (2010). Functional nanoprobes for ultrasensitive detection of biomolecules. *Chemical Society Reviews*, *39*, 4234–4243.

Srikar, R., Upendran, A., & Kannan, R. (2014). Polymeric nanoparticles for molecular imaging. *WIREs Nanomedicine and Nanobiotechnology*, *6*, 245–267.

Suginta, W., Khunkaewla, P., & Schulte, A. (2013). Electrochemical biosensor applications of polysaccharides chitin and chitosan. *Chemical Reviews*, *113*, 5458–5479.

Sun, Y.-K., & Oh, I.-H. (1996). Preparation of ultrafine $YBa_2cu_3O_{7-x}$ superconductor powders by the poly (vinyl alcohol)-assisted Sol-Gel method. *Industrial & Engineering Chemistry Research*, *35*, 4296–4300.

Sun, Y.-P., Meziani, M. J., Pathak, P., & Qu, L. (2005). Polymeric nanoparticles from rapid expansion of supercritical fluid solution. *Chemistry*, *11*, 1366–1373.

Tanwar, A. S., Meher, N., Adil, L. R., & Iyer, P. K. (2020). Stepwise elucidation of fluorescence based sensing mechanisms considering picric acid as a model analyte. *Analyst*, *145*, 4753–4767.

Tanwar, A. S., Parui, R., Garai, R., Chanu, M. A., Iyer, P. K., & Dual. (2022). "Static and dynamic" fluorescence quenching mechanisms based detection of TNT via a cationic conjugated polymer. *ACS Measurement Science Au*, *2*, 23–30.

Tanwar, A. S., Patidar, S., Ahirwar, S., Dehingia, S., & Iyer, P. K. (2019). "Receptor free" inner filter effect based universal sensors for nitroexplosive picric acid using two polyfluorene derivatives in the solution and solid states. *Analyst*, *144*, 669–676.

Tao, H., Wu, T., Aldeghi, M., Wu, T. C., Aspuru-Guzik, A., & Kumacheva, E. (2021). Nanoparticle synthesis assisted by machine learning. *Nature Reviews Materials*, *6*, 701–716.

Tao, J., Chow, S. F., & Zheng, Y. (2019). Application of flash nanoprecipitation to fabricate poorly water-soluble drug nanoparticles. *Acta Pharmaceutica Sinica B*, *9*, 4–18.

Tokay, B., & Erdem-Şenatalar, A. (2011). *Nanotanelerden silikalit-1 sentezinin mekanizmasının araştırılması*, *7*.

Tomić, M., Šetka, M., Vojkůvka, L., & Vallejos, S. (2021). VOCs sensing by metal oxides, conductive polymers, and carbon-based materials. *Nanomaterials*, *11*, 552.

Tsai, W.-C., & Wang, Y. (2019). Progress of supercritical fluid technology in polymerization and its applications in biomedical engineering. *Progress in Polymer Science*, *98*, 101161.

Verma, M. L., Sukriti., Dhanya, B. S., Saini, R., Das, A., & Varma, R. S. (2022). Synthesis and application of graphene-based sensors in biology: A review. *Environmental Chemistry Letters*, *20*, 2189–2212.

Wang, Y., Li, P., Truong-Dinh Tran, T., Zhang, J., & Kong, L. (2016). Manufacturing techniques and surface engineering of polymer based nanoparticles for targeted drug delivery to cancer. *Nanomaterials.*, *6*, 26.

Wang, Y., Feng, L., & Wang, S. (2019). Conjugated polymer nanoparticles for imaging, cell activity regulation, and therapy. *Advanced Functional Materials*, *29*, 1806818.

Wei, M., Gao, Y., Li, X., & Serpe, M. J. (2017). Stimuli-responsive polymers and their applications. *Polymer Chemistry*, *8*, 127–143.

Wen, Y., Yuan, J., Ma, X., Wang, S., & Liu, Y. (2019). Polymeric nanocomposite membranes for water treatment: A review. *Environmental Chemistry Letters*, *17*, 1539–1551.

Wu, C., Szymanski, C., & McNeill, J. (2006). Preparation and encapsulation of highly fluorescent conjugated polymer nanoparticles. *Langmuir: The ACS Journal of Surfaces and Colloids*, *22*, 2956–2960.

Wu, P.-J., Chen, J.-L., Chen, C.-P., & Chan, Y.-H. (2013). Photoactivated ratiometric copper (II) ion sensing with semiconducting polymer dots. *Chemical Communications*, *49*, 898–900.

Xiao, R., Cho, S. I., Liu, R., & Lee, S. B. (2007). Controlled electrochemical synthesis of conductive polymer nanotube structures. *Journal of the American Chemical Society*, *129*, 4483–4489.

Yang, D., Wang, J., Cao, Y., Tong, X., Hua, T., Qin, R., et al. (2023). Polyaniline-based biological and chemical sensors: Sensing mechanism, configuration design, and perspective. *ACS Applied Electronic Materials*, *5*, 593–611.

Yang, H., Duan, C., Wu, Y., Lv, Y., Liu, H., Lv, Y., Xiao, D., Huang, F., Fu, H., & Tian, Z. (2013). Conjugated polymer nanoparticles with Ag^+-sensitive fluorescence emission: a new insight into the cooperative recognition mechanism. *Particle & Particle Systems Characterization*, *30*, 972–980.

Yoon, H., Kim, J.-H., Lee, N., Kim, B.-G., & Jang, J. (2008). A novel sensor platform based on aptamer-conjugated polypyrrole nanotubes for label-free electrochemical protein detection. *Chembiochem: A European Journal of Chemical Biology*, *9*, 634–641.

Yu, J.-C., Chen, Y.-L., Zhang, Y.-Q., Yao, X.-K., Qian, C.-G., Huang, J., Zhu, S., Jiang, X.-Q., Shen, Q.-D., & Gu, Z. (2014). PH-responsive and near-infrared-emissive polymer nanoparticles for simultaneous delivery, release, and fluorescence tracking of doxorubicin *in vivo*. *Chemical Communications*, *50*, 4699–4702.

Zaporotskova, I. V., Boroznina, N. P., Parkhomenko, Y. N., & Kozhitov, L. V. (2016). Carbon nanotubes: sensor properties: A review. *Modern Electronic Materials*, *2*, 95–105.

Zehra, N., Dutta, D., Malik, A. H., Ghosh, S. S., & Iyer, P. K. (2018). Fluorescence resonance energy transfer-based wash-free bacterial imaging and antibacterial application using a cationic conjugated polyelectrolyte. *ACS Applied Materials & Interfaces*, *10*, 27603–27611.

Zhang, C., Lv, X., Zhang, Z., Qing, H., & Deng, Y. (2015). Fluorescence resonance energy transfer based biosensor for rapid and sensitive gene-specific determination. *Analytical Letters*, *48*, 2423–2433.

Zhao, M., Leggett, E., Bourke, S., Poursanidou, S., Carter-Searjeant, S., Po, S., Carmo, M. P. d, Dailey, L. A., Manning, P., Ryan, S., Urbano, L., Green, M., & Rakovich, A. (2021). Theranostic near-infrared-active conjugated polymer nanoparticles. *ACS Nano*, *15*, 8790–8802.

Zhao, X., Qiu, X., Xue, H., Liu, S., Liang, D., Yan, C., Chen, W., Wang, Y., & Zhou, G. (2023). Conjugated and non-conjugated polymers containing two-electron redox dihydrophenazines for lithium-organic batteries. *Angewandte Chemie International Edition*, *62*, 202216713.

Zhou, W., & Wang, Z. L. (2007). Scanning microscopy for nanotechnology. *Techniques and Applications*.

Zhu, S., Song, Y., Shao, J., Zhao, X., & Yang, B. (2015). Non-conjugated polymer dots with crosslink-enhanced emission in the absence of fluorophore units. *Angewandte Chemie International Edition*, *54*, 14626–14637.

CHAPTER

Polymeric smart structures

11

Magdalena Mieloszyk
Institute of Fluid-Flow Machinery, Polish Academy of Sciences, Gdansk, Poland

11.1 Introduction

Structural Health Monitoring (SHM) systems are installed on a variety of real structures (civil engineering, marine, and aircraft) for the purpose of increasing the safety level of the structure and reducing the financial outlays related to necessary technical inspections, reparations, etc. Such systems contain a variety of sensors applied for measurements of different parameters. One good example is the Wind and Structural Health Monitoring System installed on the Tsing Ma Bridge in Hong Kong. It contains a variety of sensors, for example, strain gauges, optical sensors, GPS-based position sensors, and accelerometers (Chan et al., 2006).

In SHM systems, five diagnostic levels can be determined:

1. Detection
2. Localization
3. Identification (size and type)
4. Prediction
5. Self-healing (optional)

Detection is related to the possibility of finding damage in the structure. In cases where structure is damaged, the next step is to localize the degraded area. More detailed analysis performed in the selected region allows for determining size (e.g., area, length) and/or type of damage (e.g., delamination, crack, and corrosion). Information about the actual health state of the structure, in combination with operational loading and environmental conditions, as well as the material and geometrical properties of the analyzed object, are used in the prediction level. It aims to determine the time for further safe operation of the damaged structure. Self-healing is available only for specific structures whose material has been modified. Under the assumed conditions, these structures repair themselves through the activation of embedded components, such as microcapsules with an activator (Yan et al., 2023).

Polymeric smart structures can be described in relation to the SHM levels. Sensors integrated with the structure are not only applied to the measurements of different parameters, but the achieved values in combination with a database (measured characteristics during normal operation as well as data from the numerical analysis) can be used in decision-making processes. Depending on the complexity of the smart structure, different maximal SHM levels can be achieved.

Sensory Polymers. DOI: https://doi.org/10.1016/B978-0-443-13394-7.00011-2

From the variety of sensors, optical sensors were chosen as they can be embedded into the material structure of an element, and their influence on the mechanical properties of the structure is limited. Polymeric elements containing embedded sensors can be treated as smart structures because SHM systems based on the sensor array can be applied for decision-making processes related to their future operation (continuation or stop). The most sophisticated smart structures could also have self-healing abilities.

In this chapter, the polymeric smart structures with embedded fiber Bragg grating (FBG) sensors will be presented and discussed. Due to their features like small size, high temperature and corrosion resistance, no calibration requirements, and high multiplexing capabilities, FBG sensors can be permanently integrated with a polymeric structure and used as a part of the SHM system. FBG sensors are mostly applied for the measurement of strain and temperature (Udd & Spillman, 2011). FBG sensors' sensitivity between temperature and strain changes is equal to 11.2 pm/°C and 1.3 pm/μm, respectively. The values are given for silica optical fibers and a Bragg wavelength of 1550 nm (Mieloszyk, Andrearczyk, et al., 2020). Due to cross-sensitivity on temperature and strain, for the purpose of strain measurements, an appropriate temperature compensation method should be used. FBG sensors can also be used in damage detection and localization based on vibration-based methods (Mieloszyk & Ostachowicz, 2017a).

FBG sensors can be embedded into polymeric structures during a variety of manufacturing processes. In the next part of the chapter, some examples of smart structures manufactured using standard and additive manufacturing (AM) methods will be presented.

Polymeric smart structures can be understood as separated structures with integrated SHM systems or smaller elements having sensing properties. Such elements can be integrated with larger structures that were manufactured earlier and have been operated, for example, bridges. Examples of such objects are pressure sensors that are based on the FBG sensor technique (Hong et al., 2019; Zhang et al., 2017). A sensing platform can be applied for monitoring plantar pressure distribution (Zhang et al., 2017). The idea of developing polymeric smart structures with embedded FBG sensors is continuous monitoring of the behavior of elements during their manufacturing and subsequent operation (tests) processes. An example of a real civil engineering structure that fulfills such requirements is the West Mill Bridge (UK). It contains a fiber-reinforced polymer (FRP) deck with embedded FBG sensors and can monitor the traffic (weight and speed of a vehicle) (Gebremichael et al., 2005).

The possessed information related to real strain, loading, or other parameters is useful information for future design (modifications) of analyzed objects as well as to increase the safety of the operation of the structure with the SHM system. The numerical simulation of the behavior of FRP structures under operational conditions is still challenging due to the complexity of FRP materials; thus, an essential part of the analysis of FRP properties is empirical investigation. It results in the development of a variety of SHM methods that are permanently integrated with polymers to create smart structures. Pure polymers or FRP materials with embedded sensors (mostly fiber optic sensors) (Mieloszyk et al., 2021) and piezoelectric transducers (Meyer et al., 2019) form polymeric smart structures. In such elements, sensors are applied to detect damage (Mieloszyk et al., 2021; Ning et al., 2014) and determine the real loading conditions and their influence on the element's durability under operational conditions (Gebremichael et al., 2005). Strain sensors (also called FBG sensors) are very useful in SHM systems because they play a multifunctional role in assessing structural integrity. The concept of a complex SHM system is based on the combination of information possessed by both the smart composite structure and its finite element method model. Such an approach increases the safety of structures.

A more complex polymeric smart structure contains not only the sensor(s) used for the determination of its state but also an actuator. An example is an FRP element with embedded shape memory alloy (SMA) wires and FBG sensors. The aim of such a solution is to provide active control of the mechanical parameters of a composite structure. SMA actuators embedded in FRP material were applied for vibration control of the analyzed element. Their activation changed the natural frequency of the FRP cantilever beam. Simultaneously, the measurements from embedded FBG sensors were applied to determine the actual value of the natural frequency of the analyzed element (Ho et al., 2013). Another example of a smart structure is a smart wing with SMA actuators. Due to the small thickness of the outer FRP skin of the wing, FBG sensors were attached to it instead of being embedded inside. Due to SMA actuators integrated with the wing structure, it can change its shape. The aim of using FBG sensors was to monitor the shape of the wing and the detection and localization of ice (simulated as an additional mass) on the surface of the skin (Mieloszyk et al., 2011).

11.2 Polymeric smart structures

Fig. 11.1 presents the development of a smart structure, including embedded fiber optics with FBG sensors, from simple samples to real structures. The simplest sample was manufactured from a pure polymer. The more complex sample contains reinforcement in the form of fibers (e.g., carbon or glass) and is manufactured using AM or standard methods, for example, the infusion method. The

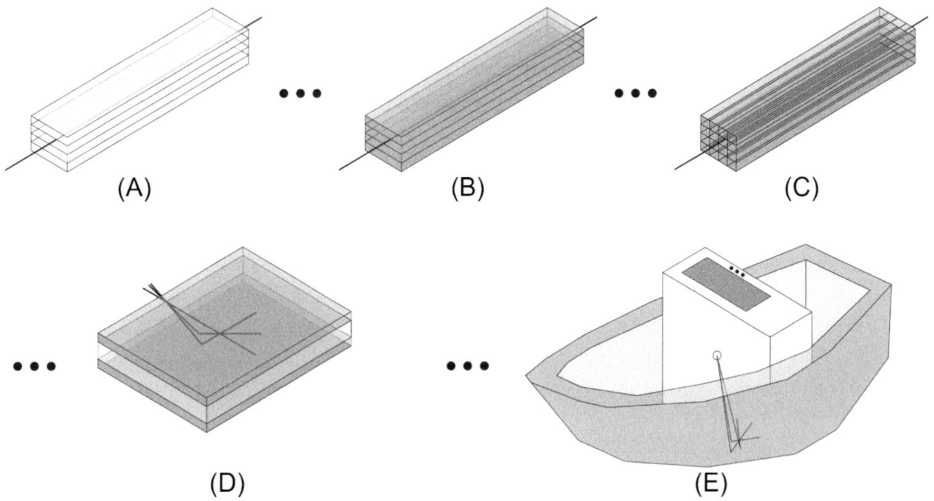

FIGURE 11.1 Smart structures.

Development of polymeric smart structures: (A) polymer sample, (B) FRP sample, (C) AM FRP sample, (D) complex element, and (E) real structure.

aim of adding reinforcement material is to increase the final element's strength while maintaining its low weight. The advantages of FRP materials are their high strength-to-weight ratio, ease of installation (connecting neighboring elements using adhesive), resistance to many environmental factors, chemical compounds, and fatigue loads (Tuwair et al., 2016). It results in FRP material applications in many branches of industry, for example, aviation (Keršienė et al., 2016), marine (Cucinotta et al., 2017; Mieloszyk et al., 2021), and civil engineering (Al-Ramahee et al., 2017; Gebremichael et al., 2005). Because of their properties, FRP materials can replace metal. It is especially important in those industrial branches (e.g., aviation) where the structure weight is related to operational costs (e.g., fuel consumption). For example, in the past, the most popular material in aviation was aluminum, while currently, the Boeing 787 contains more than 50% of its volume of composite materials (Groves, 2018). The wide popularity of FRP materials results in their operation in a variety of environmental and loading conditions. Among many FRP materials, the most popular are carbon fiber reinforcement polymers (CFRP) and glass fiber-reinforced polymers (GFRP).

FRP materials contain two main components: fiber reinforcement and a polymer matrix. Both of them have different mechanical and physical properties. From the operational point of view, the polymer is the weaker component, and its selection for a particular operational condition is the most important part of the designing and manufacturing processes. The polymer matrix is sensitive to many environmental factors, like ultraviolet radiation, temperature (especially thermal fatigue due to temperature changing cyclically), moisture, and operational liquids (e.g., hydraulic fluids and deicing chemicals). Therefore, FRP elements are protected against such factors by appropriate covers or paint layers. There are two factors (temperature and moisture) that commonly occur in the majority of operational environments and can decrease the strength/durability of FRP structures. The high relative humidity (RH) level or wet environment results in a moisture (water) diffusion process into the material. It can be accelerated by the occurrence of microcracks on the FRP element material surface or the input of optical fiber with an FBG sensor that is not properly protected against moisture uptake. On such fiber optics, under certain conditions, the capillary effect can be observed (Majewska et al., 2021). Increasing the amount of moisture inside the FRP element influences its internal stress field. It can result in internal microcracks or debonding between fiber reinforcement and matrix. Another moisture influence is the reduction of the glass transition temperature of the polymer (Eslami et al., 2012; Schutte, 1994). FRP structures' operational temperatures are designed to be much lower than the glass transition temperature value of the matrix. Therefore, moisture uptake can result in unexpected degradation processes and damage to the structure. It is worth noticing that even 1% of water mass gain results in a lowering of the glass transition temperature by *ca.* 10 °C. Additionally, a decrease in the tensile strength and Young's modulus are observed. They are lower by 15% and 10% of the base values of both parameters, respectively. Moisture is especially dangerous for composite joints because it can decrease the adhesive layer's properties. Due to its small size, the FBG sensor can be embedded inside the adhesive layer and be applied for continuous measurement of the amount of moisture (Mieloszyk & Ostachowicz, 2017b).

The second parameter is temperature. Because of the operational environment of the majority of structures, scientists are mostly focused on the effects of elevated temperatures. Such temperatures may cause changes in the internal stress field, a reduction of adhesion strength in fiber/polymer matrix bonding, and changes in physical properties in the matrix (Botelho et al., 2006). However, some structures operate at sub-zero temperatures. The most popular are airplanes whose operational temperatures are in the range of −50 °C (−70 °C) (at cruising altitude) to 50 °C (at the airport).

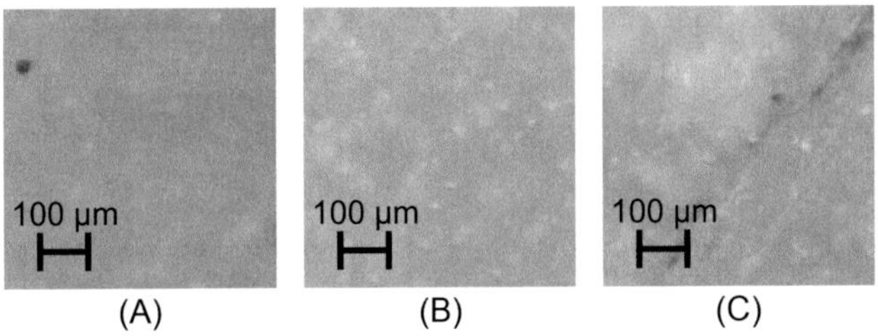

FIGURE 11.2 Polymer sample surface microstructure.

Polymer sample surface microstructure: (A) amorphous, (B) crystallized, and (C) boundary between both areas.

Sub-zero temperatures influence the structure of the polymer matrix. For example, in the amorphous structure of the polymer, a crystallization process can be observed locally. An example of the surface of a pure polymer sample that was cooled from 10 °C to −50 °C is presented in Fig. 11.2. Originally, after the end of the solidification process, the material had an amorphous structure. However, sub-zero temperature influence resulted in crystallization processes in different, unevenly distributed locations on the sample. In some places, clear boundaries between two areas with different morphologies could be observed. The material structure change resulted in a higher tensile strength of the specific samples when compared to other samples with an amorphous structure. Differences in the material structure result in changes in the internal stress distribution, which can cause microcracks to occur (Shrivastava & Nazir Hussain, 2008).

The next step is related to the development of a fabrication method that allows for reducing the amount of waste during the production process of elements with complex shapes. It results in the development of various AM methods that can be applied to both pure polymers and FRP materials. The next level is creating a more complex structure—a sandwich panel. Such elements effectively combine the lightweight and high-strength qualities of fulfillment filler with the smooth and flat surfaces of skins that allow easy installation. Such panels have higher flexural strength than solid structures with the same density. The geometrical and material properties of sandwich panels result in their high resistance to shock and impact. The final, most complex structure is a real object, for example, a composite boat hull with integrated sensors, measurement units, and post-processing software, allowing the assignment of the health condition of the structure at selected moments in time.

The smart structure description is related to pure polymers and FRP materials. Despite the fact that polymers have lower strength than metals, they are very popular in rapid prototyping. The most important reasons are related to their relatively low cost in combination with their ability to create complex structures using AM methods. It is worth highlighting that the application of 3D printing methods results in a significantly reduced amount of waste compared to standard methods when producing the final element.

Recently, AM techniques for fabricating smart polymeric structures with embedded sensors have been developed. AM technique parameters (e.g., process temperatures, layer thickness, and construction material properties) result in different problems that must be solved to develop a fully functional smart structure. Very important is the quality of the internal structure and the integrity between the sensor and the surrounding host material. Problems can be related to residual stress, strain occurrence, or shrinkage that can affect the dimensional accuracy of the final element. Rapid transition from one physical condition to another (re-melting and rapid cooling cycles) can also result in delamination between consecutive 3D printed layers, cracking, and influencing the reflected spectrum from the FBG sensor (Kousiatza & Karalekas, 2016).

AM for pure polymers can be classified as material extrusion, vat photopolymerization, powder bed fusion, binder jetting, sheet lamination, material jetting, and directed energy deposition (Dizon et al., 2018). For manufacturing FRP elements, fused deposition modeling (FDM) and selective laser sintering methods can be applied (Türk et al., 2017). The manufacturing method influences both the structure and mechanical parameters of elements.

To increase the mechanical strength of polymers, different types of reinforcement are applied. It results in the development of a variety of FRP materials with a variety of mechanical properties. FRP materials have many advantages, like a high strength-to-weight ratio, resistance to environmental factors, chemical compounds, and fatigue loads. The reinforcements can be in the form of particles and long or short fibers arranged in chosen directions. As fiber reinforcements, synthetic (e.g., glass and carbon) or natural fibers (e.g., hemp and basalt) can be used. The popularity of synthetic fibers is related to their well-known mechanical properties and behavior under a variety of environmental conditions and mechanical loadings. The problem is their non-renewability and the production process that affects the natural environment. Additionally, the person involved in the production process of both fiber and FRP composite, as well as its use, must be very careful. It is dangerous to inhale carbon dust and touch the carbon fibers themselves with an unprotected hand. Dust (containing fiber particles) may irritate the respiratory tract. Touching fibers may cause skin irritation. If the fiber penetrates the body, it may cause a wound that is difficult to heal. Therefore, people working in the production hall should wear protective clothing.

Greater interest in the impact of human activity on the environment led to the search for other solutions—natural fibers that can be extracted from a variety of plants and minerals. The main sources of vegetable fibers are agricultural waste products and plants that are intentionally cultivated for the extraction of textile fibers. Vegetable fibers are extracted mainly from leaves and bast. Natural fibers' applicability is limited by their low thermal degradation (maximal operational temperature *ca.* 200 °C). That results in a smaller set of polymers that can be used as matrices in FRP elements (Faruk et al., 2012). Natural fibers also have advantages over synthetic ones, such as production with a lower investment cost (*ca.* 50% energy used for the fabrication of glass fibers (Dicker et al., 2014)), wide availability, friendly processing (reduced wear of tools and skin irritation), renewability, and biodegradability.

There are several problems that have to be overcome before FRP with natural fibers replaces FRP elements with synthetic reinforcement. The stiffness and strength of natural fibers are comparable to those of glass fibers. However, the mechanical properties of FRP materials constructed from natural fibers are significantly lower. One of the most important issues is the chemical incompatibility between the hydrophilic natural fibers and the hydrophobic polymer matrix (Dicker et al., 2014). The degree of moisture absorption of natural fibers (e.g., jute, banana, and sisal) is equal to *ca.* 10% (Abdollahiparsa et al., 2023). It influences fiber/matrix bonding quality and can cause debonding during the operation of such elements (George et al., 2001). In the past decades, their application was mostly limited to low-load, non-critical parts (e.g., door panels containing jute fibers) (Fogorasi & Barbu, 2017). Nowadays, highly loaded structures are manufactured from biocomposites (e.g., the girder of the pedestrian bridge containing hemp and flax fibers (Abdollahiparsa et al., 2023)).

The more complex structures are sandwich panels with FRP skins and fulfillment (e.g., polyurethane foam, honeycomb) joined by adhesive. The FBG sensor array can be embedded into the skin or in a joint area between the skin and the fulfillment. The most typical sandwich structures with adhesive bonds are vessel hulls (Mieloszyk et al., 2021) and bridge decks (Gebremichael et al., 2005). The weakest part of a sandwich panel is an adhesive joint between the skin and the core. Debonding in such an area can result in the degradation of the internal part of the structure during its operation. Therefore, non-destructive techniques (NDT) applications for analyzing the quality of polymeric smart structures after their manufacturing are an important issue. For such purposes, THz spectroscopy and infrared thermography methods can be used. An example of such an application of THz spectroscopy is the evaluation of the quality of a sandwich structure with embedded arrays of FBG sensors and piezoelectric transducers. The NDT inspection allowed for the determination localizations of debondings on the adhesive joint between the skin and the core in the analyzed panel (Mieloszyk, Jurek, et al., 2020).

11.3 Optical sensors

Optical sensors can be mounted on elements made of various materials, for example, metal (Majewska et al., 2014; Mieloszyk & Ostachowicz, 2017a), composite (Mieloszyk et al., 2021), and concrete (Rodrigues et al., 2010). Effective operation of the sensor requires an appropriate mechanical connection between the fiber and the tested structure to ensure the proper transfer of measured values to light signals.

There are two types of fiber optics: multimode and single-mode. The first is applied for telecommunication signal transmission, while the second is for sensing purposes. In single-mode fiber optics, the transmitted signal distortion level is lower; therefore, such fibers are applied for manufacturing a variety of optical sensors.

External factors influence results in geometrical and physical changes in fiber optics. It is also visible in the optical signal transmitted throughout the fiber. Such effects have been observed in telecommunications since the 1970s, when fiber optics were used for the first time for data transmission. Such phenomena are unfavorable in this application and are minimized. However, such changes in the optical properties of fiber optics were the motivation for the development of fiber optic sensors. The first optical fiber strain sensor was described by Butter and Hocker in 1978 (Butter & Hocker, 1978).

Fiber optics (commonly used in sensing applications) are made of silica glass doped with GeO_2 and F (Measures, 2002). In such fiber optics, three telecommunication data transmission windows can be distinguished. They correspond to the following wavelengths: 850, 1300, and 1550 nm. The optical signal attenuation values differ among the windows and are equal to *ca.* 2, *ca.* 0.5, and *ca.* 0.2 dB/km (Measures, 2002). This results in choosing the last window as the most popular for the selection of Bragg wavelengths in the FBG sensors.

The most common classification of optical sensors is based on their location on the structure. They are called local, quasi-distributed, or distributed (Li et al., 2004). Local sensors can measure parameters at selected points. Contrary to them, in distributed sensors, the whole optical fiber is treated as a sensor. In such cases, the scattering effects (Rayleigh and Brillouin) are taken into consideration. The changes in the optical signal transmitted through the fiber optics can be converted to strain or temperature using appropriate post-processing techniques. Rayleigh scattering is linked with optical signal attenuation and can be applied for stain and temperature measurements (Souza & Tarpani, 2021). Brillouin scattering is related to the coupling of the two optical lights and is mostly applied for temperature measurement (Choi & Kwon, 2019). Quasi-distributed sensors combine the properties of both sensor types listed above. Typical quasi-distributed sensors are FBG sensors. Due to their short length (typically 10 mm), they can be treated as local sensors. However, they have multiplexing capabilities that allow manufacturing on one fiber optic line containing many FBG sensors. It results in using a large part of the fiber optics for sensing purposes. This feature is similar to those described for distributed sensors.

Optical sensors have many advantages that allow them to be applied in SHM systems. The most important can be presented in the form of a list:

- Long life (no shorter than 25 years)
- Flexibility (possibility of mounting on a variety of surfaces)
- High temperature resistance (applicability for a range of temperatures from −200 °C to 800 °C)
- High corrosion resistance
- Electromagnetic field resistance
- Non-conductive material
- Small dimensions.

The lifetime of optical sensors depends on the degradation processes of fiber optics. They are manufactured from a durable material (silica glass) that can withstand high tensile stress (elongation up to 5%) and has high corrosion resistance. The size and flexibility of optical sensors allow them to be mounted on complex surfaces and in hardly accessible locations (e.g., the circumference of a circular element, a weld). Another advantage is related to their high temperature resistance. Optical sensors can be used for strain measurements for structures under both cooling and heating conditions with an accuracy of 2 $\mu\varepsilon$. Such sensors can also be applied for temperature measurements with an accuracy of 0.1 °C. Fiber optics are characterized by their high corrosion resistance. They can be used in flammable, explosive, or chemically aggressive environments. Therefore, they can operate in a variety of corrosive environments, like hydrogen installation parts or marine structures. FBG sensors can be applied for strain measurements in CFRP parts of hydrogen tanks (Gąsior et al., 2018). The role of the CFRP layer is to ensure the mechanical strength of the container. The main cause of failure for hydrogen storage tanks is fatigue failure resulting from frequent hydrogen charging and discharging (Masuda et al., 2021). Optical sensors do not require protection against water and can be applied to offshore structures (e.g., oil platforms and wind turbines) (Mieloszyk & Ostachowicz, 2017a) or marine vessels (Majewska et al., 2014; Mieloszyk & Ostachowicz, 2017a).

Due to low signal attenuation, optical sensors can be used for measurements over very long distances (even tens of kilometers). Additionally, it is possible to mount an array of optical sensors and connect them to a measurement unit located elsewhere, while the connecting fiber optic length will be counted in kilometers. Such a feature is very important for SHM systems designed for dangerous objects (e.g., nuclear power plants and tungsten transmission pipes). Due to this, the measurement unit and operational team can be located in a safe place while the sensors are exposed to harmful environmental conditions. A good example is a measurement array that was located on the foremast of the sailing ship Dar Mlodziezy, *ca.* 26 m above the ship deck. The sensors' location was exposed to wind, water (rain), and temperature changes. The measurement unit was placed in a room with a constant temperature that was easily accessible to the researchers (Majewska et al., 2014).

The operational principle of optical sensors is based on the behavior of light (an optical signal) that propagates in fiber optics. Due to this, they are insensitive to electromagnetic fields and can operate in environments where the influence of electromagnetic fields on electric signals is too high to achieve proper readings from electric sensors. On the other side, the occurrence of optical sensors does not influence the electric sensors. Therefore, both sensor lines can be mounted close to each other. Fiber optics are non-conductive materials; therefore, the influence of mounting an array of optical sensors on a structure is minimal.

The diameter of fiber optics is comparable to the human hair diameter. Optical sensors can be both mounted on structures and embedded into their material during manufacturing processes. Their influence on the material's strength is negligible.

11.3.1 Fiber Bragg grating sensors

FBG sensors were, for the first time, applied to a real structure in 1993. It was the road bridge in Calgary (Canada), and the sensors were used for strain and temperature measurements (Measures et al., 1993). Since that time, FBG sensors have been applied for diagnostic purposes in many branches of industry, including marine (Majewska et al., 2014; Mieloszyk & Ostachowicz, 2017a), aviation (Klotz et al., 2021), and civil engineering (Gebremichael et al., 2005).

Fiber optics are dielectrics that enable the transmission of a light signal. Such elements consist of three components: core, cladding, and jacket. A scheme of a fiber optic with an FBG sensor is presented in Fig. 11.3. A core and cladding are typically made of silica glass, while a jacket is usually fabricated from polymer. The refractive index of the cladding is lower (ca. 1%) than that of the core (Peters, 2009). The core diameter is in the range of 6.5–9.5 μm, while the cladding diameter is equal to 125 μm. The outer diameter of a fiber optic depends on the jacket. Its role is to protect the brittle silica glass parts of a fiber optic against environmental factors and mechanical forces. Therefore, its diameter can be counted in millimeters.

In single-mode fiber optics, only one light ray (called the basic mode) is transmitted throughout the fiber. The light is reflected from the boundaries between the core and the cladding. In the core of a fiber optic, a permanent periodic modulation of the refractive index called a Bragg diffraction grating can be made. In such grating, the maximum reflectivity occurs for the Bragg wavelength λ_B, which is given by the following equation:

$$\lambda_B = 2n\Lambda$$

where n is the refractive index of the mode propagating in the optical fiber, and Λ is the grating period.

Bragg wavelength is used to distinguish particular FBG sensors connected in a measurement array. As presented in Fig. 11.3, a part of the signal whose wavelength is in good agreement with

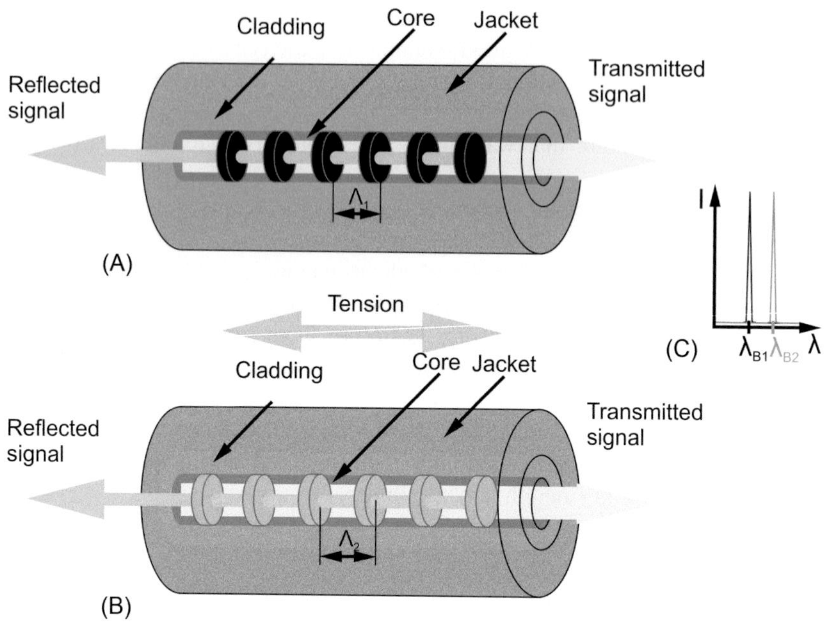

FIGURE 11.3 FBG sensor principle of operation.

Principle of operation of FBG sensor: (A) unloading condition, (B) tension loading, and (C) FBG sensor reflected spectra.

the particular FBG sensor is reflected from the grating, while the rest is still transmitted throughout the fiber optic.

FBG sensors are typically used for strain and temperature measurements. For a better understanding of the principle of operation of the FBG sensor under tensile loading, two operating states of the sensor are presented in Fig. 11.3. First (Fig. 11.3A), the sensor is presented for unloading conditions. The Bragg grating is marked in black, and its period is Λ_1. Then, a tensile force was applied to the sensor. It results in an increase in the distances between mirrors in the Bragg grating. It is presented in Fig. 11.3B. The Bragg grating under tensile loading is marked by gray, and its period is Λ_2. The Bragg wavelengths can be calculated from the relationship presented above. λ_{B1} and λ_{B2} correspond to unloading and tensile loading conditions, respectively. Because $\Lambda_2 > \Lambda_1$, the same relationship is also observed for Bragg wavelengths. Spectra from the FBG sensor under both loading conditions are presented in Fig. 11.3C. They are marked in black and gray colors—the same as used in Bragg gratings. The tensile loading results in higher values of the Bragg wavelength λ_B. In the case of compression loading, the behavior of a Bragg grating will be the opposite. Therefore, compression loading results in decreasing distances between mirrors in the grating and is observed as a lower Bragg wavelength. An analogical effect is observed for temperature measurements.

FBG sensors are manufactured with an ultraviolet laser beam of selected intensity that locally changes the refractive index in the fiber core. There are two main techniques: the phase mask method and the interferometric method. The first method is very popular due to its low complexity and high process stability. Its disadvantage is related to the possibility of using a phase mask to produce an FBG sensor with only a particular Bragg wave. Therefore, manufacturing many different sensors requires a set of phase masks. A phase mask is a special diffraction grating made on a glass plate. As a second method based on interferometry, it is possible to manufacture a variety of sensors with one laser. The laser beam is divided into two equal parts that interfere with each other. Therefore, changing the angle between the laser beams makes it possible to manufacture FBG sensors with selected Bragg wavelengths. The lower popularity of the method is related to the difficulty of obtaining a stable interference pattern throughout the entire sensor fabrication process.

FBG sensors are sensitive to changes in strain and temperature. The changes in the Bragg wavelength are linearly proportional to the changes in those measured parameters. Due to their directional properties, the strongest influences are related to the longitudinal strain (strain on the main axis of the sensor). The directional properties of the FBG sensor can be applied to determine the direction and source of an acoustic signal (Culshaw et al., 2008). The change in the maximum amplitude depends on the angle between the wave source and the main axis of the FBG sensor. It can be approximated by a cosine function where the highest values are achieved for 0° and 180° (Takeda et al., 2005; Thursby et al., 2004). Such angles are related to the main axis of the FBG sensor.

Longitudinal strains act on the periodicity of the perturbation (photoelastic effect), whereas temperature affects both the refractive index (thermo-optic effect) and the periodicity of the perturbation (thermoelastic effect) (Udd & Spillman, 2011). The influence of mechanical strain and temperature changes on an FBG sensor can be described using the relationship:

$$\frac{\Delta\lambda_B}{\lambda_B} = k\Delta\varepsilon + \alpha_f \Delta T$$

where ε means strain and T is temperature. Parameters k and α_f are the gauge factor and thermal coefficient of the FBG sensor, respectively (Shafighfard & Mieloszyk, 2022).

The gauge factor k can be presented in the form of the following formula:

$$k = \left(\frac{1}{n}\frac{\partial n}{\partial \varepsilon}\right) + \left(\frac{1}{\Lambda}\frac{\partial \Lambda}{\partial \varepsilon}\right)$$

It contains changes in the refractive index upon mechanical strain and the relative changes of the Bragg grating length related to the elastic strain occurrence (Shafighfard & Mieloszyk, 2022). The k parameter can be determined by applying a set of static forces at the end of the optical fiber and measuring the wavelength shift in the FBG sensor. The test can be realized under the assumption that Hooke's Law can be applied to determining the mechanical strain values in the fiber (Mieloszyk et al., 2020).

The second parameter (α_f) can be presented in the form of the relationship:

$$\alpha_f = \left(\frac{1}{n}\frac{\partial n}{\partial T}\right) + \left(\frac{1}{\Lambda}\frac{\partial \Lambda}{\partial T}\right)$$

It contains information about the thermal change of the refractive index of the fiber optic material and the longitudinal thermal expansion coefficient related to the change in length of the fiber optic due to temperature (Shafighfard & Mieloszyk, 2022). This parameter can be determined by exposing a particular FBG sensor to a temperature in the assumed range under a stable RH level. The parameter determination was described in detail by Mieloszyk et al. (Mieloszyk et al., 2020).

The values of both parameters are determined experimentally for free FBG sensors. For the material used in the presented investigations, the strain gauge factor of the FBG sensor was equal to 1.3 pm/με, while the thermal coefficient value was 11.2 pm/°C (Mieloszyk et al., 2020). The exact values of the parameters depend on the glass properties used in the particular fiber optic.

In the case of strain measurements, a thermal compensation is required. The most typical solution is to add an FBG sensor that will be used for temperature compensation only. It will be exposed to the same environmental conditions, but it will not be mechanically loaded. The other solution is to determine the thermal characteristics of the used FBG sensor material in a controllable environment and then, during the mechanical loading test, measure the temperature values using an additional device, for example, a thermometer. The advantage of the first method is related to having all sensors connected to one measuring device (an interrogator). The measurements are performed simultaneously. In the second method, a very important issue is avoiding time delays between two different measurement devices. Therefore, it is recommended for static or quasi-static thermal conditions.

The important feature of the FBG sensor is that it is a non-resonant sensor. Accordingly, it can be successfully applied for measuring static, quasi-static, and dynamic strains. It can also be applied for measurements of acoustic ($<$20 kHz) and ultrasonic ($>$20 kHz) waves. The non-resonant feature is an attractive advantage in acoustic-ultrasonic wave detection in comparison with a piezoelectric transducer, which is sensitive only within finite bands near its resonance frequencies (Lee et al., 2007). FBG sensors can also be applied for real-time monitoring of fatigue crack propagation (Mieloszyk, 2021). A crack development process results in mechanical wave excitation that propagates through the material of the element (Sih, 1977). Such waves can be measured by FBG sensors.

FBG strain sensors have many advantages compared to the most commonly used electrical sensors (strain gauges). A comparison of FBG strain sensors and strain gauges is collected in Table 11.1.

Contrary to FBG sensors, strain gauges installed on the oil platform were damaged by the corrosive influence of seawater after one year of exposure. FBG sensors were working properly (Ren et al., 2006). FBG sensors made on a single optical fiber can simultaneously measure several different parameters (e.g., longitudinal strain, temperature, acceleration, and displacement). The measured parameter, in this case, depends on the method of attaching the sensor to the structure or construction of the sensor for special applications. The multiplexing feature reduces the amount of cabling required and the complexity of powering the sensors. FBG sensors' utility is also linked to the possibility of manufacturing many sensors with different diffraction grating constants along one fiber. Sensors are typically distinguished based on their Bragg wavelengths. The number of sensors on one fiber optic depends on the Bragg wavelength range of the measurement unit (interrogator) and the particular application. It is related to the range of strains or temperatures that corresponds to the Bragg wavelength shift presented in Fig. 11.3. It is very important to avoid the overlap of Bragg wavelengths between two or more different sensors lying on one line during the measurements. It can happen when the assumed Bragg wavelength change range for a specific sensor is too narrow in relation to the real shifts.

Table 11.1 Comparison of FBG strain sensors and strain gauges.

Feature	FBG sensor	Strain gauge
Electromagnetic field	Immune	Sensitive
Electric conductivity	No	Yes
Explosive safety	Yes	No
Corrosion resistance	Yes	No (protection required)
Multiplexing capability/weight	Yes/small	No/high (cables)
Zero drift/calibration	No	Yes
Thermal compensation	Yes	Yes
Embedding into FRP material	Possible	Possible

It is possible to embed both strain gauges and FBG sensors into FRP material. The main difference is related to the distance in the structure between a sensor and the location of the output of the connection cable from the element material. In the case of the FBG sensor, such cable (fiber optics) length can be counted in meters. At the end of embedded strain gauges, gold-plated pins must be led out of the material. The pins are then connected to the power and measurement units. Such a strain gauge type that can be embedded into FRP material is produced, for example, by the HBK (Hottinger, Bruel, and Kjær) company.

Optical sensor arrays mounted on structures are protected against mechanical and environmental influences that can result in the degradation of the adhesive that was used to attach sensors to the element's surface. For real structures, such protection is typically based on epoxy or composite layers. For example, on the sailing ship, it prevents both the sensor array and the metal structure of the foremast from corrosion in the marine environment (Majewska et al., 2014).

The overall process of mounting FBG sensors on a structure is similar for all structural materials (e.g., metal, concrete, or FRP) and consists of the following steps:

- Determination of the location of the FBG sensor on the structure
- Marking the sensor location
- Cleaning and degreasing the area where the sensor will be glued
- Covering the gluing area with a thin layer of adhesive and the sensor placement
- Covering the gluing area (optional—applied for real structures)

The other possibility of mounting an SHM monitoring system based on FBG sensors is their embedding in the structure material. In such a method, FBG sensors' mounting steps and the building of the sensing array depend on the manufacturing process of the structure (or element). Therefore, the procedure is a modification of the fabrication process and cannot be presented in a general form. It will be briefly presented in the next section of the chapter on the examples of the three methods.

11.4 Embedded sensors

SHM methods are based on a variety of sensors mounted on structures. In this chapter, smart structures will be limited to structures with embedded FBG sensors. The embedding process strongly depends on the particular manufacturing method. However, the brittleness of optical fibers and requirements related to their localization in the structure also influence the fabrication process. During the design of smart structures, two main aspects must be taken into consideration: the quality of the reflected spectrum of the FBG sensor and the part of the element where optical fiber comes out of the structure.

Below, three manufacturing methods will be briefly described. The methods were applied for the fabrication of samples and structures analyzed in the chapter. For pure polymers, the AM method was applied, while for CFRP elements, both standard and AM methods were used. All of them were modified to embed FBG sensors.

The first AM method, multi-jet printing (MJP), can be applied to manufacturing samples from UV-curable polymers. It belongs to material jetting techniques. This method has been developed at the Institute of Fluid-Flow Machinery, Polish Academy of Sciences (IMP PAN), Poland, for the production of microturbine components (Andrearczyk et al., 2021). The most important advantages of the method are related to the minimal thickness of one layer (16 μm) and the low temperature of the process. The liquid polymer temperature in the printing head is equal to ca. 60 °C, while on the printing platform, it is much lower, and its value is ca. 40 °C, which is comparable to the human skin temperature. Due to this, there are no problems of dimensional stability linked with a polymer shrinkage occurrence. There is also no problem with residual strain occurring due to heating. Therefore, the MJP technique allows manufacturing elements without voids, whose volume fraction affects the mechanical strength of the element, as well as an embedded fiber optic sensor response to environmental factors influencing mechanical loading (Mieloszyk et al., 2020).

A scheme of the MJP printing principles is presented in Fig. 11.4. Before the manufacturing process, the shape of the sample is drawn using computer-aided design software. Then a cuboid with an inscribed sample shape is generated. During the manufacturing process, a whole volume of the cuboid is filled with polymer (build material) or casting wax material (supporting material). The materials are deposited layer by layer. The next layer can be built after the previous one is completely hardened. The final object at the end of the AM process contains the polymer element surrounded by wax. The supporting material is then removed using an oven or ultrasonic bath (Mieloszyk et al., 2022).

One of the examples of structures manufactured using the MJP technique is a compressor disk. A photograph and 3D model of the disk are presented in Fig. 11.5. Dynamical analyses were performed on the structure, and it was observed that the optimal speed range for the polymer disk is from 60,000 to 80,000 rpm. The results were compared to an aluminum disc with the same dimensions. The operation characteristics were similar for both elements. Therefore, the MJP technique can be applied for the verification of designed geometries of elements in high-speed machines. It was also observed that the mechanical properties and design of the disk require some modifications (Andrearczyk et al., 2021).

Other analyses were performed on compressor disks with improved geometry. Before their manufacturing, the thermal expansion coefficient of the used material (M3 X) was determined using an embedded FBG sensor. The tests were performed in a controllable environment (in an environmental chamber) for a range of temperatures from 20 °C to 80 °C under stable RH in a range from 10% to 90%. The selected temperatures were the operational temperature values of the compressor. The determined average thermal expansion coefficient was equal to 0.02 mm/°C and was taken into consideration during designing the final disk dimensions. The AM disk operated at

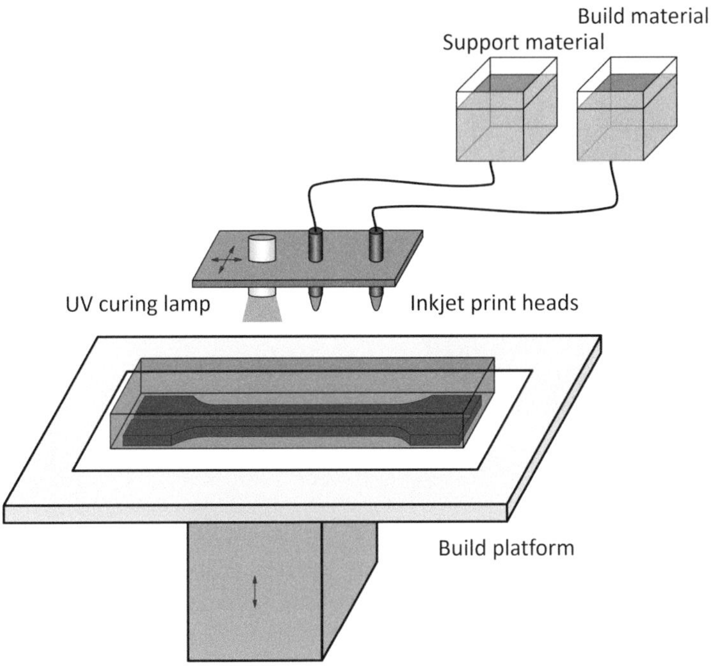

FIGURE 11.4 MJP printing principles.

MJP printing principles

From Mieloszyk, M., Majewska, K., & Andrearczyk, A. (2022). Embedded optical fibre with fibre bragg grating influence on additive manufactured polymeric structure durability. Materials, 15*(7), 2653.*

speeds ranging from 10,000 to 120,000 rpm (the disk ruptured at *ca.* 125,000 rpm) (Andrearczyk et al., 2020).

The elements and samples mentioned above were manufactured at the IMP PAN in a rapid prototyping laboratory using a 3D printer ProJet HD 3500 Max (3D Systems, Rock Hill, SC, USA). The MJP printing process is performed inside a chamber, and the process of embedding FBG sensors is based on manufacturing two separated halves of a sample with a groove (diameter equal to the outer diameter of the optical fiber). Then, the two halves were merged using the same photocurable polymer. The MJP technique modification for manufacturing polymeric samples with embedded FBG sensors is described in detail in Mieloszyk et al. (2020). The results presented in the chapter were determined for M3 X material that belongs to the category of (meth)acrylate-based photopolymer materials (Mieloszyk et al., 2022).

Another method is FDM, which is one of the most popular 3D printing methods. One of its advantages is the simplicity of the method and the lack of need to use support material. The technique can be applied to both pure polymer (thermoplastic) as well as FRP material with different reinforcement fiber types. A principle of operation of the standard FDM method is presented in Fig. 11.6A. For FRP material, two nozzles are used: one for polymer matrix and the second for fiber reinforcement. The raw material is in the form of flexible filaments (continuous fiber and

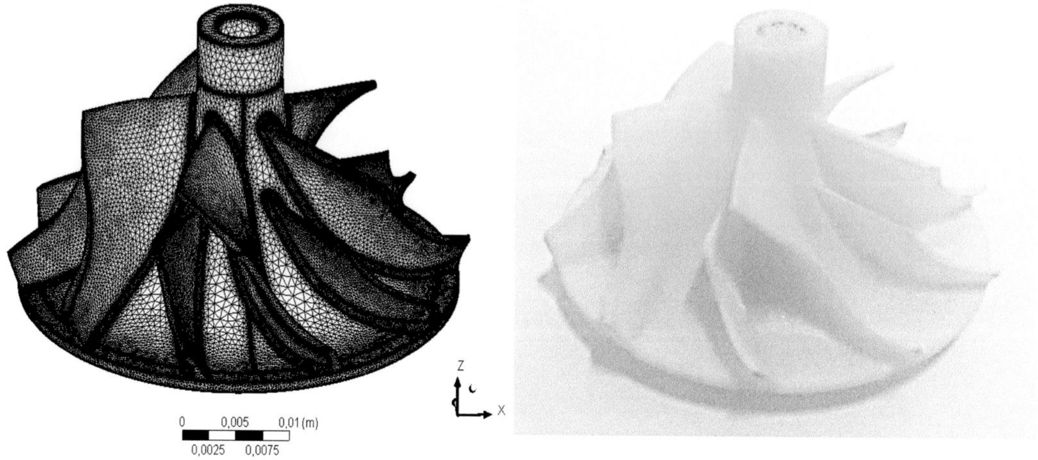

FIGURE 11.5 Model of compressor disk.

Model of compressor disk obtained by 3D scanning (left) and disk created using the MJP method (right).

From Andrearczyk, A., Konieczny, B., & Sokołowski, J. (2021). Additively manufactured parts made of a polymer material used for the experimental verification of a component of a high-speed machine with an optimised geometry—Preliminary research. Polymers, 13*(1), 137.*

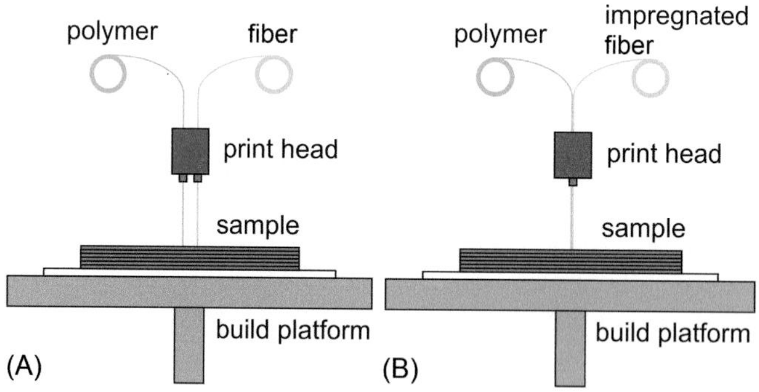

FIGURE 11.6 FDM printing principle.

FDM printing principle: (A) standard and (B) modified.

polymer). The polymer is gradually softened and melted inside the printing head. It allows connecting adjacent lines and creating a final product. Both materials are subsequently extruded through a heated nozzle onto the building platform in the form of a line. The semi-molten thermoplastic material cools rapidly, solidifies, and bonds with the neighboring rasters. The process is repeated until the final element is obtained (Kousiatza & Karalekas, 2016). In FDM parts, the formation of bonds among individual rasters of the same layer and neighboring layers is facilitated by the diffusion bonding mechanism driven by the thermal energy of the semi-molten material. The FDM technique is characterized by the accumulation of residual stresses during the layered deposition of filament. The non-uniform stresses are related to the polymer's rapid contraction during its transition from one physical state to another. Polymer is rapidly cooled on the build platform and is exposed to re-melting and rapid cooling cycles during the printing of the neighboring filaments (on the same layer and the next layer). Such a process can affect the dimensional stability and accuracy of the final element's surface roughness. It can also be observed in the form of voids, inter- or intra-layer delamination, or cracking occurrences (Kousiatza & Karalekas, 2016). The minimum thickness of the layer is *ca.* 0.254 mm (Boschetto & Bottini, 2014).

Kaunas University of Technology designed a new printing head that can be applied to the FDM method. The head is characterized by one nozzle used for both polymer and fiber reinforcement. The schematic view of the operational principle of the method is presented in Fig. 11.6B. Additionally, the continuous fiber reinforcement used is pre-impregnated before the printing process. It increases the adhesion between neighboring fiber bundles in the final element, which results in its better quality. The new method is called modified FDM, or mFDM, and is described in detail by Rimašauskas et al. (2019). The method can be applied to a variety of thermoplastic polymers (polylactic acid (PLA), acrylonitrile butadiene styrene (ABS), and polycarbonates (PC)) and carbon (Rimašauskas et al., 2019) or glass fiber (Mieloszyk et al., 2023) reinforcement. Shortly, the pre-impregnation process was based on pulling the continuous fiber throughout a bath with a thermoplastic and methylene dichloride (solvent) solution and drying with a hot air gun temperature of 220 °C (Rimašauskas et al., 2019). For the PLA, the ratio concentration is 90 g/10 g (Muna et al., 2022). After the impregnation process, the continuous fiber reinforcement can be used for printing using the mFDM method to create a composite specimen. The properties of the final products depend on impregnation process parameters, fiber reinforcement, and thermoplastic material. PLA can be used in the form of pellets, whereas ABS and PC must be mechanically crushed (Rimašauskas et al., 2019). The typical fiber content in FRP material is 18%–20%.

The mFDM method was modified to embed FBG sensors inside the samples during the printing process. The printing process is slow, and the working area is not covered. Therefore, it is possible to put the fiber in the right place on the sample material without stopping the manufacturing process. The typical thickness of one layer is 0.5 mm, which is the thickness of a continuous fiber reinforcement bundle, while the fiber optic thickness is not more than 250 μm. Therefore, the fiber optic location does not have to be taken into consideration when designing the printing process parameters. Photographs of parts of CFRP samples are presented in Fig. 11.7. The mFDM samples have characteristic edges with rounded continuous fiber (Fig. 11.7A), while the diameter of the FBG sensor is comparable to the distance between carbon fiber bundles (Fig. 11.7B).

The vacuum infusion process is a versatile process for the preparation of large composite parts due to its low cost and highly accurate manufacturing method. It belongs to liquid compaction molding processes (Gajjar et al., 2020). The most important advantages of the method are related to the possibility of manufacturing elements with complex shapes (e.g., boat hulls (Mieloszyk et al., 2021)) and high-dimensional accuracy. The curing process of the used resin (typically epoxy resin) is an exothermic reaction; therefore, a temperature increase can be observed during the process. However, it is mostly a pulse temperature change. The change in the physical state of the resin (from liquid to solid) results in the occurrence of remaining stress and can influence the shape of the embedded FBG sensor. It can be observed in the reflected spectrum distortion. The vacuum

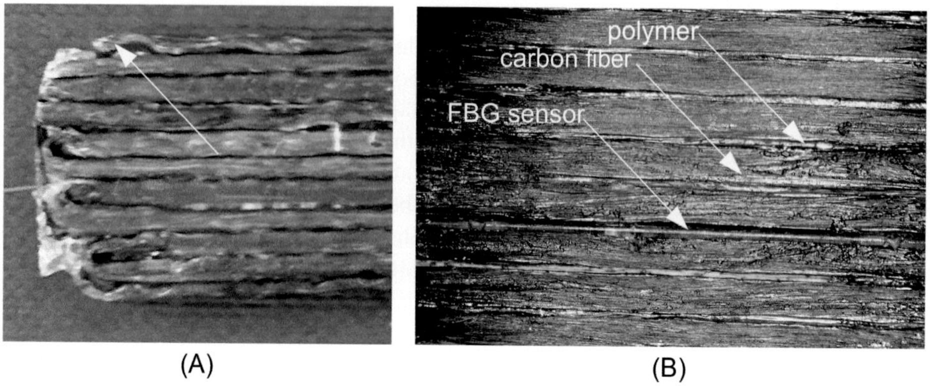

FIGURE 11.7 Structure of mFDM sample.

Structure of CFRP sample: (A) the edge of the sample with marked the end of the continuous fiber and (B) comparison of CFRP sample components and FBG sensor size.

pressure is the only driving force to consider in the infusion process to impregnate the mold cavity of the manufacturing element. Elements fabricated using the infusion method have a high fiber content (typically 60%–70% of total weight) and low porosity (less than 1%) (Gajjar et al., 2020). It is much higher than in FRP elements manufactured using the AM methods described above. However, the AM elements after the end of the process are ready-to-use, while FRP parts have to be cut to achieve final dimensions. There is also remaining resin outside the FRP element that has to be removed. The limited amount of waste and the possibility of quickly using AM elements in rapid prototyping (performing tests of designed solutions) are motivations for developing various AM methods. Not only those described above.

The infusion method can be applied to manufacturing simple FRP laminates, more complex elements like sandwich panels, or even very sophisticated structures like yacht hulls. In the process, four steps can be distinguished. They are illustrated in Fig. 11.8. First, textiles with appropriate patterns of fiber reinforcement are put on a polytetrafluoroethylene (PTFE) plate to form a stacking sequence of the former laminate. The PTFE material is chosen because it allows easy removal of the final element. During the first step, FBG sensors are placed in selected locations. Then, the whole area is covered by a special material for the purpose of equal distribution of the resin over the manufacturing element. In the next step, the vacuum chamber is built on the PTFE plate. Its edges are protected by bituminous material, and the top is created from transparent material. In appropriate locations, two pipes are mounted to ensure the appropriate resin flow during the process. Then a vacuum is generated. In the FRP elements presented in the chapter, the vacuum pressure was equal to *ca.* 0.9 bar. The last and longest step is the curing process. Its length depends on the element's complexity, the amount of the resin used, and its curing time. The samples presented in the photograph were removed from the PTFE plate after *ca.* 24 hours. Such a long time was chosen to ensure that the curing process was definitely finished. Contrary to the AM methods described above, inattentively removing samples from the working area can result in breaking optical fiber at the edge of the element. In such a case, it is impossible to use the embedded FBG sensor for measurements.

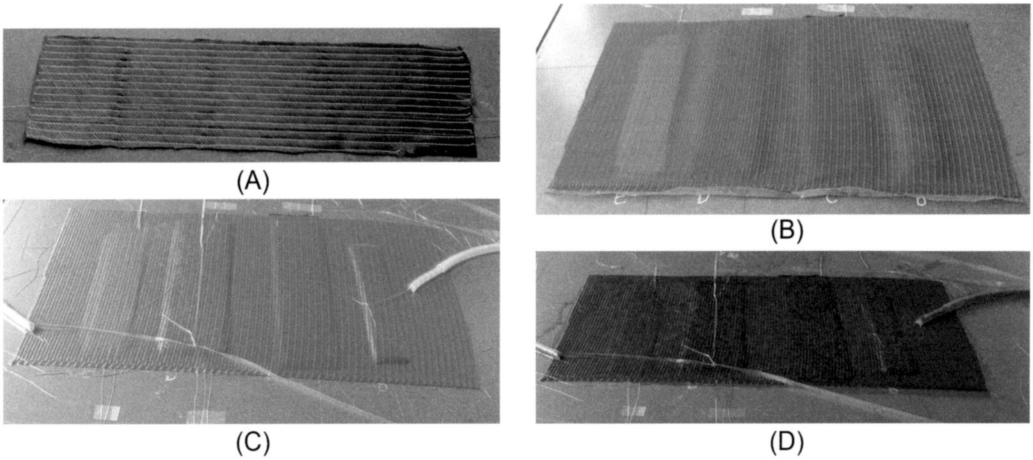

FIGURE 11.8 Infusion method.

Steps of the infusion method: (A) textile stacking sequence and FBG sensors locations, (B) covering the textiles by material, (C) the start of the resin flow, and (D) the end of the molding process.

FBG sensors can measure strain changes in the polymer material (without and with fiber reinforcement) during its manufacturing. The only limitation is the possibility of leading the fiber optic out of the production area and connecting it to the measurement unit. Such measurements are very useful in determining the remaining strain values that are a result of the curing process. Such measurements help in understanding the physical processes that occur in the material during the selected fabrication process. Knowledge about the material history is very important in predicting the remaining life length of an analyzed structure.

Embedding methods influence the spectra of embedded FBG sensors. A comparison of FBG sensor spectra for four types of smart structures is presented in Fig. 11.9. The spectra were measured using the Fiber Sensing FS4200 Portable BraggMETER. The material parameters of the samples and manufacturing method types used in their fabrication process are collected in Table 11.2. The pure polymer sample was manufactured using the MJP method. The used material (M3 X) belongs to the (meth)acrylate-based categories of photopolymer materials. The CFRP sample was manufactured using the mFDM method. The most complex structures (the sandwich panel and the boat hull) were fabricated using the infusion method.

The spectra qualities after the embedding process can be calculated using Pearson correlation. The parameter is related to the linear correlation between two variables, X and Y, and is described using the following formula (Rousseau et al., 2018):

$$P = \frac{\sum_{i=1}^{n}(x_i - \bar{x})(y_i - \bar{y})}{\sqrt{\sum_{i=1}^{n}(x_i - \bar{x})^2}\sqrt{\sum_{i=1}^{n}(y_i - \bar{y})^2}}$$

where

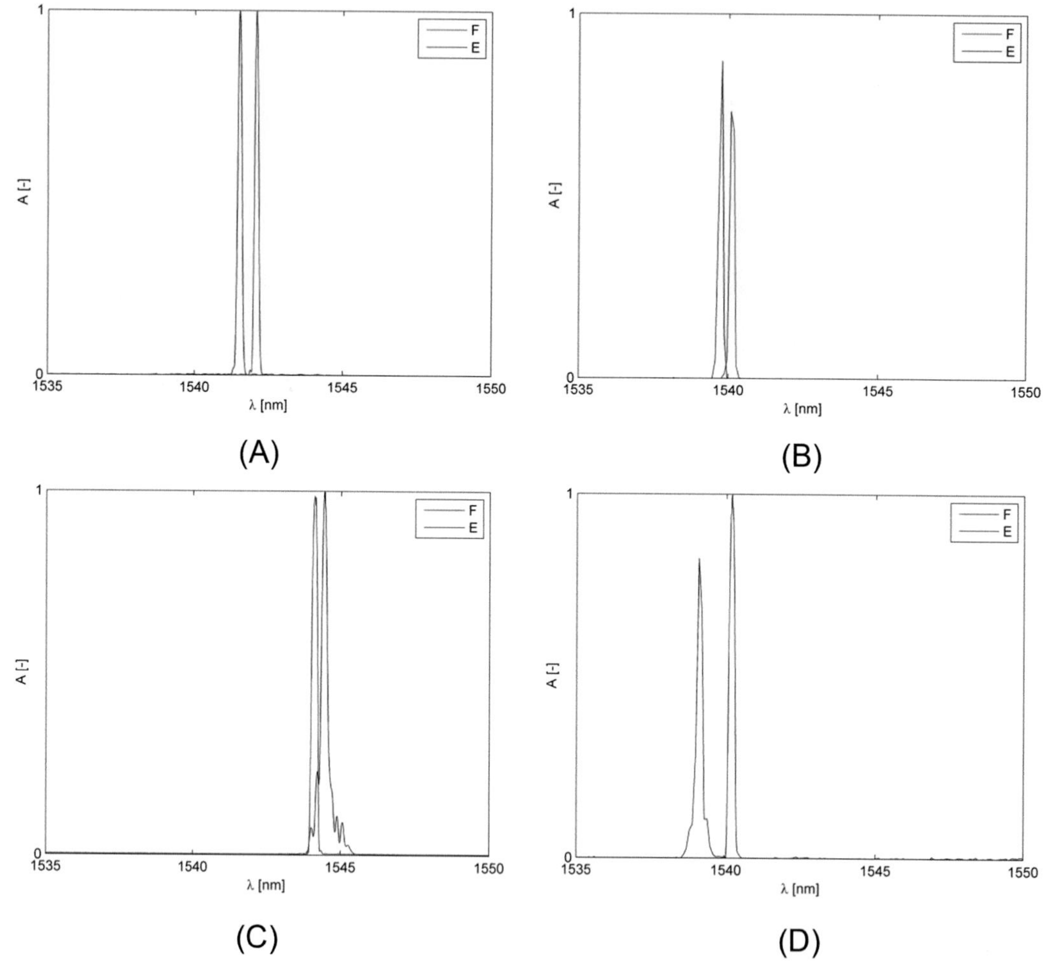

FIGURE 11.9 Spectra.

Comparison of FBG sensors' spectra: (A) polymer, (B) AM FRP, (C) sandwich structure, and (D) boat hull. *E*, Embedded; F, free.

$$\overline{x} = \frac{1}{n}\sum\nolimits_{n=1}^{n} x_i \text{ for i} = 1, \ldots, n$$

$$\overline{y} = \frac{1}{n}\sum\nolimits_{n=1}^{n} y_i \text{ for } i = 1, \ldots, n$$

where n is the analyzed signal length, and x_i and y_i are the individual signal values.

The Pearson correlation values are in a range between −1 (total negative linear correlation) and +1 (total positive linear correlation). A correlation value equal to 0 means a lack of linear correlation. The quality of the reflected spectrum of the particular embedded FBG sensor was determined

Table 11.2 Selected parameters of the analyzed samples.

	Pure polymer	CFRP	Sandwich panel/boat hull
Manufacturing method	AM	AM	Infusion
Matrix	M3 X	PLA	Vinyl ester
Fiber	–	Carbon T300B-300	SGlass
Fiber arrangement	–	0	0/45/90/ − 45
Skin	–	–	GFRP
Fulfillment	–	–	Polyvinyl chloride

Table 11.3 Pearson correlation values for selected spectra.

Sample/structure	Pearson correlation
Polymer	0.9962
CFRP	0.9275
Sandwich	0.9599
Boat	0.8783

in relation to the same sensor spectrum but under free conditions. The Pearson correlation values for the four spectra, presented in Fig. 11.9, are collected in Table 11.3. It is well visible that two parameters influenced the spectra quality: complexity of the structure (the lowest value is related to the boat—the real structure) and manufacturing method. The correlation value for the AM CFRP sample is lower than for the more complex sandwich structure. The Pearson correlation values indicated that the spectra shape changes (e.g., additional maxima with a low amplitude, Bragg wavelength amplitude decreasing, and spectrum widening) are small enough to accept the sensor for measurement application.

For better understanding the differences in embedded FBG sensor influence on FRP internal structures in relation to the manufacturing method (standard, AM), a THz spectroscopy method was applied. THz spectroscopy allows non-destructive inspection of non-conductive materials like GFRP. The measurements were performed using TPS Spectra 300 THz Pulsed Imaging and Spectroscopy (TerraView). The unit heads (emitter and receiver) were arranged in reflection mode. The angle between the heads was equal to 22°. The measurement step was equal to 0.2 mm in both directions in the xy plane, and THz signals were registered with tenfold averaging.

The method was used to examine two GFRP samples with embedded optical fibers. The differences between those samples originated from the manufacturing method used. The first was fabricated using the infusion method, while the second used the 3D printing method—mFDM. A comparison of photographs and THz spectroscopy images for the two GFRP samples is presented in Fig. 11.10. Both samples have four layers of optical fiber(s) with an FBG sensor embedded between two layers. The first sample (Fig. 11.10A) was manufactured using the standard method from glass fiber textiles. The used procedure results in differences in sample roughness. One side (the upper for THz measurements) was rough because it was covered by additional material,

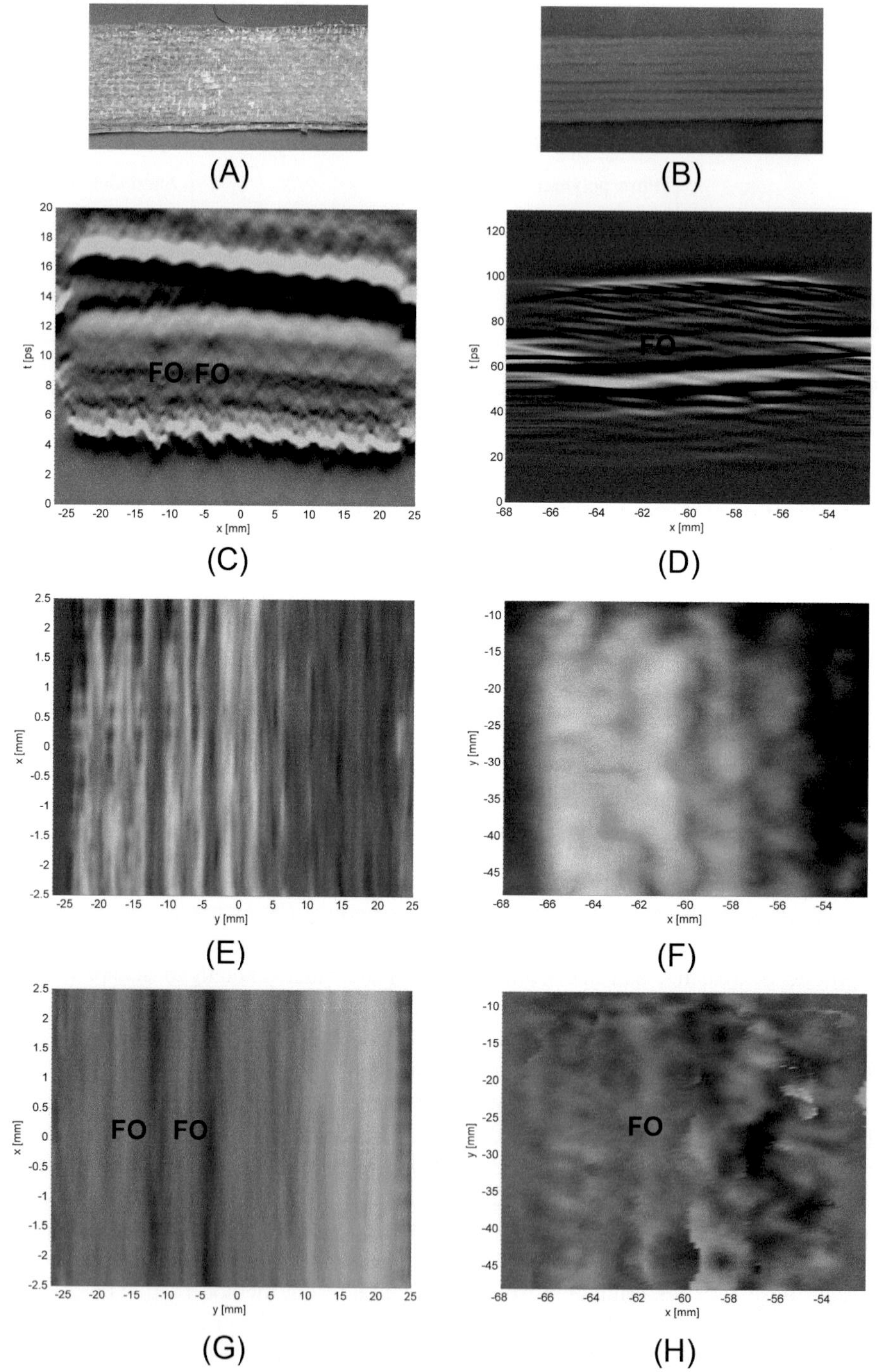

FIGURE 11.10 Comparison of GFRP.

Comparison of GFRP structures: (A) photograph of sample manufactured using infusion method, (B) photograph of AM sample, (C and D) B-scans, (E and F) C-scans for the upper surface, and (G and H) C-scans for the embedded optical fiber plane. *FO*, Fiber optic.

providing an equal distribution of resin during the manufacturing process. The second surface (the bottom for THz measurements) was smooth as the laminate during the manufacturing process was laid on a metal plate covered by PTFE. The roughness differences are well visible in the B-scan (Fig. 11.10C). The AM sample was 3D printed from continuous glass fiber on the build platform. Such a fabrication process results in layers of fiber bundles with thicknesses of *ca.* 1.3–1.5 mm (depending on the continuous fiber used) separated by gaps filled with the polymeric matrix. Due to this, it is hard to determine consecutive layers on the B-scan related to the cross-section of the sample (Fig. 11.10D). The differences in manufacturing methods are visible in the influence of the embedded optical fiber on the sample's internal structure. In the case of the infusion method, embedded optical fibers cause local deformation of the structure. The neighboring textiles did not ideally align as the optical fibers are covered by a locally thicker polymer layer (called a resin pocket) that was transported along the fibers during the fabrication process. The optical fiber influences are visible in the B-scan (Fig. 11.10C) as local curvatures on the layer boundary (between the 3rd and the 4th counting from the top surface) and as a locally higher roughness of the top surface. They are also visible as two darker parallel lines in the C-scan (Fig. 11.10E), presenting the upper surface of the sample. Contrary to this, a fiber optic is completely invisible in the AM sample cross-section (Fig. 11.10D). Also, it is hard to determine the consecutive layers. The embedded optical fiber location is observable in the C-scan (Fig. 11.10F) as a locally lighter rectangle—a bundle of reinforcement fiber in the area where the sample is locally thicker. Regardless of its small diameter, embedded optical fiber caused the local movement of consecutive layers of reinforcement fiber bundles. The last comparison is related to C-scans determined for the plate where the embedded optical fiber is located. The images are presented for standard and AM fabricated samples in Fig. 11.10G and H, respectively. The fact that the standard manufacturing method results in laminate with a comparative thickness of consecutive layers allows for presenting the exact location of the fibers—visible as two parallel dark lines. Contrary to this, the structure achieved by using the AM method contains several stacks of reinforcement fiber bundles. Therefore, the embedded optical fiber influence is more local as it moves only one reinforcement fiber bundle. The structural differences result in different responses of elements after exposition to loading. Additionally, the fracture mechanisms will not be the same for both materials. Therefore, the application of AM methods for creating smart structures requires additional experimental and theoretical analyses related to the mechanical properties of the particular structure. The elements' properties will depend on the AM process type as well as on the parameters used in the particular fabrication process.

The second important issue is related to the element edge in the location where optical fiber comes out of the host material. In very thin samples, additional protection is not needed. The user has to be careful during handling to not break the optical fiber in this area. In the case of complex elements (e.g., sandwich panels or real structures), the appropriate solution is designed depending on the used material's geometrical and mechanical parameters as well as the operational loading and environment. Examples of input/output parts for complex structures are presented in Fig. 11.11. The sandwich sample geometry was locally changed to avoid bending of the optical fibers, and the fibers were covered by an additional thick polymeric coating. Such protection was good enough to avoid damage to optical fibers and connect the sample to the interrogator during measurements in laboratory conditions. The real structure has the optical fiber ends permanently fixed to the hull structure. Additionally, the connectors are hidden under the deck of the boat. The place can be covered by a special lid that protects it against environmental influence when the SHM system is not used.

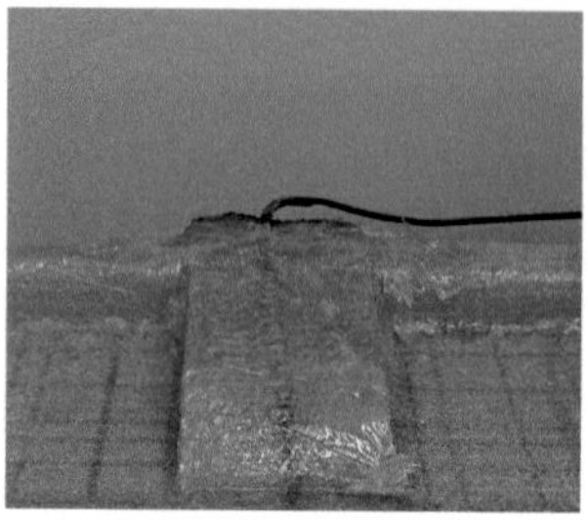
(A)

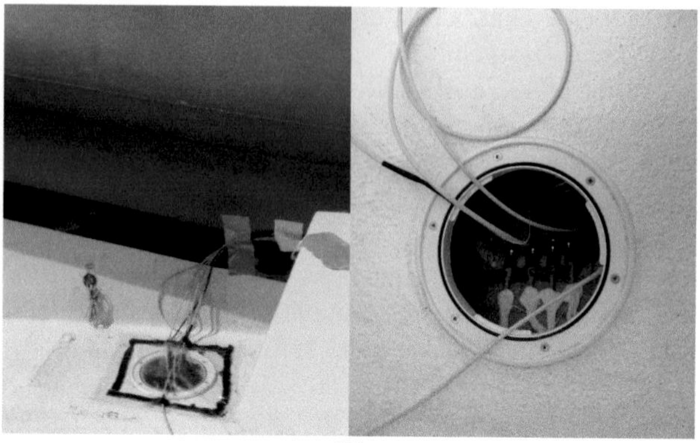
(B)

FIGURE 11.11 Examples of input/output parts in complex structures.

Examples of input/output parts in complex structures: (A) sandwich sample and (B) boat.

Table 11.4 Measurement parameters of smart structures.

Structure	Parameter		
Polymer sample	T	S	–
FRP sample	T	S	–
AM FRP sample	T	S	–
Complex element	T	S	D
Real structure	T	S	D

D, *Damage detection;* S, *strain;* T, *temperature.*

11.4.1 Applications

In Fig. 11.1, schematic representatives of polymeric smart structures were presented. Their complexity is not only related to their shape and used materials but also to their utility in SHM systems. The parameters that can be measured by the smart structures are collected in Table 11.4. For simple samples, only temperature and strain can be measured. They can be applied, for example, to determine the temperature influence on the analyzed material (strain response) or to determine the tensile strain in samples during tensile tests. They can also be treated as smart sensors that can be integrated with a real structure (e.g., a bridge) in the chosen location. The complex element contains an array of sensors that can also be applied for the localization of impact loading, and damage detection is an effect of such action. In the case of a real structure, the FBG sensor array is part of the SHM system and can be applied for damage detection during the operation of the structure.

11.4.2 Simple smart structures

As simple smart structures, the small samples with embedded optical fiber parallel to the main axis of the sample will be understood. From the list of structures presented in Fig. 11.1 and Table 11.4, the description will be related to the polymer sample and AM FRP sample. Nowadays, world strategy is related to the reduction of production processes' influence on the environment, so AM methods were chosen as the elements of manufacturing processes.

For comparison purposes, two sample types were selected: pure photocurable polymer (M3 X) and CFRP (PLA matrix and carbon fiber T300B-300). The polymer sample was manufactured using the MJP technique. The manufacturing technology is described in detail by Mieloszyk (Mieloszyk et al., 2020). The FRP samples were manufactured using the mFDM method. The details of the method were presented by Rimasauskas (Rimašauskas et al., 2019).

The geometrical parameters of the samples are collected in Table 11.5. Originally, the pure polymer material sample had the dogbone shape, but in the table, only the part related to the measurement area (the embedded FBG sensor location) was presented.

The thermal tests were performed in an environmental chamber called MyDiscovery DM600C (Angelantoni Test Technologies Srl, Massa Martana, Italy). The measurements were performed at temperatures ranging from 10 °C to 50 °C, with a 5 °C step between consecutive temperatures. The heating process was conducted at a very low speed (5 °C for 900 seconds) to ensure quasi-static conditions during the process. The measurements were performed for five stable values of RH level: 20%, 40%, 60%, 80%, and 95%. During the investigation, the samples were kept on a lattice shelf to allow them to expand in all directions. The temperature in the chamber was measured using an FBG temperature probe. The FBG sensor measurements were performed using an interrogator si425–500 (Micron Optics, Atlanta, GA, USA) with a measurement frequency of 1 Hz.

The thermal strain for each stable temperature step was determined using the relationship:

$$\varepsilon(T) = \frac{1}{n}\sum_{i=1}^{n} \varepsilon_n(T_i) \text{for i} = 1, \ldots, 9$$

where temperature T_i is the temperature at the chosen point for a stable temperature condition.

The strain values determined for temperature changes for pure polymer and CFRP materials are presented in Figs. 11.12 and 11.13, respectively. For both materials, the influence of RH is well visible, and a relationship can be determined. For polymers, the strain values increase with the RH levels due to the moisture uptake into the sample material. Contrary to this, for CFRP, the RH levels increased due to a strain decrease. Probably, moisture is captured by the voids between neighboring carbon fibers in the AM structure.

Table 11.5 Geometrical parameters of polymer and FRP sample.

Parameter	Pure polymer	CFRP
Length [mm]	70	150
Thickness [mm]	4.20	2
Width [mm]	12	15

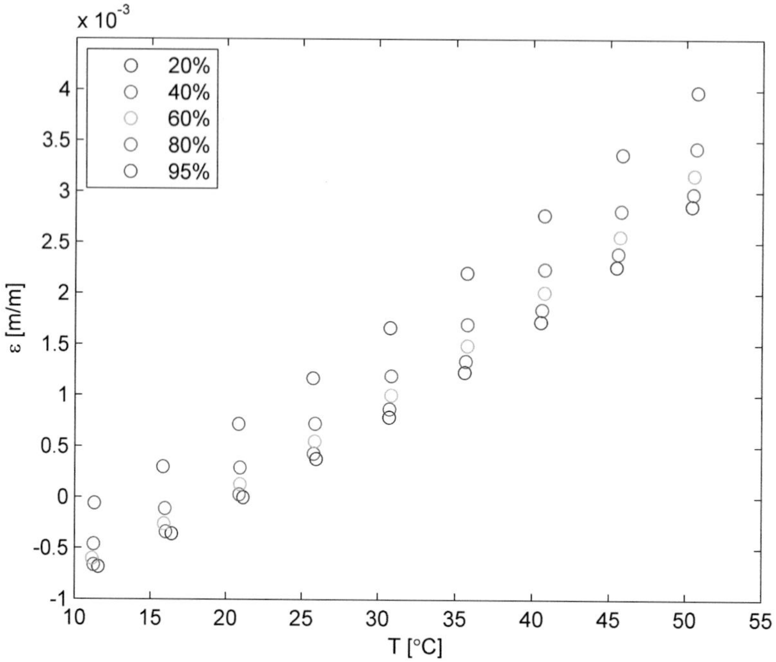

FIGURE 11.12 Strain in polymer.

Strain measured by FBG sensor embedded into AM polymer.

The strain curves for CFRP material show a characteristic behavior for temperatures higher than 45 °C. According to experimental investigations performed on pure PLA, its glass transition temperature (T_g) value was equal to 65 °C. Such value can also be expected for PLA material reinforced by carbon fibers. The main difference probably relates to the carbon fiber preparation process before the AM. The additional filler (carbon fiber) has taken space out of PLA chains due to a lack of interaction between them and influences on cross-linking in the polymer. It results in a lower T_g value and storage modulus of CFRP material compared with pure PLA (Cao et al., 2016). In the explanation of the CFRP material behavior, not only T_g temperature should be taken into consideration. The other parameter is the deflection temperature (HDT). It is a measurement of the resistance of a polymer to deformation (the stiffness of the material) when subjected to stress loading at elevated temperatures. The HDT value in a polymer material is determined by the level of crystallinity (morphological structure) and the presence of reinforcing agents such as fillers and plasticizers. Pure PLA has an HDT value of *ca.* 55 °C (Wu et al., 2019). The HDT value is related to the T_g. However, a low HDT value can also be a result of manufacturing parameters that cause a slower crystallization rate during the process (Cristea et al., 2020). However, the beginning HDT of the particular CFRP material was determined to be 42 °C.

The temperature influences the fracture mechanism observed in the CFRP sample during the tensile test, as well as the maximal tensile strength values. A comparison of strain values measured using the FBG sensor embedded in the sample and the extensometer from the tensile machine is

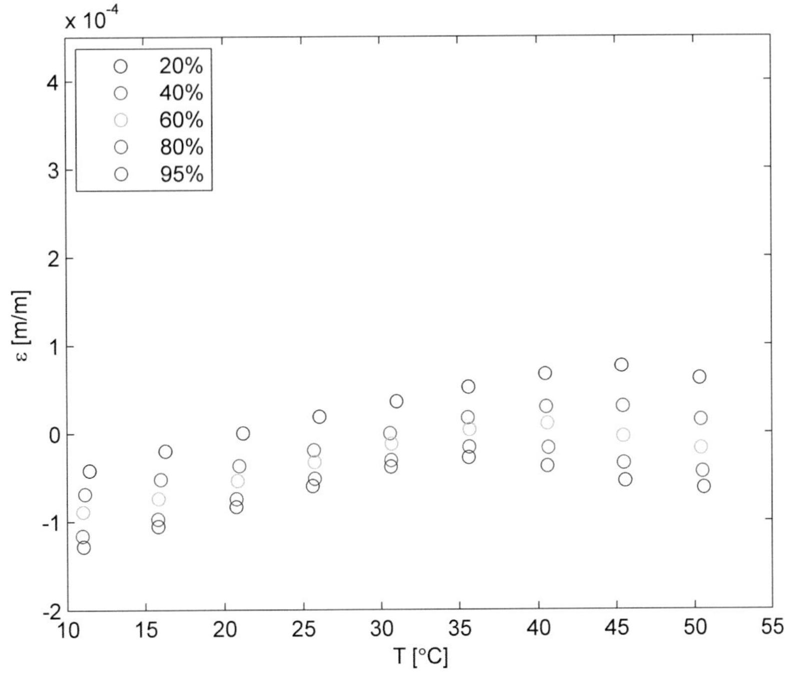

FIGURE 11.13 Strain in CFRP.

Strain measured by FBG sensor embedded into AM CFRP.

presented in Fig. 11.14. The analyses were performed on intact samples (after manufacturing) and after the thermal tests described in detail above. The tensile tests were performed on the universal static-dynamic testing machine HT-9711-25 (Hung Ta, Korea). The mean tensile strength values were *ca.* 170 MPa for both sample groups. The strain curve shapes were comparable during the tensile loading for both sample types. The main differences occur at the end of the tests. The smallest was for the sample after manufacturing. It was equal to 0.5×10^{-3} m/m and was determined at the end of the tensile test. Both the extensometer and FBG sensor stopped measurements at the same time because the sample and the embedded fiber optic were broken at the same moment in time. Contrary to this, the measurements performed on the sample after the thermal test were stopped for the FBG sensor when it was first pulled out and then broken. It is observed in the form of the consecutive tensile strain increasing and the appearance of the compression strain. However, the sample was able to withstand the tensile load within the next *ca.* 80 seconds of the test. FBG sensors were connected to the measurement unit (interrogator) from one side of the fiber optic. Because the embedded FBG sensors were not damaged, it will be possible to use the second end of the fiber optic to connect the FBG sensor and perform other measurements.

A comparison of the photographs of two samples after the tensile test, together with a zoom of the fracture, is presented in Fig. 11.15. A difference in the breaking process is well visible. Exposition of the sample at elevated temperatures results in changes in the fracture mechanism of

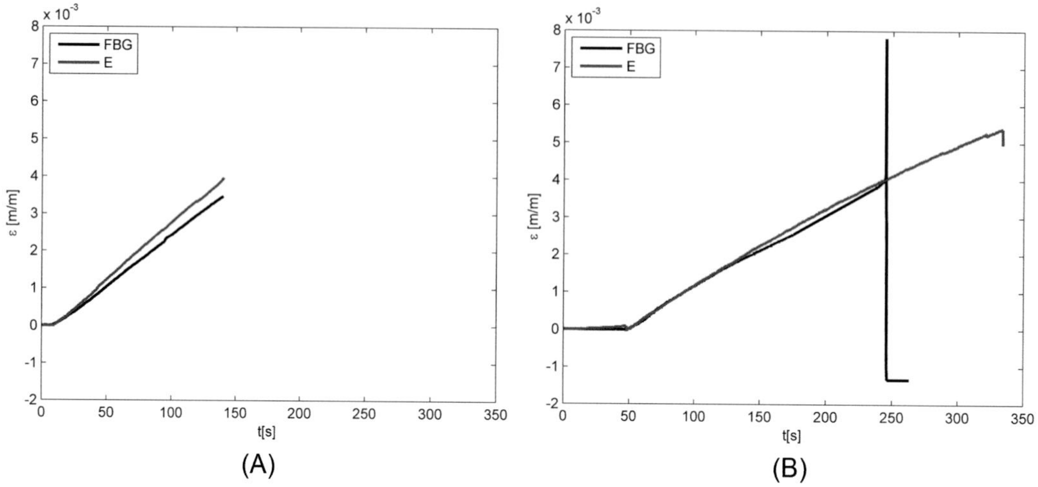

FIGURE 11.14 Tensile strain.

Comparison of tensile strain determined during the tensile test for sample: (A) after manufacturing and (B) after exposition to elevated temperatures, E—extensometer, FBG—FBG sensor.

CFRP material. The sample after the manufacturing process was separated into two parts. The carbon fibers in the bundles were broken at different heights (distances from the fracture location). Contrary to this, the sample after the thermal test remained whole. The degradation process is based on pulling out whole fiber boundless. It can be concluded that the elevated temperatures improved the boundary between carbon fiber bundles and the PLA matrix.

Such differences in material behavior do not occur for pure polymer (M3 X) material that was tested similarly. The mean tensile strength value was *ca.* 49 MPa for both sample groups. Similarly, like in the CFRP material, embedded FBG sensors were used for strain measurements, and the strain values were compared to those measured by the extensometer. The highest difference between strains measured using both methods was *ca.* 20% and was determined at the end of the tensile test. A photograph showing an example of the broken region of the sample is presented in Fig. 11.16. The sample contained both embedded and attached FBG sensors. In M3 X samples, the embedded fiber optics remained intact after the end of the tensile test. In each case, fiber optics still connected the broken parts of the sample. Only a triangular part of the sample that was ripped out during the test was separated.

It is worth noticing that FBG sensors were able to perform continuous measurements under the high loading that occurs during tensile tests. Therefore, the connection between an FBG sensor and the sample material (both pure polymers and CFRP) had very good quality.

In real structures, FRP simple elements are connected to each other, for example, in the form of adhesive joints. As it was mentioned earlier, the adhesive bonds can be damaged due to the moisture uptake from the surrounding environment. Even 1% of moisture affects the strength of an adhesive bond layer. To increase the safety of the used element, a smart polymer structure can be applied. In such a solution, the FBG sensor is embedded inside the adhesive layer and can be used

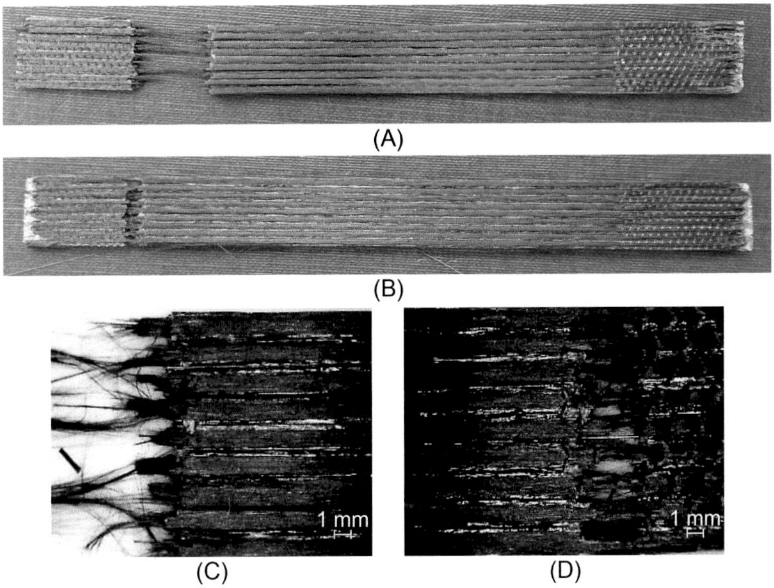

FIGURE 11.15 CFRP fracture.

Samples after the tensile test: (A and C) after manufacturing and (B and D) after exposition to elevated temperatures.

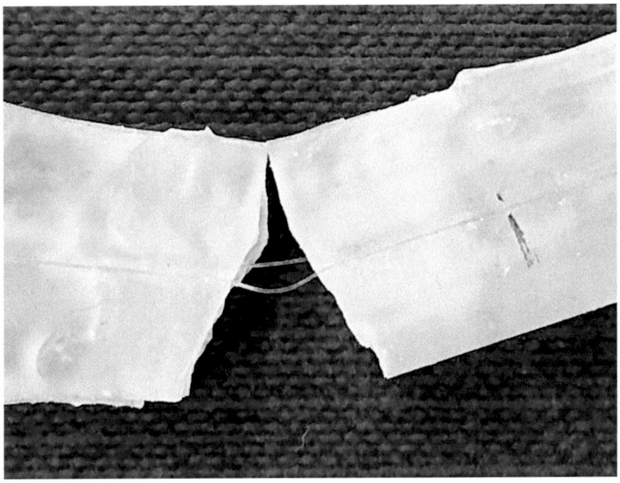

FIGURE 11.16 Polymer fracture.

Pure polymer sample after the tensile test.

for continuous monitoring of the amount of moisture inside the joint. In the solution presented by Mieloszyk and Ostachowicz (Mieloszyk & Ostachowicz, 2017b), FBG sensors were able to detect a small amount of moisture concentration (determined as 1%–3% of sample weight). The analyzed GFRP joint dimensions were 50 mm × 250 mm × 2 mm. The calibrated smart structure was able to determine the actual amount of moisture after both the drying and soaking processes.

11.4.3 Real structure—fast patrol boat

The real structure presented in this chapter is a fast patrol boat with a total length of 7 m. It has a deep V-shaped hull that has a sandwich structure; the parameters are collected in Table 11.2. The details about its manufacturing are presented in the paper (Mieloszyk et al., 2021).

Fig. 11.17 contains photographs presenting the real smart structure together with the analyzed area. This part of the boat was chosen as a possible place that can be damaged in the harbor or due to an impact with an object floating on the surface of a water body. The analyzed damage is a delamination of the hull structure close to the embedded FBG sensor array. As it is a quite big structure, the SHM system components are on the boat during its normal operational conditions. Such elements are the FBG sensors array, measurement unit for FBG sensor (interrogator Smart Scan04, Smart Fiber), and control panel integrated with the boat steering panel. It allows for measurements during the normal operation of the boat.

One of the important issues that has to be taken into consideration while performing analyses related to SHM monitoring of real complex structures is the determination of the base Bragg wavelength values for all FBG sensors in the measurement array. For simple samples, the first selected moment in time is the end of the embedding process. The second, also important, is the moment when the sample is mounted for testing. In both cases, the most important environmental parameters that must be collected are RH and outer temperature. For complex structures, like the fast

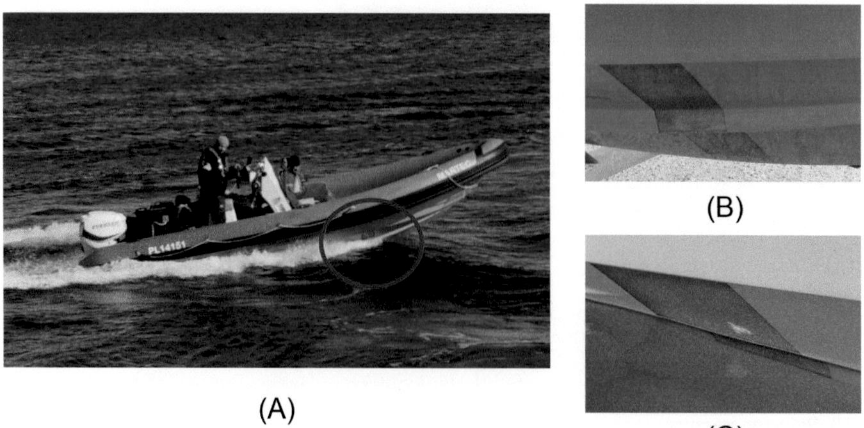

(A) (B) (C)

FIGURE 11.17 Boat.

Fast patrol boat: (A) photograph with marked analyzed area, (B) intact structure, and (C) damaged structure.

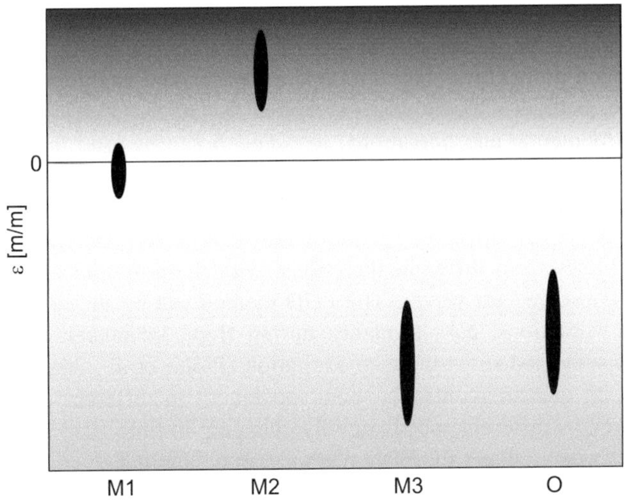

FIGURE 11.18 Strain levels of the boat.

Strain level changes in the boat after M1–M3, manufacturing steps, O—operation on the sea.

patrol boat, the procedure is not so easy. For the purpose of better understanding the complexity of the problem, strain ranges determined for all FBG sensors created the measurement array for the four chosen steps, which are presented in Fig. 11.18. In the manufacturing process, three steps can be specified. The first (denoted as M1) is the beginning of the embedding process, understood as a moment when all elements (textiles, core, and FBG sensors) are arranged. This moment corresponds to the manufacturing process stage presented in Fig. 11.8B. The second (denoted as M2) is the end of the curing process of the used polymer material. It corresponds to the last stage of the infusion method process presented in Fig. 11.8D. The occurrence of tensile strain is a typical effect of the curing process of an epoxy resin. It is observable for both simple samples and complex structures. After stage M2, the hull is finished, but it is empty. The construction of the boat required adding more elements, like a deck, engine, and steering system. Such a process (denoted as M3) is not neutral to the structure from a mechanical point of view. Instead of tension, compression is observed. Additionally, one strain range (denoted as O) was presented for the boat after its operation on the sea. The mechanical and environmental loadings also influenced the structure. Therefore, it is very important to perform not only continuous measurements using FBG sensors but also non-destructive testing (e.g., visual inspection and infrared thermography). Especially when the analyzed structure is a prototype, all information about the structure condition and its response to real loading will be used for the design purposes of new structures.

The problem related to choosing the best base condition for determining the base Bragg wavelength is more complex when a real structure does not contain an integrated SHM system that was constructed together with it. In such cases, the loading history and its influence on the material of the structure are not known. All analyses start after the end of the installation of the SHM system. A good example of the importance of the problem is a discussion presented for the FBG system installed on the foremast of the sailing ship Dar Mlodziezy (Majewska et al., 2014). The calculated strain values were compared for eight selected positions of the ship (e.g., mooring at harbors and anchoring). The higher differences among the strain levels were *ca.* 90%.

The damage detection process was based on operational modal analyses—the frequency domain decomposition (FDD) method. The method is applied to large objects, like bridges or offshore structures, under environmental and operational loadings. It can be used for analyzing signals with a high noise-to-signal ratio. The main assumption is that the structure is excited by broadband stochastic excitation like white noise (Brincker et al., 2001). In practical applications, environmental excitations like wind or waves fulfill such requirements (Carne & James, 2010). During long measurements, the wind blows from different directions, with the amplitude changing with time. An analogical situation is observed for waves. The FDD method can be successively applied for structures under different excitations, e.g., airplanes during flight maneuvers (Kocan, 2020) or civil engineering buildings subjected to earthquake excitation (Pioldi et al., 2017). In the case of a fast patrol boat, during long measurements, the boat movement in different directions in relation to waves and wind, as well as different speeds rapidly changing in time, also allows us to assume that the structure is correctly excited and the FDD method can be applied.

Because operational modal analysis is performed under the exploitation loading of a structure, it allows for achieving its real operational modal characteristics. Due to this, the method is very useful for SHM systems and damage detection in large structures.

The details of the FDD method for strain measurements (from FBG sensors), as well as a proposition of damage indexes based on such measurements, were presented by Mieloszyk and Ostachowicz (Mieloszyk & Ostachowicz, 2017a). In that case, the analyses were done for the support structure of an offshore wind turbine. The main differences between both structures are related to their operational characteristics. A wind turbine is located in one place and is affected by the environment (e.g., waves and wind). The boat structure is influenced by not only the environment but also the movement of the boat on the surface of the sea as well as hull impacts on the sea surfaces. The impacts are a normal operational procedure related to its fast speed.

The damage index, DI, that can be used for damage detection on the boat is calculated as follows:

$$DI = \int_{f1}^{f2} A(f)df$$

where f_1 and f_2 mean the beginning and end of the chosen frequency range, and A is the amplitude. The damage index is based on strain energy as a result of the loading of the structure. The frequency range during analysis was 0–1250 Hz. The damage index was calculated for two frequency ranges: the whole range and for higher frequencies above 200 Hz.

Fig. 11.19 compares frequency characteristics and damage indexes for healthy and damaged boat structures. The FDD method was applied to a set containing all the sensors embedded in the boat.

For both states, the majority of the frequency peaks are up to 200 Hz. The curve shapes in this range are similar. The number and amplitudes of excited frequencies depend on the particular excitation characteristics. The local maximum at 253 Hz occurs in both cases, but the amplitudes differ. The damage results in additional peaks (for the frequency range from 570 Hz to 598 Hz). Such frequencies are not related to the natural frequencies of the boat. However, the delamination may be between the skin and fulfillment in the sandwich structure of the hull. In such cases, impacts of

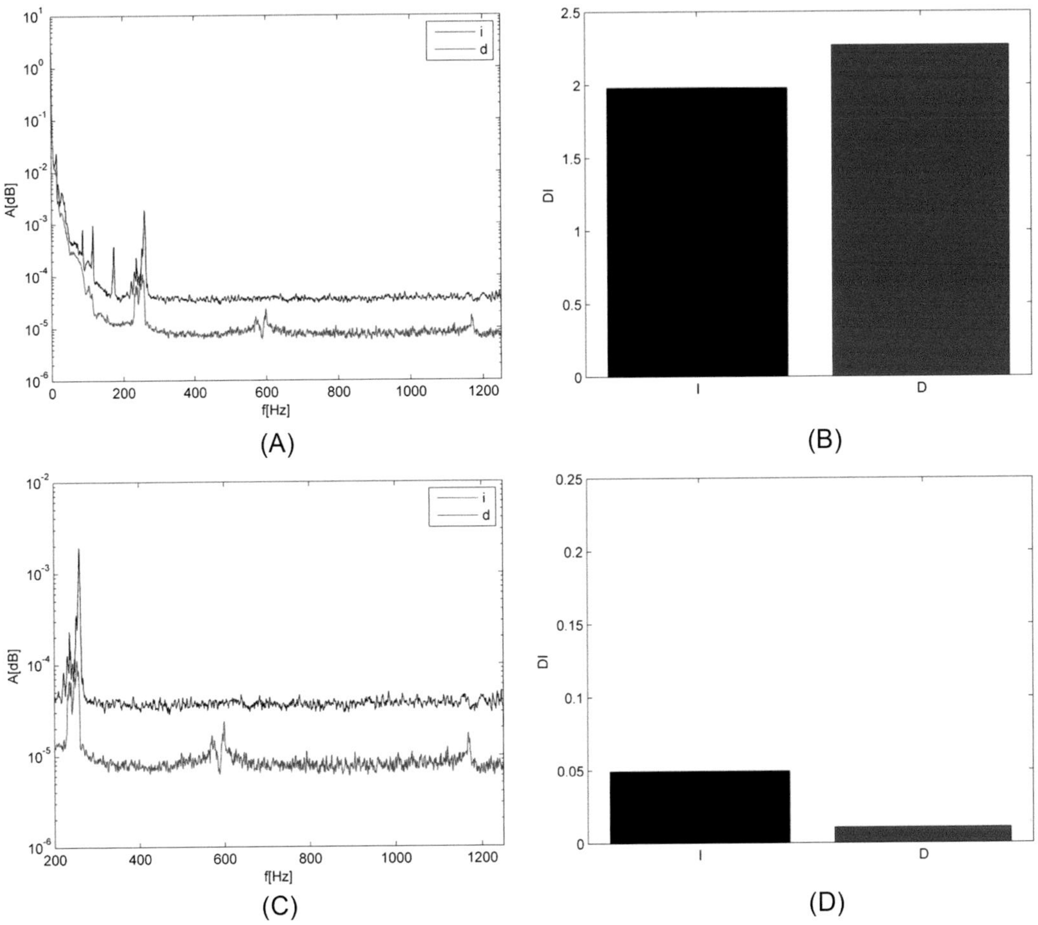

FIGURE 11.19 Damage detection.

Damage detection in the boat: (A) characteristic for frequencies 0–1250 Hz, (B) damage index for frequencies 0–1250 Hz, (C) characteristic for frequencies 200–1250 Hz, and (D) damage index for frequencies 0–1250 Hz.

the boat on the water surface (normal operational conditions) could result in a standing wave excitation in the damaged location.

Due to the frequency characteristic described above, the damage index was calculated for two ranges: the whole range and above 200 Hz. In the second range also, the peak for 253 Hz is included as damage influenced its shape. It is visible that the low-frequency part strongly influenced the DI value. Therefore, both healthy and damaged index values for the whole range are very similar. Analyses related to higher frequencies show that the DI for an intact structure is *ca.* five times smaller than for a damaged one. It seems that the strain energy for damaged structures is lower than for healthy ones. Therefore, energy dissipation occurs in the FBG sensor measurement area. For the purpose of determining the exact damage location as well as its shape and size, a non-destructive technique (e.g., infrared thermography) should be used. The hull structure stiffness is high, and it is not possible to determine the location of the delamination based on the strain measurements performed during the boat operation on the sea.

11.5 Conclusion

A variety of polymeric smart structures (from a simple sample to a real boat with a composite hull structure) were presented and discussed. Their complexity was given in relation to the SHM system levels. All measurements were performed using FBG sensors that were integrated with the polymer material. All used manufacturing methods (AM and standard) were briefly described, with particular emphasis on the FBG sensor embedding process. Also, the influence of environmental parameters on the polymer material's (without and with reinforcement) durability was presented. Two of them (temperature and moisture) were described in a more detailed way because they are very common and their influence on the polymer properties is the highest. The influence of the selected manufacturing methods on the element material structure, as well as FBG sensors and reflected spectra, was presented. For the evaluation of the quality of the spectra, the Pearson coefficient was used. Residual strain occurrence, as well as material shrinkage during the manufacturing processes, can influence the spectra and the dimensional stability of the final element. A very important issue is the place where the optical fiber comes out of an FRP element. More complex structures require better protection in this area to avoid damage to a brittle optical fiber.

Simple, smart structures are analyzed in laboratory conditions, and in that place, they can be applied for, for example, thermal strain measurement. They can also be treated as a sensing element for, for example, pressure measurement. Such an element can be integrated with a large structure that is under operation. More complex smart structures (e.g., sandwich panels) can also be applied for impact detection and localization. The most developed smart structure will contain a whole SHM system that allows the determination of the health state of the structure as well as collecting data about the real loading and environmental conditions of the structure.

Acknowledgments

This work was partially supported by the following projects:

- Analyses related to M3 X material: 'Polymeric structures with embedded FBG sensors' (2018/31/D/ST8/00463) granted by the National Science Centre, Poland.
- Analyses related to AM FRP material: 'Additive manufactured composite smart structures with embedded FBG sensors (AMCSS)' funded by the National Science Centre, Poland under M-ERA.NET 2 Call 2019, grant agreement 2019/01/Y/ST8/00075.
- Analyses related to sandwich structure and fast patrol boat: "Reliable and Autonomous Monitoring System for Maritime Structures" (MARTECII/RAMMS/1/2016), granted by the National Centre for Research and Development in Poland.

References

Abdollahiparsa, H., Shahmirzaloo, A., Teuffel, P., & Blok, R. (2023). A review of recent developments in structural applications of natural fiber-reinforced composites (NFRCs). *Composites and Advanced Materials*, *32*2634983322 1147540.

Al-Ramahee, M. A., Chan, T., Mackie, K. R., Ghasemi, S., & Mirmiran, A. (2017). Lightweight UHPC-FRP composite deck system. *Journal of Bridge Engineering*, *22*(7)04017022.

Andrearczyk, A., Konieczny, B., & Sokołowski, J. (2021). Additively manufactured parts made of a polymer material used for the experimental verification of a component of a high-speed machine with an optimised geometry—Preliminary research. *Polymers*, *13*(1), 137.

Andrearczyk, A., Mieloszyk, M., & Bagiński, P. (2020). Destructive tests of an additively manufactured compressor wheel performed at high rotational speeds. In *Advances in Manufacturing, Production Management and Process Control: Proceedings of the AHFE 2020 Virtual Conferences on Human Aspects of Advanced Manufacturing, Advanced Production Management and Process Control, and Additive Manufacturing, Modeling Systems and 3D Prototyping*, July 16–20, 2020, USA, Springer, 3030519805 (pp. 117–123).

Boschetto, A., & Bottini, L. (2014). Accuracy prediction in fused deposition modeling. *The International Journal of Advanced Manufacturing Technology*, *73*, 913–928.

Botelho, E. C., Pardini, L. C., & Rezende, M. C. (2006). Hygrothermal effects on the shear properties of carbon fiber/epoxy composites. *Journal of Materials Science*, *41*, 7111–7118.

Brincker, R., Zhang, L., & Andersen, P. (2001). Modal identification of output-only systems using frequency domain decomposition. *Smart Materials and Structures*, *10*(3), 441.

Butter, C. D., & Hocker, G. B. (1978). Fiber optics strain gauge. *Applied Optics*, *17*(18), 2867–2869.

Cao, Y., Ju, Y., Liao, F., Jin, X., Dai, X., Li, J., & Wang, X. (2016). Improving the flame retardancy and mechanical properties of poly (lactic acid) with a novel nanorod-shaped hybrid flame retardant. *RSC Advances*, *6*(18), 14852–14858.

Carne, T. G., & James, G. H., III (2010). The inception of OMA in the development of modal testing technology for wind turbines. *Mechanical Systems and Signal Processing*, *24*(5), 1213–1226.

Chan, T. H. T., Yu, L., Tam, H.-Y., Ni, Y.-Q., Liu, S., Chung, W., & Cheng, L. (2006). Fiber Bragg grating sensors for structural health monitoring of Tsing Ma bridge: Background and experimental observation. *Engineering Structures*, *28*(5), 648–659.

Choi, B.-H., & Kwon, I.-B. (2019). Damage mapping using strain distribution of an optical fiber embedded in a composite cylinder after low-velocity impacts. *Composites Part B: Engineering*, *173*107009.

Cristea, M., Ionita, D., & Maria Iftime, M. (2020). Dynamic mechanical analysis investigations of PLA-based renewable materials: How are they useful? *Materials*, *13*(22), 5302.

Cucinotta, F., Guglielmino, E., & Sfravara, F. (2017). Life cycle assessment in yacht industry: A case study of comparison between hand lay-up and vacuum infusion. *Journal of Cleaner Production*, *142*, 3822–3833.

Culshaw, B., Thursby, G., Betz, D., & Sorazu, B. (2008). The detection of ultrasound using fiber-optic sensors. *IEEE Sensors Journal*, *8*(7), 1360–1367.

Dicker, M. P. M., Duckworth, P. F., Baker, A. B., Francois, G., Hazzard, M. K., & Weaver, P. M. (2014). Green composites: A review of material attributes and complementary applications. *Composites Part A: Applied Science and Manufacturing*, *56*, 280–289.

Ryan, J., Dizon, C., Espera, A. H., Jr, Chen, Q., & Advincula, R. C. (2018). Mechanical characterization of 3D-printed polymers. *Additive Manufacturing*, *20*, 44–67.

Eslami, S., Taheri-Behrooz, F., & Taheri, F. (2012). Effects of aging temperature on moisture absorption of perforated GFRP. *Advances in Materials Science and Engineering*, *2012*.

Faruk, O., Bledzki, A. K., Fink, H.-P., & Sain, M. (2012). Biocomposites reinforced with natural fibers: 2000–2010. *Progress in Polymer Science*, *37*(11), 1552–1596.

Fogorasi, M. S., & Barbu, I. (2017). The potential of natural fibres for automotive sector-review, 2017 *IOP Conference Series: Materials Science and Engineering IOP Publishing*, *252*(1)012044.

Gajjar, T., Shah, D. B., Joshi, S. J., & Patel, K. M. (2020). Analysis of process parameters for composites manufacturing using vacuum infusion process. *Materials Today: Proceedings*, *21*, 1244–1249.

Gebremichael, Y. M., Li, W., Boyle, W. J. O., Meggitt, B. T., Grattan, K. T. V., McKinley, B., Fernando, G. F., Kister, G., Winter, D., & Canning, L. (2005). Integration and assessment of fibre Bragg grating sensors in an all-fibre reinforced polymer composite road bridge. *Sensors and Actuators A: Physical*, *118*(1), 78–85.

George, J., Sreekala, M. S., & Thomas, S. (2001). A review on interface modification and characterization of natural fiber reinforced plastic composites. *Polymer Engineering & Science*, *41*(9), 1471–1485.

Groves, R. (2018). *3.12 inspection and monitoring of composite aircraft structures. Reference Module in Materials Science and Materials Engineering: Comprehensive Composite Materials II* (pp. 300–311). Elservier.

Gąsior, P., Malesa, M., Kaleta, J., Kujawińska, M., Malowany, K., & Rybczyński, R. (2018). Application of complementary optical methods for strain investigation in composite high pressure vessel. *Composite Structures*, *203*, 718–724.

Ho, M.-P., Lau, K.-T., Au, H.-Y., Dong, Y., & Tam, H.-Y. (2013). Structural health monitoring of an asymmetrical SMA reinforced composite using embedded FBG sensors. *Smart Materials and Structures*, *22*(12) 125015.

Hong, C., Zhang, Y., & Borana, L. (2019). Design, fabrication and testing of a 3D printed FBG pressure sensor. *IEEE Access*, *7*, 38577–38583.

Keršienė, N., Raslavičius, L., Keršys, A., Kažys, R., & Žukauskas, E. (2016). Energo-mechanical evaluation of damage growth and fracture initiation in aviation composite structures. *Polymer-Plastics Technology and Engineering*, *55*(11), 1137–1144.

Klotz, T., Pothier, R., Walch, D., & Colombo, T. (2021). Prediction of the business jet Global 7500 wing deformed shape using fiber Bragg gratings and neural network. *Results in Engineering*, 9100190.

Kocan, C. (2020). A comparative study on in-flight modal identification of an aircraft using time-and frequency-domain techniques. *Journal of Vibration and Control*, *26*(21–22), 1920–1934.

Kousiatza, C., & Karalekas, D. (2016). In-situ monitoring of strain and temperature distributions during fused deposition modeling process. *Materials & Design*, *97*, 400–406.

Lee, J.-R., Tsuda, H., & Toyama, N. (2007). Impact wave and damage detections using a strain-free fiber Bragg grating ultrasonic receiver. *Ndt & E International*, *40*(1), 85–93. Available from https://doi.org/10.1016/j.ndteint.2006.07.001.

Li, H.-N., Li, D.-S., & Song, G.-B. (2004). Recent applications of fiber optic sensors to health monitoring in civil engineering. *Engineering Structures*, *26*(11), 1647–1657.

Majewska, K., Mieloszyk, M., Jurek, M., & Ostachowicz, W. (2021). Coexisting sub-zero temperature and relative humidity influences on sensors and composite material. *Composite Structures*, *260*113263.

Majewska, K., Mieloszyk, M., Ostachowicz, W., & Król, A. (2014). Experimental method of strain/stress measurements on tall sailing ships using Fibre Bragg Grating sensors. *Applied Ocean Research*, *47*, 270–283.

Masuda, S., Tomioka, J.-I., Tamura, H., & Tamura, Y. (2021). A study of decrease burst strength on compressed-hydrogen containers by drop test. *International Journal of Hydrogen Energy*, *46*(23), 12399–12406.

Measures, R. M. (2002). Fiber optics, Bragg grating sensors. *Encyclopedia of Smart Materials*.

Measures, R.M., Alavie, A.T., Maaskant, R., Ohn, M.M., Karr, S.E., Huang, S.Y., Glennie, D.J., Wade, C., Guha-Thakurta, A., Tadros, G., & Rizkalla, S. (1993). *Distributed and multiplexed fiber optic sensors III 21–29 multiplexed Bragg grating laser sensors for civil engineering* SPIE 2071.

Meyer, Y., Lachat, R., & Akhras, G. (2019). A review of manufacturing techniques of smart composite structures with embedded bulk piezoelectric transducers. *Smart Materials and Structures*, *28*(5)053001.

Mieloszyk, M. (2021). Fatigue crack propagation monitoring using fibre Bragg grating sensors. *Vibration*, *4* (3), 700–721.

Mieloszyk, M., Andrearczyk, A., Majewska, K., Jurek, M., & Ostachowicz, W. (2020). Polymeric structure with embedded fiber Bragg grating sensor manufactured using multi-jet printing method. *Measurement*, *166*108229.

Mieloszyk, M., Majewska, K., & Andrearczyk, A. (2022). Embedded optical fibre with fibre Bragg grating influence on additive manufactured polymeric structure durability. *Materials*, *15*(7), 2653.

Mieloszyk, M., Majewska, K., & Ostachowicz, W. (2021). Application of embedded fibre Bragg grating sensors for structural health monitoring of complex composite structures for marine applications. *Marine Structures*, *76*102903.

Mieloszyk, M., Majewska, K., Rimasauskiene, R., Rimasauskas, M., & Kuncius, T. (2023). An influence of temperature on additive manufactured composite with embedded fiber Bragg grating sensor unpublished content an influence of temperature on additive manufactured composite with embedded fiber Bragg grating sensor. In *Proceedings of the 14th International Workshop on Structural Health Monitoring SHM* (pp. 401–408).

Mieloszyk, M., & Ostachowicz, W. (2017a). An application of Structural Health Monitoring system based on FBG sensors to offshore wind turbine support structure model. *Marine Structures*, *51*, 65–86.

Mieloszyk, M., & Ostachowicz, W. (2017b). Moisture contamination detection in adhesive bond using embedded FBG sensors. *Mechanical Systems and Signal Processing*, *84*, 1–14.

Mieloszyk, M., Skarbek, L., Krawczuk, M., Ostachowicz, W., & Zak, A. (2011). Application of fibre Bragg grating sensors for structural health monitoring of an adaptive wing. *Smart Materials and Structures*, *20*(12)125014.

Mieloszyk, M., Jurek, M., Majewska, K., & Ostachowicz, W. (2020). Terahertz time domain spectroscopy and imaging application for analysis of sandwich panel with embedded fibre Bragg grating sensors and piezoelectric transducers. *Optics and Lasers in Engineering*, *134*106226.

Muna, I. I., Mieloszyk, M., Rimasauskiene, R., Maqsood, N., & Rimasauskas, M. (2022). Thermal effects on mechanical strength of additive manufactured CFRP composites at stable and cyclic temperature. *Polymers*, *14*(21), 4680.

Ning, X., Murayama, H., Kageyama, K., Wada, D., Kanai, M., Ohsawa, I., & Igawa, H. (2014). Dynamic strain distribution measurement and crack detection of an adhesive-bonded single-lap joint under cyclic loading using embedded FBG. *Smart Materials and Structures*, *23*(10)105011.

Peters, K. (2009). Fiber-optic sensor principles. *Encyclopedia of Structural Health Monitoring*.

Pioldi, F., Salvi, J., & Rizzi, E. (2017). Refined FDD modal dynamic identification from earthquake responses with soil-structure interaction. *International Journal of Mechanical Sciences*, *127*, 47–61.

Ren, L., Li, H.-N., Zhou, J., Li, D.-S., & Sun, L. (2006). Health monitoring system for offshore platform with fiber Bragg grating sensors. *Optical Engineering*, *45*(8), 084401.

Rimašauskas, M., Kuncius, T., & Rimašauskienė, R. (2019). Processing of carbon fiber for 3D printed continuous composite structures. *Materials and Manufacturing Processes*, *34*(13), 1528–1536.

Rodrigues, C., Félix, C., Lage, A., & Figueiras, J. (2010). Development of a long-term monitoring system based on FBG sensors applied to concrete bridges. *Engineering Structures*, *32*(8), 1993–2002.

Rousseau, R., Egghe, L., & Guns, R. (2018). *Becoming metric-wise: A bibliometric guide for researchers*. Chandos Publishing.

Schutte, C. L. (1994). Environmental durability of glass-fiber composites. *Materials Science and Engineering: R: Reports*, *13*(7), 265–323.

Shafighfard, T., & Mieloszyk, M. (2022). Experimental and numerical study of the additively manufactured carbon fibre reinforced polymers including fibre Bragg grating sensors. *Composite Structures*, *299*116027.

Shrivastava, A. K., & Nazir Hussain, M. (2008). Effect of low temperature on mechanical properties of bidirectional glass fiber composites. *Journal of Composite Materials*, *42*(22), 2407–2432.
Sih, G. C. (1977). Elastodynamic crack problems. . (4). Springer Science & Business Media.
Souza, G., & Tarpani, J. R. (2021). Using OBR for pressure monitoring and BVID detection in type IV composite overwrapped pressure vessels. *Journal of Composite Materials*, *55*(3), 423–436.
Takeda, N., Okabe, Y., Kuwahara, J., Kojima, S., & Ogisu, T. (2005). Development of smart composite structures with small-diameter fiber Bragg grating sensors for damage detection: Quantitative evaluation of delamination length in CFRP laminates using Lamb wave sensing. *Composites Science and Technology*, *65* (15–16), 2575–2587.
Thursby, G., Sorazu, B., Betz, D., Staszewski, M., & Culshaw, B. (2004). The use of fibre optic sensors for damage detection and location in structural materials. *Applied Mechanics and Materials*, *1*, 191–196.
Tuwair, H., Volz, J., ElGawady, M., Mohamed, M., Chandrashekhara, K., & Birman, V. (2016). Behavior of GFRP bridge deck panels infilled with polyurethane foam under various environmental exposure. *Structures Elsevier*, *5*, 141–151.
Türk, D.-A., Brenni, F., Zogg, M., & Meboldt, M. (2017). Mechanical characterization of 3D printed polymers for fiber reinforced polymers processing. *Materials & Design*, *118*, 256–265.
Udd, E., & Spillman, W. B., Jr (2011). *Fiber optic sensors: An introduction for engineers and scientists*. John Wiley & Sons.
Wu, F., Misra, M., & Mohanty, A. K. (2019). Studies on why the heat deflection temperature of polylactide bioplastic cannot be improved by overcrosslinking. *Polymer Crystallization*, *2*(6)e10088.
Yan, H., Xu, X., Fu, B., Fan, X., Kan, Y., & Yao, X. (2023). Constitutive model and damage of self-healing 3D braided composites with microcapsules. *Composites Communications*, *40*101586.
Zhang, Y.-F., Hong, C.-Y., Ahmed, R., & Ahmed, Z. (2017). A fiber Bragg grating based sensing platform fabricated by fused deposition modeling process for plantar pressure measurement. *Measurement*, *112*, 74–79.

CHAPTER

Sensor arrays

12

Coral Salvo Comino[1], Clara Pérez González[2] and María Luz Rodríguez Méndez[1]

[1]*Inorganic and Physical Chemistry Department, University of Valladolid, Valladolid, Spain* [2]*Materials Science Department, University of Valladolid, Valladolid, Spain*

12.1 Introduction

During the last few years, the industry has required constant renewal to meet consumer demands and current legislation. The development of new methods and devices to evaluate the quality and characteristics of the analyzed products, as well as the environmental impact, is of vital importance to improve industrial processes and control systems in several types of samples, such as biological, pharmaceutical, food, and beverages, among a wide variety of research areas (Schütze et al., 2018). The use of conventional detection materials and/or classical analysis techniques for the determination of specific compounds is becoming obsolete due to the complexity of the measurements, the cost of the equipment, the need for experts in the area, and the pretreatment of the samples. For this reason, most industrial sectors are involved in the fourth industrial revolution (Industry 4.0), based on the use of intelligent systems that allow control of all the production stages, including materials, processes, and products. Guarantying control from the raw material to the final product ensures the safety and quality of production, avoids fraud, reduces environmental impact, favors traceability, etc. (Schütze et al., 2018).

For that purpose, the use of sensor arrays coupled with chemometric tools, including chemical sensors, gas sensors, or optical sensors, has been proposed as an excellent alternative to achieve these premises due to their high reproducibility, linearity, repeatability, stability, lower limits of detection, etc. In this sense, electronic systems require combining analytical systems (usually sensors) with data processing software to obtain qualitative and quantitative information (Méndez et al., 2016; Schütze et al., 2018; Wasilewski et al., 2019). Furthermore, the use of these systems can help reduce costs and the merging of errors in analysis results. Generally, the errors are correlated with the sampling or with the sample matrix itself, but they are hardly correlated with the implementation of the devices. To ensure this characteristic, every analytical instrument must be exposed to a calibration process directly related to the expected response by traditional techniques through the implementation of a mathematical model that makes it possible to establish a validated correlation with the sample matrix analyzed (Calvini & Pigani, 2022).

Depending on the nature of the sensor developed, there is a large range of detection methodologies that can be employed to analyze complex matrixes. Some devices, such as commercial glucose detection sensors for insulin levels, can directly quantify the concentration of substances of interest

Sensory Polymers. DOI: https://doi.org/10.1016/B978-0-443-13394-7.00012-4

that are searched, while others need to be connected with data processing systems, like spectrophotometric sensors for full-range detection, where the measurement is indirect. This means that in the use of sensors that measure indirectly, there is need to have a data-reading system and an interpretation system using applied statistics. Those arrays use mathematical algorithms and software patterns for data interpretation (Munekata et al., 2023; Spinelle et al., 2017).

At the same time, the use of conducting polymers (CPs) has attracted a lot of attention due to their electrochemical properties. These polymers exhibit electrical conductivity that can be tailored through electronic doping, making them highly versatile materials. Additionally, they offer extraordinary environmental friendliness and biocompatibility, which make them ideal for a wide range of technological and scientific applications (Lange et al., 2008). These properties make conductive polymers exceptional candidates for chemical sensor modifications. This capability is crucial for the detection or determination of a wide variety of analytes. CPs not only enhance the sensitivity and selectivity of sensors but also enable the creation of devices that are both cost-effective and easy to install in situ. These advantages have significantly increased the importance of integrating conducting polymers into sensor systems in contemporary science. The potential to develop inexpensive and easy-to-deploy sensors has driven substantial research and development in this field (Parameswaranpillai & Ganguly, 2023)

In addition, CPs performance can be improved by their combination with sensitive materials. These advanced materials have synergist effects that increase electronic transfer on the sensor's surface, enhancing the quality of the responses obtained. Some examples of these materials include metals and alloys, nanomaterials, coordination compounds, porous materials, and carbonaceous materials, among others. Among them, the use of nanomaterials stands out due to their great electrocatalytic activity and high conductivity, their high surface/volume ratio, and the possibility of being modified or functionalized.

Furthermore, to improve the selectivity of the sensors, biological material can be used in the detection of analytes of interest; for instance, the immobilization of enzymes inside the conducting polymeric layers enhances the sensor's properties (Lakard, 2020; Parameswaranpillai & Ganguly, 2023). Concretely, biosensors have properties such as stability, precision, repeatability, low cost, the possibility of miniaturization, and portability, among others. These properties are in high demand. In turn, the use of nanomaterials or combinations between materials (that favor synergy in their electrocatalytic activity) and enzymes (capable of increasing their sensitivity and selectivity), applied with new preparation methodologies, will allow obtaining more efficient biosensor arrays, improving their specificity and sensitivity. In conclusion, it is expected that the scope of the advances in the use of CPs in the development of sensor arrays will be such that it allows the development of completely new methods to be implemented in several applications. It is intended that the developed devices allow the evaluation of parameters through portable, reusable, and non-contaminating automated systems, regardless of the need for pretreatment of the samples to carry out the measurements in real-time.

Due to the great variety of polymeric sensors reported in the bibliography during the last 20 years, it has been possible to develop a long number of technologies for producing multisensor systems based on arrays of sensors (Adhikari & Majumdar, 2004; Lange et al., 2008). For that purpose, the selection of materials with different reactivity and the ability to detect simultaneously dissimilar analytes is the principal challenge. Definitely, the more common polymers used in the development of sensitive sensors are based on CPs. These kinds of polymers can generate a measurable signal by changing their physical properties when they are exposed to certain chemicals. CPs are materials that offer a high variety and flexibility in their structures. This is the principal reason why CPs have received interesting uses in the development of e-noses, tongues, and eyes. Among the high variability of CPs, a selection of them for further development has been made in this chapter.

12.2 Conducting polymers

Synthetic polymers that can transfer electrical current are known as conductive polymers (CPs). These materials' inherent qualities or alterations are what give these polymers their conductivity. One of the primary drivers behind the intense research and development of these materials is their high application potential in chemical and biological sensors because they show conjugated bonds. That is, they present an alternating sequence of single and double bonds in their chemical structure. These conjugated bonds allow the delocalization of electrons along the polymer chain, providing special electrical and optical properties. Obtaining CPs can be achieved in different ways, through the conjugation of chains of unsaturated carbon atoms, carbon atom heteroatoms, or even heteroatom chains (Kovarski, 1998).

Electrical conductivity in polymers is generally low due to their nonconductive structure. However, some polymers can be modified to exhibit electrically conductive properties. These conductive polymers fall into two main categories: intrinsically conductive polymers and doped conductive polymers.

The term intrinsically conducting polymers (ICPs) refers to organic polymers that possess the inherent ability to conduct electricity. This conductivity is facilitated through extended conjugation of π electrons along the polymer chain. Notable examples of these polymers include polypyrrole (PPy), polythiophene (PTh), and polyaniline. The carbon atoms in the main chains of these polymers exhibit sp^2 hybridization (), which is essential for forming covalent bonds with the carbons of the branched chains. This specific hybridization leaves an unbonded p_z orbital *via* a π bond, resulting in an alternating distribution of double and single bonds along the polymer chain. This extended π-conjugation pathway is what imparts the conductive properties to the polymer.In contrast, extrinsically conductive polymers achieve their electronic conductivity through the incorporation of conductive materials into the polymer matrix. This inclusion can involve various conductive fillers, such as metals or carbon-based materials, which enhance the overall conductivity of the polymer composite (Hameed et al., 2023; Nasajpour-Esfahani et al., 2023). By combining these materials, extrinsically conductive polymers can be engineered to meet specific conductivity requirements for various applications, expanding their utility in fields such as electronics, sensing, and energy storage.

On the other hand, doped conductive polymers are nonconductive polymers that are modified by adding doping agents. Dopants are chemicals that donate or accept electrons, leading to the generation of charge carriers and increasing the electrical conductivity of the polymer. Due to their morphology, CPs reveal nearly negligible conductivity in the neutral (uncharged) condition. For this reason, insulating or semiconducting polymeric materials with main chains that alternate single and multiple bonds can be converted into conductive materials by doping (p or n). The conductivity of polymers follows a complex process that depends on preparation and doping. Nonconductive polymers behave like insulating materials; that is, they show an occupied valence band and an unoccupied conduction band, separated by a large gap. In these, the energy separation between bands is higher than in the case of semiconductors (Sierra-Padilla et al., 2021; Verma et al., 2023).

The main difference between CPs and insulating polymers is due to the existence of doping agents that alter the number of electrons in the different bands. P-type dopants take out electrons from the valence band; in consequence, the molecule turns positively charged, while n-type dopants increase the number of electrons in the conduction band, leaving the molecule negatively charged (Kaiser et al., 1995). That means the intrinsic conductivity arises from the generation of charge carriers upon oxidizing (p-doping) or reducing (n-doping) their conjugated chain. With the creation of a radical cation, the neutral polymer is oxidized in the initial stage. This phenomenon results in localized deformations along the polymeric chains, which generate localized electronic states and give rise to the so-called polaron (cationic radical). According to the deformation between the conducting band and the valence band, intermediate electronic states are created. Moreover, the elimination of a second electron favors the elimination of an electron from the polaron, resulting in the creation of a bipolaron (a pair of polarons with opposite spins, once the chain becomes saturated with polarons). The polymer can then undergo reoxidation, producing more charge carriers (Lange et al., 2008).

Finally, once high doping levels are reached in the polymer chains, the new bands begin to overlap, generating half-filled bands through which electrons can flow freely, giving the polymer the ability to conduct electricity. The doping of polymers can modify aspects such as color or volume. The processes described above are represented in Fig. 12.1.

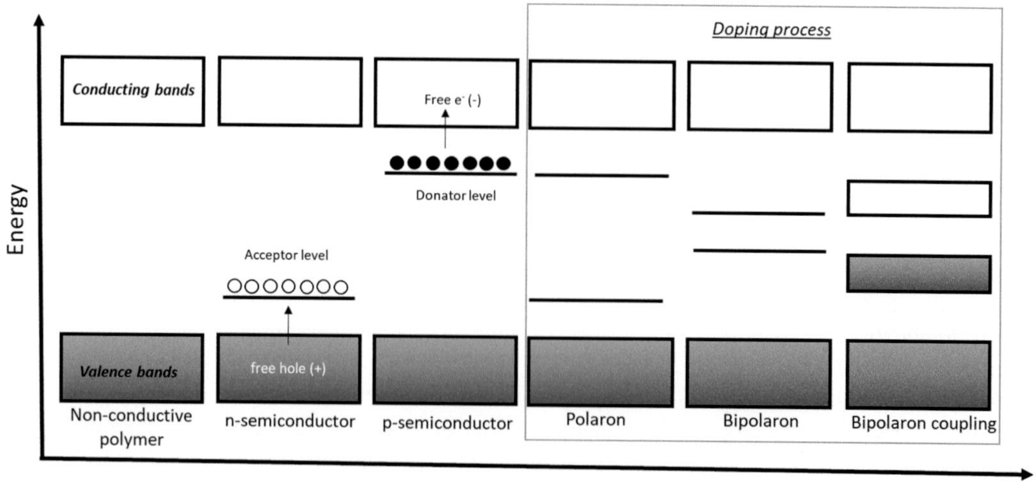

FIGURE 12.1

Schematic representation of electronic transference processes for conducting polymer formation.

The first high-conduction polymer was polyacetylene (PA). This polymer was discovered by Shirakawa, MacDiarmid, and Heeger in 1977. It is characterized by its linear structure, composed of repeating units of acetylene (−C≡C−). The triple bond between the carbon atoms in the main chain of the polymer gives it unique properties, such as high electrical and optical conductivity (Shirakawa et al., 1977). The discovery of PA provided a new perspective into the area of organic conductive materials and served as an inspiration for the creation of other conductive polymers like polyaniline (PAni), PTh, and PPy.

Since that time, conductive polymers have been the subject of intensive research and development to find novel chemical structures, synthesis methods, and uses for a wide range of applications. Throughout this chapter, due to the wide variety of conductive polymers such as PPy, polyindole, PAni, PTh, polyfuran, polyphenylene, polyfluorene, polypyrene, polycarbazole, and their derivatives, copolymers, analogs, and composites, this section will be limited to using the more stable CPs (mentioned above) that are commonly used in sensor development.

12.2.1 Polyethylene dioxythiophene:poly(styrene sulfonate)

Polyethylene dioxythiophene (PEDOT) is an intrinsic conducting polymer (ICP) derived from PThs. Its chemical and physical properties, such as its high electrical conductivity, excellent chemical and thermal stability, as well as its transparency, flexibility, and biocompatibility, make this polymer a perfect candidate for the development of sensors in a wide range of applications. PEDOT has a highly conjugated chemical structure that favors the transfer of electrons through the polymer matrix. The most common form of this polymer is in aqueous dispersion, which is compatible with some coating or printing techniques to develop sensors. The formation of a stable solution is based on the conjugation of PEDOT (positively charged) with a negatively charged saturated poly(styrene sulfonate) (PSS), which has excellent water solubility. This conjugated polymer (PEDOT:PSS), its derivatives, or their composites make it a stable aqueous dispersion that can be commercialized. However, other polymer structures, such as solid films or emerging PEDOT:PSS hydrogels, can be used in the development of sensors (Zhang et al., 2021; 2022). Stretchability, flexibility, viscosity, and a variable modulus of elasticity are some characteristics of PEDOT gel-based sensors that help them adhere to specific surfaces. Flexible electrodes, strain/pressure sensors, and a range of substrates are current examples of how gel application in wearable sensors has advanced (Zhang et al., 2022; Zhao et al., 2022). Despite having inferior mechanical qualities, PEDOT-based fibers have drawn a lot of attention in recent years because of their lightweight, innate flexibility, favorable skin compatibility, and ease of manufacture. The most common processes for developing PEDOT:PSS fibers are wet spinning and electrospinning (Zhao et al., 2022). The development of PEDOT-based films is another technique used in the creation of flexible sensor systems using flexible and portable conductive polymeric films. To do this, various factors like the manufacturing process choice, the chemical makeup of the additional elastomer, and the presence of copolymers have been assessed during the past few decades (Su et al., 2022).

One of the main advantages of this CP is its high conductivity (up to 4000 S), which makes it an excellent candidate for the development of sensors fabricated on a large scale and with cheap technology (Zhang et al., 2022). A precise and sensitive detection of electrical changes in reaction to changes in the environment or mechanical forces is made possible by PEDOT:PSS's high electrical conductivity. This enables the monitoring of changing conditions in real-time. Additionally, because of its natural adaptability and flexibility, it can be incorporated into a wide range of devices of all sizes and shapes, advancing the convenience and usefulness of sensor technology. These sensors are stretchable and can monitor most of the mechanical deformations involving extravagant movements, reducing the negative reactions of electronic devices (Fan et al., 2019; Zhang et al., 2021).

In conclusion, PEDOT:PSS-based sensor arrays represent an important development in sensor technology. Applications for environmental monitoring, health and safety, energy management, and industrial automation are driven by its benefits, which range from electrical conductivity to flexibility and sensing capacities. These sensor systems continue to play a crucial role in technological innovation with every advancement in the study of conductive materials (Fan et al., 2019).

12.2.2 Polyaniline

PAni is a conductive polymer that has gained significant attention in several fields. It is a versatile organic polymer with excellent electrical and optical properties, making it suitable for many applications. However, despite the fact that it is easily synthesized, cheaply produced, and shows thermal stability and environmental value, it presents drawbacks due to its low solubility in the solvents normally used in the industry, and it is unstable at processing temperatures (MacDiarmid et al., 1991).

The structure of this polymer varies with its oxidation state. For that reason, PAni can be easily synthesized through the chemical oxidation of aniline monomers. The resulting polymer exhibits a range of oxidation states, which can be controlled by adjusting the reaction conditions. In the oxidative polymerization reaction, an oxidizing agent such as ammonium persulfate, ferric trichloride, potassium dichromate, or hydrogen peroxide, among others, is used in an acid medium (the presence of anions from the acid used during polymerization is necessary because PAni behaves like an insulating material at a pH higher than 3). This oxidation-reduction behavior gives PAni its conductivity and makes it an excellent candidate for electronic and optoelectronic devices. The conducting properties of PAni arise from the delocalization of π-electrons along the polymer backbone. This allows for the flow of charge and the transport of electrons. The conductivity can further be enhanced by doping the polymer with dopant molecules (such as sulfuric acid, chlorhydric acid, or metal salts), which introduce additional charge carriers into the polymer matrix (Stejskal & Prokeš, 2020; Yang et al., 2023).

PAni has been investigated for several applications based on the development of sensors, energy storage devices, corrosion protection coatings, and actuators, among others. Its sensitivity to chemical and environmental changes makes it an attractive material for gas sensors, biosensors, and environmental monitoring systems. Nowadays, there has been a growing interest in the development of PAni-based sensor arrays. These devices consist of interconnected PAni sensors that can communicate and share information, enabling the monitoring of large areas or complex systems. In the design of these sensor arrays, it is necessary to evaluate aspects such as the selection of suitable substrate materials, the choice of the types of PAni sensors according to the variables to be measured, the optimization of manufacturing methods, and the implementation of efficient techniques of data communication. Moreover, it has been explored the benefits of developing nanocomposites based on PAni and other polymeric matrixes and electroactive materials such as nanoparticles, phthalocyanines, carbon nanotubes, or graphene (Al-Haidary et al., 2021; Dhand et al., 2011; Kashyap et al., 2019; Oliveira et al., 2022; Sen et al., 2016).

The use of aniline-conductive polymers in the development of sensor arrays is due to their ability to vary their electrical conductivity depending on the external stimuli to which they are subjected. That is, in the presence of certain analytes or chemical variations, the structure and conductivity of the polymer can be altered, which makes it possible to quantify the conductivity variations. That means that when the interaction between the polymer and the analyte occurs, a structural variation of the polymer is generated, which translates into electronic changes at the molecular level. Once the interaction has occurred, the electron transport capacity and its electrical conductivity are modified accordingly.

Due to its low cost, versatility, and processing method, which enable its construction in a variety of forms and sizes, adjusting to the needs of the sensors and their usage, sensor arrays based on PAni are of significant interest. In addition, they exhibit excellent detection performance for a variety of analytes, including metal ions and organic molecules.

Moreover, the ability to chemically modify or incorporate sensitive materials into the polymeric matrix, which results in an improvement in properties like sensitivity and selectivity, is one of the most notable benefits of using PAni as matrices for the development of sensors (Al-Haidary et al., 2021).

To improve its performance and expand its possible applications, research into PAni and its uses is always looking at new synthesis techniques, doping schemes, and device topologies. PAni continues to be an attractive material for future developments in a variety of disciplines thanks to its distinctive mix of electrical conductivity, processability, and environmental responsiveness (Aksnes, 2009).

12.2.3 Polypyrrole

PPy is a conjugated polymer that belongs to the group of intrinsically conductive polymers (ICPs). PPy is synthesized from the polymerization of pyrrole monomers. The synthesis process can be carried out in a solution, vapor phase, or solid state. The most common is the chemical initiation by oxidizing agents, such as persulfates or metal chlorides. With this method, the size range of PPy particles can be easily controlled. Despite being an easy synthesis method, the deposition of PPy in the electrodes by this method has shown poor adherence. To solve this problem, the electrochemical polymerization of PPy through different solvents, such as acetonitrile or water, can be applied to control the thickness and morphology of the immobilized layer (Stejskal & Prokeš, 2020).

However, despite having conductive properties due to the presence of carbon-nitrogen single and double bonds, which allow charge movement, its conductivity is lower compared to other ICPs, such as PAni or PEDOT. One of the solutions to increase its conductivity and its solubility in usual solvents consists of doping the matrix with ions such as polyelectrolytes, anionic surfactants, or even dyes (Bayat et al., 2021; Mahun et al., 2020). The concentration of the dopant during the polymerization process and the preparation technique have a significant impact on PPy solubility. It has been demonstrated that the solubility will rise with the concentration of the dopant because the dopant counterions surrounding the PPy chains diminish the inter- and intra-molecular interactions (John et al., 2018). In addition, one of its characteristics is the variation of the tonality of the polymer depending on the degree of doping. In its natural state, PPy is black, but depending on the degree and type of doping, it can fluctuate. This property makes it very useful in applications based on electronics or optical sensors (Quijada, 2020; Håkansson et al., 2006).

As previously specified, the introduction of sensitive materials or nanomaterials has generated a breakthrough in the development of polymeric sensor arrays. It is important to remember that PPy is susceptible to moisture and can break down in natural environments. Encapsulation techniques have been created, or additional materials have been mixed to increase its stability. The combination of PPy with nanoparticles or other nanomaterials (nanocomposites) provides great improvements in the development of sensor arrays in terms of sensitivity and selectivity (Jain et al., 2017).

For example, a notable improvement in the electrical conductivity of sensors has been demonstrated after the integration of nanoparticles in PPy as doping agents. In addition, PPy nanocomposites present an increase in the active surface area due to the dispersion of the nanoparticles, which favors the sensitivity of the sensors. In the same way, the integration of sensitive material or specific nanoparticles increases the specificity and detection capacity of the sensors because they respond to specific reactions, changes in temperature, pH, etc., which favors the detection of specific analytes (Jain et al., 2017).

12.3 Statistical analysis and modeling

The application of chemometrics is the last step in the development of sensor arrays, as it plays a crucial role in the analysis and interpretation of the complex data generated by these sensors. Sensor arrays consist of multiple sensors that respond to various chemical compounds or analytes present in a sample, producing a large number of data responses that have to be processed to determine the performance of the systems developed. Nowadays, scientists can conduct a wide range of mathematical and statistical analyses and find solutions to complex problems thanks to the development of statistical software. As a result, the use of multivariate statistical approaches as well as complex analytical techniques has significantly expanded.

The chemometric approach associated with the processing of signals produced by sensor arrays usually comprises three phases of analysis. First, descriptive studies are carried out to determine the ability of the system to find differences between the analyzed samples. Among them, the use of unsupervised techniques stands out thanks to their automatic approach, in which no external information is provided for the development of the mathematical model. Subsequently, depending on the purpose of the application of each sensor array, classification or regression methodologies can be applied. To do this, a wide range of statistical methods have been developed to create reliable models for the prediction of categories of unknown samples. The choice of the correct mathematical algorithm will, therefore, be the key step in the development of these mathematical models and will, therefore, require a high level of understanding of both the sensor signals and the samples to be analyzed.

The following sections describe some of the most widely applied chemometric methods in the development of sensor arrays, going from descriptive methods to the development of prediction models that can be applied in the analysis of samples at an industrial level.

12.3.1 Descriptive methods

One of the main objectives of the use of sensor arrays is the discrimination of samples based on their global composition. Principal component analysis (PCA) and hierarchical cluster analysis (HCA) are the most often used approaches for studying patterns and hidden relationships among samples in circumstances when the relationship between data and grouping is still not obvious. These methods usually focus on larger chemical datasets, bioactive ingredients, and functional characteristics.

12.3.1.1 Hierarchical cluster analysis

In statistics and data mining, HCA is a common technique for cluster analysis. It is a method for classifying objects or data points into clusters based on how similar they are to one another. The dendrogram, a type of tree-like structure created by the analysis, contains nested clusters that are used to create a hierarchical representation of the data.

There are two primary methods of HCA. Agglomerative HCA is the first type. Hierarchically, individual data points are first clustered separately, and the closest pairings of clusters are then repeatedly combined, progressing up the dendrogram until every single data point is part of a single cluster, and then the merging procedure is repeated.

The second HCA type is known as Divisive Hierarchical Clustering. Unlike agglomerative clustering, divisive HCA begins with all data points grouped on a single cluster, which will be divided repeatedly into several smaller clusters, moving downward in the dendrogram until each data point forms a separate cluster (Bridges, 1966; Köhn & Hubert, 2015).

This discrimination method has been used in several applications for liquid and gas samples, such as the discrimination of wine samples based on their chemical composition as a tool to corroborate the specificity of a PPy and PAni-based sensor array (Geană et al., 2020; Riul et al., 2004). HCA was also used for the analysis of oil samples by processing the data obtained on cyclic voltammetry, and the result showed that the clustering obtained by the dendrogram had more accuracy than other methodologies, such as factor analysis (Men et al., 2013).

The collected data of a portable e-nose based on a unique chemical surface acoustic wave array built out of seven polymers coating the sensing area of a gold electrode was analyzed using two-way hierarchical clustering. Gases and polymers with similar chemical properties were categorized into the same family based on the analysis procedure. The development of a portable e-nose using this analysis method may have real-world applications for gas recognition and detection, according to the results (Hao et al., 2010).

In summary, HCA is a valuable method for identifying patterns and grouping similar data points into clusters. HCA provides useful insights into the underlying structure and relationships within datasets, assisting in numerous sectors of study and decision-making processes by building a hierarchical representation of the data.

12.3.1.2 Principal component analysis

PCA is a popular statistical method for data visualization and dimensionality reduction. It is a technique that converts high-dimensional data into a lower-dimensional space while retaining the key structures and patterns from the original data.

This technique aims to find the principal components of the system (PCs), which are orthogonal (uncorrelated) linear combinations of the original data variables. These PCs are arranged in a clear hierarchy, with the first PC capturing the most important data variance, the second PC capturing the next most important variance, and so on. It is possible to represent the data in a reduced-dimensional space by choosing just a few of the top PCs, making it simpler to see and analyze large datasets. (Bro & Smilde, 2014).

PCA can also be used for data visualization by plotting the data points in the reduced-dimensional space based on their principal components, giving form to the Scores plot. This representation gives a better understanding of the underlying structure of the data and helps to identify clusters, outliers, or trends that may not be apparent in the original high-dimensional space (Greenacre et al., 2022).

PCA is one of the methods more widely applied in sensor arrays processing data, being considered one of the main tools of many systems. The advantages of PCA are simpler complex data, the discovery of patterns and correlations between variables, and a lower risk of overfitting in machine learning models. This methodology has been used as the discrimination methodology for the quality control of organoleptic properties of food products (Guadarrama et al., 2000), to discriminate volatile organic biomarkers (Castro et al., 2011), to compensate for the drift of gas sensors (Ziyatdinov et al., 2010), and even to evaluate the selectivity of conductive polymer sensors (Nguyen et al., 2000).

In summary, PCA is a powerful technique for dimensionality reduction, visualization, and data exploration that enables the extraction of the most important data from complex datasets. By preserving the essential patterns, PCA facilitates data analysis and interpretation, leading to more efficient and accurate decision-making processes, making it one of the best tools for sensor data processing.

12.3.2 Classification methods

Classification models play a crucial role in data processing and analysis, offering numerous benefits that aid in extracting meaningful insights and making informed decisions. Some of the main classification tools used on sensor arrays are explained in this chapter.

12.3.2.1 Linear discriminant analysis

An example of a supervised learning algorithm in machine learning is linear discriminant analysis (LDA). It is typically employed for classification and dimensionality reduction tasks. The objective of LDA is to find a linear combination of characteristics that maximizes class separation while minimizing data dimensionality. LDA works under the assumption that the data comes from different classes and that the features have a normal distribution.

LDA has several advantages, including an effective dimensionality reduction while retaining the most discriminative information between classes. An improved classification performance by emphasizing the separation between classes allows for better classification accuracy compared to using all original features. Moreover, the linear combination of features obtained from LDA allows for easy interpretation and understanding of the discriminating factors between classes (Robinson, 2012).

LDA has been commonly used in many applications, such as face recognition, pattern recognition, and flavor analysis, where the goal is to classify data into different categories based on its features (Debashis Ghosh, 2009). An example of the application of LDA on sensor arrays was the development of an e-tongue based on an array of six sensors, including PAni and PPy as the sensing materials for cava wine, where their responses were preprocessed and used as input data for an LDA model, allowing the classification of the samples depending on the aging process they underwent (Cetó et al., 2014).

12.3.2.2 Support vector machine

A strong and popular supervised machine learning technique used for classification tasks is called support vector machine (SVM). SVM performs particularly well when attempting to categorize data points into one of two groups in binary classification issues. SVM can, however, be enhanced to tackle multi-class classification issues using a variety of methods, including one-vs-all or one-vs-one.

The main goal of SVM is to find the optimal hyperplane in the feature space that divides the data points of various classes. A subspace known as a hyperplane has one less dimension than the original feature space. A hyperplane is a line in two-dimensional space, but it can also be a plane or a hyperplane in higher dimensions.

The gap between the nearest data points from each class and the hyperplane is referred to as the margin. Support vectors are the data points that define the margin and are located closest to the hyperplane.

SVM is relatively robust to outliers in the data due to its focus on the support vectors, which are the critical data points for defining the hyperplane. Furthermore, SVM can handle nonlinearly separable data and perform well in challenging classification problems due to the usage of various kernel functions (Suthaharan, 2016).

SVM has found applications in several fields, such as text classification, image recognition, medical diagnosis, and finance, where it is used for tasks involving binary or multi-class classification (Tiwari, 2013). An example of its use in sensor arrays based on polymers was its use combined with PCA (PCA-SVM) as a tool for classification to identify metallic ions (potassium, calcium, and magnesium) in liquids (Lu et al., 2021). In other works, potentiometric bioe-tongue formed by nine polymeric sensors used SVM as the classification tool to develop mathematical models for the classification of milk samples with high accuracy obtained by nonlinear approaches (Pérez-González et al., 2021).

12.3.3 Regression methods

Regression models are fundamental tools in data processing and analysis. Their ability to identify relationships, make predictions, and quantify effects makes them one of the most interesting tools in quimiometric analysis.

12.3.3.1 Partial least squares

Partial least squares (PLS) is a statistical method that combines elements of PCA and multiple regression to model the relationships between predictor variables and response variables. It is widely used in multivariate data analysis, especially in situations where there are many predictor variables and potential collinearity among them.

Creating a set of latent variables, often referred to as components, with the highest possible covariance between the predictor variables (X) and the response variables (Y) is the main objective of PLS analysis. Unlike traditional regression methods that use all predictor variables simultaneously, PLS constructs these components in a way that maximizes the explained variance in both X and Y spaces.

PLS can be used for both regression and classification tasks, depending on the nature of the response variables. PLS is used to predict continuous response variables in regression, but it can also predict categorical response variables in classification (Vlascici et al., 2008).

When dealing with high-dimensional data with potential multicollinearity among predictor variables, PLS is especially helpful. It can handle situations where traditional regression methods may suffer from instability or poor performance due to multicollinearity. Moreover, PLS reduces the dimensionality of the data while maintaining the highest amount of variance for predicting the response variables, a feature that is especially beneficial when dealing with a large number of predictors (Mehmood et al., 2012).

One of the most common problems in the use and application of sensors in the industry is the limited number of samples that can be used to train recognition systems; however, PLS can perform well even when the sample size is smaller than the number of predictor variables, making it one of the most used techniques in data processing for sensor arrays. An example of their ability to manage a large number of variables was its application by Fitzgerald et al., where signals obtained from 64 polymeric sensors were used to determine volatile profiles and were transformed into a response pattern by least squares regression (Fitzgerald et al., 2016).

12.3.3.2 Support vector machine regression

Finally, a SVM can also be used for regression proposals. Support vector machine regression (SVMR) is a supervised machine learning algorithm for solving regression problems that, in contrast to conventional linear regression, seeks to identify a nonlinear function that accurately captures the underlying patterns in the data.

Finding the hyperplane in a higher-dimensional space that best captures the relationship between the predictor factors and the continuous response variable is the goal of SVMR. The decision boundary that optimizes the distance between the data points and the hyperplane is known as the hyperplane, subject to a user-defined tolerance, or ε (epsilon). This ε-tube around the hyperplane contains the majority of the data points, and data points outside this tube incur a penalty proportional to their distance from the hyperplane (Brereton & Lloyd, 2010).

Similar to classification tasks, the data is preprocessed to ensure that all features have the same scale, typically by standardizing or normalizing them. This step is crucial to prevent any feature from dominating the optimization process due to its larger magnitude. Moreover, SVM regression can also handle nonlinear data using kernel functions, the same way SVM classification does. The original feature space is transformed into a higher-dimensional space *via* the kernel core function, where a linear hyperplane can be created that nonlinearly fits the data, making it useful for a variety of regression applications (Brereton & Lloyd, 2010).

This approach has been highly used in the development of sensor arrays when the predictor linear relationships with the sensor array data could not be found or had low correlation parameters. By choosing the appropriate core functions, authors have been able to reach high correlation values up to 0.99 in some cases (Pérez-González et al., 2023).

12.4 Sensor array systems

Sensor systems offer major advantages over traditional laboratory methods as they are portable, versatile, and affordable (Bunney et al., 2017; Zhu et al., 2015). In consequence, sensors and biosensors are ideal tools to complement or even substitute classical chemical analysis. However, there is a need to develop improved sensors able to detect new chemical compounds with lower limits of detection, better sensitivity, are cheap, portable, and fast enough to be used at-line or online, with no sample preparation required. In this frame, nanotechnology and nanoscience can provide new strategies to develop chemical sensors and biosensors with improved characteristics using new materials, nanomaterials (nanoparticles, nanocarbons, nanotubes, nanocomposites, and nanostructured films), and methodologies (He & Hwang, 2016; Maduraiveeran & Jin, 2017; Manikandan et al., 2018; Pérez-López & Merkoçi, 2011; Sinha et al., 2018; Zhu et al., 2015).

Nanomaterials or sensitive materials such as CPs have enormous potential in the development of biosensors due to their potential to amplify electrochemical signals, improving both sensitivity and selectivity. In addition, nanomaterials can act as effective immobilization matrices for enzymes that can induce unique performance characteristics in biosensors in terms of sensitivity and specificity (Gómez-Arribas et al., 2018; Pérez-López & Merkoçi, 2011; Wang & Duncan, 2017).

In the last few years, a new class of instruments entitled electronic tongues (e-tongues), electronic noses (e-noses), and electronic eyes (e-eyes) have been developed that operate in an analogous manner to the human sense and can be applied to the analysis of complex liquids, gases, and optical samples (Chovin et al., 2004; Méndez et al., 2016). As shown in Fig. 12.2, these sensor systems typically include a sampling mechanism to store samples during analysis, a sensor array for collecting signals from the interactions of sensing materials with different analytes, and a computer to store and interpret data (Munekata et al., 2023). The interaction between the electrodes and the compounds, which results in physical or chemical changes in the modified layer, is the basis for these devices. Different signals are produced by these modifications, and the kinds of transduction mechanisms used are connected with these signals. The more often used transducers use electrical measures (such as changes in current, voltage, resistance, or impedance), changes in mass or

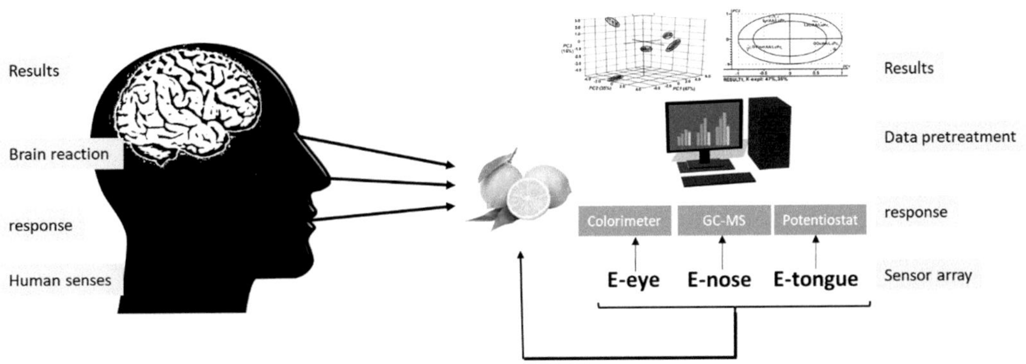

FIGURE 12.2

Schematic representation of electronic systems detection in comparison with the human sense.

temperature, or measurements of optical properties based on light absorption, polarization, fluorescence, etc. (Persaud, 2016).

Comparing sensor arrays to traditional analysis methods reveals a number of benefits, like constant real-time monitoring of numerous variables or parameters. They enable the use of numerous sensors that are dispersed across a large area, resulting in higher coverage and sampling capacity. Researchers can obtain data and carry out analysis using sensor arrays because of their low cost and low power consumption. They can also be remotely controlled and monitored. Finally, it should be highlighted that these devices allow integration with digital technologies and have the capacity to transmit data across communication networks. For that reason, they can be specially developed to satisfy certain standards, depending on the application. All of this makes it easier to gather, process, and analyze huge amounts of data, which results in quicker and more accurate decision-making (Di Rosa et al., 2017; Windmiller & Wang, 2013).

High sensitivity, rapid response, good repeatability, and reversibility to a wide range of chemicals are essential characteristics of the sensor array systems. Moreover, the ability to be exposed to various environments, the capability to develop compact devices, and the capacity to operate at specific temperatures are other necessities to improve the device configuration (Persaud, 2016).

Multivariate statistical approaches are typically applied to electronic sense system data to acquire meaningful information from the sample analysis; also, the combination of data from various complementary techniques can provide a more precise understanding of complex samples. According to the processing stage at which fusion occurs, the combining of data derived from various analytical approaches is referred to as "data fusion" and can be categorized into different levels (Buratti et al., 2018). This makes it possible to provide global sample information instead of information on individual components. Pattern recognition algorithms that combine these systems are very effective analytical tools due to their low cost, speed, and accuracy. In addition to being suitable for in- and out-line measurements, sensor arrays are also very useful for tracking final product quality and monitoring sample processing (Tan & Xu, 2020).

12.4.1 Electronic noses (gas analysis)

Persaud and Dodd conceptualized the electronic nose (e-nose) term in 1982. This intelligent sensing system was described as a device that consists of an electrochemical sensor array with partial specificity and a suitable pattern recognition system capable of identifying simple or complex odors (Persaud & Dodd, 1982). From the beginning of the design of such systems to the present, the development method of e-noses has been greatly optimized. The main advantage of these electronic systems in comparison with human noses is their capability for testing toxic gases that are not possible for humans to detect due to their dangerousness or poisonousness. Furthermore, the human nose, despite being trained to smell and qualitatively analyze certain gases, shows limitations in discriminating, classifying, and quantifying specific parameters (Tan & Xu, 2020).

The principle of e-noses is similar to the human sense; that is, to detect specific molecules (with a certain size and shape), a receptor is necessary. This receptor recognizes the specific molecule and sends the signal to the brain, which can identify it based on previous training. The analysis of specific volatile molecules in a complex matrix can be provided by using a gas sensor array that mimics the human olfaction sense. The e-nose idea essentially relies on the combination of the electronic fingerprint obtained for the specific gases, measured with the sensor array, and a pattern recognition engine to carry out the sample discrimination and/or classification.

The scheme that e-noses follow as a detection system is composed of sample delivery, detection, and computing systems (see Fig. 12.3). At first, the sample delivery system makes it possible to create a sample or volatile chemical headspace, which is the portion that will be studied. The e-nose's sensing device is then sent to this headspace by the system. The detection system, which consists of a collection of sensors, will come into contact with the volatile compounds once they have been removed, making this element of the instrument reactive. The sensors respond to contact by changing their electrical properties. Finally, a particular response is recorded and sent to the digital value whenever the sensors detect any compound (Persaud, 2016).

Sensors involved in the development of e-noses can be composed of a high variety of materials, such as metal oxide semiconductor (MOS) field-effect MOS transistors, mass-sensitive sensors, CPs, solid electrolyte sensors, and optic fiber sensors, among others (Aouadi et al., 2020). The most common gas sensor systems are metal oxide gas sensors, instead of their high temperature working limitations (from 300°C to 550°C). This is a drawback because they require high energy consumption since

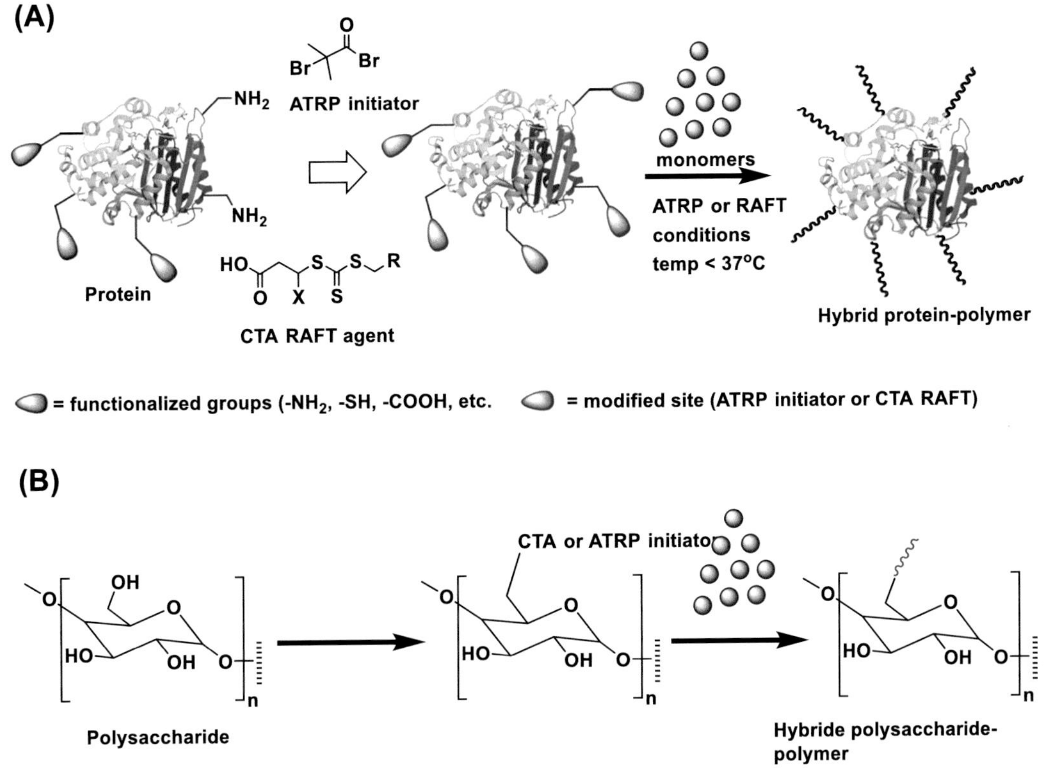

FIGURE 12.3

Schematic representation of electronic nose system.

the reaction rate at the oxide surface is too slow at lower temperatures. To solve this problem, the use of CPs in the development of gas sensors allows us to end this problem. The most often utilized monomers for sensor coatings are PAni, PPy, and PThs such as PEDOT, due to their excellent conducting properties. The main disadvantages of CPs, on the other hand, are their extreme sensitivity to environmental humidity and how that affects the effectiveness of their gas-detecting systems.

12.4.2 Electronic tongues (fluid analysis)

According to IUPAC, an e-tongue is an analytical device that includes an array of nonselective chemical sensors with partial specificity for dissimilar analytes, which work together with pattern recognition software and are capable of recognizing, discriminating, and classifying simple and complex matrixes (Vlasov et al., 2005).

The human sense of taste has been a fundamental tool in evaluating food quality for millennia. However, with the advancements in technology, e-tongues have emerged as potent alternatives to human sensory evaluation. As mentioned above, these systems consist of a sensor array that reacts to specific chemical components in a sample, enabling rapid and accurate taste analysis. Among the various types of sensors, polymeric sensors have shown remarkable potential due to their versatility, cost-effectiveness, and ease of fabrication (Tahara & Toko, 2013).

E-tongues operate based on the principle of chemical sensing, where the interaction between a sample and the sensing elements generates a response that can be quantified and analyzed. The array of sensors in the e-tongue provides a pattern of responses, which is then processed using pattern recognition algorithms to identify and differentiate taste attributes (del Valle et al., 2013) (see Fig. 12.4). Polymeric sensors utilizing polymeric materials as the sensing element, such as PAni, PEDOT, and PPy, can be engineered to respond selectively to specific taste compounds, allowing

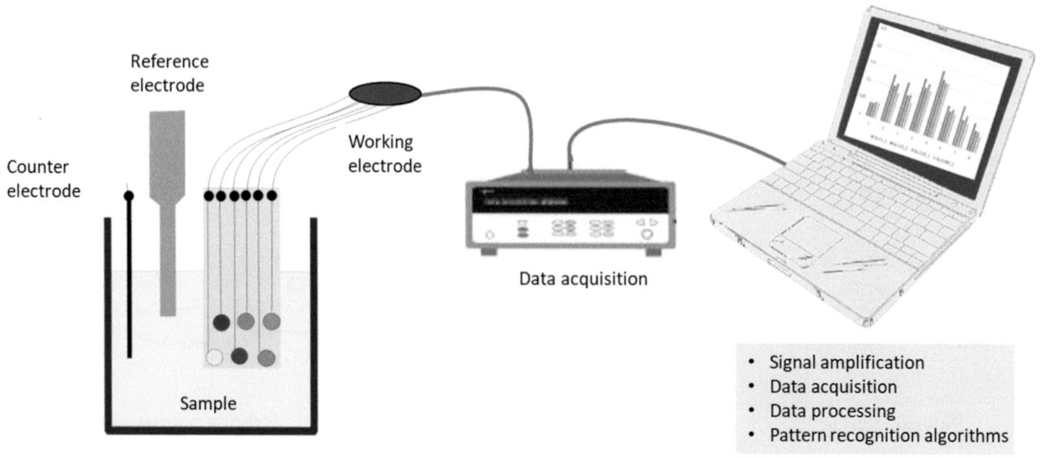

FIGURE 12.4

Schematic representation of electronic tongue system.

for the detection of various taste attributes, such as sweet, salty, sour, bitter, and umami, based on their chemical composition.

The electrochemical parameters of the system (substrate, electrolyte, and analyte) play a key role in determining the electrochemical detection technique that will be used to study the reactions occurring on the electrode surface. The working electrode serves as a transducer for the reaction; the counter electrode keeps the solution in contact with the electrode's surface and permits current to be applied to the working electrode; and the reference electrode has a stable equilibrium potential and is used to measure the potential between the other two electrodes. On the other hand, there are electrochemical detection systems that employ two electrodes, such as potentiometric sensors or impedimetric sensors that employ a counter electrode (Chambers IV, 2019; Vasilescu et al., 2019).

To guarantee a good electrochemical performance, the selection of the working electrode is essential; therefore, as has been reported, the modification of the surface with materials and nanomaterials, as well as with biological material, plays a really important role. Polymeric sensors utilizing polymeric materials as the sensing element, such as PAni, PEDOT, and PPy, can be engineered to respond selectively to specific taste compounds, allowing for the detection of various taste attributes, such as sweet, salty, sour, bitter, and umami, based on their chemical composition (Vahdatiyekta et al., 2022).

A large variety of samples can be determined using CP sensor arrays, including beverages, pharmaceutical and clinical applications, residual water, and pollutants, among many other analytes (Sierra-Padilla et al., 2021). The fabrication of polymeric sensors for e-tongues involves several techniques, including thin-film deposition, molecular imprinting, and polymer blending. Each method offers distinct advantages and challenges in terms of sensitivity, selectivity, and reproducibility. At the end of this chapter, a brief review of the latest advances in the development of e-tongues based on electrochemical sensor arrays using CPs is provided.

12.4.3 Electronic eyes (optical analysis)

The mental reaction to the visible spectrum of light reflected or emitted by an object is called color. This reaction is brought about by the interaction of light with retinal receptors, which produces stimulation that is transmitted to the brain *via* the optic nerve. Ambient lighting, as well as the item, affects how colors are perceived. Because of this, color analysis is crucial for categorizing products, including beverages and food, among many others (Munekata et al., 2023). The e-eye is an analytical tool made to simulate human visual sensitivity and collect color and aspect-related data from samples, enabling an objective assessment of an object's color characteristics. E-eye sensors are typically based on computer vision systems, colorimetric techniques, or spectrophotometry (see Fig. 12.5). The e-eye is a detection technology that gathers color information by identifying and analyzing samples. A color space, a mathematical representation that can link color coordinates to each observed color, is used to describe color. Therefore, using the appropriate tools can guarantee the most accurate color description. Hardware-oriented color spaces, human-oriented color spaces, and instrument-oriented color spaces are the three different categories of color spaces (Buratti et al., 2018; Calvini & Pigani, 2022). The sensors based on this system rely on fundamental perceptual characteristics to understand colors. Important components of this chromatic perception include hue, which enables one to discriminate between colors like red, green, and blue; saturation, which shows the strength and purity of a color; and brightness, which determines how light or dark a color is.

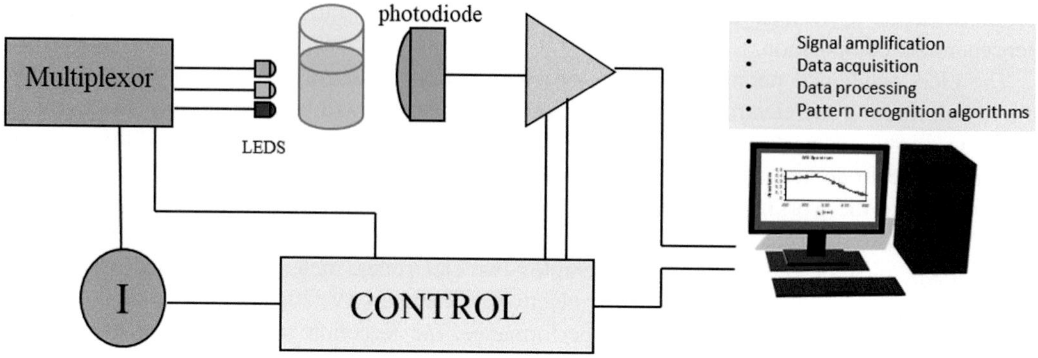

FIGURE 12.5

Schematic representation of electronic eye system based on spectrophotometry technique.

These sensors' functioning procedure starts with the capture of light using a specific lens. This light is converted into electrical signals, which are then digitalized to allow a computer or other processor to process them. Various levels of analysis are applied to the generated digital images. These sensors are based on scientific models for chromatic analysis, such as the CIELAB (CIELab*) color space, which aims to represent colors in a perceptually consistent way (Calvini & Pigani, 2022). Electronic vision sensors use additional crucial techniques to analyze colors and visual data. For instance, spectrophotometry uses the wavelength of light to determine how much of it an object absorbs or reflects. This method is essential for figuring out how much light a certain substance absorbs and its color makeup. On the other hand, the discipline of computer vision focuses on creating the systems and algorithms that enable computers to comprehend and interpret images. These principles are used by electronic vision sensors to recognize forms, objects, and, of course, colors in collected images (Munekata et al., 2023; Orlandi et al., 2019).

In several applications, including machine-vision and robotics, e-eyes based on CPs are used to detect and process light similarly to how human eyes do. CPs can be used to create machine-vision sensors because of their distinctive optical and electrical features. As light detectors, they can gather light energy and transform it into an electrical signal that can be analyzed and processed to produce visual data. When compared to other conventional materials used in optics and electronics, their key advantages are their lightweight, flexibility, and low production costs (Orlandi et al., 2019).

Despite the fact that CPs based e-eyes are a developing technology, there are still issues with resolution, sensitivity, and response time. However, research and development keep enhancing and expanding their capabilities.

12.5 Conducting polymer array system applications

Sensor arrays equipped with polymeric sensors find applications across various industries, such as the premature diagnosis of diseases (Behera et al., 2019; Liu et al., 2019) and the assessment of food and beverage industries for quality control and flavor profiling of food products, detection of adulterants and contaminants in food and beverages, or the evaluation of sensory attributes in fruits, wine, coffee, and tea (Chen et al., 2022; Escuder-Gilabert & Peris, 2010; Nategh et al., 2021; Tan & Xu, 2020), but also in other products such as meat or fish (Munekata et al., 2023). In the pharmaceutical industry, this type of technology has been used for the analysis and detection of traces of medicines to improve patient compliance and the control of pharmaceutical formulations. Finally, these systems have also been applied for the detection of pollutants in water and have been extremely useful for the detection of toxins and the remains of agricultural waste (Kirsanov et al., 2021).

12.5.1 Polyaniline-based sensor arrays

PAni-based sensor arrays find applications in diverse fields, including food and beverage analysis, environmental monitoring, and quality control. In the food industry, they are employed for organoleptic analysis and discrimination of flavors, ensuring product consistency and quality. However, in environmental monitoring, PAni sensors can detect and identify pollutants and gases, aiding in pollution control and safety assessments (Vahdatiyekta et al., 2022).

One of the advantages of PAni-based sensors is their relative cost-effectiveness compared to some traditional sensors. PAni is relatively inexpensive and can be easily synthesized in large quantities. This makes PAni-based sensor arrays promising for commercial applications where cost and scalability are essential factors (Wang et al., 2020).

Sensors modified with PAni have been used to analyze different complex samples such as oils (Stella et al., 2000), wines (Cetó et al., 2014; Parra et al., 2006), fruit juices (Mondal et al., 2019), teas (Kulikova et al., 2019; Lvova et al., 2003), and samples of contaminated water (Facure et al., 2017), among others.

Rañola et al. developed an array of sensors based on chemiresistors using CPs based on a potentiostatic electrodeposition of PPy, PAni, and poly(3-methylthiophene) assembled for its application in distinguishing between coconut oil products. These sensors exhibited reversible and repeatable responses based on their electrical resistance, with a relatively short response time demonstrating different selectivity when it came in contact with the headspace of coconut oil samples.

Impedimetric e-tongues based on CPs have also been applied to determine differences between standard solutions. Riul et al. developed an e-tongue by coating interdigitated electrodes with various materials such as PAni, stearic acid, PPy, and their different combinations. The e-tongue developed demonstrated the capability to detect basic flavors such as sweet, salt, acids, and bitter solutions with LODs of 5 mM. Moreover, the e-tongue exhibited the ability to differentiate between nonelectrolyte compounds like sucrose and discriminate between different salts, such as NaCl and KCl solutions, without the need for any pre-conditioning procedures (Riul et al., 2003).

Furthermore, more complex e-tongues constructed using ultra-thin films have demonstrated potential for food quality control. For example, an e-tongue incorporating PPy, PAni, and ruthenium pyridine in thin nanostructured films by Langmuir–Blodgett on the surface of a gold-interdigitated electrode was able to detect differences between HCl, NaCl, quinine, and sucrose at very low concentrations in comparison with the human taste threshold (Ferreira et al., 2003).

These techniques have allowed the development of very complex sensor arrays capable of determining the presence of highly specific compounds such as daidzein (Wang et al., 2016). Wan et al. developed a voltammetric sensor array where PAni and multi-wall carbon nanotubes (MWCNTs) were deposited vertically on the electrode surface, controlling their disposition using Langmuir–Blodgett technology, achieving lower detection limits (8×10^{-8} mol/L), and a wider linear range (1.0×10^{-7}–9.0×10^{-6} mol/L) compared to similar works.

Similarly, an e-nose was created by depositing thin conductive polymer films onto interdigitated electrodes, which were then connected to a personal computer through a data acquisition board. This allowed for the early identification of microbial activity in oranges for quality control. After just 24 hours of incubation, or at an early stage of biodeterioration, significant responses were obtained, making it possible to effectively separate fruits of good and poor quality and implement corrective measures in storage facilities (Sen et al., 2016).

Other approaches in the development of sensor arrays based on the use of PAni have been its combination with electroactive enhancers such as metallic nanoparticles. For example, gold and silver nanoparticles have been proven to enhance the detection capability of sensors built with gold-interdigitated microelectrodes using PPy and PAni-assembled nanocomposites to detect geosmin and 2-methylisoborneol traces in water samples with concentrations from 25 to 500 ng/L (Migliorini et al., 2020). Other applications include the analysis of ketone levels in the medical field as a biological marker of fat burning. Using PAni and a combination of silver nanoparticles and PAni, breath-acetone sensor arrays for tracking ketosis have been created. At room temperature, the sensing behaviors of gas sensors were examined in relation to 100 parts per million of acetone. When compared to pure PAni, the AgNPs-PAni film demonstrated superior overall acetone sensing behavior in terms of sensing response, highlighting the enhanced properties of sensor arrays, including metallic nanoparticles ().

Furthermore, these sensors can also be combined with biological materials to enhance the selectivity of the system. Electropolymerization has facilitated the development of integrated polymer films comprising PAni and DNA. These sensors possess intrinsic redox activity, enabling their characterization for the determination of single antioxidants and the classification of samples of tea from different brands. Notably, the incorporation of DNA in this work serves to stabilize the oxidized PAni while also reducing the number of individual sensors required for antioxidant classification (Kulikova et al., 2019; Wilson et al., 2012).

12.5.2 Polypyrrole-based sensor arrays

As has been reported previously in this chapter, PPy is widely used in the development of sensor arrays due to its excellent electrochemical properties, making it suitable for use in industry applications (Ameer & Adeloju, 2005).

PPy-based e-tongues have been applied in multiple fields. Apetrei used this technology to analyze the organoleptic properties of extra-virgin olive oils and to assess the quantification of polyphenols (Apetrei & Apetrei, 2013; Constantin Apetrei, 2012). In this work, the e-tongue was exposed to oil emulsions to perform cyclic voltammetry to evaluate the redox properties of the electroactive compounds present in the matrix. By analyzing the oxidation and reduction peaks, it was possible to discriminate between the samples analyzed, demonstrating that the methodology employed was able to overcome the challenges of measuring oily samples, such as their low conductivity and solubility in typical solvents as well as their viscosity. Based on this work, samples with a high fatty acid content, such as milk, have also been analyzed and classified by their fat content (Arrieta et al., 2023). Similarly, this system was applied in the detection of adulterations in coffee samples where the composition was modified with concentrations of 20%, 10%, 5%, and 2% of roasted corn and soybean, achieving correlation values of 0.97 by PLS (Arrieta et al., 2019).

Additionally, PPy e-noses have been used to analyze extra-virgin olive oils and coffee samples. Extra-virgin oils were analyzed with a PPy-based voltammetric electrode as sensing components. Sample emulsions were analyzed by the e-nose, and each sensor exhibited a distinct electrochemical signal, offering a high level of cross-selectivity. The acquired signals were subjected to PCA and PLS-DA to enable the six extra-virgin olive oils tested to be discriminated against based on their degree of bitterness (Constantin Apetrei, 2012).

Other studies describe the creation of an e-nose based on nanocomposites made by combining PPy with ZnO, In_2O_3, and ZnO/In_2O_3 nanoparticles. The sensor array consisted of six sensing units that were subjected to seventeen samples of coffee. The coffee samples were successfully separated according to their quality level using dendrograms' Euclidian distances and PCA. As a result, the sensor array created using PPy nanocomposites shows potential as a substitute instrument for classifying and assessing the quality of coffee samples instead of traditional analysis (Andre et al., 2021).

Other studies successfully distinguished how the use of different oak barrels (based on their origins and toasting levels) affected the development of organoleptic properties in wine samples (Parra et al., 2006). PPy-modified electrodes did not show the oxidation peaks associated with antioxidants due to the overlapping of the first oxidation pick between the polyphenols of the antioxidant and the doped PPy, while the second oxidation that should be studied could not be achieved because of the instability of the polymer film. However, wine composition still possesses electroactive molecules and ions, allowing the detection of pH changes and changes in antioxidant levels using the voltammetric e-tongue as a global information-providing tool. In a separate study (Geană et al., 2020), cyclic voltammetry was used as a potential tool for a fast way of fingerprinting the chemical composition of white wines using miniaturized disposable screen-printed carbon electrodes (SPCEs), in which CNTs were immobilized and 1-decanesulfonic acid sodium salt-doped PPy. By quimiometric analysis, high discrimination accuracies were achieved for the different sensors developed, with a 92.31% for MWCNT/SPCE, an 86.72% for SWCNT/SPCE, and finally a 78.17% for PPy/DSA/SPCE for all samples analyzed.

Bio-electronic tongues have also been used for this purpose. Garcia et al. developed a voltammetric bioe-tongue based on glucose oxidase and tyrosinase using PPy doped with gold nanoparticles as the electron mediator (Garcia-Hernandez et al., 2019). The combination of the polymer with metallic nanoparticles improved the performance of the sensors by increasing the electroactive surface area. The PCA results showed that wines were clustered based on their total polyphenol content and alcoholic degree, while musts were mainly grouped by their sugar content. These results enlighten the influence of the enzymes on the experiment results, allowing a much higher sensitivity and selectivity of the system.

Bioe-noses have also been applied in the food industry as a method of safety and quality control. Oh et al. developed a highly stable bioe-nose capable of detecting liquid and gaseous traces of cadaverine, one of the main molecules produced during spoilage processes in the meat and fish industries. In this work, a highly selective trace amine receptor (TAAR13c) was used to construct a functionalized bioe-nose based on a nickel (Ni)-decorated carboxylated PPy nanoparticle (cPPyNP). The oriented assembly of TAAR13c-embedded electrode transistors with Ni/cPPyNPs was used to create the biosensors. The bioe-nose demonstrated exceptional stability in both the aqueous and gas phases, together with great performance in terms of selectivity and sensitivity for the detection of cadaverine. Furthermore, the system's development made it possible to measure tainted, genuine food samples with efficiency (Oh et al., 2019).

While PPy-based sensor arrays show significant potential, there are challenges that need to be addressed. The stability of PPy under different environmental conditions, especially humidity and temperature fluctuations, can impact the long-term performance and reliability of the sensors. Moreover, the cross-sensitivity of PPy to different analytes may pose challenges in discriminating complex samples with overlapping signals.

12.5.3 Polyethylene dioxythiophene-based sensor arrays

PEDOT-based sensor arrays exhibit fast response and recovery times, making them suitable for real-time monitoring applications. This rapid response allows for quick detection and analysis of samples, making PEDOT e-tongues, noses, and eyes a valuable tool for process control and quality assurance.

Several studies have examined the use of e-tongue electrodes modified by depositing PEDOT on their surfaces in the wine industry. Pigani et al. developed a voltammetric e-tongue composed of three electrodes modified with PEDOT and a composite material of gold and platinum nanoparticles embedded in a PEDOT film. The effectiveness of the array was studied by analyzing different samples of white wines by differential pulse voltammetry and developing a classification model with good accuracy for all models developed (Pigani et al., 2008). Further studies demonstrated the ability of the system to analyze other types of wines. In this case, nine different types of red wines from an Italian winery from the same crop were analyzed using a PEDOT-modified electrode. The e-tongue applied in this work was able to identify wine samples containing off-average values for specific compounds such as oxide sulfide, color intensity, and polyphenol derived from differences produced during the aging or wood fining processes (Pigani et al., 2009, 2011).

Parallel to these studies, different works have been carried out in which these technologies are used as a control method for the ripening process of the grapes. To track the evolution of grape ripening, an e-tongue based on two voltammetric sensors, one poly-ethylendioxythiophene-modified Pt electrode and an unmodified carbon electrode, was designed. To evaluate the effectiveness of the device, samples of three varieties of grapes were collected to produce the musts that were subjected to an electrochemical. Following a preliminary study using PCA, calibration models using PLS were used to estimate physicochemical parameters that determine the ripening state, such as sugar and polyphenol content (Pigani et al., 2018). Further work has been able to merge the information provided by these e-tongues with the data obtained by e-eyes, allowing more accurate studies (Orlandi et al., 2019).

Moreover, PEDOT-based sensor arrays have also been applied to the assessment of juice composition. An e-tongue using amperometric electrochemistry, consisting of a Pt electrode modified with PEDOT, was employed in the blind analysis of fruit juices (Martina et al., 2007). The obtained data was used in the analysis to determine the discriminating capabilities of the e-tongue in terms of its sensitivities and specificities by using PLS-DA quimiometric analysis, which was able to tell apart juices derived from the same fruit commercialized by different brands. Similarly, Hong et al.

were able to determine the presence of adulterations in tomato juice depending on the level of adulteration employed using an e-nose (Hong et al., 2014).

In addition to the food field, sensor arrays with PEDOT-modified sensors have been widely used in the environmental area. For example, there have been studies where e-tongues have been employed as a new approach to identify trace quantities of organophosphate pesticides (OPs) in water samples with exceptional stability, reproducibility, and sensitivity, enabling the discrimination of OPs even at nanomolar concentrations (Facure et al., 2017). The e-tongue was constructed using sensing units created by reduced graphene oxide when conductive polymers are present (PEDOT-PSS and PPy) and gold nanoparticles.

Qualitative and quantitative detection of metallic ions in aqueous environments has also been an area of interest in sensor array development (Magro et al., 2019). An impedimetric sensor array of functionalized PEDOT:PSS developed by Lu et al. was able to determine K(I), Ca(II), and Mg(II) at concentration ranges of 1–10 mM (Lu et al., 2021). Other works reported the development of piezoresistive sensing arrays modified with PEDOT:PSS layers for Pb monitoring in water. These studies reported that the systems developed had sensitivities to Pb concentrations in a linear range of 0.01–1000 ppm with very low limits of detection of 5 ppb and high specificity (Yen & Lai, 2020).

Environmental pollutants based on volatile organic compounds have also been studied by e-noses, as they are supposed to pose a great risk to human health even in very low concentrations. Conti et al. developed a TiO_2-modified sensor array based on conductive polymers, including PEDOT:PSS, deposited on the surface of gold-interdigitated electrodes. Electrical impedance spectroscopy was used to collect the residence data of their behavior, coupled with a PCA, which allowed them to discriminate between increasing concentrations of the pollutants (Conti et al., 2021).

Moreover, PEDOT is a biocompatible material, which makes it suitable for use in applications involving biological samples or in vivo measurements. Its biocompatibility allows for the analysis of biological fluids, such as saliva or urine, in medical and healthcare applications. For example, humidity sensor arrays have been developed utilizing conductive ink PEDOT:PSS for their application in wireless urine detection in incontinence materials (Boehler et al., 2019).

Silva et al. developed a system based on PEO-PEDOT:PSS humidity sensors for breath monitoring, a widely used analysis method for the early diagnosis or treatment of several diseases. Several sensors with different ratios of the conductive polymer PEDOT:PSS with a polymer matrix based on polyethylene oxide were designed in this work. The test results show strong reversibility, repeatability, and reproducibility with standard deviations of 2.6% and a wide relative humidity detection range (6%–92%). It also demonstrated how the system developed was able to determine differences in breath humidity on real samples in up to 2 seconds of intervals (Assunção da Silva et al., 2021).

Similar works have shown the application of PEDOT-modified sensors in no breath analysis for asma monitoring, where the duration of breath was overcome by designing and implementing two measuring procedures that reached LODs of about 6 ppb on the working level of the actual hand-held diagnostic device used in the market nowadays (Pantalei et al., 2013).

Despite their promising potential, sensor arrays with polymeric sensors face several challenges and limitations. These include sensor drift, limited stability, cross-sensitivity, and difficulty in identifying complex taste interactions. Advancements in material science, nanotechnology, and data analysis techniques are expected to address the current challenges and propel e-tongues with polymeric sensors into new realms of applications and accuracy.

Sensor arrays with polymeric sensors have demonstrated great promise as efficient and reliable tools for taste analysis. By mimicking the human sense of taste, smell, and vision, these devices have found applications in diverse industries, benefiting both consumers and manufacturers. However, ongoing research and technological developments are necessary to overcome the current limitations and unlock the full potential of these devices in the future. As technology continues to evolve, sensor arrays are expected to play an increasingly vital role in ensuring product quality, safety, and consumer satisfaction across various domains.

12.6 Conclusion

Researchers have become interested in electronic devices that simulate the senses of smell, vision, and taste, and their applications have grown to benefit a wide range of industries. The primary advantage of electronic noses, electronic eyes, and/or electronic tongues is their versatility in combining multivariate methodologies to evaluate, characterize, and discriminate a variety of products.

The fourth industrial revolution, which is defined by an increase in automation and online monitoring *via* said electronic systems, is the one that is now influencing how the industrial sector is evolving. Progress is still required, though. For this, conductive polymers have been incorporated into the design of such devices in numerous studies. The electrical conductivity of CPs initially attracted interest in them; however, it has been demonstrated that additional properties, such as optical or chemical ones, broaden their multifunctionality and increase the range of CPs' uses in the development of sensor arrays. Numerous potential applications exist for the use of modified or unmodified CPs as receptor material or as one of the elements of the receptor layer in sensors. Furthermore, multiple receptor devices might be integrated to improve the poor selectivity.

Most industrial tests for analysis, detection, and quantification are carried out using traditional chemical analysis. However, these analysis methods can be improved by combining different technological disciplines that allow the development of new, faster, more precise, and more economical techniques. In this way, the development of new analysis methods based on intelligent systems will allow the application of new strategies in a wide variety of sectors, improving their competitiveness and sustainability. According to a trend report from the Deloitte company, nanotechnology will soon be the basis of the entire industry. However, the use of nanotechnology in the field of polymeric sensors and its implementation at an industrial level remains immature at a scientific level and therefore still represents an opportunity for new developments. In fact, it is expected that in the near future, the use of sensor arrays based on nanomaterials and CPs will have an important industrial impact, presenting optimal capabilities. To achieve this, the fundamental objective consists of the implementation of artificial intelligence methodologies using noninvasive and highly sensitive methods. The theoretical basis consists of training these systems (through supervised and unsupervised methodologies) to identify patterns and make decisions with the ability to classify and discriminate data without the need for human intervention. The following are the bases for the suggested benefits, which center on addressing the more significant challenge facing researchers: the transference of small-scale technology to the industrial level following the deployment of automatable and portable devices that can measure in real-time without the need for sample pretreatment:

- Ensure better control and characterization of raw materials and final products.
- Comprehensively control the evolution of the production process.
- Obtain a high-precision comparative tool that validates market behavior.
- Ensure environmental sustainability by using a durable and reliable device that requires no maintenance.
- Reduce costs in detection systems through simple, noninvasive systems.

The majority of the research only offered laboratory-scale data, but after they were combined, excellent improvements in terms of selectivity, sensitivity, and selectivity were discovered. The stage of development of each of these technologies varies, though. While e-eye systems still mostly consist of laboratory-designed equipment, suggesting the need to improve its technological development toward commercial applications, e-noses and e-tongues are more sophisticated technologies that have entered the commerce sector with portable devices.

References

Adhikari, B., & Majumdar, S. (2004). Polymers in sensor applications. *Progress in Polymer Science.*, *29*(7), 699–766. Available from https://doi.org/10.1016/j.progpolymsci.2004.03.002.

Aksnes, A. (2009). *Sensors for environment, health and security*. Springer.

Al-Haidary, Q. N., Al-Mokaram, A. M., Hussein, F. M., & Ismail, A. H. (2021). Development of polyaniline for sensor applications: A review. *Journal of Physics: Conference Series.*, *1853*(1), 012062. Available from https://doi.org/10.1088/1742-6596/1853/1/012062.

Ameer, Q., & Adeloju, S. B. (2005). Polypyrrole-based electronic noses for environmental and industrial analysis. *Sensors and Actuators, B: Chemical*, *106*(2), 541–552. Available from https://doi.org/10.1016/j.snb.2004.07.033.

Andre, R. S., Campaner, K., Facure, M. H. M., Mercante, L. A., Bogusz, S., & Correa, D. S. (2021). Nanocomposite-based chemiresistive electronic nose and application in coffee analysis. *ACS Food Science & Technology*, *1*(8), 1464–1471. Available from https://doi.org/10.1021/acsfoodscitech.1c00173.

Aouadi, B., Zaukuu, J. L. Z., Vitális, F., Bodor, Z., Fehér, O., Gillay, Z., Bazar, G., & Kovacs, Z. (2020). Historical evolution and food control achievements of near infrared spectroscopy, electronic nose, and electronic tongue—Critical overview. *Sensors (Switzerland).*, *20*(19), 1–42. Available from https://doi.org/10.3390/s20195479, https://www.mdpi.com/1424-8220/20/19/5479/pdf.

Apetrei, I. M., & Apetrei, C. (2013). Voltammetric e-tongue for the quantification of total polyphenol content in olive oils. *Food Research International.*, *54*(2), 2075–2082. Available from https://doi.org/10.1016/j.foodres.2013.04.032.

Arrieta, A., Barrera, I., & Mendoza, J. (2023). Application of a polypyrrole sensor array integrated into a smart electronic tongue for the discrimination of milk adulterated with sucrose. *International Journal of Technology.*, *14*(1), 90–99. Available from https://doi.org/10.14716/ijtech.v14i1.5613.

Arrieta, A. A., Arrieta, P. L., & Mendoza, J. M. (2019). Analysis of coffee adulterated with roasted corn and roasted soybean using voltammetric electronic tongue. *Acta Scientiarum Polonorum, Technologia Alimentaria.*, *18*(1), 35–41. Available from https://doi.org/10.17306/J.AFS.2019.0619.

Assunção da Silva, E., Duc, C., Redon, N., & Wojkiewicz, J. L. (2021). Humidity sensor based on PEO/PEDOT:PSS blends for breath monitoring. *Macromolecular Materials and Engineering*, *306*(12), 2100489. Available from https://doi.org/10.1002/mame.202100489.

Bayat, M., Izadan, H., Santiago, S., Estrany, F., Dinari, M., Semnani, D., Alemán, C., & Guirado, G. (2021). Study on the electrochromic properties of polypyrrole layers doped with different dye molecules. *Journal of Electroanalytical Chemistry.*, *886*, 115113. Available from https://doi.org/10.1016/j.jelechem.2021.115113.

Behera, B., Joshi, R., Anil Vishnu, G. K., Bhalerao, S., & Pandya, H. J. (2019). Electronic nose: A non-invasive technology for breath analysis of diabetes and lung cancer patients. *Journal of Breath Research.*, *13*(2). Available from https://doi.org/10.1088/1752-7163/aafc77, https://iopscience.iop.org/article/10.1088/1752-7163/aafc77/pdf.

Boehler, C., Aqrawe, Z., & Asplund, M. (2019). Applications of PEDOT in bioelectronic medicine. *Bioelectronics in Medicine.*, *2*(2), 89–99. Available from https://doi.org/10.2217/bem-2019-0014.

Brereton, R. G., & Lloyd, G. R. (2010). Support vector machines for classification and regression. *Analyst.*, *135*(2), 230–267. Available from https://doi.org/10.1039/b918972f.

Bridges, C. C. (1966). Hierarchical cluster analysis. *Psychological Reports*, *18*(3), 851–854. Available from https://doi.org/10.2466/pr0.1966.18.3.851.

Bro, R., & Smilde, A. K. (2014). Principal component analysis. *Analytical Methods*, *6*(9), 2812–2831. Available from https://doi.org/10.1039/c3ay41907j.

Bunney, J., Williamson, S., Atkin, D., Jeanneret, M., Cozzolino, D., Chapman, J., Power, A., & Chandra, S. (2017). The use of electrochemical biosensors in food analysis. *Current Research in Nutrition and Food Science*, *5*(3), 183–195. Available from https://doi.org/10.12944/CRNFSJ.5.3.02, http://www.foodandnutritionjournal.org/download/4661.

Buratti, S., Malegori, C., Benedetti, S., Oliveri, P., & Giovanelli, G. (2018). E-nose, e-tongue and e-eye for edible olive oil characterization and shelf life assessment: A powerful data fusion approach. *Talanta*, *182*, 131–141. Available from https://doi.org/10.1016/j.talanta.2018.01.096.

Calvini, R., & Pigani, L. (2022). Toward the development of combined artificial sensing systems for food quality evaluation: A review on the application of data fusion of electronic noses, electronic tongues and electronic eyes. *Sensors.*, *22*(2), 577. Available from https://doi.org/10.3390/s22020577.

Castro, M., Kumar, B., Feller, J. F., Haddi, Z., Amari, A., & Bouchikhi, B. (2011). Novel e-nose for the discrimination of volatile organic biomarkers with an array of carbon nanotubes (CNT) conductive polymer nanocomposites (CPC) sensors. *Sensors and Actuators B: Chemical.*, *159*(1), 213–219. Available from https://doi.org/10.1016/j.snb.2011.06.073.

Cetó, X., Capdevila, J., Puig-Pujol, A., & del Valle, M. (2014). Cava wine authentication employing a voltammetric electronic tongue. *Electroanalysis*, *26*(7), 1504–1512. Available from https://doi.org/10.1002/elan.201400057.

Chambers IV, E. (2019). Analysis of sensory properties in foods: A special issue. *Foods.*, *8*(8), 291. Available from https://doi.org/10.3390/foods8080291.

Chen, J., Yang, C., Yuan, C., Li, Y., An, T., & Dong, C. (2022). Moisture content monitoring in withering leaves during black tea processing based on electronic eye and near infrared spectroscopy. *Scientific Reports.*, *12*(1). Available from https://doi.org/10.1038/s41598-022-25112-6, 20721.

Chovin, A., Garrigue, P., Vinatier, P., & Sojic, N. (2004). Development of an ordered array of optoelectrochemical individually readable sensors with submicrometer dimensions: Application to remote electrochemiluminescence imaging. *Analytical Chemistry*, *76*(2), 357–364. Available from https://doi.org/10.1021/ac034974w.

Constantin Apetrei. (2012). Novel method based on polypyrrole-modified sensors and emulsions for the evaluation of bitterness in extra virgin olive oils. *Food Research International*, *48*(2), 673–680. Available from https://doi.org/10.1016/j.foodres.2012.06.010.

Conti, P. P., Andre, R. S., Mercante, L. A., Fugikawa-Santos, L., & Correa, D. S. (2021). Discriminative detection of volatile organic compounds using an electronic nose based on TiO2 hybrid nanostructures. *Sensors and Actuators, B: Chemical, 344*. Available from https://doi.org/10.1016/j.snb.2021.130124, https://www.journals.elsevier.com/sensors-and-actuators-b-chemical.

Debashis Ghosh, A. J. (2009). Modern Multivariate Statistical Techniques: Regression, Classification, and Manifold Learning. *Biometrics*, *65*(3), 990–991.

del Valle, M., Cetó, X., & Gutiérrez-Capitán, M. (2013). *Bioelectronic tongues. Portable biosensing of food toxicants and environmental pollutants* (pp. 339–372). Spain: CRC Press. Available from http://www.tandfebooks.com/doi/book/10.1201/b15589, 10.1201/b15589.

Dhand, C., Das, M., Datta, M., & Malhotra, B. D. (2011). Recent advances in polyaniline based biosensors. *Biosensors and Bioelectronics.*, *26*(6), 2811–2821. Available from https://doi.org/10.1016/j.bios.2010.10.017.

Di Rosa, A. R., Leone, F., Cheli, F., & Chiofalo, V. (2017). Fusion of electronic nose, electronic tongue and computer vision for animal source food authentication and quality assessment – A review. *Journal of Food Engineering*, *210*, 62–75. Available from https://doi.org/10.1016/j.jfoodeng.2017.04.024, http://www.sciencedirect.com/science/journal/02608774.

Escuder-Gilabert, L., & Peris, M. (2010). Review: Highlights in recent applications of electronic tongues in food analysis. *Analytica Chimica Acta*, *665*(1), 15–25. Available from https://doi.org/10.1016/j.aca.2010.03.017.

Facure, M. H. M., Mercante, L. A., Mattoso, L. H. C., & Correa, D. S. (2017). Detection of trace levels of organophosphate pesticides using an electronic tongue based on graphene hybrid nanocomposites. *Talanta*, *167*, 59–66. Available from https://doi.org/10.1016/j.talanta.2017.02.005, https://www.journals.elsevier.com/talanta.

Fan, X., Nie, W., Tsai, H., Wang, N., Huang, H., Cheng, Y., Wen, R., Ma, L., Yan, F., & Xia, Y. (2019). PEDOT:PSS for flexible and stretchable electronics: Modifications, strategies, and applications. *Advanced Science.*, *6*(19), 1900813. Available from https://doi.org/10.1002/advs.201900813.

Ferreira, M., Riul, A., Wohnrath, K., Fonseca, F. J., Oliveira, O. N., & Mattoso, L. H. C. (2003). High-performance taste sensor made from Langmuir-Blodgett films of conducting polymers and a ruthenium complex. *Analytical Chemistry*, *75*(4), 953–955. Available from https://doi.org/10.1021/ac026031p.

Fitzgerald, J. E., Zhu, J., Bravo-Vasquez, J. P., & Fenniri, H. (2016). Cross-reactive, self-encoded polymer film arrays for sensor applications. *RSC Advances.*, *6*(86), 82616–82624. Available from https://doi.org/10.1039/c6ra13874h, http://pubs.rsc.org/en/journals/journalissues.

Garcia-Hernandez, C., Garcia-Cabezon, C., Martin-Pedrosa, F., & Rodriguez-Mendez, M. L. (2019). Analysis of musts and wines by means of a bio-electronic tongue based on tyrosinase and glucose oxidase using polypyrrole/gold nanoparticles as the electron mediator. *Food Chemistry*, *289*, 751–756. Available from https://doi.org/10.1016/j.foodchem.2019.03.107.

Geană, E.-I., Ciucure, C. T., Artem, V., & Apetrei, C. (2020). Wine varietal discrimination and classification using a voltammetric sensor array based on modified screen-printed electrodes in conjunction with chemometric analysis. *Microchemical Journal*, *159*, 105451. Available from https://doi.org/10.1016/j.microc.2020.105451.

Gómez-Arribas, L. N., Benito-Peña, E., Hurtado-Sánchez, M. D. C., & Moreno-Bondi, M. C. (2018). Biosensing based on nanoparticles for food allergens detection. *Sensors (Switzerland).*, *18*(4). Available from https://doi.org/10.3390/s18041087, http://www.mdpi.com/1424-8220/18/4/1087/pdf.

Greenacre, M., Groenen, P. J. F., Hastie, T., D'Enza, A. I., Markos, A., & Tuzhilina, E. (2022). Principal component analysis. *Nature Reviews Methods Primers.*, *2*(1). Available from https://doi.org/10.1038/s43586-022-00184-w, https://www.nature.com/nrmp/.

Guadarrama, A., Rodríguez-Méndez, M. L., De Saja, J. A., Ríos, J. L., & Olías, J. M. (2000). Array of sensors based on conducting polymers for the quality control of the aroma of the virgin olive oil. *Sensors and Actuators, B: Chemical*, *69*(3), 276–282. Available from https://doi.org/10.1016/S0925-4005(00)00507-4.

Håkansson, E., Lin, T., Wang, H., & Kaynak, A. (2006). The effects of dye dopants on the conductivity and optical absorption properties of polypyrrole. *Synthetic Metals*, *156*(18–20), 1194–1202. Available from https://doi.org/10.1016/j.synthmet.2006.08.006.

Hameed, N., Capricho, J. C., Salim, N., & Thomas, S. (2023). *Multifunctional epoxy resins: Self-healing, thermally and electrically conductive resins*. In *Ser. Engineering Materials)*, (1st ed). Springer.

Hao, H. C., Tang, K. T., Ku, P. H., Chao, J. S., Li, C. H., Yang, C. M., & Yao, D. J. (2010). Development of a portable electronic nose based on chemical surface acoustic wave array with multiplexed oscillator and readout electronics. *Sensors and Actuators B: Chemical.*, *146*(2), 545–553. Available from https://doi.org/10.1016/j.snb.2009.12.023.

He, X., & Hwang, H. M. (2016). Nanotechnology in food science: Functionality, applicability, and safety assessment. *Journal of Food and Drug Analysis.*, *24*(4), 671–681. Available from https://doi.org/10.1016/j.jfda.2016.06.001, http://www.elsevier.com/journals/journal-of-food-and-drug-analysis/1021-9498#.

Hong, X., Wang, J., & Qiu, S. (2014). Authenticating cherry tomato juices—Discussion of different data standardization and fusion approaches based on electronic nose and tongue. *Food Research International.*, *60*, 173–179. Available from https://doi.org/10.1016/j.foodres.2013.10.039.

Jain, R., Jadon, N., & Pawaiya, A. (2017). Polypyrrole based next generation electrochemical sensors and biosensors: A review. *TrAC Trends in Analytical Chemistry.*, *97*, 363–373. Available from https://doi.org/10.1016/j.trac.2017.10.009.

John, J., Saheeda, P., Sabeera, K., & Jayalekshmi, S. (2018). Doped polypyrrole with good solubility and film forming properties suitable for device applications. *Materials Today: Proceedings.*, *5*(10), 21140–21146. Available from https://doi.org/10.1016/j.matpr.2018.06.512.

Kaiser, A. B., Subramaniam, C. K., Gilberd, P. W., & Wessling, B. (1995). Electronic transport properties of conducting polymers and polymer blends. *Synthetic Metals*, *69*(1–3), 197–200. Available from https://doi.org/10.1016/0379-6779(94)02415-u.

Kashyap, R., Kumar, R., Kumar, M., Tyagi, S., & Kumar, D. (2019). Polyaniline nanofibers based gas sensor for detection of volatile organic compounds at room temperature. *Materials Research Express.*, *6*(11), 1150d3. Available from https://doi.org/10.1088/2053-1591/ab4e43.

Kirsanov, D., Mukherjee, S., Pal, S., Ghosh, K., Bhattacharyya, N., Bandyopadhyay, R., Jendrlin, M., Radu, A., Zholobenko, V., Dehabadi, M., & Legin, A. (2021). A pencil-drawn electronic tongue for environmental applications. *Sensors.*, *21*(13), 4471. Available from https://doi.org/10.3390/s21134471.

Kovarski, A. (1998). *Molecular Dynamics of Additives in Polymers* (1st ed.). CRC Press.

Kulikova, T. N., Porfireva, A. V., Vorobev, V. V., Saveliev, A. A., Ziyatdinova, G. K., & Evtugyn, G. A. (2019). Discrimination of tea by the electrochemical determination of its antioxidant properties by a polyaniline – DNA – Polyphenazine dye modified glassy carbon electrode. *Analytical Letters*, *52*(16), 2562–2582. Available from https://doi.org/10.1080/00032719.2019.1618321.

Lakard, B. (2020). Electrochemical biosensors based on conducting polymers: A review. *Applied Sciences.*, *10* (18). Available from https://doi.org/10.3390/app10186614, 6614.

Lange, U., Roznyatovskaya, N. V., & Mirsky, V. M. (2008). Conducting polymers in chemical sensors and arrays. *Analytica Chimica Acta*, *614*(1), 1–26. Available from https://doi.org/10.1016/j.aca.2008.02.068.

Köhn, H., & Hubert, L. J. (2015). Hierarchical cluster analysis. *Wiley StatsRef Stat Ref Online*, 1–13. https://doi.org/10.1002/9781118445112.stat02449.pub2.

Liu, B., Huang, Y., Kam, K. W. L., Cheung, W.-F., Zhao, N., & Zheng, B. (2019). Functionalized graphene-based chemiresistive electronic nose for discrimination of disease-related volatile organic compounds. *Biosensors and Bioelectronics: X*, *1*, 100016. Available from https://doi.org/10.1016/j.biosx.2019.100016.

Lu, T., Al-Hamry, A., Talbi, M., Zhang, J., Adiraju, A., Hou, M., & Kanoun, O. (2021). Germany Functionalized PEDOT:PSS based sensor array for determination of metallic ions in smart agriculture. In 6th International Conference on Nanotechnology for Instrumentation and Measurement, *NanofIM 2021*, Institute of Electrical and Electronics Engineers Inc. Available from http://ieeexplore.ieee.org/xpl/mostRecentIssue.jsp?punumber = 9737336, https://doi.org/10.1109/NanofIM54124.2021.9737340.

Lvova, L., Legin, A., Vlasov, Y., Cha, G. S., & Nam, H. (2003). Multicomponent analysis of Korean green tea by means of disposable all-solid-state potentiometric electronic tongue microsystem. *Sensors and Actuators, B: Chemical*, *95*(1–3), 391–399. Available from https://doi.org/10.1016/S0925-4005(03)00445-3.

MacDiarmid, A. G., Manohar, S. K., Masters, J. G., Sun, Y., Weiss, H., & Epstein, A. J. (1991). Polyaniline: Synthesis and properties of pernigraniline base. *Synthetic Metals*, *41*(1–2), 621–626. Available from https://doi.org/10.1016/0379-6779(91)91145-z.

Maduraiveeran, G., & Jin, W. (2017). Nanomaterials based electrochemical sensor and biosensor platforms for environmental applications. *Trends in Environmental Analytical Chemistry.*, *13*, 10–23. Available from https://doi.org/10.1016/j.teac.2017.02.001.

Magro, C., Mateus, E. P., Raposo, M., & Ribeiro, A. B. (2019). Overview of electronic tongue sensing in environmental aqueous matrices: Potential for monitoring emerging organic contaminants. *Environmental Reviews.*, *27*(2), 202–214. Available from https://doi.org/10.1139/er-2018-0019, http://pubs.nrc-cnrc.gc.ca/cgi-bin/rp/rp2_desc_e?er.

Mahun, A., Abbrent, S., Bober, P., Brus, J., & Kobera, L. (2020). Effect of structural features of polypyrrole (PPy) on electrical conductivity reflected on 13C ssNMR parameters. *Synthetic Metals*, *259*. Available from https://doi.org/10.1016/j.synthmet.2019, 116250.

Manikandan, V. S., Adhikari, B. R., & Chen, A. (2018). Nanomaterial based electrochemical sensors for the safety and quality control of food and beverages. *Analyst.*, *143*(19), 4537–4554. Available from https://doi.org/10.1039/c8an00497h, http://pubs.rsc.org/en/journals/journal/an.

Martina, V., Ionescu, K., Pigani, L., Terzi, F., Ulrici, A., Zanardi, C., & Seeber, R. (2007). Development of an electronic tongue based on a PEDOT-modified voltammetric sensor. *Analytical and Bioanalytical Chemistry*, *387*(6), 2101–2110. Available from https://doi.org/10.1007/s00216-006-1102-1.

Mehmood, T., Liland, K. H., Snipen, L., & Sæbø, S. (2012). A review of variable selection methods in partial least squares regression. *Chemometrics and Intelligent Laboratory Systems.*, *118*, 62–69. Available from https://doi.org/10.1016/j.chemolab.2012.07.010.

Men, H., Zhang, C., Ning, K., & Chen, D. (2013). Detection of edible oils based on voltammetric electronic tongue. *Research Journal of Applied Sciences, Engineering and Technology.*, *5*(4), 1197–1202. Available from https://doi.org/10.19026/rjaset.5.4837.

Méndez, M. L. R., De Saja, J. A., Medina-Plaza, C., & García-Hernández, C. (2016). *Chapter 26 – Electronic tongues for the organoleptic characterization of wines. BT – Electronic noses and tongues in food science.* Elsevier Inc. Available from 10.1016/B978-0-12-800243-8/00026-3.

Migliorini, F. L., Teodoro, K. B. R., dos Santos, D. M., Fonseca, F. J., Mattoso, L. H. C., & Correa, D. S. (2020). Electrospun nanofibers versus drop casting films for designing an electronic tongue: Comparison of performance for monitoring geosmin and 2-methylisoborneol in water samples. *Polymers for Advanced Technologies*, *31*(9), 2075–2082. Available from https://doi.org/10.1002/pat.4930, http://onlinelibrary.wiley.com/journal/10.1002/(ISSN)1099-1581.

Mondal, D., Paul, D., & Mukherji, S. (2019). Conducting polymer coated filter paper based disposable electronic tongue. In 2018 Proceedings of the International Conference on Sensing Technology, 7–12. Available from https://doi.org/10.1109/ICSensT.2018.8603573, http://ieeexplore.ieee.org/xpl/conferences.jsp.

Munekata, P. E. S., Finardi, S., de Souza, C. K., Meinert, C., Pateiro, M., Hoffmann, T. G., Domínguez, R., Bertoli, S. L., Kumar, M., & Lorenzo, J. M. (2023). Applications of electronic nose, electronic eye and electronic tongue in quality, safety and shelf life of meat and meat products: A review. *Sensors.*, *23*(2). Available from https://doi.org/10.3390/s23020672, http://www.mdpi.com/journal/sensors.

Nasajpour-Esfahani, N., Dastan, D., Alizadeh, A., Shirvanisamani, P., Rozati, M., Ricciardi, E., Lewis, B., Aphale, A., & Toghraie, D. (2023). A critical review on intrinsic conducting polymers and their applications. *Journal of Industrial and Engineering Chemistry*, *125*, 14–37. Available from https://doi.org/10.1016/j.jiec.2023.05.013.

Nategh, N. A., Dalvand, M. J., & Anvar, A. (2021). Detection of toxic and non-toxic sweet cherries at different degrees of maturity using an electronic nose. *Journal of Food Measurement and Characterization.*, *15*(2), 1213–1224. Available from https://doi.org/10.1007/s11694-020-00724-6, http://rd.springer.com/journal/11694.

Nguyen, T. A., Kokot, S., Ongarato, D. M., & Wallace, G. G. (2000). The use of cyclic voltammetry and principal component analysis for the rapid evaluation of selectivity of conductive polymer sensors. *Electroanalysis*, *12*(2), 89–95. Available from http://onlinelibrary.wiley.com/journal/10.1002/(ISSN)1521-4109, 10.1002/(SICI)1521–4109(200002)12:2 < 89::AID-ELAN89 > 3.0.CO;2-2.

Oh, J., Yang, H., Jeong, G. E., Moon, D., Kwon, O. S., Phyo, S., Lee, J., Song, H. S., Park, T. H., & Jang, J. (2019). Ultrasensitive, selective, and highly stable bioelectronic nose that detects the liquid and gaseous cadaverine. *Analytical Chemistry*, *91*(19), 12181–12190. Available from https://doi.org/10.1021/acs.analchem.9b01068, http://pubs.acs.org/journal/ancham.

Oliveira, G. P., Barboza, B. H., & Batagin-Neto, A. (2022). Polyaniline-based gas sensors: DFT study on the effect of side groups. *Computational and Theoretical Chemistry.*, *1207*, 113526. Available from https://doi.org/10.1016/j.comptc.2021.113526.

Orlandi, G., Calvini, R., Foca, G., Pigani, L., Simone, G. V., & Ulrici, A. (2019). Data fusion of electronic eye and electronic tongue signals to monitor grape ripening. *Talanta*, *195*, 181–189. Available from https://doi.org/10.1016/j.talanta.2018.11.046.

Pantalei, S., Zampetti, E., Bearzotti, A., De Cesare, F., & Macagnano, A. (2013). Improving sensing features of a nanocomposite PEDOT:PSS sensor for NO breath monitoring. *Sensors and Actuators B: Chemical.*, *179*, 87–94. Available from https://doi.org/10.1016/j.snb.2012.10.015.

Parameswaranpillai, J., & Ganguly, S. (2023). *Introduction to polymer composite-based sensors* (pp. 1–21). Elsevier BV. Available from 10.1016/b978-0-323-98830-8.00006-0.

Parra, V., Arrieta, Á. A., Fernández-Escudero, J.-A., Rodríguez-Méndez, M. L., & De Saja, J. A. (2006). Electronic tongue based on chemically modified electrodes and voltammetry for the detection of adulterations in wines. *Sensors and Actuators B: Chemical*, *118*(1–2), 448–453. Available from https://doi.org/10.1016/j.snb.2006.04.043.

Parra, V., Arrieta, A. A., Fernández-Escudero, J. A., García, H., Apetrei, C., Rodríguez-Méndez, M. L., & Saja, J. A. D. (2006). E-tongue based on a hybrid array of voltammetric sensors based on phthalocyanines, perylene derivatives and conducting polymers: Discrimination capability towards red wines elaborated with different varieties of grapes. *Sensors and Actuators, B: Chemical.*, *115*(1), 54–61. Available from https://doi.org/10.1016/j.snb.2005.08.040.

Pérez-González, C., Salvo-Comino, C., Martin-Pedrosa, F., Dias, L., Rodriguez-Perez, M. A., Garcia-Cabezon, C., & Rodriguez-Mendez, M. L. (2021). Analysis of milk using a portable potentiometric electronic tongue based on five polymeric membrane sensors. *Frontiers in Chemistry.*, *9*. Available from https://doi.org/10.3389/fchem.2021.706460.

Pérez-González, C., Salvo-Comino, C., Martín-Pedrosa, F., García-Cabezón, C., & Rodríguez-Méndez, M. L. (2023). Bioelectronic tongue dedicated to the analysis of milk using enzymes linked to carboxylated-PVC membranes modified with gold nanoparticles. *Food Control*, *145*. Available from https://doi.org/10.1016/j.foodcont.2022.109425, https://www.journals.elsevier.com/food-control.

Pérez-López, B., & Merkoçi, A. (2011). Nanomaterials based biosensors for food analysis applications. *Trends in Food Science & Technology*, *22*(11), 625–639. Available from https://doi.org/10.1016/j.tifs.2011.04.001.

Persaud, K. (2016). *Electronic noses and tongues in the food industry* (pp. 1–12). Elsevier BV. Available from 10.1016/b978-0-12-800243-8.00001-9.

Persaud, K., & Dodd, G. (1982). Analysis of discrimination mechanisms in the mammalian olfactory system using a model nose. *Nature*, *299*(5881), 352–355. Available from https://doi.org/10.1038/299352a0.

Pigani, L., Culetu, A., Ulrici, A., Foca, G., Vignali, M., & Seeber, R. (2011). Pedot modified electrodes in amperometric sensing for analysis of red wine samples. *Food Chemistry*, *129*(1), 226–233. Available from https://doi.org/10.1016/j.foodchem.2011.04.046.

Pigani, L., Foca, G., Ionescu, K., Martina, V., Ulrici, A., Terzi, F., Vignali, M., Zanardi, C., & Seeber, R. (2008). Amperometric sensors based on poly(3,4-ethylenedioxythiophene)-modified electrodes: Discrimination of white wines. *Analytica Chimica Acta*, *614*(2), 213–222. Available from https://doi.org/10.1016/j.aca.2008.03.029.

Pigani, L., Foca, G., Ulrici, A., Ionescu, K., Martina, V., Terzi, F., Vignali, M., Zanardi, C., & Seeber, R. (2009). Classification of red wines by chemometric analysis of voltammetric signals from PEDOT-modified electrodes. *Analytica Chimica Acta*, *643*(1–2), 67–73. Available from https://doi.org/10.1016/j.aca.2009.03.040.

Pigani, L., Vasile Simone, G., Foca, G., Ulrici, A., Masino, F., Cubillana-Aguilera, L., Calvini, R., & Seeber, R. (2018). Prediction of parameters related to grape ripening by multivariate calibration of voltammetric signals acquired by an electronic tongue. *Talanta*, *178*, 178–187. Available from https://doi.org/10.1016/j.talanta.2017.09.027.

Quijada, C. (2020). Special issue: Conductive polymers: Materials and applications. *Materials*, *13*(10), 2344. Available from https://doi.org/10.3390/ma13102344.

Riul, A., De Sousa, H. C., Malmegrim, R. R., Dos Santos, D. S., Carvalho, A. C. P. L. F., Fonseca, F. J., Oliveira, O. N., & Mattoso, L. H. C. (2004). Wine classification by taste sensors made from ultra-thin films and using neural networks. *Sensors and Actuators, B: Chemical*, *98*(1), 77–82. Available from https://doi.org/10.1016/j.snb.2003.09.025.

Riul, A., Soto, A. M. G., Mello, S. V., Bone, S., Taylor, D. M., & Mattoso, L. H. C. (2003). An electronic tongue using polypyrrole and polyaniline. *Synthetic Metals*, *132*(2), 109–116. Available from https://doi.org/10.1016/S0379-6779(02)00107-8.

Robinson, S. M. (2012). *Briefs in Optimization*. Springer.

Schütze, A., Helwig, N., & Schneider, T. (2018). Sensors 4.0 – smart sensors and measurement technology enable Industry 4.0. *Journal of Sensors and Sensor Systems*, *7*(1), 359–371. Available from https://doi.org/10.5194/jsss-7-359-2018.

Sen, T., Mishra, S., & Shimpi, N. G. (2016). Synthesis and sensing applications of polyaniline nanocomposites: A review. *RSC Advances.*, *6*(48), 42196–42222. Available from https://doi.org/10.1039/c6ra03049a, http://pubs.rsc.org/en/journals/journalissues.

Shirakawa, H., Louis, E. J., MacDiarmid, A. G., Chiang, C. K., & Heeger, A. J. (1977). Synthesis of electrically conducting organic polymers: Halogen derivatives of polyacetylene, (CH)x. *Journal of the Chemical Society, Chemical Communications* (16), 578–580. Available from https://doi.org/10.1039/C39770000578.

Sierra-Padilla, A., García-Guzmán, J. J., López-Iglesias, D., Palacios-Santander, J. M., & Cubillana-Aguilera, L. (2021). E-tongues/noses based on conducting polymers and composite materials: Expanding the possibilities in complex analytical sensing. *Sensors.*, *21*(15). Available from https://doi.org/10.3390/s21154976, https://www.mdpi.com/1424-8220/21/15/4976/pdf.

Sinha, A., Dhanjai., Jain, R., Zhao, H., Karolia, P., & Jadon, N. (2018). Voltammetric sensing based on the use of advanced carbonaceous nanomaterials: A review. *Microchimica Acta.*, *185*(2), 89. Available from https://doi.org/10.1007/s00604-017-2626-0.

Spinelle, L., Gerboles, M., Kok, G., Persijn, S., & Sauerwald, T. (2017). Review of portable and low-cost sensors for the ambient air monitoring of benzene and other volatile organic compounds. *Sensors (Switzerland).*, *17* (7). Available from https://doi.org/10.3390/s17071520, http://www.mdpi.com/1424-8220/17/7/1520/pdf.

Stejskal, J., & Prokeš, J. (2020). Conductivity and morphology of polyaniline and polypyrrole prepared in the presence of organic dyes. *Synthetic Metals*, *264*, 116373. Available from https://doi.org/10.1016/j.synthmet.2020.116373.

Stella, R., Barisci, J. N., Serra, G., Wallace, G. G., & De Rossi, D. (2000). Characterization of olive oil by an electronic nose based on conducting polymer sensors. *Sensors and Actuators, B: Chemical*, *63*(1), 1–9. Available from https://doi.org/10.1016/S0925-4005(99)00510-9.

Su, Z., Jin, Y., Xiao, Y., Zheng, H., Yang, Z., Wang, H., & Li, Z. (2022). Excellent rate capability supercapacitor based on a free-standing PEDOT:PSS film enabled by the hydrothermal method. *Chemical Communications.*, *58*(33), 5088–5091. Available from https://doi.org/10.1039/d2cc00427e.

Suthaharan, S. (2016). Machine learning models and algorithms for big data classification: Thinking with examples for effective learning. *Integrated Series in Information Systems.*, *36*. Available from https://doi.org/10.1007/978-1-4899-7641-3.

Tahara, Y., & Toko, K. (2013). Electronic tongues–A review. *IEEE Sensors Journal*, *13*(8), 3001–3011. Available from https://doi.org/10.1109/JSEN.2013.2263125.

Tan, J., & Xu, J. (2020). Applications of electronic nose (e-nose) and electronic tongue (e-tongue) in food quality-related properties determination: A review. *Artificial Intelligence in Agriculture.*, *4*, 104–115. Available from https://doi.org/10.1016/j.aiia.2020.06.003.

Tiwari, M. D. (2013). Communications in computer and information science: Preface. *Communications in Computer and Information Science*, *276*, V–VI. Available from https://doi.org/10.1007/978-3-642-37463-0, http://www.springer.com/series/7899.

Vahdatiyekta, P., Zniber, M., Bobacka, J., & Huynh, T.-P. (2022). A review on conjugated polymer-based electronic tongues. *Analytica Chimica Acta*, *1221*, 340114. Available from https://doi.org/10.1016/j.aca.2022.340114.

Vasilescu, A., Fanjul-Bolado, P., Titoiu, A. M., Porumb, R., & Epure, P. (2019). Progress in electrochemical (bio)sensors for monitoring wine production. *Chemosensors*, *7*(4), 66. Available from https://doi.org/10.3390/chemosensors7040066.

Verma, A., Gupta, R., Verma, A. S., & Kumar, T. (2023). A review of composite conducting polymer-based sensors for detection of industrial waste gases. *Sensors and Actuators Reports.*, *5*, 100143. Available from https://doi.org/10.1016/j.snr.2023.100143.

Vlascici, D., Pica, E. M., Fagadar-Cosma, E., Cosma, V., & Bizerea, O. (2008). Thiocyanate and fluoride electrochemical sensors based on nanostructurated metalloporphyrin systems. *Journal of Optoelectronics and Advanced Materials.*, *10*(9), 2303–2306. Available from http://joam.inoe.ro/index.php.

Vlasov, Y., Legin, A., Rudnitskaya, A., Di Natale, C., & D'Amico, A. (2005). Nonspecific sensor arrays ("electronic tongue") for chemical analysis of liquids (IUPAC Technical Report). *Pure and Applied Chemistry.*, *77*(11), 1965–1983. Available from https://doi.org/10.1351/pac200577111965.

Wang, L., Li, Y., Wang, Q., Zou, L., & Ye, B. (2016). The construction of well-aligned MWCNTs-PANI Langmuir–Blodgett film modified glassy carbon electrode and its analytical application. *Sensors and Actuators B: Chemical.*, *228*, 214–220. Available from https://doi.org/10.1016/j.snb.2016.01.030.

Wang, Y., & Duncan, T. V. (2017). Nanoscale sensors for assuring the safety of food products. *Current Opinion in Biotechnology*, *44*, 74–86. Available from https://doi.org/10.1016/j.copbio.2016.10.005, http://www.elsevier.com/locate/copbio.

Wang, Y., Liu, A., Han, Y., & Li, T. (2020). Sensors based on conductive polymers and their composites: A review. *Polymer International.*, *69*(1), 7–17. Available from https://doi.org/10.1002/pi.5907, http://onlinelibrary.wiley.com/journal/10.1002/(ISSN)1097-0126.

Wasilewski, T., Migoń, D., Gębicki, J., & Kamysz, W. (2019). Critical review of electronic nose and tongue instruments prospects in pharmaceutical analysis. *Analytica Chimica Acta*, *1077*, 14–29. Available from https://doi.org/10.1016/j.aca.2019.05.024.

Wilson, J., Radhakrishnan, S., Sumathi, C., & Dharuman, V. (2012). Polypyrrole-polyaniline-Au (PPy-PANi-Au) nano composite films for label-free electrochemical DNA sensing. *Sensors and Actuators, B: Chemical*, *171–172*, 216–222. Available from https://doi.org/10.1016/j.snb.2012.03.019.

Windmiller, J. R., & Wang, J. (2013). Wearable electrochemical sensors and biosensors: A review. *Electroanalysis*, *25*(1), 29–46. Available from https://doi.org/10.1002/elan.201200349.

Yang, D., Wang, J., Cao, Y., Tong, X., Hua, T., Qin, R., & Shao, Y. (2023). Polyaniline-based biological and chemical sensors: Sensing mechanism, configuration design, and perspective. *ACS Applied Electronic Materials.*, *5*(2), 593–611. Available from https://doi.org/10.1021/acsaelm.2c01405.

Yen, Y. K., & Lai, C. Y. (2020). Portable real-time detection of Pb(II) using a CMOS MEMS-based nanomechanical sensing array modified with PEDOT:PSS. *Nanomaterials.*, *10*(12), 1–11. Available from https://doi.org/10.3390/nano10122454, https://www.mdpi.com/2079-4991/10/12/2454/pdf.

Zhang, W., Su, Z., Zhang, X., Wang, W., & Li, Z. (2022). Recent progress on PEDOT-based wearable bioelectronics. *VIEW*, *3*(5). Available from https://doi.org/10.1002/VIW.20220030, https://onlinelibrary.wiley.com/journal/2688268x.

Zhang, X., Yang, W., Zhang, H., Xie, M., & Duan, X. (2021). PEDOT:PSS: From conductive polymers to sensors. *Nanotechnology and Precision Engineering.*, *4*(4), 045004. Available from https://doi.org/10.1063/10.0006866.

Zhao, Q., Liu, J., Wu, Z., Xu, X., Ma, H., Hou, J., Xu, Q., Yang, R., Zhang, K., Zhang, M., Yang, H., Peng, W., Liu, X., Zhang, C., Xu, J., & Lu, B. (2022). Robust PEDOT:PSS-based hydrogel for highly efficient interfacial solar water purification. *Chemical Engineering Journal.*, *442*, 136284. Available from https://doi.org/10.1016/j.cej.2022.136284.

Zhu, C., Yang, G., Li, H., Du, D., & Lin, Y. (2015). Electrochemical sensors and biosensors based on nanomaterials and nanostructures. *Analytical Chemistry*, *87*(1), 230–249. Available from https://doi.org/10.1021/ac5039863.

Ziyatdinov, A., Marco, S., Chaudry, A., Persaud, K., Caminal, P., & Perera, A. (2010). Drift compensation of gas sensor array data by common principal component analysis. *Sensors and Actuators B: Chemical*, *146*(2), 460–465. Available from https://doi.org/10.1016/j.snb.2009.11.034.

SECTION

2

Lab-on-a-chip and sensory devices

CHAPTER

Polymers in sensory and lab-on-a-chip devices

13

Samar Damiati

Department of Chemistry, University of Sharjah, Sharjah, United Arab Emirates

13.1 Introduction

Nature is a toolkit that offers many puzzle pieces that can be used to build simple sensing devices with high performance. Many efforts have been made to mimic natural structures and behaviors using biological materials such as nucleic acids (DNA and RNA), sugars, lipids, and proteins, as well as nonbiological materials such as metals, graphene, and carbon nanotubes. In this sense, polymers could be natural, semisynthetic, or synthetic substances based on the origin of the source and composed of repeated monomer units as the basic building blocks that are created by the polymerization process. Polymers are widely employed in biomimetics, drug delivery systems, scaffolds in tissue engineering, and medical devices.

Natural biopolymers are produced by or derived from living organisms, such as plants, animals, and microbes, and consist of monomeric units that are covalently linked in chains to form large, low-to-high-molecular-weight molecules. The primary sources of a wide range of natural polymers (e.g., cellulose, starch, chitosan, collagen, alginate, and microbial polyesters) are derived from renewable resources (Kaplan, 1998a, 1998b). On the other side, synthetic polymers are classified into two classes: biodegradable and nonbiodegradable polymers. Synthetic biodegradable polymers such as polylactic acid (PLA), polyglycolic acid, polycaprolactone, and polylactic acid-co-glycolic acid (PLGA) are obtained from bio-based products and play important roles in clinical applications. Synthetic nonbiodegradable polymers, such as Teflon, nylon, polyethylene, polyester, and epoxy, are derived from petroleum oil and synthesized by the chemical polymerization of bio-monomers. Most synthetic polymers are nonbiodegradable in nature because they are resistant to environmental degradation processes. However, semisynthetic polymers (e.g., cellulose rayon and cellulose nitrate) are mostly derived from natural polymers and modified chemically. Indeed, they can be made by combining natural and synthetic materials (Damiati et al., 2022a; Gutierrez Cisneros et al., 2021).

Polymers have attracted the interest of researchers because of their unique properties involving elasticity, toughness, transparency, processing, excellent bending ability, being lightweight but comparatively stiff and strong, lack of heat and electrical conductivity, shape memory, and low cost (Brinson & Catherine Brinson, 2008a, 2008b). Furthermore, the physical properties of polymers, such as molecular weight, charges, composition, degree of polymerization, and crystallinity of material, as well as their ability to act as anchors by self-assembling on surfaces via functional end or side groups, play a dominant role in utilizing polymers in health and environmental control

Sensory Polymers. DOI: https://doi.org/10.1016/B978-0-443-13394-7.00013-6

sectors (Neoh et al., 2015). One field that widely uses polymers is the development of microfluidic devices. In polymer science, the synthesis, characterization, and implementation of stimuli-responsive polymers support the development of advanced microfabrication techniques that carry out multiple analytic investigations on the same platform. Recent developments in microfluidic chips exploit novel polymeric materials in various applications involving the fabrication of microfluidic sensors, designing well-controlled microenvironments for manipulating fluids, the development of effective drug delivery systems, and the construction of artificial cells and biodegradable scaffolds for tissue engineering (Damiati et al., 2022a,b).

This book chapter describes the integration of microfluidic technology and polymers in device fabrication, biomimicry, and sensing technology (Fig. 13.1). First, an overview of the basics of microfluidic technology is discussed. Next, the selection of polymers and manufacturing techniques to construct microfluidic platforms are highlighted. Since applications of microfluidics and polymers are widely exploited, selected studies focus on the fabrication of polymeric microfluidic devices, the formation of polymeric vesicles, and the functionalization of sensing platforms.

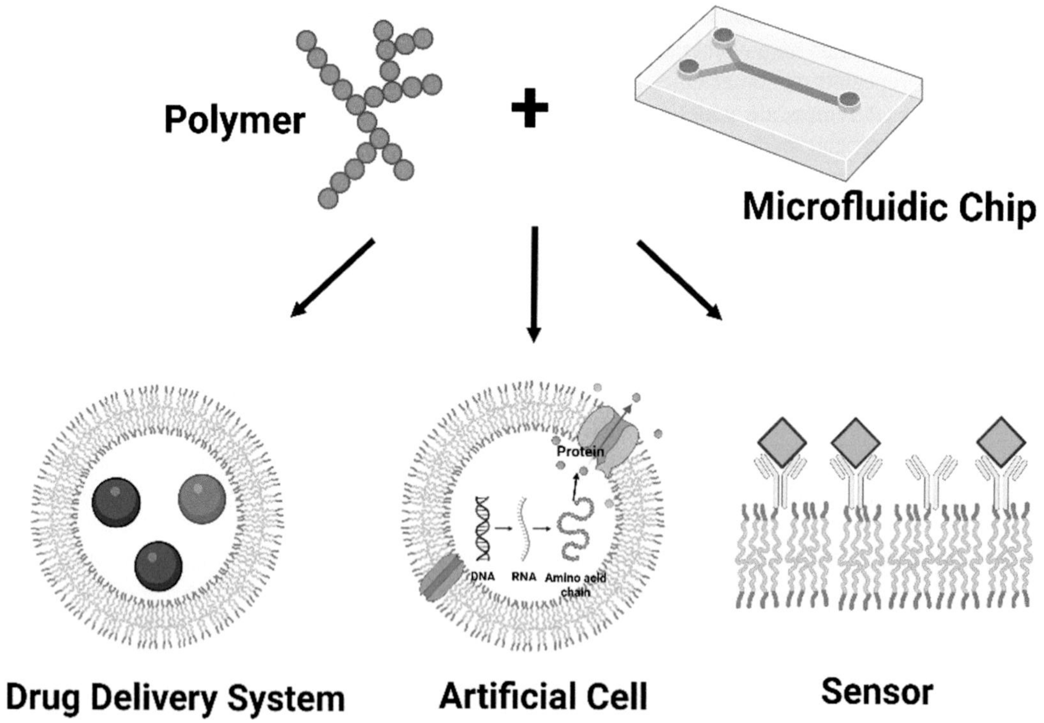

FIGURE 13.1

Combining polymers and microfluidic technology for various applications (created with Biorender.com).

13.2 Basic principles of microfluidic technology

Microfluidic devices, also called lab-on-a-chip or micrototal analysis systems, allow the integration of various biological, chemical, and biochemical reactions on a single microchip. Since its development in the early 1980s, microfluidic technology has exhibited several advantages, such as reducing the consumption of reagents/samples to picoliters, minimizing waste generation, allowing continuous oxygen and nutrient supply, allowing the integration of multiple steps on the same chip, being fast, easy to use, and low-cost (Convery & Gadegaard, 2019; Damiati et al., 2022a). Further, it provides accurate results without the requirement of centralized laboratories and specialized equipment. The construction of these miniaturized devices permits real-time observation and in situ investigation outside of a laboratory environment. Further, an important advantage that a microfluidic chip offers is the ability to perform basic functions involving mixing, preparation, filtration, separation, heating, and detection on the same platform. The functionality of a microfluidic chip is based on its design, which is composed of a network of microchannels that allows controlling the flow behavior of small volumes of fluids in microchambers with different geometries hollowed out through the chip. Controlling the geometry allows the subdivision of microchannels into multiple functional units involving pumps, mixers, reactors, valves, actuators, and sensors. Thus, the fabrication of a microfluidic chip is not just a simple arrangement of microchannels but a more sophisticated assembly of different pieces to perform specific functions (Andersson et al., 2001; Damiati, 2019; Mitrogiannopoulou et al., 2023).

In microfluidic channels, fluid dynamics can be classified into either laminar or turbulent flow. In laminar flow, the fluid moves in parallel with the direction of the moving fluid, while in turbulent flow, the fluid chaotically flows without distinct streamlines, which allows mixing and irregular fluctuations. The Reynolds number (R_e) characterizes the flow of a fluid through a microchannel:

$$R_e = \frac{pVD_h}{\mu}$$

where ρ is the flow density, v is the fluid average velocity, D_h is the hydraulic diameter, and μ is the fluid viscosity. The low R_e is usually obtained because of the micrometer dimensions of microfluidic channels. If R_e is extremely low on a microfluidic chip, laminar flow can be achieved. Whereas, at high R_e, turbulent flow without distinct streamlines occurs (Fig. 13.2). There are two

FIGURE 13.2

Schematic representation of the laminar and turbulent flow regimes in a microchannel. The Reynolds number (Re) demonstrates the physical characteristics of the fluid flow. If $R_e < 2300$, fluids exhibit laminar flow, while if $R_e > 4000$, fluids move in all three dimensions without correlation with the flow direction and exhibit turbulent flow. The transition region shares the features of laminar and turbulent flow.

common methods to allow fluid actuation through microfluidic channels: pressure-driven flow and electrokinetic flow. In pressure-driven flow, fluids are pumped through the chip via positive displacement pumps. In fluid dynamics, the no-slip condition assumes that the velocity profile of the fluids at the walls must be zero at the solid/liquid boundary. This generates a parabolic velocity profile within the microchannel. Pressure-driven flow is an inexpensive and quite reproducible approach with the possibility of integrating micropumps (Lauga et al., 2007; Saliba et al., 2018; Convery & Gadegaard, 2019; Zhang et al., 2023). The second technique for pumping fluids is electro-osmotic pumping, which is based on the movement of uncharged fluid relative to a stationary charged surface when an electric field is applied across the microchannel. In contrast to pressure-driven flow, diffusion nonuniformities can be avoided by using electrokinetic flow, which generates a blunt velocity profile (Wang et al., 2009).

13.3 Selection of a polymer to fabricate a microfluidic chip

The ability of microfluidics to scale down systems to micro-sized platforms is associated with the availability of polymeric and nonpolymeric materials and the automation of manufacturing via different microfabrication techniques. Formerly, glass and silicon were the first materials used to fabricate microfluidic chips using standard integrated circuit (IC) processes such as photolithography and surface micro-matching. Further progress in microfluidic technology has introduced novel polymeric and plastic materials that complement the original microfluidic devices (Alrifaiy et al., 2012). However, choosing the material depends on the desired application, reaction type, and the requirements of high temperature, pressure, or utilization of strong solvents or corrosive fluids. This section focuses on soft polymers and hard polymers as traditional materials to fabricate microfluidic chips (Fig. 13.3).

13.3.1 Rigid polymers

Rigid polymers are commercially available, and their use to fabricate microfluidic platforms is significantly increasing due to their simplicity of processing. Rigid thermoplastic polymers are synthetic polymers, and most are high-molecular-weight polymers consisting of chains interacting through weak van der Waals forces, stronger dipole–dipole interactions, hydrogen bonding, or stacking of aromatic rings (Haino, 2013). Thermoplastics are elastic and flexible near or above their glass transition temperature (T_g) and, thus, can be readily molded and reshaped via heating without any change in their chemical structure. Various properties make thermoplastics suitable substrates for microfluidic applications, including low manufacturing costs, being lightweight, optical transparency, good mechanical rigidity, organic solvent resistivity, acid/base resistivity, low water absorbance, and disposal advantages. The most commonly used thermoplastics in microfluidic fabrication for biological applications are poly (methyl methacrylate) (PMMA), polycarbonate (PC), polystyrene (PS), cyclo-olefin-copolymer (COC), and cyclo-olefin polymer (COP). Table 13.1 and 15.1 summarizes the thermoplastic properties that make them a good choice to be used in microfluidic applications. Still, other thermoplastics cannot be used to fabricate

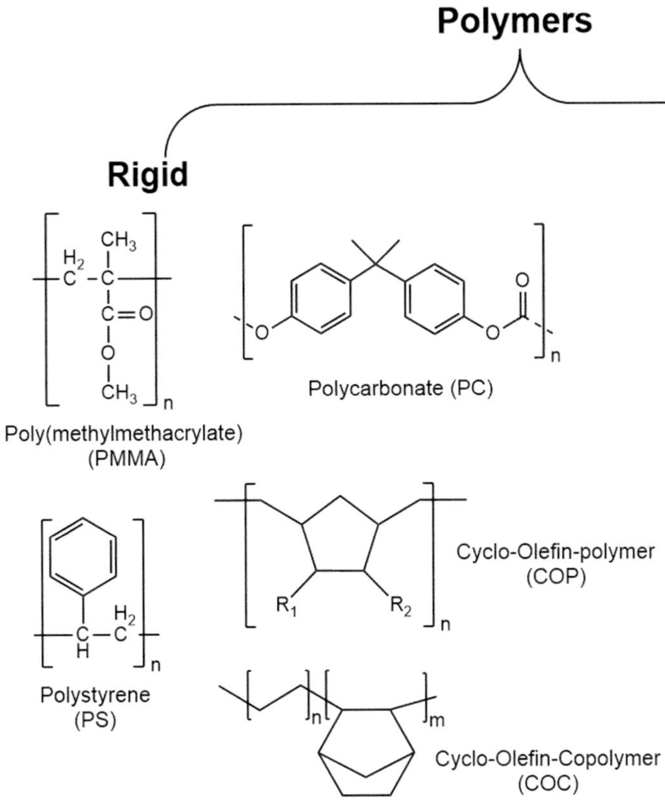

FIGURE 13.3

Molecular structures of several rigid and soft polymers commonly used in microfluidic chip fabrication

microfluidics for biological applications due to lack of biocompatibility, minimal optical transparency, dimensional instability, and difficulties in microfluidic fabrication, such as acrylonitrile butadiene styrene (Tsao, 2016; Gencturk et al., 2017; Shakeri et al., 2022). The creation of open microchannels on polymer substrates is possible based on two approaches: mold-based and non-mold-based. Then, fabricated open microchannels need to be sealed with polymers or other materials to form a closed platform through various bonding methods (Juang & Chiu, 2022). There are different fabrication options for manufacturing thermoplastic microfluidics, and they are mainly classified into two categories: rapid prototyping and replication methods. Rapid prototyping methods such as micromachining, laser machining, and 3D printing are low-cost methods that establish proof-of-concept for researchers at small-scale production levels without micro-mold fabrication. Replication methods such as hot embossing/imprinting, roller imprinting, injection molding, and thermoforming methods are the optimum choices for the mass production of thermoplastic chips (Foret et al., 2013; Tsao, 2016; Juang & Chiu, 2022).

Table 13.1 Summary of physical, chemical, and optical properties of commonly used hard and soft polymers in microfluidics.

Polymers	Hard polymers (thermoplastics)				Soft polymer
	PMMA	PC	PS	COC/COP	PDMS
Mechanical property	Rigid	Rigid	Rigid	Rigid	Elastomer
Thermal property (°C)	140 ~ 150	100 ~ 125	90 ~ 100	70 ~ 155	~80
T_g (°C)	105	145	95	65–170	−150
Solvent resistance	Good	Good	Poor	Excellent	Poor
Acid/base resistance	Good	Good	Good	Good	Poor
Optical transmissivity (visible range)	Excellent	Excellent	Excellent	Excellent	Excellent
Optical transmissivity (UV range)	Good	Poor	Poor	Excellent	Good
Autofluorescence	Low	High	High	Low	Low
Biocompatibility	Good	Good	Good	Good	Good

13.3.1.1 Rapid prototyping methods

13.3.1.1.1 Micromachining

Traditional techniques such as micro-cutting and micro-drilling produce manually cut thermoplastic substrates. The micromachining mechanical process removes unwanted parts and typically produces auxiliary components such as chassis. However, the development of low-cost rapid prototyping techniques such as digital craft cutters and computer numerically controlled (CNC) milling has improved the creation of microchannels on thin thermoplastic films. CNC micromachining is a fast method that produces precise tools that offer the possibility of generating open channels without any need for micro-mold fabrication or cleanroom facilities and typically produces microfluidic chips with channel dimensions larger than 100 μm. However, this simple method has limitations that involve microchannel resolution, surface roughness, and carrying microscratches.

13.3.1.1.2 Laser Micromachining

This is another non-mold-based technique that involves laser cutting, laser milling, laser engraving, laser drilling, and laser etching. It is a popular method to manufacture thermoplastics via the ablation process by exposing a high-energy laser beam to the substrate surface, which absorbs photon energy and thus removes materials due to melting or vaporization. The ablation process can create various components of the microfluidic device, such as channels, chambers, columns, and ports. Pieces cut by using a laser beam can have width channels down to 5 μm. Although the produced microfluidic chip is scratch-free because of its noncontact technology, surface roughness could exist because of residual decomposed polymers.

13.3.1.1.3 This prototyping technique is 3D Printing

This technology has recently become a popular method for manufacturing polymer microfluidic devices due to its advantageous features, such as low cost, easy design of complex 3D structures with multi-materials, and no requirement for a clean room. Indeed, robust connection ports, integration of sensors, and cell culture on microfluidic chips can be achieved. The most established 3D printing processes are fused deposition modeling, stereolithography, selective laser sintering, and multi-jet modeling. Each technology is suitable for different applications and has its own advantages and disadvantages. Choosing between these various 3D printing methods depends on the final desired application. However, the limitations of 3D printing are related to microchannel sizes and some final steps of laborious fabrication (Damiati et al., 2022a; Juang & Chiu, 2022).

13.3.1.2 Replication methods

13.3.1.2.1 Hot embossing

This mold-based technique is an medium-volume replication method that generates good quantities of thermoplastic chips at a rate of 10–30 min/cycle, depending on the heating/cooling conditions (Tsao, 2016). To produce the desired structure, the polymer is heated above its T_g to make it pliable and then pressed via a glass or metal stamp. The optimization of the heating temperature, pressure, embossing time, and demolding temperature is important to avoid the deformation of embossed structures (Foret et al., 2013).

13.3.1.2.2 Injection molding

It is an ideal choice for high-volume production in commercial manufacturing as the replication cycle can be completed within seconds and large numbers of devices can be produced per day. It is a mold-based method that relies on feeding melted polymers into a cavity mold, allowing the liquid polymer to set and solidify, and subsequently separating and collecting new platforms from the mold (Li et al., 2020; Juang & Chiu, 2022).

13.3.2 Soft polymers

The fabrication of microfluidic chips using soft polymers is relatively straightforward. The creation of self-standing structures or structural patterns such as molds, photomasks, and stamps can be employed via simple soft lithography techniques. Polydimethylsiloxane (PDMS) is a well-known member of the soft polymer family. PDMS is widely used in the mass production of microfluidics due to the ease of fabrication of PDMS devices. Furthermore, PDMS is inert, flexible, nontoxic, optically transparent, and biocompatible. On the contrary, PDMS still possesses several drawbacks for microfluidic applications, such as hydrophobic molecule absorption, solvent swelling, and a lack of mechanical rigidity (Tsao, 2016; Gencturk et al., 2017; Giri & Tsao, 2022). However, versatile front-end fabrication capabilities make thermoplastic polymers a more favorable choice than PDMS for commercial production.

13.3.2.1 Soft lithography

This is probably the most common rapid prototyping technique for microfluidic fabrication that processes soft materials such as polymers. Thus, the term "soft" is used, which is related to a wide range of elastomeric materials. However, PDMS is reported as the most widely used mechanically soft material for soft lithography applications. Briefly, the first step in the fabrication of a mask via soft lithography is designing a mold that has the desired structural and geometrical features with graphics software; then, the design is transferred to a photomask such as chrome-coated glass. The produced micropatterned photoresist film can then be used to fabricate a microfluidic chip by filling the master mold with PDMS. After degassing and cooling to room temperature, the PDMS mold can be peeled off the mask. Subsequently, inlets and outlets for tubing connections can be made. Further, to close the microfluidic chip and form a complete channel, the PDMS mold bonds to another PDMS piece, glass slide, silicon wafer, or other substrate material (Li & Wang, 2012; Rein et al., 2023).

13.4 The role of polymers in microfluidics

Many techniques benefit from the precise, small-volume fluidic control offered by microfluidic chips, including drug carrier synthesis, DNA amplification, cell—cell interaction, genetic analysis, clinical chemistry, cell-based assays, single-cell analysis, diagnostics, and capillary electrophoresis. All these techniques are attributed to the full control of a miniaturized, experimental environment through the concepts of fluid behavior, laminar flow, and fluid mixing (Lee, 2020). The next sections highlight applications of polymers in microfluidics, either as chassis, to generate polymeric particles that act as drug delivery carriers or artificial cells, or to fabricate sensors.

13.4.1 Polymeric drug delivery systems

Traditional drug delivery systems have several drawbacks, such as poor bioavailability and being unable to achieve sustained release. On the contrary, novel drug delivery systems control the release of the therapeutic agents at a specified rate to achieve the desired response and maximum efficacy. There are different polymeric-based drug delivery systems, such as nanoparticles and polymersomes (Fig. 13.4). The artificial polymeric vesicles can encapsulate hydrophilic drugs in the aqueous core and hydrophobic drugs within the hydrophobic membrane. Hence, microfluidic

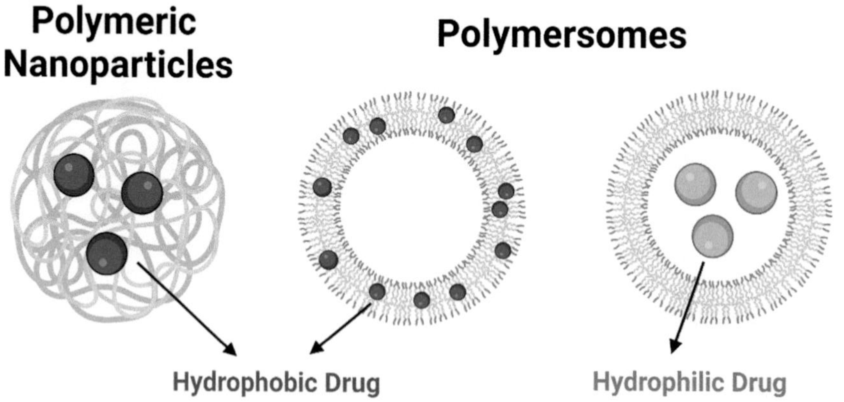

FIGURE 13.4

Schematic illustration of polymeric-based drug delivery systems (created with Biorender.com).

technology enables the production of highly controlled drug delivery systems ranging from the nanoscale to the macroscale (Adepu & Ramakrishna, 2021; Damiati et al., 2018a; Karnik et al., 2008). Microfluidics can scale down the fluidic environment, control drug delivery system size and shape, and control physical properties that, in turn, control drug release time and particle circulation lifetime and improve target specificity. Indeed, encapsulating drugs into lipid or polymer particles reduces the toxic effect of drugs, enhances the therapeutic effect of drugs, and improves the absorption of poorly water-soluble drugs (Ahn et al., 2018). However, microfluidic chips offer various methods to fabricate drug delivery systems, such as droplet-based, hydrodynamic flow focusing (HFF), and chaotic flow approaches.

Droplet-based microfluidics is a popular carrier production method that offers unique opportunities for precise control of nano- and micro-scale particle synthesis ranging from $\sim$1 nm to $\sim$500 μm that can be used in drug encapsulation. Droplets can be generated by using active or passive methods. Active microfluidic chips use external energy sources for mixing and separation, while passive microfluidic chips generate droplets without external actuation. Droplet microfluidic techniques generate discrete droplets in an immiscible phase by combining shear stress with the interfacial tension between the continuous aqueous stream and the immiscible carrier phase just before droplet generation in a microchannel. Thus, the ability of microfluidic chips to produce droplets is exploited to produce particles (Bayareh et al., 2020; Damiati et al., 2018a).

Self-assembling drug delivery systems can be easily generated in microfluidic systems via mixing and diffusion-controlled reactions (Fig. 13.5). HFF is generated by introducing two or multiple streams of fluids with different velocities side by side. The simplest device design for flow focusing contains two or three inlets, where the main channel contains the polymer solution with or without the drug and is sheathed by side fluids. Further, the chip design allows the rapid mixing of reagents to provide a homogeneous reaction environment. To enhance mixing in microfluidics, flow disruptive patterns, for example, a staggered herringbone mixer (SHM), can passively mix fluids within a microchannel. This chaotic flow method relies on designing an array of herringbone grooves on one or more surfaces of a channel to induce turbulent mixing within a continuous flow. Compared with the HFF approach at equivalent flow rate ratios, the SHM platform increases the surface area between two flowing fluids, which improves diffusional mixing performance (Ahn et al., 2018; Damiati et al., 2018b; Stroock et al., 2002).

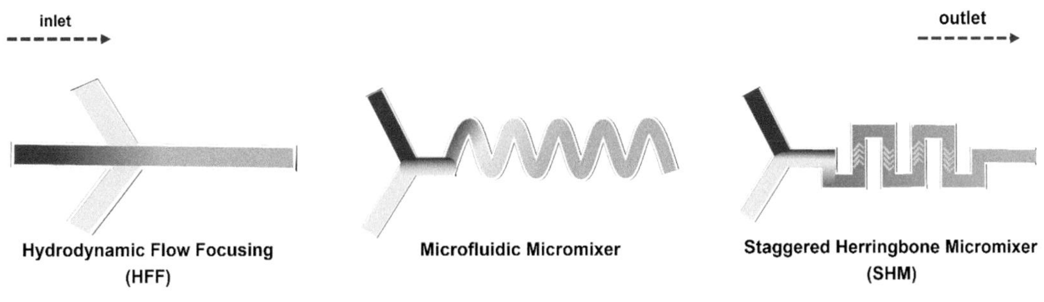

FIGURE 13.5

Schematic designs of microfluidic mixers.

The physicochemical properties of drug delivery systems, including their size, shape, and composition, significantly affect the drug release profile. For example, nanoparticles (1–100 nm) can easily cross the blood-brain barrier, while microparticles can easily be recognized by the immune system (Champion et al., 2007; Sharma et al., 2010; Saliba et al., 2018). Size uniformity is a key factor that influences the stability of polymeric particles; therefore, microfluidics is used to produce optimized drug carriers with high monodispersity and precisely tunable structures. Recently, combining modern technologies is attracting more attention to improve production processes. For example, artificial intelligence and microfluidics can effectively fabricate polymeric particles. PLGA particles are produced by different glass microfluidic systems, either in the form of single or multiple microparticles with or without indomethacin, a nonsteroidal antiinflammatory drug. Polymer concentrations and the flow rates of dispersed and aqueous phases in microfluidics were experimentally and computationally assessed for the effect of droplet/particle sizes. Microfluidic chips allowed the production of PLGA microparticles with high monodispersity and precisely tunable structures, while in silico models using an artificial neural network allowed accurate size prediction of the PLGA droplet/particle. Such a strategy allows rapid production of size-tunable polymeric particles at a desired size, which can be further used to load water-insoluble, toxic drugs (Damiati et al., 2020a; Damiati & Damiati, 2021).

Reproducibility is an issue in bulk particle generation methods as it does not provide control of mixing, but this limitation can be overcome by using an SHM microfluidic device, as shown by Chiesa et al. (2019). The synthesis of curcumin-loaded nanoparticles is performed by using the natural polymers chitosan and sodium tripolyphosphate as cross-linker molecules. The developed system exhibited a reproducible strategy to induce electrostatic interactions between the polymer and its cross-linker due to the precise control of mixing. Further, particles showed high encapsulation efficacy for curcumin and slow release via the diffusion mechanism. Karnik et al. (2008) fabricated a PDMS microfluidic device using a standard micro-molding process to synthesize poly(lactide-co-glycolide)-b-poly(ethylene glycol)(PLGA-PEG) nanoparticles via HFF. PLGA-PEG is a block copolymer composed of PLGA, which provides a biodegradable and biocompatible matrix for the loaded and controlled release of drugs, while PEG provides a long circulation half-life and stealth properties for immune evasion. The study proved that microfluidics can generate more homogeneous polymeric nanoparticles compared with bulk synthesis.

13.4.2 Polymeric artificial cells

The field of synthetic biology is significantly growing and interacts with various disciplines such as chemistry, pharmacology, engineering, and computer science. It builds bridges between biological and artificial systems, or between several biological communities, to establish mechanical systems. Biomimetics copies and imitates natural systems in a simpler form to overcome the limitations of complex biological systems. The engineering of custom cells/organs requires an understanding of the native structure and function of biological cells. Collecting biological materials such as DNA, RNA, proteins, and lipids is the starting point in biomimicry, followed by combining materials with nonbiological components, such as combining metal with a functional seminatural model. These new artificial systems can perform a well-known function in a modified manner or a new function that does not exist in nature. Indeed, artificial models facilitate the discovery of new drugs by mimicking human physiology and diseases (Damiati, 2018d; Damiati et al., 2018b, 2022a).

The cell is the basic unit of life because it is a highly advanced microreactor that performs fundamental functions to carry out vital processes. Constructing human-made cells independently or as part of tissue can be carried out by exploiting the bottom-up approach in synthetic biology. Artificial cells have micrometer-sized structures and mimic the morphological and functional characteristics of biological cells. The interior of a natural cell is compartmentalized by a membrane that separates the interior from the surrounding environment and acts as a barrier between these two distinct areas. Most studies on models of cell-like vesicles are based on the self-assembly of polymers or phospholipids to generate polymersomes and liposomes, respectively. Both have cell-like bilayer membrane structures, which constitute a real cell (Damiati, 2019).

It is known that the choice of raw materials mainly depends on the application of artificial cells. Hence, polymers with excellent chemical versatility are commonly used to construct artificial cells based on polymersomes (Fig. 13.6). Furthermore, polymersomes exhibit higher stability and encapsulation efficacies than liposomes. The construction of cell-like vesicles composed of amphiphilic block copolymers depends on the relative length of the hydrophobic and hydrophilic chains. Furthermore, controlling the chemical composition of the chosen polymer significantly improves membrane thickness, permeability, elasticity, and mechanical stability. The continuous progress in microfluidics allows high-throughput polymersome fabrication with desired components, tunable

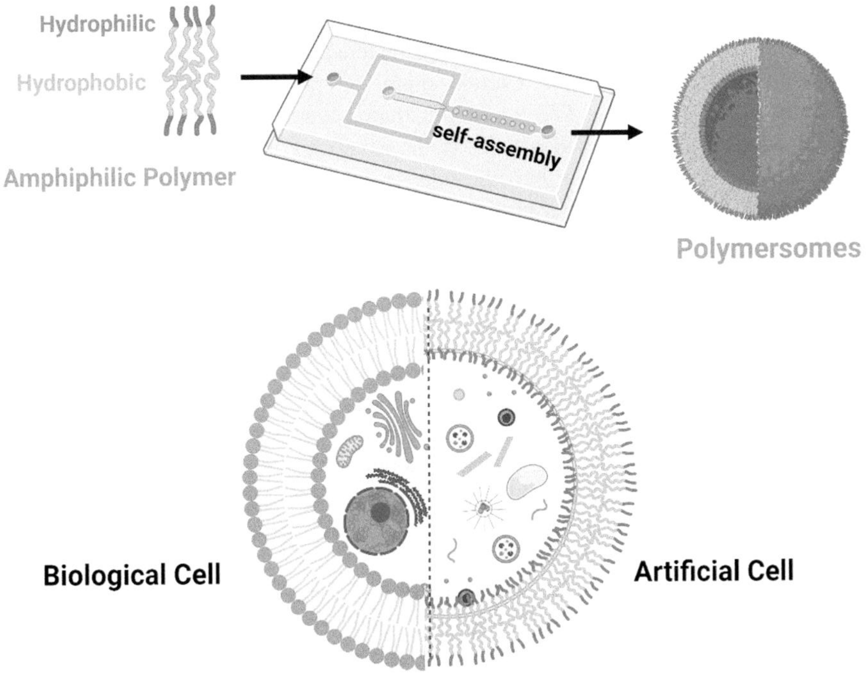

FIGURE 13.6

Schematic illustration of self-assembly of polymers into polymersomes, which is an artificial cell model that mimics the natural cell in a simple form (created with Biorender.com).

structures, and high uniformity (Chiara & deMello, 2016; Buddingh' & van Hest, 2017; Ai et al., 2020).

Microfluidic platforms that are usually applied in synthetic biology are classified into the two categories previously described: droplet-based microfluidics and channel-based microfluidics (Ai et al., 2020; Damiati et al., 2018a). Droplet-based microfluidics enables the generation of millions of miniaturized compartments, such as droplets and giant vesicles, in a highly controlled manner that can be used as artificial cells to mimic the membrane structure and conduct biological assays. Compared with traditional bulk methods, microfluidics constructs high-throughput, size-controllable monodisperse compartments with consistent physical and chemical properties and is appropriate for artificial cell fabrication (Li et al., 2018; Tan et al., 2023). In a study conducted by Weitz and coworkers to bring droplets to life, a cytoskeletal protein was expressed in biocompatible polymersomes that mimic artificial cells. The developed strategy showed the ability of microfluidics to fabricate highly monodisperse polymersomes with high encapsulation efficiency for active biomaterials that are required for protein expression. Polymersomes composed of the diblock copolymer poly(ethylene glycol)-block-poly(lactic acid)(PEG-b-PLA) and PLA homopolymer were prepared. The PEG-b-PLA serves as the amphiphile, while PLA serves as an extra coating layer to enhance stability. A glass capillary microfluidic device was used to generate water−oil−water (W/O/W) double-emulsion drops that act as a template to produce the monodisperse polymersomes. To express a membrane-related bacterial protein, MreB, an aqueous mixture of the *Escherichia coli* ribosomal extract containing all molecules that are required for transcription-translation machinery, was mixed with the MreB DNA plasmid and then injected into the microfluidic channel to form the innermost drop of the polymersome. Protein expression was evaluated by monitoring the fluorescence intensity signal over time. The maximum fluorescent MreB production was reached after 2 hours, and the MreB proteins formed patches in the aqueous core of the polymersomes with some protein attached to the membrane. Further, the release of the MreB proteins into the exterior environment was triggered by a negative osmotic shock that formed pores in the membrane (Martino et al., 2012).

Eukaryotic cells typically consist of multiple inner compartments attributed to the multiple organelles in the cellular compartment, which enable multiple processes to be accomplished. Encapsulation of multiple active molecules in single capsules with multiple compartments could reduce the risk of cross-contamination. Hence, many attempts were made to construct artificial cells with a cell-like compartmentalized architecture. Microfluidic devices showed the ability to generate multiple-compartment polymersomes and polymersome-in-polymersome structures (Shum et al., 2011). Raghavan and coworkers fabricated an approach that allows the construction of well-controlled multicompartment capsules (MCCs) with different quantities of inner compartments using a water−gas microfluidic setup. All capsules were constructed via electrostatic complexation between two natural biopolymers: the anionic alginate and the cationic chitosan. Various payloads were encapsulated in each inner compartment, including enzymes, colloidal particles, and microbial cells, while preserving native functions. A successful bacterial cascade process was also achieved where a change occurred in one MCC compartment and was transduced into a response in an adjacent compartment (Lu et al., 2017). It is more common to generate spherical particles via microfluidic devices, but nonspherical particles can also be generated, which is favorable since biological cells are not limited to a perfectly spherical shape and have various shapes and morphologies. With the advantages of using microfluidic technology to tune microchannel geometry and adjust flow rates for different phases, nonspherical polymeric particles can be generated with controlled morphology and size. For example, Shum et al. used glass capillary microfluidics to fabricate nonspherical polymersomes with multiple compartments to allow the encapsulation of multiple actives in separate compartments. A stable core-shell structure originated from a W/O/W double-emulsion system containing amphiphilic diblock copolymers and PEG–PLA stabilized in the oil shells. The PEG–PLA at the W/O and the O/W interfaces are attracted toward each other to form the membranes. Subsequently, neighboring inner droplets adhere to one another, and this results in the formation of multicompartment polymersomes. Also, they showed the ability of the shell to respond to osmotic shock by monitoring the release of an encapsulated fluorescent dye (Shum et al., 2008). One study reported on the importance of microfluidic design for the generation of stable polymeric architectures. The creation of alginate microgels was performed using four different types of glass microfluidic chips with flow-focusing or coflowing droplet generators. The generated alginate beads exhibited various architectures, including individual monodisperse or polydisperse beads, small clusters, and multicompartment systems. The effects of several parameters on microgel morphologies and dispersity were investigated at fixed polymer concentrations, including microfluidic design, flow rates, surfactant concentration, and the use of internal or external gelation. Hence, these fabricated mimetic models in the form of polymeric-based vesicles can be further used in several applications, including drug encapsulation, cell-like structures, and tissue engineering (Damiati, 2020b).

The second category, channel-based microfluidics, allows real-time observation of slight changes in cell behavior. The incorporation of microwells, hydrodynamic traps, and various channel configurations into microfluidic channel-based devices allows the creation of cellular arrays in which soluble stimuli can be controlled precisely. Coupling microfluidics with various techniques can provide more information about the multicomponent signals regulating cellular functions. For example, coupling microfluidic platforms with time-lapse microscopy assists in the investigation of dynamic properties, such as proliferation kinetics and gene expression. Further, coupling PCR into microfluidic systems improves integrated genomic analysis (Underhill et al., 2012). Microfluidics minimizes the gap between in vivo and in vitro models and may replace in vivo animal testing in the future. The combination of basic cell-based techniques and microfluidic 3D cell culture allows multiple steps involving the cultivation and investigation of viable cells on a single platform to be carried out. Microfluidic devices enable high control of in vitro culture environments of cells and tissues by regulating nutrient exchange and oxygen diffusion. There are three categories of microfluidic-based 3D cell cultures based on the substrates used for fabrication: glass/silicon-based, polymer-based, and paper-based platforms. As highlighted in the previous section, PDMS and PMMA have various applications, and hence they are also exploited in cell and tissue cultures. In microfluidic devices, it is common to seed cells within a hydrogel network that acts as a scaffold to permit oxygen diffusion and nutrient transportation (Li et al., 2012). Natural polymers such as collagen, fibrinogen, and elastin are used as native extracellular matrix (ECM) and as gel-supported 3D cell cultures in microfluidic platforms. For instance, van Duinen et al. (2017) developed a robust, long-term platform to culture endothelial cells. To mimic the microenvironment of vasculature in vivo, the microvessels were cultured against a 3D collagen matrix for the cells to adhere to. To assess the barrier function, the diffusion of fluorescent dextrans across the vessel wall was monitored, and real-time quantification of the permeability exhibited successful mimicking of in vivo vasculature functionality. However, the main drawback of hydrogel-supported 3D cell cultures is that the gel matrix may prevent the cells from culturing at enough density, which contributes to the formation of interstitial spaces. Chen et al. (2016) reported on the creation of an artificial liver in a drop via a controlled assembly of hepatic cells in biocompatible 3D core-shell hydrogel scaffolds. Droplet-based microfluidics were used to fabricate many individual monodisperse mimicking microtissues and then reconstitute the material to produce an organ-in-a-droplet artificial liver. In the constructed water−water−oil (W/W/O) double emulsions, the inner phase is a cell culture medium, while the middle phase is an alginate solution. Alginate was used due to its excellent features in that it does not interfere with cell survival or function. CO-culturing of hepatocytes and fibroblasts in the droplet core and shell, respectively, allowed homotypic and heterotypic cell−cell interactions that cannot be achieved in a 2D culture system. The functionality of the developed system was assessed by measuring albumin secretion and urea synthesis over time, which are two key biomarkers of the liver. The thousands of microtissues expressed high-level formation, and each maintained enhanced liver-specific functions.

13.4.3 Polymers with sensory properties or as a matrix in sensors

Microfluidic sensors are a promising technology that can be exploited in the health sector and environmental applications due to their characteristics of rapid, inexpensive, small size, and sensitive detection with high yield. Furthermore, microfluidic sensors offer a special advantage in that they can manipulate fluid samples without the need for pretreatment steps because the sample is processed on the chip and direct operation is performed. In addition to using polymers as chassis for microfluidic devices, they can also be used as sensing materials. Polymers are prominent in sensing devices because they can be tuned through appropriate synthetic or modification techniques. Hence, polymer-based sensors are good candidates for microfluidic applications due to their electrical properties, which are related to dielectric and electrical conductivity with a variety of structural features and fabrication processes (Waters et al., 1997; Radhakrishnan et al., 2022; Mitrogiannopoulou et al., 2023).

Microfluidic sensors can identify biological elements, such as cells, microorganisms, proteins, and enzymes, and chemical substances within a gas, liquid, or solid medium. The efficiency of sensor devices is evaluated via selectivity, which is related to the ability to capture a specific analyte in a mixture containing various target molecules, and sensitivity, which is the ability to detect the limit or the minimal quantity of the analyte (Fig. 13.7) (Ruiz et al., 2018). In response to biological or chemical compounds or to biochemical interactions, quantified signals can be measured using acoustic, optical, fluorescent, electrochemical, or magnetic methods. The efficiency of the sensing platform relies on the careful assembly of several components to build an efficient interface matrix. In this sense, smart polymers offer promising materials that can be applied to sensing devices. Sensory or smart polymers can respond to different stimuli and rearrange themselves in diverse conformations. The terms "smart" and "sensory" are used since these polymers themselves can detect and quantify the target substance and exhibit a specific response. Sensory polymers are classified into three classes based on stimulus (chemical or physical), mechanism (reversibly or irreversibly), and response (visual or electrochemical). Different sources can cause polymer responses, including temperature, pH, humidity, light, magnetic and electric fields, and biological and chemical molecules. After stimulation, polymers change their physicochemical properties, such as color,

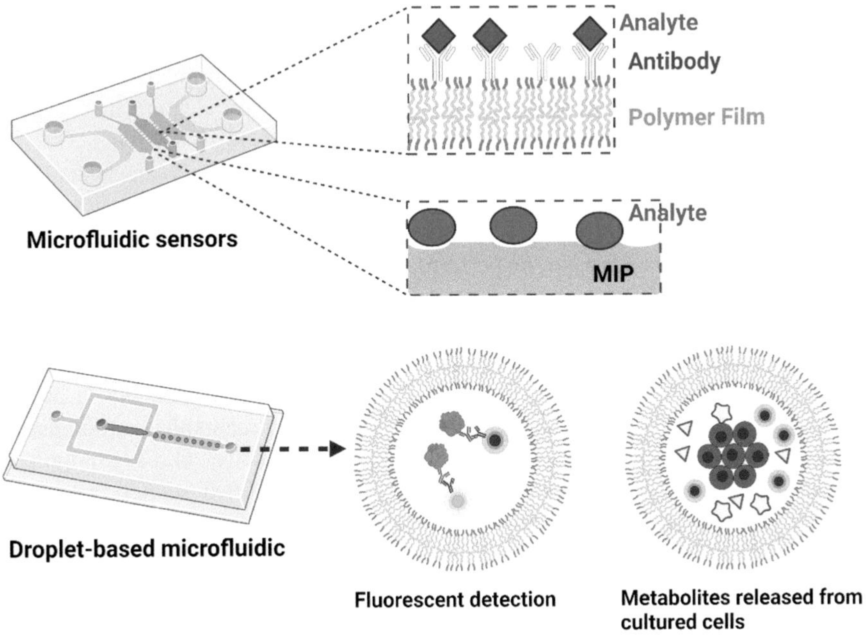

FIGURE 13.7

Schematic diagram of a polymer-based sensor and droplet-based microfluidic to detect different analytes either captured onto polymer film or trapped in polymersomes (created with Biorender.com).

fluorescence, solubility, shape, electrical conductivity, or luminescence (Basak & Bandyopadhyay, 2021; García et al., 2022; Damiati, 2022).

There is a large variety of polymers that can be used as sensory materials, such as acrylic polymers, molecularly imprinted polymers (MIPs), conductive polymers, polymeric nanocomposites, and hybrid polymers. Acrylic polymers are classic, versatile polymers prepared with esters of acrylic or methacrylic acid and acrylamide derivatives. Acrylic polymers are widely used as sensory materials because they offer an important advantage in that different sensory units can be anchored chemically to their acrylic motif structures (Ruiz et al., 2018).

MIPs are synthetic polymers with high selectivity for a target analyte and are ideally used in separation processes. MIPs can mimic biorecognition elements such as antibodies and receptors. MIPs are constructed by combining the polymer with the target species, which acts as a template. After the polymerization process and removal of the target species, hollow spaces are created and act as receptor sites for the target analyte (Scriba, 2016; Moulahoum et al., 2021).

Conductive polymers are environmentally friendly organic materials with unique electrical and optical characteristics. These polymers are readily assembled into supramolecular structures by alternating simple and multiple bonds and synthesized by using simple electro-polymerization processes. Examples of conducting polymers are fluorene derivatives, polyaniline, and polypyrrole (Nezakati et al., 2018).

Polymeric nanocomposites and hybrid polymers are two-phase structures formed by organic/inorganic phases in different ways. Polymeric nanocomposites are prepared by mixing nanoscale inorganic moieties with charged polymers, while hybrid polymers are formed of two or more different types of molecules, but at least one is a polymer, and the molecules are linked covalently to produce a single-phase structure (Kickelbick, 2003; Dai et al., 2017).

Exploiting smart polymers in microfluidic sensors improves the early diagnosis of many diseases. The development of polymer-based microfluidic devices for sensing applications facilitates the processing, analysis, and detection of an analyte in small sample volumes ranging from μL to pL. Moreover, several process steps can be performed on a single platform; for example, serial coupling of cell separation methods can be conducted before cell detection or investigation. The reduction in the cost of the analysis is attributed to the reduction in material consumption and step integration. Recently, digital and droplet microfluidics have also been used as routine methods for single-cell sequencing. In these systems, the sample solution is segmented into a droplet, which results in the generation of thousands of discrete reaction compartments that can be used for high-throughput screenings in drug testing or single-cell analysis (Ruiz et al., 2018; Berlanda et al., 2021; García et al., 2022). The next section highlights some reported studies that used different polymers as a matrix layer or chassis for microfluidic devices in various applications.

The detection of circulating tumor cells (CTCs) plays a vital role in early-stage cancer diagnosis, monitoring of cancer development, and individualized cancer therapy. A 3D flow cell composed of acrylate-based photopolymer exploited commercial screen-printed electrodes for the rapid detection of hepatic oval cells. Multiwall carbon nanotube electrodes with a chitosan film were functionalized to act as a scaffold for the immobilization of oval cell marker antibodies. The continuous-flow platform allowed samples to be exposed continuously to the sensor surface, which increased the capability of capturing the cells. Indeed, it has been shown that the chitosan insulation layer improves sensor stability and peak shape. The electrochemical measurements confirmed the efficiency and selectivity of the assembled sensing platform (Damiati et al., 2018c). Another study to detect cancer cells by quantifying the fluorescence signal was reported by Wang et al. (2018). They fabricated a hyaluronic acid-functionalized chitosan nanofiber-integrated microfluidic platform to capture CTCs from lung cancer. The counted cells under the fluorescence microscope showed the high efficiency of the developed system for isolation and detection of CTC.

Son et al. (2017) developed a PDMS microfluidic device and fluorescent microbead biosensors for in situ detection of growth factors secreted by primary hepatocytes. The design of the microfluidic device aimed to separate cell culture chambers from sensing channels using a permeable hydrogel barrier. Streptavidin-coated PS particles without and with fluorescent dye were used as capture microbeads and detection microbeads, respectively, and functionalized with specific antibodies. After seeding hepatocytes in the cell culture chamber, cell-secreted factors diffused across the hydrogel barrier to the sensing chamber. Subsequently, the growth factors were bound to capture or detect microbeads, which resulted in bead aggregation and generated fluorescence signals. Such an application shows the possibility of using several polymers for different on-chip purposes.

Nagabooshanam et al. (2020) developed a microfluidic affinity sensor to selectively detect chlorpyrifos, an organophosphate pesticide. The MIP sensor was electrochemically synthesized using the conductive polymer PPy (polypyrrole) as a polymeric matrix and chlorpyrifos as a template molecule on a gold microelectrode. The electrochemical sensor showed high selectivity and stability while using only a tiny amount of the sample (2 μL). Such an approach may improve the fabrication of sensitive sensors that are applicable in the detection of pesticides in fruit and vegetable samples, which are in high demand due to serious issues related to public health.

Mei et al. (2023) reported on building a nanofiber-based microfluidic sensor for in situ and real-time sweat analysis using MIP with built-in redox probes for specific detection of cortisol. Polyethylene terephthalate (PET), which is a thermoplastic polymer, was used as the substrate and consisted of two layers: a MIP-modified electrode for sensing and a nanofiber-based microfluidic for spontaneous sweat pumping, which act as the bottom and upper layers, respectively. To accurately capture cortisol molecules, the sensor was provided with specific imprinted cavities by decorating the MIP membrane with Prussian blue nanoparticles on carbon electrodes via electropolymerization. Indeed, the sensor was integrated with a flexible circuit board to monitor cortisol electrochemically. The developed approach allowed selective, sensitive, rapid, and stable detection of cortisol in a wide detection range.

To save patients' lives in cases of emergency bleeding and in patients who suffer from a high risk of thrombosis, a point-of-care microfluidic channel-based device was developed by Ban and colleagues (2022) that allowed on-site detection of fibrinogen concentration. In this approach, a reaction strip, wicking strip, and protection flake are embedded in the microfluidic device. The reaction is started by converting fibrinogen to fibrin on the reaction strip, resulting in the formation of blood clots. Then, the plastic protection flake that separates the reaction strip and the wicking strip is removed to allow unclotted blood flow. The fibrinogen concentration was inversely related to the rate of blood flow along the wicking strip. The developed platform allows direct measurement of the fibrinogen level in whole blood in 5 minutes without disruption by using other molecules or factors such as Ca, lipids, or tissue plasminogen activators.

13.4.4 Polymers in cell culture

Cell cultures are fundamental to cell biology, biochemistry, pharmacokinetics, drug discovery and development, and tissue engineering. The classic cell culture method is 2D cell culture, which involves culturing cells under uniform conditions, including nutrient supply, gas exchange, and waste removal. Indeed, it allows for easy monitoring and the collection of cells for further use. However, the 2D cell culture model is quite different from the biological system, and owing to its limitations, such as not mimicking the natural structures of tissues or tumors, noticeable differences arise compared to physiological conditions in cell morphology, behavior, differentiation, division method, polarity, cell-to-cell interaction, and interaction between cellular and extracellular environments. These limitations have prompted the creation and development of 3D cell culture, which is closer to mimicking the in vivo environment than classic 2D cell culture (Fig. 13.8). Optimization of cell culture conditions can improve our understanding of cancer biology and enhance investigations of biomarkers and targeted therapies. One of the most typical models for 3D culture is the 3D cell spheroid, a self-assembled spherical cell aggregate that establishes cell–cell and cell–substrate interactions in vitro to mimic the complex intracellular microenvironment. Furthermore, 3D spheroids exhibit histological properties that maintain higher levels of tissue-specific function for long periods of time in vitro (Kapałczyńska et al., 2018; Shao et al., 2020). The combination of polymers and microfluidics has led to pioneering advances in cell culture, offering high control and precision in studying cellular behavior. This promising duo provides a versatile platform with various

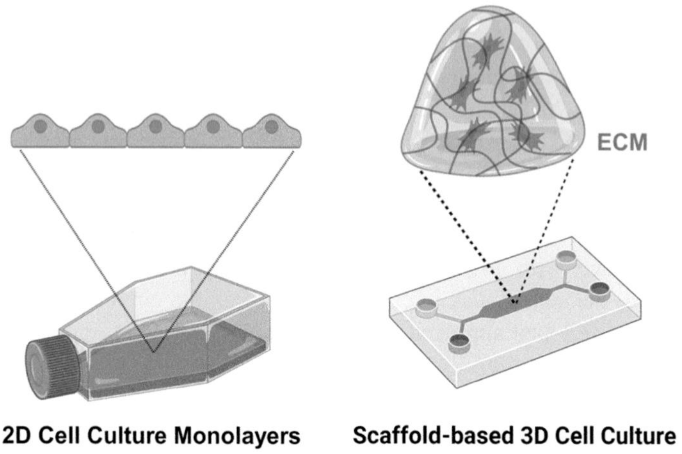

FIGURE 13.8

Schematic illustration of 2D vs. 3D cell culture models (created with BioRender.com).

advantages, enhancing traditional aspects of cell culture and offering more promising applications across various domains of biomedical research.

In the drug industry, microfluidics and polymers enable the acceleration of discovery processes by providing high-throughput platforms for screening potential drug candidates and facilitating the parallelized testing of multiple compounds. Polymer-based microfluidic devices enable the precise manipulation of fluid flow and allow the creation of microenvironments conducive to investigating cell responses to various drug concentrations. This high level of control contributes to identifying and assessing the efficacy and toxicity of drug candidates more efficiently than traditional methods. Furthermore, the encapsulation of drugs into polymeric particles plays a crucial role in the development of drug delivery systems. The controlled release of drugs into the 3D cell culture microfluidic system enables researchers to monitor the real-time responses of cells to pharmaceutical compounds, providing insights into drug kinetics and cellular interactions.

Many microfluidic chips have been fabricated and used to prepare spheroids for drug testing. Zhao et al. (2017) tested the anticancer drug doxorubicin on HepG2 spheroids using a PDMS microfluidic sidewall-attached droplet array. In a specific drug concentration range, the cell survival rate of spheroids was significantly higher than that of 2D cell culture, and the obtained results showed the possibility of applying the developed system to massive and high-throughput drug screening. Chen et al. prepared HepG2 spheroids using a PDMS microfluidic device and then generated a platform used to conduct high-throughput antitumor drug screening (Chen, Wu, et al., 2016). They found that the effect of 5-fluorouracil in 2D cell culture and 3D tumor spheroids was very different under the same concentrations and times of drug treatment. This difference was attributed to the difficulty of drug molecules penetrating the center of the 3D tumor spheroids. Another PDMS microfluidic spheroid culture device with a concentration gradient generator was developed by Lim & Park (2018). The model enabled colon cells to form spheroids and grow in the presence of cancer drug gradients. The cell viability decreased continuously with the increase in irinotecan concentration in each channel. The efficacy of gradient drug concentrations was sufficient to present the developed model as a potential high-throughput screening platform for screening the efficacy of cancer drugs. Another important study done by Mulholland et al. (2018) allowed extensive anticancer drug screening before treatment with glioma spheroids in a PDMS chip. This model highlights automated drug screening based on microfluidic spheroids, saving time and money. The system could create repeatable drug concentration gradients across a set of spheroids without requiring an external fluid drive. Additionally, it could deliver a series of drug concentrations to multiple-sized spheroids and replicate each concentration.

13.4.5 Polymers for tissue engineering

Tissue engineering is an interdisciplinary field that combines biological, engineering, and materials sciences to create functional tissues and organs for transplantation or to improve the body's regenerative processes. In recent years, the powerful combination of polymers and microfluidics has led to the development of sophisticated platforms and techniques to advance the capabilities of tissue engineering. This combination offers high control over the cellular microenvironment, enhances tissue regeneration, and opens new opportunities for personalized medicine.

Polymeric materials serve as the backbone in many tissue engineering applications due to their tunable properties, biocompatibility, and ability to mimic the ECM. The ECM is the natural scaffold, a connective network that provides structural and biochemical support to surrounding cells. Researchers can create artificial scaffolds that guide cell behavior, proliferation, and differentiation by exploiting polymers that closely resemble the ECM. For example, it is common to use hydrogels in tissue engineering due to their high water content and flexibility. Furthermore, the mechanical properties of hydrogels can be adjusted to match those of specific tissues, making them an ideal choice for various applications.

When microfluidic devices and polymers are integrated into tissue engineering, many unique advantages can be offered. A 3D environment that resembles the natural tissue matrix can be created, promoting cell adhesion and allowing the transportation of nutrients and waste (Damiati et al., 2022a; Li et al., 2012). Microfluidics involves the precise manipulation of small volumes of fluids in microscale channels, enabling precise control over the cellular microenvironment and the generation of microvascular networks within polymeric scaffolds. All these features promote the formation of functional blood vessels and provide dynamic regulation of biochemical and physical reactions. This level of control is crucial for mimicking the complexity of native tissues and addresses a major challenge in large-scale tissue engineering (Hesh et al., 2020).

Microfluidic chips can be designed to reassemble the complex vascular networks found in tissues. This is important to ensure an adequate supply of nutrients and oxygen to cells within engineered tissues and to optimize temperature and pH (Sakolish et al., 2016). By mimicking physiological conditions, scientists can improve the functionality and viability of 3D tissue-on-a-chip models, making them more clinically relevant.

Microfluidics has been applied to construct various biomimetic tissue models, such as skin, heart, lung, liver, kidney, and bone. In bone tissue engineering, biodegradable polymers are extensively used in combination with microfluidic systems to provide precise delivery of growth factors and nutrients to support osteogenesis. Furthermore, it allows the creation of scaffolds with controlled porosity and mechanical properties, promoting the formation of bone tissue with enhanced strength and functionality. Most of the developed models enabled the investigation of drug uptake, metabolism, and removal.

Osteogenesis-on-a-chip has been developed by Bahmaee and colleagues (2020). The two important features of bone, the 3D environment, and fluid shear stresses were included in the developed model. They used polymerized high-internal-phase emulsion materials as polymeric scaffolds to construct the model, which supports proliferation, differentiation, and ECM production of human embryonic stem cell-derived mesenchymal progenitor cells over long periods of up to 3 weeks. Indeed, cells exhibited a positive response to both chemical and mechanical stimulation of osteogenesis. An intermittent flow profile, which includes rest periods, significantly enhances differentiation and matrix formation when compared to static and continuous-flow conditions.

An attempt has been made by Esch et al. (2012) to create microporous, polymeric membranes that match the unique 3D shapes of the gut epithelium. The integrated membrane with the microfluidic chip can be either flat or contain controllable 3D shapes that, when populated with colon cells, mimic key aspects of the intestinal epithelium. The proposed technique uses a partially polymerized photoresist (SU-8) and offers an opportunity to mimic a variety of different barrier tissues.

Several microsystems have reported culturing liver tissues for functionality and hepatotoxicity tests due to the importance of the liver. Schütte et al. (2011) created an organ-like liver 3D coculture of hepatocytes and an endothelial cell-based microfluidic system with integrated electrodes for the assembly of liver sinusoids by dielectrophoresis. This chip allowed liver toxicity testing and liver sinusoid assembly.

Since microfluidics can address the inherent complexity of cellular systems, these devices present promising platforms to investigate the intricacies of cell-to-cell variability, drug-cell interactions, and their relevance to cancer therapy. Montanez-Sauri et al. (2011) designed a PDMS microfluidic 3D compartmentalized model to monitor a critical step in breast cancer progression. In the developed model, they cocultured human mammary fibroblasts with epithelial cells to promote a transition from ductal carcinoma in situ to invasive ductal carcinoma in vitro.

However, applications of polymers and microfluidics in tissue engineering have shown remarkable progress, but challenges remain. Constructing the fully functional integration and vascularization of large, engineered tissues for transplantation is a complex task that requires further optimization. Moreover, for clinical use, microfluidic systems in manufacturing tissue-engineered constructs need further investigations to ensure scalability and reproducibility.

13.4.6 Polymers for organ-on-chips

Organ-on-chips are microfluidic devices used to mimic physiological organ systems, combining the advantages of in vitro and in vivo models. Organ-on-chip technology serves as a preclinical model that allows multicellular assemblage and tissue–tissue interaction and mimics numerous human pathologies. Since polymers are versatile materials that play a crucial role in the design and fabrication of organ-on-chips, combining polymers and microfluidics improves the development of microenvironments for specific organs. These microscale systems aim to mimic the structural and functional characteristics of organs, providing researchers with a more physiologically relevant platform for cell culture studies. Organ-on-chips emerge as revolutionary technology overcoming the limitations of classic in vitro cell cultures, unable to mimic the complex microenvironment and physiological conditions of living organs. Moreover, organ-on-chips are a promising alternative to animal models associated with ethical concerns and high costs (Damiati et al., 2022a). Hence, this technology combines the advantages of in vitro and in vivo models. These microfluidic devices aim to copy the key features of human organs by culturing cells on a chip, with each chip design replicating specific organ functions, such as the beating of a heart or the filtration of kidneys, providing a more physiologically relevant platform for experimentation.

The success of organ-on-chip technology relies on the selection of suitable materials and designs to fabricate these microscale devices. Polymers such as PDMS and various hydrogels form the structural foundation of these chips and offer a diverse range of properties that make them ideal for mimicking the structural and mechanical characteristics of human tissues. Any polymer used in organ-on-chip systems must be biocompatible to support cell growth and function and can create a microenvironment that closely mimics the ECM of specific organs in microfluidic systems. Hence, hydrogels are commonly used in this regard due to their high water content and soft, pliable nature, mimicking the mechanical properties of tissues. Different organs exhibit varying degrees of stiffness and elasticity, and polymers can be engineered to mimic these characteristics. The recreation of specific tissue stiffness is controlled by the tunable mechanical properties of polymers, thus facilitating the growth and function of cells in a more organ-like manner. Indeed, microfluidic channels within organ-on-chip enable the continuous perfusion of nutrients and cell culture media. This dynamic fluid flow mimics the in vivo circulatory systems of organs, ensuring a constant supply of essential nutrients and waste removal. Moreover, the stiffness of materials is a crucial consideration in the design of organ-on-a-chip systems, especially when membrane flexibility is employed to improve the physiological relevance of the cultured environment. The substrate stiffness serves as a passive element influencing the cells and as an active component within the on-chip microphysiological environment (Dalsbecker et al., 2022). Devices such as lung-on-a-chip and intestine-on-a-chip are recognized for employing cyclical stretching to simulate breathing and peristalsis, respectively (Huh, 2015; Kim et al., 2012). Similarly, heart-on-chip utilizes the flexibility of PDMS for diaphragm layers to assess the contraction of heart cells during culture (Abulaiti et al., 2020). These organ-on-chip platforms have wide applications, such as drug testing, disease modeling, and understanding organ-organ interactions. Polymers facilitate the microfabrication of organ-on-chips using techniques such as soft lithography and 3D printing. These methods allow for the precise patterning of microstructures, channels, and compartments on the chip, enabling the mimicry of organ architectures.

Communication among the organ systems of different organs in the human body is a complex phenomenon that is challenging to replicate in traditional experimental setups. Hence, organ-on-chip platforms aim to understand organ-organ interactions by linking multiple chips to simulate systemic responses. This approach provides a more comprehensive understanding of how organs interconnect and influence each other's functions.

Heart-on-a-chip platforms, which miniaturize the structure and functionality of a human heart into a microfluidic device, are created by incorporating human cells, employing microfabrication techniques, and integrating advanced sensors. These platforms offer the potential to transform drug testing, disease modeling, and personalized treatment, ushering in a new era of more efficient interventions for cardiovascular diseases. Zimmermann et al. (2002) were the first to engineer heart tissue by arranging rat neonatal cardiomyocytes with gel around silicon tubes. Subsequently, studies continued to improve the first model. For example, Schaaf et al. (2011) added hiPSC cardiomyocytes to silicone-based force sensors to enhance clinical relevance and enable noninvasive contractile force measurement. Heart-on-a-chip platforms, which miniaturize the structure and functionality of a human heart into a microfluidic device, are created by incorporating human cells, employing microfabrication techniques, and integrating advanced sensors. These platforms offer the potential to transform drug testing, disease modeling, and personalized treatment, ushering in a new era of more efficient interventions for cardiovascular diseases (Kieda et al., 2023). Another heart-on-a-chip platform was developed to investigate the complex 3D cell–cell and cell–matrix interactions and recapitulate the physiologic mechanical environment experienced by cells in the native myocardium. The chip is composed of an array of suspended posts designed to contain gels loaded with cells, along with a pneumatic actuation system that imparts consistent uniaxial cyclic strains to the 3D cell constructs throughout the culture. Utilizing this device, mature and exceptionally functional micro-engineered cardiac tissues were produced from both neonatal rat and human-induced hiPSC cardiomyocytes. This outcome strongly indicates the reliability and effectiveness of the engineered cardiac microenvironment. This beating heart-on-a-chip showed superior cardiac differentiation, electrical, and mechanical coupling. Indeed, mechanical stimulation facilitated early spontaneous synchronous beating and improved contractile capability in response to electric pacing (Marsano et al., 2016).

The liver is a large and complex organ that performs a multitude of metabolomic, regulatory, and immune functions. These include the regulation of blood sugar and ammonia levels, as well as the synthesis of hormones (Wiśniewski et al., 2016). The liver performs various functions because its structure is composed of different types of cells, such as HCs, Kupffer cells, hepatic stellate cells, and liver sinusoid endothelial cells, with each cell type responsible for specific functions and connected to each other through autocrine and paracrine signaling (Deng et al., 2019). Liver stiffness plays a key role in the regulation of hepatic responses in both healthy and diseased states. Many hepatic cells require a stiff environment. To improve our understanding of the effect of varying stiffness on liver cells, in vitro models were designed to mimic the liver stiffness corresponding to various stages of disease progression to clarify the role of individual cellular responses (Dalsbecker et al., 2022). Natarajan et al. (2015) reported on the culturing of primary rat hepatocytes on various PDMS-based substrates to investigate the effect of substrate stiffness on the behavior of the cells. Cells cultured on a soft PDMS substrate retained a functional and differentiated phenotype for a longer time than cells cultured on a stiffer PDMS substrate or tissue-culture plastic surface. Hence, stiffness variation allowed for mimicking healthy and fibrotic liver tissue. It is known that any new drug needs to undergo testing in animals before being used in human clinical studies. The main challenge is that animals and humans have fundamental differences in their physiologies and drug metabolism pathways. Hence, it is found that a liver-on-a-chip, which mimics the human liver in vitro, is a good alternative that can provide more information about the pharmacokinetic and pharmacodynamic behavior of a drug in humans. Jang (2019) used a commercial liver chip consisting of species-specific primary hepatocytes interfaced with liver sinusoidal endothelial cells, cultured under physiological fluid flow. Microfluidic channels of the liver chip were coated with a species-specific combination of ECM composed of a mixture of collagen and fibronectin. The developed model detected various phenotypes of liver toxicity, such as hepatocellular injury, cholestasis, and fibrosis, and species-specific toxicities when treated with tool compounds.

Microfluidics also allows the integration of multiple tissues together to perform a more complex multitissue system that closely simulates the microenvironment between organs in the body. Since multiorgan systems contain different cell types being cocultured in the microfluidic chips, the optimization becomes more challenging. Compartmentalization of cells into separated compartments via permeable membranes can overcome this obstacle. Compartmentalization enables the supply of cell-specific media to each compartment while still permitting paracrine interactions between the compartments (Leung et al., 2022). For example, a human liver-kidney model was developed to test organ-organ interactions, where the metabolism of drugs by living tissue may influence their toxicity toward one another. Microphysiological systems, composed of human-derived cell/tissue cultures in continuously perfused chambers, were designed to mimic organ and tissue function and response to injury. The role of hepatic metabolism in aristolochic acid I, a nephrotoxin and human carcinogen, inducing kidney injury was extensively studied. Liver-on-a-chip was linked to kidney-on-a-chip that was cultured with human hepatocytes and human proximal tubular epithelial cells, respectively. Two natural polymers were used: Matrigel and collagen. Hepatocytic cells were seeded into the microfluidic chamber with Matrigel, while renal proximal tubular epithelial cells were cultured with collagen to form a tubular structure. The obtained results recorded an increase in the metabolic activity of human hepatocyte-specific metabolism of aristolochic acid I, which increased the cytotoxicity toward human kidney proximal tubular epithelial cells, leading to the formation of aristolactam adducts and the release of kidney injury biomarkers (Chang, 2017). Another model by Skardal et al. (2017) developed a three-tissue organ-on-a-chip platform composed of the liver, heart, and lung to assess the response of inter-organs to drugs. They examined the response of drugs in single, dual (liver and cardiac), and triple (liver, cardiac, and lung) organoid systems. The obtained results proved that drug responses depend on inter-tissue interactions, highlighting the functional correlation and compatibility of multiple organs and the side effects associated with candidate drugs. More interestingly, Zhang et al. (2009) designed a multi-channel 3D microfluidic cell culture system with compartmentalized microenvironments to mimic four organs: liver, lung, kidney, and adipose tissue. In the microfluidic system, gelatin microspheres were mixed with cells to create a cell-specific microenvironment, and growth factors were supplemented in the common medium to optimize cellular functions. The fabricated model exhibited limited cross-talk between cell culture compartments, similar to biological conditions.

Although organs-on-a-chip devices have advanced significantly, with great potential in the future and various models launched from research laboratories finding a place in pharmaceutical and biotech companies such as GSK, Roche, and Pfizer for drug testing in the pharmaceutical industry, it is a new technology and still faces several challenges in the area of scalable tissue/organ development, requiring further advancements. These limitations can be overcome by offering more standardization and improving compatibility with imaging, robotics, mass production, and analytical instruments. Indeed, making these platforms user-friendly would contribute to their wider adoption by nonspecialist end-users (Kieda et al., 2023; Leung et al., 2022).

13.5 Conclusion

The development of a microfluidic device has various aspects and depends on the final desired application. Selecting the optimum polymer and optimum fabrication method is a key factor in constructing a functional miniaturized system. The use of polymers in microfluidic manufacturing, biomimetics, drug delivery systems, and sensors is not limited to their ease of functionalization, modification, biocompatibility, longer lifetimes, and cost-effective production. Despite the growing development in microfluidic fabrication, it has not yet reached its full potential in terms of applications in real life. Thus, collaboration between engineers and biologists is in high demand to achieve the benefits of combining microfluidics and polymers in synthetic biology, diagnosis, and therapeutics. Microfluidics is classified into continuous-flow microfluidics and segmented-flow microfluidics. Both approaches are used to produce drug delivery systems in different forms and different sizes, ranging from the microscale to the nanoscale. Microfluidics-mediated drug delivery carriers offer a promising strategy to generate monodisperse polymeric particles with narrow size distribution and high encapsulation capacity. Furthermore, exploiting polymers, spheroids, and microfluidics offers wide applications in the fields of drug development and toxicity screening due to its simplicity, reproducibility, and close approach to physiological tissues. The ability to construct artificial cells in simpler forms, such as polymersomes and multicompartmentalized polymeric vesicles, offers the ability to build cell-like and organ-like architectures. These microfluidic-based artificial cells are life-like systems and can improve our understanding of natural systems, allowing us to investigate new cell functions and connect living and nonliving materials to perform new functions that do not exist in the biological world. A wide range of microfluidic biosensors have been developed with the continuous development of polymer materials. Polymers are broadly exploited in sensing applications, not only because of their high degree of flexibility in designing microfluidic chips. Polymer-based sensors offer rapid detection, high selectivity, and sensitivity either when polymers are used directly, such as in the case of smart polymers, or when immobilizing specific antibodies onto polymer film. Moreover, organs-on-a-chip is a promising candidate that bridges the gap between in vitro and in vivo by supporting biological research and high-throughput applications. Finally, the microfluidics field still faces some challenges due to its interdisciplinary nature; therefore, it requires the synergistic collaboration of scientists from different backgrounds and with different areas of expertise to overcome its limitations, allowing us to successfully apply microfluidics more frequently in our lives.

References

Abulaiti, M., Yalikun, Y., Murata, K., et al. (2020). Establishment of a heart-on-a-chip microdevice based on human iPS cells for the evaluation of human heart tissue function. *Scientific Reports*, *10*, 19201.

Adepu, S., & Ramakrishna, S. (2021). Controlled drug delivery systems: Current status and future directions. *Molecules*, *26*(19), 5905.

Ahn, J., Ko, J., Lee, S., Yu, J., Kim, Y. T., & Jeon, N. L. (2018). Microfluidics in nanoparticle drug delivery: From synthesis to pre-clinical screening. *Advanced Drug Delivery Reviews*, *128*, 29–53. Available from https://doi.org/10.1016/j.addr.2018.04.001.

Ai, Y., Xie, R., Xiong, J., & Liang, Q. (2020). Microfluidics for biosynthesizing: From droplets and vesicles to artificial cells. *Small (Weinheim an der Bergstrasse, Germany)*, *16*(9), 201903940. Available from https://doi.org/10.1002/smll.201903940.

Alrifaiy, A., Lindahl, O. A., & Ramser, K. (2012). Polymer-based microfluidic devices for pharmacy, biology and tissue engineering. *Polymers*, *4*(3), 1349–1398.

Andersson, H., van der Wijngaart, W., Nilsson, P., Enoksson, P., & Stemme, G. (2001). A valve-less diffuser micropump for microfluidic analytical systems. *Sensors and Actuators B: Chemical*, *72*(3), 259–265. Available from https://doi.org/10.1016/s0925-4005(00)00644-4.

Bahmaee, H., Owen, R., Boyle, L., et al. (2020). Design and evaluation of an osteogenesis-on-a-chip microfluidic device incorporating 3D cell culture. *Frontiers in Bioengineering and Biotechnology*, *8*, 557111.

Ban, Q., Zhang, Y., Li, Y., Cao, D., Ye, W., Zhan, L., Wang, D., & Wang, X. (2022). A point-of-care microfluidic channel-based device for rapid and direct detection of fibrinogen in whole blood. *Lab Chip*, *22*, 2714–2725.

Basak, S., & Bandyopadhyay, A. (2021). Tethering smartness to the metal containing polymers – Recent trends in the stimuli-responsive metal containing polymers. *Journal of Organometallic Chemistry*, *956*, 122129. Available from https://doi.org/10.1016/j.jorganchem.2021.122129.

Bayareh, M., Ashani, M. N., & Usefian, A. (2020). Active and passive micromixers: A comprehensive review. *Chemical Engineering and Processing. Process Intensification*, *147*, 107771.

Berlanda, S. F., Breitfeld, M., Dietsche, C. L., & Dittrich, P. S. (2021). Recent advances in microfluidic technology for bioanalysis and diagnostics. *Analytical Chemistry*, *93*(1), 311–331. Available from https://doi.org/10.1021/acs.analchem.0c04366.

Brinson, H. F., & Brinson, L. C. (2008a). *Characteristics, applications and properties of polymers*. Springer Science and Business Media LLC. Available from 10.1007/978-0-387-73861-1_3.

Brinson, H. F., & Catherine Brinson, L. (2008b). *Characteristics, applications and properties of polymers. Polymer engineering science and viscoelasticity: An introduction* (pp. 55–97). Boston, MA: Springer US.

Buddingh', B. C., & van Hest, J. C. M. (2017). Artificial cells: Synthetic compartments with life-like functionality and adaptivity. *Accounts of Chemical Research*, 769–777.

Champion, J. A., Katare, Y. K., & Mitragotri, S. (2007). Particle shape: A new design parameter for micro- and nanoscale drug delivery carriers. *JCR*, *121*, 3–9.

Chang, S. Y., et al. (2017). Human liver–kidney model elucidates the mechanisms of aristolochic acid nephrotoxicity. *JCI Insight*, *2*, e95978.

Chen, K. J., Wu, M. X., Guo, F., et al. (2016). Rapid formation of size-controllable multicellular spheroids via 3D acoustic tweezers. *Lab on a Chip*, *16*, 2636.

Chen, Q., Utech, S., Chen, D., Prodanovic, R., Lin, J. M., & Weitz, D. A. (2016). Controlled assembly of heterotypic cells in a core-shell scaffold: Organ in a droplet. *Lab on a chip*, *16*(8), 1346–1349. Available from https://doi.org/10.1039/c6lc00231e.

Chiara, M., & deMello, A. J. (2016). Droplet-based microfluidics for artificial cell generation: A brief review. *Interface Focus*, *6*, 4.

Chiesa, E., Greco, A., Riva, F., Tosca, E. M., Dorati, R., Pisani, S., Modena, T., Conti, B., & Genta, I. (2019). Staggered herringbone microfluid device for the manufacturing of chitosan/tpp nanoparticles: Systematic optimization and preliminary biological evaluation. *International Journal of Molecular Sciences*, *20*(24), 6212.

Convery, N., & Gadegaard, N. (2019). 30 years of microfluidics. *Micro and Nano Engineering*, *2*, 76–91.

Dai, J., Zhang, T., Zhao, H., & Fei, T. (2017). Preparation of organic-inorganic hybrid polymers and their humidity sensing properties. *Sensors and Actuators B: Chemical*, *242*, 1108–1114. Available from https://doi.org/10.1016/j.snb.2016.09.139.

Dalsbecker, P., Adiels, C. B., & Goksör, M. (2022). Liver-on-a-chip devices: The pros and cons of complexity. *The American Journal of Physiology-Gastrointestinal and Liver Physiology*, *323*(3), G188–G204.

Damiati, S., Kompella, U. B., Damiati, S. A., & Kodzius, R. (2018a). Microfluidic devices for drug delivery systems and drug screening. *Genes*, *9*(2), 103.

Damiati, S., Mhanna, R., Kodzius, R., & Ehmoser, E.-K. (2018b). Cell-free approaches in synthetic biology utilizing microfluidics. *Genes*, *9*(3), 144.

Damiati, S., Peacock, M., Leonhardt, S., Damiati, L., Baghdadi, M. A., Becker, H., Kodzius, R., & Schuster, B. (2018c). Embedded disposable functionalized electrochemical biosensor with a 3D-printed flow cell for detection of hepatic oval cells (HOCs). *Genes*, *9*, 89.

Damiati, S. (2018d). *Can we rebuild the cell membrane? Artmann G., Artmann A., Zhubanova A., & Digel I. (Eds.),* Biological, physical and technical basics of cell engineering. Singapore: Springer.

Damiati, S. (2019). New opportunities for creating man-made bioarchitectures utilizing microfluidics. *Biomedical Microdevices*, *21*(3), 62.

Damiati, S. A., Rossi, D., Joensson, H. N., & Damiati, S. (2020a). Artificial intelligence application for rapid fabrication of size-tunable PLGA microparticles in microfluidics. *Arabia Scientific Reports*, *10*(1). Available from https://doi.org/10.1038/s41598-020-76477-5.

Damiati, S. (2020b). In situ microfluidic preparation and solidification of alginate microgels. *Macromol. Res.*, *28*, 1046–1053.

Damiati, L. A., Damiati, S. A., & Damiati, S. (2022a). Developments in the use of microfluidics in synthetic biology. In Singh Vijai (Ed.), *New Frontiers and applications of synthetic biology* (pp. 423–435). Academic Press.

Damiati, L. A., El-Yaagoubi, M., Damiati, S. A., Kodzius, R., Sefat, F., & Damiati, S. (2022b). Role of polymers in microfluidic devices. *Polymers.*, *14*(23), 5132.

Damiati, S. (2022). *Acoustic biosensors for cell research.. Thouand G. (Ed.), Handbook of cell biosensors..* Springer Cham.

Damiati, S. A., & Damiati, S. (2021). Microfluidic synthesis of indomethacin-loaded PLGA microparticles optimized by machine learning. *Frontiers in Molecular Biosciences*, *8*, 677547.

Deng, J., Wu, S., Zhao, H., & Cai, D. (2019). Engineered liver-on-a-chip platform to mimic liver functions and its biomedical applications: A review. *Micromachines.*, *10*(676), 1–26.

van Duinen, V., van den Heuvel, A., Trietsch, S. J., et al. (2017). 96 perfusable blood vessels to study vascular permeability in vitro. *Scientific Reports*, *7*, 18071.

Esch, M. B., Sung, J. H., Yang, J., et al. (2012). On chip porous polymer membranes for integration of gastrointestinal tract epithelium with microfluidic ‘body-on-a-chip’ devices. *Biomedical Microdevices*, *14*, 895–906.

Foret, F., Smejkal, P., & Macka, M. (2013). Miniaturization and microfluidics. In S. Fanali, P. R. Haddad, C. F. Poole, P. Schoenmakers, & D. Lloyd (Eds.), *Liquid chromatography* (pp. 453–467). Elsevier.

García, J., García, F., Reglero Ruiz, J., Vallejos, S., & Trigo-López, M. (2022). *Sensory polymers. Smart polymers: principles and applications* (pp. 12–50). Berlin, Boston: De Gruyter.

Gencturk, E., Mutlu, S., & Ulgen, K. O. (2017). Advances in microfluidic devices made from thermoplastics used in cell biology and analyses. *Biomicrofluidics*, *11*(5), 051502.

Giri, K., & Tsao, C. W. (2022). Recent advances in thermoplastic microfluidic bonding. *Micromachines*, *13* (3), 486.

Gutierrez Cisneros, C., Bloemen, V., & Mignon, A. (2021). Synthetic, natural, and semisynthetic polymer carriers for controlled nitric oxide release in dermal applications: A review. *Polymers.*, *13*(5), 760.

Haino, T. (2013). Molecular-recognition-directed formation of supramolecular polymers. *Polymer Journal*, *45*, 363–383.

Hesh, C. A., Qiu, Y., & Lam, W. A. (2020). Vascularized microfluidics and the blood–endothelium interface. *Micromachines*, *11*(1), 18.

Huh, D. (2015). A human breathing lung-on-a-chip. *Annals ATS.*, *12*, S42–S44.

Jang, K. J., et al. (2019). Reproducing human and cross-species drug toxicities using a liver-chip. *Science Translational Medicine*, *11*, eaax5516.

Juang, Y.-J., & Chiu, Y.-J. (2022). Fabrication of polymer microfluidics: An overview. *Polymers.*, *14*(10), 2028.

Kapałczyńska, M., Kolenda, T., Przybyła, W., et al. (2018). 2D and 3D cell cultures – A comparison of different types of cancer cell cultures. *Archives of Medical Science: AMS*, *14*(4), 910–919.

Kaplan, D. L. (1998a). *Introduction to biopolymers from renewable resources* (pp. 1–29). Springer Science and Business Media LLC. Available from 10.1007/978-3-662-03680-8_1.

Kaplan, D. L. (1998b). Introduction to biopolymers from renewable resources. In D. L. Kaplan (Ed.), *Biopolymers from renewable resources. Macromolecular systems—Materials approach.* Berlin, Heidelberg: Springer.

Karnik, R., Gu, F., Basto, P., Cannizzaro, C., Dean, L., Kyei-Manu, W., Langer, R., & Farokhzad, O. C. (2008). Microfluidic platform for controlled synthesis of polymeric nanoparticles. *Nano Letters*, *8*(9), 2906–2912. Available from https://doi.org/10.1021/nl801736q.

Kickelbick, G. (2003). Concepts for the incorporation of inorganic building blocks into organic polymers on a nanoscale. *Progress in Polymer Science*, *28*, 83–114.

Kieda, J., Shakeri, A., Landau, S., et al. (2024). Advances in cardiac tissue engineering and heart-on-a-chip. *J Biomed Mater Res*, *112*, 492–511. Available from https://doi.org/10.1002/jbm.a.37633.

Kim, H. J., Huh, D., Hamilton, G., & Ingber, D. E. (2012). Human gut-on-a-chip inhabited by microbial flora that experiences intestinal peristalsis-like motions and flow. *Lab on a Chip*, *12*, 2165–2174.

Lauga, E., Brenner, M., & Stone, H. (2007). Microfluidics: The no-slip boundary condition. In C. Tropea, A. L. Yarin, & J. F. Foss (Eds.), *Springer handbook of experimental fluid mechanics*. Berlin, Heidelberg: Springer Handbooks. Springer.

Lee, C. S. (2020). Grand challenges in microfluidics: A call for biological and engineering action. *Frontiers in Sensors*, *1*, 583035.

Leung, C. M., de Haan, P., Ronaldson-Bouchard, K., et al. (2022). A guide to the organ-on-a-chip. *Nature Reviews Methods Primers*, *2*, 33.

Li, C. W., & Wang, G. J. (2012). MEMS manufacturing techniques for tissue scaffolding devices. In S. Bhansali, & A. Vasudev (Eds.), *In Woodhead Publishing series in biomaterials, MEMS for biomedical applications* (pp. 192–217). Woodhead Publishing.

Li, W., Zhang, L., Ge, X., Xu, B., Zhang, W., Qu, L., Choi, C.-H., Xu, J., Zhang, A., Lee, H., & Weitz, D. A. (2018). Microfluidic fabrication of microparticles for biomedical applications. *Chemical Society reviews*, *47*(15), 5646–5683. Available from https://doi.org/10.1039/c7cs00263g.

Li, X., Valadez, A. V., Zuo, P., & Nie, Z. (2012). Microfluidic 3D cell culture: Potential application for tissue-based bioassays. *Bioanalysis*, *4*(12), 1509–1525. Available from https://doi.org/10.4155/bio.12.133.

Li, Y., Motschman, J. D., Kelly, S. T., & Yellen, B. B. (2020). Injection molded microfluidics for establishing high-density single cell arrays in an open hydrogel format. *Analytical Chemistry*, *92*(3), 2794–2801. Available from https://doi.org/10.1021/acs.analchem.9b05099.

Lim, W., & Park, S. A. (2018). Microfluidic spheroid culture device with a concentration gradient generator for high-throughput screening of drug efficacy. *Molecules*, *23*(12), 3355.

Lu, A. X., Oh, H., Terrell, J. L., Bentley, W. E., & Raghavan, S. R. (2017). A new design for an artificial cell: Polymer microcapsules with addressable inner compartments that can harbor biomolecules, colloids or microbial species. *Chemical Science*, *8*(10), 6893–6903. Available from https://doi.org/10.1039/c7sc01335c.

Marsano, A., Conficconi, C., Lemme, M., et al. (2016). Beating heart on a chip: A novel microfluidic platform to generate functional 3D cardiac microtissues. *Lab on a Chip*, *16*, 599–610.

Martino, C., Kim, S. H., Horsfall, L., Abbaspourrad, A., Rosser, S. J., Cooper, J., & Weitz, D. A. (2012). Protein expression, aggregation, and triggered release from polymersomes as artificial cell-like structures. *Angewandte Chemie – International Edition*, *51*(26), 6416–6420. Available from https://doi.org/10.1002/anie.201201443.

Mei, X., Yang, J., Yu, X., Peng, Z., Zhang, G., & Li, Y. (2023). Wearable molecularly imprinted electrochemical sensor with integrated nanofiber-based microfluidic chip for in situ monitoring of cortisol in sweat.

Sensors and Actuators B: Chemical, 381, 133451. Available from https://doi.org/10.1016/j.snb.2023.133451.

Mitrogiannopoulou, A.-M., Tselepi, V., & Ellinas, K. (2023). Polymeric and paper-based lab-on-a-chip devices in food safety: A review. *Micromachines, 14*, 986.

Montanez-Sauri, S. I., Sung, K. E., Puccinelli, J. P., Pehlke, C., & Beebe, D. J. (2011). Automation of three-dimensional cell culture in arrayed microfluidic devices. *Journal of Laboratory Automation, 16*(3), 171–185.

Moulahoum, H., Ghorbanizamani, F., Zihnioglu, F., & Timur, S. (2021). Tracking and treating: molecularly imprinted polymer-based nanoprobes application in theranostics. In D. Adil (Ed.), *Molecular imprinting for nanosensors and other sensing applications* (pp. 45–68). Turkey: Elsevier. Available from 10.1016/B978-0-12-822117-4.00003-4.

Mulholland, T., McAllister, M., Patek, S., et al. (2018). Drug screening of biopsy-derived spheroids using a self-generated microfluidic concentration gradient. *Scientific Reports, 8*, 14672.

Nagabooshanam, S., Roy, S., Deshmukh, S., Wadhwa, S., Sulania, I., Mathur, A., Krishnamurthy, S., Bharadwaj, L. M., & Roy, S. S. (2020). Microfluidic affinity sensor based on a molecularly imprinted polymer for ultrasensitive detection of chlorpyrifos. *ACS Omega, 5*(49), 31765–31773. Available from https://doi.org/10.1021/acsomega.0c04436.

Natarajan, V., Berglund, E. J., Chen, D. X., & Kidambi, S. (2015). Substrate stiffness regulates primary hepatocyte functions. *RSC Advances, 5*, 80956–80966.

Neoh, K. G., Wang, R., & Kang, E. T. (2015). Surface nanoengineering for combating biomaterials infections. In L. Barnes, & I. R. Cooper (Eds.), *Biomaterials and medical device – Associated infections* (pp. 133–161). Woodhead Publishing.

Nezakati, T., Seifalian, A., Tan, A., & Seifalian, A. M. (2018). Conductive polymers: Opportunities and challenges in biomedical applications. *Chemical Reviews, 118*(14), 6766–6843. Available from https://doi.org/10.1021/acs.chemrev.6b00275, http://pubs.acs.org/journal/chreay.

Radhakrishnan, S., Mathew, M., & C., S. (2022). Microfluidic sensors based on two-dimensional materials for chemical and biological assessments. *Materials Advances, 3*, 1874–1904.

Rein, C., Toner, M., & Sevenler, D. (2023). Rapid prototyping for high-pressure microfluidics. *Scientific Reports, 13*, 1232.

Ruiz, J. R., Sanjuán, A., Vallejos, S., García, F., & García, J. (2018). Smart polymers in micro and nano sensory devices. *Chemosensors, 6*(2). Available from https://doi.org/10.3390/chemosensors6020012.

Sakolish, C. M., Esch, M. B., Hickman, J. J., et al. (2016). Modeling barrier tissues in vitro: Methods, achievements, and challenges. *EBioMedicine, 5*, 30–39.

Saliba, J., Daou, A., Damiati, S., Saliba, J., El-Sabban, M., & Mhanna, R. (2018). Development of microplatforms to mimic the in vivo architecture of CNS and PNS physiology and their diseases. *Genes, 9*(6), 285.

Schaaf, S., Mewe, M., Eder, A., et al. (2011). Human engineered heart tissue as a versatile tool in basic research and preclinical toxicology. *PLoS ONE, 6*(10), e26397.

Schütte, J., Hagmeyer, B., Holzner, F., et al. (2011). Artificial microorgans—A microfluidic device for dielectrophoretic assembly of liver sinusoids. *Biomedical Microdevices, 13*, 493–501.

Scriba, G. K. E. (2016). Chiral recognition in separation science – An update. *Journal of Chromatography. A, 1467*, 56–78.

Shakeri, A., Jarad, N. A., Khan, S., & Didar, T. F. (2022). Bio-functionalization of microfluidic platforms made of thermoplastic materials: A review. *Analytica Chimica Acta, 1209*, 339283. Available from https://doi.org/10.1016/j.aca.2021.339283.

Shao, C., Chi, J., Zhang, H., Fan, Q., Zhao, Y., & Ye, F. (2020). Development of cell spheroids by advanced technologies. *Advanced Materials Technologies, 2020*, 2000183.

Sharma, H. S., Hussain, S., Schlager, J., Ali, S. F., & Sharma, A. (2010). Influence of nanoparticles on blood-brain barrier permeability and brain edema formation in rats. *Acta Neurochirurgica. Supplement*, *106*, 359–364.

Shum, H. C., Kim, J. W., & Weitz, D. A. (2008). Microfluidic fabrication of monodisperse biocompatible and biodegradable polymersomes with controlled permeability. *Journal of the American Chemical Society*, *130* (29), 9543–9549.

Shum, H. C., Zhao, Y.-jin, Kim, S.-H., & Weitz, D. A. (2011). Multicompartment polymersomes from double emulsions. *Angewandte Chemie*, *123*(7), 1686–1689. Available from https://doi.org/10.1002/ange.201006023.

Skardal, A., Murphy, S. V., Devarasetty, M., et al. (2017). Multi-tissue interactions in an integrated three-tissue organ-on-a-chip platform. *Scientific Reports*, *7*(1), 8837.

Son, K. J., Gheibi, P., Stybayeva, G., Rahimian, A., & Revzin, A. (2017). Detecting cell-secreted growth factors in microfluidic devices using bead-based biosensors. *Microsystems & Nanoengineering*, *3*(1). Available from https://doi.org/10.1038/micronano.2017.25.

Stroock, A. D., Dertinger, S. K. W., Ajdari, A., Mezić, I., Stone, H. A., & Whitesides, G. M. (2002). Chaotic mixer for microchannels. *Science*, *295*(5555), 647–651. Available from https://doi.org/10.1126/science.1066238.

Tan, S., Ai, Y., Yin, X., et al. (2023). Recent Advances in Microfluidic Technologies for the Construction of Artificial Cells. *Advanced Functional Materials*, *33*(45).

Tsao, C. W. (2016). Polymer microfluidics: simple, low-cost fabrication process bridging academic lab research to commercialized production. *Micromachines*, *7*(12), 225.

Underhill, G. H., Peter, G., Chen, C. S., & Bhatia, S. N. (2012). Bioengineering methods for analysis of cells in vitro. *Annual Review of Cell and Developmental Biology*, *28*(1), 385–410. Available from https://doi.org/10.1146/annurev-cellbio-101011-155709.

Wang, M., Xiao, Y., Lin, L., Zhu, X., Du, L., & Shi, X. (2018). A microfluidic chip integrated with hyaluronic acid-functionalized electrospun chitosan nanofibers for specific capture and nondestructive release of CD44-overexpressing circulating tumor cells. *Bioconjugate Chemistry*, *29*(4), 1081–1090. Available from https://doi.org/10.1021/acs.bioconjchem.7b00747.

Wang, X., Cheng, C., Wang, S., & Liu, S. (2009). Electroosmotic pumps and their applications in microfluidic systems. *Microfluidics and Nanofluidics*, *6*(2), 145–162. Available from https://doi.org/10.1007/s10404-008-0399-9.

Waters, L. C., Jacobson, S. C., Kroutchinina, N., Khandurina, J., Foote, R. S., & Ramsey, J. M. (1997). Microchip device for cell lysis, multiplex PCR amplification, and electrophoretic sizing. *Analytical Chemistry*, *70*, 158–162.

Wiśniewski, J. R., Vildhede, A., Norén, A., & Artursson, P. (2016). In-depth quantitative analysis and comparison of the human hepatocyte and hepatoma cell line HepG2 proteomes. *Journal of Proteome Research*, *136*, 234–247.

Zhang, C., Zhao, Z., Abdul Rahim, N. A., van Noort, D., & Yu, H. (2009). Towards a human-on-chip: culturing multiple cell types on a chip with compartmentalized microenvironments. *Lab on a Chip*, *9*, 3185–3192.

Zhang, H., Yang, J., Sun, R., Han, S., Yang, Z., & Teng, L. (2023). Microfluidics for nano-drug delivery systems: From fundamentals to industrialization. *Acta Pharmaceutica Sinica B*, *13*(8), 3277–3299. Available from https://doi.org/10.1016/j.apsb.2023.01.018.

Zhao, S. P., Ma, Y., Lou, Q., Zhu, H., Yang, B., & Fang, Q. (2017). Three-dimensional cell culture and drug testing in a microfluidic sidewall-attached droplet array. *Analytical Chemistry*, *89*(19), 10153–10157.

Zimmermann, W. H., Schneiderbanger, K., Schubert, P., et al. (2002). Tissue engineering of a differentiated cardiac muscle construct. *Circulation Research*, *90*(2), 223–230.

Further reading

Damiati, S. (2018). Can we rebuild the cell membrane? In G. Artmann, A. Artmann, A. Zhubanova, & I. Digel (Eds.), *Biological, physical and technical basics of cell engineering*. Singapore: Springer.

Damiati, S. (2022c). Acoustic biosensors for cell research. In G. Thouand (Ed.), *Handbook of cell biosensors*. Cham: Springer.

CHAPTER

Gas sensors

14

Jian Zhang and Xiao Huang
Institute of Advanced Materials (IAM), Nanjing Tech University (Nanjing Tech), Nanjing, P. R. China

14.1 Introduction

Synthetic polymeric materials are widely used in a great variety of applications, such as packaging, adhesives, lubricants, and electrical insulators. Especially due to their intrinsic electrostatic and chemical characteristics, polymers have gained a lot of attention for various sensing applications (Ruiz et al., 2018). For example, the designed sensory polymers could be used for the early detection and diagnosis of different diseases, the quick detection of harmful gases or explosives, and the detection of pesticides or heavy metallic cations (Sanjuán et al., 2018). For these purposes, extensive research has been conducted based on various types of polymers for different application scenarios.

Sensory polymers are a kind of polymer that can exhibit a response when they are in contact with a target species. The selective interaction that gives rise to the response relies on the recognition of the target species, or guest analyte, by the receptor motifs of the polymer, and it must be followed by a transduction process giving rise to an easily measurable change. The sensor polymers can be classified by stimulus, mechanisms, or responses. Based on different stimuli, two large groups of stimuli—chemical stimuli (gas species, metal, or nonmetal ions) and physical stimuli (temperature, pressure, etc.)—can be used to distinguish different sensory polymers. Besides, according to different sensing mechanisms, sensory polymers can also be labeled as reversible mechanism-based sensory polymers or irreversible mechanism-based sensory polymers. A great variety of polymers, including conjugated polymers (CPs), conducting polymers, acrylic polymers, polymers with different morphological structures, or polymers modified with different ligands, can be used as sensory materials. Among them, a particular type of polymer in analytical chemistry is CPs, which are polyunsaturated compounds with alternating single and double bonds along the polymer chain, offering a myriad of opportunities to couple analyte receptor interactions as well as nonspecific interactions into observable responses (McQuade et al., 2000). Compared with small-molecule elements, CPs can amplify the signal from a recognition event. When an analyte binds locally to a functional receptor of CPs, the backbone and electronic properties of the entire CP are affected, which leads to amplified signal transduction over devices using small-molecule elements (Kyokunzire et al., 2023; Sugiyasu & Swager, 2007). Besides, conducting polymers, such as polypyrrole (PPy), polyaniline (PANI), polythiophene (PT), and their derivates, have also been used as

Sensory Polymers. DOI: https://doi.org/10.1016/B978-0-443-13394-7.00014-8

the active layers for gas sensing since the 1980s. The sensors made of conducting polymers have many advantages, such as the higher sensitivity, the short response time, and the lower operating temperature. In particular, conducting polymers are easy to synthesize through chemical or electrochemical processes, and their molecular chain structure can be modified conveniently by copolymerization or structural derivations for selective gas detection. In addition to the above-described, two main types of polymers, acrylic polymers and various porous organic polymers (POP), have also exhibited potential applications in gas sensing, which will be detailed in the third section.

Considering the nature of the generated responses, the sensory polymers can be classified by the most relevant response, including colorimetric, fluorescence, and conductive sensory polymers. Colorimetric sensing relies on changes in a polymer's light absorption properties, which have attracted growing attention due to their notable advantages, especially in rapid and on-site detection. In particular, colorimetric sensing could achieve fast detection by the naked eye without complex instruments and require non-professionals to operate (Lee et al., 2010). As for fluorescence sensing, except for the inherent sensitivity, diverse transduction schemes based upon changes in intensity, energy transfer, wavelength, and lifetime can be offered. By using CPs as sensing materials, the amplification effect resulting from efficient energy migration can lead to enhanced fluorescence sensitivity (Sun et al., 2015). Conductive sensors display changes in electrical conductivity in response to an analyte interaction, which is a particularly natural sensory scheme for electrically conductive polymers. These measurements can be easily achieved by connecting two adjacent electrodes with a conductive polymer film. When an analyte with different concentrations is brought into contact with the conductive polymer, the conductivity varies accordingly due to the changes in carrier density and/or carrier mobility.

This book chapter is organized as follows: The following section will give a description of gas sensing mechanisms based on three main transduction modes, which are fluorescence sensing, colorimetric sensing, and conductive sensing. In each of them, the detailed sensing mechanism based on different characteristics of polymers and gas species is also introduced, such as the electron exchange mechanism, resonance energy transfer mechanism, and charge-hopping mechanism. In the third section, the design and application of polymeric gas sensing materials for fluorescence sensing, colorimetric sensing, and conductive sensing are analyzed in detail. Then, the device construction methods for polymer gas sensors are also introduced, covering the ink-jet printing method, various traditional polymerization techniques, and novel solid-phase synthesis methods. After that, considering the long-standing problems of selectivity, the advancements in polymer sensor arrays are also detailed. Finally, future perspectives and challenges are briefly provided.

14.2 Sensory mechanism

A gas-sensing process involves changes in various physical and/or chemical properties of the polymer. Based on different physicochemical characteristics, sensory polymers can respond to different stimuli through a specific sensing mechanism to generate a certain response, such as a change in fluorescence intensity, color, and conductivity. This section will give a brief description of the main sensing mechanisms based on different response types.

14.2.1 Fluorescence

Fluorescence sensing is usually based on either the turn-on mechanism or the turn-off mechanism, that is, through the enhanced or quenched fluorescence that is induced by external target molecules (Gheitarani et al., 2023). In a turn-on system, a target binding event changes the electron density along the polymer backbone or the conformation of the polymer chain, which leads to enhanced fluorescence intensity. For the turn-off mode, a binding event can cause the fluorescence of the polymer to be effectively quenched through non-radiative relaxation pathways in a polymer chain or through the inter-molecule aggregation of chains. There are several mechanisms responsible for fluorescence sensing, including photo-induced electron transfer, resonance energy transfer (RET), electron exchange, and intermolecular charge transfer.

In a photoinduced electron transfer (Fig. 14.1A), the fluorescent polymer in its excited state as the donor, for example, transfers electrons to the detecting species with an acceptor characteristic. After the electron transfer, a complex is formed between the electron donor and the electron acceptor, which then returns to the ground state without light emission, but in some cases, exciplex emission can be observed. Finally, extra electrons on the acceptor are returned to the electron donor. During this process, the energy gap between the LUMO of the polymer and that of the gas as the acceptor is the thermodynamic driving force for this electron transfer process.

The RET mechanism has also been employed to develop gas sensors. In this system, an initially excited donor polymer returns to the ground state because of the energy transfer to an acceptor analyte, while the transferred energy excites the acceptor to its excited state (Fig. 14.1B). RET is due to a long-range dipolar interaction between the donor and acceptor, which depends on the relative orientation of donor and acceptor dipoles, the extent of overlap in the emission spectrum of the donor and the absorption spectrum of the acceptor, and the distance between the donor and the acceptor.

As for the electron exchange mechanism (Fig. 14.1C), the electron transfer occurs from the LUMO of the excited donor polymer to the acceptor analyte and also from the HOMO of the acceptor to the donor. Therefore, the acceptor is left in an excited state, and the donor becomes quenched. The electron exchange is a short-range phenomenon that largely depends on the spatial overlap of the orbitals of the donor polymer and the acceptor analyte, predominantly over RET when the spectral overlap is small.

14.2.2 Colorimetric

Visible colorimetric sensing is convenient and useful since the signal can be seen by the "naked eye," avoiding using expensive or cumbersome equipment. Such an event is triggered when a binding event causes a change in the conjugation length of a polymer backbone or conformation and thus changes the wavelength of the light absorbed by the polymer. For colorimetric gas sensing, there are various types of sensing mechanisms, including ring opening and closing reactions, change of functional groups, ligand exchange, and so on (Cho et al., 2021). The mechanism of the ring opening and closing reactions is mainly based on the conjugated pi-system. It is believed that electrons in a conjugated pi-system can capture specific photons when resonating along p-orbitals at a specific distance. In other words, the energy of photons captured is affected by the length and number of conjugated pi-systems, which can shift the absorption spectra of the sensing polymer. As

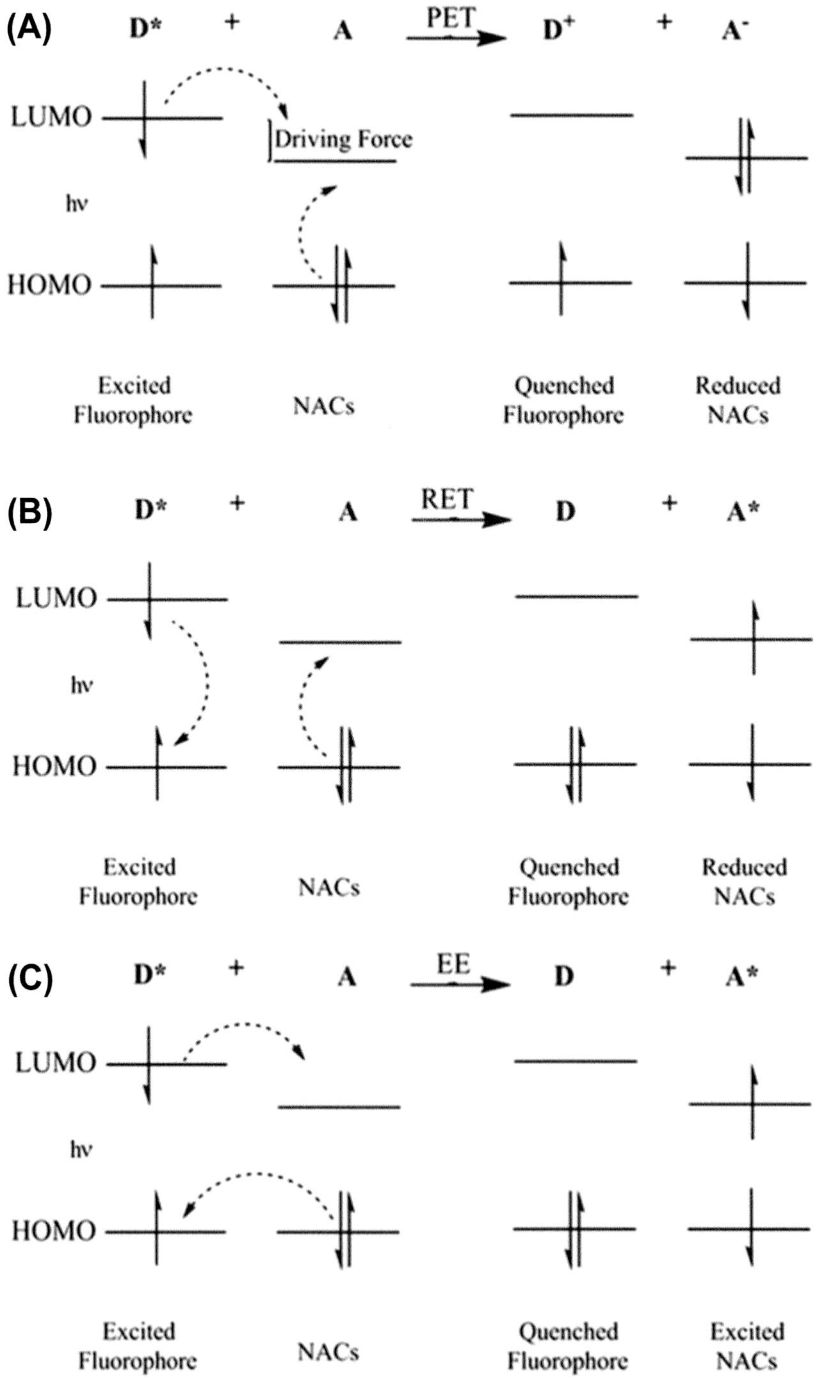

FIGURE 14.1 Molecular orbital schematic illustration for fluorescence sensing mechanism.

(A) photoinduced electron transfer, (B) resonance energy transfer, and (C) energy exchange (Sun et al., 2015).

From Sun, X., Wang, Y., & Lei, Y. (2015). Fluorescence based explosive detection: From mechanisms to sensory materials. Chemical Society Reviews, 44*(22), 8019–8061. Available from https://doi.org/10.1039/C5CS00496A.*

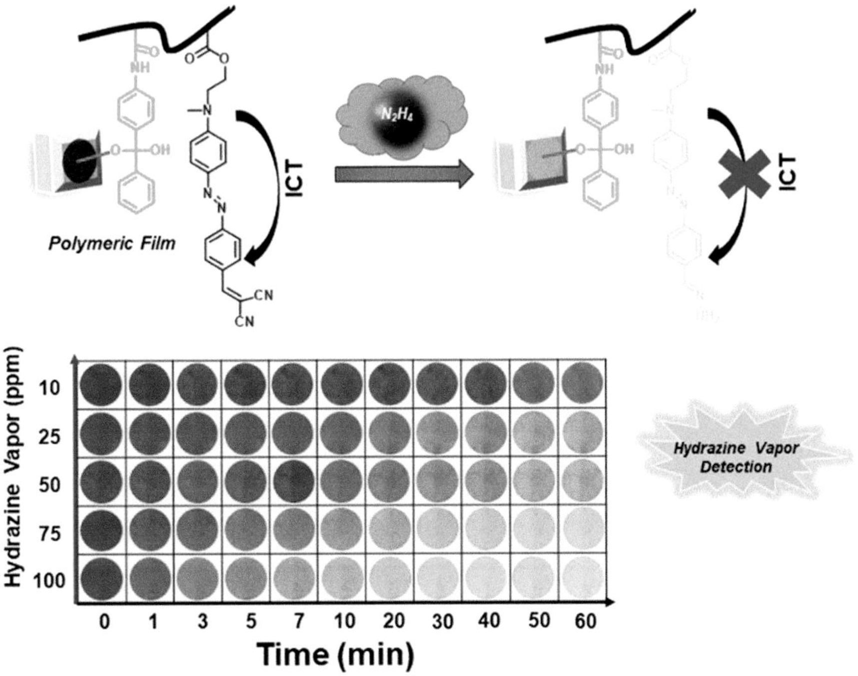

FIGURE 14.2 Colorimetric polymer sensors.

Illustration of vapor-phase hydrazine detection by polymeric film (Sharma et al., 2021).

From Sharma, R., Jung, S. H., & Lee, H. I. (2021). Thin polymeric films for real-time colorimetric detection of hydrazine vapor with parts-per-million sensitivity. ACS Applied Polymer Materials, 3*(12), 6632–6641. Available from: https://doi.org/10.1021/acsapm.1c01292.*

shown in Fig. 14.2, when hydrazine vapor attacks the α-position of the dicyano group of the polymer, the intramolecular charge transfer (ICT) process can be suppressed, which leads to the apparent color change of the polymeric film (Sharma et al., 2021). Similarly, if ligands are modified or substituted by gases in the surroundings, the state of electrons could be changed, resulting in the absorption of different wavelengths of light, which is called spin crossover.

14.2.3 Conductive sensing

Conductive sensing is based on the changes in the electrical conductivity of polymers upon gas exposure. The conductivity of polymers could decrease or increase depending on the acceptor or donor characteristics of the gas species, as well as their redox activity. To further detail the gas sensing mechanism, a p-type conductive polymer is taken as an example. When the conducing polymers are doped with an oxidizing agent, the electrons in the conductive polymers will transfer to the oxidizing agent, which can result in a decreased electron density and then an increased hole concentration, giving rise to a p-type conducing polymer (Fig. 14.3A). Upon exposure to an oxidizing gas such as NO_2,

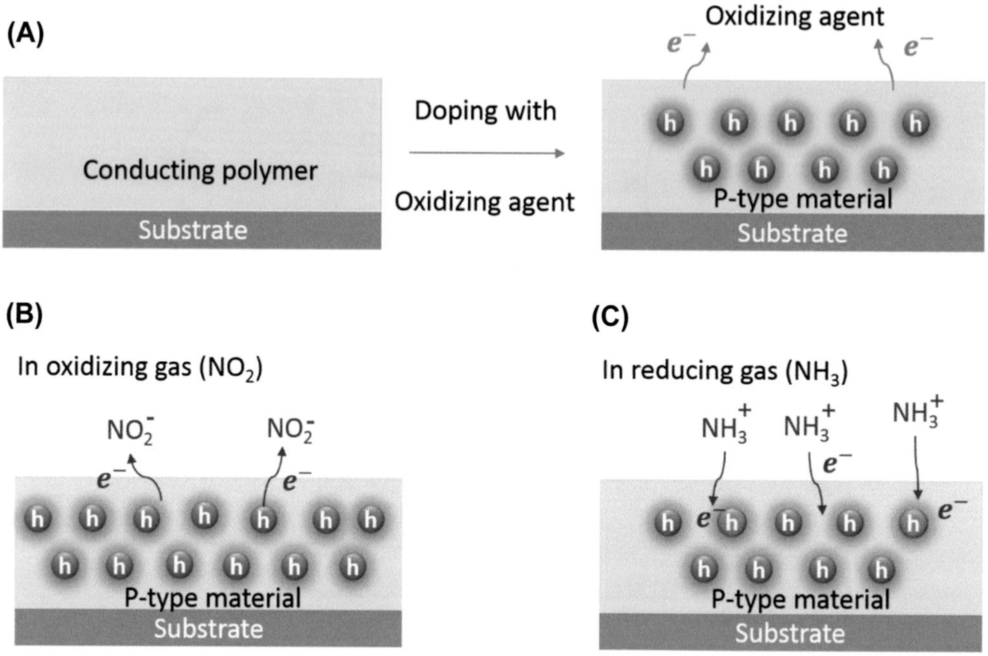

FIGURE 14.3 Conductive sensing mechanism of p-type conducting polymer.

(A) conducting polymer doping with oxidizing agent, (B) the conducting polymer in oxidizing gas (NO_2), and (C) the conducting polymer in reducing gas (NH_3) (Liu et al., 2022).

From Liu, X., Zheng, W., Kumar, R., Kumar, M., & Zhang, J. (2022). Conducting polymer-based nanostructures for gas sensors. Coordination Chemistry Reviews, 462, *214517. Available from: https://doi.org/10.1016/j.ccr.2022.214517.*

the NO_2 adsorbs the electron from the p-type conductive polymers. This results in a further increase in the hole concentration and a decrease in resistance (Fig. 14.3B). When a reducing gas such as ammonia interacts with the conductive polymers, electrons are donated to the p-type conductive polymers, and then the hole concentration of the conductive polymer is reduced due to the electron-hole combination. This further increases the sensor resistance (Fig. 14.3C). To improve the sensing performance of conductive polymers, p-n heterojunctions composed of inorganic materials and conductive polymers are usually constructed; the sensing mechanism is mainly based on the modulation of the space charge region after exposure to the target gas. The detailed sensing mechanism of such heterojunctions can be referred to in other related review papers (Liu et al., 2022).

14.3 Design and application of polymeric gas sensing materials

The basic design of a polymer-based sensor is to connect a molecular recognition site with a signaling unit, triggering a change in the measurable properties of the polymer material, such as conductivity, luminescence, and color change. To achieve improved sensing performance, the affinity between the receptor site and the target analyte (binding constant and selectivity) and the transduction efficiency of the binding event should both be considered. Therefore, the design principles for polymeric gas sensing materials should be based on the detection mode and the detection mechanism.

A great number of polymers can be used as gas-sensing materials, including CPs, conductive polymers, acrylic polymers, POP, and polymers modified with different ligands. Taking CO_2-responsive polymers as an example, the design principle of polymer sensing materials is briefly analyzed. In CO_2-responsive polymers, the functional groups in the polymer chains change from a neutral to a charged state upon exposure to CO_2. Among them, amidines, guanidines, and amines are the most widely adopted functional groups in these polymers. Compared with the amidine and guanidine groups, polymers containing the tertiary amine group have received special attention due to their simple synthesis. More importantly, tertiary amines with a moderate base (pK_{aH} = 6–7) could lead to good switchability between neutral and charge states, which are easier to protonate and deprotonate. It is well known that higher pK_{aH} leads to an increased degree of protonation, but the reversible reaction becomes difficult and is not suitable for CO_2 sensing. It was concluded that amidine with a weak base exhibited a low degree of protonation, and guanidine with a superhigh base was challenging to deprotonate, both of which are not benefiting from CO_2 sensing. As for primary and secondary amines, N-H bonds can react with CO_2 in the presence of water, forming carbamate salt. However, the formation reaction of carbamate salt cannot be reversed unless at a higher temperature, which demands extra heat energy. Till now, various amine-functionalized polymers have been employed for CO_2 detection. For example, through the proton hopping mechanism, poly(N-[3-(dimethylamino)propyl] methacrylamide) (pDMAPMAm) exhibited its capability in room-temperature CO_2 detection. However, the fabricated polymer sensor suffered from a nonlinear and irreversible response to CO_2 at different concentrations due to the saturation of amine sites (Shahrbabaki et al., 2022). To address this issue, 2-N-morpholinoethyl methacrylate with a lower pK_{aH} was employed to composite with DMAPMAm through a free radical polymerization reaction, which exhibited highly sensitive and reversible CO_2 sensing with excellent selectivity (Shahrbabaki et al., 2023). In this section, we summarize the recent progress in the application of polymeric gas sensing materials from the perspective of different sensing modes, including fluorescent sensing, colorimetric sensing, and conductive sensing. The molecular structures of selected polymers and the corresponding sensing mechanisms are also discussed.

14.3.1 Fluorescent polymers

Based on different fluorescent sensing mechanisms, various types of fluorescent polymer gas sensing platforms have been developed. For instance, porphyrinated polymers, which exhibited distinct fluorescent properties, were electrospun into nanofibrous membranes for gaseous HCl detection. Their thermal stability in harsh environments was further improved by being hybridized with polyimide. Because of the large amount of available surface area and, hence, good gas accessibility, the nanofibrous membrane of porphyrinated polyimide showed extremely high sensitivity and a fast response to HCl gas. Upon exposure to HCl gas, the structure of the porphyrin ring was changed from a free base to a protonated form. This structural change increased the resonance interaction between phenyl groups and the porphyrin nucleus, resulting in the photoluminescence quenching (Lv et al., 2010). In another example, by using 2-(acetoacetoxy)ethyl methacrylate (AAEMA) as the monomer (Fig. 14.4A), the fluorescent

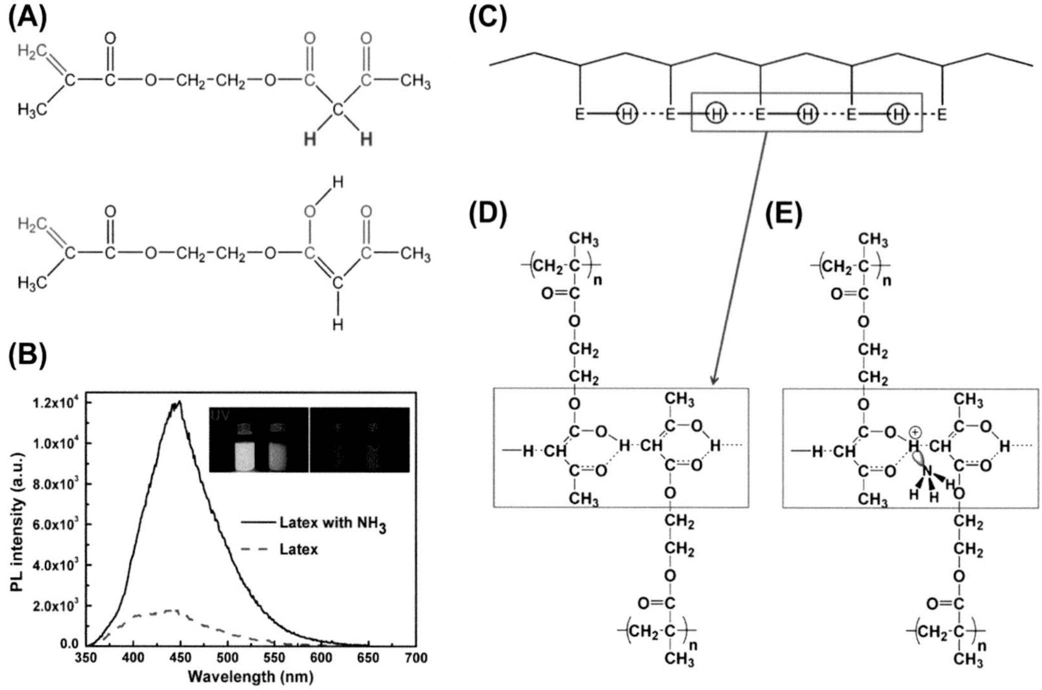

FIGURE 14.4 The molecular structure of fluorescence polymer and its sensing mechanism.

(A) chemical structure of monomer AAEMA, (B) fluorescence spectra and images of PAAEMA latex and its ammonia mixture, (C) cascade proton transfer among β-dicarbonyl units, (D) aggregate of β-dicarbonyl, and (E) ammonia and β-dicarbonyl aggregates complex (He et al., 2012).

From He, J., Zhang, T. Y., & Chen, G. (2012). Ammonia gas-sensing characteristics of fluorescence-based poly(2-(acetoacetoxy)ethyl methacrylate) thin films. Journal of Colloid and Interface Science, *373(1), 94–101. Available from: https://doi.org/10.1016/j.jcis.2011.11.074.*

PAAEMA latexes prepared by the miniemulsion polymerization method tended to form a complex with ammonia, emitting significant visible fluorescence under UV light (Fig. 14.4B). It should be noted that AAEMA is a commercially available function monomer bearing β-dicarbonyl moiety, and in the polymerized form, they tend to entangle with each other to support proton transfer (Fig. 14.4C and D). The NH_3 group induced strong fluorescence in the PAAEMA latexes, mainly originating from the supramolecular complex formed between NH_3 and β-dicarbonyl aggregates (Fig. 14.4E), which exhibited a strong visible fluorescence due to the electron donor effect. (He et al., 2012)

Based on the RET mechanism, thin blend films composed of a conjugated semiconducting polymer (fluorene polymer) and a conveniently chosen pH indicator dye (bromocresol green) were investigated for acid vapor sensing. Upon exposure to HCl, the emission of fluorene polymers was greatly quenched due to the energy transfer process. It should be noted that the fabricated thin blend films showed a fast response and excellent repeatability after consecutive exposure-recovery cycles at low and moderate acid concentrations, but these features could not be observed at higher acid vapor concentrations.

Therefore, resonant and non-resonant contributions to the observed photoluminescence should be considered simultaneously (Guillén et al., 2016).

Based on a protonation-induced charge transfer mechanism, a fluorescent triazine-based covalent organic polymer (COP) sensor was synthesized, which exhibited stable fluorescence and a sensitive HCl/NH_3 response in both the powder state and solvent state. The fluorescence emission of the polymer was weakened with red shifts after the immersion in HCl. After injection of NH_3, the fluorescence of the polymer was recovered via deprotonation. Interestingly, the microporous COP film also showed a similar fluorescence response to HCl/NH_3 with high sensitivity and good reversibility, which could serve as a solid-state optical probe for continuous and quantitative detection of HCl and NH_3 gases (Xu et al., 2018).

Both turn-on and turn-off fluorescence sensing strategies can be employed with the same sensing platform. For example, two-dimensional (2D) fluorescence chemosensors based on PANI derivatives were constructed, which showed excellent selectivity and sensitivity to various metal ions and HCl gas. Owing to the ICT effect, fluorescence quenching of the polymer occurred when reacting with the target species (Lu, Qian, et al., 2023). The HCl-doped 2D PANI could be further utilized for sensing triethylamine, which changed the doping state of HCl-2D PANI and thus revived its fluorescence as a turn-on probe (Lu, Hu, et al., 2023).

14.3.2 Colorimetric polymers

A number of polymer-based colorimetric probes for the detection of gaseous toxics were fabricated to produce reliable and practical analytical methods. As described before, based on a nanofibrous membrane of porphyrinated polyimide, rapid fluorescence quenching can be observed upon exposure to HCl gas. Furthermore, the vivid color variations of the nanofibrous membrane in response to different concentrations of HCl gas can also be easily witnessed, confirming that the porphyrinated polyimide nanofibrous membrane is an applicable material for colorimetric gas sensing (Lv et al., 2010). As a commonly reported polymer, PANI can also be employed for colorimetric ammonia sensors because the protonation of the PANI backbone by HCl forms an emeraldine salt type of PANI in green color, and such PANI chains are deprotonated and transform into a blue emeraldine base in the presence of ammonia (Fig. 14.5A). Based on this, various solid-state PANI-based colorimetric sensors have been developed. To improve the sensitivity of these sensors, a thin emeraldine salt type of PANI film was coated on each cellulose nanofiber of filter paper through in situ polymerization of aniline molecules (Fig. 14.5B), which exhibited a reversible color change of PANI from green to blue within 5 min in the presence of gaseous ammonia (Fig. 14.5C) (Gu & Huang, 2013). In another example, the manufactured polydiacetylene (PDA)/polyurethane nanofibers were also employed for colorimetric gas sensors. Upon exposure to volatile organic compound (VOC) gases, the color was changed from blue to red according to the nonpolar and aprotic classification of the solvent (Kim et al., 2019).

Through the incorporation of an optical pH indicator, a polymer-based colorimetric sensor for formaldehyde gas has been developed. Commonly, the formaldehyde sensing mechanism involves a nucleophilic addition reaction between formaldehyde and an amine group. The formation of a Schiff base accompanied by a pH shift can lead to color change using pH indicator dyes. Based on the above-described sensing mechanism, spherical and uniform poly (styrene-co-maleic anhydride)/polyethyleneimine core-shell nanoparticles were prepared and then impregnated with various pH indicator dyes to construct

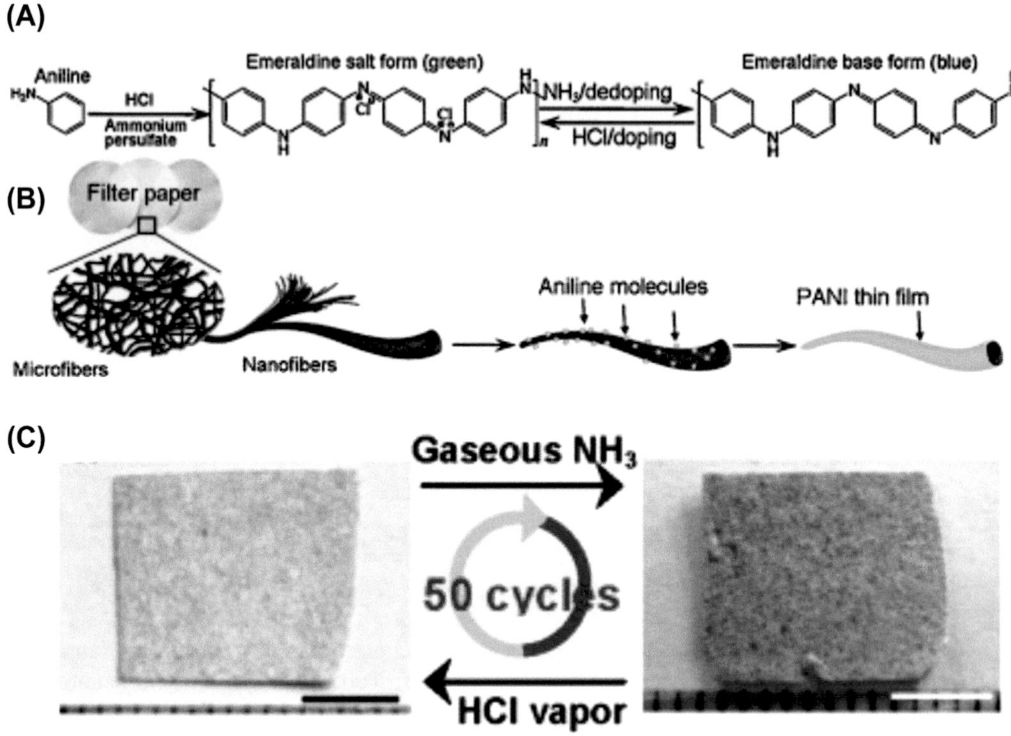

FIGURE 14.5 The PANI-based colorimetric gas sensors.

(A) schematic illustration of color change mechanism of PANI, (B) the fabrication process of the PANI-deposited filter paper, and (C) the reversible colorimetric detection of gaseous ammonia using the PANI-deposited filter paper, scale bars: 5 mm (Gu & Huang, 2013).

From Gu, Y., & Huang, J. (2013). Colorimetric detection of gaseous ammonia by polyaniline nanocoating of natural cellulose substances. Colloids and Surfaces A: Physicochemical and Engineering Aspects, 433, *166–172. Available from: https://doi.org/10.1016/j.colsurfa.2013.05.016.*

colorimetric formaldehyde gas sensors. Upon gas exposure, the colors of methyl red, bromocresol purple, and 4-nitrophenol changed to yellow, red, and gray, respectively, allowing sensitive and rapid naked-eye detection of 0.5 ppm formaldehyde. It was also found that the colorimetric response was dependent on the relative humidity. That was because the color change related to the pH change was induced by the reaction between the amine group of the colorimetric sensor and formaldehyde. With increasing humidity, instead of the reaction between formaldehyde and the amine group of sensors, formaldehyde reacted with water to form methylene glycol, which then reduced the Schiff base and the extent of pH change (Park et al., 2020). Considering that the concentration of water vapor in the testing environment is usually significantly higher than that of target gas analytes, the water vapor will present a huge challenge to most gases. To address this issue, various methods can be adopted to mitigate the humidity interference, such as using desiccants to reduce humidity levels, using the matrices of substrates for humidity regulation, and using humidity sensors to signal compensation.

Although colorimetric polymer sensors have unique strengths in collecting chemical information from the environment, the humidity effect should be rationally treated based on different sensing mechanisms. The specific method was summarized in the review paper (Yu et al., 2021).

For the realization of the colorimetric textile sensor with the characteristics of eco-friendliness and washfastness, a UV-induced photografting method based on cotton fabrics was introduced. The fabricated sensors exhibited distinctive color changes under acidic gas, in addition to outstanding durability and reusability after several washing/drying cycles (Park et al., 2022). Similarly, a sensitive and selective colorimetric ammonia sensor based on a simple pH-indicator immobilized electrospun core-shell bi-polymer nanofiber mat was fabricated, which can detect ammonia concentrations as low as 0.5 ppm with a fast response time of 10 s. Especially due to the core-shell structure, the sensor showed high selectivity against various interferences such as VOCs, basic liquids, and aerosols. Moreover, the detection range can be modulated and extended by adjusting the ratios of dual polymers without compromising the other performance parameters of the sensor (Wu & Selvaganapathy, 2023). Recently, due to their vivid color display and the capability of manipulating optical properties, stimulus-responsive chromogenic materials have attracted great attention. For instance, facile protonation-induced self-assembly of bipyridyl containing linear diacetylene (DA) structure followed by generation of UV-irradiation-promoted blue-phase PDA was demonstrated, which served as a colorimetric stimuli-responsive material for ammonia detection. Upon exposure to ammonia, blue-phase PDA displayed an excellent, naked-eye-detectable colorimetric sensory response with a color transition from brilliant blue to red (Baek et al., 2023).

14.3.3 Conductive polymers

Because of their versatile nanostructure, easy synthesis, and good environmental stability, conductive polymers have attracted great attention for room-temperature sensing. Since the early 1980s, PANI, PPy, PT, poly(3,4-ethylene dioxythiophene) (PEDOT), poly(phenylene vinylene) (PPV), and their derivatives have been used as the active layers of gas sensors. Generally, conductive polymers with typical conjugated structures exhibit semiconducting characteristics. Upon exposure to target gas species, they can operate as electron donors or acceptors, altering the concentration of charge carriers and hence the electrical conductivity or resistance of conducting polymer sensors. Despite this, because of the lower conductivities and the strong adsorption energy of VOCs or water molecules, their gas sensing applications are limited by their lower sensitivity, decreased stability, and poor selectivity. Therefore, many efforts have been made to improve the sensing properties of conductive polymers, including morphological modulation, redox doping, and functionalization.

Considering that the sensing properties of conductive polymers are closely related to their morphological structures, much attention has been paid to preparing zero-dimensional (0D), one-dimensional (1D), 2D, and three-dimensional (3D) nanostructured conductive polymers to improve the sensing properties, meeting the requirements of various sensor configurations. The 0D conductive polymers were usually prepared in micelles. For instance, through chemical oxidation polymerization of aniline in a micellar solution of dodecylbenzene sulfonic acid, conductive nanoparticles with average sizes of 20–30 nm were successfully prepared, which exhibited high stability and processability (Han et al., 2002). In another research work, the microemulsion polymerization method was also employed to synthesize PPy nanoparticles with diameters ranging from 2 to 8 nm, in which octyltrimethylamine bromide was used as a template and ferric chloride acted as an oxidant (Jang & Oh, 2002). In addition to the above-described microemulsion polymerization method, PANI nanoparticles with a diameter of 30–80 nm were successfully synthesized by using the interfacial polymerization method, in which aniline monomer and oxidizer was dissolved in the organic phase and the ionic liquid, respectively. Therefore, the polymerization reaction occurred at the interface, and PANI nanoparticles were collected at the interface (Gao et al., 2004). Till now, there have been only a few reports about 0D conductive polymer-based gas sensors. Most 0D conductive polymers are selectively anchored on the surface of other host materials to provide special functionality. For example, by grafting 0D conductive polymers on the surface of silica nanoparticles, the detection of trinitrotoluene based on the fluorescence quenching effect was realized (Feng et al., 2010).

Concerning 1D conductive polymers, both template and template-free protocols have been employed for the synthesis. Initially, the template polymerization method was proposed to prepare micro- and nano-tubes and wire arrays with adjustable length and diameter, which require a porous medium as the template to guide the formation of 1D conducting polymers, including anodic aluminum oxide templates (Back et al., 2011; Blaszczyk-Lezak et al., 2016) and particle track-etched membranes (Delvaux et al., 2000). To date, a group of 1D conductive polymers, such as PANI, PPy, poly(3-methylthiphene), PEDOT, and PPV, have already been synthesized by the template polymerization method. In addition to the template method, the template-free method can also be adopted to prepare 1D conductive polymers via a self-assembly process involving non-covalent interactions such as van der Waals forces, hydrogen bonding, and electrostatic interactions. Besides, electrospinning was confirmed to be an efficient method to fabricate 1D conducting polymers with the aid of electrostatic forces. Moreover, the diameter of a nanofiber can be simply adjusted. At present, micro- and nano-fibers of bare PANI, PPy, and poly(3-hexyl-thiophene)/PEO have been prepared using this method. Commonly, the as-prepared 1D conductive polymers have a unique morphology, small diameter, and high surface-to-volume ratio that allow the charge carriers to deplete or accumulate rapidly. Therefore, more and more research works centered on 1D conductive polymer-based gas sensors have been widely reported. For instance, by using the electrochemical polymerization method, intertwined PANI nanowire-based gas sensors exhibited an excellent response to 0.5 ppm NH_3 and ethanol in the ambient environment (Wang et al., 2004). In another example, instead of using the traditional homogeneous aqueous solution of aniline, acid, and oxidant, the interfacial polymerization method was adopted to prepare PANI nanofibers with diameters below 100 nm. Compared with conventional PANI, the prepared PANI nanofibers showed enhanced sensing performance because of the higher surface area, even though the film thickness was higher than that of a conventional PANI film. It was also found that the response increased with decreasing film thickness since only the outermost surface interacts with the vapor molecules, and a thicker film has a more inactive sensing area (Huang et al., 2004).

Due to their ultrathin thickness and larger specific surface area, 2D conductive polymers have also exhibited great potential in gas sensing and can be readily prepared by various methods. Among them, the electrochemical deposition method was regarded as an efficient method to deposit conducting polymer films on a conductive substrate (Gribkova et al., 2020), which can simplify the sensing fabrication process. By varying the amount of charge passing through the electrochemical cell, the film thickness can be adjusted. In addition, the spin-coating method and layer-by-layer assembly technique can also be employed to prepare 2D conductive polymer thin films. Recently, a scalable method for the production of crystalline PANI films with an area of 50 cm (Sanjuán et al., 2018) and adjustable thickness was developed by using the air-water interface and monolayer surfactant as the template. The as-prepared 2D PANI displayed superior chemiresistive sensing toward ammonia (30 ppb) and VOCs (10 ppm) (Zhang et al., 2019). Moreover, given that oxidative polymerization was applicable to the synthesis of other conducting polymers, it was anticipated that crystalline 2D thin films of PPy, PT, and their analogs could also be obtained by using this method. In addition to 2D conducting polymer thin films, 3D conductive polymers with interconnected 3D networks and highly continuous porous structures were also expected to obtain better sensing performances, which are usually prepared by the sacrificial template method (Cassagneau & Caruso, 2002), the electrospinning method (Kweon et al., 2018), or the combination of the vapor deposition polymerization method with freezing technology (Bai et al., 2007). For example, by using a template method, PAIN inverse opals with a 3D ordered structure of interconnected voids were successfully prepared, which exhibited higher sensitivity and faster response to dry gas, ethanol vapor, HCl, and NH_3 compared to thin film-based sensors, attributing to the porous structure with the larger specific surface area that facilitates gas diffusion and gas adsorption (Yang and Liau, 2010). Similarly, by using the template-assisted method, the hierarchical PPy nanotubes assembled into 3D urchins and 3D flower-like nanostructure were prepared (Fig. 14.6A), which were mounted on the low-cost and highly flexible polyvinylidene fluoride membrane substrate for the fabrication of flexible sensors. Due to the higher surface area and porous structure (Fig. 14.6B), the PPy urchin-based flexible NH_3 sensor exhibited a higher response value than that of PPy flowers and PPy granules (Fig. 14.6C) (Lawaniya et al., 2023).

In addition to the dimensional, structural, and morphological effects, introducing extra phases into conductive polymers also has a significant influence on the sensing properties due to the synergistic effect. Therefore, conductive polymers composited with various inorganic components, including metal nanoparticles, nanocarbons, and metal oxides, were widely reported to improve the sensing properties of bare conductive polymers. As a kind of catalyst with high electrocatalytic activity and good stability, metal nanoparticles based on Pd, Pt, Au, Ag, and Cu are composited with conductive polymers through different approaches, which can significantly change the electrical properties of the newly formed system, improving the gas adsorption ability and accelerating charge transfer between conductive polymers and gas molecules. In 2005, research work about PANI/Pd nanocomposites synthesized by the reflux method was reported. Upon exposure to different aliphatic alcohol vapors, the sensors based on PANI/Pd exhibited higher sensitivity, excellent selectivity, and a faster response to methanol vapor (Athawale et al., 2006). After that, different kinds of metal/conductive polymer nanocomposites were widely investigated. Among them, the effect of loading concentration, size, shape, and type of metal nanomaterials on sensing performance has become a research hotspot. For instance, through solution reduction of Pd ions and subsequent gas-phase polymerization of pyrrole, PPy/Pd nanocomposites were fabricated for NH_3 detection, which showed a higher response compared to that of the bare PPy because the well-distributed Pd nanoparticles provide more active sites and then influence the charge transfer between PPy and NH_3 (Hong et al., 2010). In another example, PANI fibers decorated with different concentrations of Au nanoparticles were prepared by using a simple ultrasound mixing method. The response of nanocomposites to 1000 ppm CO reached its maximum with a lower detection limit of 33 ppm when the concentration of Au was 2.5%, which was attributed to the high catalytic properties of Au nanoparticles and the positive charge transfer to PANI by Au nanoparticles after exposure to CO gas. It was believed that the size of the Au nanoparticles and their uniform distribution over the PANI materials played an important role in modulating the sensing properties (Nasresfahani et al., 2020). To sum up, the reason for the loaded metal nanoparticles to improve the sensing properties of conducting polymers is due to the following reasons: First, the

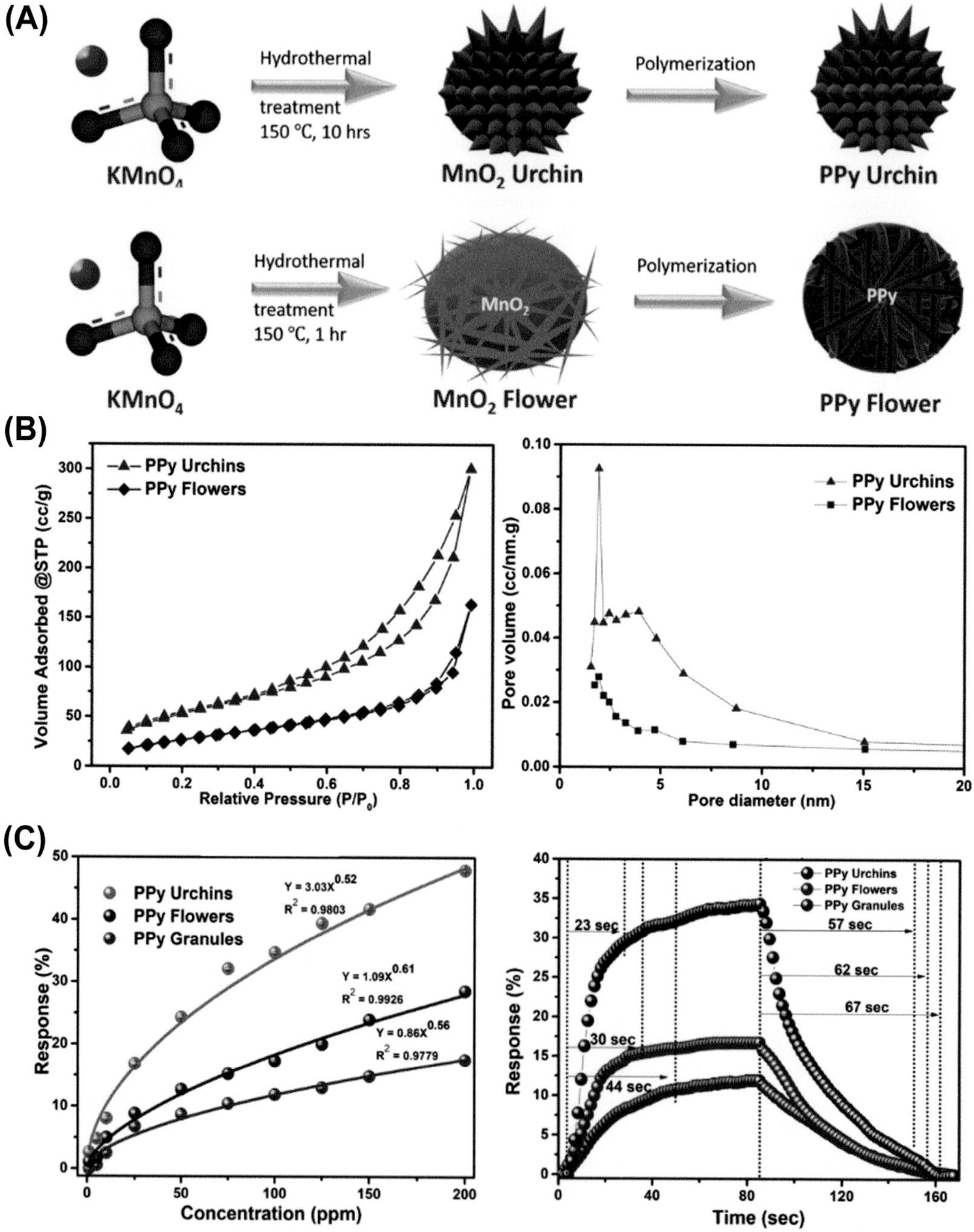

FIGURE 14.6 The morphological structure and the sensing properties of as-prepared PPy.

(A) schematic of synthesis procedure of PPy urchins and PPy flowers, (B) N_2 adsorption-desorption isotherms and pore size distribution curves of PPy urchins and flowers, and (C) comparison of gas sensing performances of different PPy nanostructures (Lawaniya et al., 2023).

From Lawaniya, S. D., Kumar, S., Yu, Y., & Awasthi, K. (2023). Ammonia sensing properties of PPy nanostructures (urchins/flowers) towards low-cost and flexible gas sensors at room temperature. Sensors and Actuators B: Chemical, 382. *Available from: https://doi.org/10.1016/j.snb.2023.133566.*

conductivity was changed upon the introduction of metal nanoparticles. Second, the introduced metal nanoparticles exhibited a chemical affinity for special gas species and then enhanced the selectivity of nanocomposites-based gas sensors. Last, the larger specific surface area of nanocomposites facilitated gas adsorption (Yan et al., 2020).

Apart from the loading of metal nanoparticles, nanocarbons are also a good choice to be combined with conductive polymers for gas sensing due to their higher environmental stability and excellent mechanical and electrical properties. Among them, carbon nanotubes and graphene nanocarbons have received special attention. In 2011, core-shell structures composed of single-walled carbon nanotubes (SWCNT) and PANI nanofibers were successfully prepared, which exhibited a huge change in resistance value within 120 s upon exposure to 100 ppb NH_3, while the bare PANI needed 1000 s to respond at the identical gas concentration. The largely enhanced sensing performance of PANI-SWCNT nanocomposites was attributed to the higher surface area and the higher conductivity (Liao et al., 2011). To further improve the sensing properties of the core-shell PANI/SWCNT nanocomposites, different types of heterojunction were constructed at the interfaces between PANI and CNTs. The p-type PANI/CNTs showed a higher response to NO_2 with a response time of 5.2 s and a detection limit of 16.7 ppb, while the n-type PANI/CNTs exhibited a remarkably higher response to NH_3 with a response time of 1.8 s and a detection limit of 6.4 ppb. The higher sensing properties of n-type PANI/CNTs were attributed to the higher specific surface area, the high permeability, and the construction of extra p-n heterojunctions (Zhang et al., 2020). Because of its antiaggregation ability and larger specific surface area, graphene was also widely adopted as a composite with conducting polymers for higher sensing performances. For example, by using the soft and hard dual templates approach, dual mesoporous PPy/graphene nanosheets were prepared with small and large mesopore layers laid on the top and bottom sides of the graphene nanosheet, respectively, which exhibited a higher sensing response to NH_3 with a lower detection limit of 200 ppb than that of both single-mesoporous PPy/graphene nanocomposites and non-mesoporous PPy/graphene nanocomposites (Qin et al., 2020).

It is well known that sensors based on bare conductive polymers have some inherent drawbacks, including limited sensitivity, instability, and reversibility. Fortunately, metal oxide semiconductors have shown great potential in gas sensing owing to their higher sensitivity and fast response/recovery kinetics. However, the bare metal oxide semiconductor-based sensors still have some drawbacks, such as the higher operating temperature and the poor selectivity. Therefore, many efforts have been made in the past decades to incorporate metal oxides into conductive polymers for gas sensing, including PPy/ZnO, PANI/Co_3O_4, PPy-WO_3, PPy-SnO_2, PANI-Fe_2O_3, and PANI/ZnO. Commonly, a series of techniques, such as sol-gel, self-assembly, evaporative deposition, electrochemical polymerization, vapor-phase synthesis, hydrothermal, solvothermal, template-assisted, electrospinning, photochemical, inclusion, solid-state, plasma, chemical, and electrical processes, can be used for the fabrication of conductive polymers-metal oxide nanocomposites thin films (Zegebreal et al., 2023), which exhibited higher gas sensing performances than those of bare conductive polymers or bare metal oxides. Taking PANI-metal oxide nanocomposites as an example, PANI/TiO_2 nanocomposites were prepared for NH_3 detection, which showed higher sensitivity, faster responsivity, better stability, and shorter recovery time compared to mono-phase sensors (Tai et al., 2007). After that, different kinds of metal oxides, such as SnO_2, ZnO, and WO_3 Nb_2O_5, were employed to composite with PANI for gas detection, all of which exhibited enhanced sensing performance due to the formation of heterojunction and the increased specific surface area. More examples of metal oxide-conductive polymers nanocomposites-based gas sensors are detailed in the cited review papers (Liu et al., 2022; Wong et al., 2020; Yan et al., 2020; Zegebreal et al., 2023). The comparison of polymer-based gas sensors under three different transduction modes is summarized in Table 14.1.

Table 14.1 Summary of polymer-based sensors toward different gases under three different transduction modes.

Order	Sensing material	Sensing mode	Gas	LOD (ppm)	t_{res} (s)	Ref.
1	Porphyrinated polyimide	Fluorescent	HCl	1.25	10	Lv et al. (2010)
2	Poly(2-(acetoacetoxy)ethyl methacrylate)	Fluorescent	NH_3	54	80	He et al. (2012)
3	Poly(9,9-di-n-octyl-2,7-fluorene) doped with bromocresol green	Fluorescent	HCl	9	32	Guillén et al. (2016)
4	Polymethacrylate	Fluorescent	NH_3	200	120	Xu et al. (2018)
5	Triazine-based COP	Fluorescent	HCl, NH_3	–	–	Lu, Qian, et al. (2023)
6	2D polyaniline derivatives	Fluorescent	TEA	10	20	Gu and Huang (2013)
7	Poly(N,N-dimethylacrylamide-co-azo-dicyano-co-N-(4-benzoylphenyl)acrylamide)	Colorimetric	N_2H_4	10	–	Sharma et al. (2021)
8	Natural cellulose substance coated with polyaniline films	Colorimetric	NH_3	100	30	Kim et al., (2019)
9	Polydiacetylene/polyurethane	Colorimetric	VOCs	100	–	Park et al. (2020)
10	Poly(styrene-co-maleic anhydride)/polyethyleneimine core-shell nanoparticles	Colorimetric	HCHO	0.5	60	Yu et al. (2021)
11	Protonated Bipyridyl-containing polydiacetylene	Colorimetric	NH_3	110	–	Han et al. (2002)
12	2D polyaniline	Conductive	NH_3	0.03	–	Zhang et al. (2019)
13	Urchin-like polypyrrole	Conductive	NH_3	1	23	Lawaniya et al. (2023)
14	Pd-polyaniline	Conductive	CH_3OH	1	8	Hong et al. (2010)
15	Au-polyaniline	Conductive	CO	33	180	Nasresfahani et al. (2020)
16	Polyaniline/carbon nanotube	Conductive	NH_3	0.0065	1.8	Zhang et al. (2020)
17	Dual-mesoporous polypyrrole/graphene	Conductive	NH_3	0.2	100	Qin et al. (2020)
18	Polyaniline/TiO_2	Conductive	NH_3	23	–	Tai et al. (2007)

14.3.4 Other polymers

Except for the above-described polymers, POP has also drawn widespread attention due to the availability of plentiful building blocks and their tunable structures, porosity, and functions (Salaris et al., 2023), which could be used as a promising platform for detecting various gas species. The sensing mechanism mainly includes the RET, donor-acceptor electron transfer, absorption competition quenching, the inner filter effect, ground-state complex formation, and coordination or hydrogen-bond interactions. Based on different sensing mechanisms, POP-based sensors can also be divided into various categories, including fluorescence sensors, colorimetric sensors, and chemiresistive sensors, which usually show rapid response, good selectivity, high sensitivity, and excellent sensing reproducibility, attributing to the interactions between the sensing moieties of POPs and the analytes, the amplification effect of polymeric skeletons, the porous structures, the strong covalent bond connection, etc.

Various types of POPs have been synthesized in the past two decades, including covalent organic frameworks (COFs), amorphous conjugated microporous polymers (CMPs), COPs, porous aromatic frameworks (PAFs), metal-organic frameworks (MOFs), hyper-crosslinked polymers, and polymers of intrinsic microporosity. Although POPs were reported earlier than the 2010s, their sensing applications were not well studied until 2012. In 2012, a strategy for the construction of molecular detection systems with CMPs was reported. The condensation of a carbazole derivative, TCB, leads to the synthesis of TCB-CMPs with blue luminescence and a large surface area, which showed enhanced sensitivity and rapid detection upon exposure to arene vapors. TCP-CMPs displayed prominent fluorescence enhancement in the presence of electron-rich arene vapors and drastic fluorescence quenching in the presence of electron-deficient arene vapors, which are attributed to the microporous conjugated network of the sensing material (Liu et al., 2012). Since then, different types of POPs have been widely employed for various sensing applications.

COFs are a class of crystalline polymer materials with a periodic network and inherent porosity that are usually connected by reversible covalent bonds. For instance, based on triazatruxene derivatives, two 2D COFs with a regular honeycomb lattice were synthesized, which showed a rapid fluorescence-on and fluorescence-off nature toward electron-rich and electron-deficient arene vapors. After being exposed to electron-rich arene vapors, the enhanced fluorescent intensity can be recovered to a normal level after 5 min in an ambient atmosphere, which indicates that the as-prepared COFs can be used for detecting electron-rich arene vapors (Xie et al., 2015). In another example, based on the changes in fluorescence emission and color, tetraphenylethene-based 2D COF exhibited a fast response and a high sensitivity to HCl gas, which could be recovered upon exposure to ammonia vapor, further indicating that COFs are suitable for the sensing of toxic gases (Cui et al., 2019). Although numerous studies have shown that hydrogen bonds between the 2D COFs and NH_3 have a major influence on the luminescence system, the luminescent mechanism of the 2D COF is still in its infancy. Therefore, to further detail the sensing mechanism between luminescent COFs and indoor pollutant NH_3, the density functional theory and time-dependent density function theory were proposed, which revealed that the luminescence sensing mechanism was strongly affected by the hydrogen bond between COF and NH_3. Upon exposure to NH_3, the hydrogen bond in the S_1 state was enhanced with a reduced fluorescence rate coefficient of the COF, which means that the COF also showed potential application in the detection of NH_3 (Yang et al., 2019). Another 2D COF, constructed from an environmentally sensitive fluorophore, can undergo concerted and adaptive structural transitions upon adsorption of gas and vapors, which is capable of detecting various VOCs rapidly and reliably even for non-polar hydrocarbon gas under humid conditions. Because of the adaptive guest inclusion, the host-guest interactions were strengthened, which can also facilitate the differentiation of organic vapors by their polarity and shapes (Wei et al., 2022). Recently, to achieve H_2 sensing, imine-based COF nanosheets composited with Pd nanoparticles were developed. Because of the continuous π-electron conjugation and the stability in acidic conditions, Schiff-base-linked $(C_{12}H_8N_2O_5S)_n$ was chosen, in which -SO_3H groups were believed to play a vital role in stabilizing the Pd nanoparticles. The as-prepared COF-based nanocomposites exhibited a shorter response and recovery time to H_2, which are significantly lower than those of most Pd-decorated H_2 sensors (Krishnaveni et al., 2023).

As a subfamily of POPs, CMPs have attracted increasing attention due to their unique pore sizes. They are a class of fascinating amorphous porous materials consisting of aromatic compounds with conjugated bonds. The uniqueness of CMPs is derived from the fully conjugated backbones and the inherent micropores, which can be employed for various analyte sensing, such as arene vapors, explosives, metal ions, and small molecules. For instance, by using an acceptor doping strategy, the fluorescence of a native polyphenylene network can be well controlled, which could endow these doped porous polymers with fluorescence sensing ability for VOC detection (Bonillo et al., 2016). In another example, based on a novel core of tetraphenyl-5,5-dioctylcyclopentadiene (TPDC), conjugated POP-based samples in three different forms (powder, solution in organic solvents, and aqueous dispersion of nanoparticles) are successfully prepared and employed for hydrogen and carbon dioxide adsorption and nitroaromatic sensing, respectively. It was confirmed that TPDC-based polymers could be set as potential multifunctional materials for sensing because of their characteristics of porosity, gas adsorption ability, solubility, and strong fluorescence sensing capabilities (Bandyopadhyay et al., 2015).

PAFs are porous materials with covalent bonds and aromatic rings as the main components, which possess a higher specific surface area and excellent chemical stability due to covalent bonding and rigid aromatic ring structures. In 2009, the porous aromatic framework PAF-1 was first synthesized, which not only has a higher surface area but also has higher thermal and hydrothermal stability than porous MOFs. More importantly, PAF-1 was demonstrated to have high uptake capacities for hydrogen and carbon dioxide as well as benzene and toluene vapors, which exhibited promising performance in sensing applications (Ben et al., 2009). Subsequently, through the assembly of luminescent building blocks of tetra(4-dihydroxyborylphenyl) germanium and 2,3,6,7,10,11-hexahydroxytriphenylene, 3D crystalline PAF-15 with a higher fluorescence quantum yield was synthesized, which exhibited a higher luminescence quenching effect after exposure to hazardous and explosive molecules such as nitrobenzene, 2,4-dinitrotoluene, and 2,4,6-trinitrotoluene. This quenching effect makes PAF-15 a promising sensing material (Yuan et al., 2012).

MOFs, also known as porous coordination polymers (PCPs), have also emerged as promising candidates for gas sensing due to their adjustable structure, large surface area, and high affinity for gas. Considering that MOFs can actively react with gases owing to the high catalytic activities of metals and ligands and that the target gases can be easily adsorbed by the numerous pores, various MOFs have been explored for gas sensing. Despite this, MOFs are usually inherently insulating because the regularly arranged organic ligands block the conducting path. Commonly, MOFs serve as a template to obtain diverse derivatives with various porous structures for gas sensing. Besides, various methods, such as compositional control, catalyst encapsulation, and light activation, were adopted to further improve the sensing properties of MOFs. Till now, the demonstration of bare MOF-based gas sensors with semiconducting characteristics is still scarce. For instance, 2D $Cu_3(HITP)_2$ (HITP = 2,3,6,7,10,11-hexaiminotriphenylene) was synthesized with a bulk conductivity of 0.2 S cm^{-1}. Through simple drop casting of $Cu_3(HITP)_2$ dispersions, the as-fabricated 2D MOF-based sensing device is capable of detecting sub-ppm levels of ammonia vapor (Campbell, Sheberla, et al., 2015). In another example, a sensor array constructed from three structurally analogous, but chemically distinct, conductive 2D MOFs can reliably distinguish five categories of VOCs, which are composed of HHTP (2,3,6,7,10,11-hexahydroxytriphenylene), HITP, and Cu or Ni metal centers (Campbell, Liu, et al., 2015). However, the rigid ligands in these MOFs lead to a lack of conductivity and pore structure tunability. Therefore, the exploration of novel MOFs with a tunable microstructure and excellent conductivity is attractive but remains challenging. It was believed that the use of bandgap engineering techniques and an enhanced understanding of the electrically conducting mechanism could facilitate the discovery of novel MOFs for gas sensing. Recently, to achieve the tunability characteristics of conductivity and pore structure, a series of isoreticular semiconductive MOFs (Co-pyNDI, Ni-pyNDI, and Zn-pyNDI) were constructed by assembling conformationally flexible naphthalene diimide bipyrazole-based linkers (denoted as pyNDI) with corresponding metal ions, which exhibited through-space charge transfer due to the formation of pi-stacking columns. Benefiting from tunable metal coordination geometry (octahedral geometry Co^{2+}, square planar Ni^{2+}, and tetrahedral geometry Zn^{2+}) and conformationally flexible ligand, NDI-based MOFs featured not only the pi-stacking dependent conductivity tunability but also the optimized pore structure and surface function by varying the ligand conformation. The gas-sensing results revealed that Co-pyNDI and Zn-pyNDI were promising active materials for selective ammonia and acetone chemiresistive sensing (Xue et al., 2023), which could provide insights into designing PCP semiconductors with specific functions by synergistically regulating conductivity and pore environment. Compared with bulk PCPs, 2D PCPs tend to have more accessible active sites that facilitate more interactions between active sites and gas molecules, which then improves gas sensing performance. In view of this, 2D ultrathin $Co[Ni(CN)_4]$ PCPs nanosheets prepared by a simple bottom-up method were used for NH_3 sensing, which possessed planar square structure (50 nm) and ultrathin thickness (2.5 nm). To further improve the sensing properties of bare 2D $Co[Ni(CN)_4]$ nanosheets, carbon quantum dots were incorporated onto the surface of 2D nanosheets. It was found that the fabricated nanocomposites exhibited unparalleled NH_3 sensitivity with fast response and excellent selectivity (Zhang et al., 2023).

Except for the previously introduced POPs, COPs, as a class of micro- or meso-porous organic materials, are usually amorphous and have no long-range crystalline structure, which has also been reported in the gas-sensing field. For instance, by using imine as a linkage between two triazine-cored precursors, micro-spherical COP was synthesized, which exhibited solid-state luminescent properties because of the N-rich core architecture. Moreover, morphologically oriented COP exhibited a reversible chromogenic response toward noxious HCl gas with better gas adsorption and diffusion ability (Prakash and Masram, 2020). In another example, a hydrazide-based COP with pyridine functionalities was also employed to fabricate an efficient chemosensor for the detection of H_2S gas, which showed a higher selectivity for H_2S gas with a fast response and recovery rate. After the adsorption of H_2S gas, the proton transportation process inside the COP backbone was restricted, and then the resistance of COP was increased accordingly (Maiti et al., 2020). Recently, the effect of N-contained groups on the sensing properties of COP was detailed by embedding primary amine, secondary amine, tertiary amine, and quaternary ammonium in four COPs through reasonable designation of building blocks. The sensing results indicated that primary amine, secondary amine, and tertiary amine were excellent NH_3 adsorption sites and beneficial for NH_3 sensing, while quaternary ammonium was more suitable for NO_2 adsorption and sensing. It was confirmed by DFT calculation and in situ UV-vis spectra that NH_3 interacted with primary amine, secondary amine, and tertiary amine through hydrogen bonding mode, and NO_2 interacted with quaternary nitrogen atoms through coordination (Zhu et al., 2023). In addition to the above discussions, there is still a lot of room for further developments of POP-based sensing applications, such as the improvement of sensing performance in sensitivity and selectivity, the exploration of inherently sensing mechanisms, and the development of novel sensing models.

14.4 Device construction methods for polymer gas sensors

Solution-based techniques have been considered the simplest technologies for fabricating polymer gas sensors, including spin-coating, drop casting, spray coating, and solution shearing. Spin coating is the most commonly used technique for polymer sensor fabrication due to its simplicity and low cost. Spin-coated devices with different thicknesses and coverage can be obtained by controlling the rotational speed, spin-coating time, and the viscosity of the solution. However, the spin-coating method has several drawbacks, including material waste, restriction on small-scale substrates, and high film thickness. Drop-casting is a simple, well-established, waste-free, low-cost, and low-energy consumption method that is widely used in laboratories. Despite this, the thickness cannot be precisely controlled, which leads to inconsistency among different sensor devices. Spray coating is a convenient, cost-effective, and industrially scalable deposition technique suitable for depositing polymers on a 3D surface during low-temperature processing to create electronics. Commonly, the atomization process is controlled by N_2 air pressure or an ultrasonic technique, in which tiny droplets are formed and directly sprayed onto the target substrate to form a very thin, dense, and stable film. The surface morphology of the film can be controlled by adjusting several parameters, including air pressure, viscosity, solvent properties, the distance between the nozzle and substrate, spray time, and spray temperature. It has been confirmed that the spray coating method can produce high-quality polymer thin films on various substrates. The solution shear-coating method is also a simple, efficient, low-cost, and scalable technique for fabricating polymer thin films for sensing. During the shearing process, the polymer solution is dropped on a heated substrate, followed by dragging a shearing plate, and then a polymer thin film is obtained after solvent evaporation.

Similar to spray coating, the ink-jet printing method is an attractive printing technique, especially for the controlled solution deposition of functional materials in small quantities in conformity with specific patterns, which is operated by ejecting single drops of ink through very fine nozzles at ambient temperature without needing vacuum conditions or direct contact. The essential requirements for the ink-jet printing method are a suitable viscosity and surface tension. Compared with the other construction methods, such as electrochemical deposition, spin coating, and dipping coating, the pattern capability of this method and its compatibility with a wide range of substrates are outstanding advantages. Till now, ink-jet printing methods have been widely investigated for chemical sensing, mainly using conducting polymers such as PANI and PT. For instance, dodecylbenzene sulfonate-doped PANI nanoparticles were inkjet printed onto a screen-printed carbon paste electrode, which allowed the rapid fabrication of PANI-based sensors that were not easily achievable in the past (Crowley et al., 2008). In another work, PT polymers with a variety of side chains, end groups, and secondary polymer chains were selectively deposited onto the array of transduction electrodes by using the inkjet printing method, which demonstrated higher sensitivity and excellent selectivity for the detection and discrimination of VOCs, indicating the huge potential application of this method (Li et al., 2007). In addition to single-phase conducting polymers, the ink-jet printing method can also be employed to fabricate nanocomposite-based gas sensors. For instance, based on polymer-carbon black composites, ink-jet-manufactured chemical sensors exhibited relatively higher sensing properties for VOC gases, which are comparable to those of the casting sensing devices. Despite this, to achieve better sensing properties, the ink deposition method still needs further investigation and optimization (Loffredo et al., 2009). Besides, screen printing and 3D printing have also been considered promising printing technologies to fabricate polymer sensor devices. Although several techniques, including solution-based methods and printing technologies, were employed to fabricate polymer sensing films, the suitable selection of solvents and the pretreatment of substrates are very vital for the reliability and efficacy of sensing devices.

Molecularly imprinted polymers (MIPs) are crosslinked polymers that are formed in the presence of template molecules (Akgönüllü et al., 2023; Nguyen et al., 2023; Saylan et al., 2019; Zhao et al., 2023). Removal of the template creates binding cavities with affinity and selectivity for the template molecules, as shown in Fig. 14.7. This templated imprinting process has the advantage of different binding selectivity by using different templates. On the other hand, like other synthetic polymers, MIPs can also be rapidly prepared with excellent thermal, chemical, and mechanical stabilities (Shimizu and Stephenson, 2010). To synthesize MIPs, different reagents are required,

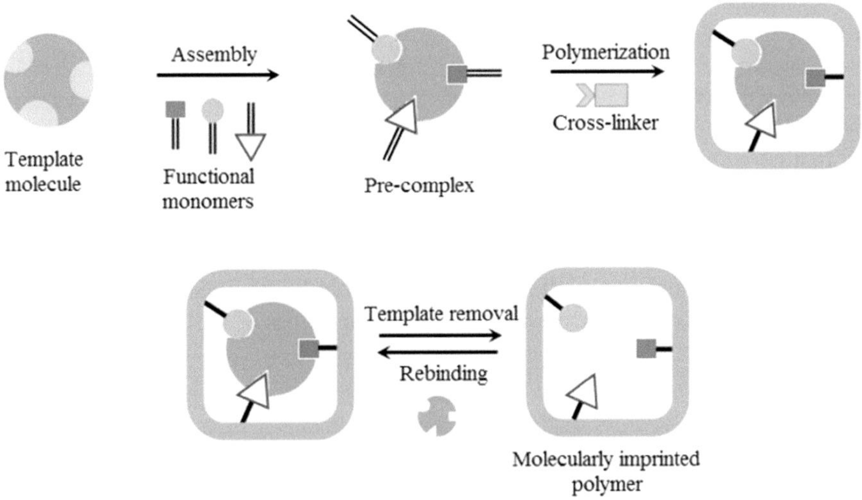

FIGURE 14.7 Molecularly imprinted polymer

. Scheme of the principle of molecularly imprinted polymer preparation (Saylan et al., 2019).

From Saylan, Y., Akgönüllü, S., Yavuz, H., Ünal, S., & Denizli, A. (2019). Molecularly imprinted polymer based sensors for medical applications. Sensors, 19*(6). Available from: https://doi.org/10.3390/s19061279.*

including a functional monomer, template, cross-linker, and polymerization initiator. Their ratio with respect to one another greatly influences the specific interaction between the polymer and the template and, subsequently, the binding capacity and imprinting factor of the resulting MIP. Besides, different types of polymerization, initiators, and solvents also play a vital role in the whole process. Till now, various polymerization techniques have been employed to synthesize MIPs, including thermal polymerization, emulsion polymerization, electropolymerization, electrospinning, and photopolymerization. Here, different polymerization methods will be introduced, not limited to the synthesis of MIPs. The thermal polymerization method is one of the most commonly used methods, is relatively straightforward, and allows for the creation of large batches of materials cost-effectively. The major drawbacks of this method are the time-consuming, tedious grinding and sieving procedures that lead to a large loss of product and the generation of a heterogeneous mixture of micro-scaled particles. Within this method, different initiators can be used, such as azobisisobutyronitrile, benzoyl peroxide, and $FeCl_30.6H_2O$. To avoid generating bulk polymers, another popular approach named emulsion polymerization is usually adopted to create polymers with various dimensions by optimizing the reaction conditions. In this method, surfactant molecules are needed. Commonly, the monomers act as oil phases that are shielded from the water phase by the surfactant and undergo cross-linking inside a microreactor, leading to more homogenous polymers with a tunable dimension. Another interesting approach, named electropolymerization, has also gained increasing attention due to the high control of the layer thickness and the direct grafting onto the electrode. However, the main drawbacks associated with electropolymerization are the possible low degree of cross-linking, the limited choice of electro-active monomers, and the difficult upscaling of the fabrication process. Another polymerization technique named photopolymerization was also widely used for polymer production. This approach is a technique that uses the energy of a light source to initiate a polymerization reaction. Commonly, photoinitiators, photosensitive functional polymers, photocross-linkable polymers, and reversible addition-fragmentation chain-transfer agents are all needed. The specific advantages of this approach are similar to those of thermally induced methods, which could allow the detection of different targets by using different polymers. However, this approach is not suitable for the detection of target species that are sensitive to irradiation with high-energy light sources. The above-mentioned techniques are frequently used for polymer production. Recently, novel approaches for MIP synthesis have been developed. Among them, solid-phase synthesis of MIP has proven to be an effective method that exhibited higher industrial potential. By using an automated synthesis protocol, a wide variety of targets were successfully imprinted (Poma et al., 2013), which have also been employed for the fabrication of electro-responsive nanoMIPs for glucose recognition (Garcia-Cruz et al., 2020).

14.5 Strategies to improve gas sensing performance

As described before, based on different sensing mechanisms, fluorescence polymers or colorimetric polymers can all detect various gas species. Moreover, according to the rational design of molecule structure, the selective detection of target gas species can be partially achieved. Despite this, selectivity is still a key point when detecting a specific gas species in complex environments. To achieve this aim, except for the rationally designed molecular structure, the exploration of novel polymer materials or the construction of sensor arrays should also be developed. Besides, to achieve higher sensing performance in a specific testing environment, such as acid gas or alkaline gas, the rational selection of synthesis methods for polymer sensing materials and the integration of polymers with suitable substrates should be simultaneously considered for the higher sensing performance. Therefore, based on the previously described sensing mechanism and the recent development of polymer sensors for fluorescence sensing, colorimetric sensing, and conductivity sensing, the strategies to improve the gas sensing performance of polymers will be discussed as follows:

In recent years, fluorescent sensor arrays have shown outstanding advantages in multiple gas molecule detection, which can detect different gas species in complex environments with higher selectivity. To detect lung cancer-related VOCs (p-xylene, styrene, isoprene, and hexanal), a fluorescent sensor array based on a cross-responsive mechanism was proposed. It consisted of seven elements: porphyrin, porphyrin derivatives, and three chemically responsive dyes. Moreover, the sensory array system with a special gas chamber, FLUENT software, and the data collection and processing system were also designed. The proposed method could 100% discriminate between the four VOCs by extracting the characteristic matrix of spectra. Combined with an artificial neural network, the correct rate of quantitative detection was up to 100%, and the sensing device still has effective responses at concentrations below 50 ppb. It was believed that the fluorescent cross-responsive sensor array could have potential clinical and practical value in the future (Lei et al., 2015). In another example, based on a series of structurally analogous perylene diimides (PDIs), a fluorescent sensor array was also constructed to differentiate gas-phase amines. Through tuning the specific end groups of PDIs, the molecular electron-donating ability and polarity can be adjusted. The fabricated sensor array assembled from different PDI structures exhibited different sensitivities to a variety of amines, which made it possible to realize fingerprint detection and selective differentiation of different types of trace amine vapors in the ppm range. Moreover, the fabricated sensor array showed a distinct higher sensitivity to the selected amines, with a lower detection limit of ppt level for aniline and ppb level for the other amines (Hu et al., 2018). In another research work, to achieve the highly discriminative analysis of 20 clinically and environmentally relevant volatile, small organic molecules, four coordinated, non-planar mono-boron complexes and four relevant polymers were first prepared. After that, eight fluorescent films based on the as-prepared polymers were integrated on different substrates to form the fluorescent sensor array. The sensing results revealed that the fabricated sensor array exhibited unprecedented discriminating capability toward different kinds of volatile small organic molecules. Moreover, for the specific signal molecule of lung cancer, n-pentane, the detection limit can reach 3.7 ppm with a 1 s response time and excellent stability after several repeated tests. Because of the tetrahedral structure of the boron centers in the polymers, the molecular channels are produced in fluorescent films, which are the key points for fast and reversible sensing. Additionally, the fluorescent films with additional selectivity were attributed to the polarity of the microchannels (Qi et al., 2018). Except for the construction of fluorescent sensor arrays, a kind of novel polymer was also developed for higher sensitivity. For instance, benefiting from its central tetraphenylethylene core with aggregation-induced emission effect and its highly electro-active branches, a dendrimer was fabricated into highly-luminescence CMPs sensing films, which exhibited superior sensitivity to VOCs. More importantly, by establishing a kind of 2D fluorescence sensor array, 18 types of VOC vapors can be precisely distinguished by linear discriminant analysis (Liu et al., 2020), which exhibited great potential application in VOC vapor discrimination.

A colorimetric sensor array composed of seven MIPs was shown to accurately identify seven different aromatic amines by using a dye-displacement strategy. It should be noted that the molecularly imprinted strategy enabled the rapid and rational preparation of polymers with excellent selectivity for sensory arrays. Moreover, a dye-displacement strategy provided an easily measurable colorimetric response. In another research work, by using the drop-casting technique, eight DA monomers with amphiphilic or bolaamphiphilic characteristics were coated onto a filter paper surface and converted to PDAs by UV irradiation. The PDA sensor array was capable of distinguishing 18 distinct VOCs in the vapor phase with higher reproducibility and discriminating ability by using color change patterns, as measured by RGB values and statistically analyzed by principal component analysis (Eaidkong et al., 2012). Another simple and low-cost colorimetric sensor array was also prepared for the discrimination of gaseous amines, which was achieved by combining chromogenic sensing dye with different polarities of polymer nanoparticles. By simply varying the mixing ratio of polymer nanoparticles with two different polarities, the sensor array enables the reliable polarity-based discrimination of closely related primary aliphatic amines. Moreover, the fabricated sensor array exhibited the characteristics of a visually recognizable gas polarity-dependent colorimetric response, high sample discrimination ability, and high fabrication reproducibility (Soga et al., 2013).

A conducting polymer sensor array consisting of PPy with different thicknesses was employed to analyze the diabetic patient's breath. A significant difference between the diabetic patient and the control person was shown. Throughout the experimental trials, the portable gas analyzing system has shown the possibility of a potential diagnostic tool for the case of diabetes mellitus (Yu et al., 2005). In another work, nine regioregular PT-based polymers with different side chains, end-groups, and copolymers were integrated into a single chip, which was fabricated by a custom inkjet printing method, demonstrating the higher sensitivity and excellent selectivity for the detection and discrimination of VOCs. Due to the different chemical structures of the polymers, the conductivity of the polymers increased or decreased after exposure to VOCs. Through principle component analysis, different VOCs can be discriminated against (Li et al., 2007). To further increase the discriminability of the sensing array, a conducting polymer film-based sensor array composed of PANI, PPy, and modified poly-3-methylthiophene was designed. Through statistical analysis of the data retrieved from the responses of 20 different polymer sensors to the vapors of nine volatile organic solvents, the discrimination ability of various sensor subsets was evaluated. Moreover, the sensing array optimization procedure was also proposed to achieve the required analyte recognition reliability, in which individual specificity factors were extracted to avoid the time-consuming exhaustive search routine (Kukla et al., 2009). To create a cross-reactive conducting polymer nanowire-based sensor array with each individual sensor having partial selectivity, distinct catalytic nanoparticles were incorporated into each sensing element, resulting in a unique response of each sensor when exposed to a specific analyte. As a proof of concept, for each sensing element, three types of catalytic nanoparticles (CuO, Mn_2O_3, and Ag) were deposited on PANI, respectively, for the simultaneous classification and quantification of three chemical species (ascorbic acid, dopamine, and hydrogen peroxide). Based on the chemiresistive signal responses from the sensor array, a principle component analysis algorithm was applied to the single analyte measurements. Afterward, the data points were mapped onto the 2D plot, which allows for easy visualization of the mapping of the data points (Song and Choi, 2015). To detect a low concentration range of VOCs, poly(3,4-ethylene-dioxythiphene)-poly(styrene sulfonate) (PEDOT:PSS) nanowires functionalized with different self-assembled monolayers (SAMs) were also developed. Especially different from conventional nanofabrication techniques, a cost-effective nanoscale soft lithography method was adopted to fabricate a gas sensor array, which consisted of well-ordered sub-100 nm-wide conducting polymer nanowires. Due to the effect of side chains and functional groups in SAMs, different sensitivities to different targeted analytes were achieved, leading to a clear separation toward ten VOCs (Jiang et al., 2018). With the development of polymer-based sensor arrays, accurate identification of complex gas environments has been partially achieved. Based on the different sensing mechanisms, the toxic gases and the biomarker molecules can be effectively distinguished even in the presence of several interference gases. Despite this, there are still many opportunities and challenges. First, the deeper application of mathematical statistics and data analysis is still worth exploring. Second, mixing transduce mode could be an efficient method to solve the problem of signal crossing. Third, the specific response value should be further enhanced, particularly when the interference gas concentration is much larger than that of the characteristic gas. Last but not least, the integration of sensor arrays on flexible substrates with biocompatible materials should be explored for wider applications.

14.6 Conclusions and perspectives

The present book chapter focuses on the recent development of polymer-based gas sensors from the perspective of three different transduction modes, covering the different sensing mechanisms, the design and application of sensing materials, and the device fabrication methods, as well as the strategies to improve the sensing performances. Recent advances in polymer-based sensors published in academic studies demonstrate that these sensors are rapidly approaching real-life applications. Despite this, the realization of the practical application still has a long way to go. The main challenges of polymer gas sensors in terms of selectivity have already been analyzed. Considering the long-standing problems with selectivity, the construction of sensor arrays could be an effective method. It should be noted that the construction of sensor arrays should consider the chip construction method simultaneously. On the other hand, the exploration of novel polymers with specific function groups or other outstanding characteristics (a higher specific surface area, a tunable pore size, etc.) can also be helpful for the improvement of sensing properties. Generally, the field of polymer gas sensors can be further developed and exploited, as described below.

First, considering the easy-processing properties of polymers, the integration of polymer materials on flexible substrates for more extensive application in human health real-time detection should be further developed, such as in human motion, breath analysis, and virus detection.

Second, based on the previously described three different transduction modes, including fluorescence, colorimetry, and conductive sensing, the mixed transduction mode could be helpful for complex environment detection with multiple analytes.

Third, to achieve long-term structural stability of polymers, the covalent stabilization of supramolecular assemblies of responsive polymers (shell or core cross-linking strategies) can be utilized.

Finally, to meet the requirements of industrial applications, further exploration of novel polymers should be developed rather than standing still. Besides, to reduce the distance between fundamental research and practical applications, the production method for sensory devices should also be improved to obtain cheap and portable sensory devices that can be used by non-professionals.

References

Akgönüllü, S., Kılıç, S., Esen, C., & Denizli, A. (2023). Molecularly imprinted polymer-based sensors for protein detection. *Polymers*, *15*(3), 629. Available from https://doi.org/10.3390/polym15030629.

Athawale, A. A., Bhagwat, S. V., & Katre, P. P. (2006). Nanocomposite of Pd-polyaniline as a selective methanol sensor. *Sensors and Actuators, B: Chemical*, *114*(1), 263–267. Available from https://doi.org/10.1016/j.snb.2005.05.009.

Back, J. W., Lee, S., Hwang, C. R., Chi, C. S., & Kim, J. Y. (2011). Fabrication of conducting PEDOT nanotubes using vapor deposition polymerization. *Macromolecular Research*, *19*(1), 33–37. Available from https://doi.org/10.1007/s13233-011-0111-x.

Baek, S., Khazi, M. I., & Kim, J. M. (2023). Colorimetric and fluorometric ammonia sensor based on protonated bipyridyl-containing polydiacetylene. *Dyes and Pigments*, *215*. Available from https://doi.org/10.1016/j.dyepig.2023.111254, http://www.journals.elsevier.com/dyes-and-pigments/.

Bai, H., Li, C., Chen, F., & Shi, G. (2007). Aligned three-dimensional microstructures of conducting polymer composites. *Polymer*, *48*(18), 5259–5267. Available from https://doi.org/10.1016/j.polymer.2007.06.071, http://www.journals.elsevier.com/polymer/.

Bandyopadhyay, S., Pallavi, P., Anil, A. G., & Patra, A. (2015). Fabrication of porous organic polymers in the form of powder, soluble in organic solvents and nanoparticles: A unique platform for gas adsorption and efficient chemosensing. *Polymer Chemistry*, *6*(20), 3775–3780. Available from https://doi.org/10.1039/c5py00235d, http://pubs.rsc.org/en/Journals/JournalIssues/PY.

Ben, T., Ren, H., Ma, S., Cao, D., Lan, J., Jing, X., Wang, W., Xu, J., Deng, F., Simmons, J. M., Qiu, S., & Zhu, G. (2009). Targeted synthesis of a porous aromatic framework with high stability and exceptionally high surface area. *Angewandte Chemie International Edition*, *48*(50), 9457–9460. Available from https://doi.org/10.1002/anie.200904637.

Blaszczyk-Lezak, I., Desmaret, V., & Mijangos, C. (2016). Electrically conducting polymer nanostructures confined in anodized aluminum oxide templates (AAO. *Express Polymer Letters*, *10*(3), 259–272. Available from https://doi.org/10.3144/expresspolymlett.2016.24.

Bonillo, B., Sprick, R. S., & Cooper, A. I. (2016). Tuning photophysical properties in conjugated microporous polymers by comonomer doping strategies. *Chemistry of Materials*, *28*(10), 3469–3480. Available from https://doi.org/10.1021/acs.chemmater.6b01195, http://pubs.acs.org/journal/cmatex.

Campbell, M. G., Liu, S. F., Swager, T. M., & Dincă, M. (2015). Chemiresistive sensor arrays from conductive 2D metal–organic frameworks. *Journal of the American Chemical Society*, *137*(43), 13780–13783. Available from https://doi.org/10.1021/jacs.5b09600.

Campbell, M. G., Sheberla, D., Liu, S. F., Swager, T. M., & Dincă, M. (2015). Cu 3 (hexaiminotriphenylene) 2: An electrically conductive 2D metal-organic framework for chemiresistive sensing. *Angewandte Chemie International Edition*, *54*(14), 4349–4352. Available from https://doi.org/10.1002/anie.201411854.

Cassagneau, T., & Caruso, F. (2002). Semiconducting polymer inverse opals prepared by electropolymerization. *Advanced Materials*, *14*(1), 34–38. Available from https://doi.org/10.1002/1521-4095(20020104).

Cho, S. H., Suh, J. M., Eom, T. H., Kim, T., & Jang, H. W. (2021). Colorimetric sensors for toxic and hazardous gas detection: A review. *Electronic Materials Letters*, *17*(1). Available from https://doi.org/10.1007/s13391-020-00254-9, http://www.springerlink.com/content/1738-8090/.

Crowley, K., O'Malley, E., Morrin, A., Smyth, M. R., & Killard, A. J. (2008). An aqueous ammonia sensor based on an inkjet-printed polyaniline nanoparticle-modified electrode. *Analyst*, *133*(3), 391–399. Available from https://doi.org/10.1039/b716154a, http://pubs.rsc.org/en/journals/journal/an.

Cui, F. Z., Xie, J. J., Jiang, S. Y., Gan, S. X., Ma, D. L., Liang, R. R., Jiang, G. F., & Zhao, X. (2019). A gaseous hydrogen chloride chemosensor based on a 2D covalent organic framework. *Chemical Communications*, *55*(31), 4550–4553. Available from https://doi.org/10.1039/c9cc01548e, http://pubs.rsc.org/en/journals/journal/cc.

Delvaux, M., Duchet, J., Stavaux, P. Y., Legras, R., & Demoustier-Champagne, S. (2000). Chemical and electrochemical synthesis of polyaniline micro- and nano-tubules. *Synthetic Metals*, *113*(3), 275–280. Available from https://doi.org/10.1016/S0379-6779(00)00226-5.

Eaidkong, T., Mungkarndee, R., Phollookin, C., Tumcharern, G., Sukwattanasinitt, M., & Wacharasindhu, S. (2012). Polydiacetylene paper-based colorimetric sensor array for vapor phase detection and identification of volatile organic compounds. *Journal of Materials Chemistry*, *22*(13), 5970–5977. Available from https://doi.org/10.1039/c2jm16273c, http://www.rsc.org/Publishing/Journals/jm/index.asp.

Feng, J., Li, Y., & Yang, M. (2010). Conjugated polymer-grafted silica nanoparticles for the sensitive detection of TNT. *Sensors and Actuators B: Chemical*, *145*(1), 438–443. Available from https://doi.org/10.1016/j.snb.2009.12.056.

Gao, H., Jiang, T., Han, B., Wang, Y., Du, J., Liu, Z., & Zhang, J. (2004). Aqueous/ionic liquid interfacial polymerization for preparing polyaniline nanoparticles. *Polymer*, *45*(9), 3017–3019. Available from https://doi.org/10.1016/j.polymer.2004.03.002.

Garcia-Cruz, A., Ahmad, O. S., Alanazi, K., Piletska, E., & Piletsky, S. A. (2020). Generic sensor platform based on electro-responsive molecularly imprinted polymer nanoparticles (e-NanoMIPs). *Microsystems & Nanoengineering*, *6*(1). Available from https://doi.org/10.1038/s41378-020-00193-3.

Gheitarani, B., Golshan, M., Safavi-Mirmahalleh, S. A., Salami-Kalajahi, M., Hosseini, M. S., & Alizadeh, A. A. (2023). Fluorescent polymeric sensors based on N-(rhodamine-G) lactam-N′-allyl-ethylenediamine and 7-(allyloxy) − 2H-chromen-2-one for Fe3 + ion detection. *Colloids and Surfaces A: Physicochemical and Engineering Aspects*, *656*. Available from https://doi.org/10.1016/j.colsurfa.2022.130473, http://www.elsevier.com/locate/colsurfa.

Gribkova, O. L., Kabanova, V. A., & Nekrasov, A. A. (2020). Electrodeposition of thin films of polypyrrole-polyelectrolyte complexes and their ammonia-sensing properties. *Journal of Solid State Electrochemistry*, *24*(11–12), 3091–3103. Available from https://doi.org/10.1007/s10008-020-04766-0.

Gu, Y., & Huang, J. (2013). Colorimetric detection of gaseous ammonia by polyaniline nanocoating of natural cellulose substances. *Colloids and Surfaces A: Physicochemical and Engineering Aspects*, *433*, 166–172. Available from https://doi.org/10.1016/j.colsurfa.2013.05.016.

Guillén, M. G., Gámez, F., Lopes-Costa, T., Cabanillas-González, J., & Pedrosa, J. M. (2016). A fluorescence gas sensor based on Förster resonance energy transfer between polyfluorene and bromocresol green assembled in thin films. *Sensors and Actuators B: Chemical*, *236*, 136–143. Available from https://doi.org/10.1016/j.snb.2016.06.011.

Han, M. G., Cho, S. K., Oh, S. G., & Im, S. S. (2002). Preparation and characterization of polyaniline nanoparticles synthesized from DBSA micellar solution. *Synthetic Metals*, *126*(1), 53–60. Available from https://doi.org/10.1016/S0379-6779(01)00494-5.

He, J., Zhang, T. Y., & Chen, G. (2012). Ammonia gas-sensing characteristics of fluorescence-based poly(2-(acetoacetoxy)ethyl methacrylate) thin films. *Academic Press Inc., Hong Kong Journal of Colloid and Interface Science*, *373*(1), 94–101. Available from https://doi.org/10.1016/j.jcis.2011.11.074, http://www.elsevier.com/inca/publications/store/6/2/2/8/6/1/index.htt.

Hong, L., Li, Y., & Yang, M. (2010). Fabrication and ammonia gas sensing of palladium/polypyrrole nanocomposite. *Sensors and Actuators B: Chemical*, *145*(1), 25–31. Available from https://doi.org/10.1016/j.snb.2009.11.057.

Hu, Y., Zhou, Z., Zhao, F., Liu, X., Gong, Y., Xiong, W., & Sillanpää, M. (2018). Fingerprint detection and differentiation of gas-phase amines using a fluorescent sensor array assembled from asymmetric perylene diimides. *Scientific Reports*, *8*(1), 10277. Available from https://doi.org/10.1038/s41598-018-28556-x.

Huang, J., Virji, S., Weiller, B. H., & Kaner, R. B. (2004). Nanostructured polyaniline sensors. *Chemistry – A European Journal*, *10*(6), 1314–1319. Available from https://doi.org/10.1002/chem.200305211, http://www.interscience.wiley.com.

Jang, J., & Oh, J. H. (2002). Novel crystalline supramolecular assemblies of amorphous polypyrrole nanoparticles through surfactant templating. *Chemical Communications*, *2*(19), 2200–2201. Available from https://doi.org/10.1039/b207744m.

Jiang, Y., Tang, N., Zhou, C., Han, Z., Qu, H., & Duan, X. (2018). A chemiresistive sensor array from conductive polymer nanowires fabricated by nanoscale soft lithography. *Nanoscale*, *10*(44), 20578–20586. Available from https://doi.org/10.1039/c8nr04198a.

Kim, M. O., Khan, M. Q., Ullah, A., Duy, N. P., Zhu, C., Lee, J. S., & Kim, I. S. (2019). Development of VOCs gas sensor with high sensitivity using colorimetric polymer nanofiber: A unique sensing method. *Materials Research Express*, *6*(10). Available from https://doi.org/10.1088/2053-1591/ab42a5, https://iopscience.iop.org/article/10.1088/2053-1591/ab42a5/pdf.

Krishnaveni, V., Esclance DMello, M., Sahoo, P., Thokala, N., Bakuru, V. R., Vankayala, K., Basavaiah, K., & Kalidindi, S. B. (2023). Palladium-nanoparticle-decorated covalent organic framework nanosheets for effective hydrogen gas sensors. *ACS Applied Nano Materials*. Available from https://doi.org/10.1021/acsanm.3c01806, https://pubs.acs.org/journal/aanmf6.

Kukla, A. L., Pavluchenko, A. S., Shirshov, Y. M., Konoshchuk, N. V., & Posudievsky, O. Y. (2009). Application of sensor arrays based on thin films of conducting polymers for chemical recognition of volatile organic solvents. *Sensors and Actuators, B: Chemical.*, *135*(2), 541–551. Available from https://doi.org/10.1016/j.snb.2008.09.027.

Kweon, O. Y., Lee, S. J., & Oh, J. H. (2018). Wearable high-performance pressure sensors based on three-dimensional electrospun conductive nanofibers. *Asia Materials*, *10*(6), 540–551. Available from https://doi.org/10.1038/s41427-018-0041-6, http://www.nature.com/am/index.html.

Kyokunzire, P., Jeong, G., Shin, S. Y., Cheon, H. J., Wi, E., Woo, M., Vu, T. T., & Chang, M. (2023). Enhanced nitric oxide sensing performance of conjugated polymer films through incorporation of graphitic carbon nitride. *International Journal of Molecular Sciences*, *24*(2), 1158. Available from https://doi.org/10.3390/ijms24021158.

Lawaniya, S. D., Kumar, S., Yu, Y., & Awasthi, K. (2023). Ammonia sensing properties of PPy nanostructures (urchins/flowers) towards low-cost and flexible gas sensors at room temperature. *Sensors and Actuators B: Chemical*, *382*, 133566. Available from https://doi.org/10.1016/j.snb.2023.133566.

Lee, K., Povlich, L. K., & Kim, J. (2010). Recent advances in fluorescent and colorimetric conjugated polymer-based biosensors. *Analyst*, *135*(9), 2179–2189. Available from https://doi.org/10.1039/c0an00239a, http://pubs.rsc.org/en/journals/journal/an.

Lei, J. C., Hou, C. J., Huo, D. Q., Luo, X. G., Bao, M. Z., Li, X., Yang, M., & Fa, H. B. (2015). A novel device based on a fluorescent cross-responsive sensor array for detecting lung cancer related volatile organic compounds. *Review of Scientific Instruments*, *86*(2). Available from https://doi.org/10.1063/1.4907628, http://scitation.aip.org/content/aip/journal/rsi.

Li, B., Santhanam, S., Schultz, L., Jeffries-EL, M., Iovu, M. C., Sauvé, G., Cooper, J., Zhang, R., Revelli, J. C., Kusne, A. G., Snyder, J. L., Kowalewski, T., Weiss, L. E., McCullough, R. D., Fedder, G. K., & Lambeth, D. N. (2007). Inkjet printed chemical sensor array based on polythiophene conductive polymers. *Sensors and Actuators, B: Chemical*, *123*(2), 651–660. Available from https://doi.org/10.1016/j.snb.2006.09.064.

Liao, Y., Zhang, C., Zhang, Y., Strong, V., Tang, J., Li, X. G., Kalantar-Zadeh, K., Hoek, E. M. V., Wang, K. L., & Kaner, R. B. (2011). Carbon nanotube/polyaniline composite nanofibers: Facile synthesis and chemosensors. *Nano Letters*, *11*(3), 954–959. Available from https://doi.org/10.1021/nl103322b.

Liu, H., Wang, Y., Mo, W., Tang, H., Cheng, Z., Chen, Y., Zhang, S., Ma, H., Li, B., & Li, X. (2020). Dendrimer-based, high-luminescence conjugated microporous polymer films for highly sensitive and selective volatile organic compound sensor arrays. *Advanced Functional Materials*, *30*(13). Available from https://doi.org/10.1002/adfm.201910275, http://onlinelibrary.wiley.com/journal/10.1002/(ISSN)1616-3028.

Liu, X., Xu, Y., & Jiang, D. (2012). Conjugated microporous polymers as molecular sensing devices: Microporous architecture enables rapid response and enhances sensitivity in fluorescence-on and fluorescence-off sensing. *Journal of the American Chemical Society*, *134*(21), 8738–8741. Available from https://doi.org/10.1021/ja303448r.

Liu, X., Zheng, W., Kumar, R., Kumar, M., & Zhang, J. (2022). Conducting polymer-based nanostructures for gas sensors. *Coordination Chemistry Reviews*, *462*, 214517. Available from https://doi.org/10.1016/j.ccr.2022.214517.

Loffredo, F., De Girolamo Del Mauro, A., Burrasca, G., La Ferrara, V., Quercia, L., Massera, E., Di Francia, G., & Della Sala, D. (2009). Ink-jet printing technique in polymer/carbon black sensing device fabrication. *Sensors and Actuators B: Chemical*, *143*(1), 421–429. Available from https://doi.org/10.1016/j.snb.2009.09.024.

Lu, Q., Hu, Z., zhang, D., Xu, F., & Xia, J. (2023). 2D polyaniline derivatives as turn-on fluorescent probe for efficient triethylamine detection at room temperature. *Talanta*, *265*, 124868. Available from https://doi.org/10.1016/j.talanta.2023.124868.

Lu, Q., Qian, J., Xu, F., He, G., Liu, Y., & Xia, J. (2023). Synthesis of 2D fluorescent polyaniline derivatives as multifunctional fluorescent chemosensor. *Journal of Polymer Science*, *61*(3), 211–222. Available from https://doi.org/10.1002/pol.20220503.

Lv, Y. Y., Wu, J., & Xu, Z. K. (2010). Colorimetric and fluorescent sensor constructing from the nanofibrous membrane of porphyrinated polyimide for the detection of hydrogen chloride gas. *Sensors and Actuators, B: Chemical*, *148*(1), 233–239. Available from https://doi.org/10.1016/j.snb.2010.05.029.

Maiti, S., Mandal, B., Sharma, M., Mukherjee, S., & Das, A. K. (2020). A covalent organic polymer as an efficient chemosensor for highly selective H2S detection through proton conduction. *Chemical Communications*, *56*(65), 9348–9351. Available from https://doi.org/10.1039/d0cc02704a, http://pubs.rsc.org/en/journals/journal/cc.

McQuade, D. T., Pullen, A. E., & Swager, T. M. (2000). Conjugated polymer-based chemical sensors. *Chemical Reviews*, *100*(7), 2537–2574. Available from https://doi.org/10.1021/cr9801014.

Nasresfahani, S., Zargarpour, Z., Sheikhi, M. H., & Ana, S. F. N. (2020). Improvement of the carbon monoxide gas sensing properties of polyaniline in the presence of gold nanoparticles at room temperature. *Synthetic Metals*, *265*, 116404. Available from https://doi.org/10.1016/j.synthmet.2020.116404.

Nguyen, V. B. C., Ayankojo, A. G., Reut, J., Rappich, J., Furchner, A., Hinrichs, K., & Syritski, V. (2023). Molecularly imprinted co-polymer for class-selective electrochemical detection of macrolide antibiotics in aqueous media. *Sensors and Actuators B: Chemical*, *374*. Available from https://doi.org/10.1016/j.snb.2022.132768, https://www.journals.elsevier.com/sensors-and-actuators-b-chemical.

Park, J. J., Kim, Y., Lee, C., Kook, J. W., Kim, D., Kim, J. H., Hwang, K. S., & Lee, J. Y. (2020). Colorimetric visualization using polymeric core-shell nanoparticles: Enhanced sensitivity for formaldehyde gas sensors. *Polymers*, *12*(5). Available from https://doi.org/10.3390/POLYM12050998, https://www.mdpi.com/2073-4360/12/5/998.

Park, Y. K., Oh, H. J., Lee, H. D., Lee, J. J., Kim, J. H., & Lee, W. (2022). Facile and eco-friendly fabrication of a colorimetric textile sensor by UV-induced photografting for acidic gas detection. *Journal of Environmental Chemical Engineering*, *10*(5), 108508. Available from https://doi.org/10.1016/j.jece.2022.108508.

Poma, A., Guerreiro, A., Whitcombe, M. J., Piletska, E. V., Turner, A. P. F., & Piletsky, S. A. (2013). Solid-phase synthesis of molecularly imprinted polymer nanoparticles with a reusable template-"plastic antibodies". *Advanced Functional Materials*, *23*(22), 2821–2827. Available from https://doi.org/10.1002/adfm.201202397.

Prakash, K., & Masram, D. T. (2020). Chromogenic covalent organic polymer-based microspheres as solid-state gas sensor. *Journal of Materials Chemistry*, *8*, 9201–9204. Available from https://doi.org/10.1039/D0TC02129F.

Qi, Y., Xu, W., Kang, R., Ding, N., Wang, Y., He, G., & Fang, Y. (2018). Discrimination of saturated alkanes and relevant volatile compounds via the utilization of a conceptual fluorescent sensor array based on organoboron-containing polymers. *Chemical Science*, *9*(7), 1892–1901. Available from https://doi.org/10.1039/C7SC05243J.

Qin, J., Gao, J., Shi, X., Chang, J., Dong, Y., Zheng, S., Wang, X., Feng, L., & Wu, Z. S. (2020). Hierarchical ordered dual-mesoporous polypyrrole/graphene nanosheets as bi-functional active materials for high-performance planar integrated system of micro-supercapacitor and gas sensor. *Advanced Functional Materials*, *30*(16). Available from https://doi.org/10.1002/adfm.201909756, http://onlinelibrary.wiley.com/journal/10.1002/(ISSN)1616-3028.

Ruiz, J. A. R., Sanjuán, A. M., Vallejos, S., García, F. C., & García, J. M. (2018). Smart polymers in micro and nano sensory devices. *Chemosensors*, *6*(2). Available from https://doi.org/10.3390/chemosensors6020012, http://www.mdpi.com/2227-9040/6/2/12/pdf.

Salaris, N., Haigh, P., Papakonstantinou, I., & Tiwari, M. K. (2023). Self-assembled porous polymer films for improved oxygen sensing. *Sensors and Actuators B: Chemical*, *374*, 132794. Available from https://doi.org/10.1016/j.snb.2022.132794.

Sanjuán, A. M., Reglero Ruiz, J. A., García, F. C., & García, J. M. (2018). Recent developments in sensing devices based on polymeric systems. *Reactive and Functional Polymers*, *133*, 103–125. Available from https://doi.org/10.1016/j.reactfunctpolym.2018.10.007.

Saylan, Y., Akgönüllü, S., Yavuz, H., Ünal, S., & Denizli, A. (2019). Molecularly imprinted polymer based sensors for medical applications. *Sensors*, *19*(6), 1279. Available from https://doi.org/10.3390/s19061279.

Shahrbabaki, Z., Farajikhah, S., Ghasemian, M. B., Oveissi, F., Rath, R. J., Yun, J., Dehghani, F., & Naficy, S. (2023). A flexible and polymer-based chemiresistive CO2 gas sensor at room temperature. *Advanced Materials Technologies*, *8*(10), 2201510. Available from https://doi.org/10.1002/admt.202201510.

Shahrbabaki, Z., Oveissi, F., Farajikhah, S., Ghasemian, M. B., Jansen-van Vuuren, R. D., Jessop, P. G., Yun, J., Dehghani, F., & Naficy, S. (2022). Electrical response of poly (N-[3-(dimethylamino)propyl] methacrylamide) to CO2 at a long exposure period. *ACS Omega*, *7*(26), 22232–22243. Available from https://doi.org/10.1021/acsomega.2c00914.

Sharma, R., Jung, S. H., & Lee, H. I. (2021). Thin polymeric films for real-time colorimetric detection of hydrazine vapor with parts-per-million sensitivity. *ACS Applied Polymer Materials*, *3*(12), 6632–6641. Available from https://doi.org/10.1021/acsapm.1c01292, pubs.acs.org/journal/aapmcd.

Shimizu, K. D., & Stephenson, C. J. (2010). Molecularly imprinted polymer sensor arrays. *Current Opinion in Chemical Biology*, *14*(6), 743–750. Available from https://doi.org/10.1016/j.cbpa.2010.07.007.

Soga, T., Jimbo, Y., Suzuki, K., & Citterio, D. (2013). Inkjet-printed paper-based colorimetric sensor array for the discrimination of volatile primary amines. *Analytical Chemistry*, *85*(19), 8973–8978. Available from https://doi.org/10.1021/ac402070z.

Song, E., & Choi, J. W. (2015). Multi-analyte detection of chemical species using a conducting polymer nanowire-based sensor array platform. *Sensors and Actuators, B: Chemical.*, *215*, 99–106. Available from https://doi.org/10.1016/j.snb.2015.03.039.

Sugiyasu, K., & Swager, T. M. (2007). Conducting-polymer-based chemical sensors: Transduction mechanisms. *Bulletin of the Chemical Society of Japan*, *80*(11), 2074–2083. Available from https://doi.org/10.1246/bcsj.80.2074United, http://www.jstage.jst.go.jp/article/bcsj/80/11/2074/_pdf.

Sun, X., Wang, Y., & Lei, Y. (2015). Fluorescence based explosive detection: From mechanisms to sensory materials. *Chemical Society Reviews*, *44*(22), 8019–8061. Available from https://doi.org/10.1039/C5CS00496A.

Tai, H., Jiang, Y., Xie, G., Yu, J., & Chen, X. (2007). Fabrication and gas sensitivity of polyaniline–titanium dioxide nanocomposite thin film. *Sensors and Actuators B: Chemical*, *125*(2), 644–650. Available from https://doi.org/10.1016/j.snb.2007.03.013.

Wang, J., Chan, S., Carlson, R. R., Luo, Y., Ge, G., Ries, R. S., Heath, J. R., & Tseng, H. R. (2004). Electrochemically fabricated polyaniline nanoframework electrode junctions that function as resistive sensors. *Nano Letters*, *4*(9), 1693–1697. Available from https://doi.org/10.1021/nl049114p.

Wei, L., Sun, T., Shi, Z., Xu, Z., Wen, W., Jiang, S., Zhao, Y., Ma, Y., & Zhang, Y.-B. (2022). Guest-adaptive molecular sensing in a dynamic 3D covalent organic framework. *Nature communications*, *13*, 7936. Available from https://doi.org/10.1038/s41467-022-35674-8.

Wong, Y. C., Ang, B. C., Haseeb, A., Baharuddin, A. A., & Wong, Y. H. (2020). Conducting polymers as chemiresistive gas sensing materials: A review. *Journal of the Electrochemical Society*, *167*, 037503. Available from http://dx.doi.org/10.1149/2.0032003JES.

Wu, R., & Selvaganapathy, P. R. (2023). Porous biocompatible colorimetric nanofiber-based sensor for selective ammonia detection on personal wearable protective equipment. *Sensors and Actuators B: Chemical*, *393*, 134270. Available from https://doi.org/10.1016/j.snb.2023.134270.

Xie, Y. F., Ding, S. Y., Liu, J. M., Wang, W., & Zheng, Q. Y. (2015). Triazatruxene based covalent organic framework and its quick-response fluorescence-on nature towards electron rich arenes. *Journal of Materials Chemistry C*, *3*(39), 10066–10069. Available from https://doi.org/10.1039/c5tc02256h, http://pubs.rsc.org/en/journals/journalissues/tc.

Xu, N., Wang, R. L., Li, D. P., Zhou, Z. Y., Zhang, T., Xie, Y. Z., & Su, Z. M. (2018). Continuous detection of HCl and NH3 gases with a high-performance fluorescent polymer sensor. *New Journal of Chemistry*, *42*(16), 13367–13374. Available from https://doi.org/10.1039/c8nj02344a, http://pubs.rsc.org/en/journals/journal/nj.

Xue, Z., Zheng, J.-J., Nishiyama, Y., Yao, M.-S., Aoyama, Y., Fan, Z., Wang, P., Kajiwara, T., Kubota, Y., Horike, S., Otake, K.-I., & Kitagawa, S. (2023). Fine pore-structure engineering by ligand conformational control of naphthalene diimide-based semiconducting porous coordination polymers for efficient chemiresistive gas sensing. *Angewandte Chemie*, *135*(2), e202215234. Available from https://doi.org/10.1002/ange.202215234.

Yan, Y., Yang, G., Xu, J. L., Zhang, M., Kuo, C. C., & Wang, S. D. (2020). Conducting polymer-inorganic nanocomposite-based gas sensors: A review. *Science and Technology of Advanced Materials*, *21*(1), 768–786. Available from https://doi.org/10.1080/14686996.2020.1820845, http://www.tandfonline.com/loi/tsta20#.V0KNDU1f3cs.

Yang, L. Y., & Liau, W. B. (2010). Environmental responses of polyaniline inverse opals: Application to gas sensing. *Synthetic Metals*, *160*(7–8), 609–614. Available from https://doi.org/10.1016/j.synthmet.2009.12.016.

Yang, Y., Zhao, Z., Yan, Y., Li, G., & Hao, C. (2019). A mechanism of the luminescent covalent organic framework for the detection of NH 3. *New Journal of Chemistry*, *43*(23), 9274–9279. Available from https://doi.org/10.1039/c9nj00243j.

Yu, J., Wang, D., Tipparaju, V. V., Tsow, F., & Xian, X. (2021). Mitigation of humidity interference in colorimetric sensing of gases. *ACS Sensors*, *6*(2), 303–320. Available from https://doi.org/10.1021/acssensors.0c01644, http://pubs.acs.org/journal/ascefj.

Yu, J. B., Byun, H. G., So, M. S., & Huh, J. S. (2005). Analysis of diabetic patient's breath with conducting polymer sensor array. *Sensors and Actuators, B: Chemical*, *108*(1–2), 305–308. Available from https://doi.org/10.1016/j.snb.2005.01.040.

Yuan, Y., Ren, H., Sun, F., Jing, X., Cai, K., Zhao, X., Wang, Y., Wei, Y., & Zhu, G. (2012). Targeted synthesis of a 3D crystalline porous aromatic framework with luminescence quenching ability for hazardous and explosive molecules. *The Journal of Physical Chemistry C*, *116*(50), 26431–26435. Available from https://doi.org/10.1021/jp309068x.

Zegebreal, L. T., Tegegne, N. A., & Hone, F. G. (2023). Recent progress in hybrid conducting polymers and metal oxide nanocomposite for room-temperature gas sensor applications: A review. *Sensors and Actuators A: Physical*, *359*. Available from https://doi.org/10.1016/j.sna.2023.114472, https://www.journals.elsevier.com/sensors-and-actuators-a-physical.

Zhang, D., Luo, Y., Huang, Z., Tang, M., Sun, J., Wang, X., Wang, X., Wang, Y., Wu, W., & Dai, F. (2023). Ultrathin coordination polymer nanosheets modified with carbon quantum dots for ultrasensitive ammonia sensors. *Journal of Colloid and Interface Science*, *630*, 776–785. Available from https://doi.org/10.1016/j.jcis.2022.10.059.

Zhang, T., Qi, H., Liao, Z., Horev, Y. D., Panes-Ruiz, L. A., Petkov, P. S., Zhang, Z., Shivhare, R., Zhang, P., Liu, K., Bezugly, V., Liu, S., Zheng, Z., Mannsfeld, S., Heine, T., Cuniberti, G., Haick, H., Zschech, E., Kaiser, U., ... Feng, X. (2019). Engineering crystalline quasi-two-dimensional polyaniline thin film with enhanced electrical and chemiresistive sensing performances. *Nature Communications*, *10*(1). Available from https://doi.org/10.1038/s41467-019-11921-3, http://www.nature.com/ncomms/index.html.

Zhang, W., Cao, S., Wu, Z., Zhang, M., Cao, Y., Guo, J., Zhong, F., Duan, H., & Jia, D. (2020). High-performance gas sensor of polyaniline/carbon nanotube composites promoted by interface engineering. *Sensors*, *20*(1), 149. Available from https://doi.org/10.3390/s20010149.

Zhao, G., Zhang, Y., Sun, D., Yan, S., Wen, Y., Wang, Y., Li, G., Liu, H., Li, J., & Song, Z. (2023). Recent advances in molecularly imprinted polymers for antibiotic analysis. *Molecules (Basel, Switzerland)*, *28*(1), 335. Available from https://doi.org/10.3390/molecules28010335.

Zhu, J. L., Ma, D. F., Jia, Y. N., Zhao, Y. Q., Tao, L. M., & Niu, F. (2023). Nitrogen configuration dependent gas sensing performance based on nitrogen contained covalent organic polymers and frameworks. *SSRN, China SSRN*. Available from https://doi.org/10.2139/ssrn.4436070, https://www.ssrn.com/index.cfm/en/.

CHAPTER

Humidity sensors

15

Daniela M. Correia[1], Ana S. Castro[1], Liliana C. Fernandes[2], Carmen R. Tubio[3] and Senentxu Lanceros-Méndez[2,3,4]

[1]Centre of Chemistry, University of Minho, Braga, Portugal [2]Physics Centre of Minho and Porto Universities (CF-UM-UP), and Laboratory of Physics for Materials and Emergent Technologies, LapMET, University of Minho, Braga, Portugal [3]BCMaterials, Basque Centre for Materials and Applications, UPV/EHU Science Park, Leioa, Spain [4]IKERBASQUE, Basque Foundation for Science, Bilbao, Spain

15.1 Introduction

15.1.1 Relevance of humidity sensing

The current technological evolution/revolution related to the digitalization of society and the economy requires the development of sensors and actuators with improved performance (Javaid et al., 2022). The sensing humidity technology is essential in this context as it enables precise monitoring and control of the environment's humidity/moisture levels, supporting improved well-being, operational efficiency, and quality in various domains (Arman Kuzubasoglu, 2022).

The development of humidity sensors is of great importance, as these types of sensors play an important role in safety and comfort, ensuring overall health and safety, improving air quality and energy efficiency, providing product quality information, and contributing to weather monitoring. Humidity sensing is relevant in a large variety of areas, ranging from industrial processes to biomedicine and food production and transportation, among others (Ku & Chung, 2023). Thus, the search for better/optimal materials to be implemented in humidity sensing devices is continuous, aiming for higher stability, sensitivity, durability, and linearity, as well as faster response and recovery times and low hysteresis (Arman Kuzubasoglu, 2022; Ku & Chung, 2023). Additionally, with the growing need for more environmentally friendly approaches, more natural and/or sustainable materials are also being explored (Li et al., 2021). Fig. 15.1 summarizes the main areas of application of humidity sensing technologies.

As relevant examples, humidity sensors are employed in ventilation and air conditioning (HVAC) systems, regulating indoor humidity levels and ensuring comfort and energy efficiency (Bamodu et al., 2018). Also, in buildings and homes, this type of sensor helps maintain air quality and prevent mold growth (Soussi et al., 2022). In the area of agriculture, these sensors are essential for monitoring and controlling humidity to support plant growth (Kirci et al., 2022). In the food industry, humidity sensors ensure proper moisture levels during processing and storage (Aguiar et al., 2022), and in the pharmaceutical manufacturing field, laboratories and medical settings allow for control and maintenance within specific levels of the surrounding environment (Kumar & Jha, 2017). The ability to

Sensory Polymers. DOI: https://doi.org/10.1016/B978-0-443-13394-7.00015-X

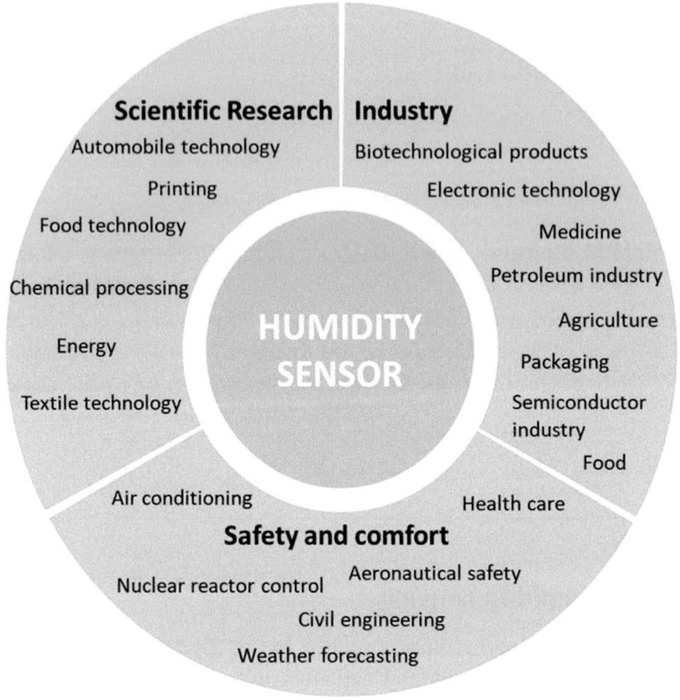

FIGURE 15.1

Main applications and areas of humidity sensor technology (based on Arman Kuzubasoglu, 2022).

accurately measure and control humidity levels is essential in multiple areas and environments to ensure optimal conditions, product quality, and overall well-being (Razjouyan et al., 2020).

15.1.2 Humidity sensing mechanisms

The main mechanisms for humidity sensing follow relatively straightforward principles, where an active material, which works as the core of the humidity sensor, produces a physical signal, typically in terms of electrical variations, in the presence of water molecules (Montes-García & Samorì, 2023). Other physical signals, including color variations, have also been applied to humidity sensors (Fernandes et al., 2019a). This sensitivity to water can rely on different principles, such as hydrogen bonds (Ali et al., 2021; Turetta et al., 2022), electrostatic (Zhao et al., 2021; Zhu et al., 2021), hydrophilic/hydrophobic interactions (Park et al., 2018; Sun et al., 2023), or intramolecular contacts (Korotcenkov, 2023; Rahman et al., 2022), among others (Zhan et al., 2023; Zhang et al., 2020).

As such, humidity-sensing devices typically measure one or more of the following parameters:

- Relative humidity (RH) is among the most commonly employed humidity sensors and relates to the amount of moisture in the air compared to the maximum amount of air moisture held at a certain temperature. Typically, RH sensors present the measurement in the form of a percentage, ranging from 0% for no RH to 100% for total saturation (Ma et al., 2023);

- Absolute humidity measures the real content of water vapor in the air, and the usual measuring unit is represented in grams per cubic meter (g/m^3), representing the mass of the water vapor per volume of air (Manickam et al., 2010);
- The dew point is the temperature at which air saturation occurs and, consequently, moisture starts to condense as dew/fog. Some humidity sensors measure this temperature, thus determining the probability of condensation occurring (Yan et al., 2022);
- Specific humidity measures the mass of water vapor present in a given mass of air, thus being typically expressed in grams of water vapor per kilogram of air (g/kg) (Kren & Anthes, 2021);
- Mixing ratio: similarly, to the specific humidity, mixing ratio is a measurement of the mass of water vapor pressure, in this case, in relation to the mass of dry air in a given operating volume. It is expressed in grams of water vapor per kilogram of dry air (g/kg) (Di Girolamo et al., 2020);
- Vapor pressure measures the partial pressure exerted by water vapor in the surrounding air. It is measured in units of pressure, such as pascals (Pa) or millibars (mbar) (Lazik et al., 2019).

The performance of humidity sensors is mostly determined by their structure, including at the nano- or micro-scopic level in the active material. This structure can be determined by the pore size and thickness of the active layer, as well as its stability, uniform morphology, and distance in the electrode layer. A porous structure in the active layer is typically more implemented as it is shown to be more effective in the sensor design as opposed to a compact structure due to its higher surface area and, consequently, higher contact area between the active material and the water molecules (Korotcenkov, 2019). The electrode layer design also has a large impact on sensor accuracy and performance. Thus, electrode placement, geometry, shielding, and protection/reactivity to the surroundings are parameters that need to be taken into consideration. In terms of electrode geometry, the spacing, separation, and design are parameters that can have a great impact on sensor performance. Additionally, the sensor's shape is very important, as, for example, sensors with serpentine electrodes or spiral electrodes tend to present a higher sensitivity, but they present lower yield rates and more complicated configurations. Ultimately, interdigitated electrodes have proven to provide the best balance between manufacturing yield, sensitivity, and accuracy (Rivadeneyra et al., 2016).

Humidity sensors can be based on various technologies, with the sensor design related to the specific measurement technology. Depending on the measuring mechanism/principle, humidity sensors can be classified into different categories, including capacitive (Ali et al., 2021; Boudaden et al., 2018), resistive (Hajian et al., 2020), optical (Mehrabani et al., 2013; Rao et al., 2021), thermal (Mack et al., 2013; Zhu et al., 2022), and gravimetric (Muckley et al., 2017; Zheng et al., 2019), among others. Similarly, they can also be categorized by the sensitive material type as polymer, carbon-, metal oxide-, or composite-based humidity sensors. In the following, the most common operating mechanisms in humidity sensors are briefly described (Arman Kuzubasoglu, 2022).

The most common mechanisms in humidity sensors are capacitive and resistive response variations (Najeeb et al., 2018). Capacitive humidity sensors are based on the principle that the electrical capacitance of the sensing/active material is altered by humidity variations in the working environment (Boudaden et al., 2018). These sensors are typically based on a moisture-absorbing dielectric material in which a humidity variation leads to a capacitance change. This capacitance change is then converted into a humidity reading (Kim et al., 2021). On the other hand, resistive humidity sensors work on the principle that some materials, such as polymers or ceramics, suffer changes in their electrical resistance with humidity variations (Ma et al., 2023). They operate similarly to capacitive humidity sensors, following the principle that the sensor's active materials can absorb and release water vapor. This water vapor affinity causes changes in the materials' resistance, and this variation determines the humidity level in the surrounding environment (Hajian et al., 2020). Color humidity sensors are particularly interesting as they give a qualitative visual output when the active material is exposed to humidity variations based on specific chemical reactions that occur, leading to a color variation (Mendes-Felipe et al., 2021). Thermal conductivity humidity sensors measure changes in the thermal conductivity of the active material due to humidity variations (Fernandes et al., 2019a). Finally, gravimetric humidity sensors directly measure humidity by weighing the moisture content of a sample of air (Muckley et al., 2017).

It is important to note that the different humidity sensor technologies are characterized by different characteristics regarding their response time, calibration necessities, accuracy, and surrounding environment sensitivity. As such, depending on the specific application, the sensor should be selected accordingly for optimal performance (Ma et al., 2023). These sensing mechanisms are based on different humidity-sensitive materials that can be integrated with, for example, stretchable polymeric substrates and a variety of conductive materials to create wearable humidity sensors. Numerous studies have also concentrated on the optimization of humidity-sensitive materials, aiming for the maximum performance of humidity sensors (Arman Kuzubasoglu, 2022). Ideally, the active material must obey several specifications, such as long operating time, low cost, stability and reproducibility, a high response to water vapor and a low cross-sensitivity to other gases, and uniform and strong binding to the substrate's surface (Korotcenkov, 2019).

15.2 Polymer-based composites for humidity sensing

The humidity sensor response is commonly based on variations in the physical characteristics, mainly the electrical properties, of the sensitive material resulting from the adsorption and/or desorption of water molecules from the surrounding environment (Fratoddi et al., 2016).

During sensor fabrication, several parameters need to be adjusted and the materials used in its design selected according to the specific application requirements, including the need to identify compatible materials with the application, suitable substrates, and appropriate fabrication methods (Lazarova et al., 2021). In this context, material selection emerges as a critical step in developing sensing devices.

As previously mentioned, some materials detect humidity by physically absorbing water molecules into the sensing layer's porous structure, while others detect humidity through the chemical adsorption of hydroxyl ions and water molecules onto the binding sites of the sensing materials (Sajid et al., 2022). Accordingly, humidity sensors have been developed using a variety of different materials in different geometries and morphologies, namely ceramics (Blank et al., 2016; Chou et al., 1999), polymers (Cheng et al., 2023; Guan et al., 2022), carbon-based materials (Ali et al., 2016; Anichini, 2020), metal–organic frameworks (Garg et al., 2021; Wu et al., 2022), composites materials (Fernandes et al., 2019b; Wu et al., 2021), 2D materials (He et al., 2018; Owji et al., 2021), 3D materials (Yu et al., 2018; Zhang et al., 2022), piezoelectric materials (Gu et al., 2016; Sun et al., 2018), nanoparticles (Kumar et al., 2020; Parthibavarman et al., 2011), nanofibers (Presti et al., 2018; Zeng et al., 2010), crystals (Xuan et al., 2011; Zhang et al., 2018), and optical materials (Ascorbe et al., 2017; Chen et al., 2021).

In recent years, polymer-based materials have gained special interest in the development of humidity sensors due to their easy processing into different morphologies and shapes, low cost, and compatibility with most additive and traditional device integration techniques (Sajid et al., 2022). Another advantage of polymer-based sensors is the possibility of modifying their chemical properties and tuning their reactivity, biocompatibility, flexibility, and resistance to degradation (Hu & Liu, 2010). For this purpose, a wide variety of polymers have been processed with different properties, such as conducting and insulating polymers, copolymers, supramolecular materials, and their composites with different organic and inorganic materials (Sajid et al., 2022).

Depending on their origin, polymers are classified as natural or synthetic. Polymers of natural origin, commonly known as biopolymers, are macromolecules derived from animals or plants. Due to their sustainable nature, these polymers have been extensively studied and applied in different industries, such as food, pharmaceuticals, and cosmetics (Malhotra et al., 2015; Ogaji et al., 2012; Alves et al., 2020). The key advantages of biopolymers include their abundance in nature and non-toxicity. Additionally, they are suitable for chemical modification, are biodegradable, and are biocompatible (Kulkarni Vishakha et al., 2012; Ezzat et al., 2020). Some notable biopolymers are polysaccharides (cellulose, chitin, chitosan, alginate, starch, or lignin) and proteins (gelatin, gluten, silks, or collagen) (Balaji et al., 2018; Cui et al., 2020).

In this context, Mahadeva et al. (2011) produced a biodegradable and flexible temperature sensor based on cellulose and polypyrrole (PPy). The combination of these materials resulted in an environmentally friendly, low-cost, and flexible temperature sensor. Voznesenskiy et al. (2013) developed integrated optical humidity sensors using chitosan waveguide films. The sensor shows a response time of 2 seconds and a dynamic range of 38%–40% RH. It does not use the functional material as a coating but rather as a waveguide. According to Yun et al. (2020), carbonized lignin-conductive particles and sodium alginate (CL/SA) composite films exhibit ultra-high sensitivity, reproducibility, and stability. The maximum response of CL/SA films with a ratio of 3:1 was 502895.40% under 97% of RH. Further, the developed composite films as humidity sensors present ultrahigh sensitivity, stable repeatability, and low hysteresis in a range of humidity between 11% and 97%, with the ability to monitor different human respiration rates (Yun et al., 2020).

Synthetic polymers are widely used components in everyday life systems and devices. Polypropylene and polyethylene (PE), as well as polyesters and polyethylene terephthalate, are present in numerous products (Scrivens & Jackson, 2000). These polymers are synthesized for a wide range of applications, including electronics, energy generation and storage, and biomedical solutions (Cao & Uhrich, 2019; Englert et al., 2018; Rashidi et al., 2014). They offer advantageous properties such as excellent mechanical properties and thermal stability, surpassing most natural polymers (Sionkowska, 2011). Several synthetic polymers have gained prominence in the fabrication of electronic systems, including humidity sensors. These polymers include poly(vinyl alcohol) (PVA) (Yun et al., 2020), polyvinylidene fluoride (PVDF) (Arularasu et al., 2020), poly(lactic acid) (PLA) (Parangusan et al., 2021), poly(ethylene glycol) (PEG) (Su et al., 2019), and polydimethylsiloxane (Yang et al., 2021). Some polymers, such as PPy (Su & Huang, 2007), polyaniline (PANI)

(Kotresh et al., 2016), and poly(3,4-ethylenedioxythiophene) (PEDOT), also exhibit conductive properties, making them particularly useful for electronic devices (Feig et al., 2018; Morais et al., 2018). In particular, the significantly larger exposed surface area, reversible electrical qualities brought on by the redox behavior, mechanical flexibility, and ease of manufacture of nanosized and micro-sized conjugated polymer particles have been proven to enhance material response compared to conventional materials (Rivadeneyra et al., 2014).

Furthermore, both natural and synthetic polymers can be combined with different fillers to improve their mechanical, thermal, or chemical properties, as well as to tune their electrical properties (Landel & Nielsen, 1993). These fillers include a variety of inorganic and organic fillers such as clays (Erdoğan & Karakişla, 2021), metals and ceramics (Yaseen et al., 2021; Asgerov et al., 2019; Ganbold et al., 2023), carbon black (Kim et al., 2021), graphene (Lee et al., 2019; Lim et al., 2016), ionic liquids (ILs) (Fernandes et al., 2019b; Serra et al., 2022), and carbon nanotubes (CNTs) (Ahmad et al., 2021; Fei et al., 2014), which are used to create hybrid composite materials.

Several techniques have been explored for the fabrication of wearable humidity sensors, including spin and dip coating (Karimov et al., 2010; McGovern et al., 2005), casting (Power et al., 2010; Bakar et al., 2022), freeze-drying (Yang et al., 2021), immersion (Bakar et al., 2022), screen printing (Lim et al., 2013), and electrospinning (Wang et al., 2021). The major drawbacks associated with these techniques rely on their time-consuming processes, material waste, and limited scalability (Osman & Lu, 2023).

In recent years, 3D printing methods, also known as additive manufacturing (AM), have gained prominence with respect to conventional methods due to their advantages, such as the ability to provide customized designs, short processing times, waste reduction, process scalability, and increased material efficiency (Huang et al., 2013). The diversity of printing methods and materials makes 3D printing technology a versatile, efficient, and powerful technological tool for advanced manufacturing (Huang et al., 2013; Truby & Lewis, 2016). Three categories of printing can be distinguished: (1) vat photopolymerization, which uses specific light (e.g., UV light) to selectively cure liquid photopolymer in the vat by light-activated layer-by-layer polymerization to form the target entity (an example of this type of manufacturing is stereolithography—SLA) (Turcin et al., 2019). (2) With respect to the inkjet method, it is based on droplets of material that are selectively deposited to form the target entity. This method is represented by techniques such as inkjet printing (Rivadeneyra et al., 2014) and binder jetting (Osman & Lu, 2023). (3) In the extrusion method, the material is extruded through a nozzle due to a mechanical force to form a continuous filament, selectively deposited to build the target entity. This category includes direct ink writing (Yan et al., 2021) and electrohydrodynamic direct writing (Wang et al., 2019), among others. However, AM technology cannot yet fully compete with conventional manufacturing, in particular with respect to mass production, due to some limitations, such as size, imperfections, and cost (Huang et al., 2013).

Several studies have been conducted to produce optimized humidity sensors by employing different manufacturing techniques. Table 15.1 summarizes studies concerning both conventional and printing systems used in the development of humidity sensors.

Kotresh et al. (2016) studied the properties of a composite material consisting of PANI−carboxymethyl cellulose (PANI−CMC), a synthetic polymer derived from a natural polymer, for the development of moisture sensors. The composite was manufactured by spin coating. A humidity sensing response of about 99% was obtained with a variation of impedance of three orders

Table 15.1 Polymer-based humidity sensors: materials, processing technologies, sensing principles, and sensing main characteristics.

Polymeric material	Sensing material (filler)		Fabrication technologies	Sensor type	Sensitivity	Relative humidity (%)	Response/recovery time (seconds)	References
PANI/CMC	—		Spin coating	Capacitance	NA	25–75	10/90	Kotresh et al. (2016)
PANI/PEDOT:PSS	—		Inkjet printing	Resistive	200%	16–98	NA	Morais et al. (2018)
Cellulose–PPy (CP)	—		Spin coating	Capacitive	NA	30–90	NA/418	Mahadeva et al. (2011)
Chitosan	—		Spin coating	Integrated-optical	—	38–40	2/NA	Voznesenskiy et al. (2013)
CL/SA	—		Solvent casting	Resistive	502895.40%	11–97	300/NA	Yun et al. (2020)
PVA/PVP/PEG	Ceramic	TiC (1–3 wt.% ratios were synthesized)	Electrospinning	Capacitance	60 nF	30–90	17/5	Yaseen et al. (2021)
PVDF	Metal	ZnO	Spin coating	Resistive	3417 Ω	5–98	30/51	Arularasu et al. (2020)
PLA/PANI		ZnO	Electrospinning	NA	NA	20–90	85/120	Parangusan et al. (2021)
PEG		Gold nanoparticles	Inkjet printing	Resistive	NA	1.8–95	1.2/3	Su et al. (2019)
PVDF		LiCl	Drop coating	Capacitive	12.678.96 pF/%RH	25–95	17.7/21	Ganbold et al. (2023)
PVDF	ILs	[Bmim][$FeCl_4$] (5, 10, and 20 wt.%)	Solvent casting	Resistive	2.66 MΩ, 1.06 MΩ, and 156 kΩ	35/90	NA	Fernandes et al. (2019b)
PVDF-HFP		[Bmim][$FeCl_4$], [Bmim][$Co(SCN)_4$], and [Bmim][$CoCl_4$]	Solvent casting	NA	$-2245 \pm 240\Omega$/RH	30/85	NA	Serra et al. (2022)
PEO	Carbon nanotubes	MWCNT and CuO (1% and 3%)	Electrospinning	Resistive/capacitive	3798.2%/53837.6%	30–90	3/22 and 20/11	Ahmad et al. (2021)
PVA		CNTs	Chemical vapor deposition (CVD)	Resistive	3.4 MΩ	11–95	NA	Fei et al. (2014)
Agarose	Graphene	GO/AgNWs	Extrusion	Resistive	30 Ω m	10–75	~1/NA	Lee et al. (2019)
PVA		TA with GO	Solution casting	Resistive	NA	40–90	NA	Lim et al. (2016)
Polyimide	Carbon black 0.01 wt.%/0.05 wt.%		Spin coating	Resistive	10.13%/15.21%	20–80	78/NA	Kim et al. (2021)

of magnitude in the RH range of 25%−75%. The developed sensor presented a response time of 10 seconds and a recovery time of 90 seconds at 75% RH. Moreover, the stability of the sensor was proven to be over 1 month. The advantage of this sensor lies in its easy processability and cost-effectiveness.

Designing advanced and cost-effective materials with improved moisture sensing represents a major challenge. Thus, advanced material compositions have been developed. As an example, humidity sensors have been developed by electrospinning based on three synthetic polymers: PVA, polyvinylpyrrolidone (PVP), and PEG reinforced with titanium carbide (TiC) as filler. This sensor exhibited high efficiency and a RH response in the 30%−90% range, and incorporating TiC further increased the sensing capability. Among the compositions tested, PVA-PEG-PVP-TiC:1% exhibited the best time response (17 seconds) and recovery (5 seconds), demonstrating high efficiency, simple processing, and cost-effectiveness (Yaseen et al., 2021).

Additionally, Ahmad et al. (2021) developed a humidity sensor based on composite nanofibers by electrospinning. The composites consisted of polyethylene oxide (PEO), oxidized multi-wall CNTs (MWCNT), and copper oxide nanoparticles (CuO) with 1% and 3% filler content (i.e., PEO-CuO-MWCNC: 1%, and PEO-CuO-MWCND: 3%). These low-cost and easy-to-manufacture moisture sensors exhibited high sensitivity for two different concentrations. PEO−MWCNT−CuO:1% showed a sensitivity value of 3798.2%, with response and recovery times of 3 seconds and 22 seconds, respectively, whereas PEO−MWCNT−CuO: 3% achieved a sensitivity of 53837.6%, with response and recovery times of 20 seconds and 11 seconds for 30%−90% RH.

Su et al. (2019) developed a highly responsive humidity sensor by printing gold nanoparticles (GNPs) grafted with a PEG hygroscopic polymer. Hygroscopic polymers exhibit changes in electrical conductivity when exposed to water vapor, making them suitable for humidity detection applications. The monitoring of the humidity changes in the surrounding environment was performed by measuring the electrical resistance of the printed thin films. The resistance of the GNP-based humidity sensor also shows temperature dependence. At higher temperatures, the GNP thin film absorbs more water, leading to lower resistance and maintaining the slope of the moisture resistance constant with the humidity variation. This consistent coefficient for the resistance/humidity correlation at different temperatures indicates that the observed resistance variation is primarily associated with the amount of water absorbed (Su et al., 2019).

Within the RH range from 1.8% to 95%, the sensor demonstrated fast responses to humidity changes with response and recovery times of ≤ 1.2 and ≤ 3 seconds, revealing the potential of the printed GNP humidity sensor for applications (Su et al., 2019).

ILs also show a strong potential for the development of humidity sensors due to their ionic nature. These salts are entirely composed of cations and anions, with a wide range of possible cations and anion combinations, allowing the development of ILs with distinct functionalities and characteristics tailored for a specific application (Welton, 2018; Correia et al., 2020).

Printable composite films based in 1-butyl 3-methylimidazolium tetrachloroferrate ([Bmim][$FeCl_4$])/PVDF have been developed to be applied as humidity sensors, as represented in Fig. 15.2, displaying the materials a linear resistance variation with the RH variation from 35% to 90% and a humidity sensitivity increase with increasing IL content (Fig. 15.2) (Fernandes et al., 2019b).

Other imidazolium-based ILs comprising the same cation but different cation chain lengths and anions types, e.g., tetrachlorocobaltate ([$CoCl_4$]$^{2-}$), tetrachloroferrate ([$FeCl_4$]$^{-}$), and tetrathiocyanatocobaltate ([$Co(SCN)_4$]$^{2-}$), have been used in humidity sensor development. The different ILs

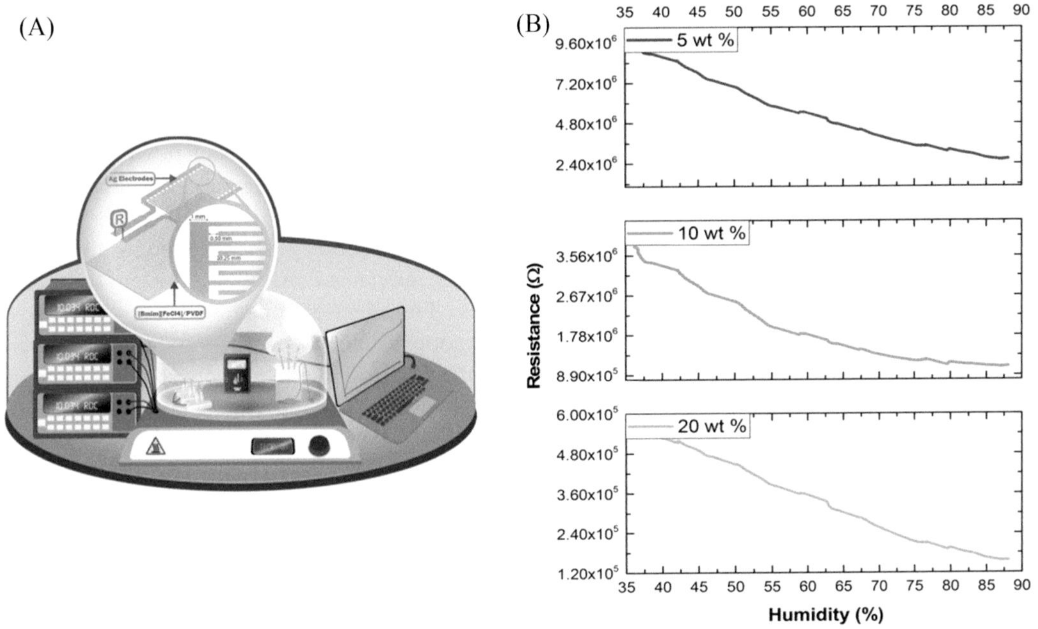

FIGURE 15.2

(A) Schematic illustration of the humidity sensitivity measurements. (B) Humidity sensor sensitivity for the [Bmim][$FeCl_4$]/PVDF with different contents of [Bmim][$FeCl_4$] (5, 10, and 20 wt.%).

Reprinted with permission from Fernandes, L., et al. (2019b). Highly sensitive humidity sensor based on ionic liquid–polymer composites. ACS Applied Polymer Materials. 1(10): 2723–2730.

were incorporated into a poly(vinylidene fluoride-*co*-hexafluoropropylene) (PVDF-HFP) matrix, and their potential as humidity sensors was evaluated through the analysis of the impedance variation as a function of RH between 30% and 85%, showing the composite bis(1-butyl-3-methylimidazolium) tetrachlorocobaltate $[Bmim]_2[CoCl_4]$/PVDF-HFP the highest humidity sensing response (-2245 ± 240 Ω/RH) (Serra et al., 2022).

Poly(ILs) (PILs) have also been employed in the development of humidity sensors (Nie et al., 2019; Wang et al., 2016). PEVIm-Br PIL exhibited a high sensitivity in the RH range from 11% to 98%, a short recovery and response time, good repeatability, and small hysteresis. As a result of the interaction between the PIL and the water molecules, a decrease in impedance values occurred with increasing RH (Wang et al., 2016). Cross-linked PILs prepared through in situ thiol-ene click polymerization have also been reported to be used in low-humidity detection. Combining hydrophobic components, the structure of PILs can be adjusted to optimize the humidity-sensing properties of the sensors. The low humidity sensing properties of the as-prepared sensors were investigated, and the possible sensing mechanism was discussed. Both characteristics, PILs strong hydrophilicity and cross-linked structure, promote the development of a stable sensing film with a response of 48.0 for a RH between 5% and 35% with a recovery/response time of 19 seconds/1 seconds, respectively, and a low humidity hysteresis (Yu et al., 2024).

15.3 Polymer-based humidity sensors and user cases

As mentioned above, polymer-based humidity sensors have received great attention, as the polymer characteristics allow to improve mechanical flexibility and solution processability, allowing to address the requirements of traditional or new-generation applications in terms of flexibility or integration into portable and wearable systems, specific characteristics highly demanded in many fields, such as electronics, robotics, and biomedicine, among others.

The advances and performance of polymer-based humidity sensory devices are closely related to the improvement and integration of several aspects: advanced materials, design strategies, transduction modes, and manufacturing technologies. Transduction modes include capacitive, resistive, optical, and surface acoustic waves, among others, where resistive and capacitive are the most used in practical applications. Furthermore, the practical use of sensors requires the control of several specifications, such as operating life, sensitivity, response/recovery time, and RH range, among others. In addition, many applications require the use of wireless monitoring to obtain real-time information. Therefore, significant innovations in all these requirements are required to implement polymer-based humidity sensors in applications including packaging, structural health monitoring, biomedicine, clothing, electronic skin, and industrial processes. Some user cases are briefly outlined below.

Humidity sensing devices have attracted great attention in human body-related detection with great application prospects in the field of medical devices, such as respiratory, speech recognition, skin health monitoring, and diapers, among others. A large number of studies are related to respiratory behavior detection, where monitoring can be performed by detecting the humidity of exhaled gases (Cai et al., 2021; Xu et al., 2022).

Fig. 15.3 illustrates some of these humidity-sensing devices. J. P. Serra et al. (2022) developed a device for real-time monitoring of human breathing through humidity sensing. The sensing material was based on PVDF-HFP with different types of imidazolium ILs and different magnetic anions (Fig. 15.3A). The suitability of the material for humidity sensing was evaluated by analyzing the impedance variation with varying RH from 30% to 85%, and the highest RH sensing response was obtained for the $[Bmin]_2[CoCl_4]$/PVDF-HFP sample. Also, the composite was attached to a cylindrical tube due to its mechanical flexibility, and the real-time monitoring of human breathing (inhale to exhale) was evaluated. Fig. 15.3A displays the prototype and the sensor response in terms of impedance variation. Zhou et al. (2020) developed a humidity sensor to monitor human respiration in real-time with a fast response and low power assumption. It is based on confining conductive polymers poly(3,4-ethylenedioxythiophene) polystyrene sulfonate (PEDOT:PSS) into 1D nanowires to facilitate the absorption and desorption of water molecules. The humidity sensor shows high sensitivity (5.46%) and ultrafast response (0.63 seconds) when exposed to varying humidity between 0% and 13%. Interestingly, different respiration patterns can be distinguished, including normal, fast, and deep respiration. Lu et al. (2021) developed a humidity sensor for asthma detection via monitoring of the respiration rate in real-time. The system was developed using multilayer graphene into electrospun polyamide 66 nanofibers. The flexible and noncontact system exhibits high humidity sensitivity and is capable of monitoring human respiration and asthma detection. The system is connected to a mobile phone to send an alert signal in case of an emergency. Another example was

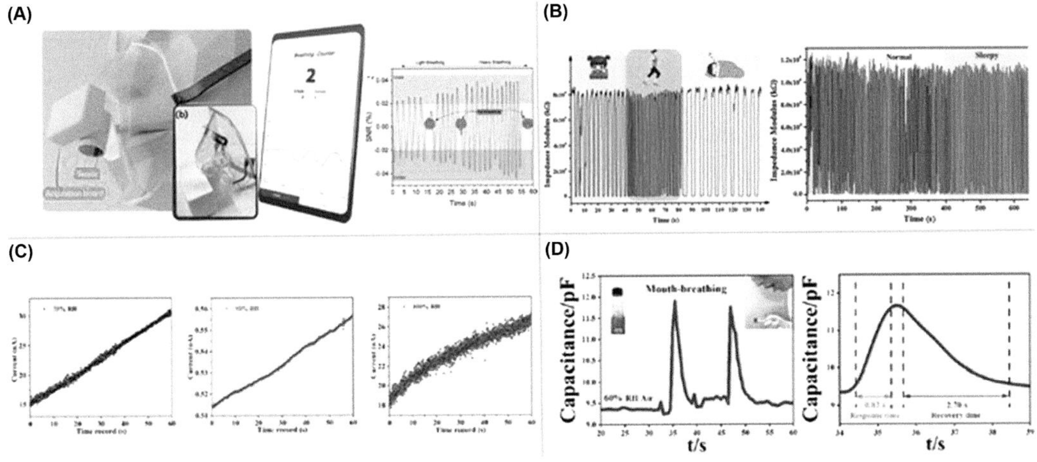

FIGURE 15.3

(A) Breathing monitoring prototype, which includes the breathing mask with the electronic system and humidity sensor (left), and corresponding response of the $[Bmim]_2[CoCl_4]$/PVDF-HFP composite (right). (B) Three respiration patterns recorded by the MPOSS-PIL humidity sensor (left), and breath monitoring for 10 min by the MPOSS-PIL humidity sensor (right). (C) Response of nanocomposite film to humidity within 60 s. Changes of current from low RH to high RH with time. (D) Mouth-breathing response of the noncontact humidity sensor (left), and the response/recovery time (right).

(A) Adapted with permission from Serra, J.P., et al. (2022). Humidity sensors based on magnetic ionic liquids blended in poly (vinylidene fluoride-co-hexafluoropropylene). ACS Applied Polymer Materials, 5*(1), 109–119. (B) Adapted with permission from Dai, J., et al. (2020). Design strategy for ultrafast-response humidity sensors based on gel polymer electrolytes and application for detecting respiration.* Sensors and Actuators B: Chemical, 304, *127270. (C) Adapted with permission from Wen, L., et al. (2022). Humidity-/sweat-sensitive electronic skin with antibacterial, antioxidation, and ultraviolet-proof functions constructed by a cross-linked network.* ACS Applied Materials & Interfaces, 14*(50), 56074–56086. (D) Adapted with permission from Wang, Y., et al. (2020). Flexible capacitive humidity sensors based on ionic conductive wood-derived cellulose nanopapers.* ACS Applied Materials & Interfaces, 12*(37), 41896–41904.*

reported by Dai et al. (2020), where a humidity sensor based on gel polymer electrolytes is presented. The obtained sensors demonstrate ultrafast response and recovery times, which are very promising for monitoring respiration (Fig. 15.3B).

Another application area of humidity sensors is in the area of electronic skin (e-skin), which can imitate the features and functions of natural skin (Trung et al., 2017; Chen et al., 2022). Wen et al. (2022) developed a silver-loaded nanocomposite film (PVA/carboxymethyl starch (CMS)/vanillin/nanoAg), which can be used as humidity/sweat sensors and wearable electronics with additional antibacterial and antioxidant activities. The nanocomposite film responds to humidity within 60 seconds (Fig. 15.3C) and records the electric signals of human joint movements and skin sweating with a response range of 0%–140% to strain at 93% RH. Wang et al. (2020) developed a flexible capacitive humidity sensor based on ionic-conductive wood-derived cellulose nanopapers

(Fig. 15.3D). The sensors showed a $>10^4$ times increase in the sensing signal over the 7%–94% RH range at a frequency of 20 Hz, while many reported humidity sensors with high sensitivity often show a working range limited to high RH levels. Interestingly, the sensors can be used for breath detection and noncontact skin humidity sensing.

Another interesting application related to human body detection is the moisture detection of a commonly used baby diaper. Flexible CdS/polyacrylamide nanocomposite for humidity sensing was reported by Chaudhary et al. (2022) and used for the moisture detection of a baby's diaper. The obtained sample shows a high sensitivity with good linearity in the humidity range from 50% to 95% RH, being 306.47 nF/%RH in highly humid regions. Similarly, MWCNT/poly-L-lysine (PLL) composite films were produced to detect RH changes in baby diapers, spatial humidity distribution, and human exhaled breath. Results show that the response is improved 600 times by PLL modification at 91.5% compared to unmodified MWCNT (Draper et al., 2022).

Other growing application areas include smart packaging based on sensing technologies to control the quality of stored products such as pharmaceuticals, cosmetics, and foods, among others. Monitoring the humidity environment in the packaging is essential to controlling the quality of the products, where the use of low-cost manufacturing processes and wireless communication approaches is desired. In this field of packaging, many efforts have been focused on the development of environmentally sustainable humidity sensing devices. For example, the use of wheat gluten protein has been investigated as a potential eco-friendly humidity-monitoring polymer for food packaging (Bibi et al., 2016). The results indicated a good sensitivity of the natural hydrophilic polymer at a high RH >70%. However, in this case, a rigid interdigital capacitor is used, which limits practical applications. Therefore, polymer-flexible substrates have been implemented. Gopalakrishnan et al. (2022) developed a low-cost sensor for wireless monitoring of moisture levels, where the sensor is fabricated on a metallized paper through laser ablation of the aminated aluminum on the parchment substrate. The frequency change of the sensor is observed to be linear within the 0%–85% RH range with a sensitivity of −87 kHz/RH. Syrový et al. (2019) developed humidity sensing devices with appropriate oxygen barrier properties for packaging, and high impedance change in response to RH changes. Samples are based on self-standing cellulose nanofibril (CNF) films and biocomposites with PEG, which was selected as a plasticizer to increase the ductility of CNF films. The humidity sensors formed by printing electrodes on CNF films exhibited a high impedance change, four orders of magnitude, in response to RH variations from 20% to 90%. Exploring new sensing materials and transduction modes led to the development of optical colorimetric sensors (Bumbudsanpharoke et al., 2019) based on photonic cellulose nanocrystals (Fig. 15.4A). The film can be used as a reversible humidity optical indicator in intelligent packaging devices for products that undergo deterioration under high moisture and electronic parts to monitor the quality during transportation or storage.

Additional applications include real-time humidity monitoring in agriculture and industry. In agriculture, humidity sensors can be used to monitor soil moisture, irrigate plants, and control water use. For example, Yin et al. (2021) developed a schematic of a wearable humidity sensor for real-time monitoring of plants in controlled environments (Yin et al., 2021). By integrating graphene in a gold-based film thermistor placed on a flexible polyimide sheet, a flexible and conformable sensor was produced and coupled to the leaf surface of plants. It allows us to obtain information related to the effect of irrigation and fertilization on the transpiration of

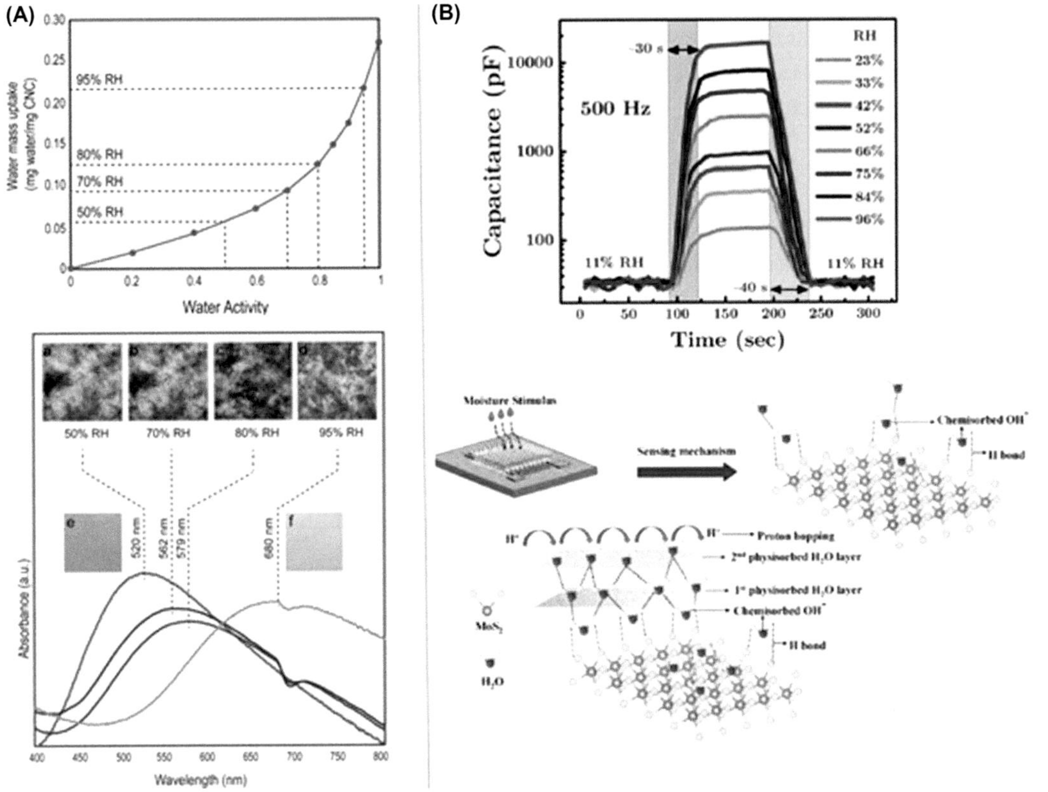

FIGURE 15.4

(A) Absorption isotherm of the iridescent CNC film, and absorbance spectra of the reflection colors from the CNC film under varied humidity exposures. (B) Response of the sensor after switching from 11%RH to different RH levels, and schematic of the moisture sensing mechanism at low and high RH environment.

(A) Adapted with permission from Bumbudsanpharoke, N., et al. (2019). Optical response of photonic cellulose nanocrystal film for a novel humidity indicator. International Journal of Biological Macromolecules, 140, *91–97. (B) Adapted with permission from Siddiqui, M.S. (2022). Highly sensitive few-layer MoS2 nanosheets as a stable soil moisture and humidity sensor. Sensors and Actuators B: Chemical, 365, Article 131930. https://doi.org/10.1016/j.snb.2022.131930.*

plants. In other work, capacitive-type humidity and soil moisture sensors are fabricated using MoS_2 nanosheets as the sensing material (Siddiqui, 2022). Capacitive-type humidity and soil moisture sensors are fabricated employing MoS_2 nanosheets as the sensing material. As shown in Fig. 15.4B, the sensors show capacitance changes when exposed to variations of RH (11–96%), and soil water content (up to 53% levels).

In industrial operating conditions, the monitoring of the humidity level is necessary, as it can affect industrial equipment and interfere with the detection performance of gas sensors, among others. The water vapor that exists in the air could cause interference in gas sensors, which are of great importance in the fields of environmental protection and human health, among others. Therefore, significant advances have been made in humidity sensors for industrial applications. For example, a polymer-based capacitive sensor was developed for process air and nonreactive gases within humidity ranges below 10% RH (Majewski et al., 2016).

In summary, advances in a large number of applications can be expected, as the implementation of smart sensing capabilities is required in the scope of digitalization to address several of the current global challenges. The next generation of humidity-sensing devices will require technological advances for improved manufacturing, integration, and sustainability. Furthermore, wireless communication capabilities and coupling with energy-harvesting devices for self-powered systems will be necessary.

15.4 Conclusions and future remarks

Humidity sensors have strongly evolved over the years due to their strong potential and importance for different fields, including industry, agriculture, and health monitoring. New demands in fields including electronics, robotics, and biomedicine, among others, are imposing novel requirements in materials, sensor responses, and integration geometries. For their development, natural and synthetic polymers have been combined with specific fillers, aiming to improve humidity sensing responses.

Besides the strong efforts devoted to the material types used, namely matrix type, different fillers, and material design, the use of natural polymers is not properly explored. Furthermore, due to their large potential, ILs need to be further studied, as their use led to the development of humidity sensors based on green chemistry without the use of nanoparticles, some of which are toxic. Additionally, improvements in polymer-based sensor performance in terms of repeatability, durability, and recovery time are needed.

Despite studies concerning printing technologies that have been presented, it is relevant to notice that the compatibility of many of the developed sensing materials with printing technologies still needs to be further explored to improve sensor implementation in lightweight, flexible, portable, and wearable systems.

Acknowledgments

This work was supported by the Portuguese Foundation for Science and Technology (FCT) in the framework of the Strategic Funding UID/FIS/04650/2020 and UID/QUI/00686/2020. The authors are grateful for funds through FCT under the project 2022.05932.PTDC, and grant SFRH/BD/145345/2019 (L.C.F.). D.M.C. thanks FCT—for the contract under the Stimulus of Scientific Employment, Individual Support 2020.02915.CEECIND (10.54499/2020.02915.CEECIND/CP1600/CT0029). This study forms part of the Advanced Materials program and was supported by MCIN with funding from European Union NextGenerationEU (PRTR-C17.I1) and by the Basque Government under the IKUR program. Funding from the Basque Government Industry Departments under the ELKARTEK program is also acknowledged.

References

Aguiar, M. L., et al. (2022). Real-time temperature and humidity measurements during the short-range distribution of perishable food products as a tool for supply-chain energy improvements. *Processes*, *10*. Available from https://doi.org/10.3390/pr10112286.

Ahmad, W., et al. (2021). Highly sensitive humidity sensors based on polyethylene oxide/CuO/multi walled carbon nanotubes composite nanofibers. *Materials*, *14*(4), 1037.

Ali, S., et al. (2021). Enhanced capacitive humidity sensing performance at room temperature via hydrogen bonding of cyanopyridone-based oligothiophene donor. *Chemosensors*, *9*. Available from https://doi.org/10.3390/chemosensors9110320.

Ali, S., et al. (2016). All-printed humidity sensor based on graphene/methyl-red composite with high sensitivity. *Carbon*, *105*, 23–32.

Alves, T. F., et al. (2020). Applications of natural, semi-synthetic, and synthetic polymers in cosmetic formulations. *Cosmetics*, *7*(4), 75.

Anichini, C., et al. (2020). Ultrafast and highly sensitive chemically functionalized graphene oxide-based humidity sensors: Harnessing device performances via the supramolecular approach. *ACS Applied Materials & Interfaces*, *12*(39), 44017–44025.

Arman Kuzubasoglu, B. (2022). Recent studies on the humidity sensor: A mini review. *ACS Applied Electronic Materials*, *4*(10), 4797–4807.

Arularasu, M., et al. (2020). PVDF/ZnO hybrid nanocomposite applied as a resistive humidity sensor. *Surfaces and Interfaces*, *21*, 100780.

Ascorbe, J., et al. (2017). Recent developments in fiber optics humidity sensors. *Sensors*, *17*(4), 893.

Asgerov, E., et al. (2019). The effect of percolation electrical properties in hydrated nanocomposite systems based on polymer sodium alginate with a filler in the form nanoparticles ZrO_2–3mol% Y_2O_3. *Advanced Physics Research*, *1*, 70–80.

Bakar, N. A. A., et al. (2022). Effect of different deposition techniques of PCDTBT: PC71BM composite on the performance of capacitive-type humidity sensors. *Synthetic Metals*, *285*, 117020.

Balaji, A. B., et al. (2018). Natural and synthetic biocompatible and biodegradable polymers. *Biodegradable and biocompatible polymer composites*, *286*, 3–32.

Bamodu, O., et al. (2018). Indoor environment monitoring based on humidity conditions using a low-cost sensor network. *Energy Procedia*, *145*, 464–471.

Bibi, F., et al. (2016). Wheat gluten, a bio-polymer layer to monitor relative humidity in food packaging: Electric and dielectric characterization. *Sensors and Actuators A: Physical*, *247*, 355–367.

Blank, T., Eksperiandova, L., & Belikov, K. (2016). Recent trends of ceramic humidity sensors development: A review. *Sensors and Actuators B: Chemical*, *228*, 416–442.

Boudaden, J., et al. (2018). Polyimide-based capacitive humidity sensor. *Sensors*, *18*. Available from https://doi.org/10.3390/s18051516.

Bumbudsanpharoke, N., et al. (2019). Optical response of photonic cellulose nanocrystal film for a novel humidity indicator. *International Journal of Biological Macromolecules*, *140*, 91–97.

Cai, C., et al. (2021). Sensitive and flexible humidity sensor based on sodium hyaluronate/MWCNTs composite film. *Cellulose*, *28*, 6361–6371.

Cao, Y., & Uhrich, K. E. (2019). Biodegradable and biocompatible polymers for electronic applications: A review. *Journal of Bioactive and Compatible Polymers*, *34*(1), 3–15.

Chaudhary, P., et al. (2022). Design and development of flexible humidity sensor for baby diaper alarm: Experimental and theoretical study. *Sensors and Actuators B: Chemical*, *350*, 130818.

Chen, L., et al. (2022). Stretchable and transparent multimodal electronic-skin sensors in detecting strain, temperature, and humidity. *Nano Energy*, *96*, 107077.

Chen, M.-Q., et al. (2021). 3D printed castle style Fabry-Perot microcavity on optical fiber tip as a highly sensitive humidity sensor. *Sensors and Actuators B: Chemical*, *328*, 128981.

Cheng, Y., et al. (2023). A flexible, sensitive and stable humidity sensor based on an all-polymer nanofiber film. *Materials Letters*, *330*, 133268.

Chou, K.-S., Lee, T.-K., & Liu, F.-J. (1999). Sensing mechanism of a porous ceramic as humidity sensor. *Sensors and Actuators B: Chemical*, *56*(1–2), 106–111.

Correia, D.M., et al., Ionic Liquid–Polymer Composites: A New Platform for Multifunctional Applications. 2020. 30(24): p. 1909736.

Cui, C., et al. (2020). Recent progress in natural biopolymers conductive hydrogels for flexible wearable sensors and energy devices: Materials, structures, and performance. *ACS Applied Biomaterials*, *4*(1), 85–121.

Dai, J., et al. (2020). Design strategy for ultrafast-response humidity sensors based on gel polymer electrolytes and application for detecting respiration. *Sensors and Actuators B: Chemical*, *304*, 127270.

Di Girolamo, P., et al. (2020). Water vapor mixing ratio and temperature inter-comparison results in the framework of the Hydrological Cycle in the Mediterranean Experiment—Special Observation Period 1. *Bulletin of Atmospheric Science and Technology*, *1*(2), 113–153.

Draper, B., et al. (2022). Reducing liver disease-related deaths in the Asia-Pacific: the important role of decentralised and non-specialist led hepatitis C treatment for cirrhotic patients. *The Lancet Regional Health–Western Pacific, 20*.

Englert, C., et al. (2018). Pharmapolymers in the 21st century: Synthetic polymers in drug delivery applications. *Progress in Polymer Science*, *87*, 107–164.

Erdoğan, M. K., & Karakişla, M. (2021). Preparation of a clay composite containing poly (o-toluidine) and halloysite, and examining of its performance as a humidity sensor. *Düzce Üniversitesi Bilim ve Teknoloji Dergisi*, *9*(2), 521–534.

Ezzat, H. A., et al. (2020). Application of natural polymers enhanced with ZnO and CuO as humidity sensor. *NRIAG Journal of Astronomy and Geophysics*, *9*(1), 586–597.

Fei, T., et al. (2014). Humidity switching properties of sensors based on multiwalled carbon nanotubes/polyvinyl alcohol composite films. *Journal of Applied Polymer Science*, *131*(1).

Feig, V. R., Tran, H., & Bao, Z. (2018). Biodegradable polymeric materials in degradable electronic devices. *ACS Central Science*, *4*(3), 337–348.

Fernandes, L., et al. (2019b). Highly sensitive humidity sensor based on ionic liquid–polymer composites. *ACS Applied Polymer Materials*, *1*(10), 2723–2730.

Fernandes, L. C., et al. (2019a). Ionic-liquid-based printable materials for thermochromic and thermoresistive applications. *ACS Applied Materials & Interfaces*, *11*(22), 20316–20324.

Fratoddi, I., et al. (2016). Role of nanostructured polymers on the improvement of electrical response-based relative humidity sensors. *Sensors and Actuators B: Chemical*, *225*, 96–108.

Ganbold, E., et al. (2023). Highly sensitive interdigitated capacitive humidity sensors based on sponge-like nanoporous PVDF/LiCl composite for real-time monitoring. *ACS Applied Materials & Interfaces*, *15*(3), 4559–4568.

Garg, A., et al. (2021). Metal-organic framework MOF-76 (Nd): Synthesis, characterization, and study of hydrogen storage and humidity sensing. *Frontiers in Energy Research*, *8*, 604735.

Gopalakrishnan, S., et al. (2022). Wireless humidity sensor for smart packaging via one-step laser-induced patterning and nanoparticle formation on metallized paper. *Advanced Electronic Materials*, *8*(7), 2101149.

Gu, L., Zhou, D., & Cao, J. C. (2016). Piezoelectric active humidity sensors based on lead-free NaNbO3 piezoelectric nanofibers. *Sensors*, *16*(6), 833.

Guan, X., et al. (2022). A flexible humidity sensor based on self-supported polymer film. *Sensors and Actuators B: Chemical*, *358*, 131438.

Hajian, S., et al. (2020). Development of a fluorinated graphene-based resistive humidity sensor. *IEEE Sensors Journal*, *20*(14), 7517–7524.
He, P., et al. (2018). Fully printed high performance humidity sensors based on two-dimensional materials. *Nanoscale*, *10*(12), 5599–5606.
Hu, J., & Liu, S. (2010). Responsive polymers for detection and sensing applications: Current status and future developments. *Macromolecules*, *43*(20), 8315–8330.
Huang, S. H., et al. (2013). Additive manufacturing and its societal impact: A literature review. *The International Journal of Advanced Manufacturing Technology*, *67*, 1191–1203.
Javaid, M., et al. (2022). Understanding the adoption of Industry 4.0 technologies in improving environmental sustainability. *Sustainable Operations and Computers*, *3*, 203–217.
Karimov, K. S., et al. (2010). Ag/PEPC/NiPc/ZnO/Ag thin film capacitive and resistive humidity sensors. *Journal of Semiconductors*, *31*(5), 054002.
Kim, J., et al. (2021). Capacitive humidity sensor based on carbon black/polyimide composites. *Sensors*, *21*, 1974. Available from https://doi.org/10.3390/s21061974.
Kirci, P., Ozturk, E., & Celik, Y. (2022). A novel approach for monitoring of smart greenhouse and flowerpot parameters and detection of plant growth with sensors. *Agriculture*, *12*. Available from https://doi.org/10.3390/agriculture12101705.
Korotcenkov, G. (2023). Paper-based humidity sensors as promising flexible devices: State of the art: Part 1. General consideration. *Nanomaterials*, *13*. Available from https://doi.org/10.3390/nano13061110.
Korotcenkov, G. (2019). *Handbook of humidity measurement: Methods, materials and technologies, Vol. 2: Electronic and electrical humidity sensors.*.
Kotresh, S., et al. (2016). Humidity sensing performance of spin coated polyaniline–carboxymethyl cellulose composite at room temperature. *Cellulose*, *23*, 3177–3186.
Kren, A. C., & Anthes, R. A. (2021). Estimating error variances of a microwave sensor and dropsondes aboard the Global Hawk in hurricanes using the three-cornered hat method. *Journal of Atmospheric and Oceanic Technology*, *38*(2).
Ku, C.-A., & Chung, C.-K. (2023). Advances in humidity nanosensors and their application: review. *Sensors*, *23*. Available from https://doi.org/10.3390/s23042328.
Kulkarni Vishakha, S., Butte Kishor, D., & Rathod Sudha, S. (2012). Natural polymers—A comprehensive review. *International Journal of Research in Pharmaceutical and Biomedical sciences*, *3*(4), 1597–1613.
Kumar, N., & Jha, A. (2017). Temperature excursion management: A novel approach of quality system in pharmaceutical industry. *Saudi Pharmaceutical Journal*, *25*(2), 176–183.
Kumar, P., et al. (2020). Fabrication of leaf shaped SnO_2 nanoparticles via sol–gel route and its application for the optoelectronic humidity sensor. *Materials Letters*, *278*, 128451.
Landel, R. F., & Nielsen, L. E. (1993). *Mechanical properties of polymers and composites*. CRC press.
Lazarova, K., et al. (2021). Flexible and transparent polymer-based optical humidity sensor. *Sensors*, *21*(11), 3674.
Lazik, D., et al. (2019). A new principle for measuring the average relative humidity in large volumes of non-homogenous gas. *Sensors*, *19*. Available from https://doi.org/10.3390/s19235073.
Lee, Y., et al. (2019). A conducting composite microfiber containing graphene/silver nanowires in an agarose matrix with fast humidity sensing ability. *Polymer*, *164*, 1–7.
Li, Z., et al. (2021). Green and sustainable cellulose-derived humidity sensors: A review. *Carbohydrate Polymers*, *270*, 118385.
Lim, D.-I., Cha, J.-R., & Gong, M.-S. (2013). Preparation of flexible resistive micro-humidity sensors and their humidity-sensing properties. *Sensors and Actuators B: Chemical*, *183*, 574–582.
Lim, M.-Y., et al. (2016). Poly (vinyl alcohol) nanocomposites containing reduced graphene oxide coated with tannic acid for humidity sensor. *Polymer*, *84*, 89–98.

Lu, L., et al. (2021). Flexible noncontact sensing for human–machine interaction. *Advanced Materials*, *33* (16), 2100218.

Ma, Z., Fei, T., & Zhang, T. (2023). An overview: Sensors for low humidity detection. *Sensors and Actuators B: Chemical*, *376*, 133039.

Mack, S., Hussein, M. A., & Becker, T. (2013). Tracking the thermal induced vapor transport across foam microstructure by means of micro-sensing technology. *Journal of Food Engineering*, *116*(2), 344–351.

Mahadeva, S. K., Yun, S., & Kim, J. (2011). Flexible humidity and temperature sensor based on cellulose–polypyrrole nanocomposite. *Sensors and Actuators A: Physical*, *165*(2), 194–199.

Majewski, J. (2016). Polymer-based sensors for measurement of low humidity in air and industrial gases. *Electrical Review* (8), 74–77.

Malhotra, B., Keshwani, A., & Kharkwal, H. (2015). Natural polymer based cling films for food packaging. *International Journal of Pharmacy and Pharmaceutical Sciences*, *7*(4), 10–18.

Manickam, V., et al. (2010). Electrolytic sensor for trace level determination of moisture in gas streams. *Measurement*, *43*(10), 1636–1643.

McGovern, S. T., Spinks, G. M., & Wallace, G. G. (2005). Micro-humidity sensors based on a processable polyaniline blend. *Sensors and Actuators B: Chemical*, *107*(2), 657–665.

Mehrabani, S., et al. (2013). Hybrid microcavity humidity sensor. *Applied Physics Letters*, *102*(24), 241101.

Mendes-Felipe, C., et al. (2021). Photocurable temperature activated humidity hybrid sensing materials for multifunctional coatings. *Polymer*, *221*, 123635.

Montes-García, V., & Samorì, P. (2023). Humidity sensing with supramolecular nanostructures. *Advanced Materials*, 2208766, **n/a**(n/a).

Morais, R. M., et al. (2018). Low cost humidity sensor based on PANI/PEDOT: PSS printed on paper. *IEEE Sensors Journal*, *18*(7), 2647–2651.

Muckley, E. S., et al. (2017). Multi-mode humidity sensing with water-soluble copper phthalocyanine for increased sensitivity and dynamic range. *Scientific Reports*, *7*(1), 9921.

Najeeb, M. A., Ahmad, Z., & Shakoor, R. A. (2018). Organic thin-film capacitive and resistive humidity sensors: A focus review. *Advanced Materials Interfaces*, *5*(21), 1800969.

Nie, J., et al. (2019). New insights on the fast response of poly(ionic liquid)s to humidity: The effect of free-ion concentration. 9(5): 749.

Ogaji, I.J., E.I. Nep, and J.D. Audu-Peter, Advances in natural polymers as pharmaceutical excipients. 2012.

Osman, A., & Lu, J. (2023). 3D printing of polymer composites to fabricate wearable sensors: A comprehensive review. *Materials Science and Engineering: R: Reports*, *154*, 100734.

Owji, E., et al. (2021). 2D materials coated on etched optical fibers as humidity sensor. *Scientific Reports*, *11* (1), 1771.

Parangusan, H., et al. (2021). Humidity sensor based on poly (lactic acid)/PANI–ZnO composite electrospun fibers. *RSC Advances*, *11*(46), 28735–28743.

Park, H., et al. (2018). Enhanced moisture-reactive hydrophilic-PTFE-based flexible humidity sensor for real-time monitoring. *Sensors*, *18*. Available from https://doi.org/10.3390/s18030921.

Parthibavarman, M., Hariharan, V., & Sekar, C. (2011). High-sensitivity humidity sensor based on SnO2 nanoparticles synthesized by microwave irradiation method. *Materials Science and Engineering: C*, *31*(5), 840–844.

Power, A. C., Betts, A. J., & Cassidy, J. F. (2010). Silver nanoparticle polymer composite based humidity sensor. *Analyst*, *135*(7), 1645–1652.

Presti, D. L., Massaroni, C., & Schena, E. (2018). Optical fiber gratings for humidity measurements: A review. *IEEE Sensors Journal*, *18*(22), 9065–9074.

Rahman, S. A., et al. (2022). Highly sensitive and stable humidity sensor based on the Bi-layered PVA/graphene flower composite film. *Nanomaterials*, *12*. Available from https://doi.org/10.3390/nano12061026.

Rao, X., et al. (2021). Review of optical humidity sensors. *Sensors*, *21*. Available from https://doi.org/10.3390/s21238049.

Rashidi, H., Yang, J., & Shakesheff, K. M. (2014). Surface engineering of synthetic polymer materials for tissue engineering and regenerative medicine applications. *Biomaterials Science*, *2*(10), 1318–1331.

Razjouyan, J., et al. (2020). Wellbuilt for wellbeing: Controlling relative humidity in the workplace matters for our health. *Indoor Air*, *30*(1), 167–179.

Rivadeneyra, A., et al. (2014). Design and characterization of a low thermal drift capacitive humidity sensor by inkjet-printing. *Sensors and Actuators B: Chemical*, *195*, 123–131.

Rivadeneyra, A., et al. (2016). Printed electrodes structures as capacitive humidity sensors: A comparison. *Sensors and Actuators A: Physical*, 244.

Sajid, M., et al. (2022). Progress and future of relative humidity sensors: A review from materials perspective. *Bulletin of Materials Science*, *45*(4), 238.

Scrivens, J. H., & Jackson, A. T. (2000). Characterisation of synthetic polymer systems. *International Journal of Mass Spectrometry*, *200*(1–3), 261–276.

Serra, J. P., et al. (2022). Humidity sensors based on magnetic ionic liquids blended in poly(vinylidene fluoride-*co*-hexafluoropropylene). *ACS Applied Polymer Materials*, *5*(1), 109–119.

Siddiqui, M. S. (2022). Highly sensitive few-layer MoS_2 nanosheets as a stable soil moisture and humidity sensor. *Sensors and Actuators B: Chemical*, *365*, 131930. Available from https://doi.org/10.1016/j.snb.2022.131930.

Sionkowska, A. (2011). Current research on the blends of natural and synthetic polymers as new biomaterials. *Progress in Polymer Science*, *36*(9), 1254–1276.

Soussi, M., et al. (2022). Comprehensive review on climate control and cooling systems in greenhouses under hot and arid conditions. *Agronomy*, *12*. Available from https://doi.org/10.3390/agronomy12030626.

Su, C.-H., et al. (2019). Highly responsive PEG/gold nanoparticle thin-film humidity sensor via inkjet printing technology. *Langmuir: The ACS Journal of Surfaces and Colloids*, *35*(9), 3256–3264.

Su, P.-G., & Huang, L.-N. (2007). Humidity sensors based on TiO_2 nanoparticles/polypyrrole composite thin films. *Sensors and Actuators B: Chemical*, *123*(1), 501–507.

Sun, C., et al. (2018). Development of a highly sensitive humidity sensor based on a piezoelectric micromachined ultrasonic transducer array functionalized with graphene oxide thin film. *Sensors*, *18*(12), 4352.

Sun, Y., et al. (2023). Hydrophobic multifunctional flexible sensors with a rapid humidity response for long-term respiratory monitoring. *ACS Sustainable Chemistry & Engineering*, *11*(6), 2375–2386.

Syrový, T., et al. (2019). Wide range humidity sensors printed on biocomposite films of cellulose nanofibril and poly (ethylene glycol). *Journal of Applied Polymer Science*, *136*(36), 47920.

Truby, R. L., & Lewis, J. A. (2016). Printing soft matter in three dimensions. *Nature*, *540*(7633), 371–378.

Trung, T. Q., et al. (2017). Transparent, stretchable, and rapid-response humidity sensor for body-attachable wearable electronics. *Nano Research*, *10*, 2021–2033.

Turcin, I., et al. (2019). *Sweat glands module with integrated sensors designed for additive manufacturing. MATEC Web of Conferences*. EDP Sciences.

Turetta, N., et al. (2022). High-performance humidity sensing in π-conjugated molecular assemblies through the engineering of electron/proton transport and device interfaces. *Journal of the American Chemical Society*, *144*(6), 2546–2555.

Voznesenskiy, S., et al. (2013). Integrated-optical sensors based on chitosan waveguide films for relative humidity measurements. *Sensors and Actuators B: Chemical*, *188*, 482–487.

Wang, D., et al. (2021). Electrospinning of flexible poly (vinyl alcohol)/MXene nanofiber-based humidity sensor self-powered by monolayer molybdenum diselenide piezoelectric nanogenerator. *Nano-micro Letters*, *13*(1), 13.

Wang, L., et al. (2016). Highly chemoresistive humidity sensing using poly(ionic liquid)s. *Chemical Communications*, *52*(54), 8417–8419.

Wang, S., et al. (2019). The preparation of graphene/polyethylene oxide/sodium dodecyl sulfate composite humidity sensor via electrohydrodynamic direct-writing. *Journal of Physics D: Applied Physics*, *52*(17), 175307.

Wang, Y., et al. (2020). Flexible capacitive humidity sensors based on ionic conductive wood-derived cellulose nanopapers. *ACS Applied Materials & Interfaces*, *12*(37), 41896–41904.

Welton, T. (2018). Ionic liquids: A brief history. *Biophysical Reviews*, *10*(3), 691–706.

Wen, L., et al. (2022). Humidity-/sweat-sensitive electronic skin with antibacterial, antioxidation, and ultraviolet-proof functions constructed by a cross-linked network. *ACS Applied Materials & Interfaces*, *14* (50), 56074–56086.

Wu, J., et al. (2021). High performance humidity sensing property of Ti3C2Tx MXene-derived Ti3C2Tx/K2Ti4O9 composites. *Sensors and Actuators B: Chemical*, *326*, 128969.

Wu, K., Fei, T., & Zhang, T. (2022). Humidity sensors based on metal–organic frameworks. *Nanomaterials*, *12*(23), 4208.

Xu, Z., et al. (2022). Self-powered multifunctional monitoring and analysis system based on dual-triboelectric nanogenerator and chitosan/activated carbon film humidity sensor. *Nano Energy*, *94*, 106881.

Xuan, R., et al. (2011). Magnetically assembled photonic crystal film for humidity sensing. *Journal of Materials Chemistry*, *21*(11), 3672–3676.

Yan, F.-J., et al. (2021). Direct ink write printing of resistive-type humidity sensors. *Flexible and Printed Electronics*, *6*(4), 045007.

Yan, J., et al. (2022). A novel dew point measurement system based on the thermal effect of humidity sensitive thin film. *Measurement*, *187*, 110248.

Yang, Y., et al. (2021). Performance of the highly sensitive humidity sensor constructed with nanofibrillated cellulose/graphene oxide/polydimethylsiloxane aerogel via freeze drying. *RSC Advances*, *11*(3), 1543–1552.

Yaseen, M., et al. (2021). Preparation of titanium carbide reinforced polymer based composite nanofibers for enhanced humidity sensing. *Sensors and Actuators A: Physical*, *332*, 113201.

Yin, S., et al. (2021). A field-deployable, wearable leaf sensor for continuous monitoring of vapor-pressure deficit. *Advanced Materials Technologies*, *6*(6), 2001246.

Yu, Y., et al. (2018). A fast response − recovery 3D graphene foam humidity sensor for user interaction. *Sensors*, *18*(12), 4337.

Yu, Y., et al. (2024). Humidity sensors based on cross-linked poly(ionic liquid)s for low humidity sensing. *Sensors and Actuators B: Chemical*, *399*, 134840.

Yun, X., et al. (2020). Fabricating flexibly resistive humidity sensors with ultra-high sensitivity using carbonized lignin and sodium alginate. *Electroanalysis*, *32*(10), 2282–2289.

Zeng, F.-W., et al. (2010). Humidity sensors based on polyaniline nanofibres. *Sensors and Actuators B: Chemical*, *143*(2), 530–534.

Zhan, L., et al. (2023). Humidity visualization through a simple thermally activated delayed fluorescent emitter: The role of hydrogen bonding. *Chemical Engineering Journal*, *454*, 140182.

Zhang, D., et al. (2018). Facile fabrication of high-performance QCM humidity sensor based on layer-by-layer self-assembled polyaniline/graphene oxide nanocomposite film. *Sensors and Actuators B: Chemical*, *255*, 1869–1877.

Zhang, Y., et al. (2020). Design and fabrication of a novel humidity sensor based on ionic covalent organic framework. *Sensors and Actuators B: Chemical*, *324*, 128733.

Zhang, Y., et al. (2022). High performance humidity sensor based on 3D mesoporous Co_3O_4 hollow polyhedron for multifunctional applications. *Applied Surface Science*, *585*, 152698.

Zhao, G., et al. (2021). Chiral nematic coatings based on cellulose nanocrystals as a multiplexing platform for humidity sensing and dual anticounterfeiting. *Small (Weinheim an der Bergstrasse, Germany)*, *17* (50), 2103936.

Zheng, Z., et al. (2019). Development of a highly sensitive humidity sensor based on the capacitive micromachined ultrasonic transducer. *Sensors and Actuators B: Chemical*, *286*, 39–45.

Zhou, C., et al. (2020). Rapid response flexible humidity sensor for respiration monitoring using nano-confined strategy. *Nanotechnology*, *31*(12), 125302.

Zhu, C., et al. (2022). High-luminescence electrospun polymeric microfibers in situ embedded with CdSe quantum dots with excellent environmental stability for heat and humidity wearable sensors. *Nanomaterials*, *12*. Available from https://doi.org/10.3390/nano12132288.

Zhu, P., et al. (2021). Electrostatic self-assembly enabled flexible paper-based humidity sensor with high sensitivity and superior durability. *Chemical Engineering Journal*, *404*, 127105.

CHAPTER

pH sensors

16

Lisa Rita Magnaghi[1,2], Camilla Zanoni[1], Giancarla Alberti[1] and Raffaela Biesuz[1,2]
[1]*Department of Chemistry, University of Pavia, Pavia, Italy* [2]*INSTM, Pavia Research Unit, Firenze, Italy*

16.1 Introduction

What is the first chemistry lesson from middle school to university? pH. Which key parameter influences and regulates almost every process in our body, world, and life? pH. But what is actually pH?

16.1.1 The advent of pH

Sørensen first proposed the pH definition in 1909 as the negative logarithm of the hydrogen ion concentration, expressed in molarity. In 1924 this first definition was made more rigorous, referring to hydrogen ion activity, which is the effective concentration of the hydrogen ion in solution, which can be considered equal to the ion concentration only moving toward highly diluted solutions in which less strong interactions between ions take place but can be quite different in concentrated solutions (Buck et al., 2002; Magnaghi, Alberti, et al., 2022; Magnaghi, Zanoni, Bellotti, et al., 2022; Magnaghi, Zanoni, Bancalari, et al., 2022). Even the symbol pH is the result of a stepwise evolution since in the first paper Sørensen used the symbol p_H, but soon several variations appeared, such as p_{H^+}, Ph, and pH, which finally became the predominant form, especially for its official adoption by the Journal of Biological Chemistry in the decade 1910–19 (Jensen, 2004).

Frankly speaking, the introduction of pH was actually not promptly accepted by the scientific community at the beginning of the 20th century, mainly because it might lead to misunderstandings when speaking about differences in acidity. As a matter of fact, as Clarke stated in 1928, "It is unfortunate that a mode of expression so well adapted to the treatment of various relations should conflict with a mental habit. $[H^+]$ represents the hydrogen ion concentration, the quantity usually thought of in conversation when we speak of increases or decreases in acidity. pH varies inversely as $[H^+]$. This is confusing" (Jensen, 2004). In the end, this parameter and its symbol have been accepted and experienced worldwide diffusion a few years after their first proposal.

16.1.2 The issue with pH measurements

The acidity or alkalinity of a given medium, that is, the pH, represents a pivotal property for all the chemical, biochemical, and physical processes occurring in that medium, and thus its measurement

Sensory Polymers. DOI: https://doi.org/10.1016/B978-0-443-13394-7.00016-1

represents a crucial task (Magnaghi, Alberti, et al., 2022; Magnaghi, Zanoni, Bellotti, et al., 2022; Magnaghi, Zanoni, Bancalari, et al., 2022). As everyone knows, or at least should know, traditionally, pH is measured electrochemically using glass electrodes or, in the most recent 50 years, using solid-state ion-sensitive field-effect transistor (ISFET) sensors. The main pros of potentiometric pH measurements are accuracy, immediate readout, and a wide range of measurable pHs (Magnaghi, Alberti, et al., 2022; Magnaghi, Zanoni, Bellotti, et al., 2022; Magnaghi, Zanoni, Bancalari, et al., 2022). Let us think about that: is there any other analytical method that ensures more than 10 orders of magnitude of the measurable range of concentrations? Unfortunately, these methods also have some cons such as the use of delicate and expensive electrodes that must be handled by trained personnel, the need for large sample volumes, and frequent calibrations (Magnaghi, Alberti, et al., 2022; Magnaghi, Zanoni, Bellotti, et al., 2022; Magnaghi, Zanoni, Bancalari, et al., 2022).

For those applications in which robustness, simplicity, and applicability on-site or for small volumes are preferred over extreme accuracy and rapidity or a wide range of measures, alternative tools have been proposed for pH assessment. pH papers or strips are undoubtedly the most famous and oldest, being paper impregnated by litmus first used to test acidity in the early 14th century, which means six centuries before the definition of pH (Magnaghi et al., 2023; Steinegger et al., 2020). From that time onward, a wide variety of pH papers have been proposed and mass-produced to test the acidity of aqueous solutions, and this cheap, simple, and quick method has become a must-have in laboratory equipment, even if their analytical performances are obviously not even comparable with glass electrode's ones (Magnaghi, Alberti, et al., 2022; Magnaghi, Zanoni, Bellotti, et al., 2022; Magnaghi, Zanoni, Bancalari, et al., 2022; Magnaghi et al., 2023). Other attractive methods for pH measurements rely both on optical methods through colorimetry, UV-vis spectroscopy, and fluorescence and on other detection approaches.

16.1.3 The explosion of sensors and the statistics of pH sensing

We already talked about the litmus test, which should actually be considered the first sensor developer in history, and fate has it that it is indeed a pH sensor! In fact, a chemical sensor is generally defined as a device that transforms chemical information, ranging from the concentration of a specific sample component to total composition analysis, into an analytically useful signal (Hulanicki et al., 1991). Chemical sensors contain two basic functional units: a receptor part, where the chemical information is transformed into a form of energy, and a transducer part, capable of transforming the energy carrying the chemical information into a useful analytical signal. Besides all the possible definitions, the pivotal feature we must stress is that, when talking about sensors, we talk about devices, which means something physical and tangible that can include different materials and should be handled in several ways, always including a solid support on which the detection occurs; for this reason, the employment of the word "sensor" for reactions or other processes occurring in solution or in vapor phase without a solid sensing platform should be considered inaccurate.

Since the mid-1980s, there has been an explosion of research interests in sensor development, as witnessed by the overwhelming number of papers published during this period, mainly driven by commercial need but also by the clear advantages these devices could offer compared to conventional analytical methods in terms of low cost, scalability, robustness, customization opportunities, and the possibility to be adapted for specific applications (in situ analysis, small volumes, and monitoring, among others) or untrained staffs and instrument-less and even power-less locations

(Regan et al., 2019). In such a wide panorama, pH has often been taken into account as chemical information to be detected and transduced by sensing devices. Just to give an idea, searching for "pH sensor" on the Web of Science Database found almost 50,000 papers that fit within the research topic. Being this book devoted to polymeric sensors, we will focus our attention on pH sensors that include the employment of polymeric materials, either as solid supports or even as receptors themselves; furthermore, except for the introduction and general descriptions, we will limit our research to papers published in the last five years (since 2019) to provide a more detailed and comprehensive description of the current trends and applications of polymers in pH sensing devices.

16.1.4 Polymers' role in pH sensors

The term "polymer" refers to a large class of materials consisting of many small molecules, named monomers, that are linked together to form long chains; polymers' molecules have a very high molecular weight (between 10,000 and 1,000,000 g/mol) and consist of several structural units usually bound together by covalent bonds (Namazi, 2017). Polymers may appear as solids, solutions, gels, nanoparticles (NPs), or films, and most of the materials recently developed can be included in this class (Alberti et al., 2021). Polymers are generally obtained through a chemical reaction of monomers that show the ability to react with another molecule, from the same type or another type, to form the polymer chain; whether this process occurs in nature, natural polymers are obtained, whereas synthetic polymers are man-made (Namazi, 2017).

In the field of sensor development, polymers provide some unique features, such as being adaptable to specific tasks through their appropriate modification or synthesis, behaving either as support for functional immobilization or directly changing their own properties upon external inputs, allowing the detection of the analytes (Alberti et al., 2021). Furthermore, they show some additional properties such as chemical and physical resistance, flexibility, biocompatibility, or biodegradability, and, in some cases, even low-cost, easy, and scalable preparation (Alberti et al., 2021).

A last general distinction must be made between natural and synthetic polymers since both have been used for sensing applications, but in different contexts and with different purposes, according to their key features, which are briefly described here below. Natural polymers derived from animals or plants are widely employed in aliments, pharmaceuticals, cosmetics, and chemistry; they ensure low cost, tunability, large availability, biodegradability, and biocompatibility, but they may be degraded by microorganisms, their synthesis is rarely reproducible, their hydration degree is variable, and they are generally more prone to external contamination (Alberti et al., 2021; Gajre Kulkarni et al., 2012). Opposite synthetic polymers, or man-made polymers, are defined as polymers that are artificially produced in laboratories, typically from petroleum oil in a controlled environment, and are made up of carbon—carbon bonds as their backbone (Shrivastava & Shrivastava, 2018). Unlike natural polymers, they have become popular only in the last 125 years; they are mainly made by the carbon-based backbone; they show identically repeated units, but chain length might vary upon synthesis conditions; last but not least, they do represent an environmental concern, to a different extent depending on the polymer; they are not intrinsically biocompatible or biodegradable, but they offer a much higher resistance to chemical or physical stress and extended recyclability if correctly manipulated and sorted (Alberti et al., 2021; Shrivastava & Shrivastava, 2018).

Specifically referring to pH sensors, polymers have been used either as support for pH-sensitive molecules, thanks to their unique chemical, physical, and mechanical properties, or directly as receptors, exploiting their reactivity toward external inputs and consequent modification of a specific property. As will be deeply discussed in the next section, combining the unique features of polymeric materials with the ubiquitous role of pH, countless sensing devices have been proposed in response to thousands of analytical issues emerging in various fields, ranging from biomedical assays to food quality assessment, from environmental control to wearable medical devices, and others.

16.2 Discussion

One of the hardest tasks in writing a review or a book chapter is to identify the proper classification for the topics to be discussed; in this case, talking about polymers, the most common classification relies on the type of polymer under investigation, but we would discard this option and prefer a more "analytical" approach. Polymer-based pH sensors will be described below according to the detection technique, starting with colorimetric sensors and moving on to other types such as fluorescent devices, electrochemical sensors, optical fibers, and so on.

Another issue that needs to be faced is that pH is somehow involved in almost every industrial, biological, and environmental process, and thus thousands of sensing devices for different purposes may hide a pH sensing ability that is not always easy to identify or not even stressed by the authors. For this reason, we did our best to summarize not only strictly speaking pH sensors but also to hint at some specific applications, especially the most common ones, but the discussion might not be complete.

16.2.1 Polymer-based colorimetric pH sensors

As we already discussed in the introduction, litmus papers deserve to be considered the progenitor of colorimetric pH sensors, and we can surely affirm that the main limitations of these devices guided the development of colorimetric pH sensors (Wencel et al., 2014). Since the 1980s, when chemical sensors started to diffuse in different fields, colorimetric or optical pH sensors have experienced worldwide fame, resulting in thousands of devices cited or discussed in several reviews (Lin, 2000; McDonagh et al., 2008; Steinegger et al., 2020; Wencel et al., 2014). Focusing our attention on the most recent papers (2019–23), it seems that this kind of approach is no longer on the front burner, but there are still a few interesting novel devices in which polymeric materials have a pivotal role either in overcoming litmus test drawbacks, such as lack of accuracy and reversibility and dye release, or in fulfilling the requirements for given applications. In the following paragraph, these materials will also be referred to as "halochromic," being materials that change their color according to pH modifications.

The first group of devices that will be presented is composed of polymer-based colorimetric sensors or sensor arrays aimed at pH measurement in a wide range, at least 3 pH units, listed in

Table 16.1; these devices have some key features in common, apart from exploiting polymeric support and relying on colorimetric detection, since the authors tried to conjugate applicability in a wide pH range with resistant support, reusability, and a strong reduction of dye leaching. To achieve this goal, both man-made and natural polymers have been tested for this type of device, and covalent immobilization is usually preferred to prevent dye leaching when present (Chalitangkoon & Monvisade, 2021; Magnaghi et al., 2023; Schoolaert et al., 2022). A different alternative is to replace pH-sensitive dyes with polymeric materials providing intrinsic halochromic properties, such as polyanilines (PANI) (Gicevicius et al., 2019; Ko et al., 2020; Nazari et al., 2023).

As far as wide pH range applicability is concerned, this goal has been achieved either by using more than one pH indicator with different pK_a values to cover a wider range or by tuning the pH-sensing moiety's properties with proper additives. As for the first strategy, Schoolaert and coauthors successfully anchored on biocompatible films both alizarin and naringenin, contained in grapefruit seed extract, in a single halochromic film that changed from brown to blue/violet in a pH range between 5 and 10 (Schoolaert et al., 2022). An even wider pH interval was reached by Chalitangkoon and coworkers by combining phenol red and rosolic acid grafted on chitosan films in two different pH-sensitive films (Chalitangkoon & Monvisade, 2021); the widest applicability has been achieved by Magnaghi et al., who proposed a 3x4 sensor array in which 12 pH indicators, each one with one or two pK_a values, were immobilized on a plastic copolymer to assess the pH in a range from 1 to 13 (Magnaghi et al., 2023). Moving to the second strategy, PANI has been exploited as an tunable halochromic species in composite films, achieving between 6 and 10 pH units of applicability (Gicevicius et al., 2019; Ko et al., 2020; Nazari et al., 2023), but also common pH indicators, such as alizarin and sulfonephthaleins, were found to show a tuneable pH range of detection depending on the chemical environment, even if in this case the extent of the measurable

Table 16.1 Polymer-based colorimetric pH sensors for wide pH range measurements.

References	Polymeric material	pH-sensing moiety	pH range
Chalitangkoon and Monvisade (2021)	Chitosan (CS)	Phenol red and rosolic acid	4–10
Gicevicius et al. (2019)	Polyaniline-o-phenylenediamine (PANI-oPD) copolymers	PANI	3–9
Ko et al. (2020)	Polyaniline- polyethylene glycol (PANI-PEG)	PANI	2–8
Magnaghi et al. (2023)	Ethylene vinyl alcohol (EVOH)	12 commercial pH indicators	1–13
Nazari et al. (2023)	Polyaniline-pectin (PANI-PEC) composite film	PANI	2–12
Roy and Rhim (2021)	Carboxymethyl Cellulose-Glycerol-Agar (CMC-Gly-Ag) composite films	Alizarin and grapefruit seed extract	5–10
Schoolaert et al. (2022)	2-n-butyl-2-oxazoline (B) and 2-ethyl-2-oxazoline (E) statistical copolymers' nanofibers	Alizarin Yellow R	3–4, 7–10
Yam et al. (2021)	Polylactic acid-polyethylene glycol (PLA-PEG) composite films	Bromocresol purple	4–7

pH range is much lower than in the other devices (Schoolaert et al., 2022; Yam et al., 2021). A summary of the performances of these devices regarding measurable pH range is depicted in Fig. 16.1, in which each device is labeled according to the numbers reported in the first column of Table 16.1.

A last remark should be made about the analytical characterization of the proposed devices, which is actually scarce or even missing at all in most of them. Most papers provide several characterizations of the innovative halochromic material and just show the color change with the pH, without proposing any method for an actual measurement but only claiming its applicability as a pH sensor (Chalitangkoon & Monvisade, 2021; Gicevicius et al., 2019; Ko et al., 2020; Roy & Rhim, 2021; Schoolaert et al., 2022; Yam et al., 2021). Opposite Nazari and coworkers developed a linear equation for pH measurements from films' UV-vis absorbance at a given wavelength in the entire pH range, and Magnaghi et al. relied on a step-wise multivariate approach for quantitatively assessing pH in synthetic samples (Magnaghi et al., 2023; Nazari et al., 2023).

Opposite to the generalized devices so far discussed, not developed for a specific application but just as *proof-of-concept* pH-sensing tools, pH sensors have been successfully exploited for very specific applications, ranging from gaseous analyte detection to biological monitoring and from biomedical assays to food quality assessment. Applicability over a wide pH range generally does not represent a key requirement for such devices, while other features may be strictly demanded, such as biocompatibility, glaring color transitions, quick kinetics, or other aspects. For instance, weak acids or base vapors, such as NH_3, can be detected by a proper pH sensor, exploiting their ability to provoke protonation or deprotonation of a suitable pH-sensing dye, but for this specific target, the polymeric support must ensure high permeability and a surface area-to-volume ratio to allow target absorption and detection (Khattab et al., 2019; Pakolpakçıl & Draczyński, 2021). Completely different requirements must be fulfilled, for instance, in seawater pH monitoring, as deeply described by Staudinger and coauthors: in this case, materials that provide long-term stability, low cross-sensitivity toward T and ionic strength, an optimal pH range of detection (typically

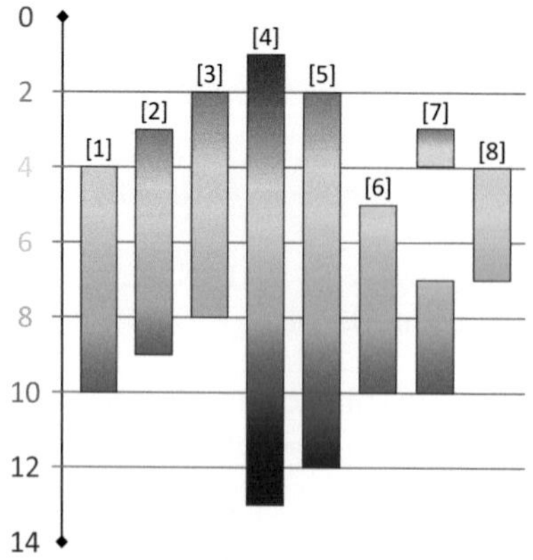

FIGURE 16.1 Comparison between devices in Table 16.1 about measurable pH ranges.

Each device in Table 16.1 is represented as a colored box covering the measurable pH range and labeled according to the numbers reported in the first column of Table 16.1.

7–8.5), and compatibility with compact and low-cost readout devices are generally preferred, and polyurethane hydrogels can be considered suitable candidates for this type of application (Staudinger et al., 2019). Again, different features are appreciated for the other specific applications to which the next sections are devoted.

16.2.1.1 pH-sensitive materials at work for food monitoring

While colorimetric pH sensors might be considered outdated in more recent literature, the development and fine-tuning of pH-sensitive materials for food freshness/spoilage monitoring represent a hot topic in the last few years due to the increasing demand for fresh and safe food products provoked both by population growth, lifestyle change, and globalization (Almasi et al., 2022). Because of the strong interest in this topic, a large number of reviews dedicated to food freshness indicators or sensors, or, generally speaking, intelligent packaging solutions, have been recently published, among which only those specifically referring to polymer-based sensors will be cited. It must be underlined that most reviews focus more on the receptor, that is, the pH-sensitive dye, while the choice of the solid support, preparation, and anchoring method and, even more, the effect of dye addition on the material's properties are seldom discussed (Almasi et al., 2022; Balbinot-Alfaro et al., 2019; Luo et al., 2022; Pirayesh et al., 2023).

Despite that, a list of the most interesting recent polymer-based pH sensors tuned for food monitoring is provided in Table 16.2, and a brief discussion mostly focused on the polymeric materials will be presented here, trying to provide an alternative viewpoint with respect to the already published papers. It must be underlined that the presented list is surely incomplete, but, being both colorimetric polymer-based sensors and food freshness sensors in wide and interdisciplinary fields, it is quite hard to identify all the papers published on this topic. As already discussed for NH_3 sensors, food freshness indicators are meant to detect analytes in low concentration and vapor phase; therefore, high permeability toward vapors is usually demanded; furthermore, polymer porosity, dye loading capacity, leaching properties, and transparency should be considered in selecting the proper support (Almasi et al., 2022). Last but not least, low cytotoxicity, biocompatibility, and easier waste management have increased the scientific and industrial interest in bio-based polymers (Kurek et al., 2019), but, on the other side, man-made polymeric materials still ensure better mechanical properties, higher resistance toward humidity and degradation, and higher scalability (Medina-Jaramillo et al., 2017).

Having roughly in mind the highly demanding requirements for this application, it is easy to understand why a large variety of polymers are currently under investigation in search of the best one, which is actually still to be found. Synthetic but biocompatible polymers commonly employed in the packaging field, such as methacrylate (Ham et al., 2023), EVOH (Magnaghi, Alberti, et al., 2021; Magnaghi, Capone, et al., 2021; Magnaghi, Zanoni, Alberti, et al., 2022a, 2022b; Magnaghi, Alberti, et al., 2022; Magnaghi, Zanoni, Bellotti, et al., 2022; Magnaghi, Zanoni, Bancalari, et al., 2022), photoinduced electron transfer (PET) (Lee et al., 2019), PS (Lyu et al., 2019), PDMS (Liu, Gurr, et al., 2020), PVOH (Wang, Feng, et al., 2022; Weston et al., 2020), and LDPE (Wells et al., 2019), have been proposed aiming at achieving the required barrier and mechanical properties and proposing a scalable device for industrial applications. For this kind of material, a colorimetric sensitive unit has been either covalently bound in the polymer chain to prevent dye leaching (Magnaghi, Alberti, et al., 2021; Magnaghi, Capone, et al., 2021; Magnaghi, Alberti, et al., 2021; Magnaghi, Zanoni, Alberti, et al., 2022a, Magnaghi, Zanoni, Alberti, et al., 2022b; Magnaghi,

Table 16.2 Polymer-based halochromic materials for food quality monitoring.

References	Polymeric material	pH-sensing moiety	Target food
Ham et al. (2023)	2-Hydroxyethyl methacrylate, [2-(methacryloyloxy)ethyl] trimethylammonium chloride and polyaniline (HEMA-MAETC-PANI) hydrogel	Bromothymol blue	Pork
Koshy et al. (2021)	Starch (S) and carbon dots (CD) film	*Clitoria ternatea* flower extract	Pork
Kurek et al. (2019)	Chitosan (CS) or carboxymethyl cellulose (CMC) films	Blueberry and red grape skin pomace extracts	Chicken
Hazarika et al. (2023)	Cellulose nanofibers and cellulose acetate nanocomposite film	*Melastoma malabathricum* extract	Meat and fish
Lee et al. (2019)	Inner film: Polyether-block-amide (PEBA) Outer film: polyethylene terephthalate (PET)	8 commercial pH indicators	Chicken
Liu et al. (2020)	Polydimethylsiloxane (PDMS) film	Chlorophenolred − dodecanoic acid (CPR − DA)	Beef and fish
Liu et al. (2020)	Polyaniline and tetraphenylethylene (PANI-TPE) composite films	PANI (colorimetric) and TPE (fluorescent)	Fish
Lyu et al. (2019)	Polystyrene and branched polyethylenimine (PS-bPEI) composite films	Bromothymol blue /tetrabutylammonium ion-paired dye	Kimchi
Magnaghi, Alberti, et al. (2021), Magnaghi, Capone, et al. (2021)	Ethylene vinyl alcohol (EVOH)	6 pH indicators	Chicken, fish and milk
Magnaghi, Zanoni, Alberti, et al. (2022a), Magnaghi, Alberti, et al. (2022), Magnaghi, Zanoni, Bellotti, et al. (2022), Magnaghi, Zanoni, Bancalari, et al. (2022)	Ethylene vinyl alcohol (EVOH)	o-Cresol red	Chicken and fish
Magnaghi, Zanoni, Alberti, et al. (2022b)	Ethylene vinyl alcohol (EVOH)	Bromocresol purple	Milk
Magnaghi, Alberti, et al. (2022), Magnaghi, Zanoni, Bellotti, et al. (2022), Magnaghi, Zanoni, Bancalari, et al. (2022)	Carboxymethyl cellulose (CMC) and cellulose (Cell)	o-Cresol red	Chicken
Mary et al. (2020)	Starch (S) film	Butterfly pea flower extract	Prawns
Sutthasupa et al. (2021)	Alginate-methylcellulose (ALG-MC) hydrogel	Bromothymol blue	Pork

Table 16.2 Polymer-based halochromic materials for food quality monitoring. ***Continued***

References	Polymeric material	pH-sensing moiety	Target food
Teixeira et al. (2022)	Cellulose acetate and glycerol or triethyl citrate (CA-Gly or CA-TEC) composite films	Açai extract	Shrimps
Wang, Wang, Sun, et al. (2022)	Polyvinyl alcohol and methylcellulose (PVOH-MC) composite film	Black wolfberry extract	Chicken and shrimps
Wells et al. (2019)	Low-density polyethylene (LDPE)	Bromophenol blue-coated hydrophilic silica	Fish
Weston et al. (2020)	Polyvinyl alcohol (PVOH) or agarose (AG) film	Red cabbage extract	Milk

Alberti, et al., 2022; Magnaghi, Zanoni, Bellotti, et al., 2022; Magnaghi, Zanoni, Bancalari, et al., 2022) or physically entrapped in the polymer bulk, making the synthetic procedure easier but the final material more prone to dye release and possible food contamination (Ham et al., 2023; Lee et al., 2019; Liu, Gurr, et al., 2020; Lyu et al., 2019; Wang, Wang, Sun, et al., 2022; Wells et al., 2019; Weston et al., 2020). PANI has also found a few applications in this field, both as a support and sensing moiety (Liu, Wang, et al., 2020) and as an additive to increase matrix viscosity and ion transport properties (Ham et al., 2023).

Moving to bio-based polymers, the casting deposition technique is generally the most common preparation procedure since it allows to obtain suitable films with a simple procedure, without the need for specific instrumentations, and to easily immobilize dyes by adding the proper amount of aqueous solution during film preparation. Cellulose derivatives (Kurek et al., 2019; Teixeira et al., 2022), chitosan (Kurek et al., 2019), starch (Koshy et al., 2021; Mary et al., 2020), and others are commonly exploited for this purpose, usually adding proper plasticizers to improve the final film. Opposite, it must be underlined that such films generally lack reproducibility, mechanical, and barrier properties and resistance and might result in dye leaching, apart from being a casting technique quite hard to scale up. Bio-based polymers have also been exploited in the form of hydrogels or nanofibrous materials to try to overcome solution-casting drawbacks (Hazarika et al., 2023; Sutthasupa et al., 2021).

16.2.1.2 pH sensors for biomedical applications

As already hinted for food monitoring devices and also in the case of biomedical applications, it turned out to be quite laborious to find the papers of interest because, in many cases, researchers from non-chemical backgrounds are actually using polymeric materials and/or pH-sensing devices without being conscious of that and without specifically claiming it, making it harder to provide a complete list of the most recent publications.

Among all, the employment of polymeric materials for wound monitoring and healing is the hottest topic in this field. While the application of polymers in wound healing has been widely reviewed in the last few years (Adamu et al., 2022; Psarrou et al., 2023; Shah, Sohail, et al., 2019; Sheokand et al., 2023; Ye et al., 2022), we will focus our attention on those devices conjugating healing properties with a pH-sensing approach to monitor wound status during the recovery process. Despite most of these devices usually exploiting textiles as preferable support, recent papers relying on both man-made and natural polymers are listed in Table 16.3.

As already discussed in the introduction, one of the main advantages of working with polymeric materials is the possibility of choosing the most suitable material for a specific application from a wide variety of polymers with many different features. In this case, an ideal and effective wound dressing should be biocompatible, elastic, maintain the moisture around the wound, provide efficient protection from infections, be durable against external stress, and possess good mechanical properties (Psarrou et al., 2023); almost all the recent literature has concentrated on exploiting either electrospun nanofibers or hydrogels. In the first case, nanofibrous materials have been often selected for their resemblance to the extracellular matrix, increased surface-to-volume ratio, porousness, and capacity to encapsulate or load medications (Adamu et al., 2022); such fibers can be obtained both from man-made (Adamu et al., 2022; Arafat et al., 2021; Bazbouz & Tronci, 2019; Zhang et al., 2022) and natural polymers (Adamu et al., 2022; Brooker & Tronci, 2023; Kurečič et al., 2018; Shah, Sohail, et al., 2019; Sheokand et al., 2023; Ye et al., 2022). In the second case, the choice falls on hydrogels for their soft texture, stretching properties, and ability to create and maintain a moist environment (Psarrou et al., 2023); differently from electrospun nanofibers, natural polymers are usually preferred for the preparation of these materials (Adamu et al., 2022; Psarrou et al., 2023; Shah, Sohail, et al., 2019; Sheokand et al., 2023; Wang et al., 2020; Ye et al., 2022;

Table 16.3 Polymer-based optical pH-sensitive materials for wound monitoring.

References	Polymeric material	pH-sensing moiety
Arafat et al. (2021)	Polyvinyl alcohol and polyacrylic acid (PVA-PAA) electrospun nanofibers	Bromothymol blue
Bazbouz and Tronci (2019)	Poly(methyl methacrylate-co-methacrylic acid) (PMMA-co-MAA) electrospun fibers	Bromothymol blue
Brooker and Tronci (2023)	Drop-cast and electrospun collagen-based materials	Bromothymol blue
Dong et al. (2022)	Agar and glycerol composite films	Bromothymol blue
Kurečič et al. (2018)	Electrospun cellulose acetate (CA)-based nanofibers	Bromocresol green
Wang et al. (2020)	Chitosan (CS)-based hydrogel	Bromothymol blue
Xie et al. (2023)	Polyvinyl alcohol and dextran (PVA-DX) hydrogel	Bromothymol blue
Zhang et al. (2022)	Polylactide and polyacrylonitrile (PLA-PAN) electrospun nanofibers	Phenol red
Zheng et al. (2021)	CD-doped polyacrylamide and quaternary ammonium chitosan (PAM-QCS) hydrogels	Phenol red
Zheng, Zhong, et al. (2023)	CD-doped polydimethylsiloxane PDMS hydrogel	Phenol red
Zhu et al. (2022)	Oxidized sodium alginate and carboxymethyl chitosan (OSA-CMCS) hydrogels	Bromothymol blue

Zheng et al., 2021; Zhu et al., 2022), even if with some exceptions (Xie et al., 2023; Zheng, Zhong, et al., 2023). Finally, the casting technique has found a place even in this field, even if it is less common than in others previously described, due to the stricter requirements the materials should satisfy to be used for wound monitoring (Brooker & Tronci, 2023; Dong et al., 2022).

16.2.2 Polymeric sensors for fluorescence-based pH detection

While colorimetric detection of pH changes based on absorption spectra has a long tradition that goes back several centuries, the employment of pH-dependent modifications in fluorescence emission has been recently investigated in the development of pH probes and sensors (Di Costanzo & Panunzi, 2021). On one hand, it is widely reported in the literature that fluorescence detection is significantly more sensitive than absorbance and can thus detect lower analyte concentrations requiring a minimum amount of receptor, but, specifically in the case of pH sensors, the chemical environment in which these materials are to be used must also be carefully evaluated since ions other than hydrogen protons, such as metal ions, might also affect fluorescence emission and consequently interfere in pH sensing (Skorjanc et al., 2021). Other differences that must be considered are the detection technique, which is generally far simpler for colorimetric devices than fluorescent ones but more selective for the latter; oxygen cross-sensitivity that hampers different fluorescent devices; and background signal, which differently affects both the techniques and other features (Di Costanzo & Panunzi, 2021; Skorjanc et al., 2021).

In describing fluorescence pH sensors, several aspects should be taken into account: first, the fluorescence mechanism involved in pH detection, such as PET, Förster Resonance Energy Transfer, Aggregation Induced Emission, Intramolecular Charge Transfer, and Excitated-State Intramolecular Proton Transfer. Being this chapter focused on polymer-based pH sensors, it is out of our purpose to explain in detail these mechanisms, and we suggest the readers investigate dedicated literature for further information (Di Costanzo & Panunzi, 2021; Skorjanc et al., 2021; Sun et al., 2015).

Another pivotal feature to be defined is which kind of modifications pH provokes in the fluorescence spectrum: on the one hand, *on-off* and *off-on* sensors undergo fluorescence intensity decrease (quenching) or increase (enhancement) upon pH modifications, either toward higher or lower pH values (Di Costanzo & Panunzi, 2021; Skorjanc et al., 2021); on the other hand, pH variations can push modifications in the fluorescence spectrum such as shifts involving both maximum emission (ratiometric sensors) and excitation's wavelengths (Di Costanzo & Panunzi, 2021; Skorjanc et al., 2021). Both of these approaches have been widely exploited in pH sensing and present specific pros and cons: fluorescence quenching or enhancement is much more common in fluorophores but is more prone to suffering from interferences, while fluorophores resulting in fluorescence shifts are more robust but less common, and they also require more complex instrumentations, especially for sensors that can be stimulated at two different wavelengths (Di Costanzo & Panunzi, 2021).

Last, the fluorophore plays a crucial role in fluorescence-based pH detection. In the case of polymer-based sensors, the pH-sensitive moiety can be either a small pH-sensitive fluorophore integrated within the polymer scaffold, protonatable monomers, or chemical functionalities directly incorporated in the polymer chains (Skorjanc et al., 2021). In the first case, the synthetic pathway is generally easier and cheaper; physical entrapment or different approaches for fluorophores embedding within the polymer matrix can be exploited; protonation or deprotonation mechanisms are usually reversible; but fluorophores solubility in the medium of interest should be considered since it is strongly related to their leaching and, consequently, to the possibility of repeated use (Skorjanc et al., 2021). The most common fluorophores are xanthene dyes, derivatives of 8-hydroxypyrene-1,3,6-trisulfonic acid (HPTS), naphthalimides, and (aza)-BODIPY dyes (Dalfen et al., 2019). Conversely, when a pH-sensitive fluorescent moiety is one of the monomers, a higher synthetic effort is required to pay for reusability, stability, and a strong reduction in dye leaching (Skorjanc et al., 2021).

Keeping these aspects in mind, in the following subsections, the most interesting devices recently proposed will be described, discussing both the principles behind their development and, when possible, their applicability.

16.2.2.1 Polymeric pH-sensitive fluorescent devices

In Table 16.4, the most interesting polymer-based fluorescent pH sensors published between 2019 and 2023 are listed, together with the polymeric material selected, the fluorophore, pH sensing mode, and, when available, possible application. We would like to first stress that, differently from colorimetric devices, for which the operating principles have been widely known for decades and the novelty stands mainly in the application, fluorescence-based sensors' operating principles are still under investigation, and thus only few devices are developed aiming at a specific application, while most of them simply represent a *proof-of-concept* to demonstrate the applicability of a given operating principle toward pH sensing. In this regard, a few applications involve NH_3 detection in the vapor phase (Petropoulou et al., 2020), CO_2 measurement (Cascales et al., 2022; Laysandra et al., 2021; Pfeifer et al., 2020), food spoilage monitoring (Li et al., 2023; Liu, Wang, et al., 2020), or biomedical applications (Cascales et al., 2022; Laysandra et al., 2021; Uzair et al., 2020).

Several approaches have been tested regarding fluorophore selection, ranging from the simpler employment of commercially available pH-sensitive fluorescent dyes (Cascales et al., 2022; Duong et al., 2020; Uzair et al., 2020), moving to their modification to improve their efficiency, linearity, or reversibility (Ermakova et al., 2023; Horak et al., 2020; Pfeifer et al., 2020; Tariq et al., 2022; Xu et al., 2022), or their covalent linkage to the polymer chain (Pablos et al., 2022; Ulrich et al., 2019) or other solid supports (Petropoulou et al., 2020) to improve leaching properties and reusability. Also, non-conventional fluorophores have been tested, such as diazaoxatriangulenium dyes, which distinguish themselves for showing an unusually long luminescence-decay time that allows back fluorescence elimination through time-resolved measurements (Dalfen et al., 2019). However, there are also alternative approaches, not relying on commercial fluorophores or existing dye modifications, such as using luminescent quantum dots (Laysandra et al., 2021) or NPs (Reyes-Cruzaley et al., 2021) or, finally, directly incorporating luminescent polymeric materials in the final matrix, as already discussed in the case of PANI for colorimetric devices (Kalisz et al., 2023; Li et al., 2023; Liu, Wang, et al., 2020).

As far as pH-dependent fluorescence spectra modification, fluorescence quenching or enhancement upon acidification or basification definitely represents the most common sensing mode, especially in the case of commercial fluorescent pH indicators or derivatives (Cascales et al., 2022; Dalfen et al., 2019; Horak et al., 2020; Kalisz et al., 2023; Laysandra et al., 2021; Pablos et al., 2022; Petropoulou et al., 2020; Pfeifer et al., 2020; Reyes-Cruzaley et al., 2021; Tariq et al., 2022; Uzair et al., 2020; Xu et al., 2022). In a few cases, a shift in fluorescence emission's maximum intensity is observed: Duong and coworkers exploit Coumarin 6 (C6) and Nile Blue A (NB) to

Table 16.4 Polymer-based fluorescent pH sensors.

References	Polymeric material	pH-sensing moiety	pH sensing mode	Application
Cascales et al. (2022)	Poly(propyl methacrylate) (PPMA)	HPTS	Fluorescence quenching in the presence of CO_2	Wearable transcutaneous CO_2 sensor
Dalfen et al. (2019)	Polyurethane (PU) hydrogel (Hydromed D1 and D4)	Diazaoxatriangulenium dyes (DAOTA)	Fluorescence quenching upon basification	pH measurement (pH range 3–7)
Duong et al. (2020)	Nafion (Nf) or polyurethane (PU) hydrogel	Coumarin 6 (C6) and Nile Blue A (NB)	Fluorescence emission shift and visible color transition	pH measurement (pH range C6: 4.5–7.5, NB: 9–12)
Ermakova et al. (2023)	Agarose matrix (sol–gel technique)	Substituted quinoxalines	Fluorescence emission shift and visible color transition	pH measurement (pH range 1–5)
Horak et al. (2020)	Dioctylsebacate-plasticized polyvinylchloride (PVC) film	Benzimidazo[1,2-a] quinolines-based derivatives	Fluorescence enhancement upon basification	pH measurement (pH range 5–8)
Kalisz et al. (2023)	Polyvinylchloride (PVC) and polycaprolactone (PCL) composite electrospun nanofibers	Polydiacetylenes (PDAs)	Fluorescence enhancement upon basification and visible color fibers transition	pH measurement (pH range 7–8.5 or 8–9.5)
Laysandra et al. (2021)	*n*-butyl acrylate and *N*-(hydroxymethyl) acrylamide composite film	N-doped graphene quantum dots (NGQDs)	Fluorescence quenching upon basification	Skin-like wearable pH sensor (pH range 3–10)
Li et al. (2023)	Polyvinylalcohol (PVOH) film	Lignin	Fluorescence emission shift	Food spoilage monitoring (pH range 9–12)
Lyu et al. (2019)	Polyaniline (PANI)	PANI and tetraphenylethylene (TPE)	Fluorescence emission shift and visible color transition	Food spoilage monitoring
Pablos et al. (2022)	N-vinylpyrrolidone, butyl acrylate and ethyl methacrylate copolymer	Covalently bound naphthalimide groups	Fluorescence quenching upon acidification	pH measurement (pH < 3)
Petropoulou et al. (2020)	Cellulose acetate (CA) electrospun fibers	Rhodamine B-functionalized Core-Shell Ferrous NP	Fluorescence quenching upon basification	pH measurement and NH_3 detection (pH range 8–13)
Pfeifer et al. (2020)	Ethylcellulose (EC) film	π-extended BODIPY dyes	Fluorescence quenching upon basification	CO_2 sensor (pH range 9–12)

(Continued)

Table 16.4 Polymer-based fluorescent pH sensors. ***Continued***

References	Polymeric material	pH-sensing moiety	pH sensing mode	Application
Reyes-Cruzaley et al. (2021)	Agarose hydrogel	Ag nanocluster capped with hyperbranched polyethyleneimine (PEI)	Fluorescence quenching upon basification	pH measurement (pH range 3–11)
Tariq et al. (2022)	Polyvinyl alcohol and glutaraldehyde (PVA-GA) matrix or polyurethane (PU) hydrogel (Hydromed D4)	Perylene-based amine fluorophores	Fluorescence quenching upon basification	pH measurement (pH range PVA-GA:6–8, D4: 5–8)
Ulrich et al. (2019)	Poly(dimethylsiloxane) (PDMS) and poly(2-hydroxyethyl acrylate) (PHEA) composite film	Pyranine	Fluorescence enhancement (λ_{exc} = 460 nm) and quenching (λ_{exc} = 406 nm) upon basification	pH measurement (pH range 5–9)
Uzair et al. (2020)	Diacrylated polyethylene glycol (PEG) hydrogel	Bromocresol green (BCG) and bromothymol blue (BTB)	Fluorescence quenching upon basification and visible color transition	pH imaging on orthopedic implants (pH range BCG: 3–5, BTB: 5–8)
Xu et al. (2022)	1,8-Naphthimide-based polymer and polyvinyl alcohol (PVOH) composite electrospun nanofiber	1,8-Naphthimide derivatives	Fluorescence quenching upon basification	pH measurement (pH range 4–10)

detect pH changes between, respectively, 4.5 and 7.5 for C6 and 9 and 12 for NB, relying both on emission spectra modification and visible color change (Duong et al., 2020); also, Ermakova and coauthors applied similar principles using substituted quinoxalines to measure acidic pH values ($pH < 5$) conjugating emission shift and color transition (Ermakova et al., 2023), and a similar result was obtained also coupling PANI and tetraphenylethylene as both halochromic and luminescent pH-sensitive moieties (Liu, Wang, et al., 2020). The visible color transition was also observed for polydiacetylenes in the device proposed by Kalisz and colleagues, who were also able to tune the apparent protonation constant and thus the operative pH range of their poly(vinyl chloride) (PVC)-PCL-based nanofibrous material, changing the monomeric chromoionophore (Kalisz et al., 2023). Obviously, a color change was also observed in preliminary characterizations by Uzair and colleagues in their X-ray excited luminescence chemical imaging device in which the pH-sensitive probes are commercial pH indicators, namely, bromocresol green and bromothymol blue (BTB), but, in this case, the color transition is no longer visible in the final application, that is, once the sensitive films are deposited on the surface of orthopedic implants in patients (Uzair et al., 2020). Finally, Ulrich and coauthors opted for quite complex pH detection in their pyranine-based sensor: increasing pH value, this fluorophore shows an enhancement in fluorescence emission upon excitation at 460 nm and a parallel fluorescence quenching moving to 406 nm as excitation wavelength; exploiting both mechanisms and relying on fluorescence ratio, better pH assessment is achieved, but more complex instrumentation and laborious measurement procedure is necessary (Ulrich et al., 2019).

Another key aspect to be discussed involves the operative pH range, which is always fundamental, especially in looking for a practical application for these devices; as already shown for colorimetric sensors, an overview of operative pH intervals for the sensors is presented in Table 16.4 and is depicted in Fig. 16.2, in which each device is labeled according to the numbers reported in the first column of Table 16.4. As for optical devices, also in the case of fluorescent probes, a small pH range is detectable when using one single fluorescent dye at a time (Cascales et al., 2022; Duong et al., 2020; Kalisz et al., 2023; Pablos et al., 2022; Tariq et al., 2022; Uzair et al., 2020), but the operative pH range can be expanded either by coupling more than one pH-sensitive moiety (Dalfen et al., 2019; Ermakova et al., 2023; Horak et al., 2020; Pfeifer et al., 2020; Xu et al., 2022) or exploiting polymeric or nano-scaled unconventional fluorophores (Laysandra et al., 2021; Li et al., 2023; Liu, Wang, et al., 2020; Petropoulou et al., 2020; Reyes-Cruzaley et al., 2021). It must be underlined that, so far, fluorescent devices allow the monitoring of smaller pH ranges than colorimetric sensor arrays previously described, measurable even because most of them are focused on specific applications rather than generalized ones.

Last but not least, the most wide-ranging polymeric materials have found an application in this kind of pH sensor, both man-made and natural materials, prepared by the casting or sol–gel technique but also by electrospinning and other deposition methods. It must be again underlined that most of these devices are actually presented as *proof-of-concept* to demonstrate their hypothetical applicability, but the focus of the research is mainly on the fluorescent probe rather than on the sensor itself. This aspect allows researchers to select the easier or simpler polymeric material for the sensor preparation without actually worrying about environmental friendliness, cost, and scalability,

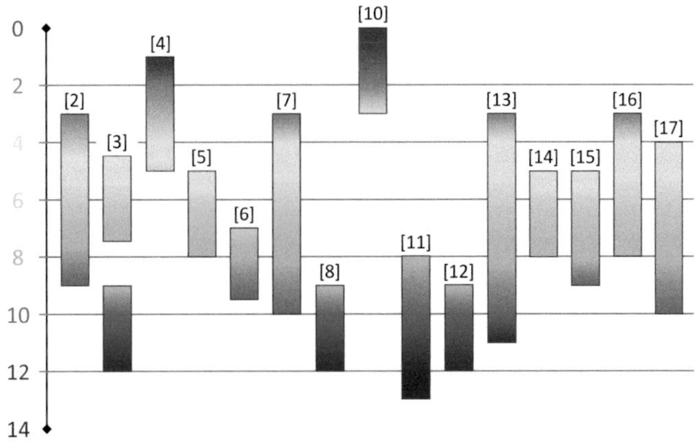

FIGURE 16.2 Comparison of measurable pH ranges for devices in Table 16.4.

Each device in Table 16.4 is represented as a colored box covering the measurable pH range and labeled according to the numbers reported in the first column of Table 16.4.

or, generally speaking, about the factual requirement for real application. We do believe that this is the reason why man-made polymers are still so common in these devices (Cascales et al., 2022; Dalfen et al., 2019; Duong et al., 2020; Horak et al., 2020; Kalisz et al., 2023; Laysandra et al., 2021; Li et al., 2023; Liu, Wang, et al., 2020; Pablos et al., 2022; Tariq et al., 2022; Ulrich et al., 2019; Uzair et al., 2020; Xu et al., 2022), while they have experienced an almost complete replacement in other applicative sensors previously described, and also commercial polymers are widely employed (Dalfen et al., 2019; Duong et al., 2020; Tariq et al., 2022) while bio-based polymers or even materials from agricultural waste (Ermakova et al., 2023; Petropoulou et al., 2020; Pfeifer et al., 2020; Reyes-Cruzaley et al., 2021) are still rarely employed, especially comparing with other fields in sensors development. Furthermore, in most cases, the principles behind polymer selection are not even described, membranes or film preparation are quickly and roughly described, and, in conclusion, the solid support does not represent the focus of the publication.

16.2.2.2 Polymer-based fluorescent probes for pH measurements in solution

Before concluding the section devoted to fluorescence-based pH sensors, we have to say that, despite falling out of our previous definition of "sensor" as a physical device, in this field the word "sensor" is often used to describe fluorescent probes working in solution. In our humble opinion, this custom is not correct, but, being widespread worldwide and trying to provide a useful literature analysis to researchers working in this field, a brief description of polymer-based fluorescent probes meant to measure pH in solution will be all the same.

Polymer dots, micelles, or NPs have been recently investigated as tuneable and stable platforms to develop fluorescent pH-sensitive probes following similar approaches to those described for physical devices; also in this case, polymeric platforms can either host fluorophores, both covalently bound or physically sorbed or entrapped (Chen et al., 2019; Du et al., 2023; Ismail et al., 2020; Nguyen et al., 2020; Shah, Tahir, et al., 2019; Wang, Feng, et al., 2022; Wu et al., 2019; Xue et al., 2019) work themselves as pH-sensitive moiety (Bao & Liu, 2022; Deb et al., 2022; Roy et al., 2022; Ou et al., 2019). Moreover, because these probes are meant to work in solution, more complex detection techniques are usually proposed compared to physical devices in which fluorescence enhancement/quenching is generally preferred. As an example, the intensity ratio between two different emission peaks, either sharing the same excitation wavelength (Chen et al., 2019; Deb et al., 2022; Wang, Feng, et al., 2022; Xue et al., 2019) or occurring at different λ_{exc} (Bao & Liu, 2022), can be exploited to assess pH with generally higher analytical performances. Or, in some cases, a single system allows the contemporary detection of different physical and chemical parameters such as pH and specific metal ions' concentrations like Cu^{2+} (Deb et al., 2022; Roy et al., 2022; Shah, Tahir, et al., 2019) or other analytes (Bao & Liu, 2022; Shah, Tahir, et al., 2019).

To conclude, the fluorescent probes recently proposed are surely interesting, and some of them might find an application in tangible sensing devices, even if a lot of work still has to be done to achieve a possible application.

16.2.3 Polymeric electrochemical pH sensors

The section on electrochemical sensors strongly brings us back to the origins of pH measurement: potentiometry and glass electrode, which are at the basis of the most widely used technique for assessing the H^+ activity in a medium. As already pointed out in the introduction, the pros of such a lucky and successful sensor are well known, as are the drawbacks. Just to remember, glass electrodes suffer from the intrinsic fragility of the sensing membrane, the need for an inner electrolyte solution together with that of constant re-calibration, the presence of a junction potential that varies with the change of external conditions, the need for a storage solution, and the impossibility of miniaturization (Avolio et al., 2022).

As underlined by Avolio and its group, in the previous decades, a lot of efforts have dealt with alternative pH sensing devices focused on high precision measurements (up to 0.001 pH units), mainly based on spectrophotometric devices. These devices were suitable for continuous monitoring, strongly required, for instance, in natural and pristine water assessments, but nevertheless, they were more expensive and complex compared to potentiometric sensors and had a longer sampling time. The presence on the market of ISFETs-based pH probes is an example of solid-state sensors, providing a cheap, robust, and miniaturizable alternative for pH measurements (Avolio et al., 2022).

The most recent papers dealing with electrochemical techniques still reflect the need to partially solve the glass electrode limitations and to face other concerns arising when conductive polymeric substances are introduced in electronic devices for pH sensing. The most recent papers are presented in Table 16.5, together with prominent features, while the pH linear range explored by the different devices is depicted in Fig. 16.3 to ease the comparison. In search of a more general trend, we can claim that the recent literature on the subject is focused on miniaturization for wearable or implantable pH sensors, possibly assessing pH together with other analytes of interest, which seems to be the frontier of electrochemical-based devices. The last contributions on this subject are reported in Table 16.6, with the same format as the previous one, and to facilitate the comparison, the pH linear range is reported in Fig. 16.4.

The electrochemical techniques are based on the measurement of electrical parameters, such as conductivity or resistivity, impedance, potential, and current, that are registered in an electrochemical cell. The nature and the number of electrodes, two or three, obviously depend on the technique.

Conductometric devices, where two simple electrodes (Pt, platinated Pt, and glassy carbon) are present, correlate the change in conductivity/resistivity of an active material to the concentration of the analyte, that is, H^+, in pH measurements. In voltammetric techniques, the analytical signal is the current generated by the redox reaction flowing between the electrodes when the potential is swept in an appropriate range. Three electrodes are present: the working electrode (WE), where the redox reaction of interest takes place; the reference electrode (RE), which guarantees a constant potential independent of the current intensity passing through the cell; and the auxiliary electrode, which does not undergo the electrochemical reactions studied but ensures the flow of current through the electrochemical cell. In this case, for pH measurements, H^+ can be correlated to the peak current or peak potential of an electroactive compound. In potentiometric sensors, the current through electrodes is zero, while the parameter of interest is the electromotive force (EMF) that sets up in a two-electrode cell, composed of a WE and a RE. In this latter case, the pH of the sample is calculated by comparing the EMF measured in the sample with a standard buffer solution of known pH, following the Nernst equation. The peculiarity of ion-selective electrodes, such as the glass electrode, is that the response is Nernstian even if no redox reaction is involved. The WE is in any case sensitive to H^+ concentration, either directly as in the ISE or indirectly through the reaction that takes place at the interface of the WE.

Table 16.5 Polymeric electrochemical pH sensors.

References	Materials	Detection technique	pH range	Sensitivity
Bao et al. (2019)	Nanoparticles (PFV/PSMA NPs) and dopamine	Fluorescence intensity	5–9	
Bhat et al. (2020)	Poly(2-hydroxyethyl methacrylate) functionalized with N-(2-aminoethyl) methacrylate (AEMA),	Impedance spectroscopy	7.35–7.45 HEPES buffer	
Chajanovsky et al. (2021)	Multi-walled carbon nanotubes (MWCNT) and PANI	Conductimetric	2.6–3.8 10.2–12.0	2.3 kΩ/pH 3.6 kΩ/pH
Choi et al. (2019)	WO_3 nanofibers and chloromethylated triptycene poly (ether sulfone) (Cl-TPES)	MOSFET	6.9–8.94	−377.5 mV/pH Artificial sw + HCO_3^-
Choi et al. (2021)	Poly(vinyl chloride) (PVC) and poly (decyl methacrylate) membrane	Potentiometry	4.8–13.1 1.9–9.2	−55.0 mV/pH -55.7 mV/pH
Choudhury et al. (2023)	MoS_2-PANI modified screen-printed carbon electrode	Potentiometry	4–8 Artificial sweat	−70.4 mV/pH
Drago et al. (2022)	PANI nanojunctions embedded within single gold nanowires	Impendence	2.0–9.0	Synthetic urine
Morshed et al. (2020)	Poly-2-Aminobenzonitrile (PABN)	Potentiometry	1–13 buffer	−52 mV/pH
Mu et al. (2021)	Indium trioxide layers on PET	Potentiometry	4–9	−50.974 mV/pH
Noriega-Navarro et al. (2020)	Carboxylic acid graphene oxide (COOH@GO) and poly[2-(diethylamino)ethyl methacrylate] (poly[DEAEMA])	Potentiometry	2.4–11.7	−47.4 mV/pH
Pfattner et al. (2019)	Poly(tetrathienoacene-diketopyrrolopyrrole) (PTDPPTFT4)	Dual gate field effect transistor, organic field-effect transistor OFET	2–12	−41 ± 7 mV buffer solution
Saikrithika and Kumar (n.d.)	PANI(4-Cl)-modified graphitized mesoporous carbon (GMC)	Potentiometric	2–11	−58 mV/pH
Sato et al. (2021)	PEI (poly(ethyleneimine)) + poly (styrenesulfonate) sodium salt (PSS) + Poly(1-vinylimidazole) (PVI) multilayer film	Cyclic voltammetry (CV)	5–6	
Sinar et al. (2020)	Nylon-66 filter paper coated with polydimethylsiloxane (PDMS)	Capacitance	4–10	Tunable
Sisodia et al. (2021)	2-(methylthio)phenol glassy carbon	Voltammetry, Peak Potential	4–9	−51 mV/pH
Zdrachek et al. (2023)	PEDOT-C14 layer on a glassy carbon	Potentiometry	5–9	
Zea et al. (2021)	PANI, polypyrrole (PPy) and Poly (styrenesulfonate) sodium salt (PSS)	Potentiometry	3–10	−81.2 mV/pH

Table 16.5 Polymeric electrochemical pH sensors. *Continued*				
References	**Materials**	**Detection technique**	**pH range**	**Sensitivity**
Žutautas et al. (2023)	A laser-induced graphene (LIG) modified with chitosan and polyfolate	Potentiometry	6–9	−27.9 mV/pH (1) −30.2 mV/pH (2)
Žutautas et al. (2022)	Chitosan on graphite electrode and polyfolate	Potentiometry, pH dopamine and glucose sensing	6–9	44 mV/pH

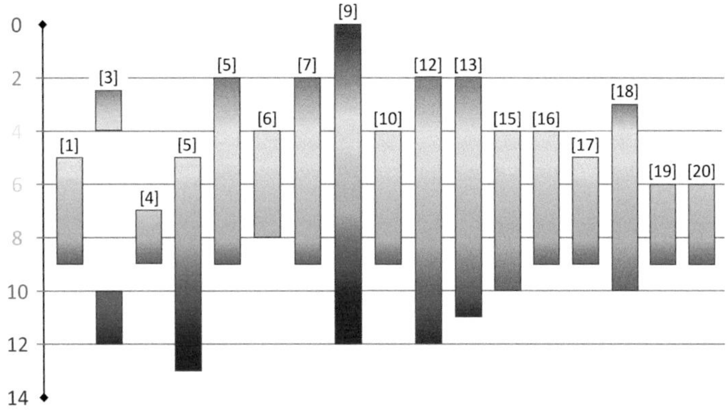

FIGURE 16.3 Comparison of measurable pH ranges for electrochemical devices in Table 16.5.

Each device in Table 16.5 is represented as a colored box covering the measurable pH range and labeled according to the numbers reported in the first column of Table 16.5.

Among other detection strategies, we must mention the metal oxide field effect transistor and ISFET, which indeed result in amplified potential measures. The main feature is the substitution of the glass electrode with solid-state sensors, where the gate electrode is modified (or substituted) by a thin layer of an insulating material (Si_3N_4, Al_2O_3, Y_2O_3, and ZrO_2) (Avolio et al., 2022). When in contact with solutions of different pH, the protonation/deprotonation process takes place on the insulator layer and sets up the electrostatic field at the gate, modulating a current that flows into the FET (Avolio et al., 2022). ISFET was presented for the first time in the 1970s and successfully commercialized in the following decades as an alternative to the glass electrode, being cheap, scalable, and miniaturizable. It is not the place here to describe them; readers can find papers, reviews, and books dedicated to the topic, but we mention it since this detection strategy and its variation were exploited in some cases described below. Other nonpotentiometric transistor-based sensors present an organic semiconductor film, such as pentacene, PANI, and poly(3-hexythiophene), which connects source and drain (Organic semiconductor Field Effect Transistors (OFET)). The current between the source and drain depends on the potential of the gate and the characteristics of the

Table 16.6 Polymeric electrochemical wearable pH sensors.

References	Materials	Detection technique	pH range/ samples	Sensibility
Choudhury et al. (2023)	MoS_2-PANI	Potentiometry	4–8 Artificial sweat	−70.4 mV/pH
Dervisevic et al. (2023)	Polymeric microneedle array coated with PANI	Potentiometry	4.0–8.6 Ex vivo mouse	−62.9 mV/pH
Dulay et al. (2021)	Polypyrrole conductive polymer	Potentiometry	4–9 In vivo intramuscular, ex vivo rabbit blood	−46.4 mV/pH
Jose et al. (2022)	Polyester woven fabric plus a. PEDOT:PSS b. Carbon alizarine c. Graphene PANI	a. Conductometry b. Voltammetry c. Potentiometry	4–9.5	−45mV/pH
Laffitte and Gray (2022)	Flexible screen printable PANI textile	Potentiometry	3–10 Buffer solution	−27.9 mV/pH −42.6 mV/pH
Lee et al. (2021)	Flexible microneedles exopyxilosane coved by PANI and poly(3-methoxypropyl acrylate) (PMC3A)	Potentiometry	3–7 Pig skin, rat artery	−94 mV/pH
Mariani et al. (2021)	IrO_x particles embedded in a PEDOT:PSS thin film.	Voltammetry	3–11 Synthetic wound exudate	−59(4) uA/pH
Mugo et al. (2023)	Polydimethylsiloxane (PDMS) microneedle platform coated with a conductive PDMS/carbon nanotube (CNT)/cellulose nanocrystal (CNC) composite (PDMS/CNT/CNC@PDMS)	Cyclic voltammetry	4.25–10 (pH range) andepinephrine, dopamine, and lactate analysis	
Possanzini et al. (2020)	PEDOT:PSS and Bromothymol blue on threads	Amperometric	4.5–7 (pH range) and Cl^- analysis in artificial sweat	
Xu et al. (2023)	Tannic acid-Ag-carbon nanotube-polyaniline (TA-Ag-CNT-PANI) composite hydrogel	Potentiometry	4–8 (pH range) and tyrosine analysis in sweat	−71mV/pH
Yoon et al. (2020)	Carbon fiber thread (CFT) electrodes coated with self-healing polymers (SHP)	Potentiometry	4–7 Buffer solution and fluids	−58.9 mV/pH
Zhao et al. (2023)	3D polyaniline (3D PANI)	Potentiometry	4–9 Sweat	69.33 mV/pH

semiconductor, which vary as a result of the adsorption of analytes and are, therefore, proportional to the concentration of the latter.

Closing the circle on the polymer-based electrochemical sensors for pH measurement, conducting polymers (CPs) have been introduced, besides inorganic pH sensing oxides, firstly because their

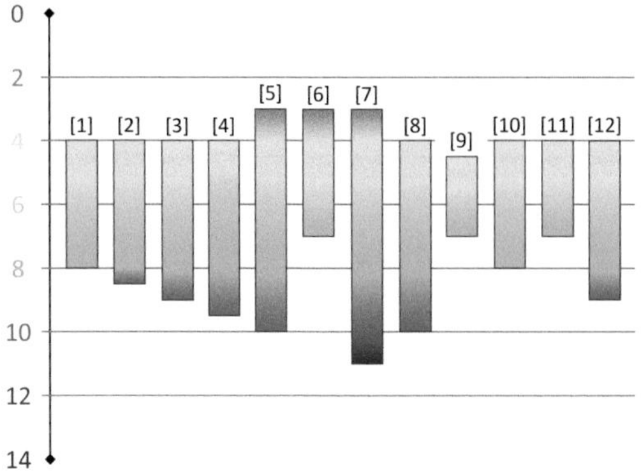

FIGURE 16.4 Comparison of measurable pH ranges for wearable electrochemical devices in Table 16.6.

Each device in Table 16.6 is represented as a colored box covering the measurable pH range and labeled according to the numbers reported in the first column of Table 16.6.

ion exchangeability serves well for the potentiometric measure, and secondly for their excellent versatility, robustness, and low cost (Avolio et al., 2022). Definitely, the CP conformation results are less rigid and more deformable than the crystal lattices of conventional conductors. It is worth underscoring that CP electrical conductivity lies in the range of semiconductors or even metals (Gualandi et al., 2021). This property derives from the conjugated Π system of carbon atoms extended along the whole length of the polymer chain, a consequence of the sp_2 hybridized polymer backbone. The charge carrier species are free to move along the chain, and electricity can easily flow through them. CPs can be doped, modified with conduction bands, or stabilized by mixing with oppositely charged polyions. In addition, CPs' mechanical features depend on their chemical composition and can be varied by means of specific additives. The main advantages of CPs can be summarized as high and controllable conductivity, excellent stability from chemical and electrochemical points of view, good optical transparency, and good biocompatibility (Gualandi et al., 2021).

16.2.3.1 Polyaniline-based devices

Among CPs, one of the most popular polymers is PANI, often prepared by direct chemical oxidation of aniline on the electrode substrate. Its response mechanism to pH is a consequence of reversible protonation/deprotonation when present in its half-oxidized emeraldine base (EB). Indeed, PANI exhibits three oxidation states: the fully reduced leucoemeraldine base (LB), the half-oxidized EB, and the fully oxidized pernigraniline base. Due to the presence of amine and imine groups, only EB can be reversibly protonated to form conductive emeraldine salt, as reported in Fig. 16.5 (Ogoshi et al., 2011). This reversible reaction is accompanied by two protons transfer, which is the origin of the Nernstian response for PANI-based potentiometric pH sensors (Tang et al., 2022).

As it will be discussed, a plenitude of possible configurations, directly with PANI or its derivatives, as well as hybrid materials in combination with PANI, have been proposed; see Table 16.5 and Table 16.6. From an analytical point of view, among the most recent contributions involving

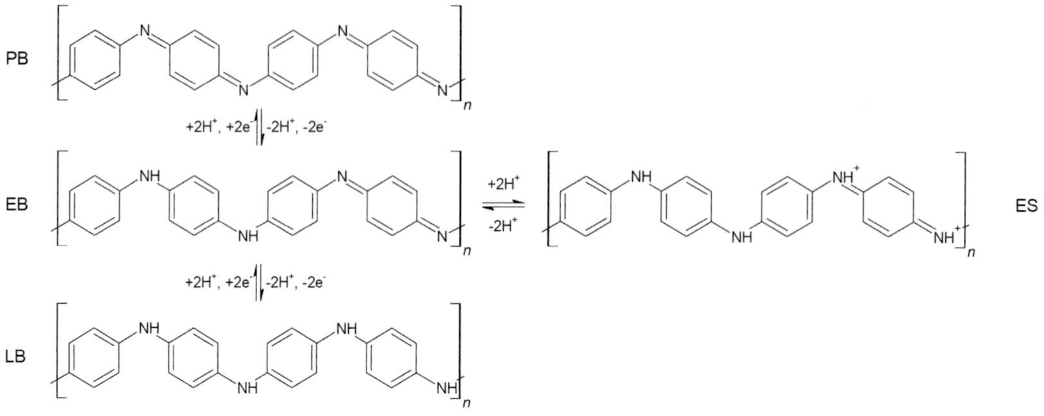

FIGURE 16.5 Polyaniline chemistry.

Redox equilibria between fully oxidized pernigraniline base (PB), half–oxidized emeraldine base (EB) and fully reduced leucoemeraldine base (LB) and protonation-deprotonation reactions between EB and protonated emeraldine salt (ES).

PANI, the proposed devices show very different performances, and some of them, as pointed out in the previous section, should be solely considered a first *proof-of-concept* that requires a more thorough investigation to obtain a kind of working prototype on real samples. This is the case of the sensor presented by Chajanovsky and coauthors, based on multi-walled CNT aniline copolymerized in the presence of dodecylbenzene sulfonic acid to obtain the sensing material for conductimetric pH measures. The device has a linear response in two extreme pH ranges where it is hard, in perspective, to find applicability; equally, the same device could be promising for phenol and aminophenol detection (Chajanovsky et al., 2021). A peculiar arrangement was proposed by Drago et al., where PANI nanojunctions are embedded within single gold nanowires for impedimetric pH measures in the range of 2.0–9.0, also tested on synthetic urine samples (Drago et al., 2022). A modified PANI polymer, poly-2-Aminobenzonitrile, is the sensing unit proposed by Morshed and coauthors that seems promising even if tested only on buffer solutions but in a wide pH range, from 1 to 13 (Morshed et al., 2020). Similar performance was described by Zea and coworkers for the sensitive multi-material composed of PANI and polypyrrole, integrated into a polyelectrolyte, poly(sodium 4-styrenesulfonate) (PSS), printed by inkjet printing, operating in the range 3–10 with a good sensitivity of −81.2 mV/pH (Zea et al., 2021).

As for wearable devices, reported in Table 16.6, Choudhury and coworkers present a hybrid material of MoS_2-PANI on a modified screen-printed carbon electrode; even here, a reasonable pH range, from 4 to 8, guarantees a possible application to biological fluid measures. Indeed, the electrode was tested on buffer solutions and artificial sweat, with, in this case, a not uncommon super-Nernstian response (Choudhury et al., 2023). Other examples seem promising candidates for wearable electronics. For instance, Zhao and coworkers succeeded in developing a three-dimensional PANI sensor for pH measures in real-time in the range 4–9, testing the device in commercial artificial perspiration samples (Zhao et al., 2023). Furthermore, in a recent study conducted by Xu and their colleagues, the researchers introduced an innovative composite hydrogel consisting of tannic acid, silver NPs, carbon nanotubes, and PANI, denoted as TA-Ag-CNT-PANI hydrogel. This hydrogel demonstrates exceptional performance within the pH range of 4 to 8, making it particularly well-suited for the accurate determination of tyrosine levels in sweat (Xu et al., 2023). We also mention Laffitte Group's textile screen printable PANI-based device, which could possibly be suited as a wearable sensor even if tested only in buffer solution (Laffitte & Gray, 2022). A wide test on different devices is finally presented by Jose et al., and the most efficient out of three was the potentiometric pH measure by a graphene electrode with PANI, tested only on buffers, even if in a pH range from 3 to 10 (Jose et al., 2022).

As for the attempts of transdermal devices, that is undoubtedly a frontier for researchers; either polymeric microneedles are covered in with PANI, producing a coated polymeric microneedle array operating in the range 4.0–8.6 (Dervisevic et al., 2023), or exopyxilosane-based flexible microneedles are covered with PANI and then poly(3-methoxypropyl acrylate) (PMC3A) (Lee et al., 2021). Tested in ex vivo mice, pig skin, and rat arteries, respectively, they presently seem far from real development and practical application.

16.2.3.2 Poly(3,4-ethylenedioxythiophene)-based devices

Another attractive CP successfully employed as a CP is poly(3,4-ethylenedioxythiophene) (PEDOT), owing to its interesting features and intrinsic electrical conductivity. Nevertheless, to be suitable for use in sensors, the conductivity of polymers must be at an appropriate level and adjustable. Indeed, too low or too high conductivity limits the sensor's performance because of the low signal-to-noise ratio. PEDOT is typically insoluble in water or common organic solvents, but when synthesized in the presence of PSS, it gives a stable, dark-blue-colored aqueous dispersion containing both polymers (PEDOT:PSS). This waterborne dispersion is commercially available and exploited for its solution-fabrication capability and miscibility into functional films by drop-casting, spin-coating, and spray-coating. It is worth mentioning that in 2000, the three discoverers of PEDOT:PSS CPs won the Nobel Prize in Chemistry, and since then, its hybrid composites with metal/metal oxide NPs, insulating polymers, carbon materials, and others have been not only extensively studied but widely applied in electronics, including electrochemical and/or electronic chemosensors. As underlined by Gao and colleagues, these research activities have produced thousands of peer-reviewed papers and patents (Gao et al., 2021).

It is also of interest to spend some time on the mechanism that allows pH measures based on PEDOT:PSS polymers. Under acidic conditions, PEDOT chains are uniformly distributed along the PSS polymer chains, and under a similar condition, a continuous electrical connection between PEDOT segments is guaranteed. Nonetheless, moving from an acidic to a basic solution, negatively charged hydroxyl groups are formed that interrupt the homogeneous distribution of PEDOT along the PSS polymer chains. A further increase in basicity results in the formation of a new hydrophobic phase of PEDOT that provokes a significant decrease in polymer conductivity. The conductivity of PEDOT:PSS used in the electrochemical sensors can be improved with other pH-sensitive materials acting as charge transfer layers of the WE. PANI, for example, does not exhibit high conductivity, as underlined above, and has pH-dependent oxidation states, making it suitable for multicomponent materials for pH sensing (Gao et al., 2021).

The most interesting applications recently proposed that PEDOT:PSS are in the field of portable sensors, and indeed, more than transdermal devices, it is a topic that deserves a deeper insight with several proposals that seem promising and at an advanced stage of development.

Wearable sensing technologies are attractive both for academic and industrial interest due to the driving force of market demand and their potentially great impact on real life. We can say that this sensor's area has received a new impulse, maybe more than others, from technical advancements not only in sensors but also in energy-harvesting and energy-storage devices. As a consequence, the request for these wearable biosensors is in constant evolution, with increasing demand for personalized medical and health care, together with specific requests from the fitness world, as underlined by Possanzini and colleagues (Possanzini et al., 2020).

The detection of biochemical markers like dopamine, adrenaline, cortisol, glucose, lactate, phenolic compounds, and electrolytes has been proposed as complementary tools to evaluate human health, physical exertion, fatigue, and mental accuracy. Presently, it seems evident that the most advanced and mature technological frontier focuses on fully textile sensor devices as solutions that show required features such as flexibility, portability, non-invasiveness, and lightweightness for continuous human body monitoring. Indeed, textile sensors are mostly eco-friendly and can rely on a well-established manufacturing background (Possanzini et al., 2020).

Moreover, it seems possible to integrate into a fabric an array of multiple sensors to detect different analytes in a sort of textile multi-sensor platform, defined by Possanzini and colleagues as "a sort of lab-on-fabric device." Human perspiration is indeed a biofluid containing many analytes, such as lactate, glucose, and urea, and electrolytes like sodium, chloride, calcium, and potassium ions, that move from the bloodstream to the skin surface through sweat. It appears clear that non-invasive sampling outside the body can be exploited to achieve information on human health, taking advantage of the already existing iontophoresis technique that results in stimulated, on-demand sweating. It is out of the scope to go into a more detailed description; we want only to highlight these cutting-edge materials that deal with electrochemical transduction that can be amperometric, potentiometric, or mediated by an organic electrochemical transistor (Possanzini et al., 2020).

Indeed, the proposal by Possanzini and coworkers is about sensing thread sensors based on the semiconducting polymer PEDOT:PSS, taking advantage of this biocompatible dispersion that results in a soft, flexible material able to work in the aqueous medium. The sensing features are given by the functionalization with silver/silver chloride (Ag/AgCl) NPs and BTB dye, which allow the simultaneous assessment of Cl^- concentration and pH detection, respectively. The single threads are woven or sewn into the same fabric without requiring gates or REs for detection. The two-terminal thread sensors behave as an electrochemically gated device both for Cl^- and pH detection. The robustness of the potentiometric-like transduction mechanism takes advantage of the highly simple and feasible geometry without the need for a RE. Chloride and pH simultaneous detection allows for an overview of hydration status, fatigue, alkalosis, and physiological conditions. Correlation with Cl^- concentrations in serum and urine demonstrated the feasibility of sweat analysis as well as those of pH that range from 4.7 to 6.6, but values up to pH 9 have been found in patients affected by cystic fibrosis (Possanzini et al., 2020).

In the same research group that developed the previous thread sensor and based on a similar architecture, Mariani and collaborators propose a fully textile device studied for a smart bandage. The idea is the real-time monitoring of wound pH since pH value is widely known to be correlated with the healing stages, meaning that possibly the wound status becomes accessible without disturbing the wound bed. The device is realized by integrating a sensing layer, including the two-terminal pH sensor made of a PEDOT:PSS thin film and iridium oxide particles, and an absorbent layer ensuring the delivery of a continuous wound exudate flow across the sensor area. The reversible response with a sensitivity of (59 ± 4) μA pH^{-1}, the operative pH range, in this case from 6 to 9, makes the device medically relevant and suitable for the application. Moreover, the bandage results were not significantly affected by the presence of the most common chemical interferents and temperature gradients from 22°C to 40°C. The relevant aspect, always evaluating the proposal under an analytical chemistry lens, is that for the smart bandage for pH sensing, Mariani and coworkers assess not only the analytical performances but also validate the device for flow analysis using synthetic wound exudate (Mariani et al., 2021).

16.2.3.3 Other semiconducting polymers-based devices

We are pleased to describe a proposal that differs from all the others reported here. Sinar and coworkers presented a disposable, thin, biocompatible device for monitoring the acidity of body fluids. It is defined by the authors as a hydrogel-activated circular discrete interdigitated capacitive (D-IDC) biosensor: it is made of a thin disk-shaped chitosan/polyvinyl alcohol hydrogel layer deposited on the multi-terminal circular D-IDC electrode, forming distinct concentric circular conductive traces covering the active sensing area. When exposed to the target solution, introduced centrally, the hydrogel transducer swells in a radial direction. The capacitive readings from the D-IDC are related to the dimensional swelling of the hydrogel, not to the incremental increase in total capacitance. It means that each electrode pair is assigned to a specific environmental pH. Moreover, the D-IDC design converts the output into a simple interrogation process of identifying binary outputs (0 or 1) over several discrete pH intervals. In principle, it can be modulated for a specific target in terms of sensitivity and pH range. To our knowledge, it seems that any further application or implementation has so long been presented by the same authors, but the D-IDC biosensor remains, compared to other examples of the described panorama, an extremely original contribution, in any case, brought to a good stage of development (Sinar et al., 2020).

Among other CPs, the contribution of Bhat and colleagues deserves mention since it is a rare example of DOE application and, as underlined elsewhere, a topic dear to these authors. The CP is a methacrylate derivative functionalized with N-(2-aminoethyl) methacrylate (AEMA). The detection is based on impedance spectroscopy, but the device was tested only on HEPES buffer in the range 7.35−7.45 (Bhat et al., 2020). A pH-sensitive polymer, poly[2-(diethylamino)ethyl methacrylate] (poly[DEAEMA]), was presented by Noriega-Navarro's group, which selected the best candidate from nine different compositions based on an orthogonal study but limited to three pH values for the measurements (Noriega-Navarro et al., 2020).

Choi and collaborators present nanofibers based on tungsten oxide (WO_3) with chloromethylated triptycene poly(ether sulfone) (Cl-TPES) binder suitable for MOSFET signal transduction, but, even with a super Nernstian response and tested on artificial seawater, exhibit a linear range from 6.9 to 8.94. From the same research group, a more robust device is later presented that is based on PVC-based ISEs with new poly(decyl methacrylate) membranes, tested only in buffer solutions in a wide range of about 8 pH unit and an almost Nernstian slope (Choi et al., 2021).

The peculiarity of Yoon's group's proposal, based on a carbon fiber thread coated with a self-healing polymer, is indeed the fast autonomous healing of the polymer at room temperature. For us, it seems more interesting that the pH sensor cable can be knitted into a headband integrated with wireless electronics and measure pH in small volumes of real human fluid samples, including urine, saliva, and sweat, and the results were similar to those of a commercial pH meter (Yoon et al., 2020). In two different papers by Žutautas and coworkers, the devices are based on chitosan combined with polyfolate in a laser-induced graphene electrode or pyrolytic graphite electrode and exhibit similar performance in the pH range from 6 to 9 with sub-Nernstian sensitivity (Žutautas et al., 2022, 2023). Finally, Pfattner and colleagues employed the semiconductor poly(tetrathienoacene-diketopyrrolopyrrole) (PTDPPTFT4) in the bottom-gate device SiO_2-based OFET signal transduction, which operates in a wide pH range from 2 to 12, but outside the buffer solution it was not tested (Pfattner et al., 2019).

We can conclude that the efforts presented in this chapter are, in many cases, improvements or modifications of previous devices or sensing strategies. In other cases, they are brilliant proofs of concepts that, unfortunately, are never tested in real samples. It is true that multi-sensing is a challenging perspective, and there is a strong demand from different fields in that sense, but in this overview, sometimes the fact that other protogenic species are detected by the proposed device sounds more like scarce selectivity than an original objective. There are also some exceptions that, in our opinion, deserve the space and attention we have dedicated.

16.2.4 Polymeric optical fibers pH sensors

Among the alternatives to potentiometric pH determination, methods based on polymeric optical fibers (POF) deserve some insight, even if their employment for this purpose is still very limited in comparison to polymer-based colorimetric pH sensors or other electrochemical devices. So far, the potential applications of POF sensors are certainly compelling. Conversely, it seems premature to say how groundbreaking they will be in the light of recent literature. Possibly, the reason for that is related to the nature of these devices. Optical fiber-based sensors require a laser source, sometimes high-power or tunable lasers. The analytical signal exploits the effects caused by the light traveling through the optical fiber system, like modulation of intensity, spectrum change, and reflection λ. Very often, the fabrication process requires specific, expensive facilities/technology to obtain devices with reliable properties. As will be commented on, all these drawbacks currently limit many of the proposed applications to remaining *proof-of-concepts*.

POF was produced by DuPont in the United States early in the 1960s, and Mitsubishi Rayon in Japan began commercializing this technology shortly thereafter, as mentioned by Macedo and coauthors (Macedo et al., 2022). In Fig. 16.6, only to give an idea of how they differ from silica fibers, the dimensional comparison between polymer optical and silica optical fibers, in the single mode and multimode, is shown. It seems quite clear that POFs could offer a good platform for low-cost sensing. From one side, their large core size makes it possible to exploit several light phenomena of reflection, absorption, or coupling of evanescent waves. On the other side, it is also possible to modify the surface of a POF to develop sensors that detect superficial refractive index (RI) change through the SPR phenomenon.

POFs are very ductile products that can be produced using discontinuous manufacturing techniques (that require at least two different steps to run successively), but also continuous manufacturing, which includes continuous extrusion, photochemical polymerization, coextrusion, dry spinning, and melt spinning (Macedo et al., 2022). The possibility of exploring multi-material fabrication was more recently explored to improve POFs, as will be described below (Booth et al., 2021; Wang, Wang, Su, et al., 2022). With respect to silica fibers, POFs are capable of transmitting data over relatively short distances due to their higher power loss, but they exhibit all polymers' excellent properties: low material and fabrication costs, lightweight, high flexibility, elasticity, and

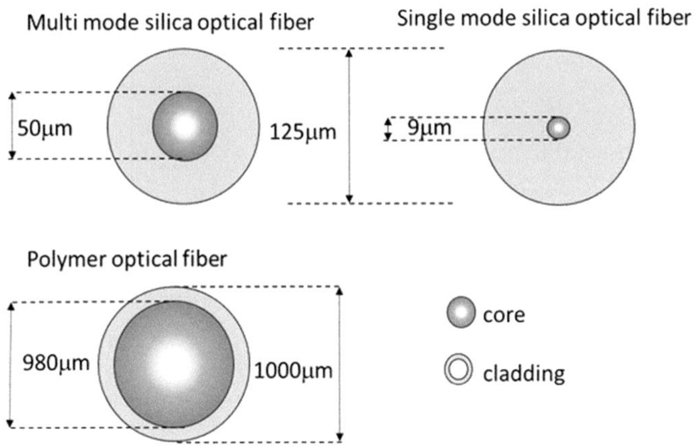

FIGURE 16.6 Dimensional comparison between silica and polymer optical fibers.

General dimensions of multimode (in the top left) and single-mode (in the top right) silica optical fibers compared to polymer optical fibers (in the bottom part).

impact resistance. The possibility of miniaturization, multiplexing, the capability of remote sensing, and high versatility, together with that of being easily integrated into existing medical devices, made POFs good candidates for sensors to be employed in biology applications and biomedicine (Tang et al., 2021).

Coming back to our focus, the possible applications of POF sensors for pH monitoring arouse great interest since, in principle, they are of interest in extremely different fields. Without going into details, we cite only a few very different examples: First, pH control is strategic in in vitro cell culture since any variation outside of the range of 6.5 to 7.7 diminishes the cell growth. Again, it was found that in diabetic foot ulcers, the chronic wound exudates pH and can indicate a healing rate or infection with early intervention, which is helpful for patients in reducing economic costs. The pH control around tissues with abnormalities is critical since the extracellular pH, which is moderately acidic (≈ 6.7), increases the process of malignant proliferation in tumor cells. Moving to a quite different field, fetal acidosis is commonly defined as a low umbilical pH or a high umbilical base deficit. With a low umbilical pH (with a threshold for cut-off varying between 7.2 and 7.0), acidosis is correlated with neonatal morbidity and mortality (Tang et al., 2021). All these examples can be suitable for a pH measurement that is difficult to access by instruments that are not microscopic, like a POF sensor. In principle, such devices may be integrated into wound dressings, fetal scalp electrodes, or biopsy needles for real-time diagnosis or prognosis. This is the reason why, very often, the pH range explored with POF sensors is around physiological values.

Regarding what was proposed by literature in the last 5 years, the number of papers is rather limited, and the main features are summarized in Table 16.7 while in Fig. 16.7 a comparison between the operative pH ranges is presented, confirming the trend of focusing in the neutral region, which is of high interest for biomedical applications.

In light of Table 16.7, we can divide pH sensors on the basis of the sensing mechanism, and almost all contributions represent an attempt to increase the performance of a previous POF based

Table 16.7 POFs pH sensors recently proposed (2019–23).

References	Nature of POF	pH-sensing mechanism	pH range
Booth et al. (2021)	Polyethylene(PE)/graphite	Potentiometry	5–8
Gong et al. (2019)	Polymers microarray	Ratiometric fluorescence	6–8
Gong et al. (2020)	Polymers microarray	Ratiometric fluorescence	6–8
Janting et al. (2019)	Polymethyl methacrylate (PMMA) + hydrogel	Bragg grating	5–7
Khanikar and Singh (2019)	PANI PVA	Transmitted power variation	2–9
Lee et al. (2023)	Acrylamide (AAM) or N-isopropylacrylamide (NIPAM) + N,N′-methylenebisacrylamide (BIS)	Ratiometric fluorescence	6.6–8.0
Macedo et al. (2022)	Recycled PMMA and PET	Rhodamine B fluorescence	4–6.6
Saha et al. (2023)	Polyacrylic acid, PAA/poly-allylamine hydrochloride (PAH)	Bragg grating	4–7
Tang et al. (2021)	Ethyl cellulose and silica prepared via a sol–gel reaction	Refractive index	4.5–12.5
Wang, Wang, Su, et al. (2022)	PVA	Long-period fiber grating, LPFG	3.1–6.2

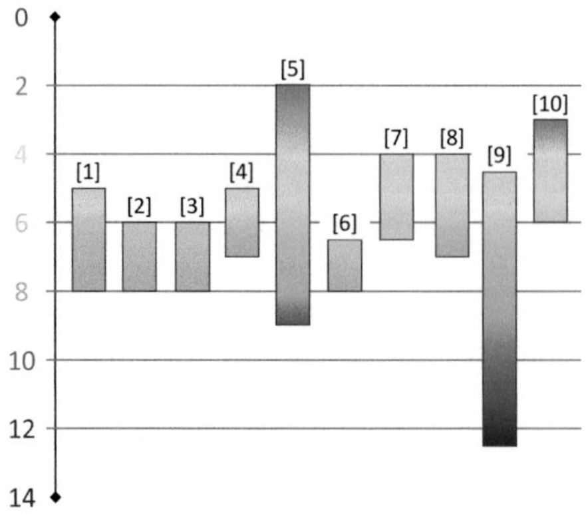

FIGURE 16.7 Comparison of measurable pH ranges for devices in Table 16.7.

Each device in Table 16.7 is represented as a colored box covering the measurable pH range and labeled according to the numbers reported in the first column of Table 16.7.

on the same concept. As for POFs based on fluorescent compounds or indicator dyes, the pH is measured by monitoring optical signals such as absorption, fluorescence intensity, or lifetime. Fluorescent pH indicators can be immobilized on an optical fiber in a variety of ways, including covalent binding and sol−gel technology (Lee et al., 2023). The versatility of POFs in the dye-based sensor is confirmed by the possibility of employing an extrusion method to produce low-cost POFs from discarded transparent plastics, as proposed by Macedo and coauthors (Macedo et al., 2022). In this case, the fiber is obtained at the desired diameter via a screw rotation control during the extrusion step. For the pH sensor assessment, PET and poly(methyl methacrylate) (PMMA) POFs are doped with Rhodamine B; the fluorescence spectrum of Rhodamine B changes after submersion of the fiber in solution with different pH values, with an increase in the fluorescence intensity as the pH decreases. Of this contribution remains the idea of using recycled materials, which makes sense only for massive production and is certainly extremely limited in the case of fiber construction (Macedo et al., 2022).

Despite high sensitivity, in current applications, these sensors typically suffer from problems such as dye-leaching and photobleaching that cause a reduction in the sensitivity and long-term stability (Tang et al., 2021); often, authors tried to partially overcome these drawbacks, as done by Gong and coworkers in two different papers (Gong et al., 2019, 2020), moving from the fluorescence spectrum of a single fluorophore-based pH sensor to a more robust ratiometric-type approach. In the first case, 5(6)-carboxyfluorescein is the reactive fluorophore, and 5,10,15,20-tetrakis-(pentafluorophenyl)-21H,23H-porphine palladium(II) is the reference dye (Gong et al., 2019); in the second case, instead, porphyrin, whose emission was also insensitive to pH, was employed as a reference (Gong et al., 2020). Undoubtedly, this trick is helpful in avoiding loss of sensitivity due to sensor concentration that decreases with time due to leaching, photobleaching, and fluctuations in the intensity of the light source.

Another question that deserves attention is related to the origin of the optical signal. Indeed, all sensors based on fluorescent pH indicators, including POF sensors, cover a dynamic range of maximally 3 pH units, driven by the pK_a value of the chromogenic unit (Lee et al., 2023). Thus, usually, the indicator for a fluorescence pH sensor is chosen with a pK_a that matches the requirements of the specific application. In the case of Lee and coauthors (Lee et al., 2023), they exploited the performance of a cellulose nanofiber (CNF) sensing unit that has a pK_a of 7.6 and exhibits an intense fluorescence spectrum, which decreases with increasing pH. The limitation of the pH range is partially overcome by synthesizing a library of fibers, where the type of monomer, in their case, AAM or NIPAM, the amount of crosslinker, BIS, triethylamine, and photopolymerization time are varied while CNF is kept constant. In this way, even if ostensibly there is a precise rationale behind the experimental setup, the authors succeeded in selecting three different fibers: AAM/BIS, NIPAM/BIS (3:1 or 2:1), and NIPAM/BIS (8:1 or 5:1). While AAM/BIS probes showed a linear response over a range of pH 6.8−7.8, not affected by the different ratio exploited, by changing the monomeric unit into NIPAM, the microenvironment of the fiber becomes more lipophilic, and the higher the NIPAM ratio, with respect to BIS, the less the CNF becomes prone to deprotonation. Since the apparent pK_a values increase, the response range, despite using the same molecule entrapped in different polymers, moves to a higher pH with NIPAM/BIS (3:1 or 2:1), showing a linear response to pH between pH 7.5 and 9.0, or even more with NIPAM/BIS (8:1 or 5:1) between pH 8.2 and 9.8. After the entrapment in a solid phase, when AAM is employed, the apparent pK_a value of CNF is almost not affected, close to the value of 7.6 of the soluble ligand, while, substituting AAM with NIPAM, pK_a becomes 8.4 and 9.2, further increasing the NIPAM ratio. This effect was already observed in the case of dyes entrapped in EVOH, discussed in the previous section (Magnaghi, Alberti, et al., 2021; Magnaghi et al., 2023; Magnaghi, Capone, et al., 2021; Magnaghi, Zanoni,

Alberti, et al., 2022a, Magnaghi, Zanoni, Alberti, et al., 2022b; Magnaghi, Alberti, et al., 2022; Magnaghi, Zanoni, Bellotti, et al., 2022; Magnaghi, Zanoni, Bancalari, et al., 2022), but, as suggested by the authors, it is also possible to modulate the apparent pK_a to expand the operative range of the dye, changing the nature and the properties of the polymer (Lee et al., 2023).

A similar attempt to improve the POF sensor is described by Gong and coauthors in two similar papers already hinted at above (Gong et al., 2019, 2020). Their idea is to employ a polymer microarray constituted of 121 different polymers (each printed in quadruplicate), with each polymer printed with three different concentrations of 5(6)-carboxyfluorescein in a 32 $\times$ 48 array. From the high-throughput screening, the optimal polymeric substrates capable of trapping functional ratiometric fluorescence-based pH sensors with the best performance can be easily selected (Gong et al., 2019). In this case, the screening leads to the identification of poly(methyl methacrylate-co-2-(dimethylamino) ethyl acrylate) (PA101), which allowed, via dip coating, the attachment of fluorescent pH sensors onto the tips of optical fibers. This strategy was still employed and improved, with the difference of using a hydrogel-based optical fiber pH sensor fabricated in situ employing a hydrogel as an immobilization matrix for the pH indicator. The polymer-microarray platform allows the identification of the hydrogel with optimal dye-trapping abilities. At the same time, as in the previous case, the pH responsivity, sensitivity, pH precision, time response, and reversibility were investigated (Gong et al., 2020). It is worth mentioning that the employment of such platforms, which is a clear and common aspect of most recent approaches, not only here in optical fiber synthesis, could be even more successful if the polymer's choice, ratio, and conditions followed a design of experiments strategy. As it happens in both of these cases, most of the sensors on the platform are absolutely blind, and the choice of the best solution seems to be discretionary and not based on quantitative criteria, as would be possible (Gong et al., 2019; Gong et al., 2020).

As mentioned at the beginning of the chapter, there is considerable interest in developing dye-free POFs for monitoring pH via measurement of the RI or the mechanical properties of a pH-sensitive film. The polymer fiber Bragg grating (PFBG) sensors are an attempt in which pH is determined via monitoring strain induced by swelling of a pH-sensitive hydrogel film deposited on the fiber (Tang et al., 2021). In the PFBG, the beam of light passes through the fiber, and the core dimension is made to reflect a particular wavelength of light, named λ_B, while transmitting all the others. When hydrogels, made of cross-linked polymers with a very hydrophilic structure, cover an OF, a shift λ_B can be produced. Indeed, if the composite fiber is put in contact with a solution where a species protonates or deprotonates, the hydrogel undergoes a phase transition and consequently a dramatic volumetric change. The swelling/shrinking mechanism makes the gel capable of working as a mechanical transducer, transferring/releasing strain to the PFBG as the pH changes, which is detected by corresponding shifts in Bragg reflection wavelength (Tang et al., 2021).

A similar device is presented by Janting and coauthors: a PMMA fiber is coated with hydrogel obtained by dissolving in 2,2-dimethoxy-2-phenylacetophenone directly in mixtures of 2-hydroxyethylmethacrylate, methacrylic acid, and ethylene glycol dimethacrylate (Janting et al., 2019). The sensor exhibits a linear response in the pH 5–7 range. The estimated sensitivity (found by a linear fit of the six data points) is 73 (2) pm/pH. From a strictly analytical point of view, it seems an unrealistic value if the device is lowered to real experimental conditions. A comparison of the sensitivities with these Bragg grating devices has been recently presented by Saha and coauthors (Saha et al., 2023).

Conversely, the same authors mentioned above present a polymer coating of polyacrylic acid/poly-allylamine hydrochloride (PAA/PAH) whose reflection properties are modulated by the layer's thickness; it must be underlined and requires a careful setup. Still, the variation of resonance wavelength (λ_B) with the pH values is almost linear from 4 to 7, with a sensitivity of 10.41 nm/pH. In comparison with the previous device, it seems to have a reasonable practical value (Saha et al., 2023).

A variation of the same approach, not involving the Bragg gratings, is proposed by Tang and coauthors as a hybrid organic-inorganic composite film: a silica matrix is coated by cross-linked ethyl cellulose; the hydrogen bonds between the two phases are affected by the pH of a solution in contact with it, leading to RI change of the film. The developed device exhibits a reversible response in the pH range of 4.5−12.5 and a linear correlation between transmitted light power and pH, with a resolution of 0.02 pH units (Tang et al., 2021).

Among this series of POF sensors, the contribution presented by Khanikar and Singh appears to be the most convincing, or simply, the device is presented in an advanced stage of its development. The sensor is obtained in a cleaved segment of silica fiber, and the core is coated with a PANI-polyvinyl alcohol (PVA) composite solution. Once dried, it constitutes the sensitive pH region in contact with the sample. The signal is the transmitted power registered at the end of the fiber when the laser source of wavelength 630 nm is launched through it, which is linear with a pH in the range of 2−9. From the analytical point of view, all the validation steps are presented, and sensing ability, time response, reproducibility, stability, and reusability are addressed. Also, the effect of temperature and ionic strength, almost always not evaluated, besides measurements of real samples, is presented (Khanikar & Singh, 2019).

As a final trend that emerges in the most recent publications, we signal the integrated sensor for simultaneous measurements. For instance, Wang and coauthors present simultaneous real-time measurements of temperature, liquid level, humidity, and pH (Wang, Wang, Su, et al., 2022), and Booth and coauthors present a platform for sensing pH and neurometabolic lactate (Booth et al., 2021).

In the first case, such a peculiar combination could be of interest in a very restricted field, such as monitoring chemical waste liquid storage tanks. The authors present three different, separated sensors, while the signals are simultaneously processed in the same unit. The synthesis of the sensors themselves, all based on polarization-maintaining fibers (PMFs), is rather complicated, with several steps exploiting different strategies that seem unrealistic to reproduce. As an example, long-period fiber grating is employed for liquid level monitoring, while for humidity, PMFs are coated with PEDOT/PSS. PMFs are instead coated with indium tin oxide film by electrostatic self-assembly. Over this film, another is made by PAH, PAA, and PSS with high permeability covered with PANI/PMMA as the protective layer obtained by the laser-assisted process (Wang, Wang, Su, et al., 2022).

In principle, the proposal described by Booth and coworkers seems promising: in the same fiber, they insert different channels, functionalizing one electrode as a potentiometric pH sensor, another as an amperometric lactate sensor, and a third with no functionalization to act as a control (a bare carbon electrode). The separate fibers have satisfactory performance. The potentiometric one, based on a carbon electrode coated with iridium oxide, shows almost Nernstian behavior within the pH range of 5 and 8. Also, the enzymatic electrode works in a satisfactory way. However, when measuring simultaneously with the fibers simultaneously present in the same device, sensitivities decrease with respect to those of separate fibers, especially for pH sensing, indicating that the electronics and methods for fabricating the sensors on the same fiber require further optimization (Booth et al., 2021).

A final comment, from the analytical point of view, is related to the validation process. Most of the POF sensors are not only early proof of concept, but the pH was tested exclusively in clean buffered solutions. Under these conditions, the temperature effect, number of cycles, response time, and hysteresis are often assessed, but we are all well aware that performances can drop dramatically when moving into the real world. In contrast, Booth and their colleagues have conducted measurements on tangible samples (Booth et al., 2021), specifically focusing on in vivo lactate levels within mouse brain tissue. Similarly, Gong et al. employed the fiber in ex vivo ovine lung tissue, utilizing it as a tumor model for comparative analysis against results obtained with a glass electrode (Gong et al., 2019, 2020). The same validation is presented by Khanikar and Singh with a wide series of liquid samples, from tap water to human saliva (Khanikar & Singh, 2019), and by Tang and coauthors reporting a measure in human serum artificially acidified (Tang et al., 2021).

To conclude, as already observed in the literature, POF sensors show promising results for pH determinations, but all evidence suggests how far we are from widespread use in the pH sensor field. These authors also recommend "how it would be beneficial to develop intensity-based sensors that can be monitored using low-cost instrumentation," which, as already mentioned above, is often a limit in these POF-based prototypes (Tang et al., 2021). We can only agree with them.

16.2.5 Polymer-based pH sensors relying on other sensing mechanisms

This final section will be devoted to pH-responsive polymeric materials relying on different operating principles than the most common so far discussed (optical and electrochemical mechanisms). Despite being far less widespread than the other approaches and, in many cases, far from real applicability, we do believe some interesting hints can be drawn from these unconventional approaches.

A significant number of papers have been recently published on polymers showing pH-dependent swelling/shrinking behavior that can, in turn, be exploited first for unusual pH measurements (Lavine et al., 2020, 2021) but also for interesting applications mainly associated with biomedical applications like in vivo local pH monitoring (Corsi et al., 2022; Kiridena et al., 2023; Wijayaratna et al., 2021) or pH-dependent drug release (Molina et al., 2022). Trying to get to the root of the operating principles of these sensors, pH-dependent swelling or shrinking is commonly achieved by inserting in the polymer chains weak acids or bases, either as monomers or as covalently bound pendants. Depending on the medium pH, these moieties can be found in their neutral or positively/negatively charged form, thus impacting the charge density of the bulk polymer and the repulsion between chains that results in material swelling or shrinking (Lavine et al., 2020). This effect is generally higher in low-ionic strength solutions due to the absence or scarcity of shielding of the charges by solvent ions, but several efforts have been carried out to overcome this limitation (Lavine et al., 2020).

When a carboxylic acid is exploited to achieve this behavior, polymeric materials present their shrunk or collapsed form at an acidic pH, that is, when neutral carboxylic acid ($-COOH$) is predominant, while exhibiting a gradual swelling that increases the pH due to the formation of negatively charged carboxylate anions ($-COO^-$). Being this chemical function widely present in common monomers, natural materials, or polymeric additives and building blocks, the carboxylic acid-associated pH-dependent swelling mechanism is undoubtedly the most common one (Bacheller et al., 2021; Eken et al., 2023; Kiridena et al., 2023; Lavine et al., 2020, 2021; Molina et al., 2022; Wijayaratna et al., 2021; Yue et al., 2020). An opposite mechanism is observed using

N-containing weak basses, such as amines or pyridines, that present swollen conformation at acidic pH due to the presence of repulsing protonated N atoms ($-NH^+$), while they shrink upon pH increasing and consequent deprotonation (Liu et al., 2019; Wu et al., 2022). These opposite chemical functions were both exploited by Corsi and coauthors in an interesting way: Two oppositely charged polyelectrolyte layers, namely PAH and poly-methacrylic-acid, containing, respectively, amine and carboxylic functions, were layer-by-layer deposited on inert support and used for in vivo pH measurements between 4 and 7.5. At a lower pH, the composite material swells due to the repulsion associated with positively charged protonated amino groups, but as the pH increases, carboxylic acid moieties deprotonate, and negatively charged carboxylate ions reduce the repulsion by interacting with $-NH^+$ and provoking polymer shrinkage. Being this device meant for biomedical applications, no investigations above pH 7, which means involving amine deprotonation, were performed (Corsi et al., 2022).

Identifying the best detection technique to measure swelling or shrinkage is definitely a hard task. Microscopy obviously represents the straightforward technique to detect a change in the conformation, but, despite being extremely useful to characterize the materials, it is unsuitable for sensing purposes, and, for this reason, it has been exploited only for those *proof-of-concept* not developed for a specific application (Eken et al., 2023; Bacheller et al., 2021). Opposite, fluorescence offers several solutions to monitor the spatial movement of the polymer chains, both labeling the monomers with fluorophores that undergo self-quenching (Corsi et al., 2022) or exploiting polymers autofluorescence (Eken et al., 2023) or aggregation-induced fluorescence quenching (Liu et al., 2019). Moving to biomedical applications, these sensitive polymers have also been labeled with radiopaque markers at their ends to allow the device's length measurement, associated with swelling or shrinking and thus local pH, by planar radiography (Kiridena et al., 2023; Wijayaratna et al., 2021). Other less common proposed detection techniques are turbidity measurements (Lavine et al., 2020, 2021), cyclic voltammetry (Molina et al., 2022), direct measurement by Vernier calliper (Wu et al., 2022), or weighing (Yue et al., 2020).

Another novel trend in the biomedical field is represented by pH sensors based on Surface-Enhanced Raman Spectroscopy (SERS), which recently turned out to be interesting candidates for in vivo pH monitoring in different biological samples. In the last years, SERS has experienced increasing popularity in biomedical monitoring assays thanks to its peculiar features such as structural specificity based on vibrational fingerprints, low detection limit, multiplexed detection, low autofluorescence, and resistance to photobleaching (Park et al., 2019; Quinn et al., 2019; Skinner et al., 2021; Zhang et al., 2021). Implantable SERS sensors, composed of plasmonic substrates encapsulated in biocompatible polymeric materials, gained even bigger attention, especially for enzymatic reaction monitoring (Quinn et al., 2019). As far as pH detection is concerned, SERS-based pH sensors are composed of a SERS substrate, commonly Au or Ag nano-scaled materials, and a pH-sensitive molecule both embedded in a biocompatible polymeric film (Zhu et al., 2020). Some thiol-containing weak acids have been applied as a pH probe since these dual-function molecules ensure a sensitive pH response, a high Raman signal, and a strong affinity with the metal NPs (Zhu et al., 2020). 4-Mercaptobenzoic acid is the most common pH probe (Park et al., 2019; Skinner et al., 2021, 2023; Zhang et al., 2021), but also 4-mercaptopyridine has been evaluated (Quinn et al., 2019; Zhu et al., 2020). Unfortunately, in the recent papers on this topic, emphasis is mainly placed on the SERS substrate and pH probe, while very short and cursory discussions are presented on the polymeric materials.

16.3 Conclusions and perspectives

This chapter aims to present the readers with a complete and wide overview of the most recent trends in polymer-based pH sensors. To fulfill the scope, a brief introduction to pH, concerning its history and its current role in analytical chemistry, was necessary, together with a compendium of the main information about sensors and polymeric sensors. As already hinted above, we limited our research to sensing devices, defined as physical devices with sensing properties, and we excluded all those probes that somehow exert recognition or measurement but only in solution, often erroneously called sensors.

Moving specifically to polymeric sensors, we opted for an analytical classification dividing the different sensors according to the detection technique required, specifically colorimetric, fluorescent, electrochemical, optical fibers, and other unusual mechanisms. Focusing only on recent advances, two different trends can be observed. On one hand, "old" types of pH sensors, namely colorimetric and electrochemical, still raise the interest of academia focused on the development of highly specific devices for real applications; the most prominent examples are smart labels for food degradation, wound monitoring devices and smart bandages, wearable sensors for rapid and early diagnosis, and even implantable sensors. On the other hand, fluorescent devices, some particular and complex electrochemical ones, optical fibers, and sensors exploiting unusual mechanisms represent innovations themselves, and thus a lot of work has yet to be done to clarify both the working principle and the achievable analytical performances, as well as to develop real-world applicable devices, rather than the *proof-of-concept* so far proposed.

Having this in mind and trying to offer a valuable take-home message to the readers, polymeric pH sensors represent an interesting and extremely wide field, in which both theoretical and fundamental studies on innovative working principles or detection techniques and highly practical optimization of low-cost and scalable devices for specific applications are welcomed. Furthermore, considering the ubiquitous role of pH in our lives and the almost limitless variety of polymeric materials available, we can assume that so far we are just scratching the surface, and a lot of interesting solutions are still to be found and investigated.

References

Adamu, B. F., Gao, J., Gebeyehu, E. K., Beyene, K. A., Tadesse, M. G., & Liyew, E. Z. (2022). Self-responsive electrospun nanofibers wound dressings: The future of wound care. *Advances in Materials Science and Engineering*, *2022*, 2025170. Available from https://doi.org/10.1155/2022/2025170.

Alberti, G., Zanoni, C., Losi, V., Magnaghi, L. R., & Biesuz, R. (2021). Current trends in polymer based sensors. *Chemosensors*, *9*(108). Available from https://doi.org/10.3390/chemosensors9050108, https://www.mdpi.com/2227-9040/9/5/108.

Almasi, H., Forghani, S., & Moradi, M. (2022). Recent advances on intelligent food freshness indicators: An update on natural colorants and methods of preparation. *Food Packaging and Shelf Life*, *32*, 100839. Available from https://doi.org/10.1016/j.fpsl.2022.100839.

Arafat, M. T., Mahmud, M. M., Wong, S. Y., & Li, X. (2021). PVA/PAA based electrospun nanofibers with pH-responsive color change using bromothymol blue and on-demand ciprofloxacin release properties.

Journal of Drug Delivery Science and Technology, *61*, 102297. Available from https://doi.org/10.1016/j.jddst.2020.102297.

Avolio, R., Grozdanov, A., Avella, M., Barton, J., Cocca, M., De Falco, F., Dimitrov, A. T., Errico, M. E., Fanjul-Bolado, P., Gentile, G., Paunovic, P., Ribotti, A., & Magni, P. (2022). Review of pH sensing materials from macro- to nano-scale: Recent developments and examples of seawater applications. *Critical Reviews in Environmental Science and Technology*, *52*(6), 979–1021. Available from https://doi.org/10.1080/10643389.2020.1843312.

Bacheller, S., Dianat, G., & Gupta, M. (2021). Synthesis of pH-responsive polymer sponge coatings and free-standing films via vapor-phase deposition. *ACS Applied Polymer Materials*, *3*(12), 6366–6374. Available from https://doi.org/10.1021/acsapm.1c01151.

Balbinot-Alfaro, E., Craveiro, D. V., Lima, K. O., Gouveia Costa, H. L., Lopes, D. R., & Prentice, C. (2019). Intelligent packaging with pH indicator potential. *Food Engineering Reviews*, *11*(4), 235–244. Available from https://doi.org/10.1007/s12393-019-09198-9.

Bao, B., Yang, Z., Liu, Y., Xu, Y., Gu, B., Chen, J., Su, P., Tong, L., & Wang, L. (2019). Two-photon semiconducting polymer nanoparticles as a new platform for imaging of intracellular pH variation. *Biosensors and Bioelectronics*, *126*, 129–135. Available from https://doi.org/10.1016/j.bios.2018.10.027.

Bao, L., & Liu, S. (2022). A dual-emission polymer carbon nanoparticles for ratiometric and visual detection of pH value and bilirubin. *Spectrochimica Acta – Part A: Molecular and Biomolecular Spectroscopy*, *267*, 120513. Available from https://doi.org/10.1016/j.saa.2021.120513.

Bazbouz, M. B., & Tronci, G. (2019). Two-layer electrospun system enabling wound exudate management and visual infection response. *Sensors (Switzerland)*, *19*(5), 991. Available from https://doi.org/10.3390/s19050991.

Bhat, A., Amanor-Boadu, J. M., & Guiseppi-Elie, A. (2020). Toward impedimetric measurement of acidosis with a pH-responsive hydrogel sensor. *ACS Sensors*, *5*(2), 500–509. Available from https://doi.org/10.1021/acssensors.9b02336.

Booth, M. A., Gowers, S. A. N., Hersey, M., Samper, I. C., Park, S., Anikeeva, P., Hashemi, P., Stevens, M. M., & Boutelle, M. G. (2021). Fiber-based electrochemical biosensors for monitoring pH and transient neurometabolic lactate. *Analytical Chemistry*, *93*(17), 6646–6655. Available from https://doi.org/10.1021/acs.analchem.0c05108.

Brooker, C., & Tronci, G. (2023). A collagen-based theranostic wound dressing with visual, long-lasting infection detection capability. *International Journal of Biological Macromolecules*, *236*, 123866. Available from https://doi.org/10.1016/j.ijbiomac.2023.123866.

Buck, R. P., Rondinini, S., Covington, A. K., Baucke, F. G. K., Brett, C. M. A., Camoes, M. F., Milton, M. J. T., Mussini, T., Naumann, R., Pratt, K. W., Spitzer, P., & Wilson, G. S. (2002). Measurement of pH. Definition, standards, and procedures (IUPAC Recommendations 2002). *Pure and Applied Chemistry. Chimie Pure et Appliquee*, *74*(11), 2169–2200. Available from https://doi.org/10.1351/pac200274112169, https://doi.org/10.1351/pac200274112169.

Cascales, J. P., Li, X., Roussakis, E., & Evans, C. L. (2022). A patient-ready wearable transcutaneous CO2 Sensor. *Biosensors*, *12*(5), 333. Available from https://doi.org/10.3390/bios12050333.

Chajanovsky, I., Cohen, S., Shtenberg, G., & Suckeveriene, R. Y. (2021). Development and characterization of integrated nano-sensors for organic residues and pH field detection. *Sensors*, *21*(17), 5842. Available from https://doi.org/10.3390/s21175842.

Chalitangkoon, J., & Monvisade, P. (2021). Synthesis of chitosan-based polymeric dyes as colorimetric pH-sensing materials: Potential for food and biomedical applications. *Carbohydrate Polymers*, *260*, 117836. Available from https://doi.org/10.1016/j.carbpol.2021.117836.

Chen, P., Ilyas, I., He, S., Xing, Y., Jin, Z., & Huang, C. (2019). Ratiometric pH sensing and imaging in living cells with dual-emission semiconductor polymer dots. *Molecules (Basel, Switzerland)*, *24*(16), 2923. Available from https://doi.org/10.3390/molecules24162923.

Choi, K. R., Chen, X. V., Hu, J., & Bühlmann, P. (2021). Solid-contact pH Sensor with covalent attachment of ionophores and ionic sites to a poly(decyl methacrylate) matrix. *Analytical Chemistry*, *93*(50), 16899–16905. Available from https://doi.org/10.1021/acs.analchem.1c03985.

Choi, S. J., Savagatrup, S., Yo Kim, J. H., Lang, T. M., & Swager. (2019). Precision pH sensor based on WO3 nanofiber-polymer composites and differential amplification. *ACS Sensors*, *4*(10), 2593–2598. Available from https://doi.org/10.1021/acssensors.9b01579.

Choudhury, S., Deepak, D., Bhattacharya, G., McLaughlign, J., & Roy, S. S. (2023). MoS2-polyaniline based flexible electrochemical biosensor: Toward pH monitoring in human sweat. *Macromolecular Materials and Engineering*, *308*(8), 2300007. Available from https://doi.org/10.1002/mame.202300007.

Corsi, M., Paghi, A., Mariani, S., Golinelli, G., Debrassi, A., Egri, G., Leo, G., Vandini, E., Vilella, A., Dähne, L., Giuliani, D., & Barillaro, G. (2022). Bioresorbable nanostructured chemical sensor for monitoring of pH level in vivo. *Advanced Science*, *9*(22), 2202062. Available from https://doi.org/10.1002/advs.202202062.

Dalfen, I., Dmitriev, R. I., Holst, G., Klimant, I., & Borisov, S. M. (2019). Background-free fluorescence-decay-time sensing and imaging of pH with highly photostable diazaoxotriangulenium dyes. *Analytical Chemistry*, *91*(1), 808–816. Available from https://doi.org/10.1021/acs.analchem.8b02534.

Deb, M., Hassan, N., Chowdhury, D., Hussain Sanfui, M. D., Roy, S., Bhattacharjee, C., Majumdar, S., Chattopadhyay, P. K., & Singha, N. R. (2022). Nontraditional redox active aliphatic luminescent polymer for ratiometric pH sensing and sensing-removal-reduction of Cu(II): Strategic optimization of composition. *Macromolecular Rapid Communications*, *43*(19), 2200317. Available from https://doi.org/10.1002/marc.202200317.

Dervisevic, M., Dervisevic, E., Esser, L., Easton, C. D., Cadarso, V. J., & Voelcker, N. H. (2023). Wearable microneedle array-based sensor for transdermal monitoring of pH levels in interstitial fluid. *Biosensors and Bioelectronics*, *222*, 114955. Available from https://doi.org/10.1016/j.bios.2022.114955.

Di Costanzo, L., & Panunzi, B. (2021). Visual pH sensors: From a chemical perspective to new bioengineered materials. *Molecules (Basel, Switzerland)*, *26*(10), 2952. Available from https://doi.org/10.3390/molecules26102952.

Dong, M., Sun, X., Li, L., He, K., Wang, J., Zhang, H., & Wang, L. (2022). A bacteria-triggered wearable colorimetric band-aid for real-time monitoring and treating of wound healing. *Journal of Colloid and Interface Science*, *610*, 913–922. Available from https://doi.org/10.1016/j.jcis.2021.11.146.

Drago, N. P., Choi, E. J., Shin, J., Kim, D. H., & Penner, R. M. (2022). A nanojunction pH sensor within a nanowire. *Analytical Chemistry*, *94*(35), 12167–12175. Available from https://doi.org/10.1021/acs.analchem.2c02606.

Du, N., Zhang, G., Hou, P., Zhang, H., & Guan, R. (2023). Ratiometric fluorescent probe based on non-conjugated polymer dots for pH measurements in ordinary Portland cement-based materials. *Microchimica Acta*, *190*(4), 119. Available from https://doi.org/10.1007/s00604-023-05691-5.

Dulay, S., Rivas, L., Miserere, S., Pla, L., Berdún, S., Parra, J., Eixarch, E., Gratacós, E., Illa, M., Mir, M., & Samitier, J. (2021). In vivo Monitoring with micro-implantable hypoxia sensor based on tissue acidosis. *Talanta*, *226*, 122045. Available from https://doi.org/10.1016/j.talanta.2020.122045.

Duong, H. D., Shin, Y., & Rhee, J. I. (2020). Development of novel optical pH sensors based on coumarin 6 and nile blue A encapsulated in resin particles and specific support materials. *Materials Science and Engineering C*, *107*, 110323. Available from https://doi.org/10.1016/j.msec.2019.110323.

Eken, G. A., Huang, Y., Guo, Y., & Ober, C. (2023). Visualization of the pH response through autofluorescent poly(styrene-alt-N-maleimide) polyelectrolyte brushes. *ACS Applied Polymer Materials*, *5*(2), 1613–1623. Available from https://doi.org/10.1021/acsapm.2c02066.

Ermakova, E. V., Bol'shakova, A. V., & Bessmertnykh-Lemeune, A. (2023). Dual-responsive and reusable optical sensors based on 2,3-diaminoquinoxalines for acidity measurements in low-pH aqueous solutions. *Sensors*, *23*(6), 2978. Available from https://doi.org/10.3390/s23062978.

Gajre Kulkarni, V., Butte, K., & Rathod, S. (2012). Natural polymers – A comprehensive review. *International Journal of Research in Pharmaceutical and Biomedical Sciences*, *3*(4), 1597–1613.

Gao, N., Yu, J., Tian, Q., Shi, J., Zhang, M., Chen, S., & Zang, L. (2021). Application of PEDOT:PSS and its composites in electrochemical and electronic chemosensors. *Chemosensors*, *9*(4), 79. Available from https://doi.org/10.3390/chemosensors9040079.

Gicevicius, M., Kucinski, J., Ramanaviciene, A., & Ramanavicius, A. (2019). Tuning the optical pH sensing properties of polyaniline-based layer by electrochemical copolymerization of aniline with o-phenylenediamine. *Dyes and Pigments*, *170*, 107457. Available from https://doi.org/10.1016/j.dyepig.2019.04.002.

Gong, J., Tanner, M. G., Venkateswaran, S., Stone, J. M., Zhang, Y., & Bradley, M. (2020). A hydrogel-based optical fibre fluorescent pH sensor for observing lung tumor tissue acidity. *Analytica Chimica Acta*, *1134*, 136–143. Available from https://doi.org/10.1016/j.aca.2020.07.063.

Gong, J., Venkateswaran, S., Tanner, M. G., Stone, J. M., & Bradley, M. (2019). Polymer microarrays for the discovery and optimization of robust optical-fiber-based pH sensors. *ACS Combinatorial Science*, *21*(5), 417–424. Available from https://doi.org/10.1021/acscombsci.9b00031.

Gualandi, I., Tessarolo, M., Mariani, F., Possanzini, L., Scavetta, E., & Fraboni, B. (2021). Textile chemical sensors based on conductive polymers for the analysis of sweat. *Polymers*, *13*(6), 894. Available from https://doi.org/10.3390/polym13060894.

Ham, M., Kim, S., Lee, W., & Lee, H. (2023). Fabrication of printable colorimetric food sensor based on hydrogel for low-concentration detection of ammonia. *Biosensors*, *13*(1), 18. Available from https://doi.org/10.3390/bios13010018.

Hazarika, K. K., Konwar, A., Borah, A., Saikia, A., Barman, P., & Hazarika, S. (2023). Cellulose nanofiber mediated natural dye based biodegradable bag with freshness indicator for packaging of meat and fish. *Carbohydrate Polymers*, *300*, 120241. Available from https://doi.org/10.1016/j.carbpol.2022.120241.

Horak, E., Babić, D., Vianello, R., Perin, N., Hranjec, M., & Steinberg, I. M. (2020). Photophysical properties and immobilisation of fluorescent pH responsive aminated benzimidazo[1,2-a]quinoline-6-carbonitriles. *Spectrochimica Acta – Part A: Molecular and Biomolecular Spectroscopy*, *227*, 117588. Available from https://doi.org/10.1016/j.saa.2019.117588.

Hulanicki, A., Glab, S., & Ingman, F. (1991). Chemical sensors definitions and classification. *Pure and Applied Chemistry*, *63*(9), 1247–1250. Available from https://doi.org/10.1351/pac199163091247, https://doi.org/10.1351/pac199163091247.

Ismail, S. R., Bryaskova, R. G., Georgiev, N. I., Philipova, N. D., Bakov, V. V., Uzunova, V. P., Tzoneva, R. D., & Bojinov, V. B. (2020). Design and synthesis of fluorescent shell functionalized polymer micelles for biomedical application. *Polymers for Advanced Technologies*, *31*(6), 1365–1376. Available from https://doi.org/10.1002/pat.4866.

Janting, J., Pedersen, J. K. M., Woyessa, G., Nielsen, K., & Bang, O. (2019). Small and robust all-polymer fiber bragg grating based pH sensor. *Journal of Lightwave Technology*, *37*(18), 4480–4486. Available from https://doi.org/10.1109/JLT.2019.2902638.

Jensen, W. B. (2004). The symbol for pH. *Journal of Chemical Education*, *81*(1), 1–21. Available from https://doi.org/10.1021/ed081p21, https://doi.org/10.1021/ed081p21.

Jose, M., Mylavarapu, S. K., Bikkarolla, S. K., Machiels, J., Sankaran, K. J., McLaughlin, J., Hardy, A., Thoelen, R., & Deferme, W. (2022). Printed pH sensors for textile-based wearables: A conceptual and experimental study on materials, deposition technology, and sensing principles. *Advanced Engineering Materials*, *24*(5), 2101087. Available from https://doi.org/10.1002/adem.202101087.

Kalisz, J., Zarębska, J., Kijeńska-Gawrońska, E., Maksymiuk, K., & Michalska, A. (2023). Colorimetric and fluorimetric pH sensing using polydiacetylene embedded within PVC-PCL nanofibers. *Electroanalysis*, *35*(6), e202200497. Available from https://doi.org/10.1002/elan.202200497.

Khanikar, T., & Singh, V. K. (2019). PANI-PVA composite film coated optical fiber probe as a stable and highly sensitive pH sensor. *Optical Materials*, *88*, 244–251. Available from https://doi.org/10.1016/j.optmat.2018.11.044.

Khattab, T. A., Dacrory, S., Abou-Yousef, H., & Kamel, S. (2019). Development of microporous cellulose-based smart xerogel reversible sensor via freeze drying for naked-eye detection of ammonia gas. *Carbohydrate Polymers*, *210*, 196–203. Available from https://doi.org/10.1016/j.carbpol.2019.01.067.

Kiridena, S. D., Wijayaratna, U. N., Levon, E., Moschella, P., Pirrallo, R. G., Tzeng, T. R. J., & Anker, J. N. (2023). X-ray visualized sensors for peritoneal dialysis catheter infection. *Advanced Functional Materials*, *33*(31), 2204899. Available from https://doi.org/10.1002/adfm.202204899.

Ko, Y., Jeong, H. Y., Kwon, G., Kim, D., Lee, C., & You, J. (2020). pH-responsive polyaniline/polyethylene glycol composite arrays for colorimetric sensor application. *Sensors and Actuators, B: Chemical*, *305*, 127447. Available from https://doi.org/10.1016/j.snb.2019.127447.

Koshy, R. R., Koshy, J. T., Mary, S. K., Sadanandan, S., Jisha, S., & Pothan, L. A. (2021). Preparation of pH sensitive film based on starch/carbon nano dots incorporating anthocyanin for monitoring spoilage of pork. *Food Control*, *126*, 108039. Available from https://doi.org/10.1016/j.foodcont.2021.108039.

Kurečič, M., Maver, T., Virant, N., Ojstršek, A., Gradišnik, L., Hribernik, S., Kolar, M., Maver, U., & Kleinschek, K. S. (2018). A multifunctional electrospun and dual nano-carrier biobased system for simultaneous detection of pH in the wound bed and controlled release of benzocaine. *Cellulose*, *25*(12), 7277–7297. Available from https://doi.org/10.1007/s10570-018-2057-z.

Kurek, M., Hlupić, L., Ščetar, M., Bosiljkov, T., & Galić, K. (2019). Comparison of two pH responsive color changing bio-based films containing wasted fruit pomace as a source of colorants. *Journal of Food Science*, *84*(9), 2490–2498. Available from https://doi.org/10.1111/1750-3841.14716.

Laffitte, Y., & Gray, B. L. (2022). Potentiometric pH sensor based on flexible screen-printable polyaniline composite for textile-based microfluidic applications. *Micromachines*, *13*(9), 1376. Available from https://doi.org/10.3390/mi13091376.

Lavine, B. K., Kaval, N., Oxenford, L., Kim, M., Dahal, K. S., Perera, N., Seitz, R., Moulton, J. T., & Bunce, R. A. (2021). Synthesis and characterization of N-isopropylacrylamide microspheres as pH sensors. *Sensors*, *21*(19), 6493. Available from https://doi.org/10.3390/s21196493.

Lavine, B. K., Pampati, S. R., Dahal, K. S., Kim, M., Perera, U. D. T. N., Benjamin, M., & Bunce, R. A. (2020). Swellable copolymers of N-isopropylacrylamide and alkyl acrylic acids for optical pH sensing. *Molecules (Basel, Switzerland)*, *25*(6), 1408. Available from https://doi.org/10.3390/molecules25061408.

Laysandra, L., Kurniawan, D., Wang, C. L., Chiang, W. H., & Chiu, Y. C. (2021). Synergistic effect in a graphene quantum dot-enabled luminescent skinlike copolymer for long-term pH detection. *ACS Applied Materials and Interfaces*, *13*(50), 60413–60424. Available from https://doi.org/10.1021/acsami.1c18077.

Lee, K., Park, H., Baek, S., Han, S., Kim, D., Chung, S., Yoon, J. Y., & Seo, J. (2019). Colorimetric array freshness indicator and digital color processing for monitoring the freshness of packaged chicken breast. *Food Packaging and Shelf Life*, *22*, 100408. Available from https://doi.org/10.1016/j.fpsl.2019.100408.

Lee, K. J., Capon, P. K., Ebendorff-Heidepriem, H., Keenan, E., Brownfoot, F., & Schartner, E. P. (2023). Influence of the photopolymerization matrix on the indicator response of optical fiber pH sensors. *Sensors and Actuators B: Chemical*, *376*, 132999. Available from https://doi.org/10.1016/j.snb.2022.132999.

Lee, W., Jeong, S.-H., Lim, Y.-W., Lee, H., Kang, J., Lee, H., Lee, I., Han, H.-S., Kobayashi, S., Tanaka, M., & Bae, B.-S. (2021). Conformable microneedle pH sensors via the integration of two different siloxane polymers for mapping peripheral artery disease. *Science Advances*, *7*, 6290. Available from https://www.science.org.

Li, Y., Shen, Q., Li, S., & Xue, Y. (2023). High quantum-yield lignin fluorescence materials based on polymer confinement strategy and its application as a natural ratiometric pH sensor film. *Industrial Crops and Products*, *194*, 116384. Available from https://doi.org/10.1016/j.indcrop.2023.116384.

Lin, J. (2000). Recent development and applications of optical and fiber-optic pH sensors. *Trends in Analytical Chemistry*, *19*(9), 541–552. Available from https://doi.org/10.1016/S0165-9936(00)00034-0, https://www.sciencedirect.com/science/article/abs/pii/S0165993600000340.

Liu, B., Gurr, P. A., & Qiao, G. G. (2020). Irreversible spoilage sensors for protein-based food. *ACS Sensors, 5*(9), 2903–2908. Available from https://doi.org/10.1021/acssensors.0c01211.

Liu, L., Zhang, Q., Wang, J., Zhao, L., Liu, L., & Lu, Y. (2019). Rapid and specific assays for intracellular pH fluctuations based on pH-dependent disassembly-assembly of conjugated polymer nanoparticles. *Sensors and Actuators, B: Chemical, 297*, 126801. Available from https://doi.org/10.1016/j.snb.2019.126801.

Liu, X., Wang, Y., Zhu, L., Tang, Y., Gao, X., Tang, L., Li, X., & Li, J. (2020). Dual-mode smart packaging based on tetraphenylethylene-functionalized polyaniline sensing label for monitoring the freshness of fish. *Sensors and Actuators, B: Chemical, 323*, 128694. Available from https://doi.org/10.1016/j.snb.2020.128694.

Luo, X., Zaitoon, A., & Lim, L. T. (2022). A review on colorimetric indicators for monitoring product freshness in intelligent food packaging: Indicator dyes, preparation methods, and applications. *Comprehensive Reviews in Food Science and Food Safety, 21*(3), 2489–2519. Available from https://doi.org/10.1111/1541-4337.12942.

Lyu, J. S., Choi, I., Hwang, K. S., Lee, J. Y., Seo, J., Kim, S. Y., & Han, J. (2019). Development of a BTB − /TBA + ion-paired dye-based CO2 indicator and its application in a multilayered intelligent packaging system. *Sensors and Actuators, B: Chemical, 282*, 359–365. Available from https://doi.org/10.1016/j.snb.2018.11.073.

Macedo, L., Junior, R. W. M. P., Frizera, A., Pontes, M. J., & Leal-Junior, A. (2022). An alternative to discarded plastic: A report of polymer optical fiber made from recycled materials for the development of biosensors. *Optical Fiber Technology, 72*, 103001. Available from https://doi.org/10.1016/j.yofte.2022.103001.

Magnaghi, L. R., Alberti, G., Milanese, C., Quadrelli, P., & Biesuz, R. (2021). Naked-eye food freshness detection: Innovative polymeric optode for high-protein food spoilage monitoring. *ACS Food Science and Technology, 1*(2), 165–175. Available from https://doi.org/10.1021/acsfoodscitech.0c00089.

Magnaghi, L. R., Alberti, G., Zanoni, C., Guembe-Garcia, M., Quadrelli, P., & Biesuz, R. (2023). Chemometric-assisted litmus test: One single sensing platform adapted from 1–13 to narrow pH ranges. *Sensors, 23*(3), 1696. Available from https://doi.org/10.3390/s23031696.

Magnaghi, L. R., Capone, F., Alberti, G., Zanoni, C., Mannucci, B., Quadrelli, P., & Biesuz, R. (2021). EVOH-based pH-sensitive optode array and chemometrics: From naked-eye analysis to predictive modeling to detect milk freshness. *ACS Food Science and Technology, 1*(5), 819–828. Available from https://doi.org/10.1021/acsfoodscitech.1c00065.

Magnaghi, L. R., Zanoni, C., Alberti, G., Quadrelli, P., & Biesuz, R. (2022a). Freshness traffic light for fish products: Dual-optode label to monitor fish spoilage in sales packages. *ACS Food Science and Technology, 2*(6), 1030–1038. Available from https://doi.org/10.1021/acsfoodscitech.2c00097.

Magnaghi, L. R., Zanoni, C., Alberti, G., Quadrelli, P., & Biesuz, R. (2022b). Towards intelligent packaging: BCP-EVOH@ optode for milk freshness measurement. *Talanta, 241*. Available from https://doi.org/10.1016/j.talanta.2022.123230.

Magnaghi, L. R., Zanoni, C., Bancalari, E., Saadoun, J. H., Alberti, G., Quadrelli, P., & Biesuz, R. (2022c). pH-sensitive sensors at work on poultry meat degradation detection: From the laboratory to the supermarket shelf. *AppliedChem, 2*(3), 128–141. Available from https://doi.org/10.3390/appliedchem2030009.

Magnaghi, L. R., Zanoni, C., Bellotti, D., Alberti, G., Quadrelli, P., & Biesuz, R. (2022d). Quick and easy covalent grafting of sulfonated dyes to CMC: From synthesis to colorimetric sensing applications. *Polymers, 14*(19). Available from https://doi.org/10.3390/polym14194061.

Magnaghi, L. R., Alberti, G., Pazzi, B. M., Zanoni, C., & Biesuz, R. (2022e). A green-PAD array combined with chemometrics for pH measurements. *New Journal of Chemistry, 46*(40), 19460–19467. Available from https://doi.org/10.1039/d2nj03675d.

Mariani, F., Serafini, M., Gualandi, I., Arcangeli, D., Decataldo, F., Possanzini, L., Tessarolo, M., Tonelli, D., Fraboni, B., & Scavetta, E. (2021). Advanced wound dressing for real-time pH monitoring. *ACS Sensors, 6*(6), 2366–2377. Available from https://doi.org/10.1021/acssensors.1c00552.

Mary, S. K., Koshy, R. R., Daniel, J., Koshy, J. T., Pothen, L. A., & Thomas, S. (2020). Development of starch based intelligent films by incorporating anthocyanins of butterfly pea flower and TiO2and their applicability as freshness sensors for prawns during storage. *RSC Advances*, *10*(65), 39822–39830. Available from https://doi.org/10.1039/d0ra05986b.

McDonagh, C., Burke, C. S., & MacCraith, B. D. (2008). Optical chemical sensors. *Chemical Reviews*, *108*(2), 400–422. Available from https://doi.org/10.1021/cr068102g, https://doi.org/10.1021/cr068102g.

Medina-Jaramillo, C., Ochoa-Yepes, O., Bernal, C., & Famá, L. (2017). Active and smart biodegradable packaging based on starch and natural extracts. *Carbohydrate Polymers*, *176*, 187–194. Available from https://doi.org/10.1016/j.carbpol.2017.08.079, https://www.sciencedirect.com/science/article/pii/S0144861717309554.

Molina, B. G., Vasani, R. B., Jarvis, K. L., Armelin, E., Voelcker, N. H., & Alemán, C. (2022). Dual pH- and electro-responsive antibiotic-loaded polymeric platforms for effective bacterial detection and elimination. *Reactive and Functional Polymers*, *181*, 105434. Available from https://doi.org/10.1016/j.reactfunctpolym.2022.105434.

Morshed, M., Wang, J., Gao, M., & Wang, Z. (2020). Poly-2-amino-benzonitrile: A wide dynamic pH linear responding material. *Journal of Molecular Structure*, *1222*, 128891. Available from https://doi.org/10.1016/j.molstruc.2020.128891.

Mu, B., Cao, G., Zhang, L., Zou, Y., & Xiao, X. (2021). Flexible wireless pH sensor system for fish monitoring. *Sensing and Bio-Sensing Research*, *34*, 100465. Available from https://doi.org/10.1016/j.sbsr.2021.100465.

Mugo, S. M., Robertson, S. V., & Lu, W. (2023). A molecularly imprinted electrochemical microneedle sensor for multiplexed metabolites detection in human sweat. *Talanta*, *259*, 124531. Available from https://doi.org/10.1016/j.talanta.2023.124531.

Namazi, H. (2017). Polymers in our daily life. *BioImpacts*, *7*(2), 73–74. Available from https://doi.org/10.15171/bi.2017.09.

Nazari, S., Khiabani, M. S., Mokarram, R. R., Hamishehkar, H., Chisti, Y., & Tizchang, S. (2023). Optimized formulation of polyaniline-pectin optical film sensor for pH measurement. *Materials Science and Engineering B: Solid-State Materials for Advanced Technology*, *294*, 116517. Available from https://doi.org/10.1016/j.mseb.2023.116517.

Nguyen, T. H., Sun, T., & Grattan, K. T. V. (2020). Novel coumarin-based pH sensitive fluorescent probes for the highly alkaline pH region. *Dyes and Pigments*, *177*, 108312. Available from https://doi.org/10.1016/j.dyepig.2020.108312.

Noriega-Navarro, R., Castro-Medina, J., Escárcega-Bobadilla, M. V., & Zelada-Guillén, G. A. (2020). Control of ph-responsiveness in graphene oxide grafted with poly-DEAEMA via tailored functionalization. *Nanomaterials*, *10*(4), 614. Available from https://doi.org/10.3390/nano10040614.

Ogoshi, T., Hasegawa, Y., Aoki, T., Ishimori, Y., Inagi, S., & Yamagishi, T. A. (2011). Reduction of emeraldine base form of polyaniline by pillar[5]arene based on formation of poly(pseudorotaxane) structure. *Macromolecules*, *44*(19), 7639–7644. Available from https://doi.org/10.1021/ma2016979.

Ou, J., Tan, H., Chen, Z., & Chen, X. (2019). FRET-based semiconducting polymer dots for pH sensing. *Sensors (Switzerland)*, *19*(6), 1455. Available from https://doi.org/10.3390/s19061455.

Pablos, J. L., Hernández, E., Catalina, F., & Corrales, T. (2022). Solid fluorescence pH sensors based on 1,8-naphthalimide copolymers synthesized by UV curing. *Chemosensors*, *10*(2), 73. Available from https://doi.org/10.3390/chemosensors10020073.

Pakolpakçil, A., & Draczyński, Z. (2021). A facile design of colourimetric polyurethane nanofibrous sensor containing natural indicator dye for detecting ammonia vapour. *Materials*, *14*(22), 6949. Available from https://doi.org/10.3390/ma14226949.

Park, J. E., Yonet-Tanyeri, N., Ende, E. V., Henry, A. I., White, B. E. P., Mrksich, M., & Van Duyne, R. P. (2019). Plasmonic microneedle arrays for in situ sensing with surface-enhanced raman spectroscopy (SERS). *Nano Letters*, *19*(10), 6862–6868. Available from https://doi.org/10.1021/acs.nanolett.9b02070.

Petropoulou, A., Kralj, S., Karagiorgis, X., Savva, I., Loizides, E., Panagi, M., Krasia-Christoforou, T., & Riziotis, C. (2020). Multifunctional gas and pH fluorescent sensors based on cellulose acetate electrospun fibers decorated with rhodamine B-functionalised core-shell ferrous nanoparticles. *Scientific Reports*, *10* (1). Available from https://doi.org/10.1038/s41598-019-57291-0.

Pfattner, R., Foudeh, A. M., Chen, S., Niu, W., Matthews, J. R., He, M., & Bao, Z. (2019). Dual-gate organic field-effect transistor for pH sensors with tunable sensitivity. *Advanced Electronic Materials*, *5*(1). Available from https://doi.org/10.1002/aelm.201800381.

Pfeifer, D., Russegger, A., Klimant, I., & Borisov, S. M. (2020). Green to red emitting BODIPY dyes for fluorescent sensing and imaging of carbon dioxide. *Sensors and Actuators, B: Chemical*, *304*. Available from https://doi.org/10.1016/j.snb.2019.127312.

Pirayesh, H., Park, B. D., Khanjanzadeh, H., Park, H. J., & Cho, Y. J. (2023). Cellulosic material-based colorimetric films and hydrogels as food freshness indicators. *Cellulose*, *30*(5), 2791–2825. Available from https://doi.org/10.1007/s10570-023-05057-3.

Possanzini, L., Decataldo, F., Mariani, F., Gualandi, I., Tessarolo, M., Scavetta, E., & Fraboni, B. (2020). Textile sensors platform for the selective and simultaneous detection of chloride ion and pH in sweat. *Scientific Reports*, *10*(1). Available from https://doi.org/10.1038/s41598-020-74337-w.

Psarrou, M., Mitraki, A., Vamvakaki, M., & Kokotidou, C. (2023). Stimuli-responsive polysaccharide hydrogels and their composites for wound healing applications. *Polymers*, *15*(4). Available from https://doi.org/10.3390/polym15040986.

Quinn, A., You, Y. H., & McShane, M. J. (2019). Hydrogel microdomain encapsulation of stable functionalized silver nanoparticles for SERS pH and urea sensing. *Sensors (Switzerland)*, *19*(16). Available from https://doi.org/10.3390/s19163521.

Regan, F., Worsfold, P., Poole, C., Townshend, A., & Miró, M. (2019). *Sensors | overview*☆ (pp. 172–178). Oxford: Academic Press. Available from https://www.sciencedirect.com/science/article/pii/B9780124095472145408, https://doi.org/10.1016/B978-0-12-409547-2.14540-8.

Reyes-Cruzaley, A. P., Ochoa-Terán, A., Tirado-Guízar, A., Félix-Navarro, R. M., Alonso-Núñez, G., & Pina-Luis, G. (2021). A fluorescent PET probe based on polyethyleneimine-Ag nanoclusters as a reversible, stable and selective broad-range pH sensor. *Analytical Methods*, *13*(22), 2495–2503. Available from https://doi.org/10.1039/d1ay00302j.

Roy, J. S. D., Chowdhury, D., Sanfui, M. H., Hassan, N., Mahapatra, M., Ghosh, N. N., Majumdar, S., Chattopadhyay, P. K., Roy, S., & Singha, N. R. (2022). Ratiometric pH sensing, photophysics, and cell imaging of nonaromatic light-emitting polymers. *ACS Applied Bio Materials*, *5*(6), 2990–3005. Available from https://doi.org/10.1021/acsabm.2c00297.

Roy, S., & Rhim, J. W. (2021). Fabrication of carboxymethyl cellulose/agar-based functional films hybridized with alizarin and grapefruit seed extract. *ACS Applied Bio Materials*, *4*(5), 4470–4478. Available from https://doi.org/10.1021/acsabm.1c00214.

Saha, N., Brunetti, G., Armenise, M. N., & Ciminelli, C. (2023). A compact, highly sensitive pH sensor based on polymer waveguide Bragg grating. *IEEE Photonics Journal*, *15*(2). Available from https://doi.org/10.1109/JPHOT.2023.3242820.

Saikrithika, S., & Kumar, A. S. (n.d.). A selective voltammetric pH sensor using graphitized mesoporous carbon/polyaniline hybrid system. Available from: https://doi.org/10.1007/s12039-021-01908-3S.

Sato, K., Sato, F., Kumano, M., Kamijo, T., Sato, T., Zhou, Y., Korchev, Y., Fukuma, T., Fujimura, T., & Takahashi, Y. (2021). Electrochemical quantitative evaluation of the surface charge of a poly(1-vinylimidazole) multilayer film and application to nanopore pH sensor. *Electroanalysis*, *33*(6), 1633–1638. Available from https://doi.org/10.1002/elan.202100041.

Schoolaert, E., Merckx, R., Becelaere, J., Rijssegem, S., Hoogenboom, R., & De Clerck, K. (2022). Eco-friendly colorimetric nanofiber design: Halochromic sensors with tunable pH-sensing regime based on

2-ethyl-2-oxazoline and 2-n-butyl-2-oxazoline statistical copolymers functionalized with alizarin yellow R. *Advanced Functional Materials*, *32*(1), 2106859. Available from https://doi.org/10.1002/adfm.202106859, https://doi.org/10.1002/adfm.202106859.

Shah, S. A., Sohail, M., Khan, S., Minhas, M. U., de Matas, M., Sikstone, V., Hussain, Z., Abbasi, M., & Kousar, M. (2019). Biopolymer-based biomaterials for accelerated diabetic wound healing: A critical review. *International Journal of Biological Macromolecules*, *139*, 975–993. Available from https://doi.org/10.1016/j.ijbiomac.2019.08.007.

Shah, T. U. H., Tahir, M. H., & Liu, H. (2019). Polyethylene glycol-modified cystamine for fluorescent sensing. *Journal of Materials Science*, *54*(1), 313–322. Available from https://doi.org/10.1007/s10853-018-2867-2.

Sheokand, B., Vats, M., Kumar, A., Srivastava, C. M., Bahadur, I., & Pathak, S. R. (2023). Natural polymers used in the dressing materials for wound healing: Past, present and future. *Journal of Polymer Science*, *61*, 1389–1414. Available from https://doi.org/10.1002/pol.20220734.

Shrivastava, A., & Shrivastava, A. (2018). *1 – Introduction to plastics engineering. Plastics Design Library* (pp. 1–16). William Andrew Publishing. Available from https://www.sciencedirect.com/science/article/pii/B9780323395007000010, https://doi.org/10.1016/B978-0-323-39500-7.00001-0.

Sinar, D., Mattos, G. J., & Knopf, G. K. (2020). Disposable hydrogel activated circular interdigitated sensor for monitoring biological fluid pH. *IEEE Sensors Journal*, *20*(24), 14624–14631. Available from https://doi.org/10.1109/JSEN.2020.3011797.

Sisodia, N., Miranda, M., McGuinness, K. L., Wadhawan, J. D., & Lawrence, N. S. (2021). Intra- and intermolecular sulf- hydryl hydrogen bonding: Facilitating proton transfer events for determination of pH in sea water. *Electroanalysis*, *33*(3), 559–562. Available from https://doi.org/10.1002/elan.202060332.

Skinner, W. H., Chung, M., Mitchell, S., Akidil, A., Fabre, K., Goodwin, R., Stokes, A. A., Radacsi, N., & Campbell, C. J. (2021). A SERS-active electrospun polymer mesh for spatially localized pH measurements of the cellular microenvironment. *Analytical Chemistry*, *93*(41), 13844–13851. Available from https://doi.org/10.1021/acs.analchem.1c02530.

Skinner, W. H., Robinson, N., Hardisty, G. R., Fleming, H., Geddis, A., Bradley, M., Gray, R. D., & Campbell, C. J. (2023). SERS microsensors for pH measurements in the lumen and ECM of stem cell derived human airway organoids. *Chemical Communications*, *59*(22), 3249–3252. Available from https://doi.org/10.1039/d2cc06582g.

Skorjanc, T., Shetty, D., & Valant, M. (2021). Covalent organic polymers and frameworks for fluorescence-based sensors. *ACS Sensors*, *6*(4), 1461–1481. Available from https://doi.org/10.1021/acssensors.1c00183.

Staudinger, C., Strobl, M., Breininger, J., Klimant, I., & Borisov, S. M. (2019). Fast and stable optical pH sensor materials for oceanographic applications. *Sensors and Actuators B: Chemical*, *282*, 204–217. Available from https://doi.org/10.1016/j.snb.2018.11.048, https://www.sciencedirect.com/science/article/pii/S0925400518320021.

Steinegger, A., Wolfbeis, O. S., & Borisov, S. M. (2020). Optical sensing and imaging of pH values: Spectroscopies, materials, and applications. *Chemical Reviews*, *120*(22), 12357–12489. Available from https://doi.org/10.1021/acs.chemrev.0c00451, https://doi.org/10.1021/acs.chemrev.0c00451.

Sun, X., Wang, Y., & Lei, Y. (2015). Fluorescence based explosive detection: From mechanisms to sensory materials. *Chemical Society Reviews*, *44*(22), 8019–8061. Available from https://doi.org/10.1039/c5cs00496a.

Sutthasupa, S., Padungkit, C., & Suriyong, S. (2021). Colorimetric ammonia (NH3) sensor based on an alginate-methylcellulose blend hydrogel and the potential opportunity for the development of a minced pork spoilage indicator. *Food Chemistry*, *362*. Available from https://doi.org/10.1016/j.foodchem.2021.130151.

Tang, Y., Zhong, L., Wang, W., He, Y., Han, T., Xu, L., Mo, X., Liu, Z., Ma, Y., Bao, Y., Gan, S., & Niu, L. (2022). Recent advances in wearable potentiometric pH sensors. *Membranes*, *12*(5), 504. Available from https://doi.org/10.3390/membranes12050504.

Tang, Z., Gomez, D., He, C., Korposh, S., Morgan, S. P., Correia, R., Hayes-Gill, B., Setchfield, K., & Liu, L. (2021). A U-shape fibre-optic pH sensor based on hydrogen bonding of ethyl cellulose with a sol-gel matrix. *Journal of Lightwave Technology*, *39*(5), 1557–1564. Available from https://doi.org/10.1109/JLT.2020.3034563.

Tariq, A., Garnier, U., Ghasemi, R., Lefevre, J. P., Mongin, C., Brosseau, A., Audibert, J. F., Pansu, R., Dauzères, A., & Leray, I. (2022). Perylene based PET fluorescent molecular probes for pH monitoring. *Journal of Photochemistry and Photobiology A: Chemistry*, *432*. Available from https://doi.org/10.1016/j.jphotochem.2022.114035.

Teixeira, S. C., de Oliveira, T. V., Silva, R. R. A., Ribeiro, A. R. C., Stringheta, P. C., Rigolon, T. C. B., Pinto, M. R. M. R., & de Fátima Ferreira Soares, N. (2022). Colorimetric indicators of açaí anthocyanin extract in the biodegradable polymer matrix to indicate fresh shrimp. *Food Bioscience*, *48*. Available from https://doi.org/10.1016/j.fbio.2022.101808.

Ulrich, S., Osypova, A., Panzarasa, G., Rossi, R. M., Bruns, N., & Boesel, L. F. (2019). Pyranine-modified amphiphilic polymer conetworks as fluorescent ratiometric pH sensors. *Macromolecular Rapid Communications*, *40*(21). Available from https://doi.org/10.1002/marc.201900360.

Uzair, U., Johnson, C., Beladi-Behbahani, S., Rajamanthrilage, A. C., Raval, Y. S., Benza, D., Ranasinghe, M., Schober, G., Tzeng, T. R. J., & Anker, J. N. (2020). Conformal coating of orthopedic plates with X-ray scintillators and pH indicators for X-ray excited luminescence chemical imaging through tissue. *ACS Applied Materials and Interfaces*, *12*(47), 52343–52353. Available from https://doi.org/10.1021/acsami.0c13707.

Wang, D., Wang, X., Sun, Z., Liu, F., & Wang, D. (2022). A fast-response visual indicator film based on polyvinyl alcohol/methylcellulose/black wolfberry anthocyanin for monitoring chicken and shrimp freshness. *Food Packaging and Shelf Life*, *34*. Available from https://doi.org/10.1016/j.fpsl.2022.100939.

Wang, H., Zhou, S., Guo, L., Wang, Y., & Feng, L. (2020). Intelligent hybrid hydrogels for rapid in situ detection and photothermal therapy of bacterial infection. *ACS Applied Materials and Interfaces*, *12*(35), 39685–39694. Available from https://doi.org/10.1021/acsami.0c12355.

Wang, J., Wang, L., Su, X., Xiao, R., Gao, D., Kang, C., Fang, X., & Zhang, X. (2022). Simultaneous real-time measurements of temperature, liquid level, humidity, and pH by ZnSe/Co nanostructure-coated polymer films. *ACS Applied Nano Materials*. Available from https://doi.org/10.1021/acsanm.2c03334.

Wang, X., Feng, Y., Liu, J., Cheng, K., Liu, Y., Yang, W., Zhang, H., & Peng, H. (2022). Fluorescein isothiocyanate-doped conjugated polymer nanoparticles for two-photon ratiometric fluorescent imaging of intracellular pH fluctuations. *Spectrochimica Acta – Part A: Molecular and Biomolecular Spectroscopy*, *267*. Available from https://doi.org/10.1016/j.saa.2021.120477.

Wells, N., Yusufu, D., & Mills, A. (2019). Colourimetric plastic film indicator for the detection of the volatile basic nitrogen compounds associated with fish spoilage. *Talanta*, *194*, 830–836. Available from https://doi.org/10.1016/j.talanta.2018.11.020.

Wencel, D., Abel, T., & McDonagh, C. (2014). Optical chemical pH sensors. *Analytical Chemistry*, *86*(1), 15–29. Available from https://doi.org/10.1021/ac4035168.

Weston, M., Phan, M. A. T., Arcot, J., & Chandrawati, R. (2020). Anthocyanin-based sensors derived from food waste as an active use-by date indicator for milk. *Food Chemistry*, *326*. Available from https://doi.org/10.1016/j.foodchem.2020.127017.

Wijayaratna, U. N., Kiridena, S. D., Adams, J. D., Behrend, C. J., & Anker, J. N. (2021). Synovial fluid pH sensor for early detection of prosthetic hip infections. *Advanced Functional Materials*, *31*(37). Available from https://doi.org/10.1002/adfm.202104124.

Wu, C. Y., Chen, J. R., & Su, C. K. (2022). 4D-printed pH sensing claw. *Analytica Chimica Acta*, *1204*. Available from https://doi.org/10.1016/j.aca.2022.339733.

Wu, J., Xu, B., Liu, Z., Yao, Y., Zhuang, Q., & Lin, S. (2019). The synthesis, self-assembly and pH-responsive fluorescence enhancement of an alternating amphiphilic copolymer with azobenzene pendants. *Polymer Chemistry*, *10*(29), 4025–4030. Available from https://doi.org/10.1039/c9py00634f.

Xie, X., Lei, Y., Li, Y., Zhang, M., Sun, J., Zhu, M. Q., & Wang, J. (2023). Dual-crosslinked bioadhesive hydrogel as NIR/pH stimulus-responsiveness platform for effectively accelerating wound healing. *Journal of Colloid and Interface Science*, *637*, 20–32. Available from https://doi.org/10.1016/j.jcis.2023.01.081.

Xu, L., Liu, X., Jia, J., Wu, H., Xie, J., & Jia, Y. (2022). Electrospun nanofiber membranes from 1,8-naphthimide-based polymer/poly(vinyl alcohol) for ph fluorescence sensing. *Molecules (Basel, Switzerland)*, *27* (2). Available from https://doi.org/10.3390/molecules27020520.

Xu, Z., Qiao, X., Tao, R., Li, Y., Zhao, S., Cai, Y., & Luo, X. (2023). A wearable sensor based on multifunctional conductive hydrogel for simultaneous accurate pH and tyrosine monitoring in sweat. *Biosensors and Bioelectronics*, *234*. Available from https://doi.org/10.1016/j.bios.2023.115360.

Xue, Y., Wan, Z., Ouyang, X., & Qiu, X. (2019). Lignosulfonate: A convenient fluorescence resonance energy transfer platform for the construction of a ratiometric fluorescence pH-sensing probe. *Journal of Agricultural and Food Chemistry*, *67*(4), 1044–1051. Available from https://doi.org/10.1021/acs.jafc.8b05286.

Yam, N. J., Rusli, A., Hamid, Z. A. A., Abdullah, M. K., & Marsilla, K. I. K. (2021). Halochromic poly (lactic acid) film for acid base sensor. *Journal of Applied Polymer Science*, *138*(13). Available from https://doi.org/10.1002/app.50093.

Ye, W., Qin, M., Qiu, R., & Li, J. (2022). Keratin-based wound dressings: From waste to wealth. *International Journal of Biological Macromolecules*, *211*, 183–197. Available from https://doi.org/10.1016/j.ijbiomac.2022.04.216.

Yoon, J. H., Kim, S. M., Park, H. J., Kim, Y. K., Oh, D. X., Cho, H. W., Lee, K. G., Hwang, S. Y., Park, J., & Choi, B. G. (2020). Highly self-healable and flexible cable-type pH sensors for real-time monitoring of human fluids. *Biosensors and Bioelectronics*, *150*. Available from https://doi.org/10.1016/j.bios.2019.111946.

Yue, Y., Luo, H., Han, J., Chen, Y., & Jiang, J. (2020). Assessing the effects of cellulose-inorganic nanofillers on thermo/pH-dual responsive hydrogels. *Applied Surface Science*, *528*. Available from https://doi.org/10.1016/j.apsusc.2020.146961.

Zdrachek, E., Forrest, T., & Bakker, E. (2023). Symmetric cell for improving solid-contact pH electrodes. *Analytica Chimica Acta*, *1239*. Available from https://doi.org/10.1016/j.aca.2022.340652.

Zea, M., Texidó, R., Villa, R., Borrós, S., & Gabriel, G. (2021). Specially designed polyaniline/polypyrrole ink for a fully printed highly sensitive pH microsensor. *ACS Applied Materials and Interfaces*, *13*(28), 33524–33535. Available from https://doi.org/10.1021/acsami.1c08043.

Zhang, X., Lv, R., Chen, L., Sun, R., Zhang, Y., Sheng, R., Du, T., Li, Y., & Qi, Y. (2022). A multifunctional janus electrospun nanofiber dressing with biofluid draining, monitoring, and antibacterial properties for wound healing. *ACS Applied Materials and Interfaces*, *14*(11), 12984–13000. Available from https://doi.org/10.1021/acsami.1c22629.

Zhang, Y., Gallego, I., Plou, J., Pedraz, J. L., Liz-Marzán, L. M., Ciriza, J., & García, I. (2021). SERS monitoring of local pH in encapsulated therapeutic cells. *Nanoscale*, *13*(34), 14354–14362. Available from https://doi.org/10.1039/d1nr03969e.

Zhao, Y., Yu, Y., Zhao, S., Zhu, R., Zhao, J., & Cui, G. (2023). Highly sensitive pH sensor based on flexible polyaniline matrix for synchronal sweat monitoring. *Microchemical Journal*, *185*. Available from https://doi.org/10.1016/j.microc.2022.108092.

Zheng, K., Tong, Y., Zhang, S., He, R., Xiao, L., Iqbal, Z., Zhang, Y., Gao, J., Zhang, L., Jiang, L., & Li, Y. (2021). Flexible bicolorimetric polyacrylamide/chitosan hydrogels for smart real-time monitoring and promotion of wound healing. *Advanced Functional Materials*, *31*(34). Available from https://doi.org/10.1002/adfm.202102599.

Zheng, X. T., Zhong, Y., Chu, H. E., Yu, Y., Zhang, Y., Chin, J. S., Becker, D. L., Su, X., & Loh, X. J. (2023). Carbon dot-doped hydrogel sensor array for multiplexed colorimetric detection of wound healing.

ACS Applied Materials and Interfaces, *15*(14), 17675–17687. Available from https://doi.org/10.1021/acsami.3c01185.

Zhu, G., Cheng, L., Liu, G., & Zhu, L. (2020). Synthesis of gold nanoparticle stabilized on silicon nanocrystal containing polymer microspheres as effective surface-enhanced raman scattering (SERS) substrates. *Nanomaterials*, *10*(8), 1–9. Available from https://doi.org/10.3390/nano10081501.

Zhu, Z., Wang, L., Peng, Y., Chu, X., Zhou, L., Jin, Y., Guo, H., Gao, Q., Yang, J., Wang, X., Long, Z., Ge, Y., Lu, S., & Wang, B. (2022). Continuous self-oxygenated double-layered hydrogel under natural light for real-time infection monitoring, enhanced photodynamic therapy, and hypoxia relief in refractory diabetic wounds healing. *Advanced Functional Materials*, *32*(32). Available from https://doi.org/10.1002/adfm.202201875.

Žutautas, V., Jelinskas, T., & Pauliukaite, R. (2022). A novel sensor for electrochemical pH monitoring based on polyfolate. *Journal of Electroanalytical Chemistry*, *921*. Available from https://doi.org/10.1016/j.jelechem.2022.116668.

Žutautas, V., Trusovas, R., Sartanavičius, A., Ratautas, K., Selskis, A., & Pauliukaite, R. (2023). A sensor for electrochemical pH monitoring based on laser-induced graphene modified with polyfolate. *Chemosensors*, *11*(6), 329. Available from https://doi.org/10.3390/chemosensors11060329.

CHAPTER

Temperature sensors

17

Yosuke Mizuno
Faculty of Engineering, Yokohama National University, Yokohama, Japan

17.1 Introduction

This chapter explores the realm of temperature sensing via polymer optical fibers (POFs) (Fasano et al., 2016; Husdi et al., 2004; Kaino et al., 1981; Koike & Asai, 2009; Kuzyk, 2006; Minakawa et al., 2015; Peters, 2011; Rao et al., 1991; Ziemann et al., 2002; Zubia & Arrue, 2001), but before we venture into the specifics, let us set the stage with a wider perspective. In the contemporary era, we are witnessing a surge in societal challenges stemming from the degradation and damage inflicted by natural disasters on infrastructure. These infrastructures, encompassing buildings, tunnel walls, bridges, dams, and pipelines, are often relics from periods of high economic growth. As these structures age and succumb to the forces of nature, the need for monitoring systems, such as optical fiber sensors, escalates. These sensors are instrumental in tracking various conditions of these structures, including temperature, strain, pressure, and vibration.

Optical fiber sensors are a diverse group, broadly categorized into three types: point, multipoint, and distributed (Brogan & Walt, 2005; Hartog, 2017; Lee, 2003; Leung et al., 2015; Motil et al., 2016). Point sensors are limited to sensing at a single location on the optical fiber, typically where the fiber Bragg grating (FBG) is inscribed or at the optical fiber's end face. Sensing strategies based on multimode interference also fall into this category. Multipoint sensors, on the other hand, offer the ability to sense at multiple locations along the optical fiber, with measurements using multiple FBGs serving as a classic example. Contrastingly, distributed sensors break these boundaries, offering sensing capabilities at any location along the optical fiber. These sensors employ techniques based on Rayleigh (Chen et al., 2012; Froggatt & Moore, 1998; Koyamada et al., 2009; Lu et al., 2019; Pastor-Graells et al., 2016; Sang et al., 2008; Shatalin et al., 1998; Song et al., 2014; Yan et al., 2017), Brillouin (Denisov et al., 2016; Ding et al., 2014; Garus et al., 1996; Horiguchi & Tateda, 1989; Hotate, 2000; Kim et al., 2015; Kurashima et al., 1993; Minardo et al., 2016; Mizuno et al., 2008; Zou et al., 2009), and Raman scattering (Bolognini & Hartog, 2013; Han et al., 2005; Hwang et al., 2010; Saxena et al., 2015; Wang, Wu, et al., 2017). Each of these methods brings its own unique strengths and weaknesses to the table, and none can claim absolute superiority over the others.

In this chapter, our exploration commences with an introduction to distributed temperature sensors based on Brillouin scattering in POFs. These sensors hold a significant advantage as they eliminate blind spots in measurement. Our journey then advances to the realm of temperature sensing using FBGs inscribed in POFs. The final stage of our exploration introduces us to temperature sensing based on multimode interference in POFs.

Sensory Polymers. DOI: https://doi.org/10.1016/B978-0-443-13394-7.00017-3

17.2 Brillouin-based techniques

We begin with an introduction to distributed temperature-sensing techniques based on Brillouin scattering in POFs (Mizuno et al., 2021). However, given the inherent dependency of Brillouin scattering on both temperature and strain, a narrative solely centered on temperature would be somewhat constrained. Consequently, this section will also encompass strain sensing, running in parallel with our discourse on temperature sensing. We appreciate your understanding of this comprehensive approach.

Historically, the evolution of distributed optical fiber sensors has predominantly leveraged silica glass optical fibers, a mainstay in high-speed communication. Yet, these glass optical fibers present challenges. They are susceptible to damage, necessitate careful handling, and fracture at a strain of a few percent, rendering the measurement of larger strains or temperatures in such conditions unfeasible. To surmount these hurdles, contemporary research and development initiatives are pivoting toward distributed optical fiber sensors employing POFs.

POFs bring to the table a plethora of advantages over their glass counterparts, including a substantial core diameter, exceptional flexibility capable of withstanding strains exceeding 100% in certain instances, cost-effective installation, straightforward fiber interconnection, and enhanced safety. The deployment of distributed optical fiber sensors using POFs not only simplifies handling but also significantly broadens the spectrum of strains that can be addressed. Moreover, POFs offer a unique advantage—the capacity to imbue the sensor with a "memory" function. This function is attributed to the propensity of POFs to undergo plastic deformation in response to substantial strains or temperature changes nearing the glass-transition temperature, thereby enabling the POF itself to retain the magnitude and location of strains or temperature changes (Husdi et al., 2004; Minakawa et al., 2015).

Leveraging this property gives rise to an innovative concept. Instead of perpetually installing costly analysis equipment, as is the case with glass optical fibers, one can merely embed POFs in structures and conduct a roving inspection of numerous structures with a single analysis device after an event like an earthquake. From a societal perspective, this approach could drastically curtail the cost of optical fiber sensor technology, broadening its application from large-scale edifices to general individual residences and potentially aiding in the reduction of the duration of evacuation life for residents postearthquakes.

In this section, we will elucidate the sensing characteristics of Brillouin scattering in POFs and the demonstration of distributed strain and temperature sensing using POFs. We will also explore its high spatial resolution and speed enhancement, all aimed toward the realization of distributed optical fiber sensors equipped with memory functions.

17.2.1 Fundamentals of polymer optical fibers

Optical fibers are known for their scattering phenomena, such as Rayleigh and Raman scattering, which form the basis for strain and temperature measurements. While sensors that utilize these phenomena are relatively straightforward to implement, their accuracy and stability have been somewhat compromised due to their reliance on the intensity of scattered light. Consequently, there has been a growing interest in distributed sensors that exploit Brillouin scattering, an interaction between ultrasonic waves and light within optical fibers.

Brillouin scattering (Agrawal, 2019) involves a Doppler effect triggered by ultrasonic waves in the optical fiber, leading to a known decrease in frequency, termed the Brillouin frequency shift (BFS). As the BFS changes in proportion to the magnitude of strain or temperature applied to the optical fiber, it becomes possible to determine these quantities by measuring the BFS. This method is particularly notable for its high stability and accuracy, as it relies on frequency rather than the intensity of scattered light.

Let us now turn our attention to POFs (Fasano et al., 2016; Husdi et al., 2004; Kaino et al., 1981; Koike & Asai, 2009; Kuzyk, 2006; Minakawa et al., 2015; Peters, 2011; Rao et al., 1991; Ziemann et al., 2002; Zubia & Arrue, 2001). POFs, composed of polymer materials for both the core and cladding, are characterized by their significantly larger diameter than silica glass optical fibers. Despite the large diameter of the bare fiber, which ranges from 500 μm to 1 mm, POFs are flexible and resistant to breakage caused by bending or pulling. The core diameter is also substantial, ranging from 50 μm to 980 μm, which eliminates the need for high alignment accuracy for coupling between POFs or between POFs and optical elements. This facilitates low-cost connector processing and other end-face treatments.

Currently, most POF products are step-index POFs with polymethyl methacrylate (PMMA) as the core material. The optical propagation loss in the visible light band is a few hundred dB/km, while it is extremely high in the communication wavelength band and unsuitable for use. In contrast, graded-index POFs with fully fluorinated PMMA, expanding the low-loss wavelength band to the communication wavelength band, are called perfluorinated graded-index (PFGI)-POFs (Koike & Asai, 2009). The optical propagation loss at wavelengths of 850 nm and 1310 nm is about 10 dB/km, and even at the communication wavelength of 1550 nm, the loss is suppressed to about 250 dB/km. Currently, PFGI-POFs with core diameters ranging from 50 μm to 120 μm are commercially available.

17.2.2 Brillouin characterization

Observing Brillouin scattering in POFs requires two key elements. The first is the ability to use relatively inexpensive optical devices in the communication wavelength band, which have been developed in the field of optical communication technology. In particular, narrow-linewidth semiconductor lasers and erbium-doped fiber amplifiers are crucial. The second requirement is that the core diameter be as small as possible. This is because the Brillouin scattering signal becomes stronger as the power density is higher. Fortunately, the PFGI-POF, with a core diameter of 50 μm, possesses both features. In fact, the POFs in which Brillouin scattering has been experimentally observed so far are limited to PFGI-POFs (Mizuno & Nakamura, 2010a).

At the 1550 nm band and room temperature, while the BFS of a silica fiber is about 10.8 GHz, the BFS of a PFGI-POF (hereafter simply referred to as a POF) is about 2.8 GHz, as shown in Fig. 17.1 (Mizuno & Nakamura, 2010a). This is due to the lower speed of sound in a POF (softer) compared to silica fiber. Also, the coefficients of the strain and temperature dependence of the BFS are, respectively, +580 MHz/% and +1.18 MHz/K for a silica fiber, and −121.8 MHz/% (relatively small strains of less than about 2%) and −4.1 MHz/K (or −3.2 MHz/K depending on the structure) for a POF (Minakawa et al., 2014; Mizuno & Nakamura, 2010b). Both coefficients have the opposite sign, and their absolute values are 0.2 times and 3.5 times, respectively. This is due to the unique strain and temperature dependence of Young's modulus of

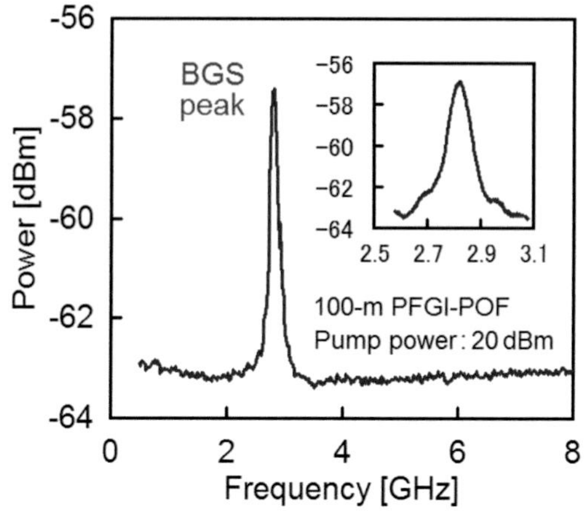

FIGURE 17.1 Brillouin gain spectrum observed in a POF.

The inset shows the magnified view around the peak. *POF,* Polymer optical fiber.

From Mizuno, Y., & Nakamura, K. (2010). Experimental study of Brillouin scattering in perfluorinated polymer optical fiber at telecommunication wavelength. Applied Physics Letters, *97(2). Available from https://doi.org/10.1063/1.3463038.*

a POF, and it indicates that Brillouin scattering in a POF, which is less dependent on strain, can be used for high-precision temperature sensing.

One of the features of a POF is that it can apply to large strains exceeding a few percent, but the dependence of the BFS in a POF on large strains is special (Hayashi et al., 2012; Hayashi, Minakawa, et al., 2014; Mizuno, Matsutani, et al., 2018). First, the large strain dependence of the Brillouin gain spectrum (BGS) is shown in Fig. 17.2A. The input light power was 26 dBm, and the pulling speed was 200 μm/s. When the applied strain was 0%, the BFS was at 2.85 GHz, shifted to the low-frequency side in the range of 0%−2.3%, and shifted to the high-frequency side in the range of 2.3% or more. When a strain of 10% or more was applied, the scattered light power at around 2.8 GHz gradually attenuated. When a strain of 7.3% or more was applied, a new peak appeared around 3.2 GHz, and its power increased with increasing applied strain. When the applied strain was 31%, the power of the two Brillouin signal peaks was about the same, and at 60%, the original peak completely disappeared. The relatively small peak that always exists at 2.85 GHz is a scattered signal from the input end side where no strain is applied (6 cm), which is not essential. For the two peaks around 2.8 GHz and 3.2 GHz, the large strain dependence of the BFS is shown in Fig. 17.2B. The original peak showed nonlinear dependence. However, it was difficult to accurately measure the BFS when the applied strain was 20% or more because it was buried in the small peak at 2.85 GHz. On the other hand, the new peak around 3.2 GHz showed almost no strain dependence.

Next, Fig. 17.2C shows a photograph of the side of the POF when a strain of about 7.3% was applied at 200 μm/s. A part that became thin in stages was observed. In this thin part, the speed of sound increased due to the change in core structure, and the BFS changed steeply to about 3.2 GHz ("BFS hopping"). As the applied strain increased, the thin part expanded along the POF, and the

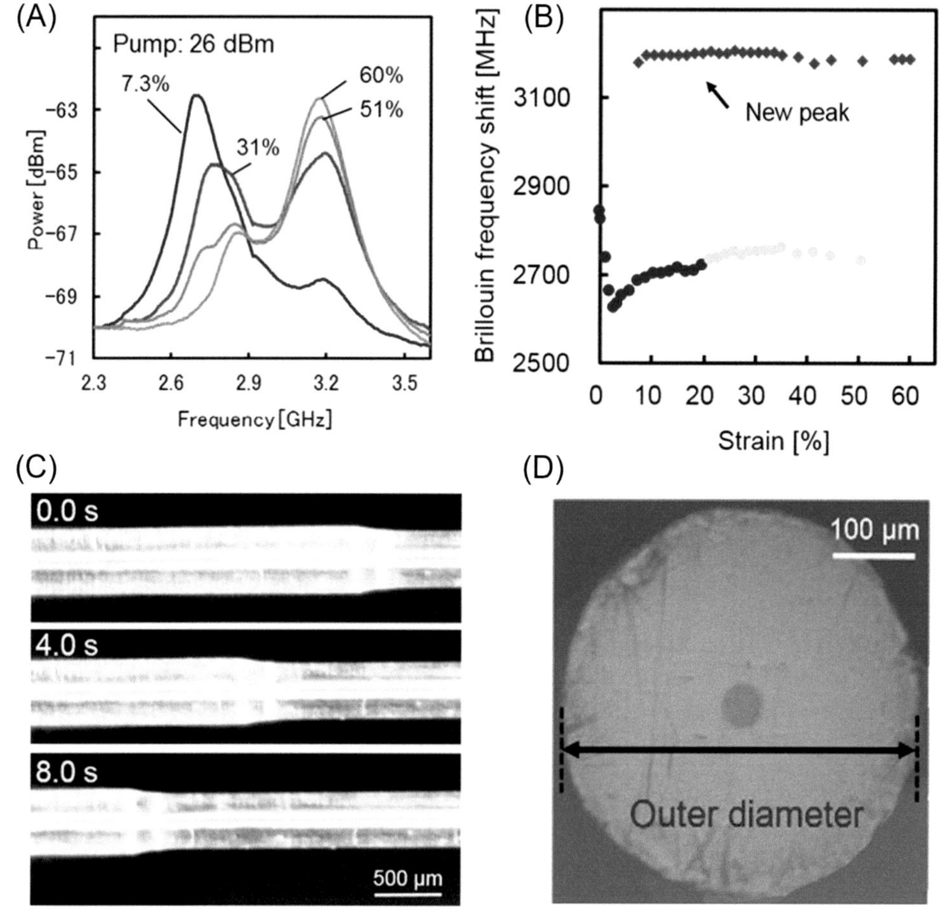

FIGURE 17.2 Response of the POF to large strain.

(A) Strain dependence of the BGS, where % represents the magnitude of strain relative to the initial POF length. (B) Strain dependence of the BFS. (C) The process of POF thinning. (D) Cross-sectional view of the POF after thinning. *BGS*, Brillouin gain spectrum; *POF*, polymer optical fiber; *BFS*, Brillouin frequency shift.

From Hayashi, N., Minakawa, K., Mizuno, Y., & Nakamura, K. (2014). Brillouin frequency shift hopping in polymer optical fiber. Applied Physics Letters, 105*(9). Available from https://doi.org/10.1063/1.4895041.*

entire POF became thin with a strain of 60%. At that time, the outer diameter of the thin part was kept almost constant. This is why the BFS of the new peak does not depend on strain, and only its power depends on strain. Fig. 17.2D shows a cross-sectional photograph of the thin part. From the change in outer diameter, it is estimated that the core diameter of the thin part is about 0.85 times that of the case without strain. Finally, after applying a strain of 60%, when the strain was released, the entire length of the POF was kept thin due to plastic deformation. When the BFS strain and temperature dependence coefficients were measured again for this POF, both were linear, and the coefficients were −65.6 MHz/% and −4.04 MHz/K, respectively. These values were about 0.5 times and about 1.3 times the reported values in conventional PFGI-POFs, respectively. Therefore by using Brillouin scattering in the thinned POF, it is possible to enhance further the characteristic value of the POF, which is high-precision temperature sensing that is less dependent on strain.

In addition to this, the BFS in a tapered POF (Hayashi, Fukuda, et al., 2014; Mizuno et al., 2017; Ujihara et al., 2015), the response of the BFS to even larger strains exceeding 130% (Mizuno, Matsutani, et al., 2018), the temperature dependence of BFS strain dependence (and vice versa) (Minakawa et al., 2017), and the humidity and pressure dependences of the BFS (Mizuno, Lee, Hayashi, et al., 2018; Schreier et al., 2018) have also been clarified.

17.2.3 Distributed sensing

The field of distributed strain and temperature sensors utilizing Brillouin scattering in optical fibers has seen the proposal of various methodologies. These include time-domain, frequency-domain, and correlation-domain approaches (Garus et al., 1996; Horiguchi & Tateda, 1989; Hotate, 2000; Kurashima et al., 1993; Minardo et al., 2016; Mizuno et al., 2008). However, the capacity for real-time distributed measurement was previously confined to configurations that necessitated light injection from both ends of the sensing fiber (Dong et al., 2014; Minardo et al., 2014). The flexibility of sensor installation within structures was compromised, and a single break in the sensing fiber could halt the operation of the entire system. These challenges were addressed with the advent of high-speed Brillouin optical correlation-domain reflectometry (BOCDR) (Mizuno et al., 2008).

The experimental setup of the ultrahigh-speed BOCDR (Mizuno et al., 2016) is depicted in Fig. 17.3. The optical setup mirrors that of the conventional BOCDR, achieving distributed strain and temperature measurement by sweeping the "correlation peak"—formed by frequency-modulating the output of a 1550 nm band laser—along the sensing fiber. It is recognized that the spatial resolution inversely correlates with the product of the modulation amplitude and frequency, while the measurable length of the fiber (measurement range) inversely correlates with the modulation frequency. This implies a trade-off relationship between the measurement range and spatial resolution of BOCDR. For a detailed understanding of the operating principle of BOCDR, we refer readers to the relevant literature (Lee et al., 2016; Mizuno et al., 2009, 2016; Mizuno et al., 2008, 2010; Mizuno, Lee, & Nakamura, 2018).

Conventionally, the sampling rate was restricted to less than 20 Hz, as the BGS was detected in the frequency domain using the frequency sweep function of the electrical spectrum analyzer (Mizuno et al., 2009). In contrast, the high-speed BOCDR employs a voltage-controlled oscillator to detect the BGS in the time domain, thereby enabling a sampling rate of several hundred Hz. In addition, the use of phase-detection technology to derive the BFS from the BGS allows for a sampling rate exceeding 100 kHz. We have already achieved notable advancements, including the enhancement of the signal-to-noise ratio, the expansion of the strain dynamic range, and the real-time detection of propagating mechanical waves (Mizuno et al., 2016).

We now turn our attention to experiments utilizing a POF as the sensing fiber for BOCDR (Hayashi, Minakawa, et al., 2015; Hayashi, Mizuno, et al., 2014; Hayashi, Mizuno, et al., 2015; Lee et al., 2017; Mizuno, Lee, et al., 2019). In the case of a POF, a specialized noise removal method is required. Initially, we conducted a measurement of the temperature distribution.

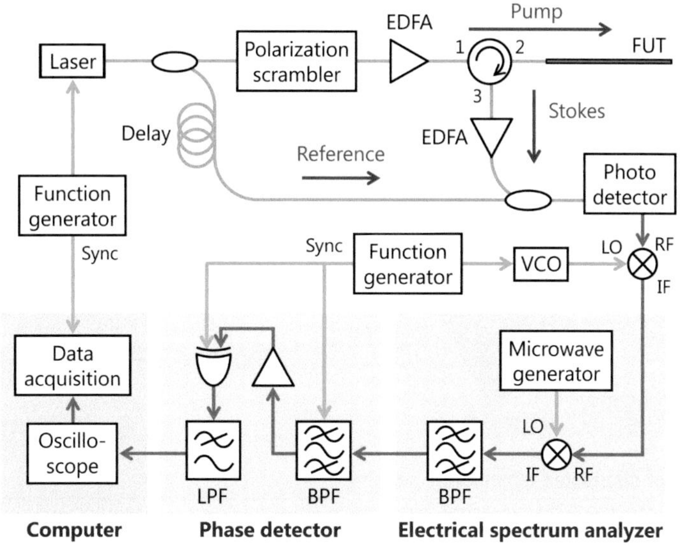

FIGURE 17.3 Experimental setup of ultrahigh-speed BOCDR.

BOCDR, Brillouin optical correlation-domain reflectometry; *BPF*, band-pass filter; *EDFA*, erbium-doped fiber amplifier; *FUT*, fiber under test; *IF*, intermediate frequency; *LO*, local oscillator; *LPF*, low-pass filter; *RF*, radio frequency; *VCO*, voltage-controlled oscillator.

From Mizuno, Y., Hayashi, N., Fukuda, H., Song, K. Y., & Nakamura, K. (2016). Ultrahigh-speed distributed Brillouin reflectometry. Light: Science and Applications, 5. *Available from https://doi.org/10.1038/lsa.2016.184.*

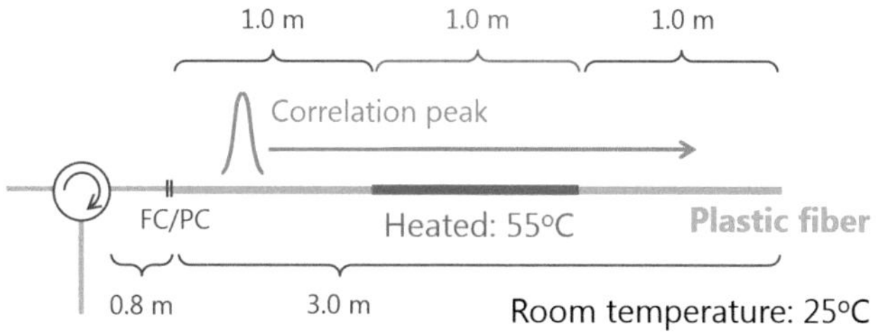

FIGURE 17.4 Structure of the sensing fiber.

A 1.0 m section of the 3.0 m POF was heated. *POF*, Polymer optical fiber.

From Mizuno, Y., Lee, H., Hayashi, N., & Nakamura, K. (2019). Noise suppression technique for distributed Brillouin sensing with polymer optical fibers. Optics Letters, *44(8), 2097–2100. Available from https://doi.org/10.1364/OL.44.002097.*

Fig. 17.4 illustrates the configuration of the sensing fiber. A 1.0 m section of the 3.0 m POF was heated to 55°C (with room temperature at 25°C). The modulation amplitude of the laser was set at 0.4 GHz, and the modulation frequency was swept within the range of 13.515–13.590 MHz. This corresponds to a measurement range of 8.2 m and a theoretical

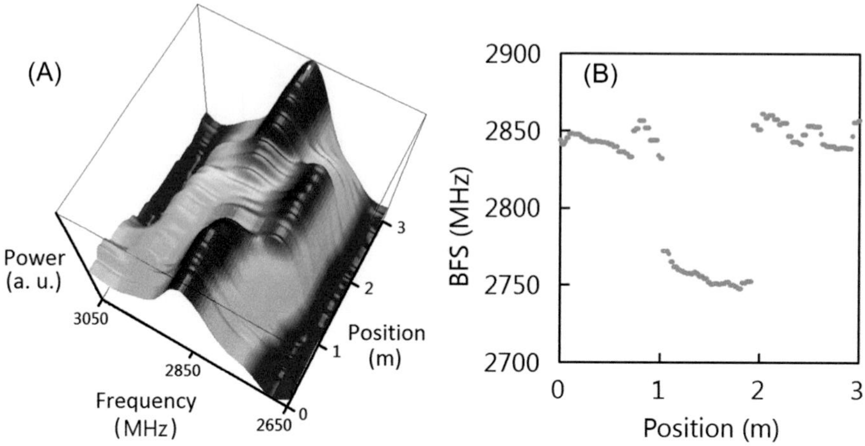

FIGURE 17.5 Results of distributed temperature sensing.

Distributions of (A) the BGS and (B) the BFS. *BGS*, Brillouin gain spectrum; *BFS*, Brillouin frequency shift.

From Mizuno, Y., Lee, H., Hayashi, N., & Nakamura, K. (2019). Noise suppression technique for distributed Brillouin sensing with polymer optical fibers. Optics Letters, 44*(8), 2097–2100. Available from https://doi.org/10.1364/OL.44.002097.*

spatial resolution of 0.65 m. The sampling rate at a single point was 42 Hz, with 84 measurement points and a single distribution measurement time of 2 s. Fig. 17.5A and B present the measurement results of the BGS and BFS distributions, respectively. The BFS in the heated section decreased by an amount corresponding to a temperature change of 30°C, thereby validating the distributed temperature sensing.

We also demonstrated the detection of dynamic strain applied to a 1.0 m section of the 3.0 m POF. The measurement range, spatial resolution, and sampling rate were consistent with the above. After preapplying a static tensile strain of 0.3%, a sinusoidal dynamic strain of $\pm 0.3\%$ was applied at 2 Hz. Fig. 17.6A and B display the time variation of the measured BGS and BFS, respectively. The BFS fluctuated by an amount corresponding to a strain of $\pm 0.3\%$ at 2 Hz, thereby validating the detection of dynamic strain.

In this section, we have introduced the sensing characteristics of Brillouin scattering in a PFGI-POF and provided examples of real-time distributed sensing of strain and temperature using a POF with the high-speed BOCDR method. This is geared toward the realization of distributed strain and temperature sensors with a "memory" function. At the 1550 nm band, the strain and temperature dependence coefficients of the BFS (about 2.8 GHz) in a POF are opposite in sign compared to a silica fiber, and their absolute values are 0.2 times and 3.5 times, respectively. This suggests that Brillouin scattering in a POF, which is less dependent on strain, can be harnessed for high-sensitivity temperature sensing. We also discussed the BFS hopping phenomenon that occurs when a large strain exceeding a few percent is applied to a POF. Last, we applied the high-speed BOCDR method to a POF and demonstrated the measurement of temperature distribution and the detection of dynamic strain. Future challenges include the implementation of memory functions and the improvement of measurement accuracy.

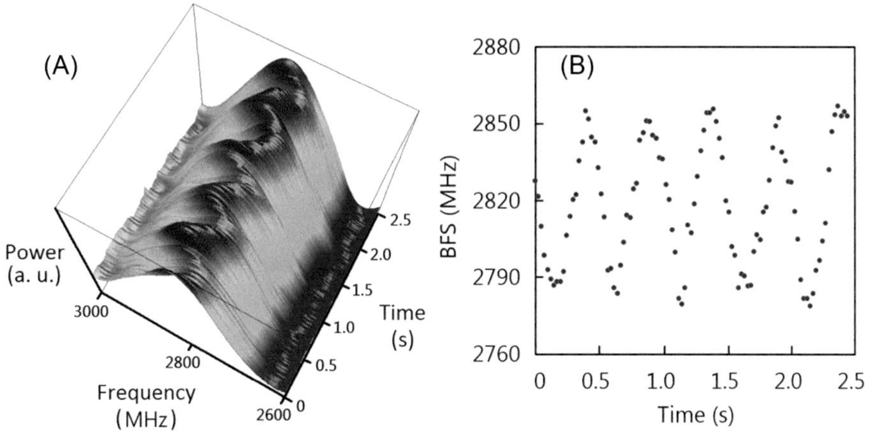

FIGURE 17.6 Results of dynamic strain detection.

Temporal variations of (A) the BGS and (B) the BFS. *BGS*, Brillouin gain spectrum; *BFS*, Brillouin frequency shift.

From Mizuno, Y., Lee, H., Hayashi, N., & Nakamura, K. (2019). Noise suppression technique for distributed Brillouin sensing with polymer optical fibers. Optics Letters, 44*(8), 2097–2100. Available from https://doi.org/10.1364/OL.44.002097.*

17.3 Fiber-Bragg-grating-based techniques

FBGs are localized reflectors embedded within a segment of an optical fiber, designed to reflect specific light wavelengths. Their wide-ranging applications in the realms of communication and sensing have established them as one of the most essential optical devices (Farahi et al., 1990; Kashyap, 1999; Othonos & Kalli, 1996; Wang, Liu, et al., 2017; Wang, Zhao, et al., 2017; Woyessa et al., 2016). When incorporated into single-mode fibers (SMFs), FBGs reflect a singular light wavelength, referred to as the Bragg wavelength, denoted by $\lambda_B = 2\, n_{eff}\Lambda$. Here, n_{eff} represents the effective refractive index of the fiber core, and Λ signifies the grating pitch. Reflections of higher-order diffraction also occur at wavelengths that are submultiples of the original. Conversely, when FBGs are inscribed in multimode fibers (MMFs), they exhibit multiple reflection peaks in the vicinity of the fundamental Bragg peak. These multiple peaks are a result of the coupling between different linearly polarized modes and can be exploited for the discriminative sensing of various physical parameters.

When inscribing FBGs into silica glass fibers, techniques such as the phase mask method and the two-beam interference method have been employed. These methods leverage the property that the refractive index of the core of silica optical fibers doped with germanium changes upon exposure to ultraviolet light. However, it was challenging to apply the same techniques to inscribe FBGs into standard POFs due to their lack of photosensitivity (Carroll et al., 2007; Johnson et al., 2010; Webb, 2015; Xiong et al., 1999). One solution to this problem was the development of a technique for inscribing FBGs into POFs using femtosecond laser irradiation. This method has made FBG inscription into POFs relatively easy, leading to significant advancements in POF-FBG research (Ishikawa et al., 2017; Ishikawa et al., 2018; Koerdt et al., 2016;

Lacraz et al., 2015, 2016; Leal-Junior et al., 2019; Leal-Junior, Theodosiou, et al., 2018; Min et al., 2018; Mizuno et al., 2019a; 2019b; Mizuno, Ishikawa, et al., 2019; Singal, 2017; Stajanca et al., 2016; Theodosiou, Komodromos, et al., 2017; Theodosiou, Lacraz, et al., 2016, 2017; Theodosiou, Polis, et al., 2016; Zheng et al., 2018).

As detailed in the previous section, in addition to widely used PMMA-POFs, there are PFGI-POFs that can be used in the communication wavelength band. In this section, we would like to introduce two techniques related to temperature sensing using FBGs inscribed in PFGI-POFs. One technique involves discriminative sensing of temperature and strain using multiple Bragg wavelengths in PFGI-POFs, which are multimode fibers. The other technique takes advantage of the flexibility of POFs and involves controlling the temperature sensitivity of FBGs by twisting the POF.

17.3.1 Potential of discriminative sensing

The Bragg wavelength's temperature dependence in a PFGI-POF-FBG was thoroughly scrutinized at 1550 nm (Mizuno, Ishikawa, et al., 2019). Utilizing a femtosecond laser irradiation method, an FBG was etched into a 1.2 m section of PFGI-POF. This FBG covered a span of approximately 2 mm. The PFGI-POF (GigaPOF − 50SR, Chromis Fiberoptics) consisted of three layers: a core (diameter: 50 μm; refractive index: ~1.35), a cladding (diameter: 70 μm, refractive index: ~1.34), and an overcladding (diameter: 490 μm). The core and cladding layers were made up of doped and undoped amorphous fluoropolymer (CYTOP), respectively. The optical propagation loss stood at ~0.25 dB/m at 1550 nm, with a numerical aperture of ~0.19. The FBG was directly etched, without the need to remove the overcladding layer, using a femtosecond laser system (High Q femtoREGEN, High Q Laser) operating at 517 nm. The pulse duration was 220 fs, the repetition rate was 1 kHz, and the pulse energy was ~100 nJ. The POF was affixed to an air-bearing translation system capable of high-precision, high-accuracy two-axis motion. A long-working-distance objective (×50) was mounted on a third axis, and the laser beam was focused and directed into the fiber. Precise synchronization of the laser pulse repetition rate and the stage motion facilitated a plane-by-plane grating inscription with a specified length and an index-modulation value.

Fig. 17.7 depicts the experimental setup used to measure the temperature dependence of the FBG-reflected spectrum. All optical paths, except the POF, were silica SMFs. The output from an ASE source was channeled into the POF, and the FBG-reflected light was routed via an optical circulator to an optical spectrum analyzer. The POF was housed in a thermostatic chamber. Due to the chamber's insufficient temperature accuracy, a thermocouple was positioned near the POF-FBG to calibrate the temperature. One end of the POF was connected to a silica SMF outside the chamber using a butt-coupling technique, and the other end was cut at an 8° angle to minimize Fresnel reflection. It is worth noting that the subsequent experimental results were highly reproducible, provided the fibers' alignment in the setup was preserved.

First, we examined the optical spectrum of the light reflected by the FBG, as depicted in Fig. 17.8A. Despite the overlap with the spectrum of light Fresnel-reflected primarily at the SMF-to-POF boundary, a clear observation was made at approximately 1560 nm. Fig. 17.8B provides a magnified view of the FBG-reflected spectrum around its peak points. The spectrum exhibited multiple peaks and dips, a consequence of the POF's multimode nature.

Following this, we measured the spectra of the FBG-reflected light while varying the temperature from 27.5°C to 60.1°C, as illustrated in Fig. 17.9A. At 27.5°C, three peaks were discernible at

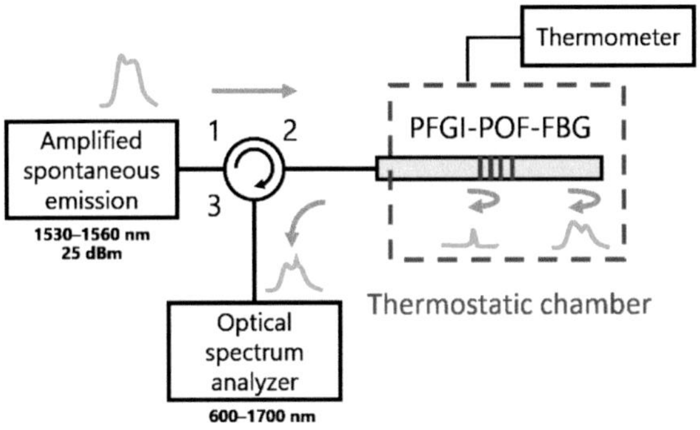

FIGURE 17.7 Experimental setup.

The temperature dependence of the Bragg wavelength of the PFGI-POF-FBG was measured. *FBG*, Fiber Bragg grating; *PFGI-POF*, perfluorinated graded-index polymer optical fiber.

From Mizuno, Y., Ishikawa, R., Lee, H., Theodosiou, A., Kalli, K., & Nakamura, K. (2019). Potential of discriminative sensing of strain and temperature using perfluorinated polymer FBG. IEEE Sensors Journal, 19*(12), 4458–4462. Available from https://doi.org/10.1109/jsen.2019.2900464.*

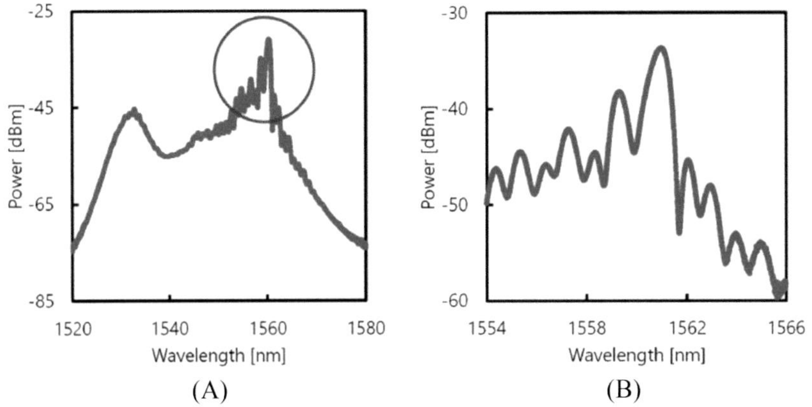

FIGURE 17.8 Measured spectrum of the FBG-reflected light.

(A) Wide rage and (B) magnified view of the red-circled part in (A), around the FBG-induced peaks. *FBG*, Fiber Bragg grating.

From Mizuno, Y., Ishikawa, R., Lee, H., Theodosiou, A., Kalli, K., & Nakamura, K. (2019). Potential of discriminative sensing of strain and temperature using perfluorinated polymer FBG. IEEE Sensors Journal, 19*(12), 4458–4462. Available from https://doi.org/10.1109/jsen.2019.2900464.*

1557.7, 1558.7, and 1560.4 nm within this range. As the temperature rose, all these peaks shifted toward a longer wavelength, albeit with differing dependence coefficients. Tracing specific peaks in this broad temperature range was challenging due to the overlapping of some peaks. Fig. 17.9B

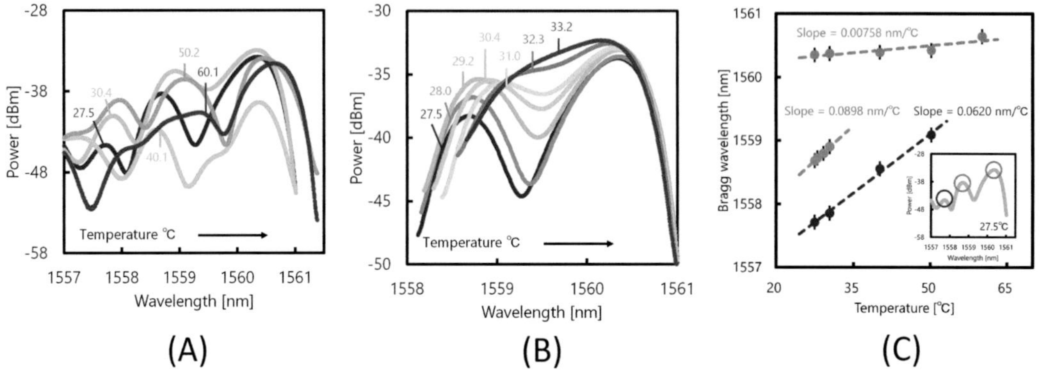

FIGURE 17.9 Temperature dependence of the POF-FBG-reflected spectrum.

(A) Spectral change when temperature increased from 27.5°C to 60.1°C. (B) Spectral change when temperature increases from 27.5°C to 33.2°C. (C) Central wavelengths of the three spectral peaks are plotted as functions of temperature. The error bars were calculated using the standard deviations of 20 measurements. The dashed lines are linear fits. *FBG*, Fiber Bragg gratings; *POF*, polymer optical fiber.

From Mizuno, Y., Ishikawa, R., Lee, H., Theodosiou, A., Kalli, K., & Nakamura, K. (2019). Potential of discriminative sensing of strain and temperature using perfluorinated polymer FBG. IEEE Sensors Journal, 19*(12), 4458–4462. Available from https://doi.org/10.1109/jsen.2019.2900464.*

shows the spectral dependence on temperature in the narrow range from 27.5°C to 33.2°C. The temperature dependence of the peak at 1558.7 nm was almost linear until it became indistinguishable at approximately 33°C due to overlap with the peak at 1560.4 nm.

Fig. 17.9C presents the temperature dependence of the central wavelengths of the three peaks. The temperature-dependence coefficients of the peaks at 1557.7 nm and 1558.7 nm were 0.062 nm/°C and 0.090 nm/°C (with coefficients of determination R^2: 0.995 and 0.993), respectively. The latter is approximately eight times that of a silica SMF-FBG (0.011 nm/°C) and nearly identical to that of a PMMA-POF-FBG (0.088 nm/°C). Conversely, the peak at 1560.4 nm remained almost constant as the temperature rose from 27.5°C to around 50°C. When the temperature increased to approximately 60°C, the peak exhibited a clear upward shift, and the linear fit in the wide temperature range yielded a dependence coefficient of 0.0076 nm/°C ($R^2 = 0.777$), which is significantly smaller than those of the other peaks.

Last, as depicted in Fig. 17.10A, we measured the strain dependence of the FBG-reflected spectrum, which included two peaks: one temperature-dependent (at 1558.7 nm) and the other largely temperature-independent (at 1560.4 nm). Strains ranging from 0% to 0.5% were applied to a 10 cm long section encompassing the FBG. As the strain increased, both peaks shifted toward a longer wavelength. Fig. 17.10B shows the strain dependence of the central wavelengths of the two peaks. Another peak occasionally appeared near the two peaks, but we traced the initial two peaks through continuous observation. The dependence was almost linear for both peaks, and the coefficients were nearly identical (11.2 nm/% at 1558.7 nm and 11.3 nm/% at 1560.4 nm; these values are smaller than previously reported values, but this is not unusual considering that the previous report estimated the Bragg wavelength using unique demodulation methods; their R^2 values were both

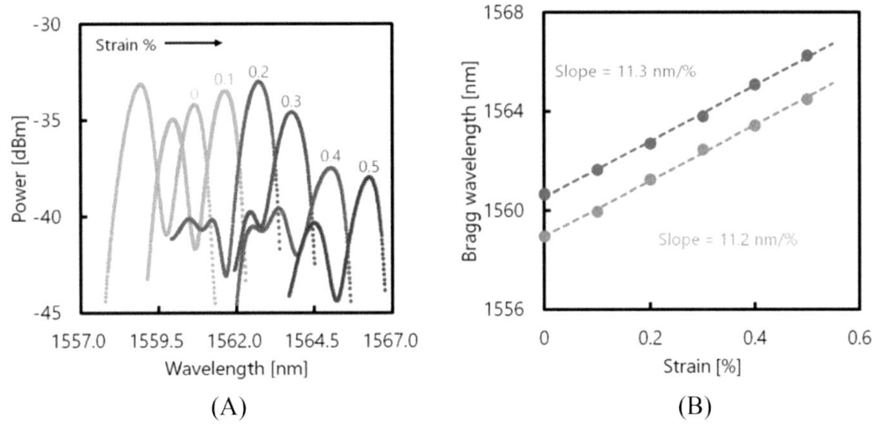

FIGURE 17.10 Measured strain characteristics.

(A) Strain dependence of the POF-FBG-reflected spectrum. (B) Central wavelengths of the two peaks are plotted as functions of strain. The dashed lines are linear fits. *FBG*, Fiber Bragg gratings; *POF*, polymer optical fiber.

From Mizuno, Y., Ishikawa, R., Lee, H., Theodosiou, A., Kalli, K., & Nakamura, K. (2019). Potential of discriminative sensing of strain and temperature using perfluorinated polymer FBG. IEEE Sensors Journal, 19*(12), 4458–4462. Available from https://doi.org/10.1109/jsen.2019.2900464.*

approximately 0.998). These results suggest that highly accurate discrimination of strain and temperature could potentially be achieved by simultaneously employing the multiple peaks (i.e., the temperature-dependent and strain-dependent peak and the temperature-independent but strain-dependent peak) of the POF-FBG-reflected spectrum. It is worth noting that the spectral peak at 1557.7 nm had a strain sensitivity of 12.2 nm/%, which is slightly higher than those of the other two peaks, but it suffered from a relatively low R^2 value of 0.971. This indicates a trade-off relationship between the temperature range and the measurement precision. One advantage of this method is its straightforward calculation procedure (i.e., strain information can be directly obtained), in which we do not need to use matrix-based discrimination of strain and temperature. The achievable temperature range is limited by the overlap of two spectral peaks, but it can be extended to up to approximately 50°C if a reduced measurement precision is permissible.

17.3.2 Sensitivity control through twisting

The ability to modulate the strain and temperature sensitivities of a POF-FBG without incurring significant costs offers advantages for both high-precision measurements and the discriminative sensing of strain and temperature. A cost-effective approach to achieving this control involves twisting the POF-FBG. Given the inherent flexibility of POFs compared to their more brittle silica counterparts, greater degrees of twist can be applied, thereby enhancing the tunability of strain and temperature coefficients. In this subsection, we explore the impact of twisting on the spectral characteristics of PFGI-POF-FBG and its sensitivities to strain and temperature (Mizuno et al., 2019b).

We employed the same PFGI-POF-FBG sample and experimental apparatus as described in Section 5.1. An ASE source served as the broadband light source, and an optical spectrum analyzer with a 2.0 nm resolution was used to capture the FBG-reflected spectrum. Lorentzian demodulation techniques were utilized for stable Bragg wavelength measurements (Mizuno et al., 2019a). As illustrated in Fig. 17.11, a 100-mm section containing the FBG at its midpoint was secured, and twists were applied at one end. Strain evaluations were conducted by pulling the opposite end, while temperature dependencies were assessed by placing the section on a heating element. The ambient temperature was maintained at 19°C.

Initially, we focused on understanding how the application of twists affects the FBG-reflected spectrum. Fig. 17.12A presents the spectra with and without a twist of 150 turns/m, normalized to a peak power of 1. Notably, the Bragg wavelength remained virtually unchanged despite the significant twist. When the twist was removed, the spectrum reverted to its original state with high repeatability, indicating the POF's elastic behavior at a twist level of 150 turns/m. Fig. 17.12B plots the Bragg wavelength as a function of applied twist, revealing negligible changes. This is consistent with the expectation that the grating pitch remains largely unaffected by twisting when the length of the 100-mm section is held constant. Fig. 17.12C and D further demonstrate that the spectral power begins to diminish at approximately 160 turns/m and is virtually obliterated by noise at 165 turns/m. This suggests that mechanical distortions and significant optical losses occur beyond this twist threshold, marking the transition from elastic to plastic behavior in the PFGI-POF.

Subsequent experiments were designed to measure twist-dependent strain sensitivity. Fig. 17.13A and B present the strain-induced spectral shifts and their linear relationship with a coefficient of 13.6 nm/%. When twists were introduced, as shown in Fig. 17.13C and D, the strain coefficient was reduced to 9.9 nm/%. Additional data points were collected at various twist levels and plotted in Fig. 17.14, revealing a decrease in strain sensitivity as the twist increased. The calculated linear fitting yielded a dependence coefficient of −34.2 pm/%/turns/m, likely attributable to the twist-induced reduction in the core's refractive index strain dependence.

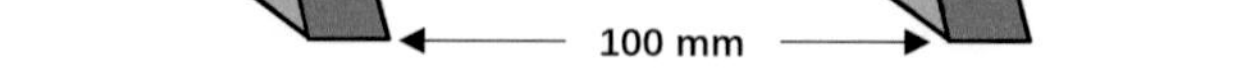

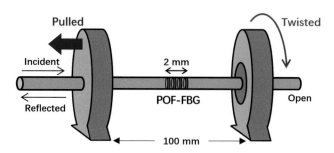

FIGURE 17.11 Setup for applying twists to a POF-FBG.

A 100-mm section containing the FBG at its midpoint was secured. *FBG*, Fiber Bragg grating; *POF*, polymer optical fiber.

From Mizuno, Y., Ma, T., Ishikawa, R., Lee, H., Theodosiou, A., Kalli, K., & Nakamura, K. (2019). Twist dependencies of strain and temperature sensitivities of perfluorinated graded-index polymer optical fiber Bragg gratings. Applied Physics Express, 12*(8), 082007. Available from https://doi.org/10.7567/1882-0786/ab3013.*

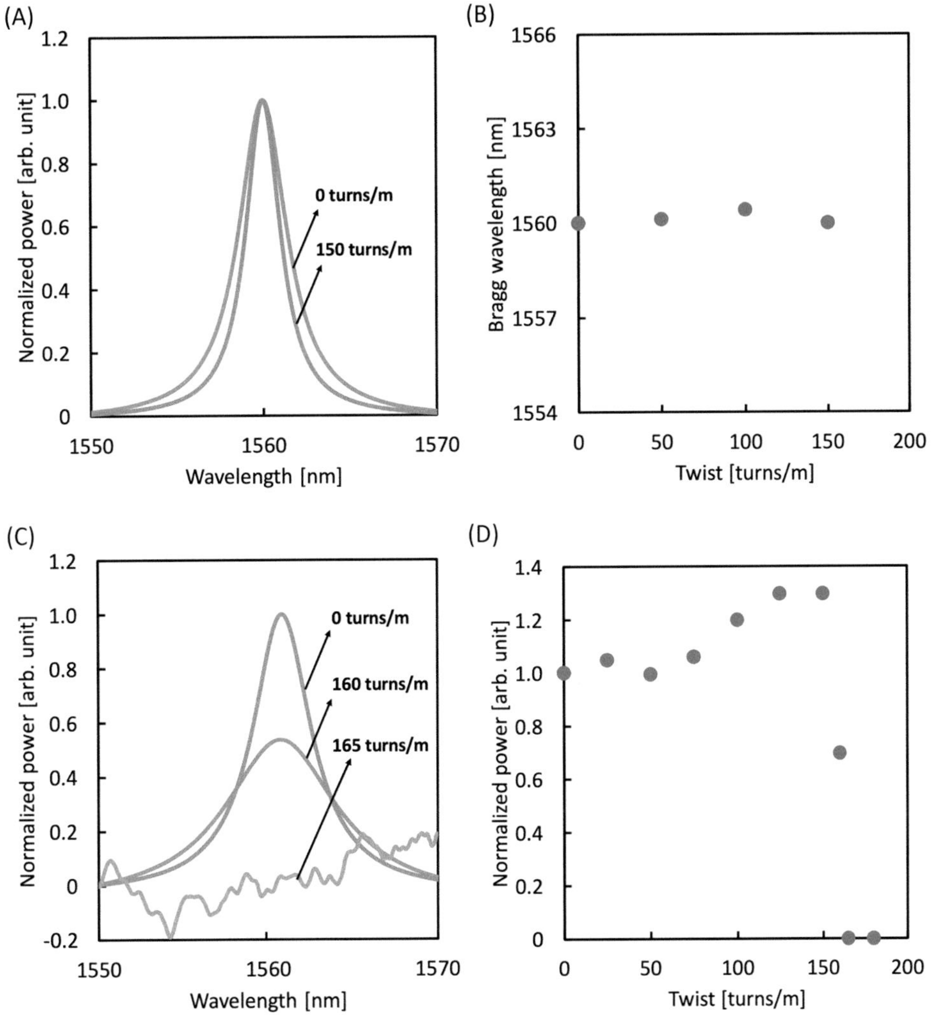

FIGURE 17.12 Measured dependences of the FBG-reflected spectra on twist.

(A) FBG-reflected spectra measured when no twist was applied and when a twist of 150 turns/m was applied. The vertical axis was normalized so that the peak powers became 1. (B) Bragg wavelength dependence on twist. (C) FBG-reflected spectra were measured when no twist was applied and when twists of 160 and 165 turns/m were applied. The vertical axis was normalized using the spectrum with no twist. Lorentzian fitting was not performed for the data at 165 turns/m. (D) Normalized peak power dependence on the twist. *FBG*, Fiber Bragg grating.

Last, we turned our attention to examining the twist-dependent temperature sensitivity. Fig. 17.15A and B depict the temperature-induced spectral shifts and their quasilinear relationship with a coefficient of −30.0 pm/°C. Upon introducing twists, as shown in Fig. 17.15C and D, the

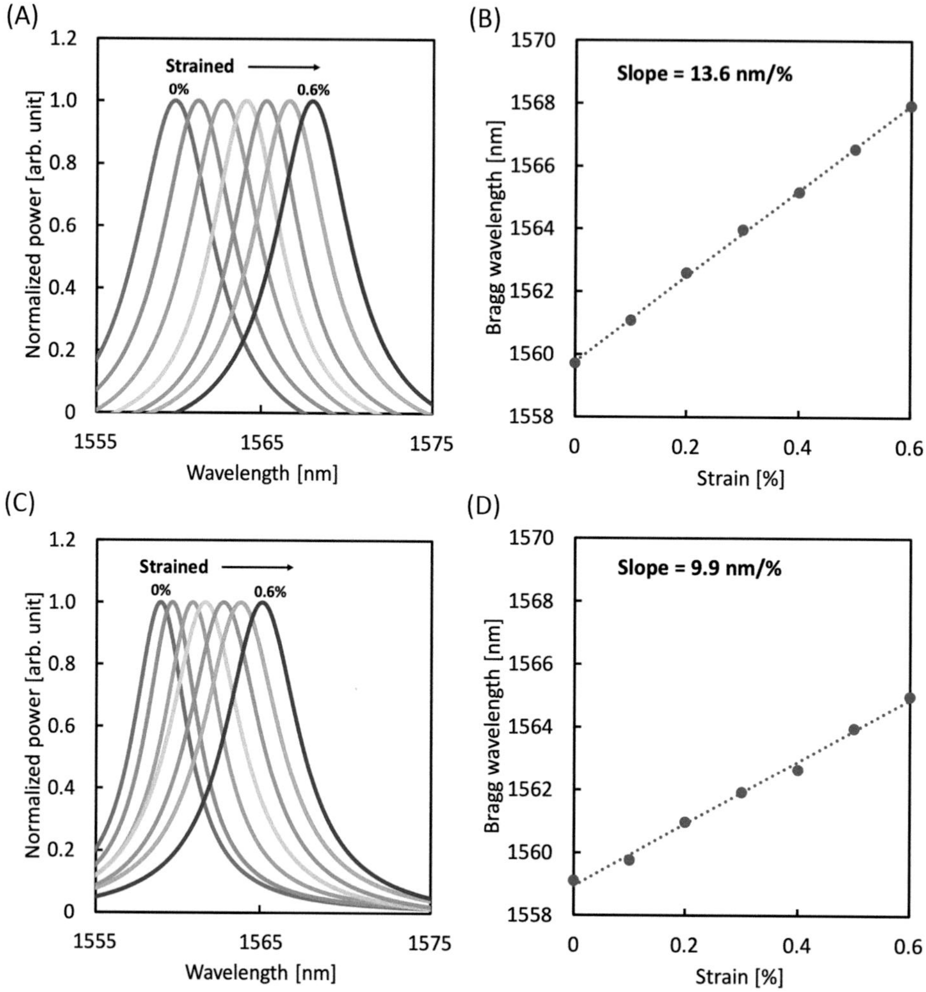

FIGURE 17.13 Strain dependences of the FBG-reflected spectra at different twists.

(A) Strain dependence of the FBG-reflected spectrum when no twist was applied. Step: 0.1%. (B) Bragg wavelength dependence on strain when no twist was applied. (C) Strain dependence of the FBG-reflected spectrum when a twist of 100 turns/m was applied. Step: 0.1%. (D) Bragg wavelength dependence on strain when a twist of 100 turns/m was applied. The dotted lines in (B) and (D) are linear fits. *FBG*, Fiber Bragg gratings.

temperature coefficient escalated to approximately −91.1 pm/°C. Fig. 17.16 plots this temperature sensitivity against applied twist, revealing an increase in sensitivity with increasing twist. Although the dependence exhibited a nonlinear increase, it was linearly fitted to roughly evaluate the dependence coefficient, which was found to be approximately 0.39 pm/°C/turns/m. By applying a twist of 150 turns/m, the temperature sensitivity can be enhanced to approximately three times that of an

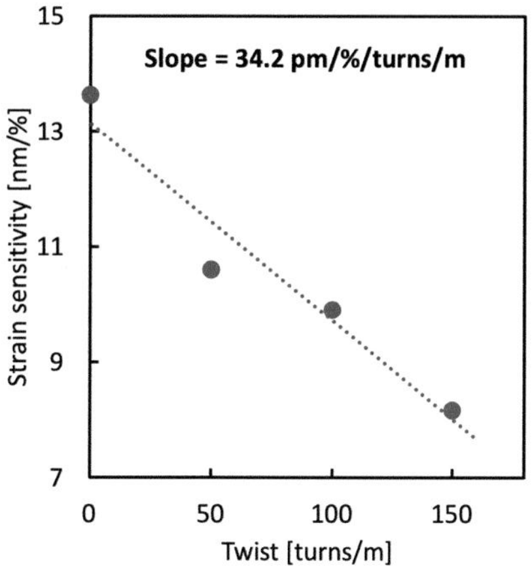

FIGURE 17.14 Strain sensitivity plotted as a function of twist.

The dotted line is a linear fit.

From Mizuno, Y., Ma, T., Ishikawa, R., Lee, H., Theodosiou, A., Kalli, K., & Nakamura, K. (2019). Twist dependencies of strain and temperature sensitivities of perfluorinated graded-index polymer optical fiber Bragg gratings. Applied Physics Express, 12*(8), 082007. Available from https://doi.org/10.7567/1882-0786/ab3013.*

untwisted POF-FBG. This behavior is likely explained by the assumption that the absolute value of the negative coefficient of the refractive index dependence on temperature is augmented by the applied twist.

Thus the comprehensive experimental data affirm that the strain and temperature sensitivities of PFGI-POF-FBGs can be effectively modulated through the application of twists. This method offers a cost-efficient avenue for altering FBG characteristics and is particularly well-suited for flexible POFs, which can accommodate significant twists. These results indicate that by twisting the PFGI-POF-FBG, one can achieve highly sensitive temperature sensing while simultaneously reducing strain sensitivity. Furthermore, the potential for discriminative sensing of strain and temperature using both twisted and untwisted PFGI-POF-FBGs becomes feasible. This can be particularly advantageous when employing a matrix-based method for multiparameter sensing.

17.4 Multimode interference-based techniques

One of the most cost-effective and straightforward approaches to temperature sensing using POFs involves the utilization of multimode interference-based sensors. These sensors are typically constructed using a single-mode-multimode-single-mode (SMS) structure, where an MMF is flanked by two SMFs (Frazão et al., 2011; Kumar et al., 2003; Liu & Wei, 2007; Mehta et al., 2003; Tripathi et al., 2009). Given that the majority of commercially available POFs are MMFs, they are ideally suited to serve as the central fiber in SMS-based sensor configurations.

To extend the upper limit of measurable strain, a large-strain sensor employing a 0.16 m-long PMMA-based step-index POF as the MMF was developed (Huang et al., 2012). This sensor

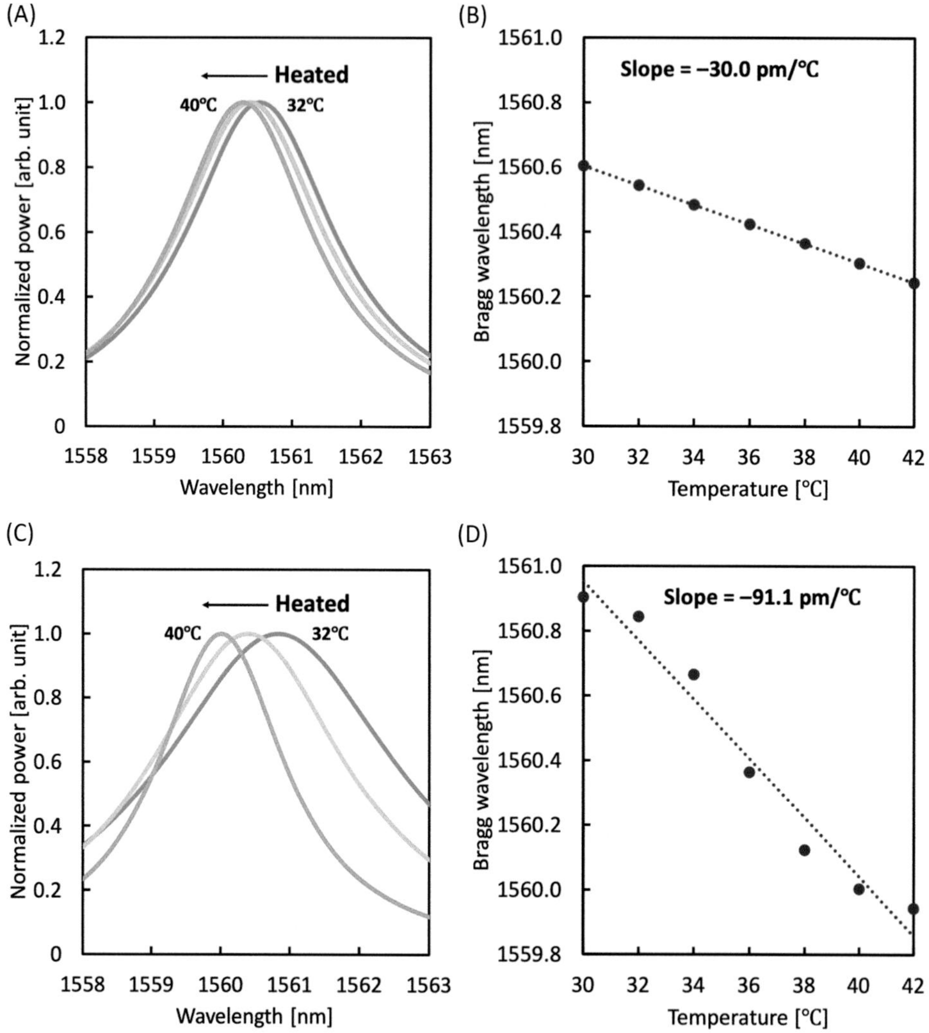

FIGURE 17.15 Temperature dependences of the FBG-reflected spectra at different twists.

(A) Temperature dependence of the FBG-reflected spectrum when no twist was applied. Step: 4°C. (B) Bragg wavelength dependence on temperature when no twist was applied. (C) Temperature dependence of the FBG-reflected spectrum when a twist of 150 turns/m was applied. Step: 4°C. (D) Bragg wavelength dependence on temperature when a twist of 150 turns/m was applied. The dotted lines in (B) and (D) are linear fits. *FBG*, Fiber Bragg gratings.

demonstrated strain and temperature sensitivities of −1.72 pm/με and +56.8 pm/°C, respectively, at a wavelength of 1570 nm. However, the high propagation loss of PMMA-based POFs at telecommunication wavelengths (> 100 dB/m) inherently restricts the maximum sensing range to just a

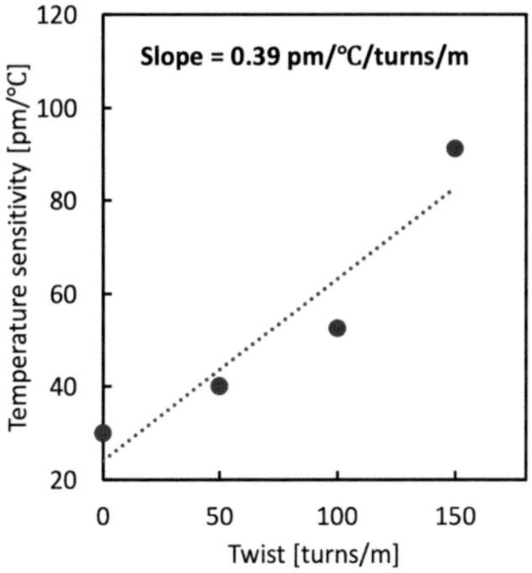

FIGURE 17.16 Temperature sensitivity (i.e., absolute value of the temperature coefficient of the Bragg wavelength) plotted as a function of twist.

The dotted line is a linear fit.

From Mizuno, Y., Ma, T., Ishikawa, R., Lee, H., Theodosiou, A., Kalli, K., & Nakamura, K. (2019). Twist dependencies of strain and temperature sensitivities of perfluorinated graded-index polymer optical fiber Bragg gratings. Applied Physics Express, *12(8), 082007. Available from https://doi.org/10.7567/1882-0786/ab3013.*

few tens of meters. Additionally, the capability for discriminative measurement of strain and temperature is crucial for practical applications. While this could theoretically be achieved through the combined use of multimode interference and another physical phenomenon like Brillouin scattering, such scattering has not yet been observed in PMMA-based POFs. In contrast, PFGI-POFs are unique in that they exhibit a relatively low loss of approximately 0.25 dB/m at 1550 nm (or ~0.05 dB/m at 1300 nm), and Brillouin scattering has been experimentally verified in these fibers, as outlined in Section 17.4.

In this section, we initially focus on the implementation and performance evaluation of strain and temperature sensors based on an SMS structure incorporating a PFGI-POF, operating at a wavelength of 1300 nm. We employ three 1 m-long PFGI-POFs with core diameters of 50, 62.5, and 120 μm for this study. Notably, when the core diameter is 62.5 μm, the strain sensitivity reaches −112 pm/με. This value is approximately 13 times greater than that observed in a silica GI-MMF, based on a rudimentary calculation. Subsequently, we explore the performance of temperature sensors using 1 m-long PFGI-POFs across a broad temperature range, extending from room temperature to temperatures approaching the glass transition point. The temperature sensitivity increases with rising temperature, peaking at +202 nm/°C at approximately 72°C. This behavior is unique to POFs with relatively low glass-transition temperatures. Furthermore, utilizing the Fresnel reflection at the distal open end of the PFGI-POF, we introduce a single-end-access sensor configuration for both strain and temperature measurements. The experimental results confirm that the sensitivities remain largely consistent regardless of the configuration, whether single- or two-end access, provided the length of the PFGI-POF remains constant. This single-end-access configuration offers greater flexibility for sensor integration into various structures.

17.4.1 Fundamental characterization

The SMS structure is composed of an MMF that is flanked by identical SMFs at both ends. Light is introduced into the MMF through the leading SMF at the first SMF/MMF interface (Frazão et al., 2011; Kumar et al., 2003; Liu & Wei, 2007; Mehta et al., 2003; Tripathi et al., 2009). Due to the difference in spot sizes between the fundamental modes of the MMF and the SMF, multiple modes within the MMF are excited and propagate with their individual propagation constants. At the second SMF/MMF interface, the resultant field coupled to the outgoing SMF is governed by the relative phase differences among these MMF modes. When axial alignment is maintained at the first SMF/MMF boundary, only the axially symmetric modes within the MMF are excited. Detailed calculations reveal that the optical spectrum in the outgoing SMF is influenced by physical variables, such as strain and temperature, which affect both the propagation constants and the length of the MMF. These changes can be quantitatively assessed by monitoring shifts in spectral power or the location of spectral peaks or dips.

The experimental setup is illustrated in Fig. 17.17 (Numata et al., 2014). We utilized three 1 m-long PFGI-POFs with varying core diameters of 50, 62.5, and 120 μm. These PFGI-POFs were butt-coupled to silica SMFs. A swept-source laser (SSL) served as the light source, operating as a broadband source with a central wavelength of 1320 nm and a bandwidth of 110 nm. After polarization adjustment, the laser output was launched into the PFGI-POF, and the transmitted light spectrum was analyzed using an optical spectrum analyzer. Both strain and temperature variations were applied across the entire length of the PFGI-POF, with the ambient room temperature maintained at approximately 27°C.

Fig. 17.18 presents the optical spectra before and after transmission through a PFGI-POF with a 62.5-μm core diameter. A distinct spectral dip was observed around 1330 nm in the transmitted spectrum.

The strain-dependent behavior of this spectral dip is depicted in Fig. 17.19A. As the applied strain increased, the spectral dip shifted toward shorter wavelengths. Fig. 17.19B plots this wavelength shift as a function of applied strain, revealing an almost linear relationship with a coefficient of −1118 nm/% (equivalent to −112 pm/με). This value is approximately 13 and 7.7 times greater than those observed in a silica GI-MMF and a PMMA-based POF, respectively. Similarly, as the

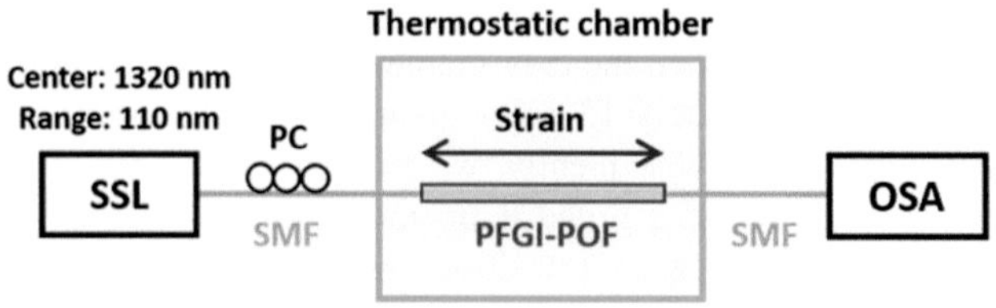

FIGURE 17.17 Schematic of experimental setup.

OSA, Optical spectrum analyzer; *PC*, polarization controller; *PFGI-POF*, perfluorinated graded-index polymer optical fiber; *SMF*, single-mode fiber; *SSL*, swept-source laser.

From Numata, G., Hayashi, N., Tabaru, M., Mizuno, Y., & Nakamura, K. (2014). Ultra-sensitive strain and temperature sensing based on modal interference in perfluorinated polymer optical fibers. IEEE Photonics Journal, 6*(5). Available from https://doi.org/10.1109/JPHOT.2014.2352637.*

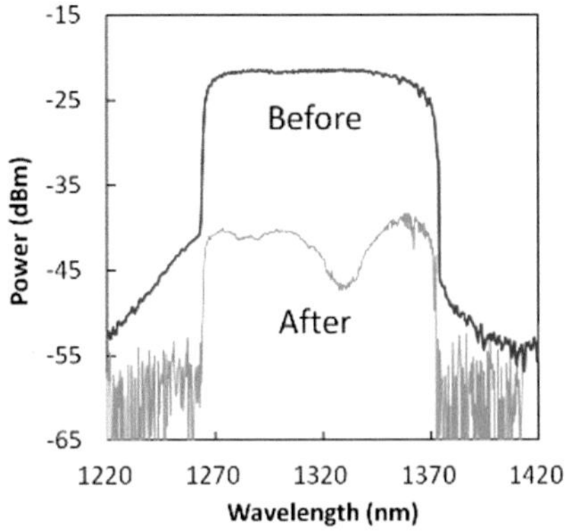

FIGURE 17.18 Measured optical spectra.

Before and after transmission through the POF. *POF*, Polymer optical fiber.

From Numata, G., Hayashi, N., Tabaru, M., Mizuno, Y., & Nakamura, K. (2014). Ultra-sensitive strain and temperature sensing based on modal interference in perfluorinated polymer optical fibers. IEEE Photonics Journal, 6*(5). Available from https://doi.org/10.1109/JPHOT.2014.2352637.*

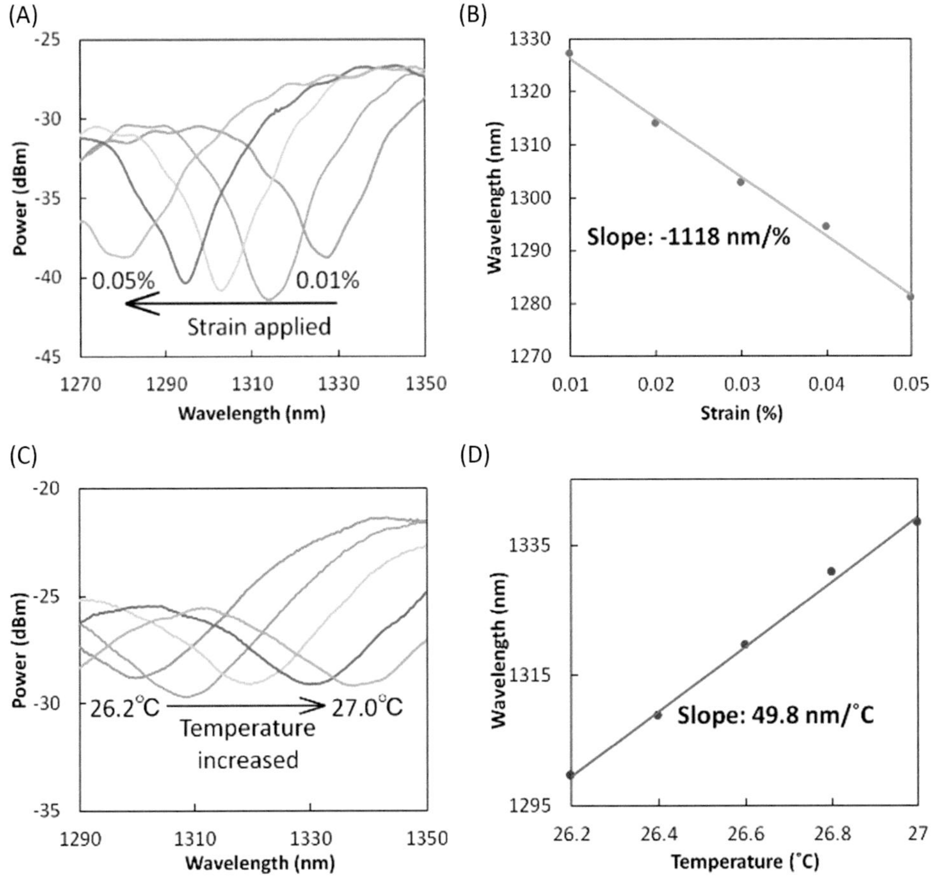

FIGURE 17.19 Measurement results for the POF with 62.5-μm core.

(A) Spectral dependence on strain, (B) dip wavelength versus strain, (C) spectral dependence on temperature, and (D) dip wavelength versus temperature. *POF*, Polymer optical fiber.

temperature rose, the spectral dip shifted toward longer wavelengths, as shown in Fig. 17.19C. Fig. 17.19D illustrates the temperature-dependent behavior of the dip wavelength, yielding a coefficient of +49.8 nm/°C, which is over 1800 and 100 times greater than those in a silica GI-MMF and a PMMA-based POF, respectively.

Subsequent measurements were conducted using PFGI-POFs with 50-μm and 120-μm core diameters. The resulting strain and temperature coefficients were +3.42 pm/με and −4.71 nm/°C for the 50-μm core, and −8.21 pm/με and +0.74 nm/°C for the 120-μm core. Although these values exceeded those in silica GI-MMFs with corresponding core diameters, they were not as pronounced as those in the PFGI-POF with a 62.5-μm core. This discrepancy is partially attributed to the closer proximity of the critical wavelength (Tripathi et al., 2009) for the 62.5-μm core to the measured wavelength region (~1300 nm) compared to the 50-μm and 120-μm cores.

Considering that SMS-based sensors using silica GI-MMFs with the same core diameters do not have large strain and temperature coefficients, the exceptionally high strain and temperature sensitivities observed in PFGI-POFs, particularly those with a 62.5-μm core, are likely not a consequence of the fiber structure but rather stem from the unique properties of its perfluorinated polymer material. In other words, the propagation constant of each mode and the length of perfluorinated polymer exhibit a significantly higher dependency on strain and temperature than those in other MMFs, whether composed of glass or polymer.

17.4.2 Temperature sensitivity enhancement

In this subsection, we extend our investigation to the temperature sensitivity enhancement of the SMS sensors using 1 m-long PFGI-POFs with core diameters of 50, 62.5, and 120 μm (Numata et al., 2015). The experimental setup largely mirrors that depicted in Fig. 17.17, with the entire length of the PFGI-POF placed in a thermostatic chamber to regulate ambient temperature. A notable modification involves the light source; an ultra-wideband source emitting super-continuum light with an output spectrum ranging from 1100 to 1760 nm (pumped at 1550 nm) was employed to facilitate comprehensive spectral measurements. The room temperature for these experiments was set at 20°C.

Fig. 17.20 presents the optical spectra before and after transmission through a PFGI-POF with a 50-μm core diameter at room temperature. The pretransmission spectrum displayed a relatively flat profile across the 1100–1760 nm range, with a peak at 1550 nm corresponding to the pump frequency for super-continuum generation. Posttransmission, the spectrum exhibited not only a 10–20 dB uniform loss but also wavelength-dependent losses, manifesting as distinct spectral dips and peaks. These features are attributed to modal interference and are sensitive to the ambient temperature of the POF.

Our study then focused on the temperature-dependent behavior of these spectral dips across a wide temperature range (Fig. 17.21A). The data were discontinuously plotted for two reasons: (1) the time required for temperature stabilization at each measurement point was approximately 10 minutes, necessitating separate measurements at 10°C intervals, and (2) certain peaks were obscured near the 1550-nm pump peak. As the temperature increased, all dip wavelengths decreased. The dependence coefficient at room temperature at 1300 nm was approximately −5.3 nm/°C, which moderately aligns with previously reported values (Numata et al., 2014).

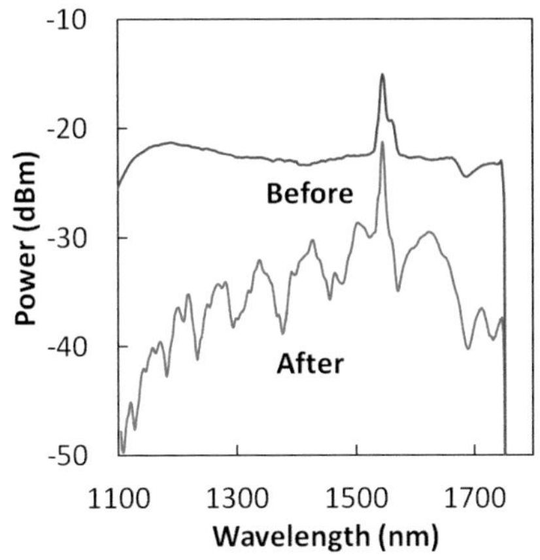

FIGURE 17.20 Optical spectra measured within a wider wavelength range.

Before and after transmission through the POF. *POF*, Polymer optical fiber.

From Numata, G., Hayashi, N., Tabaru, M., Mizuno, Y., & Nakamura, K. (2015). Drastic sensitivity enhancement of temperature sensing based on multimodal interference in polymer optical fibers. Applied Physics Express, *8(7), 072502. Available from https://doi.org/10.7567/apex.8.072502.*

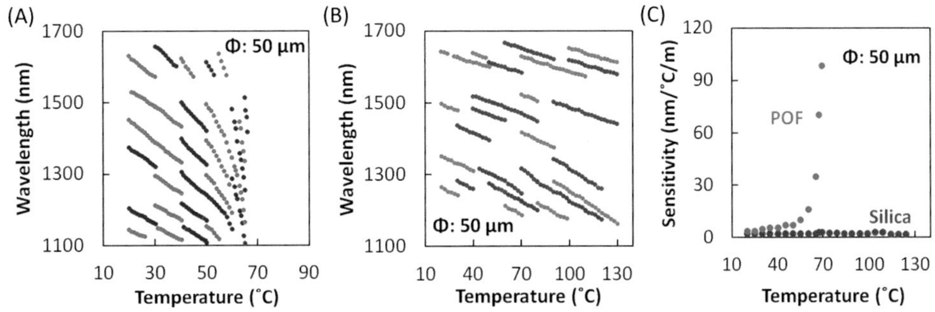

FIGURE 17.21 Measured results for the fibers with 50-μm core diameter.

Dip-wavelength dependence on wide-range temperature in (A) the PFGI-POF and (B) the silica GI-MMF. Different colors were used to clarify different dips. (C) Temperature sensitivity (absolute value of the temperature-dependence coefficients) at 1300 nm in the PFGI-POF and the silica GI-MMF plotted as functions of temperature. *PFGI-POF*, Perfluorinated graded-index polymer optical fiber; *GI-MMF*, graded-index multimode fiber.

Interestingly, the coefficients gradually increased beyond 50°C, peaking at −85 nm/°C at 67°C—approximately 16 times the room temperature value.

To validate that this behavior is unique to POFs, we conducted similar experiments using a silica GI-MMF with a 50-μm core diameter (Fig. 17.21B). The results confirmed that the gradual change in temperature-dependent coefficients is indeed specific to POFs, likely due to the partial phase transition of the polymer material. Fig. 17.21C plots the temperature sensitivities for both PFGI-POF and silica GI-MMF, highlighting the abrupt increase in sensitivity at 67°C for the PFGI-POF in contrast to the near-constant sensitivity in the silica GI-MMF.

We conducted additional experiments using PFGI-POFs with core diameters of 62.5 μm and 120 μm, as illustrated in Fig. 17.22A–D. In both cases, an increase in temperature led to a corresponding increase in the dip wavelengths. This variation, both in absolute value and sign, has been previously reported (Numata et al., 2014) and is attributed to the structural influence on the critical wavelengths (Tripathi et al., 2009), which are shorter than 1100 nm for the 62.5-μm-core PFGI-POF and longer than 1700 nm for the 120-μm-core PFGI-POF. At room temperature and a wavelength of 1300 nm, the coefficients were +7.7 nm/°C for the 62.5-μm-core PFGI-POF and +1.1 nm/°C for the 120-μm-core PFGI-POF. The discrepancies between these values and those previously reported (Numata et al., 2014) are likely due to variations in manually prepared end-surfaces and lot-to-lot nonuniformity, which result in different excited modes. At high temperatures, the maximal values obtained at 1300 nm were +202 nm/°C for the 62.5-μm-core PFGI-POF and +85.6 nm/°C for the 120-μm-core PFGI-POF, representing increases of approximately 26 and

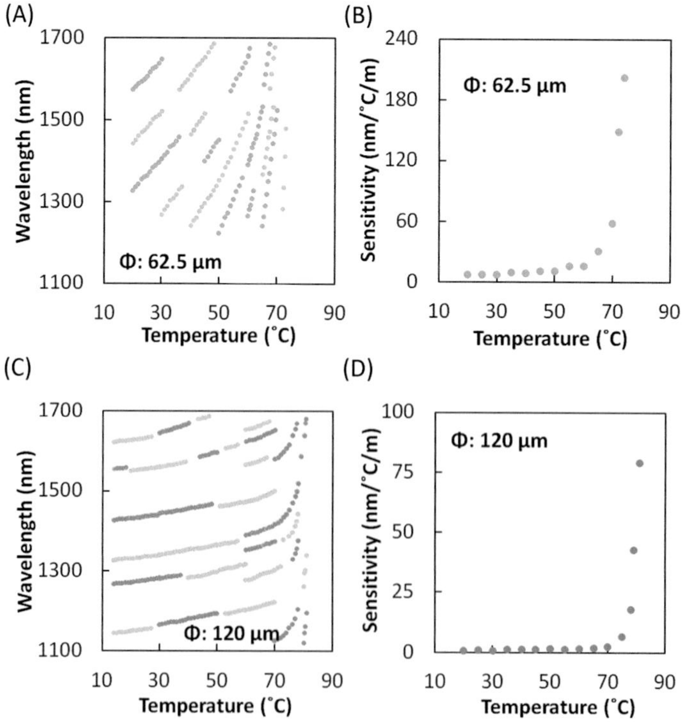

FIGURE 17.22 Measured results for the PFGI-POFs.

Temperature dependences of (A) the dip wavelengths and (B) the coefficient in the PFGI-POF with 62.5-μm core diameter, and (C, D) those in the PFGI-POFs with 120-μm core diameter. *PFGI-POF*, Perfluorinated graded-index polymer optical fiber.

78 times over their room temperature values, respectively. Notably, the temperature sensitivity of +202 nm/°C at 1300 nm, achieved at approximately 72°C using the 62.5-μm-core PFGI-POF, exceeds the largest previously reported value (Liu & Wei, 2007) by a factor of over 7000. It is important to note that while the dependence is nonlinear, the wavelength of a specific dip correlates one-to-one with temperature, offering potential applications in temperature sensing. However, caution is advised in system design due to expected variations in measurement error with temperature. Experimental data revealed that the temperatures at which the highest temperature sensitivities were obtained—specifically, the absolute values of the temperature-dependence coefficients—were approximately 69°C, 72°C, and 82°C for PFGI-POFs with core diameters of 50 μm, 62.5 μm, and 120 μm, respectively. This suggests a positive correlation between the highest operable temperature and the core diameter of the POF.

Thus in this subsection, our findings indicate that PFGI-POFs exhibit unique temperature-dependent behaviors, particularly at elevated temperatures. The temperature sensitivity of PFGI-POFs, especially those with a 62.5-μm core diameter, significantly increases with temperature, reaching up to +202 nm/°C at approximately 72°C. This value surpasses any previously reported measurements and represents a groundbreaking advancement in the field of modal-interference-based temperature sensors. We also demonstrated that this behavior is unique to POFs with relatively low glass-transition temperatures. Besides, considering that the phase-transition temperature of polymers can be controlled by adding plasticizers and by copolymerizing/blending different materials, we expect that such an ultrahigh temperature sensitivity will be achievable not only around 70°C–80°C but also around somewhat arbitrary temperatures. Note that the temperature sensitivity at room temperature has also been shown to be enhanced by annealing (Kawa et al., 2017b).

17.4.3 Single-end-access configuration

Significant advancements have been made in developing SMS-based strain and temperature sensors. However, most of the sensors reported in the literature utilize two-end-access systems, where both ends of the MMF are physically connected to SMFs, as depicted in Fig. 17.23A. From a practical standpoint, single-end-access configurations, illustrated in Fig. 17.23B, offer greater flexibility for embedding sensors into various structures. In this subsection, we introduce a single-end-access configuration for SMS-based strain and temperature sensors, employing PFGI-POFs as MMFs to achieve high sensitivities (Kawa et al., 2017a). Our experimental results demonstrate that the strain and temperature sensitivities remain largely consistent, irrespective of whether a single- or two-end-access configuration is used.

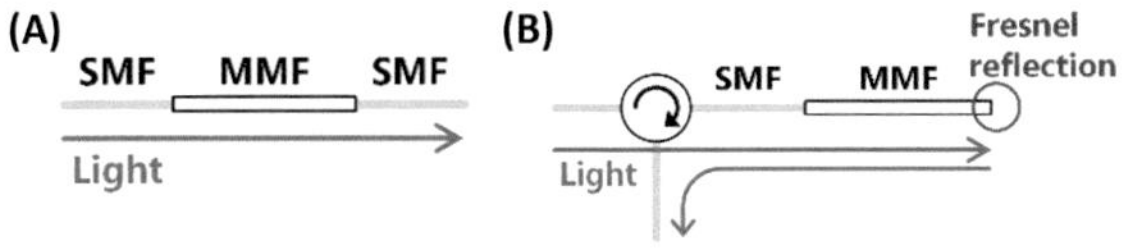

FIGURE 17.23 Two conceptual setups for optical fiber sensors.

(A) Transmissive configuration and (B) reflectometric configuration.

We utilized PFGI-POFs with a core diameter of 62.5 μm and a length of 0.7 m as the MMFs. Fig. 17.24A and B presents the actual experimental setups for transmissive and reflectometric configurations, respectively. In both setups, an SSL with a 20-kHz sweep rate, a central wavelength of 1320 nm, and a bandwidth of 110 nm was employed. The laser output was first directed through a polarization controller to mitigate polarization-dependent spectral fluctuations before being injected

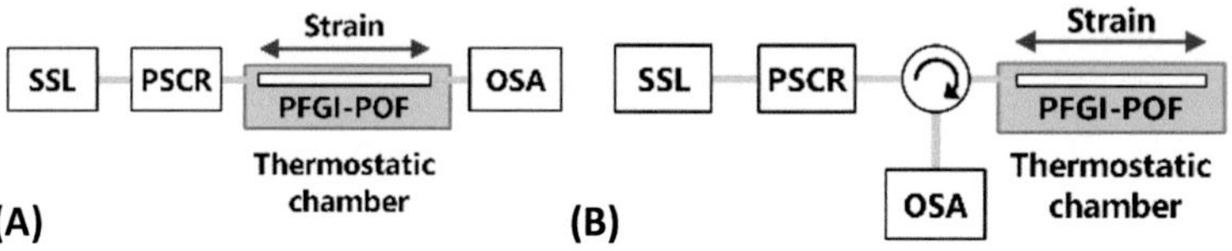

FIGURE 17.24 Actual experimental setups.

(A) Transmissive configuration and (B) reflectometric configuration. *OSA*, Optical spectrum analyzer, *PSCR*, polarization scrambler, *SSL*, swept-source laser. The yellow lines indicate silica single-mode fibers.

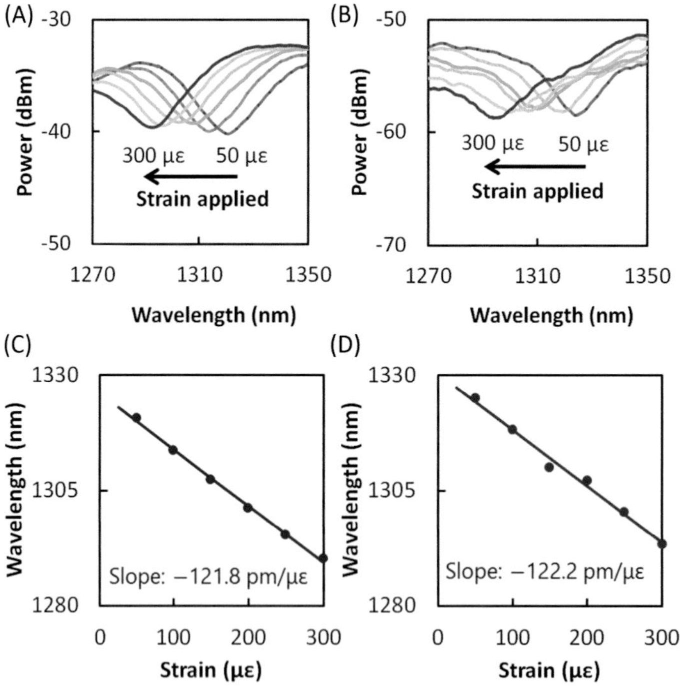

FIGURE 17.25 Measured strain dependences.

Measured spectral dependences on strain in (A) transmissive configuration and (B) reflectometric configuration. The dip wavelength depends on strain in (C) transmissive configuration and (D) reflectometric configuration.

into the POF. In the transmissive configuration, both ends of the POF were butt-coupled to silica SMFs, and the transmitted light spectrum was monitored using an optical spectrum analyzer. Conversely, in the reflectometric configuration, one end of the POF was butt-coupled to a silica SMF, while the other end was left open and polished flat; the reflected light spectrum was monitored similarly. It is worth noting that the optical circulator used in the reflectometric setup had an operating wavelength range of 1280–1340 nm and insertion losses of approximately 0.5 dB for both directions. The room temperature during the experiments was approximately 28°C.

Fig. 17.25A and B displays the measured strain-dependent spectral dips for both configurations. A single, relatively broad dip was observed, which shifted to shorter wavelengths as the applied strain increased. The spectral power in the reflectometric configuration was approximately 20 dB lower than that in the transmissive configuration, a difference attributable to the Fresnel reflection loss at the POF end and additional POF propagation loss. Fig. 17.25C and D plot the dip wavelengths as functions of strain, revealing nearly linear dependencies with slopes of −1218 nm/% (transmissive) and −1222 nm/% (reflectometric). These values are in moderate agreement with previously reported data (Numata et al., 2014), and any discrepancies are likely due to imperfections in the end faces and structural nonuniformities in the POF.

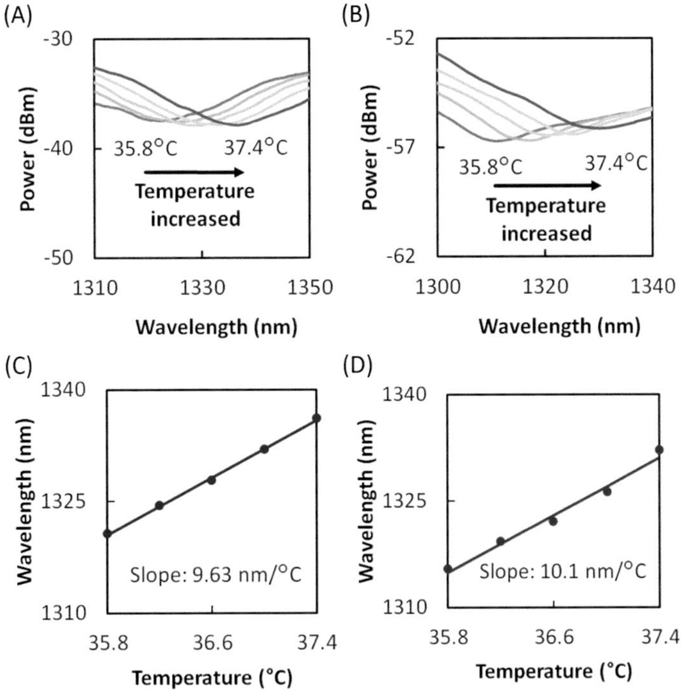

FIGURE 17.26 Measured temperature dependences.

Measured spectral dependences on the temperature in (A) transmissive configuration and (B) reflectometric configuration. The dip wavelength depends on the temperature in (C) transmissive configuration and (D) reflectometric configuration.

Fig. 17.26A and B shows the temperature-dependent spectral dips for both configurations. The dips shifted to shorter wavelengths as the temperature increased. Fig. 17.26C and D plot the dip wavelengths as functions of temperature, indicating temperature-dependence coefficients of 9.63 nm/°C (transmissive) and 10.1 nm/°C (reflectometric). These values are consistent with those previously reported (Numata et al., 2014).

This subsection introduces a single-end-access configuration for SMS-based strain and temperature sensors, utilizing Fresnel reflection at the distal open end of the MMF (PFGI-POF). We achieved high strain and temperature sensitivities of −122.2 pm/με and 10.1 nm/°C, respectively, comparable to those obtained in two-end-access configurations. This development holds significant promise for enhancing the ease of deployment and versatility of SMS-based fiber-optic sensors. One consideration for future work is that for POF lengths exceeding several dozen meters, additional propagation loss may result in the reflected spectrum being obscured by noise. This issue could potentially be mitigated by affixing a mirror at the distal end of the POF.

17.5 Conclusions and perspectives

In this chapter, we have presented a comprehensive review of temperature (and strain) sensing methodologies that employ PFGI-POFs. These fibers offer significant advantages for sensing applications, particularly in the measurement of large strains. Unlike silica glass fibers, PFGI-POFs are highly sensitive to a variety of measures, including temperature, strain, humidity, and radiation. We have specifically focused on distributed sensing based on Brillouin scattering, as well as point-type and area-type sensing through the use of FBGs and multimode interference.

Initially, we explored the sensing characteristics of Brillouin scattering in PFGI-POFs, providing examples of real-time distributed strain and temperature sensing using the high-speed BOCDR method. This approach aims to develop distributed sensors with memory functions. We found that the strain and temperature dependence coefficients of the BFS in PFGI-POFs are notably different from those in silica fibers, making them particularly useful for high-sensitivity temperature sensing. We also discussed the BFS hopping phenomenon that occurs under large strain conditions. Future challenges in this area include the implementation of memory functions and the enhancement of measurement accuracy.

We then shifted our focus to point-type temperature sensing, specifically those based on FBGs inscribed in PFGI-POFs. Our experimental results revealed that the temperature coefficient of one of the clearest peaks in the FBG-reflected spectrum was 0.09 nm/°C, which is significantly higher than those observed in silica fibers. We also demonstrated that the strain and temperature sensitivities of PFGI-POF-FBGs can be modulated by twisting the fiber, offering a cost-efficient method for altering FBG characteristics.

Last, we evaluated the performance of strain and temperature sensors based on an SMS structure incorporating PFGI-POFs. Our experiments utilized PFGI-POFs with core diameters of 50, 62.5, and 120 μm. Notably, the strain sensitivity reached −112 pm/με when the core diameter was 62.5 μm, a value approximately 13 times greater than that observed in silica fibers. We also found that the temperature sensitivity increases with rising temperature, peaking at +202 nm/°C at approximately 72°C. We then introduced a single-end-access sensor configuration that offers greater flexibility for sensor integration into various structures.

Though not directly addressed in this chapter, several alternative methods exist for measuring temperature using PFGI-POFs. One notable approach involves the utilization of long-period gratings inscribed within POFs (Theodosiou et al., 2019). In this method, the mode-conversion wavelengths exhibit temperature dependence, offering a viable mechanism for temperature detection. In addition, various interference effects, distinct from the multimode interference discussed herein, have been extensively documented in the literature (Szczerska, 2022). Such techniques provide a complementary avenue for temperature sensing. Furthermore, plasmonic effects have been harnessed in PMMA-POF-based temperature sensors (Cennamo et al., 2022). These sensors exploit the temperature-dependent shifts in surface plasmon resonance wavelengths. This principle could potentially be extended to PFGI-POF-based temperature sensors.

Another intriguing method involves the exploitation of the fiber fuse phenomenon in PFGI-POFs (Arnaldo et al., 2018; Leal-Junior et al., 2022; Leal-Junior, Frizera, Pontes, et al., 2018; Mizuno, Hayashi, Tanaka, & Nakamura, 2014; Mizuno, Hayashi, Tanaka, Nakamura, et al., 2014a, 2014b; Paixão et al., 2022). Characterized by its unique attributes compared to glass fiber fuses, including markedly slow fuse propagation, low threshold power, and the generation of a black curve composed of fluorinated graphene oxide postpropagation, this phenomenon offers distinct advantages. The transmission power of a fused POF has been reported to be over 10-fold more sensitive than its nonfused counterpart, making it a compelling option for temperature sensing (Leal-Junior, Frizera, Lee, et al., 2018). By integrating the fuse effect with multimode interference, the potential for stable, wavelength-domain temperature sensing with heightened sensitivity is considerable. This is analogous to the advancements observed in magnetic field sensing (Leal-Junior et al., 2021).

PFGI-POF-based temperature and strain sensing technologies have thus demonstrated significant advancements across various dimensions. However, several technical challenges must be addressed to broaden their applicability in practical scenarios. One such challenge is the limited specificity of the sensor's response to potential measures. The issue of cross-sensitivities to multiple parameters can adversely affect sensor performance, although this can be mitigated to some extent through the judicious selection of specialized polymers.

Another area of concern, particularly for applications requiring large-strain sensing, is the viscoelastic behavior of the polymers used in POFs. Factors such as temporal stress relaxation and molecular reorganization can alter the properties of Brillouin scattering over time, thereby affecting the accuracy of strain measurements. Additionally, the temperature range for standard POFs is generally limited to a maximum of approximately 80°C–100°C. While high-temperature POFs can extend this range to over 120°C, they still fall short of the high-temperature capabilities of silica glass fibers.

Furthermore, the full potential of POF-based sensing technologies remains untapped due to the current unavailability of low-attenuation single-mode POFs. The absence of such fibers restricts the application of more sensitive, coherent measurement principles in distributed sensing systems. The development of low-loss single-mode POFs would not only enhance the performance and applicability of FBG sensors but would also lower the Brillouin threshold power, thereby improving the signal-to-noise ratio in distributed Brillouin sensing systems.

Despite these challenges, it is unequivocal that POFs offer a compelling platform for the development of advanced temperature and strain sensing systems, boasting numerous advantages that traditional silica fibers cannot provide. As we look to the future, we are optimistic that POF-based sensing will emerge as a cornerstone technology, instrumental in the development of smart materials and structures. Addressing the aforementioned challenges will be crucial in realizing this potential, and we anticipate that the field will see significant advancements in the years to come.

References

Agrawal, G. P. (2019). *Nonlinear fiber optics* (6th). San Diego: Academic Press. Available from https://doi.org/10.1016/C2018-0-01168-8.

Leal, A., Jr., Frizera, A., Lee, H., Mizuno, Y., Nakamura, K., Leitão, C., Domingues, M. F., Alberto, N., Antunes, P., Andre, P., Marques, C., & Pontes, M. J. (2018). Design and characterization of a curvature sensor using fused polymer optical fibers. *Optics Letters*, *43*(11), 2539–2542. Available from https://doi.org/10.1364/OL.43.002539.

Bolognini, G., & Hartog, A. (2013). Raman-based fibre sensors: Trends and applications. *Optical Fiber Technology*, *19*(6), 678–688. Available from https://doi.org/10.1016/j.yofte.2013.08.003.

Brogan, K. L., & Walt, D. R. (2005). Optical fiber-based sensors: Application to chemical biology. *Current Opinion in Chemical Biology*, *9*(5), 494–500. Available from https://doi.org/10.1016/j.cbpa.2005.08.009.

Carroll, K. E., Zhang, C., Webb, D. J., Kalli, K., Argyros, A., & Large, M. C. J. (2007). Thermal response of Bragg gratings in PMMA microstructured optical fibers. *Optics Express*, *15*(14), 8844–8850. Available from https://doi.org/10.1364/OE.15.008844.

Cennamo, N., Prete, D. D., Arcadio, F., & Zeni, L. (2022). A temperature sensor exploiting plasmonic phenomena changes in multimode POFs. *IEEE Sensors Journal*, *22*(13), 12900–12905. Available from https://doi.org/10.1109/JSEN.2022.3178753.

Chen, T., Wang, Q., Chen, R., Zhang, B., Chen, K. P., Maklad, M., & Swinehart, P. R. (2012). Distributed hydrogen sensing using in-fiber Rayleigh scattering. *Applied Physics Letters*, *100*(19). Available from https://doi.org/10.1063/1.4712592.

Denisov, A., Soto, M. A., & Thévenaz, L. (2016). Going beyond 1000000 resolved points in a Brillouin distributed fiber sensor: Theoretical analysis and experimental demonstration. *Light: Science and Applications*, *5*. Available from https://doi.org/10.1038/lsa.2016.74.

Ding, M., Mizuno, Y., & Nakamura, K. (2014). Discriminative strain and temperature measurement using Brillouin scattering and fluorescence in erbium-doped optical fiber. *Optics Express*, *22*(20), 24706–24712. Available from https://doi.org/10.1364/OE.22.024706.

Dong, Y., Xu, P., Zhang, H., Lu, Z., Chen, L., & Bao, X. (2014). Characterization of evolution of mode coupling in a graded-index polymer optical fiber by using Brillouin optical time-domain analysis. *Optics Express*, *22*(22), 26510. Available from https://doi.org/10.1364/OE.22.026510.

Farahi, F., Webb, D. J., Jones, J. D. C., & Jackson, D. A. (1990). Simultaneous measurement of temperature and strain: Cross-sensitivity considerations. *Journal of Lightwave Technology*, *8*(2), 138–142. Available from https://doi.org/10.1109/50.47862.

Fasano, A., Woyessa, G., Stajanca, P., Markos, C., Stefani, A., Nielsen, K., Rasmussen, H. K., Krebber, K., & Bang, O. (2016). Fabrication and characterization of polycarbonate microstructured polymer optical fibers for high-temperature-resistant fiber Bragg grating strain sensors. *Optical Materials Express*, *6*(2), 649–659. Available from https://doi.org/10.1364/OME.6.000649.

Frazão, O., Silva, S. O., Viegas, J., Ferreira, L. A., Araújo, F. M., & Santos, J. L. (2011). Optical fiber refractometry based on multimode interference. *Applied Optics*, *50*(25), E184–E188. Available from https://doi.org/10.1364/AO.50.00E184.

Froggatt, M., & Moore, J. (1998). High-spatial-resolution distributed strain measurement in optical fiber with Rayleigh scatter. *Applied Optics*, *37*(10), 1735–1740. Available from https://doi.org/10.1364/AO.37.001735.

Garus, D., Krebber, K., Schliep, F., & Gogolla, T. (1996). Distributed sensing technique based on Brillouin optical-fiber frequency-domain analysis. *Optics Letters*, *21*(17), 1402–1404. Available from https://doi.org/10.1364/OL.21.001402.

Han, Y. G., Tran, T. V. A., Kim, S. H., & Lee, S. B. (2005). Multiwavelength Raman-fiber-laser-based long-distance remote sensor for simultaneous measurement of strain and temperature. *Optics Letters*, *30*(11), 1282–1284. Available from https://doi.org/10.1364/OL.30.001282.

Hartog, A. H. (2017). *An introduction to distributed optical fibre sensors. An introduction to distributed optical fibre sensors* (pp. 1–440). United Kingdom: CRC Press. Available from https://doi.org/10.1201/9781315119014.

Hayashi, N., Fukuda, H., Mizuno, Y., & Nakamura, K. (2014). Observation of Brillouin gain spectrum in tapered polymer optical fiber. *Journal of Applied Physics*, *115*(17), 173108. Available from https://doi.org/10.1063/1.4875102.

Hayashi, N., Minakawa, K., Mizuno, Y., & Nakamura, K. (2014). Brillouin frequency shift hopping in polymer optical fiber. *Applied Physics Letters*, *105*(9). Available from https://doi.org/10.1063/1.4895041.

Hayashi, N., Mizuno, Y., & Nakamura, K. (2012). Brillouin gain spectrum dependence on large strain in perfluorinated graded-index polymer optical fiber. *Optics Express*, *20*(19), 21101–21106. Available from https://doi.org/10.1364/OE.20.021101.

Hayashi, N., Mizuno, Y., & Nakamura, K. (2014). Distributed Brillouin sensing with centimeter-order spatial resolution in polymer optical fibers. *Journal of Lightwave Technology*, *32*(21), 3999–4003. Available from https://doi.org/10.1109/JLT.2014.2339361.

Hayashi, N., Mizuno, Y., & Nakamura, K. (2015). Simplified Brillouin optical correlation-domain reflectometry using polymer optical fiber. *IEEE Photonics Journal*, *7*(1). Available from https://doi.org/10.1109/JPHOT0.2014.2381650.

Hayashi, N., Minakawa, K., Mizuno, Y., & Nakamura, K. (2015). Polarization scrambling in Brillouin optical correlation-domain reflectometry using polymer fibers. *Applied Physics Express*, *8*(6), 062501. Available from https://doi.org/10.7567/apex.8.062501.

Horiguchi, T., & Tateda, M. (1989). BOTDA—Nondestructive measurement of single-mode optical fiber attenuation characteristics using Brillouin interaction: Theory. *Journal of Lightwave Technology*, *7*(8), 1170–1176. Available from https://doi.org/10.1109/50.32378.

Hotate, K. (2000). Measurement of Brillouin gain spectrum distribution along an optical fiber using a correlation-based technique-proposal, experiment and simulation. *IEICE Transactions on Electronics*, *E83-C*(3), 405–411. Available from http://www.jstage.jst.go.jp/browse/.

Huang, J., Lan, X., Wang, H., Yuan, L., Wei, T., Gao, Z., & Xiao, H. (2012). Polymer optical fiber for large strain measurement based on multimode interference. *Optics Letters*, *37*(20), 4308–4310. Available from https://doi.org/10.1364/OL.37.004308.

Husdi, I. R., Nakamura, K., & Ueha, S. (2004). Sensing characteristics of plastic optical fibres measured by optical time-domain reflectometry. *Measurement Science and Technology*, *15*(8), 1553–1559. Available from https://doi.org/10.1088/0957-0233/15/8/022.

Hwang, D., Yoon, D. J., Kwon, I. B., Seo, D. C., & Chung, Y. (2010). Novel auto-correction method in a fiber-optic distributed-temperature sensor using reflected anti-Stokes Raman scattering. *Optics Express*, *18*(10), 9747–9754. Available from https://doi.org/10.1364/OE.18.009747.

Ishikawa, R., Lee, H., Lacraz, A., Theodosiou, A., Kalli, K., Mizuno, Y., & Nakamura, K. (2017). Pressure dependence of fiber Bragg grating inscribed in perfluorinated polymer fiber. *IEEE Photonics Technology Letters*, *29*(24), 2167–2170. Available from https://doi.org/10.1109/LPT.2017.2767082.

Ishikawa, R., Lee, H., Lacraz, A., Theodosiou, A., Kalli, K., Mizuno, Y., & Nakamura, K. (2018). Strain dependence of perfluorinated polymer optical fiber Bragg grating measured at different wavelengths. *Japanese Journal of Applied Physics*, *57*(3), 038002. Available from https://doi.org/10.7567/jjap.57.038002.

Johnson, I. P., Kalli, K., & Webb, D. J. (2010). 827 nm Bragg grating sensor in multimode microstructured polymer optical fibre. *Electronics Letters*, *46*(17), 1217–1218. Available from https://doi.org/10.1049/el.2010.1595.

Kaino, T., Fujiki, M., & Nara, S. (1981). Low-loss polystyrene core-optical fibers. *Journal of Applied Physics, 52*(12), 7061–7063. Available from https://doi.org/10.1063/1.328702.

Kashyap, R. (1999). *Fiber Bragg gratings*. Elsevier.

Kawa, T., Numata, G., Lee, H., Hayashi, N., Mizuno, Y., & Nakamura, K. (2017a). Temperature sensing based on multimodal interference in polymer optical fibers: Room-temperature sensitivity enhancement by annealing. *Japanese Journal of Applied Physics, 56*(7), 078002. Available from https://doi.org/10.7567/jjap.56.078002.

Kawa, T., Numata, G., Lee, H., Hayashi, N., Mizuno, Y., & Nakamura, K. (2017b). Single-end-access strain and temperature sensing based on multimodal interference in polymer optical fibers. *IEICE Electronics Express, 14*(3), 20161239. Available from https://doi.org/10.1587/elex.14.20161239.

Kim, Y. H., Lee, K., & Song, K. Y. (2015). Brillouin optical correlation domain analysis with more than 1 million effective sensing points based on differential measurement. *Optics Express, 23*(26), 33241–33248. Available from https://doi.org/10.1364/OE.23.033241.

Koerdt, M., Kibben, S., Bendig, O., Chandrashekhar, S., Hesselbach, J., Brauner, C., Herrmann, A. S., Vollertsen, F., & Kroll, L. (2016). Fabrication and characterization of Bragg gratings in perfluorinated polymer optical fibers and their embedding in composites. *Mechatronics, 34*, 137–146. Available from https://doi.org/10.1016/j.mechatronics.2015.10.005.

Koike, Y., & Asai, M. (2009). The future of plastic optical fiber. *NPG Asia Materials, 1*(1), 22–28. Available from https://doi.org/10.1038/asiamat.2009.2.

Koyamada, Y., Imahama, M., Kubota, K., & Hogari, K. (2009). Fiber-optic distributed strain and temperature sensing with very high measurand resolution over long range using coherent OTDR. *Journal of Lightwave Technology, 27*(9), 1142–1146. Available from https://doi.org/10.1109/JLT.2008.928957.

Kumar, A., Varshney, R. K., Antony, C. S., & Sharma, P. (2003). Transmission characteristics of SMS fiber optic sensor structures. *Optics Communications, 219*(1–6), 215–219. Available from https://doi.org/10.1016/S0030-4018(03)01289-6.

Kurashima, T., Horiguchi, T., Izumita, H., Furukawa, Si, & Koyamada, Y. (1993). Brillouin optical-fiber time domain reflectometry. *IEICE Transactions on Communications, E76-B*(4), 382–390.

Kuzyk, M. G. (2006). *Polymer fiber optics: Materials, physics, and applications. Polymer Fiber Optics: Materials, Physics, and Applications* (pp. 1–405). United States: CRC Press.

Lacraz, A., Polis, M., Theodosiou, A., Koutsides, C., & Kalli, K. (2015). Femtosecond laser inscribed Bragg gratings in low loss CYTOP polymer optical fiber. *IEEE Photonics Technology Letters, 27*(7), 693–696. Available from https://doi.org/10.1109/LPT.2014.2386692.

Lacraz, A., Theodosiou, A., & Kalli, K. (2016). Femtosecond laser inscribed Bragg grating arrays in long lengths of polymer optical fibres: A route to practical sensing with POF. *Electronics Letters, 52*(19), 1626–1627. Available from https://doi.org/10.1049/el.2016.2238.

Leal-Junior, A., Díaz, C., Frizera, A., Lee, H., Nakamura, K., Mizuno, Y., & Marques, C. (2021). Highly sensitive fiber-optic intrinsic electromagnetic field sensing. *Advanced Photonics Research, 2*(1). Available from https://doi.org/10.1002/adpr.202000078.

Leal-Junior, A., Frizera, A., Lee, H., Mizuno, Y., Nakamura, K., Paixão, T., Leitão, C., Domingues, M. F., Alberto, N., Antunes, P., André, P., Marques, C., & Pontes, M. J. (2018). Strain, temperature, moisture, and transverse force sensing using fused polymer optical fibers. *Optics Express, 26*(10), 12939. Available from https://doi.org/10.1364/oe.26.012939.

Leal-Junior, A., Frizera, A., Pontes, M. J., Antunes, P., Alberto, N., Domingues, M. F., Lee, H., Ishikawa, R., Mizuno, Y., Nakamura, K., André, P., & Marques, C. (2018). Dynamic mechanical analysis on fused polymer optical fibers: Towards sensor applications. *Optics Letters, 43*(8), 1754. Available from https://doi.org/10.1364/ol.43.001754.

Leal-Junior, A., Theodosiou, A., Díaz, C., Marques, C., Pontes, M. J., Kalli, K., & Frizera-Neto, A. (2018). Polymer optical fiber Bragg gratings in CYTOP fibers for angle measurement with dynamic compensation. *Polymers*, *10*(6). Available from https://doi.org/10.3390/polym10060674.

Leal-Junior, A. G., Marques, C., Lee, H., Nakamura, K., & Mizuno, Y. (2022). Sensing applications of polymer optical fiber fuse. *Advanced Photonics Research*, *3*(6). Available from https://doi.org/10.1002/adpr.202100210.

Leal-Junior, A. G., Theodosiou, A., Diaz, C. R., Marques, C., Pontes, M. J., Kalli, K., & Frizera, A. (2019). Simultaneous measurement of axial strain, bending and torsion with a single fiber Bragg grating in CYTOP fiber. *Journal of Lightwave Technology*, *37*(3), 971–980. Available from https://doi.org/10.1109/JLT.2018.2884538.

Lee, B. (2003). Review of the present status of optical fiber sensors. *Optical Fiber Technology*, *9*(2), 57–79. Available from https://doi.org/10.1016/S1068-5200(02)00527-8.

Lee, H., Hayashi, N., Mizuno, Y., & Nakamura, K. (2016). Slope-assisted Brillouin optical correlation-domain reflectometry: Proof of concept. *IEEE Photonics Journal*, *8*(3). Available from https://doi.org/10.1109/JPHOT.2016.2562512.

Lee, H., Hayashi, N., Mizuno, Y., & Nakamura, K. (2017). Slope-assisted Brillouin optical correlation-domain reflectometry using polymer optical fibers with high propagation loss. *Journal of Lightwave Technology*, *35*(11), 2306–2310. Available from https://doi.org/10.1109/JLT.2017.2663440.

Leung, C. K. Y., Wan, K. T., Inaudi, D., Bao, X., Habel, W., Zhou, Z., Ou, J., Ghandehari, M., Wu, H. C., & Imai, M. (2015). Review: Optical fiber sensors for civil engineering applications. *Materials and Structures/Materiaux et Constructions*, *48*(4), 871–906. Available from https://doi.org/10.1617/s11527-013-0201-7.

Liu, Y., & Wei, L. (2007). Low-cost high-sensitivity strain and temperature sensing using graded-index multimode fibers. *Applied Optics*, *46*(13), 2516–2519. Available from https://doi.org/10.1364/AO.46.002516.

Lu, P., Mihailov, S. J., Coulas, D., Ding, H., & Bao, X. (2019). Low-loss random fiber gratings made with an fs-IR laser for distributed fiber sensing. *Journal of Lightwave Technology*, *37*(18), 4697–4702. Available from https://doi.org/10.1109/JLT.2019.2917389.

Mehta, A., Mohammed, W., & Johnson, E. G. (2003). Multimode interference-based fiber-optic displacement sensor. *IEEE Photonics Technology Letters.*, *15*(8), 1129–1131. Available from https://doi.org/10.1109/LPT.2003.815338.

Min, R., Ortega, B., Leal-Junior, A., & Marques, C. (2018). Fabrication and characterization of Bragg grating in CYTOP POF at 600-nm wavelength. *IEEE Sensors Letters*, *2*(3). Available from https://doi.org/10.1109/LSENS.2018.2848542.

Minakawa, K., Hayashi, N., Mizuno, Y., & Nakamura, K. (2015). Thermal memory effect in polymer optical fibers. *IEEE Photonics Technology Letters*, *27*(13), 1394–1397. Available from https://doi.org/10.1109/LPT.2015.2421950.

Minakawa, K., Hayashi, N., Shinohara, Y., Tahara, M., Hosoda, H., Mizuno, Y., & Nakamura, K. (2014). Wide-range temperature dependences of Brillouin scattering properties in polymer optical fiber. *Japanese Journal of Applied Physics*, *53*(4), 042502. Available from https://doi.org/10.7567/jjap.53.042502.

Minakawa, K., Mizuno, Y., & Nakamura, K. (2017). Cross effect of strain and temperature on Brillouin frequency shift in polymer optical fibers. *Journal of Lightwave Technology*, *35*(12), 2481–2486. Available from https://doi.org/10.1109/JLT.2017.2689331.

Minardo, A., Bernini, R., Ruiz-Lombera, R., Mirapeix, J., Lopez-Higuera, J. M., & Zeni, L. (2016). Proposal of Brillouin optical frequency-domain reflectometry (BOFDR). *Optics Express*, *24*(26), 29994–30001. Available from https://doi.org/10.1364/OE.24.029994.

Minardo, A., Bernini, R., & Zeni, L. (2014). Distributed temperature sensing in polymer optical fiber by BOFDA. *IEEE Photonics Technology Letters*, *26*(4), 387–390. Available from https://doi.org/10.1109/LPT.2013.2294878.

Mizuno, Y., Hayashi, N., Fukuda, H., Song, K. Y., & Nakamura, K. (2016). Ultrahigh-speed distributed Brillouin reflectometry. *Light: Science and Applications*, 5. Available from https://doi.org/10.1038/lsa.2016.184.

Mizuno, Y., Hayashi, N., Tanaka, H., & Nakamura, K. (2014). Spiral propagation of polymer optical fiber fuse accompanied by spontaneous burst and its real-time monitoring using Brillouin scattering. *IEEE Photonics Journal*, *6*(3). Available from https://doi.org/10.1109/JPHOT.2014.2323301.

Mizuno, Y., Hayashi, N., Tanaka, H., Nakamura, K., & Todoroki, S. I. (2014a). Observation of polymer optical fiber fuse. *Applied Physics Letters*, *104*(4). Available from https://doi.org/10.1063/1.4863413.

Mizuno, Y., Hayashi, N., Tanaka, H., Nakamura, K., & Todoroki, S. I. (2014b). Propagation mechanism of polymer optical fiber fuse. *Scientific Reports*, *4*. Available from https://doi.org/10.1038/srep04800.

Mizuno, Y., He, Z., & Hotate, K. (2009). One-end-access high-speed distributed strain measurement with 13-mm spatial resolution based on Brillouin optical correlation-domain reflectometry. *IEEE Photonics Technology Letters*, *21*(7), 474–476. Available from https://doi.org/10.1109/LPT.2009.2013643.

Mizuno, Y., Ishikawa, R., Lee, H., Theodosiou, A., Kalli, K., & Nakamura, K. (2019). Potential of discriminative sensing of strain and temperature using perfluorinated polymer FBG. *IEEE Sensors Journal*, *19*(12), 4458–4462. Available from https://doi.org/10.1109/jsen.2019.2900464.

Mizuno, Y., Lee, H., Hayashi, N., & Nakamura, K. (2018). Hydrostatic pressure dependence of Brillouin frequency shift in polymer optical fibers. *Applied Physics Express*, *11*(1), 012502. Available from https://doi.org/10.7567/apex.11.012502.

Mizuno, Y., Lee, H., Hayashi, N., & Nakamura, K. (2019). Noise suppression technique for distributed Brillouin sensing with polymer optical fibers. *Optics Letters*, *44*(8), 2097–2100. Available from https://doi.org/10.1364/OL.44.002097.

Mizuno, Y., Lee, H., & Nakamura, K. (2018). Recent advances in Brillouin optical correlation-domain reflectometry. *Applied Sciences (Switzerland)*, *8*(10). Available from https://doi.org/10.3390/app8101845.

Mizuno, Y., Ma, T., Ishikawa, R., Lee, H., Theodosiou, A., Kalli, K., & Nakamura, K. (2019a). Lorentzian demodulation algorithm for multimode polymer optical fiber Bragg gratings. *Japanese Journal of Applied Physics*, *58*(2). Available from https://doi.org/10.7567/1347-4065/aaf897.

Mizuno, Y., Ma, T., Ishikawa, R., Lee, H., Theodosiou, A., Kalli, K., & Nakamura, K. (2019b). Twist dependencies of strain and temperature sensitivities of perfluorinated graded-index polymer optical fiber Bragg gratings. *Applied Physics Express*, *12*(8), 082007. Available from https://doi.org/10.7567/1882-0786/ab3013.

Mizuno, Y., Matsutani, N., Hayashi, N., Lee, H., Tahara, M., Hosoda, H., & Nakamura, K. (2018). Brillouin characterization of slimmed polymer optical fibers for strain sensing with extremely wide dynamic range. *Optics Express*, *26*(21), 28030–28037. Available from https://doi.org/10.1364/OE.26.028030.

Mizuno, Y., & Nakamura, K. (2010a). Experimental study of Brillouin scattering in perfluorinated polymer optical fiber at telecommunication wavelength. *Applied Physics Letters*, *97*(2). Available from https://doi.org/10.1063/1.3463038.

Mizuno, Y., & Nakamura, K. (2010b). Potential of Brillouin scattering in polymer optical fiber for strain-insensitive high-accuracy temperature sensing. *Optics Letters*, *35*(23), 3985–3987. Available from https://doi.org/10.1364/OL.35.003985.

Mizuno, Y., Ujihara, H., Lee, H., Hayashi, N., & Nakamura, K. (2017). Polymer optical fiber tapering using hot water. *Applied Physics Express*, *10*(6), 062502. Available from https://doi.org/10.7567/apex.10.062502.

Mizuno, Y., Theodosiou, A., Kalli, K., Liehr, S., Lee, H., & Nakamura, K. (2021). Distributed polymer optical fiber sensors: A review and outlook. *Photonics Research*, *9*(9), 1719–1733. Available from https://doi.org/10.1364/PRJ.435143.

Mizuno, Y., Zou, W., He, Z., & Hotate, K. (2010). Operation of Brillouin optical correlation-domain reflectometry: Theoretical analysis and experimental validation. *Journal of Lightwave Technology*, *28*(22), 3300–3306. Available from https://doi.org/10.1109/JLT.2010.2081348.

Mizuno, Y., Zou, W., He, Z., & Hotate, K. (2008). Proposal of Brillouin optical correlation-domain reflectometry (BOCDR). *Optics Express*, *16*(16), 12148–12153. Available from https://doi.org/10.1364/OE.16.012148.

Motil, A., Bergman, A., & Tur, M. (2016). [INVITED] State of the art of Brillouin fiber-optic distributed sensing. *Optics and Laser Technology*, *78*, 81–103. Available from https://doi.org/10.1016/j.optlastec.2015.09.013.

Numata, G., Hayashi, N., Tabaru, M., Mizuno, Y., & Nakamura, K. (2014). Ultra-sensitive strain and temperature sensing based on modal interference in perfluorinated polymer optical fibers. *IEEE Photonics Journal*, *6*(5). Available from https://doi.org/10.1109/JPHOT.2014.2352637.

Numata, G., Hayashi, N., Tabaru, M., Mizuno, Y., & Nakamura, K. (2015). Drastic sensitivity enhancement of temperature sensing based on multimodal interference in polymer optical fibers. *Applied Physics Express*, *8*(7), 072502. Available from https://doi.org/10.7567/apex.8.072502.

Othonos, A., & Kalli, K. (1996). *Fiber Bragg gratings: Fundamentals and applications in telecommunications and sensing*. Artech House.

Paixão, T., Belo, J. H., Carvalho, A. F., Amaral, V. S., Araújo, J. P., Lee, H., Nakamura, K., Mizuno, Y., André, P., & Antunes, P. (2022). Magneto-responsive optical fiber with fuse-effect-induced fluorinated graphene oxide core. *Advanced Photonics Research*, *3*. Available from https://doi.org/10.1002/adpr.202100209.

Pastor-Graells, J., Martins, H. F., Garcia-Ruiz, A., Martin-Lopez, S., & Gonzalez-Herraez, M. (2016). Single-shot distributed temperature and strain tracking using direct detection phase-sensitive OTDR with chirped pulses. *Optics Express*, *24*(12), 13121–13133. Available from https://doi.org/10.1364/OE.24.013121.

Peters, K. (2011). Polymer optical fiber sensors—A review. *Smart Materials and Structures*, *20*(1), 013002. Available from https://doi.org/10.1088/0964-1726/20/1/013002.

Rao, B. S., Puschett, J. B., & Matyjaszewski, K. (1991). Preparation of pH sensors by covalent linkage of dye molecules to the surface of polystyrene optical fibers. *Journal of Applied Polymer Science*, *43*(5), 925–928. Available from https://doi.org/10.1002/app.1991.070430510.

Sang, A. K., Froggatt, M. E., Gifford, D. K., Kreger, S. T., & Dickerson, B. D. (2008). One centimeter spatial resolution temperature measurements in a nuclear reactor using Rayleigh scatter in optical fiber. *IEEE Sensors Journal*, *8*(7), 1375–1380. Available from https://doi.org/10.1109/JSEN.2008.927247.

Saxena, M. K., Raju, S. D. V. S. J., Arya, R., Pachori, R. B., Ravindranath, S. V. G., Kher, S., & Oak, S. M. (2015). Raman optical fiber distributed temperature sensor using wavelet transform based simplified signal processing of Raman backscattered signals. *Optics and Laser Technology*, *65*, 14–24. Available from https://doi.org/10.1016/j.optlastec.2014.06.012.

Schreier, A., Wosniok, A., Liehr, S., & Krebber, K. (2018). Humidity-induced Brillouin frequency shift in perfluorinated polymer optical fibers. *Optics Express*, *26*(17), 22307. Available from https://doi.org/10.1364/oe.26.022307.

Shatalin, S. V., Treschikov, V. N., & Rogers, A. J. (1998). Interferometric optical time-domain reflectometry for distributed optical-fiber sensing. *Applied Optics*, *37*(24), 5600–5604. Available from https://doi.org/10.1364/AO.37.005600.

Singal, T. L. (2017). *Optical fiber communications: Principles and applications. Optical fiber communications: Principles and applications* (pp. 1–450). India: Cambridge University Press. Available from http://doi.org/10.1017/9781316661505, https://doi.org/10.1017/9781316661505.

Song, J., Li, W., Lu, P., Xu, Y., Chen, L., & Bao, X. (2014). Long-range high spatial resolution distributed temperature and strain sensing based on optical frequency-domain reflectometry. *IEEE Photonics Journal*, *6*(3). Available from https://doi.org/10.1109/JPHOT.2014.2320742.

Stajanca, P., Lacraz, A., Kalli, K., Schukar, M., & Krebber, K. (2016). Strain sensing with femtosecond inscribed FBGs in perfluorinated polymer optical fibers. In *Proceedings of SPIE – The International Society for Optical Engineering*, 9899 SPIE Germany. Available from https://doi.org/10.1117/12.2225081.

Szczerska, M. (2022). Temperature sensors based on polymer fiber optic interferometer. *Chemosensors*, *10*(6), 228. Available from https://doi.org/10.3390/chemosensors10060228.

Theodosiou, A., Komodromos, M., & Kalli, K. (2017). Accurate and fast demodulation algorithm for multipeak FBG reflection spectra using a combination of cross correlation and hilbert transformation. *Journal of Lightwave Technology*, *35*(18), 3956–3962. Available from https://doi.org/10.1109/JLT.2017.2723945.

Theodosiou, A., Lacraz, A., Polis, M., Kalli, K., Tsangari, M., Stassis, A., & Komodromos, M. (2016). Modified fs-laser inscribed FBG array for rapid mode shape capture of free-free vibrating beams. *IEEE Photonics Technology Letters*, *28*(14), 1509–1512. Available from https://doi.org/10.1109/LPT.2016.2555852.

Theodosiou, A., Lacraz, A., Stassis, A., Koutsides, C., Komodromos, M., & Kalli, K. (2017). Plane-by-plane femtosecond laser inscription method for single-peak bragg gratings in multimode CYTOP polymer optical fiber. *Journal of Lightwave Technology*, *35*(24), 5404–5410. Available from https://doi.org/10.1109/JLT.2017.2776862.

Theodosiou, A., Min, R., Leal-Junior, A. G., Ioannou, A., Frizera, A., Pontes, M. J., Marques, C., & Kalli, K. (2019). Long period grating in a multimode cyclic transparent optical polymer fiber inscribed using a femtosecond laser. *Optics Letters*, *44*(21), 5346. Available from https://doi.org/10.1364/ol.44.005346.

Theodosiou, A., Polis, M., Lacraz, A., Kalli, K., Komodromos, M., & Stassis, A. (2016). Comparative study of multimode CYTOP graded index and single-mode silica fibre Bragg grating array for the mode shape capturing of a free-free metal beam. In *Proceedings of SPIE – The International Society for Optical Engineering*. 9886 SPIE Cyprus. Available from https://doi.org/10.1117/12.2230792.

Tripathi, S. M., Kumar, A., Varshney, R. K., Kumar, Y. B. P., Marin, E., & Meunier, J. P. (2009). Strain and temperature sensing characteristics of single-mode-multimode-single-mode structures. *Journal of Lightwave Technology*, *27*(13), 2348–2356. Available from https://doi.org/10.1109/JLT.2008.2008820.

Ujihara, H., Hayashi, N., Minakawa, K., Tabaru, M., Mizuno, Y., & Nakamura, K. (2015). Polymer optical fiber tapering without the use of external heat source and its application to refractive index sensing. *Applied Physics Express*, *8*(7), 072501. Available from https://doi.org/10.7567/apex.8.072501.

Wang, J., Zhao, L., Liu, T., Li, Z., Sun, T., & Grattan, K. T. V. (2017). Novel negative pressure wave-based pipeline leak detection system using fiber Bragg grating-based pressure sensors. *Journal of Lightwave Technology*, *35*(16), 3366–3373. Available from https://doi.org/10.1109/JLT.2016.2615468.

Wang, M., Wu, H., Tang, M., Zhao, Z., Dang, Y., Zhao, C., Liao, R., Chen, W., Fu, S., Yang, C., Tong, W., Shum, P. P., & Liu, D. (2017). Few-mode fiber based Raman distributed temperature sensing. *Optics Express*, *25*(5), 4907–4916. Available from https://doi.org/10.1364/OE.25.004907.

Wang, T., Liu, K., Jiang, J., Xue, M., Chang, P., & Liu, T. (2017). Temperature-insensitive refractive index sensor based on tilted moiré FBG with high resolution. *Optics Express*, *25*(13), 14900–14909. Available from https://doi.org/10.1364/OE.25.014900.

Webb, D. J. (2015). Fiber Bragg grating sensors in polymer optical fibres. *Measurement Science and Technology*, *26*(9). Available from https://doi.org/10.1088/0957-0233/26/9/092004.

Woyessa, G., Nielsen, K., Stefani, A., Markos, C., & Bang, O. (2016). Temperature insensitive hysteresis free highly sensitive polymer optical fiber bragg grating humidity sensor. *Optics Express*, *24*(2), 1206–1213. Available from https://doi.org/10.1364/OE.24.001206.

Xiong, Z., Peng, G. D., Wu, B., & Chu, P. L. (1999). Highly tunable Bragg gratings in single-mode polymer optical fibers. *IEEE Photonics Technology Letters*, *11*(3), 352–354. Available from https://doi.org/10.1109/68.748232.

Yan, A., Huang, S., Li, S., Chen, R., Ohodnicki, P., Buric, M., Lee, S., Li, M. J., & Chen, K. P. (2017). Distributed optical fiber sensors with ultrafast laser enhanced Rayleigh backscattering profiles for real-time monitoring of solid oxide fuel cell operations. *Scientific Reports*, *7*(1). Available from https://doi.org/10.1038/s41598-017-09934-3.

Zheng, Y., Bremer, K., & Roth, B. (2018). Investigating the strain, temperature and humidity sensitivity of a multimode graded-index perfluorinated polymer optical fiber with Bragg grating. *Sensors*, *18*(5), 1436. Available from https://doi.org/10.3390/s18051436.

Ziemann, O., Krauser, J., Zamzow, P. E., & Daum, W. (2002). *POF – Polymer optical fibers for data communication*. Springer.

Zou, W. W., He, Z., & Hotate, K. (2009). Complete discrim5ination of strain and temperature using Brillouin frequency shift and birefringence in a polarization-maintaining fiber. *Optics Express*, *17*(3), 1248–1255. Available from https://doi.org/10.1364/OE.17.001248.

Zubia, J., & Arrue, J. (2001). Plastic optical fibers: An introduction to their technological processes and applications. *Optical Fiber Technology*, *7*(2), 101–140. Available from https://doi.org/10.1006/ofte.2000.0355.

CHAPTER

Detection of nitroaromatic and nitramine explosives

18

Roberto J. Aguado
LEPAMAP-PRODIS Research Group, University of Girona, Girona, Girona (Catalonia), Spain

18.1 Introduction to nitroaromatic and nitramine explosives

18.1.1 From dyeing to killing

The German chemist Julius Wilbrand is credited with having undertaken the first synthesis of 2,4,6-trinitrotoluene (TNT) (Zhao, Yin, et al., 2021). It happened in 1863, and for some three decades, the only known usage of this product of toluene nitration was as a yellow dye (Lee et al., 2020). Little did Wilbrand suspect that TNT would become the most commonly used military explosive in World War II and, in general, during the 20th century (Agrawal, 2010). Less could he guess that, in a more specific context, this nitroaromatic compound would be the most typical explosive charge in antipersonnel landmines (Hussein & Waller, 2000).

Needless to say, TNT has also been used in many terrorist attacks,[1] such as the killing of 34 people at the Taba Hilton Hotel in the Sinai Peninsula in 2004 (McCarthy & Pedahzur, 2014), the 2004 Madrid train bombings that killed 193 and injured more than 2000 (Hamilos, 2007), the assassination of former Prime Minister of Lebanon Rafic Hariri in 2005 (The Guardian, 2005), the suicide bombings at the 2015 Ankara peace rally that killed 95 (Frasera, 2015), and, more recently, the 2023 Peshawar mosque building that killed eighty-four (including the perpetrator) and injured more than two-hundred people (Dawn.com, 2023). Nonetheless, it is only fair to say that TNT also has important non-lethal applications. One of them is precisely restoring soils and saving lives by de-mining (Raza et al., 2020).

18.1.2 Nitro compounds used as secondary explosives

TNT is not the only nitroaromatic compound that is widely employed as a secondary explosive. Other examples include methyl(2,4,6-trinitrophenyl)nitramide, generally known as *Tetryl*, and picric acid[2] (PA) (Toal & Trogler, 2006). Indeed, until the introduction of a relatively inexpensive

[1]Since the line between warfare and terrorism may be thin, "terrorism" in this chapter refers to non-state violence with the purpose of creating fear and/or fostering a political, ideological, or religious objective. This is not denying the fact that some states, their regular armies, and their private military companies have recently used or are using means of terrorism.
[2]IUPAC name: 2,4,6-trinitrophenol.

Sensory Polymers. DOI: https://doi.org/10.1016/B978-0-443-13394-7.00018-5

process for toluene nitration at the gates of World War I, PA was the most common filler for artillery shells. In addition, albeit not aromatic, one cannot talk about landmines without mentioning plasticized explosives based on hexahydro-1,3,5-trinitro-1,3,5-triazine, more commonly known as Royal Detonation Explosive (RDX) (Mundy et al., 2023). One of the most widely used families of explosives, Composition C, which includes the (in)famous C-4, implies RDX as the main explosive agent. Finally, considering the chemical resemblance to RDX, 1,3,5,7-tetranitro-1,3,5,7-tetrazocane, or High Melting Explosive (HMX), is less common but not less worth mentioning (Longo & Musah, 2020). HMX is of utmost importance in the manufacturing of armor-piercing warheads.

Fig. 18.1 presents the structural formula of TNT, PA, Tetryl, RDX, and HMX, along with *Jmol 14* optimized 3D structure (by MM energy minimization) in each case. Still, these are not the only nitroaromatic and/or nitramine compounds of interest when it comes to explosive detection. For instance, TNT may be accompanied by side-products corresponding to an incomplete nitration of toluene (often intentionally, as in the case of some industrial explosives), mainly 2,4-dinitrotoluene (2,4-DNT) and 2,6-dinitrotoluene (2,6-DNT). Vapors and leaks from TNT-filled landmines contain high proportions of 2,4-DNT, dinitrobenzene isomers, and nitrobenzene, resulting from the photolysis of TNT by sunlight (Mihas et al., 2007). In the environment, TNT is metabolized by some organisms to 2-amino-4,6-dinitrotoluene (2-DANT) and 4-amino-2,6-dinitrotoluene (4-ADNT) (Tauqeer et al., 2021).

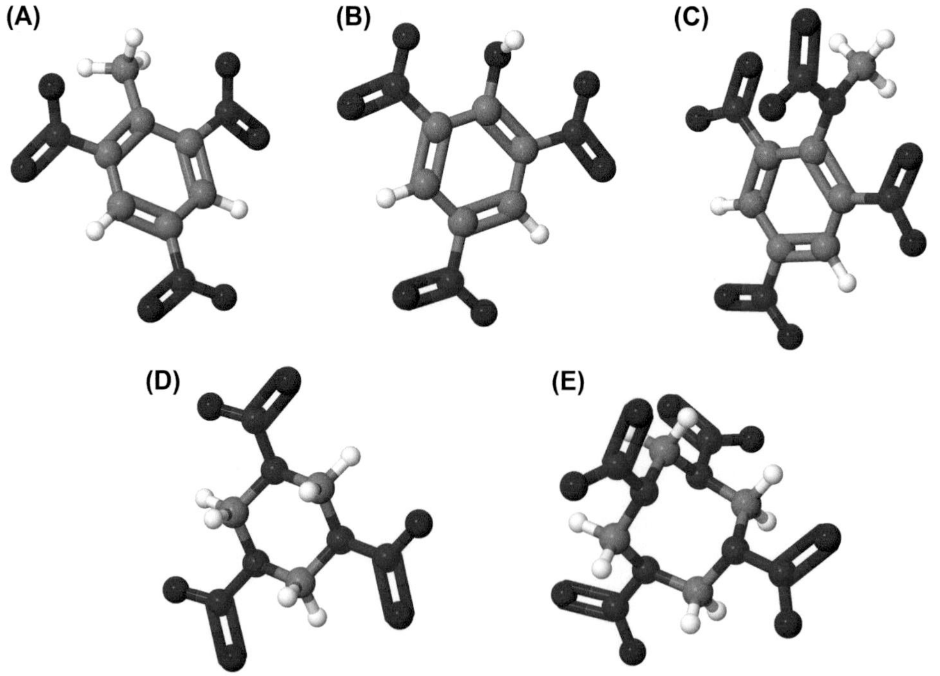

FIGURE 18.1

Ball-and-stick molecular structures of TNT (A), PA (B), Tetryl (C), RDX (D), and HMX (E).

18.1.3 The current (and persistent) threats of 2,4,6-trinitrotoluene/Royal Detonation Explosive-filled landmines

The harm caused by nitroaromatic and nitramine explosives is not restricted to their many direct victims. Landmines and unexploded charges pollute the soil and the groundwater, represent a risk to the ecosystem and its inhabitants, hamper the development of agriculture and other human activities, and heavily complicate restoration plans once the active conflict is over. In the words of the Ukrainian environmental activist Evgenia Zasiadko (McCarthy, 2022): "We're an agricultural country, and when it's not an active war, I don't know how we're going to rebuild anything because it's going to be polluted." By April 2023, the area of Ukrainian land to be contaminated by mines was estimated at 174,000 km^2 (Waterhouse James, 2023). This includes nearly 10 million hectares of arable land. Even if the war, still going on as this chapter is being written, were about to finish, the cost of de-mining operations would be as high as $37 billion (UN News, 2023).

Antipersonnel, antivehicle, and antilanding mines used in the Russo-Ukrainian war comprise PMN-1 (240 g of TNT), PMN-2 (100 g of RDX and TNT), PMN-4 (50 g of RDX and TNT), MON-50 (700 g of RDX), MON-90 (6 kg of RDX), MON-100 (2 kg of TNT), MON-200 (12 kg of TNT), OZM-72 (660 g of TNT), POM-2 (140 g of TNT), and PDM-1 (10 kg of TNT), among others (Evans & Seddon, 2022). The environmental damage produced by TNT, RDX, and their degradation products, cooccurring compounds, or metabolites has been extensively studied (Chakraborty et al., 2022; Tauqeer et al., 2021; Zhao, Yang et al., 2021). Besides nitrobenzene and the aforementioned derivatives, products from the degradation of TNT include aniline, which is highly toxic for aquatic life (Chen et al., 2023), 2-DANT, 4-DANT, and TNT itself are known to attack humic monomers and phenoloxidases in soils by oxidative coupling. And, while the adsorption of TNT on soils is deemed reversible, the adsorption of RDX is not (Chatterjee et al., 2017).

As a European, I could not help referring to a shocking war taking place on this continent. However, Ukraine and the contenders of the Yugoslav Wars, still contaminated with mines, cannot be considered the only sources of motivation. Subsaharan countries such as Angola, Mozambique, Chad, Ethiopia, Mali, and Somalia can attest to the persistent hazards of landmine fields (Cluster Munition Coalition, International Campaign to Ban Landmines, 2020; Doswald-Beck et al., 1995). Middle-East countries such as Iraq and Turkey, besides the obvious cases of Syria and Yemen, are considered to have "massive antipersonnel mine contamination." Southeast Asian countries like Thailand and Cambodia also make the list, dragging the consequences of conflicts that took place several decades ago. This is also the case in Afghanistan, with the aggravating condition of insecurity for de-mining (Cluster Munition Coalition, International Campaign to Ban Landmines, 2020).

18.2 Brief overview of the detection of secondary explosives involving polymers

18.2.1 Nitroexplosive-responsive polymers in the recent literature

Considering what has been exposed up to this point, the need for detecting nitroaromatic and nitramine explosives is amply justified. Evaluating risks in restoration plans, finding landmines and explosive remnants in clearance operations, addressing the environmental impact of leaks, performing forensic analyses, and avoiding terrorist attacks are surely not the only reasons why, but they are surely enough. This chapter explains the convenience of using polymeric materials for this important task, describes the mechanisms of response involved, and addresses the proposals and inventions that have emerged over the last decade. All in all, in order of relative frequency in the recent literature, detection and quantification of nitroexplosives by responsive polymers fall within one or more of these categories:

- Fluorescence quenching of conjugated polymers, including polymetalloles, usually involving electron transfer and/or energy transfer mechanisms;
- electrochemical detection, often implying conjugated polymers as well;
- colorimetric (naked eye) detection, frequently (but not exclusively) by the formation of Meisenheimer complexes;
- fluorescence enhancement or turn-on, a strategy that seems more common with non-polymeric probes;
- surface acoustic wave sensing, a typical approach in the first decade of the century, but with few recent advances (involving polymers) in the last few years.

The prevalence of fluorescence quenching mechanisms in this topic seems evident in Fig. 18.2. This cooccurrence diagram has been constructed with the University of Leiden's VOSViewer software, following a search in Scopus with the following query string: "(nitroaromatic OR nitramine) AND explosive AND detection AND polymer." The search was restricted from 2014 to 2023; both indexing keywords and author keywords were considered; their minimum appearance was set to 10; keywords imposed by the search string (such as "nitroaromatic") were manually excluded; and

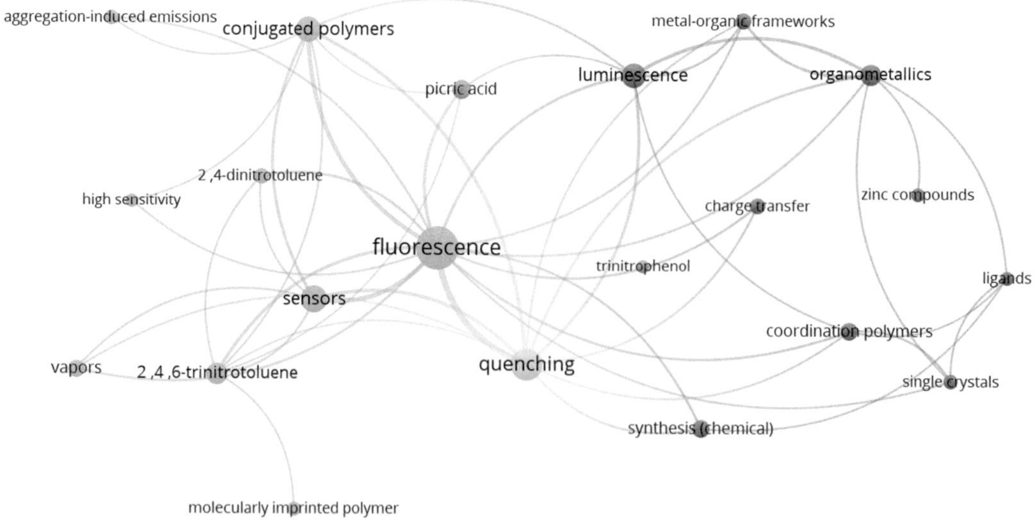

FIGURE 18.2

Cooccurrence diagram of the indexing and author keywords in the recent literature (2014–2023) on the detection of nitroexplosives involving polymers.

the number of connecting lines was limited to 50. As indicated, "fluorescence," "quenching," and "conjugated polymers" were found to be the most frequent keywords, and "charge transfer" or electron transfer is the main mechanism behind this quenching. TNT, PA, and 2,4-DNT appeared as the most common target compounds.

Some strategies to improve the sensitivity and/or selectivity of polymer-containing sensing systems may apply to several kinds of detection (fluorescent, electrochemical, and colorimetric). Three examples are quantum dot coating, molecular imprinting, and Meisenheimer complex formation. They have subsections dedicated to them within the section in which they account for more publications in the 2014–2023 period. Hence, quantum dots (QDs) are presented in the context of luminescence alterations, molecular imprinting is described within the section dedicated to electrochemical detection, and Meisenheimer complexes are addressed to explain colorimetric methods, even though some of them are colorless.

18.2.2 Complementing (not replacing) other analytical techniques

Probably, as of today, the most convincing option for the detection of explosives at trace concentrations is ion mobility spectrometry (IMS). Despite its excellent sensitivity and despite being the most widely used technique for this purpose in airports, ports, and prisons, it has a rather high rate of false positives: up to 5% for RDX and 25% for TNT in the analysis of blank soils (Fernandez-Maestre et al., 2022). Other useful analytical methods are gas chromatography coupled with mass spectrometry, energy dispersive X-ray diffraction, nuclear quadrupole resonance, neutron activation, and longwave infrared spectral reflectance. They are covered in a recent book published by Elsevier, "Counterterrorist Detection Techniques of Explosives" (Kagan & Oxley, 2022). By no means are responsive polymers intended to replace these techniques, in which all further advances are encouraged and will be welcome.

The general purpose of the proposals that are based on polymers is to offer possibilities of detection that, despite not matching the analytical performance of IMS or other specific techniques, offer great user friendliness, do not require trained staff or animals, and provide a quick response. Furthermore, with the aid of analytical devices such as spectrofluorimeters, colorimeters, particle size detectors, and voltameters, the accurate quantification of nitroexplosives in fluid or solid samples is factible and, in fact, generally reported in practically every recent work on the topic (Duraimurugan & Siva, 2016; Feng et al., 2012; Goudappagouda et al., 2020).

A typical flaw of the most basic and commercially available responsive polymers, even when using up-to-date analytical devices, lies in their sensitivity. Arguably, it is a consequence of what allows the system to be used for quantification over a broad concentration range. The response, be it color change, luminescence quenching, luminescence enhancement, or alteration of the electrical resistivity of the polymer, is generally monotonously increasing with the concentration of a colored complex, a non-fluorescent adduct, or any other product of the association between the target analyte and the polymer. Therefore, the signal-to-noise ratio will tend to be low in dilute systems. Nonetheless, researchers have come up with covalent and non-covalent modifications that attain extraordinarily high values of the Stern-Volmer constant (in the case of fluorescence quenching), the extinction coefficient (in the case of colorimetric detection), or other parameters that may express proportionality between the concentration of the nitro compound and the response.

18.3 Altering the luminescence of conjugated polymers

18.3.1 The simplest being we know

Although responsive polymers of many different kinds have been proposed for the detection of nitroexplosives, no category is more frequently found in the literature than that of conjugated polymers with luminescent properties. Fluorescent polymers are often based on fluorene units (Yang et al., 2010), fluorenone units (Mothika et al., 2018), phenylene units (Duraimurugan & Siva, 2016), anthracene units (Goudappagouda et al., 2020), triphenyl or tetraphenyl moieties (Zhou et al., 2016), or pyrene units (Gou et al., 2022), among other possibilities. Nonetheless, perhaps the most classical option in this sense is polyacetylene (Liu et al., 2001), while polymers and copolymers of fluorene and/or phenylene have probably been the most common choice as of recently (Malik et al., 2023).

The alternating double and single bonds of conjugated polymers involve a system of delocalized π electrons, owing to overlapping *p* orbitals. This has the following implications:

- They are photoluminescent since conjugated entities have great absorptivity of highly energetic radiation, and π electrons require less energy to reach an excitation state than in non-delocalized systems. Hence, even if their quantum yield is low, they can be good fluorescence emitters (Chou et al., 2005).
- They are, in general, electrically conducting polymers—a discovery that provided Heeger, MacDiarmid, and Shirakawa with the 2000 Nobel Prize in Chemistry (Shirakawa et al., 1977).
- They undergo an amplified quenching effect: a single quencher molecule can reduce or suppress the fluorescence emission of a whole polymer chain (Malik et al., 2023).

The heading of this subsection comes from Goethe: "[Light is] the simplest, most undivided, most homogeneous being that we know." The quantum mechanics of photons may be as intricate as we can make it, but the raw phenomenon of light is indeed simple and universal. In this case, detecting whether or not a polymeric material or a polymer-impregnated material emits light under certain conditions is a really quick realization. Then, the intensity of the emission and/or the extent of quenching can be quantified by means of a fluorescence spectrophotometer, or "spectrofluorometer." Spectrofluorometers may also record the excitation spectra, the photoluminescence quantum efficiency (PLQE), the fluorescence lifetime, or the radiative and non-radiative decay rates if desired (Fan et al., 2023; Zhang et al., 2020).

The broad absorption bands of conjugated polymers indicate that many different sources of radiation are eligible for excitation. During excitation, the electromagnetic radiation absorbed by the polymer promotes electrons to a π^* antibonding orbital. Transitions are of the $\pi-\pi^*$ type if electrons are promoted from a low-energy π bonding orbital, or $n-\pi^*$ transitions if they come from an *n*-orbital. The aforementioned fact that even affordable ultraviolet (UV)-A lamps suffice for excitation implies a relatively low $\pi-\pi^*$ transition energy. Obviously, $n-\pi^*$ transitions require absorbing more energy, and thus shorter wavelengths are generally necessary. Except in the case of certain chemical modifications, both the absorption due to $\pi-\pi^*$ transitions and the emission due to π^*-π radiative recombination are more intense than their $n-\pi^*$ counterparts.

Usually, but not always, the polymers used for colorimetric detection (e.g., unsubstituted polyfluorene) are colorless under conventional halogen lamps, sodium vapor lamps, incandescent tungsten light bulbs, and conventional LEDs (Abdel-Rahman et al., 2017). Nonetheless, solar radiation, cheap bulbs with significant emission in the UV region, and affordable UV-A lamps (even those commercialized as "black light") will do the job, making polyfluorene display its characteristic blue emission (Aguado, Gomes, et al., 2023). Although blue is the most typical emission color, the fluorescence of conjugated polymers can be modulated to emit yellow light, red light, green light, or light at other wavelengths by modifying their backbone with proper functional groups, such as benzothiadiazole, pyrrole, or thiophene units (Kim et al., 2018; Zheng, Duan, et al., 2023). In another context, the copolymerization of fluorene or other conjugated monomers in the presence of perylene dyes is a long-known strategy (Ego et al., 2003). These kinds of functionalizations amplify the energy difference between excitation and emission. Finally, there are also examples of successful non-covalent modification, including the incorporation of europium(III) ions so that the polymer works as a sensitizer for said lanthanide ions, resulting in their characteristic bright red emission (Turchetti et al., 2015).

18.3.2 Fluorescence quenching mediated by (photoinduced) electron transfer

Theorizing and modeling electron transfer reactions gave Rudolph A. Marcus the 1992 Nobel Prize in Chemistry (Nobel Prize Outreach, 1992). According to the Marcus theory, electron transfer in solution implies the rearrangement of solvent molecules in the vicinity of both the donor and the acceptor, not violating the restrictions of the Franck-Condon principle. Marcus proposed models to calculate not only the energy barrier between reactants and products but also the rate of electron transfer. The rate equation is of little usefulness in what pertains to the detection of nitroaromatic compounds, given that their interaction with conjugated polymers is usually strong and electron transfer is fast in the presence of a light source. Nonetheless, the photoluminescence of some polymers is sometimes selectively quenched by a certain nitroexplosive (e.g., by TNT but not by 2,4-DNT), and thus the Gibbs free energy of activation (ΔG), provided that the reactants and/or the products are uncharged, is:

$$\Delta G = \frac{1}{\lambda}\left(1 + \frac{\Delta G^{\circ}}{\lambda}\right)^2$$

where λ is the reorganization energy and ΔG° is the Gibbs free energy difference when the reactants and products are an infinite distance apart. The value of $-\Delta G^{\circ}$ is deemed the driving force of the process.

Nitroarenes are avid electron acceptors. Out of the ones mentioned to this point (including 2,4-DNT and Tetryl), TNT and PA are the strongest Lewis acids, and thus electron transfer reactions that involve them will have greater $-\Delta G^{\circ}$ values. This is the principle behind the many works on the selective detection of both (Gou et al., 2022; Xu et al., 2011), more particularly of TNT (Nguyen et al., 2023). In any case, the electron-withdrawing effect of the nitro groups ($-NO_2$) decreases the electron density of the aromatic ring, easing its association with electron-rich entities. Needless to say, this includes conjugated polymers with delocalized π electrons populating their backbone. The formation of an adduct by charge transfer is driven by the tendency to balance electron densities between the electron-rich polymer and the electron-deficient nitroaromatic compound. This transfer generally takes place from the excited state of the photoluminescent polymer (π^*) to the nitroarene. The energy required is usually supplied as photons, conveniently provided by the same light source that promotes the excitation of the conjugated polymer. Hence, charge transfer phenomena between conjugated polymers and nitro compounds often appear in the literature as "photoinduced electron transfer," abbreviated as "PET" (Martelo et al., 2019).

A simplified and generalized mechanism of fluorescence quenching by PET phenomena is schematized in Fig. 18.3. As aforementioned, the particularity of conjugated polymers is that the photoluminescence of many units of the conjugated structure, as long as it has not been intentionally interrupted during polymerization (e.g., co-polymerizing fluorene or *p*-phenylene oxide with nonconjugated monomers), can be deactivated with a single electron acceptor molecule. This is why films from conventional conjugated polymers typically have detection limits in the micromolar range, or around 0.1–1 ppm of TNT or PA (Toal & Trogler, 2006), clearly corresponding to a number of deactivated polymer units that exceed the number of nitroarene molecules in the analyte.

Even though detection in the micromolar range may suffice for soils contaminated by military production sites, it does not for groundwater, and much less for drinking water. That said, certain chemical modifications and strategies for specific supramolecular arrangements result in extraordinarily sensitive detection by fluorescence quenching, with limits of detection for TNT of 23 ppb (Xu et al., 2011) or 64 nM, that is, 15 ppb (Öztürk & Şehitoğlu, 2019). An approach involving hyperbranched conjugated polymer nanoparticles attained a detection limit of TNT in water as low as 0.8 ppb (Wu et al., 2017). Furthermore, it is one of the few cases in which quenching was selective toward TNT without the interference of PA.

The PET mechanism involving conjugated polymers and nitroarenes, as described here, generally causes static quenching as opposed to dynamic or collisional quenching (Genovese et al., 2021). Static quenching corresponds to complex formation, and the numerous binding sites along a polymer chain, along with the possibility of π-π interactions between the conjugated polymer and nitroarenes, favor it. Indeed, as evidenced by shifted emission bands, there is frequent exciplex formation between donor and acceptor (Sun et al., 2015), although not in the case of polyacetylene (Toal et al., 2007).

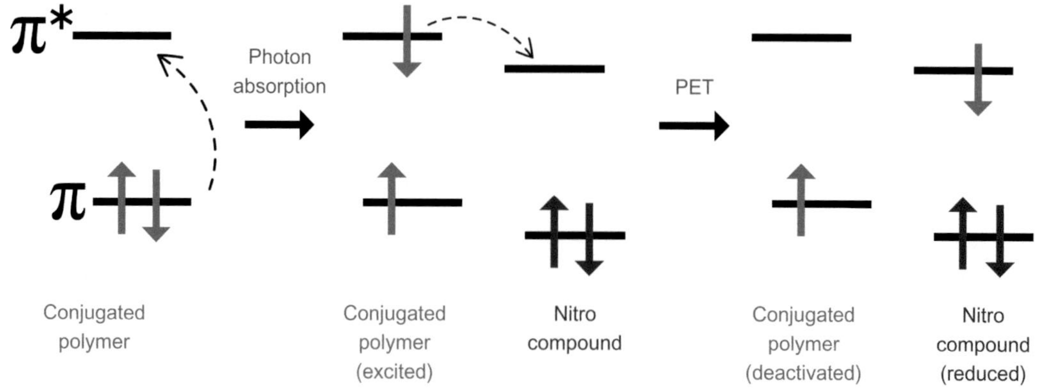

FIGURE 18.3

Jablonski-like diagram of photoluminescence quenching by photoinduced electron transfer between a conjugated polymer and a nitroexplosive.

Stating quenching certainly eases the quantification of the target nitroexplosives based on the extent of emission diminishment. Under these conditions, and considering that nitroarenes can often be regarded as ideal quenchers for conjugated polymers, the ratio of the emission intensity without quencher (I_0) to the emission intensity with quencher (I) follows a linear trend with the concentration of quencher. Hence, data from the intensity maxima in emission spectra are linearly fitted to the Stern-Volmer plot:

$$I_0/I = 1 + K_{SV}[Q]$$

where K_{SV} is often referred to as the "Stern-Volmer constant" and $[Q]$ is the molar concentration of quencher molecules (e.g., TNT). The higher the Stern-Volmer constant, the better the sensitivity of the system. In the aforementioned case with an exceptionally low detection limit (0.8 ppb), the Stern-Volmer constant was $1.2 \cdot 10^6\ M^{-1}$ (Wu et al., 2017). Improving these parameters by means of covalent and/or non-covalent modifications has recently been and keeps being a great source of motivation for the research on conjugated polymers as chemosensors. Fig. 18.4A–C displays some typical conjugated polymers in their native forms, but innovation generally comes from their modification in many different ways.

Nitramine explosives such as RDX and HMX are electron acceptors as well, although the saturated nature of their bonds excludes the possibility of $\pi-\pi$ interactions between donor and acceptor. Furthermore, while the LUMO energy of TNT has been estimated as -3.65 eV (Zhu et al., 2022) or

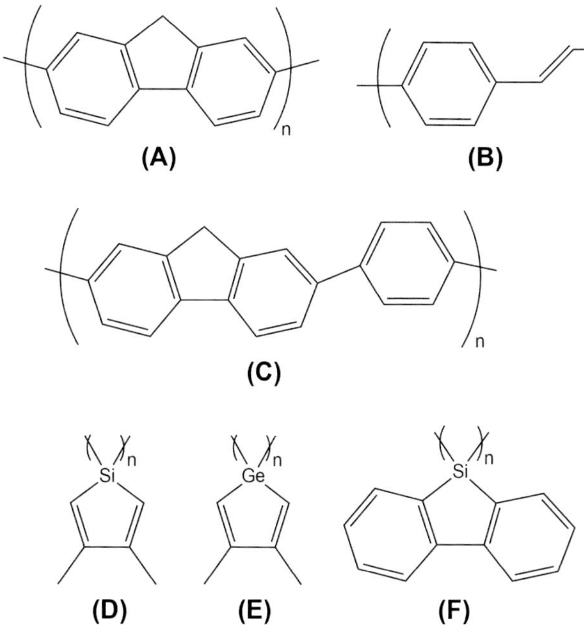

FIGURE 18.4

Structures of unsubstituted polyfluorene (A), poly(p-phenylene vinylene) (B), poly(fluorene-1,4-phenylene) (C), poly(2,3-dimethyl silole) (D), poly(2,3-dimethyl germole) (E), and polysilafluorene (F).

−3.93 eV (Noh et al., 2022), accounting for a LUMO-HOMO difference of roughly 6.0 eV, this gap is approximately 7.0 eV in the case of RDX, with a LUMO energy of −2.96 eV (Kovalev et al., 2018), and 7.3 eV in the case of HMX, with a LUMO energy of −2.25 eV (Zhu et al., 2022). All considered, electron transfer to saturated nitramines is less favored in both thermodynamic and kinetic terms. That said, the successful detection of RDX with silole-based polymers has been reported (Sanchez et al., 2009). Referring to those conjugated macromolecules requires the introduction of polymetalloles.

Polymetalloles comprise metal (or metalloid) and organic units along a conjugated polymer backbone. Metal or metalloid atoms are simultaneously linked among themselves and coordinated with conjugated organic ligands, generally substituted cyclopentadiene. Interestingly enough, the detection of nitroexplosives was probably the most popular application for polymetalloles at the beginning of the century (Sohn et al., 2003). The most common metallole units are siloles, germoles, and silafluorene units. Fig. 18.4D–E shows the simplest polymers resulting from those monomers, but, similarly to the case of organic conjugated polymers, further substitution is usual.

Poly(tetraphenylsilole-vinylene), poly(tetraphenylsilole-silafluorene-vinylene), and poly(silafluorene-vinylene) can be used for the detection and quantification of TNT, DNT, and PA by fluorescence quenching, with Stern-Volmer constants of up to 2×10^4 L/mol. Poly(silafluorene-vinylene) could also be used for tetryl and nitramine explosives such as RDX and HMX (Sanchez et al., 2007). Photoluminescent polymetalloles, such as poly(1,4-diethynylbenzene)2,3,4,5-tetraphenylsilole, poly(1,4-diethynylbenzene)2,3,4,5-tetraphenylgermole, and poly(1,4-diethynylbenzene)silafluorene, were shown to undergo fluorescence quenching in the presence of TNT, DNT, and PA (Toal & Trogler, 2006). Nonetheless, as few research groups were dedicated to their study, the popularity of polymetalloles for explosive detection decreased after the first decade of the 21st century.

18.3.3 Fluorescence quenching mediated by energy transfer

Different energy transfer mechanisms may promote fluorescence quenching, possibly coexisting with electron transfer. In the most typical case, Förster resonance energy transfer (FRET), the conjugated polymer in its excited state transfers energy to an acceptor molecule in its vicinity. Regarding the acceptor, nitrophenols such as PA are generally easier to detect by FRET-driven quenching than aprotic nitroaromatics and nitramines. The reasons behind this preference are the acidity of PA and its higher absorptivity in the UV-A region, significantly overlapping the emission spectra of different photoluminescent polymers (Khalil et al., 2020).

Even before Theodor Förster finished school, Jean B. Perrin estimated that two molecules could exchange energy if the distance between them was ~15 nm or less (Perrin, 1927). It was two decades later when Förster postulated two formidable models for energy transfer—one based on classical mechanics and another one rooted in quantum mechanics (Förster, 1948). He established that energy transfer is mediated by dipole-dipole interactions between donor and acceptor and derived the following equation for the energy transfer rate (k_{ET}):

$$k_{ET} = \frac{1}{\tau_D}\left(\frac{R_0}{R}\right)^6$$

where τ_D is the unperturbed donor lifetime, R is the distance between donor and acceptor, and R_0 is the critical distance at which the energy transfer is half-maximal. The energy transfer efficiency (ETE) is then strongly dependent on the intermolecular distance (Müller et al., 2013):

$$ETE = \frac{1}{1 + (R/R_0)^6}$$

To a slight extent and with little relevance in most cases, FRET coexists with electron transfer in fluorescence quenching phenomena because the emission intensity of the conjugated polymer (typically emitting blue light) somehow overlaps the absorption spectrum of yellow or orange nitroaromatic compounds in the visible region (2,4-DNT, TNT, PA, Tetryl, etc.).[3] Understandably, it is seldom mentioned unless intended to be a key quenching mechanism, as its influence on the photoluminescence of conjugated polymers is generally outweighed by that of electron transfer. After all, as a matter of several orders of magnitude, nitroarenes have much higher molar absorptivity in the UV-C region (<280 nm) than in the visible region, except PA (Hughes et al., 2015). Both PET and FRET were identified as significant mechanisms for quenching in a work involving truxene-based polymers as donors, resulting in higher sensitivity toward PA than toward TNT and in deviations from linearity in Stern-Volmer plots (Nailwal et al., 2022).

The FRET mechanism was also claimed to be responsible for the quenching of cadmium-coordinated N,N,N′,N′-tetrakis(4-(4H-1,2,4-triazol-4-yl)phenyl)benzene-1,4-diamine polymers by nitrophenols, including PA. Interestingly, with a concentration of 3.5×10^{-4} M, aprotic nitroaromatic compounds attained quenching efficiencies below 10%, while that of PA was 87%. K_{SV} for PA was $4.7 \times 10^4\ M^{-1}$. More recently and similarly, a zinc-coordinated polymer attained a slightly lower K_{SV} for PA, $1.6 \times 10^4\ M^{-1}$, but higher quenching efficiencies were observed for 2,4-dinitrophenol and especially for 4-nitrophenol (Dascălu et al., 2023).

In 2020, researchers from the Indian Institute of Technology (Roorkee) and the Indian Institute of Science Education and Research (Kolkata) used poly[(N,N-dimethylacrylamide)-co-(*tert*-butyl carbamate-tryptophan-ethyl methacrylate)] as a donor (Kumar et al., 2020). Unlike that of polyfluorenes (blue emission), the photoluminescence emission of this copolymer is mainly located in the UV region, granting great overlap with the absorption spectrum of nitroaromatics. Furthermore, to provide the system with selectivity, the donor polymer was paired with carbazole as an acceptor. This way, nitroaromatics competed with it and prevailed. K_{SV} values were estimated as $5.3 \times 10^3\ M^{-1}$, $5.7 \times 10^3\ M^{-1}$, and $2.9 \times 10^4\ M^{-1}$ for 2,4-DNT, TNT, and PA, respectively. This corresponded to limits of detection in the micromolar range, while ETE was roughly 35%. The same group of authors improved the sensitivity of the polymer toward nitro compounds in 2021, using a more sophisticated approach (Kumar et al., 2021). In the latter case, a dansyl moiety-containing copolymer was used as the acceptor, but its role was not that of competing with nitroarenes. Instead, quenching did not take place in the donor-acceptor pair in the absence of nitro compounds, but favorable interactions between them and the accepting copolymer mediated FRET phenomena.

[3]This is not the case of RDX, which is a white crystalline solid with strong absorptivity below 260 nm.

18.3.4 Polymer-coated quantum dots and quantum dot-doped polymers

Within the realm of fluorescence alterations, the optical properties of QDs offer interesting opportunities for the detection of nitroexplosives. The term first appeared in the literature in the mid-eighties (Reed et al., 1986), but it had been synthesized at the end of the 1970s by Alexey Ekimov. He observed the unexpected luminescent properties of nanocrystals of copper chloride and calcium selenide, which he referred to as "semiconductor microcrystals" (Ekimov & Onushchenko, 1982). Later, Ekimov and Alexander Efros postulated a theory on the "quantum confinement effects" of QDs (Ekimov et al., 1985). In parallel and independently, on the other side of the Cold War, Louis E. Brus synthesized his own "small semiconductor crystallites" and related the wavelength emitted to their particle size (Brus, 1984). As their size changes, so does the color of the light emitted when QDs are photoexcited, ranging from the visible region to infrared wavelengths. At Bells Lab, Brus mentored Moungi Bawendi, who made the reproducible synthesis of monodisperse QD suspensions possible (Murray et al., 1993). In 2023, owing to their research on QDs, Ekimov, Brus, and Bawendi were awarded the Nobel Prize in Chemistry.

Although the use of QDs exceeds by far that of a fluorescence-based probe, their size-dependent photoluminescence provides experimenters with the possibility of tuning the emission wavelength. This gives us control over the FRET phenomena described above, for example, making the emission band match the absorption peak of a certain nitroexplosive. Moreover, QDs can emit light at extremely narrow wavelengths, allowing for the differentiation between different nitro compounds based on their spectral fingerprints. Finally, due to their small size, when suspended in a liquid, gravitational forces will not outweigh Brownian motion, and they will stay stably dispersed. It should be noted that, although traditional QDs consist of semiconducting salts or oxides, this is one of the many areas affected by the discovery of graphene at the beginning of the 21st century. Graphene QDs are also called "carbon dots." They have a spherical core formed by graphene stacking and a surface that is rich in functional groups (carboxyl, carbonyl, epoxy, hydroxyl, etc.).

Many applications require the surface modification of QDs and/or their dispersion in a solid matrix, and here is where polymers can play relevant roles (Tomczak et al., 2009). On one hand, they can play the role of a transparent or at least UV-dull matrix for applications in optical sensing. Some popular polymer matrices are paper (Tian et al., 2017), polyethylene (Ushakov & Kosobudsky, 2019), and even poly(lactic acid) for additive manufacturing (Brubaker et al., 2018). On the other hand, when attached to the surface of QDs, they can offer hydrophilicity with terminal polar groups, hydrophobicity with alkyl chains, or complex formation ability with Lewis-basic or Lewis-acidic sites (Hezinger et al., 2008). Needless to say, the latter function is of utmost importance for detection purposes.

Electrospinning can achieve a good dispersion of CdSe QDs or carbon dots in thermoplastic matrices, including poly(vinyl chloride), polystyrene, and poly(methyl methacrylate). For the non-selective detection of vapors such as DNT, TNT, RDX, and triacetone triperoxide (not a nitro compound), the green fluorescent emission of the composite was progressively quenched. Out of the polymers used as matrix, polystyrene was the one that provided the highest signal-to-noise ratio, probably due to the establishment of π-π interactions with DNT and RDX. These kinds of systems, in which QDs lack specific functionalization, are easy to build and user-friendly but fall short in terms of selectivity and sensitivity.

Regarding polymer-functionalized QDs, the functional macromolecules will need to have grafting groups to attach to the interface—groups that, in most cases, will establish dative bonds toward the metal ions or functional groups of the semiconductor nanoparticles. Thiol groups, phosphine moieties, and amine groups are generally suitable for this task. As an alternative, functional groups of the polymer can be attached to a crosslinker that, in turn, is anchored on the surface of QDs. For example, it is known that nitroaromatics form complexes with polyaniline by accepting electrons from the latter. In fact, polyaniline alone suffices as a probe of different kinds, including photothermal and electrochemical, but detection limits below 1 nM (for PA) could only be achieved by using this conjugated polymer with CdTe (core)/ZnS (shell) QDs (Dutta et al., 2015). They used 2-mercaptosuccinic acid as a crosslinker between QDs and the polymer, promoting the transfer of excitons to the latter in the presence of UV radiation. Polyaniline's luminescence was ultrasensitively quenched by PA, as hinted before. The effects exerted by TNT were not addressed, but its ability to cause quenching on the same system is likely. In a more recent study, polyethyleneimine (PEI) chains were attached to carbon dots on one unit and bound to the nitroarene on a distant unit. The formation of an anionic Meisenheimer complex, not unlike in the case of polyaniline, provided graphene QDs with rapid, sensitive, and selective fluorescence quenching in the presence of TNT (Şen et al., 2022).

18.3.5 Systems based on aggregation-induced emission

In 2001, Ben Zhong Tang's group (Hong Kong University of Science and Technology) found that aggregation of 1-methyl-1,2,3,4,5-pentaphenylsilole, simply by adding water, increased its PLQE from <0.1% to 21% (Luo et al., 2001). The mechanism underlying aggregation-induced emission (AIE) is generally accepted to consist of the restriction of intramolecular motion (Zhang et al., 2023). Two examples are highlighted in Fig. 18.5: restriction of intramolecular rotation in the case of a tetraphenyl derivative and restriction of vibration in the case of cyclooctatetrathiophene. Both molecules have conjugated structures, but their vibration or rotation causes non-radiative decay, making electrons in excited states return to the ground state without photon emission. Aggregation and the subsequent motion restriction convert, to a lesser or greater extent, these non-radiative decay phenomena into radiative recombination.

The same principle can be applied to many polymers that are weak fluorophores in solution but whose aggregates have strong photoluminescence emission. When it comes to the detection of nitroexplosives, two opposite responses are then possible: turning off the emission of the aggregates or promoting a turn-on effect by contributing to aggregation. While AIE-based detection systems are often related to enhancement, fluorescence quenching phenomena still prevail for nitro compounds. In these cases, turn-off may be not only caused by PET and/or FRET, as described above, but also by disrupting the aggregates.

The fluorescent green emission from nanoaggregates of tetraphenyl derivatives with arylimidazole groups was selectively suppressed by PA, attaining K_{SV} values as high as $1.6 \times 10^6\ M^{-1}$. Both PET and FRET were postulated as quenching mechanisms. In a less conventional case of AEI luminogens, organic molecular cages with triphenylphosphine units also allowed for the selective and ultrasensitive detection of PA (Tao et al., 2022). Finally, in a more recent approach, polymers with a non-conjugated backbone (a polycarbonate backbone) but with conjugated pendant units were synthesized and proved to display AIE (Huang et al., 2023). Once again, the photoluminescence of the system was selectively quenched by PA, while aprotic nitroaromatics had little effect. Although the overlapping absorption and emission spectra are generally provided as reasons for this selectivity, the possibility of hydrogen bonding should also be taken into account, given how important

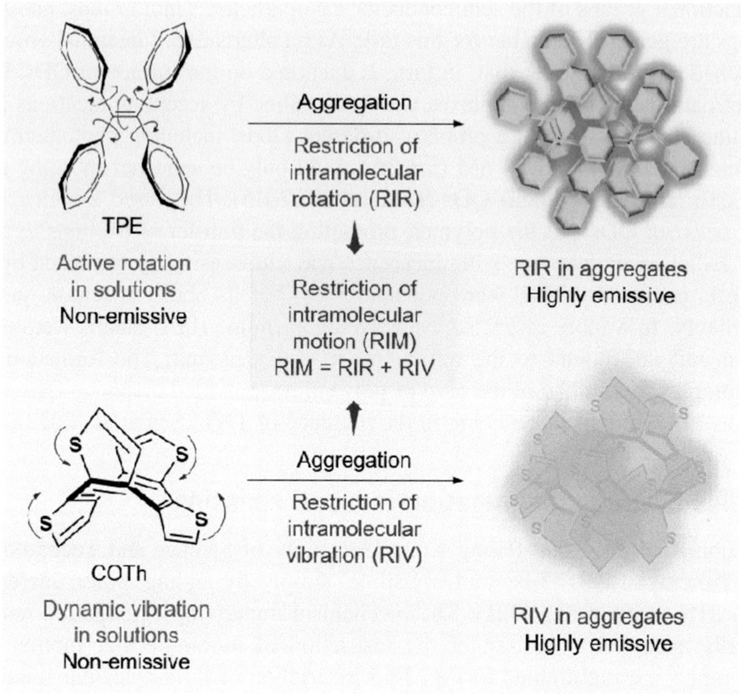

FIGURE 18.5

Aggregation-induced emission by restriction of intramolecular motion, either rotation or vibration.

From Zhang, Z., Zhu, L., Deng, Z., Zeng, J., Cai, X.-M., Qiu, Z., et al. (2023). Aggregation-induced emission biomaterials for antipathogen medical applications: Detecting, imaging and killing. *ChemRxiv. Cambridge: Cambridge Open Engage.*

interatomic or intermolecular distances are for FRET phenomena. Likewise, it should be noted that these approaches involving AIE, followed by quenching, tended to achieve better sensitivity and selectivity toward PA than those based on direct FRET-mediated quenching.

As hinted above, AIE offers some possibilities for fluorescence enhancement as well. A system involving protamine-capped gold nanoparticles, with weak photoluminescence if well-dispersed, underwent aggregation and subsequent AIE in the presence of TNT, working as a polydentate ligand for protamine chains and promoting the agglomeration of nanoparticles (Bener et al., 2022). Another turn-on proposal was based on induced polymerization, which should not be classified as AIE, but it also restricts intramolecular motion. A nitro compound could exert oxidative polymerization on conjugated units, such as 9,9-dihydridofluorene (Sanchez & Trogler, 2008).

Proposals for the detection of nitro compounds based on fluorescence turn-on are less traditional and generally less straightforward to fabricate than those based on fluorescence turn-off. Allegedly, they tend to offer two advantages: (1) more ease of detection in the absence of analytical devices, because a slight increase in brightness over a dark surface seems more evident than a slight decrease in emission from an already bright film; (2) better selectivity to a certain target, since triggering a response will usually imply having interacted with both donor and acceptor moieties. That said, turn-on systems are not necessarily more sensitive than fluorescence quenching-based systems, especially considering the recent advances that have been exposed to this point. Besides AIE and oxidative polymerization, there are several strategies that could lead to fluorescence enhancement:

- An intrinsically luminescent macromolecule has its photoluminescence quenched by PET to accept moieties that are intentionally included in the system. Then, the presence of a target nitroexplosive triggers a chemical reaction that interrupts charge transfer phenomena (Madhu et al., 2014).
- A conjugated polymer includes in its backbone electron acceptor units that inhibit its fluorescence, as in the previous case. Unlike triggering a chemical reaction, the target compound in its excited state (with a lower LUMO energy than that of the paired acceptor) works as a charge carrier, increasing the decay lifetime by more than one order of magnitude (Balan et al., 2012).

All in all, common conjugated polymers can be directly used for nitroexplosive detection, but modifying their backbone, their side chains, and/or their supramolecular structure is essential to generating innovative proposals with enhanced sensitivity and selectivity. Nonetheless, applications of polymers with high electronic delocalization do not only stand high because of their optical properties but also because of the electrical properties of many of them.

18.4 Polymers for electrochemical detection

18.4.1 The thousandfold mistake

Shirakawa and Ikeda first synthesized a conducting polyacetylene thin film by a fortuitous error, as narrated by Shirakawa himself in his 2000 Nobel Prize lecture (Shirakawa, 2000): "After a series of experiments to reproduce the error, we noticed that we had used a concentration of the Ziegler-Natta catalyst nearly one thousand times greater than that usually used." It should be noted that pristine polyacetylene is an insulator, but the introduction of charge carriers allowed electrons to move freely along the polymer chains. The accidental doping of polyacetylene led to the development of a new class of materials—conducting polymers.

Currently, the use of polyacetylene for these applications is overwhelmed by other polymers whose conductivity depends on their degree of oxidation:

- Polyaniline and its derivatives;
- polythiophene and its derivatives, mainly poly(3,4-ethylenedioxythiophene) (PEDOT);
- polypyrrole and its derivatives;
- poly(phenylenediamines):
- conducting polymetalloles such as polyselenophene;
- porphyrin polymers.

For instance, fully reduced polyaniline, that is, consisting completely of −Ph−NH− units, is named "leucoemeraldine," and it acts as an insulator. Its oxidation results in "emeraldine salt," which has $-Ph=NH^{+}-$ units that can be stabilized with an anion. Likewise, highly oxidized PEDOT (with $-S^{+}-$ in its charged thiophene units) is frequently paired with poly(styrene sulfonate) (PSS) for the dual objective of stabilization and ease of dispersion in water (Shi, Liu, et al., 2015). In general, the greater the extent to which a conjugated polymer is oxidized, the greater its conductivity. This principle allows for their use in both anodes and cathodes, although the latter tend to be more frequent than the former: the query "polyaniline cathode" yielded 1113 results in Scopus (as of August 16, 2023), while "polyaniline anode" provided 511 matches.

Once conducting polymers have been introduced, the next unavoidable question is related to the role of nitroexplosives in this scheme. They have been presented as electron acceptors and, thus, as oxidants. It is known that the electrochemical reduction of TNT in water displays three distinct peaks (one per nitro group) in cyclic voltammetry assays (Yu et al., 2017). In protic solvents, reduction implies six electrons and six protons per nitro group, totaling 18 electrons and 18 protons per TNT molecule. Hence, the reduction of 2,4-DNT will imply 12 electrons and 12 protons. In aprotic solvents, the reduction mechanism has not been fully understood as far as I am concerned, but the reversible generation of radical anions has been shown to involve one electron per nitro group (Olson et al., 2015), although further exchange can lead to diradical anions, at least in the case of 2,4-DNB (Dief et al., 2022). Fig. 18.6 shows the reduction pathways that have been proposed for 2,4-DNT in protic and aprotic systems.

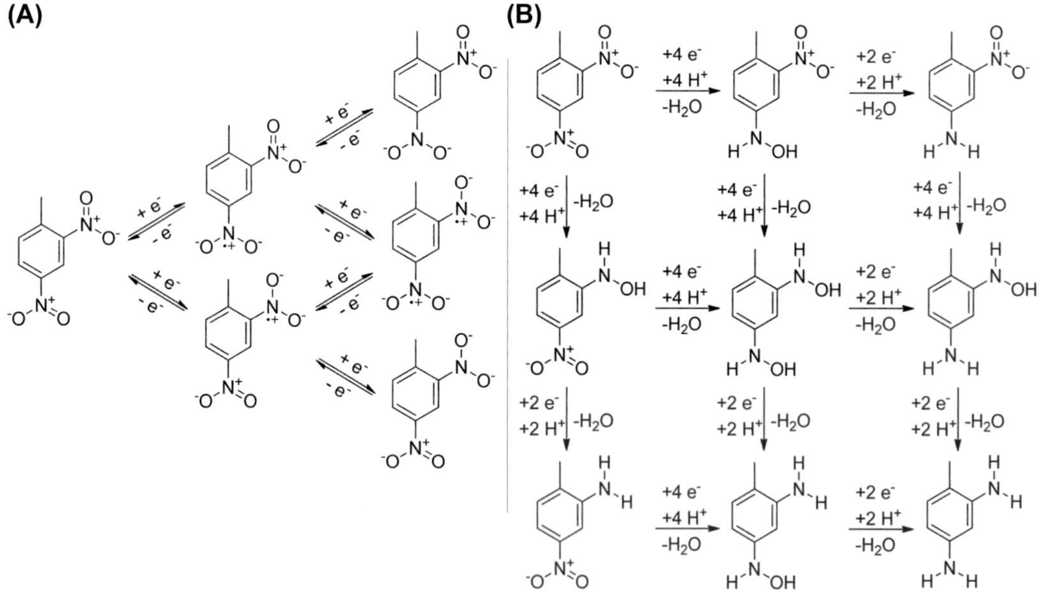

FIGURE 18.6

Reduction pathways for 2,4-DNT in aprotic solvents (A) and in water (B).

Adapted with permission from: Olson, E. J., Isley, W. C. I., Brennan, J. E., Cramer, C. J., & Bühlmann, P. (2015). Electrochemical reduction of 2,4-dinitrotoluene in aprotic and pH-buffered media. The Journal of Physical Chemistry C, 119*(23), 13088–13097. https://doi.org/10.1021/acs.jpcc.5b02840.*

The equally interesting electrochemical reduction of nitramine explosives, such as RDX and HMX, has also been studied in aprotic media (Holubowitch et al., 2020). Even in dried acetonitrile, the reduction was postulated to be driven forward by the acidic protons from trace water, which is practically unavoidable (Hui & Webster, 2011). The reduction of each nitro group then generates a hydroxide ion (one proton exchanged) or a water molecule (involving two protons), involving a gain of two electrons. The resulting nitroso groups are prone to dimerization or even polymerization, and the subsequent azodioxy moieties can be further reduced to azoxy, azo, and hydrazo groups. In total, the complete reduction of RDX or HMX implies five electrons per nitro group.

18.4.2 Reduced nitro compound, doped polymer?

In the most typical case, a nitroexplosive undergoes reduction over a conducting polymer/composite film or a polymer-coated working electrode by applying a certain potential or voltage range. Ag/AgCl is the most typical reference electrode. The electron loss from the polymer enhances its conductivity, allowing the user to record a higher current intensity. In general, the change in intensity is directly proportional to the number of exchanged electrons (Faraday's first law) and thus to the concentration of the analyte. The most popular electrochemical technique for the detection of nitro compounds is probably differential pulse voltammetry, although some alternatives include cyclic voltammetry, linear sweep voltammetry, fast amperometry, and stripping voltammetry.

Common conducting polymers, including polyaniline, polythiophene, and their derivatives, work as p-type semiconductors. This means that the majority of charge carriers are holes, while electrons are the minority carriers. PA, TNT, RDX, and electron acceptors in general are well-known dopants for p-type semiconductors (Small et al., 2012). That said, nanocomposite films and composite-coated electrodes are more commonly found among recent works (2014–2023) than unblended conducting polymers, and certain hybrid materials may display n-type behavior despite containing a p-type conducting polymer.

As one of many examples of the prevalence of composites, a carbon paste electrode modified with both PEDOT and carbon nanotubes showed increases in current intensity upon the addition of 4-nitrobenzene, attaining a detection limit of 83 nM (i.e., 10 ppb) (Xu et al., 2014). Likewise, TNT concentration exerted a linear increase in current intensity over an electrode coated with gold nanoparticles and poly(o-phenylenediamine–aniline) (Sağlam et al., 2015). In another study, electrodeposited molybdenum disulfide-poly(m-aminobenzenesulfonic acid) was shown to have its conductivity greatly enhanced by TNT in aqueous media, especially at pH 6 (Yang et al., 2016). Assays based on differential pulse voltammetry resulted in very high sensitivity, with a detection limit as low as 0.043 ppb and a linear range (current intensity—TNT concentration) of 0.1–200 ppb.

Recent contributions tend to be increasingly complex in terms of electrode fabrication. Layered double hydroxides of Co and Al were combined with silver nanoparticles and poly(o-phenylenediamine) to coat a glassy carbon electrode (Dhanasekaran et al., 2019). The current intensity increased linearly with the concentration of 4-nitrophenol, at least up to 13 μM. 4-nitrophenol was also electrochemically reduced to p-hydroxyaminophenol over $ZnFe_2O_4$-decorated polyaniline-coupled graphene nanosheets, resulting in direct proportionality between the current and the concentration of the nitro compound (Wei et al., 2021).

In some systems, interactions between conducting polymers and nitroarenes increase the dynamic resistance of the former instead of boosting their conductivity. This is the case of a recent report on reduced graphene oxide@MoS_2@PEDOT:PSS nanocomposite films (Gupta et al., 2023). Unlike the aforementioned p-type conducting polymers, these films had n-type behavior, implying that electrons were the majority carriers. Interactions with TNT disrupted the continuity of the electronic conduction paths. Fortunately, the dedoping effect in this sort of system (based on n-type semiconductors) is compatible with highly sensitive detection. The limit of detection was 0.1–0.45 ppb, and the linear range of the resistance–concentration plot was from 1.5 to 47 ppb.

Finally, although the most widely used conducting polymers are petro-based and non-biodegradable, bio-based and/or biodegradable alternatives are also plausible. Polycurcumin methacrylate, an intrinsically conducting polymer, displayed a 90.44% increase in current density in the presence of PA (Gogoi et al., 2013).

Often, these systems are alleged to be selective toward nitroaromatic compounds, at least at certain voltage values, given that both a certain degree of electron deficiency (or a certain reduction potential) and the capability to establish π-π interactions are requirements to trigger an electrochemical response at a certain order of magnitude (e.g., 10–100 μA). However, among said nitroaromatic compounds, detection is rarely selective toward one or two in particular. Furthermore, the frequency by which interference studies are performed is lower than the frequency by which selectivity towards nitroexplosives is mentioned as a feature of the electrochemical system.

18.4.3 Molecular imprinting: toward higher selectivity

Molecularly imprinting polymers (MIPs) are half a century old, but contributions in the last decade outweigh those in the first forty years, probably because computational chemistry eases the experimental design a lot. The technique was first presented by Günter Wulff in 1972 in a meeting held in Freiburg, provoking a discussion that lasted for 45 min (Wulff, 2013). The same year, Wulff and his PhD student Ali Sarhan published the paper "Use of polymers with enzyme-analogous structures for resolution of racemates" (Wulff & Sarhan, 1972). At some point, those "enzyme-analogous polymers" were renamed MIPs.

The technique usually implies the polymerization of functional monomers in the presence of the template molecule (in this case, a nitroaromatic or nitramine compound), along with radical initiators and crosslinkers, if necessary, in porogenic solvents. First, the monomers, with or without crosslinkers, form a complex with the nitro compound. Then, the functional monomers are polymerized, and the target molecule is removed (Han et al., 2022). When it comes to the detection of nitro compounds, electrochemical techniques are not the only ones whose selectivity benefits from MIPs. They can also be used to promote selective fluorescence quenching (Tanwar et al., 2022) and changes in structural color (Fan et al., 2020). However, partly because MIPs can be conveniently synthesized by electropolymerization over the working electrode of an electrochemical cell, the use of MIPs for this application (in the context of nitroaromatic and nitroamine compounds) outweighs those based on luminescence and color changes.

Electropolymerization allows experimenters to control the thickness of the MIP film. The most common alternative, bulk polymerization, may result in an uneven distribution of molecular recognition sites, their blockage, and their deterioration (Apak et al., 2023). For instance, for the specific detection of PA, a MIP layer was electrosynthesized from pyrrole and PA (template), with a KCl-containing methanol/water (1:4) solvent, and over a graphene-coated graphite electrode (Karthika et al., 2022). The >NH groups of polypyrrole were postulated to act as electron donors. In pulse voltammetry assays, the measured current at a given applied potential increased linearly with the concentration of PA. Other examples of MIPs in electrochemical detection include a copolymer of bis(2,2′-bithienyl)-(4- aminophenyl)methane (for affinity) and thiophene derivatives (for conductivity) for different nitroaromatics (Huynh et al., 2013), polyaniline for PA and TNT (Shi, Hou, et al., 2015), and PEI for the same two nitroarenes (Arman et al., 2022).

Although the use of conducting polymers such as polypyrrole and polyaniline for molecular imprinting helps enhance the electrical response, it is not a strict requisite. Applications of non-conducting MIPs for electrochemical detection were more frequent in less recent contributions, and, while they tended to be less sensitive, they did not fall short of selectivity. A non-conducting MIP formed by the bulk polymerization of ethylene glycol dimethacrylate and methacrylic acid in the presence of TNT, initiated by 2,2′-Azobisisobutyronitrile, achieved the selective detection of said nitroexplosive when coated on a screen-printed electrode (Pesavento et al., 2013). Likewise, the difficult task of separating RDX and HMX was undertaken by means of a non-conducting MIP synthesized with HMX as the template, methacrylic acid as a monomer, and trimethylolpropane trimethacrylate as a crosslinker (Wang et al., 2017).

The use of the target analyte as a template molecule is not strictly required. Trimesic acid, an innocuous compound with resemblance to TNT (but with carboxyl groups instead of nitro groups), can be used as a dummy template to obtain molecularly imprinted PEDOT that, in turn, is capable of detecting TNT without interference from nonaromatic explosives (Zheng, Ling, et al., 2023). Nonetheless, if the objective is the detection of nitramine explosives, Kemp's acid is a good molecular analog for RDX (Leibl et al., 2020). Tracking the increase in current intensity at 0.4 V allowed for a linear fit in the nanomolar range.

Fig. 18.7 schematizes the process of molecular imprinting, both using TNT (the most common target analyte among nitroexplosives) and a dummy template. Their structures have been optimized by MM energy minimization, and their van der Waals surfaces have been estimated by means of *Jmol 14*. As an example of a functional monomer, aziridine can be polymerized toward PEI. Overall, electrochemical detection with a MIP-coated working electrode offers key advantages in terms of sensitivity, selectivity, and quick response. However, it is less direct than optical techniques because it needs to apply electric potential. Furthermore, the signal is heavily dependent on the lack of electrode contamination (Karadurmus et al., 2022).

18.5 Polymers for colorimetric detection

18.5.1 The soul of nature

The most evident and visual way in which a polymeric material can indicate the presence of a nitroexplosive is color change. Selectively chromogenic, flexible, durable, water-resistant (and moisture-resistant) materials can be soaked in an aqueous sample, exposed to the vapors from a potential explosive artifact in an airport, or placed in the proximity of a landmine or a military production site. Then, if and only if the target nitro compounds are there, it adopts a different color,

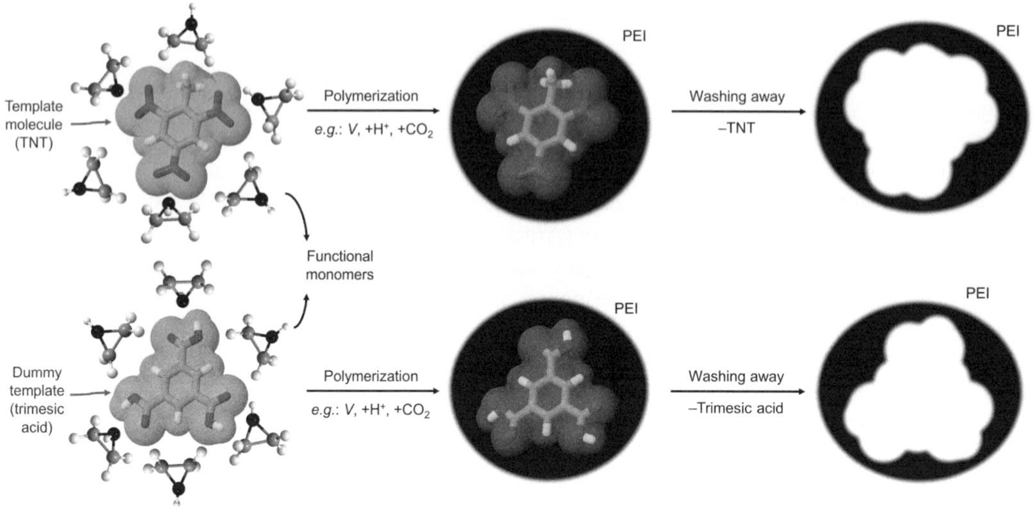

FIGURE 18.7

Polymerization of aziridine (electrochemical, acid-catalyzed or CO_2-catalyzed) over TNT and trimesic acid as template molecules, resulting in a characteristic imprint when the template molecule is washed away.

directly identifiable by the naked eye.[4] A Rudolf Steiner's quote comes to mind: "Color is the soul of nature and the entire cosmos." Light physicists would use less poetic, but not more accurate, descriptions. In any case, even if it is partially because of a social bias toward the importance of vision (Hutmacher, 2019), color (including brightness and opacity) is probably the most reliable phenomenon of our experience.

The scientific literature contains many examples of chromogenic polymers that fulfill, with more or less success, the description exposed above (Aguado et al., 2022; Lee et al., 2020; Pablos et al., 2014a, 2014b). Besides raising alarms by naked-eye inspection, they are also useful for quantification, only requiring a colorimeter. Unlike spectrofluorimeters, colorimeters are among the cheapest and easiest-to-use analytical devices. However, as of today, laboratory-like colorimeters are not even necessary. On one hand, there are portable, lightweight colorimeters that provide accurate determinations of the coordinates of a certain color space, such as CIE 1976 L*a*b* (Aguado, Mazega, et al., 2023). On the other hand, many proposals in the recent literature actually suggest smartphones for colorimetric detection. With this purpose in mind, researchers often develop their own apps (Tang et al., 2017).

There are a number of commercially available products for the colorimetric detection of nitroexplosives. One of the most traditional kits for this task, Expray, has been extensively used in forensic applications since 1991 (Plexus Scientific, 2023). It is based on a series of sequential reactions to distinguish between (1) nitroaromatics; (2) nitrate esters and nitramines; and (3) inorganic nitrates.

[4]Along this chapter, "colorimetric detection" refers to an optical response that does not need to receive radiation other than that from natural light, that involves absorption within the visible region of the electromagnetic spectrum, and that could be visualized in the absence of external equipment (although a colorimeter could be used for quantification).

As a drawback, fertilizers give way to false positives. Another successful system is the pellet-based explosive detection kit developed by the Indian Defense Research and Development Organization (Kavi, 2010). Equally worthy of mention are ChemSee's SprayView (ChemSee, 2023) and Westminster's Explosives and Narcotics Detectors (Westminster Group, 2023).

An interesting assessment of nitroexplosive-contaminated soils took place in the areas surrounding Los Alamos National Laboratory (New Mexico, United States) in 1993 (Haywood et al., 1995). For that, fluorescence quenching and chromogenic complexation complemented each other. Other than filter paper to collect the sample, none of the formulations used in those assays involved polymers, but the kind of association that produced a purple color with TNT and a red color with Tetryl can be extrapolated to stable macromolecules. For instance, instead of using dibutylamine (a liquid at room conditions), one could employ a solid polymer with secondary amino groups. That being said, covalently bound functional groups are not the only possibility. Chromogenic polymer gels with key active molecules embedded in them are full-fledged sensory polymeric materials, even if the polymer used to support those active molecules is as optically dull as sodium carboxymethyl cellulose (Thipwimonmas et al., 2021).

Colorimetric approaches usually depend on the formation of colored adducts with nitroexplosives (Meisenheimer complexes) and/or the surface plasmon resonance phenomena of some metal nanoparticles. These two mechanisms can be combined to maximize selectivity with strategies such as molecular imprinting, the use of specific antibodies, and the selection of aptamers.

18.5.2 Meisenheimer complexes of electron-donating polymers and nitroarenes

When it comes to the colorimetric detection of nitroaromatic explosives, the underlying mechanism for most proposals in the recent literature involves taking advantage of the strong electron-withdrawing capabilities of $-NO_2$ groups (Aguado et al., 2022; Aparna et al., 2018; Hughes et al., 2015). The first report on the formation of colored adducts between a nitroarene and a nucleophilic substance dates from 1886, and it is due to the Czech chemist Jaroslav Janovski (Pollitt & Saunders, 1965). However, it was Jakov Meisenheimer who, in 1902 and in what is deemed the first of his many notable achievements, figured out the correct structure of the adduct (Meisenheimer, Reactionen, 1902).

Although initially involving methoxide ions, ethoxide ions, and other relatively small chemical entities, the electron-donating part of the so-called Meisenheimer complexes has been expanded with great success to include polymeric materials. The list is not restricted to very specific macromolecules from fine chemistry experiments. Some common amino polymers, such as polyaniline (Huang et al., 2015) and PEI (Liu et al., 2011), readily form adducts with TNT. The kinetics of this complex formation differ from those of electron transfer phenomena described by the Marcus theory (Fendler & Larsen, 1972).

Poly(2-(dimethylamino)ethyl methacrylate), which is not a particularly expensive polymer, can be used for the selective colorimetric detection of TNT (Aguado et al., 2022; Pablos et al., 2014a). 2,4-DNT and dinitrobenzene isomers did not trigger a color change. Researchers from the University of Burgos noted that by copolymerizing the repeating unit of that polymer, 2-(dimethylamino)ethyl methacrylate, in the presence of non-sensing monomers, they could tailor the hydrophilic/hydrophobic balance and other properties of the resulting copolymer while keeping its responsiveness (Pablos et al., 2014a). The same can be done with (aminomethyl)styrene monomers, which also form red Meisenheimer complexes with TNT (Pablos et al., 2014b).

Amino groups, unless electronically deactivated or sterically hindered, are nucleophilic enough to form adducts with TNT and, at the same time, stable enough to allow users to store amino polymers under a broad range of conditions. Alkoxides can form Meisenheimer complexes with nitroarenes, as in the case of the first Meisenheimer complexes reported, but they are too reactive toward moisture. Hydroxyl groups are weak nucleophiles, but they offer great versatility for functionalization. Indeed, an alternative to the direct use of amino polymers is the chemical modification of polyols and polysaccharides. One of the most popular pathways to provide hydroxy polymers with amino groups is silane coupling. For instance, 3-aminopropyltriethoxysilane can be attached to cellulose paper to obtain red Meisenheimer complexes in the presence of TNT (Hughes et al., 2015).

The 3D molecular structures of a plausible adduct formation reaction, particularly involving the terminal amino group of an aminosilane coupling agent and TNT's aromatic ring, are presented in Fig. 18.8. The adduct's geometry has been computed by Chem3D Pro's energy minimization tool. The amino-TNT dative bond is postulated to include the amino's nitrogen as the electron-pair donor atom and TNT's carbon 1 (the one linked to the methyl group) as the acceptor (Gavrila et al., 2020).

In another work, 3-aminopropyltrimethoxysilane was reacted with aluminum oxide [AlOOH|Al$(OH)_2^+$] to obtain nanowells with amino groups on their surfaces (Liu et al., 2011). This surface was then coated with alternate layers of PSS and poly(allylamine hydrochloride). Surface reflectance was red-shifted as amino groups formed Meisenheimer complexes with TNT. Interestingly, PEI could also be used instead of the silane coupling agent. Other aminosilane agents known to form colored adducts with TNT are 3-aminopropyl(diethoxy)methylsilane and 3-(ethoxydimethylsilyl)propylamine (Tang et al., 2017).

Nevertheless, not all Meisenheimer complexes are colored. Nitroexplosives may form complexes with amines and other nucleophiles that are colorless and thus of restricted usefulness for

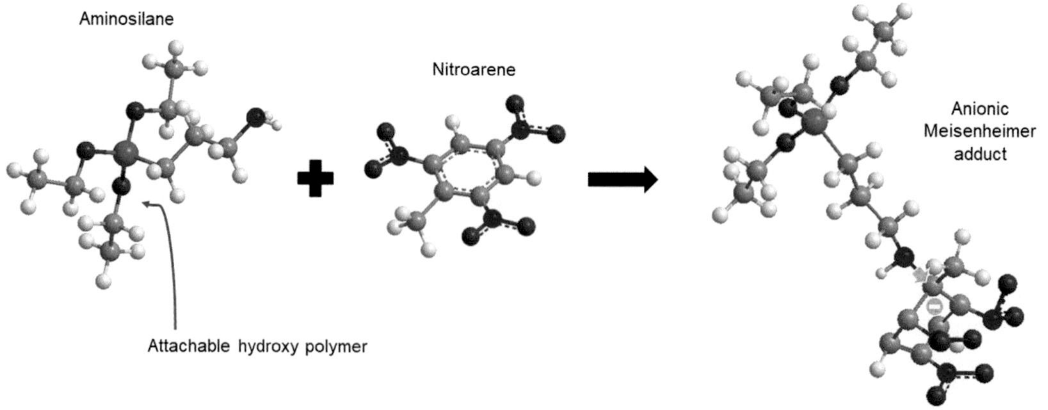

FIGURE 18.8

Formation of an anionic Meisenheimer adduct: 3D structures of a typical aminosilane coupling agent (3-Aminopropyltriethoxysilane), TNT, and the resulting complex.

colorimetric detection. That said, colorless complexes can somehow be part of a selective color-change mechanism. For instance, TNT can promote the aggregation of amino-decorated gold nanoparticles in suspension, resulting in the shifting of the absorption band assigned to surface plasmon resonance (Bai et al., 2016). Similarly, copper nanoparticles capped with PEI chains were useful for the colorimetric quantification of TNT (Aparna et al., 2018). Still, in terms of sensitivity, that system works better as a fluorescence-based probe, given that PEI-Cu nanoparticles have more absorptivity in the UV region than in the visible region.

18.5.3 Hydrolysis of nitramines and detection of decomposition products

As can be noticed from virtually any subsection so far, nitramine explosives such as RDX and HMX tend to be less frequently addressed by works on electrochemical, fluorescent, or colorimetric detection involving polymers. They are not only less popular in the scientific literature but also trickier to detect in selective ways. The fact that their LUMO-to-HOMO energy difference is greater than that of TNT, as aforementioned, has some relevance. Nonetheless, in my judgment, the greatest limitation is that they cannot establish π-π interactions with conjugated polymers, and this hinders FRET processes, electron transfer, molecular imprinting, and complex formation in general. Hence, one of the most common strategies for the qualitative and quantitative determination of nitramine explosives is hydrolyzing them in an aqueous alkaline solution and measuring the concentration of decomposition products by colorimetric means. It should be noted that this strategy is not specific to nitramines, as it has also been used for nitroaromatics, nitrate esters, and nitroalcohols. For instance, the Griess reagent can be immobilized on a polymeric substrate, such as paper, for the simultaneous identification of many different explosives (Peters et al., 2015).

The mechanism of the alkaline decomposition of RDX and HMX is not trivial. One could expect the former to release three nitrite ions per nitramine molecule and the latter to give out four nitrite ions, but that is not the case. Indeed, the alkaline hydrolysis of RDX begins like this, generating a nitrite ion and 3,5-dinitro-1,3,5-triaza-1-ene (Balakrishnan et al., 2003):

$$C_3H_6N_6O_6 + OH^- \rightarrow C_3H_5N_5O_4 + NO_2^- + H_2O$$

However, further decomposition of 3,5-dinitro-1,3,5-triaza-1-ene ($C_3H_5N_5O_4$) implies competitive reactions leading to complex intermediates, N_2O, formaldehyde, ammonia, acetate ions, etc. In light of this, researchers approaching colorimetric detection by this step cannot rely on stoichiometric calculations. Instead, they generally perform an internal standard calibration, obtaining a function of the color change with the concentration of nitramine explosive (Xie et al., 2018). In any case, there are several colorimetric systems to measure the concentration of nitrite or ammonia:

- The Griess reaction, a traditional method for nitrite quantification that dates from 1879 and that has been applied indiscriminately to nitroexplosives such as RDX, HMX, pentaerythritol tetranitrate, and trinitroglycerin (Üzer et al., 2017);
- the Berthelot reaction, as traditional as the former (Berthelot was a contemporary of Griess), but aimed at the ammonia resulting from the complete hydrolysis of RDX (Üzer et al., 2008);
- the oxidation of 3,3′,5,5′-tetramethylbenzidine (TMB), due to the nitrous acid formed in the presence of HCl, and its subsequent color change (Xie et al., 2018).

In the latter case, the role of the polymer was unusual in the context of explosive detection. It did not act as a solid matrix, as a charge carrier, or as an electron donor. Instead, the authors used polystyrene nanoparticles as a Pickering emulsifier. In this sense, two solid-stabilized emulsions, one for alkaline hydrolysis and another of acidic nature, were compartmentalized. As soon as they were mixed, TMB was converted into its yellow diimine form. The precursors of RDX and HMX exerted little interference, and the same could be said of some "common inorganic ions." Nevertheless, the interference of iron(iii) salts, whose high reduction potential makes it a likely oxidizer for TMB, was not tested.

18.5.4 Colorimetric MIPs, immunosensors, and aptasensors

As described above in the context of electrochemical cells, the use of nitroexplosives such as TNT, RDX, and HMX for molecular templating allows for their selective detection. Polymerization in the presence of the template molecule is one of the most popular approaches, and it can be done with easily processable monomers, such as methyl methacrylate and acrylamide (Fan et al., 2020). Increasing concentrations of nitroexplosives (up to 10 mM) red-shifted the structural color of the patterned particles or films. This elegant example relied on morphological changes in the solid, not implying the formation of Meisenheimer complexes. Nonetheless, some approaches combine the selectivity of MIPs with Meisenheimer complex formation and/or surface plasmon resonance (Giustina et al., 2016; Hughes et al., 2015).

Wulff spoke of "enzyme-analogous" polymers before MIPs were called MIPs (Wulff & Sarhan, 1972). Nonetheless, possibilities for the optical detection of nitroexplosives include the use of natural proteins, such as antibodies, or nucleic acids. The former gives way to immunosensors, sensing systems based on specific antigen-antibody conjugations. In a typical experiment, a host animal (e.g., mice) is immunized with the antigen of interest (e.g., TNT). When the animals' immune system then responds with specific antibodies, the serum is collected and purified to yield the desired antibodies.

For instance, a colorimetric lateral flow immunoassay for TNT was prepared by dipping strips with filter paper to wick liquids in a buffered solution containing colloidal gold and mouse anti-TNT antibody (Romolo et al., 2015). In the absence of TNT, two transversal red lines were observed. With increasing concentrations of nitroarene, the control line remained, but the color of the test line progressively faded away. The mechanism, being based on the loss of plasmon resonance, resembles that of PEI-capped gold nanoparticles (discussed above). The sensitivity of this assay, 1 ppm, was in the typical range of colorimetric detection systems, but high selectivity toward TNT was guaranteed.

While the use of antibodies for detection purposes leads to immunosensors, the use of nucleic acids leads to aptasensors. Aptamers are single-stranded DNA or RNA chains that become bound to specific targets. Typically, a large pool of random DNA or RNA sequences is subjected to iterative rounds of selection. Those sequences with the strongest association with the target molecule (once again, TNT, for example) are isolated. The nucleotide sequence of a certain anti-TNT aptamer was 5′-g gat ccg ttg ata taa aat tcc aca tat cac ata ccg agc gcg cga cgt cgt ctc act gtc ctg ctg cct ccg cgt cat ggt tga ttg tgg tgt tgg ctc-3′ (Idros et al., 2015), where the locant numbers (referring to the sugar-phosphate backbone) indicate that the sequence is expressed from the 5′ end, and the letters indicate the order of nitrogenous bases (adenine, thymine, cytosine, or guanine). That aptamer was attached to a green amine-functionalized silica surface, whose color shifted to red in the presence of TNT. Once again, these amino groups triggered the change in color by forming a Meisenheimer complex with the nitroaromatic compound.

18.6 Concluding remarks

This chapter has insisted on the need for nitroexplosive detection and classified the most relevant polymer-based systems in the recent literature (2014–2023) into three big categories: changes in photoluminescence, electrochemical detection, and naked-eye colorimetric detection. Usually, the first two categories imply the use of conjugated polymers, while colorimetric detection often involves the formation of colored Meisenheimer complexes with polymers containing amino groups. What is common to nearly all detection systems is the role that the nitro compound tends to play: that of electron acceptor. The same role can be extended to systems in which the polymers involved lack specific functional groups or high π-electron density, working as support materials such as gels, paper strips, or plastic substrates. In such cases, the materials or chemical entities triggering the response (e.g., QDs or liquid amines) are embedded in the polymeric matrix. However, this chapter focuses on applications where the chemosensing function is provided by the polymer itself, and in this context, fluorescence quenching is by far the most prevalent mechanism.

In the most typical case, the optical properties of a conjugated polymer, commonly including fluorene, metallole, and/or phenylene units, are altered in the presence of TNT and/or PA. Those optical properties frequently mean blue light emission following photoexcitation, although the emission wavelength can be modulated by chemical modification or by association with lanthanide ions, and that alteration is generally quenching. TNT usually provides the strongest response if the quenching mechanism is PET, while proposals based on FRET-mediated quenching are often selective toward PA, given the high absorptivity of the latter in the UV-A region, overlapping the emission spectrum of some conjugated polymers. The concentration of nitroexplosive is often related to quenching by means of a Stern-Volmer plot, given that static quenching tends to prevail over dynamic quenching. Additionally, AIE offers many possibilities when it comes to the emission wavelength, and this is why it often goes hand in hand with FRET phenomena. Alternatively, the material whose photoluminescence is quenched is not a polymer but QDs in dispersion, and the polymer works as a binder. Despite the advantages of fluorescence turn-on for sensing purposes, it is less straightforward and harder to attain with nitroaromatic or nitramine explosives. In general, it requires the nitroexplosive to react with another electron acceptor that is causing quenching by a PET mechanism.

While photoluminescent systems take advantage of the electron deficiency of nitroexplosives (in general) and of the absorption spectrum of PA (in particular), electrochemical detection exploits their reduction potential. Some conjugated polymers are interesting not only because of their optical properties but also because of their variable electrical conductivity, which depends on the degree of oxidation. The most common of them, including polyaniline, PEDOT, and polypyrrole, are p-type semiconductors. Then, electron-deficient molecules such as TNT, PA, and RDX work as dopants. Regarding the assay for detection, differential pulse voltammetry stands as the preferred choice. When conducting polymers are used to coat an electrode (often a glassy carbon electrode), the current intensity measured within a certain voltage range generally increases linearly with the concentration of oxidizer (nitroexplosive). As in the case of detection by fluorescence quenching, polymer-coated QDs can also be used for electrochemical detection, given the nature of semiconductor nanoparticles.

In general, sensing systems involving a fluorimeter or pulse voltammetry assays offer more sensitivity than colorimetric detection, but the latter may offer a more direct and easier interpretation, explaining its commercial success. Classical chromogenic kits for explosive detection have tended to relegate polymers to the role of support or matrix for the sample, but the same principles that allow for colored complexes between nitroaromatic molecules and small nucleophiles can be applied to nucleophilic polymers. In fact, many amino-containing polymers that form Meisenheimer complexes with TNT are relatively cheap, and since their backbone can be modified while keeping their chromogenic activity, unlike the consequences of disrupting the conjugated continuity of a conducting polymer, the properties of those polymers can be tailored in terms of hydrophilic/hydrophobic balance, film-forming capabilities, etc. Besides Meisenheimer complex formation, a colorimetric response can also be attained by the nitroexplosive-induced aggregation of polymer-capped metal nanoparticles, altering their surface plasmon resonance.

Finally, there are a number of strategies to improve selectivity that can be applied to detection systems of different kinds. One that has become quite trendy in recent literature, despite being half a century old, is molecular imprinting. In this chapter, the subsection dedicated to it belongs to electrochemical detection, given that it provides a solution to interfering problems in pulse voltammetry tests. Nonetheless, the possibility of polymerizing functional monomers around TNT or other nitroexplosives used as template molecules provides specificity to fluorescence quenching and colorimetric detection as well. Alternatively, innocuous template molecules, such as trimesic acid, can be used as "dummy" templates. Other selectivity-enhancing proposals involve proteins with quaternary structure, especially antibodies (for immunosensors) and nucleic acids (for aptasensors).

References

Abdel-Rahman, F., Okeremgbo, B., Alhamadah, F., Jamadar, S., Anthony, K., & Saleh, M. A. (2017). Caenorhabditis elegans as a model to study the impact of exposure to light emitting diode (LED) domestic lighting. *Journal of Environmental Science and Health, Part A*, *52*(5), 433–439. Available from https://doi.org/10.1080/10934529.2016.1270676.

Agrawal, J. P. (2010). *High energy materialsstatus of explosives* (pp. 69–161). Weinheim: Wiley VCH. Available from https://doi.org/10.1002/9783527628803.ch2.

Aguado, R., Rita, A., Santos, M. G., Vallejos, S., & Valente, A. J. M. (2022). Paper-based probes with visual response to vapors from nitroaromatic explosives: Polyfluorenes and tertiary amines. *Molecules (Basel, Switzerland)*, *27*(9), 2900. Available from https://doi.org/10.3390/molecules27092900.

Aguado, R. J., Gomes, B. O., Durães, L., & Valente, A. J. M. (2023). Luminescent papers with asymmetric complexes of Eu(III) and Tb(III) in polymeric matrices and suggested combinations for color tuning. *Molecules (Basel, Switzerland)*, *28*(16), 1420–3049. Available from https://doi.org/10.3390/molecules28166164.

Aguado, R. J., Mazega, A., Fiol, N., Tarrés, Q., Mutjé, P., & Delgado-Aguilar, M. (2023). Durable nanocellulose-stabilized emulsions of dithizone/chloroform in water for Hg2 + detection: A novel approach for a classical problem. *ACS Applied Materials & Interfaces*, *15*(9), 12580–12589. Available from https://doi.org/10.1021/acsami.2c22713.

Apak, R., Üzer, A., Sağlam, Ş., & Arman, A. (2023). Selective electrochemical detection of explosives with nanomaterial based electrodes. *Electroanalysis*, *35*(1), e202200175. Available from https://doi.org/10.1002/elan.202200175.

Aparna, R. S., Devi, J. S. A., Sachidanandan, P., & George, S. (2018). Polyethylene imine capped copper nanoclusters-fluorescent and colorimetric onsite sensor for the trace level detection of TNT. *Sensors and Actuators B: Chemical*, *254*, 811–819. Available from https://doi.org/10.1016/j.snb.2017.07.097, https://www.sciencedirect.com/science/article/pii/S0925400517313059.

Arman, A., Sağlam, Ş., Üzer, A., & Apak, R. (2022). Electrochemical determination of nitroaromatic explosives using glassy carbon/multi walled carbon nanotube/polyethyleneimine electrode coated with gold nanoparticles. *Talanta*, *238*, 122990. Available from https://doi.org/10.1016/j.talanta.2021.122990, https://www.sciencedirect.com/science/article/pii/S0039914021009127.

Bai, X., Xu, S., Hu, G., & Wang, L. (2016). Surface plasmon resonance-enhanced photothermal nanosensor for sensitive and selective visual detection of 2,4,6-trinitrotoluene. *Sensors and Actuators, B: Chemical*, *237*, 224–229. Available from https://doi.org/10.1016/j.snb.2016.06.093.

Balakrishnan, V. K., Halasz, A., & Hawari, J. (2003). Alkaline hydrolysis of the cyclic nitramine explosives RDX, HMX, and CL-20: New insights into degradation pathways obtained by the observation of novel intermediates. *Environmental Science & Technology*, *37*(9), 1838–1843. Available from https://doi.org/10.1021/es020959h.

Balan, B., Vijayakumar, C., Tsuji, M., Saeki, A., & Seki, S. (2012). Detection and distinction of DNT and TNT with a fluorescent conjugated polymer using the microwave conductivity technique. *The Journal of Physical Chemistry. B*, *116*(34), 10371–10378. Available from https://doi.org/10.1021/jp304791r.

Bener, M., Şen, F. B., & Apak, R. (2022). Protamine gold nanoclusters – Based fluorescence turn-on sensor for rapid determination of Trinitrotoluene (TNT). *Spectrochimica Acta Part A: Molecular and Biomolecular Spectroscopy*, *279*, 121462. Available from https://doi.org/10.1016/j.saa.2022.121462, https://www.sciencedirect.com/science/article/pii/S1386142522006114.

Brubaker, C. D., Frecker, T. M., McBride, J. R., Reid, K. R., Jennings, G. K., Rosenthal, S. J., & Adams, D. E. (2018). Incorporation of fluorescent quantum dots for 3D printing and additive manufacturing applications. *Journal of Materials Chemistry C*, *6*(28), 7584–7593. Available from https://doi.org/10.1039/C8TC02024H.

Brus, L. E. (1984). Electron–electron and electron-hole interactions in small semiconductor crystallites: The size dependence of the lowest excited electronic state. *The Journal of Chemical Physics*, *80*(9), 4403–4409. Available from https://doi.org/10.1063/1.447218.

Chakraborty, N., Begum, P., Patel, B. K., Rodriguez-Couto, S., & Shah, M. P. (2022). *Chapter 13 – Counterbalancing common explosive pollutants (TNT, RDX, and HMX) in the environment by microbial degradation* (pp. 263–310). Elsevier. Available from https://www.sciencedirect.com/science/article/pii/B9780323858397000128, 10.1016/B978-0-323-85839-7.00012-8.

Chatterjee, S., Ut Deb, S., Datta, C., Walther, D. K., & Gupta. (2017). Common explosives (TNT, RDX, HMX) and their fate in the environment: Emphasizing bioremediation. *Chemosphere*, *184*, 438–451. Available from https://doi.org/10.1016/j.chemosphere.2017.06.008, https://www.sciencedirect.com/science/article/pii/S0045653517309104.

ChemSee. (2023). SprayView explosive detection kit. https://www.chemsee.com/commercial/explosive-detection/products/sprayview/.

Chen, Z., Yang, J., Huang, D., Wang, S., Jiang, K., Sun, W., Chen, Z., Cao, Z., Ren, Y., Wang, Q., Liu, H., Zhang, X., & Sun, X. (2023). Adsorption behavior of aniline pollutant on polystyrene microplastics. *Chemosphere*, *323*, 138187. Available from https://doi.org/10.1016/j.chemosphere.2023.138187, https://www.sciencedirect.com/science/article/pii/S004565352300454X.

Chou, C. H., Hsu, S. L., Dinakaran, K., Chiu, M.-Y., & Wei, K.-H. (2005). Synthesis and characterization of luminescent polyfluorenes incorporating side-chain-tethered polyhedral oligomeric silsesquioxane units. *Macromolecules*, *38*(3), 745–751. Available from https://doi.org/10.1021/ma0479520.

Cluster Munition Coalition, International Campaign to Ban Landmines. (2020). Landmine monitor 2020. Global CWD Repository. Available from https://commons.lib.jmu.edu/cisr-globalcwd/1554.

Dascălu, M., Chibac-Scutaru, A. L., & Roman, G. (2023). Detection of nitroaromatics by a Zn(II)-containing coordination polymer derived from a 1,2,3-triazole-based tricarboxylate ligand. *Journal of Molecular Liquids*, *386*, 122457. Available from https://doi.org/10.1016/j.molliq.2023.122457, https://www.sciencedirect.com/science/article/pii/S0167732223012618.

Dawn.com. (2023). Death toll of Peshawar mosque blast revised down to 84: Police. Available from https://www.dawn.com/news/1735288. [Unpublished content].

Dhanasekaran, T., Manigandan, R., Padmanaban, A., Suresh, R., Giribabu, K., & Narayanan, V. (2019). Fabrication of Ag@Co-Al layered double hydroxides reinforced poly(o-phenylenediamine) nanohybrid for efficient electrochemical detection of 4-nitrophenol, 2,4-dinitrophenol and uric acid at nano molar level. *Scientific Reports*, *9*(1), 13250. Available from https://doi.org/10.1038/s41598-019-49595-y.

Dief, E. M., Hoffmann, N., & Darwish, N. (2022). Electrochemical Detection of Dinitrobenzene on Silicon Electrodes: Toward Explosives Sensors. *Surfaces*, *5*(1), 218–227. Available from https://doi.org/10.3390/surfaces5010015.

Doswald-Beck, L., Herby, P., & Dorais-Slakmon, J. (1995). Basic facts: The human cost of landmines. Available from https://www.icrc.org/en/doc/resources/documents/misc/57jmcy.htm.

Duraimurugan, K., & Siva, A. (2016). Phenylene(vinylene) based fluorescent polymer for selective and sensitive detection of nitro-explosive picric acid. *Journal of Polymer Science Part A: Polymer Chemistry*, *54* (24), 3800–3807. Available from https://doi.org/10.1002/pola.28270.

Dutta, P., Saikia, D., Adhikary, N. C., & Sarma, N. S. (2015). Macromolecular systems with MSA-capped CdTe and CdTe/ZnS core/shell quantum dots as superselective and ultrasensitive optical sensors for picric acid explosive. *ACS Applied Materials & Interfaces*, *7*(44), 24778–24790. Available from https://doi.org/10.1021/acsami.5b07660.

Ego, C., Marsitzky, D., Becker, S., Zhang, J., Grimsdale, A. C., Müllen, K., MacKenzie, J. D., Silva, C., & Friend, R. H. (2003). Attaching perylene dyes to polyfluorene: Three simple, efficient methods for facile color tuning of light-emitting polymers. *Journal of the American Chemical Society*, *125*(2), 437–443. Available from https://doi.org/10.1021/ja0205784.

Ekimov, A. I., Efros, A. L., & Onushchenko, A. A. (1985). Quantum size effect in semiconductor microcrystals. *Solid State Communications*, *56*(11), 921–924. Available from https://doi.org/10.1016/S0038-1098(85)80025-9.

Ekimov, A. I., & Onushchenko, A. A. (1982). Quantum size effect in the optical spectra of semiconductor microcrystals. *Soviet Physics: Semiconductors*, *16*(7), 775–778.

Evans, R., & Seddon, B. (2022). Explosive ordnance guide for Ukraine (2nd ed.). GICHD. Available from https://www.gichd.org/publications-resources/publications/explosive-ordnance-guide-for-ukraine-second-edition/.

Fan, J., Meng, Z., Dong, X., Xue, M., Qiu, L., Liu, X., Zhong, F., & He, X. (2020). Colorimetric screening of nitramine explosives by molecularly imprinted photonic crystal array. *Microchemical Journal*, *158*, 105143. Available from https://doi.org/10.1016/j.microc.2020.105143, https://www.sciencedirect.com/science/article/pii/S0026265X20310791.

Fan, X. C., Wang, K., Shi, Y. Z., Cheng, Y. C., Lee, Y. T., Yu, J., Chen, X. K., Adachi, C., & Zhang, X. H. (2023). Ultrapure green organic light-emitting diodes based on highly distorted fused π-conjugated molecular design. *Nature Photonics*, *17*(3), 280–285. Available from https://doi.org/10.1038/s41566-022-01106-8.

Frasera, S. (2015). AP news unpublished content suicide bombings kill 95 people at Ankara peace rally. Available from https://apnews.com/article/d381289f64da47dc95914372825e55c0.

Fendler, J. H., & Larsen, J. W. (1972). Thermodynamic and kinetic analysis of Meisenheimer complex formation. *The Journal of Organic Chemistry*, *37*(16), 2608–2611. Available from https://doi.org/10.1021/jo00981a019.

Feng, L., Li, H., Qu, Y., & Lü, C. (2012). Detection of TNT based on conjugated polymer encapsulated in mesoporous silica nanoparticles through FRET. *Chemical Communications*, *48*(38), 4633–4635. Available from https://doi.org/10.1039/C2CC16115J.

Fernandez-Maestre, R., Tabrizchi, M., & Meza-Morelos, D. (2022). Ion–shift reagent binding energy and the mass–mobility shift correlation in ion mobility spectrometry. *Rapid Communications in Mass Spectrometry*, *36*(20), e9360. Available from https://doi.org/10.1002/rcm.9360.

Förster, T. (1948). Zwischenmolekulare energiewanderung und fluoreszenz. *Annalen der Physik*, *437*(1–2), 55–75. Available from https://doi.org/10.1002/andp.19484370105.

Gavrila, A. M., Iordache, T. V., Lazau, C., Rotariu, T., Cernica, I., Stroescu, H., Stoica, M., Orha, C., Bandas, C. E., & Sarbu, A. (2020). Biomimetic sensitive elements for 2,4,6-trinitrotoluene tested on multi-layered sensors. *Coatings*, *10*(3). Available from https://doi.org/10.3390/coatings10030273.

Genovese, D., Cingolani, M., Rampazzo, E., Prodi, L., & Zaccheroni, N. (2021). Static quenching upon adduct formation: A treatment without shortcuts and approximations. *Chemical Society Reviews*, *50*(15), 8414–8427. Available from https://doi.org/10.1039/D1CS00422K.

Giustina, G. D., Sonato, A., Gazzola, E., Ruffato, G., Brusa, S., & Romanato, F. (2016). SPR enhanced molecular imprinted sol–gel film: A promising tool for gas-phase TNT detection. *Materials Letters*, *162*, 44–47. Available from https://doi.org/10.1016/j.matlet.2015.09.114, https://www.sciencedirect.com/science/article/pii/S0167577X1530625X.

Gogoi, B., Dutta, P., Paul, N., Dass, N. N., & Sarma, N. S. (2013). Polycurcumin acrylate and polycurcumin methacrylate: Novel bio-based polymers for explosive chemical sensor. *Sensors and Actuators B: Chemical*, *181*, 144–152. Available from https://doi.org/10.1016/j.snb.2013.01.071, https://www.sciencedirect.com/science/article/pii/S0925400513000889.

Gou, Z., Wang, A., Tian, M., & Zuo, Y. (2022). Pyrene-based monomer-excimer dual response organosilicon polymer for the selective detection of 2,4,6-trinitrotoluene (TNT) and 2,4,6-trinitrophenol (TNP). *Materials Chemistry Frontiers*, *6*(5), 607–612. Available from https://doi.org/10.1039/D1QM01574E.

Goudappagouda., Dongre, S. D., Das, T., & Babu, S. S. (2020). Dual mode selective detection and differentiation of TNT from other nitroaromatic compounds. *Journal of Materials Chemistry A*, *8*(21), 10767–10771. Available from https://doi.org/10.1039/D0TA02091E.

Gupta, J., Singhal, P., Gupta, B. K., & Rattan, S. (2023). Advanced functional rGO@MoS2@PEDOT: PSS multicomponent-based nanocomposite films for rapid and ultra-sensitive TNT detection. *Materials Today Communications*, *35*, 106316. Available from https://doi.org/10.1016/j.mtcomm.2023.106316, https://www.sciencedirect.com/science/article/pii/S2352492823010073.

Hamilos, P. (2007). The Guardian unpublished content the worst Islamist attack in European history. Available from https://www.theguardian.com/world/2007/oct/31/spain.

Han, Y., Tao, J., Ali, N., Khan, A., Malik, S., Khan, H., Yu, C., Yang, Y., Bilal, M., & Mohamed, A. A. (2022). Molecularly imprinted polymers as the epitome of excellence in multiple fields. *European Polymer Journal*, *179*, 111582. Available from https://doi.org/10.1016/j.eurpolymj.2022.111582, https://www.sciencedirect.com/science/article/pii/S0014305722005869.

Haywood, W., McRae, D., Powell, J., & Harris, B. W. (1995). An assessment of high-energy explosives and metal contamination in soil at TA-67 (12), L-site, and TA-14, Q-site. Available from https://sgp.fas.org/othergov/doe/lanl/lib-www/la-pubs/00287159.pdf.

Hezinger, A. F. E., Teßmar, J., & Göpferich, A. (2008). Polymer coating of quantum dots – A powerful tool toward diagnostics and sensorics. *Interactive Polymers for Pharmaceutical and Biomedical Applications*, *68*(1), 138–152. Available from https://doi.org/10.1016/j.ejpb.2007.05.013, https://www.sciencedirect.com/science/article/pii/S0939641107001956.

Holubowitch, N. E., Crabtree, C., & Budimir, Z. (2020). Electroanalysis and spectroelectrochemistry of nonaromatic explosives in acetonitrile containing dissolved oxygen. *Analytical Chemistry*, *92*(17), 11617−11626. Available from https://doi.org/10.1021/acs.analchem.0c01174.

Huang, S., He, Q., Xu, S., & Wang, L. (2015). Polyaniline-based photothermal paper sensor for sensitive and selective detection of 2,4,6-trinitrotoluene. *Analytical Chemistry*, *87*(10), 5451−5456. Available from https://doi.org/10.1021/acs.analchem.5b01078.

Huang, S., Wang, E., Tong, J., Shan, G. G., Liu, S., Feng, H., Qin, C., Wang, X., & Su, Z. (2023). Rational design of AIE-active biodegradable polycarbonates for high-performance WLED and selective detection of nitroaromatic explosives. *Chinese Chemical Letters*, *34*(8), 108008. Available from https://doi.org/10.1016/j.cclet.2022.108008, https://www.sciencedirect.com/science/article/pii/S1001841722010178.

Hughes, S., Dasary, S. S. R., Begum, S., Williams, N., & Yu, H. (2015). Meisenheimer complex between 2,4,6-trinitrotoluene and 3-aminopropyltriethoxysilane and its use for a paper-based sensor. *Sensing and Bio-Sensing Research.*, *5*, 37−41. Available from https://doi.org/10.1016/j.sbsr.2015.06.003, https://www.sciencedirect.com/science/article/pii/S2214180415300015.

Hui, Y., & Webster, R. D. (2011). Absorption of water into organic solvents used for electrochemistry under conventional operating conditions. *Analytical Chemistry*, *83*(3), 976−981. Available from https://doi.org/10.1021/ac102734a.

Hussein, E. M. A., & Waller, E. J. (2000). Landmine detection: The problem and the challenge. *Applied Radiation and Isotopes*, *53*(4), 557−563. Available from https://doi.org/10.1016/S0969-8043(00)00218-9, https://www.sciencedirect.com/science/article/pii/S0969804300002189.

Hutmacher, F. (2019). Why is there so much more research on vision than on any other sensory modality? *Frontiers in Psychology*, *10*. Available from https://doi.org/10.3389/fpsyg.2019.02246, https://www.frontiersin.org/articles/.

Huynh, T., Sosnowska, M., Sobczak, J. W., Chandra, B. K. C., Nesterov, V. N., D'Souza, F., & Kutner, W. (2013). Simultaneous chronoamperometry and piezoelectric microgravimetry determination of nitroaromatic explosives using molecularly imprinted thiophene polymers. *Analytical Chemistry*, *85*(17), 8361−8368. Available from https://doi.org/10.1021/ac4017677.

Idros, N., Ho, M. Y., Pivnenko, M., Qasim, M. M., Xu, H., Gu, Z., & Chu, D. (2015). Colorimetric-based detection of TNT explosives using functionalized silica nanoparticles. *Sensors*, *15*(6), 12891−12905. Available from https://doi.org/10.3390/s150612891.

Kagan, A., & Oxley, J.C. (2022). Front matter (pp. i−ii). Elsevier. Available from https://www.sciencedirect.com/science/article/pii/B9780444641045099872, https://doi.org/10.1016/B978-0-444-64104-5.09987-2.

Karadurmus, L., Bilge, S., Sınağ, A., & Ozkan, S. A. (2022). Molecularly imprinted polymer (MIP)-based sensing for detection of explosives: Current perspectives and future applications. *TrAC Trends in Analytical Chemistry*, *155*, 116694. Available from https://doi.org/10.1016/j.trac.2022.116694, https://www.sciencedirect.com/science/article/pii/S0165993622001777.

Karthika, P., Shanmuganathan, S., & Viswanathan, S. (2022). Electrochemical sensor for picric acid by using molecularly imprinted polymer and reduced graphene oxide modified pencil graphite electrode. *Proceedings of the Indian National Science Academy*, *88*(3), 263−276. Available from https://doi.org/10.1007/s43538-022-00084-3.

Kavi, P. (2010). U.S. firm keen on DRDO's explosive detection kit, pact on anvil. Available from https://pib.gov.in/newsite/PrintRelease.aspx?relid = 68492.

Khalil, I. E., Pan, T., Shen, Y., & Zhang, W. (2020). A water-stable luminescent coordination polymer for sensitive detection of nitroaromatic compounds. *Inorganic Chemistry Communications*, *120*, 108170. Available from https://doi.org/10.1016/j.inoche.2020.108170, https://www.sciencedirect.com/science/article/pii/S1387700320307607.

Kim, D., Kim, J., & Lee, T. S. (2018). Dual-signal detection of trypsin using controlled aggregation of conjugated polymer dots and magnetic nanoparticles. *Sensors and Actuators B: Chemical*, *264*, 45–51. Available from https://doi.org/10.1016/j.snb.2018.02.118, https://www.sciencedirect.com/science/article/pii/S0925400518303976.

Kovalev, I. S., Taniya, O. S., Kopchuk, D. S., Giri, K., Mukherjee, A., Santra, S., Majee, A., Rahman, M., Zyryanov, G. V., Bakulev, V. A., & Chupakhin, O. N. (2018). 1-Hydroxypyrene-based micelle-forming sensors for the visual detection of RDX/TNG/PETN-based bomb plots in water. *New Journal of Chemistry*, *42*(24), 19864–19871. Available from https://doi.org/10.1039/C8NJ03807D.

Kumar, V., Choudhury, N., Kumar, A., De, P., & Satapathi, S. (2020). Poly-tryptophan/carbazole based FRET-system for sensitive detection of nitroaromatic explosives. *Optical Materials*, *100*, 109710. Available from https://doi.org/10.1016/j.optmat.2020.109710, https://www.sciencedirect.com/science/article/pii/S0925346720300616.

Kumar, V., Saini, S. K., Choudhury, N., Kumar, A., Maiti, B., De, P., Kumar, M., & Satapathi, S. (2021). Highly sensitive detection of nitro compounds using a fluorescent copolymer-based FRET system. *ACS Applied Polymer Materials*, *3*(8), 4017–4026. Available from https://doi.org/10.1021/acsapm.1c00540.

Lee, M. G., Yoo, H. W., Lim, S. H., & Yi, G. R. (2020). Inkjet-printed low-cost colorimetric tickets for TNT detection in contaminated soil. *Korean Journal of Chemical Engineering*, *37*(12), 2171–2178. Available from https://doi.org/10.1007/s11814-020-0627-x.

Leibl, N., Duma, L., Gonzato, C., & Haupt, K. (2020). Polydopamine-based molecularly imprinted thin films for electro-chemical sensing of nitro-explosives in aqueous solutions. *Bioelectrochemistry (Amsterdam, Netherlands)*, *135*, 107541. Available from https://doi.org/10.1016/j.bioelechem.2020.107541, https://www.sciencedirect.com/science/article/pii/S1567539420300736.

Liu, Y., Mills, R. C., Boncella, J. M., & Schanze, K. S. (2001). Fluorescent polyacetylene thin film sensor for nitroaromatics. *Langmuir: The ACS Journal of Surfaces and Colloids*, *17*(24), 7452–7455. Available from https://doi.org/10.1021/la010696p.

Liu, Y., Wang, H. H., Indacochea, J. E., & Wang, M. L. (2011). A colorimetric sensor based on anodized aluminum oxide (AAO) substrate for the detection of nitroaromatics. *Sensors and Actuators B: Chemical*, *160* (1), 1149–1158. Available from https://doi.org/10.1016/j.snb.2011.09.040, https://www.sciencedirect.com/science/article/pii/S0925400511008422.

Longo, C. M., & Musah, R. A. (2020). MALDI-mass spectrometry imaging for touch chemistry biometric analysis: Establishment of exposure to nitroaromatic explosives through chemical imaging of latent fingermarks. *Forensic Chemistry*, *20*, 100269. Available from https://doi.org/10.1016/j.forc.2020.100269, https://www.sciencedirect.com/science/article/pii/S2468170920300576.

Luo, J., Xie, Z., Lam, J. W. Y., Cheng, L., Chen, H., Qiu, C., Kwok, H. S., Zhan, X., Liu, Y., Zhu, D., & Tang, B. Z. (2001). Aggregation-induced emission of 1-methyl-1,2,3,4,5-pentaphenylsilole. *Chemical Communications* (18), 1740–1741. Available from https://doi.org/10.1039/B105159H.

McCarthy, J. (2022). How Russia's invasion of Ukraine is harming water, air, soil, and wildlife. Available from https://www.globalcitizen.org/en/content/environmental-impact-of-war-in-ukraine/.

McCarthy, H. L., & Pedahzur, A. (2014). *20. The Sinai terrorist attacks. The evolution of the global terrorist threat from 9/11 to Osama bin Laden's death* (pp. 483–497). Columbia University Press. Available from https://doi.org/10.7312/hoff16898-020.

Madhu, S., Bandela, A., & Ravikanth, M. (2014). BODIPY based fluorescent chemodosimeter for explosive picric acid in aqueous media and rapid detection in the solid state. *RSC Advances*, *4*(14), 7120–7123. Available from https://doi.org/10.1039/C3RA46565A.

Malik, A. H., Habib, F., Qazi, M. J., Ganayee, M. A., Ahmad, Z., & Yatoo, M. A. (2023). A short review article on conjugated polymers. *Journal of Polymer Research*, *30*(3), 115. Available from https://doi.org/10.1007/s10965-023-03451-w.

Martelo, L. M., Marques, L. F., Burrows, H. D., & Berberan-Santos, M. N. (2019). *Explosives detection: From sensing to response* (pp. 293–320). Cham: Springer International Publishing. Available from https://doi.org/10.1007/4243_2019_9.

Meisenheimer, J., & Reactionen, Ü. (1902). Aromatischer nitrokörper. *Justus Liebigs Annalen der Chemie*, *323*(2), 205–246.

Mihas, O., Kalogerakis, N., & Psillakis, E. (2007). Photolysis of 2,4-dinitrotoluene in various water solutions: Effect of dissolved species. *Environmental Applications of Advanced Oxidation Processes*, *146*(3), 535–539. Available from https://doi.org/10.1016/j.jhazmat.2007.04.054, https://www.sciencedirect.com/science/article/pii/S0304389407005390.

Mothika, V. S., Räupke, A., Brinkmann, K. O., Riedl, T., Brunklaus, G., & Scherf, U. (2018). Nanometer-thick conjugated microporous polymer films for selective and sensitive vapor-phase TNT detection. *ACS Applied Nano Materials*, *1*(11), 6483–6492. Available from https://doi.org/10.1021/acsanm.8b01779.

Mundy, P. C., Werner, A., Singh, L., Singh, V., Mendieta, R., Patullo, C. E., Wulff, H., & Lein, P. J. (2023). Hexahydro-1,3,5-trinitro-1,3,5-triazine (RDX) causes seizure activity in larval zebrafish via antagonism of γ-aminobutyric acid type A receptor $\alpha1\beta2\gamma2$. *Archives of Toxicology*, *97*(5), 1355–1365. Available from https://doi.org/10.1007/s00204-023-03475-7.

Murray, C. B., Norris, D. J., & Bawendi, M. G. (1993). Synthesis and characterization of nearly monodisperse CdE (E = sulfur, selenium, tellurium) semiconductor nanocrystallites. *Journal of the American Chemical Society*, *115*(19), 8706–8715. Available from https://doi.org/10.1021/ja00072a025.

Müller, S., Galliardt, H., Schneider, J., Barisas, B., & Seidel, T. (2013). Quantification of Förster resonance energy transfer by monitoring sensitized emission in living plant cells. *Frontiers in Plant Science*, *4*. Available from https://doi.org/10.3389/fpls.2013.00413, https://www.frontiersin.org/articles/.

Nailwal, Y., Devi, M., & Pal, S. K. (2022). Luminescent conjugated microporous polymers for selective sensing and ultrafast detection of picric acid. *ACS Applied Polymer Materials*, *4*(4), 2648–2655. Available from https://doi.org/10.1021/acsapm.1c01905.

Nguyen, C. H. T., Nguyen, T. H., Nguyen, T. P. L., Tran, H. L., Luu, T. H., Tran, C. D., Nguyen, Q. T., Nguyen, L. T., Yokozawa, T., & Nguyen, H. T. (2023). Aerobic direct arylation polycondensation of N-perylenyl phenoxazine-based fluorescent conjugated polymers for highly sensitive and selective TNT explosives detection. *Dyes and Pigments*, *219*, 111613. Available from https://doi.org/10.1016/j.dyepig.2023.111613, https://www.sciencedirect.com/science/article/pii/S0143720823005399.

Nobel Prize Outreach. (1992). Award ceremony speech. Available from https://www.nobelprize.org/prizes/chemistry/1992/ceremony-speech/.

Noh, D., Ampadu, E. K., & Oh, E. (2022). Influence of air flow on luminescence quenching in polymer films towards explosives detection using drones. *Polymers*, *14*(3), 2073–4360. Available from https://doi.org/10.3390/polym14030483.

Olson, E. J., Isley, W. C., III, Brennan, J. E., Cramer, C. J., & Bühlmann, P. (2015). Electrochemical reduction of 2,4-dinitrotoluene in aprotic and pH-buffered media. *The Journal of Physical Chemistry C*, *119*(23), 13088–13097. Available from https://doi.org/10.1021/acs.jpcc.5b02840.

Öztürk, B. Ö., & Şehitoğlu, S. K. (2019). Pyrene substituted amphiphilic ROMP polymers as nano-sized fluorescence sensors for detection of TNT in water. *Polymer*, *183*, 121868. Available from https://doi.org/10.1016/j.polymer.2019.121868, https://www.sciencedirect.com/science/article/pii/S0032386119308742.

Pablos, J. L., Trigo-López, M., Serna, F., García, F. C., & García, J. M. (2014a). Solid polymer substrates and smart fibres for the selective visual detection of TNT both in vapour and in aqueous media. *RSC Advances*, *4*(49), 25562–25568. Available from https://doi.org/10.1039/C4RA02716G.

Pablos, J. L., Trigo-López, M., Serna, F., García, F. C., & García, J. M. (2014b). Water-soluble polymers{,} solid polymer membranes{,} and coated fibres as smart sensory materials for the naked eye detection and quantification of TNT in aqueous media. *Chemical Communications*, *50*(19), 2484–2487. Available from https://doi.org/10.1039/C3CC49260E.

Perrin, J. B. (1927). Fluorescence et induction moléculaire par résonance. *Comptes Rendus Hebdomadaires des Séances Academie Science*, *184*, 1097–1100.

Pesavento, M., D'Agostino, G., Alberti, G., Biesuz, R., & Merli, D. (2013). Voltammetric platform for detection of 2,4,6-trinitrotoluene based on a molecularly imprinted polymer. *Analytical and Bioanalytical Chemistry*, *405*(11), 3559–3570. Available from https://doi.org/10.1007/s00216-012-6553-y.

Peters, K. L., Corbin, I., Kaufman, L. M., Zreibe, K., Blanes, L., & McCord, B. R. (2015). Simultaneous colorimetric detection of improvised explosive compounds using microfluidic paper-based analytical devices (μPADs). *Anal. Methods*, *7*(1), 63–70. Available from https://doi.org/10.1039/C4AY01677G.

Plexus Scientific. (2023). Explosive detection. Available from https://www.plexsci.com/expray.

Pollitt, R. J., & Saunders, B. C. (1965). 860. The Janovsky reaction. *Journal of the Chemical Society* (0), 4615–4628. Available from https://doi.org/10.1039/JR9650004615.

Raza, A., Biswas, A., Zehra, A., & Mengesha, A. (2020). Multiple tier detection of TNT using curcumin functionalized silver nanoparticles. *Forensic Science International: Synergy*. Available from https://doi.org/10.1016/j.fsisyn.2020.08.001.

Reed, M. A., Bate, R. T., Bradshaw, K., Duncan, W. M., Frensley, W. R., Lee, J. W., & Shih, H. D. (1986). Spatial quantization in GaAs–AlGaAs multiple quantum dots. *Journal of Vacuum Science & Technology B: Microelectronics Processing and Phenomena. 0734-211X*, *4*(1), 358–360. Available from https://doi.org/10.1116/1.583331.

Romolo, F. S., Ferri, E., Mirasoli, M., D'Elia, M., Ripani, L., Peluso, G., Risoluti, R., Maiolini, E., & Girotti, S. (2015). Field detection capability of immunochemical assays during criminal investigations involving the use of TNT. *Forensic Science International*, *246*, 25–30. Available from https://doi.org/10.1016/j.forsciint.2014.10.037, https://www.sciencedirect.com/science/article/pii/S0379073814004563.

Sanchez, J. C., DiPasquale, A. G., Mrse, A. A., & Trogler, W. C. (2009). Lewis acid–base interactions enhance explosives sensing in silacycle polymers. *Analytical and Bioanalytical Chemistry*, *395*(2), 387–392. Available from https://doi.org/10.1007/s00216-009-2846-1.

Sanchez, J. C., DiPasquale, A. G., Rheingold, A. L., & Trogler, W. C. (2007). Synthesis, luminescence properties, and explosives sensing with 1,1-tetraphenylsilole- and 1,1-silafluorene-vinylene polymers. *Chemistry of Materials: A Publication of the American Chemical Society*, *19*(26), 6459–6470. Available from https://doi.org/10.1021/cm702299g.

Sanchez, J. C., & Trogler, W. C. (2008). Efficient blue-emitting silafluorene–fluorene-conjugated copolymers: Selective turn-off/turn-on detection of explosives. *Journal of Materials Chemistry*, *18*(26), 3143–3156. Available from https://doi.org/10.1039/B802623H.

Şen, F. B., Beğiç, N., Bener, M., & Apak, R. (2022). Fluorescence turn-off sensing of TNT by polyethylenimine capped carbon quantum dots. *Spectrochimica Acta Part A: Molecular and Biomolecular Spectroscopy*, *271*, 120884. Available from https://doi.org/10.1016/j.saa.2022.120884, https://www.sciencedirect.com/science/article/pii/S1386142522000324.

Sağlam, Ş., Üzer, A., Tekdemir, Y., Erçağ, E., & Apak, R. (2015). Electrochemical sensor for nitroaromatic type energetic materials using gold nanoparticles/poly(o-phenylenediamine–aniline) film modified glassy carbon electrode. *Talanta*, *139*, 181–188. Available from https://doi.org/10.1016/j.talanta.2015.02.059, https://www.sciencedirect.com/science/article/pii/S0039914015001423.

Shi, H., Liu, C., Jiang, Q., & Xu, J. (2015). Effective approaches to improve the electrical conductivity of PEDOT:PSS: A review. *Advanced Electronic Materials*, *1*(4), 1500017. Available from https://doi.org/10.1002/aelm.201500017.

Shi, L., Hou, A. G., Chen, L. Y., & Wang, Z. F. (2015). Electrochemical sensor prepared from molecularly imprinted polymer for recognition of TNT. *Polymer Composites*, *36*(7), 1280–1285. Available from https://doi.org/10.1002/pc.23032.

Shirakawa, H., Louis, E. J., MacDiarmid, A. G., Chiang, C. K., & Heeger, A. J. (1977). Synthesis of electrically conducting organic polymers: Halogen derivatives of polyacetylene, (CH). *Journal of the Chemical Society. Chemical Communications* (16), 578–580. Available from https://doi.org/10.1039/C39770000578.

Shirakawa, H. (2000). The discovery of polyacetylene film: The dawning of an era of conducting polymers. Available from https://www.nobelprize.org/prizes/chemistry/2000/shirakawa/lecture.

Small, C. E., Tsang, S. W., Kido, J., So, S. K., & So, F. (2012). Origin of enhanced hole injection in inverted organic devices with electron accepting interlayer. *Advanced Functional Materials*, *22*(15), 3261–3266. Available from https://doi.org/10.1002/adfm.201200185.

Sohn, H., Sailor, M. J., Magde, D., & Trogler, W. C. (2003). Detection of nitroaromatic explosives based on photoluminescent polymers containing metalloles. *Journal of the American Chemical Society*, *125*(13), 3821–3830. Available from https://doi.org/10.1021/ja021214e.

Sun, X., Wang, Y., & Lei, Y. (2015). Fluorescence based explosive detection: From mechanisms to sensory materials. *Chemical Society Reviews*, *44*(22), 8019–8061. Available from https://doi.org/10.1039/C5CS00496A.

Tang, N., Mu, L., Qu, H., Wang, Y., Duan, X., & Reed, M. A. (2017). Smartphone-enabled colorimetric trinitrotoluene detection using amine-trapped polydimethylsiloxane membranes. *ACS Applied Materials & Interfaces*, *9*(16), 14445–14452. Available from https://doi.org/10.1021/acsami.7b03314.

Tanwar, A. S., Parui, R., Garai, R., Chanu, M. A., & Iyer, P. K. (2022). Dual "static and dynamic" fluorescence quenching mechanisms based detection of TNT via a cationic conjugated polymer. *ACS Measurement Science Au*, *2*(1), 23–30. Available from https://doi.org/10.1021/acsmeasuresciau.1c00023.

Tao, R., Zhao, X., Zhao, T., Zhao, M., Li, R., Yang, T., Tang, L., Jin, Y., Zhang, W., & Qiu, L. (2022). Cage-confinement induced emission enhancement. *The Journal of Physical Chemistry Letters*, *13*(28), 6604–6611. Available from https://doi.org/10.1021/acs.jpclett.2c01651.

Tauqeer, H. M., Karczewska, A., Lewińska, K., Fatima, M., Khan, S. A., Farhad, M., Turan, V., Ramzani, P. M. A., Iqbal, M., Hasanuzzaman, M., & Prasad, M. N. V. (2021). *Chapter 36 - Environmental concerns associated with explosives (HMX, TNT, and RDX), heavy metals and metalloids from shooting range soils: Prevailing issues, leading management practices, and future perspectives*. Academic Press. Available from https://www.sciencedirect.com/science/article/pii/B9780128193822000363, 10.1016/B978-0-12-819382-2.00036-3.

The Guardian. (2005). Unpublished content Lebanese minister hit by car bomb. Available from https://www.theguardian.com/world/2005/jul/12/syria.lebanon.

Thipwimonmas, Y., Thiangchanya, A., Phonchai, A., Thainchaiwattana, S., Jomsati, W., Jomsati, S., Tayayuth, K., & Limbut, W. (2021). The development of digital image colorimetric quantitative analysis of multi-explosives using polymer gel sensors. *Sensors*, *21*(23). Available from https://doi.org/10.3390/s21238041.

Tian, X., Peng, H., Li, Y., Yang, C., Zhou, Z., & Wang, Y. (2017). Highly sensitive and selective paper sensor based on carbon quantum dots for visual detection of TNT residues in groundwater. *Sensors and Actuators B: Chemical.*, *243*, 1002–1009. Available from https://doi.org/10.1016/j.snb.2016.12.079, https://www.sciencedirect.com/science/article/pii/S0925400516320512.

Toal, S. J., & Trogler, W. C. (2006). Polymer sensors for nitroaromatic explosives detection. *Journal of Materials Chemistry*, *16*(28), 2871–2883. Available from https://doi.org/10.1039/B517953J.

Toal, S. J., Sanchez, J. C., Dugan, R. E., & Trogler, W. C. (2007). Visual detection of trace nitroaromatic explosive residue using photoluminescent metallole-containing polymers. *Journal of Forensic Sciences*, *52*(1), 79–83. Available from https://doi.org/10.1111/j.1556-4029.2006.00332.x.

Tomczak, N., Jańczewski, D., Han, M., & Vancso, G. J. (2009). Designer polymer–quantum dot architectures. *Progress in Polymer Science*, *34*(5), 393–430. Available from https://doi.org/10.1016/j.progpolymsci.2008.11.004, https://www.sciencedirect.com/science/article/pii/S0079670008001196.

Turchetti, D. A., Nolasco, M. M., Szczerbowski, D., Carlos, L. D., & Akcelrud, L. C. (2015). Light emission of a polyfluorene derivative containing complexed europium ions. *Physical Chemistry Chemical Physics: PCCP*, *17*(39), 26238–26248. Available from https://doi.org/10.1039/C5CP03567H.

UN News. (2023). Demining Ukraine: Bringing lifesaving expertise back home. Available from https://news.un.org/en/story/2023/07/1138477.

Ushakov, N. M., & Kosobudsky, I. D. (2019). Dielectric measurements of polymer composite based on CdS quantum dots in low density polyethylene at microwave frequencies. *Semiconductors*, *53*(16), 2162–2165. Available from https://doi.org/10.1134/S1063782619120315.

Üzer, A., Erçağ, E., & Apak, R. (2008). Spectrophotometric determination of cyclotrimethylenetrinitramine (RDX) in explosive mixtures and residues with the Berthelot reaction. *Analytica Chimica Acta*, *612*(1), 53–64. Available from https://doi.org/10.1016/j.aca.2008.02.015, https://www.sciencedirect.com/science/article/pii/S0003267008003061.

Üzer, A., Yalçın, U., Can, Z., Erçağ, E., & Apak, R. (2017). Indirect determination of pentaerythritol tetranitrate (PETN) with a gold nanoparticles − based colorimetric sensor. *Talanta*, *175*, 243–249. Available from https://doi.org/10.1016/j.talanta.2017.06.049, https://www.sciencedirect.com/science/article/pii/S0039914017306847.

Wang, J., Meng, Z., Xue, M., Qiu, L., & Zhang, C. (2017). Separation of 1,3,5,7-tetranitro-1,3,5,7-tetraazacyclooctane and 1,3,5-trinitro-1,3,5- triazacyclohexane by molecularly imprinted solid-phase extraction. *Journal of Separation Science*, *40*(5), 1201–1208. Available from https://doi.org/10.1002/jssc.201601024.

Waterhouse James. (2023). Ukraine war: The deadly landmines killing hundreds. Available from https://www.bbc.com/news/world-europe-65204053.

Wei, W., Yang, S., Hu, H., Li, H., & Jiang, Z. (2021). Hierarchically grown $ZnFe_2O_4$-decorated polyaniline-coupled-graphene nanosheets as a novel electrocatalyst for selective detecting p-nitrophenol. *Microchemical Journal*, *160*, 105777. Available from https://doi.org/10.1016/j.microc.2020.105777, https://www.sciencedirect.com/science/article/pii/S0026265X20322724.

Westminster Group. (2023). Explosives detection identification field test kit. Available from https://www.wg-plc.com/product/explosives-detection-identification-field-test-kit.

Wu, X., Hang, H., Li, H., Chen, Y., Tong, H., & Wang, L. (2017). Water-dispersible hyperbranched conjugated polymer nanoparticles with sulfonate terminal groups for amplified fluorescence sensing of trace TNT in aqueous solution. *Materials Chemistry Frontiers*, *1*(9), 1875–1880. Available from https://doi.org/10.1039/C7QM00173H.

Wulff, G. (2013). Fourty years of molecular imprinting in synthetic polymers: Origin, features and perspectives. *Microchimica Acta*, *180*(15), 1359–1370. Available from https://doi.org/10.1007/s00604-013-0992-9.

Wulff, G., & Sarhan, A. (1972). The use of polymers with enzyme-analogous structures for the resolution of racemates. AngewThe use of polymers with enzyme-analogous structures for the resolution of racemates. *Angewandte Chemistry International.*, *11*, 341–344.

Xie, Z., Ge, H., Du, J., Duan, T., Yang, G., & He, Y. (2018). Compartmentalizing incompatible tandem reactions in pickering emulsions to enable visual colorimetric detection of nitramine explosives using a smartphone. *Analytical Chemistry*, *90*(19), 11665–11670. Available from https://doi.org/10.1021/acs.analchem.8b03331.

Xu, B., Wu, X., Li, H., Tong, H., & Wang, L. (2011). Selective detection of TNT and picric acid by conjugated polymer film sensors with donor–acceptor architecture. *Macromolecules*, *44*(13), 5089–5092. Available from https://doi.org/10.1021/ma201003f.

Xu, G., Li, B., Wang, X., & Luo, X. (2014). Electrochemical sensor for nitrobenzene based on carbon paste electrode modified with a poly(3,4-ethylenedioxythiophene) and carbon nanotube nanocomposite. *Microchimica Acta*, *181*(3), 463–469. Available from https://doi.org/10.1007/s00604-013-1136-y.

Yang, J., Aschemeyer, S., Martinez, H., & Trogler, W. C. (2010). Hollow silica nanospheres containing a silafluorene–fluorene conjugated polymer for aqueous TNT and RDX detection. *Chemical Communications*, *46*(36), 6804–6806. Available from https://doi.org/10.1039/C0CC01906B.

Yang, T., Yu, R., Chen, H., Yang, R., Wang, S., Luo, X., & Jiao, K. (2016). Electrochemical preparation of thin-layered molybdenum disulfide-poly(m-aminobenzenesulfonic acid) nanocomposite for TNT detection. *Special issue in honor of Chinese Academician Prof. Hong-Yuan Chen for his 80th birthday*, *781*, 70–75. Available from https://doi.org/10.1016/j.jelechem.2016.09.009, https://www.sciencedirect.com/science/article/pii/S1572665716304544.

Yu, H. A., DeTata, D. A., Lewis, S. W., & Silvester, D. S. (2017). Recent developments in the electrochemical detection of explosives: Towards field-deployable devices for forensic science. *TrAC Trends in Analytical Chemistry*, *97*, 374–384. Available from https://doi.org/10.1016/j.trac.2017.10.007, https://www.sciencedirect.com/science/article/pii/S0165993617302674.

Zhang, Q., Wu, Y., Lian, S., Gao, J., Zhang, S., Hai, G., Sun, C., Li, X., Xia, R., Cabanillas-Gonzalez, J., & Mo, Y. (2020). Simultaneously enhancing photoluminescence quantum efficiency and optical gain of polyfluorene via backbone intercalation of 2,5-dimethyl-1,4-phenylene. *Advanced Optical Materials*, *8*(12), 2000187. Available from https://doi.org/10.1002/adom.202000187.

Zhang, Z., Zhu, L., Deng, Z., Zeng, J., Cai, X.-M., Qiu, Z., et al. (2023). *Aggregation-induced emission biomaterials for anti-pathogen medical applications: Detecting, imaging and killing*. Cambridge Open Engage. Available from https://chemrxiv.org/engage/chemrxiv/article-details/63f62ddd32cd591f125169b8.

Zhao, G., Yin, P., Staples, R., & Shreeve, J. M. (2021). One-step synthesis to an insensitive explosive: N,N′-bis((1H-tetrazol-5-yl)methyl)nitramide (BTMNA). *Chemical Engineering Journal*, *412*, 128697. Available from https://doi.org/10.1016/j.cej.2021.128697, https://www.sciencedirect.com/science/article/pii/S1385894721002953.

Zhao, W., Yang, X., Feng, A., Yan, X., Wang, L., Liang, T., Liu, J., Ma, H., & Zhou, Y. (2021). Distribution and migration characteristics of dinitrotoluene sulfonates (DNTs) in typical TNT production sites: Effects and health risk assessment. *Journal of Environmental Management*, *287*, 112342. Available from https://doi.org/10.1016/j.jenvman.2021.112342, https://www.sciencedirect.com/science/article/pii/S0301479721004047.

Zheng, C., Ling, Y., Chen, J., Yuan, X., Li, S., & Zhang, Z. (2023). Design of a versatile and selective electrochemical sensor based on dummy molecularly imprinted PEDOT/laser-induced graphene for nitroaromatic explosives detection. *Environmental Research*, *236*, 116769. Available from https://doi.org/10.1016/j.envres.2023.116769, https://www.sciencedirect.com/science/article/pii/S0013935123015736.

Zheng, Q., Duan, Z., Zhang, Y., Huang, X., Xiong, X., Zhang, A., Chang, K., & Li, Q. (2023). Conjugated polymeric materials in biological imaging and cancer therapy. *Molecules (Basel, Switzerland)*, *28*(13), 1420–3049. Available from https://doi.org/10.3390/molecules28135091.

Zhou, H., Wang, X., Lin, T. T., Song, J., Tang, B. Z., & Xu, J. (2016). Poly(triphenyl ethene) and poly(tetraphenyl ethene): Synthesis, aggregation-induced emission property and application as paper sensors for effective nitro-compounds detection. *Polymer Chemistry*, *7*(41), 6309–6317. Available from https://doi.org/10.1039/C6PY01358A.

Zhu, L., Zhu, B., Wan, Y., Deng, S., Yu, Z., Zhang, C., & Luo, J. (2022). AIEgen-based metal-organic frameworks as sensing "toolkit" for identification and analysis of energetic compounds. *Interface Control and Characterization of Energetic Materials*, *3*(4), 257–265. Available from https://doi.org/10.1016/j.enmf.2022.09.001, https://www.sciencedirect.com/science/article/pii/S2666647222000756.

CHAPTER

Sensing of metal ions and anions with fluorescent polymeric nanoparticles

19

Suban K. Sahoo, Anuj Saini, Arup K. Ghosh and Aditi Tripathi
Department of Chemistry, Sardar Vallabhbhai National Institute of Technology (SVNIT), Surat, Gujarat, India

19.1 Introduction

The history of fluorescent polymeric nanoparticles (FPNs) can be traced back to the mid-20th century, when the concept of nanotechnology was in it's nascent stages. Paul Ehrlich initiated the historical development of polymer nanoparticles after Ursula Scheffel made her initial experimental attempts. The Peter Speiser group at ETH Zurich carried out extensive research in the late 1960s and early 1970s (Khanna & Speiser, 1969; Quintanar-Guerrero et al., 1998; Khanna et al., 1970; Kreuter, 2007). The FPNs typically have dimensions ranging from 1 to 1000 of nanometers. The core structure of FPNs is composed of a polymeric material, which can be designed to provide specific functionalities such as biocompatibility, stability and controlled release properties, making them ideal for a variety of applications, especially in the fields of biomedicine (Banik et al., 2016), imaging (Zhang et al., 2014), sensing (Gale & Caltagirone, 2018; Prakash et al., 2016; Wu et al., 2017; Bayen et al., 2016), and drug delivery (Hofmann et al., 2014; Feuser et al., 2015).

This chapter presents the fundamental principles governing FPNs synthesis, encompassing various polymerization techniques and their influence on nanoparticle size, morphology, and surface properties. Understanding the impact of these parameters on FPNs performance as ion sensors is crucial for achieving superior stability, reproducibility, and facile functionalization. In the recent years, FPNs have witnessed significant advancements, presenting promising opportunities for sensitive and selective cations and anions sensing. By incorporating fluorescent probes within polymeric materials, FPNs have shown exceptional capabilities in detecting and quantifying ions with remarkable precision and sensitivity (Turos et al., 2007; Shastri, 2003; Bettencourt & Almeida, 2012). The accurate detection of cations and anions holds critical importance across various scientific domains, encompassing environmental studies, industrial processes, medical diagnostics, and pharmaceutical research (Wan et al., 2021; Prasad, 2012). In the past, multiple ion sensing methodologies have been used. However, traditional ion sensing methods have been plagued by complexities, including limited sensitivity, poor selectivity, prolonged response times, and laborious sample preparation. Consequently, the advent of FPNs as efficient ion sensors has sparked widespread interest among researchers, offering a promising alternative to overcome these challenges (Li & Liu, 2014; Elsabahy et al., 2015).

Sensory Polymers. DOI: https://doi.org/10.1016/B978-0-443-13394-7.00019-7

One of the distinguishing characteristics of FPNs is their ability to exhibit dynamic fluorescence responses upon interaction with specific cations and anions. We elucidate the mechanisms underlying fluorescence modulation, such as photo-induced electron transfer (PET) (Bissell et al., 1993), intramolecular charge transfer (ICT) (Thiagarajan et al., 2005), Förster resonance energy transfer (FRET) (Chen et al., 2013) and aggregation-induced emission (AIE) (Liu et al., 2017). These mechanisms play a pivotal role in selective and sensitive ions detection. This chapter emphasizes on the synthetic approaches, the underlying sensing mechanisms, and the uses of FPNs in different fields. This chapter also aims to inspire further research and innovation in this emerging field by addressing the difficulties and opportunities of FPNs-based sensors, ultimately resulting in the widespread use of FPNs as innovative ion sensors across a wide range of research fields.

19.2 Design and synthesis of fluorescent polymeric nanoparticles

Fluorescent polymeric nanoparticles have garnered significant attention as versatile and functional platforms for various applications, including sensing, imaging, drug delivery, and bioanalysis. The design and synthesis of FPNs involve careful consideration of their composition, size, morphology, surface functionalization, and incorporation of fluorophores. This section briefly discusses an overview of different types of fluorescent polymeric nanoparticles and the strategies employed in their design and synthesis. The different types of FPNs have been classified as core-shell nanoparticles, conjugated polymer nanoparticles, non-conjugated polymer nanoparticles, and polymer dots, which are discussed below.

19.2.1 Core–shell nanoparticles

Core–shell nanoparticles consist of a central core surrounded by a shell layer, offering control over optical properties and functionalities. The inorganic nanoparticles (e.g., quantum dots, silver/gold nanoparticles) or organic dyes can serve as the core to provide fluorescence properties, whereas polymers can be used as shell materials to encapsulate the core and provide photostability, biocompatibility, and functionalities for post-functionalization (Ghosh Chaudhuri & Paria, 2012; Zhang et al., 2005; Chen, Hsu, et al., 2015).

19.2.2 Conjugated polymer nanoparticles

Conjugated polymers possess inherent fluorescent properties are synthesized by polymerizing small aromatic molecules with extended π-conjugation and own fluorescence followed by fabricated into nanoparticles for enhanced brightness and stability. Design considerations include selecting a conjugated polymer with a high fluorescence quantum yield, an appropriate bandgap, and good solubility. To obtain conjugated polymer nanoparticles, techniques such as nanoprecipitation, emulsion polymerization, or template-assisted assembly can be used (Kang et al., 2022; MacFarlane et al., 2021).

19.2.3 Nonconjugated polymer nanoparticles

The nonconjugated polymers can also be used to synthesize fluorescent polymeric nanoparticles. These nonconjugated fluorescent polymer dots (NCPDs) are water soluble, synthetically simple in nature, nontoxic to cells, have high fluorescence quantum yield, and can be used for in vitro bioimaging. Recent research has shown that certain nonconjugated PNPs, such as poly(ethyleneimine) (PEI), can emit strong fluorescence under specific conditions despite the lack of typical chromophores (Yang et al., 2021; Saini & Sahoo, 2023).

19.2.4 Polymer dots

Polymer dots, also known as polymer nanoparticles or semiconducting polymer nanoparticles, are small fluorescent nanoparticles composed of conjugated or nonconjugated polymers. These nanoparticles contain greater than 50% volume or weight fraction and have a diameter smaller than 40 nm (Yuan et al., 2021; Bai et al., 2022).

19.3 Synthetic approaches

19.3.1 One-pot synthesis methods involving polymerization

A practical and straightforward method for creating polymeric nanoparticles is one-pot synthesis. This method uses a direct reaction in one step to generate polymers that will later be converted into nanomaterials by one or more processes, which include self-assembly, crosslinking, precipitation, and aggregation processes. The carbonyl group, a double-bonded component present in such polymers, often plays a pivotal role in crosslinking and self-assembly. Subsequently, the post-functionalization process with ligands or biomolecules further improves the stability, target specificity, and enables specific functions (Campbell et al., 2013; Tang et al., 2020; Kumar et al., 2016).

19.3.2 Dye-loaded polymeric nanoparticles

Small organic dyes encapsulated within a polymeric matrix serve as fluorophores in dye-loaded nanoparticles. For dye encapsulation, methods such as solvent evaporation, emulsion-based methods, or self-assembly can be used. These NPs are so chosen that they allow the selection of a wide range of dyes with desirable fluorescence properties and tunability (Klymchenko, 2018; Ma et al., 2011).

19.3.3 Stimuli-responsive polymeric nanoparticles

Stimuli-responsive nanoparticles can undergo reversible changes in their fluorescence properties in response to external stimuli such as pH, temperature, light, or specific analytes. For design purposes, incorporation of stimuli-responsive moieties into the polymer matrix or on the nanoparticle surface. These nanoparticles are useful in controlled release systems, environmental sensing, and responsive imaging (Zhang et al., 2018; Jia et al., 2016).

The design and synthesis of fluorescent polymeric nanoparticles offer immense possibilities for tailoring their properties and functionalities to meet specific application requirements. Through careful selection of materials, optimization of synthesis methods, and surface functionalization, researchers can develop a wide range of fluorescent polymeric nanoparticles with enhanced optical properties, stability, and target specificity. These nanoparticles continue to drive advancements for the sensing of cations and anions in various fields, including biology, environment, and nanotechnology.

19.4 Characterization techniques

FPNs are widely used in biomedical applications, including drug delivery, imaging, and ion sensing. They consist of a polymeric matrix decorated with fluorescent dyes, functional groups, or quantum dots. Understanding the physiochemical properties and behavior of FPNs in biological systems requires extensive characterization. The size and morphology of FPNs are critical aspects of their characterization. Techniques such as dynamic light scattering (DLS), scanning electron microscope (SEM), and transmission electron microscopy (TEM) are generally employed to determine the size and morphology of the nanoparticles. TEM also provides information about their shape and structure. Surface properties play an essential role in the biological performance of the nanoparticles. Zeta potential analysis is often used to measure the surface charge, which indicates the electrostatic interaction of the particles with the surrounding medium. Surface chemistry can be characterized using techniques like Fourier transform infrared spectroscopy (FTIR) and X-ray photoelectron spectroscopy (XPS). The fluorescence properties of FPNs are essential for imaging and sensing applications. Spectroscopic techniques like fluorescence spectroscopy and time-resolved fluorescence spectroscopy determine the fluorescence intensity, wavelength, and lifetime of the nanoparticles. These techniques also enable the assessment of stability and photostability. Biocompatibility and toxicity are crucial considerations for biomedical applications. In vitro cell-based assays such as the MTT assay and lactate dehydrogenase (LDH) assay can evaluate the cytotoxicity of FPNs. To investigate the biodistribution and pharmacokinetics of nanoparticles, in vivo imaging techniques such as positron emission tomography and magnetic resonance imaging (MRI) are used.

19.5 Cations and anions sensing with fluorescent polymeric nanoparticles

Metal ions have essential functions in both environmental and biological processes. Nevertheless, excessive amounts of metal ions can lead to ecological issues and pose risks to the living systems. In recent years, there has been significant progress in the development of fluorescent and colorimetric chemosensors that exhibit high selectivity and sensitivity for the real-time detection of metal ions in environmental and biological systems (Saleem & Lee, 2015; Wu et al., 2018; Aderinto & Imhanria, 2018). Similarly, anions are ubiquitous and therefore, there is a growing interest on developing fluorescence based analytical techniques for detecting anions, such as halides, nitrides, phosphates, sulfates, nitrates, and superoxides (Basabe-Desmonts et al., 2007; Ruedas-Rama et al., 2012; Goshisht & Tripathi, 2021).

In general, a fluorescent chemosensor consists of a signaling unit (light-emitting unit) and a analyte-binding unit. The analyte-binding unit selectively recognizes the target analyte. Upon analyte recognition, the fluorescence behavior of the signaling unit is altered in the form of fluorescence quenching, fluorescence enhancement and/or fluorescence red/blue-shift. Several sensing mechanisms based on electron/charge/energy transfer (such as PET, ICT, FRET etc.) are reported to explain the analyte-induced distinct fluorescence response from a fluorescent chemosensor. The signaling unit in the fluorescent chemosensors can be an organic dyad (fluorophore) or a fluorescent nanomaterial, such as semiconductor quantum dots, metal nanoclusters, carbon dots, silicon dots, polymeric nanoparticles and organic nanoparticles. One approach involves the construction of fluorescent sensors using fluorescent polymeric nanoparticles, enabling the monitoring of various metal ions and anions (Aderinto & Imhanria, 2018). This section summarizes a few examples of FPNs employed for the sensing of metal ions and anions.

Chen, Li, et al. (2015) developed a FPNs through the copolymerization of styrene, vinyl benzyl chloride, and a fluorescent vinylic crosslinking monomer called fluorescein-O,O-bis-propene (FBP) in an oil-in-water mini emulsion. The mini-emulsion was stabilized using a cationic surfactant called dodecyl trimethyl ammonium bromide. Subsequently, the surface of the nanoparticles was grafted with a ligand 1,4,7,10-tetraazacyclododecane (Cyclen). The nanoparticles exhibited an empathetic fluorescence "on-off" response to Cu^{2+} ions in an aqueous solution with a detection limit of 340 nM. The fluorescence change of FPNs by Cu^{2+} was achieved due to the FRET effect between FBP and the Cu^{2+}–Cyclen complex. The nanoparticles displayed a high selectivity towards Cu^{2+} ions (Fig. 19.1). Sequentially, the in-situ generated nanoparticle-Cu^{2+} ensemble can allow the fluorescence "off-on" detection of S^{2-} anion with a detection limit of 2.1 μM.

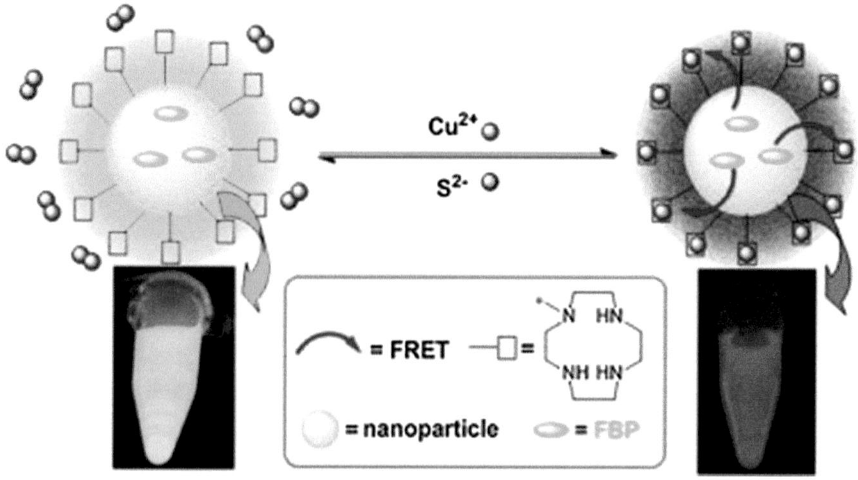

FIGURE 19.1

Scheme showing the fluorescent detection of Cu^{2+} and S^{2-} ions using the Cyclen-functionalized FPNs.

Courtesy: Chen, J., Li, Y., Zhong, W., Hou, Q., Wang, H., Sun, X., & Yi, P. (2015). Novel fluorescent polymeric nanoparticles for highly selective recognition of copper ion and sulfide anion in water. Sensors and Actuators B: Chemical, 206, *230–238.*

Yang et al. (2021) synthesized a FPNs through a one-pot reaction between hyperbranched polyethyleneimine (hPEI) and 6-hydroxy-2-naphthaldehyde (HNA). The Schiff base reaction followed by self-assembly formed the PEI-HNA FPNs, which emit at 525 nm ($\lambda_{ex} = 390$ nm) (Fig. 19.2). The fluorescence emission of PEI-HNA FPNs was quenched selectively by Cu^{2+} ions, and no interference was observed in the co-presence of other competitive metal ions. The PEI-HNA rapidly detect Cu^{2+} ions within the concentration range of 0–60 μM in just 30 seconds with a detection limit of 243 nM. The probe PEI-HNA was employed for bioimaging in living cells, which exhibited excellent cell penetrability and low toxicity. The PEI-HNA was also incorporated into filter paper, hydrogel, and nanofibrous film to create solid-phase sensors, which showed rapid response for Cu^{2+}. In another work, Liu et al. (2017) synthesized a FPNs by crosslinking hyperbranched PEI (hPEI) with formaldehyde in an aqueous medium (Fig. 19.3). The FPNs showed strong intrinsic fluorescence without the need for any external fluorescent agent. When Cu^{2+} ions adsorbed onto the surface of the FPNs caused a rapid fluorescence quenching through electron transfer process. The developed sensor was utilized to quantify copper ions in environmental water samples. Furthermore, the fluorescence response from the FPNs upon the addition of Cu^{2+} and cysteine was used to mimic the IMPLICATION logic gate. The hPEI is widely used to obtain nonconjugated FPNs because of the facile synthesis, good aqueous solubility and easy post-functionalization. The small molecular cross-linkers like glutaraldehyde, formaldehyde, citric acid, *iso*phthalaldehyde and salicylaldehyde are used for the synthesis of PEI-based FPNs. In a recent work, Sahoo and coworkers used the formaldehyde as cross-linking agent and PEI as polymeric platform to obtain the cyan-blue-emitting PEIF FPNs. Subsequently, the surface of the PEIF FPNs was post-functionalized with vitamin B_6 cofactor pyridoxal 5′-phosphate (PLP) to obtain the yellow-emitting polymeric nanoparticles PEIFPLP FPNs (Saini & Sahoo, 2023). The fluorescence emission of PEIFPLP at 552 nm ($\lambda_{ex} = 450$ nm) was quenched by Fe^{2+} and Cu^{2+} ions due to the electron transfer process from the excited FPNs to the available empty *d*-orbital of Fe^{2+} and Cu^{2+}. Using PEIFPLP FPNs, the concentration of Fe^{2+} and Cu^{2+} ions can be detected down to 31 and 17 nM, respectively. Also, the developed probe PEIFPLP was successfully employed to detect Fe^{2+} and Cu^{2+} ions in vegetables, fruits and environmental water samples.

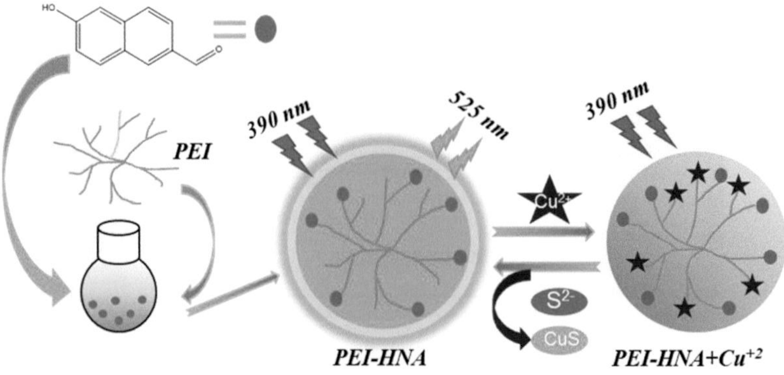

FIGURE 19.2

The synthesis of PEI-HNA and complexation with Cu^{2+}.

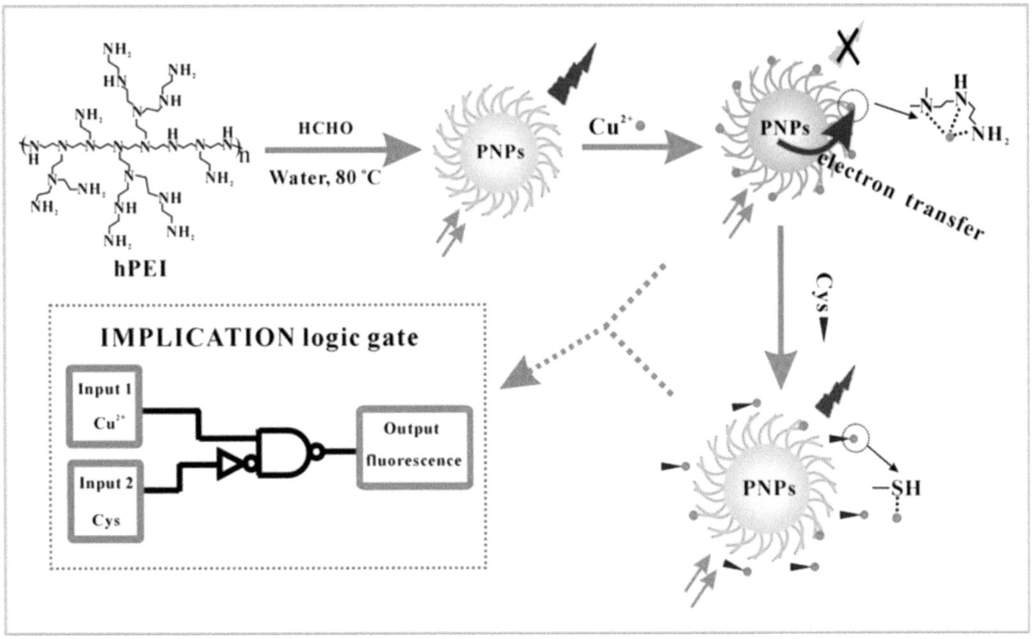

FIGURE 19.3

Scheme showing the preparation of FPNs, Cu^{2+} detection, and IMPLICATION logic gate operation by using the FPNs.

Courtesy: Liu, S.G., Li, N., Fan, Y.Z., Li, N.B., & Luo, H.Q. (2017). Intrinsically fluorescent polymer nanoparticles for sensing Cu^{2+} in aqueous media and constructing an IMPLICATION logic gate. Sensors and Actuators B: Chemical, 243, *634–641.*

The amidation reaction between PEI and citric acid followed by self-assembly of PEI formed the rice-like nonconjugated polymer dots (NPDs) (Zhang et al., 2019). The citric acid chemically reacts with the primary and secondary amines of PEI. The formation of cross-linked amide linkages self-assembled PEI to form the rice-shaped NPDs with a width of 70 nm and a length of 200 nm. The NPDs showed good dispersibility because of the hydrophilic groups present on the surface, such as hydroxyl, carboxyl, amide, and amino groups. The NPDs emit at 438 nm, when excited at 354 nm. The fluorescence emission of NPDs was selectively quenched by Cu^{2+} due to the formation of copper complexes over the surface of the dots and inner-filter effect. The NPDs fluorescence emission was also quenched by ClO^- due to the oxidation of hydroxyl groups on the surface leading to the suppression of the electron transfer process between the chromophore groups and the hydroxyl groups. The detection limit of NPDs for Cu^{2+} and ClO^- ions was reported to be 3.5 nM and 1.9 nM, respectively. The analytical novelty of NPDs was further examined by quantifying Cu^{2+} and ClO^- ions in real environmental water samples. In another approach, PEI and p-phenylenediamine (PPDA) was reacted at room temperature in polar medium to form the blue-greenish emitted polymeric nanoparticles PEIPPDA ($\lambda_{ex} = 377$ nm, $\lambda_{em} = 499$ nm) (Liu et al., 2019). The fluorescence emission of PEIPPDA was quenched selectively by Cu^{2+} with a detection limit of 11.27 nM. Zhong et al., have synthesized a hPEI-based fluorescent and biocompatible polymeric nanoparticles F-ohPEIs, and employed for the fluorescent turn-off detection of Cu^{2+} (Zhong et al., 2021). In order to minimize the intrinsic high positive charge density, hPEI was first oxidized with H_2O_2 to form the ohPEI. The oxidized hPEI (ohPEI) was cross-linked and self-assembled by formaldehyde to form the fluorescent F-ohPEIs ($\lambda_{ex} = 365$ nm, $\lambda_{em} = 452$ nm). The selective complexation of Cu^{2+} ions on the surface of the F-ohPEIs, and the energy transfer from the excited F-ohPEIs to the coordinated Cu^{2+} ions quenched the fluorescence emission at 452 nm with a detection limit of 0.013 μM.

Fluorescent nanoparticles of poly(methylmethacrylate-co-glycidylmethacrylate) were synthesized using semi-continuous emulsion polymerization by Ghezelsefloo et al. (2021) These nanoparticles were functionalized with rhodamine B ethylenediamine acrylate (RhAcL-3). Upon the addition of Fe^{2+} and Fe^{3+} ions, the RhAcL-3 nanoparticles exhibited an immediate and noticeable color change accompanied by a significant increase in the fluorescence emission at 590 nm (Fig. 19.4). The fluorescence enhancement reached approximately twofold for Fe^{2+} ions and 2.twofold for Fe^{3+} ions. This effect was attributed to the chelation of these ions with the spirolactam moiety of rhodamine B resulted to an open-ring amide form. The linear dynamic detection range of the RhAcL-3 for Fe^{2+} and Fe^{3+} ions was determined to be 4−320 μM with the detection limits of 2.63 μM and 2.5 μM, respectively.

Wang et al. (2017) developed a ratiometric fluorescent probe NP3 based on fluorescent polymeric nanoparticles for the selective detection of Hg^{2+} ions in an aqueous medium and living cells. This probe consists of two components: a reference fluorescent dye (4-ethoxy-9-allyl-1,8-naphthalimide: EANI) located in the core of the nanoparticle and the Hg^{2+}-recognition group (fluorescein derivative: AEMH-FITC) on the surface of the nanoparticles. The probe NP3 exhibited distinct dual emissions at 432 nm and 528 nm (Fig. 19.5). Hg^{2+} ions can effectively quench the fluorescence from the FITC moieties through a PET mechanism. In contrast, the fluorescence from EANI remains unchanged. This differential response leads to a ratiometric fluorescent detection of Hg^{2+} ions. The probe NP3 demonstrates favorable water dispersibility, excellent long-term photostability, and selective ratiometric response for Hg^{2+} ions. It exhibited a low detection limit of 75 nM. Additionally, the intracellular fluorescence imaging experiments revealed that NP3 possesses good cell-membrane permeability and enables the visualization of changes in Hg^{2+} ion levels within living cells. In another work, blue and red dual emission carbon dots (DDCDs) was synthesized using ethylenediamine and citric acid as precursor materials in a water-formamide binary system through a one-step hydrothermal method (Lu et al., 2019). Subsequently, the DDCDs were used to prepare ion-imprinted fluorescence polymers (IIPs) using CTAB as soft template. The blue emission of DDCDs was explicitly quenched by Cr^{3+}, while the red emission of DDCDs was selectively quenched by Pb^{2+}. This unique property of DDCDs enabled the simultaneous dual channel detection of Cr^{3+} and Pb^{2+} ions using the IIPs. The fluorescence quenching at 440 nm due to Cr^{3+} was found to be linear from 0.1 to 6.0 μM with a detection limit of 27 nM. Whereas the quenching effect at 580 nm for the red channel was linear in the range of 0.1 to 5.0 μM of Pb^{2+} with a detection limit of 34 nM. The IIPs was employed for detecting Cr^{3+} and Pb^{2+} ions in real water samples.

Zou et al. (2018) synthesized intrinsically fluorescent polymer nanoparticles (F-PNPs) using melamine and 2-hydroxy-5-methylisophthalaldehyde as precursor materials through a solvothermal method. The F-PNPs exhibited a strong yellow-green fluorescence at 542 nm (Fig. 19.6A). The

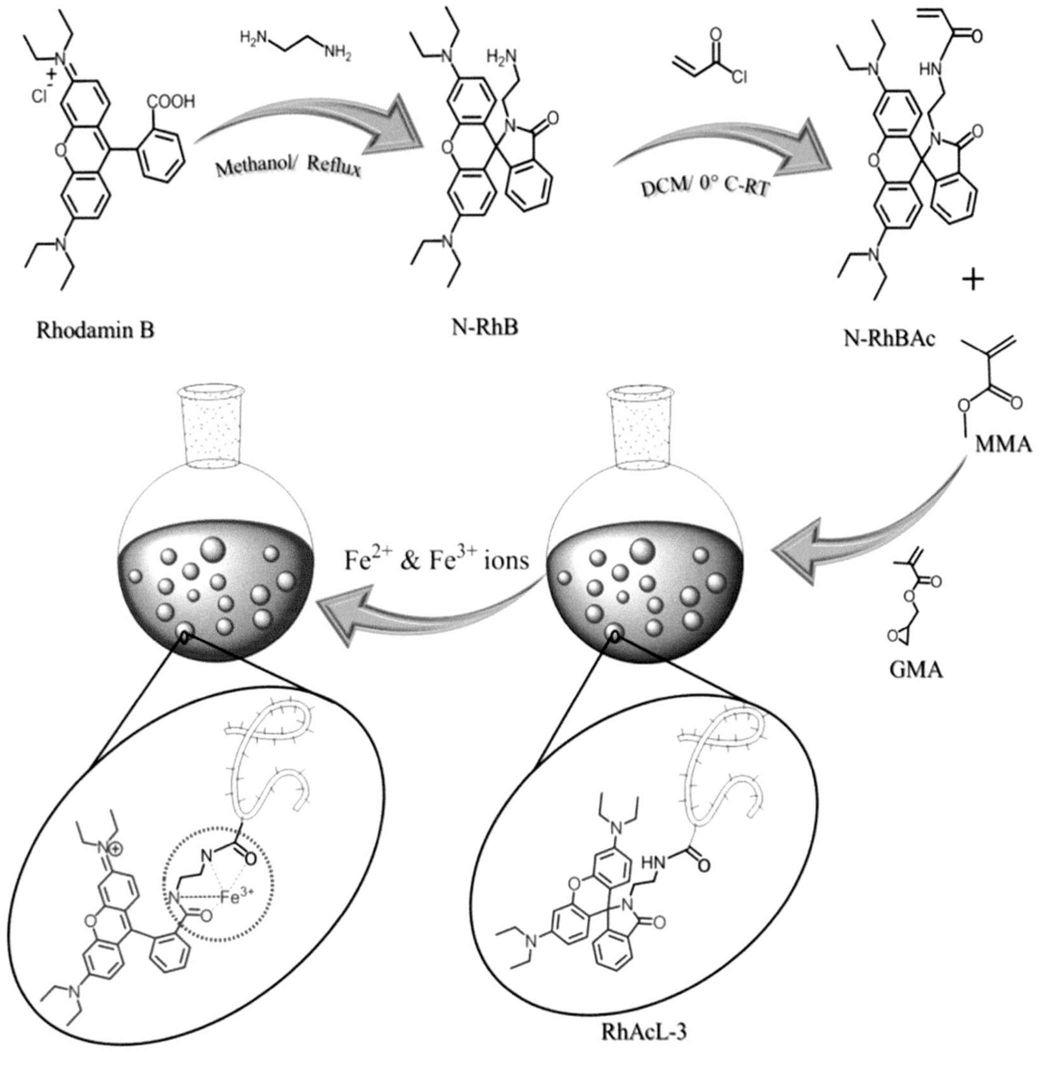

FIGURE 19.4

Schematic representation for synthesizing N-RhBAc and preparing fluorescent polymer nanoparticles (RhAcL-3) and their interaction with Fe^{2+} or Fe^{3+} ions.

Courtesy: Ghezelsefloo, S., Rad, J.K., Hajiali, M., & Mahdavian, A.R. (2021). Rhodamine-based fluorescent polyacrylic nanoparticles: A highly selective and sensitive chemosensor for Fe (II) and Fe (III) cations in water. Journal of Environmental Chemical Engineering, 9*(2), 105082.*

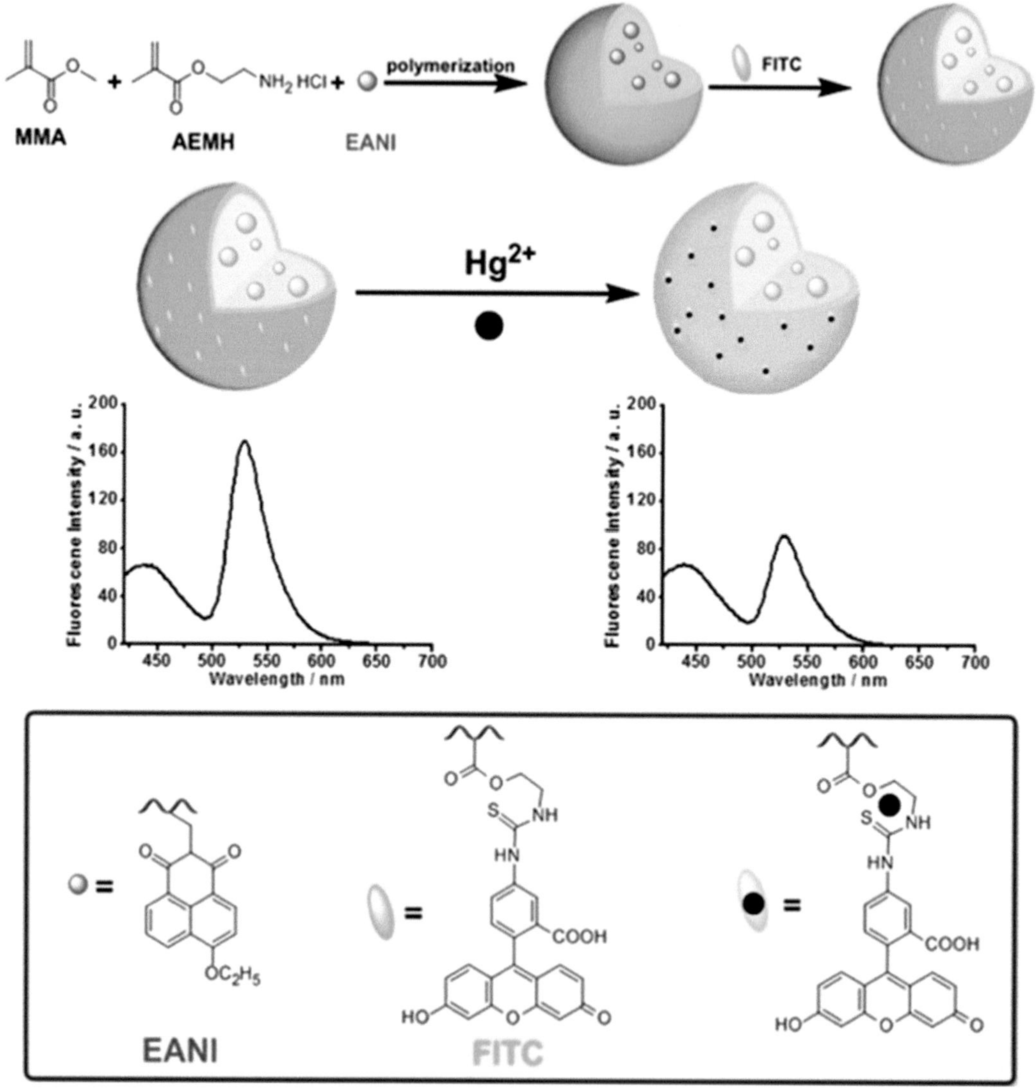

FIGURE 19.5

Scheme presented the synthesis of the probe NP3 and its application as a ratiometric fluorescent probe for detecting Hg^{2+}.

Courtesy: Wang, H., Zhang, P., Chen, J., Li, Y., Yu, M., Long, Y., Yi, P. (2017). Polymer nanoparticle-based ratiometric fluorescent probe for imaging Hg^{2+} ions in living cells. Sensors and Actuators B: Chemical, 242, *818–824.*

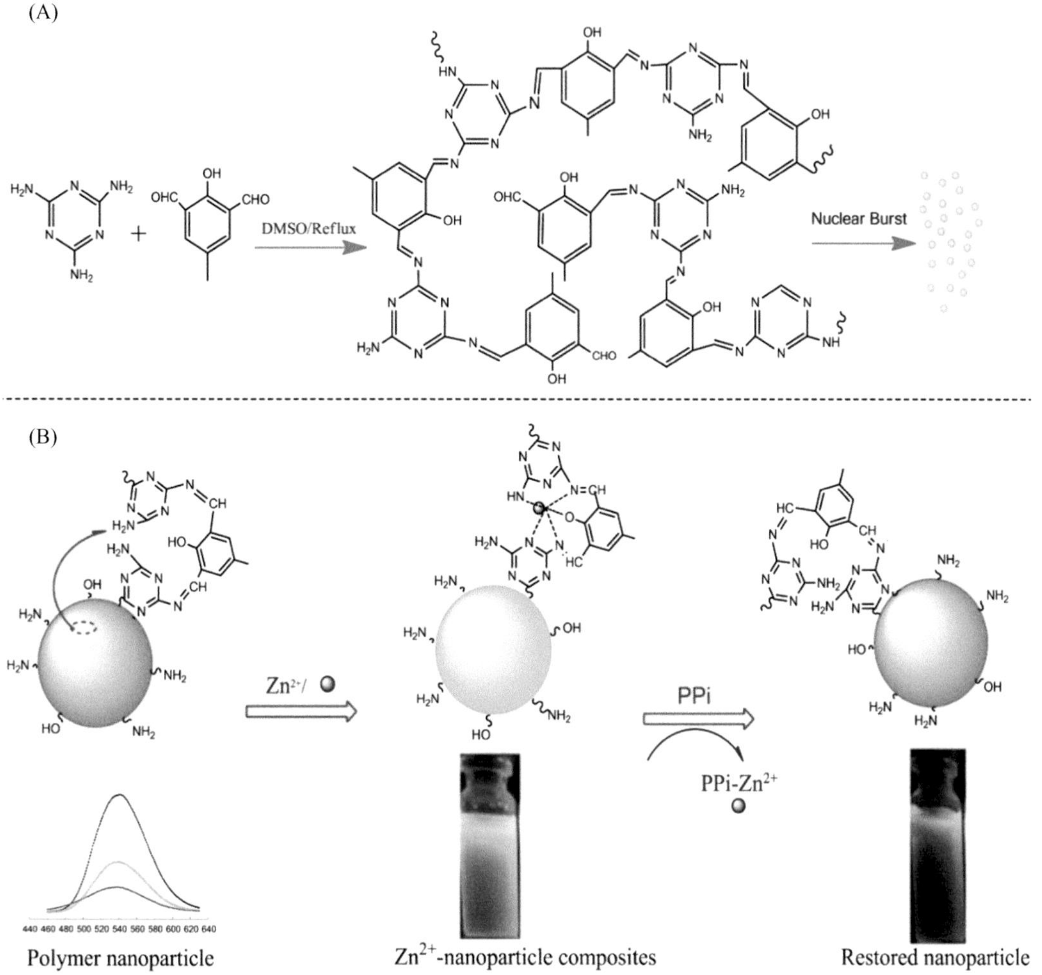

FIGURE 19.6

(A) Synthesis of F-PNPs using melamine and 2-hydroxy-5-methylisophthalaldehyde as precursor materials by a solvothermal method; (B) Schematic illustration of the detection principle of Zn^{2+} and PPi based on F-PNPs.

Courtesy: Zou, W., Gong, F., Chen, X., Cao, Z., Xia, J., Gu, T., & Li, Z. (2018). Intrinsically fluorescent and highly functionalized polymer nanoparticles as probes for the detection of zinc and pyrophosphate ions in rabbit serum samples. Talanta, 188, *203–209.*

surface of the fluorescent polymer nanoparticles contained abundant amino and hydroxyl groups, providing multiple binding sites as well as excellent water solubility and biocompatibility. When Zn^{2+} ions were added to the F-PNPs, a significant blue shift ($\Delta\lambda = 40$ nm, $\lambda_{shift} = 542$ to 502 nm) and noticeable fluorescence enhancement was observed at 502 nm. Subsequent addition of pyrophosphate (PPi) displaced Zn^{2+} ions from the coordination cavity of the fluorescent polymer nanoparticles-Zn^{2+} nanocomposites and restored the fluorescence of the F-PNPs (Fig. 19.6B). By performing fluorescence titration experiments, the detection limits of F-PNPs for Zn^{2+} and F-PNPs-Zn^{2+} ensemble for PPi were determined to be 2.75×10^{-8} M and 7.63×10^{-8} M, respectively.

A ratiometric detection of Cu^{2+} ions was achieved by developing nanohybrid (GCDs@RSPN) sensing system by loading green fluorescent carbon dots (GCDs) prepared from sunset yellow and (PEI) via hydrothermal method on the surface of red-emitting semiconducting polyfluorene-4,7-dithiophene-2,1,3-benzothiadiazole (PFDBT-5) polymer nanoparticles (RSPN) through electrostatic adsorption (Luo et al., 2023). The GCDs with abundant amino groups present over the surface of RSPN selectively recognize Cu^{2+} ions that activate the PET process and quench the fluorescence emission. With a linearity range of 0–100 μM, the probe GCDs@RSPN can be applied to detect Cu^{2+} down to 0.577 μM. Also, GCDs@RSPN coated paper strips were developed for the cost-effective visual detection of Cu^{2+}. In another work, core–shell nanoparticle-based nanohybrid sensing system was fabricated for the detection of Cu^{2+} in aqueous media (Zhang et al., 2013). The sensing system was synthesized by one-pot facile miniemulsion polymerization, where the fluorophore 4-methamino-9-allyl-1,8-naphthalimide (MANI) was incorporated covalently into the PMMA core of the particle and the ligand vinylbenzylcyclam was linked chemically onto the surface. The cyclam functionalized over the particles selectively bind with Cu^{2+}. With the addition of Cu^{2+} ions, the fluorescence emission of MANI dye in nanoparticles was quenched due to the intraparticle FRET from the dye present within PMMA core to the Cu^{2+}-cyclam complexes on the nanoparticle surface. Without any interference from other metal ions, the nanohybrid sensing system can detect Cu^{2+} down to 500 nM and over a wide pH range (pH 4–10) in water.

The semiconducting polymer nanoparticles made of poly(9,9-dioctylfluorenyl-2,7-diyl) (PFO) was assembled with 7-nitro-1,2,3-benzoxadiazole amine (NBD-A) to form the NBD@PFO nanohybrids (Wang et al., 2021). The NBD@PFO nanohybrids showed two emission bands at 420–490 nm and 490–650 nm due to the polymer nanoparticles PFO and NBD-A dye, respectively. The dual emissions observed from NBD@PFO because of the effective FRET process between PFO and NBD-A. The NBD@PFO nanohybrids were employed for ratiometric detection of H_2S and two-photon imaging of H_2S in zebrafish and living cells. Under physiological conditions, H_2S exist in the form of HS^-, and therefore, NaHS reagent was added to the nanohybrids to examine the fluorescence response. Addition of HS^- to NBD@PFO caused thiolysis of NBD-A that inhibited the intraparticle FRET that decreased the green emission NBD-A without altering the fluorescent intensity of PFO (Fig. 19.7). The NBD@PFO nanohybrids showed a good linear range for H_2S from 0.1 to 100 μM with the detection limit of 0.034 μM. On the other hand, Song et al. (2017) developed a fluorescence sensor for detecting the superoxide anion ($O_2^{\bullet -}$) using the suffocated polystyrene nanoparticles ($PS\text{-}SO_3H$)@terbium (Tb)/guanine (G) nanoscale coordination polymers ($PS\text{-}SO_3H$@Tb/G NCPs). The $PS\text{-}SO_3H$ nanoparticles contained an abundance of sulfonyl groups, which served as binding sites for the spontaneous self-assembly of Tb^{3+} and G on the $PS\text{-}SO_3H$ surface, resulting in the formation of the $PS\text{-}SO_3H$@Tb/G NCPs sensor. The fluorescence intensity ratio exhibited a linear decrease as the concentration of $O_2^{\bullet -}$ increased within the range of 10.12 nM to 6.0 μM. Impressively, this sensor achieved a low detection limit of 3.4 nM.

Semiconducting polymer dots (PDs) possess strong and tunable fluorescence, high quantum yield, excellent photostability and biocompatibility. The PDs are made of π-conjugated polymers with diameters ≤40 nm (Bai et al., 2022). Using the nano-precipitation method, the carboxyl

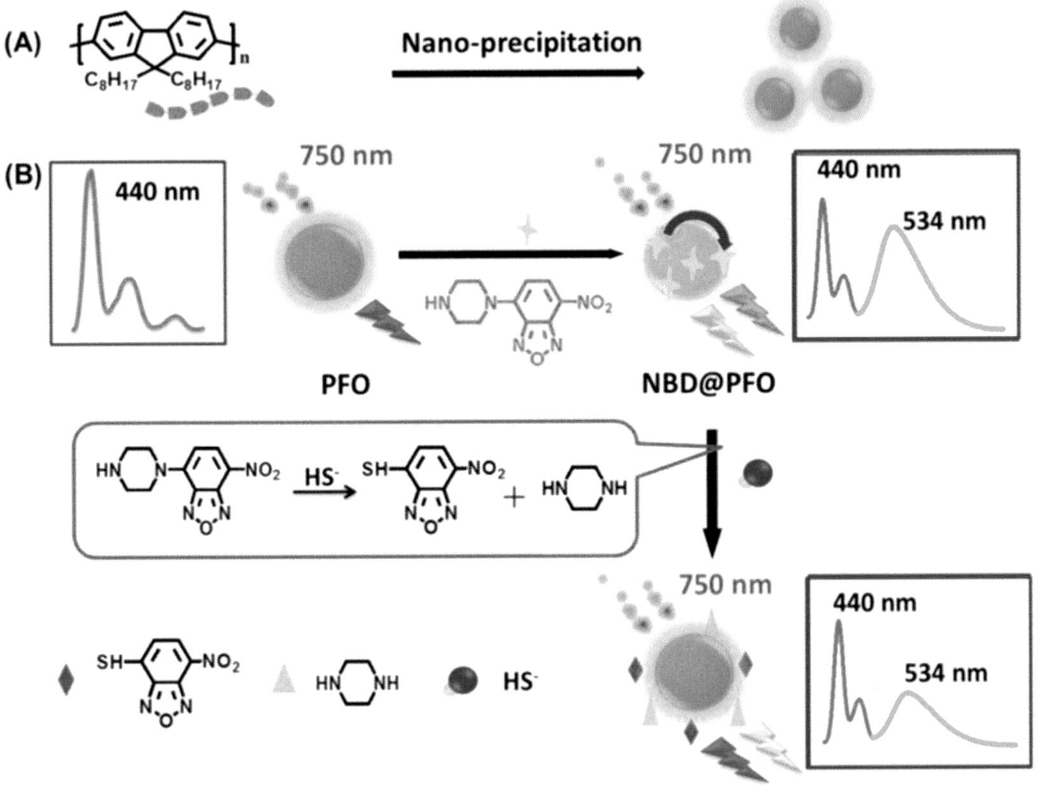

FIGURE 19.7

Scheme showing the two-photon ratiometric sensing of H_2S using the nanobybrid sensing system NBD@PFO.

Courtesy: Wang, D., He, J., & Sun, J. (2021). Two-photon ratiometric fluorescent probe based on NBD-amine functionalized semiconducting polymer nanoparticles for real-time imaging of hydrogen sulfide in living cells and zebrafish. Talanta, 228, *122269.*

functionalized poly[(9,9-dioctylfluorenyl-2,7-diyl)-alt-co-(1,4-benzo-(2,1′,3)-thiadiazole)] (PFBT) PDs were synthesized and decorated with 9-anthracenecarboxylic acid (Chabok et al., 2019). The fluorescence emission at 540 nm (λ_{ex} = 460 nm) was selectively quenched by Cu^{2+} and Fe^{3+}, and restored upon the sequential addition of histidine (His). The PDs showed a good linear range from 0.1 to 630 μM and 0.1 to 720 μM for Cu^{2+} and Fe^{3+} with a detection limit of 61.7 nM and 58.1 nM, respectively. For cascade detection, the in-situ generated PDs/Cu^{2+} assembly was used a nanoprobe for turn-on detection of His down to 79.6 nM. The practical utility of PDs/Cu^{2+} probe was examined by detecting His in blood serum and live cells. In another work, Sun et al. (2017) the PDs of PFBT was prepared by adopting nanoprecipitation method followed by assembled with the rhodamine B hydrazide (RB-hy) to obtain the nanohybrids PDs@RB-hy. The functionalized nonfluorescent RB-hy selectively bind with Cu^{2+} and open the spirolactam ring. The nanohybrids PDs@RB-hy exhibited FRET due to the presence of PFBT PDs as an energy donor and RB-hy-Cu

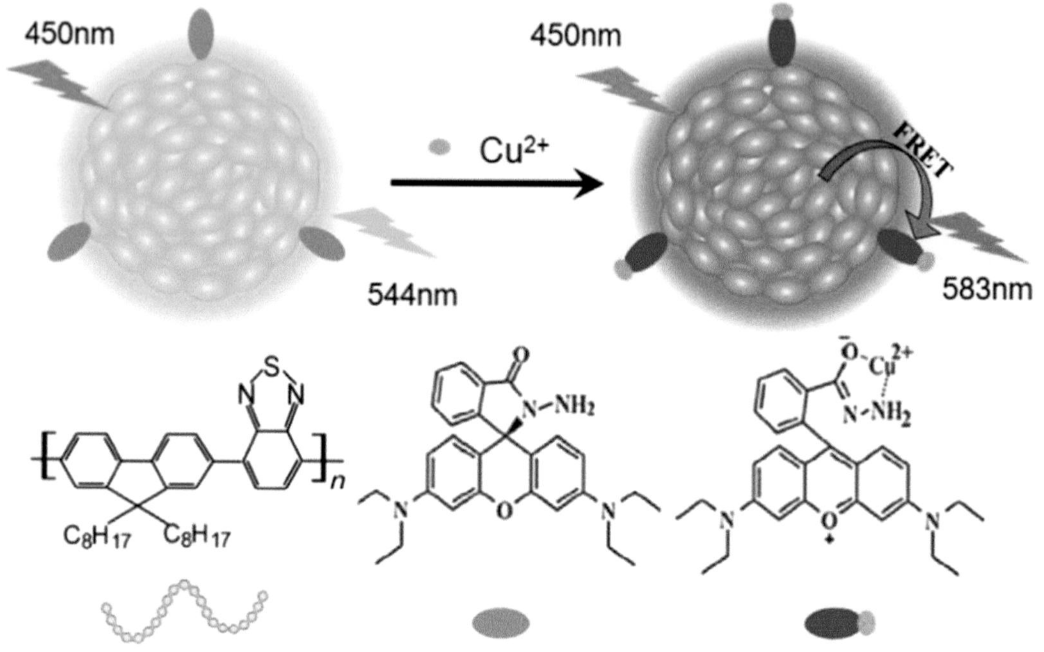

FIGURE 19.8

Schematic representation of the ratiometric fluorescent detection of Cu^{2+} using the nanohybrids PDs@RB-hy.

Courtesy: Sun, J., Mei. H., & Gao, F. (2017). Ratiometric detection of copper ions and alkaline phosphatase activity based on semiconducting polymer dots assembled with rhodamine B hydrazide. Biosensors and Bioelectronics, 91, *70–75.*

(II) complex act as an energy acceptor (Fig. 19.8). Addition of Cu^{2+} to the solution of PDs@RB-hy resulted a decrease in the fluorescence emission of PDs at 544 nm while the emission of RB-hy-Cu (II) complex enhanced at 583 nm. The ratiometric response from PDs@RB-hy allowed the detection of Cu^{2+} down to 15 nM. Subsequently, the nano-assembly PDs@RB-hy-Cu(II) was employed for the detection of alkaline phosphatase (ALP) activity down to 0.0018 U/L by using pyrophosphate (PPi) as phosphate substrates. The phosphate-functionalized PFBT PDs was prepared by doping tributyl phosphate (TBP) (Shamsipur et al., 2019). The TBP@PFBT PDs was employed for the detection of Cu^{2+}, Fe^{3+} and Cytochrome c (Cyt c). Addition of Cu^{2+} and Fe^{3+} ions to the solution of TBP@PFBT PDs resulted a fluorescence quenching at 540 nm (λ_{ex} = 460 nm) due to the complexing interaction with the phosphate groups present over the surface that caused aggregation induced fluorescence quenching. Using the fluorescence turn-off response from TBP@PFBT PDs, the concentrations of Cu^{2+} and Fe^{3+} ions can be detected down to 30 pM and 0.35 nM, respectively. Subsequently, as the Cyt c consists a Fe^{3+} coordinated porphyrin ring and Cyt c is used as an apoptotic biomarker, the TBP@PFBT PDs was successfully employed for the detection of Cyt c with a limit of detection of 32.7 pM.

There is a significant interest on the synthesis of fluorescent polydopamine (PDA) nanoparticles for potential sensing applications owing to their biocompatibility, one-step facile synthesis and

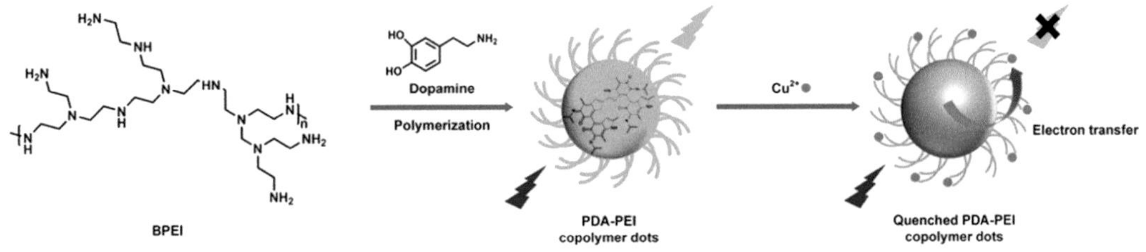

FIGURE 19.9

Schematic representation for the synthesis of PDA-PEI copolymer dots and sensing of Cu^{2+}.

Courtesy: Zhong, Z., & Jia, L. (2019). Room temperature preparation of water-soluble polydopamine-polyethyleneimine copolymer dots for selective detection of copper ions. Talanta, 197, *584–591.*

good water-solubility. The neurotransmitter dopamine (DA) undergoes self-polymerization by converting into its quinone form to obtain the melanin-mimicking polymeric material PDA, which under suitable experimental conditions can be converted into a PDA FPNs. Yin et al., synthesized the blue luminescent periodate-oxidized o-PDANPs FPNs by oxidizing and self-polymerization of DA using sodium periodate as an oxidant (Yin et al., 2018). Subsequently, the fluorescence emission of o-PDANPs FPNs was enhanced upon reducing the carbonyl groups existing in o-PDANPs by adding $NaBH_4$. The reduced r-PDANPs FPNs showed a selective fluorescent turn-off response for the detection of Fe^{3+}. The Fe^{3+} bind with the phenolic hydroxyl groups present over the surface of r-PDANPs FPNs followed by the nonradiative electron transfer quenched the fluorescence emission at 423 nm. Using r-PDANPs FPNs, the concentration of Fe^{3+} can be detected down to 0.15 μM. In another work, the DA was self-polymerized and cross-linked with branched PEI to obtain the PDA-PEI copolymer dots of diameters 9.4 ± 2.2 nm (Zhong & Jia, 2019). The PDA-PEI copolymer dots emitted at 530 nm, when excited at 380 nm. With the addition of different metal ions, the fluorescence emission of PDA-PEI was selectively quenched by Cu^{2+} due to the electron transfer process at the excited state (Fig. 19.9). The developed copolymer dots showed a linearity range from 0.0016 to 80 μM and detection limit of 1.6 nM for Cu^{2+}. The practical utility was also validated by quantifying Cu^{2+} in food and environmental water samples. Recently, the DA was self-polymerized under mild conditions in the presence of folic acid (FA) to form the blue fluorescent FA-PDA FPNs of diameter 1.9 ± 0.3 nm (Chen et al., 2023). The fluorescence emission of FA-PDA FPNs was selectively quenched by Hg^{2+} due to the energy/electron transfer process. The FA-PDA FPNs showed a linearity range from 0–18 μM with a detection limit of 0.18 μM.

19.6 Conclusions

This chapter discusses the growing interest in the synthesis and applications of fluorescent polymeric nanoparticles for cation and anion sensing due to their good aqueous solubility, excellent fluorescence, excellent photostability, tunable fluorescence, high quantum yield and biocompatibility. It provides insights into the design, synthesis, functionalization, and applications of these FPNs showcasing their potential for various analytical, environmental, and biomedical applications. The chapter begins with an overview on the design, synthesis and characterization of FPNs. The FPNs for sensing applications are designed by different approaches, such as one-pot synthesis and doping of dye within polymer core. Subsequently, the various techniques used for the characterization of the synthesized FPNs have been presented. The primary characterization methods include DLS, SEM, and TEM. Surface characterization also involves techniques such as zeta potential analysis, FTIR and XPS. FPNs have shown tremendous potential for the detection of a wide range of cations, anions and other small neutral molecules. FPNs have demonstrated a good linear range of detection with a low detection limit for several cations such as Cu^{2+}, Zn^{2+}, Fe^{2+}, Fe^{3+} and Hg^{2+}. Sensitive and specific detection of several anions such as S^{2-}, HS^{-} and ClO^{-} have also been achieved with the use of FPNs. However, the use of FPNs for the detection of anions is still a developing field and requires more research. Besides developing potential sensing based applications, future research on FPNs must also focus on developing simple, green and cost-effective techniques for producing size-controlled FPNs on a large scale for commercial applications. To summarize, by compiling the ever growing research on FPNs in this chapter, an opportunity has been presented to address the challenges involved and highlight the future scope associated with FPNs. Thus, this review aims to foster further research in the development of FPNs for ion sensing, leading to advancements in the field of detection technologies.

References

Aderinto, S. O., & Imhanria, S. (2018). Fluorescent and colourimetric 1, 8-naphthalimide-appended chemosensors for the tracking of metal ions: Selected examples from the year 2010 to 2017. *Chemical Papers, 72* (8), 1823–1851.

Bai, X., Wang, K., Chen, L., Zhou, J., & Wang, J. (2022). Semiconducting polymer dots as fluorescent probes for in vitro biosensing. *Journal of Materials Chemistry B, 10*(33), 6248–6262.

Banik, B. L., Fattahi, P., & Brown, J. L. (2016). Polymeric nanoparticles: the future of nanomedicine. *WIREs Nanomedicine and Nanobiotechnology, 8*(2), 271–299.

Basabe-Desmonts, L., Reinhoudt, D. N., & Crego-Calama, M. (2007). Design of fluorescent materials for chemical sensing. *Chemical Society reviews, 36*(6), 993–1017.

Bayen, S. P., Mondal, M. K., Naaz, S., Mondal, S. K., & Chowdhury, P. (2016). Design and sonochemical synthesis of water-soluble fluorescent silver nanoclusters for Hg^{2+} sensing. *Journal of environmental chemical engineering, 4*(1), 1110–1116.

Bettencourt, A., & Almeida, A. J. (2012). Poly (methyl methacrylate) particulate carriers in drug delivery. *Journal of Microencapsulation, 29*(4), 353–367.

Bissell, R. A., Prasanna de Silva, A., Nimal Gunaratne, H., Mark Lynch, P., Maguire, G. E., McCoy, C. P., & Samankumara Sandanayake, K. (1993). Fluorescent PET (photoinduced electron transfer) sensors. *Photoinduced Electron Transfer V*, 223–264.

Campbell, P. S., Lorbeer, C., Cybinska, J., & Mudring, A. V. (2013). One-pot synthesis of luminescent polymer-nanoparticle composites from task-specific ionic liquids. *Advanced Functional Materials, 23*(23), 2924–2931.

Chabok, A., Shamsipur, M., Yeganeh-Faal, A., Molaabasi, F., Molaei, K., & Sarparast, M. (2019). A highly selective semiconducting polymer dots-based "off-on" fluorescent nanoprobe for iron, copper and histidine detection and imaging in living cells. *Talanta, 194*, 752–762.

Chen, G., Song, F., Xiong, X., & Peng, X. (2013). Fluorescent nanosensors based on fluorescence resonance energy transfer (FRET). *Industrial & Engineering Chemistry Research, 52*(33), 11228–11245.

Chen, G.-J., Hsu, C., Ke, J.-H., & Wang, L.-F. (2015). Imaging and chemotherapeutic comparisons of iron oxide nanoparticles chemically and physically coated with poly (ethylene glycol)-b-poly (ε-caprolactone)-g-poly (acrylic acid). *Journal of Biomedical Nanotechnology*, *11*(6), 951–963.
Chen, J., Li, Y., Zhong, W., Hou, Q., Wang, H., Sun, X., & Yi, P. (2015). Novel fluorescent polymeric nanoparticles for highly selective recognition of copper ion and sulfide anion in water. *Sensors and Actuators B: Chemical*, *206*, 230–238.
Chen, L., Chen, C., Yan, Y., Yang, L., Liu, R., Zhang, J., Zhang, X., & Xie, C. (2023). Folic acid adjustive polydopamine organic nanoparticles based fluorescent probe for the selective detection of mercury ions. *Polymers*, *15*, 1892.
Elsabahy, M., Heo, G. S., Lim, S.-M., Sun, G., & Wooley, K. L. (2015). Polymeric nanostructures for imaging and therapy. *Chemical Reviews*, *115*(19), 10967–11011.
Feuser, P. E., dos Santos Bubniak, L., dos Santos Silva, M. C., da Cas Viegas, A., Fernandes, A. C., Ricci-Junior, E., Nele, M., Tedesco, A. C., Sayer, C., & de Araújo, P. H. H. (2015). Encapsulation of magnetic nanoparticles in poly (methyl methacrylate) by miniemulsion and evaluation of hyperthermia in U87MG cells. *European Polymer Journal*, *68*, 355–365.
Gale, P. A., & Caltagirone, C. (2018). Fluorescent and colorimetric sensors for anionic species. *Coordination Chemistry Reviews*, *354*, 2–27.
Ghezelsefloo, S., Rad, J. K., Hajiali, M., & Mahdavian, A. R. (2021). Rhodamine-based fluorescent polyacrylic nanoparticles: A highly selective and sensitive chemosensor for Fe (II) and Fe (III) cations in water. *Journal of Environmental Chemical Engineering*, *9*(2), 105082.
Ghosh Chaudhuri, R., & Paria, S. (2012). Core/shell nanoparticles: classes, properties, synthesis mechanisms, characterization, and applications. *Chemical Reviews*, *112*(4), 2373–2433.
Goshisht, M. K., & Tripathi, N. (2021). Fluorescence-based sensors as an emerging tool for anion detection: Mechanism, sensory materials and applications. *Journal of Materials Chemistry C*, *9*(31), 9820–9850.
Hofmann, D., Messerschmidt, C., Bannwarth, M. B., Landfester, K., & Mailänder, V. (2014). Drug delivery without nanoparticle uptake: delivery by a kiss-and-run mechanism on the cell membrane. *Chemical Communications*, *50*(11), 1369–1371.
Jia, X., Zhao, X., Tian, K., Zhou, T., Li, J., Zhang, R., & Liu, P. (2016). Novel fluorescent pH/reduction dual stimuli-responsive polymeric nanoparticles for intracellular triggered anticancer drug release. *Chemical Engineering Journal*, *295*, 468–476.
Kang, S., Yoon, T. W., Kim, G.-Y., & Kang, B. (2022). Review of conjugated polymer nanoparticles: From formulation to applications. *ACS Applied Nano Materials*, *5*(12), 17436–17460.
Khanna, S. C., Jecklin, T., & Speiser, P. (1970). Bead polymerization technique for sustained-release dosage form. *Journal of Pharmaceutical Sciences*, *59*(5), 614–618.
Khanna, S. C., & Speiser, P. (1969). Epoxy resin beads as a pharmaceutical dosage form I: Method of preparation. *Journal of Pharmaceutical Sciences*, *58*(9), 1114–1117.
Klymchenko, A.S. (2018). Dye-loaded fluorescent polymeric nanoparticles: bright platform for biosensing and bioimaging.
Kreuter, J. (2007). Nanoparticles-a historical perspective. *International Journal of Pharmaceutics*, *331*(1), 1–10.
Kumar, A., Chowdhuri, A. R., Laha, D., Chandra, S., Karmakar, P., & Sahu, S. K. (2016). One-pot synthesis of carbon dot-entrenched chitosan-modified magnetic nanoparticles for fluorescence-based Cu^{2+} ion sensing and cell imaging. *RSC Advances*, *6*(64), 58979–58987.
Li, K., & Liu, B. (2014). Polymer-encapsulated organic nanoparticles for fluorescence and photoacoustic imaging. *Chemical Society Reviews*, *43*(18), 6570–6597.
Liu, S. G., Li, N., Fan, Y. Z., Li, N. B., & Luo, H. Q. (2017). Intrinsically fluorescent polymer nanoparticles for sensing Cu^{2+} in aqueous media and constructing an IMPLICATION logic gate. *Sensors and Actuators B: Chemical*, *243*, 634–641.

Liu, X., Tang, Y., Gao, J., Luo, Q., Zeng, Z., & Wang, Q. (2019). Room-temperature synthesis of novel polymeric nanocluster with emissions and its Cu^{2+} recognition performance. *Journal of Luminescence*, *205*, 142–147.

Liu, Y., Mao, L., Liu, X., Liu, M., Xu, D., Jiang, R., Deng, F., Li, Y., Zhang, X., & Wei, Y. (2017). A facile strategy for fabrication of aggregation-induced emission (AIE) active fluorescent polymeric nanoparticles (FPNs) via post modification of synthetic polymers and their cell imaging. *Materials Science and Engineering: C*, *79*, 590–595.

Lu, H., Xu, S., & Liu, J. (2019). One pot generation of blue and red carbon dots in one binary solvent system for dual channel detection of Cr^{3+} and Pb^{2+} based on ion imprinted fluorescence polymers. *ACS Sensors*, *4*(7), 1917–1924.

Luo, F., Zhu, M., Liu, Y., Sun, J., & Gao, F. (2023). Ratiometric and visual determination of copper ions with fluorescent nanohybrids of semiconducting polymer nanoparticles and carbon dots. *Spectrochimica Acta Part A: Molecular and Biomolecular Spectroscopy*, *295*, 122574.

Ma, C., Zeng, F., Huang, L., & Wu, S. (2011). FRET-based ratiometric detection system for mercury ions in water with polymeric particles as scaffolds. *The Journal of Physical Chemistry B*, *115*(5), 874–882.

MacFarlane, L. R., Shaikh, H., Garcia-Hernandez, J. D., Vespa, M., Fukui, T., & Manners, I. (2021). Functional nanoparticles through π-conjugated polymer self-assembly. *Nature Reviews Materials*, *6*(1), 7–26.

Prakash, K., Sahoo, P. R., & Kumar, S. (2016). A substituted spiropyran for highly sensitive and selective colorimetric detection of cyanide ions. *Sensors and Actuators B: Chemical*, *237*, 856–864.

Prasad, P. N. (2012). *Introduction to nanomedicine and nanobioengineering*. John Wiley & Sons.

Quintanar-Guerrero, D., Allémann, E., Fessi, H., & Doelker, E. (1998). Preparation techniques and mechanisms of formation of biodegradable nanoparticles from preformed polymers. *Drug Development and Industrial Pharmacy*, *24*(12), 1113–1128. Available from https://doi.org/10.3109/03639049809108571.

Ruedas-Rama, M. J., Walters, J. D., Orte, A., & Hall, E. A. (2012). Fluorescent nanoparticles for intracellular sensing: a review. *Analytica Chimica Acta*, *751*, 1–23.

Saini, A. K., & Sahoo, S. K. (2023). Vitamin B6 cofactor-conjugated fluorescent polymeric nanoparticles for nanomolar detection of Cu(II) and Fe(II) and their application in edible items. *ACS Applied Nano Materials*, *6*(5), 3277–3284.

Saleem, M., & Lee, K. H. (2015). Optical sensor: a promising strategy for environmental and biomedical monitoring of ionic species. *RSC Advances*, *5*(88), 72150–72287.

Shamsipur, M., Chabok, A., Molaabasi, F., Seyfoori, A., Hajipour-Verdom, B., Shojaedin-Givi, B., Sedghi, M., Naderi-Manesh, H., & Yeganeh-Faal, A. (2019). Label free phosphate functionalized semiconducting polymer dots for detection of iron(III) and cytochrome c with application to apoptosis imaging. *Biosensors and Bioelectronics*, *141*, 111337.

Shastri, V. P. (2003). Non-degradable biocompatible polymers in medicine: Past, present and future. *Current Pharmaceutical Biotechnology*, *4*(5), 331–337.

Song, Y., Hao, J., Hu, D., Zeng, M., Li, P., Li, H., Chen, L., Tan, H., & Wang, L. (2017). Ratiometric fluorescent detection of superoxide anion with polystyrene@nanoscale coordination polymers. *Sensors and Actuators B: Chemical*, *238*, 938–944.

Sun, J., Mei, H., & Gao, F. (2017). Ratiometric detection of copper ions and alkaline phosphatase activity based on semiconducting polymer dots assembled with rhodamine B hydrazide. *Biosensors and Bioelectronics*, *91*, 70–75.

Tang, Y., Zhou, X., Xu, K., & Dong, X. (2020). One-pot synthesis of fluorescent non-conjugated polymer dots for Fe3 + detection and temperature sensing. *Spectrochimica Acta Part A: Molecular and Biomolecular Spectroscopy*, *240*, 118626.

Thiagarajan, V., Ramamurthy, P., Thirumalai, D., & Ramakrishnan, V. T. (2005). A novel colorimetric and fluorescent chemosensor for anions involving PET and ICT pathways. *Organic Letters*, *7*(4), 657–660.

Turos, E., Shim, J.-Y., Wang, Y., Greenhalgh, K., Reddy, G. S. K., Dickey, S., & Lim, D. V. (2007). Antibiotic-conjugated polyacrylate nanoparticles: new opportunities for development of anti-MRSA agents. *Bioorganic & Medicinal Chemistry Letters*, *17*(1), 53–56.

Wan, H., Xu, Q., Gu, P., Li, H., Chen, D., Li, N., He, J., & Lu, J. (2021). AIE-based fluorescent sensors for low concentration toxic ion detection in water. *Journal of Hazardous Materials*, *403*, 123656.

Wang, D., He, J., & Sun, J. (2021). Two-photon ratiometric fluorescent probe based on NBD-amine functionalized semiconducting polymer nanoparticles for real-time imaging of hydrogen sulfide in living cells and zebrafish. *Talanta*, *228*, 122269.

Wang, H., Zhang, P., Chen, J., Li, Y., Yu, M., Long, Y., & Yi, P. (2017). Polymer nanoparticle-based ratiometric fluorescent probe for imaging Hg^{2+} ions in living cells. *Sensors and Actuators B: Chemical*, *242*, 818–824.

Wu, D., Chen, L., Lee, W., Ko, G., Yin, J., & Yoon, J. (2018). Recent progress in the development of organic dye based near-infrared fluorescence probes for metal ions. *Coordination Chemistry Reviews*, *354*, 74–97.

Wu, D., Sedgwick, A. C., Gunnlaugsson, T., Akkaya, E. U., Yoon, J., & James, T. D. (2017). Fluorescent chemosensors: the past, present and future. *Chemical Society reviews*, *46*(23), 7105–7123.

Yang, J., Chen, W., Chen, X., Zhang, X., Zhou, H., Du, H., Wang, M., Ma, Y., & Jin, X. (2021). Detection of Cu^{2+} and S^{2-} with fluorescent polymer nanoparticles and bioimaging in HeLa cells. *Analytical and Bioanalytical Chemistry*, *413*, 3945–3953.

Yin, H., Zhang, K., Wang, L., Zhou, K., Zeng, J., Gao, D., Xia, Z., & Fu, Q. (2018). Redox modulation of polydopamine surface chemistry: a facile strategy to enhance the intrinsic fluorescence of polydopamine nanoparticles for sensitive and selective detection of Fe^{3+}. *Nanoscale*, *10*, 18064–18073.

Yuan, Y., Hou, W., Qin, W., & Wu, C. (2021). Recent advances in semiconducting polymer dots as optical probes for biosensing. *Biomaterials Science*, *9*(2), 328–346.

Zhang, H., Dong, X., Wang, J., Guan, R., Cao, D., & Chen, Q. (2019). Fluorescence emission of polyethylenimine-derived polymer dots and its application to detect copper and hypochlorite ions. *ACS Applied Materials & Interfaces*, *11*, 32489–32499.

Zhang, K., Liu, J., Guo, Y., Li, Y., Ma, X., & Lei, Z. (2018). Synthesis of temperature, pH, light and dual-redox quintuple-stimuli-responsive shell-crosslinked polymeric nanoparticles for controlled release. *Materials Science and Engineering: C*, *87*, 1–9.

Zhang, P., Chen, J., Huang, F., Zeng, Z., Jia Hu, P., Yi, F., Zeng., & Wu, S. (2013). One-pot fabrication of polymernanoparticle-based chemosensors for Cu^{2+} detection in aqueous media. *Polym. Chem*, *4*, 2325–2332.

Zhang, W., Shi, L., Miao, Z. J., Wu, K., & An, Y. (2005). Core–shell–corona micellar complexes between poly (ethylene glycol)-block-poly (4-vinyl pyridine) and polystyrene-block-poly (acrylic acid). *Macromolecular Chemistry and Physics*, *206*(23), 2354–2361.

Zhang, X., Zhang, X., Yang, B., Hui, J., Liu, M., Chi, Z., Liu, S., Xu, J., & Wei, Y. (2014). A novel method for preparing AIE dye based cross-linked fluorescent polymeric nanoparticles for cell imaging. *Polymer Chemistry*, *5*(3), 683–688.

Zhong, J., Wang, B., Sun, K., & Duan, J. (2021). Hyperbranched polyethylenimine–based polymeric nanoparticles: synthesis, properties, and an application in selective response to copper ion. *Colloid and Polymer Science*, *299*, 1577–1586.

Zhong, Z., & Jia, L. (2019). Room temperature preparation of water-soluble polydopamine-polyethyleneimine copolymer dots for selective detection of copper ions. *Talanta*, *197*, 584–591.

Zou, W., Gong, F., Chen, X., Cao, Z., Xia, J., Gu, T., & Li, Z. (2018). Intrinsically fluorescent and highly functionalized polymer nanoparticles as probes for the detection of zinc and pyrophosphate ions in rabbit serum samples. *Talanta*, *188*, 203–209.

CHAPTER

Protein sensors

20

Marta Guembe-García[1,2] and Ana Arnaiz[1,3]

[1]*Facultad de Ciencias, Departamento de Química Orgánica, Universidad de Burgos, Burgos, Spain* [2]*Department of Chemistry, University of Pavia, Pavia, Italy* [3]*Universidad Politécnica de Madrid, Madrid, Spain*

20.1 Introduction

In recent years, the detection and quantification of proteins have become areas of research of great relevance in fields such as medicine, biology, and biotechnology. Proteins play a crucial role in numerous biological processes, and their precise detection can provide valuable information about human health, immune response, and disease diagnosis.

Polymeric sensors for proteins have emerged as a promising tool for rapidly, sensitively, and selectively detecting and quantifying these biomolecules. These sensors are based on the specific interaction between proteins and polymers, enabling the detection of target proteins with high affinity and precision. Moreover, these types of sensor materials offer a wide range of applications, such as the detection of proteins from pathogens and the monitoring of enzymatic activity.

In addition to their selectivity and sensitivity, polymeric sensors for proteins also offer other significant advantages. On one hand, they are more cost-effective and easier to manufacture compared to other protein detection methods, such as those based on antibodies. On the other hand, polymeric sensors are highly stable and resistant, allowing them to be used in a wide range of conditions and environments, including real-time and field applications.

20.2 Protein-based polymer sensor

Polymeric sensors for proteins arise as an alternative to conventional methods of molecular recognition. Unlike techniques such as enzyme-linked immunoassay (Clark & Adams, 1977), zymography (Galis et al., 1995; Kleiner & Stetlerstevenson, 1994), high-performance liquid chromatography (Eckert & Roller, 1990), or nuclear magnetic resonance (Feeney et al., 1989), which require expensive reagents, equipment, and a complex result interpretation process, polymeric sensors offer a more accessible and user-friendly solution.

These sensors are composed of two main components: the matrix and the sensing units. The sensing unit is responsible for interacting with the analyte, in this case, the protein, and generating an analytical signal. On the other hand, the matrix acts as a support for the sensing part, and an appropriate matrix will facilitate the interaction between the sensor and the target, improving its response.

Sensory Polymers. DOI: https://doi.org/10.1016/B978-0-443-13394-7.00020-3

The main objective of polymeric sensors for proteins is to achieve molecular biorecognition, that is, to mimic the molecular recognition that occurs at the cellular level, such as antigen-antibody or enzyme-substrate interactions (Mazzotta et al., 2022). Using artificial systems presents several advantages compared to biological systems, such as high selectivity, greater stability, lower cost, versatility to adapt to different analytes and applications, and simpler synthesis (Bossi et al., 2007). Synthetic polymers themselves possess many of these characteristics, making them excellent support materials for these sensors.

Within the category of polymeric sensors for proteins, there are different types classified according to the characteristics of their matrix:

- Hydrogels.
- Conjugated polymers (CPs) and conjugated polyelectrolytes (CPEs).
- Molecularly imprinted polymers (MIPs).
- Other polymers.

In conclusion, polymeric sensors for proteins offer an efficient and cost-effective alternative to conventional protein detection methods. By harnessing the properties of synthetic polymers, these sensors enable selective and stable molecular biorecognition. With different types of matrices available, specific sensors can be designed for different analytes and applications, opening a wide range of possibilities in protein detection and quantification.

20.2.1 Hydrogels

Hydrogels were first described in the 1960s by Wichterle and Lim ("Industrial Research Associations: Some Taxation Problems," 1960) as polymers formed by a three-dimensional matrix with the main characteristic of being able to absorb and retain large amounts of water in their structure (Choi et al., 2013; Hoffman, 2002; Pinelli et al., 2020; Tang et al., 2022). The three-dimensional matrix that characterizes these materials derives from their cross-linking, which can be physical or chemical (see Fig. 20.1):

- *Chemical crosslinking* forms what is known as chemical or permanent gels, whose three-dimensional network is formed by covalent bonds.
- *Physical crosslinking* gives rise to physical or reversible gels, whose three-dimensional networks are maintained through secondary forces such as hydrogen bonds and electrostatic or hydrophobic interactions. When this physical crosslinking occurs between a polyelectrolyte and a multivalent ion of opposite charge, a physical gel known as an "ionotropic" hydrogel is formed. And if it occurs between two polyelectrolytes of opposite charges, complex coacervates or polyelectrolyte complex hydrogels form. Calcium alginate tends to form this type of hydrogel.

Regardless of the type of crosslinking, none of these materials are homogeneous; they have regions of greater swelling and regions of greater crosslinking known as clusters.

On the other hand, hydrogels can be classified based on their composition as natural, synthetic, or semi-synthetic (a mixture of both), and they are described as follows:

- *Natural hydrogels* are prepared by the chemical or physical crosslinking of natural polymers such as starch, alginate, cellulose, gelatin, proteins, or polysaccharides (see Table 20.1).

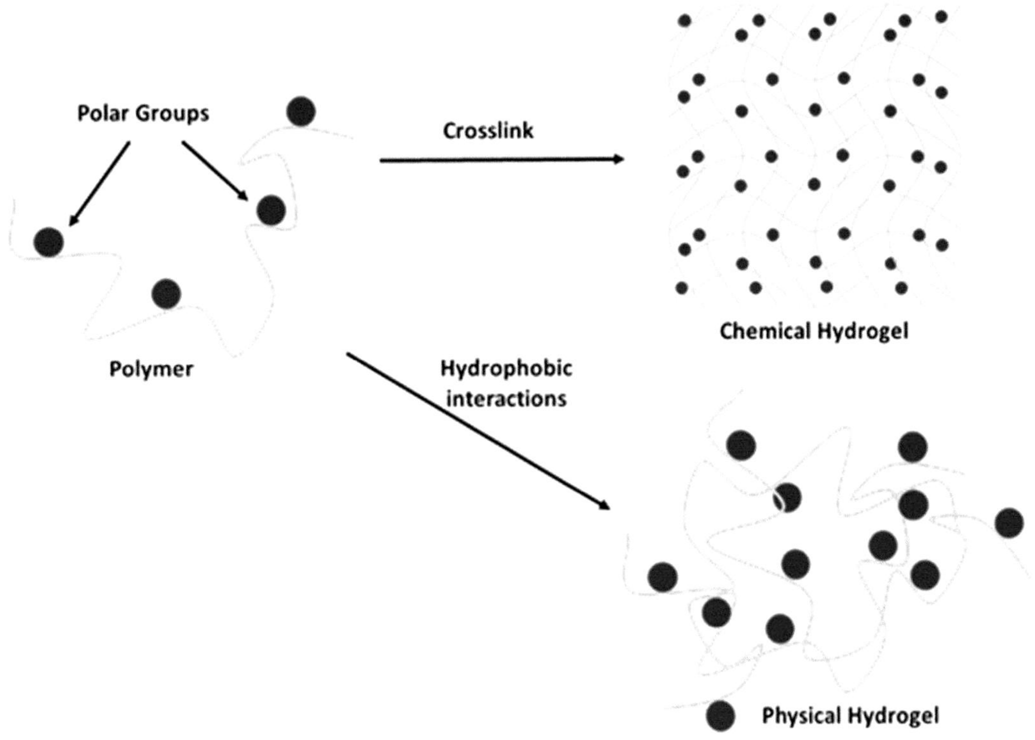

FIGURE 20.1 Chemical and physical hydrogel structure.

Structural difference between chemical and physical hydrogel.

Natural hydrogels have various biomedical applications, such as forming supports for tissues and organs, controlled drug release, tissue engineering, and regenerative medicine, due to their biocompatibility and biodegradability. Natural hydrogels are also used in food production, personal care products, and agriculture due to their low toxicity and liquid-absorbing capacity.

- *Synthetic hydrogels* are prepared by the chemical polymerization of hydrophilic monomers such as acrylates or vinyls (see Table 20.1) using crosslinking agents. Synthetic hydrogels have various applications, including tissue engineering, controlled drug release, and the manufacture of personal care products such as diapers and sanitary pads. Synthetic hydrogels also have unique properties, such as the ability to change their size and shape in response to external stimuli such as pH or temperature, making them useful for applications such as sensors and switching devices. However, synthetic hydrogels may also have higher toxicity and lower biocompatibility than natural hydrogels.
- *Semisynthetic hydrogels* refer to those hydrogels that are prepared by combining natural and synthetic polymers (see Table 20.1). These hydrogels combine the useful properties of natural polymers, such as biocompatibility and biodegradability, with the unique properties of synthetic polymers, such as responsiveness to external stimuli and the ability to change shape.

Table 20.1 Hydrogels classified according to their nature.

Natural		Semisynthetic	Synthetic	
Anionic	Hyaluronic acid	P(PEG-co-peptides)	*Polyesters*	PEG-PLA-PEG
	Alginic acid			PEG-PLGA-PEG
	Aectin	Alginate-g-(PEO-PPO-PEO)		PEG-PCL-PEG
	Carrageenan			PLA-PEG-PLA
	Chondroitin sulfate			PHB
		P(PLGA-co-serine)		P(PF-co-EG) ± acrylate end groups
		Dextran sulfate		P(PEG/PBO terephthalate
Cationic	Chitosan	Collagen-acrylate	*Other polymers*	PEG-bis-(PLA-acrylate)
				PEG ± CDs
	Polylysine	Alginate-acrylate		PEG-g-P(AAm-co-Vamine)
				PAAm
Amphipathic	Collagen and gelatin	P(HPMA-g-peptide)		P(NIPAAm-co-AAc)
				P(NIPAAm-co-EMA)
	Carboxymethyl chitin			PVAc/PVA
		P(HEMA/Matrigel)		PNVP
		Fibrin		P(biscarboxy-phenoxy-phosphazene)
Neutral	Dextran	HA-g-NIPAAm		P(MMA-co-HEMA)
	Agarose			P(AN-co-allyl sulfonate)
	Pullulan			P(GEMA-sulfate)

CD, cyclodextrin; EG, ethylene glycol; HEMA, hydroxyethyl methacrylatel; P(...), poly(...); PAAc, poly(acrylic acid); PAAm, polyacrylamide; PBO, poly(butylene oxide); PCL, polycaprolactone; PEG, poly(ethylene glycol); PEO, poly(ethylene oxide); PF, propylene fumarate; PGEMA, poly(glucosylethyl methacrylate); PHB, poly(hydroxy butyrate); PHEMA, poly(hydroxyethyl methacrylate); PHPMA, poly(hydroxypropyl methacrylamide); PLA, poly(lactic acid); PLGA, poly(lactic-co-glycolic acid); PMMA, poly(methyl methacrylate); PNIPAAm, poly(N-isopropyl acrylamide); PNVP, poly(N-vinyl pyrrolidone); PPO, poly (propylene oxide); PVA, poly(vinyl alcohol); PVAc, poly(vinyl acetate); PVamine, poly(vinyl amine).
Adapted from Hoffman, A. S. (2002). Hydrogels for biomedical applications. Advanced Drug Delivery Reviews, 54(1), 3–12. http://www.elsevier.com/locate/drugdeliv, https://doi.org/10.1016/S0169-409X(01)00239-3.

Semisynthetic hydrogels are also used in a wide variety of biomedical applications, such as forming supports for tissues and organs, controlled drug release, and tissue engineering, but they are also used in other applications such as food and personal care product production.

In general terms, hydrogels are known for their high elasticity and low mechanical resistance, which allows them to easily deform and recover their original shape. Moreover, hydrogels can be designed and modulated for use in different applications, such as controlled drug release, tissue engineering, the manufacturing of contact lenses, and as sensors for various types of molecules, including proteins. Hydrogels are highly promising materials in the fields of biotechnology and medicine due to their biocompatibility and ability to mimic the characteristics of living tissue.

20.2.1.1 Synthesis

The routes used for hydrogel synthesis can vary widely and depend on the type of crosslinking, whether physical or chemical, and the polymer composition.

In the case of *physical hydrogels*, there are some methods for synthesizing them. Warming or cooling a polymer solution to form a gel is a common method, where a warm polymer solution is allowed to cool down to room temperature to form a gel. For example, block copolymers like PEO-PPO-PEO can form hydrogels in water upon warming. Crosslinking a polymer in an aqueous solution using freeze-thaw cycles is another method, where the polymer solution is frozen, thawed, and then frozen again to create polymer microcrystals. Lowering the pH to form an H-bonded gel between different polymers in the same aqueous solution is another method, while mixing solutions of a polyanion and a polycation to form a complex coacervate gel is also a popular technique. Finally, gelling a polyelectrolyte solution with a multivalent ion of opposite charge is another way to form physical hydrogels.

Chemical hydrogels are formed by crosslinking polymers in a solid state or solution using various methods. These methods include radiation, chemical crosslinkers, multi-functional reactive compounds, copolymerization, and polymerization of a monomer within a different solid polymer. For example, radiation can be used to crosslink PEO in water. Chemical crosslinkers like glutaraldehyde or bis-epoxide can be used to treat collagen to form a gel. Copolymerization of a monomer and a crosslinker in solution can be done using a mixture of HEMA and ethylene glycol dimethacrylate. Monomers can also be copolymerized with a multidimensional macromolecule, such as bis-methacrylate-terminated polylactide-polyethyleneoxide-polylactide, which can be activated by a photosensitizer and visible light radiation to form a gel. Additionally, a monomer can be polymerized within a different solid polymer to form an interpenetrating polymer network gel, such as acrylonitrile within starch. Finally, hydrophobic polymers like PVAc or polyacrylonitrile (PAN) can be partially hydrolyzed to form hydrogels such as polyvinyl alcohol or PAN/PAAm/PAAc, respectively.

In some cases, hydrogels can also be synthesized as nanoparticles (NPs) using molecular imprinting techniques (see Section 20.3) (Garcia-Cruz et al., 2020; Sener et al., 2010).

20.2.1.2 Hydrogels for proteins

Historically, hydrogels have been used to build temperature, pressure, or pH sensors. However, in recent years, hydrogel sensors for enzymes, nucleotides, or cells have also been developed. This chapter will focus on the applications of these materials as protein sensors.

The use of hydrogels as sensors for biological molecules was initially proposed as an alternative to existing systems, as these devices had some disadvantages, such as (Jung et al., 2017):

- *Non specific interactions* at the interface between liquid and solid substrates.
- *Autofluorescence* of polymeric sensor materials.
- *Low stability of analytes* in measurement systems.

However, the use of polymer matrices based on hydrogels for the development of protein sensors can overcome these disadvantages and provide other advantages, such as:

- Greater analyte retention capacity.
- The design can be adapted to the needs of the sensor system.
- Reducing autofluorescence.

- They provide biologically compatible, aqueous environments.
- Increases specificity by avoiding non-specific interactions with other proteins.

Hydrogels for protein detection are based on a polymer matrix that encapsulates, traps, or fixes a substrate capable of interacting with the target protein. The response of the sensor will depend on the nature and properties of that substrate, providing a wide variety of responses for this type of sensor.

- *Gravimetric*: The sensor-analyte interaction results in a variation in the mass of the sensor device, either through a decrease in the material due to analyte destruction/degradation (West & Hubbell, 1999) or through an increase generated by swelling (Lim et al., 2020; Sener et al., 2010). In some cases, quartz crystal microgravimetry with dissipation (QCM-D) is used to quantify the results (Sener et al., 2010).
- *Volumetric*: In this case, interaction with the analyte generates a change in material swelling that can also be quantified as a variation in the material's volume (Wu et al., 2019).
- *Changes in refractive index*: The binding of the analyte to the sensor generates a change in the material's refractive index. This change is quantified using surface plasmon resonance (SPR) (Chou et al., 2016).
- *Electrochemical*: The reaction between the sensor device and the target generates a change in the electrochemical properties of the system (Biela et al., 2015; Garcia-Cruz et al., 2020; Phonklam et al., 2020). In some cases, this can be quantified using electrochemical impedance spectroscopy (EIS) (Biela et al., 2015).
- *Fluorescent*: The binding of the analyte to the hydrogel generates a change in the fluorescence of the system. In some cases, the binding turns off the material's previous fluorescence, creating on/off systems (Liang et al., 2015; Randriantsilefisoa et al., 2019; Shohatee et al., 2018). In others, a non fluorescent or low fluorescence system is increased by the presence of the analyte, creating off/on systems (Ebrahimi et al., 2015; Gunda et al., 2016; Sadat Ebrahimi et al., 2015).
- *Colorimetric*: Material interactions with the analyte result in changes in the color of the hydrogel (Choi et al., 2013; Ebrahimi et al., 2015; Guembe-García et al., 2020; Guembe-García et al., 2021; Gunda et al., 2016; Sadat Ebrahimi et al., 2015). Some sensors show a dual colorimetric and fluorimetric response (Ebrahimi et al., 2015; Gunda et al., 2016; Sadat Ebrahimi et al., 2015).

The most common hydrogel protein sensors are those with a colorimetric and fluorescent response. Generally, attempts are made to avoid responses that require complicated interpretation techniques or expensive instrumentation like QCM-D, EIS, or SPR.

20.2.1.3 Progress and future perspectives

Although hydrogels as protein sensors have made significant progress and demonstrated the capability to detect a wide range of proteins with diverse responses (see Table 20.2), there are still challenges to overcome. While the field has seen advancements since the creation of the first sensors of this type, meeting the necessary clinical requirements for medical use remains an ongoing endeavor.

In addition to meeting clinical standards, there are technical hurdles that need to be addressed before hydrogel-based protein sensors can be effectively employed in clinical applications. One of the key challenges lies in the commercial development of these sensors, as the current stage is still relatively nascent. There is a need for the development of a robust system capable of quantifying

Table 20.2 Examples of proteins detected by hydrogels, classified according to the sensor response.

Answer		Target
Gravimetric	Matrix metalloproteinase-1 (MMP-1) (West & Hubbell, 1999) Hepatitis B core antigen (HBcAg) (Lim et al., 2020) Lys (Sener et al., 2010)	
Volumetric		Glutathione (GSH) (Wu et al., 2019)
Changes in refractive index		Human thyroid stimulating hormone (TSH) antibody (Chou et al., 2016)
Electrochemical	Cardiac troponin T (Phonklam et al., 2020) Trypsin (Garcia-Cruz et al., 2020) Matrix metalloproteinase-9 (MMP-9) (Biela et al., 2015)	
Fluorescent	On-Off	FLAG antibody (Randriantsilefisoa et al., 2019) GSH antibody (Randriantsilefisoa et al., 2019) Vascular endothelial growth factor (VEGF) (Shohatee et al., 2018) Collagenase (Liang et al., 2015)
	Off-On	β-d-galactosidase (*Escherichia coli* K-12) (Gunda et al., 2016) α-glucosidase (gram +) (Ebrahimi et al., 2015) Elastase (gram −) (Ebrahimi et al., 2015) β-glucuronidase (β-GUS) (Sadat Ebrahimi et al., 2015)
Colorimetric	Immunoglobulin G (IgG) antibody (Choi et al., 2013) β-d-galactosidase (*Escherichia coli* K-12) (Gunda et al., 2016) α-glucosidase (gram +) (Ebrahimi et al., 2015) Elastase (gram −) (Ebrahimi et al., 2015) β-glucuronidase (β-GUS) (Sadat Ebrahimi et al., 2015) MMP activity (Guembe-García et al., 2021; Guembe-García et al., 2020)	

and translating the optical responses of these sensors simply and efficiently for clinical use. This would enable accurate and reliable protein detection in a clinical setting.

Furthermore, the long term and medium term shelf life of hydrogel-based protein sensors has not been extensively studied (Jung et al., 2017). To ensure their practical viability, it is crucial to conduct further research into the storage and stability of these materials. Understanding how the sensors maintain their performance over extended periods and under various conditions is essential for their successful implementation.

Addressing these challenges will require interdisciplinary collaborations between researchers, clinicians, and industry experts. By combining expertise in materials science, biochemistry, sensor development, and commercialization, it is possible to overcome the current limitations and pave the way for the practical application of hydrogel-based protein sensors in clinical settings.

Continued research and development efforts are necessary to advance the field of hydrogel-based protein sensors. By refining their performance, optimizing their clinical compatibility, and establishing robust commercialization strategies, hydrogel sensors have the potential to revolutionize protein detection and contribute to improved diagnostics, personalized medicine, and patient care.

20.2.2 Conjugates polymers and polyelectrolytes

CPs are polymers whose main chain is formed by a system of conjugated double bonds, as shown in Fig. 20.2. This conjugated bond system allows for greater electron mobility throughout the material structure, resulting in characteristic electrical and optical properties. When the polymer chains are short, they are referred to as oligoconjugated polymers (CO). CO synthesis is more complex than CP synthesis, but due to their higher solubility and defined structure, they are better suited for biological applications.

Polymers that are soluble in water and have side chains with charged residues are called *polyelectrolytes*. They can be classified in two ways; according to their charges or according to their main chain. In the former case, there are three types (Ambade et al., 2007):

- *Cationic polyelectrolytes* have positively charged groups, which are commonly ammonium quaternary, phosphonium quaternary, or guanidyl or protonated amino groups.

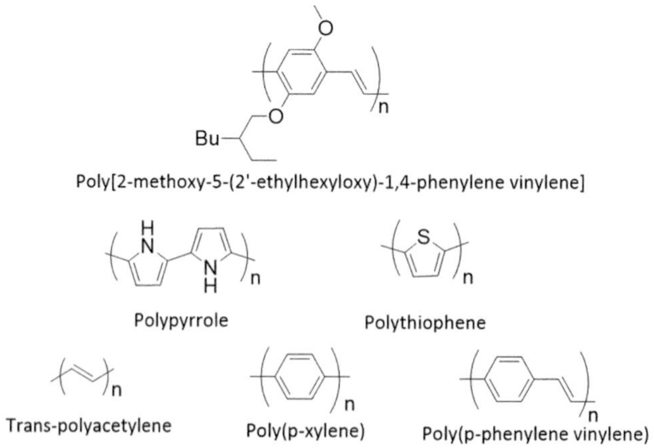

FIGURE 20.2 Example of conjugated polymers (CP).

Example of six different structures of CP.

- *Anionic polyelectrolytes* with negatively charged residues such as carboxylic acid, sulfonic, or phosphonate anionic groups, or phenolic hydroxide groups.
- *Polyanfiolites* have zwitterionic side chains (with both charges).

However, according to the nature of their main chain, two types of polyelectrolytes can be distinguished:

- Non conjugated polyelectrolytes (Fig. 20.3A).
- Conjugated polyelectrolytes (CPEs) (Fig. 20.3B).

From an application standpoint, distinguishing between CPs and polyelectrolytes can pose challenges. However, hybrid materials called CPEs hybrids have emerged as a favorable solution in many cases. These hybrids leverage the structural properties of CPs, which offer optical and electrical characteristics while incorporating charged side chains that provide the solubility typical of polyelectrolytes.

Notably, the solubility of CPs can be enhanced by introducing side chains with polyethylene glycol (PEG) residues (Zhu et al., 2011). This modification improves their solubility in aqueous environments and facilitates interactions with proteins. Additionally, non CPEs have also been explored as protein sensors (Sandanaraj et al., 2006), although their use is currently limited in scope.

CPEs hybrids combine the unique properties of CPs and polyelectrolytes, making them versatile and adaptable to various measurement needs. These materials exhibit characteristics such as high brightness, excellent photosensitivity, low toxicity, and good stability in biological media. Moreover, they offer solubility and conductivity, inherited from polyelectrolytes, in addition to other desirable features (Zhou et al., 2019).

These hybrid materials have found applications in several fundamental biological areas. They are employed as chemical or biological sensors, facilitating the detection and monitoring of proteins. Additionally, they are utilized in fluorescence imaging techniques for visualizing protein distribution and interactions. Furthermore, CPEs hybrids play a crucial role in diagnostic applications, aiding in disease detection and analysis. They have also shown potential in photodynamic, bactericidal, and tumoricidal therapies, among other therapeutic approaches.

In addition to their use as protein sensors, CPEs hybrids have broader applications. They contribute to the development of optoelectronic systems, photosynthesis research, photocatalysis, and bioenergy-related endeavors (Zhou et al., 2019).

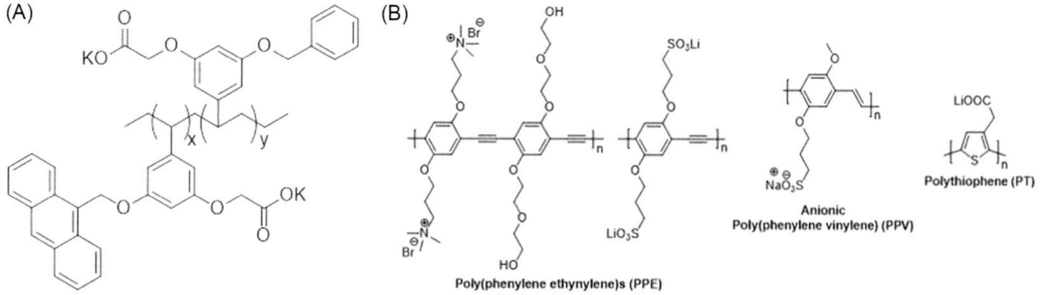

FIGURE 20.3 (A) Example of nonconjugated polyelectrolytes; (B) example of Conjugated polyelectrolytes (CPEs).

Example of different structures of (A) nonconjugated polyelectrolytes and (B) CPEs.

This chapter will specifically delve into the applications of CPEs hybrids as protein sensors, highlighting their unique capabilities in protein detection, quantification, and characterization.

20.2.2.1 Synthesis

20.2.2.1.1 Synthesis of conjugated polymers

The synthesis of CPs has come a long way since the first polymer, polyacetylene, was synthesized in 1974 (Ito et al., 1974). While the structure of conjugated bonds remains fundamental to their properties, the current synthesis methods involve *organometallic coupling reactions* for the construction of these polymers.

Suzuki coupling, Sonogashira coupling, Heck reaction, and Wittig crosscoupling are some of the commonly used methods for synthesizing CPs. These reactions enable the formation of specific bonds and the assembly of monomers to create the desired polymer structure. Additionally, olefin metathesis reactions and oxidative aromatic coupling have proven useful in synthesizing polymers similar to poly(p-phenylene vinylene) and other electron-rich aromatic polymers. In the case of polythiophenes and polypyrroles, alternative electrochemical polymerization methods can also be employed for synthesis.

While the main chain of CPs is responsible for their optical properties, the incorporation of side chains can introduce additional desirable characteristics. For instance, the addition of positively or negatively charged residues to the side chains can enhance solubility in water, resulting in the formation of CPEs. Alternatively, the use of PEG as a side chain can improve solubility and is particularly beneficial in situations where electrostatic interactions need to be avoided (Zhu et al., 2011).

Furthermore, the presence of thiol groups in side chains can promote interactions with reactive oxygen species, showcasing the versatility of CPs for various applications. The combination of main chain structure and tailored side chain modifications allows for the fine-tuning of properties and expand the range of applications for these materials.

20.2.2.1.2 Synthesis polyelectrolyte

Non CP synthesis encompasses various methods, including addition polymerization, condensation polymerization, and free radical polymerization. Unlike CPs, non CPs typically lack the distinctive optical properties associated with light absorption and emission in the visible and near-infrared ranges. However, they possess unique mechanical and thermal properties that make them valuable in applications such as plastics, adhesives, and coatings.

The synthesis of non CPs is a broad and diverse field, with methodologies tailored to specific research objectives or application requirements. Addition polymerization involves the formation of polymer chains through the repeated addition of monomers, resulting in a high-molecular-weight polymer. This process is commonly used for the synthesis of polyethylene, polypropylene, and other commercially important non CPs.

Condensation polymerization, on the other hand, involves the stepwise elimination of small molecules, such as water or alcohol, during polymer chain formation. This method is commonly employed for the synthesis of polyesters, polyamides, and other condensation polymers. By carefully selecting monomers and reaction conditions, researchers can control the molecular weight, structure, and properties of the resulting non CPs.

Free radical polymerization is a versatile technique widely used for both conjugated and non CPs. It involves the initiation, propagation, and termination of free radicals to form polymer chains. Through careful selection of monomers, initiators, and reaction conditions, researchers can achieve desired molecular weights and tailor the properties of non CPs for specific applications.

It is important to note that the synthesis of non CPs is a vast and continually evolving field, with ongoing research focused on improving polymerization techniques, exploring novel monomers, and developing advanced characterization methods. These advancements contribute to the development of innovative materials with enhanced mechanical strength, thermal stability, and compatibility with various industrial and technological applications.

20.2.2.2 Conjugated polymers and polyelectrolytes as protein sensors

Thanks to their electrical and optical properties, CPs can be used as highly sensitive optical probes in many applications. In the case of protein biodetection, the responses of these sensors can be colorimetric (Herland & Inganäs, 2007), fluorimetric (Ambade et al., 2007), or potentiometric (Das et al., 2015). The functioning of these polymers as sensors can be due to three mechanisms:

- *Mecanochromism*: That is, the optical properties of the material vary according to its structure. This change is the result of the material's interaction with the analyte, in this case, the protein. Fig. 20.4A schematically shows the functioning of this type of sensor for a colorimetric response (Charych et al., 1993; Faïd & Leclerc, 1996; Faïd & Leclerc, 1998; Ho et al., 2005), while Fig. 20.4B shows an example for an off/on or on/off fluorimetric response (Bérn Abérem et al., 2006; Ho et al., 2005; Nilsson & Inganäs, 2004; Nilsson et al., 2003; Nilsson, Olsson, et al., 2005). The same behavior can be observed in sensors with potentiometric responses (Fig. 20.4C) (Faïd & Leclerc, 1998; Kumpumbu-Kalemba & Leclerc, 2000).
- *Förster Resonance Energy Transfer (FRET)*: Another possibility is the use of FRET pairs; FRET pairs are composed of a fluorescent molecule and an inhibitor that absorbs in the region where the fluorophore emits so that when they are united or at a certain distance, fluorescence disappears. When this interaction is broken, the fluorophore recovers its fluorescence. In the case of CPEs, the material itself acts as a fluorophore, and the interaction with the target protein releases the inhibitor. In the case of fluorescent responses, there are always two possibilities of responses, on/off and off/on. In the case of *on/off* systems, part of the analyte will act as a translator/fluorophore, so its binding with the polymer causes the extinction of the material's fluorescence and the increase of the transducer's fluorescence. This is because the material's emission band overlaps the transducer's excitation band, resulting in the transducer's emission band (Kumaraswamy et al., 2004; Pinto & Schanze, 2004). In the case of *off/on* systems, the binding of the recognition molecule and the transducer to the polymer turns off the material's fluorescence. This extinction, as in the previous case, is due to an overlap of the material's emission band with the transducer/quencher absorption band. The reaction with the analyte releases the transducer that acts as an inhibitor, and the system recovers fluorescence (Feng et al., 2007; Liu et al., 2022; Yang et al., 2014). Both behaviors are schematically shown in Fig. 20.5.
- *Electronic transfer*. Like the previous mechanism, this one is also associated with a fluorescent response. The same two responses can be observed: on/off or off/on. In the case of *on/off* systems, and like in the previous case, part of the analyte acts as a transducer and turns off the

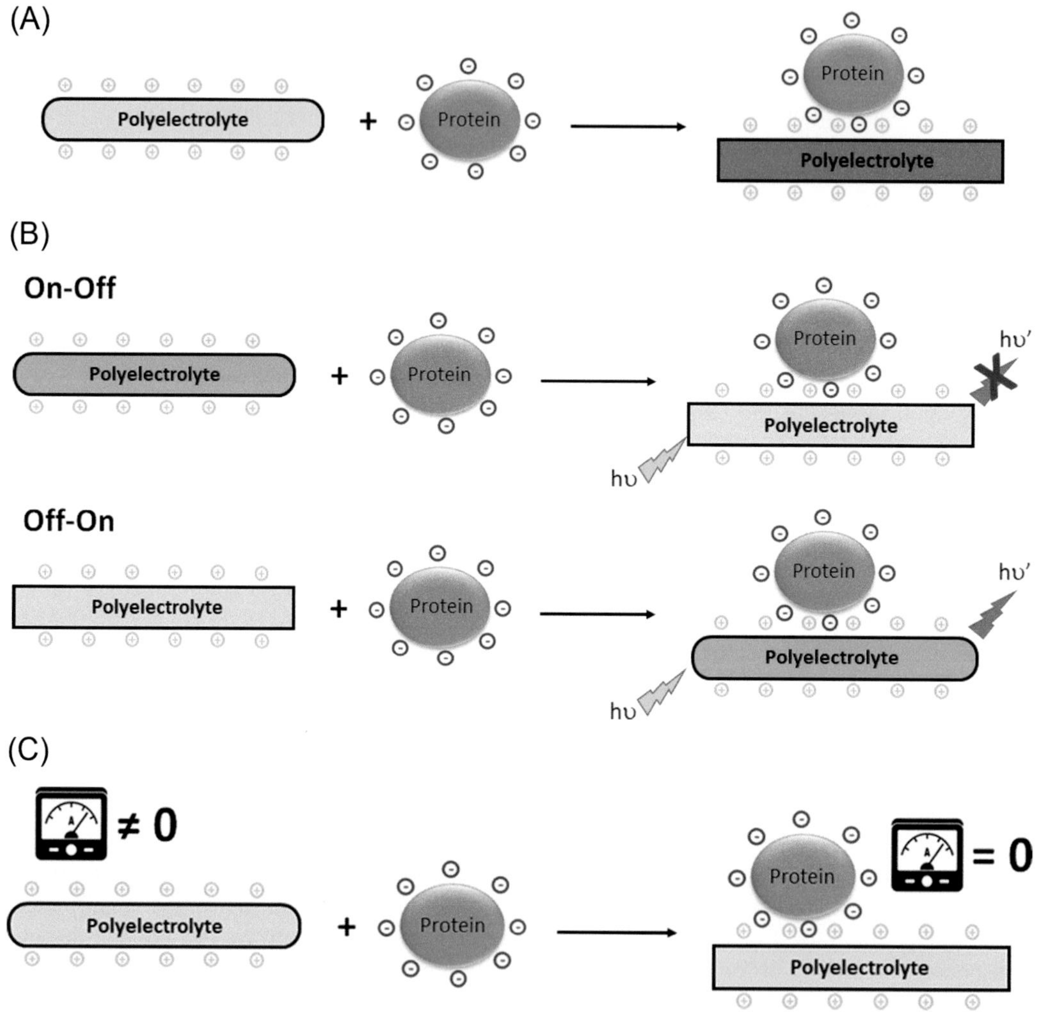

FIGURE 20.4 Schematic mecanochromism behavior of CPEs as (A) colorimetric, (B) fluorimetric, and (C) potentiometric sensors.

Mechanochromic behavior scheme of conjugated polyelectrolyte sensors, with (A) colorimetric, (B) fluorimetric, (C) potentiometric responses.

fluorescence of the CPEs. The difference between this visual response and the previous one (FRET) is the mechanism. In this case, the extinction of fluorescence is due to an electronic transfer mechanism and not to an overlap of emission-absorption bands (Fan et al., 2002). In the case of *on/off* systems, like in the case of FRET, the union between the recognition unit system and the polymer turns off its characteristic fluorescence, due in this case to an electronic transfer mechanism. Finally, the interaction between the analyte and the sensor system releases

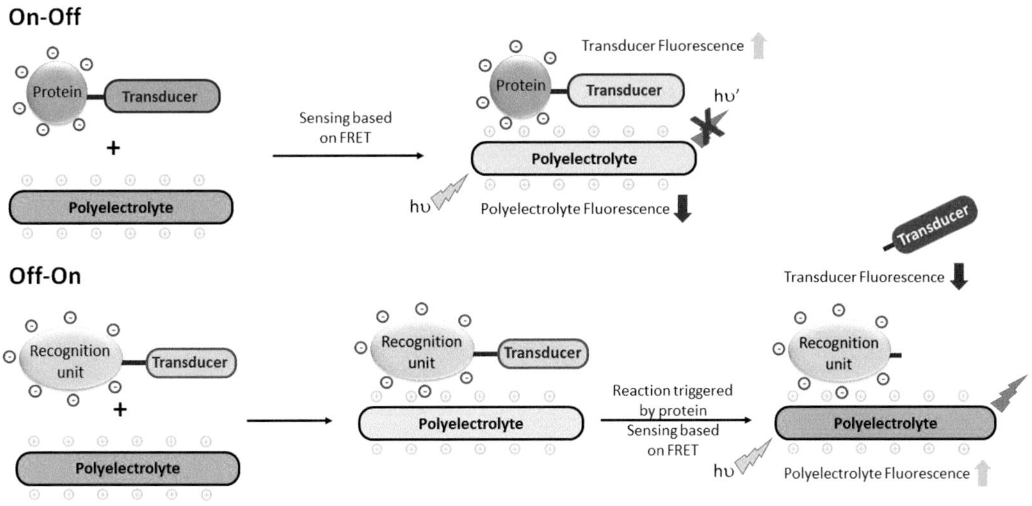

FIGURE 20.5 Schematic FRET behavior of CPEs sensors.

Scheme of the fluorescent behavior of conjugated polyelectrolyte sensors based on FRET mechanisms, both in On/Off and On/Off cases.

the transducer and restores the original fluorescence to the system (Chen et al., 1999). Both behaviors are schematically shown in Fig. 20.6.

Due to their optical properties, the fluorescent responses of these sensors, regardless of their internal mechanisms, are the most common and studied. However, it is true that these mechanisms can be confused by only observing the analytical response. From a physical point of view, the operation of the amplified fluorescent response of these systems can be explained by the Stern-Volmer equation (Lakowicz, 2006):

$$F_0/F = 1 + K_{SV}[Q]$$

where F_0 is the initial fluorescence, F is the final fluorescence, $[Q]$ is the concentration of the quencher, and K_{SV} is the Stern-Volmer constant that is calculated from the linear portion of the F_0/F versus $[Q]$ graph. The values of K_{SV} obtained can explain the type of fluorescent interaction, whether FRET or electronic transfer, that occurs in the studied system.

In the fluorescence quenching analysis, the initial fluorescence intensity (F_0) and the final fluorescence intensity (F) are measured in the presence of a quencher at various concentrations ($[Q]$). The Stern-Volmer constant (K_{SV}) is then calculated from the linear region of the plot of F_0/F versus $[Q]$. This constant provides valuable information about the nature of the fluorescent interaction occurring in the system, whether it involves FRET or electronic transfer.

By examining the variation of the F_0/F ratio as a function of the quencher concentration $[Q]$, the type of process taking place can be determined. If the F_0/F ratio shows a linear relationship with $[Q]$, it suggests that an energy transfer process, such as FRET, is occurring. On the other hand, if the F_0/F relationship does not follow a linear dependence with $[Q]$, it is more likely to involve an electron transfer process.

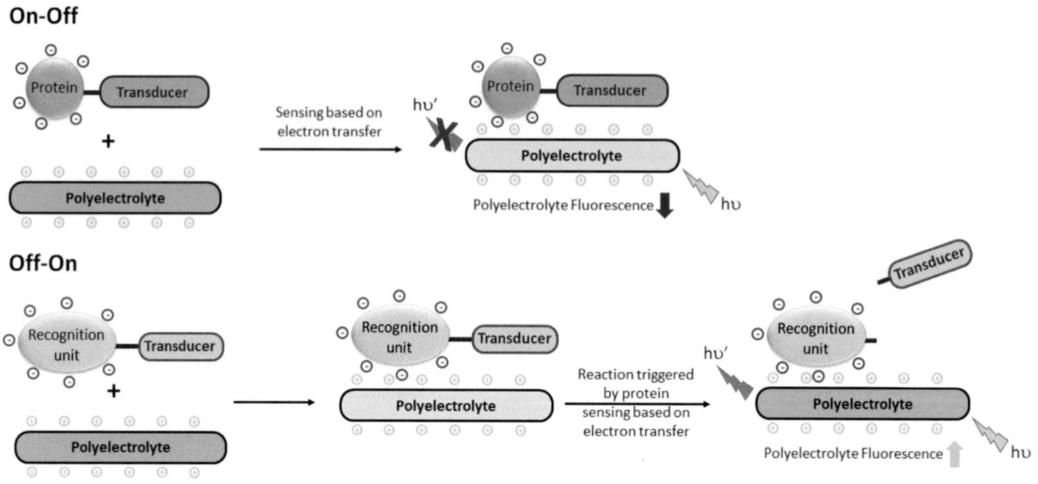

FIGURE 20.6 Schematic electronic transfer behavior of CPEs sensors.

Scheme of the fluorescent behavior of conjugated polyelectrolyte sensors based on electronic transfer mechanisms, both in On/Off and On/Off cases.

It is important to emphasize that this initial analysis indicates the type of process involved, but further detailed analysis and additional experiments may be required to confirm and fully understand the specific energy transfer mechanism at play. Additionally, in certain cases, modifications to the Stern-Volmer equation may be necessary to accommodate specific experimental conditions or more complex systems (Mátyus et al., 2006).

Due to their hydrophobic aromatic groups, these types of sensors *tend to aggregate*, which is a drawback in their application. Another issue is that protein-based sensors using CPEs may not exhibit specificity due to *non specific* hydrophobic and/or electrostatic *interactions* of these materials with other proteins or macromolecules such as DNA (Kim et al., 2005). To address these two problems, alternative strategies have been studied, which are as follows:

- To prevent *aggregation*, the use of CPEs-coated polystyrene microspheres has been proposed and has been successfully tested for two types of enzymes, proteases and phosphatase kinases (Kumaraswamy et al., 2004; Rininsland et al., 2004).
- To avoid *non-specific interactions*, non CPEs in nanoparticle format have been studied. These systems are capable of selectively responding to only metalloproteins (Sandanaraj et al., 2006). The response of these materials is also fluorescent and can occur through either FRET or electron transfer mechanisms (Basu et al., 2004).

In recent years, CPE NPs, also known as polymer dots, have been studied for the detection of proteins in vitro. Examples are shown in Table 20.3 (Bai et al., 2022).

The general goal of protein sensors is to quantify the concentration or activity of proteins. However, in the case of CPE sensors, there are also sensors that are capable of determining the structure of certain proteins (Gunda et al., 2016; Liang et al., 2015; Shohatee et al., 2018). This is

Table 20.3 Examples of proteins detected by CPE-nanoparticles/polymers dots, classified according to the sensor response.

Answer			Protein target
Fluorescent	FRET	On/Off	Kinase (Moon et al., 2007)
		Off/On	IgG (Andronico et al., 2018)
			Carcinoembryonic antigen (CEA) (Lin et al., 2012)
			Matrix Metallopeptidase 2 (MMP-2) (Yang et al., 2014)
Electrochemiluminescent	On/Off	CEA (Wang et al., 2018)	
		Cytokeratin-19-fragment (CY211) (Wang et al., 2018)	
		Neuron-specificenolase (NSE) (Wang et al., 2018)	
Potentiometric			Cytokine (IL-6) (Liu et al., 2022)

because CPEs can interact with proteins in a specific manner, leading to changes in their fluorescence properties that can provide information on protein structure. By analyzing the fluorescent response of CPE-based sensors to different protein structures, it is possible to identify specific protein conformations or structural changes that are indicative of particular disease states or cellular processes. This makes CPE-based sensors a valuable tool for both basic research and clinical applications.

20.2.2.3 Progress and future perspectives

Despite their promising optical properties and versatility, the utilization of CPE materials as protein sensors has not been extensively explored compared to other sensor supports, as evident in the literature (see Table 20.4). Although there was significant development in the early 2000s, the prominence of CPE-based polymer dots as protein sensors has somewhat diminished in the subsequent decade. At present, the focus of CPE-related publications primarily revolves around therapeutic applications and drug delivery systems rather than protein sensing.

It is essential to address this relative lack of exploration in the protein sensing domain and rekindle interest in the potential of CPE materials for this purpose. The unique combination of optical and polyelectrolyte properties in CPE hybrids makes them promising candidates for sensitive and selective protein detection. By redirecting research efforts toward the development of CPE-based protein sensors, we may uncover new opportunities and applications in fields such as biomedicine, diagnostics, and biotechnology.

Efforts to enhance the visibility of CPE materials as protein sensors can be instrumental in advancing state-of-the-art sensor technology. Researchers and scientists can investigate novel strategies to optimize the performance of CPE-based protein sensors, explore their potential in real-time and in situ applications, and identify ways to improve their sensitivity and stability. Collaborations between experts in material science, biochemistry, and sensor development can foster new insights and innovative approaches.

Table 20.4 Examples of proteins detected by CPE and non-conjugated polyelectrolyte, classified according to the sensor response.

Polymer	Answer			Protein target
Conjugated polyelectrolyte	Mechanocromismo	Colorimetric	Hemagglutin (Virus protein) (Charych et al., 1993)	
			Avidin (Faïd & Leclerc, 1998; Faïd & Leclerc, 1996)	
			Human thrombin (Ho et al., 2005)	
		Fluorescent	On/Off	Human thrombin (Ho et al., 2005)
				Calmodulin (Nilsson & Inganäs, 2004)
				Amyloid Fibril Formation (Nilsson, Herland, et al., 2005)
			Off/On	Human thrombin (Bérn Abérem et al., 2006; Ho et al., 2005)
				JR2K and JR2E (Nilsson et al., 2003)
		Potentiometric		Avidin (Faïd & Leclerc, 1998; Kumpumbu-Kalemba & Leclerc, 2000)
	FRET	Fluorescent	On/Off	Kinase (Rininsland et al., 2004)
				Papain (Pinto & Schanze, 2004)
			Off/On	Acetylcholinesterase (AChE) (Feng et al., 2007)
				Phospholipase C (PLC) (Liu et al., 2008)
				MMP-2 (Yang et al., 2014)
	Electro transfer	Fluorescent	On/Off	Cytochrome c (cyt-c) (Fan et al., 2002)
			Off/On	Avidin (Chen et al., 1999)
Non-conjugated polyelectrolyte	Electro transfer	Fluorescent	On/Off	Metalloproteins (Sandanaraj et al., 2006)

Moreover, initiatives to showcase successful applications of CPE materials as protein sensors in various contexts can inspire further interest and investment in this area. Demonstrating their practical utility and efficacy in protein detection can encourage the wider adoption of CPE materials in research, clinical settings, and industrial applications.

Overall, revitalizing the exploration of CPE materials as protein sensors can lead to valuable advancements in sensor technology and contribute to a deeper understanding of biological processes and interactions at the molecular level. With continued research and attention, CPE-based protein sensors have the potential to emerge as a prominent and valuable tool in the realm of biotechnology and medical research.

20.2.3 Molecular Imprinted polymer

MIPs are synthetic polymers with specific physical selectivity toward a particular molecule, making them artificial antibodies. The basis of their selectivity and primary characteristic is their synthesis process, which involves the use of templates that are as similar as possible to the analyte, resulting in cavities within the polymer structure that allow recognition of these molecules. Because of these cavities, MIPs can be used as sensors, where the presence/absence of target molecules in the polymer structure generates quantifiable variations in material properties. Based on the affected property, optical, electrochemical, or piezoelectric sensors can be distinguished. In the case of protein sensors, electrochemical sensors are the most studied (Mazzotta et al., 2022).

20.2.3.1 Synthesis

Regardless of the type of sensor, the synthesis process that defines these materials is the same. As shown in the diagram in Fig. 20.7, the synthesis process consists of several stages, which are as follows:

- *Choice of monomers, solvent, and template*: Regarding the choice of *monomers*, up to six monomers with specific chemical structures capable of interacting with the template, either through covalent (Hedin-Dahlström et al., 2004; Takeda & Kobayashi, 2005) or non covalent interactions (Sellergren, 1997; Wulff & Knorr, 2001) or both (Hwang & Lee, 2002), can be selected. The most commonly used monomers contain acrylic acid or methacrylic acid grafted onto PEG derivatives (Bergmann & Peppas, 2008). These monomers are very convenient for biological applications since their anionic acid groups can form non covalent complexes with the templates, favoring their final elimination. Crosslinking agents can also be added, but they are not necessary in all cases. Crosslinkers greatly affect the stiffness of the polymer structure and its swelling. Swelling can affect the dimensions/structure of cavities formed in polymers in aqueous media. Additionally, the reactivity/compatibility of these crosslinking monomers with the rest of the structural monomers should be considered. Although MIPs with low percentages of crosslinking exist, it is rare to find them with less than 10% (Lu et al., 2002), and in fact, they can reach up to 80% of the final material composition (Cormack & Elorza, 2004). Another aspect to consider in the synthesis of these materials is the choice of *solvent*, which should not interfere with the formation of the template-monomer complex. Polar solvents, like water, are commonly avoided in favor of nonpolar solvents. Historically, chloroform has been one of the most commonly used solvents (Komiyama et al., 2003), although others such as dimethyl sulfoxide (Bergmann & Peppas, 2008), toluene (Byrne et al., 2002), benzene (Komiyama et al., 2003), or even tetrahydrofuran (Oral & Peppas, 2004) have also been studied. On the other hand, in the case of the *template*, two factors should be considered: the structural similarity between the analyte-template (in some cases, the analyte itself can be used) and its stability during the synthetic process.

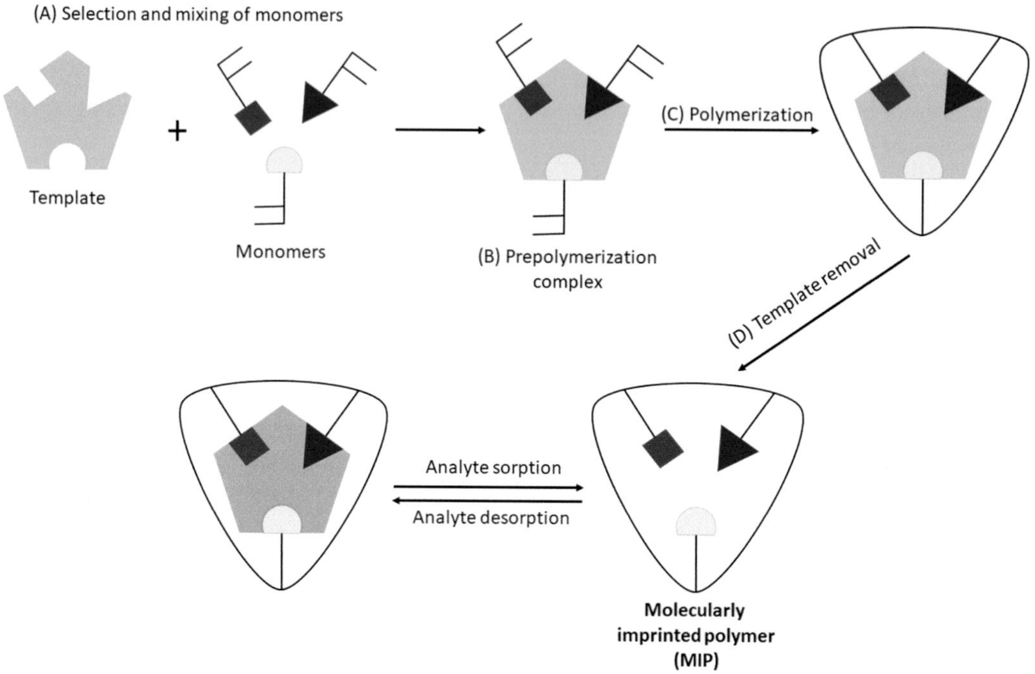

FIGURE 20.7 Scheme of MIP synthesis.

Scheme of molecularly imprinted polymer synthesis: (A) Selection and mixing of monomers; (B) Pre-polymerization and complex formation; (C) Polymerization; (D) Template removal.

- *Prepolymerization complex formation*: Next, the prepolymerization complex is formed between the monomers and the template, which will give rise to the cavities in the structure. Depending on the type of complex formed, two types of MIPs can be distinguished: covalent or noncovalent. The most common are the noncovalent ones since their synthesis is simpler (it is easier to extract the templates); however, the covalent ones are more efficient in terms of molecular recognition.
- *Polymerization*: Originally, polymerization took place through thermal initiation or photoinitiation. Currently, it is also carried out through electrochemical polymerization by oxidation of electroactive monomers or by autopolymerization in an oxygen environment, as it allows controlling the thickness of the membrane and achieving extra thin membranes.
- *Template removal*: Finally, the template is extracted so that the cavities of the material are left free and the sensor material can absorb the analyte in its structure. In most cases, the process is reversible, and the sensors can be reused. Initially, this was one of the critical steps in the synthesis of these sensors since complete removal of the templates was not easy.

20.2.3.2 Molecularly imprinted polymers for macromolecules

Initially, these materials were designed for the detection of small molecules, and their use for the detection of macromolecules such as proteins, which are already complex molecules, required an adaptation of the synthetic process. The two limitations of the conventional synthesis of MIPs using biological macromolecules are *the crosslinking of the materials* and *the stability of the macromolecules* (Bergmann & Peppas, 2008).

- *Crosslinking of the materials*, which is fundamental to maintaining the structure of the cavities, becomes a problem when working with voluminous templates derived from macromolecules. On one hand, they get trapped in the structure and cannot be properly removed, rendering the material useless as a sensor. On the other hand, crosslinking hinders the diffusion of analytes inside the polymeric matrix, increasing response times. Although in some cases the imprinting in gels, which are structurally more flexible, has been achieved, examples are very few.
- *Stability of the macromolecules*. Proteins are complex macromolecules from the point of view of stability. Their structure and functionality are seriously affected by factors such as pH, ionic strength, or solvation. A protein does not behave the same way in an aqueous (biological) medium as it does in an organic medium; in fact, in an organic medium, both its structure and activity can be altered to a greater or lesser extent. This is a crucial problem when synthesizing MIPs in organic media. If the structure of the template in the organic medium (of synthesis) is not the same as that of the final analyte in the aqueous/biological medium (work medium), it is possible that the relevant interaction will not occur and the sensor system will not function. In addition, the solubility of proteins decreases drastically in organic solvents, making their use as templates difficult.

Another problem with protein detection is cross-reactivity, which arises from the complexity of these molecules. Proteins are non-branched heteropolymers formed by sequences of between 200 and 300 amino acids. Depending on their composition, proteins acquire different types of structures, often related to their functionality. Generally, these structures are established by orienting the hydrophobic lateral residues of the amino acids toward the interior and the hydrophilic residues toward the exterior. This means that different proteins may have a sufficiently similar external "conformation" to be confused with each other, leading to cross-reactivity (Bergmann & Peppas, 2008).

The experimental difficulties in the synthesis of these materials explain why, although the first MIP sensor for proteins, specifically for glucose oxidase, was developed in 1985 (Glad et al., 1985), it was not until 2005 that these types of sensor systems began to be developed consistently. In fact, even in 2014, only 1% of developed MIPs targeted proteins. Furthermore, in all of these studies, the selected proteins had very specific characteristics (Erdőssy et al., 2016):

- Good conformational stability.
- High isoelectric point.
- High glycosylation.

These properties allowed for strong and selective electrostatic interactions to be established between negatively charged polymers formed by monomers with glycan and aminophenylboronic acid and positively charged proteins such as lysozyme or avidin (Wang et al., 2014). In other words, although the foundations for the synthesis of these sensor systems were being laid, sensors for all types of proteins could not yet be developed without restrictions.

20.2.3.2.1 Synthetic strategies

The emergence of MIPs as sensors for macromolecules has occurred within the last 15 years. Specifically, from 2008 to 2023, the annual number of publications in this field has doubled, increasing from 610 documents to 1264 in 2023, according to Scopus data from July to December 2023. This remarkable growth can be attributed to the implementation of various approaches that have successfully addressed previous challenges (Mazzotta et al., 2022). These approaches have focused on resolving issues related to template denaturation and material cross-linking.

- *Template denaturation*: To solve this problem, a compromise solution was reached; polymerization was carried out in water with soluble polymers and activators. There is controversy regarding the negative influence of water on the formation of dipole-dipole interactions and hydrogen bonds between the template and the monomers. However, the use of water as a synthesis solvent solves the initial structural and solubility problems.
- *Material crosslinking*: Several possible strategies were suggested for this problem. The first, proposed by Hjertén and his colleagues (Hjertén et al., 1997; Tong et al., 2001), was the use of slightly crosslinked polymeric hydrogels similar to those used in electrophoresis. Unfortunately, the results were not as expected; the resulting materials did not meet the quality and stability expectations of the print due to the excessive flexibility of the chain (Ge & Turner, 2008). Another possible strategy is the *printing of epitopes* (Teixeira et al., 2021). Epitopes are small amino acid chains with fewer stability and diffusion problems than complete proteins. The most successful strategy was *surface imprinting*, which has become one of the main methods for macromolecular printing. It is characterized by using materials with very low thicknesses comparable to those of the template, and the cavities are focused on the surface of the material. This arrangement facilitates the removal of templates and eliminates the problem of diffusing analytes. Finally, and in line with the previous strategy, the use of small-dimension MIPs was studied in the form of *thin films* or *NPs* for their application in electrochemical sensors (Ge & Turner, 2008; Mazzotta et al., 2022).

Printing epitopes. The complexity of proteins poses a stability problem when using them as a template in the synthesis of MIPs (Fig. 20.8). However, in the 2000s, Rachkov and Minoura suggested the use of shorter sequences epitopes (Rachkov & Minoura, 2000, 2001). This strategy presents two main advantages:

- Allows polymerization in organic media.
- The small size of these molecules facilitates their complete removal from the template at the end of the synthetic process.

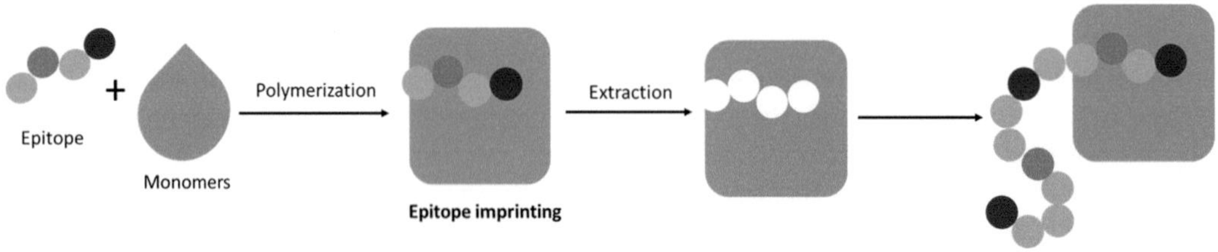

FIGURE 20.8 Epitope MIPs synthesis.

Schematic representation of the synthesis of molecularly imprinted polymers (MIP) using an epitope as a template.

The key to this strategy is the choice of epitopes. Initially, specific sequences of the protein were selected (to avoid cross-reactivity) that were found on the exterior of its structure. Nowadays, target proteins are studied in advance to select the most suitable epitope. There are different options when choosing the sequence:

- *Terminal sequences*: Choosing linear peptides contained in the C-/N-terminal sequences can be advantageous. Especially in the case of C-terminal sequences, which usually do not suffer posttranslational modifications and are thus more specific. The advantage of this strategy is that the production of these linear peptides is economical and easy (Canfarotta et al., 2018; Li et al., 2016; Li et al., 2017; Ma et al., 2017, 2019; Zhao et al., 2017).
- *Sequences extracted from the target protein*: The protein is scanned for fragments that are suitable for the experimental synthesis conditions (Altintas et al., 2019; Gupta et al., 2016, 2018). Sometimes, proteins of the same family are studied together to obtain sensors capable of detecting various proteins of the same family. The selection process itself is more complicated, but the specificity of the sensors is greater.
- *Natural epitopes*: In some cases, peptide sequences involved in biological processes recognized by known antibodies or receptors are selected (Kushwaha et al., 2019; Lu et al., 2012; Palladino et al., 2018; Tai et al., 2010). However, in some cases, it is not possible to follow this strategy because not all proteins are equally known.
- *Non-linear peptides*: Proteins generally have at least a secondary structure, so the use of cyclic peptides can provide an advantage over linear peptides in terms of specificity (Hou et al., 2021; Xu et al., 2019).
- *Epitopes, not based on peptides*: Monosaccharides or polysaccharides can be used as templates for the detection of glycoproteins. It is easier to select epitopes, but they can only be used for a specific group of proteins (Zhang et al., 2019).

Nowadays, computational analysis techniques are very useful for selecting the most accessible and specific protein sequences.

Surface printing. Also, due to the need to generate polymer nanostructures with a high area to volume ratio to maximize the number of sites printed on the surface and reduce problems related to diffusion, various strategies have been developed, which are as follows:

- *Microcontact printing*: In this technique, synthesis takes place between two glass surfaces in three steps (Fig. 20.9): (1) Protein Stamp: template proteins are absorbed onto the surface of one of the glass surfaces; (2) Polymerization: the two glass surfaces are brought into contact, the one with the absorbed proteins and the one containing the monomer solution, and the photochemical initiation of polymerization is started. The arrangement of the glass allows for microcontact printing of the material and generates the characteristic cavities on the surface; and (3) Extraction: finally, the upper glass is removed and the template proteins are eliminated. This technique allows the synthesis of MIPs without solubilizing the templates (one of the most common impediments in the synthesis of these materials), requires very small amounts of monomers, and allows multiple polymerizations to be carried out simultaneously with different monomer combinations (Chunta et al., 2016, 2018).
- *Sacrificial template support*: Although there are different variations in terms of support types and geometries, the bases of this technique are the same; protein templates are anchored onto a

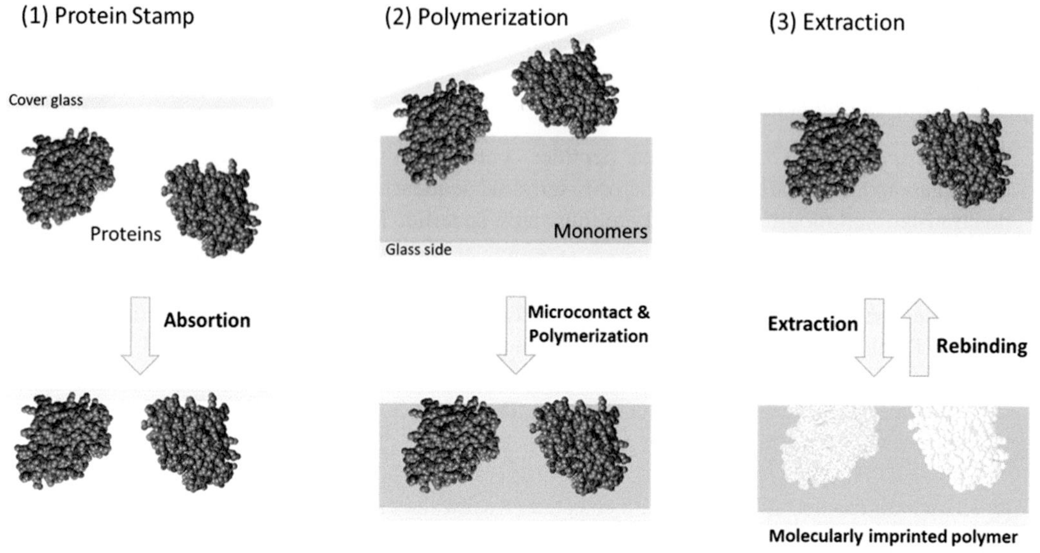

FIGURE 20.9 Microcontact printing MIPs synthesis.

Schematic representation of the synthesis of molecularly imprinted polymers (MIP) using microcontact printing.

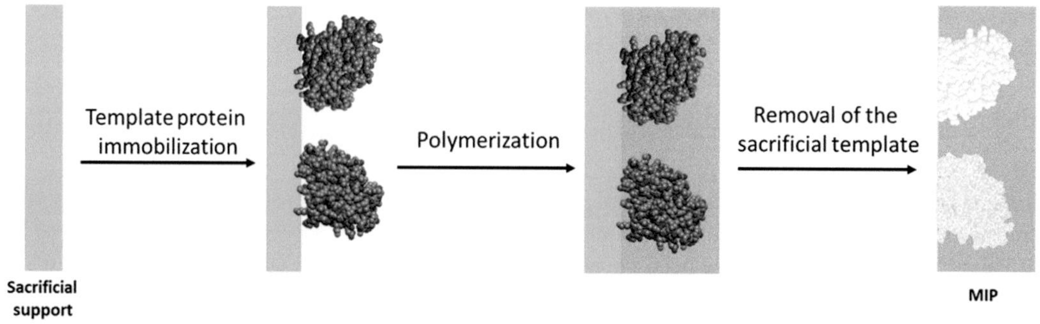

FIGURE 20.10 Sacrificial template supports MIPs synthesis.

Schematic representation of the synthesis of molecularly imprinted polymers (MIP) using sacrificial template support.

support (polymeric or not), polymer printing is carried out over them, and in some cases, polymerization is carried out directly on the final electrode. Finally, the template is removed in a non conservative manner. Fig. 20.10 shows a scheme of the process (Menaker et al., 2009; Orozco et al., 2013).

- *Soft lithography*: Synthesis is carried out on a support, where the monomer solution and the template are deposited. Next, the system is covered with a polydimethylsiloxane mold (stamp), and photopolymerization is initiated. Finally, the top mold is removed, and the template is eliminated. This process (see Fig. 20.11) allows for control of the dimensions of the material and the creation of micro- and nano-scale systems without the need for specialized equipment or high costs (Lautner et al., 2011). There is a variant of this process called two-dimensional molecular printing. In some cases, a modified stamp is used where the templates are anchored and then covered with the monomer solution necessary for the formation of the material. Finally, it is brought into contact with the support, and polymerization is initiated.

- *Use of aptamers*: To enhance selectivity, some surface imprinting techniques employ aptamers (short sequences of nucleic acids). These molecules are commonly used as "synthetic antibodies" due to their ability to selectively recognize and bind to specific molecules such as proteins, peptides, small molecules, or even whole cells. In the case of MIPs and surface imprinting, aptamers are immobilized onto the synthesis surface/electrode and become embedded in the final material after polymerization (see Fig. 20.12), thereby promoting protein recognition (Jolly et al., 2016; Kalecki et al., 2020; Liao et al., 2012).

Molecularly imprinted polymers films. The objective of this strategy is to obtain materials with a high surface-to-volume ratio. A large surface area maximizes the number of possible cavities on the material's surface, and its low thickness (resulting from its reduced volume) reduces diffusion problems. It could be said that it is a variant of surface printing strategies with the advantage of allowing the integration of materials into electrochemical measurement systems. These types of materials can be synthesized in situ, directly on the electrode, or in two ex situ steps. The two-phase synthesis allows for prior characterization and subsequent deposition of the materials onto the electrode. Various strategies have been developed for both procedures:

- *Drop-casting, or spin-and-dip coating*: It consists of a drop-casting deposit of the polymer onto the electrode. The polymer can either be pre-polymerized and melted or polymerized directly on the electrode. In this latter case, it has been found that the use of ionic liquids with an amino

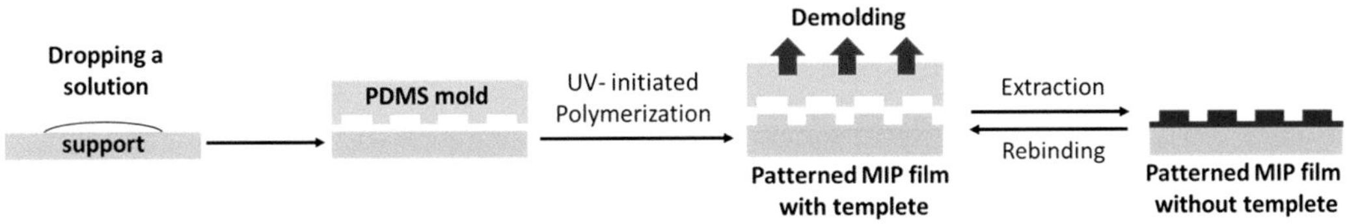

FIGURE 20.11 Soft lithography MIPs synthesis.

Schematic representation of the synthesis of molecularly imprinted polymers (MIP) using soft lithography.

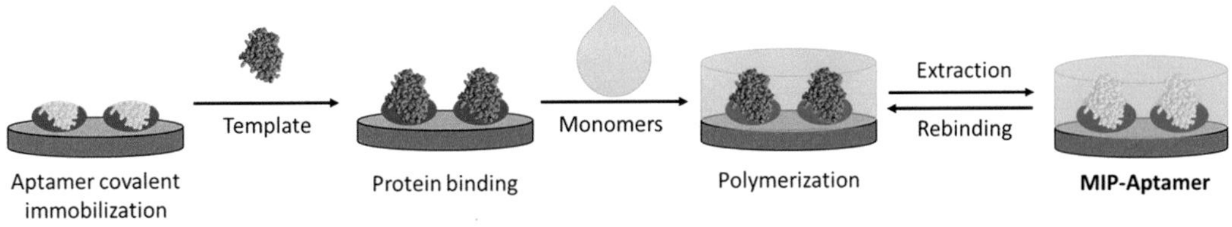

FIGURE 20.12 Use of aptamer for MIPs synthesis.

Schematic representation of the synthesis of molecularly imprinted polymers (MIP) using aptamer.

group and a vinyl group (1-{3-[(2-aminoethyl)amino]propyl}-3-vinylimidazole bromide) as a functional monomer has great advantages: it increases thermal stability, improves conductivity, is soluble in water, is biocompatible, and allows for polymerization at room temperature (contributing to the stability of the protein templates). Fig. 20.13 shows a synthesis scheme (Karimian et al., 2013; Viswanathan et al., 2012).

- *Use of thiol derivatives*: Thiol-derived monomers are combined with protein templates to generate polymer self-assembled monolayers that give rise to characteristic cavities on the material's surface (see Fig. 20.14). Generally, these syntheses are carried out in situ on a gold electrode. The system is straightforward to design but presents selectivity problems with similar molecules. However, they have had very good results with cancer markers (Wang et al., 2008; Yu et al., 2016).
- *Grafting polymerizable groups and/or initiators onto the electrode surface*: Polymerizable groups that are compatible with the templates are grafted onto the electrode surface (AU), and the in situ polymerization of the monomers and template is carried out (see Fig. 20.15). This technique is often combined with controlled/"living" radical polymerization, which allows for the control of polymer thickness (Feng et al., 2014).
- *Electropolymerization*: In previous strategies, MIPs were synthesized through radical or photochemical polymerization. An alternative approach is to use in situ electropolymerization directly on the electrode. This technique allows for control of polymer thickness (facilitating removal of the template), pore size, and synthesis conditions that are compatible with protein template stability. However, the redox activity of the templates should be considered, as it may affect the synthesis process and the final response of the sensor material. Nevertheless, this technique leads to materials with more uniform binding sites (Kalecki et al., 2020).

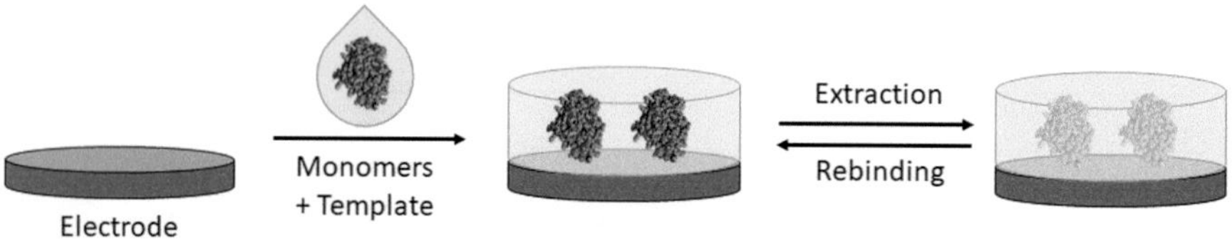

FIGURE 20.13 Drop-casting or spin-and-dip-coating MIPs synthesis.

Schematic representation of the synthesis of molecularly imprinted polymers (MIP) using drop-casting or spin-and-dip-coating.

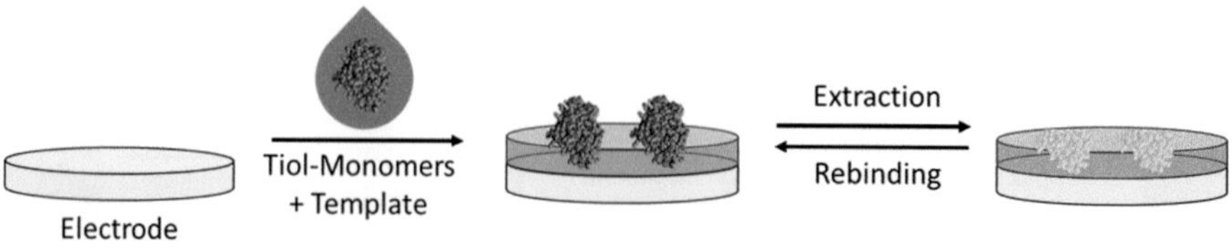

FIGURE 20.14 Thiol derivatives MIPs synthesis.

Schematic representation of the synthesis of molecularly imprinted polymers (MIP) using thiol derivatives monomers.

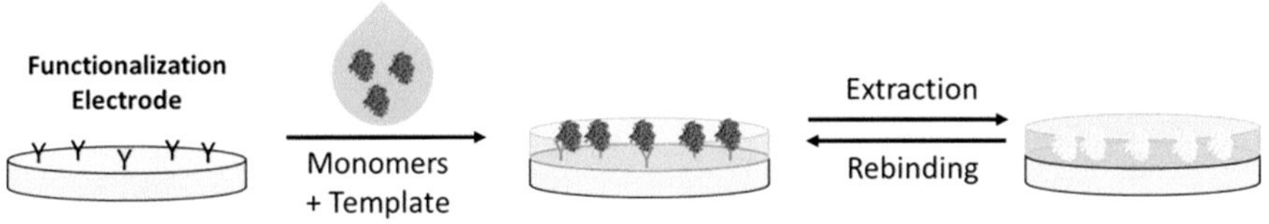

FIGURE 20.15 Grafting polymerizable groups and/or initiators onto the electrode surface MIPs synthesis.

Schematic representation of the synthesis of molecularly imprinted polymers (MIP) using grafting polymerizable groups and/or initiators onto the electrode surface.

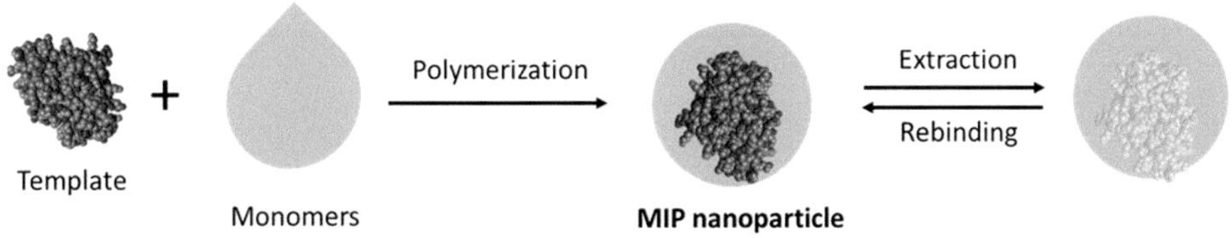

FIGURE 20.16 Nanogels MIPs synthesis.

Schematic representation of the synthesis of Nanogels molecularly imprinted polymers (MIP).

Molecularly imprinted polymer nanoparticles. Similar to the previous approach, this strategy aims to maximize the number of cavities on the material's surface while minimizing volume. Various synthetic methods have been described for these materials, such as precipitation polymerization, emulsion polymerization, nanogels, solid-phase synthesis, and surface grafting. However, the first two have not been used for protein targets.

- *Nanogels*: In this case, synthesis is carried out using acrylic monomers and small peptides (about 26 amino acids long) as templates. As a result, gel NPs are obtained with a size very similar to proteins (Fig. 20.16). The dimensions of these materials enable good interaction with the target, and the mild reaction conditions are compatible with biological molecules. The selectivity obtained is quite good (Zhu et al., 2011). The disadvantage of this method, however, is that it does not allow for large-scale production (Garcia-Cruz et al., 2020; Sener et al., 2010).
- *Solid-phase synthesis*: Solid-phase synthesis (see Fig. 20.17) is a method of peptide synthesis in which a solid support is used to grow the peptide chain, one peptide at a time. Once the sequence is constructed, it is separated from the support. To combine it with the synthesis of NP MIP, the final sequence is not separated from the support and is used as a template. After polymerization, the system can be cleaned of unpolymerized residues, and finally, the NP MIP can be extracted. The great advantage of this synthesis is that it allows for large scale production of these materials (Xu et al., 2016).
- *Core-shell grafting*: In this synthesis, (Fig. 20.18) porous NPs are used as a base, usually made of silica and externally functionalized. Surface grafts allow the templates to be fixed during the polymerization process. Polymerization occurs in a controlled manner, creating a thin layer of polymer around the nanoparticle. Finally, the templates are removed, leaving on the surface of the material the characteristic cavities of MIPs (Gao et al., 2011).

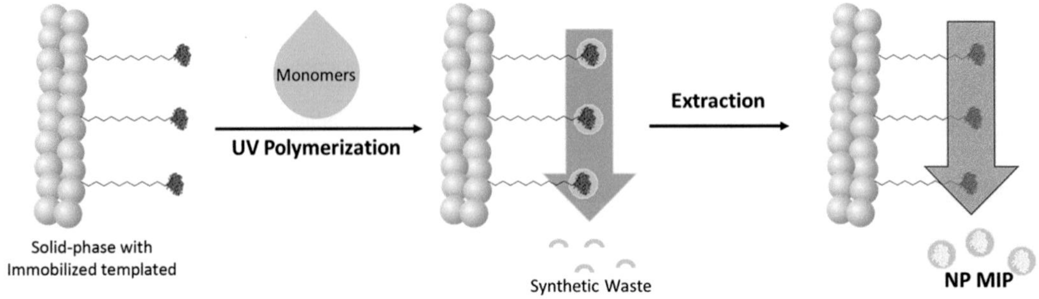

FIGURE 20.17 Solid-phase MIPs synthesis.

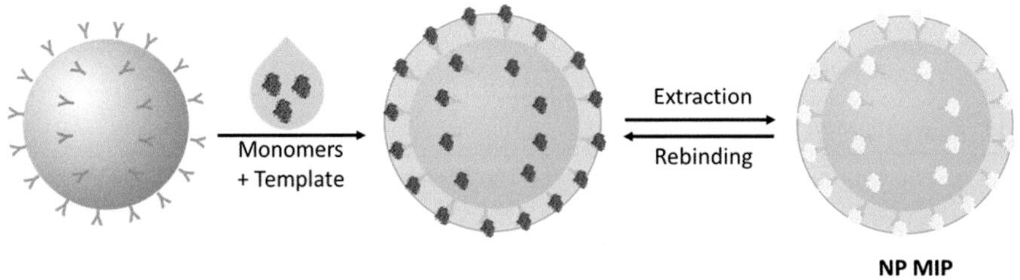

FIGURE 20.18 Core-shell grafting MIPs synthesis.

Schematic representation of the synthesis of molecularly imprinted polymers (MIP) using core-shell grafting.

MIPs in NP format are the most successful examples of nanoscale materials. They have the highest surface to volume ratio and the largest total active surface area per unit weight of polymer. In fact, they have also been used for other applications such as imaging, spectroscopy, and image processing. The only drawback of using these materials as sensors is that they cannot be directly synthesized onto electrodes, as with MIP films. This represents an extra step in their synthesis compared to previous strategies. Two approaches have been studied for this step:

- *Direct incorporation onto the electrode*: This option is only suitable for graphite or carbon paste electrodes. The MIP NPs are mixed directly with the assembled components that will form the electrode.
- *Immobilization on the electrode using polymeric membranes*: This second approach is the most versatile because it can be adapted to the needs or characteristics of the transducer. The NPs are immobilized on the electrode using polymeric materials such as agarose, chitosan, or acrylic derivatives as adhesives. The NPs are mixed with these materials and deposited by drip spin-coating onto the surface of the electrode.

20.2.3.3 Progress and future perspectives

Initially, MIPs for proteins were proposed as substitutes for conventional molecular recognition systems such as antigen-antibody, a kind of artificial antibody. However, in 2016, this goal was still far from reality. Although significant progress had been made since the first MIP for RNase, the range of proteins that could be detected with MIPs was quite limited, and half of the developed models used hemoglobin, serum albumin, or avidin as templates.

One of the main challenges that these materials faced was their routine use, which was not possible at that time despite some discrepancies between authors. In most cases, the compositional complexity of biological samples interferes with the sensing ability of these materials.

Additionally, there was no common synthesis method that could optimize the synthetic yield of these materials and their sensing ability, adaptable to diverse analytes and their needs.

Nonetheless, in recent years, the development of MIPs has grown exponentially, particularly in this field. Currently, 70% of the developed MIPs are directed toward the detection of macromolecules. In fact, the spectrum of proteins that these systems can detect is wider, even including cancer protein markers. Table 20.5 shows some examples of target proteins for MIPs.

Table 20.5 MIPs for protein targets.

Synthetic strategies			**Protein target**
Printing epitopes	Terminal sequences	C-Terminal	Human serum albumin (HSA) (Ma et al., 2017)
			Bovine Serum Albumin (BSA) (Li et al., 2017)
			Insulin (Zhao et al., 2017)
			Epidermal growth factor receptor (EGFR) (Canfarotta et al., 2018)
			Cytochrome c (Cyt c) (Ma et al., 2019)
		N-Terminal	Cyt c (Li et al., 2016)
	Sequences extracted from the target protein	Class 3 outer membrane protein allele of *Neisseria meningitidis* (Gupta et al., 2016)	
		fbpA periplasm protein of *N. meningitidis* (Gupta et al., 2018)	
		Neuron-specific enolase (NSE) (Altintas et al., 2019)	
	Natural epitopes	Anthrax-protective antigen (PA) 83 (Tai et al., 2010)	
		Proline-tRNA ligase protein of *Mycobacterium leprae* (Kushwaha et al., 2019)	
		HIV-1 glycoprotein 41 (gp41) (Lu et al., 2012)	
		Troponin T (TnT) isoform 6 (Palladino et al., 2018)	
	Non-linear peptides	Atrial natriuretic peptide (ANP) (Hou et al., 2021)	
		Glicoproteína 41 (gp41) (VIH) (Xu et al., 2019)	

(Continued)

Table 20.5 MIPs for protein targets. ***Continued***

Synthetic strategies			Protein target
	Epitopes not based on peptides		Tyrosine-phosphorylated peptides (Zhang et al., 2019)
Surface printing	Microcontact printing	Low-density lipoprotein (LDL) (Chunta et al., 2016)	
		High-density lipoprotein (HDL) (Chunta et al., 2018)	
	Sacrificial template support	BSA (Menaker et al., 2009)	
		Anti-IgG (Orozco et al., 2013)	
	Soft lithography	ExtrAvidin (EAv) (Lautner et al., 2011)	
		Lysozyme (Lys) (Lautner et al., 2011)	
	Use of aptamers	Thrombin (Liao et al., 2012)	
		Prostate-specific antigen (PSA) (Jolly et al., 2016)	
MIPs films	Drop-casting or spin- and dip-coating	Antigen-125 (CA 125) (Viswanathan et al., 2012)	
		Troponin T (TnT) (Karimian et al., 2013)	
	Use of thiol derivatives	(CEA) (Yu et al., 2016)	
		Mioglobin (Myo) (Wang et al., 2008)	
		Hemoglobin (PHb) (Wang et al., 2008)	
	Grafting polymerizable groups and/or initiators onto the electrode surface	Carcinoembryonic antigen (CEA) (Feng et al., 2014)	
		Carbohydrate antigen-199 (CA199) (Feng et al., 2014)	
	Electropolymerization	Human apurinic-apyrimidinic endonuclease-1 (APE1) (Kalecki et al., 2020)	
		Vascular endothelial growth factor (VEGF) (Kalecki et al., 2020)	
MIPs NPs	Nanogels	Lys (Sener et al., 2010)	
		Trypsin (Garcia-Cruz et al., 2020)	
	Solid-phase synthesis	Kallikrein (Xu et al., 2016)	
		Ribonuclease A (Xu et al., 2016)	
	Core-shell grafting	Bovine pancreas ribonuclease A (Gao et al., 2011)	
		Bovine hemoglobin (BHb) (Gao et al., 2011)	

In recent years, some progress has been made toward synthesizing MIPs as protein sensors. While chemical polymerization is commonly used to prepare MIPs, electropolymerization has proven to be the most effective approach for developing MIPs as protein sensors for several reasons:

- Allows in situ *polymerization* on an electrode or other widely studied transducers.

- Allows the *use of water as a synthesis solvent*, which helps maintain the natural structure of the templates.
- Buffer usage for *protein stabilization* can serve as a supporting electrolyte during polymerization.
- Does *not* require an *external initiator* that could damage the templates, unlike some chemical initiators that can act as oxidizing agents.
- *Control over the dimensions of the material*: Electropolymerization allows for control over the thickness of the synthesized materials as well as the synthesis of nanostructures through the use of sacrificial supports.
- *Confined surface deposition*: By using electrooxidation and suitable monomers, surface deposition can be controlled by generating an "active" initiator electrochemically or by creating a local pH change on the electrode surface.

The main advantage of electrochemical polymerization is that it allows for polymerization on an electrode or other widely studied transducers. Water can also be used as a synthesis solvent, which helps maintain the natural structure of the templates. The use of buffer for protein stabilization can also serve as a supporting electrolyte during polymerization, thus eliminating the requirement for an external initiator that could damage the templates, unlike some chemical initiators that act as oxidizing agents. There is also control over the dimensions of the material as well as confined surface deposition.

Despite the great advances of recent years, there are still some challenges to be overcome in the development of these materials:

- *Improving knowledge about protein-MIP recognition interfaces*: This would undoubtedly favor the development of more specific and effective sensing systems.
- *Using organic solvents*: Although not essential for electrochemical synthesis, they are sometimes necessary due to the low solubility of the starting reagents. A good alternative could be the use of ionic liquids.
- *Increasing the specificity of the MIP* for better detection limits and sensitivity. One option proposed is the development of hybrid materials.
- *Large scale production* for commercialization. The ultimate goal of these materials is their mass production and subsequent commercialization. In this case, the use of chemometric techniques could optimize the homogeneous and large scale production of these sensors.

Although there is still room for improvement, MIPs are one of the best options under development in terms of molecular recognition, particularly protein recognition.

20.2.4 Other polymers

There are additional polymers that fall outside the categories mentioned earlier, as they are not synthesized using molecular imprinting techniques, are not CPs, lack charged side groups (polyelectrolytes), and are not crosslinked (hydrogels). However, research has been conducted on their potential use as protein sensors.

The properties of these polymers are determined by the selection of monomers. Typically, a combination of hydrophilic and hydrophobic monomers is employed to create materials that are insoluble in aqueous media without the need for crosslinking while exhibiting a certain degree of swelling. These polymers can be prepared in various formats, such as films or powders, when derived from linear polymers.

Given the wide range of possibilities, various synthesis methods are available for these polymers. These methods include electropolymerization, radical polymerization (thermal or photochemical), and other techniques that enable the controlled formation of polymer chains and the incorporation of desired functional groups.

The exploration of these diverse polymer systems for protein sensing purposes allows for the development of materials with unique properties and capabilities. Researchers continue to investigate their synthesis methods, optimize their performance, and explore their potential applications in fields such as diagnostics, biotechnology, and biomedical engineering. This ongoing research aims to expand the repertoire of polymer-based protein sensors and advance their integration into various analytical and biomedical platforms.

Similarly, the responses of these sensors can be found in different types, such as colorimetric (Magnaghi, Capone, et al., 2020; Magnaghi et al., 2022; Magnaghi, Alberti, Capone, et al., 2020; Magnaghi, Alberti, et al., 2021; Magnaghi, Alberti, Quadrelli, et al., 2020; Magnaghi, Capone, et al., 2021), piezoelectric (Tsai & Lin, 2005), electrochemical (Panasyuk et al., 1999), amperometric (Sarkar, 2000), or fluorescent (off-on) of the FRET type (Arnaiz et al., 2022, 2023). Regarding the spectrum of proteins they can detect, as shown in Table 20.6, it is considerably more restricted than that of the preceding polymers. However, their applications in the food industry, specifically

Table 20.6 Other sensory polymers for protein detection according to the sensory response.

<table>
<tr><th>Type of polymer</th><th colspan="2">Answer</th><th>Target</th></tr>
<tr><td rowspan="4">Film</td><td>Colorimetric</td><td>Milk Protein degradation (Magnaghi et al., 2022; Magnaghi, Capone, et al., 2021)
Fish Protein degradation (Magnaghi, Capone, et al., 2020)
Meat Protein degradation (Magnaghi, Alberti, Capone, et al., 2020; Magnaghi, Alberti, et al., 2021; Magnaghi, Alberti, Quadrelli, et al., 2020; Magnaghi, Capone, et al., 2020)</td><td></td></tr>
<tr><td colspan="2">Piezoelectric</td><td>α-fetoprotein (AFP) (Tsai & Lin, 2005)</td></tr>
<tr><td colspan="2">Electrochemical</td><td>Rabbit IgG (Ag) (Panasyuk et al., 1999)</td></tr>
<tr><td>Amperometric</td><td>Casillan-90 (Sarkar, 2000)
Protease (Sarkar, 2000)
Hemoglobin (Sarkar, 2000)
Cyt-c (Sarkar, 2000)
α-Amylase (Sarkar, 2000)</td><td></td></tr>
<tr><td>Linear polymers</td><td>Fluorescent (FRET)</td><td>Off/On</td><td>Trypsin (Arnaiz et al., 2022)
Mpro of SARS-CoV-2 (Arnaiz et al., 2023)</td></tr>
</table>

in quality control and food safety (Arnaiz et al., 2022, 2023), have been highlighted, as have their recent applications in clinical diagnosis (Arnaiz et al., 2023).

Due to the fact that the majority of these sensors are relatively new, it is likely that in the coming years, they will be able to expand their capabilities to detect a wider range of proteins. These sensors have shown great promise as materials for protein detection and other applications, thanks to their ability to identify specific molecules with high sensitivity and precision.

20.3 Conclusions

The increasing demand for protein sensors has led to significant advancements in their development in recent years. Protein sensors based on conjugated and/or polyelectrolyte polymers experienced substantial progress in the 2000s, with applications primarily focused on imaging and diagnostic techniques in the field of biomedicine.

Hydrogels have been extensively utilized in protein detection since their introduction in the late 1990s. They have undergone continuous development; however, their current suitability for clinical use is limited due to their inherent fragility and susceptibility to degradation over time. MIPs, on the other hand, have witnessed exponential growth as versatile sensors, including for protein detection. MIPs are the most extensively studied option and can be applied to a wide range of proteins. Nevertheless, challenges remain in terms of their commercialization and large-scale industrial production.

Despite the significant progress achieved in protein sensor polymer research, several hurdles need to be addressed before their clinical application becomes feasible:

- *Meeting clinical standards*: protein sensor polymers need to comply with stringent clinical standards, including sensitivity, specificity, accuracy, and reliability. Extensive validation and testing in clinical settings are necessary to ensure their effectiveness and safety.
- *Industrial development and large scale production*: scaling up the production of protein sensor polymers to meet the demands of clinical applications is a significant challenge. Establishing cost-effective manufacturing processes and ensuring consistent quality control are crucial for their widespread adoption.
- *Long term storage and stability*: Protein sensors must maintain their performance over extended periods of storage and remain stable under various environmental conditions. Developing strategies to enhance their stability, such as optimizing storage conditions and employing appropriate packaging materials, is essential.

Addressing these challenges will pave the way for the clinical implementation of protein sensor polymers, enabling their use in diverse medical applications, including disease diagnosis, therapeutic monitoring, and personalized medicine. Continued research and collaboration between academia, industry, and regulatory bodies are crucial for overcoming these hurdles and realizing the full potential of protein sensor polymers in clinical settings.

References

Altintas, Z., Takiden, A., Utesch, T., Mroginski, M. A., Schmid, B., Scheller, F. W., & Süssmuth, R. D. (2019). Integrated approaches toward high-affinity artificial protein binders obtained via computationally simulated epitopes for protein recognition. *Advanced Functional Materials*, *29*(15), 1807332. Available from https://doi.org/10.1002/adfm.201807332.

Ambade, A. V., Sandanaraj, B. S., Klaikherd, A., & Thayumanavan, S. (2007). Fluorescent polyelectrolytes as protein sensors. *Polymer International*, *56*(4), 474–481. Available from https://doi.org/10.1002/pi.2185.

Andronico, L. A., Chen, L., Mirasoli, M., Guardigli, M., Quintavalla, A., Lombardo, M., Trombini, C., Chiu, D. T., & Roda, A. (2018). Thermochemiluminescent semiconducting polymer dots as sensitive nanoprobes for reagentless immunoassay. *Nanoscale*, *10*(29), 14012–14021. Available from https://doi.org/10.1039/c8nr03092h, http://pubs.rsc.org/en/journals/journal/nr.

Arnaiz, A., Guembe-García, M., Delgado-Pinar, E., Valente, A. J. M., Ibeas, S., García, J. M., & Vallejos, S. (2022). The role of polymeric chains as a protective environment for improving the stability and efficiency of fluorogenic peptide substrates. *Scientific Reports*, *12*(1). Available from https://doi.org/10.1038/s41598-022-12848-4.

Arnaiz, A., Guirado-Moreno, J. C., Guembe-García, M., Barros, R., Tamayo-Ramos, J. A., Fernández-Pampín, N., García, J. M., & Vallejos, S. (2023). Lab-on-a-chip for the easy and visual detection of SARS-CoV-2 in saliva based on sensory polymers. *Sensors and Actuators B: Chemical*, *379*, 133165. Available from https://doi.org/10.1016/j.snb.2022.133165.

Bai, X., Wang, K., Chen, L., Zhou, J., & Wang, J. (2022). Semiconducting polymer dots as fluorescent probes for in vitro biosensing. *Journal of Materials Chemistry B*, *10*(33), 6248–6262. Available from https://doi.org/10.1039/d2tb01385a.

Basu, S., Vutukuri, D. R., Shyamroy, S., Sandanaraj, B. S., & Thayumanavan, S. (2004). Invertible amphiphilic homopolymers. *Journal of the American Chemical Society*, *126*(32), 9890–9891. Available from https://doi.org/10.1021/ja047816a.

Bergmann, N. M., & Peppas, N. A. (2008). Molecularly imprinted polymers with specific recognition for macromolecules and proteins. *Progress in Polymer Science (Oxford)*, *33*(3), 271–288. Available from https://doi.org/10.1016/j.progpolymsci.2007.09.004.

Bérn Abérem, M., Najari, A., Ho, A. A., Gravel, J. F., Nobert, P., Boudreau, D., & Leclerc, M. (2006). Protein detecting arrays based on cationic polythiophene-DNA-aptamer complexes. *Advanced Materials*, *18*(20), 2703–2707. Available from https://doi.org/10.1002/adma.200601651.

Biela, A., Watkinson, M., Meier, U. C., Baker, D., Giovannoni, G., Becer, C. R., & Krause, S. (2015). Disposable MMP-9 sensor based on the degradation of peptide cross-linked hydrogel films using electrochemical impedance spectroscopy. *Biosensors and Bioelectronics*, *68*, 660–667. Available from https://doi.org/10.1016/j.bios.2015.01.060, http://www.elsevier.com/locate/bios.

Bossi, A., Bonini, F., Turner, A. P. F., & Piletsky, S. A. (2007). Molecularly imprinted polymers for the recognition of proteins: The state of the art. *Biosensors and Bioelectronics*, *22*(6), 1131–1137. Available from https://doi.org/10.1016/j.bios.2006.06.023.

Byrne, M. E., Park, K., & Peppas, N. A. (2002). Molecular imprinting within hydrogels. *Advanced Drug Delivery Reviews*, *54*(1), 149–161. Available from https://doi.org/10.1016/S0169-409X(01)00246-0, http://www.elsevier.com/locate/drugdeliv.

Canfarotta, F., Czulak, J., Betlem, K., Sachdeva, A., Eersels, K., Van Grinsven, B., Cleij, T. J., & Peeters, M. (2018). A novel thermal detection method based on molecularly imprinted nanoparticles as recognition elements. *Nanoscale*, *10*(4), 2081–2089. Available from https://doi.org/10.1039/c7nr07785h, http://pubs.rsc.org/en/journals/journal/nr.

Charych, D. H., Nagy, J. O., Spevak, W., & Bednarski, M. D. (1993). Direct colorimetric detection of a receptor-ligand interaction by a polymerized bilayer assembly. *Science (New York, N.Y.)*, *261*(5121), 585–588. Available from https://doi.org/10.1126/science.8342021.

Chen, L., McBranch, D. W., Wang, H.-L., Helgeson, R., Wudl, F., & Whitten, D. G. (1999). Highly sensitive biological and chemical sensors based on reversible fluorescence quenching in a conjugated polymer.

Proceedings of the National Academy of Sciences, *96*(22), 12287–12292. Available from https://doi.org/10.1073/pnas.96.22.12287.

Choi, E., Choi, Y., Nejad, Y. H. P., Shin, K., & Park, J. (2013). Label-free specific detection of immunoglobulin G antibody using nanoporous hydrogel photonic crystals. *Sensors and Actuators, B: Chemical*, *180*, 107–113. Available from https://doi.org/10.1016/j.snb.2012.03.053.

Chou, Y. N., Sun, F., Hung, H. C., Jain, P., Sinclair, A., Zhang, P., Bai, T., Chang, Y., Wen, T. C., Yu, Q., & Jiang, S. (2016). Ultra-low fouling and high antibody loading zwitterionic hydrogel coatings for sensing and detection in complex media. *Acta Biomaterialia*, *40*, 31–37. Available from https://doi.org/10.1016/j.actbio.2016.04.023, http://www.journals.elsevier.com/acta-biomaterialia.

Chunta, S., Suedee, R., & Lieberzeit, P. A. (2016). Low-density lipoprotein sensor based on molecularly imprinted polymer. *Analytical Chemistry*, *88*(2), 1419–1425. Available from https://doi.org/10.1021/acs.analchem.5b04091, http://pubs.acs.org/journal/ancham.

Chunta, S., Suedee, R., & Lieberzeit, P. A. (2018). High-density lipoprotein sensor based on molecularly imprinted polymer. *Analytical and Bioanalytical Chemistry*, *410*(3), 875–883. Available from https://doi.org/10.1007/s00216-017-0442-3, http://link.springer.de/link/service/journals/00216/index.htm.

Clark, M. F., & Adams, A. N. (1977). Characteristics of the microplate method of enzyme linked immunosorbent assay for the detection of plant viruses. *Journal of General Virology*, *34*(3), 475–483. Available from https://doi.org/10.1099/0022-1317-34-3-475.

Cormack, P. A. G., & Elorza, A. Z. (2004). Molecularly imprinted polymers: Synthesis and characterisation. *Journal of Chromatography B: Analytical Technologies in the Biomedical and Life Sciences*, *804*(1), 173–182. Available from https://doi.org/10.1016/j.jchromb.2004.02.013.

Das, S., Chatterjee, D. P., Ghosh, R., & Nandi, A. K. (2015). Water soluble polythiophenes: Preparation and applications. *RSC Advances*, *5*(26), 20160–20177. Available from https://doi.org/10.1039/c4ra16496b, http://pubs.rsc.org/en/journals/journalissues.

Ebrahimi, M. M. S., Laabei, M., Jenkins, A. T. A., & Schönherr, H. (2015). Autonomously sensing hydrogels for the rapid and selective detection of pathogenic bacteria. *Macromolecular Rapid Communications*, *36*(24), 2123–2128. Available from https://doi.org/10.1002/marc.201500485, http://www3.interscience.wiley.com/journal/117932056/grouphome.

Eckert, H., & Roller, M. (1990). Derivatizing reagents based on ferrocene for HPLC-ECD determination of peptides and proteins. *Journal of Liquid Chromatography*, *13*(17), 3399–3414. Available from https://doi.org/10.1080/01483919008049110.

Erdőssy, J., Horváth, V., Yarman, A., Scheller, F. W., & Gyurcsányi, R. E. (2016). Electrosynthesized molecularly imprinted polymers for protein recognition. *TrAC - Trends in Analytical Chemistry*, *79*, 179–190. Available from https://doi.org/10.1016/j.trac.2015.12.018, http://www.elsevier.com/locate/trac.

Faïd, K., & Leclerc, M. (1996). Functionalized regioregular polythiophenes: Towards the development of biochromic sensors. *Chemical Communications*, *24*, 2761–2762. Available from https://doi.org/10.1039/CC9960002761.

Faïd, K., & Leclerc, M. (1998). Responsive supramolecular polythiophene assemblies. *Journal of the American Chemical Society*, *120*(21), 5274–5278. Available from https://doi.org/10.1021/ja9802753.

Fan, C., Plaxco, K. W., & Heeger, A. J. (2002). High-efficiency fluorescence quenching of conjugated polymers by proteins. *Journal of the American Chemical Society*, *124*(20), 5642–5643. Available from https://doi.org/10.1021/ja025899u.

Feeney, J., Birdsall, B., Akiboye, J., Tendler, S. J. B., Jiménez Barbero, J., Ostler, G., Arnold, J. R. P., Roberts, G. C. K., Kühn, A., & Roth, K. (1989). Optimising selective deuteration of proteins for 2D 1 H NMR detection and assignment studies application to the Phe residues of Lactobacillus casei dihydrofolate reductase. *FEBS Letters*, *248*(1–2), 57–61. Available from https://doi.org/10.1016/0014-5793(89)80431-4.

Feng, F., Tang, Y., Wang, S., Li, Y., & Zhu, D. (2007). Continuous fluorometric assays for acetylcholinesterase activity and inhibition with conjugated polyelectrolytes. *Angewandte Chemie – International Edition*, *46*(41), 7882–7886. Available from https://doi.org/10.1002/anie.200701724.

Feng, X., Gan, N., Zhou, J., Li, T., Cao, Y., Hu, F., Yu, H., & Jiang, Q. (2014). A novel dual-template molecularly imprinted electrochemiluminescence immunosensor array using Ru(bpy)32 + -Silica@Poly-L-lysine-Au composite nanoparticles as labels for near-simultaneous detection of tumor markers. *Electrochimica Acta*, *139*, 127–136. Available from https://doi.org/10.1016/j.electacta.2014.07.008, http://www.journals.elsevier.com/electrochimica-acta/.

Galis, Z. S., Sukhova, G. K., & Libby, P. (1995). Microscopic localization of active proteases by in situ zymography: Detection of matrix metalloproteinase activity in vascular tissue. *FASEB Journal*, *9*(10), 974–980. Available from https://doi.org/10.1096/fasebj.9.10.7615167, https://onlinelibrary.wiley.com/journal/15306860.

Gao, R., Kong, X., Wang, X., He, X., Chen, L., & Zhang, Y. (2011). Preparation and characterization of uniformly sized molecularly imprinted polymers functionalized with core-shell magnetic nanoparticles for the recognition and enrichment of protein. *Journal of Materials Chemistry*, *21*(44), 17863–17871. Available from https://doi.org/10.1039/c1jm12414e.

Garcia-Cruz, A., Ahmad, O. S., Alanazi, K., Piletska, E., & Piletsky, S. A. (2020). Generic sensor platform based on electro-responsive molecularly imprinted polymer nanoparticles (e-NanoMIPs). *Microsystems and Nanoengineering*, *6*(1). Available from https://doi.org/10.1038/s41378-020-00193-3, http://www.nature.com/micronano/.

Ge, Y., & Turner, A. P. F. (2008). Too large to fit? Recent developments in macromolecular imprinting. *Trends in Biotechnology*, *26*(4), 218–224. Available from https://doi.org/10.1016/j.tibtech.2008.01.001.

Glad, M., Norrlöw, O., Sellergren, B., Siegbahn, N., & Mosbach, K. (1985). Use of silane monomers for molecular imprinting and enzyme entrapment in polysiloxane-coated porous silica. *Journal of Chromatography. A*, *347*(C), 11–23. Available from https://doi.org/10.1016/S0021-9673(01)95465-2.

Guembe-García, M., Peredo-Guzmán, P. D., Santaolalla-García, V., Moradillo-Renuncio, N., Ibeas, S., Mendía, A., García, F. C., García, J. M., & Vallejos, S. (2020). Why is the sensory response of organic probes within a polymer film different in solution and in the solid-state? evidence and application to the detection of amino acids in human chronic wounds. *Polymers*, *12*(6), 1249. Available from https://doi.org/10.3390/polym12061249.

Guembe-García, M., Santaolalla-García, V., Moradillo-Renuncio, N., Ibeas, S., Reglero, J. A., García, F. C., Pacheco, J., Casado, S., García, J. M., & Vallejos, S. (2021). Monitoring of the evolution of human chronic wounds using a ninhydrin-based sensory polymer and a smartphone. *Sensors and Actuators, B: Chemical*, *335*. Available from https://doi.org/10.1016/j.snb.2021.129688, https://www.journals.elsevier.com/sensors-and-actuators-b-chemical.

Gunda, N. S. K., Chavali, R., & Mitra, S. K. (2016). A hydrogel based rapid test method for detection of: Escherichia coli (E. coli) in contaminated water samples. *Analyst*, *141*(10), 2920–2929. Available from https://doi.org/10.1039/c6an00400h, http://pubs.rsc.org/en/journals/journal/an.

Gupta, N., Shah, K., & Singh, M. (2016). An epitope-imprinted piezoelectric diagnostic tool for Neisseria meningitidis detection. *Journal of Molecular Recognition*, *29*(12), 572–579. Available from https://doi.org/10.1002/jmr.2557.

Gupta, N., Singh, R. S., Shah, K., Prasad, R., & Singh, M. (2018). Epitope imprinting of iron binding protein of Neisseria meningitidis bacteria through multiple monomers imprinting approach. *Journal of Molecular Recognition*, *31*(7), e2709. Available from https://doi.org/10.1002/jmr.2709.

Hedin-Dahlström, J., Shoravi, S., Wikman, S., & Nicholls, I. A. (2004). Stereoselective reduction of menthone by molecularly imprinted polymers. *Tetrahedron, Asymmetry*, *15*(15), 2431–2436. Available from https://doi.org/10.1016/j.tetasy.2004.06.002.

Herland, A., & Inganäs, O. (2007). Conjugated polymers as optical probes for protein interactions and protein conformations. *Macromolecular Rapid Communications*, *28*(17), 1703–1713. Available from https://doi.org/10.1002/marc.200700281.

Hjertén, S., Liao, J. L., Nakazato, K., Wang, Y., Zamaratskaia, G., & Zhang, H. X. (1997). Gels mimicking antibodies in their selective recognition of proteins. *Chromatographia*, *44*(5–6), 227–234. Available from https://doi.org/10.1007/BF02466386.

Ho, H. A., Béra-Abérem, M., & Leclerc, M. (2005). Optical sensors based on hybrid DNA/conjugated polymer complexes. *Chemistry – A European Journal*, *11*(6), 1718–1724. Available from https://doi.org/10.1002/chem.200400537.

Hoffman, A. S. (2002). Hydrogels for biomedical applications. *Advanced Drug Delivery Reviews*, *54*(1), 3–12. Available from https://doi.org/10.1016/S0169-409X(01)00239-3, http://www.elsevier.com/locate/drugdeliv.

Hou, H., Jin, Y., Xu, K., Sheng, L., Huang, Y., & Zhao, R. (2021). Selective recognition of a cyclic peptide hormone in human plasma by hydrazone bond-oriented surface imprinted nanoparticles. *Analytica Chimica Acta*, *1154*. Available from https://doi.org/10.1016/j.aca.2021.338301, http://www.journals.elsevier.com/analytica-chimica-acta/.

Hwang, C. C., & Lee, W. C. (2002). Chromatographic characteristics of cholesterol-imprinted polymers prepared by covalent and non-covalent imprinting methods. *Journal of Chromatography. A*, *962*(1–2), 69–78. Available from https://doi.org/10.1016/S0021-9673(02)00559-9.

Industrial Research Associations: Some taxation problems. (1960). *Nature*, *185*(4706), 63–64. https://doi.org/10.1038/185063a0.

Ito, T., Shirakawa, H., & Ikeda, S. (1974). Simultaneous polymerization and formation of polyacetylene film on the surface of concentrated soluble Ziegler-type catalyst solution. *Journal of Polymer Science: Polymer Chemistry Edition*, *12*(1), 11–20. Available from https://doi.org/10.1002/pol.1974.170120102.

Jolly, P., Tamboli, V., Harniman, R. L., Estrela, P., Allender, C. J., & Bowen, J. L. (2016). Aptamer-MIP hybrid receptor for highly sensitive electrochemical detection of prostate specific antigen. *Biosensors and Bioelectronics*, *75*, 188–195. Available from https://doi.org/10.1016/j.bios.2015.08.043, http://www.elsevier.com/locate/bios.

Jung, I. Y., Kim, J. S., Choi, B. R., Lee, K., & Lee, H. (2017). Hydrogel based biosensors for in vitro diagnostics of biochemicals, proteins, and genes. *Advanced Healthcare Materials*, *6*(12). Available from https://doi.org/10.1002/adhm.201601475, http://onlinelibrary.wiley.com/journal/10.1002/(ISSN)2192-2659.

Kalecki, J., Iskierko, Z., Cieplak, M., & Sharma, P. S. (2020). Oriented immobilization of protein templates: A new trend in surface imprinting. *ACS Sensors*, *5*(12), 3710–3720. Available from https://doi.org/10.1021/acssensors.0c01634, http://pubs.acs.org/journal/ascefj.

Karimian, N., Vagin, M., Zavar, M. H. A., Chamsaz, M., Turner, A. P. F., & Tiwari, A. (2013). An ultrasensitive molecularly-imprinted human cardiac troponin sensor. *Biosensors and Bioelectronics*, *50*, 492–498. Available from https://doi.org/10.1016/j.bios.2013.07.013.

Kim, I. B., Dunkhorst, A., & Bunz, U. H. F. (2005). Nonspecific interactions of a carboxylate-substituted PPE with proteins. A cautionary tale for biosensor applications. *Langmuir: The ACS Journal of Surfaces and Colloids*, *21*(17), 7985–7989. Available from https://doi.org/10.1021/la051152g.

Kleiner, D. E., & Stetlerstevenson, W. G. (1994). Quantitative zymography: Detection of picogram quantities of gelatinases. *Analytical Biochemistry*, *218*(2), 325–329. Available from https://doi.org/10.1006/abio.1994.1186.

Komiyama M., Takeuchi T., Mukawa T., & Asanuma H. (2003). *Molecular imprinting: From fundamentals to applications.*

Kumaraswamy, S., Bergstedt, T., Shi, X., Rininsland, F., Kushon, S., Xia, W., Ley, K., Achyuthan, K., McBranch, D., & Whitten, D. (2004). Fluorescent-conjugated polymer superquenching facilitates highly sensitive detection of proteases. *Proceedings of the National Academy of Sciences of the United States of America*, *101*(20), 7511–7515. Available from https://doi.org/10.1073/pnas.0402367101.

Kumpumbu-Kalemba, L., & Leclerc, M. (2000). Electrochemical characterization of monolayers of a biotinylated polythiophene: Towards the development of polymeric biosensors. *Chemical Communications* (19), 1847–1848. Available from https://doi.org/10.1039/a909744i, http://pubs.rsc.org/en/journals/journal/cc.

Kushwaha, A., Srivastava, J., Singh, A. K., Anand, R., Raghuwanshi, R., Rai, T., & Singh, M. (2019). Epitope imprinting of Mycobacterium leprae bacteria via molecularly imprinted nanoparticles using multiple monomers approach. *Biosensors and Bioelectronics*, *145*, 111698. Available from https://doi.org/10.1016/j.bios.2019.111698.

Lakowicz, J. R. (2006). *Principles of fluorescence spectroscopy. Principles of Fluorescence Spectroscopy* (pp. 1–954). United States: Springer. Available from http://www.springerlink.com/openurl.asp?genre = book&isbn = 978-0-387-31278-1, https://doi.org/10.1007/978-0-387-46312-4.

Lautner, G., Kaev, J., Reut, J., Öpik, A., Rappich, J., Syritski, V., & Gyurcsányi, R. E. (2011). Selective artificial receptors based on micropatterned surface-imprinted polymers for label-free detection of proteins by SPR imaging. *Advanced Functional Materials*, *21*(3), 591–597. Available from https://doi.org/10.1002/adfm.201001753.

Li, D. Y., Zhang, X. M., Yan, Y. J., He, X. W., Li, W. Y., & Zhang, Y. K. (2016). Thermo-sensitive imprinted polymer embedded carbon dots using epitope approach. *Biosensors and Bioelectronics*, *79*, 187–192. Available from https://doi.org/10.1016/j.bios.2015.12.016, http://www.elsevier.com/locate/bios.

Li, M. X., Wang, X. H., Zhang, L. M., & Wei, X. P. (2017). A high sensitive epitope imprinted electrochemical sensor for bovine serum albumin based on enzyme amplifying. *Analytical Biochemistry*, *530*, 68–74. Available from https://doi.org/10.1016/j.ab.2017.05.006, http://www.elsevier.com/inca/publications/store/6/2/2/7/8/1/index.htt.

Liang, Y., Bar-Shir, A., Song, X., Gilad, A. A., Walczak, P., & Bulte, J. W. M. (2015). Label-free imaging of gelatin-containing hydrogel scaffolds. *Biomaterials*, *42*, 144–150. Available from https://doi.org/10.1016/j.biomaterials.2014.11.050, http://www.journals.elsevier.com/biomaterials/.

Liao, Y. J., Shiang, Y. C., Huang, C. C., & Chang, H. T. (2012). Molecularly imprinted aptamers of gold nanoparticles for the enzymatic inhibition and detection of thrombin. *Langmuir: The ACS Journal of Surfaces and Colloids*, *28*(24), 8944–8951. Available from https://doi.org/10.1021/la204651t.

Lim, S. L., Ooi, C. W., Low, L. E., Tan, W. S., Chan, E. S., Ho, K. L., & Tey, B. T. (2020). Synthesis of poly (acrylamide)-based hydrogel for bio-sensing of hepatitis B core antigen. *Materials Chemistry and Physics*, *243*. Available from https://doi.org/10.1016/j.matchemphys.2019.122578, http://www.journals.elsevier.com/materials-chemistry-and-physics.

Lin, Z., Zhang, G., Yang, W., Qiu, B., & Chen, G. (2012). CEA fluorescence biosensor based on the FRET between polymer dots and Au nanoparticles. *Chemical Communications*, *48*(79), 9918–9920. Available from https://doi.org/10.1039/c2cc35645g.

Liu, J., Liu, X., Chen, H., Yang, L., Cai, A., Ji, H., Wang, Q., Zhou, X., Li, G., Wu, M., Qin, Y., & Wu, L. (2022). Bifunctional Pdots-based novel ECL nanoprobe with qualitative and quantitative dual signal amplification characteristics for trace cytokine analysis. *Analytical Chemistry*, *94*(19), 7115–7122. Available from https://doi.org/10.1021/acs.analchem.2c01041.

Liu, Y., Ogawa, K., & Schanze, K. S. (2008). Conjugated polyelectrolyte based real-time fluorescence assay for phospholipase C. *Analytical Chemistry*, *80*(1), 150–158. Available from https://doi.org/10.1021/ac701672g.

Lu, C. H., Zhang, Y., Tang, S. F., Fang, Z. B., Yang, H. H., Chen, X., & Chen, G. N. (2012). Sensing HIV related protein using epitope imprinted hydrophilic polymer coated quartz crystal microbalance. *Biosensors and Bioelectronics*, *31*(1), 439–444. Available from https://doi.org/10.1016/j.bios.2011.11.008.

Lu, Y., Li, C., Liu, X., & Huang, W. (2002). Molecular recognition through the exact placement of functional groups on non-covalent molecularly imprinted polymers. *Journal of Chromatography. A*, *950*(1–2), 89–97. Available from https://doi.org/10.1016/S0021-9673(02)00058-4.

Ma, X. T., He, X. W., Li, W. Y., & Zhang, Y. K. (2017). Epitope molecularly imprinted polymer coated quartz crystal microbalance sensor for the determination of human serum albumin. *Sensors and Actuators, B: Chemical*, *246*, 879–886. Available from https://doi.org/10.1016/j.snb.2017.02.137.

Ma, X. T., He, X. W., Li, W. Y., & Zhang, Y. K. (2019). Oriented surface epitope imprinted polymer-based quartz crystal microbalance sensor for cytochrome c. *Talanta*, *191*, 222–228. Available from https://doi.org/10.1016/j.talanta.2018.08.079, https://www.journals.elsevier.com/talanta.

Magnaghi, L. R., Alberti, G., Capone, F., Zanoni, C., Mannucci, B., Quadrelli, P., & Biesuz, R. (2020). Development of a dye-based device to assess the poultry meat spoilage. Part II: Array on act. *Journal of Agricultural and Food Chemistry*, *68*(45), 12710–12718. Available from https://doi.org/10.1021/acs.jafc.0c03771, http://pubs.acs.org/journal/jafcau.

Magnaghi, L. R., Alberti, G., Milanese, C., Quadrelli, P., & Biesuz, R. (2021). Naked-eye food freshness detection: Innovative polymeric optode for high-protein food spoilage monitoring. *ACS Food Science and Technology*, *1*(2), 165–175. Available from https://doi.org/10.1021/acsfoodscitech.0c00089, https://pubs.acs.org/page/afsthl/about.html.

Magnaghi, L. R., Alberti, G., Quadrelli, P., & Biesuz, R. (2020). Development of a dye-based device to assess poultry meat spoilage. Part I: Building and testing the sensitive array. *Journal of Agricultural and Food Chemistry*, *68*(45), 12702–12709. Available from https://doi.org/10.1021/acs.jafc.0c03768, http://pubs.acs.org/journal/jafcau.

Magnaghi, L. R., Capone, F., Alberti, G., Zanoni, C., Mannucci, B., Quadrelli, P., & Biesuz, R. (2021). EVOH-based pH-sensitive optode array and chemometrics: From naked-eye analysis to predictive modeling to detect milk freshness. *ACS Food Science and Technology*, *1*(5), 819–828. Available from https://doi.org/10.1021/acsfoodscitech.1c00065, https://pubs.acs.org/page/afsthl/about.html.

Magnaghi, L. R., Capone, F., Zanoni, C., Alberti, G., Quadrelli, P., & Biesuz, R. (2020). Colorimetric sensor array for monitoring, modelling and comparing spoilage processes of different meat and fish foods. *Foods*, *9*(5), 684. Available from https://doi.org/10.3390/foods9050684.

Magnaghi, L. R., Zanoni, C., Alberti, G., Quadrelli, P., & Biesuz, R. (2022). Towards intelligent packaging: BCP-EVOH@ optode for milk freshness measurement. *Talanta*, *241*, 123230. Available from https://doi.org/10.1016/j.talanta.2022.123230.

Mátyus, L., Szöllosi, J., & Jenei, A. (2006). Steady-state fluorescence quenching applications for studying protein structure and dynamics. *Journal of Photochemistry and Photobiology B: Biology*, *83*(3), 223–236. Available from https://doi.org/10.1016/j.jphotobiol.2005.12.017.

Mazzotta, E., Di Giulio, T., & Malitesta, C. (2022). Electrochemical sensing of macromolecules based on molecularly imprinted polymers: Challenges, successful strategies, and opportunities. *Analytical and Bioanalytical Chemistry*, *414*(18), 5165–5200. Available from https://doi.org/10.1007/s00216-022-03981-0, https://link.springer.com/journal/216/volumes-and-issues.

Menaker, A., Syritski, V., Reut, J., Öpik, A., Horváth, V., & Gyurcsányi, R. E. (2009). Electrosynthesized surface-imprinted conducting polymer microrods for selective protein recognition. *Advanced Materials*, *21*(22), 2271–2275. Available from https://doi.org/10.1002/adma.200803597, http://www3.interscience.wiley.com/cgi-bin/fulltext/122310489/PDFSTART.

Moon, J. H., MacLean, P., McDaniel, W., & Hancock, L. F. (2007). Conjugated polymer nanoparticles for biochemical protein kinase assay. *Chemical Communications*, *46*, 4910–4912. Available from https://doi.org/10.1039/b710807a, http://pubs.rsc.org/en/journals/journal/cc.

Nilsson, K. P. R., Herland, A., Hammarström, P., & Inganäs, O. (2005). Conjugated polyelectrolytes: Conformation-sensitive optical probes for detection of amyloid fibril formation. *Biochemistry*, *44*(10), 3718–3724. Available from https://doi.org/10.1021/bi047402u.

Nilsson, K. P. R., & Inganäs, O. (2004). Optical emission of a conjugated polyelectrolyte: Calcium-induced conformational changes in calmodulin and calmodulin-calcineurin interactions. *Macromolecules*, *37*(24), 9109–9113. Available from https://doi.org/10.1021/ma048605t.

Nilsson, K. P. R., Olsson, J. D. M., Stabo-Eeg, F., Lindgren, M., Konradsson, P., & Inganäs, O. (2005). Chiral recognition of a synthetic peptide using enantiomeric conjugated polyelectrolytes and optical spectroscopy. *Macromolecules*, *38*(16), 6813–6821. Available from https://doi.org/10.1021/ma051188f.

Nilsson, K. P. R., Rydberg, J., Baltzer, L., & Inganäs, O. (2003). Self-assembly of synthetic peptides control conformation and optical properties of a zwitterionic polythiophene derivative. *Proceedings of the National Academy of Sciences of the United States of America*, *100*(18), 10170–10174. Available from https://doi.org/10.1073/pnas.1834422100.

Oral, E., & Peppas, N. A. (2004). Responsive and recognitive hydrogels using star polymers. *Journal of Biomedical Materials Research - Part A*, *68*(3), 439–447. Available from https://doi.org/10.1002/jbm.a.20076, http://onlinelibrary.wiley.com/journal/10.1002/(ISSN)1552-4965.

Orozco, J., Cortés, A., Cheng, G., Sattayasamitsathit, S., Gao, W., Feng, X., Shen, Y., & Wang, J. (2013). Molecularly imprinted polymer-based catalytic micromotors for selective protein transport. *Journal of the American Chemical Society*, *135*(14), 5336–5339. Available from https://doi.org/10.1021/ja4018545.

Palladino, P., Minunni, M., & Scarano, S. (2018). Cardiac Troponin T capture and detection in real-time via epitope-imprinted polymer and optical biosensing. *Biosensors and Bioelectronics*, *106*, 93–98. Available from https://doi.org/10.1016/j.bios.2018.01.068, http://www.elsevier.com/locate/bios.

Panasyuk, T., Nigmatullin, R., Piletsky, S., Maltceva, T., & Bryk, M. (1999). Polyvinylchloride membranes in immunosensor design. *Colloids and Surfaces A: Physicochemical and Engineering Aspects*, *149*(1–3), 539–545. Available from https://doi.org/10.1016/S0927-7757(98)00692-X, http://www.elsevier.com/locate/colsurfa.

Phonklam, K., Wannapob, R., Sriwimol, W., Thavarungkul, P., & Phairatana, T. (2020). A novel molecularly imprinted polymer PMB/MWCNTs sensor for highly-sensitive cardiac troponin T detection. *Sensors and Actuators B: Chemical*, *308*, 127630. Available from https://doi.org/10.1016/j.snb.2019.127630.

Pinelli, F., Magagnin, L., & Rossi, F. (2020). Progress in hydrogels for sensing applications: A review. *Materials Today Chemistry*, *17*, 100317. Available from https://doi.org/10.1016/j.mtchem.2020.100317.

Pinto, M. R., & Schanze, K. S. (2004). Amplified fluorescence sensing of protease activity with conjugated polyelectrolytes. *Proceedings of the National Academy of Sciences of the United States of America*, *101*(20), 7505–7510. Available from https://doi.org/10.1073/pnas.0402280101.

Rachkov, A., & Minoura, N. (2000). Recognition of oxytocin and oxytocin-related peptides in aqueous media using a molecularly imprinted polymer synthesized by the epitope approach. *Journal of Chromatography. A*, *889*(1–2), 111–118. Available from https://doi.org/10.1016/S0021-9673(00)00568-9.

Rachkov, A., & Minoura, N. (2001). Towards molecularly imprinted polymers selective to peptides and proteins. The epitope approach. *Biochimica et Biophysica Acta – Protein Structure and Molecular Enzymology*, *1544*(1–2), 255–266. Available from https://doi.org/10.1016/S0167-4838(00)00226-0.

Randriantsilefisoa, R., Cuellar-Camacho, J. L., Chowdhury, M. S., Dey, P., Schedler, U., & Haag, R. (2019). Highly sensitive detection of antibodies in a soft bioactive three-dimensional bioorthogonal hydrogel. *Journal of Materials Chemistry B*, *7*(20), 3220–3231. Available from https://doi.org/10.1039/c9tb00234k, http://pubs.rsc.org/en/journals/journal/tb.

Rininsland, F., Xia, W., Wittenburg, S., Shi, X., Stankewicz, C., Achyuthan, K., McBranch, D., & Whitten, D. (2004). Metal ion-mediated polymer superquenching for highly sensitive detection of kinase and phosphatase activities. *Proceedings of the National Academy of Sciences of the United States of America*, *101*(43), 15295–15300. Available from https://doi.org/10.1073/pnas.0406832101.

Sadat Ebrahimi, M. M., Voss, Y., & Schönherr, H. (2015). Rapid detection of Escherichia coli via enzymatically triggered reactions in self-reporting chitosan hydrogels. *ACS Applied Materials and Interfaces*, *7*(36), 20190–20199. Available from https://doi.org/10.1021/acsami.5b05746, http://pubs.acs.org/journal/aamick.

Sandanaraj, B. S., Demont, R., Aathimanikandan, S. V., Savariar, E. N., & Thayumanavan, S. (2006). Selective sensing of metalloproteins from nonselective binding using a fluorogenic amphiphilic polymer.

Journal of the American Chemical Society, *128*(33), 10686–10687. Available from https://doi.org/10.1021/ja063544v.

Sarkar, P. (2000). One-step separation-free amperometric biosensor for the detection of protein. *Microchemical Journal*, *64*(3), 283–290. Available from https://doi.org/10.1016/S0026-265X(00)00005-9.

Sellergren, B. (1997). Noncovalent molecular imprinting: Antibody-like molecular recognition in polymeric network materials. *TrAC – Trends in Analytical Chemistry*, *16*(6), 310–320. Available from https://doi.org/10.1016/S0165-9936(97)00027-7, http://www.elsevier.com/locate/trac.

Sener, G., Ozgur, E., Yilmaz, E., Uzun, L., Say, R., & Denizli, A. (2010). Quartz crystal microbalance based nanosensor for lysozyme detection with lysozyme imprinted nanoparticles. *Biosensors and Bioelectronics*, *26*(2), 815–821. Available from https://doi.org/10.1016/j.bios.2010.06.003.

Shohatee, D., Keifer, J., Schimmel, N., Mohanty, S., & Ghosh, G. (2018). Hydrogel-based suspension array for biomarker detection using horseradish peroxidase-mediated silver precipitation. *Analytica Chimica Acta*, *999*, 132–138. Available from https://doi.org/10.1016/j.aca.2017.10.033, http://www.journals.elsevier.com/analytica-chimica-acta/.

Tai, D. F., Jhang, M. H., Chen, G. Y., Wang, S. C., Lu, K. H., Lee, Y. D., & Liu, H. T. (2010). Epitope-cavities generated by molecularly imprinted films measure the coincident response to anthrax protective antigen and its segments. *Analytical Chemistry*, *82*(6), 2290–2293. Available from https://doi.org/10.1021/ac9024158.

Takeda, K., & Kobayashi, T. (2005). Bisphenol A imprinted polymer adsorbents with selective recognition and binding characteristics. *Science and Technology of Advanced Materials*, *6*(2), 165–171. Available from https://doi.org/10.1016/j.stam.2004.11.008.

Tang, Y., Zhang, X., Li, X., Ma, C., Chu, X., Wang, L., & Xu, W. (2022). A review on recent advances of Protein-Polymer hydrogels. *European Polymer Journal*, *162*, 110881. Available from https://doi.org/10.1016/j.eurpolymj.2021.110881.

Teixeira, S. P. B., Reis, R. L., Peppas, N. A., Gomes, M. E., & Domingues, R. M. A. (2021). Epitope-imprinted polymers: Design principles of synthetic binding partners for natural biomacromolecules. *Science Advances*, *7*(44). Available from https://doi.org/10.1126/sciadv.abi9884, https://www.science.org/doi/10.1126/sciadv.abi9884.

Tong, D., Hetényi, C., Bikádi, Z., Gao, J. P., & Hjertén, S. (2001). Some studies of the chromatographic properties of gels (\artificial antibodies/receptors\) for selective adsorption of proteins. *Chromatographia*, *54*(1–2), 7–14. Available from https://doi.org/10.1007/BF02491825, http://www.chromatographia.de/.

Tsai, W.-C., & Lin, I.-C. (2005). Development of a piezoelectric immunosensor for the detection of alpha-fetoprotein. *Sensors and Actuators B: Chemical*, *106*(1), 455–460. Available from https://doi.org/10.1016/j.snb.2004.09.007.

Viswanathan, S., Rani, C., Ribeiro, S., & Delerue-Matos, C. (2012). Molecular imprinted nanoelectrodes for ultra sensitive detection of ovarian cancer marker. *Biosensors and Bioelectronics*, *33*(1), 179–183. Available from https://doi.org/10.1016/j.bios.2011.12.049.

Wang, N., Feng, Y., Wang, Y., Ju, H., & Yan, F. (2018). Electrochemiluminescent imaging for multi-immunoassay sensitized by dual DNA amplification of polymer dot signal. *Analytical Chemistry*, *90*(12), 7708–7714. Available from https://doi.org/10.1021/acs.analchem.8b01610, http://pubs.acs.org/journal/ancham.

Wang, S., Ye, J., Bie, Z., & Liu, Z. (2014). Affinity-tunable specific recognition of glycoproteins via boronate affinity-based controllable oriented surface imprinting. *Chemical Science*, *5*(3), 1135–1140. Available from https://doi.org/10.1039/c3sc52986j, http://pubs.rsc.org/en/Journals/JournalIssues/SC.

Wang, Y., Zhou, Y., Sokolov, J., Rigas, B., Levon, K., & Rafailovich, M. (2008). A potentiometric protein sensor built with surface molecular imprinting method. *Biosensors and Bioelectronics*, *24*(1), 162–166. Available from https://doi.org/10.1016/j.bios.2008.04.010.

West, J. L., & Hubbell, J. A. (1999). Polymeric biomaterials with degradation sites for proteases involved in cell migration. *Macromolecules*, *32*(1), 241–244. Available from https://doi.org/10.1021/ma981296k.

Wu, R., Ge, H., Liu, C., Zhang, S., Hao, L., Zhang, Q., Song, J., Tian, G., & Lv, J. (2019). A novel thermometer-type hydrogel senor for glutathione detection. *Talanta*, *196*, 191–196. Available from https://doi.org/10.1016/j.talanta.2018.12.020, https://www.journals.elsevier.com/talanta.

Wulff, G., & Knorr, K. (2001). Stoichiometric noncovalent interaction in molecular imprinting. *Bioseparation*, *10*(6), 257–276. Available from https://doi.org/10.1023/A:1021585518592.

Xu, J., Ambrosini, S., Tamahkar, E., Rossi, C., Haupt, K., & Tse Sum Bui, B. (2016). Toward a universal method for preparing molecularly imprinted polymer nanoparticles with antibody-like affinity for proteins. *Biomacromolecules*, *17*(1), 345–353. Available from https://doi.org/10.1021/acs.biomac.5b01454, http://pubs.acs.org/journal/bomaf6.

Xu, J., Merlier, F., Avalle, B., Vieillard, V., Debré, P., Haupt, K., & Tse Sum Bui, B. (2019). Molecularly imprinted polymer nanoparticles as potential synthetic antibodies for immunoprotection against HIV. *ACS Applied Materials and Interfaces*, *11*(10), 9824–9831. Available from https://doi.org/10.1021/acsami.8b22732, http://pubs.acs.org/journal/aamick.

Yang, W., Zhang, G., Weng, W., Qiu, B., Guo, L., Lin, Z., & Chen, G. (2014). Signal on fluorescence biosensor for MMP-2 based on FRET between semiconducting polymer dots and a metal organic framework. *RSC Advances*, *4*(102), 58852–58857. Available from https://doi.org/10.1039/C4RA12478B.

Yu, Y., Zhang, Q., Buscaglia, J., Chang, C. C., Liu, Y., Yang, Z., Guo, Y., Wang, Y., Levon, K., & Rafailovich, M. (2016). Quantitative real-time detection of carcinoembryonic antigen (CEA) from pancreatic cyst fluid using 3-D surface molecular imprinting. *Analyst*, *141*(14), 4424–4431. Available from https://doi.org/10.1039/c6an00375c, http://pubs.rsc.org/en/journals/journal/an.

Zhang, G., Jiang, L., Zhou, J., Hu, L., & Feng, S. (2019). Epitope-imprinted mesoporous silica nanoparticles for specific recognition of tyrosine phosphorylation. *Chemical Communications*, *55*(67), 9927–9930. Available from https://doi.org/10.1039/c9cc03950c, http://pubs.rsc.org/en/journals/journal/cc.

Zhao, C. J., Ma, X. H., & Li, J. P. (2017). An insulin molecularly imprinted electrochemical sensor based on epitope imprinting. *Chinese Journal of Analytical Chemistry*, *45*(9), 1360–1366. Available from https://doi.org/10.1016/S1872-2040(17)61039-9, https://www.journals.elsevier.com/chinese-journal-of-analytical-chemistry.

Zhou, L., Lv, F., Liu, L., & Wang, S. (2019). Water-soluble conjugated organic molecules as optical and electrochemical materials for interdisciplinary biological applications. *Accounts of Chemical Research*, *52*(11), 3211–3222. Available from https://doi.org/10.1021/acs.accounts.9b00427, http://pubs.acs.org/journal/achre4.

Zhu, C., Yang, Q., Liu, L., Lv, F., Li, S., Yang, G., & Wang, S. (2011). Multifunctional cationic poly(p-phenylene vinylene) polyelectrolytes for selective recognition, imaging, and killing of bacteria over mammalian cells. *Advanced Materials*, *23*(41), 4805–4810. Available from https://doi.org/10.1002/adma.201102850.

CHAPTER

Detection of neutral species: unveiling new targets of interest

21

Saúl Vallejos[1] and Álvaro Miguel[1,2]

[1]*Department of Chemistry, Universidad de Burgos, Burgos, Spain* [2]*Universidad Autónoma de Madrid, Madrid, Spain*

21.1 Introduction

As defined by García Pérez et al. (2022) a smart polymer generates a response from a stimulus and through a specific mechanism. The sensory polymers are those in which the response is an alert, and therefore, they can be classified through the mechanism, the response, or the stimulus. The main subject of this chapter revolves around a specific type of chemical stimulus, namely neutral species, that is, non-ionic species. Specifically, in this chapter, species or families of chemical species are identified and defined, whose detection is highly relevant in today's society but has not been addressed using sensory polymers, at least not as extensively as the detection of other targets such as heavy metals, nitro-aromatic explosives, etc.

In general, it can be stated that detecting neutral molecules is more complex than detecting charged molecules. Even though this distinction varies depending on specific circumstances, the easiness of anions and cations detection arises from their higher mobility in solution, especially in aqueous solutions. The aquatic medium is the most effective connecting wire on our planet, although pollutants can spread through soil and air as well. The aqueous medium, whether fresh or marine, represents the origin of life, and therefore all animal and plant species inhabiting the earth depend directly on the water. Thus, all harmful substances that are discharged into the aquatic environment end up returning to humans through the intake of water, meat/fish, and vegetables (Fig. 21.1) (Liu et al., 2021). With all that said, monitoring the presence of harmful neutral compounds in water may prevent public health disorders, as it will be tackled in the following sections.

This book has dedicated several chapters to the detection of specific neutral molecules, such as nitroaromatic explosives or volatile organic compounds (VOCs). In this chapter, we will discuss different groups or families of neutral molecules whose detection is of great interest in various fields, including food safety, biomedicine, environmental monitoring, and quality control.

Furthermore, considering the potential readers of this chapter, the molecules or groups of molecules to be discussed are either emerging contaminants or substances of significant interest that may not necessarily be new or emerging but have a significant impact in the aforementioned sectors.

Emerging contaminants are chemical substances which show signs of being harmful to health and/or the environment, but are little studied and not regulated by governments or public health agencies. The laws and regulations associated with substances are developed as more information

Sensory Polymers. DOI: https://doi.org/10.1016/B978-0-443-13394-7.00021-5

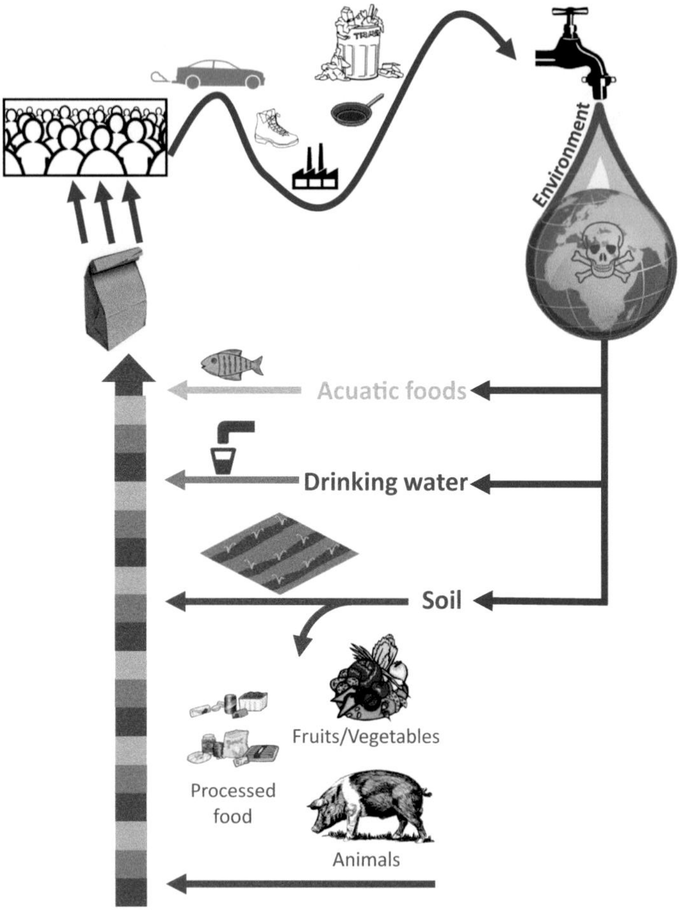

FIGURE 21.1

Graphical abstract of the cycle of emerging pollutants in the world today. The cycle shows how water incorporates pullutants produced by human activity to the food chain.

becomes available. Consequently, in the meantime, these emerging contaminants continue to be used in product formulations or different industrial treatments waiting for restrictive legislation. As a consequence, in the meantime, these emerging contaminants continue to be used in product formulations or in different industrial treatments waiting for restrictive legislation. Therefore, the definition of emerging pollutants is constant, but the list of emerging pollutants varies depending on where they are regulated and the historical moment in which they are published (Kantiani et al., 2010; Liu et al., 2021; Matei et al., 2022; Salthammer, 2020; Thomaidis et al., 2012). Nowadays, the most significant emerging water and food contaminants are shown in Table 21.1, but these are just some examples that are not intended to be limiting.

Table 21.1 Families of emerging pollutants.

Family	Examples	Uses
POPs Persistent organic pollutants (Liu et al., 2021; Nobile et al., 2020)	Polybrominated diphenyl ethers (**PBDEs**)	Flame retardant, plastics, textiles, electronic castings, circuitry.
	Hexabromocyclododecanes (**HBCDDs**)	Flame retardant, thermal insulation in the building industry.
	Polybrominated biphenyls (**PBBs**) and polychlorinated biphenyls (**PCBs**)	Flame retardant, consumer appliances, textiles, plastic foams.
	Tetrabromobisphenol A (**TBBPA**) and other phenols	Flame retardant, printed circuit boards, thermoplastics (mainly in TVs)
	Organochlorine contaminants (**OCs**)	Pesticides
	Polycyclic aromatic hydrocarbons (**PAHs**)	Produced oil deposits and by the combustion of organic matter
	PFCs Perfluorocarbons (Kantiani et al., 2010; Liu et al., 2021) • Perfluorooctanesulfonic acid (**PFOS**) • Perfluorinated carboxylic acids (**PFCAs**), as Perfluorooctanoic acid (**PFOA**) • Fluorotelomer alcohols (**PFTOHs**) • Perfluorinated sulfonamides (**PFASAs**) • Perfluorinated sulfonamide ethanols (**PFASEs**)	Industrial organic pollutants. stain repellents, textiles, paints, waxes, polishes, electronics, adhesives, omniphobic coatings and food packaging
PPCPs Pharmaceutical and personal care products (Kantiani et al., 2010)	Parabens	Personal care products
	Antibiotics	Pharmaceutical use
	Coccidiostats	Pharmaceutical use
Biotoxins	Marine biotoxins (palitoxins, spirolides) (Kantiani et al., 2010)	Produced by natural marine phytoplankton
	Mycotoxins (Yuan et al., 2021; Braun et al., 2021; Crudo et al., 2021)	Produced by various fungal species
Agricultural chemicals	Nonpolar pesticides (e.g., pyrethroids) (Llorca et al., 2017)	Insecticide, malaria control
MPs & NPs Microplastics and nanoplastics	Polyethylene, polypropylene, polyvinylchloride, polyethyelen terephthalate, polystyrene,	Packaging

It is worth noting that for many of these emerging contaminants, there is a scarcity of documented cases utilizing sensory polymers for their detection. In fact, in some cases, no existing examples have been found to date.

Therefore, the main objective of this chapter is to complement this knowledge gap and provide innovative insights for detecting emerging contaminants using sensory polymers. Various approaches and strategies for designing and developing smart polymers capable of selectively recognizing and quantifying these neutral contaminant molecules will be explored. Additionally, the current challenges and limitations in this field will be discussed, along with future perspectives for research and development of more efficient and sensitive sensory polymers.

Through this chapter, we aim to provide scientists, engineers, and professionals in the field of environmental detection and monitoring with a solid foundation and inspiration to address the challenge of detecting these emerging contaminants. By advancing our understanding of the interactions between polymers and neutral contaminant molecules, we can open up new possibilities for enhancing the accuracy, sensitivity, and selectivity of sensory polymers, thus significantly contributing to the protection and preservation of our natural environment.

The first part of the upcoming sections is devoted to the identification and definition of the targets, while the second part focuses on the most relevant detection strategies utilizing sensory polymers that have been published to date.

21.2 Persistent organic pollutants

21.2.1 Definition and contextualization

Persistent Organic Pollutants (POPs) are a class of highly toxic chemical compounds that encompass different families, from organochlorine pesticides such as DDT, industrial chemicals such as polychlorinated biphenyls (PCBs), and unintentional by-products of industrial activities like dioxins and furans. Historically, many of these compounds found extensive use in industrial, agricultural, and domestic applications. However, due to their persistence and inherent toxicity, their production and usage have been heavily regulated, with several bans in place (Devi, 2020; Fu et al., 2003; Kanan & Samara, 2018).

Their toxicity arises from their resistance to biological degradation, their propensity to accumulate in living organisms' tissues, and their ability to undergo long-range transport through air and water. These substances pose significant threats to human health and the environment, qualifying them as environmental contaminants of grave concern (Liu et al., 2021; Nobile et al., 2020).

The health impacts of POPs on human beings can be severe, manifesting as reproductive disorders, hormonal disruptions, immunological impairments, and heightened cancer risks. Furthermore, POPs exert detrimental effects on ecosystems, causing harm to wildlife and disrupting food chains (Crain et al., 2008; Gascon et al., 2013; Yilmaz et al., 2019).

To address the issue of POPs, the Stockholm Convention on Persistent Organic Pollutants was established as an international treaty. Its primary objective is to eliminate or restrict the production, use, and release of POPs into the environment. By doing so, the convention aims to safeguard human health and protect the integrity of the natural world. Additionally, the agreement promotes the development and adoption of safer and more sustainable alternatives to these hazardous substances (Bilcke, 2001; Karlaganis et al., 2008).

In this section, 7 different POPs will be analyzed from the detection and quantification point of view using sensory polymers, polybrominated diphenyl ethers (PBDEs), hexabromocyclododecanes

(HBCDDs), polybrominated biphenyls (PBBs) / polychlorinated biphenyls (PCBs), tetrabromobisphenol A (TBBPA) and other phenols, organochlorine contaminants (OCs), polycyclic aromatic hydrocarbons (PAHs) and perfluorocarbons (PFCs).

Polybrominated diphenyl ethers (PBDEs) are a group of persistent organic pollutants (POPs) widely utilized as flame retardants. These compounds consist of bromine atoms attached to a diphenyl ether backbone. PBDEs are commonly found in consumer goods such as furniture, electronics, textiles, and foam insulation materials. Their primary purpose is to enhance fire resistance and prevent rapid fire spread. Nonetheless, PBDEs exhibit persistence and bioaccumulation in the environment. They have been detected globally in the atmosphere, oceans, land, and living organisms. Prolonged exposure to PBDEs has been linked to developmental neurotoxicity, endocrine disruption, and potential carcinogenic effects (Hale et al., 2003; Hites, 2004; McDonald, 2002; Ohoro et al., 2021; Schreiber et al., 2010).

Hexabromocyclododecanes (HBCDDs) are another group of flame retardants used in diverse applications like building materials, electronics, and textiles. HBCDDs comprise various isomers, with the most prevalent being γ-HBCDD. These compounds persist in the environment and have been identified in both indoor and outdoor air, water, sediment, and biota. HBCDDs have been associated with adverse effects on human health, including disruption of thyroid hormones and developmental neurotoxicity (Al-Omran et al., 2022; Covaci et al., 2006; Morel et al., 2022).

Polybrominated biphenyls (PBBs) and polychlorinated biphenyls (PCBs) are two related classes of POPs that have been extensively employed in various industrial sectors. PBBs were primarily used as flame retardants, while PCBs served as coolants and insulating fluids in electrical equipment. Both PBBs and PCBs exhibit high persistence and can bioaccumulate in the environment. Despite restrictions and prohibitions in many countries, these substances persist in the atmosphere, water bodies, soil, and living organisms. PCBs and certain PBBs are recognized as toxic, exerting adverse effects on the immune system, reproductive system, and neurological development, and are classified as probable human carcinogens (Crisp et al., 1998; Jacobson et al., 1989; Kimbrough, 1987; Safe & Hutzinger, 1984).

Tetrabromobisphenol A (TBBPA) and other phenolic compounds are flame retardants utilized in various applications, including electronic devices, electrical equipment, and building materials. TBBPA is among the most widely used flame retardants worldwide. Phenolic compounds, including TBBPA, can leach out from products and contaminate the environment. While TBBPA has a relatively short environmental half-life, it can still accumulate in specific ecosystems. The health effects of TBBPA and other phenolic flame retardants are still being investigated, but studies suggest potential endocrine disruption (Covaci et al., 2009; Crisp et al., 1998; Kitamura et al., 2002).

Organochlorine contaminants (OCs) encompass a group of persistent organic pollutants containing chlorine atoms in their molecular structure. These compounds were extensively used in the past for purposes such as pesticides, industrial chemicals, and solvents. Examples of OCs include dichlorodiphenyltrichloroethane (DDT), hexachlorobenzene (HCB), and chlordane. OCs exhibit high persistence in the environment and can bioaccumulate in the food chain. Similar to PBDEs, they have been identified on a global scale in the atmosphere, water bodies, soil, and living organisms, even after restrictions and bans on their production and use. OCs are known for their toxic effects on humans and wildlife, with long-term exposure associated with endocrine disruption, neurotoxicity, immune system disorders, and certain cancers (Ajiboye et al., 2020; Helou et al., 2019; Martyniuk et al., 2020; Qi et al., 2022).

Polycyclic aromatic hydrocarbons (PAHs) constitute a group of organic compounds composed of fused aromatic rings. They are formed during the incomplete combustion of organic materials, such as fossil fuels, wood, and tobacco. PAHs can be released into the environment through industrial processes, vehicle emissions, and natural sources like forest fires. Some common PAHs include naphthalene, benzo[a]pyrene, and anthracene. PAHs persist in the environment and can be found in air, soil, water, and sediments. They can also adhere to particulate matter, making inhalation a significant exposure route for humans. PAHs are known carcinogens, with certain compounds posing a higher risk than others. Additionally, PAHs have been linked to developmental and reproductive toxicity, as well as adverse effects on aquatic organisms (Guo, 2021; Patel et al., 2020; Ye et al., 2022).

Perfluorocarbons (PFCs) are a class of synthetic organic compounds containing carbon atoms bonded to fluorine atoms. These compounds are characterized by their strong carbon-fluorine bonds, which make them highly stable and resistant to degradation. PFCs have been extensively used in industrial and consumer applications, including stain-resistant coatings, non-stick cookware, waterproof textiles, and fire-fighting foams. Examples of PFCs include perfluorooctanoic acid (PFOA) and perfluorooctane sulfonic acid (PFOS). PFCs persist in the environment and can accumulate in living organisms, including humans. They have been detected worldwide in the atmosphere, water, soil, and biota. PFCs raise concerns due to their potential adverse health effects, including liver toxicity, reproductive and developmental effects, and possible disruption of the immune system (Cennamo et al., 2019; Fenton et al., 2021; Gong et al., 2015; Schrenk et al., 2020).

21.2.2 Detection with sensory polymers

21.2.2.1 Polycyclic aromatic hydrocarbons

When discussing Molecularly Imprinted Polymers (MIPs) as smart polymers, the possibilities are nearly infinite, and various authors have been sharpening their ingenuity to harness the potential of these magnificent materials for the detection of PAHs. Essentially, a MIP is prepared by dissolving the template molecule (target), a cross-linking monomer (vinyl or acrylic, typically EGDMA or divinylbenzene), a functional monomer (MMA, 4-VP, etc.), and a porogen (a solvent with low dielectric constant, such as ACN, toluene, DCM, etc.) in the same solution. Different authors have fabricated MIPs in which the porogen and the template were the same species, for example, toluene (Egli, 2014; Egli et al., 2015; Xue et al., 2021). PAHs are molecules with multiple aromatic rings, such as naphthalene, anthracene, and pyrene. These species are chemically similar to toluene, which is why it was chosen as the template model. In the thesis published by Egli, (Egli, 2014) the MIPs were combined with gas chromatography-mass spectrometry (GC-Ms) in selected ion monitoring (SIM) mode as an offline sensor to characterize the absorption properties of these materials. The response of the MIPs was linear in the range of 0.1 to 120.0 μg/L and demonstrated selectivity towards PAHs in the presence of structurally or hydrophobically similar compounds, such as octane, octanol, and p-cresol. The calculated detection limits were 0.235 to 258 ng/L, while the recovery percentages ranged from 6.2% to 66.9% for naphthalene and phenanthrene, respectively.

21.2.2.2 Perfluorocarbons

Setting aside the works focusing on the detection of PFCs using MIP-type sensory polymers, (Lu et al., 2022) recent ideas have recently emerged that open up new possibilities for researchers in this field. An example of this is the approach proposed by Breshears et al., based on a microfluidic paper chip (Breshears et al., 2023). This simple device takes advantage of the competitive interactions between PFOA/PFOS, cellulose fibres, and various reagents (L-lysine, casein, and albumin). These interactions can alter the surface tension at the wetting front and, subsequently, the capillary flow rate (Fig. 21.2, left). Furthermore, this process can be recorded using a smartphone by capturing videos of the capillary action occurring in the chip, which contains a channel through which the samples flow in less than two minutes (Fig. 21.2, right). Considering the different reagents used, albumin yielded the best results, followed by casein, achieving detection limits of 10 g/μL in DI water and 1 fg/μL in effluent wastewater (processed).

One of the concerns that always surrounds sensory polymers is specificity, which is why the sensory system was tested with other non-fluorocarbon surfactants such as anionic sodium dodecyl

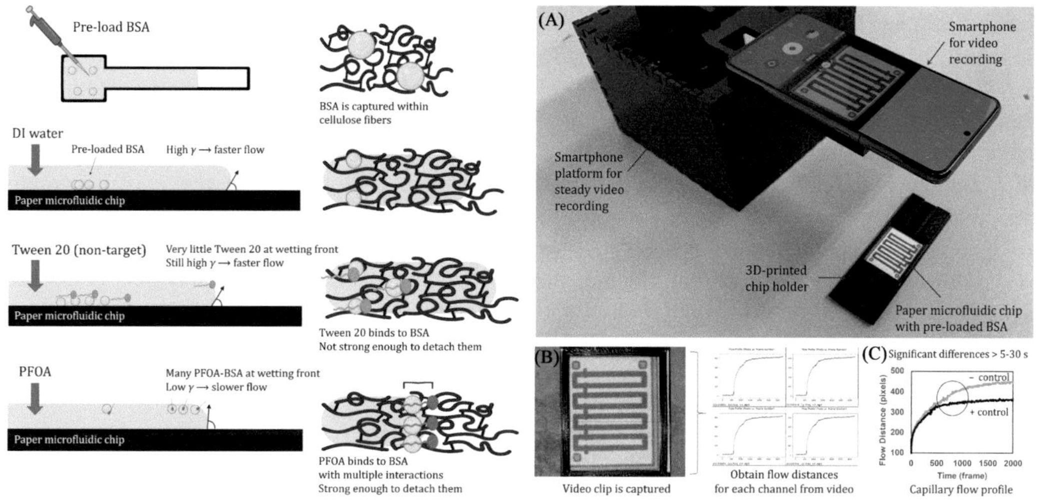

FIGURE 21.2

(Left) Schematic representation of how the molecular interactions between the reagent (e.g., BSA), paper substrate (cellulose fibers), and target molecule (e.g., PFOA or Tween 20) can affect the interfacial tension and subsequently the capillary flow rate. **(Right)** Experimental setup and procedure. (A) A smartphone is placed steadily on a plastic box to record the video of liquid flow through the paper microfluidic channels. The four-channel paper microfluidic chip is placed on a 3D-printed chip holder. 3 μL of reagent (BSA, casein, or L-lysine) solution is loaded right after the inlets and dried. (B) 3 μL of the sample solution is then loaded into each channel, and the video clip is recorded. The video clip is uploaded to Google Drive from a smartphone and automatically analyzed using a custom Python code in Google Colab. (C) A raw flow rate profile is collected for each channel. The earliest time to make a significant distinction between positive and negative controls is determined, which varied from 5 s to 30 s.

Source*: Adapted with permission from Breshears, L. E., Mata-Robles, S., Tang, Y., Baker, J. C., Reynolds, K. A., Yoon, J., & Rapid, Y. (2023). Sensitive detection of PFOA with smartphone-based flow rate analysis utilizing competitive molecular interactions during capillary action.* Journal of Hazardous Materials, 446, *130699. https://doi.org/10.1016/J.JHAZMAT.2022.130699. Copyright © Elsevier 2023 Schematic representation and experimental setup of the PFOA/PFOS paper chip sensor (Breshears et al., 2023).*

sulfate (SDS), non-ionic Tween 20, and cationic cetyltrimethylammonium bromide (CTAB). The sensory system successfully distinguished PFOA from the three surfactants with 100% accuracy. This system can be seen as a combination of two polymers (one protein and cellulose) for the preparation of a highly simple and effective sensory system.

Another strategy for detecting PFCs is based on displacement sensors, or IDAs, which can be easily implemented in a polymeric sensor. For instance, Gou et al. have developed a cationic siloxane that acts as a quencher for Erythrosin B (EB) (Gou et al., 2022), a dye that also exhibits fluorescence (Fig. 21.3). When this duo interacts with PFOS, the EB dye is released due to the strong attraction between PFOS and the cationic siloxane, resulting in a fluorescent signal that can be easily measured.

21.2.2.3 Others

To date, no bibliographic references have been found utilizing sensory polymers for the detection of hexabromocyclododecanes, polybrominated diphenyl ethers, or organochlorine contaminants. This lack of information indicates a compelling challenge for smart polymers in this field. However, MIP-based systems have been proposed for the detection of other persistent organic pollutants (POPs) such as polybrominated biphenyls, (Chen et al., 2022; Nzangya et al., 2021) polychlorinated biphenyls, (Guo et al., 2020) and tetrabromobisphenol A, (Feng et al., 2019; Wu et al., 2019; Zeng et al., 2020) employing similar strategies to those discussed in the previous paragraphs for other POPs.

The development of innovative approaches for the detection of hexabromocyclododecanes, polybrominated diphenyl ethers, and organochlorine contaminants using smart polymers is a motivating challenge. A myriad of exploration opportunities pops up that will contribute to the monitoring and control of these harmful compounds in various environments in the future.

21.3 Pharmaceutical and personal care products

21.3.1 Definition and contextualization

Pharmaceutical and personal care products (PPCPs) are a diverse group of chemical substances that include prescription and over-the-counter drugs, as well as personal care products used for hygiene, beauty, and grooming purposes. At the same time, PPCPs encompass a wide range of items commonly found in households, such as medications, cosmetics, fragrances, sunscreens, shampoos, soaps, and lotions (Kantiani et al., 2010; Nobile et al., 2020).

These products are designed to have specific physiological or cosmetic effects on the human body. Pharmaceuticals are intended for medical purposes, such as treating diseases, alleviating symptoms, or preventing illnesses. Personal care products, on the other hand, are primarily used for maintaining personal hygiene, enhancing appearance, or promoting well-being.

PPCPs are typically designed to be biologically active and effective at low concentrations. When people use these products, a certain amount cannot be absorbed by the body, and it is released to the environment through various routes, such as flushing down the toilet or washing personal care products down the drain. As a result, PPCPs can enter wastewater treatment systems or directly contaminate water bodies, soil, and air.

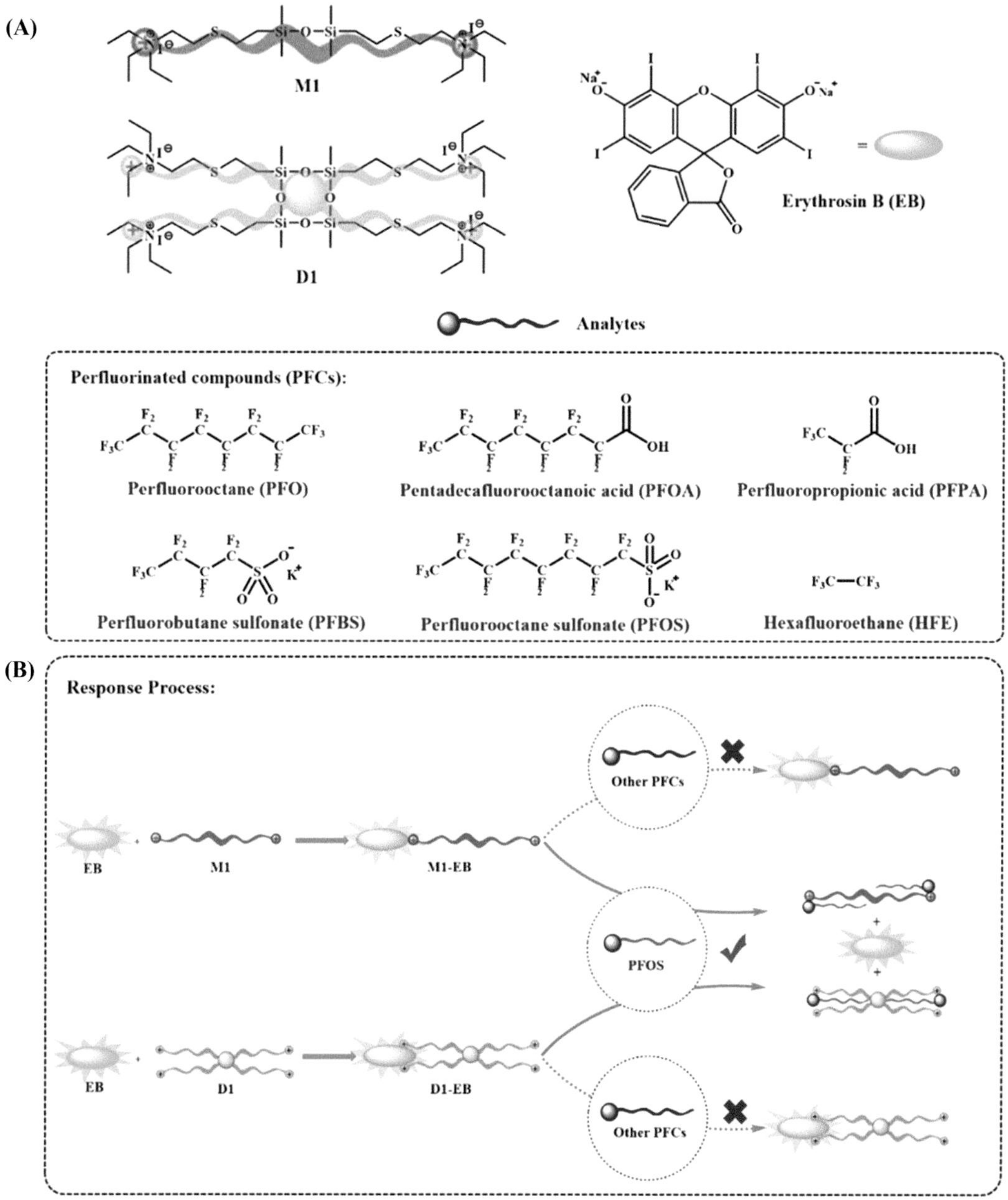

FIGURE 21.3

(A) Molecular structures of M1, D1, erythrosine B (EB), and PFCs analytes. (B) Schematic illustration of EB-siloxane system for PFCS detection.

Source: *Adapted with permission from Gou, Z., Wang, A., Zhang, X., Zuo, Y., & Lin, W. (2022). Multi-head cationic siloxane based "turn on" fluorescent system for selective detection of perfluorooctanoic sulfonate (PFOS).* Sensors Actuators B Chemical, 367, *132017. https://doi.org/10.1016/J.SNB.2022.132017. Copyright © Elsevier 2022 Molecular structures and schematic operation procedure of the EB-siloxane sensor for PFCS (Gou et al., 2022).*

Due to the widespread use and disposal of PPCPs, concerns have arisen regarding their potential impacts on the environment and human health. Some PPCPs have been detected in water sources, including rivers, lakes, and groundwater, as well as in soil and aquatic organisms. There is growing interest in understanding the fate, behaviour, and potential risks associated with PPCPs in the environment, as well as developing strategies for their proper management and mitigation.

It is important to note that the presence of PPCPs in the environment does not necessarily imply immediate harm or risk to human health. However, the continuous release and accumulation of these compounds, combined with their potential to interact with ecosystems and organisms, have prompted studies and regulations to ensure their safe use and minimize potential environmental and health impacts.

The most concerning contaminants from cosmetics are Parabens. Since some of them have been studied, and their toxicity has been proven, the industry has looked for less known substitutes such as propyl-paraben, whose use could be demonstrated to be equally risky. Estrogenic activity of parabens is one of their main hazards. They bind to estrogen receptors which could result in bioaccumulation in human breast tissues, and the development of cancer (Harvey & Darbre, 2004).

Another group of greatest concern is antibiotics, given the resistance that certain microorganisms can develop to these drugs. Within the antibiotics, several families can be differentiated, such as sulfonamides, fluoroquinolones, nitroimidazoles, penicillin, cephalosporins, tetracyclines, nitrofurans, and macrolides. In addition, there are other types of drugs that are also receiving much attention, such as salicylic acid, (Nobile et al., 2020) which causes adverse reactions to intolerants.

Finally, coccidiostats are another subgroup of pharmaceuticals and personal care products analyzed in this chapter. They are chemical compounds used to prevent and treat infections caused by parasites of the genus coccidia, which are responsible for animal diseases, especially poultry, and livestock. Exposure to coccidiostats in the environment can negatively impact aquatic organisms and soil microbiota. These compounds can be toxic to certain non-target organisms and disrupt natural ecosystems' biological processes. In addition, there is concern related to antimicrobial resistance since frequent and prolonged use of coccidiostats in animal production may contribute to the development of resistance in coccidia parasites.

21.3.2 Detection with sensory polymers

21.3.2.1 Parabens

Microporous organic networks (MONs) are a type of polymers which have been used for the detection of a specific type of PPCPs, parabens (Han et al., 2021). Chemically speaking, parabens are esters of p-hydroxybenzoic acid. The basic structure of parabens (Fig. 21.4) consists of a benzene ring substituted with a hydroxyl group (-OH) and an ester group in para (p) position.

One of the strategies for the detection of this type of compound is the extraction with a smart polymer (MONs, MIPs, polymeric ionic liquids, etc.), and the subsequent characterization with a complementary technique, such as HPLC (Han et al., 2021; López-Darias et al., 2010; Núñez et al., 2010). Fig. 21.5 graphically depicts the procedure using a microporous organic network, namely MON.

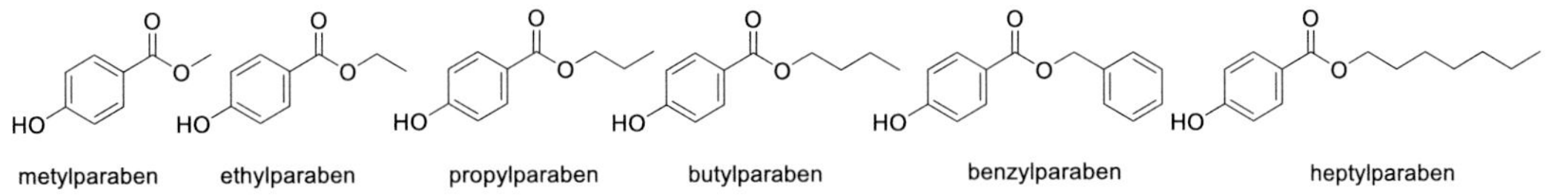

FIGURE 21.4 Most commonly used parabens in cosmetics.

Six different paraben molecules, commonly used in cosmetics.

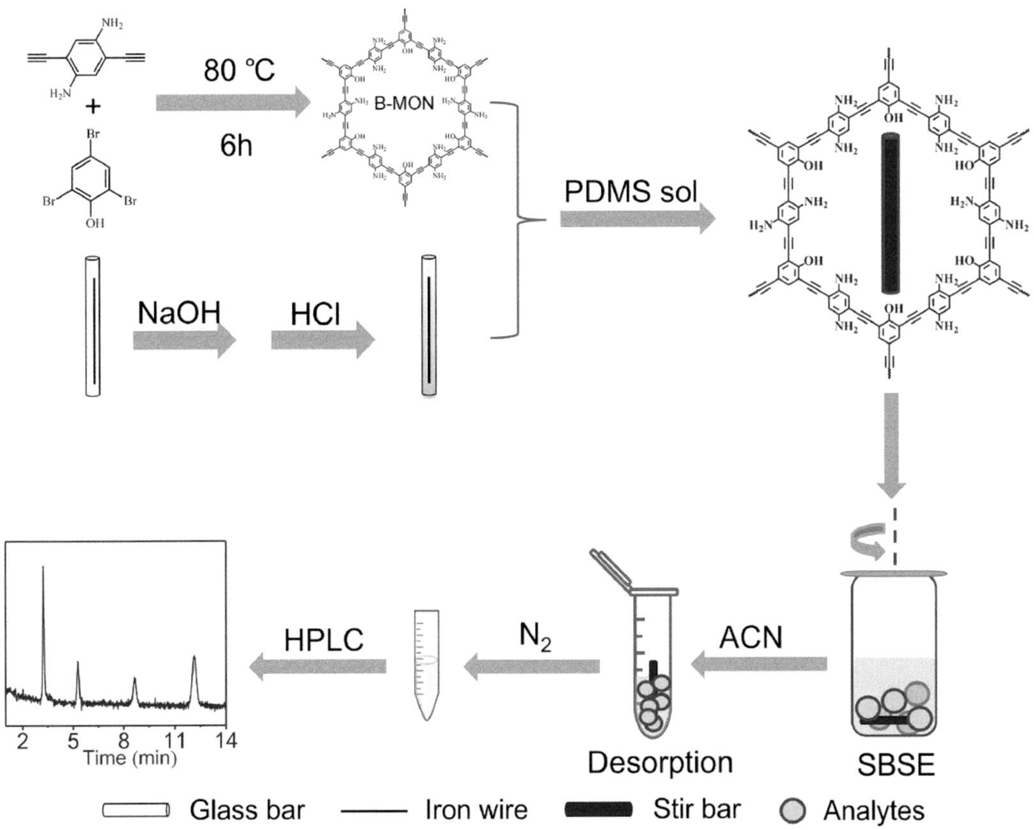

FIGURE 21.5

Illustration for the fabrication of B-MON stir bar in SBSE.

Source: *Adapted with persmission from Han, J. H., Cui, Y. Y., & Yang, C. X. (2021). Tailored amino/hydroxyl bifunctional microporous organic network for efficient stir bar sorptive extraction of parabens and flavors from cosmetic and food samples.* Journal of Chromatography, 1655, *462521. https://doi.org/10.1016/J.CHROMA.2021.462521. Copyright © Elsevier 2021 Diagram of the synthesis of microporous organic network, the adsorption/desorption process and HPLC quantification of the target molecule (Han et al., 2021).*

The detection procedure using MIPs is similar, although with some distinguishing points. Firstly, a template is synthesized using a paraben as a "molecular mold." This paraben is typically benzylparaben and serves as a template for the detection or extraction of other parabens since the binding sites on the polymer matrix are often the same. Regarding the structural polymer matrix, some of the most commonly used monomers to fabricate MIPs include methyl methacrylate, (2-hydroxyethyl methacrylate), 4-vinylpyridine, styrene, methacrylic acid, acrylic acid, ethylene glycol dimethacrylate, and N,N'-methylenebisacrylamide (Belbruno, 2019; Haupt et al., 2012). For more information about MIPs, we refer the reader to Chapter 4 of this book.

21.3.2.2 Antibiotics

Antibiotics are undoubtedly another of the PPCPs that have aroused the most interest. Chemically speaking, antibiotics are a diverse class of compounds that exhibit a wide variety of structures. There is no single chemical structure that defines all antibiotics, as they are derived from different classes of organic compounds. This fact complicates the design of sensor polymers that detect in a general way all the different antibiotic families that are depicted in Fig. 21.6:

Among all the species, tetracyclines and nitrofurans have been the chemical structures that have attracted most of the attention of researchers in sensory polymers. Its detection has been proposed with polymers such as polyfluorenes, (Malik & Iyer, 2017) cadmium(II) coordinated polymers, (Fan et al., 2020) or metal-organic frameworks (Li et al., 2021). In most cases, fluorogenic polymers are the most commonly used for the detection of this type of molecule, a class of polymers that exhibit fluorescence properties. These polymers are designed to emit light upon excitation by a specific wavelength of light. They possess fluorophores or chromophores within their molecular structure, which are responsible for the emission of light. Their unique optical properties, such as high quantum yield, tunable emission wavelengths, and high sensitivity, make them valuable tools for chemical and biological sensing, as they can selectively interact with the desired target and produce a measurable fluorescence signal.

These polymers can be synthesized through various methods, including copolymerization, post-polymerization modification, and blending with fluorophores. By incorporating specific functional groups or molecules, the fluorescence emission of the polymers can be tailored to detect specific antibiotics.

21.3.2.3 Coccidiostats

From a structural perspective, coccidiostats can belong to different classes of chemical compounds, such as sulfonamides, quinolones, amides, ionophores, and imidazole derivatives. Each class of coccidiostat possesses a characteristic chemical structure that imparts its biological properties and activity against coccidia. Therefore, the sensory polymers for coccidiostats need to be designed individually for each specific compound. In such a situation, Molecularly Imprinted Polymers (MIPs) always serve as a valid and highly pragmatic choice, as demonstrated in the case of the coccidiostat "clopidol" by preparing a polypyrrole MIP (Radi et al., 2014, 2019).

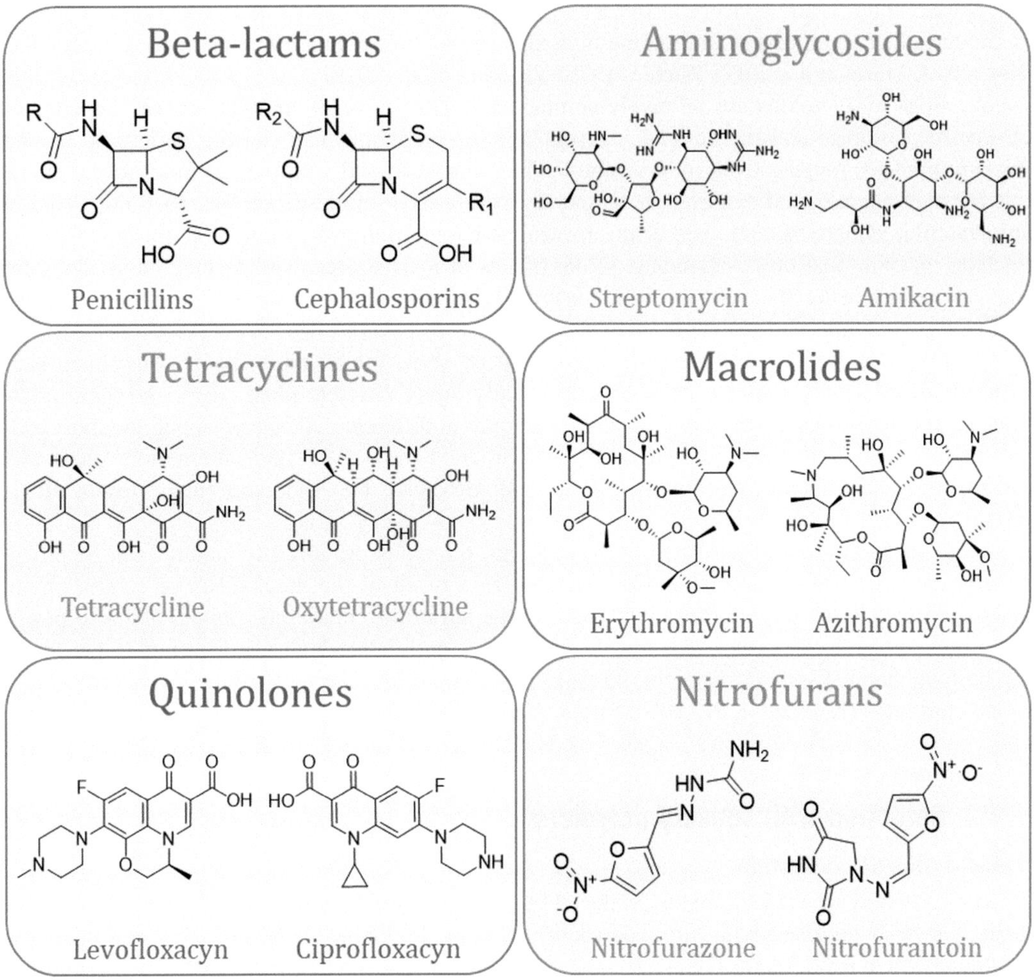

FIGURE 21.6

Families of antibiotics based on their chemical structure, and some examples. Six different antibiotic families, based on their chemical structure, namely betalactams, aminoglycosides, tetracyclines, macrolides, quinolones, and nitrofurans.

21.4 Biotoxins

21.4.1 Definition and contextualization

Biotoxins are toxic compounds produced by living organisms, such as bacteria, fungi, plants, and animals. These toxins can be natural or result from human activity and can be found in various environments, including water, soil, and food. Biotoxins are considered emerging contaminants that have sparked interest and concern in the scientific community. In this section, two types of biotoxins are described: mycotoxins and marine biotoxins (Bennett & Klich, 2003; Gerssen & Gago-Mart Í Nez, 2019; Marin et al., 2013; Ritchie, 1975; Visciano et al., 2016).

Poisonings by biotoxins are relatively common (60,000 per year), and the severity depends on the type of substance and the amount ingested. These compounds can cause a variety of adverse effects in humans, ranging from mild symptoms like diarrhoea and vomiting, to more severe symptoms such as neurological syndromes, temporary paralysis or amnesia, serious gastrointestinal or cardiovascular symptoms, and even death in extreme cases. Moreover, some biotoxins are bioaccumulative, meaning that they accumulate in the tissues of marine organisms as they move along the food chain, increasing the risk for end consumers. In this section, two types of biotoxins are described: mycotoxins and marine biotoxins

Mycotoxins are secondary metabolites produced by some microscopic filamentous fungi, such as Fusarium or Claviceps species. These fungi usually infect different plant crops, and their mycotoxins have been shown to be genotoxic and mutagenic in different cell lines. Among the most relevant mycotoxins, and those of greatest concern right now, we find alternariol, alternariol monomethyl ether and the altertoxins, all of them produced by Alternaria fungi (Braun et al., 2021).

Analogous to the toxins generated by some fungi, biotoxins produced by natural marine phytoplankton are toxic chemical compounds generated as a result of the metabolism of certain phytoplankton species in the ocean. These biotoxins can be harmful to marine organisms and also pose a risk to human health when they accumulate in shellfish and fish, which can incorporate the toxins into the food chain (Espiña & Rubiolo, 2008; Landsberg, 2002). The most studied ones are palitoxins and spirolides (Kantiani et al., 2010; Fire & Van Dolah, 2012).

21.4.2 Detection with sensory polymers

21.4.2.1 Marine biotoxins

There are different types of biotoxins produced by marine phytoplankton, such as paralytic shellfish toxins, diarrheic shellfish toxins, amnesic shellfish toxins, lipophilic shellfish toxins, and ciguatoxins, among others (Fig. 21.7).

There have been few authors who have ventured to design sensory polymers for the detection of marine biotoxins, (González et al., 2019; Jigyasa Rajput, 2022; Sanjuán et al., 2018, 2019) mainly due to the complexity of their chemical structures. A relatively accessible subgroup from the perspective of molecular recognition is the amnesic shellfish toxins, as they are all derivatives of domoic acid. Similar to previously mentioned detection strategies, such as in the case of parabens (Section 21.3), in this case, a generic molecular template of the domoic acid structure can be valid for derived compounds such as its 5' diastereomer or isomers like isodomoic acids A-H.

Jiang et al. have proposed an efficient solution for the detection of domoic acid using a polymer sensor based on polyacrylamide, polydopamine, and graphene oxide. Firstly, polyacrylamide allowed the authors to generate a molecular imprinting template, creating specific cavities where domoic acid interacts through hydrogen bonding. The signal generated by this hydrogen bonding interaction can be amplified using the polydopamine-graphene oxide tandem, as it facilitates charge

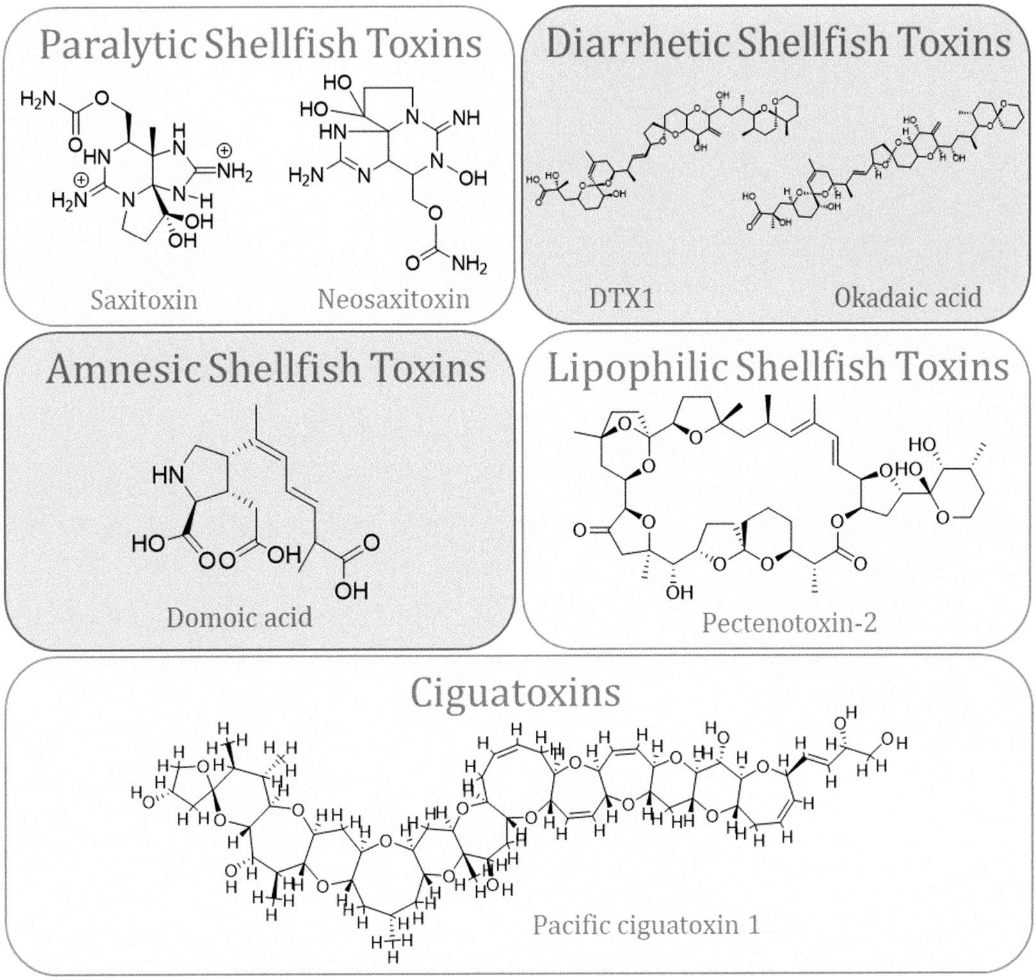

FIGURE 21.7 Families of marine biotoxins, and chemical structures of some examples.

Five marine biotoxins families, namely paralytic, diarrhoetic, amnesic and lipophilic shellfish toxins and ciguatoxins, and some structural examples.

transfer. This entire system was screen-printed onto an electrode, which was capable of achieving detection limits of 0.31 nM (Jiang et al., 2022).

21.4.2.2 Mycotoxins

There are several types of mycotoxins, each produced by different species of moulds. Some of the most well-known mycotoxins include aflatoxins, ochratoxin A, deoxynivalenol (DON), zearalenone, and fumonisins (Fig. 21.8) (Böhm et al., 2010; Soleimany et al., 2012; Solís-Cruz et al., 2017).

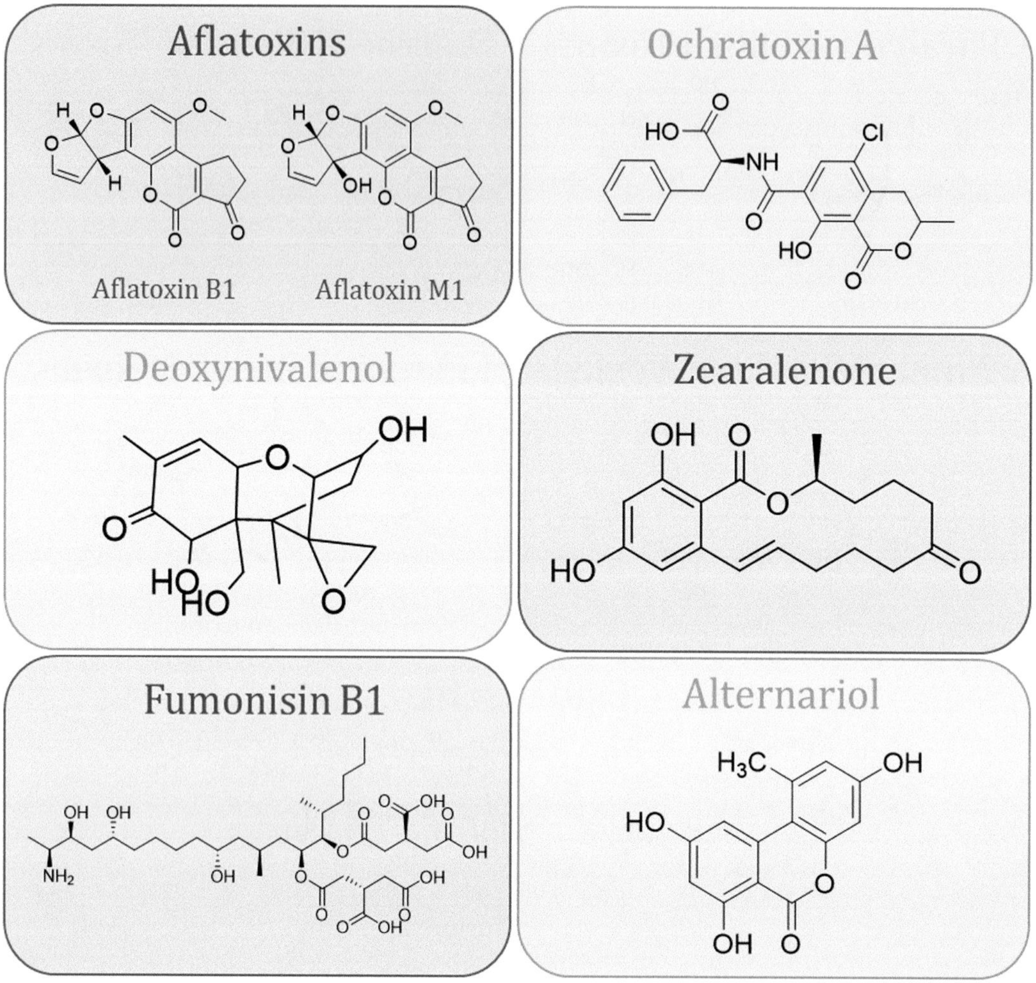

FIGURE 21.8

Families of mycotoxins, and chemical structures of some examples. Five families of mycotoxins, namely aflatoxins, ochratoxin A, deoxynivalenol, zearalenone, fumonisins, and respective structural examples.

Once again, MIPs (molecularly imprinted polymers), or the coupling of MIPs and other types of porous structures are one of the most pragmatic alternatives for the detection of these targets. Yang et al. ventured into the detection of aflatoxin B1 (AFT B1) and sterigmatocystin, using a combination of MIP and HMON (hollow structure microporous organic network). The authors combined the capability of HMONs to generate hydrophobic surfaces when fabricating the MIP, thereby increasing the cavity imprinting density. This material achieved detection limits of 4.4 and 6.7 ng/L for AFT B1 and sterigmatocystin, respectively (Yang et al., 2023).

Beyond MIPs, some authors have utilized aptamers for the detection of such targets. An aptamer is a single-stranded nucleic acid molecule (RNA or DNA) that can specifically and selectively bind to a specific target. Aptamers are selected through a process called SELEX (Systematic Evolution of Ligands by Exponential Enrichment), which involves generating a diverse library of nucleic acid molecules and iteratively amplifying and selecting those with high affinity and specificity towards the desired target. In the specific case described by Liu et al., a fluorometric aptamer-based assay (labelled with carboxyfluorescein) was employed for the recognition of ochratoxin A (OTA). The fluorescent signal increases with increasing concentrations of OTA, achieving a detection limit of 0.11 ng/mL (Liu et al., 2018).

From a more applied perspective, other researchers have developed polymer dots for sensitive monitoring of zearalenone in agricultural products. Polymer dots are polymeric nanoparticles that exhibit fluorescence properties. These particles consist of a conjugated polymer core that can be excited by light, resulting in fluorescence emission at different wavelengths. In the study published by Bu et al., polymer dots were fabricated using hydroquinone and SA as precursors and exhibited excellent colorimetric-fluorescent sensing properties, achieving a detection limit of 0.036 ng/mL (Bu et al., 2022).

21.5 Agricultural chemicals

21.5.1 Definition and contextualization

Among emerging contaminants, agricultural chemicals have gained particular attention due to their widespread use in modern agriculture and their potential adverse effects on ecosystems and human health. Agricultural chemicals, including pesticides, herbicides, fungicides, and fertilizers, are essential for ensuring high crop yields and combating agricultural pests and diseases. However, their extensive application, combined with inadequate management practices, has led to their persistence as a contaminant in various environmental compartments such as water, soil, and air (Akesson & Yates, 2003; Mann et al., 2009; Thomson, 1973).

The occurrence of agricultural chemicals as emerging contaminants poses challenges to retrieve their potential risks to ecosystems. Their transport and fate in the environment, as well as their ability to bioaccumulate in organisms, raise concerns about their long-term impacts on both terrestrial and aquatic ecosystems. Furthermore, the high probability for agricultural chemicals to enter the food chain and ultimately impact human health necessitates a comprehensive understanding of their occurrence, behaviour, and potential risks (Bouwer, 1990; Khan et al., 2022).

Efforts are underway to develop advanced analytical methods and monitoring strategies to detect and quantify these emerging contaminants in environmental matrices. Additionally, research focuses on understanding their toxicological effects, fate, and transport mechanisms to evaluate their potential risks and develop appropriate mitigation strategies. The identification of alternative, and environmentally friendly, agricultural practices and the promotion of sustainable approaches are also a key points to reduce the reliance on traditional agricultural chemicals and minimise their impact on the environment, (Luo et al., 2016; Migliorelli & Dessertine, 2018; Suthersan et al., 2016) but without giving up high production yield.

21.5.2 Detection with sensory polymers

In Table 21.2, the most relevant agricultural chemicals are presented, citing various examples of chemical structures, as well as the category and subcategory to which they belong.

Table 21.2 List of agricultural chemicals classified by chemical structure and functionality.

Category	Subcategory	Examples
Pesticides	Insecticides	Organophosphates (e.g., dichlorvos, chlorpyrifos, diazinon, malathion)
		Carbamates (e.g., carbaryl, aldicarb)
		Pyrethroids (e.g., deltamethrin, cypermethrin)
		Neonicotinoids (e.g., imidacloprid, clothianidin)
	Herbicides	Glyphosate
		Paraquat
		Atrazine
		2,4-D (2,4-dichlorophenoxyacetic acid)
	Fungicides	Azoles (e.g., propiconazole, tebuconazole)
		Strobilurins (e.g., azoxystrobin, trifloxystrobin)
		Phenols (e.g., 2-phenylphenol)
		Benzimidazoles (e.g., thiabendazole, carbendazim)
	Nematicides	Dazomet
		Metam sodium
		Oxamyl
Fertilizers	Nitrogen-based	Urea
		Ammonium nitrate
		Ammonium sulfate
	Phosphorus-based	Superphosphates
		Monoammonium phosphate
		Diammonium phosphate
	Potassium-based	Potassium chloride
		Potassium sulfate
		Potassium nitrate
Plant growth Regulators	Synthetic auxins	Indole-3-acetic acid (IAA)
		2,4-D (2,4-dichlorophenoxyacetic acid)
	Gibberellins	Gibberellic acid (GA3)
	Cytokinins	Kinetin
		Zeatin
Others	Rodenticides	Warfarin
		Bromadiolone
	Insect repellents	DEET (N,N-diethyl-meta-toluamide)
		Icaridin
	Crop curing agents	Ethylene

In this section, the design of sensory polymers for some of the agricultural chemicals defined in Table 21.2 will be analyzed. For instance, Dichlorvos, also known as DDVP (Dichlorvos Vaporizing Agent), is an organophosphate chemical compound used as an insecticide and acaricide. Pimsen et al. have developed an ingenious colorimetric method for the detection of dichlorvos (Fig. 21.9). The first key component of the colorimetric method is poly(10,12-pentacosadiynoic acid), which exhibits a blue colour, and vesicles of this polymer can be prepared. The second key component is myristoylcholine (a cationic surfactant), which can cause the blue color of the polymer to transform into an intense red colour. The third key component is acetylcholinesterase, which

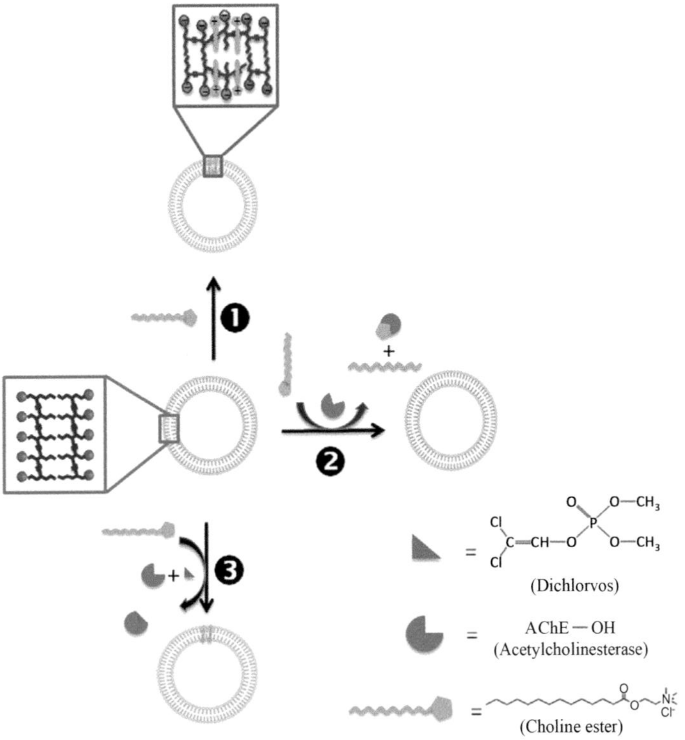

FIGURE 21.9

Schematic illustration of dichlorvos detection based on poly(PCDA) vesicles/AChE/choline ester system: ❶ colorimetric response of poly(PCDA) to choline ester; ❷ colorimetric response of poly(PCDA) to a mixture of choline ester and AChE; ❸ colorimetric response of poly(PCDA) to a mixture of choline ester, dichlorvos and AChE.

Source: *Adapted with permission from Pimsen, R., Khumsri, A., Wacharasindhu, S., Tumcharern, G., & Sukwattanasinitt, M. (2014). Colorimetric detection of dichlorvos using polydiacetylenevesicles with acetylcholinesterase and cationic surfactants. Biosensors and Bioelectronics, 62, 8–12. https://doi.org/10.1016/j.bios.2014.05.069. Copyright © Elsevier 2014 Schematic representation of dichlorvos colorimetric detection mechanism by enzymatic activity blocking of acetylcholinesterase and poly(10,12-pentacosadiynoic acid) vesicles (Pimsen et al., 2014).*

can hydrolyze myristoylcholine and thereby maintain the polymer in its blue colour state. Now, the target compound (dichlorvos) directly affects the enzymatic activity of acetylcholinesterase, making it less efficient in hydrolyzing myristoylcholine, thus allowing the latter to ultimately induce the colour change from blue to red. In this way, the authors achieved a detection limit for dichlorvos of 6.7 ppb using a UV-Vis spectrophotometer and 50 ppb visually (Pimsen et al., 2014).

A very similar strategy was pursued by Wang et al., using a coordination polymer with Cerium, in this case, aimed at the detection and quantification of organophosphorus pesticides. LODs of 0.024 μg/L were achieved (Wang et al., 2022).

Other authors, such as Bustamante et al., have proposed colorimetric dosimeters for the detection of different phenolic structures used as pesticides. The polymer contains a structural matrix formed by vinylpyrrolidone and methyl methacrylate. Additionally, it contains a small molar proportion of 0.05%–0.25% of side aniline groups provided by the monomer 4-vinylaniline. The amine group is responsible for the detection system as it can be transformed into benzene diazonium groups in the presence of sodium nitrite and hydrochloric acid, and subsequently react with the different phenols. This reaction generates an azo linkage between the polymer chains and the phenols, resulting in different colour changes depending on the chemical structure of the phenols. Specifically, 13 different colors are obtained when the polymeric material is exposed to m-cresol, 2-chlorophenol, bisphenol-A, 4-chloro-2-methylphenol, 2,4-dimethylphenol, 2-phenylphenol, 1-naphthol, 2,4-dinitrophenol, 4-chlorophenol, 1,8-dihydroxyanthraquinone, 2-nitrophenol, fenhexamid, and 2,4-dichlorophenol (Fig. 21.10). The obtained detection limits also vary depending on the detected chemical species, ranging from 20 to 800 ppb (Bustamante et al., 2019).

Chlorpyrifos, an organophosphorus insecticide, has also been the subject of investigation by several authors. Among them, Soongsong et al. stand out with their proposal of a colorimetric aptasensor that utilizes the localized surface plasmon resonance (LSPR) of gold nanoparticle (AuNP) aggregates coupled with a specific aptamer and cationic polyethylenimine (PEI). The aptasensor presents various advantages such as a simple procedure, low cost, reduced analysis time, and the absence of complicated instruments. Additionally, it offers high sensitivity, selectivity, and stability. The measurement principle relies on gold nanoparticles (AuNP), which can undergo colour changes under aggregation and dispersion conditions. In the absence of Chlorpyrifos, the negatively charged phosphate groups in the aptamer potentially interact with cationic polyethylenimine (PEI), resulting in the dispersion of AuNPs and their red coloration. However, in the presence of Chlorpyrifos, the aptamer specifically binds to this compound, consequently releasing the cationic PEI. The lack of interaction between PEI and the aptamer induces the aggregation of AuNPs, leading to a visually observable colour change from red to blue. This system achieved a LOD of 7.4 ng mL^{-1} and was further applied to real samples of tap water, grapefruit, and longan (Soongsong et al., 2021). Very similar approach was also proposed by Bala et al. for the detection of malathion, by using the polyelectrolyte polydiallyldimethylammonium chloride and gold particles (Bala et al., 2016).

Again, the detection of chlorpyrifos, along with profenofos and cypermethrin, was studied by Zhu et al. using a polymeric coating on paper that generated a colorimetric response when exposed to these 3 pesticides. This alternative provides very short response times, as the contact surface is typically high due to the porosity of the paper. The polymer used was zwitterionic, specifically poly(sulfobetaine methacrylate), and detection limits of 0.235, 4.891, and 4.053 mg/L were achieved for chlorpyrifos, profenofos, and cypermethrin, respectively (Zhu et al., 2023).

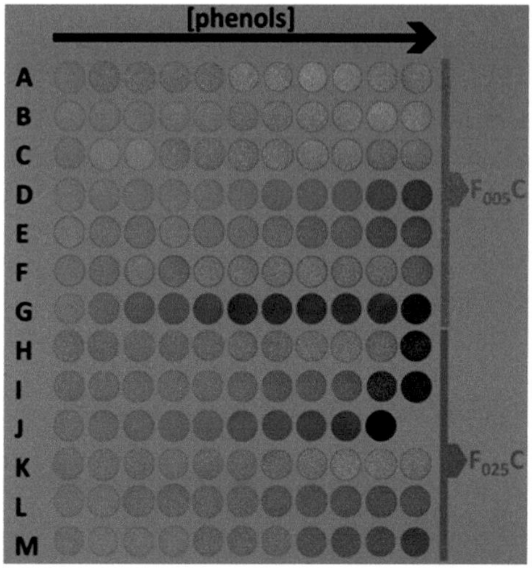

FIGURE 21.10

Figure shows the colorimetric response of sensory materials for the analyzed phenols. When the colorimetric response was not clearly visible with $F_{005}B$, a material with a higher concentration of the sensory motifs was used ($\mathbf{F}_{025}B$).

Adapted with permission from Bustamante, S. E., Vallejos, S., Pascual-Portal, B. S., Muñoz, A., Mendia, A., Rivas, B. L., García, F. C., & García, J. M. (2019). Polymer films containing chemically anchored diazonium salts with long-term stability as colorimetric sensors. Journal of Hazardous Materials, 365, *725–732. https://doi.org/10.1016/j.jhazmat.2018.11.066. Copyright © Elsevier 2019 Smart polymeric sensors with different color sensing depending on the structure of the target phenol. Thirteen different colors are present in the figure with an increasing intensity related to the concentration of phenols (Bustamante et al., 2019).*

Finally, Zhao et al. proposed the detection of the herbicide atrazine by using MIPs, specifically methacrylic acid-based MIPs. The authors propose an initial separation of the herbicide using molecularly imprinted polymers (MIPs), and then suggest two types of detection methods one visual that does not require equipment, and another using surface-enhanced Raman spectroscopy (SERS). They applied their development to real apple juice samples and achieved an extraction percentage of 93% using polyacrylic acid MIPs. The detection limits reached with visual analysis and SERS were 0.01 mg/L and 0.0012 mg/L, respectively (Zhao et al., 2019).

21.6 Micro and nano plastics

21.6.1 Definition and contextualization

Micro and nano plastics are, without a doubt, the emerging contaminant that is of greatest concern today. There is greater awareness at all levels, thanks partly to the media's dissemination in recent years. They can have different origins: they can derive from the degradation of plastics that are not correctly disposed of or recycled; they can be mixtures of polymers, functional additives and residual impurities deriving from plastic manufacturing; they can derive from the spontaneous degradation of larger plastics; they could be intentionally added to products for a specific purpose (e.g., in cosmetic products, detergents and maintenance products, fertilizers and plant protection products, medical devices, medicinal products for human and veterinary use, food complement and medical food, paints, inks and other coatings, etc.) or could be generated in households, where, for example, the use of washing machines is constantly eroding synthetic fibres of clothes (Galloway, 2015; Liu et al., 2021; Revel et al., 2018; Rocha-Santos, 2018; Shen et al., 2019).

Every year, 176,300 tonnes per year (71,800–280,600) are not intentionally emitted to EU surface waters (Hann et al., 2018). In particular, most of the releases to surface water were identified to be road tyre wear (94,000 tonnes per year) and losses of pre-production plastic pellets (41,000 tonnes per year), followed by road marking (15,000 tonnes per year) and washing of clothes (13,000 tonnes per year). According to the ECHA estimations, by 2040, the aggregate emission is forecast to exceed 640,00 tonnes and to range from 160,000 tonnes to 1.1 million tonnes. This forecast considers the uncertainty about sector-specific emissions and it is expressed as cumulative emissions (European Chemicals Agency, 2020).

MPs and NPs are present in aquatic media, soil and air, and primarily affect the gastrointestinal tract and/or the lungs. In general, the composition of microplastics can be divided into two parts: (1) one or more polymers that represent the majority of the material's formulation (polyethylene, PET, polypropylene, etc.), and (2) the rest of the components or fillers that are added in the manufacture or processing of these materials (dyes, flame retardants, plasticizers, etc.). When these plastics degrade and form MPs and NPs, they contain the functional groups of both parts, which makes them extraordinary support to interact with macromolecules such as proteins and glycoproteins through different processes, such as adsorption. These interactions are responsible for activating the macrophage response or generating the production of cytokines, that is, for activating certain mechanisms of our immune system that produce inflammation processes in the intestines and surrounding tissues.

The most recent studies focus on both aquatic (such as oysters) and terrestrial species (such as certain types of earthworms). MPs and NPs are already known to inhibit lipid metabolism in oysters, but the results with terrestrial species have been even more devastating. In soils, the adsorption capacity of MPs and NPs is combined with the already harmful effects of another family of emerging pollutants such as PFCs. In fact, the presence of MPs and NPs in soils increases the absorption capacity of PFCs in earthworms and seriously affects their reproductive capabilities. Therefore, these PMs and NPs not only pose a risk by themselves, but also enhance the harmful effect of other emerging pollutants.

21.6.2 Detection with sensory polymers

In general terms, the detection and quantification of micro and nanoplastics (MNPs) is a complex task, as graphically depicted in Fig. 21.11, in which some critical points are mentioned (Jakubowicz et al., 2021; Schwaferts et al., 2019). Being a relatively new field, it is still in active development, and there are various approaches to the detection of these emerging contaminants. Most of the methods described in the literature typically require a prior separation process (sample pretreatment by digestion and preconcentration), followed by different analytical techniques (Möller et al., 2020; Nguyen et al., 2019). Among the different analysis methods, (Schwaferts et al., 2019) microscopy (light), scanning electron microscopy, Fourier transform infrared spectroscopy, Raman spectroscopy, atomic force microscopy

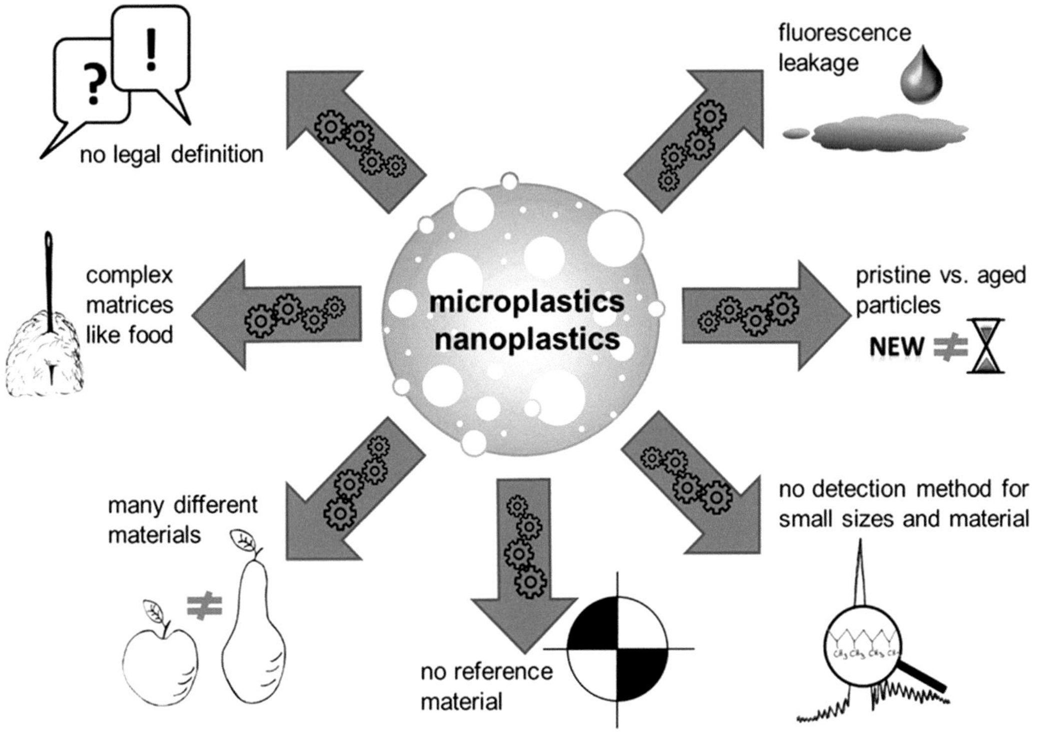

FIGURE 21.11

Challenges and limitations in micro- and nanoplastics research.

Source: *Adapted with permission from Paul, M. B., Stock, V., Cara-Carmona, J., Lisicki, E., Shopova, S., Fessard, V., Braeuning, A., Sieg, H., & Böhmert, L. (2020). Micro- and nanoplastics – current state of knowledge with the focus on oral uptake and toxicity.* Nanoscale Advances, 2*(10), 4350–4367. https://doi.org/10.1039/D0NA00539H, licensed under CC BY-NC 3.0 (Paul et al., 2020). Diagram where the main challenges and limitations in terms of micro and nanoplastics detection are presented (Paul et al., 2020).*

based infrared spectroscopy (AFM-IR), flow cytometry, matrix-assisted laser desorption/ionization time-of-flight (MALDI-TOF) mass spectrometry, pyrolysis gas chromatography–mass spectrometry (PYR-GC–Ms), and liquid chromatography–tandem mass spectrometry (LC–Ms/MS) are prominent.

However, with regard to the detection of these targets using sensory polymers, the literature on this topic is currently very limited. Nevertheless, the following paragraphs present the latest advancements in this field.

Firstly, it is important to differentiate between the detection of compounds associated with microplastics, such as PCBs, PAHs, or bisphenol A, and the detection of microplastics themselves. When we refer to the detection of the latter, there are different receptors (immobilized in polymers) specifically designed for the detection of MNPs.

One group of receptors for MNP detection includes enzymes, which have shown great efficiency in recognizing polymers such as polyethylene terephthalate (PET) and its sustainable

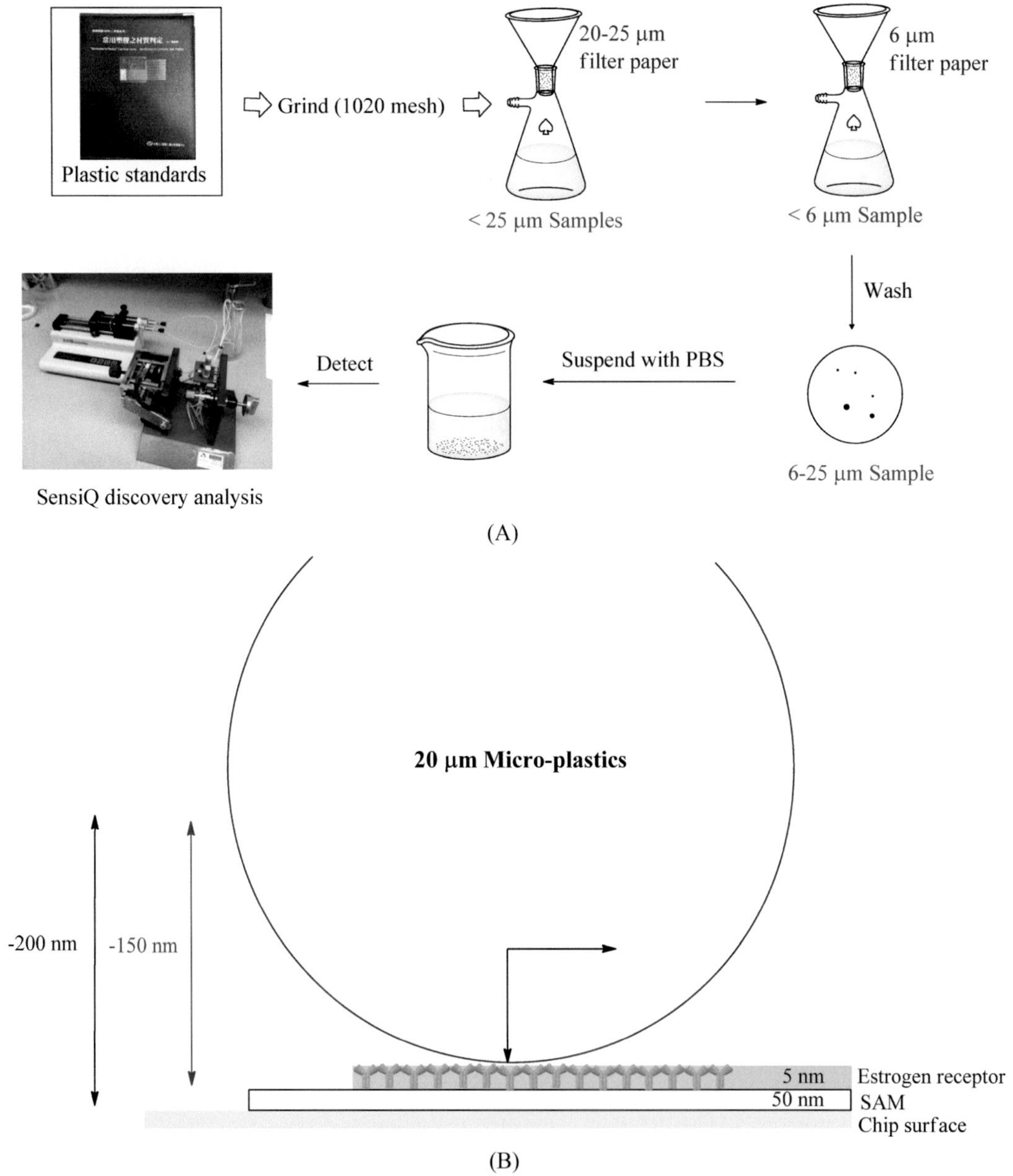

FIGURE 21.12

Graphical representation of the experimental procedure. (A) Overview of the general procedure for detecting microplastics using the surface plasmon resonance (SPR) biosensor. (B) Schematic depiction of microplastic detection utilizing estrogen receptors (ERs) immobilized on the SPR sensor. The double arrow (200 nm) corresponds to the detectable range of SPR, while the double arrow (150 nm) represents the detectable range of microplastics after subtracting the self-assembled monolayer (SAM) and estrogen receptors immobilized on the chip. The arrow inside the particle indicates the gravitational force and the flow rate of microplastics in the microfluidic system.

Source: *From Huang, C. J., Narasimha, G. V., Chen, Y. C., Chen, J. K., & Dong, G. C. (2021). Measurement of low concentration of micro-plastics by detection of bioaffinity-induced particle retention using surface plasmon resonance biosensors.* Biosensors, 11*(7), 219. https://doi.org/10.3390/bios11070219, licensed under CC BY 4.0 (Huang et al., 2021) (A) Representation of the detection mechanisms of microplastics using a surface plasmon resonance biosensor. (B) Scheme of a microplastic particle immobilized on an estrogen receptor. (Huang et al., 2021).*

alternative, polyethylene furanoate (PEF). The enzymes responsible for the degradation of these plastics are PETase and PEFase, both of which have been immobilized on different polymeric supports to carry out this bioremediation function. However, this system cannot be considered a sensory polymer in the traditional sense, as it lacks the final component of generating a response. Therefore, these sensory polymers with immobilized enzymes (referred to as "Type 1 SP" according to the terminology established by Garcia et al.) are often coupled with an electrode as a transducer element. Enzyme-based sensors tend to produce or consume protons and/or electroactive species, and the electrode is responsible for generating the response, thus completing the sensory system (Tang et al., 2023; Wilson & Hu, 2000).

Given the time and technique costs associated with the separation and concentration of microparticles in real samples, some authors opt for preparing them directly in the laboratory through crushing and filtration (Fig. 21.12). Huang et al. chose this strategy and proposed sensors based on

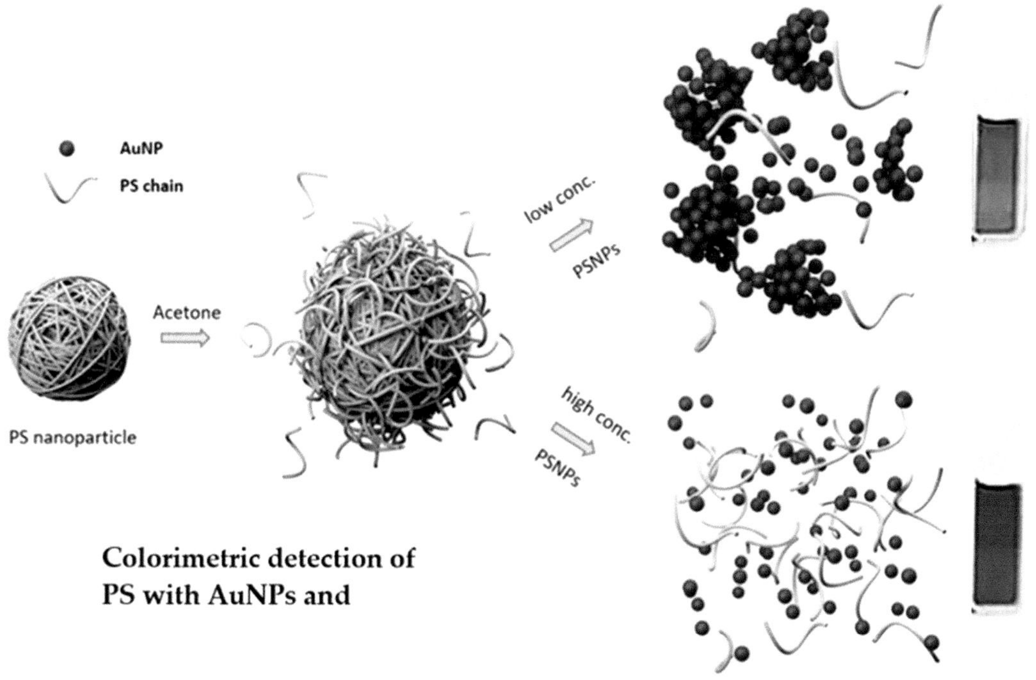

FIGURE 21.13

Graphical abstract of polystyrene nanoplastic detection in aqueous phase, a colorimetric approach with gold nanoparticles.

Source: *Adapted with permission from Hong, J., Lee, B., Park, C., & Kim, Y. A. (2022). A colorimetric detection of polystyrene nanoplastics with gold nanoparticles in the aqueous phase.* Science of the Total Environment, 850, *158058. https://doi.org/10.1016/j.scitotenv.2022.158058. Copyright © Elsevier 2022 Graphical representation of the detection of polystyrene nanoparticles. The plastic avoids gold nanoparticles agglomeration and thus triggers a color change from blue to reddish (Hong et al., 2022).*

estrogen receptors. This study focused on the mode of movement and the development of low-concentration detection for microplastics using surface plasmon resonance (SPR), a technique through which an overoccupation of biologically sourced polystyrene (PS) was found via a stronger binding force with the estrogen receptors (ER) and a longer retention time. The study was also successful for PVC and PE microplastics (Huang et al., 2021).

Another strategy that has been proven effective is the utilization of metallic nanoparticles, such as Au nanoparticles. Hong et al. recently achieved visual detection of high or low concentrations of microplastics (PS) through a colour change (red-blue), as shown in Fig. 21.13 (Hong et al., 2022).

This fact paves the way for new investigations that combine metallic nanoparticles deposited/anchored on polymeric supports, (García-Calvo et al., 2016) for the easy and visual detection of microplastics.

21.7 Summary and conclusions

Despite the inherent difficulties of detecting harmful neutral species, there is a need to produce economical, portable, and easy-to-use and interpret sensors that interact with emerging neutral pollutants. The more information available about emerging pollutants, the more restrictive the regulations become, and therefore, the importance of reliable sensors grows at the same time. The demonstrated effectiveness and versatility of smart polymers with ionic species make them interesting candidates to meet the demands of different sectors, including food safety, biomedicine, environmental monitoring, and quality control.

Due to their immense versatility, tailor-made MIPs are among the most commonly employed receptors, regardless of the target family of pollutants. Their ability to capture specific molecules can modify certain properties, such as surface tension, which is very useful for the detection of PFCs. MIPs can be used both as generic or specific receptors, either individually or in combination with characterization techniques to obtain a readable response. Some examples of species captured by MIPs, or MIPs in combination with MONs, include PBBs, PCBs, TBBPAs, parabens, coccidiostats, and mycotoxins like AFB1 or sterigmatocystin. Common combined techniques found in the literature include chromatographic techniques, such as HPLC or GC, which have been used for the detection of parabens and PHAs. Additionally, SERS or electrochemical measurements are useful for the detection and quantification of marine toxins or atrazine. Thanks to their extensive surface area, MOFs have also been used as receptors to quantify emerging pollutants through a mechanism similar to that of MIPs.

More traditional colorimetric and fluorimetric sensors have also been widely reported. By leveraging the interactions between a siloxane compound and erythrosine dye, it is possible to detect PFOs. Other common strategies include the utilization of polyfluorene, metal-coordinated polymers such as Cd, Ce, or AuNPs (gold nanoparticles), polymer dots, and zwitterionic polymers. An intriguing approach involves the use of diazonium salts through the interaction of phenols and nitrogenated aromatic rings, which enables the detection, quantification, and identification of various phenols. This is achieved thanks to the diverse range of colours generated by the smart polymer in response to different phenolic compounds.

Additional recognition elements that have been developed include aptamers, polyionic liquids, and enzymes. These elements are especially relevant for detecting biotoxins or magnetic nanoparticles (MNPs). However, to obtain a readable response, it is often necessary to combine these recognition elements with electrochemical characterization techniques, such as deposition over an electrode. This integration of different elements and techniques enhances the sensitivity and specificity of the sensor, making it capable of detecting and quantifying a wide range of analytes, including biotoxins and MNPs.

In summary, this chapter has compiled the initial efforts in detecting neutral emerging pollutants to help identify opportunities in this field. On one hand, as an increasing number of species are recognized as potentially hazardous, there is a growing need to develop smart polymers as specific sensors for them. On the other hand, research in sensing technologies is well-established, but there is still a significant journey ahead to integrate these technologies into solid-state applications effectively. Smart polymers have demonstrated their ability to enhance the robustness and efficiency of non-polymeric sensors. As a result, they represent a reliable strategy for addressing new threats that society will encounter in the future.

Acknowledgments

We gratefully acknowledge the financial support provided by the Regional Government of Castilla y León (Junta de Castilla y León), Ministry of Science and Innovation MICIN, and European Union Next Generation EU PRTR.

References

Ajiboye, T. O., Kuvarega, A. T., & Onwudiwe, D. C. (2020). Recent Strategies for Environmental Remediation of Organochlorine Pesticides. *Applied Sciences*, *10*(18), 6286. Available from https://doi.org/10.3390/APP10186286.

Akesson, N. B., & Yates, W. E. (2003). Problems relating to application of agricultural chemicals and resulting drift residues. *Annual Review of Entomology*, *9*(1), 285–318. Available from https://doi.org/10.1146/ANNUREV.EN.09.010164.001441.

Al-Omran, L. S., Stubbings, W. A., & Harrad, S. (2022). Concentrations and isomer profiles of hexabromocyclododecanes (HBCDDs) in floor, elevated surface, and outdoor dust samples from Basrah, Iraq. *Environmental Science: Processes & Impacts*, *24*(6), 910–920. Available from https://doi.org/10.1039/D2EM00133K.

Bala, R., Kumar, M., Bansal, K., Sharma, R. K., & Wangoo, N. (2016). Ultrasensitive aptamer biosensor for malathion detection based on cationic polymer and gold nanoparticles. *Biosensors and Bioelectronics*, *85*, 445–449. Available from https://doi.org/10.1016/J.BIOS.2016.05.042.

Belbruno, J. J. (2019). Molecularly imprinted polymers. *Chemical Reviews*, *119*(1), 94–119. Available from https://doi.org/10.1021/acs.chemrev.8b00171.

Bennett, J. W., & Klich, M. (2003). Mycotoxins. *Clinical Microbiology Reviews*, *16*(3), 497–516. Available from https://doi.org/10.1128/CMR.16.3.497-516.2003.

Bilcke, C. V. (2001). The stockholm convention on persistent organic pollutants. *American Journal of International Law*, *95*(3), 692–708. Available from https://doi.org/10.2307/2668517.

Bouwer, H. (1990). Agricultural chemicals and ground water quality. *Journal of Soil and Water Conservation*, *45*(2), 184–189. Available from https://doi.org/10.1111/j.1745-6592.1990.tb00324.x.

Braun, D., Eiser, M., Puntscher, H., Marko, D., & Warth, B. (2021). Natural contaminants in infant food: the case of regulated and emerging mycotoxins. *Food Control*, *123*, 107676. Available from https://doi.org/10.1016/J.FOODCONT.2020.107676.

Breshears, L. E., Mata-Robles, S., Tang, Y., Baker, J. C., Reynolds, K. A., & Yoon, J. Y. (2023). Rapid, sensitive detection of PFOA with smartphone-based flow rate analysis utilizing competitive molecular interactions during capillary action. *Journal of Hazardous Materials*, *446*, 130699. Available from https://doi.org/10.1016/J.JHAZMAT.2022.130699.

Bu, T., Bai, F., Zhao, S., Sun, X., Jia, P., He, K., Wang, Y., Li, Q., & Wang, L. (2022). Dual-modal immunochromatographic test for sensitive detection of zearalenone in food samples based on biosynthetic staphylococcus aureus-mediated polymer dot nanocomposites. *Analytical Chemistry*, *94*(14), 5546–5554. Available from https://doi.org/10.1021/ACS.ANALCHEM.1C04721.

Bustamante, S. E., Vallejos, S., Pascual-Portal, B. S., Muñoz, A., Mendia, A., Rivas, B. L., García, F. C., & García, J. M. (2019). Polymer films containing chemically anchored diazonium salts with long-term stability as colorimetric sensors. *Journal of Hazardous Materials*, *365*, 725–732. Available from https://doi.org/10.1016/J.JHAZMAT.2018.11.066.

Böhm, J., Koinig, L., Razzazi-Fazeli, E., Blajet-Kosicka, A., Twaruzek, M., Grajewski, J., & Lang, C. (2010). Survey and risk assessment of the mycotoxins deoxynivalenol, zearalenone, fumonisins, ochratoxin a, and aflatoxins in commercial dry dog food. *Mycotoxin Res*, *26*(3), 147–153. Available from https://doi.org/10.1007/S12550-010-0049-4/TABLES/3.

Cennamo, N.; Arcadio, F.; Perri, C.; Zeni, L.; Sequeira, F.; Bilro, L.; Nogueira, R.; D'Agostino, G.; Porto, G.; Biasiolo, A. (2019). Water monitoring in smart cities exploiting plastic optical fibers and molecularly imprinted polymers. the case of PFBS detection. 2019 IEEE Int. Symp. Meas. Networking, M N 2019 - Proc. https://doi.org/10.1109/IWMN.2019.8805049.

Chen, X., Yao, H., Song, D., Sun, G., & Xu, M. (2022). Extracellular chemoreceptor of deca-brominated diphenyl ether and its engineering in the hydrophobic chassis cell for organics biosensing. *Chemical Engineering Journal*, *433*, 133266. Available from https://doi.org/10.1016/J.CEJ.2021.133266.

Covaci, A., Gerecke, A. C., Law, R. J., Voorspoels, S., Kohler, M., Heeb, N. V., Leslie, H., Allchin, C. R., & De Boer, J. (2006). Hexabromocyclododecanes (HBCDs) in the environment and humans: a review. *Environmental Science & Technology*, *40*(12), 3679–3688. Available from https://doi.org/10.1021/ES0602492.

Covaci, A., Voorspoels, S., Abdallah, M. A. E., Geens, T., Harrad, S., & Law, R. J. (2009). Analytical and environmental aspects of the flame retardant tetrabromobisphenol-a and its derivatives. *Journal of Chromatography A*, *1216*(3), 346–363. Available from https://doi.org/10.1016/j.chroma.2008.08.035.

Crain, D. A., Janssen, S. J., Edwards, T. M., Heindel, J., Ho, Sm, Hunt, P., Iguchi, T., Juul, A., McLachlan, J. A., Schwartz, J., Skakkebaek, N., Soto, A. M., Swan, S., Walker, C., Woodruff, T. K., Woodruff, T. J., Giudice, L. C., & Guillette, L. J. (2008). Female reproductive disorders: the roles of endocrine-disrupting compounds and developmental timing. *Fertility and Sterility*, *90*(4), 911–940. Available from https://doi.org/10.1016/j.fertnstert.2008.08.067.

Crisp, T. M., Clegg, E. D., Cooper, R. L., Wood, W. P., Andersen, D. G., Baetcke, K. P., Hoffmann, J. L., Morrow, M. S., Rodier, D. J., Schaeffer, J. E., Touart, L. W., Zeeman, M. G., & Patel, Y. M. (1998). Environmental endocrine disruption: an effects assessment and analysis. *Environmental Health Perspectives*, *106*(1), 11–56. Available from https://doi.org/10.1289/EHP.98106S111.

Crudo, F., Aichinger, G., Mihajlovic, J., Varga, E., Dellafiora, L., Warth, B., Dall'Asta, C., Berry, D., & Marko, D. (2021). In vitro interactions of alternaria mycotoxins, an emerging class of food contaminants, with the gut microbiota: a bidirectional relationship. *Archives of Toxicology*, *95*(7), 2533–2549. Available from https://doi.org/10.1007/S00204-021-03043-X.

Devi, N. L. (2020). *Persistent Organic Pollutants (POPs): environmental risks, toxicological effects, and bioremediation for environmental safety and challenges for future research. Bioremediation of Industrial Waste for Environmental Safety* (pp. 53–76). Singapore: Springer. Available from https://doi.org/10.1007/978-981-13-1891-7_4.

Egli, S.N. (2014). Thin-film molecularly imprinted polymers for detection systems for polycyclic aromatic hydrocarbons in water, memorial universitu, https://research.library.mun.ca/6483/.

Egli, S. N., Butler, E. D., & Bottaro, C. S. (2015). Selective extraction of light polycyclic aromatic hydrocarbons in environmental water samples with pseudo-template thin-film molecularly imprinted polymers. *Analytical Methods*, *7*(5), 2028–2035. Available from https://doi.org/10.1039/C4AY02849J.

Espiña, B., & Rubiolo, J. A. (2008). Marine toxins and the cytoskeleton: pectenotoxins, unusual macrolides that disrupt actin. *FEBS J, 275*(24), 6082–6088. Available from https://doi.org/10.1111/J.1742-4658.2008.06714.X.

European Chemicals Agency. (2020). *Background document on the opinion on the annex XV report proposing restrictions on intentionally added microplastics.*

Fan, L., Wang, F., Zhao, D., Sun, X., Chen, H., Wang, H., & Zhang, X. (2020). Two cadmium(II) coordination polymers as multi-functional luminescent sensors for the detection of Cr(VI) anions, dichloronitroaniline pesticide, and nitrofuran antibiotic in aqueous media. *Spectrochimica Acta Part A, 239*, 118467. Available from https://doi.org/10.1016/J.SAA.2020.118467.

Feng, J., Tao, Y., Shen, X., Jin, H., Zhou, T., Zhou, Y., Hu, L., Luo, D., Mei, S., & Lee, Y. I. (2019). Highly sensitive and selective fluorescent sensor for tetrabromobisphenol-A in electronic waste samples using molecularly imprinted polymer coated quantum dots. *Microchem. J, 144*, 93–101. Available from https://doi.org/10.1016/j.microc.2018.08.041.

Fenton, S. E., Ducatman, A., Boobis, A., DeWitt, J. C., Lau, C., Ng, C., Smith, J. S., & Roberts, S. M. (2021). Per- and polyfluoroalkyl substance toxicity and human health review: current state of knowledge and strategies for informing future research. *Environmental Toxicology and Chemistry, 40*(3), 606–630. Available from https://doi.org/10.1002/etc.4890.

Fire, S. E., & Van Dolah, F. M. (2012). Marine Biotoxins. In A. Aguirre Alonso, R. Ostfeld, & P. Daszak (Eds.), *New Directions in Conservation Medicine: Applied Cases of Ecological Health.* Oxford University Press.

Fu, J., Mai, B., Sheng, G., Zhang, G., Wang, X., Peng, P., Xiao, X., Ran, R., Cheng, F., Peng, X., Wang, Z., & Tang, U. W. (2003). Persistent organic pollutants in environment of the Pearl River Delta, China: an overview. *Chemosphere, 52*(9), 1411–1422. Available from https://doi.org/10.1016/S0045-6535(03)00477-6.

Galloway, T. S. (2015). Micro- and nano-plastics and human health. *Marine Anthropogenic Litter*, 343–366. Available from https://doi.org/10.1007/978-3-319-16510-3_13.

García-Calvo, J., García-Calvo, V., Vallejos, S., García, F. C., Avella, M., García, J. M., & Torroba, T. (2016). Surface coating by gold nanoparticles on functional polymers: on-demand portable catalysts for suzuki reactions. *ACS Applied Materials & Interfaces, 8*(38), 24999–25004. Available from https://doi.org/10.1021/acsami.6b07746.

García Pérez, J. M., García García, F. C., Vallejos, S., Trigo, M., & Reglero-Ruiz, J. A. (2022). *Sensory polymers. Smart Polymers. Principles and Applications* (p. 12) Berlin/Boston: De Gruyter.

Gascon, M., Morales, E., Sunyer, J., & Vrijheid, M. (2013). Effects of persistent organic pollutants on the developing respiratory and immune systems: a systematic review. *Environment International, 52*, 51–65. Available from https://doi.org/10.1016/J.ENVINT.2012.11.005.

Gerssen, A., & Gago-Mart Í Nez, A. (2019). Emerging marine biotoxins. *Toxins (Basel), 11*(6), 314. Available from https://doi.org/10.3390/TOXINS11060314.

Gong, J., Fang, T., Peng, D., Li, A., & Zhang, L. (2015). A highly sensitive photoelectrochemical detection of perfluorooctanic acid with molecularly imprined polymer-functionalized nanoarchitectured hybrid of AgI–BiOI composite. *Biosens. Bioelectron, 73*, 256–263. Available from https://doi.org/10.1016/J.BIOS.2015.06.008.

González, J.Á., de la T., González, J.Á., & de la T. (2019). Sondas Fotoquímicas y Materiales Luminiscentes Para La Detección de Micotoxinas Carboxiladas.

Gou, Z., Wang, A., Zhang, X., Zuo, Y., & Lin, W. (2022). Multi-Head cationic siloxane based "turn on" fluorescent system for selective detection of perfluorooctanoic sulfonate (PFOS). *Sensors and Actuators B: Chemical, 367*, 132017. Available from https://doi.org/10.1016/J.SNB.2022.132017.

Guo, Y. (2021). Polycyclic aromatic hydrocarbons in meat. *Nature Food, 2*(12), 914. Available from https://doi.org/10.1038/s43016-021-00440-4.

Guo, Y., He, X., Huang, C., Chen, H., Lu, Q., & Zhang, L. (2020). Metal–organic framework-derived nitrogen-doped carbon nanotube cages as efficient adsorbents for solid-phase microextraction of polychlorinated biphenyls. *Analytica Chimica Acta, 1095*, 99–108. Available from https://doi.org/10.1016/J.ACA.2019.10.023.

Hale, R. C., Alaee, M., Manchester-Neesvig, J. B., Stapleton, H. M., & Ikonomou, M. G. (2003). Polybrominated diphenyl ether flame retardants in the North American environment. *Environment International*, *29*(6), 771–779. Available from https://doi.org/10.1016/S0160-4120(03)00113-2.

Han, J. H., Cui, Y. Y., & Yang, C. X. (2021). Tailored amino/hydroxyl bifunctional microporous organic network for efficient stir bar sorptive extraction of parabens and flavors from cosmetic and food samples. *Journal of Chromatography A*, *1655*, 462521. Available from https://doi.org/10.1016/J.CHROMA.2021.462521.

Hann, S., Kershaw, P., Sherrington, C., Bapasola, A., Jamieson, O., Cole, G., & Hickman, M. (2018). *Investigating Options for Reducing Releases in the Aquatic Environment of Microplastics Emitted by (but Not Intentionally Added in) Products*.

Harvey, P. W., & Darbre, P. (2004). Endocrine disrupters and human health: could oestrogenic chemicals in body care cosmetics adversely affect breast cancer incidence in women? A review of evidence and call for further research. *Journal of Applied Toxicology*, *24*(3), 167–176. Available from https://doi.org/10.1002/jat.978.

Haupt, K., Linares, A. V., Bompart, M., & Bui, B. T. S. (2012). *Molecularly imprinted polymers*. *Topics in Current Chemistry* (Vol. 325, pp. 1–28). Berlin, Heidelberg: Springer. Available from https://doi.org/10.1007/128_2011_307.

Helou, K., Harmouche-Karaki, M., Karake, S., & Narbonne, J. F. (2019). A review of organochlorine pesticides and polychlorinated biphenyls in Lebanon: environmental and human contaminants. *Chemosphere*, *231*, 357–368. Available from https://doi.org/10.1016/j.chemosphere.2019.05.109.

Hites, R. A. (2004). Polybrominated diphenyl ethers in the environment and in people: a meta-analysis of concentrations. *Environmental Science & Technology*, *38*(4), 945–956. Available from https://doi.org/10.1021/ES035082G.

Hong, J., Lee, B., Park, C., & Kim, Y. (2022). A colorimetric detection of polystyrene nanoplastics with gold nanoparticles in the aqueous phase. *Science of the Total Environment*, *850*, 158058. Available from https://doi.org/10.1016/J.SCITOTENV.2022.158058.

Huang, C. J., Narasimha, G. V., Chen, Y. C., Chen, J. K., & Dong, G. C. (2021). Measurement of low concentration of micro-plastics by detection of bioaffinity-induced particle retention using surface plasmon resonance biosensors. *Biosensors*, *11*(7), 219. Available from https://doi.org/10.3390/bios11070219.

Jacobson, J. L., Humphrey, H. E. B., Jacobson, S. W., Schantz, S. L., Mullin, M. D., & Welch, R. (1989). Determinants of polychlorinated biphenyls (PCBs), polybrominated biphenyls (PBBs), and dichlorodiphenyl trichloroethane (DDT) levels in the sera of young children. *American Journal of Public Health*, *79*(10), 1401–1404. Available from https://doi.org/10.2105/AJPH.79.10.1401.

Jakubowicz, I., Enebro, J., & Yarahmadi, N. (2021). Challenges in the search for nanoplastics in the environment—a critical review from the polymer science perspective. *Polymer Testing*, *93*, 106953. Available from https://doi.org/10.1016/J.POLYMERTESTING.2020.106953.

Jiang, M., Tang, J., Zhou, N., Liu, J., Tao, F., Wang, F., & Li, C. (2022). Rapid electrochemical detection of domoic acid based on polydopamine/reduced graphene oxide coupled with in-situ imprinted polyacrylamide. *Talanta*, *236*, 122885. Available from https://doi.org/10.1016/J.TALANTA.2021.122885.

Jigyasa Rajput, J. K. (2022). Nanomaterial-based sensors as potential remedy for detection of biotoxins. *Food Control*, *135*, 108686. Available from https://doi.org/10.1016/J.FOODCONT.2021.108686.

Kanan, S., & Samara, F. (2018). Dioxins and furans: a review from chemical and environmental perspectives. *Trends in Environmental Analytical Chemistry*, *17*, 1–13. Available from https://doi.org/10.1016/J.TEAC.2017.12.001.

Kantiani, L., Llorca, M., Sanchís, J., Farré, M., & Barceló, D. (2010). Emerging food contaminants: a review. *Anal. Bioanal. Chem*, *398*(6), 2413–2427. Available from https://doi.org/10.1007/S00216-010-3944-9.

Karlaganis, G., Marioni, R., Sieber, I., & Weber, A. (2008). The Elaboration of the 'stockholm convention' on persistent organic pollutants (POPs): a negotiation process fraught with obstacles and opportunities. *Environmental Science and Pollution Research*, *8*(3), 216–221. Available from https://doi.org/10.1007/BF02987393.

Khan, S., Naushad, M., Govarthanan, M., Iqbal, J., & Alfadul, S. M. (2022). Emerging contaminants of high concern for the environment: current trends and future research. *Environment Research*, *207*, 112609. Available from https://doi.org/10.1016/j.envres.2021.112609.

Kimbrough, R. D. (1987). Human health effects of polychlorinated biphenyls (PCBs) and polybrominated biphenyls (PBBs). *Annual Review of Pharmacology and Toxicology*, *27*, 87–111. Available from https://doi.org/10.1146/ANNUREV.PA.27.040187.000511.

Kitamura, S., Jinno, N., Ohta, S., Kuroki, H., & Fujimoto, N. (2002). Thyroid hormonal activity of the flame retardants tetrabromobisphenol A and tetrachlorobisphenol A. *Biochemical and Biophysical Research Communications*, *293*(1), 554–559. Available from https://doi.org/10.1016/S0006-291X(02)00262-0.

Landsberg, J. H. (2002). The effects of harmful algal blooms on aquatic organisms. *Reviews in Fisheries Science*, *10*(2), 113–390. Available from https://doi.org/10.1080/20026491051695.

Li, R., Wang, W., El-Sayed, E. S. M., Su, K., He, P., & Yuan, D. (2021). Ratiometric fluorescence detection of tetracycline antibiotic based on a polynuclear lanthanide metal–organic framework. *Sensors Actuators, B Chem*, *330*, 129314. Available from https://doi.org/10.1016/J.SNB.2020.129314.

Liu, Q., Chen, Z., Chen, Y., Yang, F., Yao, W., & Xie, Y. (2021). Microplastics and nanoplastics: emerging contaminants in food. *Journal of Agricultural and Food Chemistry*, *69*(36), 10450–10468. Available from https://doi.org/10.1021/acs.jafc.1c04199.

Liu, Y., Yan, H., Shangguan, J., Yang, X., Wang, M., & Liu, W. (2018). A fluorometric aptamer-based assay for ochratoxin a using magnetic separation and a cationic conjugated fluorescent polymer. *Microchimica Acta*, *185*(9), 427. Available from https://doi.org/10.1007/S00604-018-2962-8.

Llorca, M., Farré, M., Eljarrat, E., Díaz-Cruz, S., Rodríguez-Mozaz, S., Wunderlin, D., & Barcelo, D. (2017). Review of emerging contaminants in aquatic biota from latin america: 2002–2016. *Environmental Toxicology and Chemistry*, *36*(7), 1716–1727. Available from https://doi.org/10.1002/etc0.3626.

Lu, D., Zhu, D. Z., Gan, H., Yao, Z., Luo, J., Yu, S., & Kurup, P. (2022). An ultra-sensitive molecularly imprinted polymer (MIP) and gold nanostars (AuNS) modified voltammetric sensor for facile detection of perfluorooctance sulfonate (PFOS) in drinking water. *Sensors and Actuators B: Chemical*, *352*, 131055. Available from https://doi.org/10.1016/J.SNB.2021.131055.

Luo, L., Qin, L., Wang, Y., & Wang, Q. (2016). Environmentally-friendly agricultural practices and their acceptance by smallholder farmers in China—a case study in Xinxiang County, Henan Province. *Sci. Total Environ*, *571*, 737–743. Available from https://doi.org/10.1016/J.SCITOTENV.2016.07.045.

López-Darias, J., Pino, V., Meng, Y., Anderson, J. L., & Afonso, A. M. (2010). Utilization of a benzyl functionalized polymeric ionic liquid for the sensitive determination of polycyclic aromatic hydrocarbons; parabens and alkylphenols in waters using solid-phase microextraction coupled to gas chromatography–flame ionization detection. *Journal of Chromatography A*, *1217*(46), 7189–7197. Available from https://doi.org/10.1016/J.CHROMA.2010.09.016.

Malik, A. H., & Iyer, P. K. (2017). Conjugated polyelectrolyte based sensitive detection and removal of antibiotics tetracycline from water. *ACS Applied Materials & Interfaces*, *9*(5), 4433–4439. Available from https://doi.org/10.1021/acsami.6b13949.

Mann, R. M., Hyne, R. V., Choung, C. B., & Wilson, S. P. (2009). Amphibians and agricultural chemicals: review of the risks in a complex environment. *Environmental Pollution*, *157*(11), 2903–2927. Available from https://doi.org/10.1016/J.ENVPOL.2009.05.015.

Marin, S., Ramos, A. J., Cano-Sancho, G., & Sanchis, V. (2013). Mycotoxins: occurrence, toxicology, and exposure assessment. *Food and Chemical Toxicology*, *60*, 218–237. Available from https://doi.org/10.1016/J.FCT.2013.07.047.

Martyniuk, C. J., Mehinto, A. C., & Denslow, N. D. (2020). Organochlorine pesticides: agrochemicals with potent endocrine-disrupting properties in fish. *Molecular and Cellular Endocrinology*, *507*, 110764. Available from https://doi.org/10.1016/J.MCE.2020.110764.

Matei, E., Covaliu-Mierla, C. I., Ţurcanu, A. A., Râpa, M., Predescu, A. M., & Predescu, C. (2022). Multifunctional membranes—a versatile approach for emerging pollutants removal. *Membranes (Basel), 12*, 67. Available from https://doi.org/10.3390/MEMBRANES12010067.

McDonald, T. A. (2002). A perspective on the potential health risks of PBDEs. *Chemosphere, 46*(5), 745–755. Available from https://doi.org/10.1016/S0045-6535(01)00239-9.

Migliorelli, M., & Dessertine, P. (2018). Time for new financing instruments? A market-oriented framework to finance environmentally friendly practices in EU agriculture. *Journal of Sustainable Finance & Investment, 8*(1), 1–25. Available from https://doi.org/10.1080/20430795.2017.1376270.

Möller, J. N., Löder, M. G. J., & Laforsch, C. (2020). Finding microplastics in soils: a review of analytical methods. *Environmental Science & Technology, 54*(4), 2078–2090. Available from https://doi.org/10.1021/ACS.EST.9B04618/SUPPL_FILE/ES9B04618_SI_001.PDF.

Morel, C., Christophe, A., Maguin-Gaté, K., Paoli, J., Turner, J. D., Schroeder, H., & Grova, N. (2022). Head-to-Head study of developmental neurotoxicity and resultant phenotype in rats: α-hexabromocyclododecane versus valproic acid, a recognized model of reference for autism spectrum disorders. *Toxics, 10*(4), 180. Available from https://doi.org/10.3390/TOXICS10040180/S1.

Nguyen, B., Claveau-Mallet, D., Hernandez, L. M., Xu, E. G., Farner, J. M., & Tufenkji, N. (2019). Separation and analysis of microplastics and nanoplastics in complex environmental samples. *Accounts of Chemical Research, 52*(4), 858–866. Available from https://doi.org/10.1021/ACS.ACCOUNTS.8B00602/ASSET/IMAGES/LARGE/AR-2018-00602E_0004.JPEG.

Nobile, M., Arioli, F., Pavlovic, R., Ceriani, F., Lin, S. K., Panseri, S., Villa, R., & Chiesa, L. M. (2020). Presence of emerging contaminants in baby food. *Food Additives & Contaminants: Part A: Chemistry, Analysis, Control, Exposure & Risk Assessment, 37*(1), 131–142. Available from https://doi.org/10.1080/19440049.2019.1682686.

Núñez, L., Turiel, E., Martin-Esteban, A., & Tadeo, J. L. (2010). Molecularly imprinted polymer for the extraction of parabens from environmental solid samples prior to their determination by high performance liquid chromatography-ultraviolet detection. *Talanta, 80*(5), 1782–1788. Available from https://doi.org/10.1016/J.TALANTA.2009.10.023.

Nzangya, J. M., Ndunda, E. N., Bosire, G. O., Martincigh, B. S., & Nyamori, O. V. (2021). *Polybrominated diphenyl ethers (PBDEs) as emerging environmental pollutants: advances in sample preparation and detection techniques. Emerging Contaminants*. InTechOpen. Available from https://doi.org/10.5772/intechopen.93858.

Ohoro, C. R., Adeniji, A. O., Okoh, A. I., & Okoh, O. O. (2021). Polybrominated diphenyl ethers in the environmental systems: a review. *Journal of Environmental Health Science and Engineering, 19*(1), 1229–1247. Available from https://doi.org/10.1007/S40201-021-00656-3.

Patel, A. B., Shaikh, S., Jain, K. R., Desai, C., & Madamwar, D. (2020). Polycyclic aromatic hydrocarbons: sources, toxicity, and remediation approaches. *Front. Microbiol, 11*, 562813. Available from https://doi.org/10.3389/fmicb.2020.562813.

Paul, M. B., Stock, V., Cara-Carmona, J., Lisicki, E., Shopova, S., Fessard, V., Braeuning, A., Sieg, H., & Böhmert, L. (2020). Micro- and nanoplastics – current state of knowledge with the focus on oral uptake and toxicity. *Nanoscale Adv, 2*(10), 4350–4367. Available from https://doi.org/10.1039/D0NA00539H.

Pimsen, R., Khumsri, A., Wacharasindhu, S., Tumcharern, G., & Sukwattanasinitt, M. (2014). Colorimetric detection of dichlorvos using polydiacetylene vesicles with acetylcholinesterase and cationic surfactants. *Biosensors and Bioelectronics, 62*, 8–12. Available from https://doi.org/10.1016/J.BIOS.2014.05.069.

Qi, S. Y., Xu, X. L., Ma, W. Z., Deng, S. L., Lian, Z. X., & Yu, K. (2022). Effects of organochlorine pesticide residues in maternal body on infants. *Frontiers in Endocrinology (Lausanne), 13*, 890307. Available from https://doi.org/10.3389/fendo.2022.890307.

Radi, A. E., El-Naggar, A. E., & Nassef, H. M. (2014). Determination of coccidiostat clopidol on an electropolymerized-molecularly imprinted polypyrrole polymer modified screen printed carbon electrode. *Anal. Methods, 6*(19), 7967–7972. Available from https://doi.org/10.1039/c4ay01320d.

Radi, A.-E., Wahdan, T., & El-Basiony, A. (2019). Electrochemical sensors based on molecularly imprinted polymers for pharmaceuticals analysis. *Current Analytical Chemistry*, *15*(3), 219–239. Available from https://doi.org/10.2174/1573411014666180501100131.

Revel, M., Châtel, A., & Mouneyrac, C. (2018). Micro(Nano)Plastics: A threat to human health? *Current Opinion in Environmental Science & Health*, *1*, 17–23. Available from https://doi.org/10.1016/J.COESH.2017.10.003.

Ritchie, J. (1975). Mechanism of action of local-anesthetic agents and biotoxins. *British Journal of Anaesthesia*, *47*, 191–198.

Rocha-Santos, T. A. P. (2018). Editorial Overview: Micro and Nano-Plastics. *Current Opinion in Environmental Science & Health*, *1*, 52–54. Available from https://doi.org/10.1016/J.COESH.2018.01.003.

Safe, S., & Hutzinger, O. (1984). Polychlorinated biphenyls (PCBs) and polybrominated biphenyls (PBBs): biochemistry, toxicology, and mechanism of action. *Critical Reviews in Toxicology*, *13*(4), 319–395. Available from https://doi.org/10.3109/10408448409023762.

Salthammer, T. (2020). Emerging indoor pollutants. *International Journal of Hygiene and Environmental Health*, *224*, 113423. Available from https://doi.org/10.1016/J.IJHEH.2019.113423.

Sanjuán, A. M., Reglero Ruiz, J. A., García, F. C., & García, J. M. (2018). Recent developments in sensing devices based on polymeric systems. *Reactive & Functional Polymers*, *133*, 103–125. Available from https://doi.org/10.1016/J.REACTFUNCTPOLYM.2018.10.007.

Sanjuán, A.M.; Reglero Ruiz, J.A.; García, F.C.; García, J.M. (2019). Smart polymers for highly sensitive sensors and devices: micro- and nanofabrication alternatives. In Smart Polymers and Their Applications; Woodhead Publishing, pp. 607–650. https://doi.org/10.1016/B978-0-08-102416-4.00017-X.

Schreiber, T., Gassmann, K., Götz, C., Hübenthal, U., Moors, M., Krause, G., Merk, H. F., Nguyen, N. H., Scanlan, T. S., Abel, J., Rose, C. R., & Fritsche, E. (2010). Polybrominated diphenyl ethers induce developmental neurotoxicity in a human in vitro model: evidence for endocrine disruption. *Environmental Health Perspectives*, *118*(4), 572–578. Available from https://doi.org/10.1289/EHP.0901435.

Schrenk, D., Bignami, M., Bodin, L., Chipman, J. K., del Mazo, J., Grasl-Kraupp, B., Hogstrand, C., Hoogenboom, L., Leblanc, J. C., Nebbia, C. S., Nielsen, E., Ntzani, E., Petersen, A., Sand, S., Vleminckx, C., Wallace, H., Barregård, L., Ceccatelli, S., Cravedi, J. P., Halldorsson, T. I., Haug, L. S., Johansson, N., Knutsen, H. K., Rose, M., Roudot, A. C., Van Loveren, H., Vollmer, G., Mackay, K., Riolo, F., & Schwerdtle, T. (2020). Risk to human health related to the presence of perfluoroalkyl substances in food. *EFSA Journal*, *18*(9), 6223. Available from https://doi.org/10.2903/J.EFSA.2020.6223.

Schwaferts, C., Niessner, R., Elsner, M., & Ivleva, N. P. (2019). Methods for the analysis of submicrometer- and nanoplastic particles in the environment. *Trends in Analytical Chemistry*, *112*, 52–65. Available from https://doi.org/10.1016/J.TRAC.2018.12.014.

Shen, M., Zhu, Y., Zhang, Y., Zeng, G., Wen, X., Yi, H., Ye, S., Ren, X., & Song, B. (2019). Micro(nano) plastics: unignorable vectors for organisms. *Marine Pollution Bulletin*, *139*, 328–331. Available from https://doi.org/10.1016/J.MARPOLBUL.2019.01.004.

Soleimany, F., Jinap, S., Faridah, A., & Khatib, A. (2012). A UPLC–MS/MS for simultaneous determination of aflatoxins, ochratoxin A, zearalenone, DON, fumonisins, T-2 toxin and HT-2 toxin, in cereals. *Food Control*, *25*(2), 647–653. Available from https://doi.org/10.1016/J.FOODCONT.2011.11.012.

Solís-Cruz, B., Hernández-Patlán, D., Beyssac, E., Latorre, J. D., Hernandez-Velasco, X., Merino-Guzman, R., Tellez, G., & López-Arellano, R. (2017). Evaluation of chitosan and cellulosic polymers as binding adsorbent materials to prevent aflatoxin B1, fumonisin B1, ochratoxin, trichothecene, deoxynivalenol, and zearalenone mycotoxicoses through an in vitro gastrointestinal model for poultry. *Polymers*, *9*(10), 529. Available from https://doi.org/10.3390/POLYM9100529.

Soongsong, J., Lerdsri, J., & Jakmunee, J. (2021). A Facile colorimetric aptasensor for low-cost chlorpyrifos detection utilizing gold nanoparticle aggregation induced by polyethyleneimine. *Analyst*, *146*(15), 4848–4857. Available from https://doi.org/10.1039/D1AN00771H.

Suthersan, S., Quinnan, J., Horst, J., Ross, I., Kalve, E., Bell, C., & Pancras, T. (2016). Making strides in the management of "emerging contaminants". *Groundwater Monitoring & Remediation*, *36*(1), 15–25. Available from https://doi.org/10.1111/gwmr.12143.

Tang, Y., Hardy, T. J., & Yoon, J. Y. (2023). Receptor-based detection of microplastics and nanoplastics: current and future. *Biosensors and Bioelectronics*, *234*, 115361. Available from https://doi.org/10.1016/J.BIOS.2023.115361.

Thomaidis, N. S., Asimakopoulos, A. G., & Bletsou, A. A. (2012). Emerging contaminants: a tutorial mini-review. *Global NEST Journal*, *14*(1), 72–79. Available from https://doi.org/10.30955/gnj.000823.

Thomson, W. T. (1973). *Agricultural Chemicals Book IV - Fungicides*. Indianapolis: Thomson Publications.

Visciano, P., Schirone, M., Berti, M., Milandri, A., Tofalo, R., & Suzzi, G. (2016). Marine biotoxins: occurrence, toxicity, regulatory limits and reference methods. *Front. Microbiol*, *7*, 1051. Available from https://doi.org/10.3389/FMICB.2016.01051.

Wang, J., Wang, X., Wang, M., Bian, Q., & Zhong, J. (2022). Novel Ce-based coordination polymer nanoparticles with excellent oxidase mimic activity applied for colorimetric assay to organophosphorus pesticides. *Food Chem*, *397*, 133810. Available from https://doi.org/10.1016/j.foodchem.2022.133810.

Wilson, G. S., & Hu, Y. (2000). Enzyme-based biosensors for in vivo measurements. *Chemical Reviews*, *100*(7), 2693–2704. Available from https://doi.org/10.1021/CR990003Y/ASSET/IMAGES/LARGE/CR990003YF00005.JPEG.

Wu, M., Wang, X., Shan, J., Zhou, H., Shi, Y., Li, M., & Liu, L. (2019). Sensitive and selective electrochemical sensor based on molecularly imprinted polypyrrole hybrid nanocomposites for tetrabromobisphenol A detection. *Anal. Lett*, *52*(16), 2506–2523. Available from https://doi.org/10.1080/00032719.2019.1617298.

Xue, Z., Zheng, X., Yu, W., Li, A., Li, S., Wang, Y., & Kou, X. (2021). Research progress in detection technology of polycyclic aromatic hydrocarbons. *Journal of the Electrochemical Society*, *168*(5), 057528. Available from https://doi.org/10.1149/1945-7111/AC0227.

Yang, L., Wang, J., Li, C. Y., Liu, Q., Wang, J., Wu, J., Lv, H., Ji, X. M., Liu, J. M., & Wang, S. (2023). Hollow-structured molecularly imprinted polymers enabled specific enrichment and highly sensitive determination of aflatoxin B1 and sterigmatocystin against complex sample matrix. *Journal of Hazardous Materials*, *451*, 131127. Available from https://doi.org/10.1016/J.JHAZMAT.2023.131127.

Ye, Q., Xi, X., Fan, D., Cao, X., Wang, Q., Wang, X., Zhang, M., Wang, B., Tao, Q., & Xiao, C. (2022). Polycyclic aromatic hydrocarbons in bonehomeostasis. *Biomedicine & Pharmacotherapy*, *146*, 112547. Available from https://doi.org/10.1016/j.biopha.2021.112547.

Yilmaz, B., Terekeci, H., Sandal, S., & Kelestimur, F. (2019). Endocrine disrupting chemicals: exposure, effects on human health, mechanism of action, models for testing and strategies for prevention. *Reviews in Endocrine and Metabolic Disorders*, *21*(1), 127–147. Available from https://doi.org/10.1007/S11154-019-09521-Z, *2019 211*.

Yuan, S., Li, C., Zhang, Y., Yu, H., Xie, Y., Guo, Y., & Yao, W. (2021). Ultrasound as an emerging technology for the elimination of chemical contaminants in food: a review. *Trends in Food Science and Technology*, *109*, 374–385. Available from https://doi.org/10.1016/j.tifs.2021.01.048.

Zeng, L., Cui, H., Chao, J., Huang, K., Wang, X., Zhou, Y., & Jing, T. (2020). Colorimetric determination of tetrabromobisphenol A based on enzyme-mimicking activity and molecular recognition of metal-organic framework-based molecularly imprinted polymers. *Microchim. Acta*, *187*(2), 142. Available from https://doi.org/10.1007/s00604-020-4119-9.

Zhao, B., Feng, S., Hu, Y., Wang, S., & Lu, X. (2019). Rapid determination of atrazine in apple juice using molecularly imprinted polymers coupled with gold nanoparticles-colorimetric/SERS dual chemosensor. *Food Chem*, *276*, 366–375. Available from https://doi.org/10.1016/J.FOODCHEM.2018.10.036.

Zhu, J., Yin, X., Zhang, W., Chen, M., Feng, D., Zhao, Y., & Zhu, Y. (2023). Simultaneous and sensitive detection of three pesticides using a functional poly(sulfobetaine methacrylate)-coated paper-based colorimetric sensor. *Biosensors*, *13*(3), 309. Available from https://doi.org/10.3390/BIOS13030309.

SECTION 3

Research trends and challenges in polymer sensors

CHAPTER

Trends and challenges in polymer sensors 22

José M. García
Departamento de Química, Facultad de Ciencias, Universidad de Burgos, Burgos, Spain

22.1 Introduction

22.1.1 About this book

As we approach the final chapter of "Sensory Polymers: From their Design to Practical Applications," we reflect on the journey that has unfolded through the diverse and intricate landscape of sensory polymers. Beginning with the foundational understanding of these sensory materials, we have explored their multifunctional capabilities to perceive and respond to external stimuli, a theme that has been the leitmotif of our narrative. From the initial examination of their transformative role in sensing technology to the detailed discussions on various types of sensory polymers or sensing solutions, each chapter has built upon the last, creating a tapestry of knowledge that spans from molecularly imprinted polymers to the innovative designs of hybrid and composite sensors.

Delving into specific applications, we have witnessed how these materials have impacted colourimetric and fluorogenic sensing, offering sensitivity and specificity. Electrochemical sensors, biosensors, and those based on nanomaterials have demonstrated the breadth of sensory polymers' potential. In industrial and environmental contexts, we have seen their impact on gas, humidity, and pH sensing, as well as temperature measurement and explosive detection. Furthermore, the advent of smart polymeric structures and sensor arrays has signaled a leap towards complex, integrated systems, while lab-on-a-chip devices have epitomized the miniaturization and sophistication of these technologies.

Each chapter has not only showcased advancements but sometimes also the challenges that lie in the commercialization of these technologies, emphasizing the need for scalability, reproducibility, and practical integration. As we converge on the final chapter, we aim to encapsulate these insights, drawing on the comprehensive understanding we've gained to outline the future of sensory polymers. Scientists in the field are on the verge of new breakthroughs, prepared to overcome the remaining challenges and fully utilize the potential of sensory polymers in real-world applications, extending their use beyond the laboratory and into everyday life.

As the field of sensory polymer advances, this final chapter aims to delve into the future trends that are likely to shape the landscape of sensory polymer technology and identify the challenges that must be overcome to harness the full spectrum of possibilities they promise. We will explore the next frontier in sensor technology, where polymers are not just passive detectors but active participants in data collection and environmental interaction. The discussion will cover anticipated innovations in material science, potential applications in various industries, and the expected

Sensory Polymers. DOI: https://doi.org/10.1016/B978-0-443-13394-7.00022-7

integration of sensory polymers with other technological advancements such as artificial intelligence (AI) and the Internet of Things (IoT).

22.1.2 From low molecular mass chemosensors to sensory polymers

The insights regarding the historical, current, and future aspects of chemosensors, as discussed by Gunnlaugsson, Akkaya, Yoon, James, and others, are highly relevant in this concluding chapter. While readers may perceive, as they progress through this book, that challenges in the field of chemosensors have been resolved, such an assumption is incorrect. Instead, the field confronts numerous ongoing challenges that require continuous scientific exploration and innovation. First and foremost, there is an unceasing need for the development of "new" chemosensors capable of detecting analytes that are yet unknown. This imperative goes beyond creating sensors for emerging biomarkers or detecting trace pollutants in our environmental surroundings, including air and water supplies. The biological and environmental analysis field operates within a framework of increasingly stringent standards established by regulatory authorities. As a result, even though existing chemosensors may demonstrate functionality, they frequently fall short in terms of the necessary selectivity and sensitivity required for specific real-world applications. Whether the challenge necessitates the design of custom receptors tailored to unique targets or the improvement of existing sensor systems, the demand for an expanding array of chemosensors persists. In summary, our perspective as researchers anticipates the ongoing growth and development of chemosensor research. This trajectory is driven by the imperative to address emerging analytical requirements, meet regulatory guidelines, and continuously enhance the performance of chemosensors to meet the evolving demands of practical applications (Wu et al., 2017).

While these comments apply to all probes, including conventional low molecular mass chemosensors and polymer sensory systems, polymers provide numerous benefits for sensor technology, surpassing traditional chemosensors in various ways. Their advantages include being lightweight, flexible, and biocompatible, as well as offering easy functionalisation and low-cost, efficient production methods. Polymers can either directly engage in sensing reactions or serve as platforms for receptor attachment. They are adaptable for use on various substrates and can be tailored to create user-friendly sensory devices. Furthermore, modifying the chemical structure of polymers can improve crucial properties like selectivity and response time for future sensor systems. Despite efforts to enhance current sensors, only slight improvements have been seen in their market presence. To tackle these challenges effectively, a collaborative approach that includes chemistry, physics, engineering, and both data and computer science is crucial. This approach is fundamental for achieving large-scale progress and manufacturing of these technologies. In this scenario, polymeric sensory gauges are expected to play a significant role in various domains, such as in medical applications, controlling chemical reactions, identifying gases, and potentially serving as electronic noses or tongues (Alberti et al., 2021; Cichosz et al., 2018).

22.2 Challenges and trends related to polymer-based sensory materials

Sensory polymers have emerged as a transformative force in the landscape of sensing technology, marking their evolution from simple passive materials to complex systems capable of active detection and response. The historical trajectory of sensory polymers is characterized by increasing sophistication, expanding from basic chemical detection to the multifaceted and highly selective sensors we see today. Such advancements have firmly established their role in advancing the frontiers of sensor technologies, leveraging their inherent ability to exhibit pronounced changes in physical, chemical, or electrical properties when exposed to specific environmental stimuli.

The implications of sensory polymers stretch far and wide across various scientific disciplines. In materials science, they have catalyzed a new era of smart materials, demonstrating the potential for tailored, selective, and reversible interactions with external stimuli. This versatility has led to breakthroughs in biomedical sensing, where sensory polymers now play a critical role in diagnostics and continuous health monitoring, offering minimally invasive options for patient care. The environmental sector has seen significant advancements in monitoring and pollution control thanks to the sensitivity and selectivity of polymer-based detection systems. Industrial applications, too, have been revolutionised by the integration of sensory polymers in process monitoring and safety measures.

With each innovative step, sensory polymers have unlocked functionalities that once seemed beyond reach—pioneering wearable technology that synergises with personal devices for health tracking and creating resilient sensors for extreme environments. Such innovations stem from a confluence of chemistry, physics, materials science, and engineering, driving interdisciplinary research that continually pushes the boundaries of what sensory polymers can achieve.

As we examine the current landscape and predict the trajectory of sensory polymers, it's clear that their evolution has spurred technological growth but has not yet significantly enhanced people's quality of life. To this end, the challenge of integrating them into everyday objects and systems to make our surroundings safer, medical care more proactive, and technology more intimately attuned to human needs must be addressed. They must meet market requirements to be integrated into objects or devices for both the simplest use by specialised personnel and widespread use among the population. Addressing this complex challenge is crucial for realising the full potential of polymers, which have been the predominant focus of sensor construction research, a trend recognized since the early 21st century (Fig. 22.1; Adhikari & Majumdar, 2004).

The grand challenges related to polymer-based sensory materials can be summarised as follows (Adhikari & Majumdar, 2004; Alam et al., 2022; Cichosz et al., 2018; Gamboa et al., 2023; Sanjuán et al., 2018);

1. Development of new biopolymers with sensory properties, focusing on expanding the range of detectable target species, especially for biomedical applications to enhance disease detection and diagnosis.
2. Refinement of recognition mechanisms for instant identification of target components, including the search and selection of suitable materials and the creation of effective signalling mechanisms.
3. Improvement of the transduction process from the chemical interactions between the receptor and the target to a measurable property (electrical, mechanical, or optical), aiming to enhance detection characteristics.
4. Reduction of detection or response time, critical for the identification of hazardous substances like heavy metallic cations or explosives.
5. Development of solid-state pH and ion-selective sensors, offering more durable and robust alternatives to existing technologies.

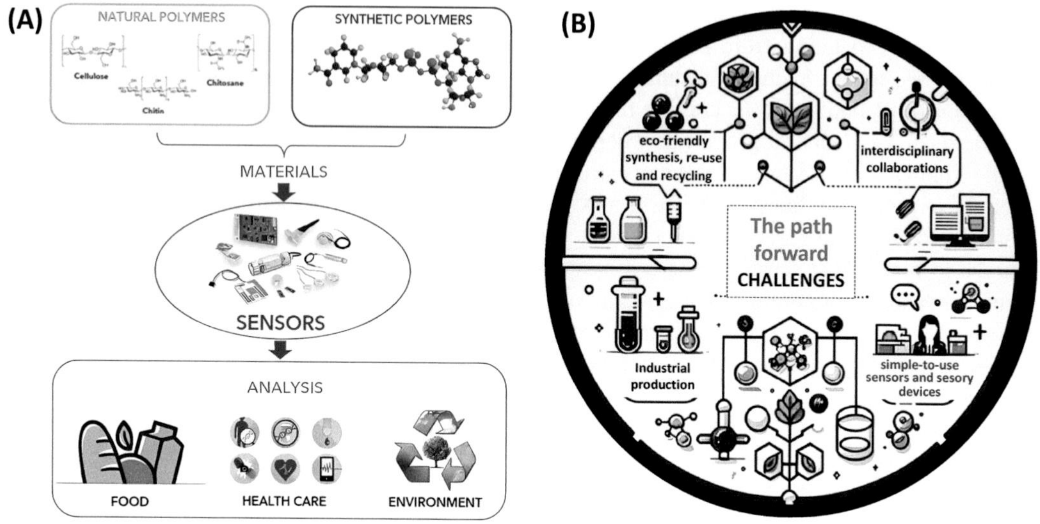

FIGURE 22.1

Sensory polymers based devices: (A) application, licensed under CC BY; and (B) market adoption challenges.

Source: *Licensed under CC BY and adapted from Alberti, G., Zanoni, C., Losi, V., Magnaghi, L. R., & Biesuz, R. (2021). Current trends in polymer based sensors.* Chemosensors, 9, *108. https://doi.org/10.3390/CHEMOSENSORS9050108 using own inputs and DALL·E.*

6. Automation of sensor manufacturing processes, employing advanced fabrication techniques such as printing or semiconductor technologies, to facilitate the production of miniaturised sensor arrays.
7. Enhancement of signal processing technologies and instrumentation to improve sensor functionality and reliability.
8. Creation of novel sensor substrates and internal electrodes for new planar and solid-state fabrication designs.
9. Development of conducting polymer hydrogels for their practical applications in biomedical sensors.
10. Improvement of production methods to commercialise affordable, portable sensory devices for non-specialised users, bridging the gap between fundamental research and real-life applications in response to societal demands.

Prioritizing by technological advancement and potential for broad impact, the trends in polymer-based sensors can be outlined as follows (Adhikari & Majumdar, 2004; Alam et al., 2022; Cichosz et al., 2018; Gamboa et al., 2023; Sanjuán et al., 2018):

1. Enhanced molecular recognition: Focusing on the improvement of receptor component immobilization through chemical modifications, this trend aims to fine-tune molecular recognition for selectivity and utilizes new materials for both transducers and chemical transduction strategies.

2. Multianalyte sensing capabilities: Innovations in sensor arrays that enable the detection of multiple analytes, which are crucial for complex applications like implantable sensors in medical diagnostics, highlight the progress in partially selective sensing technologies.
3. Sensor miniaturisation and integration: The development of compact sensor arrays with high-density, individually addressable elements facilitates advanced two-dimensional concentration mapping and signifies progress in the miniaturization and integration of sensor components.
4. Microtechniques and confined space phenomena: Advancements in the use of microtechniques and the understanding of physicochemical phenomena in dimensionally confined spaces are crucial for adapting sensor devices to handle small sample volumes, particularly in biosensor research.
5. Biochemical studies in non-aqueous media: Exploring biochemical reactions outside of aqueous environments opens new biosensing possibilities, expanding the applications of polymer-based sensors beyond traditional settings.
6. In-depth mechanism analysis: A new trend is the deepened investigation into the specific reaction mechanisms of sensory polymers, which is vital as these reactions are not fully understood, and their comprehensive analysis is essential for the development of more sophisticated sensing technologies.

With challenges and trends in mind, in the coming years, the trajectory of polymer sensor research is poised to advance through (Adhikari & Majumdar, 2004; Cichosz et al., 2018):

1. Refinement in the chemical modification techniques for the immobilization of receptor motifs.
2. Exploration and adoption of novel smart materials and the development of innovative production techniques.
3. Engineering of sensory materials capable of multi-response and distinct responses to varied stimuli.
4. In-depth investigation into the chemical and biochemical reactions occurring in aqueous, non-aqueous, and gas-phase environments.
5. Progression towards the miniaturisation and seamless integration of sensor components.
6. Marketing and impact on people's lives.

22.3 Market and industry outlook

As we delve into the realm of sensory polymers, we optimistically stand on the cusp of a potential market transformation. Despite this, it's important to acknowledge the marketing challenges that have been highlighted, along with the risk that these innovations may remain confined within research environments. Addressing these challenges head-on is crucial to transition sensory polymers from the laboratory to a market revolution, ensuring their successful commercialization and integration into everyday use.

In 2022, the global sensor market was valued at a formidable USD 204.8 billion, and it's projected to soar to an estimated USD 508.64 billion by 2032, flourishing at a compound annual growth rate (CAGR) of 8.40% from 2023 to 2032. This explosive growth trajectory spans across diverse sensor types—including biosensors, optical, RFID, image, temperature, touch, flow, pressure, and level sensors—powered by cutting-edge technologies like CMOS, MEMS, and NEMS. The sensory polymer field, poised for success, could be the linchpin in this expansion, finding its stride in key industries such as healthcare, IT/telecom, automotive, industrial, and aerospace & defence. The successful integration of sensory polymers into the market holds the promise of not just enriching these sectors but also of catalyzing a new era of innovation and economic prosperity (Sensor Market, 2023).

In a different market prospect, the market stood at over US$ 34 billion in 2020 and is projected to burgeon at a CAGR of 18% from 2021 to 2031. With an array of sensor types, including flow, image, motion, pressure, temperature and humidity, touch, and water sensors, and driven by technologies such as MEMS and CMOS, the market is poised to surpass a valuation of US$ 208 billion by 2031. This ascent reflects a rapidly evolving industry that is not only diversifying in its applications but also marking a significant imprint across various end-use industries, signaling a decade of unprecedented growth and technological breakthroughs (Smart/Intelligent Sensors Market, 2024).

Considering a specific technology market prospect, e.g., pH sensor market (glass type sensor, ISFET sensor, others), it was valued at USD 603 million in 2022 and is anticipated to reach USD 1601 million by 2029 (Arshak and Korostynska, 2007). And in the landscape of disease diagnostics, valued at $112 billion in 2021, antibody-related tests command a substantial market share of $43 billion (Dixit et al., 2022).

Within this frame, the introduction of new legislative measures presents an opportunity by significantly increasing the need for materials and equipment designed to monitor human-related hazards, such as exposure to toxic vapours and gases in work environments, contamination of water by industrial waste, and the use of pesticides in agriculture. This requirement extends into the medical sector as well, further driving the development of these essential materials. Polymers, known for their versatility and ability to be custom-modified or synthesised to address specific needs, have emerged as key components in the creation of sensory devices for these vital areas (Cichosz et al., 2018).

Recent advances in non-enzymatic optical and electrochemical biosensors for pesticide detection utilise stable recognition elements like antibodies, aptamers, and molecularly imprinted polymers (MIPs), often combined with nanomaterials for enhanced performance. These biosensors, including immunochromatographic assays and fluorescence assays, offer high sensitivity and selectivity and are ideal for on-site applications due to their simplicity and visual detection capabilities. Electrochemical biosensors provide low detection limits but are generally more complex and less suited for on-site use. Future improvements in biosensor performance are expected to leverage novel nanomaterials like metal-organic frameworks and bimetallic nanoparticles for their catalytic and optical properties, as well as aptamers and MIPs for their high stability and low cost. Sustainable challenges like simplifying sample preparation and ensuring method reproducibility must be addressed to transition these biosensors from the lab to commercial markets (Majdinasab et al., 2021).

22.4 Future prospects and research directions

This section can be conceptualized from two distinct but interconnected perspectives. The first offers a forward-looking view of polymeric sensors, examining the evolution of material usage across short-, medium-, and long-term horizons within various domains. The second perspective inverts this approach, focusing on the demands of specific application fields and identifying the requisite polymeric sensor materials tailored to these needs. Both perspectives grapple with the challenge of overlapping materials or applications, where certain contexts may demand more emphasis on specific polymers or sensor applications. To streamline the discussion for the reader, we endeavor to synthesize these two approaches, carefully curating the selection of both application fields and materials. This strategy aims to reduce redundancy while acknowledging that some repetition of materials and applications is unavoidable due to their intrinsic relevance across multiple contexts.

Regarding materials, our discussion focuses on molecularly imprinted polymers (MIPs), conducting polymers (CPs), composites and nanocomposites, and acrylic polymers. In this case, there is also an overlap between composites and nanocomposites with MIPs and CPs, as well as between the latter two. As for applications, we will concentrate on sensors for health and wearable applications, with a specific section dedicated to saliva analysis, food control and safety, environmental pollutants control, and wearable gloves and smart sensory textiles. Finally, we include three overarching sections that encompass these topics: multi-response and multi-tasking materials, response time and transduction processes, and leveraging smartphones for portable and convenient detection and recognition tasks.

22.4.1 Polymers imprinting polymers

The main strengths of MIPs, such as their robustness, reusability, and tailored specificity, are crucial for accurate analyte detection in diverse settings; however, challenges like limited dynamic range and possible cross-reactivity in complex matrices remain. They can be engineered to detect a wide array of substances, including enzymes, proteins, bacteria, viruses, metal ions, and toxins, serving as cost-effective and versatile sensors for use in electrical, chemical, optical, and electromechanical monitoring. Future developments in MIP technology aim to increase selectivity and sensitivity, broaden the spectrum of detectable substances, and leverage nanotechnology for superior sensing performance, e.g., in the antibody-related diagnostics market, valued at $43 billion in 2021, MIPs have the potential to compete with antibodies by offering cost-effective assay development and simplified reagent preparation in clinical settings, thereby standing to capture a significant market share (Alam et al., 2022; Wang & Zhang, 2023).

The hybridization of MIPs with nanotechnology opens the door to new possibilities, including addressing the specific challenges of these materials, such as, for example, the incorporation of carbon dots (CDs) and metal nanoparticles. Thus, since their discovery in 2004, carbon dots (CDs) have garnered significant attention for their physicochemical properties, including biocompatibility, optical features, and eco-friendliness, making them candidates with good potential for fluorescent probes in nanosensor applications. The integration of MIPs with green CDs addresses critical challenges in sensor development, particularly the need for improved selectivity and sensitivity towards specific analytes. This combination results in sensor platforms capable of precise and selective detection in complex matrices, including biological, food, and environmental samples. The attributes of MIPs, such as robustness and reusability, together with the advantageous properties of CDs, like water solubility and photostability, position MIP/CD-based sensors as formidable tools for a wide range of applications, promising significant advancements in the fields of biology, food safety, and environmental monitoring. Additionally, novel detection methods like lifetime changes and ratiometric sensing are poised to enhance MIP/CDs probing capabilities. Embracing green chemistry for reusable and eco-friendly sensing solutions, alongside developing techniques for micro- and macromolecule imprinting, will further broaden the application of these sensors in detecting viruses, bacteria, and proteins (Ansari & Masoum, 2021; Keçili et al., 2023; Lahcen & Amine, 2019).

Regarding metal nanoparticles, the synergy between metal/metal oxide nanoparticles (NPs) and MIPs is advancing sensor technology, offering enhanced optical, catalytic, and conductive properties for improved sensor functions, including enzyme-free biosensors and more efficient electrochemical reactions. While selectivity and stability require further enhancement, the potential for these composites in sensor development is vast, with research moving towards new applications and green synthesis methods. Future directions include reducing non-specific adsorption for greater MIP selectivity, exploring 2D materials like MXene (material composed of transition metals, carbon, and/or nitrogen, with unique properties) for heightened sensitivity, and developing MIPs for large biological entities to advance disease biomarker detection (Ait Lahcen et al., 2023).

22.4.2 Sensors based on conducting polymers

In the landscape of sensor technology, CPs stand at the forefront of several industrial needs, and researchers seek out innovations for greater efficiency and autonomy, where three primary trends emerge: miniaturization, process automation, and the development of multi-tasking devices. These polymers can pave the way for the development of sensors that operate with minimal human supervision and integrate smoothly with digital systems, surpassing traditional materials in efficiency and compatibility with computing analysis.

CPs, such as polypyrrole, polyaniline, and polythiophene, have become relevant in the development of innovative sensor technologies due to their unique electrical conductance changes in response to environmental stimuli. These materials enable the creation of highly sensitive and selective sensors for applications ranging from gas detection to biomedical diagnostics. For example, CP-based gas sensors can detect low concentrations of harmful gases by exhibiting significant changes in resistance or capacitance upon exposure, making them invaluable for air quality monitoring and industrial safety (Pascual et al., 2019; Rochat & Swager, 2013; Thomas et al., 2007). In the biomedical field, CPs are integrated into biosensors by coupling with biological recognition elements to detect disease biomarkers with high specificity. Challenges such as improving stability, reproducibility, and device integration are being addressed to enhance CP sensors' practicality.

Future research is focused on tailoring CPs through chemical modifications or hybridization with nanotechnology for higher performance, paving the way for their incorporation into portable, flexible, and wearable technologies. For example, through the incorporation of carbon nanotubes (CT). Thus, CP/CT nanocomposites, utilizing polymers like polyaniline, polypyrrole, polythiophene, and poly(*o*-phenylenediamine), exhibit enhanced chemical and electrical properties for sensor applications, thanks to the synergistic effects of CPs and CTs. These nanocomposites have been effectively used to detect a wide range of biomolecules, gases, metal ions, environmental pollutants, and in strain and pressure sensors, underscoring their versatility. However, further research is necessary to improve their sensitivity, selectivity, detection limits, recovery, and response times for broader and more efficient applications (Mostafa et al., 2022).

22.4.3 Polymer composites and nanocomposites

Polymer nanocomposites (PNCs) synergise a polymeric matrix with nanoscale fillers, enhancing mechanical, optical, and electrical properties and garnering significant interest for their performance improvements and versatility in forms like films, fibers, or coatings. These advancements position PNCs as key materials in the development of chemical and biological sensors with increased sensitivity, selectivity, and response times across various applications. Additionally, polymers' role in electronic devices extends to pristine, composite, or nanocomposite forms, offering a broad spectrum of domestic and industrial applications with unique functionalities. The conductance in polymer nanocomposites, determined by the filler networks within the matrix, showcases their potential in creating sensors for detecting physical parameters such as strain, pressure, and temperature, leveraging the stimuli-responsive nature of these conductive networks, with heightened sensitivity, selectivity, and response times, making them highly effective for a wide range of applications. Despite ongoing technical and phenomenological research and debates on their formation mechanisms, the exploration of polymer nanocomposites continues to demand extensive scientific effort to unlock their full potential and elucidate the underlying principles governing their functionalities (Shukla & Saxena, 2021).

Graphene and its derivatives, known for their excellent electrical and mechanical properties and large specific surface area, have drawn significant research interest, especially when combined with CPs like polyaniline, polypyrrole, and others, to form novel nanocomposite materials for gas sensing. These composites, benefiting from synergistic effects, outperform pure graphene or CPs alone, offering enhanced sensing capabilities at room temperature. Despite the promising results and various preparation methods like in situ polymerization and electropolymerisation enhancing gas sensor performance, the potential of graphene-conducting polymer chemiresistors remains underexplored, with limited studies addressing their responsiveness to different gases (Zamiri & Haseeb, 2020).

Building on this momentum, polymer composite-based flexible pressure sensors are advancing rapidly, with potential uses in health monitoring, electronic skin, tactile systems, and bio-sensing. These sensors, which include capacitive, piezoresistive, and piezoelectric types, are gaining attention for their high sensitivity, low-cost production, and biocompatibility. Future research aims to create new polymer materials with adjustable electrical and mechanical features, enhance sensitivity to various pressures, and integrate with other sensors for multifunctional capabilities. Efforts are also geared toward developing flexible energy sources for wearables and refining manufacturing processes to boost reliability and durability for real-world applications (Guo et al., 2023).

The integration of polymer composites with mechanochromic materials has led to the development of innovative colorimetric mechanical sensors that transform mechanical actions—like force, pressure, strain, and impact—into visually perceptible color changes. This unique blend offers a dual advantage: the polymer matrices ensure the sensors' durability and reusability under various mechanical stresses, while the mechanochromic components provide a distinct color response to different types of mechanical stimuli, enabling the quantification of forces directly by the naked eye. Such sensors are pioneering a shift away from conventional, electrically-based mechanochromic sensors, favouring a simpler, more intuitive approach to detecting and monitoring mechanical changes. As research progresses, the future of these sensors holds promise for broader applications, from construction safety to transportation, where rapid, equipment-free detection of mechanical failure or stress is crucial. This trend towards material and functional diversification in sensor design marks a significant advance in the field of mechanical sensing, emphasizing the growing importance of visual diagnostics in real-world applications (Inci et al., 2020).

22.4.4 Sensors based on acrylic polymers

Acrylic polymers have carved a niche in the research field of sensor technology, primarily due to their inherent capacity to be chemically tailored. The functionalisation of these polymers involves covalently attaching sensing groups to the polymeric backbone, allowing for the creation of materials with a broad spectrum of properties to meet specific sensing requirements. This adaptability is particularly vital in the development of sensors, where the detection and measurement of various environmental stimuli or analytes are critical.

Among the acrylic polymers, acrylamide derivatives and their copolymers, as well as methacrylic and acrylic acids, are frequently utilized. These compounds serve as versatile canvases onto which a plethora of functional groups can be grafted. The choice of functional group is pivotal as it determines the polymer's sensitivity and selectivity towards particular stimuli. For instance, the incorporation of ionic groups can render the polymer sensitive to changes in pH or electrolyte concentration, making them suitable for use in pH sensors or ion-selective electrodes. Hydrophobic groups, on the other hand, can be used to create sensors that respond to organic compounds or nonpolar substances.

The process of incorporating these functional groups into the acrylic moieties is an interplay of organic chemistry and material science. A wide array of monomers can be copolymerized to introduce the desired functionality into the acrylic polymer. This synthesis process is not merely a matter of chemical reaction but a deliberate and strategic design to achieve the optimal configuration for the intended sensing application.

The design of sensor-based materials using acrylic polymers is not without its challenges. One must carefully consider the environment in which the sensor will operate, the longevity of the sensor's response, and the potential for interference from other substances. Moreover, the physical form of the sensor—whether it be a thin film, a bead, a fiber, or a coating—must be compatible with the polymer's properties and the application's demands.

The use of acrylic polymers in sensor technology showcases the remarkable potential of synthetic polymers when combined with the power of chemical customisation. Through meticulous design and synthesis, these polymers can be transformed into highly sensitive and selective sensors, opening doors to new possibilities in environmental monitoring, medical diagnostics, and beyond (Arnaiz et al., 2022; González-Ceballos et al., 2021; Guirado-Moreno et al., 2021; Guirado-Moreno, Carreira-Barral, et al., 2023). Solid-state chemistry has recently facilitated the preparation of materials featuring chemically unstable groups under ambient conditions. This innovative approach takes advantage of the unique chemical behaviors exhibited by macromolecular environments in solid states, which significantly differ from their behaviors in solution or as low molecular mass entities (Bustamante et al., 2019).

22.4.5 Polymeric mechanochemical force sensors

Polymer mechanochemistry, leveraging mechanical forces to trigger chemical reactions, has emerged as a promising avenue for developing flexible force sensors suited for the IoT, bioelectronics, wearable electronics, and fluid-based manufacturing processes like bioprinting. Significant progress in designing mechanophores, primarily driving optical changes and moving towards reversible systems, marks a leap in application potential. However, hurdles such as achieving reproducibility, enhancing sensitivity, and shortening response times still stand.

The translation of these materials into viable applications necessitates a deeper understanding of the interactions within the polymer host network and the development of systems that allow for activation without matrix damage. Innovations like filler incorporation, double networks, or micelle-based structures in hydrogels have been proposed to overcome these challenges. Furthermore, the integration of these materials into broader technologies calls for standardised calibration and testing methods, especially for novel electrical signal outputs.

Addressing outstanding questions, such as developing reversible electronic property mechanophores and ensuring insensitivity to non-target stimuli, is crucial for advancing the field. The quest for cost-effective scale-up, the impact of manufacturing forces, and the establishment of necessary standards and infratechnology underscore the multidisciplinary effort needed to transition these sensors from laboratory to commercial success. Fig. 22.2 schematically shows the development phases of flexible force sensors using mechanosensitive polymers, from the design to commercialization (Willis-Fox et al., 2023).

22.4.6 Wearables and health

The field of biomonitoring has seen significant interest in the development of advanced wearable strain sensors due to their easy implementation, enhanced diagnostic capabilities, improved user experience, and long-term monitoring potential. These wearable sensors find applications across a spectrum of areas, including healthcare, medical diagnosis, robotics, prosthetics, virtual reality, professional sports, and entertainment. Traditional metallic foil or semiconductor strain gauges, due to their strain limitations, prove impractical for wearable applications. Ideal strain sensors should exhibit high stretchability, sensitivity, and durability for extended use without performance degradation. Over the last decade, strides in material science, nanotechnology, and microelectronics, combined with the growing demand for wearable technologies, have led to notable progress. A typical polymer nanocomposite strain sensor comprises a conductive network for signal generation and an elastomeric polymer for flexibility, stretchability, and protective insulation, requiring careful material and design choices to balance sensitivity and stretchability. Factors like electron conduction mechanisms, including disconnection, crack propagation, and tunneling, are crucial for optimizing strain sensing performance. Despite these advancements, several challenges remain. Material selection is complex, with each conductive nanomaterial presenting unique drawbacks—from limited conductivity and aggregation issues in carbon black to the high cost and poor adhesion of metal nanowires, alongside the processing difficulties of conductive polymers. Achieving high stretchability, sensitivity, and linearity in a single sensor is challenging due to the conflicting nature of these requirements. Additionally, scalability and cost-effectiveness are critical for commercialization, necessitating innovations like fully printed electronics for mass production. Integrating wearable strain sensors into a complete system with power, signal processing, and communication capabilities remains an unmet goal, calling for interdisciplinary efforts to address reliability, robustness, and power efficiency. Future developments will likely explore attractive features such as self-healing, self-powering, and biodegradability, requiring further research and collaboration between

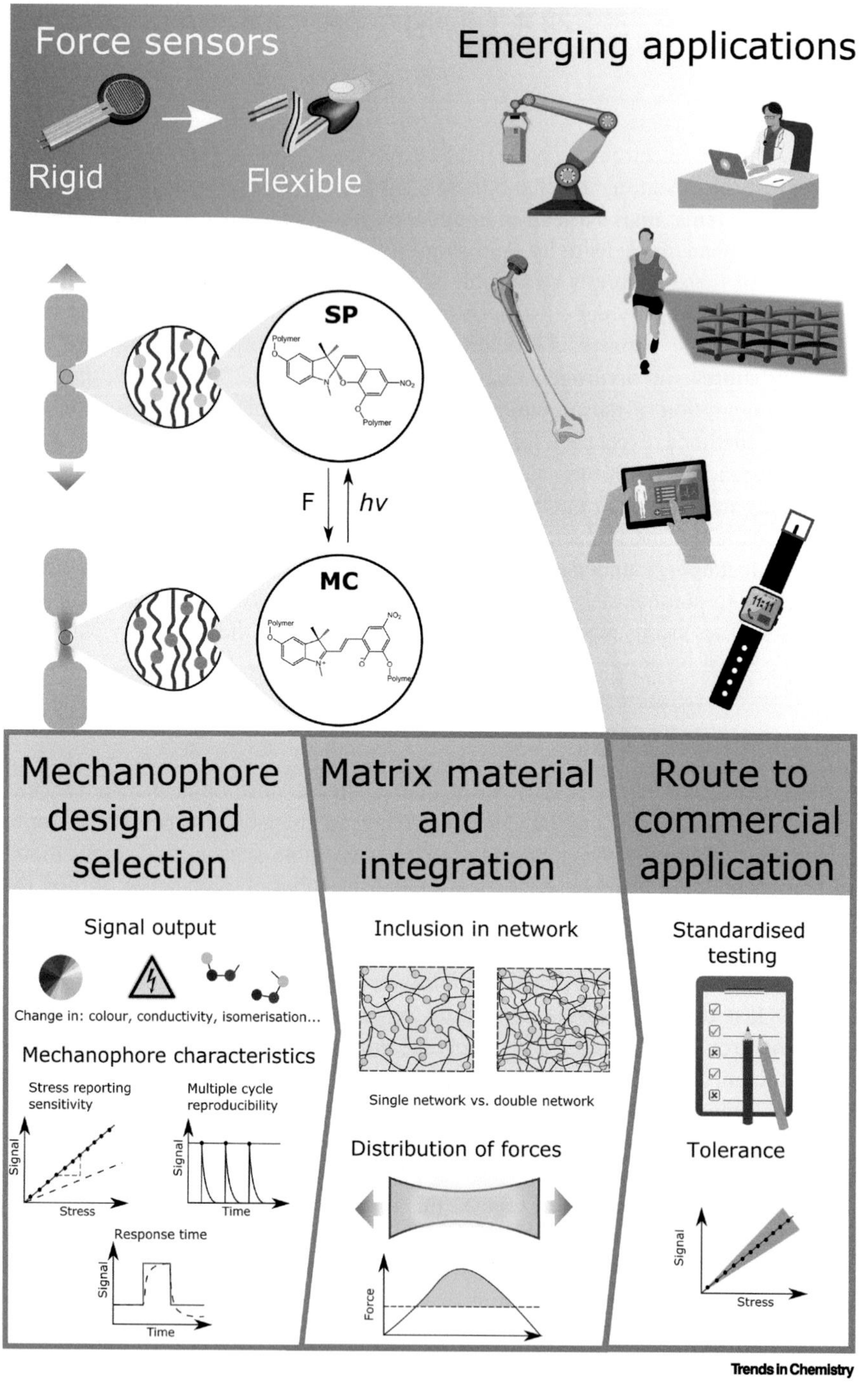

FIGURE 22.2

Flexible force sensors. Summary of the development phases of flexible force sensors using mechanosensitive polymers.

Source: *Image published under CC BY 4.0 license, from Willis-Fox, N., Watchorn-Rokutan, E., Rognin, E., & Daly, R. (2023). Technology pull: Scale-up of polymeric mechanochemical force sensors.* Trends in Chemistry, 5, *415–431. https://doi.org/10.1038/s41598-017-09755-4.*

scientists and industry professionals to surmount these obstacles (Lu et al., 2019; Palumbo & Yang, 2022; Veeramuthu et al., 2021; Wang et al., 2023; Yadav et al., 2023).

In recent years, there has been notable progress in the use of hydrogels and conducting polymer hydrogels (CPH) for diagnostic, wearable, and implantable biomedical sensors, leveraging their unique properties and advanced fabrication techniques, particularly in applications that demand flexibility and resilience, such as electronic skin (e-skin) sensors. E-skin sensors represent a growing trend in the utilization of CPHs for wearable technology and innovations in human health monitoring by closely mimicking the tactile and responsive attributes of natural skin. This burgeoning area focuses on creating sensors that are not only sensitive to changes in strain but also capable of conforming to the dynamic contours of the human body, thereby offering precise and uninterrupted monitoring capabilities. CPHs are especially suited to these applications due to their inherent high mechanical strength and stretchability, attributes that are essential for wearables that require durability and flexibility. Moreover, the physical properties of CPHs, including their stretchability, can be easily adjusted through simple modifications in their chemical composition or structure, allowing for customisation to specific sensor requirements. This tunability is a key advantage, enabling the design of sensors that can maintain high performance across a range of mechanical stresses, for durable, flexible wearables that can autonomously repair themselves, promising a new era in real-time, seamless human health monitoring (Gamboa et al., 2023; Tran et al., 2022).

Early and cost-effective cancer detection through polymeric sensors represents a pivotal advancement in healthcare, offering the potential to significantly improve patient outcomes by identifying malignancies at their most treatable stages. These sensors, with their high specificity and sensitivity, are emerging as crucial tools in the timely diagnosis and management of cancer. Thus, cancer bio- and chemosensors, especially those utilising electrochemical approaches, are increasingly recognised for their high sensitivity and selectivity in detecting trace cancer biomarkers, with MIPs showing promise due to their cost-effectiveness. Despite existing challenges in performance optimisation, MIP-based sensors are well-positioned for future advancements. Prospects include novel MIP synthesis for enhanced binding, integration with microfluidics for portable diagnostics, smartphone linkage for rapid self-testing, and aptamer hybridisation for multiplexed biomarker detection, all contributing to the potential of these sensors to revolutionise cancer diagnosis and therapy (Ben Moussa, 2023).

The same rationale applies to rare diseases, such as cystic fibrosis, whose detection and disease-following through polymeric sensors underscores a crucial leap in personalised medicine, enabling early diagnosis and treatment strategies tailored to individual needs. These innovative sensors offer a non-invasive, rapid, and cost-effective means to monitor and manage complex conditions, significantly enhancing patients' quality of life (Vallejos et al., 2018).

22.4.7 Trends in wearable glove sensors and smart sensory textiles

Wearable glove-based sensors represent a significant leap forward in the field of wearable technology, merging the practicality of gloves with advanced sensing capabilities for a wide array of applications. These sensors serve as versatile analytical tools that enable the detection of various substances, from illicit drugs and hazardous chemicals to pathogens, on different surfaces. This technology finds its utility in fields such as crime scene investigation, airport security, and disease control, offering a unique solution for analysing both dry and liquid samples by simply swiping the sensor-equipped glove over the target surface.

The integration of glove-based sensors into smart sensory textiles and garments further expands their potential, blending seamlessly with daily wear to provide continuous health monitoring without sacrificing comfort or convenience. Unlike other wearable sensors that may pose risks of infection or discomfort through prolonged use, glove-based sensors offer a non-intrusive alternative that fits naturally into the user's lifestyle. The development of these sensors heavily relies on the careful selection of glove sensory materials and conducting nanomaterials, with a focus on enhancing their functionality through various transducer modification techniques tailored for real-world applications.

Challenges in material selection, sensor design, and the integration of sensors with textiles are addressed through innovative approaches, showcasing the adaptability and potential of this technology. Moreover, the sustainability of these sensors is a crucial consideration, with studies exploring strategies for the proper disposal of used sensors in alignment with Sustainable Development Goals (SDGs).

As the technology behind wearable glove-based sensors and smart sensory textiles evolves, it promises not only to enhance public safety and health monitoring but also to introduce new dimensions of interaction with our environment. This advancement heralds a future where wearable technology becomes an indispensable part of our daily lives, offering insights and safety measures that were previously unattainable. The synergy between glove-based sensors and smart textiles underscores the potential for wearable technology to create garments that not only offer comfort and style but also serve as powerful tools for health and safety monitoring (González-Ceballos et al., 2020; Pablos, Trigo-López, et al., 2014; Pablos et al., 2015; Trigo-López et al., 2018; Tsong et al., 2023).

22.4.8 Sensing of analytes in saliva

The detection of biomarkers in saliva has emerged as a pivotal strategy for noninvasive health monitoring, offering a wealth of information about physiological and pathological states without the discomfort and invasiveness associated with blood tests. Saliva, as a diagnostic fluid, contains a wide range of analytes, including hormones, enzymes, antibodies, and nucleic acids, whose concentrations can reflect the body's health condition, making it an ideal medium for early disease detection, monitoring of health status, and personalised medicine (Cardoso et al., 2023).

Polymer chemosensors represent a significant advancement in the field of saliva-based diagnostics, offering high sensitivity, selectivity, and rapid response times for the detection of various analytes. These sensors leverage the unique properties of polymers, such as their structural versatility and ability to undergo functionalisation, to create highly specific interactions with target biomolecules. The incorporation of recognition elements within the polymer matrix enables the direct transduction of chemical interactions into measurable signals, facilitating the quantitative analysis of saliva samples. An exemplary application of this technology can be seen in the efforts to combat the SARS-CoV-2 pandemic. Traditional testing methods are invasive and require complex sample collection. A fluorogenic peptide substrate of the virus's main protease (Mpro) was covalently immobilized in a polymer to coat cellulose-based materials. These sensory labels exhibited fluorescence upon contact with saliva from COVID-19 positive patients. This approach underscored the potential of polymer chemosensors in early disease detection, offering a simple, noninvasive, and rapid testing solution (Fig. 22.3; Arnaiz et al., 2023).

However, the development of polymer chemosensors for saliva analysis faces several challenges. One of the primary obstacles is the complex composition of saliva, which contains a wide range of substances that can interfere with sensor performance. Additionally, the low concentration of certain biomarkers in saliva necessitates the development of highly sensitive sensors capable of detecting analytes at nanomolar or even picomolar levels. Furthermore, the stability of polymer chemosensors in the oral environment, where factors such as pH and temperature can vary significantly, is crucial for reliable measurements.

Recent progress in polymer chemosensor technology has led to the development of innovative approaches to overcome these challenges. Advances in polymer chemistry have enabled the synthesis of novel materials with enhanced sensitivity and selectivity for specific biomarkers. Nanotechnology has been instrumental in improving sensor performance, incorporating nanoparticles into polymer matrices, enhancing signal transduction, and enabling the detection of low-abundance analytes. Moreover, the integration of polymer chemosensors with microfluidic devices and wearable technology has opened new avenues for real-time, on-site saliva analysis.

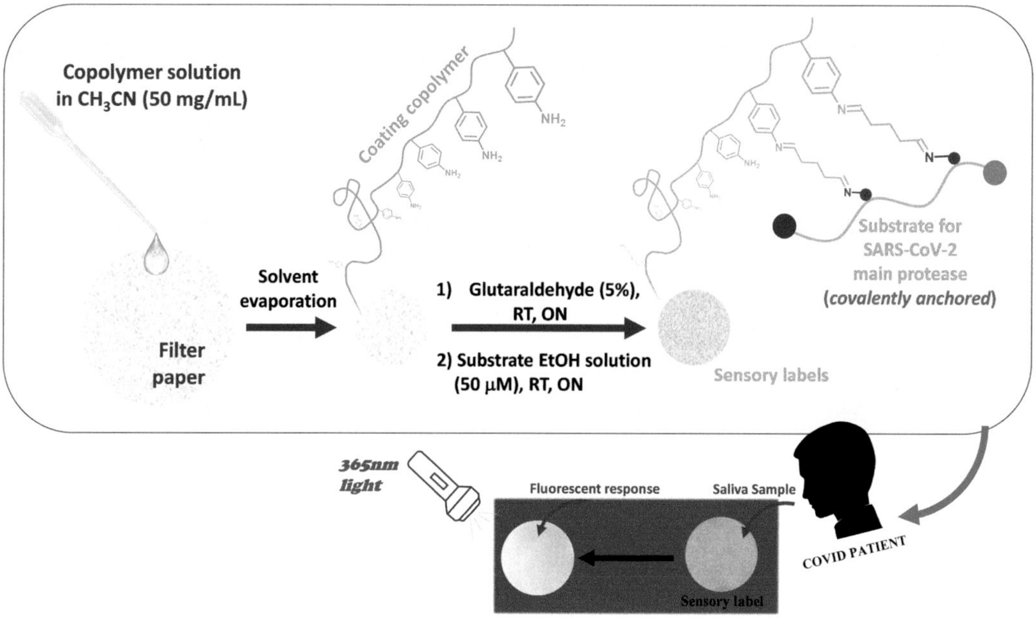

FIGURE 22.3

SARS-CoV-2 in saliva polymeric sensor. Polymeric sensory film for the detection of SARS-CoV-2 in saliva based on a peptide substrate containing a FRET pair (fluorophore and quencher) to detect the Mpro protein.

Source: *Published under CC BY 4.0 license from Arnaiz, A., Guirado-Moreno, J. C., Guembe-García, M., Barros, R., Tamayo-Ramos, J. A., Fernández-Pampín, N., García, J. M., & Vallejos, S. (2023). Lab-on-a-chip for the easy and visual detection of SARS-CoV-2 in saliva based on sensory polymers.* Sensors and Actuators B: Chemical, 379, 133165. *https://doi.org/10.1016/j.snb.2022.133165.*

Looking ahead, the future of saliva-based diagnostics with polymer chemosensors is bright, with several promising research directions emerging. The design of multiplexed sensors capable of simultaneously detecting multiple analytes in saliva could significantly enhance the utility of saliva as a diagnostic medium. The development of smart polymers that can respond to changes in analyte concentration with a proportional change in signal offers the potential for dynamic monitoring of health conditions. Additionally, the exploration of biodegradable and biocompatible polymers could address environmental concerns associated with sensor disposal and expand the applicability of these sensors in personalized medicine.

22.4.9 Food control and safety

Sensory polymers are set to improve food safety and quality control, integrating smart packaging for detecting contaminants, along with sensors that monitor temperature, humidity, gas levels, and pH to ensure the freshness and safety of food products. These polymers are central to developing advanced sensors for a range of analytes, including biological and chemical contaminants, allergens, nutritional ingredients, and potentially harmful food additives, thereby facilitating smart systems for shelf-life prediction, nutrient monitoring, and comprehensive quality assurance throughout the food supply chain (González-Ceballos et al., 2023; Guirado-Moreno et al., 2023; Oveissi et al., 2021; Vallejos et al., 2022).

Polymer chemosensors, particularly colourimetric sensors, are increasingly integrated into food packaging materials due to their user-friendliness and ability to indicate thermal histories, such as temperature fluctuations and thawing cycles, which indirectly model the freshness and safety of food. Although these sensors do not directly measure the food's condition, they provide valuable information on potential quality degradation. Moreover, humidity sensors have been developed to maintain the quality of dry packaged foods, and gas concentration changes in packaging are monitored to detect food spoilage and microbial activity. Examples of such market-available sensors include Timestrip, MonitorMark by 3 M, and the Ageless Eye by Mitsubishi, which signal key environmental changes within food packaging (González-Ceballos et al., 2022).

For instance, MIPs have been extensively studied to enhance food safety through their ability to efficiently, sensitively, and cost-effectively detect food safety hazard factors (FSHFs) such as chemicals, pathogens, and physical contaminants. Leveraging their high specificity and stability under adverse conditions, MIPs are integrating into advanced sensing platforms like electrochemical and optical sensors, extending the detection range from small molecules to pathogens and toxins, and offering portable solutions for rapid, on-site analysis by nonspecialists, marking a significant advancement in food analysis technology (Cao et al., 2019). For instance, MIP-based optical sensors represent a solution for the selective, sensitive, and cost-effective detection of pesticides in food systems, addressing the critical need for monitoring pesticide residues that pose health risks based on fluorescence, colourimetry, and spectroscopy (Fang et al., 2021). MIPs coupled with carbon nanomaterials have been exploited in the field for creating electrochemical, optical, and mass-sensitive sensors that enhance the detection of food hazards to achieve specificity and sensitivity. This innovative approach leverages the advantages of carbon nanomaterials and the high selectivity of MIPs, though it faces challenges in fully understanding the binding mechanism, optimising selectivity, stability, and the necessity for sample pretreatment due to complex food matrices (Chi & Liu, 2023).

Utilising sensory polymers for their adaptability and sensitivity, both the food industry and consumers stand to gain through increased accuracy in quality control, minimised waste, heightened safety measures, and adherence to international food safety standards. However, sensory systems in the food industry must still vie with traditional analytical methods regarding cost, performance, and reliability. Furthermore, in the food industry, where safety, quality control, and consumer trust are of paramount importance, the reluctance of producers to adopt smart labels indicating spoilage levels stems from commercial concerns. Despite the critical need for innovative sensors capable of detecting freshness, spoilage, toxicity, and overall quality, which could revolutionise the precision and reliability of food monitoring systems, commercial apprehensions hold back their widespread implementation (Oveissi et al., 2021; Pavase et al., 2018).

The detection of residual antibiotics in food is crucial for ensuring public health and regulatory compliance, given the risks associated with antibiotic pollution stemming from agricultural and veterinary practices. Luminescent sensors have gained prominence in this arena due to their ability to offer rapid, sensitive, and specific detection, positioning them as an invaluable tool in the fight against antibiotic contamination in the food supply. Among various sensor technologies, MIPs have shown promise, exhibiting selectivity and sensitivity, offering advantages such as rapid response times, cost-effectiveness, and the ability to operate without the typical sample pretreatment (Zhang et al., 2024).

22.4.10 Sensing environmental pollutants

The detection of environmental pollutants has become an increasingly critical concern, given their pervasive impact on ecosystems, human health, and global sustainability. Pollutants such as heavy metals, pesticides, volatile organic compounds, and particulate matter pose severe risks, including chronic diseases, ecological degradation, and acute toxicity. Sensing these contaminants accurately and promptly is vital for implementing effective environmental management strategies, ensuring the safety of drinking water, improving air quality, and adhering to regulatory standards aimed at protecting both the environment and public health. Advances in sensor technology, particularly those utilising novel materials and sophisticated detection mechanisms, offer the potential to revolutionise our ability to monitor and respond to environmental pollutants, facilitating more informed decisions and actions toward mitigating pollution and promoting a healthier planet.

In this context, the development of polymer nanocomposite-based electrochemical sensors is a tool for detecting environmental pollutants, offering an alternative to traditional, labour-intensive analytical methods. These sensors capitalise on the electrocatalytic activities, conductivity, and surface area of polymer nanocomposites, delivering sensitive and real-time monitoring of contaminants like heavy metals and pesticides. Challenges such as achieving uniform dispersion of nanofillers and simplifying production processes persist, with current efforts being a bridge between laboratory innovation and commercial viability. Future research is geared toward novel synthetic methods and efficient integration of nanoparticles with conductive polymers to enhance sensor performance, with biopolymer nanocomposites presenting a nontoxic, biodegradable, and biocompatible option for environmental sensing. This review has consolidated recent advancements and highlighted the potential of polymer nanocomposites in creating advanced, cost-effective, and user-friendly sensors for widespread environmental application (Tajik et al., 2021). For instance, sensory polymers can be used to detect and remove toxic cations, i.e., mercury ions, from complex aqueous environments for their sensitivity, structural stability, and tunable sensing properties. These sensors incorporate mercury-specific binding sites within their structures, which are promising for providing accurate signalling and effective separation (Pablos, Ibeas, et al., 2014; Sharma & Lee, 2022; Vallejos et al., 2011, 2017).

22.4.11 Multiresponse and multitasking materials

In the sensory polymer field, one of the most significant challenges has been the development of sensors with the capability to selectively respond to multiple stimuli simultaneously. The creation of these multi-tasking materials is not just an incremental improvement, but a transformative leap that could significantly broaden the utility of standard sensors. These advanced polymers are envisioned to possess a level of sensitivity and specificity that allows them to discern and react to a complex array of environmental changes in real-time. Such sensors could integrate multiple sensing functions within a single platform, enabling the detection of a diverse range of analytes—from temperature and pressure to chemical and biological substances—thereby streamlining processes across various industries. This innovation could lead to more comprehensive monitoring systems that can provide detailed environmental assessments, enhance safety protocols, and contribute to smarter, more responsive technology applications (Cichosz et al., 2018).

Multi-tasking in sensory polymers can be achieved through a combination of approaches: Sensory arrays: Deploying an assembly of sensors, each tailored to a specific stimulus, allows for simultaneous data collection on various environmental factors; material engineering: Crafting polymers with molecular structures that are responsive to diverse stimuli enables the detection of multiple environmental changes in a single material; hybrid materials: Integrating different reactive components within one polymer, such as metal nanoparticles for magnetic sensitivity and organic compounds for chemical reactivity, bestows multiple sensing capabilities; molecular imprinting: This method creates selective binding sites within polymers, which, when coupled with electrical or optical signals, can detect and quantify multiple specific substances; smart nanocomposites: The incorporation of nanoscale elements into polymers imparts the ability to sense and respond to various stimuli like light, pressure, and chemical interactions; cross-responsive networks: Connecting sensors with overlapping detection abilities forms a responsive network capable of identifying complex stimuli through interlinked reactions; chemometrics, data analysis and AI: Utilising advanced algorithms and AI to analyse sensor output enhances pattern recognition and differentiates between similar stimuli, effectively amplifying the sensor's discernment capabilities, thus enabling the differentiation and quantification of multiple analytes from the polymers' responses (Vallejos et al., 2011).

Similarly, sensory materials that yield varied responses to a single stimulus are essential for deepening data collection accuracy and breadth, thereby facilitating a more thorough analysis and understanding of environmental or analyte conditions (Pascual et al., 2019).

22.4.12 Improving the response time and the transduction processes

One of the paramount challenges in enhancing the functionality of sensory polymers lies in diminishing the detection or response time, particularly when it comes to identifying dangerous substances such as heavy metal ions or explosives. The rapid identification of these hazardous materials is crucial for ensuring safety and mitigating risks, making the need for swift sensor responses a critical aspect of sensor development.

Furthermore, the transduction process, which converts the chemical interaction between the sensor's receptor and the target analyte into a quantifiable signal (be it electrical, mechanical, or optical), presents a significant opportunity for advancement. By refining this conversion mechanism, sensors can achieve more precise, reliable, and sensitive detection capabilities. The focus on enhancing the transduction process is aimed not just at improving the quality of detection but also at reducing the time it takes to obtain measurable results. Additionally, modifying the microstructure can enhance critical parameters. For example, employing foaming techniques can decrease response times and improve detection and quantification limits due to the expanded specific surface area available for analyte interaction (Pascual et al., 2018).

Recent developments in sensor technology, particularly the incorporation of microfluidic fabrication techniques and quartz crystal microbalances (QCMs), have already made notable strides in shortening detection times across various sensing applications. These innovations leverage the precise control of fluid flow and the sensitive measurement of mass changes, respectively, to expedite the detection process without compromising accuracy. As research continues to evolve in this area, the integration of such advanced technologies into sensory polymer applications holds the promise of setting new benchmarks for response times and transduction efficiency in the detection of hazardous substances (Sanjuán et al., 2018).

22.4.13 Leveraging smartphones for portable and convenient detection and recognition tasks

The coupling of sensory polymers with smartphone technology is emerging as a novel approach for portable detection and recognition. The development of sensors that combine sensory polymers with smartphones leverages the latter's built-in sensors and computational power, offering a portable and immediate solution for detection needs across multiple domains, including environmental monitoring, clinical and point-of-care diagnostics, and bioanalysis. This could be considered an alternative approach to sensing devices that allow the creation of a cost-effective, quickly fabricated, and easy-to-use detection system (Guembe-García et al., 2021, 2022). Despite these advantages, the transition of sensory polymer-based smartphone sensors from lab to market again faces significant challenges. These include the previously commented optimization need for the sensory performance of sensory polymers and the limitations of current smartphone imaging and computing capabilities. However, as technology progresses, these initial hurdles are expected to be overcome, paving the way for future commercial applications (He et al., 2024).

The colourimetric method, favoured for its simplicity, speed, and broad applicability, stands out for its potential to transform these devices into versatile analytical instruments for health monitoring and safety assessments (Qian et al., 2022). For instance, the performance of colourimetric sensory polymers from the qualitative on-site visual detection can be greatly improved with the smartphone assisted on-site fine quantification, as was demonstrated in the easy nitrite analysis of processed meat with colourimetric sensory polymers and a smartphone app. The method involves capturing a photograph of a calibration color chart alongside test polymeric sensory discs using the smartphone app 'Colorimetric Tritration', allowing for self-calibration of the system under identical

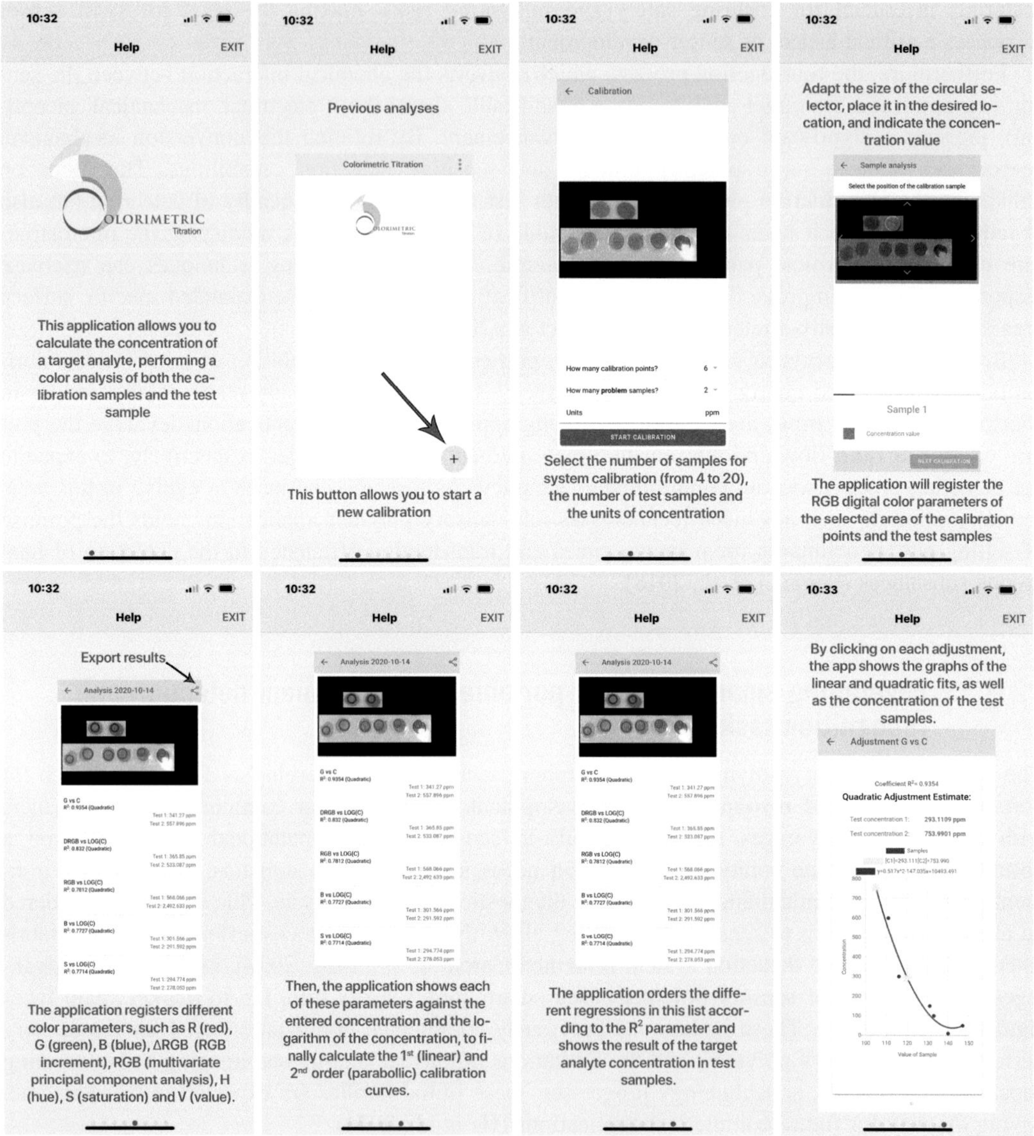

FIGURE 22.4

Smartphone's app for quantification of a target species. Colorimetric titration: sample of a smartphone's app for determining the concentration of a target species based on a colourimetric or fluorogenic sensory film. It can be free downloaded from App Store and Google Play.

Source: *Screenshots of the Colorimetric Titration App (available on Google Play. https://play.google.com/store/apps/details?id5es.inforapps.chameleon&gl5ES [Accessed February 2024]; and App Store. https://apps.apple.com/si/app/colorimetric-titration/id1533793244 [Accessed February 2024]).*

conditions for both elements. This approach ensures the method's independence from smartphone models, lighting, and distance. The app then analyses digital colour parameters and performs multiple fits to determine concentration levels based on the best R2 coefficient, demonstrating the process's adaptability and precision without specific lighting requirements (Fig. 22.4; Guembe-García et al., 2022; Guirado-Moreno, González-Ceballos, et al., 2023).

22.5 Conclusions

To sum up, any commonly utilized polymer can function as a sensor, provided it is incorporated with suitable additives or undergoes precise modifications. Sensory polymers are unlocking significant advancements due to their adaptability, processing ease, and compatibility, both as intrinsic sensing elements and through functionalisation for enhanced detection capabilities. The journey toward realising their full potential involves not only overcoming challenges in developing novel, reliable, sensitive, selective, and biocompatible materials but also in device preparation and market integration. Collaborative efforts between polymer scientists and technologists are essential to refine polymer-based sensor technologies, aiming to produce miniaturised, multifunctional sensor arrays. This integrated approach addresses the dual challenge of creating sophisticated devices that meet stringent market demands while navigating the complexities of commercialisation, setting the stage for widespread adoption in various applications.

AI disclosure

During the preparation of this work the author(s) used DALL·E in order to construct and adapt Fig. 22.1B from a reference and using own inputs and DALL·E. After using this tool/service, the author(s) reviewed and edited the content as needed and take(s) full responsibility for the content of the publication.

Acknowledgment

The financial support provided by MCIN/AEI/10.13039/501100011033 and the "European Union NextGenerationEU/PRTR" (TED2021-129419B-C21) is gratefully acknowledged.

References

Adhikari, B., & Majumdar, S. (2004). Polymers in sensor applications. *Progress in Polymer Science*, *29*, 699–766. Available from https://doi.org/10.1016/j.progpolymsci.2004.03.002.

Ait Lahcen, A., Lamaoui, A., & Amine, A. (2023). Exploring the potential of molecularly imprinted polymers and metal/metal oxide nanoparticles in sensors: recent advancements and prospects. *Microchimica Acta*, *190*, 1–25. Available from https://doi.org/10.1007/s00604-023-06030-4.

Alam, M. W., Bhat, S. I., Al Qahtani, H. S., Aamir, M., Amin, M. N., Farhan, M., Aldabal, S., Khan, M. S., Jeelani, I., Nawaz, A., & Souayeh, B. (2022). Recent progress, challenges, and trends in polymer-based sensors: a review. *Polymers*, *14*, 2164. Available from https://doi.org/10.3390/polym14112164.

Alberti, G., Zanoni, C., Losi, V., Magnaghi, L. R., & Biesuz, R. (2021). Current trends in polymer based sensors. *Chemosensors*, *9*, 108. Available from https://doi.org/10.3390/CHEMOSENSORS9050108.

Ansari, S., & Masoum, S. (2021). Recent advances and future trends on molecularly imprinted polymer-based fluorescence sensors with luminescent carbon dots. *Talanta*, *223*, 121411. Available from https://doi.org/10.1016/j.talanta.2020.121411.

Arnaiz, A., Guembe-García, M., Delgado-Pinar, E., Valente, A. J. M., Ibeas, S., García, J. M., & Vallejos, S. (2022). The role of polymeric chains as a protective environment for improving the stability and efficiency of fluorogenic peptide substrates. *Scientific Reports*, *12*, 121411. Available from https://doi.org/10.1038/s41598-022-12848-4.

Arnaiz, A., Guirado-Moreno, J. C., Guembe-García, M., Barros, R., Tamayo-Ramos, J. A., Fernández-Pampín, N., García, J. M., & Vallejos, S. (2023). Lab-on-a-chip for the easy and visual detection of SARS-CoV-2 in saliva based on sensory polymers. *Sensors and Actuators B: Chemical*, *379*, 133165. Available from https://doi.org/10.1016/j.snb.2022.133165.

Arshak, K., & Korostynska, O. (2007). State-of-the-art polymer based pH sensors. Sensors, Special Issue. Available from https://www.mdpi.com/journal/sensors/special_issues/polymer_ph_sensors.

Ben Moussa, F. (2023). Molecularly imprinted polymers meet electrochemical cancer chemosensors: A critical review from a clinical and economic perspective. *Microchemical Journal*, *191*, 108838. Available from https://doi.org/10.1016/J.MICROC.2023.108838.

Bustamante, S. E., Vallejos, S., Pascual-Portal, B. S., Muñoz, A., Mendia, A., Rivas, B. L., García, F. C., & García, J. M. (2019). Polymer films containing chemically anchored diazonium salts with long-term stability as colorimetric sensors. *Journal of Hazardous Materials*, *365*, 725–732. Available from https://doi.org/10.1016/j.jhazmat.2018.11.066.

Cao, Y., Feng, T., Xu, J., & Xue, C. (2019). Recent advances of molecularly imprinted polymer-based sensors in the detection of food safety hazard factors. *Biosensors and Bioelectronics*, *141*, 111447. Available from https://doi.org/10.1016/j.bios.2019.111447.

Cardoso, A. G., Viltres, H., Ortega, G. A., Phung, V., Grewal, R., Mozaffari, H., Ahmed, S. R., Rajabzadeh, A. R., & Srinivasan, S. (2023). Electrochemical sensing of analytes in saliva: Challenges, progress, and perspectives. *TrAC - Trends in Analytical Chemistry*, *160*, 116965. Available from https://doi.org/10.1016/J.TRAC.2023.116965.

Chi, H., & Liu, G. (2023). Carbon nanomaterial-based molecularly imprinted polymer sensors for detection of hazardous substances in food: Recent progress and future trends. *Food Chemistry*, *420*, 136100. Available from https://doi.org/10.1016/j.foodchem.2023.136100.

Cichosz, S., Masek, A., & Zaborski, M. (2018). Polymer-based sensors: a review. *Polymer Testing*, *67*, 342–348. Available from https://doi.org/10.1016/j.polymertesting.2018.03.024.

Dixit, C. K., Bhakta, S., Reza, K. K., & Kaushik, A. (2022). Exploring molecularly imprinted polymers as artificial antibodies for efficient diagnostics and commercialization: A critical overview. *Hybrid Advances*, *1*, 100001. Available from https://doi.org/10.1016/j.hybadv.2022.100001.

Fang, L., Jia, M., Zhao, H., Kang, L., Shi, L., Zhou, L., & Kong, W. (2021). Molecularly imprinted polymer-based optical sensors for pesticides in foods: Recent advances and future trends. *Trends in Food Science & Technology*, *116*, 387–404. Available from https://doi.org/10.1016/j.tifs.2021.07.039.

Gamboa, J., Paulo-Mirasol, S., Estrany, F., & Torras, J. (2023). Recent progress in biomedical sensors based on conducting polymer hydrogels. *ACS Applied Bio Materials*, *6*, 1720–1741. Available from https://doi.org/10.1021/acsabm.3c00139.

González-Ceballos, L., Carlos Guirado-Moreno, J., Utzeri, G., Miguel García, J., Fernández-Muiño, M. A., Osés, S. M., Teresa Sancho, M., Arnaiz, A., Valente, A. J. M., & Vallejos, S. (2023). Straightforward purification method for the determination of the activity of glucose oxidase and catalase in honey by extracting polyphenols with a film-shaped polymer. *Food Chemistry*, *405*, 134789. Available from https://doi.org/10.1016/j.foodchem.2022.134789.

González-Ceballos, L., Cavia, Md. M., Fernández-Muiño, M. A., Osés, S. M., Sancho, M. T., Ibeas, S., García, F. C., García, J. M., & Vallejos, S. (2021). A simple one-pot determination of both total phenolic content and antioxidant activity of honey by polymer chemosensors. *Food Chemistry*, *342*, 128300. Available from https://doi.org/10.1016/j.foodchem.2020.128300.

González-Ceballos, L., Guirado-moreno, J. C., Guembe-García, M., Rovira, J., Melero, B., Arnaiz, A., Diez, A. M., García, J. M., & Vallejos, S. (2022). Metal-free organic polymer for the preparation of a reusable antimicrobial material with real-life application as an absorbent food pad. *Food Packaging and Shelf Life*, *33*, 100910. Available from https://doi.org/10.1016/j.fpsl.2022.100910.

González-Ceballos, L., Melero, B., Trigo-López, M., Vallejos, S., Muñoz, A., García, F. C., Fernandez-Muiño, M. A., Sancho, M. T., & García, J. M. (2020). Functional aromatic polyamides for the preparation of coated fibres as smart labels for the visual detection of biogenic amine vapours and fish spoilage. *Sensors and Actuators, B: Chemical*, *304*, 127249. Available from https://doi.org/10.1016/J.SNB.2019.127249.

Guembe-García, M., González-Ceballos, L., Arnaiz, A., Fernández-Muiño, M. A., Sancho, M. T., Osés, S. M., Ibeas, S., Rovira, J., Melero, B., Represa, C., García, J. M., & Vallejos, S. (2022). Easy nitrite analysis of processed meat with colorimetric polymer sensors and a smartphone app. *ACS Applied Materials and Interfaces*, *14*, 37051–37058. Available from https://doi.org/10.1021/acsami.2c09467.

Guembe-García, M., Santaolalla-García, V., Moradillo-Renuncio, N., Ibeas, S., Reglero, J. A., García, F. C., Pacheco, J., Casado, S., García, J. M., & Vallejos, S. (2021). Monitoring of the evolution of human chronic wounds using a ninhydrin-based sensory polymer and a smartphone. *Sensors and Actuators, B: Chemical*, *335*, 129688. Available from https://doi.org/10.1016/J.SNB.2021.129688.

Guirado-Moreno, J. C., Carreira-Barral, I., Ibeas, S., García, J. M., Granès, D., Marchet, N., & Vallejos, S. (2023). Democratization of copper analysis in grape must following a polymer-based lab-on-a-chip approach. *ACS Applied Materials and Interfaces*, *15*, 16055–16062. Available from https://doi.org/10.1021/acsami.3c00395.

Guirado-Moreno, J. C., González-Ceballos, L., Carreira-Barral, I., Ibeas, S., Fernández-Muiño, M. A., Teresa Sancho, M., García, J. M., & Vallejos, S. (2023). Smart sensory polymer for straightforward Zn(II) detection in pet food samples. *Spectrochimica Acta—Part A: Molecular and Biomolecular Spectroscopy*, *284*, 121820. Available from https://doi.org/10.1016/J.SAA.2022.121820.

Guirado-Moreno, J. C., Guembe-García, M., García, J. M., Aguado, R., Valente, A. J. M., & Vallejos, S. (2021). Chromogenic anticounterfeit and security papers: an easy and effective approach. *ACS Applied Materials and Interfaces*, *13*, 60454–60461. Available from https://doi.org/10.1021/acsami.1c19228.

Guo, W. T., Tang, X. G., Tang, Z., & Sun, Q. J. (2023). Recent advances in polymer composites for flexible pressure sensors. *Polymers (Basel)*, *15*(9), 2176. Available from https://doi.org/10.3390/POLYM15092176.

He, X., Ji, W., Xing, S., Feng, Z., Li, H., Lu, S., Du, K., & Li, X. (2024). Emerging trends in sensors based on molecular imprinting technology: Harnessing smartphones for portable detection and recognition. *Talanta*, *268*, 125283. Available from https://doi.org/10.1016/j.talanta.2023.125283.

Inci, E., Topcu, G., Guner, T., Demirkurt, M., & Demir, M. M. (2020). Recent developments of colorimetric mechanical sensors based on polymer composites. *Journal of Materials Chemistry C*, *8*, 12036–12053. Available from https://doi.org/10.1039/D0TC02600J.

Keçili, R., Hussain, C. G., & Hussain, C. M. (2023). Emerging trends in green carbon dots coated with molecularly imprinted polymers for sensor platforms. *TrAC—Trends in Analytical Chemistry*, *166*, 117205. Available from https://doi.org/10.1016/J.TRAC.2023.117205.

Lahcen, A. A., & Amine, A. (2019). Recent advances in electrochemical sensors based on molecularly imprinted polymers and nanomaterials. *Electroanalysis*, *31*, 188–201. Available from https://doi.org/10.1002/elan.201800623.

Lu, Y., Biswas, M. C., Guo, Z., Jeon, J. W., & Wujcik, E. K. (2019). Recent developments in bio-monitoring via advanced polymer nanocomposite-based wearable strain sensors. *Biosensors and Bioelectronics*, *123*, 167–177. Available from https://doi.org/10.1016/j.bios.2018.08.037.

Majdinasab, M., Daneshi, M., & Marty, J. L. (2021). Recent developments in non-enzymatic (bio)sensors for detection of pesticide residues: Focusing on antibody, aptamer and molecularly imprinted polymer. *Talanta*, *232*, 122397. Available from https://doi.org/10.1016/j.talanta.2021.122397.

Mostafa, M. H., Ali, E. S., & Darwish, M. S. A. (2022). Recent developments of conductive polymers/carbon nanotubes nanocomposites for sensor applications. *Polymer-Plastics Technology and Materials*, *61*, 1456–1480. Available from https://doi.org/10.1080/25740881.2022.2069038.

Oveissi, F., Nguyen, L. H., Giaretta, J. E., Shahrbabaki, Z., Rath, R. J., Apalangya, V. A., Yun, J., Dehghani, F., & Naficy, S. (2021). *Sensors for food quality and safety. Food Engineering Innovations Across the Food Supply Chain* (pp. 389–410). Australia: Elsevier. Available from https://doi.org/10.1016/B978-0-12-821292-9.00010-8.

Pablos, J. L., Ibeas, S., Muñoz, A., Serna, F., García, F. C., & García, J. M. (2014). Solid polymer and metallogel networks based on a fluorene derivative as fluorescent and colourimetric chemosensors for Hg(II). *Reactive and Functional Polymers*, *79*, 14–23. Available from https://doi.org/10.1016/j.reactfunctpolym.2014.02.009.

Pablos, J. L., Trigo-López, M., Serna, F., García, F. C., & García, J. M. (2014). Solid polymer substrates and smart fibres for the selective visual detection of TNT both in vapour and in aqueous media. *RSC Advances*, *4*, 25562–25568. Available from https://doi.org/10.1039/C4RA02716G.

Pablos, J. L., Vallejos, S., Muñoz, A., Rojo, M. J., Serna, F., García, F. C., & García, J. M. (2015). Solid polymer substrates and coated fibers containing 2,4,6-trinitrobenzene motifs as smart labels for the visual detection of biogenic amine vapors. *Chemistry - A European Journal*, *21*, 8733–8736. Available from https://doi.org/10.1002/CHEM.201501365.

Palumbo, A., & Yang, E. H. (2022). *Current trends on flexible and wearable mechanical sensors based on conjugated polymers combined with carbon nanotubes. Conjugated Polymers for Next-Generation Applications, Volume 1: Synthesis, Properties and Optoelectrochemical Devices* (pp. 361–399). United States: Elsevier. Available from https://doi.org/10.1016/B978-0-12-823442-6.00008-8.

Pascual, B. S., Vallejos, S., Ramos, C., Sanz, M. T., Ruiz, J. A. R., García, F. C., & García, J. M. (2018). Sensory polymeric foams as a tool for improving sensing performance of sensory polymers. *Sensors (Switzerland)*, *18*, 4378. Available from https://doi.org/10.3390/S18124378.

Pascual, B. S., Vallejos, S., Reglero Ruiz, J. A., Bertolín, J. C., Represa, C., García, F. C., & García, J. M. (2019). Easy and inexpensive method for the visual and electronic detection of oxidants in air by using vinylic films with embedded aniline. *Journal of Hazardous Materials*, *364*, 238–243. Available from https://doi.org/10.1016/J.JHAZMAT.2018.10.039.

Pavase, T. R., Lin, H., Shaikh, Qua, Hussain, S., Li, Z., Ahmed, I., Lv, L., Sun, L., Shah, S. B. H., & Kalhoro, M. T. (2018). Recent advances of conjugated polymer (CP) nanocomposite-based chemical sensors and their applications in food spoilage detection: A comprehensive review. *Sensors and Actuators, B: Chemical*, *273*, 1113–1138. Available from https://doi.org/10.1016/J.SNB.2018.06.118.

Qian, S., Cui, Y., Cai, Z., & Li, L. (2022). Applications of smartphone-based colorimetric biosensors. *Biosensors and Bioelectronics: X*, *11*, 100173. Available from https://doi.org/10.1016/J.BIOSX.2022.100173.

Rochat, S., & Swager, T. M. (2013). Conjugated amplifying polymers for optical sensing applications. *ACS Applied Materials and Interfaces*, *5*, 4488–4502. Available from https://doi.org/10.1021/am400939w.

Sanjuán, A. M., Reglero Ruiz, J. A., García, F. C., & García, J. M. (2018). Recent developments in sensing devices based on polymeric systems. *Reactive and Functional Polymers*, *133*, 103–125. Available from https://doi.org/10.1016/j.reactfunctpolym.2018.10.007.

Sensor Market. (2023). Precedence Research. Available from https://www.precedenceresearch.com/sensor-market. [Accessed February 2024].

Sharma, R., & Lee, H.-il (2022). Recent advances in polymeric chemosensors for the detection and removal of mercury ions in complex aqueous media. *Journal of Macromolecular Science, Part A*, *59*, 389–402. Available from https://doi.org/10.1080/10601325.2022.2054348.

Shukla, P., & Saxena, P. (2021). Polymer nanocomposites in sensor applications: A review on present trends and future scope. *Chinese Journal of Polymer Science*, *39*, 665–691. Available from https://doi.org/10.1007/s10118-021-2553-8.

Smart/Intelligent Sensors Market Outlook, Trends, Analysis 2031. Available from https://www.transparencymarketresearch.com/smart-intelligent-sensor-market.html. [Accessed February 2024].

Tajik, S., Beitollahi, H., Nejad, F. G., Dourandish, Z., Khalilzadeh, M. A., Jang, H. W., Venditti, R. A., Varma, R. S., & Shokouhimehr, M. (2021). Recent developments in polymer nanocomposite-based electrochemical sensors for detecting environmental pollutants. *Industrial and Engineering Chemistry Research*, *60*, 1112–1136. Available from https://doi.org/10.1021/acs.iecr.0c04952.

Thomas, S. W., Joly, G. D., & Swager, T. M. (2007). Chemical sensors based on amplifying fluorescent conjugated polymers. *Chemical Reviews*, *107*, 1339–1386. Available from https://doi.org/10.1021/cr0501339.

Tran, V., Lee., Nguyen, T. N., & Lee, D. (2022). Recent advances and progress of conducting polymer-based hydrogels in strain sensor applications. *Gels*, *9*, 12. Available from https://doi.org/10.3390/GELS9010012.

Trigo-López, M., Muñoz, A., Mendía, A., Ibeas, S., Serna, F., García, F. C., & García, J. M. (2018). Palladium-containing polymers as hybrid sensory materials (water-soluble polymers, films and smart textiles) for the colorimetric detection of cyanide in aqueous and gas phases. *Sensors and Actuators, B: Chemical*, *255*, 2750–2755. Available from https://doi.org/10.1016/j.snb.2017.09.089.

Tsong, J. L., Robert, R., & Khor, S. M. (2023). Emerging trends in wearable glove-based sensors: A review. *Analytica Chimica Acta*, *1262*, 341277. Available from https://doi.org/10.1016/J.ACA.2023.341277.

Vallejos, S., Estévez, P., Ibeas, S., Muñoz, A., García, F. C., Serna, F., & García, J. M. (2011). A selective and highly sensitive fluorescent probe of Hg^{2+} in organic and aqueous media: The role of a polymer network in extending the sensing phenomena to water environments. *Sensors and Actuators, B: Chemical*, *157*, 686–690. Available from https://doi.org/10.1016/j.snb.2011.05.041.

Vallejos, S., Hernando, E., Trigo, M., García, F. C., García-Valverde, M., Iturbe, D., Cabero, M. J., Quesada, R., & García, J. M. (2018). Polymeric chemosensor for the detection and quantification of chloride in human sweat. Application to the diagnosis of cystic fibrosis. *Journal of Materials Chemistry B*, *6*, 3735–3741. Available from https://doi.org/10.1039/c8tb00682b.

Vallejos, S., Reglero, J. A., García, F. C., & García, J. M. (2017). Direct visual detection and quantification of mercury in fresh fish meat using facilely prepared polymeric sensory labels. *Journal of Materials Chemistry A*, *5*, 13710–13716. Available from https://doi.org/10.1039/C7TA03902F.

Vallejos, S., Trigo-López, M., Arnaiz, A., Miguel, Á., Muñoz, A., Mendía, A., & García, J. M. (2022). From classical to advanced use of polymers in food and beverage applications. *Polymers (Basel)*, *14*, 4954. Available from https://doi.org/10.3390/POLYM14224954.

Veeramuthu, L., Venkatesan, M., Benas, J. S., Cho, C. J., Lee, C. C., Lieu, F. K., Lin, J. H., Lee, R. H., & Kuo, C. C. (2021). Recent pProgress in conducting polymer composite/nanofiber-based strain and pressure sensors. *Polymers (Basel)*, *13*, 4281. Available from https://doi.org/10.3390/POLYM13244281.

Wang, L., Wang, H., Wan, Q., & Gao, J. (2023). Recent development of conductive polymer composite-based strain sensors. *Journal of Polymer Science*, *61*, 3167–3185. Available from https://doi.org/10.1002/pol.20230200.

Wang, L., & Zhang, W. (2023). Molecularly imprinted polymer (MIP) based electrochemical sensors and their recent advances in health applications. *Sensors and Actuators Reports*, *5*, 100153. Available from https://doi.org/10.1016/J.SNR.2023.100153.

Willis-Fox, N., Watchorn-Rokutan, E., Rognin, E., & Daly, R. (2023). Technology pull: scale-up of polymeric mechanochemical force sensors. *Trends in Chemistry*, *5*, 415–431. Available from https://doi.org/10.1016/j.trechm.2023.02.005.

Wu, D., Sedgwick, A. C., Gunnlaugsson, T., Akkaya, E. U., Yoon, J., & James, T. D. (2017). Fluorescent chemosensors: The past, present and future. *Chemical Society Reviews*, *46*, 7105–7123. Available from https://doi.org/10.1039/C7CS00240H.

Yadav, A., Yadav, N., Wu, Y., RamaKrishna, S., & Hongyu, Z. (2023). Wearable strain sensors: state-of-the-art and future applications. *Materials Advances*, *4*, 1444–1459. Available from https://doi.org/10.1039/d2ma00818a.

Zamiri, G., & Haseeb, A. S. M. A. (2020). Recent trends and developments in graphene/conducting polymer nanocomposites chemiresistive sensors. *Materials (Basel)*, *13*, 3311. Available from https://doi.org/10.3390/MA13153311.

Zhang, Z., Zhang, H., Tian, D., Phan, A., Seididamyeh, M., Alanazi, M., Ping Xu, Z., Sultanbawa, Y., & Zhang, R. (2024). Luminescent sensors for residual antibiotics detection in food: Recent advances and perspectives. *Coordination Chemistry Reviews*, *498*, 215455. Available from https://doi.org/10.1016/j.ccr.2023.215455.

Index

Note: Page numbers followed by "*f*" and "*t*" refer to figures and tables, respectively.

C

D

E

F

G

I

J

K

L

M

N

O

P

Printed in the United States
by Baker & Taylor Publisher Services